U0156898

nature

The Living Record of Science

《自然》学科经典系列

总顾问：李政道（Tsung-Dao Lee）

英方总主编：Sir John Maddox
Sir Philip Campbell

中方总主编：路甬祥

生命科学的进程 II
PROGRESS IN LIFE SCIENCES II

（英汉对照）

主编：许智宏

外语教学与研究出版社 · 麦克米伦教育 · 《自然》旗下期刊与服务集合

FOREIGN LANGUAGE TEACHING AND RESEARCH PRESS · MACMILLAN EDUCATION · NATURE PORTFOLIO

北京 BEIJING

图书在版编目 (CIP) 数据

生命科学的进程. II：英汉对照／许智宏主编. —— 北京：外语教学与研究出版社，2021.4
（《自然》学科经典系列／路甬祥等总主编）
ISBN 978-7-5213-2511-9

Ⅰ. ①生… Ⅱ. ①许… Ⅲ. ①生命科学－文集－英、汉 Ⅳ. ①Q1-53

中国版本图书馆 CIP 数据核字 (2021) 第 056346 号

出 版 人　徐建忠
项目统筹　章思英
项目负责　刘晓楠　顾海成
责任编辑　刘晓楠
责任校对　王　菲
封面设计　孙莉明　高　蕾
版式设计　孙莉明
出版发行　外语教学与研究出版社
社　　址　北京市西三环北路 19 号（100089）
网　　址　http://www.fltrp.com
印　　刷　北京华联印刷有限公司
开　　本　787×1092　1/16
印　　张　61.5
版　　次　2021 年 6 月第 1 版 2021 年 6 月第 1 次印刷
书　　号　ISBN 978-7-5213-2511-9
定　　价　568.00 元

购书咨询：（010）88819926　电子邮箱：club@fltrp.com
外研书店：https://waiyants.tmall.com
凡印刷、装订质量问题，请联系我社印制部
联系电话：（010）61207896　电子邮箱：zhijian@fltrp.com
凡侵权、盗版书籍线索，请联系我社法律事务部
举报电话：（010）88817519　电子邮箱：banquan@fltrp.com
物料号：325110001

《自然》学科经典系列

（英汉对照）

总顾问：李政道（Tsung-Dao Lee）

英方总主编：Sir John Maddox
Sir Philip Campbell

中方总主编：路甬祥

英方编委：
Philip Ball
Arnout Jacobs
Magdalena Skipper

中方编委（以姓氏笔画为序）：
万立骏
朱道本
许智宏
武向平
赵忠贤
滕吉文

生命科学的进程

（英汉对照）

主编：许智宏

审稿专家 （以姓氏笔画为序）

王 昕	王秀娥	王晓晨	王敏康	冯兴无	邢 松	吕 扬
刘 武	刘京国	江丕栋	孙 军	李素霞	杨 志	杨茂君
吴新智	张 旭	张德兴	陈继征	陈新文	林圣龙	昌增益
金 侠	周筠梅	赵凌霞	胡 荣	袁 峥	莫 韫	顾孝诚
崔 巍	梁前进	董 为				

翻译工作组稿人 （以姓氏笔画为序）

| 王耀杨 | 刘 明 | 关秀清 | 李 琦 | 何 铭 | 蔡 迪 |

翻译人员 （以姓氏笔画为序）

王耀杨	毛晨晖	尹　金	田晓阳	吕　静	刘振明	刘皓芳
孙玉诚	苏　慧	杜　丽	李　响	李　梅	吴　彦	张玉光
张立召	张锦彬	周志华	郑建全	荆玉祥	高如丽	郭　娟
崔娅铭	彭丽霞	董培智	曾菊平			

校对人员 （以姓氏笔画为序）

王　菲	王　敏	王帅帅	王阳兰	王晓萌	王晓蕾	王赛儿
孔凌楠	田晓阳	代　娟	丛　岚	冯　琛	乔萌萌	刘　明
刘丛丛	齐文静	阮玉辉	孙　娟	孙瑞静	苏　慧	李　四
李　梅	李　琦	李　景	李红菊	吴　茜	吴　彦	何　铭
张　帆	张玉光	张世馥	周玉凤	周平博	赵广宇	赵凤轩
姜　薇	顾海成	郭　琴	唐　颖	崔天明	葛云霄	董　为
韩玲俐	曾芃斐	蔡　迪	管　冰	潮兴娟	潘卫东	

Contents
目 录

1969

1970

1971

VI

Volume II

New Evidence on the Antiquity of Piltdown Man

K. P. Oakley and C. R. Hoskins

Editor's Note

The famous skull and jawbone of a fossil human excavated near Piltdown in southern England in 1912 had been the focus of much attention by anthropologists. By the 1940s, the combination of modern-looking skull and ape-like jaw seemed out of tune with discoveries from Asia and Africa, showing fossil humans with small brains and human-like jaws. Here Kenneth Oakley and Randall Hoskins used a new method of dating bones mixed up in the same deposit (such as found at Piltdown), by measuring the amount of fluorine that had seeped into the bone. The Piltdown bones showed relatively little fluorine, suggesting a relatively recent date. At this stage, however, nobody was prepared to admit that the Piltdown remains were fraudulent.

FEW, if any, fossils have given rise to more controversy than the remains discovered by Charles Dawson in gravel at Piltdown, near Fletching, Sussex, and described in 1913 by Sir Arthur Smith Woodward as representing a new genus and species of man, *Eoanthropus dawsoni*. The geological age of "Piltdown man" was a matter of dispute from the first, moreover, this problem has latterly become linked with the question as to whether the thick but essentially human cranial bones and the remarkably ape-like lower jaw (and canine tooth) might be a chance association and represent two creatures of different geological ages. The present investigation, using the fluorine method for determination of the relative antiquity of fossil bones, has shown that the cranial bones and jaw-bone are of the same age, at the earliest Middle, more probably early Upper, Pleistocene.

Earlier Evidence

The Piltdown gravel is a thin remanié deposit in a terrace about 50 ft. above the River Ouse, containing fossil mammals of two distinct ages: a derived Villafranchian series ("Upper Pliocene" of earlier authors, now classed as Lower Pleistocene), and a later Pleistocence group, in part at least contemporary with the gravel. The "contemporary" group was classed by Dawson and Woodward[1] as "early Pleistocene", but according to more recent authorities it is not earlier than Middle Pleistocene.

To which of these two groups does *Eoanthropus* belong? This might be thought a simple question to decide by comparison of the states of preservation of the various remains. Yet examination of the specimens from this very point of view has led different authorities to diametrically opposed conclusions. In fact, anatomists regarding

2

皮尔当人年代的新证据

奥克利，霍斯金斯

编者按

这件著名的化石人类的头骨和颌骨，是于1912年在英格兰南部皮尔当附近发掘得到的，一直以来都得到了古人类学家的广泛关注。它具有类似现代人类的头骨和类似猿类的颌骨。到20世纪40年代，人们发现它与在亚洲和非洲发现的化石人类不同，这些化石人类的脑较小，而颌骨更像人类。本文中肯尼思·奥克利和兰德尔·霍斯金斯使用一种全新的方法，即通过测量渗透入骨的氟含量，对同一沉积物（例如在皮尔当发现的沉积物）中掺杂在一起的骨进行年代测定。皮尔当人骨的含氟量相对较低，表明其处于相对较近的年代。然而直到此时，还没有人会想到皮尔当化石是个骗局。

几乎没有化石比查尔斯·道森在萨塞克斯的弗莱彻附近的皮尔当砾石层中发现的化石引发过更多的争议，1913年阿瑟·史密斯·伍德沃德爵士将这些化石描述成一种人类新属和新种的代表——道森曙人。皮尔当人的地质年代从一开始就存在争议，此外，最近这个难题已经开始变得与另一个问题相关，即这些厚的但本质上属于人类的颅骨和明显类似猿类的下颌骨（和犬齿）是否只是偶然共生在一起，并代表了两种不同地质年代的生物？目前使用氟年代测定法对骨化石的相对年代进行了测定，结果表明颅骨和颌骨属于同一年代，即中更新世的最早期，更可能是上更新世早期。

早期证据

皮尔当砾石层是位于乌斯河之上大约50英尺的一处阶地上的经过再沉积的薄层沉积物，含有两个不同年代的哺乳动物化石：维拉方期的一个次生系列（早期作者将其命名为"上新世晚期"，现在被列入下更新世）和更新世晚期群，其至少部分与砾石层属于同一年代。道森和伍德沃德[1]将这"同年代"的群划为"更新世早期"，但是根据更为新近的主流观点，它们所处的时代并不早于中更新世。

那么曙人属于这两个群中的哪一个呢？人们认为这是一个简单的问题，可以通过比较不同化石的保存状况而确定。然而仅仅从这一点来检验标本，不同的专家就得到了完全相反的结论。事实上，认为该下颌骨和犬齿属于类人猿的解剖学家表示

3

the mandible and canine tooth as anthropoid have felt free to place these with the Villafranchian group, and to dismiss the cranium as later Pleistocene and referable to *Homo sapiens* (cf. Marston[2]). This hypothesis is permissible in so far as mammalian remains of two ages (for example, "*Stegodon*", Villafranchian; and *Castor fiber*, post-Villafranchian) occurred with *Eoanthropus* at the base of the dark gravel (ref. 3, pp. 83–85), but it would be erroneous to claim that none of the cranial fragments was closely associated with the mandible. Although the fragments of skullcap were for the most part recovered loose on spoil heaps, Dawson states (ref. 1, p. 121) that Smith Woodward dug out a small portion of the occipital bone "from within a yard of the point where the jaw was discovered, and at precisely the same level".

In reading their first paper, Dawson and Woodward (ref. 1, pp. 123, 143) stated their view that the human skull and mandible, being practically unrolled, were contemporary with the gravel and therefore probably "Pleistocene". However, in the discussion which followed, Sir Arthur Keith[4] argued that the skull should be assigned to the "Pliocene" group. He was influenced, he said, by the fact that in the Heidelberg jaw, of early Pleistocene date, the region of the chin was essentially human, whereas the Piltdown mandible showed simian characters. E. T. Newton[5] said that the mineralized condition of the skull bones pointed to their being of "Pliocene" age. In replying to the discussion, Dawson admitted this possibility, for he said (ref. 1, p. 151): "... the occurrence of certain Pliocene specimens in a considerably rolled condition, suggested a difference as to age, but not to the extent of excluding the possibility of their being coeval. The rolled specimens may have entered the stream further up the river than the human remains. ... Then again the skull might have been surrounded by some colloid material [that is, clay] which preserved it in its passage from some earlier deposit." But in the second paper (ref. 3, p. 86) Dawson wrote: "Putting aside the human remains and those of the beaver, the remains all point to a characteristic land fauna of the Pliocene age; and though all are portions of hard teeth, they are rolled and broken. The human remains on the other hand, although of much softer material, are not rolled, and the remains of beaver are in a similar condition. It would therefore seem that the occurrence of these two individuals belongs to one of the periods of reconstruction of this gravel."

In 1935, Dr. A. T. Hopwood[6] reconsidered the evidence, and, largely on the basis of the state of preservation of the specimens, concluded that *Eoanthropus* belonged to the Villafranchian or Lower Pleistocene faunal assemblage. He pointed out that the absence of indications of rolling was unreliable as a criterion, for one of the teeth of "*Stegodon*" (*Elephas* cf. *planifrons*), undoubtedly a member of the derived group, was practically unrolled. The evidence then available, he said, justified the statement that "Piltdown Man is the oldest human fossil yet discovered" (ref. 6, p. 57).

In recent years many anthropologists have held the view that, so long as its date remained uncertain, this material was better placed in a suspense account, for the anatomical features of *Eoanthropus* (assuming the material to represent one creature) are wholly

可以把这些样本归入维拉方群中去，并且排除了其颅骨所处年代为更新世晚期而将其列为智人的可能性（参见马斯顿[2]）。如果仅就目前为止与曙人一起出现于暗色砾石层底部（参考文献 3，第 83~85 页）的两个不同年代的哺乳动物化石（例如，维拉方期的"剑齿象"以及后维拉方期的欧亚河狸）而言，这一假说是可以成立的，但是如果声称没有一个颅骨破片与下颌骨存在密切关系的话，那么这一假说就是错误的。尽管后来修复的大部分颅骨顶部破片都曾散布在废石堆上，但是道森仍强调（参考文献 1，第 121 页），史密斯·伍德沃德从"距离发现颌骨的地点一码之内，并且正是在同一高度上"挖出了小部分枕骨。

在读到道森和伍德沃德的第一篇论文（参考文献 1，第 123 页和第 143 页）时，他们在文中陈述了自己的观点，即实际上未被滚磨过的人类头骨和下颌骨与砾石层是同一时代的，因此可能是"更新世"时期的。然而在接下来的讨论中，阿瑟·基思[4]爵士认为，头骨应该属于"上新世"时期，这是由于更新世早期的海德堡下颌骨的颏部本质上是人类的颏，而皮尔当人的下颌骨却显示出了猿的特征。牛顿[5]认为头骨的矿化情况表明它们是"上新世"时期的生物。在对这一讨论进行回复时，道森承认了这一可能性，因为他指出（参考文献 1，第 151 页）："……处于被显著滚磨过的状态的某些上新世时期标本的出现，提示了年代的差异，但是并不排除它们属于同时代生物的可能性。这些被滚磨过的标本可能在比人类化石更上游的位置进入溪流之中……然后头骨可能又被一些胶体材料（即黏土）包围，这种材料将标本从较早沉积物的搬运过程中保留下来。"但是第二篇论文中（参考文献 3，第 86 页），道森写道："先撇开人类化石和河狸化石不谈，其他所有的化石都指向一种特征性的上新世时期的陆地动物群；尽管所有的化石都是坚硬牙齿的一部分，但它们都被滚磨过并且是破碎的。另一方面，尽管人类化石都是些硬度小得多的材料，但是它们都没有被滚磨过，河狸化石的保存状况也与之相似。因此这两个个体的出现时期似乎属于该砾石层的一个再堆积时期。"

1935 年，主要基于这些样本的保存状况，霍普伍德[6]博士重新考虑了这一证据，他推断曙人属于维拉方期或下更新世期的动物群组合。他指出将缺乏滚磨迹象作为一个标准是不可靠的，因为"剑齿象"（平额象相似种）的一枚牙齿完全没有被滚磨过，而它毫无疑问是次生组群的成员之一。他认为现有的证据可用于证明"皮尔当人是至今发现的最古老的人类化石"的论述（参考文献 6，第 57 页）。

最近几年许多人类学家所持有的观点是，只要没有确定标本所属的年代，那么最好将其暂且搁置，因为曙人（假设这些材料代表一种生物）的解剖特征与在远东

contrary to what discoveries in the Far East and in Africa have led us to expect in an early Pleistocene hominid. However, if one could at least be certain of the contemporaneity of the parts associated under the name *Eoanthropus*, and of their precise geological age, the number of possible interpretations would be much reduced.

The Fluorine-dating Method

It has long been known that fossil bones accumulate fluorine in course of time. The major constituent of bones and teeth is hydroxyapatite, which acts as a natural trap for wandering fluorine ions. Bones absorb fluorine from the ground-water, and it becomes fixed in their substance as fluorapatite by a process of ionic interchange. Owing to the porous texture of bones (and teeth), this progressive alteration is not confined to the surface, but usually proceeds more or less uniformly throughout the body of the material. It was suggested at the British Association meeting in 1947[7] that the percentage distribution of this element in the various bones and teeth from the Piltdown gravel might reflect their relative ages sufficiently clearly to throw some light on the major problem. Later, in response to a request from the Keeper of Geology at the British Museum, the Government Chemist agreed to undertake experimental work on the fluorine analysis of fossil bones.

After preliminary trials, Mr. R. H. Settle and one of us (C. R. H.), assisted by Mr. E. C. W. Maycock, adapted published methods of analysis to the exact determination of fluorine in very small samples of bone. A description of the analytical technique and a discussion on the limits of accuracy will be presented in a report on "fluorine-dating" to be published later in the *Bulletin of the British Museum (Nat. Hist.)*, and it is not necessary in this article to refer other than briefly to these matters.

The method was based on that originally devised by Willard and Winter[8] and afterwards modified by various authors. The fluorine in the bone was separated as hydrofluosilicic acid by distillation with perchloric acid in presence of a few beads of soft glass to serve as a source of silica. Suitable aliquots of concentrated distillate were titrated with dilute thorium nitrate solution, using Alizarin S as indicator. The solution of thorium nitrate was standardized against solutions of known fluorine content under identical conditions, particularly with regard to titrating to the same stage in the colour change of the indicator. In the majority of cases, the amount of fluorine in the distillate was sufficient for at least three aliquots to be titrated, and the average fluorine content, after making a small correction for a blank determination, was adopted. There was generally close agreement between the fluorine contents calculated from titrations of different aliquots from a given distillation.

Where possible, at least 20 mgm. of bone was used for fluorine determination; but in several cases it was necessary to rely on samples of the order of 5 mgm. The errors of analysis naturally increase as the weight of sample decreases, but it is believed that with

和非洲的发现完全相反，而那些发现让我们对更新世早期的人科有所期待。然而，如果有人能够至少确定与曙人相关的同时代部分以及它们确切的地质年代的话，那么就会大大减少可能性的解释的数量了。

氟年代测定法

很久以前人们就知道，随着时间的推移，氟会在骨化石中累积。骨和牙齿的主要成分是羟磷灰石，它是游离氟离子的天然收集器。骨从地下水中吸收氟，然后氟通过离子交换过程在骨中以氟磷灰石的形态固定下来。由于骨（和牙齿）的多孔结构，这一渐进性的变化并不局限于表面，而是通常在整个材料体大致均匀地发生。1947 年的英国科学促进会会议 [7] 上提出，这一元素在皮尔当砾石层中发现的各种骨和牙齿中的百分比分布情况可能反映了它们所处的相对年代，这对阐明主要问题会有帮助。后来，应大英博物馆的地质学管理人的请求，政府部门的化验师同意承担对骨化石进行氟分析的实验工作。

初步尝试之后，在梅科克先生的协助下，塞特尔先生和我们中的成员之一霍斯金斯将已发表的分析方法进行调整，以适用于准确测定非常小的骨样本中的氟。对这种分析技术的描述及对其在准确性方面的局限性的讨论将在关于"氟年代测定法"的报道中进行陈述，这篇报道不久将发表在《大英博物馆公报（自然史）》上，因此在这篇文章里只简单地介绍一下这种方法。

这一方法建立在威拉德和温特 [8] 最初的设计基础之上，后来经过了不同研究者的修改。通过几粒软质玻璃珠作为硅的来源，将骨骼中的氟通过高氯酸的蒸馏作用分离为氟硅酸。使用茜素 S 作为指示剂，将浓缩的蒸馏液适当地分为几等份，用稀释的硝酸钍溶液进行滴定。在相同条件下，用已知的含氟溶液对硝酸钍溶液进行标准化，尤其是在指示剂颜色变化的指导下滴定到相同阶段。大多数情况下，蒸馏物中的氟含量足以滴定至少三等份，在对空白测定进行微小的矫正后，就可以采用平均氟含量了。根据不同等份蒸馏物得到的滴定溶液计算出的氟含量之间，通常具有良好的一致性。

如果可能的话，进行氟测定至少需要用 20 毫克的骨；但是有些情况下，依样本情况该值可能以 5 毫克为单位向下浮动。分析误差随样本重量的减少而自然增加，

sample weights of 5 mgm. and upwards the error in the adopted values is not greater than ±0.1 percent of fluorine. For sample weights less than 5 mgm. the error may be ±0.2 percent of fluorine.

Approximate estimations of the iron and phosphate contents of the bones were made on residues of the samples remaining after the fluorine determinations. Colorimetric methods were used in each case (thioglycollic acid for iron, and ammonium molybdate followed by reduction of phosphomolybdate with stannous chloride for phosphate).

The fluorine-dating method was first applied to the Galley Hill skeleton[9]. Briefly, it was shown that indigenous fossil bones in the Middle Pleistocene terrace gravels at Swanscombe contain around 2 percent fluorine, those from Upper Pleistocene deposits in the same region around 1 percent, and post-Pleistocene bones not more than 0.3 percent; while the Galley Hill skeleton, although found *in* the Middle Pleistocene gravels, proved to contain only about 0.3 percent fluorine, and was therefore clearly an intrusive burial, at earliest end-Pleistocene. The Swanscombe skull bones, on the other hand, discovered in these gravels by Mr. A. T. Marston in 1935–36, showed the expected 2 percent fluorine.

Series of bones from other sites have been analysed, and the results show that the method, although limited in scope, is useful for differentiating fossil bones of diverse antiquity when they occur mixed together, provided that the specimens compared have similar matrices. It cannot be used to determine the relative antiquity of bones from widely separated localities. The method was ideally suited to the Galley Hill problem. There seemed reasonable hope that it would help to resolve the Piltdown enigma. Accordingly, in October 1948, Mr. W. N. Edwards, Keeper of Geology in the British Museum, authorized the sampling of *Eoanthropus* and associated mammalian bones and teeth. For the most part the samples were obtained by applying a dental drill to broken or worn edges of the specimens until a small but sufficient quantity of bone powder had been cored out. Where possible, powder from several drill holes in each specimen was mixed in order to ensure a representative sample. In view of the ferruginous nature of the deposits, it was thought advisable to determine the iron content of all the samples, but we found that there is no appreciable correlation in the Piltdown material between fluorine content and iron impregnation.

In the case of coarsely porous bone, it is sometimes difficult to obtain a sample which is completely free from silt contamination. The fluorine content of a contaminated portion of a bone will obviously be misleadingly low. It was therefore decided to determine the phosphate content of all samples, and to express the fluorine value of each sample as the percentage ratio of fluorine to phosphate (as P_2O_5). This procedure facilitates comparison of the fluorine contents of bones in which there has been variable contamination.

但是如果使用 5 毫克或更多量样本的话，就认为采用的氟含量测定值误差不大于 ±0.1%。而对于样本重量少于 5 毫克的检测，氟含量的误差可能达到 ±0.2%。

在测定完氟含量之后，根据剩余的样本残渣近似估算骨中铁和磷酸盐含量。每个样本的铁和磷酸盐含量估算都使用比色法（铁通过巯基乙酸进行比色，磷酸盐通过钼酸铵以及随后的氯化亚锡对磷钼酸盐的还原作用进行比色）完成。

氟年代测定法首先应用于伽力山骨架 [9]。简单地说，斯旺斯孔布的中更新世阶地砾石层中的本地的骨化石含有约 2% 的氟，而同一地区的上更新世沉积物化石骨的氟含量约为 1%，晚更新世骨的氟含量则不高于 0.3%。尽管伽力山骨架发现于中更新世时期的砾石层**之中**，但是事实证明这些样本的氟含量只有约 0.3%，因此很明显这是发生在更新世末的早期阶段的一处侵入型埋藏。另一方面，1935 年 ~ 1936 年间，马斯顿在这些砾石层发现的斯旺斯孔布头骨氟含量为 2%，这与预期的结果一致。

对其他遗址的一系列骨也进行了分析，结果表明尽管使用范围有局限性，但是当许多样本混合在一起同时出现、并且假如用来进行比较的样本具有相似的基质的话，那么用这种方法来区分不同年代的骨化石还是很有效的。这种方法不能用来确定出土地点相差很远的骨的相对年代关系，而对于解决伽力山问题则是非常理想的，因此人们希望这种方法有助于解决皮尔当之谜的想法也是合理的。相应地，大英博物馆的地质学管理人爱德华兹先生于 1948 年 10 月批准了对曙人以及共生的哺乳动物骨和牙齿进行采样。采样方法为使用牙钻来钻取样本破碎的或磨损的边缘，直到获取少量但足以用来进行分析的骨粉为止。如果有可能，对从每份样本的不同钻孔得到的骨粉进行混合以保证采样具有代表性。鉴于沉积物含铁的这一性质，有人认为确定所有样本的含铁量是可取的，但我们发现皮尔当材料的氟含量和铁浸渍之间并没有明显的相关性。

对于粗糙多孔的骨，有时很难得到完全没有受土壤污染的样本。显然，骨的污染部分的低含氟量会产生误导。因此我们决定测定所有样本的磷酸盐含量，然后将每个样本的氟值表达为氟与磷酸盐（以 P_2O_5 的形式）的百分比。这一步骤使得比较存在不同程度污染的骨的含氟量容易很多。

Application to the Piltdown Material

Every available bone and tooth from the Piltdown gravel and from neighbouring deposits was analysed, including seventeen samples of the *Eoanthropus* material. The results are shown in the accompanying table. For comparative purposes the mammalian remains have been grouped according to known or probable age. Their colour, degree of rolling or other physical states have been ignored in making this age classification, except in the case of some of the subfossil or recent specimens. The teeth of *Mastodon arvernensis*, *Elephas* cf. *planifrons* (= "*Stegodon*" auctt.) and "*Rhinoceros*" cf. *etruscus* cannot be younger than Lower Pleistocene. All the rest of the Pleistocene material from the gravel is either certainly later (*Cervus elaphus*, *Castor fiber*), or possibly later (*Hippopotamus* sp., *Equus* sp., *Cervus* sp.), than Villafranchian. Any such post-Villafranchian elements might theoretically be of first interglacial age (when according to Hopwood's terminology the "Middle Pleistocene" fauna began to appear); of great interglacial age (that is, Middle Pleistocene of all authors), or of last interglacial age (that is, early Upper Pleistocene of some authors, late Middle Pleistocene of others). Since distinctively Cromerian forms are absent and as there are indications that the deposit has been repeatedly re-worked, the post-Villafranchian material is probably partly Middle and partly Upper Pleistocene. A problem of classification typical of this site is presented by the so-called bone implement from Piltdown (*P*. 18). It is part of the femur of an elephant, judged on the basis of size to be that of a member of the *Elephas meridionalis–antiquus* group. It is conceivably Cromerian (first interglacial), more probably later, but can safely be classed as "possibly Middle or Upper Pleistocene".

Analyses of fossil materials from Piltdown

	Fluorine %	P_2O_5 %	$\dfrac{\%F}{\%P_2O_5} \times 100$	Iron %
Eoanthropus I				
P.1 L. parieto-frontal (*E*. 590)†	0.1 (2)*	21	0.5	7
P.2 L. temporal (*E*. 591)	0.4	18	2.2	7
P.3 R. parietal (*E*. 592)	0.3	17	1.8	6
P.4 Occipital (*E*. 593)	0.2	28	0.7	5
P.5 R. mandibular ramus (*E*. 594)	0.2 (5)	20	1.0	6
P.17 Canine (*E*. 611)	<0.1	27	0.4	Trace
P.42 Molar (rm₁) (*E*. 594)	<0.1	23	0.4	Trace
Eoanthropus II				
P.30 R. frontal (*E*. 646)	0.1	13	0.8	12
P.31 Occipital (*E*. 647)	0.1 (2)	17	0.6	17
P.32 Molar (lm₁) (*E*. 648)	0.4 (2)	30	1.3	Trace
Other Mammalian Remains **Lower Pleistocene**				
P.6 Molar, *Mastodon* cf. *arvernensis* (*E*. 595)	1.9	23	8.3	5
P.7 Molar, *Elephas* cf. *planifrons* (*E*. 596)	2.7 (2)	33	8.2	3
P.8 Molar, *Elephas* cf. *planifrons* (*E*. 597)	2.5	34	7.4	1
P.23 Molar, *Elephas* cf. *planifrons* (*E*. 620)	3.1	39	7.9	4
P.25 Molar, *Mastodon arvernensis* (*E*. 622)	2.3	36	6.4	4
P.26 Premolar, "*Rhinoceros*" cf. *etruscus* (*E*. 623)	2.0	24	8.3	6

应用到皮尔当材料

我们对每件从皮尔当砾石层及附近沉积物中得到的骨和牙齿都进行了分析，包括 17 件曙人材料样本。分析结果见附表。为了进行比较，根据已知或者可能的年代对哺乳动物化石进行了分组。除了亚化石样本或近世的样本外，在进行年代分类时，都忽略了样本的颜色、滚磨程度或其他物理状态。阿维尔乳齿象、平额象相似种（一般作者所认为的"剑齿象"）和艾特鲁斯克"古犀"相似种的牙齿都不会晚于下更新世时期。源自砾石层的所有其余的更新世材料，其年代或者可以肯定比维拉方期更晚（马鹿、欧亚河狸），或者可能比维拉方期晚（河马未定种、马未定种、鹿未定种）。理论上，任何这种后维拉方期的成员都可能生存于第一间冰期（根据霍普伍德的说法，当时正是"中更新世"动物群开始出现的时候），大间冰期（即所有作者所说的中更新世）或者末次间冰期（即有些作者所谓的上更新世早期，而另一些作者认为应该说是中更新世晚期）的。由于具有特征性的克劳默间冰期动物种类并不存在，并且有迹象表明此处堆积物曾经多次再沉积，所以后维拉方期的材料可能一部分是中更新世时期的，另一部分则是上更新世时期的。这一遗址典型的分类问题可以由所谓的皮尔当骨制品来说明 (*P*. 18)。该样本是一头大象的股骨的一部分，根据大小来判断，应该是南方象-古象群的成员之一的股骨。它可能属于克劳默间冰期（第一间冰期）或者更晚的时期，但将其分类到"可能为中更新世或上更新世"时期是没有问题的。

皮尔当人化石材料分析

	含氟量 %	P_2O_5%	$\dfrac{\%F}{\%P_2O_5} \times 100$	铁含量 %
曙人 I				
P.1 左顶骨–额骨 (*E*. 590) †	0.1 (2)*	21	0.5	7
P.2 左颞骨 (*E*. 591)	0.4	18	2.2	7
P.3 右顶骨 (*E*. 592)	0.3	17	1.8	6
P.4 枕骨 (*E*. 593)	0.2	28	0.7	5
P.5 右下颌支 (*E*. 594)	0.2 (5)	20	1.0	6
P.17 犬齿 (*E*. 611)	<0.1	27	0.4	痕量
P.42 臼齿（第一右下白齿）(*E*. 594)	<0.1	23	0.4	痕量
曙人 II				
P.30 右额骨 (*E*. 646)	0.1	13	0.8	12
P.31 枕骨 (*E*. 647)	0.1 (2)	17	0.6	17
P.32 臼齿（第一左下白齿）(*E*. 648)	0.4 (2)	30	1.3	痕量
其他哺乳动物化石 下更新世				
P.6 臼齿，阿维尔乳齿象相似种 (*E*. 595)	1.9	23	8.3	5
P.7 臼齿，平额象相似种 (*E*. 596)	2.7 (2)	33	8.2	3
P.8 臼齿，平额象相似种 (*E*. 597)	2.5	34	7.4	1
P.23 臼齿，平额象相似种 (*E*. 620)	3.1	39	7.9	4
P.25 臼齿，阿维尔乳齿象 (*E*. 622)	2.3	36	6.4	4
P.26 前臼齿，艾特鲁斯克"古犀"相似种 (*E*. 623)	2.0	24	8.3	6

Continued

	Fluorine %	P_2O_5 %	$\dfrac{\%F}{\%P_2O_5} \times 100$	Iron %
Possibly Middle and Upper Pleistocene				
P.9 Molar, *Hippopotamus* sp. (*E.* 598)	0.1 (3)	37	0.3	3
P.10 Premolar, *Hippopotamus* sp. (*E.* 599)	1.1 (3)	29	3.8	5
P.11 Antler, *Cervus elaphus* (*E.* 600)	1.5 (3)	28	5.4	3
P.12 Metatarsal, *Cervus* sp. (*E.* 601)	0.1	27	0.4	4
P.13 Molar, *Equus* sp. (*E.* 602)	0.4 (3)	25	1.6	2
P.14 Molar, *Castor fiber* (*E.* 603)	0.4	30	1.3	3
P.18 Femur, *Elephas* cf. *antiquus* (*E.* 615)	1.3 (3)	30	4.3	2
P.19 Indet. bone from basal clay (*E.* 616)	1.4	33	4.2	1
P.21 Incisor, *Castor fiber* (*E.* 618)	0.1	27	0.4	10
P.22 Mandible, *Castor fiber* (*E.* 619)	0.3	18	1.7	6
P.24 Fragment of enamel, *Elephas* sp. indet. (*E.* 621)	0.8 (3)	36	2.2	1
Holocene or Pleistocene				
P.36 Tibia, *Cervus* sp. (*E.* 1383)	<0.1	35	0.3	1
P.37 Caprine molar (*E.* 1384)	0.3 (3)	22	1.4	2
P.39 Bovine long-bone (*E.* 1385)	0.1 (3)	30	0.3	9
P.40 Indet. bone (sub-fossil) (*E.* 1386)	0.1	30	0.3	Trace
P.41 Indet. bone (sub-fossil) (*E.* 1387)	0.3	42	0.7	2
Holocene (Recent)				
P.33 Fragment of fresh bone from soil	<0.1	33	0.3	Trace
P.34 Pelvis, *Bos taurus*	<0.1	24	0.4	4
P.35 Metatarsal, *Bos taurus* (*E.* 1388)	<0.1	27	0.4	2
P.38 Ungual phalange, *Bos taurus* (*E.* 1389)	0.3	32	0.9	5

* Where more than one determination of fluorine content has been made indicated by the number in brackets, the value recorded is the average.

† The register numbers of specimens in the Department of Geology, British Museum (Nat. Hist.), are given in brackets after the description.

In attempting to interpret the analytical results, it is important to note that there is no significant difference in the rate of absorption of fluorine by bone and by dentine. The fluorine content of the mandibular ramus of *Eoanthropus*, for example, ranged from less than 0.1 to 0.3 percent, while that of a molar embedded in this jaw-bone was less than 0.1 percent. If there is differential absorption, the slight advantage is with bone. There are indications that enamel is more resistant to absorption of fluorine than dentine. The analyses of teeth, with two exceptions, were based on samples which were either wholly dentine, or which included a substantial proportion of dentine. The two exceptions were samples of enamel (*P.* 6, *P.* 24).

On the evidence of their state of preservation, the molar and premolar of *Hippopotamus* were placed by Hopwood (ref. 6, p. 49) with *Eoanthropus* in the Lower Pleistocene group. He noted, however, that *Hippopotamus* had never been recorded in a Red Crag association; so that on general grounds it would appear that these teeth more probably belong to the later group[10]. The molar, the fluorine content of which is closely comparable with that of *Eoanthropus*, shows unique preservation (*P.* 9). Whereas its enamel (0.1 percent fluorine) is practically unaltered, its dentine is stained blackish-brown throughout and contains 0.05 percent fluorine. An X-ray powder diffraction photograph showed that this blackened

	含氟量%	P_2O_5%	$\dfrac{\%F}{\%P_2O_5} \times 100$	铁含量%
可能为中更新世和上更新世				
P.9 臼齿，河马未定种 (E. 598)	0.1 (3)	37	0.3	3
P.10 前臼齿，河马未定种 (E. 599)	1.1 (3)	29	3.8	5
P.11 鹿角，马鹿 (E. 600)	1.5 (3)	28	5.4	3
P.12 距骨，鹿未定种 (E. 601)	0.1	27	0.4	4
P.13 臼齿，马未定种 (E. 602)	0.4 (3)	25	1.6	2
P.14 臼齿，欧亚河狸 (E. 603)	0.4	30	1.3	3
P.18 股骨，古象相似种 (E. 615)	1.3 (3)	30	4.3	2
P.19 底部黏土中的未确定的骨 (E. 616)	1.4	33	4.2	1
P.21 门齿，欧亚河狸 (E. 618)	0.1	27	0.4	10
P.22 下颌骨，欧亚河狸 (E. 619)	0.3	18	1.7	6
P.24 珐琅质破片，象不能鉴定种 (E. 621)	0.8 (3)	36	2.2	1
全新世或更新世				
P.36 胫骨，鹿未定种 (E. 1383)	<0.1	35	0.3	1
P.37 山羊臼齿 (E. 1384)	0.3 (3)	22	1.4	2
P.39 牛亚科长骨 (E. 1385)	0.1 (3)	30	0.3	9
P.40 未确定的骨（亚化石）(E. 1386)	0.1	30	0.3	痕量
P.41 未确定的骨（亚化石）(E. 1387)	0.3	42	0.7	2
全新世（近世）				
P.33 土壤中的新鲜骨破片	<0.1	33	0.3	痕量
P.34 骨盆，家牛	<0.1	24	0.4	4
P.35 距骨，家牛 (E. 1388)	<0.1	27	0.4	2
P.38 有蹄类指骨，家牛 (E. 1389)	0.3	32	0.9	5

* 这里对氟含量进行了不止一次的测定，括号中的数字表明测定次数，记录的数值是平均值。

† 描述后在括号中给出的是大英博物馆（自然分馆）地质学系登记的样本编号。

在解释分析结果的过程中，有一点很重要，那就是要注意到骨和牙本质对氟的吸收率并无显著差异。例如，曙人的下颌支的含氟量从不足 0.1% 到 0.3% 不等，而附着在这具下颌骨中的臼齿的含氟量却小于 0.1%。如果说吸收情况有差异的话，那么骨应该稍具优势。有迹象表明珐琅质比牙本质对氟吸收的抗性更强。除了两次例外，牙齿的分析都是基于具有完整牙本质的样本，或者包含牙本质重要部分的样本，两次例外情况是指分析的珐琅质样本（P. 6，P. 24）。

根据样本的保存状况提供的证据，霍普伍德（参考文献 6，第 49 页）将河马的臼齿、前臼齿与曙人一起放到了下更新世群中。但是他认为，河马从来没有在与红岩共生的记录中出现过，因此一般情况下这些牙齿更可能属于较晚的群[10]。该臼齿样本的含氟量与曙人的非常接近，表明其独特的保存情况（P. 9）。然而其珐琅质（含氟量为 0.1%）实际上并没有遭受蚀变，而其牙本质全部被染成了黑褐色，含氟量为 0.05%。X 射线粉晶衍射照片显示，这一变黑的牙本质由羟磷灰石构成，还可

dentine consists of hydroxyapatite with slight admixture, possibly, of calcium sulphate. Analysis indicated 7 percent iron. A probable interpretation of the specimen is that at some stage of fossilization the hydroxyapatite prisms in the dentine became coated by iron sulphide which inhibited fluorine absorption. The teeth of *Eoanthropus* are in striking contrast to this *Hippopotamus* tooth, and indeed to all the associated fossil animal teeth. Drill holes in the canine and in the molars of *Eoathropus* revealed—most unexpectedly—that below an extremely thin ferruginous surface stain their dentine was pure white, apparently no more altered than the dentine of recent teeth from the soil.

Comparison of the fluorine values of the specimens attributed to *Eoanthropus* and of the bones and teeth the geological ages of which are certain leaves little doubt that: (1) all the specimens of *Eoanthropus*, including the remains of the second skull found two miles away, are contemporaneous; (2) *Eoanthropus* is, at the earliest, Middle Pleistocene.

There can no longer be any question of *Eoanthropus* belonging to the Villafranchian group; but whether it is Middle Pleistocene or later is arguable. That the figures scarcely provide any differentiation between *Eoanthropus* and recent bones requires some explanation; but at least it serves to emphasize the probably enormous time-gap separating the former from the Lower Pleistocene material.

The wide range of fluorine content in the post-Villafranchian material (0.1–1.5 percent) is consistent with the suggestion that this age group is composite. Although none of these specimens is markedly water-worn, the gravels could have been reconstructed at several dates without the component materials travelling far. It is interesting to recall that Dawson (ref. 3, p. 86) considered that the remains of beaver (*Castor*) were the only fossils from the Piltdown pit which could be counted as contemporary with *Eoanthropus*. This is precisely the conclusion which one would draw from the fluorine results.

The eroded (and afterwards worked) fragment of elephant femur (*P.* 18, with 1.3 percent fluorine was not found *in situ*, but appears to have come from clay below the gravel, where it would surely have been a derivative from an older deposit. The red-deer antler (*P.* 11, with 1.5 percent fluorine) was found some distance from the skull site, and may have been preserved in a patch of gravel also representing an earlier phase of the Pleistocene. The fluorine results are, in fact, so consistent with the known or probable relative ages of the mammalian fossils in the Piltdown mélange, that it now appears justifiable to regard *Eoanthropus* and *Castor fiber* as the latest elements in the mixture, and to ascribe them to the period immediately preceding the final re-arrangement of the gravel, since which time free fluorine ions have apparently been remarkably deficient in the ground-water.

Geological Evidence: Conclusion

From the palaeontological data alone, it is not possible to decide whether the final settlement of the Piltdown gravel took place during Middle or early Upper Pleistocene

能混合了少许的硫酸钙。分析表明其铁含量为 7%。对该样本可能的解释是，石化作用的某一阶段，牙本质中的羟磷灰石棱晶表面覆盖上了一层硫化铁，这就阻止了牙本质对氟的吸收。曙人的牙齿与该河马牙齿形成了鲜明的对比，实际上它与所有相关的动物牙齿化石都截然不同。非常出乎意料地是，曙人犬齿和臼齿上的钻孔揭示出，在非常薄的一层铁锈斑点表面之下，它们的牙本质是洁白的，这显然与从土壤中发现的近代牙齿的牙本质完全一样。

对归于曙人的样本的氟值与地质学年代已确定的骨和牙齿的氟值进行的比较几乎可以确信：（1）包括两英里外发现的第二件头骨化石在内的所有曙人标本都是同一时代的；（2）曙人所处的年代最早是中更新世。

任何关于曙人是否属于维拉方群的疑问都不复存在了，但是它属于中更新世还是更晚年代则还在争论之中。那些无法提供曙人和最近骨之间存在任何差异的数值尚需解释，但它至少强调了把曙人与下更新世材料分开的可能是巨大的时间差。

后维拉方期材料中的氟含量变化范围很广（0.1%~1.5%），这与这一年代的群体是混合体的推测是一致的。尽管这些样本没有明显出现被水冲磨过的现象，但是这一砾石层可能在几个时期、在组成的材料并未搬运很远的情况下重新堆积的。有趣的是，道森（参考文献 3，第 86 页）认为河狸化石是皮尔当坑发现的唯一可以视为与曙人同时期的化石，而这也正是我们可以从氟含量的计算结果推导出来的结论。

被侵蚀的（后来被加工过的）大象股骨破片（*P.* 18，氟含量为 1.3%）并不是在原位发现的，而似乎是从砾石层之下的黏土中发现的，由此可以推断出，它肯定是一处更古老的堆积物的衍生物。马鹿鹿角（*P.* 11，氟含量为 1.5%）是在距离发现头骨的地点的不远处发现的，它可能在代表更新世较早阶段的一片砾石层中保存过。事实上，得到的氟含量的结果与皮尔当记录中的哺乳动物化石已知或可能的相对年代如此一致，因此现在将混合物中的曙人和欧亚河狸当作年代最晚的成分并且将它们归于砾石层发生最后一次沉积之前的那个时期是合理的，因为很明显，该时期地下水非常缺乏自由氟离子。

地质学证据：结论

单单依据古生物学资料就决定皮尔当砾石层的最终沉积是否发生在中更新世时

times.

It was pointed out by Clement Reid[11] that these gravels rest on a low plateau surface (100–120 ft. above O.D.) which was extensively developed in the Weald, but which was nowhere covered by marine deposits of the period of the submergence represented by the Goodwood raised beach (135 ft. above O.D.). In any event, it seems unlikely that this surface existed before the base levelling associated with the close of the great interglacial period. Furthermore, if the Piltdown gravel is viewed as a river terrace deposit, its position about fifty feet above the River Ouse[12] places it in a group grading with the Main Monastirian sea-level. From the temperate character of even the latest faunal elements in the Piltdown faunal mélange, it is probable that *Eoanthropus* lived under interglacial conditions, although the final resorting of the gravels may have been brought about by periglacial soil-flow (solifluxion). The question of the precise geological age of the Piltdown gravel is open to further inquiry, but taking the balance of available evidence, *Eoanthropus* may be provisionally referred to the last warm interglacial period (Riss–Würm interglacial; that is, early Upper Pleistocene, although here it should be noted that some authorities count Riss–Würm as Middle Pleistocene).

We wish to record our thanks to the Government Chemist, Dr. G. M. Bennett, and to the Keeper of Geology, British Museum, Mr. W. N. Edwards, for their co-operation, and for permission to publish the relevant portions of this communication.

(**165**, 379-382; 1950)

Kenneth P. Oakley: Department of Geology, British Museum (Natural History).
C. Randall Hoskins: Department of the Government Chemist.

References:

1. Dawson, C., and Woodward, A. S., *Quart. J. Geol. Soc. London*, **69**, 117 (1913).

2. Marston, A. T., *Geol. Assoc. London, Circular*, No. 483, 1 (1946).

3. Dawson, C., and Woodward, A. S., *Quart. J. Geol. Soc. London*, **70**, 82 (1914).

4. Keith, A., *Quart. J. Geol. Soc. London*, **69**, 148 (1913).

5. Newton, E. T., *Quart, J. Geol. Soc. London*, **69**, 151 (1913).

6. Hopwood, A. T., *Proc. Geol. Assoc. London*, **46**, 46 (1935).

7. Oakley, K. P., *Advancement of Science*, **4**, 336 (1948).

8. Willard, H. H., and Winter, O. B., *Indust. Eng. Chem. (Anal. Edit.)*, **5**, 7 (1933).

9. Oakley, K. P., and Montagu, M. F. A., *Bull. Brit. Mus. (Nat. Hist.), Geol.*, **1**, 2 (1949).

10. Curwen, E. C., "Archaeology of Sussex", 38 (London: Methuen, 1937).

11. Reid, C., *Quart. J. Geol. Soc. London*, **69**, 149 (1913).

12. Edmunds, F. H., *Abs. Proc. Geol. Soc. London*, No. 1457, 39 (1950).

期或者上更新世早期是不可能的。

克莱门特·里德[11] 指出这个砾石场覆盖在一片低的高原表面上（海拔 100 英尺～120 英尺），这一表面在威尔德是广泛形成的，但是该表面并没有被以古德伍德上升的海滩为代表的淹没时期的海洋性沉积物（海拔 135 英尺）所覆盖。无论如何，这一表面似乎都不可能存在于与大间冰期结束相关的基准面夷平之前。此外，如果皮尔当砾石层被视为一片河流阶地堆积物，那么它位于乌斯河[12] 之上 50 英尺处的位置就会将其放置于一个依最终间冰期极相期海平面逐级变化的群体了。即使从皮尔当混杂动物群中最年轻动物成员的生活温度特征来看，曙人都是可能生活在间冰期环境下的，尽管砾石层的最终形成可能是借助于冰川边缘的泥流（泥流作用）完成的。皮尔当砾石层的准确地质学年代问题还需要进一步的研究，但是考虑到已得到的证据，我们认为可以暂时将曙人归于最后的温暖间冰期（里斯–玉木间冰期，即上更新世初期，但是这里需要说明的一点是，有些专家将里斯–玉木间冰期当作中更新世）。

我们对政府化验师贝内特博士和大英博物馆地质学管理人爱德华兹先生表示感谢，感谢他们提供的合作以及允许我们将这部分信息的相关内容进行发表。

（刘皓芳 翻译；林圣龙 审稿）

Ape or Man?

R. Broom and J. T. Robinson

Editor's Note

The validation of Dart's Taung skull, *Australopithecus*, as representative of an intermediate stage between ape and human, rather than an ape, was largely due to Robert Broom and his fossil hominid discoveries starting in 1936. Here he and colleague Robinson summarize their work to date, on *Pleisanthropus* (now *Australopithecus*) and the more robust form *Paranthropus*. They admit that controversy still raged between those who saw the fossils as almost human, and those who regarded them more as apes, with the intermediate "ape-man" designation being seen by some as a kind of compromise. Broom and Robinson say that all those who had actually seen the fossils came away with the feeling that they had seen something intermediate between ape and man.

IN 1925, Prof. Dart announced the discovery of a new type of higher primate that seemed to be somewhat intermediate between ape and man. This was the skull of the Taungs child which he called *Australopithecus africanus*. For some years there was considerable dispute between those of us who regarded the little skull as that of a being closely allied to the ancestor of man, and those who considered it only a variety of ape, allied to the chimpanzee, which by parallelism had come to resemble man in many characters.

Since 1936 we have made a large number of new discoveries, and this group of higher primates is now known by many skulls and skeletal remains of adults of a considerable variety of forms which we think should be placed in different genera and species.

In the past two years we have discovered a number of nearly complete skulls which give us a new picture of the origin of man. When the only known skulls of our so-called "ape-men" had brains of between 450 and 650 c.c., those who said they were only apes and had no close relationship to man seemed to have some case. But now that we have skulls with brains of 750, 800 and possibly 1,000 c.c., we seem to be dealing with beings that have some claim to be called human.

Recently we have found five skulls with fair-sized brains. One is a child skull with a brain of perhaps 600 c.c., and one that of a child with a brain of about 700 or 750 c.c. Then we have two nearly complete female skulls with brain cases which we estimate at about 800 or 850 c.c. Each skull lacks the occiput, but is otherwise nearly complete although somewhat crushed. It will take some time to reconstruct an uncrushed skull; but we seem very safe in considering that it had a brain of more than 800 c.c.

是猿类还是人类？

布鲁姆，鲁滨逊

编者按

达特的汤恩头骨被确定为南方古猿——一种介于猿类与人类之间的中间类型的代表——而不是猿类，这在很大程度上要归因于罗伯特·布鲁姆的研究以及他从1936年开始发现的人科化石。本文中布鲁姆和他的同事鲁滨逊将之前关于迩人（现在为南方古猿）以及更为粗壮的傍人的工作进行了总结。一些学者认为这些化石接近于人类，而另一些学者则认为它们更像猿类，布鲁姆和鲁滨逊承认这种激烈的争论长期存在着，在一些人看来，中间类型"猿人"的命名是一种折中的办法。布鲁姆和鲁滨逊称所有那些实际看过化石的人都会有这样一种感觉：他们看到的是介于猿类和人类之间的某种中间类型。

1925年，达特教授宣布发现了一种新型高等灵长类动物，这种灵长类动物似乎是介于猿类与人类之间的某种中间类型。这便是他称为南方古猿非洲种的汤恩幼儿的头骨。多年以来人们对此一直存在着很大的争议：我们这部分人认为这个小头骨所属的生物类型与人类祖先具有密切的亲缘关系，而另外一些人则认为它只是猿类的一个变种，与黑猩猩有亲缘关系；而黑猩猩经过平行进化，在许多方面已具有与人类相似的特征。

自1936年以来，我们已经取得了大量的新发现。通过许多成年头骨和骨残骸，使人们认识了这类高等灵长类动物，这些成年头骨和骨残骸的类型多种多样，我们认为应该将其归为不同的属和种。

在过去的两年中，我们已经发现了大量基本完整的头骨，为我们描绘了一幅人类起源的新图景。当只知道我们所谓的"猿人"头骨的脑量在450毫升～650毫升之间时，那些说它们只是猿类并与人类没有任何密切亲缘关系的人似乎还有一些根据，但是现在我们发现了脑量达到750毫升、800毫升甚至可能为1,000毫升的头骨，那么我们正在讨论的生物似乎就是一些人声称的人类。

最近，我们发现了5件具有较大脑量的头骨。一件是年幼个体的头骨，其脑量约为600毫升，另一件年幼个体头骨的脑量约为700毫升或750毫升。之后我们得到了两件几乎完整的女性头骨，我们估计其脑量约为800毫升或850毫升。每件头骨都缺少枕骨，但除此以外，尽管稍微有些破碎基本上还是完整的。复原出一件未破碎的头骨尚需一些时间，但是我们可以肯定地认为其脑量大于800毫升。

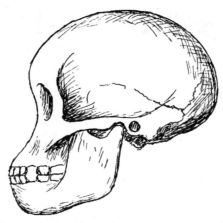

Fig. 1. Attempted restoration of seven-year-old male *Paranthropus crassidens* ($\frac{1}{3}$ natural size). The skull is almost complete, but the top is much crushed down on the base. The mandible is drawn from the jaw of another child. As restored the brain would be more than 700 c.c.

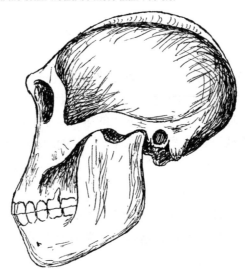

Fig. 2. Restoration of skull of female *Paranthropus crassidens* ($\frac{1}{3}$ natural size). The skull is satisfactorily known except the occiput, but is considerably crushed. The mandible is perfect, but is that of another individual. As restored the brain would be more than 800 c.c.

In structure the skulls are very interesting. The face is flat and broad. Above the eyes is a transverse supraorbital ridge; but it is more slender than in Pekin man, and a little like that of *Pithecanthropus*. There is practically no brow. The temporal muscles must have been relatively very large, and they passed up to the top of the skull as in male gorillas, and were separated by a bony sagittal crest rising about 12 mm. above the general surface of the skull. The ear region and the glenoid are almost typically human, and there is a large mastoid process and well-marked mastoid notch.

The palate and teeth are essentially human, and the incisors and canines would be accepted by most dentists as human teeth.

We have a perfect female mandible with all the teeth in perfect condition and a number

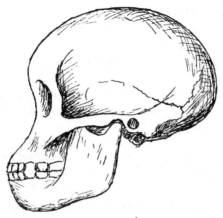

图 1. 对7岁男性傍人粗齿种的尝试复原图（真实尺寸的 $\frac{1}{3}$）。该头骨基本完整，但是颅底前部有破碎。下颌骨是根据另一个幼儿的颌骨画出来的。复原之后，该头骨脑量超过700毫升。

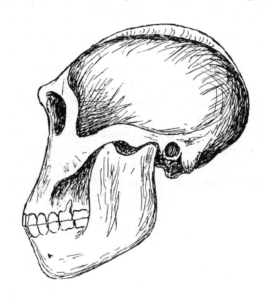

图 2. 女性傍人粗齿种头骨的复原图（真实尺寸的 $\frac{1}{3}$）。除枕骨外，人们对该头骨有了充分的认识，但是其破碎程度相当严重。下颌骨保存得很好，但属于另外一个个体。复原之后得到该头骨的脑量应超过800毫升。

这些头骨的结构非常引人注意。面部扁平而宽阔。眼上方是一个横向的眶上嵴，但是比北京人的更加纤细，与爪哇猿人的有点像。几乎没有眉毛。其颞肌肯定相对比较大，一直延伸到头骨的顶部，就像在雄性大猩猩中一样，高于头骨全表面约12毫米的一个矢状脊将颞肌分开。耳区和下颌关节盂几乎是典型的人类类型，并且有一个大的乳突和明显的乳突切迹。

腭和牙齿实质而言是人类的，大部分牙科专家认为切牙和犬齿也属于人类的牙齿。

我们有一件完整的女性下颌骨，所有牙齿都保存得很好，还有许多其他颌骨也

21

of other jaws in fairly good condition. One striking feature is that all the jaws have some indication of a chin, and one has a typical human chin with a well-marked mental prominence. If we did not have the whole symphysis, which is very thick, and the molar teeth, which are certainly those of *Paranthropus*, we might suspect this jaw to be that of a *Homo*.

Then we have a very fine pelvis. It has a large, wide ilium, practically human and not at all ape-like, with an ischium that is not human and which also differs considerably from that of the Sterkfontein ape-man *Plesianthropus*.

When Dart described the Taungs ape-man the large majority of men of science in the northern hemisphere were of opinion that it was an anthropoid ape allied to the chimpanzee and gorilla. One went so far as to call it a "dwarf gorilla". But soon one or two came to the conclusion that, if an anthropoid ape, it is the one known that is nearest to man. But most anatomists apparently took little interest in the skull.

In 1936 the senior author started a search for an adult skull, as the child skull did not seem to be understood; and a fairly good skull of an adult allied to the Taungs child was found on August 17 at Sterkfontein. This was made the type of *Plesianthropus transvaalensis*. Two years later another type of skull was found at Kromdraai, and named *Paranthropus robustus*. Some anatomists in Britain and America assumed it as probable that the two adult skulls were merely adults of the Taungs species. Then in November 1948 another new type was discovered at Swartkrans and called *Paranthropus crassidens*. Even now there are, we believe, many who consider that all our types are at most only species of *Australopithecus*, and a few still hold that they are all anthropoid apes. A considerable number of men of science from America and Europe have visited South Africa to examine our specimens, and we think all who have taken this trouble have gone back convinced that at least we have the remains of beings that were much more human in structure than any known living or fossil apes.

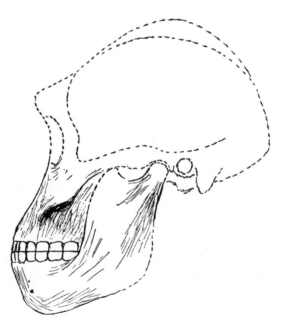

Fig. 3. Restoration of skull of male Swartkrans ape-man *Paranthropus crassidens* ($\frac{1}{3}$ natural size). Only lower jaw and face are known. As restored the brain would be about 1,000 c.c., and thus well within the human range.

保存得相当完好。所有颌骨都有一个显著的特征，即具有一些颏的迹象；其中一个具有典型的人颏特征，带有明显的颏凸。如果没有粗壮的联合部以及那些肯定属于傍人属的臼齿，我们可能会怀疑这个颌骨属于人属。

之后我们发现了一个很好的骨盆。它有一个大而宽的髂骨，该髂骨实际上是人类的而根本不类似猿类的；它还有一个坐骨，该坐骨不是人类的，与斯泰克方丹猿人——迩人的坐骨差异也很大。

当达特描述汤恩猿人时，北半球科学界的大多数人认为它是一种与黑猩猩和大猩猩具有亲缘关系的类人猿。有人甚至称它为"矮小大猩猩"。但是不久之后，有一两个人得出结论称，如果是类人猿的话，那么它就是当前已知的最接近人类的类人猿了。但是大多数解剖学家显然对这个头骨没有什么兴趣。

由于大家似乎并没有弄明白年幼个体的头骨，1936 年本文第一作者开始了对成年个体头骨的寻找；8 月 17 日在斯泰克方丹发现了一件保存相当好的与汤恩幼儿有亲缘关系的成年头骨。认为是德兰士瓦迩人。两年后在克罗姆德拉伊又发现了另一种类型的头骨，并将其命名为粗壮傍人。英国和美国的一些解剖学家认为这两个成年头骨可能仅仅是汤恩种类的成年个体。之后在 1948 年 11 月，在斯瓦特克朗斯发现了另外一种新类型，将其命名为傍人粗齿种。我们相信即使在现在，仍有许多人认为我们发现的所有类型至多不过是南方古猿的一些种类，还有一些人仍然认为它们都是类人猿。来自美国和欧洲科学界的许多人士都到访过南非来查验我们的标本，我们认为所有不辞辛苦这么做的人回去之后都至少会相信，我们拥有的生物化石在结构上比任何已知的现存猿类或化石猿类都更接近于人类。

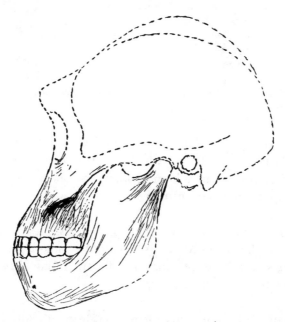

图 3. 男性斯瓦特克朗斯猿人——傍人粗齿种头骨的复原图（真实尺寸的 $\frac{1}{3}$）。只有下颌骨和面部是已知的。复原之后得到的该头骨脑量约为1,000 毫升，因而完全处于人类脑量的范围之内。

Opinions still differ considerably. Some have argued that all our ape-men are true human beings. Some—a very few, we believe—consider they are anthropoid apes. Probably most prefer to wait, and call them ape-men.

Plesianthropus has a female with a brain of about 500 c.c., and this seems to us a brain scarcely large enough to entitle it to human status—though it may be in the human line.

Paranthropus crassidens we now know fairly well. It has a brain of about 800–850 c.c. in the female, and perhaps more than 1,000 c.c. in the male. Even in the female it has a large bony sagittal crest. It has milk teeth differing so much from those of *Plesianthropus* and *Australopithecus* as to seem to confirm the correctness of placing it in a distinct genus. Its pelvis is also different in important details from those of both *Plesianthropus* and man.

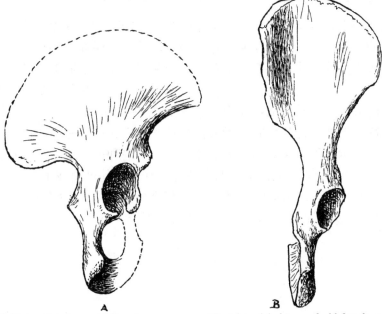

Fig. 4. (*A*) Right pelvic bones of Swartkrans ape-man; (*B*) right pelvic bones of old female orang. Both one-third natural size, and similarly orientated in plane of ilium.

The question now arises—must *Paranthropus crassidens*, with a brain well in the human range, be called an "early man"? It is clearly an ape-man that has by a mutation developed a large brain—possibly in structure a human brain, and one might argue that it is an early type of man—perhaps the ancestor of man.

Again, the teeth seem to solve our problem. *Australopithecus* and *Plesianthropus* have remarkable milk pre-molar teeth. Man has exactly the same type. But *Paranthropus* has a more primitive type, so that it seems more probable that man (*Homo*) has evolved from a *Plesianthropus*-like type than from a *Paranthropus*.

Possibly we are correct in assuming that there lived in South Africa a million or perhaps two million years ago a family of higher primates, not closely related to the living

24

目前观点仍存在很大的分歧。有些人认为，我们所有的猿人都是真正的人类。有一些人认为它们只是类人猿，我们相信这部分人只占极少数。可能大多数人更倾向于观望，并暂时称它们为猿人。

迩人标本中有一件女性头骨，其脑量约为 500 毫升，在我们看来，这样的脑量尚不足以称其为人类——尽管它可能处于人类世系中。

现在我们对傍人粗齿种已经了解得非常清楚了。其女性的脑量约为 800 毫升～850 毫升，男性可能大于 1,000 毫升。即使女性也有巨大的矢状脊。它的乳齿与迩人和南方古猿的差别如此之大，因此似乎可以确认将其归为一个不同的属是正确的。它的骨盆在很多重要细节上也与迩人和人类的不同。

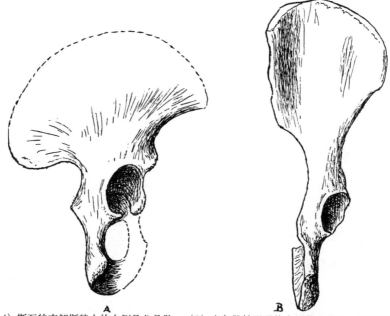

图 4. (A) 斯瓦特克朗斯猿人的右侧骨盆骨骼；(B) 老年雌性猩猩的右侧骨盆骨骼。这两幅图都为真实尺寸的 $\frac{1}{3}$，并且同样都是从髂骨平面看到的视图。

现在问题产生了——脑量正好处于人类范围内的傍人粗齿种一定要称为"早期人类"吗？很明显，猿人有一个由于发生突变而发育形成的脑——在结构上可能是人类的脑，甚至有些人可能会认为它是一种早期人类的类型——或许是人类的祖先。

那么，牙齿似乎能帮助我们解决问题。南方古猿和迩人具有显著的乳前白齿。人类也拥有相同类型的牙齿。但是傍人的牙齿类型更为原始，因此人类（人属）似乎更可能是从一种类似迩人的类型进化而来的，而不是从傍人进化而来的。

我们的以下推测可能是正确的：即一百万年或可能两百万年前在南非居住着高等灵长类动物家族，它们与现存的类人猿没有密切的亲缘关系，但有可能是从一种

25

anthropoids, but perhaps evolved from a very early anthropoid or even a pre-anthropoid by a different line, that this line early became bipedal and soon used the hands for tools and weapons, and that one branch of this family, about Upper Pliocene times, gave rise to man. There is no doubt the family varied greatly, and the safest conclusion to which we can at present come is that of the writer in *The Times* of July 28, who says: "It seems clear that the *Australopithecinae* as revealed up to now were 'almost man', and their presence in the Transvaal a million years ago strongly reinforces the possibility that something which we should recognize as man was first evolved in Africa".

(**166**, 843-844; 1950)

R. Broom and J. T. Robinson: Transvaal Museum, Pretoria.

非常早的类人猿或甚至是从一种不同世系的前类人动物进化而来的，这一世系很早就成为了两足动物并且不久就会用双手使用工具和武器，大概在上新世晚期这一家族的一个分支演化成了人类。毫无疑问，这个家族发生了巨大的变异，目前我们能够得出的最保险的结论正如 7 月 28 日《泰晤士报》上的一个作者所说的："似乎很清楚的是，目前为止的研究表明南方古猿亚科'几乎就是人类'，一百万年前它们在德兰士瓦的存在大大增强了这样的可能性；我们应该认定为人类的这一生物最早是从非洲进化而来的。"

（刘皓芳 翻译；冯兴无 审稿）

Evidence for the Pauling–Corey α-helix in Synthetic Polypeptides

W. Cochran and F. H. C. Crick

Editor's Note

In 1951 Linus Pauling and Robert Corey proposed that the molecular chains of proteins are commonly compacted into helical structures called α-helices. Here William Cochran and Francis Crick at the Cavendish Laboratory of Cambridge University provide evidence of these structures in a synthetic peptide (a kind of model protein), using X-ray crystallography. This and related work at the same time confirmed the hypothesis of Pauling and Corey. The α-helix is now known to be one of the key elements in the "secondary structure" of proteins, by which means the long polypeptide chain is folded into a compact, functional shape.

WE have calculated, in collaboration with Dr. V. Vand[1], the Fourier transform (or continuous structure factor) of an atom repeated at regular intervals on an infinite helix. The properties of the transform are such that it will usually be possible to predict the general character of X-ray scattering by any structure based on a regular succession of similar groups of atoms arranged in a helical manner. In particular, the type of X-ray diffraction picture given by the synthetic polypeptide poly-γ-methyl-L-glutamate, which has been prepared in a highly crystalline form by Dr. C. H. Bamford and his colleagues in the Research Laboratories, Courtaulds, Ltd., Maidenhead, is so readily explained on this basis as to leave little doubt that the Pauling–Corey α-helix[2], or some close approximation to it, exists in this polypeptide. Pauling and Corey[2] have already shown this correspondence in the equatorial plane; it is shown here that the correspondence extends over the whole of the diffraction pattern.

We quote here the value of the transform which applies when the axial distance between successive turns of the helix is P, the axial distance between the successive atoms lying on the helix is p, and the structure so formed is repeated exactly in an axial distance c. (For the latter condition to be possible, P/p must be expressible as the ratio of whole numbers.) In this case, the transform is restricted to planes in reciprocal space which are perpendicular to the axis of the helix, and occur at heights $\zeta = l/c$, where l is an integer. In crystallographic nomenclature, these are the layer lines corresponding to a unit cell of length c. On the lth such plane the transform has the value:

$$F\left(R,\psi,\frac{l}{c}\right) = f \sum_n J_n(2\pi Rr)\exp\left[in\left(\psi + \frac{\pi}{2}\right)\right]. \qquad (1)$$

(R, ψ, ζ) are the cylindrical co-ordinates of a point in reciprocal space, f is the atomic scattering factor, and J_n is the Bessel function of order n; r is the radius of the helix on which the set of atoms lies, the axes in real space being chosen so that one atom lies at

合成多肽中存在鲍林—科里α螺旋的证据

科克伦，克里克

编者按

1951年，莱纳斯·鲍林和罗伯特·科里提出蛋白质的分子链通常会形成被称为α螺旋的紧密的螺旋结构。在本文中，剑桥大学卡文迪什实验室的威廉·科克伦和弗朗西斯·克里克利用X射线晶体学方法，给出了这种结构在合成多肽（蛋白质的一种形式）中存在的证据。此项工作以及同一时期的相关工作证实了鲍林和科里的假说。现在α螺旋被认为是蛋白质"二级结构"的关键要素之一，通过这种方式，长的多肽链被折叠成了一个紧凑的、具有功能的形状。

通过与范特博士合作[1]，我们已经计算出在无限螺旋结构中按照固定间距重复排列的单个原子的傅里叶变换（或连续结构因子）。该变换的性质是：对于任意基于相似原子基团连续规则排列成螺旋形式的结构，傅里叶变换通常都可以推测出该结构的X射线散射结果所具有的总体特征。尤其是，合成多肽聚-γ-甲基-L-谷氨酸酯所给出的X射线衍射图样，在上述理论基础上可以很容易地得到解释，该多肽中无疑存在鲍林–科里α螺旋或者某种与之极为类似的结构，这种高度晶态的多肽由位于梅登黑德的考陶尔兹有限公司研究实验室的班福德博士及其同事合成。鲍林和科里[2]已经指出了在赤道平面上的这种对应关系；这里要说明的是，这种对应关系能够拓展到整个衍射图中。

在这个变换中，我们设位于螺旋中相邻两圈之间的轴距为 P，位于螺旋中相邻原子之间的轴距为 p，而这样形成的结构精确地按轴距 c 周期性重复排列。（为保证后一情况成立，P/p 必须用整数比来表示。）在这种情况下，变换只限于与螺旋轴垂直的倒易空间中的平面，而且出现在高度 $\zeta = l/c$ 处，其中 l 为整数。用晶体学术语来讲，它们是对应于长度为 c 的单位晶胞的层线。在第 l 个这种晶面上，变换的数值为：

$$F\left(R, \psi, \frac{l}{c}\right) = f \sum_n J_n(2\pi Rr) \exp\left[in\left(\psi + \frac{\pi}{2}\right)\right] \tag{1}$$

(R, ψ, ζ) 为倒易空间中一点的柱坐标，f 为原子的散射因子，J_n 为 n 阶贝塞尔函数；r 为原子组所在螺旋的半径，选取正空间中的轴使得一个原子位于 $(r, 0, 0)$。对于

29

(r, 0, 0). For a given value of l, the sum in equation (1) is to be taken over all integer values of n which are solutions of the equation,

$$\frac{n}{P} + \frac{m}{p} = \frac{l}{c},$$

(2)

m being any integer[1].

Thus only certain Bessel functions contribute to a particular layer line. This is illustrated in the accompanying table for the case of poly-γ-methyl-L-glutamate, for which Pauling and Corey[2] suggested $P = 5.4$ A., $p = 1.5$ A. and $c = 27$ A. The first column lists the number, l, of the layer line, while the second gives the orders (n) of the Bessel functions which contribute to it (for simplicity only the lowest two values of n are given for each layer line).

Value of l for the layer line	Lowest two values of n allowed by theory		Observed average strength of layer line(ref. 4)
0	0	±18	strong
1	−7	+11	
2	+4	−14	*weak
3	−3	+15	very weak
4	+8	−10	
5	+1	−17	medium
6	−6	+12	
7	+5	−13	
8	−2	+16	weak
9	±9		
10	+2	−16	weak
11	−5	+13	
12	+6	−12	
13	−1	+17	very weak
14	−8	+10	
15	+3	−15	
16	−4	+14	
17	+7	−11	
18	0	+18	medium
19	−7	+11	
20	+4	−14	
21	−3	+15	
22	+8	−10	
23	+1	−17	trace
24	−6	+12	
25	+5	−13	
26	−2	+16	trace
27	±9		
28	+2	−16	trace

Layers not described are absent.

*(1012), the reflexion having the smallest value of R, is absent.

给定的 l 值，方程（1）中的求和将取遍下列方程式中 n 的所有整数解：

$$\frac{n}{P} + \frac{m}{p} = \frac{l}{c} \qquad (2)$$

其中 m 为任意整数[1]。

因此对于某一特定层线而言，只有特定的贝塞尔函数有贡献。下表以聚-γ-甲基-L-谷氨酸酯为例对此进行了说明，鲍林和科里[2]认为 $P=5.4\,Å$，$p=1.5\,Å$，$c=27\,Å$。表中第一列为层线编号 l，第二列为对其有贡献的贝塞尔函数的阶数 n（为简明起见，对于每一层线只给出最小的两个 n 值）。

层线编号 l	理论范围内两个最小 n 值			层线的平均观测强度（参考文献 4）
0	0		±18	强
1	−7		+11	
2	+4		−14	* 弱
3	−3		+15	很弱
4	+8		−10	
5	+1		−17	中
6	−6		+12	
7	+5		−13	
8	−2		+16	弱
9		±9		
10	+2		−16	弱
11	−5		+13	
12	+6		−12	
13	−1		+17	很弱
14	−8		+10	
15	+3		−15	
16	−4		+14	
17	+7		−11	
18	0		+18	中等
19	−7		+11	
20	+4		−14	
21	−3		+15	
22	+8		−10	
23	+1		−17	痕量
24	−6		+12	
25	+5		−13	
26	−2		+16	痕量
27		±9		
28	+2		−16	痕量

没有描述的层是不出现的。

* 有最小 R 值的反射的（1012）层不出现。

Now there is, of course, more than one set of atoms in the polypeptide, but for all of them, P, p and c are the same, although r is different. The basis of our prediction is that a reflexion will be absent if the contribution of all sets of atoms to it is very small, and that on the average it will be strong if all sets of atoms make a large contribution.

It is a property of Bessel functions of higher order, illustrated in the graph, that they remain very small until a certain value of $2\pi Rr$ is reached, and that this point recedes from the origin as the order increases. Now, whatever the precise form of the chain, the value of r for any atom cannot be greater than about 8 A. because of the packing of the chains. This sets a limit to the value of $2\pi Rr$ within the part of the transform covered by the observed diffraction picture ($R < 0.3$ A.$^{-1}$ for $l \neq 0$). No set of atoms can make an appreciable contribution to the amplitude of a reflexion occurring on a layer line with which only high-order Bessel functions are associated, because $2\pi Rr$ comes within the very low part of the curve in the graph.

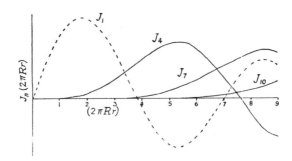

The march of higher-order Bessel functions(with J_1 added dashed)

We should therefore predict that layer lines to which only high-order Bessel functions contribute would be weak or absent, and that those to which very low orders contribute would be strong.

These predictions are strikingly borne out by the experimental data[4] summarized in the last column of the table. The significant Bessel functions involved in the first twenty-eight layer lines are shown in the second column, and, as will be seen, only layer lines associated with a function of order 4 or less are represented. This limiting value of 4 is less than might have been expected, and this fact suggests that the contribution of the side-chains to any reflexion is small, probably due to their large thermal motion.

In addition, the theory predicts (as can also be shown by a simpler approach) that meridional reflexions can occur only on layer lines which involve Bessel functions of order zero; that is, at reciprocal spacings of multiples of $1/1.5$ A.$^{-1}$. This had previously been pointed out by Perutz[3] when reporting the strong meridional 1.5 A. reflexion.

We have therefore no doubt that the structure of poly-γ-methyl-L-glutamate is based on a helix of eighteen residues in five turns and 27 A., or a helix which approximates to this

当然，由于多肽分子中有不止一组原子，就总体而言虽然它们的 r 是不同的，但 P、p 和 c 都是相同的。我们所做推测的基础是，如果所有原子组对于反射的贡献都很小，反射就不会出现，一般情况下，若所有原子组都有明显贡献则反射会很强。

图中呈现出较高阶贝塞尔函数的一个性质，即它们在达到一个特定的 $2\pi Rr$ 值之前其数值一直很小，且该点会随着阶数增大而远离原点。无论链以何种精确的形式存在，任何原子的 r 值都会因链的堆积而无法超过 8 Å 左右。$2\pi Rr$ 的值在变换部分内被限定在一个区间内，这一部分涵盖于观测到的衍射图像之中（$R<0.3$ Å^{-1}，$l \neq 0$）。对于只与高阶贝塞尔函数有关的层线上发生的反射，所有原子组都不能对其强度做出显著的贡献，因为 $2\pi Rr$ 处于图中曲线较低的那部分。

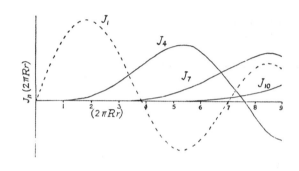

较高阶贝塞尔函数的演变（J_1 为虚线）

因此我们可以推测，那些只由高阶贝塞尔函数贡献的层线将会很弱甚至不出现，而那些由低阶贝塞尔函数贡献的层线则会很强。

由表中最后一列汇总的实验数据[4]可知，上述推测得到了惊人的证实。第二列数据为与前 28 条层线有关的有效贝塞尔函数，并且正如我们将要看到的那样，只有与阶数小于或等于 4 的函数有关的层线才会被呈现出来。4 这个极限值可能比预期值要小，这就意味着侧链对于任何反射的贡献都较小，这可能归因于它们大规模的热运动。

此外根据理论推测（通过一个较简便方法也能表明）只有在与零阶贝塞尔函数有关的层线上，即处于倍数为 1/1.5 Å^{-1} 的倒易空间时，才有可能产生子午线反射。佩鲁茨[3]在报告子午线 1.5 Å 的强反射时曾指出过这一点。

因此我们确信聚-γ-甲基-L-谷氨酸酯的结构是基于一条由 18 个残基旋转 5 圈而形成的轴距为 27 Å 的螺旋，或者是非常近似于该结构的一条螺旋。由于鲍林和科

very closely. As the structure proposed by Pauling and Corey[2] satisfies these conditions and is also stereochemically very satisfactory, it seems to us highly probable that it is correct.

We should like to thank Dr. Bamford and his colleagues for allowing us to quote their experimental results in advance of publication, and Sir Lawrence Bragg and Dr. M. Perutz for the stimulus which their interest in this work has provided.

(**169**, 234-235; 1952)

W. Cochran: Crystallographic Laboratory, Cavendish Laboratory, Cambridge.

F. H. C. Crick: Medical Research Council Unit for the Study of the Molecular Structure of Biological Systems, Cavendish Laboratory, Cambridge, Dec. 14.

References:

1. Cochran, W., Crick, F. H. C., and Vand. V. (to be published).

2. Pauling, L., and Corey, R. B., *Proc. U.S. Nat. Acad. Sci.*, 37, 241 (1951).

3. Perutz, M. F., *Nature*, 167, 1053 (1951).

4. Bamford, C. H., Brown, L., Elliott, A., Hanby, W. E., and Trotter, I. F. (to be published).

里 [2] 提出的结构满足这些条件，并且从立体化学的角度看也是令人满意的，因此我们认为这个结论极有可能是正确的。

班福德博士及其同事们准许我们引用他们尚未发表的实验结果，劳伦斯·布拉格爵士和佩鲁茨博士对这项研究的关注给我们提供了动力，我们对此表示感谢。

（王耀杨 翻译；顾孝诚 审稿）

A Structure for Deoxyribose Nucleic Acid

J. D. Watson and F. H. C. Crick

Editor's Note

This short paper is probably the most famous Nature has ever published. By the early 1950s it was becoming clear that the genes responsible for inherited traits reside on the biochemical polymer DNA. Watson and Crick used measurements from X-ray crystallography, along with chemical reasoning, to propose a molecular structure for DNA in which two polymer strands are intertwined in a double helix. The result was striking not just for its beauty and elegance but because, as the researchers laconically say, "the specific pairing we have postulated immediately suggests a possible copying mechanism for the genetic material"—that is, a means by which DNA is replicated. This marked the beginning of modern genomics.

WE wish to suggest a structure for the salt of deoxyribose nucleic acid (D.N.A.). This structure has novel features which are of considerable biological interest.

A structure for nucleic acid has already been proposed by Pauling and Corey[1]. They kindly made their manuscript available to us in advance of publication. Their model consists of three intertwined chains, with the phosphates near the fibre axis, and the bases on the outside. In our opinion, this structure is unsatisfactory for two reasons: (1) We believe that the material which gives the X-ray diagrams is the salt, not the free acid. Without the acidic hydrogen atoms it is not clear what forces would hold the structure together, especially as the negatively charged phosphates near the axis will repel each other. (2) Some of the van der Waals distances appear to be too small.

Another three-chain structure has also been suggested by Fraser (in the press). In his model the phosphates are on the outside and the bases on the inside, linked together by hydrogen bonds. This structure as described is rather ill-defined, and for this reason we shall not comment on it.

We wish to put forward a radically different structure for the salt of deoxyribose nucleic acid. This structure has two helical chains each coiled round the same axis (see diagram). We have made the usual chemical assumptions, namely, that each chain consists of phosphate diester groups joining β-D-deoxyribofuranose residues with 3′, 5′ linkages. The two chains (but not their bases) are related by a dyad perpendicular to the fibre axis. Both chains follow right-handed helices, but owing to the dyad the sequences of the atoms in the two chains run in opposite directions. Each chain loosely resembles Furberg's[2] model No. 1; that is, the bases are on the inside of the helix and the phosphates on the outside. The configuration of the sugar and the atoms near it is close to Furberg's "standard

脱氧核糖核酸的结构

沃森，克里克

编者按

这篇简短的文章可能是《自然》有史以来发表的最著名的文章。在 20 世纪 50 年代初期，人们越来越清楚地认识到，具有遗传特征的基因存在于生化多聚体 DNA 上。沃森和克里克利用 X 射线晶体学的测量结果和化学推理，提出了 DNA 的一种分子结构模型，即 DNA 分子是由两条聚合链缠绕而成的双螺旋结构。这一结果之所以令人震惊，不只是因为双螺旋结构的优雅和美丽，而且正如研究者所指出的，"我们提出的这种特定配对原则立即揭示了遗传物质一种可能的复制机制。"——也就是说，DNA 复制的一种方式。它标志着现代基因组学的开端。

我们希望提出脱氧核糖核酸（DNA）盐的一种结构。这种结构的新特性具有重要的生物学意义。

鲍林和科里已经提出了核酸的一种结构 [1]。他们在原稿发表之前慷慨地将其提供给我们。他们的模型由三条彼此缠绕的链组成，磷酸基团位于长链的中心轴附近，而碱基则位于外侧。在我们看来，该结构不太令人满意，理由有两点：（1）我们认为，给出 X 射线图的物质是盐而不是游离酸。在没有酸性氢原子的情况下，使该结构维系在一起的力并不明确，尤其是中心轴附近带负电荷的磷酸基团会彼此排斥。（2）某些范德华距离显得过小。

弗雷泽提出了另外一种三链式结构（即将发表）。在他的模型中，磷酸基团位于外侧而碱基位于内侧，通过氢键连接在一起。他对这种结构的描述非常不清楚，因此我们不会对其作任何评论。

我们要提出的是一种完全不同的脱氧核糖核酸盐的结构。这种结构有两条螺旋链，围绕同一中心轴相互缠绕（见示意图）。我们采用了一般的化学假设，即每条链由磷酸二酯基团组成，这些基团通过 3′、5′ 连接与 β–D– 脱氧呋喃核糖残基连接起来。两条链（而不是它们的碱基）通过一个与中心轴垂直的二分体相连接。两条链都是右手螺旋，不过由于成对出现，两条链上的原子顺序方向相反。各条链大致类似于富尔贝里 [2] 的一号模型，也就是说，碱基位于螺旋的内侧，而磷酸基团位于外侧。糖及其邻近原子的构型类似于富尔贝里的"标准构型"，糖分子大致上垂直于与之相连的碱基。每条链在 z 轴方向上，每隔 3.4 Å 就有一个残基。我们假定同一链上的

configuration", the sugar being roughly perpendicular to the attached base. There is a residue on each chain every 3.4 A. in the z-direction. We have assumed an angle of 36° between adjacent residues in the same chain, so that the structure repeats after 10 residues on each chain, that is, after 34 A. The distance of a phosphorus atom from the fibre axis is 10 A. As the phosphates are on the outside, cations have easy access to them.

The structure is an open one, and its water content is rather high. At lower water contents we would expect the bases to tilt so that the structure could become more compact.

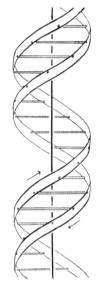

This figure is purely diagrammatic. The two ribbons symbolize the two phosphate-sugar chains, and the horizontal rods the pairs of bases holding the chains together. The vertical line marks the fibre axis.

The novel feature of the structure is the manner in which the two chains are held together by the purine and pyrimidine bases. The planes of the bases are perpendicular to the fibre axis. They are joined together in pairs, a single base from one chain being hydrogen-bonded to a single base from the other chain, so that the two lie side by side with identical z-co-ordinates. One of the pair must be a purine and the other a pyrimidine for bonding to occur. The hydrogen bonds are made as follows: purine position 1 to pyrimidine position 1; purine position 6 to pyrimidine position 6.

If it is assumed that the bases only occur in the structure in the most plausible tautomeric forms (that is, with the keto rather than the enol configurations) it is found that only specific pairs of bases can bond together. These pairs are: adenine (purine) with thymine (pyrimidine), and guanine (purine) with cytosine (pyrimidine).

In other words, if an adenine forms one member of a pair, on either chain, then on these assumptions the other member must be thymine; similarly for guanine and cytosine. The sequence of bases on a single chain does not appear to be restricted in any way. However, if only specific pairs of bases can be formed, it follows that if the sequence of bases on one chain is given, then the sequence on the other chain is automatically determined.

It has been found experimentally[3,4] that the ratio of the amounts of adenine to thymine, and the ratio of guanine to cytosine, are always very close to unity for deoxyribose nucleic acid.

It is probably impossible to build this structure with a ribose sugar in place of the deoxyribose, as the extra oxygen atom would make too close a van der Waals contact.

The previously published X-ray data[5,6] on deoxyribose nucleic acid are insufficient for a rigorous test of our structure. So far as we can tell, it is roughly compatible with the

相邻残基间的夹角为 36°，因此每条链上每 10 个残基，即 34 Å，就出现一次螺旋结构重复。磷原子与中心轴之间的距离为 10 Å。由于磷酸基团位于外侧，阳离子很容易接近它们。

这种结构尚未完全确定，其含水量相当高。在含水量低时，我们预计碱基会发生倾斜从而结构变得更加紧凑。

这种结构的新颖之处在于两条链通过嘌呤与嘧啶碱基连接在一起。碱基平面垂直于中心轴。碱基成对连接在一起，一条链上的一个碱基与另一条链上的一个碱基通过氢键结合，因此两个并排的碱基具有相同的 z 轴坐标。要产生这种氢键，其中一个碱基必须是嘌呤，另一个则必须是嘧啶。嘌呤的 1 位和嘧啶的 1 位之间以及嘌呤的 6 位和嘧啶的 6 位之间形成氢键。

如果假定碱基处于互变异构结构中最可能的构型（也就是说，处于酮式而不是烯醇式构型），可以发现，只有特定的碱基对可以结合在一起。这些碱基对是：腺嘌呤（嘌呤）与胸腺嘧啶（嘧啶），鸟嘌呤（嘌呤）与胞嘧啶（嘧啶）。

这仅仅是个示意图。两条丝带代表两条磷酸–糖链，而横线代表将链连接起来的碱基对。垂直线标示出中心轴。

换句话说，根据上述假定，如果一个碱基对中某一链上是一个腺嘌呤，那么另一链上与之对应的必定是一个胸腺嘧啶；鸟嘌呤和胞嘧啶也是类似情况。一条单链上的碱基顺序看来是不受任何限制的，但是，如果只有特定的碱基之间能够形成碱基对，那么如果给定某条链上碱基的顺序，另一条链的碱基顺序就自然确定了。

已经通过实验发现 [3,4]，对于脱氧核糖核酸而言，腺嘌呤与胸腺嘧啶数量的比值以及鸟嘌呤和胞嘧啶数量的比值总是非常接近 1 的。

如果用核糖代替脱氧核糖，很可能无法建立这种结构，因为多出来的氧原子将导致范德华接触过于紧密。

以前发表的关于脱氧核糖核酸的 X 射线数据 [5,6] 还不足以对我们的结构进行严格的检验。就我们所知，我们的结构与实验数据粗略吻合，不过在更多精确的结果

experimental data, but it must be regarded as unproved until it has been checked against more exact results. Some of these are given in the following communications. We were not aware of the details of the results presented there when we devised our structure, which rests mainly though not entirely on published experimental data and stereochemical arguments.

It has not escaped our notice that the specific pairing we have postulated immediately suggests a possible copying mechanism for the genetic material.

Full details of the structure, including the conditions assumed in building it, together with a set of co-ordinates for the atoms, will be published elsewhere.

We are much indebted to Dr. Jerry Donohue for constant advice and criticism, especially on interatomic distances. We have also been stimulated by a knowledge of the general nature of the unpublished experimental results and ideas of Dr. M. H. F. Wilkins, Dr. R. E. Franklin and their co-workers at King's College, London. One of us (J. D. W.) has been aided by a fellowship from the National Foundation for Infantile Paralysis.

(**171**, 737-738; 1953)

J. D. Watson and F. H. C. Crick: Medical Research Council Unit for the Study of the Molecular Structure of Biological Systems, Cavendish Laboratory, Cambridge, April 2.

References:

1. Pauling, L., and Corey, R. B., *Nature*, 171, 346 (1953); *Proc. U.S. Nat. Acad. Sci.*, 39, 84 (1953).

2. Furberg, S., *Acta Chem. Scand.*, 6, 634 (1952).

3. Chargaff, E., for references see Zamenhof, S., Brawerman, G., and Chargaff, E., *Biochim. et Biophys. Acta*, 9, 402 (1952).

4. Wyatt. G. R., *J. Gen. Physiol.*, 36, 201 (1952).

5. Astbury, W. T., Symp. Soc. Exp. Biol. 1, Nucleic Acid, 66 (Camb. Univ. Press, 1947).

6. Wilkins, M. H. F., and Randall, J. T., *Biochim. et Biophys. Acta*, 10, 192 (1953).

核查之前，我们还必须把它视为未经证实的。一些核查工作将在后面的通讯文章中给出。在我们设计结构时，我们还不清楚该工作结果的细节，我们的结构主要（但不是完全）依赖于已发表的实验数据和立体化学证据。

我们还注意到，我们提出的这种特定配对原则立即揭示了遗传物质一种可能的复制机制。

该结构的完整细节，包括建立此结构所用的假设条件以及一套原子坐标，将在其他地方发表。

我们非常感谢杰里·多诺霍博士长期以来的建议和批评，尤其是关于原子间距离的问题。我们还从伦敦国王学院的威尔金斯博士、富兰克林博士及其同事们未发表的实验结果与观点所包含的一般属性的知识中获得了启发。我们中的一员（沃森）由国家小儿麻痹症基金会资助。

（王耀杨 翻译；王晓晨 审稿）

Molecular Structure of Deoxypentose Nucleic Acids

M. H. F. Wilkins *et al.*

Editor's Note

James Watson and Francis Crick's famous paper describing the structure of DNA was accompanied by this contribution from their collaborator Maurice Wilkins and his coworkers at King's College in London. It describes the evidence from X-ray diffraction that the DNA molecule has a helical structure, a crucial aspect of the picture presented by Watson and Crick. Indeed, the researchers mention the "reasonable agreement" with that model. The analysis in this paper draws on the experimental work of Rosalind Franklin and Raymond Gosling, described in the following paper.

WHILE the biological properties of deoxypentose nucleic acid suggest a molecular structure containing great complexity, X-ray diffraction studies described here (cf. Astbury[1]) show the basic molecular configuration has great simplicity. The purpose of this communication is to describe, in a preliminary way, some of the experimental evidence for the polynucleotide chain configuration being helical, and existing in this form when in the natural state. A fuller account of the work will be published shortly.

The structure of deoxypentose nucleic acid is the same in all species (although the nitrogen base ratios alter considerably) in nucleoprotein, extracted or in cells, and in purified nucleate. The same linear group of polynucleotide chains may pack together parallel in different ways to give crystalline[1-3], semi-crystalline or paracrystalline material. In all cases the X-ray diffraction photograph consists of two regions, one determined largely by the regular spacing of nucleotides along the chain, and the other by the longer spacings of the chain configuration. The sequence of different nitrogen bases along the chain is not made visible.

Oriented paracrystalline deoxypentose nucleic acid ("structure *B*" in the following communication by Franklin and Gosling) gives a fibre diagram as shown in Fig. 1 (cf. ref. 4). Astbury suggested that the strong 3.4-A. reflexion corresponded to the internucleotide repeat along the fibre axis. The ~34 A. layer lines, however, are not due to a repeat of a polynucleotide composition, but to the chain configuration repeat, which causes strong diffraction as the nucleotide chains have higher density than the interstitial water. The absence of reflexions on or near the meridian immediately suggests a helical structure with axis parallel to fibre length.

脱氧戊糖核酸的分子结构

威尔金斯等

编者按

詹姆斯·沃森和弗朗西斯·克里克在他们所发表的著名文章中描述了 DNA 的结构，这一科研成果也有他们的合作者——伦敦国王学院的莫里斯·威尔金斯及其同事们的贡献。这篇文章描述了由 X 射线衍射所得到的证据，即 DNA 分子的螺旋结构——这是由沃森和克里克描绘的图景中的重要方面。的确，这些研究者们提到其结构和螺旋模型具有"合理的一致性"。这篇文章中的分析部分是依据罗莎琳德·富兰克林和雷蒙德·戈斯林的实验工作得到的，这将在下一篇文章中提到。

尽管脱氧戊糖核酸的生物学性质暗示其分子结构极为复杂，但这里所描述的 X 射线衍射研究（参见阿斯特伯里的研究[1]）却表明其基本分子构型非常简单。这篇通讯文章的目的是为了初步地描述一些实验证据，证明多核苷酸链的构型为螺形，并且以这种形式存在于自然状态中。对这一工作更完整的记述不久即将发表。

不管是在提取的或者细胞内的核蛋白中，还是处于纯化的核酸盐状态下，所有物种中的脱氧戊糖核酸都具有相同的结构（尽管含氮碱基的比例有显著差异）。一组相同的线性的多核苷酸链能够以不同方式平行排列在一起，形成晶态[1-3]、半晶态或类晶态物质。在所有情况下，X 射线衍射照片都包含两个区域，其中一个区域很大程度上取决于沿着这条链排列规则的核苷酸间距，另一个区域则取决于长链构型的更大排列间距。沿着链延伸方向，不同含氮碱基的排列顺序目前还无法观测到。

图 1（参见参考文献 4）显示的是脱氧戊糖核酸的有序类晶态（富兰克林与戈斯林在后面的通讯文章中所说的"B 型结构"）所给出的纤维衍射图。阿斯特伯里提出，强烈的 3.4Å 反射对应于沿纤维轴方向核苷酸之间的重复间距。但是，约为 34Å 的层线不是由于多核苷酸组分的重复，而是对应于链状构型的重复，正是这种链状构型的重复造成了强烈衍射，因为核苷酸链具有比间隙水更高的密度。在子午线及其附近没有出现反射，直接说明了脱氧戊糖核酸是螺旋轴平行于纤维轴方向的螺旋结构。

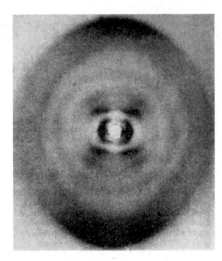

Fig. 1 . Fibre diagram of deoxypentose nucleic acid from *B. coli* Fibre axis vertical

Diffraction by Helices

It may be shown[5] (also Stokes, unpublished) that the intensity distribution in the diffraction pattern of a series of points equally spaced along a helix is given by the squares of Bessel functions. A uniform continuous helix gives a series of layer lines of spacing corresponding to the helix pitch, the intensity distribution along the nth layer line being proportional to the square of J_n, the nth order Bessel function. A straight line may be drawn approximately through the innermost maxima of each Bessel function and the origin. The angle this line makes with the equator is roughly equal to the angle between an element of the helix and the helix axis. If a unit repeats n times along the helix there will be a meridional reflexion (J_0^2) on the nth layer line. The helical configuration produces side-bands on this fundamental frequency, the effect[5] being to reproduce the intensity distribution about the origin around the new origin, on the nth layer line, corresponding to C in Fig. 2.

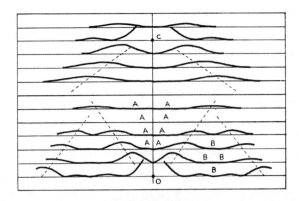

Fig. 2. Diffraction pattern of system of helices corresponding to structure of deoxypentose nucleic acid. The squares of Bessel functions are plotted about 0 on the equator and on the first, second, third and fifth layer lines for half of the nucleotide mass at 20 A. diameter and remainder distributed along a radius, the mass at a given radius being proportional to the radius. About C on the tenth layer line similar functions are plotted for an outer diameter of 12 A.

44

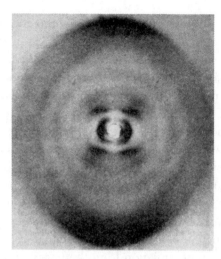

图 1. 大肠杆菌的脱氧戊糖核酸的纤维图纤维轴垂直于纸面

由螺旋产生的衍射

有文章指出 [5] 沿螺旋等间距排列的一系列点所产生的衍射图案，其强度分布可以由贝塞尔函数的平方给出（斯托克斯也有此观点但未发表）。一条均一连续的螺旋会给出一系列层线，层线间距对应于螺旋的螺距。第 n 层线的强度分布正比于 J_n 的平方，J_n 为 n 阶贝塞尔函数。通过每个贝塞尔函数最内部的极大值点和原点可以近似地画出一条直线。这条直线与赤道的夹角近似等于螺旋的一个元件与螺旋轴的夹角。如果一个单元沿着螺旋重复 n 次，就会在第 n 层线出现一个子午反射（J_0^2）。螺旋构型在此基频上则会产生边频带，其效应 [5] 是在第 n 层线上围绕新原点复制出原点周围的强度分布，对应于图 2 中的 C。

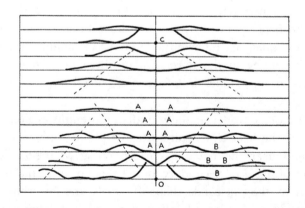

图 2. 对应于脱氧戊糖核酸结构的螺旋体系的衍射图案。贝塞尔函数的平方值在赤道线上围绕 0 作图，而在第 1、2、3 和 5 层线上按照直径 20 Å 的核苷酸质量的一半计算，其余部分则沿径向分布，因为在一给定半径下的质量均与该半径成正比。第 10 层线的 C 点附近的函数图是根据 12 Å 外径的核苷酸画出的。

We will now briefly analyse in physical terms some of the effects of the shape and size of the repeat unit or nucleotide on the diffraction pattern. First, if the nucleotide consists of a unit having circular symmetry about an axis parallel to the helix axis, the whole diffraction pattern is modified by the form factor of the nucleotide. Second, if the nucleotide consists of a series of points on a radius at right-angles to the helix axis, the phases of radiation scattered by the helices of different diameter passing through each point are the same. Summation of the corresponding Bessel functions gives reinforcement for the innermost maxima and, in general, owing to phase difference, cancellation of all other maxima. Such a system of helices (corresponding to a spiral staircase with the core removed) diffracts mainly over a limited angular range, behaving, in fact, like a periodic arrangement of flat plates inclined at a fixed angle to the axis. Third, if the nucleotide is extended as an arc of a circle in a plane at right-angles to the helix axis, and with centre at the axis, the intensity of the system of Bessel function layer-line streaks emanating from the origin is modified owing to the phase differences of radiation from the helices drawn through each point on the nucleotide. The form factor is that of the series of points in which the helices intersect a plane drawn through the helix axis. This part of the diffraction pattern is then repeated as a whole with origin at C (Fig. 2). Hence this aspect of nucleotide shape affects the central and peripheral regions of each layer line differently.

Interpretation of the X-Ray Photograph

It must first be decided whether the structure consists of essentially one helix giving an intensity distribution along the layer lines corresponding to J_1, J_2, J_3 ..., or two similar co-axial helices of twice the above size and relatively displaced along the axis a distance equal to half the pitch giving J_2, J_4, J_6 ..., or three helices, etc. Examination of the width of the layer-line streaks suggests the intensities correspond more closely to J_1^2, J_2^2, J_3^2 than to J_2^2, J_4^2, J_6^2 ... Hence the dominant helix has a pitch of ~34 A., and, from the angle of the helix, its diameter is found to be ~20 A. The strong equatorial reflexion at ~17 A. suggests that the helices have a maximum diameter of ~20 A. and are hexagonally packed with little interpenetration. Apart from the width of the Bessel function streaks, the possibility of the helices having twice the above dimensions is also made unlikely by the absence of an equatorial reflexion at ~34 A. To obtain a reasonable number of nucleotides per unit volume in the fibre, two or three intertwined coaxial helices are required, there being ten nucleotides on one turn of each helix.

The absence of reflexions on or near the meridian (an empty region AAA on Fig. 2) is a direct consequence of the helical structure. On the photograph there is also a relatively empty region on and near the equator, corresponding to region BBB on Fig. 2. As discussed above, this absence of secondary Bessel function maxima can be produced by a radial distribution of the nucleotide shape. To make the layer-line streaks sufficiently narrow, it is necessary to place a large fraction of the nucleotide mass at ~20 A. diameter. In Fig. 2 the squares of Bessel functions are plotted for half the mass at 20 A. diameter, and the rest distributed along a radius, the mass at a given radius being proportional to the radius.

现在我们将用物理学概念简要分析一下重复单元或核苷酸的形状与大小对衍射图案的一些影响。第一，如果核苷酸由具有圆形对称性的单元组成，且其对称轴平行于螺旋的中心轴，那么整个衍射图案就会受到核苷酸的形状因子的影响。第二，如果核苷酸是由一系列点组成，且这些点位于与螺旋轴成直角的半径方向上，那么通过每个点的不同直径的螺旋所散射出来的辐射的相位角就是相同的。相应的贝塞尔函数的加和就会使最内部的极大值获得增强，一般来说，还会由于相位差而抵消所有其他极大值。这种螺旋体系（对应于去除了中心部分的螺旋式楼梯）主要在有限的角度范围内产生衍射，这样的体系实际上很类似于与中心轴有固定夹角的平板的周期性排列。第三，如果核苷酸在与螺旋中心轴垂直的平面内伸展成一段圆弧，且其圆心位于轴上，那么，从原点发射出的贝塞尔函数层线条纹系统的强度就会由于通过核苷酸上每一点的各条螺旋所产生的辐射具有相位差而改变。形状因子与螺旋线和通过螺旋轴的平面相交所得的一系列点有关。这部分衍射图案将会以 C 为原点（图 2）整体重复。因此，核苷酸形状的这一因素对于每一层线的中心和外围区域会产生不同的影响。

对 X 射线照片的解释

首先必须确定，该结构本质上是由一条螺旋组成而使得沿层线的强度分布与 J_1、J_2、J_3⋯相对应，还是由两条类似的同轴螺旋组成，其大小为上述单条螺旋的两倍，相应地沿着螺旋轴每半个螺距间隔处的强度分布变成了 J_2、J_4、J_6⋯，或者甚至是由三条螺旋组成，诸如此类。对层线条纹宽度的检测表明，强度分布更接近 J_1^2、J_2^2、J_3^2，而不是 J_2^2、J_4^2、J_6^2⋯。因此，绝大部分螺旋的螺距大约是 34 Å，并且通过螺旋角可以得到其直径大约是 20 Å。约 17 Å 处的强烈的赤道反射表明，螺旋的最大直径约为 20 Å，并为六方堆积且很少相互穿插。除了贝塞尔函数条纹的宽度这一证据之外，在约 34 Å 处没有赤道反射出现也能说明尺寸上两倍于单条螺旋是不太可能的。但要使螺旋纤维中每单位体积里核苷酸的数目比较合理，又需要两条或三条相互缠绕成同轴螺旋，因为单条螺旋每一周有 10 个核苷酸。

这种螺旋结构的直接结果就是在子午线处及其附近没有反射（图 2 中 AAA 处的空白区域）。在照片上，赤道处及其附近也有一个相对空白的区域，对应于图 2 中的 BBB 区域。如同上面讨论过的，核苷酸放射状分布的外形可以导致次级贝塞尔函数极大值的缺失。要使层线条纹变得足够窄，就必须把核苷酸的大部分质量置于直径为 20 Å 处。图 2 中贝塞尔函数的平方是根据直径为 20 Å 处的核苷酸质量的一半来作图的，其余部分则沿径向分布，在给定半径下的质量与该半径成正比。

On the zero layer line there appears to be a marked J_{10}^2, and on the first, second and third layer lines, $J_9^2 + J_{12}^2$, $J_8^2 + J_{12}^2$, etc., respectively. This means that, in projection on a plane at right-angles to the fibre axis, the outer part of the nucleotide is relatively concentrated, giving rise to high-density regions spaced c. 6 A. apart around the circumference of a circle of 20 A. diameter. On the fifth layer line two J_5 functions overlap and produce a strong reflexion. On the sixth, seventh and eighth layer lines the maxima correspond to a helix of diameter $\sim$12 A. Apparently it is only the central region of the helix structure which is well divided by the 3.4-A. spacing, the outer parts of the nucleotide overlapping to form a continuous helix. This suggests the presence of nitrogen bases arranged like a pile of pennies[1] in the central regions of the helical system.

There is a marked absence of reflexions on layer lines beyond the tenth. Disorientation in the specimen will cause more extension along the layer lines of the Bessel function streaks on the eleventh, twelfth and thirteenth layer lines than on the ninth, eighth and seventh. For this reason the reflexions on the higher-order layer lines will be less readily visible. The form factor of the nucleotide is also probably causing diminution of intensity in this region. Tilting of the nitrogen bases could have such an effect.

Reflexions on the equator are rather inadequate for determination of the radial distribution of density in the helical system. There are, however, indications that a high-density shell, as suggested above, occurs at diameter $\sim$20 A.

The material is apparently not completely paracrystalline, as sharp spots appear in the central region of the second layer line, indicating a partial degree of order of the helical units relative to one another in the direction of the helix axis. Photographs similar to Fig. 1 have been obtained from sodium nucleate from calf and pig thymus, wheat germ, herring sperm, human tissue and T_2 bacteriophage. The most marked correspondence with Fig. 2 is shown by the exceptional photograph obtained by our colleagues, R. E. Franklin and R. G. Gosling, from calf thymus deoxypentose nucleate (see following communication).

It must be stressed that some of the above discussion is not without ambiguity, but in general there appears to be reasonable agreement between the experimental data and the kind of model described by Watson and Crick (see also preceding communication).

It is interesting to note that if there are ten phosphate groups arranged on each helix of diameter 20 A. and pitch 34 A., the phosphate ester backbone chain is in an almost fully extended state. Hence, when sodium nucleate fibres are stretched[3], the helix is evidently extended in length like a spiral spring in tension.

Structure *in vivo*

The biological significance of a two-chain nucleic acid unit has been noted (see preceding communication). The evidence that the helical structure discussed above does, in fact, exist in intact biological systems is briefly as follows:

在第 0 层线，似乎出现了显著的 J_{10}^2，而在第 1、2 和 3 层线则分别出现了显著的 $J_9^2 + J_{12}^2$、$J_8^2 + J_{12}^2$ 等。这意味着，在与纤维轴垂直的平面上的投影图中，核苷酸外围部分相对比较浓缩，从而形成了与直径 20 Å 的圆周相距约 6 Å 处环绕的高密度分布区。在第 5 层线上，两个 J_5 函数重叠并产生一个强反射。第 6、7 和 8 层线上的极大值对应于直径约为 12 Å 的螺旋。很显然，只有螺旋结构的中间区域才按照 3.4 Å 的间距很好地排列，核苷酸的外围部分相互重叠形成一条连续的螺旋。这意味着，在螺旋体系的中间区域，含氮碱基就像一摞便士硬币[1]般地排布着。

第 10 层线以上明显缺乏反射。样品的无序性会导致第 11、12 和 13 层线上贝塞尔函数条纹比在第 9、8 和 7 层线上更宽。由于这个原因，位于较高层线上的反射就会比较难观测到。核苷酸的形状因子也可能会导致这一区域反射强度降低。含氮碱基的倾斜可能也会产生这种效应。

赤道上的反射强度更低，有点不足以确定螺旋体系中密度的径向分布。不过如前所述，有迹象表明在直径约为 20 Å 处有一个高密度壳层。

被检测的样品显然不是完全的类晶态，因为在第 2 层线的中心区域出现了明显的斑点，这意味着在螺旋中心轴方向上各个螺旋单元彼此间的排布表现出一定程度的有序性。利用其他来源（如来自于小牛或猪的胸腺、麦芽、鲱鱼的精子，人体组织或 T_2 噬菌体）的核酸钠盐也都得到了类似于图 1 的照片。我们的同事富兰克林和戈斯林利用小牛胸腺中的脱氧戊糖核酸盐得到了异常漂亮的衍射照片（参见后一篇通讯文章），该照片和图 2 具有非常显著的一致性。

必须强调的是，上述讨论中或许不无含糊之处，但是大体上看，实验数据和沃森与克里克所描述的那种模型（参见前一篇通讯文章）之间具有合理的一致性。

有趣的是，如果在直径为 20 Å、螺距为 34 Å 的螺旋的每一圈内安置 10 个磷酸基团，那么磷酸酯骨架就几乎处于完全伸展的状态。因此，当使核酸钠盐长链伸展开时，螺旋就会明显变长[3]，如同受到牵拉的螺旋形弹簧。

在体内的结构

人们注意到了双链核酸单元具有重要的生物学意义（参见前面的通讯文章）。实际上，前面所论述的螺旋结构在完整无损的生物体内确实存在，证据可以简述如下：

Sperm heads. It may be shown that the intensity of the X-ray spectra from crystalline sperm heads is determined by the helical form-function in Fig. 2. Centrifuged trout semen give the same pattern as the dried and rehydrated or washed sperm heads used previously[6]. The sperm head fibre diagram is also given by extracted or synthetic[1] nucleoprotamine or extracted calf thymus nucleohistone.

Bacteriophage. Centrifuged wet pellets of T_2 phage photographed with X-rays while sealed in a cell with mica windows give a diffraction pattern containing the main features of paracrystalline sodium nucleate as distinct from that of crystalline nucleoprotein. This confirms current ideas of phage structure.

Transforming principle (in collaboration with H. Ephrussi-Taylor). Active deoxypentose nucleate allowed to dry at ~60 percent humidity has the same crystalline structure as certain samples[3] of sodium thymonucleate.

We wish to thank Prof. J. T. Randall for encouragement; Profs. E. Chargaff, R. Signer, J. A. V. Butler and Drs. J. D. Watson, J. D. Smith, L. Hamilton, J. C. White and G. R. Wyatt for supplying material without which this work would have been impossible; also Drs. J. D. Watson and Mr. F. H. C. Crick for stimulation, and our colleagues R. E. Franklin, R. G. Gosling, G. L. Brown and W. E. Seeds for discussion. One of us (H. R. W.) wishes to acknowledge the award of a University of Wales Fellowship.

(**171**, 738-740; 1953)

M. H. F. Wilkins: Medical Research Council Biophysics, Research Unit.
A. R. Stokes and H. R. Wilson: Wheatstone Physics Laboratory, King's College, London, April 2.

References:

1. Astbury. W. T., Symp. Soc. Exp. Biol., 1, Nucleic Acid (Cambridge Univ. Press, 1947).

2. Riley, D. P., and Oster, G., *Biochim. et Biophys. Acta*, 7, 526 (1951).

3. Wilkins, M. H. F., Gosling, R. G., and Seeds, W. E., *Nature*, **167**, 759 (1951).

4. Astbury, W. T., and Bell, F. O., Cold Spring Harb. Symp. Quant. Biol., 6, 109 (1938).

5. Cochran, W., Crick, F. H. C., and Vand, V., *Acta Cryst.*, 5, 581 (1952).

6. Wilkins, M. H. F., and Randall, J. T., *Biochim. et Biophys. Acta*, 10, 192 (1953).

精子头部　有数据表明，晶态的精子头部的 X 射线谱的强度是由图 2 中那样的螺旋形式函数决定的。经离心分离的鲑鱼精液给出的衍射图案与先前使用干燥并且重新水合的精子头部或经漂洗的精子头部得到的图案是一样的[6]。提取的或人工合成[1]的鱼精蛋白或者提取的小牛胸腺核组蛋白也给出了精子头部的纤维衍射图。

噬菌体　将 T_2 噬菌体离心得到的湿的片状沉淀物封闭在带有云母窗的样品盒中，它给出的衍射图像具有类晶态核酸钠盐的主要特征，这与结晶态核蛋白的特征迥然不同。这一结果确认了目前关于噬菌体结构的观点。

转化要素（与伊弗鲁西–泰勒合作）　在约 60% 湿度下干燥得到具有活性的脱氧戊糖核酸盐，与胸腺核酸钠盐的某些样品[3]具有相同的晶体结构。

我们要衷心感谢兰德尔教授对我们的鼓励，感谢查加夫教授、西格纳教授、巴特勒教授、沃森博士、史密斯博士、哈密顿博士、怀特博士以及怀亚特博士，没有他们所提供的各种材料我们就不可能做这项工作，我们还要感谢沃森博士和克里克先生的激励，以及我们的同事富兰克林、戈斯林、布朗和西兹所提供的讨论。我们中的一员（威尔逊）对威尔士大学所授予的奖学金表示感谢。

（王耀杨 翻译；顾孝诚 审稿）

Molecular Configuration in Sodium Thymonucleate

R. E. Franklin and R. G. Gosling

Editor's Note

This paper reports the X-ray crystallographic work on DNA by Rosalind Franklin and Raymond Gosling at King's College London, which furnished James Watson and Francis Crick with the crucial information for deducing how DNA is structured. Franklin and Gosling realise not only that DNA is helical but also that the sugar and base groups, part of the molecular building blocks, must point inwards towards the helix axis, contrary to an earlier helical model proposed by Linus Pauling. Franklin, who died from ovarian cancer only five years later, has been considered unfairly sidelined in the discovery of DNA's structure, not least because of the way she was dismissively described by James Watson in his account of the story in the book *The Double Helix*.

SODIUM thymonucleate fibres give two distinct types of X-ray diagram. The first corresponds to a crystalline form, structure *A*, obtained at about 75 percent relative humidity; a study of this is described in detail elsewhere[1]. At higher humidities a different structure, structure *B*, showing a lower degree of order, appears and persists over a wide range of ambient humidity. The change from *A* to *B* is reversible. The water content of structure *B* fibres which undergo this reversible change may vary from 40–50 percent to several hundred percent of the dry weight. Moreover, some fibres never show structure *A*, and in these structure *B* can be obtained with an even lower water content.

The X-ray diagram of structure *B* (see photograph) shows in striking manner the features characteristic of helical structures, first worked out in this laboratory by Stokes (unpublished) and by Crick, Cochran and Vand[2]. Stokes and Wilkins were the first to propose such structures for nucleic acid as a result of direct studies of nucleic acid fibres, although a helical structure had been previously suggested by Furberg (thesis, London, 1949) on the basis of X-ray studies of nucleosides and nucleotides.

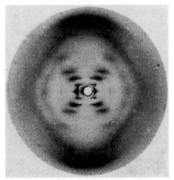

Sodium deoxyribose nucleate from calf thymus. Structure *B*

胸腺核酸钠盐的分子构型

富兰克林，戈斯林

编者按

这篇文章报道了伦敦国王学院的罗莎琳德·富兰克林和雷蒙德·戈斯林对 DNA 进行的 X 射线晶体学研究，该项研究为詹姆斯·沃森和弗朗西斯·克里克推断 DNA 结构提供了关键信息。富兰克林和戈斯林认识到不仅 DNA 是螺旋结构的，而且核糖和碱基基团作为分子构建基元的一部分，必须向内指向螺旋轴。这与莱纳斯·鲍林早先提出的螺旋模型相反。仅仅 5 年后富兰克林就死于卵巢癌，人们认为她在 DNA 结构的发现中被不公正地边缘化了，尤其是詹姆斯·沃森在其叙述 DNA 发现故事的《双螺旋》一书中曾对她做了轻蔑的排斥性描述。

胸腺核酸钠盐的纤维会产生两种不同类型的 X 射线图。第一种对应于在相对湿度大约为 75% 时得到的 A 型结构晶体形态，对它的研究在别处有详细的描述[1]。在更高湿度条件下出现的另一种 B 型结构有序性较低，能在较大范围的环境湿度下出现并且保持构型不变。从 A 型结构到 B 型结构的变化是可逆的。经历这种可逆变化后的 B 型结构纤维，其含水量可为干重的 40%~50% 到百分之几百不等。此外，某些纤维根本不会出现 A 型结构，但在这些纤维中可以得到含水量很低的 B 型结构。

B 型结构的 X 射线图（见照片）非常突出地表现了螺旋结构的代表性特征，这些特征是由斯托克斯（未发表）以及克里克、科克伦和范特等人在本实验室首先得到的。斯托克斯和威尔金斯基于对核酸纤维的直接研究，首先提出核酸有此结构，而早先富尔贝里（参见其博士论文，伦敦，1949 年）基于对核苷与核苷酸的 X 射线研究也提出过一种螺旋结构。

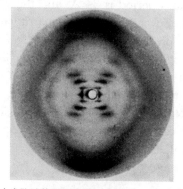

小牛胸腺的脱氧核糖核酸钠盐，B 型结构

While the X-ray evidence cannot, at present, be taken as direct proof that the structure is helical, other considerations discussed below make the existence of a helical structure highly probable.

Structure B is derived from the crystalline structure A when the sodium thymonucleate fibres take up quantities of water in excess of about 40 percent of their weight. The change is accompanied by an increase of about 30 percent in the length of the fibre, and by a substantial re-arrangement of the molecule. It therefore seems reasonable to suppose that in structure B the structural units of sodium thymonucleate (molecules on groups of molecules) are relatively free from the influence of neighbouring molecules, each unit being shielded by a sheath of water. Each unit is then free to take up its least-energy configuration independently of its neighbours and, in view of the nature of the long-chain molecules involved, it is highly likely that the general form will be helical[3]. If we adopt the hypothesis of a helical structure, it is immediately possible, from the X-ray diagram of structure B, to make certain deductions as to the nature and dimensions of the helix.

The innermost maxima on the first, second, third and fifth layer lines lie approximately on straight lines radiating from the origin. For a smooth single-strand helix the structure factor on the nth layer line is given by:

$$F_n = J_n(2\pi rR)\exp i\, n(\psi + \tfrac{1}{2}\pi),$$

where $J_n(u)$ is the nth-order Bessel function of u, r is the radius of the helix, and R and ψ are the radial and azimuthal co-ordinates in reciprocal space[2]; this expression leads to an approximately linear array of intensity maxima of the type observed, corresponding to the first maxima in the functions J_1, J_2, J_3, etc.

If, instead of a smooth helix, we consider a series of residues equally spaced along the helix, the transform in the general case treated by Crick, Cochran and Vand is more complicated. But if there is a whole number, m, of residues per turn, the form of the transform is as for a smooth helix with the addition, only, of the same pattern repeated with its origin at heights mc^*, $2mc^*$ ... etc. (c is the fibre axis period).

In the present case the fibre-axis period is 34 A. and the very strong reflexion at 3.4 A. lies on the tenth layer line. Moreover, lines of maxima radiating from the 3.4-A. reflexion as from the origin are visible on the fifth and lower layer lines, having a J_5 maximum coincident with that of the origin series on the fifth layer line. (The strong outer streaks which apparently radiate from the 3.4-A. maximum are not, however, so easily explained.) This suggests strongly that there are exactly 10 residues per turn of the helix. If this is so, then from a measurement of R_n the position of the first maximum on the nth layer line (for $n \leqslant 5$), the radius of the helix, can be obtained. In the present instance, measurements of R_1, R_2, R_3 and R_5 all lead to values of r of about 10 A.

目前虽然 X 射线研究的证据还不能直接证明该结构为螺旋形，但下面讨论中对其他方面的思考使螺旋结构的存在大有可能。

当胸腺核酸钠盐的纤维吸收了超过其自身重量 40% 的水分时，晶态的 A 型结构就变为 B 型结构。这一变化还伴随着纤维长度 30% 的增加，以及实质性的分子重排。由此我们有理由认为在 B 型结构中，胸腺核酸钠盐的结构单元（分子群中的分子）基本上不受邻近分子的影响，因为每个结构单元都被一层水分子层屏蔽。每个单元都可以独立自由地采取不依赖于邻近分子的最低能量状态的构型，另外考虑到结构中长链分子的性质，其总体形状很有可能就是螺旋形的 [3]。如果我们采纳螺旋结构的假说，那么根据 B 型结构的 X 射线图，我们立刻就可以对螺旋的实质与尺度做出某些推论。

在第 1、2、3 和第 5 层线上最内部的极大值近似地分布在从原点辐射出来的一条直线上。对于一个平滑的单链螺旋来说，位于第 n 层线上的结构因子可以由下式给出：

$$F_n = J_n(2\pi r R)\exp i\, n(\psi + \tfrac{1}{2}\pi)$$

其中，$J_n(u)$ 是 u 的 n 阶贝塞尔函数，r 是螺旋的半径，而 R 和 ψ 分别是倒易空间中的径向坐标和方位角坐标 [2]。该表达式导致所观测到的强度极大值形成近似线性排列的形式，这些极大值对应于 J_1、J_2、J_3 等函数的第一个极大值。

如果我们在考虑沿着螺旋等间距排布的一系列残基，而不是把它们当作一条平滑的螺旋来处理，那么克里克、科克伦和范特所采用的变换通常情况下会更复杂。不过，要是每周螺旋中的残基数 m 是整数，那么变换的形式就只是在一条平滑螺旋的基础上，在高度为 mc^*、$2mc^*$… （c 为纤维轴周期）等处为新原点重复添加相同的图案就行了。

在目前情况下，纤维轴周期为 34 Å，而在 3.4 Å 处出现的极强的反射位于第 10 层线上。此外，由 3.4 Å 反射所产生的极大辐射线与来自原点的相似，可以在第 5 层线和序号更低的层线上看到，这些辐射线具有 J_5 极大值，这与第 5 层线上原有辐射线的极大值是一致的。（不过，靠外的明显是从 3.4 Å 极大值辐射而来的较强条纹还不易解释。）这有力地暗示着螺旋的每一周刚好有 10 个残基。如果确实如此，那么通过测量第 n 层线上第一极大值的位置 R_n（$n \leqslant 5$），就能够得出螺旋的半径。在目前的例子中，根据测量得到的 R_1、R_2、R_3 和 R_5 推算出的螺旋半径 r 的值都大约是 10 Å。

Since this linear array of maxima is one of the strongest features of the X-ray diagram, we must conclude that a crystallographically important part of the molecule lies on a helix of this diameter. This can only be the phosphate groups or phosphorus atoms.

If ten phosphorus atoms lie on one turn of a helix of radius 10 A., the distance between neighbouring phosphorus atoms in a molecule is 7.1 A. This corresponds to the P . . . P distance in a fully extended molecule, and therefore provides a further indication that the phosphates lie on the outside of the structural unit.

Thus, our conclusions differ from those of Pauling and Corey[4], who proposed for the nucleic acids a helical structure in which the phosphate groups form a dense core.

We must now consider briefly the equatorial reflexions. For a single helix the series of equatorial maxima should correspond to the maxima in $J_0(2\pi r R)$. The maxima on our photograph do not, however, fit this function for the value of r deduced above. There is a very strong reflexion at about 24 A. and then only a faint sharp reflexion at 9.0 A. and two diffuse bands around 5.5 A. and 4.0 A. This lack of agreement is , however, to be expected, for we know that the helix so far considered can only be the most important member of a series of coaxial helices of different radii; the non-phosphate parts of the molecule will lie on inner co-axial helices, and it can be shown that, whereas these will not appreciably influence the innermost maxima on the layer lines, they may have the effect of destroying or shifting both the equatorial maxima and the outer maxima on other layer lines.

Thus, if the structure is helical, we find that the phosphate groups or phosphorus atoms lie on a helix of diameter about 20 A., and the sugar and base groups must accordingly be turned inwards towards the helical axis.

Considerations of density show, however, that a cylindrical repeat unit of height 34 A. and diameter 20 A. must contain many more than ten nucleotides.

Since structure B often exists in fibres with low water content, it seems that the density of the helical unit cannot differ greatly from that of dry sodium thymonucleate, 1.63 gm./cm.^{3} [1,5], the water in fibres of high water-content being situated outside the structural unit. On this basis we find that a cylinder of radius 10 A. and height 34 A. would contain thirty-two nucleotides. However, there might possibly be some slight inter-penetration of the cylindrical units in the dry state making their effective radius rather less. It is therefore difficult to decide, on the basis of density measurements alone, whether one repeating unit contains ten nucleotides on each of two or on each of three co-axial molecules. (If the effective radius were 8 A. the cylinder would contain twenty nucleotides.) Two other arguments, however, make it highly probable that there are only two co-axial molecules.

First, a study of the Patterson function of structure A, using superposition methods, has indicated[6] that there are only two chains passing through a primitive unit cell in this

既然极大值形成线性排列是 X 射线图最突出的特征之一，那么我们应当得出结论：该分子中在晶体学方面最重要的部分一定位于具有这一直径的螺旋上。它们只可能是磷酸基团或者磷原子。

如果 10 个磷原子分布在半径为 10 Å 的螺旋的一周上，那么分子中相邻磷原子之间的距离就是 7.1 Å。这对应于完全伸展的分子中的磷原子之间的距离，从而为磷酸基团位于结构单元的外侧提供了进一步的证明。

因此我们的结论不同于鲍林和科里 [4]，他们认为在核酸的螺旋结构中磷酸基团形成了致密的内核。

现在我们必须简要地考虑一下赤道反射。对于单链螺旋来说，赤道处极大值系列应该对应于 $J_0(2\pi rR)$ 的极大值。不过，如果 r 取前面推算得到的数值，我们照片上的极大值并不符合这一函数关系。在约 24 Å 附近有一极强的反射，然后在 9.0 Å 处只有一个微弱且尖锐的反射，另外在约 5.5 Å 和 4.0 Å 附近有两个弥散的反射带。不过，这种不一致是意料之中的，因为我们知道，迄今所考虑的螺旋只可能是一系列不同半径的同轴螺旋中最重要的一个成员而已。分子中的非磷酸盐部分将位于靠内部的同轴螺旋上，并且可以证明，尽管这些部分不会明显影响层线上的最内部的极大值，但它们可以造成破坏或移动赤道极大值以及其他层线上靠外围的其他极大值的效果。

因此，如果结构是螺旋形的，我们就会发现磷酸基团或磷原子位于一个直径约为 20 Å 的螺旋上，因而，糖和碱基基团必定只能转向内部朝向螺旋轴。

但是，考虑到密度，一个高为 34 Å、直径为 20 Å 的圆柱状重复单元中必定包含远不止 10 个核苷酸。

由于 B 型结构经常存在于含水量低的纤维中，螺旋单元的密度似乎不会和干的胸腺核酸钠盐的密度（1.63 mg/cm³）[1,5] 相差很多，因为在含水量较高的纤维中，水分子是位于结构单元外部的。基于这些考虑，我们发现一个半径为 10 Å、高为 34 Å 的圆柱体中能容纳 32 个核苷酸。不过，处于干燥状态时，圆柱状单元之间也有可能存在某些轻微的相互穿插，使其有效半径稍小。因此，单纯靠密度测量，难以断定包含 10 个核苷酸的一个重要单元是在两个或三个同轴分子的每个分子上。（如果有效半径为 8 Å，相应地圆柱中应该包含 20 个核苷酸。）不过，另外两个论据使得只有两个同轴分子的看法显得颇有可能。

首先，通过叠加法对 A 型结构的帕特森函数进行的研究表明 [6]，在此结构中，只有两条链穿过一个素晶胞。由于 $A \rightleftharpoons B$ 的可逆变换很容易实现，那么 B 型结构中

structure. Since the $A \rightleftharpoons B$ transformation is readily reversible, it seems very unlikely that the molecules would be grouped in threes in structure B. Secondly, from measurements on the X-ray diagram of structure B it can readily be shown that, whether the number of chains per unit is two or three, the chains are not equally spaced along the fibre axis. For example, three equally spaced chains would mean that the nth layer line depended on J_{3n}, and would lead to a helix of diameter about 60 A. This is many times larger than the primitive unit cell in structure A, and absurdly large in relation to the dimensions of nucleotides. Three unequally spaced chains, on the other hand, would be crystallographically non-equivalent, and this, again, seems unlikely. It therefore seems probable that there are only two co-axial molecules and that these are unequally spaced along the fibre axis.

Thus, while we do not attempt to offer a complete interpretation of the fibre-diagram of structure B, we may state the following conclusions. The structure is probably helical. The phosphate groups lie on the outside of the structural unit, on a helix of diameter about 20 A. The structural unit probably consists of two co-axial molecules which are not equally spaced along the fibre axis, their mutual displacement being such as to account for the variation of observed intensities of the innermost maxima on the layer lines; if one molecule is displaced from the other by about three-eighths of the fibre-axis period, this would account for the absence of the fourth layer line maxima and the weakness of the sixth. Thus our general ideas are not inconsistent with the model proposed by Watson and Crick in the preceding communication.

The conclusion that the phosphate groups lie on the outside of the structural unit has been reached previously by quite other reasoning[1]. Two principal lines of argument were invoked. The first derives from the work of Gulland and his collaborators[7], who showed that even in aqueous solution the —CO and —NH$_2$ groups of the bases are inaccessible and cannot be titrated, whereas the phosphate groups are fully accessible. The second is based on our own observations[1] on the way in which the structural units in structures A and B are progressively separated by an excess of water, the process being a continuous one which leads to the formation first of a gel and ultimately to a solution. The hygroscopic part of the molecule may be presumed to lie in the phosphate groups ($(C_2H_5O)_2PO_2Na$ and $(C_3H_7O)_2PO_2Na$ are highly hygroscopic[8]), and the simplest explanation of the above process is that these groups lie on the outside of the structural units. Moreover, the ready availability of the phosphate groups for interaction with proteins can most easily be explained in this way.

We are grateful to Prof. J. T. Randall for his interest and to Drs. F. H. C. Crick, A. R. Stokes and M. H. F. Wilkins for discussion. One of us (R. E. F.) acknowledges the award of a Turner and Newall Fellowship.

(**171**, 740-741; 1953)

的分子就不大可能是以三个分子为一组。其次，根据对 B 型结构的 X 射线图的测量，可以很容易地看出，无论每个单元中有两条链还是三条链，这些链都不可能是沿着纤维轴等间距排布的。比如说，三条等间距排布的链意味着第 n 层线由 J_{3n} 决定，这将会得出一个直径约为 60 Å 的螺旋。这比 A 型结构中的素晶胞要大好几倍，相对于核苷酸的尺寸来说更是大得不合情理。另一方面，三条不等间距排布的链在晶体学上是不等效的，因此看起来也不大可能。由此看来，可能只有两个同轴分子，并且它们是沿着纤维轴不等间距排布的。

因此，我们虽然并未对 B 型结构的纤维图提供一个完整详尽的解释，但还是可以陈述如下结论：该结构很可能是螺旋形的。磷酸基团位于结构单元的外侧，在直径大约为 20 Å 的螺旋上。结构单元很可能是由两个同轴分子沿纤维轴不等间距排布形成的，它们之间的相互位移正好可以解释层线上最内部极大值的观测强度的差异。如果一个分子与另一个分子间的位移约为纤维轴周期的 $\frac{3}{8}$，那就可以解释为何第 4 层线极大值缺失和第 6 层线的强度非常微弱。因此，我们的基本观点与沃森和克里克在前面的通讯文章中提出的模型没有矛盾。

此前已经依据其他一些完全不同的推论 [1] 得到了磷酸基团位于结构单元外侧这一结论。这主要借助于两条主要论据。第一条论据来自于格兰德及其同事的工作 [7]。他们证明，即使是在水溶液中，碱基中的羰基和氨基也是不可及的，并且都无法进行滴定。而磷酸基团却是完全可及的。第二条论据是基于我们自己的观测 [1]：即 A 型结构和 B 型结构中的结构单元如何逐步被过量的水分开。在这一连续过程中，首先形成凝胶，最终成为溶液。可以假定分子中的吸水部分位于磷酸基团处（$(C_2H_5O)_2PO_2Na$ 和 $(C_3H_7O)_2PO_2Na$ 都是高度吸湿的 [8]），那么对上述过程最简单的解释就是这些基团位于结构单元的外侧。此外，磷酸基团可以很容易地与蛋白质发生相互作用，这一现象也可以非常容易地以此方式作出解释。

我们要感谢兰德尔教授的关注，以及克里克博士、斯托克斯博士和威尔金斯博士等人与我们讨论。我们中的一员（富兰克林）对特纳和纽沃尔研究基金的支持表示感谢。

（王耀杨 翻译；顾孝诚 审稿）

Rosalind E. Franklin and R. G. Gosling: Wheatstone Physics Laboratory, King's College, London, April 2.

References:

1. Franklin, R. E., and Gosling, R. G. (in the press).

2. Cochran, W., Crick, F. H. C., and Vand, V., *Acta Cryst.*, 5, 501 (1952).

3. Pauling, L., Corey, R. B., and Bransom, H. R., *Proc. U.S. Nat. Acad. Sci.*, 37, 205 (1951).

4. Pauling, L., and Corey, R. B., *Proc. U.S. Nat. Acad. Sci.*, 39, 84 (1953).

5. Astbury, W. T., Cold Spring Harbor Symp. on Quant. Biol., 12, 56 (1947).

6. Franklin, R. E., and Gosling, R. G. (to be published).

7. Gulland, J. M., and Jordan, D. O., Cold Spring Harbor Symp. on Quant. Biol., 12, 5(1947).

8. Drushel, W. A., and Felty, A. R., *Chem. Zent.*, 89, 1016 (1918).

Genetical Implications of the Structure of Deoxyribonucleic Acid

J. D. Watson and F. H. C. Crick

Editor's Note

The short paper by James Watson and Francis Crick the previous month, postulating a structure for DNA, only hinted at what this structure implied for genetics and heredity. Here the two researchers expand on what they have in mind. They show that the two strands of the double helix are held together by weak bonds called hydrogen bonds, which specifically bind each of the four "nucleotide bases" with a complementary partner that has a well-fitting shape. "If the actual order of the bases on one of the pair of chains were given, one could write down the exact order of the bases on the other one", they say. So each strand may act as a template for assembling its complement, enabling replication that passes on the genetic material encoded in the sequence of bases.

THE importance of deoxyribonucleic acid (DNA) within living cells is undisputed. It is found in all dividing cells, largely if not entirely in the nucleus, where it is an essential constituent of the chromosomes. Many lines of evidence indicate that it is the carrier of a part of (if not all) the genetic specificity of the chromosomes and thus of the gene itself. Until now, however, no evidence has been presented to show how it might carry out the essential operation required of a genetic material, that of exact self-duplication.

We have recently proposed a structure[1] for the salt of deoxyribonucleic acid which, if correct, immediately suggests a mechanism for its self-duplication. X-ray evidence obtained by the workers at King's College, London[2], and presented at the same time, gives qualitative support to our structure and is incompatible with all previously proposed structures[3]. Though the structure will not be completely proved until a more extensive comparison has been made with the X-ray data, we now feel sufficient confidence in its general correctness to discuss its genetical implications. In doing so we are assuming that fibres of the salt of deoxyribonucleic acid are not artefacts arising in the method of preparation, since it has been shown by Wilkins and his co-workers that similar X-ray patterns are obtained from both the isolated fibres and certain intact biological materials such as sperm head and bacteriophage particles[2,4].

The chemical formula of deoxyribonucleic acid is now well established. The molecule is a very long chain, the backbone of which consists of a regular alternation of sugar and phosphate groups, as shown in Fig. 1. To each sugar is attached a nitrogenous base, which can be of four different types. (We have considered 5-methyl cytosine to be equivalent to cytosine, since either can fit equally well into our structure.) Two of the possible bases—adenine and guanine—are purines, and the other two—thymine and cytosine—are pyrimidines. So far as is known, the sequence of bases along the chain is irregular. The monomer unit, consisting of phosphate, sugar and base, is known as a nucleotide.

脱氧核糖核酸结构的遗传学意义

沃森，克里克

编者按

在本文发表的一个月前，詹姆斯·沃森和弗朗西斯·克里克发表了一篇短文，提出了 DNA 的一种结构，当时仅仅暗示了这种结构对于遗传学及遗传的意义。本文中两位研究者进一步阐述了他们的想法。他们指出，双螺旋的两条链是由一种叫做氢键的弱键连接在一起的，其中氢键将四个"核苷酸碱基"与具有良好拟合形状的互补核苷酸碱基特异性地结合在一起。他们指出"如果给出其中一条链的确切的碱基顺序，那么我们就能准确写出另一条链的碱基顺序"。因此，每条链都可以作为其互补链的模板，使得由碱基序列编码的遗传物质能够复制下去。

活细胞中脱氧核糖核酸（DNA）的重要性是无可争议的。在一切分裂着的细胞中，DNA 大部分（如果不是全部）存在于细胞核中，是染色体的重要组成成分。许多证据表明，DNA 是部分（如果不是全部）染色体遗传特异性或者说基因本身的载体。但是，到目前为止还没有证据能够指出它是如何完成遗传物质所必需的精确的自我复制的。

最近，我们提出了脱氧核糖核酸盐的一种结构 [1]，如果这一结构是正确的，那它立即揭示了 DNA 自我复制的机制。与此同时，伦敦国王学院的研究人员得到的 X 射线的数据对我们的结构也提供了定性的支持 [2]，该数据与以前提出过的各种结构都不一致 [3]。虽然在对 X 射线数据进行更广泛深入的比较之前我们提出的结构还不能被完全证实，但是现在我们有足够的信心认为，讨论其遗传学意义是基本正确的。为此，我们认为样品的制备方法并不会使脱氧核糖核酸盐的纤维结构出现人为假象，这是因为威尔金斯及其同事们已经发现，分离纯化的纤维与某些完整无损的生物材料，如精子头部和噬菌体颗粒等，能够得到类似的 X 射线衍射图样 [2,4]。

现在我们已经完全确定了脱氧核糖核酸的化学组成。其分子是一条很长的链，糖和磷酸基团规则交替排列组成了它的骨架，如图 1 所示。每一个糖分子与一个含氮碱基相连，碱基有四种不同的类型。（我们认为 5–甲基胞嘧啶与胞嘧啶是等同的，因为它们同样都适合于我们的结构中碱基所在位置。）碱基中有两种属于嘌呤，它们是鸟嘌呤和腺嘌呤，另外两种属于嘧啶，它们是胸腺嘧啶和胞嘧啶。就目前所知，分子长链上的碱基序列是无规则的。由磷酸基团、糖和碱基组成的单体单元被称为核苷酸。

The first feature of our structure which is of biological interest is that it consists not of one chain, but of two. These two chains are both coiled around a common fibre axis, as is shown diagrammatically in Fig. 2. It has often been assumed that since there was only one chain in the chemical formula there would only be one in the structural unit. However, the density, taken with the X-ray evidence[2], suggests very strongly that there are two.

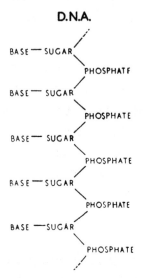

D.N.A.

Fig. 2. This figure is purely diagrammatic. The two ribbons symbolize the two phosphate-sugar chains, and the horizontal rods the pairs of bases holding the chains together. The vertical line marks the fibre axis.

Fig. 1. Chemical formula of a single chain of deoxyribonucleic acid.

The other biologically important feature is the manner in which the two chains are held together. This is done by hydrogen bonds between the bases, as shown schematically in Fig. 3. The bases are joined together in pairs, a single base from one chain being hydrogen-bonded to a single base from the other. The important point is that only certain pairs of bases will fit into the structure. One member of a pair must be a purine and the other a pyrimidine in order to bridge between the two chains. If a pair consisted of two purines, for example, there would not be room for it.

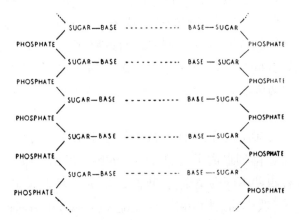

Fig. 3. Chemical formula of a pair of deoxyribonucleic acid chains. The hydrogen bonding is symbolized by dotted lines.

我们这种具有生物学意义的结构的首要特点，在于它是由双链而不是由单链组成。两条链围绕一个共同的中心轴缠绕在一起，如图2所示。此前人们一直认为，既然DNA在化学组成上只是一条链，那么其结构单元也应该只有一条链。然而，X射线衍射结果中的密度值[2]有力地表明它有两条链。

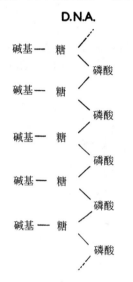

图 1. 单链脱氧核糖核酸的化学式

图 2. 这仅仅是个示意图。两条丝带代表两条磷酸-糖链，而横线代表将链连接起来的碱基对。垂直线标示出中心轴。

另外一个具有重要生物学意义的特点是两条链结合在一起的方式。双链是通过碱基之间的氢键结合在一起的，如图3所示。碱基之间配对结合在一起，一条链上的一个碱基与另一条链上的一个碱基通过氢键结合。值得注意的是，在这个结构中只有特定的碱基对可以匹配这个结构。为了将两条链桥接起来，碱基对中的一个是嘌呤则另一个必须是嘧啶。如果一个碱基对是由两个嘌呤组成的，那将没有空间容纳这个碱基对。

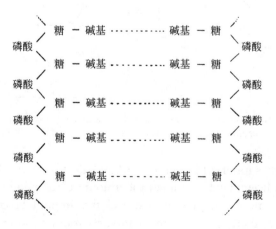

图 3. 双链脱氧核糖核酸的化学式。虚线代表氢键。

We believe that the bases will be present almost entirely in their most probable tautomeric forms. If this is true, the conditions for forming hydrogen bonds are more restrictive, and the only pairs of bases possible are:

adenine with thymine;
guanine with cytosine.

The way in which these are joined together is shown in Figs. 4 and 5. A given pair can be either way round. Adenine, for example, can occur on either chain; but when it does, its partner on the other chain must always be thymine.

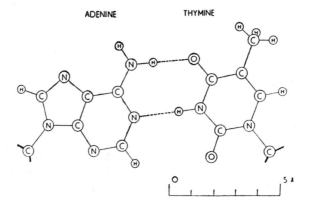

Fig. 4. Pairing of adenine and thymine. Hydrogen bonds are shown dotted. One carbon atom of each sugar is shown.

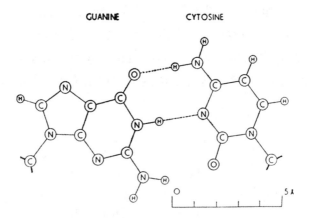

Fig. 5. Pairing of guanine and cytosine. Hydrogen bonds are shown dotted. One carbon atom of each sugar is shown.

This pairing is strongly supported by the recent analytical results[5], which show that for all sources of deoxyribonucleic acid examined the amount of adenine is close to the amount of thymine, and the amount of guanine close to the amount of cytosine, although the cross-ratio (the ratio of adenine to guanine) can vary from one source to another. Indeed, if the sequence of bases on one chain is irregular, it is difficult to explain these analytical

我们认为，碱基几乎全部以其最可能出现的互变异构体形式存在。如果事实确实如此，那么形成氢键的条件将会受到更多的限制，可能的碱基配对就会只有两种：

腺嘌呤和胸腺嘧啶；
鸟嘌呤和胞嘧啶。

它们的结合方式如图4和图5所示。对于给定的一个碱基对，正反配对都是可以的。例如，腺嘌呤可以出现在任意一条链上，但是另一条链上与它配对的一定是胸腺嘧啶。

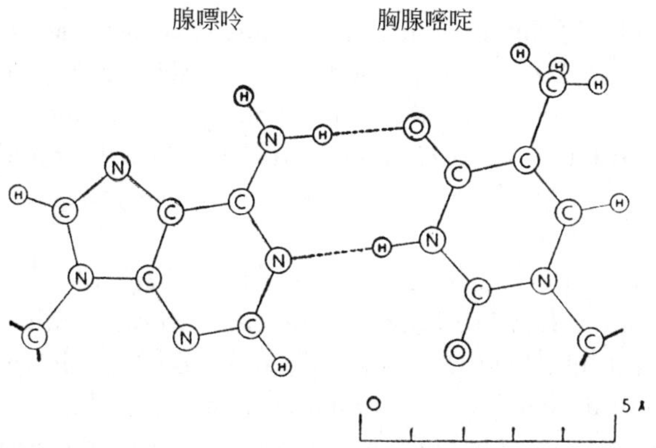

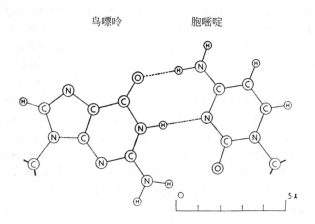

图 4. 腺嘌呤和胸腺嘧啶之间的配对。虚线代表氢键。图中只标出了每个糖基中的一个碳原子。

鸟嘌呤   胞嘧啶

图 5. 鸟嘌呤和胞嘧啶之间的配对。虚线代表氢键。图中只标出了每个糖基中的一个碳原子。

最近的分析结果有力地支持了这种配对方式 [5]，这些结果显示，虽然交叉比率（腺嘌呤与鸟嘌呤的比率）随样品来源的不同而不同，但在被检测的各种来源的脱氧核糖核酸中，腺嘌呤的量总是接近于胸腺嘧啶的量，鸟嘌呤的量总是接近于胞嘧啶的量。实际上，如果一条链上的碱基序列是无规则的，那么除非用我们提出的这种配对原则，

results except by the sort of pairing we have suggested.

The phosphate-sugar backbone of our model is completely regular, but any sequence of the pairs of bases can fit into the structure. It follows that in a long molecule many different permutations are possible, and it therefore seems likely that the precise sequence of the bases is the code which carries the genetical information. If the actual order of the bases on one of the pair of chains were given, one could write down the exact order of the bases on the other one, because of the specific pairing. Thus one chain is, as it were, the complement of the other, and it is this feature which suggests how the deoxyribonucleic acid molecule might duplicate itself.

Previous discussions of self-duplication have usually involved the concept of a template, or mould. Either the template was supposed to copy itself directly or it was to produce a "negative", which in its turn was to act as a template and produce the original "positive" once again. In no case has it been explained in detail how it would do this in terms of atoms and molecules.

Now our model for deoxyribonucleic acid is, in effect, a *pair* of templates, each of which is complementary to the other. We imagine that prior to duplication the hydrogen bonds are broken, and the two chains unwind and separate. Each chain then acts as a template for the formation on to itself of a new companion chain, so that eventually we shall have *two* pairs of chains, where we only had one before. Moreover, the sequence of the pairs of bases will have been duplicated exactly.

A study of our model suggests that this duplication could be done most simply if the single chain (or the relevant portion of it) takes up the helical configuration. We imagine that at this stage in the life of the cell, free nucleotides, strictly polynucleotide precursors, are available in quantity. From time to time the base of a free nucleotide will join up by hydrogen bonds to one of the bases on the chain already formed. We now postulate that the polymerization of these monomers to form a new chain is only possible if the resulting chain can form the proposed structure. This is plausible, because steric reasons would not allow nucleotides "crystallized" on to the first chain to approach one another in such a way that they could be joined together into a new chain, unless they were those nucleotides which were necessary to form our structure. Whether a special enzyme is required to carry out the polymerization, or whether the single helical chain already formed acts effectively as an enzyme, remains to be seen.

Since the two chains in our model are intertwined, it is essential for them to untwist if they are to separate. As they make one complete turn around each other in 34 A., there will be about 150 turns per million molecular weight, so that whatever the precise structure of the chromosome a considerable amount of uncoiling would be necessary. It is well known from microscopic observation that much coiling and uncoiling occurs during mitosis, and though this is on a much larger scale it probably reflects similar processes on a molecular level. Although it is difficult at the moment to see how these processes occur without

否则很难解释这些分析结果。

在我们的模型中，磷酸基–糖骨架是完全规则的，在这种结构中任何碱基对序列都是合适的。由此可以断定，在大分子中可能存在许多种不同的碱基排列方式，因此，这似乎表明精确的碱基序列就是携带遗传信息的密码。如果双链中一条链的碱基序列是已知的，那么我们就可以准确写出另一条链的碱基序列，因为碱基之间的配对是特异的。因此，一条链是另一条链的互补链，正是这个特点向我们揭示了脱氧核糖核酸分子是如何自我复制的。

此前关于自我复制的讨论常常涉及模板或者模具的概念。人们认为，或者是模板直接自我复制，或者是模板产生一个"负链"，此"负链"反过来又作为模板再产生一条原来的"正链"。但是任何论述都没能详细解释复制过程在原子和分子水平上是如何进行的。

实际上，我们的脱氧核糖核酸模型可以看作一**对**模板，它们之间是互补的。我们猜想，在复制之前氢键断裂，两条链解开并彼此分离。然后分别以每条链为模板，合成一条与之互补的新链，最终我们就得到了**两**对链，而在复制之前只有一对链。此外，碱基对的顺序将会被完全准确地复制下来。

深入研究我们的模型可以发现，如果单条链（或者单条链上的相应部分）为螺旋构型，复制就可以非常简单地进行。我们猜想，在细胞周期中的这个阶段，胞内存在大量游离核苷酸（严格来讲应该是多核苷酸的前体）。游离核苷酸的碱基不时地通过氢键与链上的某个碱基相结合。我们现在可以推断，只有当形成上述结构时这些单体才可能聚合形成一条新链。这应该是合理的，因为从空间上来看，除非结合到初始链上的核苷酸能形成我们所提出的结构的核苷酸，否则，它们是不可能"结晶"到初始链上并相互接近而连在一起形成一条新链的。至于是否需要一种专一性的酶来催化聚合过程，或者已形成的单股螺旋链是否可以有效地发挥酶的作用，这还需要进一步的研究。

在我们的模型中两条链是相互缠绕的，因此，要分离它们就必须解开螺旋。两条链每 34 Å 缠绕一周，按此计算，分子量为 100 万的 DNA 约有 150 个螺旋，因此，无论染色体具有怎样精确的结构，其复制过程中一定会发生大量解链。根据显微镜观察的结果，人们已经知道在有丝分裂过程中发生大量的螺旋化和解螺旋化，尽管这是在一个更大的尺度上，但是它很可能反映出在分子水平上也存在类似的过程。虽然此刻还没有弄清各个过程的关联性，很难确定它们是如何发生的，但是我们认

everything getting tangled, we do not feel that this objection will be insuperable.

Our structure, as described[1], is an open one. There is room between the pair of polynucleotide chains (see Fig. 2) for a polypeptide chain to wind around the same helical axis. It may be significant that the distance between adjacent phosphorus atoms, 7.1 A., is close to the repeat of a fully extended polypeptide chain. We think it probable that in the sperm head, and in artificial nucleoproteins, the polypeptide chain occupies this position. The relative weakness of the second layer-line in the published X-ray pictures[3a,4] is crudely compatible with such an idea. The function of the protein might well be to control the coiling and uncoiling, to assist in holding a single polynucleotide chain in a helical configuration, or some other non-specific function.

Our model suggests possible explanations for a number of other phenomena. For example, spontaneous mutation may be due to a base occasionally occurring in one of its less likely tautomeric forms. Again, the pairing between homologous chromosomes at meiosis may depend on pairing between specific bases. We shall discuss these ideas in detail elsewhere.

For the moment, the general scheme we have proposed for the reproduction of deoxyribonucleic acid must be regarded as speculative. Even if it is correct, it is clear from what we have said that much remains to be discovered before the picture of genetic duplication can be described in detail. What are the polynucleotide precursors? What makes the pair of chains unwind and separate? What is the precise role of the protein? Is the chromosome one long pair of deoxyribonucleic acid chains, or does it consist of patches of the acid joined together by protein?

Despite these uncertainties we feel that our proposed structure for deoxyribonucleic acid may help to solve one of the fundamental biological problems—the molecular basis of the template needed for genetic replication. The hypothesis we are suggesting is that the template is the pattern of bases formed by one chain of the deoxyribonucleic acid and that the gene contains a complementary pair of such templates.

One of us (J. D. W.) has been aided by a fellowship from the National Foundation for Infantile Paralysis (U.S.A.).

(**171**, 964-967; 1953)

J. D. Watson and F. H. C. Crick: Medical Research Council Unit for the Study of the Molecular Structure of Biological Systems, Cavendish Laboratory, Cambridge.

References:

1. Watson, J. D., and Crick, F. H. C., *Nature*, 171, 737 (1953).

2. Wilkins, M. H. F., Stokes, A. R., and Wilson, H. R., *Nature*, 171, 738 (1953). Franklin, R. E., and Gosling, R. G., *Nature*, 171, 740 (1953).

3. (*a*) Astbury, W. T., Symp. No. 1 Soc. Exp. Biol., 66 (1947). (*b*) Furberg, S., *Acta Chem. Scand.*, 6, 634 (1952). (*c*) Pauling, L., and Corey, R. B., *Nature*, 171, 346 (1953);

为这一障碍并非是不可克服的。

正如前文所述 [1]，我们提出的结构是尚未确定的。在一对多核苷酸链之间有一定的空间（见图 2），可使一条多肽链围绕同一螺旋轴缠绕。相邻磷原子之间的距离为 7.1 Å，与一条完全伸展的多肽链中重复单元之间的距离非常接近，这可能很重要。我们认为，在精子头部和人工合成的核蛋白中，多肽链很可能占据了这个位置。在已经发表的 X 射线衍射图片中二级衍射线相对较弱的现象 [3a,4] 与我们的这一观点粗略吻合。蛋白质的功能可能是控制螺旋和解螺旋，协助维持多核苷酸单链的螺旋构型，或者行使其他非特异性的功能。

我们的模型也可以解释其他的一些现象。比如，自发突变可能是由于某个碱基偶尔变为它的一种可能性较小的互变异构体形式。此外，减数分裂期间同源染色体之间的配对可能依赖于特定碱基之间的配对。我们将在另外的文章中详细讨论这些观点。

目前，应当将我们提出的关于脱氧核糖核酸复制的基本方案看作一种推测。即便这种推测是正确的，从我们此前的论述中也能清楚地看到，在能详细描述遗传物质复制的全景之前，还有许多问题需要我们去探索。多核苷酸的前体是什么？什么使得配对的双链解螺旋并相互分离？蛋白质的确切作用是什么？染色体到底是一条很长的脱氧核糖核酸双链，还是由通过蛋白质连接起来的许多个脱氧核糖核酸双链的片段组成的？

尽管这些问题还没有确定的答案，但是我们认为我们提出的脱氧核糖核酸的结构可能有助于解决一个最基本的生物学问题——遗传物质复制所需模板的分子基础。我们提出的假说是，复制所需的模板就是一条脱氧核糖核酸链上形成的碱基序列模式，而基因则包括了一对互补的模板链。

我们中的一员（沃森）由国家小儿麻痹症基金会资助。

（郑建全 翻译；王晓晨 审稿）

Proc. U.S. Nat. Acad. Sci., **39**, 84 (1953). (*d*) Fraser, R. D. B. (in preparation).

4. Wilkins, M. H. F., and Randall, J. T., *Biochim. et Biophys. Acta*, **10**, 192 (1953).

5. Chargaff, E., for references see Zamenhof, S., Brawerman, G., and Chargaff, E., *Biochim. et Biophys. Acta*, **9**, 402 (1952). Wyatt, G. R., *J. Gen. Physiol.*, **36**, 201 (1952).

Piltdown Man

Editor's Note

The discovery of a modern-looking skull with a primitive jaw at Piltdown in southern England in 1912 caused a sensation. As the years passed, *Eoanthropus dawsoni*—the name given to this "missing link"—looked increasingly out of step with discoveries from Africa and Asia, showing early hominids with primitive skulls and modern-looking teeth. This news report in *Nature* is an account of a paper by J. S. Weiner and colleagues (*Bulletin of the British Museum (Natural History)* 2, No. 3, 1953) in which the Piltdown finds are comprehensively revealed as forgeries. The skull was admittedly ancient, although still of a modern human. The jaw came from a modern ape, the teeth abraded mechanically. Both had been stained to look old.

A CRITICAL re-examination of all the Piltdown specimens was undertaken this year by the British Museum (Natural History) in conjunction with the Department of Anatomy, University of Oxford. The results, published over the names of Dr. J. S. Weiner, Dr. K. P. Oakley and Prof. W. E. Le Gros Clark, demonstrate clearly that the jaw and the canine tooth are in fact deliberate frauds (*Bull. Brit. Mus. (Nat. Hist.)*, **2**, No. **3**; 1953). The Piltdown brain-case must still be regarded as a genuine fossil (an Upper Pleistocene variety of modern man). The jawbone and tooth which had been falsely associated with this human cranium were so entirely out of character that, even when allowance was made for different parts of the human skull having evolved at different rates, it became increasingly difficult as time went on to reconcile them with the evidence of human evolution obtained elsewhere. The removal of the "ape-jawed Piltdown man (*Eoanthropus*)" from the fossil record thus actually clarifies the problem of human ancestry.

In their remarkably flat wear the molar teeth in the Piltdown jawbone rèsemble man rather than ape. Microscopical examination of these flatly worn surfaces showed that they are due to artificial abrasion and not to natural wear. The canine is a young tooth with incomplete root, so the fact that it has been worn right down to the pulp cavity, as in an aged individual, is only explicable by artificial abrasion.

Re-determination of the fluorine content of all the specimens, using an improved technique, has shown that whereas the brain-case contains the same amount of fluorine as some local Upper Pleistocene fossils, the jawbone and the canine contain no more than modern specimens. The organic content of all the specimens has also been determined. The jawbone and the canine contain as much nitrogen as fresh bones and teeth; but the brain-case contains only the small quantity expected in Upper Pleistocene fossils.

皮尔当人

编者按

1912 年在英国南部皮尔当发现一件具有原始下颌骨、看起来像现代人类的头骨，这一发现引起一场轰动。随着时间的流逝，道森曙人——为这个"缺失环节"所取的名字，看起来与出自非洲和亚洲的发现越来越不一致，道森曙人展现的早期人科动物具有原始的头骨和看起来像现代人类的牙齿。《自然》上的这篇消息报道是对韦纳及其同事的论文（《大英博物馆公报（自然史）》第 2 卷，第 3 期；1953 年）的一个说明，揭示了皮尔当发现完全是造假。虽然该头骨属于现代人类，但不可否认它是相当古老的。其颌骨来自于现代猿类，牙齿经过机械打磨。为了看起来古老一些，它们都经过了人工染色。

今年大英博物馆（自然史分馆）联合牛津大学解剖学系共同对所有的皮尔当样本进行了批判性的重新审查。以韦纳博士、奥克利博士和勒格罗·克拉克教授的名义发表的审查结果明确地证实了颌骨以及犬齿实际上是刻意伪造的（《大英博物馆公报（自然史）》，第 2 卷，第 3 期；1953 年）。该皮尔当脑颅一定仍被当作一个真正的化石（一种上更新世时期的现代人类物种）。即使认同人类头骨的不同部分是以不同速率进化的，颌骨和牙齿与该人类颅骨虚假地关联在一起完全不相称，以至于随着时间的流逝，想要使它们与在其他地方获得的人类进化证据相吻合变得越来越困难。因而从化石记录中将"具有猿类颌的皮尔当人（曙人）"去除，实际上就能阐明人类祖先的问题了。

皮尔当人下颌骨上磨损面非常平坦的臼齿，与人类的相似，而非与猿类的相似。对这些平坦的磨损面进行的显微镜检验表明，它们是人工磨损形成的而非自然磨损形成的。犬齿是一颗年轻的牙齿，具有不完整的牙根，所以它表现出来的像在一个老年个体中见到的磨损径直向下到牙髓腔的情况只能通过人工磨损来解释了。

使用一种改进的技术对所有样本的氟含量进行了重新检测，结果表明尽管脑颅与当地上更新世一些化石的氟含量相同，但颌骨和犬齿的氟含量却不超过现代样本。所有样本的有机物含量也已得到检测。颌骨和犬齿的含氮量与新鲜骨和牙齿的相同，但是脑颅中的含量很少，正如在上更新世化石中预期的那样。

Finally, evidence has been found which indicates that the jawbone was artificially stained by iron salts and potassium dichromate prior to its being "discovered" by the excavators. Some who were at first sceptical about the association of the human brain-case with an ape-like jaw and canine tooth became convinced that it was a genuine association, and not a fortuitous one, when it was announced that remains of another individual had been found at a second site. Tests applied to the two pieces of skull bone alleged to be from the second site show that whereas one piece is fossilized and probably belongs to the skull from the first site, the other piece is comparatively modern and has been artificially stained in an attempt to match it in colour. A molar tooth reputed to be from this second site shows signs of artificial abrasion and almost certainly came from the original "Piltdown mandible".

The tests and other investigations reported above were carried out jointly in the Departments of Geology and Mineralogy of the British Museum, in the Department of Anatomy and the Clarendon Laboratory, University of Oxford, and in the Department of the Government Chemist, London.

These investigations serve to confirm the doubts expressed by certain men of science during the controversy which raged for some years after the original discovery in 1912. For example, exactly forty years ago, Prof. D. Waterston, then professor of anatomy in King's College, London, after superimposing tracings of the radiograms of the Piltdown mandible and that of a chimpanzee, wrote (*Nature*, **92**, 319; 1913): "The similarity of the specimens brought out in this way is very striking, for the outlines are practically identical.... The cranial fragments of the Piltdown skull, on the other hand, are in practically all their details essentially human. If that be so it seems to me to be as inconsequent to refer the mandible and the cranium to the same individual as it would be to articulate a chimpanzee foot with the bones of an essentially human thigh and leg". A similar view was expressed by Mr. G. Miller, jun., in the United States after systematic comparison of the casts of the Piltdown fossils with the corresponding bones of men and apes.

(**172**, 981-982; 1953)

最终，已发现的一些证据表明颌骨在被挖掘者"发现"之前，就已用铁盐和重铬酸钾进行了人工染色。当宣布在别的地点发现了另一个体的残骸时，起初对人类脑颅与类似猿类的颌骨和犬齿之间的关联表示怀疑的那些人开始相信这是一个真实的而非偶然的关联。对宣称是来自第二地点的两件头骨进行的检测表明，尽管其中一件头骨是石化的并且可能是来自于第一地点的头骨，但另一件却相对现代些，已进行了人工染色，以使该头骨与其他头骨在颜色上匹配。一颗号称是来自第二地点的臼齿表现出了人工磨损的迹象，几乎肯定它来自于最初的"皮尔当下颌骨"。

以上报道的这些检测和其他调查是由大英博物馆地质学与矿物学部、牛津大学解剖学系和克拉伦登实验室、伦敦政府化学部联合完成的。

这些调查证实了在 1912 年最初发现之后争议盛行的几年里一些科学工作者提出的质疑。例如，正是 40 年前，时为伦敦国王学院解剖学系教授的沃特斯顿在将皮尔当下颌骨的射线照相扫描图与黑猩猩的叠加之后，写道（《自然》，92 卷，319 页；1913 年）："以这种方式呈现的这些样本的相似性很明显，因为其轮廓几乎一样……另一方面，皮尔当头骨的颅骨破片几乎在所有细节方面实质上都是人类的。如果果真如此，那么在我看来推断该下颌骨和颅骨属于同一个体是不合理的，正如将一只黑猩猩的脚与实质上是人类的大腿和小腿的骨头合成一个关节一样不合情理"。美国的小米勒先生在对皮尔当化石模型和相应的人类及猿类的骨进行了系统比较之后，表达了相似的观点。

（刘皓芳 翻译；吴新智 胡荣 审稿）

Chemical Examination of the Piltdown Implements

K. P. Oakley and J. S. Weiner

Editor's Note

Following their revelations this same year that the supposedly ancient pre-human remains from Piltdown in southern England were forgeries, Joseph Sidney Weiner and Kenneth Oakley here add a further twist. Not only had all the bones been stained to look old, but also had some of the supposed flint artefacts too. Staining might have had the purpose of hardening bone, the better to preserve it—but there is no possible reason for staining flint artefacts except to deceive. These revelations were a bitter blow to some established names in anthropology, who had fallen for the deception. The identity of the forger has, however, never been conclusively proven.

IN the report[1] on the main results of our re-examination of the Piltdown material, we gave reasons for regarding the chromate staining of the *mandible* as indicating a deliberate attempt to match a modern bone with the mineralized cranial fragments. The actual composition of this bone (3.9 percent nitrogen, less than 0.03 percent fluorine) suffices to prove its modernity; but the chromate staining, combined with the artificially abraded appearance of the molars, indicates that it is not only modern but also fraudulent.

In case there is any lingering doubt that the Piltdown finds are in part fraudulent, we think that one other fact now brought to light should be published immediately. Suspecting that some of the so-called implements[2] reported from the site might have been "doctored", we asked Mr. E. T. Hall, of the Clarendon Laboratory, Oxford, to test the composition of their surface stains by means of his X-ray spectrographic method of analysis. He has reported to us that the stains on these flints are entirely ferruginous, with one notable exception. The triangular flint (Reg. No. E. 606) recovered *in situ* from the layer immediately overlying the skull horizon[3] is chromate stained. When this stain is removed in acid the flint appears greyish-white. It is indistinguishable from a mechanically broken piece of flint such as one might encounter on the surface of any ploughed field in "Chalkland".

Whereas a bone might have been dipped in a solution of potassium dichromate with the sole purpose of trying to harden it, a flint would only have been treated in that way by a forger requiring it to be of a certain colour.

(**172**, 1110; 1953)

K. P. Oakley: Department of Geology, British Museum (Nat. Hist.), London, S.W.7.

对皮尔当工具进行的化学检验

奥克利，韦纳

编者按

这一年，韦纳和奥克利先是揭露了发现于英格兰南部皮尔当的所谓古代前人类化石是一个骗局，随后他们又对此做了进一步的验证。不仅所有的骨经过了染色处理以使其看起来陈旧，而且一些之前被猜测是燧石的石器也经过了同样的处理。给骨染色可能是为了使其硬化以便于保存，但是给燧石石器染色的原因就只能是为了欺骗了。这次揭露事件使一些著名的人类学家受到沉重的一击，并因这个骗局而声名扫地。但是伪造者的身份至今仍未得到充分证实。

通过对皮尔当样品重新进行检验，我们认为对**下颌骨**进行铬酸盐染色表明伪造者蓄意使这件现代的骨与矿化的颅骨破片相匹配，在这份包含了主要结果的检验报告中[1]，我们给出了得出这样结论的依据。该骨骼的实际组分（3.9% 的氮，不到 0.03% 的氟）足以证实其现代性；但是铬酸盐染色以及臼齿表面受到过人工磨损表明，该样本不仅是现代的，而且还带有欺诈性。

为了避免有人怀疑皮尔当发现带有部分欺诈性，我们认为应该立即将新发现的另一个事实公之于众。由于怀疑报告发现于该遗址的所谓的工具[2] 可能被"做过手脚"，所以我们请求牛津大学克拉伦登实验室的霍尔先生用 X 射线光谱分析方法对其表面的染料进行了检验。他在报告中称，这些燧石表面的染料全部含铁，但是有一个显著的例外。在头骨所在层位[3] 紧上方的地层原位发掘出的三角形燧石（登记编号 E.606）是铬酸盐染色。当用酸将这种染料去除时，燧石呈现出灰白色。这与燧石由于机械断裂而形成的碎片的颜色非常相似，这在任意一块"白垩质土地"上被犁过的田野中都可以见到。

用重铬酸钾溶液浸泡骨可能只是为了使其硬化，但是用这种方法处理燧石的目的则只有一个，即伪造者为了使其呈现出某种特定的颜色。

（刘皓芳 崔娅铭 翻译；吴新智 审稿）

J. S. Weiner: Department of Anatomy, University of Oxford. Nov. 24.

References:

1. Weiner, J. S., Oakley, K. P., and Clark, W. E. Le Gros, *Bull. Brit. Mus. (Nat. Hist.)*, Geol. Ser., **2**, No. 3 (1953).

2. Doubts about their genuineness were expressed in 1949 in a handbook of the British Museum (Nat. Hist.), "Man the Tool-Maker", 1st edit., pp. 69-70.

3. Dawson, C., and Woodward, A. S., *Quart. J. Geol. Soc. Lond.*, **69**, 122, footnote 1, pl. xvi, fig. 2 (1913).

Pithecanthropus, Meganthropus and the Australopithechinae[*]

G. H. R. von Koenigswald

Editor's Note

The flourish of African hominid discovery in the 1930s and '40s came after the discovery of the Pithecanthropines in China and Java. Here Gustav Heinrich Ralph von Koenigswald—an expert on these Asian forms—compares the Asian record with the newly discovered Australopithecines from Africa. He disputes claims that the Australopithecines were capable of technology, showing that their brains were small and they had many apelike features. The Pithecanthropines looked much more modern and had larger brains relative to their body mass. Later work vindicated this view: the Pithecanthropines are now all assigned to *Homo erectus*, the first human of modern aspect. Presciently, von Koenigswald suspected that Australopithecines and Pithecanthropines shared a common ancestry: but in 1954 no trace of such ancestry existed.

A VERY important group of higher Primates has been discovered in Africa, the Australopithecinae[1]. They include *Australopithecus*, *Plesianthropus* and *Paranthropus* from South and most probably "*Meganthropus africanus*"[2] from East Africa. The position of *Telanthropus* from the *Plesianthropus* layers of Swartkrans is still under dispute.

The large amount of material now at hand leaves no doubt that the Australopithecinae are members of the Hominidae. With this group they share the upright gait, which can be concluded from the pelvis, used by *Plesianthropus* and *Paranthropus*, and several important peculiarities of the skull and the dentition. The *Australopithecus* from Makapansgat owes the name "*prometheus*" to Dart's suggestion that he knew the use of fire, which, however, was disputed by Broom[3]. Dart has also suggested that certain battered ungulate humeri from the same site are evidence of a "predatory implemental technique"[4]; however, exactly the same type of damage is caused by hyaenas[5], which are also known from Makapansgat[6].

The oldest known, undoubtedly human beings, are the *Pithecanthropi* of Asia, including *Pithecanthropus*, *Sinanthropus* and *Meganthropus*. Do the Australopithecinae represent an ancestral type? Are they really older and more primitive?

First of all, the known Australopithecinae are, geologically, not older than *Meganthropus* and *Pithecanthropus modjokertensis* (not *Pithecanthropus robustus*) from the Lower Pleistocene of Java[7]. In at least two of the South African sites, Kromdraai and Swartkrans, the horse,

---

[*] Summary of a lecture given at University College, London, on March 12.

爪哇猿人、魁人与南方古猿亚科*

孔尼华

编者按

继在中国和爪哇发现猿人之后，20 世纪 30 年代和 40 年代在非洲出现了发现人科的高潮。本文中，研究亚洲猿人类群的专家古斯塔夫·海因里希·拉尔夫·孔尼华将亚洲的猿人与在非洲新发现的南方古猿亚科进行比较。结果显示南方古猿亚科的脑量小并且具有许多类似猿类的特征，他对南方古猿亚科能掌握技术这一说法表示怀疑。而爪哇猿人看起来更加现代并且相对于它们的体重来说，其脑量较大。后来的研究证明了这一看法：现在，爪哇猿人都被归属于直立人，它们是最早的有现代人模样的人类。孔尼华很有预见性地推测南方古猿亚科和猿人亚科拥有共同的祖先，但是在 1954 年还没有找到这一祖先存在的任何痕迹。

在非洲出土了一组非常重要的高等灵长类类群，即南方古猿亚科 [1]。它们包括来自南非的南方古猿、迩人和傍人，也很有可能包括来自东非的"非洲魁人" [2]。在斯瓦特克朗斯的迩人层位出土的远人的分类地位尚存在争议。

现在手头上的大量材料毫无疑问地表明南方古猿亚科属于人科。它们和这个群体都具有直立行走的特征（这一点可以根据迩人和傍人的骨盆推断出），头骨和齿系有着一些重要的特点。达特认为马卡潘斯盖的南方古猿会使用火，因此将其命名为"普罗米修斯"，而布鲁姆则对这一点表示了异议 [3]。达特还认为某些来自同一遗址的被磨损的有蹄类肱骨可以作为"捕食工具技术"的证据 [4]；然而，马卡潘斯盖 [6] 的证据表明，鬣狗 [5] 可以造成同种类型的损伤。

已知最古老的、确定无疑的人类是亚洲的猿人，包括爪哇猿人、中国猿人和魁人。那么南方古猿亚科是否代表了一种祖先类型？它们真的更古老更原始吗？

首先，从地质学角度而言，已知的南方古猿亚科并不比爪哇下更新世的魁人和莫佐克托猿人（不是粗壮猿人）更古老 [7]。至少在南非的克罗姆德拉伊和斯瓦特克

---

* 3 月 12 日在伦敦大学学院发表的一篇演讲的摘要。

83

Equus, is found, which is the best guide fossil for the Pleistocene, and at Sterkfontein and Swarkrans we find an extinct carnivore, *Lycyaena*, which even in Europe existed until the Villafranchian[8]. There is no definite proof that any of the other sites is really of Pliocene age.

The best-known hominid from the Lower Pleistocene of Java is *Pithecanthropus modjokertensis*. This form has retained the "simian gap", which is a very primitive characteristic which might be expected in an early human forerunner; for Remane, in a very careful study, has shown that even in the dentition of modern man there are peculiarities due to his forerunners having much better developed canines[9]. This gap has already disappeared in the full-grown Australopithecinae—although it seems to be present in the deciduous dentition of *Australopithecus africanus*; but while this *Pithecanthropus* in spite of the primitive dentition possesses a typical human entrance to the nose with a well-developed nasal septum, in the Australopithecinae—I quote from a description of *Paranthropus crassidens*—"except that the inner parts of the premaxillaries pass considerably further back, the resemblance of this region to that of the chimpanzee is considerable"[10].

The incisors in *Pithecanthropus* are well developed. In the Australopithecinae they are much reduced, in some cases even below the minimum values known for modern man. Minimum values for central incisors are, for *Paranthropus crassidens* and *Homo sapiens*, upper 7.5 and 8.0 mm., lower 5.0 and 4.6 mm., respectively. The relative values are much higher, as the average length for the first lower molar in *crassidens* is 15.0 mm., in modern man 11.1 mm. The canines are of about the same size in both groups.

In modern man the first lower premolar is only a little smaller than the second (in some Eskimos we still find the reverse), the average being 0.3 mm. But it is 1.3 mm. in *Paranthropus crassidens*. In *Sinanthropus* the first premolar is largest, also in *Meganthropus*, and generally in all anthropoids. This surely is the primitive condition. As a whole, in the Australopithecinae the incisor–canine–premolar group is more reduced not only as in *Pithecanthropus* but also more than in modern man. In connexion with this reduction, the dental arcade shows a typical shortening, which has been demonstrated by Le Gros Clark[11].

In spite of all these specializations, the roots of the premolars in *Paranthropus* do not differ in number from those observed generally in the anthropoids. In the lower jaw the second premolar has two roots, and in the upper jaw the first and the second both have three. In *Pithecanthropus* the lower premolar in question is single-rooted, the first upper has three and the second two roots. In modern man two roots in the second lower premolar are extremely rare (one case was observed among 2,089 specimens), while three separate roots occur in the first upper premolar in 3 percent, three roots in the second in only 0.3 percent, two in 3.5 percent of specimens[12].

朗斯这两处遗址发现了马属动物，这是更新世时期最好的指示性化石了，而在斯泰克方丹和斯瓦特克朗斯，我们则发现了一种已灭绝的食肉动物狼鬣狗，这种动物即使在欧洲也一直生存到维拉方期 [8]。还没有确凿的证据表明其他任何遗址真正属于上新世。

爪哇下更新世时期最著名的人科是莫佐克托猿人。它们保留了"猿的齿隙"，这是一种可能存在于早期人类祖先的非常原始的特征。雷马内通过一次非常细致的研究发现，即使在现代人的齿系中，也存在由于其祖先具有发育良好得多的犬齿而产生的一些特点 [9]。尽管这种齿隙在南方古猿非洲种的乳齿齿系中可能存在，但在发育完全的南方古猿亚科中已经消失了。另外这种爪哇猿人尽管其齿系是原始的，却具有典型的人类的鼻腔入口和发育完好的鼻中隔，而在南方古猿亚科中，"除了前颌骨的内部部分大幅向后延伸外，这一区域与黑猩猩的还是具有很大的相似性的"（引自对傍人粗齿种的描述）[10]。

爪哇猿人的门齿发育良好。南方古猿亚科门齿的发育情况则差得多，有些情况下，甚至低于现代人已知的最小值。傍人粗齿种和智人的上中门齿的最小值分别是 7.5 毫米和 8.0 毫米，下中门齿的最小值分别为 5.0 毫米和 4.6 毫米。相关数值要高得多，就如傍人粗齿种的第一下白齿的平均长度是 15.0 毫米，而现代人的是 11.1 毫米。两个群体的犬齿大小基本相同。

现代人的第一下前白齿只比第二下前白齿略小一点（在某些爱斯基摩人中，我们还发现了相反的情况），平均值是 0.3 毫米，但是傍人粗齿种的平均值是 1.3 毫米。中国猿人的第一前白齿最大，魁人也是，所有的类人猿一般也如此。因此这肯定是原始状态。总体上来说，南方古猿亚科的门齿－犬齿－前白齿群不仅像爪哇猿人的一样大大减小了，而且比现代人减小的还要多。齿弓也相应显示出典型的缩短现象，勒格罗·克拉克已证实这一点 [11]。

尽管存在以上这些特殊之处，傍人的前白齿齿根数量与在类人猿中观察到的普遍数目并无差异。下颌骨的第二前白齿有 2 个齿根，而上颌骨的第一前白齿和第二前白齿都有 3 个齿根。此处讨论的爪哇猿人下前白齿是单根的，而第一上前白齿有 3 个齿根，第二上前白齿有 2 个齿根。在现代人中，第二下前白齿有 2 个齿根的情况极为少见（2,089 个标本中只观察到 1 例），而第一上前白齿有 3 个独立齿根的情况发生的比例为 3%，第二上前白齿有 3 个齿根的情况发生的比例仅为 0.3%，有 2 个齿根的比例为 3.5%[12]。

In the relative size of the molars, the increase in size from the first to the last lower molar, and in the upper jaw the second molar being larger than the first, *Pithecanthropus* shows the same primitive conditions as the Australopithecinae and the anthropoids. There is, however, a tendency for the last upper molar in *Paranthropus crassidens* to be enlarged; this is not found either in the Hominidae (in modern man this molar is often completely reduced) or in the anthropoids.

The deciduous dentition is of great value for the determination of the phylogenetic position. In *Sinanthropus* as in *Pithecanthropus modjokertensis* the first lower deciduous molar has two main cusps—as also the chimpanzee and the orang; in the first form the entoconid is absent, as is common in the anthropoids, in the second very faint. Complete absence of the entoconid in modern man is found in about 2 percent of specimens only. In the Australopithecinae this cusp is not only developed to the same degree as in modern man (in *Australopithecus africanus* this point is not so clear), but there is also a tendency towards a complete molarization of this tooth, which reaches a maximum in *Plesianthropus transvaalensis* for the upper and in *Paranthropus robustus* for the lower molar; in the latter form, except for its size, this tooth is indistinguishable from the second deciduous molar. There is a certain amount of variability in modern man; but extreme conditions like these have not been observed in a vast collection of modern teeth at our disposal (Bolk Collection, Amsterdam.)

This comparison already reveals so many features which must be regarded as a sign of over-specialization that they exclude the Australopithecinae from the ancestorship not only of *Pithecanthropus* but also of *Homo sapiens*.

A greater resemblance exists between the Australopithecinae and the imperfectly known *Meganthropus* from Java; so much so that recently Robinson has included the Javanese species in the same group[2]. He shows that, especially in the formation of the symphysis, there is a very great resemblance, and he mentions the occurrence of a mental spine in a certain specimen of *Paranthropus crassidens*. We are looking forward with great interest to his final publication.

There are, however, some differences, in which *Meganthropus* is more primitive. There is the basic pattern of the premolars and the first molar, which still have an undivided ridge between protoconid and metaconid, conditions which I could not detect in any of the South African species. The same is true of the presence of a paraconid on the second lower deciduous molar. The first lower premolar is of a more oval outline in the Australopithecinae, whereas in *Meganthropus* it is more angular, as in *Sinanthropus* and Heidelberg man.

The dental arch is probably very similar in both forms, as is evident from a new find, discovered last year[13]. This specimen, however, is crushed, and therefore not reliable.

There is not a single feature in the original *Meganthropus* jaw fragment which cannot be

爪哇猿人在以下方面显示出了与南方古猿亚科和类人猿同样的原始特征，包括臼齿的相对大小，从第一颗下臼齿到最后一颗下臼齿逐渐增大，以及上颌第二臼齿比第一臼齿大。但是傍人粗齿种的最后一颗上臼齿有增大的趋势，在人科（现代人的这一臼齿经常是完全缺少的）和类人猿中都没有发现这种现象。

乳齿齿系的研究对于确定物种的演化发展地位非常有价值。中国猿人和莫佐克托猿人的第一下乳臼齿具有 2 个主齿尖，在黑猩猩和猩猩中也是如此。第一种类型没有下内尖，这点在类人猿中很常见，在第二种类型中，下内尖很弱。观察的样本中，现代人的下内尖完全消失的情况仅为约 2%。南方古猿亚科的这一齿尖不仅达到了与现代人相同的发育程度（南方古猿非洲种中这一点尚未确定），而且这颗牙齿也具有完全臼齿化的趋势，这种趋势在德兰士瓦迩人的上臼齿和粗壮傍人的下臼齿中达到了最大的程度。后一类型中，除了牙齿的大小，很难依据其他特征将这颗牙与第二乳臼齿区分开来。现代人虽然存在一定程度的变异，但是在我们研究的大量现代牙齿中并没有观察到像这样的极端情况（博尔克收编，阿姆斯特丹）。

这一比较已经揭示了如此多的特征，它们必定被认为是过度特化的迹象，这足以将南方古猿亚科从爪哇猿人和智人的祖先中排除出去了。

南方古猿亚科与尚不完全清楚的爪哇魁人之间存在更大的相似性，它们之间的相似性如此之多，以至于最近鲁滨逊将爪哇的这个物种纳入了与南方古猿亚科相同的类群之中 [2]。他指出二者具有非常大的相似性，尤其是在（纤维软骨）骨联合形成方面，他还提到了傍人粗齿种的一些标本中出现了颏棘。我们正怀着极大的兴趣期待他最终发表的结果。

但是也有一些差异表明魁人更加原始。它们具有最基本的前臼齿和第一臼齿模式，在下原尖和下后尖之间有尚未分开的脊，这是我在其他南非物种中都没有发现的。第二下乳磨牙有一个下前尖。南方古猿亚科的第一下前臼齿的轮廓比较椭圆，而魁人的第一前臼齿有较多的棱角，在中国猿人和海德堡人中也是如此。

去年的一个新发现表明这两种类型的齿弓可能非常相似 [13]，但是这个样本已经压碎了，所以结果并不可靠。

最初发现的魁人颌骨破片上没有一个特征能与其他原始人类的颌骨匹配。南方

matched with other primitive human jaws. In the corresponding part of the *Australopithecus* mandible the two-rooted second premolar must already be regarded as a sign that it belongs to a different category. In fossil human jaws this condition has never been observed; in modern man it occurs in less than 0.02 percent of specimens. When more is known of *Meganthropus*, it will probably be found that there are more differences of significance.

However, there can be no doubt that *Meganthropus* and the Australopithecinae have a common ancestor. Such a form must have still had the "simian gap"—*vide Pithecanthropus*—relatively large canines and incisors, large first lower premolars, and a primitive deciduous dentition. Therefore, it will not fall into the group of the Australopithecinae *s. str.*, and for that reason I cannot agree with Robinson, who claims that the Australopithecinae are "ancestral to he euhominid group" (see ref. 2, p. 37).

Within this latter group there is a very strong tendency towards a reduction of the dentition, and parallel with it an exaggerated increase in brain capacity (Table 1). The same mesiodistal lengths are for *Plesianthropus transvaalensis* 44.0 mm., and for *Paranthropus crassidens* (average) no less than 48.4 mm., respectively.

Table 1. Mesiodistal length of the three lower molars

Lower Pleistocene:	*Meganthropus*	44.0 mm.
	Pithecanthropus modjokertensis	39.5 mm.
Middle Pleistocene:	*Sinanthropus pekinensis* (G)	37.7 mm.
	Homo heidelbergensis	36.5 mm.
Upper Pleistocene:	*Homo neanderthalensis* (Spy II)	34.0 mm.
Recent:	*Homo sapiens* (White, average)	32.5 mm.

Also in regard to the brain capacity the conditions seem fundamentally different. The capacity of the Australopithecinae is a controversial question; Broom gives estimates up to more than 1,000 c.c.[14], and Schepers even calculates "cephalization coefficients" (see Schepers, ref. 14). Let us not forget that only a single skull can be measured, *Plesianthropus* V, and that here the capacity is only 482 c.c.

Most probably in their absolute brain capacity the Australopithecinae bridge the gap which exists between the largest brain of the anthropoids (gorilla with 585 c.c.) and the smallest of the Hominidae (*Pithecanthropus erectus* II with 775 c.c.). There is, however, this important difference: the greatest brain capacity in the Australopithecinae goes together with the largest molars, whereas in man it goes with the smallest molars. We have tried to express this relation in a brain-molar coefficient[15]—brain capacity in cubic centimeters divided by the mesiodistal length of the three upper molars in millimetres, both measurements taken on the same skull—the results of which are given in Table 2.

古猿下颌骨相应部分的第二前臼齿具有 2 个齿根，这肯定早已被认为是它属于不同分类单元的标志。在化石人类的颌骨中，则从未见到过这种情况，它在现代人中的发生比例低于 0.02%。在对魁人有了更多的了解之后，可能会发现更多有意义的差异。

然而，毫无疑问魁人和南方古猿亚科具有共同的祖先。这一类型肯定还具有"猿的齿隙"（参看爪哇猿人）、相对较大的犬齿和门齿、大的第一下前臼齿，以及原始的乳齿齿系。因此不应该将其归到狭义的南方古猿亚科组群中，这就是我不能同意鲁滨逊意见的原因，他认为南方古猿亚科是"真正人类群体的祖先"（见参考文献 2，第 37 页）。

表 1 中年代较晚的组群齿系减小的趋势非常明显，与之相对应，脑量增加也十分明显。德兰士瓦迩人和傍人粗齿种（平均值）的近中远中径分别是 44.0 毫米和不低于 48.4 毫米。

表 1. 三颗下臼齿的近中远中径

下更新世	魁人	44.0 毫米
	莫佐克托猿人	39.5 毫米
中更新世	中国猿人北京种（G）	37.7 毫米
	海德堡人	36.5 毫米
上更新世	尼安德特人 (Spy II)	34.0 毫米
近期	智人（白种人，均值）	32.5 毫米

至于脑量，情况似乎从根本上不同。南方古猿亚科的脑量尚存争议，布鲁姆估计其大于 1,000 毫升 [14]，舍佩尔斯甚至计算出了"脑发育系数"（见舍佩尔斯，参考文献 14）。但希望大家不要忘记，只能测量迩人 V 号这唯一一件头骨，其脑量只有482 毫升。

南方古猿亚科很有可能在绝对脑量方面填补了类人猿的最大脑量（大猩猩是585 毫升）和人科的最小脑量（直立猿人 II 的脑量是 775 毫升）之间存在的空缺。然而，这里有一个重要的差异，即南方古猿亚科的最大脑量与最大臼齿是同时出现的，而在人类中最大脑量却与最小臼齿一起出现。我们曾经尝试用脑 – 臼齿系数来表示这种关系 [15]，即用以立方厘米为单位的脑量除以以毫米为单位的三颗上臼齿的近中远中径，这两个数据都是对同一头骨测量得到的，结果见表 2。

Table 2. Brain–molar Coefficients

(The numbers in brackets indicate the number of skulls measured)

Chimpanzee (7)	10.9–13.9	Average: 11.8
Orang (16)	9.8–12.4	10.9
Gorilla (10)	9.0–13.0	11.7
Plesianthropus V	13.4	
Pithecanthropus modjokertensis	25.8	
Pithecanthropus erectus	24.3	
Sinanthropus pekinensis XI	34.2	
Homo neanderthalensis (Steinheim)	36.7	
Homo sapiens (Europeans, 10)	33.5-57.1	47.3

As we expected, the modern anthropoids form a homogeneous group. The values for *Pithecanthropus* are estimates: a capacity of 1,000 c.c. was estimated by Weidenreich for the skull of *modjokertensis*, and for *erectus* we have used a dentition, not yet described, from a different individual. In Table 2 the anthropoids are sharply separated from the Hominidae; *Plesianthropus* is not intermediate, since it comes within the range of the anthropoids.

This coefficient has still another advantage. If we take a maximum of about 15 for the anthropoids, then the Australopithecinae with their large molars will still remain within the anthropoid group, even with a brain capacity of more than 700 c.c.

Two specimens of the large *Plesianthropus* from Swartkrans have a well-developed sagittal crest[15], which is not found in the smaller species. Apart from the fact that this must be taken as a sign of a limited brain capacity, it is to be noted that a sagittal crest in the anthropoids depends upon the absolute size of the species. It is absent from the small gibbon, very rare in male chimpanzees, normal in male oranges and practically always present in the male and often in the large female gorillas. That this structure appears among the Australopithecinae under the same conditions is a very important parallel to the anthropoids, and a significant difference from the Hominidae.

Thus the Australopithecinae apparently are a group of the Hominidae, which most probably did not rise much above the "anthropoid" level. The reduction of the dentition only affects the face; their increase in brain capacity is slight and depends upon the absolute size of the species; the possession of a sagittal crest in the large specimen parallels the development of the same structure in the anthropoids. It seems that towards the end of the Pliocene period the early Hominidae were separated into several branches—Australopithecinae in Africa, *Gigantopithecus* (and undescribed forms) in China, Pithecanthropi in Asia—and that only one of them, the Pithecanthropi, by a harmonious reduction of the whole dentition and—this is the most important point—by an exaggerated and accelerated increase of the brain capacity, gave rise to the Hominidae, of which group we are the most human members.

(**173**, 795-797; 1954)

表 2. 脑 – 臼齿系数

（括号中的数字表示测量的头骨编号）

黑猩猩 (7)	10.9~13.9	均值：11.8
猩猩 (16)	9.8~12.4	10.9
大猩猩 (10)	9.0~13.0	11.7
迩人 V	13.4	
莫佐克托猿人	25.8	
直立猿人	24.3	
中国猿人北京种 XI	34.2	
尼安德特人（施泰因海姆）	36.7	
智人（欧洲人，10）	33.5~57.1	47.3

正如我们所期望的，现代类人猿形成了一个同质类群。爪哇猿人的估计值如下：魏登瑞根据莫佐克托猿人的头骨估计其脑量为 1,000 毫升，而对于直立猿人，我们则使用另外一个个体尚未被描述过的齿系来进行估算。表 2 中，类人猿明显与人科分离开了；迩人并不是介于中间，而是落在了类人猿的范围之内。

这一系数还有另外一个优点。如果我们对类人猿取最大值（大约 15），那么即使南方古猿亚科的脑量超过 700 毫升，这些具有大臼齿的南方古猿亚科也将依然位列于类人猿类群内。

斯瓦特克朗斯的两个大型迩人样本具有发育完好的矢状脊[15]，这一特征还没有在较小的物种中发现过。这肯定可以看作有限脑量的标志，此外还应该说明的是，类人猿的矢状脊取决于物种个体的绝对大小。小长臂猿没有这种矢状脊，在雄性黑猩猩中这种情况也很少见，而雄性猩猩中则很常见，实际上，雄性大猩猩总存在矢状脊，大型雌性大猩猩也常具有矢状脊。在同等条件下，南方古猿亚科中出现的这种结构与类人猿非常一致，而与人科则具有显著差异。

因此，很明显南方古猿亚科是人科的一个组群，很有可能在发生上它们并不比"类人猿"高等很多。齿系的减小只会影响到面部；它们脑量增加的幅度微小且依赖于物种个体的绝对大小；大型样本所具有的矢状脊与类人猿中同样结构的发育情况一致。可能在接近上新世末期时，早期人科分化成了几个支系——非洲的南方古猿亚科、中国的巨猿（以及其他没有描述过的类型）、亚洲的猿人，其中只有一种，即猿人的全齿系匀称地减小，另外最重要的一点是，其脑量以一种惊人的速度加速变大，由此产生了人科，在这一组群中，我们人类是最主要的成员。

（刘皓芳 崔娅铭 翻译；吴新智 审稿）

G. H. R. von Koenigswald: University of Utrecht.

References:

1. Gregory, W. K., and Hellman, M., *Ann. Transvaal Mus.*, **19**, 339 (1939).

2. Robinson, J. T., *Amer. J. Phys. Anthrop.*, **11**, 1 (1953).

3. Broom, R., "Finding the Missing Link", 74 (London, 1950).

4. Dart, R. A., *Amer. J. Phys. Anthrop.*, 7, 1 (1949).

5. Zapfe, H., *Palaeobiologica*, 7, 111 (1939).

6. Toerien, M. J., *S. Afric. J. Sci.*, **48**, 293 (1952).

7. von Koenigswald, G. H. R., Rep. Int. Geol. Congr. Great Britain 1948, part IX, 59 (1950).

8. del Campana, D., *Palaeontologia Italica*, **19**,189 (1913).

9. Remane, A., *Z. f. Anat. u. Entwicklungsgesch.*, **82**, 391 (1927).

10. Broom, R., and Robinson, J. T., *Transvaal Mus. Mem.*, **6**, 1 (1952).

11. Clark, W. E. Le Gros, *J. Roy. Anthrop. Inst.*, **80**, 37 (1952).

12. Visser, J. B., Diss. Zurich, 1 (1948).

13. Marks, P., *Madjalah Ilum Alam Untuk Indonesia*, **109**, 26 (1953).

14. Broom, R., Robinson, J. T., and Schepers, *Transvaal Mus. Mem.*, **4**, 1 (1950).

15. von Koenigswald, G. H. R., *Proc. Kon. Ned. Akad. Wetensch.*, B, **56**, 403 and 427 (1953): **57**, 85 (1954).

Structural Changes in Muscle during Contraction: Interference Microscopy of Living Muscle Fibres

A. F. Huxley and R. Niedergerke

Editor's Note

Here physiologists Andrew Huxley and Rolf Niedergerke from the University of Cambridge show that muscle fibres shorten as a result of the sliding between two sets of filaments containing the proteins myosin and actin. The researchers use interference microscopy to show that the width of thick, myosin-containing filaments called "*A*-bands" remain constant during muscle contraction, implying that the thin, actin-containing filaments of the "*I*-band" slide into the *A*-band when muscle shortens. The paper was published back-to-back with a manuscript by English biologists Hugh Huxley and Jean Hanson from the Massachusetts Institute of Technology, who used light microscopy to independently arrive at similar results. Together these studies lead to the "sliding filament" model of muscle, which is still widely accepted today.

IN spite of the numerous investigations which have been made into the changes of the striations of muscle when it contracts, there is little agreement at the present day on either the nature or the significance of these changes. Several factors contribute to this unsatisfactory position. The only contractions that could be studied in living muscle by the earlier workers[1] were the slow waves that occur in freshly isolated insect fibres; the broad striations and small diameter of these fibres are favourable for the interpretation of the microscope image, but the local nature of the contractions and the difficulty of applying passive stretch make it impossible to say whether the changes seen in the striations are accompaniments of "activation", of tension development or of shortening. Observations on fixed material, whether with visible light or with the electron microscope, are also subject to this limitation as well as to the uncertainties in the effect of the fixative. Isolated fibres from frog muscle, however, give satisfactory propagated twitches and tetani, in which activation is complete very early after the first stimulus[2], while tension develops more slowly even if the contraction is isometric, and changes of length, both during stimulation and in the resting muscle, can be controlled by holding the tendon ends. This preparation therefore provides a basis for correlating visible changes with the sequence of events which take place during a contraction, and Buchthal and his colleagues[3] have endeavoured to exploit this possibility. Their conclusions are, however, open to the objection that they used the ordinary light microscope, which cannot be expected to provide a reliable image of unstained striations (alternate bands of high and low refractive index) the repeat distance of which is 2–3 μ, in a fibre of 50–100 μ diameter.

肌肉收缩时的结构变化：
使用干涉显微镜观察活体肌纤维

赫胥黎，尼德格克

编者按

在这篇文章中，剑桥大学的生理学家安德鲁·赫胥黎和罗尔夫·尼德格克指出，肌纤维缩短是由包含肌球蛋白和肌动蛋白的两套细丝间的滑动造成的。研究者利用干涉显微镜发现包含肌球蛋白丝的粗的 A 带的宽度在肌肉收缩过程中保持恒定，这暗示着当肌肉缩短时包含肌动蛋白丝的细的"I 带"滑入 A 带。紧随这篇文章之后的是来自麻省理工学院的英国生物学家休·赫胥黎和琼·汉森完成的一篇来稿，他们利用光学显微镜独立地得到了相同的结论。综合这些研究结果所得出的肌纤维"滑动细丝"模型至今仍被广泛认可。

尽管人们已经对肌肉收缩过程中肌肉条纹的变化进行了大量的研究，但是目前关于这些变化的本质或其重要性几乎没有取得一致的意见。这种并不令人满意的局面是由许多因素造成的。早期的研究人员 [1] 只能从新鲜分离的昆虫肌纤维的慢波中研究活体肌肉的收缩；这些肌纤维的条纹宽、直径小，适合用显微镜进行观察，但是由于其收缩存在局域性以及被动拉伸困难，研究人员无法分辨出条纹的变化是与纤维的"激活"、拉伸还是与缩短相伴随。对于固定材料的观察，无论用光学显微镜还是电子显微镜，也受这种限制的影响，而且固定效果也会带来不确定性。然而，从青蛙肌肉中分离出来的肌纤维则具有良好的颤搐和强直性收缩的能力，在这种肌纤维中，第一次刺激后激活能够很快完成 [2]，而拉伸出现得更加缓慢（即使肌肉是等张收缩），并且在刺激过程中和静止肌肉中肌纤维长度的变化都可以通过固定肌腱的两端来控制。因此，这一准备工作为肉眼可见的变化同收缩过程中事件发生的顺序关联起来提供了根据，布克塔尔及其同事们在这方面作了尝试 [3]。但是他们的结论容易受到质疑，因为他们使用的普通光学显微镜不能提供一张未染色的条纹（具有高折射率和低折射率的交替带）的可靠图片，在直径为 50 微米 ~100 微米的纤维中，这些条纹的重复距离为 2 微米 ~3 微米。

The phase-contrast microscope is equally unsuitable for a specimen of these dimensions; but an interference microscope in which the reference beam does not traverse the specimen would be expected to give a satisfactory "optical section" of the fibre. An instrument of this kind was therefore built, the optical components being made to our specification by Messrs. R. and J. Beck. The basic principle was first described by Smith[4] (see also Huxley[5]), but further developments were incorporated to allow a water-immersion objective of n.a. 0.9 to be used. Gross refraction effects due to the cylindrical shape of the fibre were abolished by adding serum albumin to the Ringer's solution to bring its refractive index close to the average value for the fibre contents. A solid cone of illumination, n.a. 0.5–0.6, was always employed. Under these conditions, the fibre was completely invisible with ordinary light, but with the interference arrangement an excellent image of the striations (and also of sarcoplasm, nuclei and granules) was obtained. The contrast between A-(higher refractive index) and I-bands could be controlled or reversed by changing the background path-difference between the two beams; the measured widths of the bands were independent of this adjustment (Fig. 1). The fibre was photographed on moving film by a series of ten flashes from a discharge tube at intervals of about 20 msec., and could be stimulated by pulses of current synchronized with these flashes.

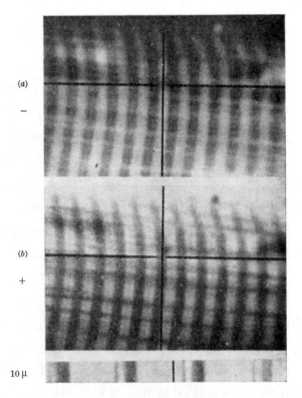

(a)

–

(b)

+

10 µ

Fig. 1. Muscle fibre in negative (a) and positive (b) contrast. A-bands (higher refractive index) light in (a), dark in (b). Note that the threads of sarcoplasm have a refractive index slightly higher than even the A-bands. Sarcomere length, 3.0 µ.

Passive stretch. The sarcomere length s could be changed by passive stretch or release from

　　相差显微镜同样不适合观察这种尺度的样品，而干涉显微镜则有可能给出一个令人满意的肌纤维"光学切片"，其参考光束并不穿透样品。因此，我们制造了这种类型的仪器，理查德·贝克先生和约瑟夫·贝克先生按照我们的要求制造了光学元件。史密斯首先描述了其基本原理[4]（也参考了赫胥黎的工作[5]），但我们对其进行了进一步改造，添加了一个数值孔径为 0.9 的水浸物镜。通过向林格液中添加血清白蛋白，以使溶液的折射率接近肌纤维含量的平均值来消除圆柱形肌纤维所带来的总折射效应。通常使用数值孔径为 0.5~0.6 的圆锥形聚光镜。在此条件下，用普通光源完全观察不到肌纤维，但是使用相干模式则可以获得极好的条纹图像（还有肌浆、细胞核以及颗粒体的图像）。可以通过改变两个光束之间的背景路径差来控制或者颠倒 A 带（折射率较高）和 I 带之间的明暗对比，这种调节不会影响带的测量宽度（图 1）。通过一个放电管以每隔约 20 毫秒闪光 10 次的速度进行系列闪拍，将纤维图像照到移动的底片上，信号可以通过与这些闪拍同步的脉冲电流而得到加强。

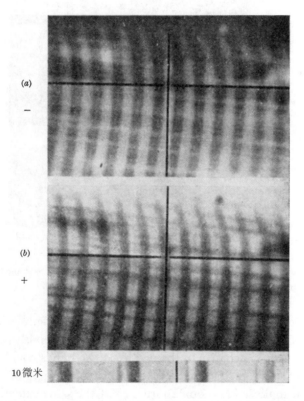

图 1. 肌纤维的负反差（a）和正反差（b）图片。（a）图中亮的部分和（b）图中暗的部分是 A 带（折射率较高）。请注意肌浆丝的折射率甚至比 A 带略高。肌节长度为 3.0 微米。

被动拉伸　通过被动拉伸或者松弛，肌节长度 s 能够从大约 2.0 微米变化到

about 2.0 to 4.2 μ, the value at the extended length in the body being 2.5 μ[6]. Almost the whole of this change of length took place in the *I*-bands (Fig. 2). The measured width of each *A*-band remained constant at 1.4–1.5 μ except for a fall to about 1.3 μ as *s* was reduced in the range 2.5–2.0 μ; but this fall may well not be real, as its amount is less than the resolving power of the optical system. When a fibre was stretched rapidly (20–30 percent in 5 msec.) the new ratio of *A*- to *I*-band widths appeared to be established without delay (less than 2 msec.).

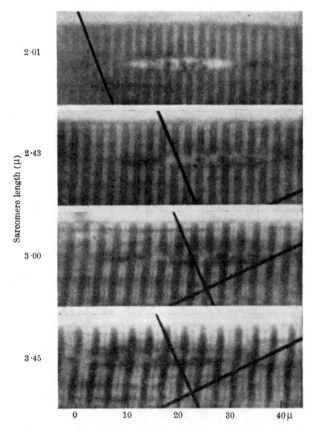

Fig. 2. Passive stretch of a muscle fibre. Positive contrast (*A*-bands dark). Sarcomere lengths[6] indicated beside the photographs. Almost all the change of length is in the *I*-bands (light).

Isometric twitches. Fibres were stimulated at a wide range of lengths with the tendon ends held stationary, and the twitch tensions, measured simultaneously with the *RCA* 5734 transducer, were normal. No change in the widths of the bands could be detected, except that when slight shortening of the region of the fibre in the field of view took place the changes were similar to those in isotonic shortening of the same extent.

Isotonic contractions (Fig. 3). Fibres were photographed during twitches and short tetani under isotonic conditions, with various initial lengths up to 3.2 μ per sarcomere. As in passive shortening, it was the *I*-bands that became narrower, the band-width of *A* being constant down to a sarcomere length of $s = 2.5$ μ and falling only slightly down to $s = 2.0$ μ.

98

4.2 微米，在体内延伸长度值为 2.5 微米 [6]。几乎全部的肌节长度的变化都产生于 I 带（图 2）。每个 A 带的测量宽度均恒定维持在 1.4 微米~1.5 微米，只有 s 降到 2.5 微米~2.0 微米时 A 带的测量宽度才会降到 1.3 微米左右；但是这种下降很可能是不真实的，因为它的数值低于光学系统的分辨率。当肌纤维被快速拉伸时（5 毫秒内拉伸 20%~30%），A 带与 I 带宽度新的比值也会即时确立（延迟在 2 毫秒以内）。

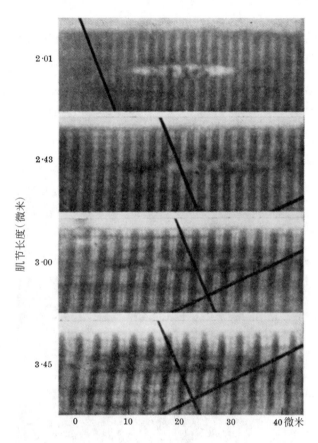

图 2. 肌纤维的被动拉伸。此图为正反差成像（A 带为暗的部分）。肌节长度在图旁标注出来。几乎所有长度的改变都产生于 I 带（亮的部分）。

等长颤搐　固定两端肌腱并刺激肌纤维使其处于不同长度，同时用 RCA 5734 换能器检测它的颤搐张力，发现其均处于正常范围。但是检测不到条带宽度的变化，只有当视野内肌纤维区域发生轻微缩短时，这样的变化才与同等程度的等张缩短中的变化类似。

等张收缩（图 3）　在等张条件下，对处于颤搐和短促的强直性收缩过程中具有不同起始长度的肌纤维（肌节长度长达 3.2 微米）进行了拍照。和被动缩短时一样，I 带变窄，当肌节长度 s 缩短到 2.5 微米时，A 带的宽度维持恒定；当肌节长度 s 缩短到 2.0 微米时，A 带的宽度略微下降。在进一步的收缩中，用电影摄影研究了持续

On further shortening, studied largely by ciné photography of slow contractions induced by constant-current stimulation, *A*-band width decreased definitely in all cases; but there were additional phenomena which were not the same in every experiment. The following sets of changes were observed on several occasions:

(*a*) Striations became extremely faint on shortening beyond a sarcomere length of about 1.8 μ (cf. Speidel[7]).

(*b*) The dense band narrowed to about half the sarcomere length, after which both bands narrowed in proportion.

(*c*) At *s* = 1.8 μ, a very narrow dense band was visible at the centre of the former *A*-band, and on shortening to *s* = 1.7 μ additional dense lines appeared midway between these (cf. Jordan's observations on stained preparations[8]).

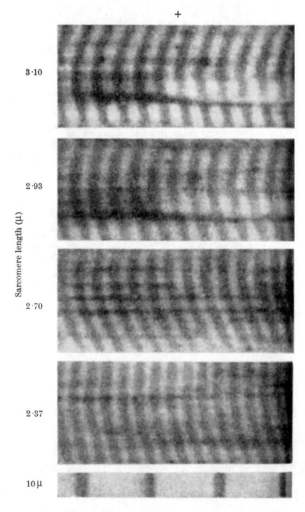

Fig. 3. Muscle fibre during a short isotonic tetanus. Positive contrast (*A*-bands dark). As in passive stretch (Fig. 2), the *A*-bands remain of almost constant width.

的电流刺激所引发的慢速收缩，在所有实验中 A 带宽度显著变窄；不过在各个实验组中也存在另外一些不同的实验现象。以下是多次观察到的变化：

(a) 当缩短超出一个肌节长度（约 1.8 微米）时，条纹会变得非常模糊（参见斯派德尔[7]）。

(b) 致密条带收缩到约半个肌节长度后，两条带成比例缩短。

(c) 当 s=1.8 微米时，在前面的 A 带的中心可以看见一条非常窄的致密条带，当缩短到 s=1.7 微米时，在两个致密条带中间又会出现一条致密的线（参见乔丹对染色样品的观察[8]）。

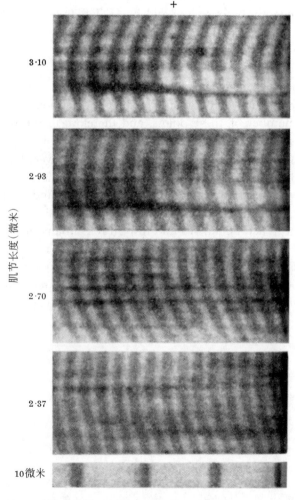

图 3. 肌纤维在短时间内等张强直性收缩时的照片。此图为正反差成像（A 带为暗的部分）。当被动拉伸时（图 2），A 带的宽度几乎维持不变。

The similarity of the changes during passive shortening and during isotonic contraction, and the absence of change during isometric twitches, show that the changes in the ratio of widths of the *A*- and *I*-band depend simply on the length of the fibre, and are unaffected by "activation" or by tension development as such. The approximate constancy of *A*-band width under a wide range of conditions (including shortening within the physiological range) agrees with the observations of Krause and of Engelmann[1], and also with those of H. E. Huxley and J. Hanson on separated myofibrils reported in the accompanying communication, though it is in conflict with the results of Buchthal *et al.*[3]. The natural conclusion, that the material which gives the *A*-bands their high refractive index and also their birefringence is in the form of submicroscopic rods of definite length, was put forward by Krause, and receives strong support from the observations reported here. The identification of this material as myosin[9], and the existence of filaments (presumably actin) extending through the *I*-bands and into the adjacent *A*-bands, as shown in many electron microscope studies, makes very attractive the hypothesis that during contraction the actin filaments are drawn into the *A*-bands, between the rodlets of myosin. (This point of view was reached independently by ourselves and by H. E. Huxley and Jean Hanson in the summer of 1953. It has already been mentioned by one of those authors[10] and is further discussed by them in the accompanying article.)

If a relative force between actin and myosin is generated at each of a series of points in the region of overlap in each sarcomere, then the tension per filament should be proportional to the number of these points, and therefore to the width of this zone of overlap. If the myosin rods are 1.5 μ long and the actin filaments 2.0 μ, the isometric tetanus tension should fall linearly as the fibre is stretched over the range of sarcomere lengths from 2.0 to 3.5 μ; this is in fair agreement with observation[11]. In arthropod striated muscle, there is a wide range of sarcomere lengths *in situ*, and narrowness of striation appears to be correlated with high speed of contraction[12]. This would be expected if the relative velocity between actin filaments and myosin rods in any one zone of overlap were the same for muscles of different sarcomere lengths, since the number of sarcomeres shortening in series per unit length is inversely proportional to sarcomere length. On this basis it would also be expected that the muscle with longer sarcomeres would be capable of producing a greater tension, but we are not aware of any experimental results on this point.

(173, 971-973; 1954)

A. F. Huxley and R. Niedergerke: Physiological Laboratory, University of Cambridge.

References:

1. Krause, W., "Die motorischen Endplatten der qnergestreiften Muskelfasern" (Hahn, Hannover, 1869). Engelmann. T. W., *Pflüg. Arch. ges. Physiol.*, **23**, 571 (1880). Rollett, A., *Denkschr. Akad. Wiss. Wien*, **58**, 41 (1891). Hürthle, K., *Pflüg. Arch. ges. Physiol.*, **126**, 1 (1909).

2. Hill, A. V., *Proc. Roy. Soc.*, B, **136**, 399 (1949).

3. Buchthal, F., Knappeis, G. G., and Lindhard, J., *Skand. Arch Physiol.*, **73**, 163 (1936). Buchthal, F., and Knappeis, G. G., *Acta Physiol. Scand.*, **6**, 123 (1943).

4. Smith, F. H., prov. Pat. Spec. No. 21996 (1947); complete Pat. Spec. No. 639014 (1950).

被动缩短和等张收缩过程中变化的相似性以及等长颤搐过程中没有变化都表明
A 带与 I 带宽度之比的改变仅仅取决于肌纤维的长度，而不受"激活"或拉伸这类
因素的影响。A 带宽度在多种条件下（包括在生理范围内的缩短）几乎都保持恒定，
这虽然与布克塔尔等人 [3] 的实验结果矛盾，但是与克劳斯和恩格尔曼 [1] 的观察结果
一致，也与赫胥黎和汉森在随后的文章中对分离出的肌原纤维的研究结果一致。本
文观察到的现象强有力地支持了克劳斯提出的结论，即 A 带的高折射率和双折射性
是由长度一定的亚显微棒状物质造成的。这种物质被证明是肌球蛋白 [9]，而且在许
多电子显微镜研究中都观察到有一种细丝（可能是肌动蛋白）贯穿 I 带并进入相邻的
A 带中。基于这些发现，我们提出了一个很有吸引力的假设，即在收缩过程中，肌动
蛋白丝被牵拉进肌球蛋白棒形结构间的 A 带。（这一观点由我们组和赫胥黎及琼·汉
森在 1953 年夏天分别独立提出。在随后的文章中他们进行了深入的探讨，并且他们
中有一位作者已经提及过这一观点 [10]。）

如果肌动蛋白与肌球蛋白之间的相对作用力产生于每个肌节中重叠部分的一系
列接触点，那么每根细丝的张力应该与这些接触点的数量成正比，并因此与重叠区
域的宽度成正比。如果肌球蛋白棒长 1.5 微米，肌动蛋白丝长 2.0 微米，那么当肌纤
维拉伸到超过肌节长度达 2.0 微米 ~3.5 微米时，等长强直性收缩张力应该呈线性下
降，这与实验结果非常一致 [11]。在节肢动物体内，横纹肌的肌节长度差别很大，并
且窄条纹似乎与高收缩速度有关 [12]。由于每单位长度上连续缩短的肌节数目与肌节
长度成反比，我们可以猜想，对于具有不同肌节长度的肌肉，任何重叠区域中肌动
蛋白丝和肌球蛋白棒间的相对速度是相同的。在这个基础上，我们也可以预期肌肉
的肌节越长，肌肉所能产生的张力也应该越大，不过关于这一点我们还不知道任何
实验结果。

（张锦彬 翻译；周筠梅 审稿）

5. Huxley, A. F., *J. Physiol.*, **117**, 52*P* (1952).

6. Sandow, A., *J. Cell. Comp. Physiol.*, **9**, 37 (1936).

7. Speidel, C. C., *Amer. J. Anat.*, **65**, 471 (1939).

8. Jordan, H. E., *Amer. J. Anat.*, **55**, 117 (1934).

9. Hasselbach, W., *Z. Naturforsch.*, **8b**, 449 (1953). Hanson, J., and Huxley, H. E., *Nature*, **172**, 530 (1953).

10. Huxley, H. E., *Biochim. Biophys. Acta*, **12**, 387 (1953).

11. Ramsey, R. W., and Street, S. F., *J. Cell. Comp. Physiol.*, **15**, 11 (1940).

12. Jasper, H. H., and Pezard, A., *C. R. Acad. Sci., Paris*, **198**, 499 (1934).

105

Changes in the Cross-striations of Muscle during Contraction and Stretch and Their Structural Interpretation

H. Huxley and J. Hanson

Editor's Note

In this paper, British biologists Hugh Huxley and Jean Hanson from the Massachusetts Institute of Technology use light microscopy to establish a sliding filament model of muscle contraction. The two researchers had previously shown that myosin is located in the thick filaments of the "*A*-band", whilst actin is located in the thin filaments of the "*I*-band". Here they photograph striated muscle fibres contracting on microscope slides and establish that the width of the *A*-band remains constant. They also show that the driving force behind muscle contraction is the splitting (hydrolysis) of molecules of ATP.

IN recent papers[1-3], we have described evidence concerning the location and arrangement of the two principal structural proteins, actin and myosin, in striated muscle at rest length. This evidence indicates that myosin is located in the anisotropic or *A*-bands, in the form of longitudinal filaments about 110 A. in diameter, spaced out in a hexagonal array 440 A. apart; these filaments are continuous from end to end of the *A*-band, and appear to be responsible for its high density and birefringence. Actin, on the other hand, is present in both the *A*-bands and the relatively isotropic or *I*-bands, in the form of filaments about 40 A. in diameter; these extend from the *Z*-lines, through the *I*-bands, and into the *A*-bands, where they lie between the myosin filaments and terminate on either side of the *H*-zone; the myosin filaments seem to have a somewhat greater thickness in this zone. Hasselbach[4] has reached similar conclusions about the location of actin and myosin, though his concept of the details of their arrangement is different from ours. We shall now describe evidence that during stretch, and during contraction down to about 65 percent of rest length, the length of the *A*-bands remains constant within the limits of accuracy of our measurements (5–10 percent), the changes in length of the muscle being taken up by changes in the length of the *I*-bands alone; further shortening beyond the point where the *I*-bands vanish (about 65 percent of the rest length) is accompanied by the formation of contraction bands where the *A*-bands have come into contact with the *Z*-lines. These changes appear to take place by a process in which actin filaments slide out of or into the parallel array of myosin filaments in the *A*-bands; as shortening proceeds, the actin filaments fold up in the *A*-band, and this folding continues after the *I*-bands have been fully retracted. The myosin filaments remain at constant length until forced to shorten by excessive contraction of the sarcomeres.

肌肉收缩与拉伸过程中的横纹变化及结构上的解释

赫胥黎，汉森

编者按

在这篇文章中，来自麻省理工学院的英国生物学家休·赫胥黎和琼·汉森通过光学显微镜建立了肌肉收缩的滑动细丝模型。之前这两位研究者指出肌球蛋白位于"A带"的粗肌丝中，而肌动蛋白位于"I带"的细肌丝中。本文中他们通过对显微镜载玻片上的横纹肌纤维收缩过程进行拍照，证明了A带的宽度保持恒定。他们也指出肌肉收缩的动力是ATP分子裂解（水解）。

在最近的几篇文章中 [1-3]，我们描述了关于两个主要的结构蛋白——肌动蛋白和肌球蛋白——在静息长度的横纹肌中的位置和排布方式的证据。这个证据表明肌球蛋白位于各向异性带或 A 带中，以直径约为 110 埃的纵向细丝形式存在，组成相互之间间隔 440 埃的六边形。从 A 带的这一端到那一端，这些细丝是连续的，使得 A 带具有高密度性和双折射性的特点。另一方面，肌动蛋白在 A 带和相对呈现为各向同性带或 I 带中都有分布。肌动蛋白相互聚集成直径约 40 埃的细丝，这些细丝从 Z 线开始伸展，穿过 I 带，进入 A 带。肌动蛋白丝位于肌球蛋白丝之间并终止于 H 区任何一边。在 H 区中，肌球蛋白丝的密度似乎更高。关于肌动蛋白和肌球蛋白的位置，哈塞尔巴赫 [4] 也得到类似的结论，不过这一结论在肌球蛋白与肌动蛋白的具体排布方式上与我们的模型有一定的差异。现在我们将给出证据，即在肌肉拉伸和收缩到静息长度的 65% 的过程中，在我们测量的精度范围内（5% ~10%），A 带的长度始终保持不变；而肌纤维长度的变化都源自 I 带长度的改变。当肌纤维继续缩短到 I 带消失时（约为静息长度的 65%）会伴随形成收缩带，A 带与 Z 线在此相接触。这些变化似乎是由肌动蛋白丝滑出或滑进平行排布于 A 带的肌球蛋白丝而引起的：当肌肉收缩时，肌动蛋白丝折叠进 A 带中，在 I 带完全收缩之后这个过程还会继续进行；而肌球蛋白丝的长度保持不变，直到肌节过度收缩时才被迫缩短。

Previous work on the changes in cross-striation accompanying stretch or contraction (reviewed by Jordan[5] and Buchthal, Knappeis and Lindhard[6]) has given results which in general we consider to be unreliable, for the following reasons. Observations made on intact muscle fibres in conventional light microscopes are liable to be misleading because of optical artefacts due to the thickness of the fibres. (This difficulty has been surmounted by the technique described by A. F. Huxley and R. Niedergerke in the accompanying paper.) Furthermore, normal contraction is so rapid that the changes taking place *during* the process are difficult to see and record. If fixed and sectioned material is used, it is possible to avoid optical artefacts and the necessity for rapid observation, but other kinds of artefacts are introduced. In spite of all these considerations, a number of workers, notably Speidel[7], have given accounts of changes of band pattern during stretch and contraction which we recognize as generally correct; but they do not establish the details of the changes with the precision necessary for satisfactory interpretation.

In order to avoid optical artefacts, we have used isolated myofibrils about 2 μ in diameter prepared by blending glycerol-extracted rabbit psoas muscle[8]. They are admirable objects for high-resolution microscopy in phase-contrast illumination or polarized light, and will contract when treated with adenosine triphosphate[9]. This contraction is a much slower process than contraction *in vivo*, and therefore provides favourable circumstances for detailed observation of the band changes taking place. The evidence that the mechanism of contraction in glycerol-extracted muscle treated with adenosine triphosphate is similar to that of normal contraction in living muscle has already been adequately discussed by Szent-Györgyi[8] and Weber and Portzehl[10]. We have also devised a simple technique for stretching isolated fibrils during observation. A suspension of fibrils, mounted as a very thin layer on a slide under a coverslip, is examined in the microscope until a fibril is found with one end embedded in a fibre fragment adhering to the coverslip, and its other end in a fragment attached to the slide. Movement of the coverslip in the appropriate direction will then produce the desired stretch or will permit the fibril to shorten if adenosine triphosphate is present. Very small movements can be produced with great ease by gentle pressure on the edge of the coverslip, for the thin layer of liquid provides smooth and highly viscous lubrication.

Photographs used for measurement were taken on microfile film at a magnification of ×370 or ×550, and an apochromatic phase-contrast objective of n.a. 1.15 was employed.

Contraction

We have studied and obtained photographic records of the details of contraction in the following systems:

以前曾有人对肌肉拉伸或收缩过程中的横纹变化进行过研究（参见乔丹[5]以及布克塔尔、纳培斯和林哈德[6]的综述），并取得了一些结果，但总体上我们认为这些结果是不可靠的，原因如下：由于肌纤维本身的厚度，使用普通光学显微镜对完整的肌纤维进行观察很容易被光学假象误导。（这一难题已经被赫胥黎和尼德格克在相关文章中所述的技术克服。）此外，肌肉的收缩是一个非常迅速的过程，人们很难对这个过程中发生的变化进行观察和记录。如果将样品固定并切片后再观察，虽然可以避免光学假象和需要快速观察的问题，但又引入了其他假象。尽管存在这些难题，许多研究人员特别是斯派德尔[7]对肌肉拉伸和收缩过程中条带样式的变化还是给出了解释，并且给出了我们认为大体正确的结论。但是，由于他们没有描述出具有足够精度的变化细节，所以未给出令人满意的解释。

为了避免出现光学假象，我们用混合甘油抽提的方法从兔腰大肌[8]中分离出直径约2微米的肌原纤维。这种纤维保持了收缩活性（在加入腺苷三磷酸后可以收缩[9]），非常适合用高分辨率相差显微镜或偏振光显微镜来观察。与体内收缩相比，这种收缩是一个较为缓慢的过程，因此为详细观察发生的条带变化提供了有利的条件。森特–哲尔吉[8]、韦伯和波策尔[10]已经充分论证了通过甘油抽提的肌肉在加入腺苷三磷酸后收缩的机制和活体肌肉正常收缩机制相似的证据。在观察中我们还设计了一种简单的拉伸分离的纤维的方法。我们首先将肌原纤维悬浮液滴到载玻片上，然后盖上盖玻片，使之铺展成很薄的一层。在显微镜下观察，寻找一端连在盖玻片上而另一端连在载玻片上的肌原纤维。往适宜的方向移动盖玻片就会产生期望的拉伸，或者在腺苷三磷酸存在的情况下肌原纤维将缩短。由于液体的薄层具有平滑和高黏的润滑作用，因此轻压盖玻片的边缘就可以很容易地产生小的位移。

所有的图片都是使用微缩胶片进行拍摄的，放大倍数为370倍或550倍。使用的是数值孔径为1.15的消色相差物镜。

收　缩

在下列系统中我们对肌肉收缩过程的细节进行了研究，并获得了一些图片记录：

(1) Fibrils contracting freely at room temperature (about 22 °C.) in 5×10^{-4} M adenosine triphosphate, 0.1 M potassium chloride, 10^{-3} M magnesium chloride, pH 7.0.

(2) Fibrils contracting freely at room temperature in a series of steps achieved by irrigating them with a succession of small amounts of a very dilute solution of adenosine triphosphate (5×10^{-6} M), irrigation being stopped as soon as the required degree of shortening had taken place.

(3) Fibrils contracting freely at a low temperature (about 2 °C.) in adenosine triphosphate in the almost complete absence of magnesium ions, when shortening takes place very slowly.

(4) Fibrils contracting as in (1) but held at both ends so that shortening is controlled.

Cinephotography was used for system 1; but in systems 2–4 it was possible to take "still" photographs on fine-grain film. In the first three systems the fibrils were contracting against virtually zero load and showed identical changes of band pattern. System 4 provided information about isometric contraction, and it was found that the changes of band pattern differed in some details from those recorded during contraction against zero load.

During contraction of single fibrils against zero load, we have observed the following changes in band patterns (illustrated in Figs. 1–4). The I-bands shorten from a resting length of approximately 0.8 μ until they disappear completely. During this process, the length of the A-bands remains constant at approximately 1.5 μ. Changes of density within the A-band do, however, occur. The H-zone, originally of low density, first becomes indistinguishable from the rest of the A-band, and is then replaced (at about 85 percent rest length) by a narrow zone which is denser than the rest of the A-band. At a slightly shorter sarcomere length (about 80 percent rest length), a very dense line becomes visible at either end of the A-band. The overall density of the A-bands decreases as the fibril diameter increases during shortening. When the I-bands disappear at about 65 percent rest length, contraction bands form at the lines of contact of adjacent A-bands. It is interesting to note that contraction down to 65 percent of the rest length covers the usual range of physiological shortening. With the further shortening which can usually be produced in isolated fibrils, the contraction bands become denser; during this process (or in some cases just before the I-bands disappear) the dense zone in the middle of the sarcomere splits into two lines which merge with the incoming contraction bands at approximately 30 percent rest length.

110

（1）室温（约 22℃）下，在含有 5×10^{-4} 摩尔 / 升腺苷三磷酸、0.1 摩尔 / 升氯化钾和 10^{-3} 摩尔 / 升氯化镁的 pH 值为 7.0 的溶液中肌原纤维自由收缩。

（2）连续使用少量低浓度（5×10^{-6} 摩尔 / 升）的腺苷三磷酸冲洗肌原纤维，当纤维缩短到目的长度时停止冲洗，通过这一系列步骤能获得室温下肌原纤维的自由收缩。

（3）在低温（约 2℃）、几乎没有镁离子的条件下，加入腺苷三磷酸后观察肌原纤维的自由收缩（此时肌原纤维的收缩会非常缓慢）。

（4）肌原纤维收缩，如同系统（1），但要固定肌原纤维的两端以便控制收缩。

我们在系统（1）中使用了电影摄影技术，但在系统（2)~(4）中则可以用微粒胶片拍摄"静态"照片。在前三个实验中，肌原纤维实际上都是在零负载的条件下进行自由收缩，因此呈现出一致的条带样式的变化。系统（4）提供了关于等长收缩的信息，其呈现出的条带样式的变化与零负载条件下条带样式的变化在一些细节上有所不同。

我们观察到单个肌原纤维在零负载下收缩时，其条带样式呈现出下列变化，如图 1~图4 所示。I 带从约为 0.8 微米的静息长度逐渐缩短直至它们完全消失，而 A 带的长度则始终保持恒定（约为 1.5 微米）。但在这个过程中，A 带的密度确实发生了变化。A 带中原本密度较低的 H 区在收缩后与 A 带中其他区域的界限开始变得不明显。当收缩至约为静息长度的 85% 时，H 带被一条很窄的比 A 带其他部位密度都大的带所取代。当肌节的长度收缩到约为静息长度的 80% 时，在 A 带的两端会出现一条高密度的线。肌原纤维的直径随着收缩过程逐渐变大，而 A 带的总密度呈逐渐减小的趋势。此外，I 带收缩至约为静息长度的 65% 时会消失，在与相邻的 A 带接触的地方会出现一条新的收缩带。有趣的是，我们注意到，达到静息长度的 65% 的收缩覆盖了生理状态下肌肉收缩的通常范围。分离得到的肌原纤维还可以进一步缩短，这使得收缩带的密度变得更高。在这个过程中（或者在 I 带恰好要消失前的一些情况下），肌节中部的密度区域分裂成两条带，当收缩到约为静息长度的 30% 时，这两条带会与新形成的收缩带合并。

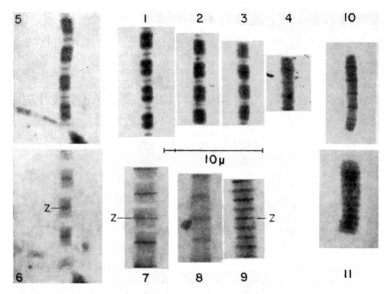

Myofibrils photographed in phase contrast. Magnification, 4,000 ×. Photographs of extracted fibrils are printed so as to give adequate contrast, and the fibrils are in fact much less dense than they appear here.
Figs. 1–4. The same four sarcomeres of one fibril photographed during contraction induced by adenosine triphosphate from rest length down to 50 percent rest length, when contraction bands have formed.
Figs. 5 and 6. Stretched fibril (115 percent rest length) before (Fig. 5) and after (Fig. 6) extraction of myosin.
Figs. 7, 8 and 9. Fibrils after extraction of myosin. Fig. 7: rest length. Fig. 8: 90 percent rest length. Fig. 9: 75 percent rest length.
Figs. 10 and 11. Fibril with contraction bands (50 percent rest length) before (Fig. 10) and after (Fig. 11) extraction of myosin.

Fibrils prepared from muscle which was allowed to shorten to equilibrium length (~80 percent rest length) before glycerol extraction usually lack the *H*-zones characteristic of rest-length fibrils, and have correspondingly shorter sarcomere lengths. Suspensions of untreated fibrils also include some specimens with sarcomere lengths down to about 65 percent of the rest length. Contraction presumably took place while the muscle was still excitable before glycerol extraction. Such fibrils exhibit the same characteristic band patterns and band lengths as those recorded for each sarcomere length during contraction induced by adenosine triphosphate.

During isometric contraction of isolated fibrils in adenosine triphosphate (achieved by holding the ends of the fibrils) the lengths of the *A*- and *I*-bands do not change. However, a narrow dark zone appears in the centre of the *A*-band (as in free contraction) as though some translation of material within the sarcomere were taking place, presumably accompanied by stretch of a series elastic component. This phenomenon is not observed in fibrils that are attached along their whole length to the coverslip or slide.

Stretch

We have found that isolated fibrils can stretch by two different processes, depending on whether the fibril has been "plasticized"[10], or whether it is being extended while in rigor or

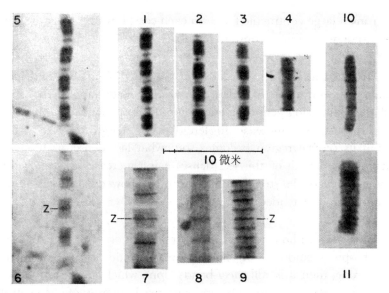

使用相差显微镜对肌原纤维进行 4,000 倍放大拍摄。为了获得适当的对比度，我们将提取的肌原纤维的照片打印了出来。实际观察到的肌原纤维的密度比图中所示的要低一些。

图 1~ 图 4. 在腺苷三磷酸的诱导下，同一肌原纤维的 4 个肌节在不同收缩程度时的照片，由静息长度一直收缩到静息长度的 50%（此时收缩带已经形成）。

图 5 和图 6. 肌球蛋白被抽提前（图 5）和被抽提后（图 6）拉伸的肌原纤维（静息长度的 115%）。

图 7、图 8 和图 9. 肌球蛋白被抽提后的肌原纤维。图 7：静息长度；图 8：静息长度的 90%；图 9：静息长度的 75%。

图 10 和图 11. 肌球蛋白被抽提前（图 10）和被抽提后（图 11）具有收缩带的肌原纤维（静息长度的 50%）。

在甘油抽提前，从缩短到平衡长度（约为静息长度的 80%）的肌肉中制备出来的肌原纤维通常缺少静息长度的肌原纤维特征性的 H 区。相应地，其肌节长度也较短；未处理过的肌原纤维的悬浮液中也含有长度只有静息长度 65% 的肌节。收缩可能发生在甘油抽提前肌肉仍具有应激性时。这些肌原纤维显示了同样的特征性条带样式和条带长度，这与腺苷三磷酸诱导的收缩中各肌节长度的记录结果一致。

分离得到的肌原纤维在腺苷三磷酸的诱导下发生等长收缩（通过固定肌原纤维的两端来实现）时，其 A 带和 I 带的长度并没有发生改变。但是，在 A 带的中央会出现一个窄的暗区（与自由收缩类似）。这似乎表明在肌节中发生了某种物质的移动，并且很可能是伴随着串联弹性组分的拉伸。在全长都被连在盖玻片或载玻片上的纤维中未观察到这一现象。

拉　伸

我们发现可以通过两种不同的途径拉伸分离得到的肌原纤维，这取决于肌原纤维是否已经被"塑料化"[10]，或者是否在强直状态下或正施加很强的收缩力时被拉伸。

while it is exerting a large contractile force. In each case, however, only the *I*-bands change in length; the *A*-bands remain at constant length.

If fibrils are stretched at 2°C., in the absence of magnesium ions and in the presence of rather high concentrations of adenosine triphosphate (about 10^{-2} *M*)—conditions which favour the plasticizing action of adenosine triphosphate rather than its contracting effect[10]—then the *I*-bands increase in length. The length of the *A*-band remains unchanged, but its central region becomes somewhat less dense, as though the *H*-zone were lengthening; the length of this less dense region increases as stretch proceeds. This process is perfectly reversible: stretched fibrils can be allowed to shorten until contraction bands form, and then re-extended.

If fibrils are stretched without any previous treatment, or in the presence of 10^{-4} *M* adenosine triphosphate and 10^{-3} *M* magnesium chloride at room temperature, when contraction is strong, then it is still the *I*-bands alone which increase in length. However, this increase is now not accompanied by any decrease in density of the central part of the *A*-band; on the other hand, the *Z*-lines now become appreciably fainter in spite of the fact that the diameter of the fibril is decreasing. (In the first type of stretch the *Z*-lines remain normal.) This suggests that some stretch is occurring in the region of the *Z*-line. This type of extension is also reversible.

We believe that the second type of stretch extends the series elastic component (Hill[11]), whereas the first type of stretch produces semi-plastic extension as in relaxed muscle.

The appearance of stretched muscle has also been investigated by examining fibrils prepared from fibres which were stretched immediately after removal from the rabbit. Such preparations contain a very high proportion of fibrils with long *I*-bands and the characteristic long zone of low density in the middle of each *A*-band (Fig. 5). We have made several hundred measurements of the lengths of *A*- and *I*-bands on such fibrils and can detect no significant difference in the lengths of *A*-bands from those found in fibrils with the shorter *I*-bands typical of rest length. These observations provide some evidence that the band pattern changes associated with stretch in whole living muscle fibres are similar to those seen in isolated fibrils which have been plasticized. Much more powerful evidence for this is described in the accompanying paper by A. F. Huxley and R. Niedergerke, to whom we are indebted for early reports of their results.

Myosin Extraction after Stretch and after Contraction

We have extracted myosin from fibrils of different sarcomere lengths by methods similar to those described previously[3]. In these earlier studies a dark zone always remained in the centre of the *A*-band of rest-length fibrils after extraction, although the density of the rest of the *A*-band was reduced to that of the *I*-band. Electron microscope studies showed that the dark band still contained the thicker filaments which elsewhere in the *A*-band had been removed. Hasselbach[4] has shown that this dark band disappears during prolonged myosin extraction, and we have found that this process can be accelerated by using 0.1 *M*

但不论哪种情况，只有 I 带的长度发生了变化，A 带的长度保持不变。

如果在 2 ℃的不含有镁离子但含有相当高浓度的腺苷三磷酸（约 0.01 摩尔/升）溶液中（这种条件下腺苷三磷酸会促进肌原纤维的塑料化而不会使其发生收缩）拉伸肌原纤维，那么 I 带的长度会增加[10]。A 带的长度保持不变，但中央区域的颜色变浅，好像 H 区伸长了一样。随着拉伸的进行，颜色变浅的区域的长度也随之增加。这一过程是完全可逆的：拉伸过的肌原纤维还可以缩短直到形成收缩带，之后还可以重新伸展。

如果没有预先对肌原纤维进行处理，或者在室温下含有 10^{-4} 摩尔/升腺苷三磷酸和 10^{-3} 摩尔/升氯化镁的溶液中就进行拉伸（此时肌原纤维会发生强烈收缩），那么仍然只有 I 带的长度增加。但此时这种增加并没有伴随 A 带中央区域的密度减小。另一方面，尽管肌原纤维的直径变小，Z 线却明显变得有点模糊（在第一种拉伸情况下，Z 线并不会发生变化）。这表明 Z 线所在的区域也被拉伸。这一过程也是可逆的。

我们认为第二种类型的拉伸拉长了串联弹性组分（参见希尔[11]），而第一种类型的拉伸使得肌原纤维出现了像松弛肌肉中那样的半塑料式的拉长。

通过检测刚从兔子中分离出来就立即进行拉伸的肌原纤维，我们也研究了肌肉拉伸的现象。如图 5 所示，我们发现很大比例的肌原纤维中都含有长的 I 带，同时在每条 A 带中央都具有特征性的长的低密度区。我们对这类肌原纤维中的 A 带和 I 带的长度进行了数百次测量，结果发现这些肌原纤维中 A 带的长度与在静息长度下具有典型的短的 I 带的肌原纤维中 A 带长度相比没有明显的不同。这些现象证明了在生物活体内的肌纤维中和在塑料化后分离得到的肌原纤维中肌小节受到拉伸时条带样式的变化是相同的。关于这一点更有力的证据参见赫胥黎和尼德格克的相关文章，他们实验结果的早期报告让我们很受益。

在拉伸和收缩后抽提肌球蛋白

采用与前述类似的方法[3]，我们将肌节长度不同的肌原纤维中的肌球蛋白抽提出来。虽然其余 A 带的密度减小到与 I 带近似，但是在早期的研究中人们还是可以发现抽提后静息长度下 A 带的中央存在一个暗区。利用电子显微镜进行观察可以发现，A 带中其他区域的粗丝都消失了，唯独暗带中仍有保留。哈塞尔巴赫[4]研究表明延长肌球蛋白抽提过程会使暗带消失。我们发现使用 pH 值为 7.0 的含有 0.1 摩尔/升焦磷酸盐和 10^{-3} 摩尔/升氯化镁的溶液可以加速这一过程。相应地，在这篇文章中，

pyrophosphate, $10^{-3} M$ magnesium chloride, pH 7.0. Accordingly, in the present studies we have employed this method for extracting all the myosin.

The appearance of fibrils extracted at different degrees of stretch and contraction is shown in Figs. 6–9 and 11. The "ghost" fibril consists of a faint backbone structure (which we believe consists largely of actin) the density of which is about the same as that of the original I-bands; the Z-lines are also still visible. In fibrils at rest length, an apparent gap is observed in place of the original H-zone (Fig. 7). In stretched fibrils, where there was originally a longer zone of low density in the centre of the A-band, the length of the gap is correspondingly greater (Fig. 6). The ghost fibrils are, however, still structurally continuous; stretched fibrils shorten spontaneously to a little less than rest length during extraction unless stuck to slide or coverslip; they may be reversibly extended again. This can be done with great ease, and it is apparent that only a very weak force opposes such a stretch; the gap elongates in the process, but the length of the material extending from the Z-line to the edge of the gap remains constant; that is, no stretching of an elastic component in the I-band region now occurs.

The ghosts obtained from contracted fibrils in which the H-zones had just disappeared have no gaps in the centres of the sarcomeres (Fig. 8). Fibrils which had shortened until dark lines appeared in the middle of the A-bands retain these lines after thorough extraction of myosin (Fig. 9); these dark lines are also present in ghosts from more strongly contracted fibrils (Fig. 11).

None of these "ghosts" will contract when treated with adenosine triphosphate.

Electron Microscopy

We have made a preliminary examination in the electron microscope of thin sections of stretched, contracted and myosin-extracted muscles prepared by the methods described elsewhere[2,12], and we have obtained results which, so far as they go, are in complete agreement with those obtained by light microscopy. This technique does not readily permit of reliable measurements of the lengths of sarcomeres or bands, for, apart from fixation artefacts, an unknown amount of compression is always present in even the best thin sections. In sections of stretched muscle, the majority of the A-bands show the characteristic long central region of low density (also observed by Philpott and Szent-Györgyi[13]); we find that the secondary array of thin (40 A.) filaments[2] is absent from this zone. In stretched muscle which has been subjected to myosin extraction the thick (110 A.) primary filaments are absent, and gaps apparently exist between the groups of thin (40 A.) filaments associated with successive Z-lines. In sections of glycerol-extracted muscle contracted in adenosine triphosphate, we have observed the same variety of band patterns seen in the light microscope. Up to the point where the I-bands disappear, the primary array of thick filaments remains apparently unchanged; and when contraction bands have been formed, the primary filaments between them are still straight. Our fixation of the secondary filaments has not yet been sufficiently good to allow us to describe adequately any changes, apart from translation into the A-bands, that may have taken place in them

我们采用这种方法来抽提所有的肌球蛋白。

图 6～图 9 以及图 11 显示了抽提后处于不同程度的拉伸和收缩状态的肌原纤维。"鬼影"肌原纤维包含一个非常模糊的骨架结构（我们认为主要由肌动蛋白组成），其密度大约和之前的 I 带相同。与此同时，Z 线仍然清晰可见。在处于静息长度的肌原纤维中，原先是 H 区的地方出现了一个明显的间隙（图 7），而在被拉伸的肌纤维中，在 A 带中央原本是长的低密度区的地方，相应地出现了更大的间隙（图 6）。然而鬼影纤维在结构上仍然是连续的，被拉伸的肌原纤维在抽提时会自动缩短到比静息长度稍微小一点的长度（除非肌原纤维粘在载玻片或盖玻片上）。它们还可以被再次拉长。这很容易做到，显然仅微弱的力就能阻碍这样的拉伸；并且伴随着拉伸我们可以观察到 H 带中的间隙逐渐变大，但从 Z 线到这一间隙边缘的距离则保持不变，也就是说，此时 I 带中的弹性组分没有被拉伸。

从 H 区刚刚消失了的处于收缩状态的肌原纤维中得到的鬼影纤维在肌节的中央没有间隙（图 8）。那些能缩短到暗线处（存在于 A 带中间）的肌原纤维在经过对肌球蛋白的彻底抽提后，这些线仍然存在（图 9）。来自收缩程度更强烈的肌原纤维的鬼影纤维中也可以观察到这些暗线（图 11）。

这些"鬼影"纤维在腺苷三磷酸的作用下都不会产生收缩反应。

电子显微镜

我们使用电子显微镜分别对拉伸、收缩以及抽提肌球蛋白后的肌肉薄切片进行了初步检测，制备方法在其他文章中有所提及 [2,12]，并且得到了与用光学显微镜观察完全一致的结果。不过，由于固定过程中会产生假象，以及即使在最薄的切片中样品也会不可避免地发生不同程度的收缩，因此这种方法不能可靠地测量肌节或条带的长度。在对拉伸的肌肉切片进行观察时，我们发现绝大多数 A 带具有特征性的长的低密度中央区域（菲尔波特和森特–哲尔吉也观察到相同的结果 [13]），并且在这个区域内没有直径为 40 埃的细肌丝 [2]。在抽提掉肌球蛋白后被拉伸的肌肉中，直径为 110 埃的粗肌丝也消失了，并且在细肌丝和 Z 线之间出现了明显的间隙区域。同样的，在对甘油抽提后由腺苷三磷酸诱导而收缩的肌原纤维的切片进行观察时，我们也得到了与使用光学显微镜观察到的相一致的条带样式。直到 I 带消失，粗肌丝的排布并未明显改变。当收缩带形成后，粗肌丝仍然是直的。受固定方法所限，我们只能观察到细丝在肌肉收缩过程中相对于 A 带发生了平移，但无法对这一过程中的其他细微变化

during contraction. During stretch of living muscle, however, the approximately 400-A. axial period seems to remain unchanged; a similar result was obtained by low-angle X-ray diffraction studies[1].

Conclusions

We believe that most of these changes in the cross-striations of muscle during stretch and contraction may be adequately described in terms of the following fairly simple model: The backbone of the muscle fibril is made up of actin filaments which extend from the Z-line up to one side of the H-zone, where they are attached to an elastic component (*not* the series elastic component) which for convenience we will call the S-filaments. The S-filaments provide continuity between the set of actin filaments associated with one Z-line and that associated with the next. The series elastic component is provided either by the actin filaments themselves or, more probably, by their mode of attachment to the Z-line. Myosin filaments extend from one end of the A-band, through the H-zone, to the other end of the A-band, and their length is unaltered by stretch or by contraction down to the point where the sarcomere length is equal to the length of the A-band; when contraction beyond this point takes place, the ends of the myosin filaments fold up and contraction bands form. Thus myosin and actin filaments lie side by side in the A-band and, in the absence of adenosine triphosphate, cross-linkages will form between them; the S-filaments are attached to the myosin filaments in the centre of the A-band by some more permanent cross-linkages.

In this model, plastic stretch takes place when the actin filaments are partly withdrawn from the A-band, leaving a long lighter central region and stretching the S-filaments in the process. Only the I-bands and the H-zones increase in length, the length of A-band remaining constant. This process would be inhibited by cross-linkages between the actin and myosin filaments; there is good evidence[10] that muscles are only readily extensible when such linkages are absent. When they are present, stretch would take place by extension of the series elastic component; in our model, this would again lead to an increase in the length of the I-band, but in this case no change in the length of the H-zone would take place. Contraction takes place in this model when the actin filaments are drawn into the A-band (until the H-zone is filled up) and are then folded up in some way to produce more extensive shortening. Thus, when the model is allowed to shorten, only the I-bands decrease in length until adjacent A-bands are pulled into contact with the Z-lines. When contraction is isometric, translation of actin filaments into the A-bands is accompanied by stretch of the series elastic component. It will be seen that in both stretch and contraction the behaviour of this model reproduces the observed behaviour of muscle quite faithfully.

A possible driving force for contraction in this model might be the formation of actin-myosin linkages when adenosine triphosphate, having previously displaced actin from myosin, is enzymatically split by the myosin. In this way, the actin filaments might be drawn into the array of myosin filaments in order to present to them as many active

进行充分的阐述。然而，在对活体肌细胞进行拉伸实验时可以观察到长度约为 400 埃的轴心区并未发生变化。通过小角度的 X 射线衍射实验 [1] 也可以得到类似的结果。

结　论

我们认为用下列非常简单的模型就可以描述肌肉拉伸和收缩过程中横纹所呈现出的变化：肌原纤维的骨架由肌动蛋白丝组成，它们从 Z 线开始伸展到 H 带边缘，在此它们与一种弹性组分相连（**不是串联弹性组分**），为简便起见称之为 S 细丝。这些 S 细丝使得相邻 Z 线的肌动蛋白丝相互连接起来。串联弹性组分可能是由肌动蛋白丝本身提供的，更有可能是由于它们与 Z 线连接的方式不同而产生的。肌球蛋白丝从 A 带的一端开始伸展，穿过 H 区直到 A 带的另一端结束，并且其长度在肌肉拉伸或收缩到肌节长度和 A 带长度相等的时候都不会改变；只有当肌节的长度缩短到比 A 带还短时，肌球蛋白的末端才会折叠起来形成收缩带。因此，肌球蛋白和肌动蛋白丝在 A 带中是紧挨着的，在没有腺苷三磷酸的情况下，它们之间会形成桥连；同时 S 细丝也会通过某种永久性的桥连与 A 带中央的肌球蛋白丝相连。

在这个模型中，当肌动蛋白丝从 A 带中部分滑出时会产生可塑性拉伸，使 H 区伸长并且颜色变浅，S 细丝也伸长。只有 I 带和 H 区的长度增加，A 带本身的长度保持恒定。肌动蛋白丝与肌球蛋白丝之间的桥连可以抑制该过程，并且有充分的证据表明只有未发生这样的桥接时肌肉才易于拉伸 [10]。在发生桥接时，拉伸可以通过串联弹性组分的拉长来完成。在我们的模型中，这会再次导致 I 带长度增加，但在实验中我们发现 H 区的长度并未发生变化。在我们的模型中，肌肉收缩时肌动蛋白丝向 A 带中滑入，直到完全进入 H 区，然后以某种方式折叠以产生更大规模的收缩。因此当肌肉收缩时只有 I 带的长度缩短，直到相邻的 A 带与 Z 线重合。在等长收缩中，伴随着肌动蛋白丝向 A 带中滑入，串联弹性组分也相应地被拉伸。无论肌肉拉伸还是收缩，上述模型都和观察到的肌肉的行为吻合得很好。

关于该模型中肌肉收缩的动力，我们认为其可能主要源自肌动蛋白与肌球蛋白之间桥接的形成。当腺苷三磷酸被肌球蛋白酶解时，肌动蛋白从肌球蛋白中移出，按照这个途径肌动蛋白丝可能被拉入到肌球蛋白束之中以便暴露出尽可能多的活性基团来形成肌动球蛋白。另外，如果肌动蛋白的结构允许足够多的活性基团暴露给

groups for actomyosin formation as possible; furthermore, if the structure of actin is such that a greater number of active groups could be opposed to those on the myosin by, for example, a coiling of the actin filaments, then even greater degrees of shortening could be produced by essentially the same mechanism. The model will remain contracted as long as the splitting of adenosine triphosphate continues; it will relax if the splitting stops and more adenosine triphosphate diffuses in, breaking actomyosin linkages and allowing the muscle to be re-extended. However, our results by no means require that contraction and relaxation be brought about in this way; and indeed, in the light of recent studies on actin[14,15], it would not be surprising if other processes are also involved. Furthermore, our results cannot exclude the possibility that repetitive configurational changes take place within the myosin filaments during contraction, unaccompanied by any overall change in the length of those filaments, and that these changes somehow bring about the observed movement of the actin filaments into the A-band.

These results will be described in greater detail elsewhere. We are much indebted to Prof. Francis O. Schmitt for his encouragement of this work, and to the Commonwealth Fund and the Rockefeller Foundation for their support.

(**173**, 973-976; 1954)

H. Huxley and J. Hanson: Department of Biology, Massachusetts Institute of Technology, Cambridge, Massachusetts.

References:

1. Huxley, H. E., *Proc. Roy. Soc.*, B, **141**, 59 (1953).

2. Huxley, H. E., *Biochim. Biophys. Acta*, **12**, 387 (1953).

3. Hanson, J., and Huxley, H. E., *Nature*, **172**, 530 (1953).

4. Hasselbach, W., *Z. Naturforsch.*, 8b, 449 (1953).

5. Jordan, H. E., *Amer. J. Anat.*, **27**, 1 (1920); "Physiological Reviews", **13**, 301 (1933).

6. Buchthal, F., Knappeis, G. G., and Lindhard, J., *Skand. Arch. Physiol.*, **73**, 163 (1936).

7. Speidel, C. C., *Amer. J. Anat.*, **65**, 471 (1939).

8. Szent-Györgyi, A., "Chemistry of Muscular Contraction" (2nd edit., Academic Press, N. Y., 1951).

9. Hanson, J., *Nature*, **169**, 530 (1952).

10. Weber, H. H., and Portzehl, H., "Advances in Protein Chemistry", 7, 161 (1952); "Progress in Biophysics" (in the press); we are indebted to Prof. Weber and to the Editors for allowing us to read this article.

11. Hill, A. V., *Proc. Roy. Soc.*, B, **137**, 273 (1950).

12. Hodge, A. J., Huxley, H. E., and Spiro, D., *J. Histo. and Cyto. Chem.*, **2**, 54 (1954).

13. Philpott, D., and Szent-Györgyi, A., *Biochim. Biophys. Acta*, **12**, 128 (1953).

14. Straub, F. B., and Feuer, G., *Biochim. Biophys. Acta*, **4**, 455 (1950).

15. Tsao, T.-C., *Biochim. Biophys. Acta*, **11**, 236 (1953).

肌球蛋白的活性基团（比如通过肌动蛋白丝的自身卷曲），那么便可以通过同样的机制进行更进一步的收缩。只要腺苷三磷酸不断地被水解，这种收缩过程就将一直持续下去。当腺苷三磷酸停止水解时，肌肉将松弛，更多的腺苷三磷酸扩散进入纤维，破坏肌动球蛋白的桥接，从而允许肌肉再度延伸。不过，我们并不认为肌肉收缩和松弛一定按照上述机制发生，事实上，根据最近的一些关于肌动蛋白的研究结果，即使肌肉的收缩还包含其他的过程也不足为奇 [14,15]。此外，我们不排除在收缩过程中肌球蛋白丝的构象发生重复改变的可能性，而并不伴随那些细丝总长度的变化，也不排除这些构象变化在一定程度上可能导致所观察到的肌动蛋白丝进入 A 带的运动。

我们会在别处对上述实验结果进行更详尽的描述。我们非常感谢弗朗西斯·施米特教授对本工作的鼓励，也非常感谢联邦基金和洛克菲勒基金会对我们工作的资助。

（张锦彬 翻译；周筠梅 审稿）

Structure of Vitamin B$_{12}$

D. C. Hodgkin *et al.*

Editor's Note

Vitamin B$_{12}$ was first identified in 1948 as the active ingredient in liver extracts used to treat pernicious anaemia. It was later manufactured by fermentation of bacteria, but chemists wondered whether it might be synthesized directly by chemical means. This required knowledge of the molecular structure of vitamin B$_{12}$. As a very complicated organic molecule, it posed a fearsome challenge to the usual method of structure determination, X-ray crystallography. That challenge is met in this paper by Dorothy Hodgkin of Oxford University and her colleagues, which was a key factor in Hodgkin's 1964 Nobel Prize in chemistry. The results show where every atom sits, enabling chemists in the USA and Switzerland to synthesize the compound between 1965 and 1972.

SINCE our communication in August 1955[1], we have carried our refinement of the structure of vitamin B$_{12}$ a critical stage further. Four more calculations of the electron density distribution have been made, one for the wet B$_{12}$ crystals, two for air-dried B$_{12}$ crystals and one for the hexacarboxylic acid.

In Fig. 1 are shown the electron density peaks over all the atoms of the B$_{12}$ molecule as it appears from our present calculations in the wet crystals. These peaks are rather lower than those observed for the hexacarboxylic acid owing to the smaller number of X-ray reflexions given by the larger molecule in proportion to its size. But they serve to show the relative positions of all the atoms in space of this very complex molecule within probably as little as 0.3 A. Similar evidence has been obtained for air-dried B$_{12}$; for both crystals the reliability indices in the latest structure factor calculations, 36.1 and 30.6 percent, respectively, are now low enough to indicate that the solution we have reached is likely to be essentially correct.

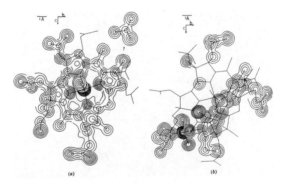

(a)   (b)

Fig. 1. Electron density-levels from the three-dimensional electron density distribution for wet B$_{12}$ crystals, calculated with terms phased on 93 atoms.

Here, and in Fig. 4, the contours are drawn in the sections of the calculated distribution parallel with the *a* place, passing at or near the atomic positions. Owing to the complexity of the molecule, complete contours for every atom are not given and the figure is divided into two parts. (*a*) shows the cobalt-containing nucleus and cyanide group; (*b*) the benziminazole, sugar, phosphate and propanolamine groups, the side chains being divided between the two. The acetamide chain on ring *B* is inset. The contour interval is 1 $e/$A.^{3}

维生素B₁₂的结构

霍奇金等

编者按

1948 年，科学家们首次发现了维生素 B₁₂，它是从肝脏提取物中分离得到的活性成分，用来治疗恶性贫血。后来又利用细菌发酵方法生产出维生素 B₁₂，但化学家一直尝试通过化学方法直接合成这种化学物质。这就需要了解维生素 B₁₂ 的分子结构。作为一种非常复杂的有机分子，利用常规的 X 射线晶体衍射分析法测定其结构是极其具有挑战性的。这篇文章的作者牛津大学的多萝西·霍奇金及其同事们就遇到了这样的挑战，这也是霍奇金获得 1964 年诺贝尔化学奖的关键因素。维生素 B₁₂ 分子中每个原子位置的确定为美国和瑞士的化学家在 1965 年～1972 年间合成该化合物提供了依据。

自从 1955 年 8 月发表维生素 B₁₂ 的结构以来 [1]，我们又对其进行了修正，并取得了关键性进展。我们已经获得了四种样品的电子密度分布的计算结果，包括一种含水维生素 B₁₂ 晶体、两种无水维生素 B₁₂ 晶体和一种六羧酸晶体。

图 1 所示的是含水维生素 B₁₂ 晶体分子上所有原子的电子密度峰，该图是依照我们目前对含水维生素 B₁₂ 晶体的计算结果绘制的。这些峰值比在六羧酸晶体上观察到的强度低得多。因为分子越大，X 射线反射的数量就越少，与分子的大小成比例。但是这些峰值表明这个非常复杂的分子上所有原子的空间相对位置可能仅有 0.3 Å。在无水维生素 B₁₂ 晶体上也得到了类似的结果。在最新的结构因子计算中，含水和无水两种晶体可靠因子值分别为 36.1% 和 30.6%，数据目前已经非常低，足以表明我们所使用的方法基本正确。

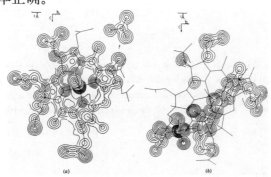

图 1. 含水维生素 B₁₂ 的电子密度分布水平图，数据来自三维的电子密度分布，通过计算 93 个原子的相位项获得。

图 1 和图 4 显示的轮廓是计算获得的部分电子密度分布图，该图平行于 a 平面（经过原子或者邻近原子位置）。因为这个分子比较复杂，所以每个原子的完整轮廓不能一次性描绘出来，而是分成了两个部分：(a) 显示的是含钴核心和氰基基团；(b) 显示的是苯并咪唑、糖、磷酸酯和丙醇胺基团，侧链则被分割到这两部分中。B 环的乙酰胺侧链是内嵌的。图中的轮廓间隔为：1 $e/Å^3$。

The molecule that appears is very beautifully composed, not far from spherical in form, with all the more chemically reactive groups on its surface. It is drawn in projection in Fig. 2. It is built around the two planes of the benziminazole nucleus and the central cobalt-containing nucleus, which are nearly at right angles to each other. The ribose ring turns in a position nearly normal to the benziminazole group, which permits its easy linking through the phosphate, propanolamine, and propionic acid residues to ring *D* of the planar group. The benziminazole nucleus is packed in on either side by the propionamide side-chains in the extended, staggered configuration, attached to rings *A* and *B*; the side-chain on ring *C*, in the gauche configuration, lies above the ribose ring. All the acetamide residues project from the opposite side of the nucleus to the propionamide residues, towards the cyanide group. It is an interesting point that one of them—that on ring *B*—swings round the carbon-carbon single bond from a position directed away from, to one in contact with, the cyanide group, when the crystals are removed from their mother liquor; this permits the rather closer packing of the molecules found in the air-dried crystals. Throughout the molecule, as shown by Figs. 1 and 2, the atomic positions found conform in a most convincing way with the stereochemical rules established by the study of simpler molecules.

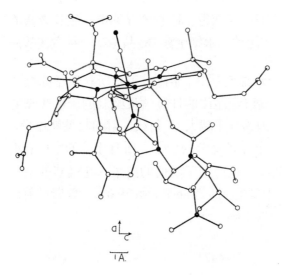

Fig. 2. The atomic positions found in the molecule of vitamin B$_{12}$. These are shown as derived in the wet crystals, projected on the *b* plane. To distinguish different parts of the molecule, bonds within the cobalt containing nucleus, benzimidazole and cyanide groups are shown in black, together with the cobalt and phosphorus atoms and the nitrogen atoms of the nucleus, benzimidazole, propanolamine and cyanide groups.

It now seems reasonably certain that we should formulate vitamin B$_{12}$, C$_{63}$H$_{88}$O$_{14}$N$_{14}$PCo, as in I, and the hexacarboxylic acid as in II, with six double bonds in the inner ring of the central nucleus; these can form a resonating system by intervention of the cobalt atom as illustrated by III. Some evidence in support of this has been obtained by theoretical calculations[2], by spectroscopic measurements[3] and from further chemical studies[4]. Our own evidence for this formulation is the geometrical form of the nucleus, which is illustrated by

维生素 B₁₂ 分子的组成非常巧妙，其造型近似球状，并且所有的化学反应活性基团都分布在其表面上。图 2 所示为投影图。它由两个面构成，分别是苯并咪唑核心和中间含钴核心，这两个面相互间夹角接近为直角。核糖环转向特定的位置以接近苯并咪唑基团，这样利于磷酸酯、丙醇胺和丙酸残基与 D 环平面的基团相连接。苯并咪唑核心要么堆积在由 A 环和 B 环上的丙酰胺侧链形成的伸展且相互交错的构型的同侧，要么堆积在由位于核糖环上面的 C 环丙酰胺侧链形成的邻位交叉构型的一侧。所有的乙酰胺残基都是沿着氰基方向从核心相反一侧投向丙酰胺残基。令人感兴趣的一点是，从母液中分离晶体时，B 环上的乙酰胺残基会绕着碳 – 碳单键旋转，从定向地远离氰基的位置转到与氰基相近的位置，这使得无水晶体形成更紧凑的分子结构。图 1 和图 2 所示为维生素 B₁₂ 整体分子结构，确定的原子位置符合那些根据较简单分子的研究建立起来的立体化学规则，因此这个结果是非常令人信服的。

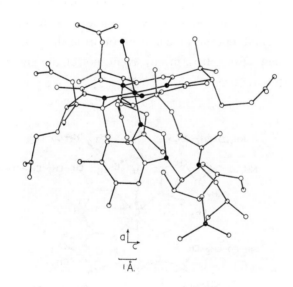

图 2. 含水维生素 B₁₂ 晶体中发现的原子位置。这些位置都是通过投射到 b 平面上获得的。为了区分该分子的不同部分，用黑实线表示含钴核心、苯并咪唑和氰基基团以及含钴核心、苯并咪唑、丙醇胺和氰基上的氮原子。

现在我们推导出维生素 B₁₂ 的化学式为 $C_{63}H_{88}O_{14}N_{14}PCo$，如（I）所示，这看起来是很合理的；六羧酸的化学式如（II）所示，其中央核的内环含有 6 个双键。这些双键可以形成烯键共轭系统并与钴原子发生络合，如（III）所示。通过理论计算 [2]、光谱测量 [3] 与更进一步的化学研究 [4] 获得的一些证据能够支持这个结论。我们用于支持这个化学式的证据来源于核心的几何学形状，如图 3 所示，此外，这三种

Fig. 3, and which is closely similar in all three crystals studied. The interatomic distances shown in this figure are derived from the latest calculations on the hexacarboxylic acid and are not individually very precise; in other parts of the molecule distances differing by as much as 0.16 A. from accepted bond-lengths have been found. But it is notable that in the region N$_6$–N$_{18}$ *all* the bond-lengths are shorter than normal single bond distances; their average value, 1.36 A., is similar to that found in the phthalocyanines[5]. The positions of C$_{29}$ and C$_{30}$, one or other of which would be expected to be out of the plane of the central ring system if only five double bonds were present, are even more significant. In the hexacarboxylic acid they appeared at all stages of the refinement to be within 0.2 A. of this plane, but in the B$_{12}$ crystals themselves their positions were at first more confused, particularly C$_{29}$. As a critical test, accordingly, this atom was omitted from the phasing calculations used for the latest electron distribution calculated for air-dried B$_{12}$. It has now appeared, as Fig. 4 shows, as a small but clearly defined peak at a site which is closely equivalent to that found in the hexacarboxylic acid along the line Co to C$_8$. There is still some departure from planar character in this region of the inner nucleus in both wet and air-dried crystals, but this seems to affect N$_{10}$, C$_9$ and C$_8$ as much as C$_{29}$. If real, it may be a consequence of a small amount of compression from the benziminazole nucleus, which is in contact with these atoms. It is, in any event, not of the geometrical form that would be expected if C$_8$ were reduced.

(I)

晶体的研究结果都极为相似。图中显示的是通过最新的计算方法获得的六羧酸分子中原子间的距离，但单个数据不是很准确；这个分子中其他部分的间距与已经被发现并被接受的键长相比最多相差 0.16 Å。但值得注意的是在 N$_6$~N$_{18}$ 区域内**所有**的键长都比正常单键的键长短。它们的平均键长为 1.36 Å，类似于在酞菁中发现的键长 [5]。如果只存在 5 个双键，那么 C$_{29}$ 和 C$_{30}$ 这两个位置不管哪一个都会偏离中心环系统平面，甚至比其他位置偏离得更为明显。对于六羧酸，在整个修正过程中，这两个位置总会在这个平面 0.2 Å 范围内显现。但是对于维生素 B$_{12}$ 晶体，它们的位置起初非常模糊，尤其是 C$_{29}$。因此，在进行关键实验的相位计算时，我们忽略这个原子，然后再计算无水的维生素 B$_{12}$ 最新电子分布。如图 4 所示，现在我们可以很清楚地看到，沿着钴和 C$_8$ 的连线，有一个细小但又清晰可辨的峰，与六羧酸上发现的近乎完全一致。在含水和无水的结晶中，内部核心区域的平面特征仍然存在着一定的偏离，但对 N$_{10}$、C$_9$ 和 C$_8$ 的影响程度与 C$_{29}$ 差不多。如果真是这样，将会产生的一个结果是与这些原子相连接的苯并咪唑核心会出现少量压缩现象。无论如何，如果 C$_8$ 不存在的话，那么这个分子也就不会形成几何学的形状了。

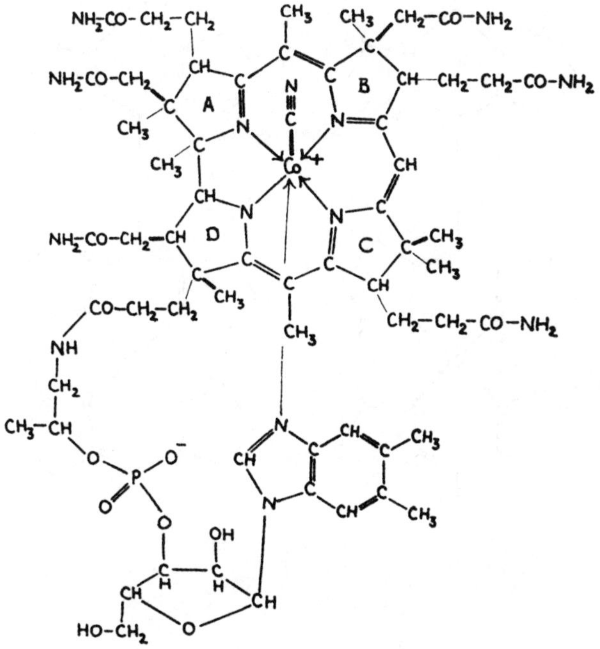

(I)

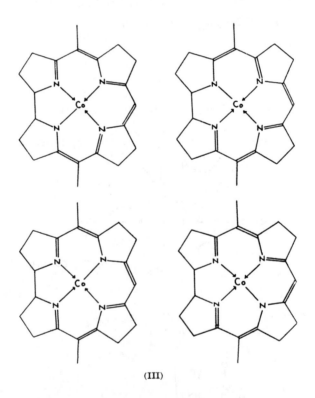

(II)

(III)

(II)

(III)

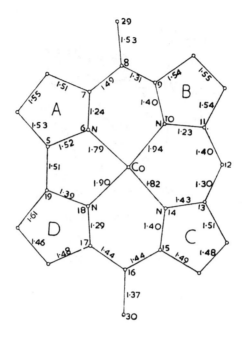

Fig. 3. Interatomic distances found in the inner nucleus of the hexacarboxylic acid, derived from nine refinements of the calculated electron density distribution.

Fig. 4. Electron density peaks over the nucleus in air-dried vitamin B$_{12}$, derived from the three-dimensional electron density distribution calculated with terms phased on 95 atoms (Princeton data).

The crystal structures of the different B$_{12}$ compounds are still not completely solved; many of the water molecules which occupy space between the vitamin molecules have been placed, but not all. Still further small adjustments are to be expected on continued

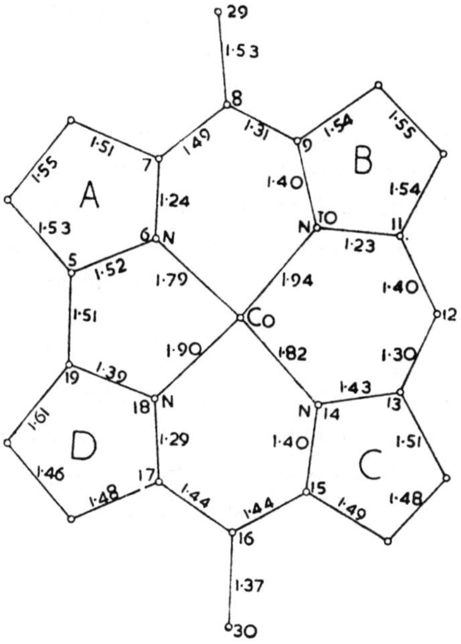

图 3. 六羧酸中内部核心内的原子间距。对计算得到的电子密度分布共进行了 9 次修正。

图 4. 无水的维生素 B₁₂ 核心上面的电子密度峰。数据来自三维的电子密度分布，是对分子中 95 个原子
的相位项进行计算得到的（数据来自普林斯顿）。

目前对不同维生素 B₁₂ 化合物的晶体结构还不完全清楚。位于维生素分子间隙
的许多结晶水分子都已确定，但并不是全部。我们仍然期望在所测量的原子位置的

refinement of the atomic positions observed. We hope that future calculations will add still greater precision and certainty to our knowledge of the structure of the molecule.

The latest series of calculations on wet B$_{12}$ were carried out on the University of Manchester electronic computer through the generous co-operation of Dr. D. W. J. Cruickshank, Miss Diana Pilling and Mr. J. F. P. Donavan, of the University of Leeds. Those on air-dried B$_{12}$ and on the hexacarboxylic acid were made on the National Bureau of Standards automatic computer in Los Angeles, supported by the United States Office of Naval Research.

Again we acknowledge the very generous support given this research by the Nuffield Foundation, and materials and help given us by Dr. E. Lester Smith, Glaxo Laboratories, Ltd., and Dr. K. Folkers, Merck Laboratories. Two of us (J. P. and J. K.) received grants from the Department of Scientific and Industrial Research.

(**178**, 64-66; 1956)

D. C. Hodgkin, F.R.S., J. Kamper, M. Mackay and J. Pickworth: Oxford.

K. N. Trueblood: Los Angeles.

J. G. White: Princeton.

References:

1. *Nature*, **176**, 325 (1955).

2. Orgel, L. E. (private communication).

3. Beaven, G. H., and Johnson, E. A., *Nature*, **176**, 1264 (1955).

4. Todd, A. R. (private communication).

5. Robertson, J. M., *J. Chem. Soc.*, 1195 (1936). Robertson, J. M., and Woodward, I., *ibid.*, 219 (1937).

持续修正方面作进一步的细小调整。我们也希望未来的计算会更加精确和明确，使我们对这个分子的结构更加了解。

最近这些基于含水维生素 B₁₂ 的一系列计算工作是在曼彻斯特大学的电子计算机上进行的，也是在与利兹大学的克鲁克香克博士、黛安娜·皮林女士和多纳文先生的大力合作下进行的。基于无水维生素 B₁₂ 和六羧酸计算的工作是在位于洛杉矶的美国国家标准局的自动计算机上进行的，其经费由美国海军研究办公室支付。

我们还要感谢纳菲尔德基金会对这项研究工作的大力支持，感谢葛兰素化学药品公司的莱斯特·史密斯博士提供了实验材料并给予帮助，也感谢默克实验室的福克斯博士的帮助。此外，我们中的两人（珍妮·皮克沃思和珍妮弗·坎珀）也得到了科学及工业研究署的资助。

（刘振明 翻译；吕扬 审稿）

A Three-dimensional Model of the Myoglobin Molecule Obtained by X-ray Analysis

J. C. Kendrew *et al.*

Editor's Note

Although X-ray diffraction had been used since the beginning of the twentieth century to analyse the structure of crystalline materials such as common salt, it had not yielded much information about the properties of materials involved in living things. Progress was made only after the Second World War in Europe. This paper reports the structure of the molecule called myoglobin, used by animals of all kinds for storing oxygen. The principal authors are John Kendrew and David Phillips. Their success in producing the first structure of a real-life protein depended on a technique developed by Max Perutz involving the attachment of various heavy atoms such as mercury to the protein molecule being studied. Kendrew and Perutz received the Nobel Prize for Chemistry in 1962.

MYOGLOBIN is a typical globular protein, and is found in many animal cells. Like haemoglobin, it combines reversibly with molecular oxygen; but whereas the role of haemoglobin is to transport oxygen in the blood stream, that of myoglobin is to store it temporarily within the cells (a function particularly important in diving animals such as whales, seals and penguins, the dark red tissues of which contain large amounts of myoglobin, and which have been our principal sources of the protein). Both molecules include a non-protein moiety, consisting of an iron–porphyrin complex known as the haem group, and it is this group which actually combines with oxygen; haemoglobin, with a molecular weight of 67,000, contains four haem groups, whereas myoglobin has only one. This, together with about 152 amino-acid residues, makes up a molecular weight of 17,000, so that myoglobin is one of the smaller proteins. Its small size was one of the main reasons for our choice of myoglobin as a subject for X-ray analysis.

In describing a protein it is now common to distinguish the primary, secondary and tertiary structures. The *primary structure* is simply the order, or sequence, of the amino-acid residues along the polypeptide chains. This was first determined by Sanger using chemical techniques for the protein insulin[1], and has since been elucidated for a number of peptides and, in part, for one or two other small proteins. The *secondary structure* is the type of folding, coiling or puckering adopted by the polypeptide chain: the α-helix and the pleated sheet are examples. Secondary structure has been assigned in broad outline to a number of fibrous proteins such as silk, keratin and collagen; but we are ignorant of the nature of the secondary structure of any globular protein. True, there is suggestive evidence, though as yet no proof, that α-helices occur in globular proteins, to an extent which is difficult to gauge quantitatively in any particular case. The *tertiary structure* is the way in which the

利用X射线分析获得肌红蛋白分子的三维模型

肯德鲁等

编者按

尽管早在 20 世纪初 X 射线衍射就已用于分析食盐等晶体物质的结构，但是对于生物体相关物质的性质并没有给出太多信息。直到第二次世界大战之后，这种情况才在欧洲有所改善。这篇文章报道了肌红蛋白（即各种动物用以储存氧气的物质）分子的结构。本文的主要作者是约翰·肯德鲁和戴维·菲利普斯。他们能够成功得到第一个生物体内蛋白质的结构，是借助了马克斯·佩鲁茨所开发的技术，即令汞等重原子附着于所研究的蛋白质分子上。肯德鲁和佩鲁茨获得了 1962 年的诺贝尔化学奖。

肌红蛋白是一种典型的球状蛋白质，存在于多种动物细胞之中。如同血红蛋白一样，它可以与氧分子可逆地结合；不过血红蛋白的作用是在血液流动过程中传输氧气，而肌红蛋白则是将氧气临时储存于细胞之中（这对鲸、海豹和企鹅等潜水动物来说是一种特别重要的功能，富含大量肌红蛋白的暗红色组织是我们获得这种蛋白质的主要来源）。两种分子中都含有非蛋白部分，由被称为血红素分子的铁 – 卟啉络合物组成，而实际上与氧结合的正是这个血红素分子。分子量为 67,000 的血红蛋白中含有四个血红素分子，而肌红蛋白中只含有一个血红素分子。这个血红素分子再加上大约 152 个氨基酸残基，总分子量合为 17,000，因此肌红蛋白是一种较小的蛋白质。我们选择肌红蛋白作为 X 射线分析对象的主要理由之一就是它的尺寸小。

在描述蛋白质时，通常要识别其一级、二级和三级结构。**一级结构**是指多肽链中氨基酸残基的排列顺序。桑格首先借助化学手段确定了胰岛素蛋白的一级结构 [1]，此后又有很多种多肽物质的一级结构被阐明，另外有一两种其他小蛋白质的部分一级结构也被解析。**二级结构**是指多肽链采取的折叠、卷曲或皱折的类型，例如 α 螺旋和折叠片。丝蛋白、角蛋白与胶原蛋白等多种纤维蛋白的二级结构已经有了大概轮廓，但我们对任何球状蛋白的二级结构的本质一无所知。当然，有一些尚未被证实的暗示性证据表明，α 螺旋存在于球状蛋白中，但是在各种具体事例中其含量还难以定量地确定。**三级结构**是指三维蛋白质分子的折叠和卷曲的多肽链在空间中形成的排布方式。要完全解释蛋白质的化学和物理性质，必须了解上述三个层次的结构，

folded or coiled polypeptide chains are disposed to form the protein molecule as a three-dimensional object, in space. The chemical and physical properties of a protein cannot be fully interpreted until all three levels of structure are understood, for these properties depend on the spatial relationships between the amino-acids, and these in turn depend on the tertiary and secondary structures as much as on the primary.

Only X-ray diffraction methods seem capable, even in principle, of unravelling the tertiary and secondary structures. But the great efforts which have been devoted to the study of proteins by X-rays, while achieving successes in clarifying the secondary (though not yet the tertiary) structures of fibrous proteins, have hitherto paid small dividends among the metabolically more important globular, or crystalline, proteins. Progress here has been slow because globular proteins are much more complicated than the organic molecules which are the normal objects of X-ray analysis (not counting hydrogens, myoglobin contains 1,200 atoms, whereas the most complicated molecule the structure of which has been completely determined by X-rays, vitamin B_{12}, contains 93). Until five years ago, no one knew how, in practice, the complete structure of a crystalline protein might be found by X-rays, and it was realized that the methods then in vogue among protein crystallographers could at best give the most sketchy indications about the structure of the molecule. This situation was transformed by the discovery, made by Perutz and his colleagues[2], that heavy atoms could be attached to protein molecules in specific sites and that the resulting complexes gave diffraction patterns sufficiently different from normal to enable a classical method of structure analysis, the so-called "method of isomorphous replacement", to be used to determine the relative phases of the reflexions. This method can most easily be applied in two dimensions, giving a projection of the contents of the unit cell along one of its axes. Perutz attached a p-chloromercuri-benzoate molecule to each of two free sulphydryl groups in haemoglobin and used the resulting changes in certain of the reflexions to prepare a projection along the y-axis of the unit cell[3]. Disappointingly, the projection was largely uninterpretable. This was because the thickness of the molecule along the axis of projection was 63 A. (corresponding to some 40 atomic diameters), so that the various features of the molecule were superposed in inextricable confusion, and even at the increased resolution of 2.7 A. it has proved impossible to disentangle them[4]. It was clear that further progress could only be made if the analysis were extended to three dimensions. As we shall see, this involves the collection of many more observations and the production of three or four different isomorphous replacements of the same unit cell, a requirement which presents great technical difficulties in most proteins.

The present article describes the application, at low resolution, of the isomorphous replacement method in three dimensions to type A crystals of sperm whale myoglobin[5]. The result is a three-dimensional Fourier, or electron-density, map of the unit cell, which for the first time reveals the general nature of the tertiary structure of a protein molecule.

Isomorphous Replacement in Myoglobin

No type of myoglobin has yet been found to contain free sulphydryl groups, so that the method of attaching heavy atoms used by Perutz for haemoglobin could not be employed.

因为这些性质取决于氨基酸分子之间的空间关系，而这些空间关系则不仅取决于蛋白质的一级结构，同样也取决于二级结构和三级结构。

一般而言，也只有 X 射线衍射的方法可以揭示二级和三级结构。尽管在阐明纤维蛋白二级结构（三级结构还不清楚）时 X 射线衍射方法取得了成功，但对于了解那些在新陈代谢过程中作用更为重要的球状蛋白或结晶蛋白，虽然投入了巨大的努力却仍然所获甚少。这一领域的进展缓慢，是由于球状蛋白分子比 X 射线分析的一般对象——有机分子结构复杂很多（不考虑氢原子，肌红蛋白中还有 1,200 个原子，而采用 X 射线方法完全确定结构的最复杂的分子——维生素 B_{12} 仅有 93 个原子）。五年前还没有人知道如何实际利用 X 射线方法来获得结晶蛋白的完整结构，而当时蛋白质晶体学家常用的方法最多只能对分子结构给出极为粗略的描述。佩鲁茨及其同事 [2] 的发现改变了这种情况。他们发现，重原子可以结合到蛋白质分子中特定的位置上，所得到的复合物给出了明显不同于普通蛋白质的衍射图案，这使得我们可以用一种名为"同晶置换方法"的经典结构分析方法确定其相对相位。这一方法可以很容易地用于二维分析，给出晶胞沿着某一轴向的投影图。佩鲁茨把对氯汞苯甲酸分子连接到血红蛋白分子中两个游离的巯基上，并利用某些反射的变化来获得晶胞沿 y 轴方向的投影图 [3]。令人遗憾的是，得到的投影图在很大程度上无法得到解释。这是因为分子在投影轴方向上的厚度为 63 Å（相当于大约四十个原子的直径），因此分子中的各种信息彼此重叠，导致了无法解释的混乱状况 [4]，即使将分辨率提高到 2.7 Å 也还是一样。很明显，只有取得新的进展后，才能将这种分析方法扩展到三维分析。就如同我们即将看到的，这将涉及更多观测结果的收集与同一晶胞的 3~4 种同晶置换晶体的制备支持，上述需求对于大多数蛋白质来说意味着巨大的技术困难。

本文介绍的是在低分辨率下研究抹香鲸肌红蛋白 [5]A 型晶体三维结构时同晶置换方法的应用。所得到的结果是晶胞的三维傅里叶谱图，或电子密度图，这是首次揭示蛋白质分子三级结构的一般特征。

肌红蛋白的同晶置换

迄今为止还没有发现任何一种肌红蛋白中具有游离的巯基，因此佩鲁茨将重原子结合到血红蛋白上的方法无法应用。最终，我们通过将肌红蛋白与重金属离子

Eventually, we were able to attach several heavy atoms to the myoglobin molecule at different specific sites by crystallizing it with a variety of heavy ions chosen because they might be expected, on general chemical grounds, to possess affinity for protein side-chains. X-ray, rather than chemical, methods were used to determine whether combination had taken place, and, if so, whether the ligand was situated predominantly at a single site on the surface of the molecule. Among others, the following ligands were found to combine in a way suitable for the present purpose: (i) potassium mercuri-iodide and auri-iodide; (ii) silver nitrate, potassium auri-chloride; (iii) p-chloromercuri-benzene sulphonate; (iv) mercury diammine ($Hg(NH_3)^{2+}$, prepared by dissolving mercuric oxide in hot strong ammonium sulphate), p-choro-aniline; (v) p-iodo-phenylhydroxylamine. Each group of ligands combined specifically at a particular site, five distinct sites being found in all. The substituted phenylhydroxylamine is a specific reagent for the iron atom of the haem group[6], and may be assumed to combine with that group; in none of the other ligands have we any certain knowledge of the mechanism of attachment or of the chemical nature of the site involved.

Methods of X-ray Analysis

Type A crystals of myoglobin are monoclinic (space group $P2_1$) and contain two protein molecules per unit cell. Only the $h0l$ reflexions are "real", that is, can be regarded as having relative phase angles limited to 0 or π, or positive or negative signs, rather than general phases; when introduced into a Fourier synthesis, these reflexions give a projection of the contents of the cell along its y-axis. In two dimensions the analysis followed lines[7] similar to that of haemoglobin. First, the heavy atom was located by carrying out a so-called difference-Patterson synthesis; if all the heavy atoms are located at the same site on every molecule in the crystal, this synthesis will contain only one peak, from the position of which the x-and z-co-ordinates of the heavy atom can be deduced, and the signs of the $h0l$ reflexions determined. These signs were cross-checked by repeating the analysis for each separate isomorphous replacement in turn; we are sure of almost all of them to a resolution of 4 A., and of most to 1.9 A. Using the signs, together with the measured amplitudes, we may, finally, compute an electron-density projection of the contents of the unit cell along y; but, as in haemoglobin and for the same reasons, the projection is in most respects uninterpretable (even though here the axis of projection is only 31A.). On the other hand, knowledge of the signs of the $h0l$ reflexions to high resolution enabled us to determine the x- and z-co-ordinates of all the heavy atoms with some precision. This was the starting point for the three-dimensional analysis now to be described.

In three dimensions the procedure is much more lengthy because all the general reflexions hkl must be included in the synthesis, and more complicated because these reflexions may have any relative phase angles, not only 0 or π. Furthermore, we need to know all three co-ordinates of the heavy atoms; the two-dimensional analysis gives x and z, but to find y is more difficult, and details of the methods used will be published elsewhere, including among others two proposed by Perutz[8] and one proposed by Bragg[9]. Finally, a formal ambiguity enters into the deduction of general phase angles if only one isomorphous

形成共晶，使几种重原子分别结合到肌红蛋白分子的几个特定位置上，选取重金属离子的依据是，按照一般的化学原理来考虑，可以预期它们与蛋白质分子侧链间具有亲和力。我们采用 X 射线衍射方法，而不是化学方法，来确定结合是否发生，并且如果结合的话，是否绝大多数重金属离子位于分子表面某一特定的位置上。在实验中，我们发现下列配体可以按上述要求方式与肌红蛋白结合：(i) 碘汞酸钾和碘金酸钾；(ii) 硝酸银，氯金酸钾；(iii) 对氯汞苯磺酸；(iv) 二氨合汞离子 $(Hg(NH_3)^{2+}$，将氧化汞溶解于热的浓硫酸铵中即可制得)，对氯苯胺；(v) 对碘苯羟胺。每组配体与某一特定的位置结合，总计有五个不同的位置。取代后的苯基羟胺对于血红素分子[6]中的铁原子来说是一种特效试剂，可以假定它是与血红素分子相结合的。对于其他配体来说，我们对其与肌红蛋白结合的机制和相关位置的化学特征都一无所知。

X 射线分析方法

肌红蛋白的 A 型晶体属于单斜晶系（空间群 $P2_1$），每个晶胞中含有 2 个蛋白质分子。只有 $h0l$ 反射线是"真正有用的"，也就是说，它们的相对相位角取值仅限于 0 或 π，于是有或正或负的符号，而不同于一般的相位；对这些反射线进行傅里叶合成后可以得到晶胞内分子沿 y 轴方向的投影图。二维分析的过程与对血红蛋白的分析过程类似[7]。首先，通过名为差值帕特森合成的方法对重原子进行定位。如果不同晶体分子中，所有重原子都位于相同位置，该合成结果将只包含一个峰位置，从峰的位置可以推算出重原子的 x 坐标和 z 坐标，并确定 $h0l$ 反射线的正负号。依次对每种同晶置换晶体重复进行同样的分析，即可得到正负号校验结果。我们确信几乎所有信息分辨率均能达到 4 Å，其中大多数能达到 1.9 Å。利用正负号与测得的结构振幅数据，我们可以计算出晶胞内沿 y 轴方向的电子密度投影图。但是与血红蛋白一样，由于相同的理由，所得的投影在很大程度上是无法解释的（尽管我们这里的投影轴仅有 31 Å）。另一方面，对高分辨率下 $h0l$ 反射线正负号的了解，使我们能够以一定的精度来确定所有重原子的 x 坐标和 z 坐标。这就是我们将要介绍的三维分析的起点。

在三维情况下，分析过程要漫长得多，因为进行合成时必须包括所有一般的 hkl 反射线，再加上这些反射线可能具有不仅限于 0 和 π 的任意的相对相位角，所以分析过程变得更为复杂。此外，我们需要知道重原子的三个坐标位置，缺一不可。二维分析给出了 x 坐标和 z 坐标，但是确定 y 坐标要困难得多。所用方法的细节将在其他地方发表，其中包括佩鲁茨[8]提出的两种方法和布拉格[9]提出的一种方法。最后，如果只有一种同晶置换晶体，那么对一般相角的推断就会出现形式上的不确定性。

replacement is available; this can be resolved by using several replacements[10], such as are available in the present case. Once the phases of the general reflexions have been determined, one can carry out a three-dimensional Fourier synthesis which will be a representation of the electron density at every point in the unit cell.

Before such a programme is embarked upon, however, the resolution to be aimed at must be decided. The number of reflexions needed, and hence the amount of labour, is proportional to the cube of the resolution. To resolve individual atoms it would be necessary to include at least all terms of the series with spacings greater than 1.5 A.—some 20,000 in all; and it is to be remembered that the intensities of all the reflexions would have to be measured for *each* isomorphous derivative. Besides this, introduction of a heavy group may cause slight distortion of the crystal lattice; as the resolution is increased, this distortion has an increasingly serious effect on the accuracy of phase determination. In the present stage of the analysis the most urgent objective was an electron-density map detailed enough to show the general layout of the molecule—in other words, its tertiary structure. If the α-helix, or something like it, forms the basis of the structure, we need only work to a resolution sufficient to show up a helical chain as a rod of high electron density. For this purpose we require only reflexions with spacings greater than about 6 A. ; in all there are some 400 of these, of which about 100 are $h0l$'s already investigated in the two-dimensional study. The Fourier synthesis described here is computed from these 400 reflexions only, and is in consequence blurred; besides this, it is distorted by an unknown amount of experimental error, believed to be small but at the moment difficult to estimate. Thus while the general features of the synthesis are undoubtedly correct, there may be some spurious detail which will require correction at a later stage.

The Three-dimensional Fourier Synthesis

The synthesis was computed in 70 min. on the EDSAC Mark I electronic computer at Cambridge (as a check, parts of the computation were repeated on DEUCE at the National Physical Laboratory). It is in the form of sixteen sections perpendicular to y and spaced nearly 2 A. apart; these must be piled on top of one another to represent the electron density throughout the cell, containing two myoglobin molecules together with associated mother liquor (which amounts to nearly half the whole). Unfortunately, the synthesis cannot be so represented within the two-dimensional pages of a journal; furthermore, if the sections are displayed side by side, they give no useful idea of the structure they represent. The examples reproduced in Fig. 1 illustrate some of the more striking features.

如果像本文实例一样使用多种同晶置换晶体[10]，就可以克服这个问题。一旦确定了一般反射线的相位，就可以进行三维傅里叶综合分析，获得结果代表了晶胞内每一点的电子密度。

不过，在开始进行这些处理之前，必须先确定需要达到的分辨率。所需反射线的数量以及与此对应的工作量正比于分辨率的立方。要分辨出每个原子，必须至少包含所有间距大于 1.5 Å 的反射线，总共约有 20,000 个。另外我们还必须记得对**每一种**同晶衍生物测定其全部反射线的强度。除此之外，重金属元素的引入还可能导致晶格的轻微变形。随着分辨率的提高，这种变形将对相位确定的精确性产生越来越严重的影响。对于现阶段的分析工作而言，最紧迫的任务是得到一张足以显示出分子的一般轮廓（即分子的三级结构）的电子密度图。如果构成分子结构的基础是 α 螺旋或者类似的结构，那么我们只需要分辨率达到让螺旋链以一根高电子密度链的形式显现出来即可。为了达到这个目的，我们就只需要那些间距超过 6 Å 的反射线，这总共约有 400 个，而且其中约有 100 个是 *h0l* 反射线，它们在研究二维情况时已经探讨了。这里所描述的傅里叶合成，是只对这 400 个左右的反射线进行计算得到的，因此是模糊的。此外，处理过程还受到某种大小未知的实验误差干扰，虽然我们认为它的影响不大，但是目前还难以估计这一误差。因此，尽管结果中的一般特征无疑是正确的，但其中可能还会存在细节问题，需要在下一步的研究中给予校正。

三维傅里叶合成

使用剑桥的 EDSAC（电子延迟存储自动计算机）Mark I 型电子计算机，70 分钟内完成合成计算（在国家物理实验室的 DEUCE 即通用电子数字计算机上对部分计算进行了重复检验）。获得结果是垂直于 *y* 轴的相互间隔大约 2 Å 的 16 个断层面。必须将这些断层面彼此叠放起来才能表示整体晶胞的电子密度，这包括两个肌红蛋白分子和其间结合的母液（约占全部的一半）。不幸的是，合成的结果无法在杂志的二维纸面上进行展示。另外，要将那些断层面合并排列，它们也不能真实地显示它们代表的结构。图 1 中展示的断层面图表现了一些很显著的特征。

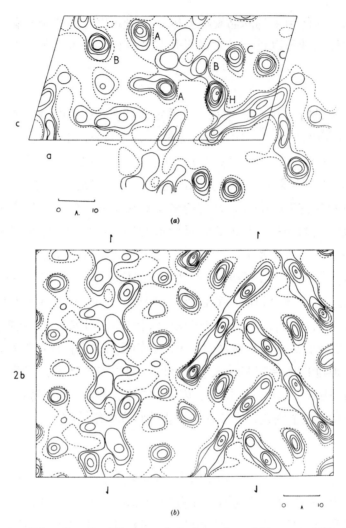

Fig. 1. (*a*) Section of three-dimensional Fourier synthesis of type *A* myoglobin at *y* = −1/8*b*. *A–D*, polypeptide chains; *H*, haem group. (*b*) Section parallel to [20$\bar{1}$] at *x*=0, showing polypeptide chain *A* (on the right).

A first glance at the synthesis shows that it contains a number of prominent rods of high electron density; these usually run fairly straight for distances of 20, 30 or 40 A., though there are some curved ones as well. Their cross-section is generally nearly circular, their diameter about 5 A., and they tend to lie at distances from their neighbours of 8–10 A. (axis to axis). In some instances two segments of rod are joined by fairly sharp corners. Fig. 1*a* shows several rods—three of them (*A*, *B* and *C*) cross the plane of the section almost at right angles, while one (*D*) lies nearly in that plane. *D* is part of a nearly straight segment of chain about 40 A. long, of which some 20 A. is visible in this section. It seems virtually certain that these rods of high density are the polypeptide chains themselves—indeed, it is hard to think of any other features of the structure which they could possibly be. Their circular cross-section is what would be expected if the configuration were helical, and the electron density along their axes is of the right order for a helical arrangement such as the

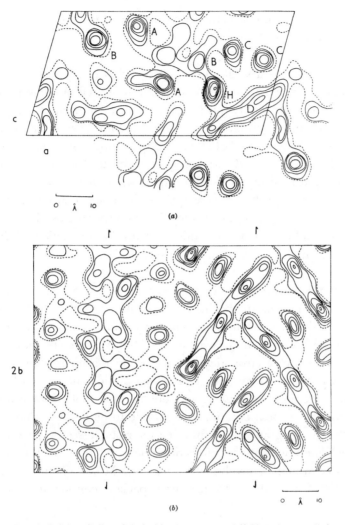

图 1. (*a*) *A* 型肌红蛋白进行三维傅里叶合成时位于 $y = -1/8b$ 处的断层面。*A~D* 代表不同的多肽链，*H* 代表血红素分子。(*b*) 位于 $x = 0$ 处与 [20$\bar{1}$] 平行的断层面，显示了多肽链 *A*（右边）。

  初步考察合成的结果，可以看出其中包含了许多明显的高电子密度链。它们通常会比较直地延伸 20 Å、30 Å 或 40 Å 的距离，尽管其中也有一些弯曲的链。它们的断层面一般近于圆形，直径约为 5 Å，它们倾向于以 8 Å~10 Å 的间距（轴与轴之间）排列。在某些地方，电子密度链的两个片断是以很尖锐的拐角连接。图 1*a* 显示了几个高电子密度链，其中三条（*A*、*B* 和 *C*）几乎垂直地穿过该断层面，而另一条（*D*）基本是平躺在该断层面上。*D* 是一段长约 40 Å 近乎笔直的链中的一部分，在这个断层面中可以看到其中约 20 Å 的部分。基本上可以确定，这些高电子密度链就是多肽链，事实上也很难认为它们会是结构中的其他部分。对于螺旋构型来说，就像预期的那样，其截面是圆形的，而且沿着轴向的电子密度分布也符合像 α 螺旋那样的螺

α-helix. The various rods in the structure are intertwined in a very complex manner, the nature of which we shall describe later.

Another prominent feature is a single disk-shaped region of high electron density which reaches a peak value greater than at any other point in the cell. A section through this disk is shown at H in Fig. 1a. We identify this feature as the haem group itself, for the following reasons: (i) the haem group is a flat disk of about the same size; (ii) its centre is occupied by an iron atom and therefore has a higher electron density than any other point in the whole molecule; (iii) a difference-Fourier projection of the p-iodo-phenylhydroxylamine derivative shows that, at least in y-projection, the position of the iodine atom is near that of our group; this is what we should expect, since this reagent specifically combines with the haem group; (iv) the orientation of the disk corresponds, as closely as the limited resolution of the map allows one to determine it, with the orientation of the haem group deduced from measurements of electron spin resonance[5,11].

We cannot understand the structure of the molecules in the crystal unless we can decide where one ends and its neighbours begin. In a protein crystal the interstices are occupied by mother liquor, in this case strong ammonium sulphate. the electron density of which is nearly equal to the average for the whole cell. Hence it is to be expected that in the intermolecular regions the electron density will be near average (the density of coiled polypeptide chains is much above average, and that of side-chains well below). It should also be fairly uniform; these regions should not be crossed by major features such as polypeptide chains. Using these criteria, it is possible to outline the whole molecule with minor uncertainties. It was gratifying to find that the result agreed very well, in projection, with a salt-water difference-Fourier projection made as part of the two-dimensional programme (for the principles involved, see ref. 12). Moreover, the dimensions of the molecule agreed closely with those deduced from packing considerations in various types of unit cell.

The Myoglobin Molecule

We are now in a position to study the tertiary structure of a single myoglobin molecule separated from its neighbours. Fig. 2 illustrates various views of a three-dimensional model constructed to show the regions of high electron density in the isolated molecule. Several points must be noticed. First, the model shows only the general distribution of dense regions. The core of a helical polypeptide chain would be such a region; but if the chain were pulled out, into a β-configuration, for example, its mean density would drop to near the average for the cell and the chain would fade out at this resolution. Similarly, side-chains should, in general, scarcely show up, so that the polypeptide rods in the model must be imagined as clothed in an invisible integument of side-chains, so thick that neighbouring chains in reality touch. Third, features other than polypeptide chains may be responsible for some of the regions of high density; patches of adsorbed salt, for example. Fourth, the surface chosen to demarcate a molecule cannot be traced everywhere with certainty, so it is possible that the molecule shown contains parts of its neighbours, and correspondingly lacks part of its own substance.

旋形排布的顺序。结构中的各种链以极为复杂的方式缠绕在一起，我们将在后面描述这一方式的特征。

另一个显著特征是存在一个单独的圆盘形高电子密度区域，其电子密度比晶胞内其他任何位置都高。图 1a 中 H 显示的就是穿过该圆盘的一个断层面。我们认为这一特征应该是血红素分子才有的：(i) 血红素分子具有平面结构；(ii) 中心被一个铁原子占据，因此电子密度比整个分子中其他任何部位都要高；(iii) 对碘苯羟胺衍生物的差值傅里叶投影图中显示，至少在沿 y 轴方向的投影图中，碘原子的位置是在这一基团附近，既然该试剂特异性地与血红素分子结合，那这一结果就和我们预期的是一致的；(iv) 在电子密度图有限的分辨率允许的范围内，我们得到的该电子密度图最准确的取向与利用电子自旋共振方法测量后推断出来的血红素分子的取向是一致的 [5,11]。

要理解晶体中分子的结构，必须先确定哪里是分子的末端位置和哪里是分子的起点位置。在蛋白质晶体结构中，空隙被母液即浓的硫酸铵溶液占据，其电子密度与整个晶胞的平均值近乎相等。因此可以预计，分子间区域的电子密度接近平均值（卷曲的多肽链的密度大大高于平均值，而侧链的密度则明显低于平均值）。并且这些区域的分布应该是非常均匀的，且不应与多肽链的主要区域交叉。根据这些原则就有可能以较小的不确定性勾画出分子的轮廓。令人欣慰的是，我们发现投影图中的结果与作为二维程序（其中所涉及的原理请参见参考文献 12）一部分的盐水差值傅里叶投影图吻合得非常好。此外，分子的尺寸与对各种类型晶胞进行填充所得出的结果十分吻合。

肌红蛋白分子

现在我们可以开始研究与其邻近分子隔开的单个肌红蛋白分子的三级结构。图 2 给出了三维分子模型的不同视角，这一模型显示了单个分子中的高电子密度区域。有几点是必须要注意的。首先，模型显示的只是高电子密度区域的大致分布。那些螺旋形多肽链的核心应该是这样的区域，但是如果多肽链伸展开，例如转变为 β 构型，那它们的平均电子密度将降低到晶胞电子密度的平均值附近，此时这些多肽链在该分辨率下就看不到了。与此类似，侧链基本上很难显示出来，因此要将模型中的多肽链想象成被一层不可见的侧链外壳所覆盖，这一外壳非常厚以至于邻近的链实际上是彼此接触的。第三，除多肽链之外的其他部分也可能对某些高密度区域有贡献，例如吸附盐分形成的斑点。第四，所选取的作为分子边界的表面并不是处处准确，因此显示出的分子中有可能包含其邻近分子的部分，同样也可能缺少其自身的某些部分。

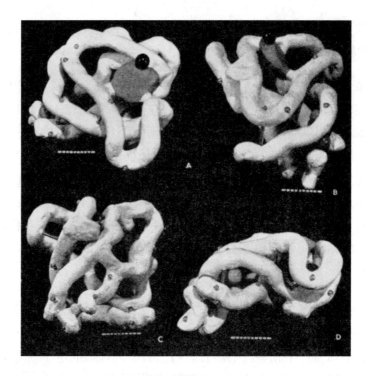

Fig. 2. Photographs of a model of the myoglobin molecule. Polypeptide chains are white; the grey disk is the haem group. The three spheres show positions at which heavy atoms were attached to the molecule (black: Hg of p-chloro-mercuri-benzene-sulphonate; dark grey: Hg of mercury diammine; light grey: Au of auri-chloride). The marks on the scale are 1 A. apart.

Making due allowance for these difficulties, we may note the main features. It is known[13] that myoglobin has only one terminal amino-group: it is simplest to suppose that it consists of a single polypeptide chain. This chain is folded to form a flat disk of dimensions about 43 A. × 35 A. × 23 A. Within the disk chains pursue a complicated course, turning through large angles and generally behaving so irregularly that it is difficult to describe the arrangement in simple terms; but we note the strong tendency for neighbouring chains to lie 8–10 A. apart in spite of the irregularity. One might loosely say that the molecule consists of two layers of chains, the predominant directions of which are nearly at right angles in the two layers. If we attempt to trace a single continuous chain throughout the model, we soon run into difficulties and ambiguities, because we must follow it around corners, and it is precisely at corners that the chain must lose the tightly packed configuration which alone makes it visible at this resolution (an α-helix, for example, cannot turn corners without its helical configuration being disrupted). Also, there are several apparent bridges between neighbouring chains, perhaps due to the apposition of bulky side-chains. The model is certainly compatible with a single continuous chain, but there are at least two alternative ways of tracing it through the molecule, and it will not be possible to ascertain which (if either) is correct until the resolution has been improved. Of the secondary structure we can see virtually nothing directly at this stage. Owing to the corners, the chain cannot be in helical configuration through out; in fact, the total length of chain in the model is 300 A., whereas an α-helix of 152 residues would be only 228 A.

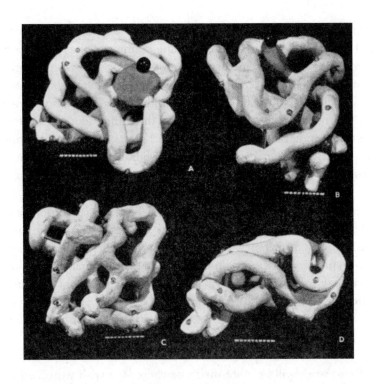

图 2. 肌红蛋白分子模型的照片。白色代表多肽链，灰色圆盘代表血红素分子。三个球指示出重原子在
分子表面结合的位置（黑色的是对氯汞苯磺酸中的汞，深灰色的是二氨合汞离子中的汞，浅灰色的是氯
金酸中的金）。图中标尺均代表 1 Å。

　　适当考虑这些问题之后，我们就可以关注主要的特征了。我们已经知道 [13] 肌红
蛋白中只有一个末端氨基，因此很容易推断出它只包含一条多肽链。这条链通过折
叠形成了尺寸为 43 Å×35 Å×23 Å 的扁圆盘。在圆盘内链经过大角度的转折缠绕出
一条很复杂的线，由于这些缠绕基本上都过于无规则而难以用简单的词汇描述其排
布情况。尽管有如此的不规则性，但我们还是可以看到一个明显的趋势即邻近链是
以 8 Å~10 Å 的间隔排列的。不太严格地说，分子是由两层链组成的，两层中链的主
要排列方向接近垂直。如果我们试图追踪单独一条贯穿模型的连续的链，那么我们
很快就会陷入困难和迷茫之中。因为我们进行追踪时必须绕过拐角，而恰恰是在拐
角处，这条链一定会丧失其紧密排列的构型，而现有的分辨率条件下只有紧密排列
的构型是可见的（例如 α 螺旋结构，它在不破坏螺旋构型的前提下是无法转过拐角
的）。另外，邻近链之间有明显的几处桥接，可能是由于大侧链的相互靠近。这一模
型无疑是与一条单独连续链相吻合的，但是至少存在两种贯穿追踪整个分子的方式，
而且要是不改进分辨率的话就不大可能确定究竟哪一种方式是正确的（如果其中某
一种是正确的话）。实际上目前我们还不能直接看到任何二级结构的信息。由于拐角
的存在，链不可能始终保持螺旋构型。实际上，模型中链的总长度为 300 Å，而一

long. The 300 A. might correspond, for example, to 70 percent α-helix and 30 percent fully extended chain, but of course intermediate configurations are probably present, too. The haem group is held in the structure by links to at least four neighbouring chains; nevertheless, one side of it is readily accessible from the environment to oxygen and to larger reagents such as p-iodo-phenylhydroxylamine (in the difference Fourier projection of this complex, referred to above, the position of the iodine atom indicates that the ligand is attached to the outside of the group). Clearly, however, the model cannot at present be correlated in detail with what we know of the chemistry of myoglobin; this must await further refinement.

Perhaps the most remarkable features of the molecule are its complexity and its lack of symmetry. The arrangement seems to be almost totally lacking in the kind of regularities which one instinctively anticipates, and it is more complicated than has been predicated by any theory of protein structure. Though the detailed principles of construction do not yet emerge, we may hope that they will do so at a later stage of the analysis. We are at present engaged in extending the resolution to 3 A., which should show us something of the secondary structure; we anticipate that still further extensions will later be possible— eventually, perhaps, to the point of revealing even the primary structure.

Full details of this work will be published elsewhere. We wish to record our debt to Miss Mary Pinkerton for assistance of all kinds; to the Mathematical Laboratory, University of Cambridge, for computing facilities on the EDSAC; to Dr. J. S. Rollett and the National Physical Laboratory for similar facilities on the DEUCE; to Mrs. Joan Blows and Miss Ann Mansfield for assistance in computing; for fellowships to the U.S. Public Health Service (H. W.), the Merck Fellowship Board (R. G. P.), the U.S. National Science Foundation (R. G. P. and H. M. D.), and the Rockefeller Foundation (H. M. D.); and to Sir Lawrence Bragg for his interest and encouragement. Finally, we wish to express our profound gratitude to the Rockefeller Foundation, which has actively supported this research from its earliest beginnings.

(**181**, 662-666; 1958)

J. C. Kendrew, G. Bodo, H. M. Dintzis, R. G. Parrish and H. Wyckoff: Medical Research Council Unit for Molecular Biology, Cavendish Laboratory, Cambridge.
D. C. Phillips: Davy Faraday Laboratory, The Royal Institution, London.

References:

1. Sanger, F., and Tuppy, H., *Biochem. J.*, **49**, 481(1951). Sanger, F., and Thompson, E. O. P., *ibid.*, **53**, 353, 366 (1953).

2. Green, D. W., Ingram, V. M., and Perutz, M. F., *Proc. Roy. Soc.*, A, **225**, 287 (1954).

3. Bragg, W. L., and Perutz, M. F., *Proc. Roy. Soc.*, A, **225**, 315 (1954).

4. Dintzis, H. M., Cullis, A. F., and Perutz, M. F. (in the press).

5. Kendrew, J. C., and Parrish, R. G., *Proc. Roy. Soc.*, A, **238**, 305 (1956).

6. Jung, F., *Naturwiss.*, **28**, 264 (1940). Keilin, D., and Hartree, E. F., *Nature*, **151**, 390 (1943).

7. Bluhm, M. M., Bodo, G., Dintzis, H. M., and Kendrew, J. C. (in the press).

条由152个残基组成的α螺旋的长度只有228 Å。300 Å的链长可能对应于多种构型，例如70% α螺旋构型和30%完全伸展的构型，当然中间构型很可能也是存在的。血红素分子通过与至少四条邻近链的连接而被固定于结构中。不过，它的一面很容易与环境中的氧或对碘苯羟胺这样的较大试剂接触（在上面提到过的复合物的差分傅里叶投影图中，碘原子的位置表明配体结合到了血红素分子的外表面）。不过很明显，目前该模型在细节上还不能与我们所知的肌红蛋白的化学性质结合起来，这有待于下一步的改进。

也许分子最显著的特征就是具有复杂性和缺乏对称性。看起来，排列形式几乎完全不具有人们通常所期望的规律性，而且它比任何蛋白质结构理论所预测的还要复杂。尽管还没有找到详细的构造原理，但我们希望能在下一阶段的分析中给出。目前我们正设法将分辨率提高到3 Å，这样我们将能看到某些二级结构的信息。我们期待未来还能有更进一步的提高，最终也许能达到揭示出一级结构的程度。

这一工作的完整细节将发表在其他地方。在这里我们要感谢为我们提供了各种帮助的玛丽·平克顿小姐，感谢允许我们在EDSAC上进行计算的剑桥大学数学实验室，感谢允许我们在DEUCE上进行同样计算的罗利特博士和英国国家物理实验室，感谢琼·布洛斯夫人和安·曼斯菲尔德小姐帮助我们进行计算。我们还要对美国公共卫生署（威科夫）、默克基金委员会（帕里什）、美国国家科学基金会（帕里什和丹特齐斯）和洛克菲勒基金会（丹特齐斯），对劳伦斯·布拉格爵士所给予的关注和鼓励表示感谢。最后，我们要向从一开始就给我们这项研究提供积极帮助的洛克菲勒基金会致以深切的谢意。

（王耀杨 翻译；吕扬 审稿）

8. Perutz, M. F., *Acta Cryst.*, **9**, 867(1956).

9. Bragg, W. L. (in the press).

10. Bokhoven, C., Schoone, J. C., and Bijvoet, J. M., *Acta Cryst.*, **4**, 275 (1951).

11. Ingram, D. J. E., and Kendrew, J. C., *Nature*, **178**, 905 (1956).

12. Bragg, W. L., and Perutz, M. F., *Acta Cryst.*, **5**, 277(1952).

13. Schmid, K., *Helv. Chim. Acta*, **32**, 105 (1949). Ingram, V. M. (unpublished work).

Sexually Mature Individuals of *Xenopus laevis* from the Transplantation of Single Somatic Nuclei

Gurdon *et al.*

Editor's Note

Here British developmental biologist John Gurdon and colleagues describe the production of the first sexually mature cloned animals, specifically frogs. Gurdon transferred nuclei from embryonic tadpole cells into unfertilized frog eggs, which went on to develop normally. A few years later he cloned frogs from adult cells, demonstrating irrefutably that nuclei from mature, differentiated cells could have their developmental clock reprogrammed. Gurdon's work paved the way for future somatic-cell nuclear-transfer experiments, including the arrival of Dolly, the first mammal to be cloned from an adult cell, nearly 40 years later. Today, somatic-cell nuclear-transfer experiments are still shedding light on the plasticity of the genome, as well as offering a route for the production of stem cells.

A method of testing the potentialities of nuclei from embryonic cells has been described by Briggs and King[1]. The method consists of transferring a nucleus from an embryonic cell into an enucleated and unfertilized egg of the same species. King and Briggs[2], who have performed their experiments on *Rana pipiens*, found that normal tadpoles resulted from eggs with transplanted nuclei in about 35 percent of cases in which the nuclei were taken from blastulae, but in only about 6 percent of cases in which the nuclei were taken from late gastrula endoderm; they have not reported normal development from nuclei of post-neurulae or later stages.

We have performed similar experiments on *Xenopus laevis*. In this species we do not enucleate the eggs, because the female pronucleus participates in the development of only a very few of the transplanted eggs; these cases can be recognized by the use of a nuclear marker[3], and are excluded from our results. The marker is introduced by taking donor nuclei from a stock of mutant individuals having only one nucleolus in all cells, as opposed to wild-type individuals in which all cells contain potentially two nucleoli. Large series of observations and measurements show no differences in the embryonic development, viability, growth-rate, and fertility between mutant individuals and their wild-type full sibs. The host eggs are taken from the wild-type stock and the female pronucleus introduces an additional nucleolar organizer when it participates in development. The origin of the nuclei in transplant-tadpoles can be interpreted with certainty by knowing the ploidy and number of nucleoli.

We have found that the great majority of normal tadpoles resulting from transferred nuclei

152

源于单个体细胞核移植的非洲爪蟾性成熟个体

格登等

编者按

本文中，英国发育生物学家约翰·格登和他的同事们描述了第一批性成熟克隆动物——特别是成蛙的产生。格登将处于胚胎时期的蝌蚪细胞的细胞核移植到未受精的蛙卵中，该蛙卵能继续正常发育。数年后，他用成体细胞获得了克隆蛙，这无可辩驳地证明了来自成熟的、已发生分化的细胞的细胞核可以重新编排其发育时钟。格登的工作为后来的体细胞核移植实验铺平了道路，其中包括在近40年之后由成体细胞克隆而来的第一只哺乳动物——多利羊的诞生。目前，体细胞核移植实验依然清楚地表明基因组具有可塑性，同时也成为获得干细胞的一条途径。

布里格斯和金[1]描述过一种可以用来检测胚胎细胞的细胞核潜能的方法。该方法是将胚胎细胞中的细胞核移植到同种生物未受精的去核卵细胞中。金和布里格斯[2]曾经用美洲豹蛙来做这个实验，他们发现采用来自囊胚期细胞的细胞核进行移植的实验中约有35%的个体可以发育成正常的蝌蚪，但是采用来自原肠胚后期内胚层细胞的细胞核进行移植的实验中只有约6%的个体可以发育成蝌蚪，他们没有报道来自神经胚后期或更晚时期的细胞核能够正常发育。

我们利用非洲爪蟾进行了类似的实验。对于这一物种我们并没有将卵细胞去核，因为卵原核仅参与了一部分移植卵的发育。通过应用细胞核标记能够证明这些情况[3]，而且我们的结果不包括这些情况。我们用来引入标记的供体细胞核来自所有细胞中只含有一个核仁的突变体库，与之相对应的野生型个体的所有细胞含两个潜在的核仁。大量的观察和测量结果表明，突变个体及其野生型全同胞在胚胎发育、生存能力、生长速度和繁殖力方面并无差别。实验中所用的受体卵细胞来自于野生型个体，卵原核参与发育时会引入一个额外的核仁形成区。核移植蝌蚪体内细胞核来源一定可以通过获知其染色体组倍数以及核仁的数目来确定。

我们发现通过核移植得到的正常蝌蚪中绝大多数都能进行正常的变态发育。年

153

pass through metamorphosis normally. The older frogs appear to be sexually mature males and females; the rate of growth and sexual differentiation is similar in transplant-embryos and controls. In order to test the capacity of gametogenesis in transplant-individuals, we hope shortly to breed from these. Immature oocytes were found in one transplant-frog which died two months after metamorphosis.

As with *Rana pipiens*, the proportion of normal development obtainable from transferred nuclei decreases with increasing age of the donor embryo. *Xenopus laevis* gives a smaller proportion of normal development from blastula nuclei than *Rana pipiens*, but on the other hand normal development can be obtained from much later donor stages; a prehatching tadpole (Nieuwkoop[4] stage 32, endoderm) was the most advanced stage of donor from which a normal individual was obtained, but this was accidentally killed shortly before metamorphosis. We have also a few frogs from nuclei which have been transferred twice (serial transfers).

Table 1 gives the numbers of metamorphosed frogs that have been obtained from different germ-layers and developmental stages of donor embryos. When transplanting endoderm nuclei from post-neurula stages care has been taken to avoid using nuclei from the region of the presumptive germ-cells.

Table 1. Metamorphosed Frogs which Resulted from Transplanted Nuclei

Donors			Transplants		
			Metamorphosed frogs		
Stage	Germ-layer	Total transfers	No.	Percentage of total	Percentage of normal late blastulae
Mid and late blastula	Ectoderm	565	17	3.0	12
	Endoderm	159	12	7.5	16
Early gastrula	Ectoderm	94	7	7.5	40
	Endoderm	287	12	4.2	15
Late gastrula	Ectoderm	26	2	7.7	33
	Endoderm	185	11	5.9	16.5
Neural folds	Ectoderm	9	0	0	0 out of 1
	Mesoderm	325	4	1.2	6
	Endoderm	137	7	5.1	14
Tail bud	Endoderm	163	4	2.5	9
Muscular response	Endoderm	287	1	0.35	3
Pre-hatching tadpoles	Endoderm	357	0	0	0 out of 70
Total		2,594	77		

When considering our results it is important to appreciate that the normal development of eggs with transferred nuclei may mean either that the nuclei were undifferentiated and totipotent, or that they were differentiated, but were able to return to a totipotent state as a result of developing again in the cytoplasm of an uncleaved egg.

154

长个体似乎已经发育为性成熟的雄性和雌性个体。核移植胚胎与对照组胚胎的生长速率与性别的分化类似。为了检测核移植个体的配子形成能力，我们希望这些个体能够快速繁殖。我们曾经在一只核移植蛙体内发现未发育成熟的卵母细胞，这只蛙在变态发育两个月后死亡。

与美洲豹蛙的实验结果一样，核移植后能进行正常发育的个体所占比例随着供体胚胎年龄的增加而减小。与美洲豹蛙相比，采用非洲爪蟾的囊胚期细胞核进行核移植得到的能正常发育的个体所占比例则相对较低。但另一方面，我们用发育时期更晚一些的胚胎细胞作供体进行核移植后能获得正常发育的个体，而孵化前的蝌蚪（尼乌科普 [4]，32 发育期，内胚层）是能获得正常个体的最晚时期的供体，但是这一个体在即将进行变态发育前意外死亡了。我们也有一些成体蛙，它们的核已经被移植了两次（即连续移植）。

表 1 给出了通过移植处于不同胚层和不同发育阶段的供体胚胎获得的变态发育的蛙的数量。在对神经胚后期的内胚层细胞核进行移植时应注意避免使用那些位于将来可能发育成生殖细胞的区域的细胞核。

表 1. 核移植获得的变态发育的蛙

供体			移植个体			
				变态发育的蛙数量		
发育时期	胚层	移植总数		数目	占总数百分比	能发育到囊胚晚期个体占总数百分比
囊胚中晚期	外胚层	565		17	3.0	12
	内胚层	159		12	7.5	16
原肠早期	外胚层	94		7	7.5	40
	内胚层	287		12	4.2	15
原肠晚期	外胚层	26		2	7.7	33
	内胚层	185		11	5.9	16.5
神经褶期	外胚层	9		0	0	0（总数为 1）
	中胚层	325		4	1.2	6
	内胚层	137		7	5.1	14
尾芽期	内胚层	163		4	2.5	9
肌肉效应期	内胚层	287		1	0.35	3
孵化前蝌蚪	内胚层	357		0	0	0（总数为 70）
总数		2,594		77		

在我们的实验结果中有一点是十分重要的，即带有移植核的卵细胞能够正常发育，这可能意味着用来移植的核是尚未分化的而且是全能的，或者是已经分化了的细胞核在尚未分裂的卵细胞质中重新发育而使得细胞的全能性得到恢复。

Our results may be summarized as follows: (1) It is possible to transplant nuclei without impairing their ability to bring about normal development. (2) Some nuclei are capable of giving normal development very shortly before the organ of which they are part becomes functional; normal development was obtained from presumptive somite nuclei nine hours before the first muscular responses. (3) Normal development results from nuclei of more advanced donor stages in *Xenopus* than in *Rana pipiens*. (4) A number of monozygotic frogs have been obtained from single donors.

It may now become possible to breed from our sexually mature individuals which have developed from single somatic nuclei, and to test their genetic qualities.

We gratefully acknowledge the technical assistance of Miss A. Jewkes. This work has been made possible by contributions from the British Empire Cancer Campaign (M. F. and T. R. E.) and the Medical Research Council (J. B. G.).

(**182**, 64-65; 1958)

J. B. Gurdon, T. R. Elsdale and M. Fischberg: Department of Zoology and Comparative Anatomy, University Museum, Oxford, May 9.

References:

1. Briggs, R., and King, T. J., *J. Exp. Zool.*, **122**, 485 (1953).

2. King, T. J., and Briggs, R., Cold Spring Harbor Symp., **21**, 271 (1956).

3. Elsdale, T. R., Fischberg, M., and Smith, S., *Exp. Cell Res.* (in the press).

4. Nieuwkoop, P. D., and Faber, J., "Normal Table of *Xenopus laevis*" (1956).

我们的结果总结如下：（1）在不损害实现其正常发育能力的条件下进行细胞核移植是可能的；（2）在胚胎发育过程中，即将进行部分功能分化的器官的细胞核具有使核移植个体正常发育的能力；在第一次肌肉效应前9小时，我们用即将发育为体节的细胞核进行移植获得的个体仍能正常发育；（3）与美洲豹蛙相比，用发育时期更晚的爪蟾胚胎细胞核进行移植后获得的个体仍能正常发育；（4）我们在单个细胞核移植的实验中，获得了一些同卵双生的成蛙。

现在，我们有可能对由单个体细胞核发育来的性成熟爪蟾个体进行繁殖，并检验它们的遗传学特性。

我们诚挚地感谢朱克斯小姐的技术协助。同时这项工作也是得到了大英帝国癌症运动组织（菲施贝格和埃尔斯代尔）以及医学研究理事会（格登）的支持才得以实现的。

（苏慧 翻译；刘京国 审稿）

A New Fossil Skull from Olduvai

L. S. B. Leakey

Editor's Note

Louis Leakey almost single-handedly made Africa the site of choice for palaeontological exploration from the 1930s to the present. He was born in Kenya, educated at Cambridge, England, and appointed director of the Coryndon Museum in Nairobi in 1945. In 1959, he reported the discovery of a new fossil skull from the Olduvai Gorge, a part of the East African Rift Valley. His first estimate of the age of the skull was 600,000 years, based largely on geological evidence, but this has since been corrected to 1.75 million years. Apart from his original work, Leakey founded a dynasty of African paleontologists. The discovery reported here was made by his wife Mary. His second son Richard Leakey became a later collaborator and was joined by his wife Meave and eventually his daughter Louise.

ON July 17, at Olduvai Gorge in Tanganyika Territory, at Site *FLK*, my wife found a fossil hominid skull, at a depth of approximately 22 ft. below the upper limit of Bed I. The skull was in the process of being eroded out on the slopes, and it was only because this erosion had already exposed part of the specimen that the discovery was possible. Excavations were begun on the site the following day and continued until August 6. As a result, an almost complete skull of a hominid was discovered. This skull was found to be associated with a well-defined living floor of the Oldowan, pre-Chelles–Acheul, culture.

Upon the living floor, in addition to Oldowan tools and waste flakes, there were the fossilized broken and splintered bones of the animals that formed part of the diet of the makers of this most primitive stone-age culture. It has not yet been possible to study the fauna found on this living floor; but it can be said that it includes birds, amphibians, reptiles such as snakes and lizards, many rodents and also immature examples of two genera of extinct pigs, as well as antelope bones, jaws and teeth.

It is of special importance to note that whereas the bones of the larger animals have all been broken and scattered. The hominid skull was found as a single unit within the space of approximately one square foot by about six inches deep. Even fragile bones like the nasals are preserved. The expansion and contraction of the bentonitic clay, upon which the skull rested and in which it was partly embedded, had resulted, over the years, in its breaking up into small fragments which have had to be pieced together. The bones, however, are not in any way warped or distorted. A large number of fragments still remain to be pieced together.

This very great difference between the condition of the hominid skull and that of the

来自奥杜威的新的头骨化石

利基

来自奥杜威的新的头骨化石

利基

编者按

路易斯·利基几乎是仅凭一己之力就使非洲成为了从 20 世纪 30 年代到现在普遍选择的古生物学研究之地。利基出生于肯尼亚，后就读于英国剑桥大学，1945 年被任命为内罗毕科里登博物馆的馆长。1959 年他报道在奥杜威峡谷（东非大裂谷的一段）发现了一件新的头骨化石。最初他主要依据地质学方面的证据，估算出该头骨的年代距今约为 60 万年，不过现已更正为 175 万年。除了他自己进行创造性的研究工作外，利基的家人也成为这一领域的权威，成就了一个非洲古生物学的家族研究。本篇报道中的发现是由他的妻子玛丽完成的。后来，他的二儿子理查德·利基也加入到这项工作之中，接着理查德的妻子梅亚维也加入进来，最后理查德的女儿路易丝也成为这项研究中的一员。

7 月 17 日，在坦噶尼喀地区奥杜威峡谷的 *FLK* 遗址中第 I 层上界之下深约 22 英尺的地方，我的妻子发现了一件人科化石的头骨。在斜坡上的头骨正处于被侵蚀出来的过程之中，正是因为受到了侵蚀令标本有一部分已经被暴露出来，使得发现它成为可能。次日我们便开始对这个遗址进行挖掘，并一直持续到 8 月 6 日。结果我们发现了一件近乎完整的人科的头骨。发现该头骨与奥杜威文化（前舍利–阿舍利文化）的一个清晰可辨的生活面具有紧密的联系。

在这个生活面中，除了奥杜威文化的工具及废弃的破片之外，还有已石化的破碎的或裂成破片的动物的骨。这些动物是这个最为原始的石器时代文化创造者们的部分食物。现在还不可能对在这个生活面上发现的动物群进行研究；但可以说这个动物群包括鸟、两栖动物、爬行动物（比如蛇与蜥蜴）、许多啮齿动物，还有两类已灭绝的猪的未成年个体，以及羚羊的骨、颌骨及牙齿。

有一点特别重要，那就是我们注意到稍大的动物的骨全破碎了，散落四处；但这一人科的头骨却是一个独立的实体，埋在约 1 平方英尺、6 英寸深的空间内。甚至像鼻骨这样易碎的骨都被保存了下来。头骨位于膨润土之上，且部分包埋在膨润土之中。多年来，由于膨润土的胀缩作用，导致头骨破裂成了小破片，于是我们不得不把它们拼在一起。不过，骨没有发生任何翘曲或变形。至今仍有大量破片有待拼凑起来。

在这一生活面上，人科动物头骨的状态与动物的骨的状态（所有动物的骨全是

159

animal bones on the same living floor (all of which had been deliberately broken up) seems to indicate clearly that this skull represents one of the hominids who occupied the living site; who made and used the tools and who ate the animals. There is no reason whatever, in this case, to believe that the skull represents the victim of a cannibalistic feast by some hypothetical more advanced type of man. Had we found only fragments of skull, or fragments of jaw, we should not have taken such a positive view of this.

It therefore seems that we have, in this skull, an actual representative of the type of "man" who made the Oldowan pre-Chelles–Acheul culture.

This skull has a great many resemblances to the known members of the sub-family of Australopithecinae. Some scientists recognize only one genus, namely, *Australopithecus*, and treat Broom's *Paranthropus* as a synonym; others consider that the demonstrable differences are of such a nature that both genera are valid. Personally, having recently re-examined all the material of the two genera, in Johannesburg and Pretoria, I accept both as valid.

The Olduvai skull is patently a member of the sub-family Australopithecinae, and in certain respects it recalls the genus *Paranthropus*. In particular, this is the case in respect of the presence of the sagittal crest, the great reduction in the size of the canines and the incisors, the relatively straight line of these teeth at the front of the palate, the position of the nasal spines and the flatness of the forehead. In certain other characters, the new skull resembles more closely the genus *Australopithecus*, for example in respect of the high cranial vault, the deeper palate and the reduction of the upper third molars to a size smaller than the second, all of which are features to be found in *Australopithecus* but not in *Paranthropus*.

The very close examination and direct comparisons which I have personally made in South Africa have convinced me that, on the basis of our present state of knowledge, the new skull from Olduvai, while clearly a member of the Australopithecinae, differs from both *Australopithecus* and *Paranthropus* much more than these two genera differ from each other.

I am not in favour of creating too many new generic names among the Hominidae; but I believe that it is desirable to place the new find in a separate and distinct genus. I therefore propose to name the new skull *Zinjanthropus boisei*. This generic name derives from the word "Zinj", which is the ancient name for East Africa as a whole, which is the specific name is in honour of Mr. Charles Boise, whose constant encouragement and financial help ever since 1948 have made this and other important discoveries possible. I would also like to acknowledge the generous help received, from time to time, from the Wenner-Gren Foundation and the Wilkie Trust.

The following is the preliminary diagnosis of the new genus and the new species:

人为蓄意弄碎的）之间的差别非常大，这似乎清楚地表明该头骨代表了占据这块生存地的人科中的一员；他们制造并使用工具，以动物为食。既然如此，我们没有理由去相信这块头骨代表的是一些假想的进化水平更高的人类同类相食的受害者。即使我们只发现头骨破片或颌骨破片，我们也不应该采纳上述看法。

因此，该头骨似乎确实代表了一种"人"，他们创造了奥杜威前舍利–阿舍利文化。

该头骨与南方古猿亚科已知成员的头骨相比有许多相似点。一些科学家认为仅存在南方古猿属这一个属，并把布鲁姆的傍人属当作同物异名；而其他科学家认为两个属之间显而易见的差异性表明这两个属都是成立的。最近我在约翰内斯堡与比勒陀利亚重新考察了这两个属的全部资料，个人认为应该分作两个属。

显然奥杜威头骨属于南方古猿亚科的成员，且在某些方面与傍人属相似。这尤其表现在以下方面：具有矢状脊，犬齿与门齿的大小减小很多，这些牙齿在腭前基本呈直线排列，鼻棘的位置，前额扁平。这件新头骨的某些其他特点更近似于南方古猿属，例如：头骨穹隆高，腭较深，第三上臼齿的大小减小到比第二臼齿还要小。所有这些特点都可以在南方古猿属中发现，而在傍人属中则找不到。

我亲自在南非进行了十分仔细的检查，并进行了直接对比，在我们目前知识水平基础上，我确信来自奥杜威的新头骨显然是南方古猿亚科的成员，其与南方古猿属及傍人属的差异度远远大于这两个属之间的差异度。

我不赞同在人科之下创立太多新属名；但我认为应该将这件新发现的头骨归为一个独立、不同的属。因此我建议命名这一新头骨为鲍氏东非人。属名来源于单词"Zinj"，是整个东非的古名，而种名是为了纪念查尔斯·鲍伊斯，自 1948 年来，是他的不断鼓励与资金支持才使得这个重要的发现及其他重要的发现成为可能。我还要对不时收到的来自温纳–格伦基金会与威尔基信托基金会的慷慨资助表示感谢。

以下是新属与新种的初步鉴定：

Zinjanthropus gen. nov. :

Genotype: a young male with third molars not yet in wear and sutures relatively open, from *FLK* I, Olduvai.

A new genus of the Hominidae, sub-family Australopithecinae, which exhibits the following major differences from the genera *Australopithecus* and *Paranthropus*:

(*a*) in males a nuchal crest is developed as a continuous ridge across the occipital bone;

(*b*) the inion, despite the great evidence of muscularity, is set lower (when the skull is in the Frankfurt plane) than in the other two genera;

(*c*) the posterior wall of the occipital bone rises more steeply to form, with the parietals, a very high-vaulted posterior region of the skull;

(*d*) the foramen magnum is less elongate and has a more horizontal position than in *Australopithecus* (in the crushed skulls of *Paranthropus* it is not possible to be quite sure of the plane of the foramen magnum);

(*e*) the presence of a very massive horizontal ridge or torus above the mastoids. This is much more marked than the normal type of supra-mastoid crest;

(*f*) the mastoids are more similar to those seen in present-day man, both in size and shape;

(*g*) the presence of a strong wide shelf above the external auditory meatus, posterior to the jugal element of the temporal bone;

(*h*) the shape and form of the tympanic plate, whether seen in *norma lateralis* or in *norma basalis*. In this character the new skull has similarities with the Far Eastern genus *Pithecanthropus*;

(*i*) the very great pneumatosis of the whole of the mastoid region of the temporal bones, which even invades the squamosal elements;

(*j*) the massiveness of the jugal element of the temporal bone relative to the total size of the temporal bone;

(*k*) the way in which the parietals rise almost vertically behind the squamous elements of the temporal before bending over to become a dome;

(*l*) the relative thinness of the parietals in comparison with the occipitals and the temporals;

东非人新属：

正型标本：年轻男性，第三臼齿还没有磨损，骨缝相对不闭合，出土于奥杜威 *FLK* 遗址第 I 层。

人科南方古猿亚科的一个新属与南方古猿属和傍人属相比，其表现出来的主要不同点如下：

(*a*) 男性的项脊发育成一个横跨枕骨的连续骨脊；

(*b*) 枕外隆凸点，尽管存在强有力的证据证明其肌肉强壮，但当将这一头骨置于法兰克福平面的位置时此测点低于其他两个属的；

(*c*) 枕骨后壁向上陡倾，与顶骨一起形成拱起程度很高的头骨后区；

(*d*) 枕骨大孔延伸得不长，相比于南方古猿它处于更加水平的位置上（在压碎了的傍人头骨上，不可能确定枕骨大孔所在的平面）；

(*e*) 在乳突之上有十分巨大的水平脊或圆枕。这比正常类型的乳突上脊要明显得多；

(*f*) 乳突的大小与形状都更类似于现在的人类所呈现的乳突；

(*g*) 在外耳道上方，颞骨颧突后面出现了强而宽阔的猿板；

(*h*) 不论是从侧面看还是从底部看，鼓板的形态与形状都与远东爪哇猿人属相似；

(*i*) 颞骨的整个乳突区气窦很大，甚至侵入鳞部；

(*j*) 相对于颞骨的总体尺寸，颞骨颧突很硕壮；

(*k*) 在颞骨的鳞部之后，顶骨几乎是垂直升起，之后弯曲形成一个圆顶；

(*l*) 与枕骨及颞骨相比，顶骨相对较薄；

(*m*) the very prominent and keeled anterior margin of the crests on the frontal bone for the anterior segment of the temporal muscles in the region of the post-orbital constriction (even the most muscular male *Paranthropus* exhibits nothing comparable);

(*n*) the very unusual position of the nasion, which is on the most anterior part of the skull, instead of being behind and below the glabella region;

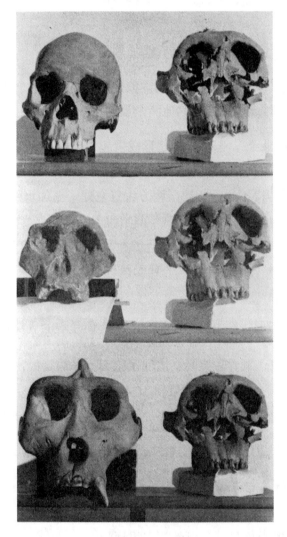

Fig. 1. *Above*: The new skull compared with the skull of an Australian aboriginal. Note the very long face, the architecture of the malar region, the unusual nasal bones, the torus above the mastoid, the sagittal and nuchal crests. *Middle*: The new skull compared with a cast of the most complete adult of *Australopithecus*. Note the difference in the size and shape of the face, the shape of the tympanic plate, the low position of the inion, the huge mastoid, as well as the difference in the shape of the malar region and the supra-orbital area. *Below*: The new skull seen next to that of a gorilla.

(m) 眶后缩狭区供颞肌前段附着的额骨上的脊的前缘十分突出，呈龙骨状突起（甚至肌肉最强健的男性傍人也没有可与之比较的结构）；

(n) 鼻根点位置尤为与众不同，是在头骨的最前部，而不是在眉间点区的后下方；

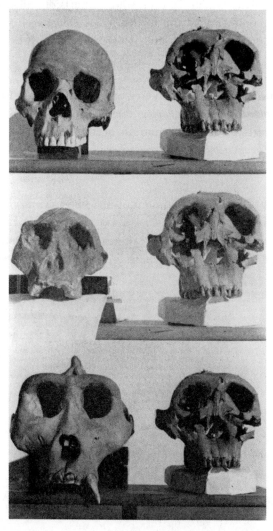

图 1. **上**：新头骨与澳大利亚土著人头骨的比较。注意其很长的脸、颧骨区的构造、与众不同的鼻骨、乳突上方的圆枕骨、矢状脊与项脊。**中**：新头骨与最完整的成年南方古猿的铸模型比较。注意面部形状和大小的差异、鼓板形状、位置低的枕外隆凸点、巨大的乳突，以及颊部与眶上部形状的差异。**下**：新头骨与大猩猩的头骨的比较。

(*o*) the very great absolute and also relative width of the inter-orbital area, with which may be associated the shape of the nasal bones, which are much wider at the top than at their inferior margin;

(*p*) the whole shape and position of the external orbital angle elements of the frontal bone;

(*q*) the very deep palate which is even more markedly like that of *Homo* than in *Australopithecus*, and is quite unlike the form seen is *Paranthropus*, except in respect of the more or less straight canine–incisor line which has already been commented on, as a character recalling *Paranthropus*;

(*r*) the conformation of the malar–maxillary area of the cheek. In all known members of the genera *Australopithecus* and *Paranthropus* there is a buttress of bone which runs down from the malar towards the alveolar margin of the maxilla in about the region of the fourth premolar; in *Zinjanthropus* this buttress is wholly absent and the form of architecture of this region is that which is found in *Homo*;

(*s*) the very great area of muscle attachment on the inferior margin of the malars;

(*t*) the relatively greater reduction of the canines in comparison with the molar–premolar series than is seen even in *Paranthropus*; where it is a marked character.

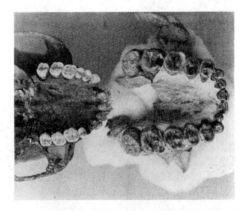

Fig. 2. The palate of the new skull compared with that of an East African native.

Zinjanthropus boisei sp. nov.

A species of *Zinjanthropus* in which the males are far more massive than the most massive male *Paranthropus*. The face is also excessively long. Males have a sagittal crest, at least posteriorly. Upper third molars smaller than the second.

The above is only a preliminary diagnosis of the genus *Zinjanthropus* species *boisei*. It is recognized that, if and when further material is found, the diagnosis will need both enlarging and possibly modifying.

(o) 眶间区的绝对宽度、相对宽度都相当大，可能与鼻骨的形状有关，其顶部比其下缘宽得多。

(p) 额骨眼眶外侧角部分的位置与整体形状；

(q) 腭很深，其与人属的相似程度明显更甚于与南方古猿属的相似程度，并完全不同于在傍人属中所见到的形式，除了犬齿–门牙排列大致呈直线，这一特点与傍人属的特点相似，这在上文已经提到过；

(r) 脸颊的颧骨–上颌骨部位的构造。南方古猿属与傍人属的所有已知成员都有壁柱状的构造，它从颧骨一直延伸到大约是第四前臼齿部位的上颌骨牙槽边缘；东非人完全没有这种壁柱状的构造，该部位的结构与人属完全相同；

(s) 颧骨下缘肌肉附着的面积很大；

(t) 与臼齿–前臼齿齿系相比，犬齿相对减小的程度甚至比傍人的还要大；这是一个显著的特点。

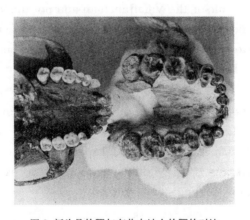

图 2. 新头骨的腭与东非本地人的腭的对比

鲍氏东非人新种：

男性东非人比最魁梧的男性傍人还要魁梧得多。面部也相当长。男性有矢状脊，至少在后部是这样。第三上臼齿比第二臼齿小。

上述仅是对鲍氏东非人的初步鉴定。我们认识到，如果发现了进一步的材料，将需要对鉴定要点进行补充并可能进行修改。

The whole question of generic value is one which is relative. There are some who maintain that *Australopithecus* and *Paranthropus* are not generically distinct, and who will wish to treat *Zinjanthropus* as a third, but less specialized, species of a single genus; but the differences seem to be too great for this.

I must now turn to the absolute and relative geological age of the new skull. As stated earlier, *Zinjanthropus* comes from Olduvai Gorge, about 22 ft. below the upper limit of Bed I. It was found in association with tools of the Oldowan culture, on a living floor and with associated fauna.

In the past it has been customary to regard Olduvai Bed I as a part of the Middle Pleistocene, not differentiation it from Bed II. During the last few years, however, detailed excavations at sites *BK* II, *SHK* II and *HWK* II have shown that there is a constant and well-marked break between the top of Bed I and the base of Bed II. It is incidentally on this clearly defined land surface that Chellean Stage I living sites are found.

There has also been found a great deal of new faunal evidence, and it is now clear that the fauna of Olduvai Bed I is the same as that of Omo, and that both are generally of the same age as that of Taungs. In other words, it is now necessary to regard Olduvai Bed I as representing the upper half of the Villafranchian and not the lower part of the Middle Pleistocene. So far as relative dating is concerned, it now seems clear that in the Far East the Djetis beds belong to the Middle, rather than to the Lower, Pleistocene, so that the new Olduvai skull would be older than the oldest *Pithecanthropus*.

In South Africa, the deposits at Taungs and Sterkfontein are now regarded as belonging to the upper part of the Lower Pleistocene; they must therefore be regarded as generally contemporary with Olduvai Bed I. The Makapan beds are a little younger, in all probability, while Swartkrans is of Middle Pleistocene age, as are the upper beds at Sterkfontein which are now yielding stone tools.

With the Taungs child, therefore, and the *Australopithecus* fossils from the lower beds at Sterkfontein, the new find represents one of the earliest Hominidae, with the Olduvai skull as the oldest yet discovered maker of stone tools.

属的价值问题是比较而言的。有些人坚持认为南方古猿属与傍人属不是两个区别明显的属，他们想要将东非人看成是与上二者同一个属的第三个较欠特化的种。但这样做的话，它们之间的差异似乎太大了。

现在我必须开始讨论新头骨的绝对地质年龄与相对地质年龄。正如前文所述，东非人来自奥杜威峡谷第 I 层上界之下深约 22 英尺的地方。在同一生活面上还发现了属于奥杜威文化的工具以及动物群。

过去一直习惯把奥杜威第 I 层作为中更新世的一部分，没有把它与第 II 层进行区分。不过最近几年，在遗址 BK II、SHK II 及 HWK II 所进行的详细的挖掘显示，在第 I 层的顶部与第 II 层的底部之间存在一个稳定和明显的间断。在这个界限清楚的地表上偶然发现了舍利文化第 I 期的生命活动遗址。

人们还发现了新动物群存在的大量证据，现在已经清楚奥杜威第 I 层的动物群与奥莫的动物群相同，而且这二者的年代与汤恩头骨的年代大体相同。换言之，现在应该用奥杜威第 I 层来代表维拉方期的上半部分，而不是中更新世的下部。就相对年代而言，似乎可以明确地认为远东的哲蒂斯地层属于中更新世，而不是下更新世，因此奥杜威新头骨比最古老的猿人属还要古老。

在南非，现在认为汤恩与斯泰克方丹的堆积物属于下更新世的晚期；因此通常认为它们与奥杜威第 I 层的时代相同。很可能马卡潘地层更年轻一点，而斯瓦特克朗斯和斯泰克方丹的上部地层（现已出土了石器）都属于中更新世。

因此，与汤恩小孩及来自斯泰克方丹下部地层的南方古猿化石一样，新发现的头骨代表了最早的人科中的一员，奥杜威头骨代表了迄今为止被发现的最早的石器制造者。

The following approximate measurements will indicate the size of the new specimen.

Length from inion to glabella	about	174 mm.
Greatest breadth at supra-mastoid torus	,,	138 mm.
Greatest breadth of brain case on squamosal element of the temporal bones	,,	118 mm.
Height (in Frankfurt plane) from basion to a point vertically above it in the sagittal plane	,,	98 mm.
External orbital angle width	,,	122 mm.
Inter-orbital width	,,	32.5 mm.
Post-orbital width	,,	88 mm.
Palate-length from front of incisors to a line joining back of third molars	,,	84 mm.
Palate-width at second molars	,,	82 mm.
Palate-width at third premolars	,,	62 mm.
Length of molar-premolar series	,,	72 mm.

Teeth measurements:

$M3$: 21×16 mm.; $M2$: 21×17 mm.; $M1$: 18×15.5 mm.; $PM4$: 18×12 mm.; $PM3$: 17×11.5 mm.; C: 9.5×9 mm.; $C2$: 7×7 mm.; $C1$(both damaged but about 10×8 mm.).

(**184**, 491-493; 1959)

L. S. B. Leakey: Coryndon Museum, Nairobi.

以下的测量近似值表明了这一新标本的大小。

枕骨隆凸点到眉间的长度	大约 174 毫米
圆枕上乳突处的最大宽度	大约 138 毫米
在颞骨鳞部脑颅的最大宽度	大约 118 毫米
当颅骨处于法兰克福平面时，从枕骨大孔前缘点到在矢状面上垂直于它的一点的高度	大约 98 毫米
眼眶外角的宽度	大约 122 毫米
眶间宽度	大约 32.5 毫米
眶后宽度	大约 88 毫米
从门齿前面到第三臼齿背面连接线的腭长度	大约 84 毫米
在第二臼齿的腭宽度	大约 82 毫米
在第三前臼齿的腭宽度	大约 62 毫米
臼齿－前臼齿齿系的长度	大约 72 毫米

牙齿的测量值：
$M3$：21×16 毫米；$M2$：21×17 毫米；$M1$：18×15.5 毫米；$PM4$：18×12 毫米；$PM3$：17×11.5 毫米；C：9.5×9 毫米；$C2$：7×7 毫米；$C1$（都损坏了，但大约是 10×8 毫米）。

（田晓阳 翻译；吴新智 胡荣 审稿）

The Affinities of the New Olduvai Australopithecine

Editor's Note

The shift of focus from southern to eastern Africa in the search for human origins was so abrupt that it can be dated precisely—to 15 August 1959, when *Nature* published Louis Leakey's discovery of a super-robust australopithecine from Olduvai Gorge. Coming at the end of decades of fruitless search, Leakey named his find *Zinjanthropus boisei*—or "Nutcracker Man", for its enormous teeth. Reviewing the find in *Nature* the following May, John T. Robinson—Broom's longtime co-author on many papers about South African australopithecines—suggested that *Zinjanthropus* was not as distinct as Leakey had proposed, suggesting that it might be referred to *Paranthropus*. Robinson's comments stuck—Nutcracker Man is now known as *Paranthropus boisei*.

Responding to J. T. Robinson's critique, Leakey responded in characteristically robust style. Although "Zinj" was similar to *Paranthropus*, it was, Leakey maintained, sufficiently different to be deserving of generic rank. This was clearly more than a trivial academic spat over names: it had taken decades for Leakey to discover a significant fossil hominid, and Robinson might have been worried about a newcomer in a field previously dominated by Broom and himself, and sought to put the interloper in his place. Such battles were not the first to have happened in palaeoanthropology, and they would not be the last.

DR. L. S. B. Leakey recently reported in *Nature*[1] the discovery of an essentially complete hominid skull without mandible from site *FLK* at Olduvai Gorge. He regards the specimen as an australopithecine that differs more from either of the two known genera, *Australopithecus* and *Paranthropus*, than these two differ from each other. He therefore erected the genus *Zinjanthropus* to accommodate the new specimen.

The description consists largely of a list of twenty "major" differences between the new specimen and the two previously known genera. Some of these points cannot be dealt with as they are briefly stated in terms which are not useful by themselves for comparison. Others, for example, continuous nuchal crest, enlargement of the mastoid area, prominent temporal lines, large areas for masseter attachment, etc., do not each represent a separate difference; all are reflexions of the fact that this specimen is a little larger and more muscular than the known South African specimens of *Paranthropus*. In at least one case (nuchal crest) comparison with the latter specimens is not possible since there is no known male specimen with the relevant area preserved. The remainder are not real differences and will be dealt with briefly:

新型奥杜威南方古猿亚科的亲缘关系

编者按

当《自然》杂志刊载了路易斯·利基在奥杜威峡谷发现超级粗壮的南方古猿亚科后，关注人类起源的目光由南非迅速转移到了东非，这一转变非常突然，甚至可以精确到 1959 年 8 月 15 日这一天。在经过数十年徒劳的探索后，利基将其发现的南方古猿命名为鲍氏东非人，又由于其具有硕大的牙齿因此被称为"胡桃钳人"。次年 5 月，在回顾《自然》杂志上发表的成果时，约翰·鲁滨逊（曾与布鲁姆合作发表过多篇关于南方古猿的论文）认为，这个东非人并没有利基所认为的那样特别，而可能与傍人有关。鲁滨逊的观点后来被证实是正确的，胡桃钳人现在被称为鲍氏傍人。

利基以其特有的粗壮风格对鲁滨逊的批评做出了回应。利基坚持认为，尽管"东非人"与傍人类似，但它们仍然存在非常大的差异从而不能将二者归为同一个属。这显然不仅仅是一个微不足道的关于命名的学术争论：利基耗费了数十年的时间发现了这一具有重要意义的人科化石，而鲁滨逊可能对这个闯入原本由布鲁姆和他自己主导的领域的外来者感到很烦恼，并想挫挫闯入者的傲气。在古人类学领域，这类纷争不是第一次发生，也不会是最后一次。

利基博士最近在《自然》杂志 [1] 上报道称在奥杜威峡谷的 *FLK* 遗址发现了一件人科的头骨，该头骨基本完整，但没有下颌骨。他认为该标本属于南方古猿亚科，但不同于两个已知属（南方古猿属和傍人属）中的任何一个，它与这两个属之间的差异要大于这两个属彼此间的差异。因此他为这件新标本建立了东非人这个新属。

关于这件新标本的描述主要包括新标本和那两个先前已知的属之间的 20 个"较大的"差异。其中一些差异表述得过于简单，我们无法进行探讨，并且这些差异本身也不利于进行比较。其他的如连续的项脊、乳突骨区的增大、凸出的颞线、咀嚼肌附着区的增大等等，并不是每一项都代表独立的区分特征，所有这些现象都反映了一个事实：这件标本比已知的南非傍人标本稍大一些且肌肉更强健。此外，至少有一项（项脊）是不可能与后一种标本进行比较的，因为没有已知男性标本的相关部位被保存下来。余下的部分没有实质性的差别，因此只作简短的探讨：

(1) The inion is lower relative to the Frankfort plane: In both *Australopithecus* and *Paranthropus* the base of the external occipital protuberance is almost exactly in the Frankfort plane, as seems to be the case with the Olduvai specimen.

(2) The posterior wall of the occipital bone rises more steeply to form, with the parietals, a very high-vaulted posterior region of the skull: This is also the case in *Paranthropus*, less so in *Australopithecus*. *Paranthropus* has an almost spheroidal brain-case which is relatively low and narrow anteriorly but steep-sided and higher posteriorly. *Australopithecus* has a brain-case more nearly like that of a dolichocephalic modern hominine. The Olduvai specimen has the *Paranthropus* type of brain-case.

(3) The form of the tympanic plate is different: Dr. Leakey demonstrated this point to me on the specimen to make clear his meaning. This feature is variable in *Paranthropus* and includes an instance of close resemblance to the Olduvai specimen. The observed range of variation is illustrated in Fig. 1. Being thus variable, it is in any event a feature of low phyletic valence.

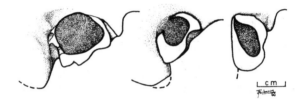

Fig. 1. Variation of the tympanic bone at the lateral end of the external auditory meatus in *Paranthropus* from Swartkrans (*SK* 48, *SK* 52 and *SK* 848). The Olduvai specimen resembles in this feature the condition illustrated in the middle diagram.

(4) The very unusual position of the nasion—almost coinciding with glabella: This is true of *Paranthropus* and in at least some cases of *Australopithecus* also. In *Paranthropus*, as in the Olduvai specimen, the nasals are relatively very wide near the nasion.

(5) The very great absolute and also relative width of the inter-orbital area. The proportionate width of the inter-orbital area to that between the external orbital angles is 26.6 in the Olduvai specimen; 26.2(*SK* 846), 25.3(*SK* 48) in two specimens of *Paranthropus*. A single specimen of *Australopithecus* gave a value of 24.1, while a random sample of eight modern human skulls gave a range of 23.4–30.5. No taxonomic significance can therefore be attached to this point on the present evidence.

(6) The very deep palate which is even more markedly like that of *Homo* than *Australopithecus*, and is quite unlike the form seen in *Paranthropus*: In *Homo*, the degree of vaulting of the hard palate is variable, but the difference in depth at the incisive fossa and the back of the palate is normally relatively slight. This is true also of *Australopithecus*. The palate of *Paranthropus* differs in that it slopes more markedly; that is, there is always an appreciable difference between the anterior and posterior depths—in some cases there

174

（1）相对于眶耳平面，枕外隆凸点较低：南方古猿与傍人的枕外隆凸点底部几乎都在眶耳平面上，这似乎与奥杜威标本是一样的。

（2）枕骨后壁向上陡倾，与顶骨形成拱起程度很高的头骨后区：傍人的情况也是如此，而南方古猿的拱起程度则欠缺一些。傍人的头盖骨类似于球体，其前部相对较低而窄，但侧面陡，后部较高。南方古猿的头盖骨与脸长的现代人类的更为相似。奥杜威标本的头盖骨类型与傍人的一样。

（3）鼓板的形态不同：利基博士利用标本清楚地论述了这一观点。这一特征在南方古猿中是变化的，且存在与奥杜威标本非常类似的情况。所观察到的变化范围如图 1 所示。就因为存在这样的变化，在任何情况下，这个特征在种系发生上的价值都不高。

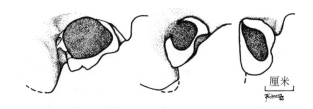

图 1. 源于斯瓦特克朗斯的傍人外耳道侧端耳鼓骨处的差异（*SK* 48、*SK* 52 和 *SK* 848），奥杜威标本的特征与本图中间所示的情形类似。

（4）鼻根点的位置几乎与眉间点一致，这点非常罕见：傍人的情况就是如此，至少有些南方古猿也是如此。像奥杜威标本一样，傍人的鼻骨在鼻根点附近相对很宽。

（5）眶间区的绝对宽度和相对宽度都非常大。奥杜威标本眶间区对外眶角的宽度比为 26.6；傍人的两个标本是 26.2（*SK* 846）与 25.3（*SK* 48）。单个南方古猿标本的比值为 24.1，而随机抽样出来的 8 个现代人类头骨的样本的比值范围为 23.4~30.5。因此就现有证据看这一点不具有分类学上的意义。

（6）腭很深，明显更类似于人属的腭，而不是南方古猿属的腭，并且与傍人属中所见的腭十分不同：人属的硬腭的拱形程度是有差异的，但门齿窝与腭后端的深度差异通常相对较小。南方古猿属的情况也是如此。傍人属的腭的不同点在于其倾斜更为明显；也就是说，前部与后部间的深度总是有明显的差异——有时前部一点也不深，有时前部明显很深。奥杜威标本的特征是：与前部深度相比后部深度相对

is no depth at all anteriorly, in others the anterior depth is appreciable. The Olduvai specimen shares this characteristic of relatively much greater posterior compared to anterior depth, reflecting the *Paranthropus* condition, not that of *Homo* and *Australopithecus*.

(7) The shape and arrangement of the zygomatic process of the maxilla: The Olduvai specimen has a relatively poorly developed zygomatic process, the lower border of which passes almost horizontally from the zygo-maxillary suture to the main body of the maxilla. In some specimens of *Paranthropus* the process passes from the zygo–maxillary suture downward and medial–ward at a fairly sharp angle like an inverted flying buttress. However, *SK* 52, of the same dental age as the Olduvai specimen, has a trace only of the buttress and *SK* 846 (mature adult) has no trace of it, as in the Olduvai specimen.

(8) The relatively greater reduction of the canines in comparison with the molar–premolar series than is seen even in *Paranthropus*: As the marked change of proportion between anterior and cheek teeth occurs in *Paranthropus* and the Olduvai specimen between canine and P^3, the degree of reduction can be measured by the ratio between the modules of these two teeth. The ratio for the Olduvai specimen is 64.9 while that for three specimens of *Paranthropus* ranges from 61.8 to 78.8.

(9) In the species diagnosis Leakey refers to the presence of a sagittal crest "at least posteriorly". The reason for this statement is not clear, since the crest occupies a position identical to that in *Paranthropus*—on roughly the middle third of the distance between glabella and inion. In neither form is the sagittal crest known to reach as far back as the planum occipitale or even the posterior part of the sagittal suture.

It seems to me that the most fruitful approach to an understanding of the australopithecine skull, within the framework of hominid structure, is in terms of diet and the nature of the dentition. *Australopithecus*, as an omnivore eating at least a moderate amount of flesh, has a dental apparatus very similar to that of the older hominines. The relative sizes of the teeth along the tooth row are closely similar, as is the arrangement of the dental arcades. The anterior teeth, especially the canines, are well developed, but the post-canine ones only moderately so. The skulls are also similarly constructed except for the relatively small cranium, compared to the face, of *Australopithecus*.

In *Paranthropus* the situation is very different. Here the anterior teeth, set in an almost straight line across the front of the palate, are appreciably reduced in size compared to those of *Australopithecus*, while the post-canine teeth are appreciably larger. This difference in arrangement and proportion is very striking and is characterized by the very small canine tooth set firmly against a massive premolar. All this must clearly have functional meaning. The massive crushing and grinding teeth, the relatively unimportant anterior teeth and frequent damage to the enamel which could only have been caused by small, very hard particles (presumably grit), and the large size of the animal suggest a predominantly vegetarian diet. Concentration of heavy dental function in the post-canine region has strongly modified the skull architecture. The massive cheek teeth, especially

较大，所反映出的是傍人属的特征，而非人属或南方古猿属的特征。

（7）上颌骨颧突的走向与形状：奥杜威标本的上颌骨颧突发育相对较差。它的下缘几乎水平地从颧骨－上颌骨骨缝延伸到上颌骨的主体。一些傍人标本的上颌骨颧突从颧骨－上颌骨骨缝向下和沿中间成锐角穿过，犹如一个倒转的飞拱。然而，尽管 SK 52 与奥杜威标本的齿龄相同，但仅有少许壁柱状构造的痕迹，而 SK 846（成熟个体）和奥杜威标本一样没有壁柱状构造的痕迹。

（8）与臼齿－前臼齿齿系相比，犬齿明显减小，甚至比傍人减小得还明显：前齿与颊齿间的大小比例在傍人和奥杜威标本（犬齿与第三上前臼齿间）中发生了明显变化，因此退化的程度可以通过测量这两颗牙齿模数之间的比率获得。奥杜威标本的比率是 64.9，而 3 个傍人标本的比率从 61.8 到 78.8 不等。

（9）在种一级的鉴定上，利基认为"至少在后部"出现了矢状脊。这一条所依据的理由是不清晰的，因为奥杜威标本矢状脊的分布位置与傍人的一致，大约位于眉间点与枕外隆凸点之间的中部三分之一处。已知的矢状脊没有一种形式是向后延伸远至枕平面或甚至到矢状缝的后部。

在我看来，在人科系统框架内了解该南方古猿头骨的最有效的方法是分析南方古猿的食性和齿系之间的特征。南方古猿作为一种杂食动物至少会吃适量的肉，其牙齿构造十分类似于较早的人亚科成员。同一齿列中牙齿的相对大小非常相似，沿齿弓排列。前齿尤其是犬齿发育完好，但是犬齿后的牙齿仅为中度发育。头骨的构造也非常相似，只是南方古猿的颅骨与面部相比相对较小而已。

傍人的情况则非常不同。与南方古猿相比，沿着腭前部几乎呈直线横向排列的前齿明显变小，而颊齿明显较大。这种差异在排列和比例上是非常明显的，这表现为非常小的犬齿牢牢地紧挨着粗大的前臼齿。所有这些特征一定具有非常重要的功能学意义。粗大的用来压碎和研磨食物的牙齿、相对不重要的前齿、只可能是由十分坚硬的小颗粒（可能是粗砂）所引起的常见的牙釉质的损坏以及庞大的体型，都表明以素食为主。牙齿的主要功能集中在后犬齿区，这对头骨的结构有很大的影响。粗大的颊齿尤其是臼齿的齿根系统发育得很完善，因此生长在粗壮的骨头上并且腭的后部相对较深。由于咀嚼力很大、腭很厚、颌骨特别是支撑颊齿的地方粗大，因

the molars, have strongly developed root systems and hence are set in heavy bone, and the back of the palate is relatively deep. Owing to the heavy chewing forces, the palate is thick, the jaws are massive, especially where supporting the cheek teeth, the zygomatic arches are strongly built, the circumorbital bone is strong in the stressed areas, the strongly stressed pillars on either side of the pyriform aperture are thick and the medial and lateral pterygoid plates, especially the latter, are large and strong. However, the large dental battery requires heavy musculature to operate it—hence further changes from the *Australopithecus* and hominine pattern. The areas of origin and insertion of these muscles are sturdy; hence the large surface area of the mandibular ramus, the massive zygomatic arch—which also is widely spaced from the brain-case to allow for the very large temporal muscle—large medial and lateral pterygoid plates, clearly defined temporal lines, sagittal crest and apparent prominence of the supraorbital tori. The large muscles associated with the large and chunky animal result in well-defined muscular ridges on the occiput and the prominence of the mastoid area—assisted by the large temporal muscles which necessitate appreciable lateral extension of the zygomatic arch. It is probable also that the large temporal muscles have influenced the shape of the brain-case anteriorly to help produce the low and narrow shape characteristic of *Paranthropus*. As is usual in relatively massive skulls where large surface area is necessary, some reduction of weight has occurred by increased pneumatism. The mastoid region (not the mastoid process alone) is strongly cellular and the air spaces in the skull base, the maxillae and the frontal bone are all very large.

The curious and characteristic features of the *Paranthropus* skull, which parallel some of those of the gorilla skull and mark it off so clearly from that of *Australopithecus*, are all functionally related and determined largely by the specialized diet of this creature as compared to all other known hominids. These functional and structural differences indicate clear adaptational differences between the two forms and hence also differences of evolutionary direction. In modern systematics, adaptational differences of this order and nature are normally accorded generic rank.

The features of the Olduvai specimen fit the *Paranthropus* pattern as here outlined very accurately—the differences being almost entirely of the sort shown to be variable within a single population by the sample of nearly 200 specimens of *Paranthropus* now in the collections of the Transvaal Museum. The chief feature of the Olduvai form which does not fall within the observed range of variation of this collection is size. The best size comparison is with the only male specimen with good teeth which has P^3–M^1 preserved in sequence and is of the same dental age as the Olduvai specimen. The latter is only 8.4 percent larger. The average percentage difference for five skull and dental series dimensions compared to those of a fully adult female skull is 17.4 percent. As Schultz has shown in a number of papers, intra-specific variation in measurable primate anatomical characters can often greatly exceed the above values.

The validity of separate specific status is not clear on the basis of the single specimen, and it is perhaps wisest to leave it as distinct. In the light of the above analysis, however,

此颧弓构造结实，围眶骨受力区坚硬，梨状口任一边上的强受力柱粗，中间的翼状板与侧面的翼状板特别是后者又大又结实。但是粗大的齿列需要强大的肌肉组织来控制，因此傍人与南方古猿及人亚科的头骨结构之间产生了进一步的差异。这些肌肉的起点区与附着区非常强健；因此，下颌支的表面积较大，颧弓较粗大（它与头盖骨之间的宽距能允许宽大的颞肌存在），中间和侧面的翼状板大，颞线、矢状脊界线清楚，眶上圆枕明显凸出。体形大而强壮的动物所特有的发达肌肉导致枕骨的肌肉脊清晰以及乳突区凸起，并伴有发达的颞肌，这使得颧弓明显侧向伸展。发达的颞肌可能已经影响了傍人头盖骨前部的形状，并致使其变得低而窄。通常情况下，肌肉在头骨上附着的面积增大必然需要相应地增大头骨的体积，因此通过增加头骨中的气室来减轻头骨的重量。乳突区（不仅仅是乳突）呈多孔状，头骨基部、上颌骨和额骨中的气室都非常大。

傍人头骨具有奇特而典型的特征，与一些大猩猩头骨非常相近但与南方古猿头骨明显不同。通过与所有其他已知的人科成员相比发现，这些特征在功能上是密切相关的，并且主要是由其特定的食性决定的。这些功能上与结构上的不同清晰地表明了两种类型之间的适应性差异，因此在进化方向上也是有差异的。在现代分类学上，适应性演化差异和特性具有属一级的分类意义。

奥杜威标本的形态符合傍人属的特征，此处概括得已非常准确——即奥杜威标本所反映的形态差异基本没有超出德兰士瓦博物馆现有藏品中约 200 个傍人标本所体现的形态变异范围。奥杜威标本明显超出藏品所观测的变异范围的特征是它的大小。若进行大小的比较，就必须是一个男性标本，并且该标本的牙齿从第三上前白齿－第一上白齿依次保存完好，且齿龄与奥杜威标本相同。后者仅大 8.4%。和完全成熟的女性标本的头骨相比，5 件头骨与齿系大小的平均百分数差为 17.4%。正如舒尔茨在很多文章里所叙述的那样，可测量的灵长类动物解剖特性的种内变化值经常大大超出上述值。

由于奥杜威标本只有一件，因此无法确定种一级的区别，也许最明智的方法是把奥杜威标本作为一个独立的种。但是综上所述，把它放到独立的属的位置似乎既

separate generic status seems unwarranted and biologically unmeaningful. I therefore propose that the name of the Olduvai form be *Paranthropus boisei* (Leakey).

J. T. Robinson

(**186**, 456-458; 1960)

*   *   *

THE exact taxonomic label that should be applied to the skull that I have named *Zinjanthropus*, from Olduvai, and which I described in *Nature* of August 15, 1959, seems to me relatively unimportant at the moment. Inevitably, different scientific workers have different ideas of what characters justify specific, generic, and even superfamilial rank. After all, this is purely a question of artificial labels.

Dr. Robinson and I agree that *Zinjanthropus boisei* is closely related to the Australopithecinae; we agree that it has certain resemblances to *Paranthropus*, and we disagree mainly in that he believes the differences to be insufficient to justify separate generic rank, while I think they do.

It is hard enough to reach agreement among zoologists on the taxonomic status of living primates, under conditions in which we possess the skull, skeleton, skin and viscera for study, and it will always be much more difficult to do so when we have only fossils to guide us. I can only say that the very considerable additional work that I have done on the *Zinjanthropus* skull since my preliminary report in *Nature* has greatly strengthened my view that it is entirely different from *Australopithecus* and *Paranthropus*, differing from both these genera more than they do from each other.

I do not feel that any useful purpose would be served by entering into a long discussion with Dr. Robinson in *Nature* at present, since the more detailed study of the Olduvai skull which is now in hand will not, I hope, be too long delayed. However, Dr. Robinson makes certain statements which may mislead those who read them, unless I comment on them. I will therefore do so as briefly as possible.

First of all, whereas in *Paranthropus* and *Australopithecus* (as Dr. Robinson says) the external occipital protuberance lies more or less on the Frankfort plane, in *Zinjanthropus* it lies below it.

Robinson's description of the brain case of *Paranthropus* as "almost spheroidal", but also "relatively low and narrow anteriorly but steep-sided and higher posteriorly", does not seem to make sense, for the two statements seem to cancel each other out. In any event, such a combined description does not fit the brain case of *Zinjanthropus*.

没有实质的根据又没有生物学意义。因此我建议将利基发现的奥杜威标本更名为鲍氏傍人。

鲁滨逊

（田晓阳 翻译；董为 审稿）

*　　*　　*

在 1959 年 8 月 15 日的《自然》杂志上，我对我所命名的来自奥杜威的东非人头骨进行了描述，此刻对我来说，应该附于此头骨准确的分类学标签并不是最重要的。然而不可避免的是，不同的科学工作者对于用什么样的特征界定种、属甚至是超科的分类阶元持有不同的观点。毕竟，这纯粹是一个人为分类的问题。

鲁滨逊博士和我都认为鲍氏东非人与南方古猿亚科有很近的亲缘关系；我们也都认为其与傍人属有某些相似之处，而我们观点的分歧主要是，他认为这些差异不足以将其归为一个独立的属，而我认为可以。

即使在有头骨、骨骼、皮肤以及内脏用于研究的条件下，让动物学家们对现存的灵长类动物进行分类并达成共识都是很难的，更何况当我们只能以化石为依据的时候，这种共识就更难达成了。我只能说，自从在《自然》杂志发表了那篇最初的报道之后，我又对这一东非人头骨进行了大量的工作，这些工作都极大地支持了我的观点，即它与南方古猿属和傍人属是完全不同的，它与这两个属之间的差异要大于这两个属彼此之间的差异。

我觉得，现在与鲁滨逊博士在《自然》杂志上展开长篇的讨论并不会产生任何有用的效果，因为我不希望现在手头上对奥杜威头骨进行更为详细的研究工作因此而被耽搁得太久。然而，鲁滨逊博士发表的一些陈述可能会误导读者，所以我不得不发表一下评论。因此我会尽可能简短地进行一些说明。

首先，（正如鲁滨逊博士所说）傍人和南方古猿的枕外隆凸点大体位于眶耳平面上，然而在东非人中其位于该平面之下。

鲁滨逊对傍人脑颅的描述是"类似于球体"，但"其前部相对较低而窄，但侧面陡，后部较高"，这似乎不具有任何意义，因为这两个陈述是自相矛盾的。任何情况下，它们都不能共同用来描述东非人的脑颅。

Robinson illustrates the range of the tympanic plate (see in profile), in *Paranthropus*. None of these three illustrations closely resembles the tympanic plate of *Zinjanthropus*, although the one to the left appears to be rather closer than the middle one.

Without knowing the points at which Robinson measures inter-orbital width and external orbital width, I cannot comment upon his comparisons of his *Paranthropus* figures with mine for *Zinjanthropus*.

As to the morphology of the palate, I do not know upon what evidence Robinson is basing his statement, since I have published no measurements of the palatal depth in *Zinjanthropus*. I must repeat, however, quite categorically, that the morphology of the *Zinjanthropus* palate in no way resembles that of *Paranthropus*.

I cannot accept Robinson's statement that the zygomatic process of the maxilla in *Zinjanthropus* is "relatively poorly developed"; I would say rather, as I have said before, that it is developed in an entirely different morphological manner from *Paranthropus*.

As regards the position of the sagittal crest, in *Zinjanthropus* it ceases to be a crest and divides into two temporal lines well behind the line drawn vertically through the ear when the skull is on the Frankfort plane. In *Paranthropus* (in all the published photographs), the sagittal crest extends a long way forward of such a vertical line through the ear when the skull is on the Frankfort plane, and it is therefore wrong to say, as Robinson does, that "the crest occupies a position identical to that in *Paranthropus*".

Finally, I do not understand the significance of a comparison of "the ratio between the modules of these two teeth", that is, the canine and P^3. Robinson has defined a module as the sum of the length and breadth of a tooth divided by two, and I am at a complete loss to understand how the ratios of modules can have any significance. It must be obvious that one can have on one band a canine tooth 16 mm. long and 6 mm. wide (module equals 11), and a premolar 9 mm. long and 7 mm. wide (module equals 8), while in another specimen one could have a canine which measured 11 mm. × 11 mm., and a premolar which was only 5 mm. long and 11 mm. wide, yet the ratio of the modules in the two sets of teeth would be identical, but completely without significance.

In any event, the ratio between the canine and the premolar alone cannot have any bearing upon the relation of the canine size to the total molar-premolar series, unless the premolar bears a constant relation to the total post canine series.

I therefore repeat my statement that in *Zinjanthropus* there is a relatively greater reduction of the canines in comparison with the total molar-premolar series than is seen in *Paranthropus*, and maintain that Robinson has in no way disproved this statement.

I agree with Robinson that we need to study the *Australopithecus* skull structure and

鲁滨逊描绘了傍人的鼓板变异范围（见侧面图）。然而这三个图示均不与东非人的鼓板相类似，尽管左边那个看起来比中间那个更接近一些。

因为不知道鲁滨逊测量眼窝间宽度和眼窝外宽度的测量点，所以对于他得到的傍人数值与我的东非人数值进行的比较，我不作出任何评论。

至于腭的形态，我不知道鲁滨逊的陈述基于什么证据，因为我从未发表过任何关于东非人腭厚度的测量结果。但是我必须重申的是，我非常确定东非人的腭形态与傍人的绝对没有相似之处。

鲁滨逊认为东非人的上颌骨颧突"发育相对较差"，我不认同这一说法；我想说的是，正如我之前已经说过的，它是以一种与傍人完全不同的形态学方式发育的。

至于矢状脊的位置，在东非人中它不再是一道脊，而是分成了两条颞线，当头骨位于眶耳平面时，这两条颞线恰好位于垂直经过耳所画出的直线的后面。在傍人中（在所有已发表的照片中），矢状脊延伸了很长一段距离，当头骨位于眶耳平面时，其位于过耳垂直线的前方，因此鲁滨逊所说的"矢状脊的分布位置与傍人的是一致的"是错误的。

最后，我不理解对"这两颗牙齿模数之间的比率"进行比较的意义何在，即对犬齿和第三上前白齿进行的比较。鲁滨逊将这种模数定义为一枚牙齿的长度和宽度的总和除以 2，我完全不能理解这种模数的比值到底有什么意义。很显然，一个个体可能具有一枚长 16 毫米、宽 6 毫米的犬齿（模数等于 11）以及一枚长 9 毫米、宽 7 毫米的前白齿（模数等于 8），然而别的标本可能具有长 11 毫米、宽 11 毫米的犬齿以及一枚仅仅长 5 毫米、宽 11 毫米的前白齿，尽管这两套牙齿的模数比值相同，但完全没有意义。

无论如何，对犬齿大小与全部白齿－前白齿齿系的关系来说，单个的犬齿和前白齿间的比值并不具备任何意义，除非前白齿与全部的后犬齿齿系间具有某种恒定的关系。

因此，我重申一下我的观点，就全部白齿－前白齿齿系的比较而言，在东非人中犬齿的退化比在傍人中看到的退化相对来说幅度更大，并且我认为鲁滨逊无法反驳我的这一观点。

鲁滨逊认为，我们需要依据其食性和生活方式研究南方古猿的头骨结构和齿系，

dentition in terms of diet and mode of life, and I shall certainly do so as far as *Zinjanthropus* is concerned in my fuller report.

L. S. B. Leakey

(**186**, 458; 1960)

J. T. Robinson: Transvaal Museum, Pretoria.
L. S. B. Leakey: Coryndon Museum, Nairobi, Kenya.

Reference:
1. *Nature*, **184**, 491 (1959).

对此我表示赞同，并且在对东非人进行更完整的报道中，我肯定会这么做。

利基

（刘皓芳 翻译；董为 审稿）

Four Adult Haemoglobin Types in One Person

C. Baglioni and V. M. Ingram

Editor's Note

By the 1960s the molecular structure of haemoglobin was well known: normal adult haemoglobin contains two identical α polypeptide chains and two identical β chains. German American biologist Vernon M. Ingram had used electrophoresis to show how amino-acid substitutions could produce disease-causing haemoglobin variants. Here Ingram, with Corrado Baglioni from the Massachusetts Institute of Technology, describes the analysis of a blood sample from a single patient containing four different adult haemoglobin types. Three were abnormal, containing various amino-acid substitutions, and the researchers use this information to make assumptions about how the haemoglobin molecule is assembled.

THE normal adult type of human haemoglobin (A) is a single molecular species with the chemical constitution $\alpha_2^A\beta_2^A$ (ref. 1). Such a molecule contains two identical α polypeptide chains and two identical β chains. It has been shown that the genetically controlled abnormal human haemoglobins previously examined contain an abnormal α or an abnormal β chain, carrying an amino-acid substitution[2,3]; for example, lysine for glutamic acid in the β chain of haemoglobin C, $\alpha_2^A\beta_2^C$ (ref. 4). We wish to report the results of chemical studies on four adult haemoglobin components[5] found in a single individual, D. K. P.; her haemoglobins turned out to be normal in both peptide chains, abnormal in one or other chain and abnormal in both chains, respectively. Apart from their genetic interest, these findings allow us to make certain deductions about the mode of assembly of the haemoglobin molecule.

The blood sample from this patient, D. K. P., became available to us through the kindness of Dr. L. M. Tocantins and Miss J. Atwater of Philadelphia. They have already described[5] the occurrence of four haemoglobins in approximately equal amounts in this blood, together with their clinical findings. The haemoglobins are electrophoretically distinct at pH 8.6; they have been called[5] haemoglobins A, G, C and X, which is the order of migration, with A as the fastest component. Electrophoretic comparison with authentic specimens is the basis for naming the first three haemoglobins; haemoglobin X was unknown and the slowest haemoglobin yet described. The haemoglobin G was found by these authors[5] to behave identically with a specimen of G from Dr. Schneider. It may therefore be identical with the haemoglobin G of Lehmann and Edington[6] and with the G_{Ib} (for G_{Ibadan}) recently described by Shooter et al.[7], and found by them to be abnormal in the α chain. On the other hand, D. K. P.'s haemoglobin G is different from the G of Schwartz et al.[3,8] which carries a substitution of glycine for glutamic acid in the β peptide chain.

一人体内的四种成人血红蛋白

巴廖尼，英格拉姆

编者按

20 世纪 60 年代，血红蛋白的分子结构已为人们所熟知。正常成人血红蛋白包括两条相同的 α 多肽链和两条相同的 β 多肽链。美籍德裔生物学家弗农·英格拉姆运用电泳方法进行的研究显示氨基酸替换可以产生多种致病的血红蛋白变异体。本文中英格拉姆与来自麻省理工学院的科拉多·巴廖尼合作，对血液中存在四种不同的成人血红蛋白的一名患者的血样进行了分析，结果发现其中三种血红蛋白是存在各种不同的氨基酸替换的异常血红蛋白，并在此基础上提出了血红蛋白分子组装机制的假说。

正常成人型人血红蛋白 A 是化学组成为 $\alpha_2^A\beta_2^A$（参考文献 1）的单分子化合物，这个分子包括两条相同的 α 多肽链和两条相同的 β 多肽链。以前的研究表明，由遗传因素引起的人异常血红蛋白中往往存在一条异常的携带氨基酸替换的 α 链或 β 链[2,3]。例如血红蛋白 C，$\alpha_2^A\beta_2^C$（参考文献 4）的 β 链上一个谷氨酸就被一个赖氨酸所取代。我们对一名女性（D. K. P.）体内发现的四种成人血红蛋白组分[5]进行了化学分析，发现其中有的血红蛋白的两条链都正常，有的一条链发生了变异，而有的两条链都发生了变异。这些发现除了具有遗传意义外，还可以让我们据此对血红蛋白分子的组装模式作出确切的推论。

费城的托坎廷斯医生和阿特沃特小姐慷慨地为我们提供了患者 D. K. P. 的血液样本，他们叙述了该患者的临床表现，并指出该患者血液样本中存在着四种几乎等量的血红蛋白[5]。由于这四种血红蛋白在 pH 值为 8.6 时电泳迁移率不同，它们分别被命名为血红蛋白 A、G、C 和 X[5]，字母表示蛋白迁移顺序，A 表示迁移速度最快的蛋白组分。其中，前三种血红蛋白是根据与各自迁移率相同的标准蛋白样本来命名的，而 X 组分是一个未知的血红蛋白，在目前已知的所有血红蛋白中迁移最慢。他们还发现[5]血红蛋白 G 的性质与施奈德医生所研究的 G 样本是一致的。因此，它很可能与莱曼和埃丁顿[6]描述的血红蛋白 G 以及舒特等人[7]最近描述的血红蛋白 G$_{Ib}$（即 G$_{伊巴丹}$）是同一种物质，其 α 链发生异常。另一方面，这一血红蛋白 G 组分又不同于施瓦茨等人[3,8]描述的血红蛋白 G，后者 β 链上的一个谷氨酸残基被甘氨酸残基所取代。

The original characterization of haemoglobins A, G, C and X was done by paper electrophoresis[5]. We have been able to confirm the presence of the four components by starch-gel electrophoresis[9] at pH 8.6 in 0.02 M veronal buffer (see Fig. 1). The four proteins were also separated by electrophoresis on a starch block[10] in the same buffer. The relative amounts (percent) recovered were: A, 35; G, 27; C, 23; X, 15.

Hb A

Hb AGCX

Fig. 1. Starch-gel electrophoresis of D. K. P.'s haemoglobin at pH 8.6 in 0.05 M veronal buffer

Each haemoglobin was digested with trypsin and fingerprinted[11]. Haemoglobin A appeared to be completely normal (Fig. 2) and haemoglobin C showed only the expected absence[4] of peptide 4 and its replacement by two new peptides, C-4a and C-4b (C-4b = haemoglobin C tryptic peptide 4b). These peptides were isolated, hydrolysed and their amino-acid content analysed. The results confirmed the glutamic acid to lysine substitution previously reported for haemoglobin C[4]. It appears, therefore, that D. K. P.'s haemoglobin C is authentic haemoglobin C.

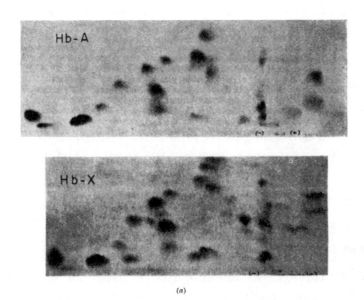

Hb-A

Hb-X

(a)

在之前的研究中，已经通过纸电泳法对血红蛋白 A、G、C 和 X 的初步性质进行了分析 [5]。我们采用淀粉胶电泳 [9]（pH 值为 8.6 的含 0.02 摩尔 / 升巴比妥的缓冲液）的方法来证实这四种血红蛋白的存在（见图 1）。在同样的缓冲体系下进行淀粉阻滞电泳 [10]，也能够成功地将这四种血红蛋白分离开。可以看出它们的相对含量分别是：A，35%；G，27%；C，23%；X，15%。

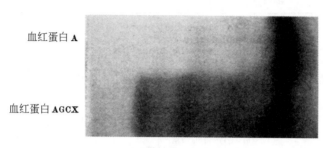

图 1. D. K. P. 血红蛋白的淀粉胶电泳图（pH 值为 8.6 的含 0.05 摩尔 / 升巴比妥缓冲液）

用胰蛋白酶消化各个血红蛋白后进行指纹印迹分析 [11]，结果表明血红蛋白 A 可能是完全正常的（图 2），而血红蛋白 C 出现了我们所预期的结果，即肽段 4 缺失并被两个新肽段 C-4a 和 C-4b 所取代（C-4b 代表血红蛋白 C 的胰蛋白酶消化肽段 4b）。将这些肽段分离、水解后进行氨基酸分析，结果表明其肽链上的一个谷氨酸被一个赖氨酸所取代，这也证实了之前的一篇关于血红蛋白 C 的报道 [4]。因此，D. K. P. 的 C 组分可能就是血红蛋白 C。

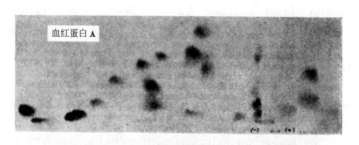

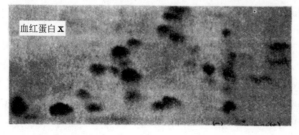

(a)

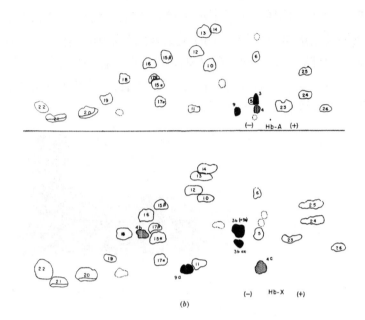

Fig. 2. Photographs (a) and tracings (b) of fingerprints (ref. 11) of haemoglobins A and X. Electrophoresis was in the usual buffer but on a cooled brass plate at about 30 V./cm. for 2.5 hr. Ascending chromatography is in pyridine: isoamyl alcohol: water—35:35:30 by volume (ref. 18). The methionine residue in G-3b is often partially oxidized to the corresponding sulphoxide peptide with a different chromatographic mobility. Such an oxidized peptide is referred to as "G-3b ox". Peptides A-4, C-4a and C-4b are marked by a vertical shading. Peptides A-3, G-3b, G-3b ox and G-9a are marked in black.

Fingerprints[11] of D. K. P.'s haemoglobin G were more difficult to interpret. The fingerprinting technique had to be modified to make it more sensitive. Isolated peptide bands from paper ionophoresis at pH 6.4 are compared side by side by a second ionophoresis on Whatman No. 3*MM* paper in a pH 4.7 buffer. It eventually appeared that in haemoglobin G, two normal peptides—A-3 and A-9—had been replaced by three new peptides. No other changes have so far been detected.

The following relationship for these five peptides was established by means of quantitative amino-acid analyses of the isolated peptides with the Spinco model Moore and Stein automatic analyser, by end-group determinations with the dinitrophenol procedures[12], by stepwise degradation with the Edman–Sjøquist procedure[13] and by fingerprinting[11] of some of the peptides after their digestion with elastase[14].

The haemoglobin A peptides A-3 and A-9 differ only in that A-9 has an additional N-terminal lysine residue; we may write A-9 = Lys-(A-3). The Lys-(A-3) bond in such a peptide would be attacked by trypsin only slowly; hence yields of peptides A-9 and A-3 were found to be variable.

(1) In D. K. P.'s haemoglobin G, an asparagine residue in peptide A-3 (and A-9) has been replaced by lysine; hence a new trypsin-sensitive bond has appeared which results in the

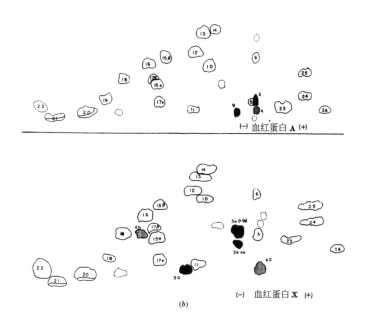

图 2. 血红蛋白 A 和血红蛋白 X 的指纹印迹图谱(参考文献 11)的原图(*a*)和摹图(*b*)。在冷却的铜板上，于常规缓冲液中 30 伏特／厘米恒压电泳 2.5 小时。之后在嘧啶：异戊醇：水（体积比 35∶35∶30）（参考文献 18）的混合溶液中完成上行层析。由于 G-3*b* 肽段上的甲硫氨酸残基常常被部分氧化为亚砜，因此其层析样点迁移率会不同，我们将此氧化肽段表示为"G-3*b* ox"。用竖线区域来表示肽段 A-4、C-4*a* 和 C-4*b*。用黑色区域来表示肽段 A-3、G-3*b*、G-3*b* ox 和 G-9*a*。

相比之下，D. K. P. 的血红蛋白 G 的指纹印迹 [11] 结果解析起来要更困难一些。我们需要对指纹印迹技术进行一些改进以提高其灵敏度。首先，利用纸电泳的方法在 pH 值为 6.4 时将多肽段初步分离，然后将分离开的多肽条带并排转移到 3*MM* 号沃特曼滤纸上，之后在 pH 值为 4.7 的缓冲液中进行二次电泳分离。结果表明血红蛋白 G 的 A-3 和 A-9 这两条正常肽段被三个新的肽段取代了。除此之外没发现其他变化。

之后，我们通过下列方法来研究上述五种肽段间的相互关系。采用特种设备公司生产的穆尔和斯坦自动分析仪对分离肽段进行定量氨基酸分析，用二硝基酚法 [12] 测定末端基团，利用埃德曼–舍奎斯特法 [13] 逐步降解肽段，并在弹性蛋白酶消化后采用指纹印迹法 [11] 分析其中的部分肽段 [14]。

血红蛋白 A 的肽段 A-3 和 A-9 之间的区别只是后者的 N 末端多了一个赖氨酸残基，可以将这一关系表示为 A-9＝Lys-(A-3)。胰蛋白酶只能缓慢地攻击这个肽段中 Lys-(A-3) 之间的肽键，因此消化产生的 A-9 和 A-3 的量并不恒定。

（1）D. K. P. 的血红蛋白 G 的肽段 A-3（和 A-9）上的一个天冬酰胺残基被一个赖氨酸残基所取代，从而增加了一个胰蛋白酶敏感位点，于是就产生了一个额外的

appearance of an additional tryptic peptide. Therefore we can write:

$$A\text{-}3 \rightarrow G\text{-}3a + G\text{-}3b$$
$$A\text{-}9 \rightarrow G\text{-}9a + G\text{-}3b \ (= G\text{-}9b).$$

Of course, peptide G-9a = Lys-(G-3a), as illustrated in Fig. 3.

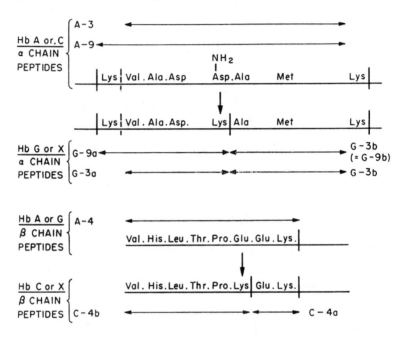

Fig. 3. Diagrammatic representation of the peptides obtained by the action of trypsin on the α- or β-chains of haemoglobins A, G, C, X. See text for the explanation of the relationship between peptide A-3 and A-9 and between G-3a and G-9a. In the lower part of the figure it can be seen that peptides A-4 and C-4b are at the N-terminus of the β-chains (ref. 19) of haemoglobins A and C. Peptide A-9 contains 29 amino-acid residues, peptide G-9a has 8.

(2) It so happens that peptide G-3b contains methionine, which in the course of fingerprinting is to some extent oxidized to the sulphoxide. The resultant peptide "G-3b ox" is definitely identifiable, since it is separated in chromatography. Peptides A-3 and A-9 (and for that matter A-25) show the same oxidation phenomenon.

Haemoglobin X was also digested and fingerprinted[11]. It is clear from the results (Figs. 2 and 3) that haemoglobin X contains the abnormal peptides characteristic of both haemoglobin C and G. Peptide A-4 (β chain) is replaced by the two haemoglobin C peptides C-4a and C-4b. Peptides A-3 and A-9 (α chain) are replaced by the three haemoglobin G peptides G-3a, G-9a and G-3b. Haemoglobin X is therefore a naturally occurring adult haemoglobin which is abnormal in both its α and its β chains. Such doubly abnormal haemoglobins have been prepared in vitro by Itano and Robinson[15] using dissociation and re-association of suitable mixtures of singly abnormal human haemoglobins.

胰蛋白酶水解片段。我们可以这样表述：

$$A\text{-}3 \rightarrow G\text{-}3a + G\text{-}3b$$
$$A\text{-}9 \rightarrow G\text{-}9a + G\text{-}3b \ (= G\text{-}9b)$$

当然，如图 3 所示，肽段 G-9a 就是 Lys-(G-3a)。

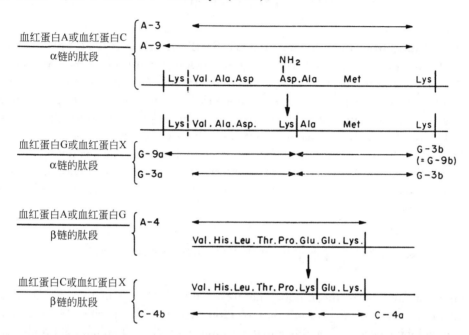

图 3. 血红蛋白 A、G、C 和 X 的 α 链或 β 链经胰蛋白酶消化后得到的肽段的分析图。肽段 A-3 和 A-9 的关系以及肽段 G-3a 和 G-9a 的关系说明见正文。我们可以从靠下方的图上看出，肽段 A-4 和 C-4b 位于血红蛋白 A 和血红蛋白 C 的 β 链的 N 末端（参考文献 19）。肽段 A-9 含有 29 个氨基酸残基，肽段 G-9a 含有 8 个氨基酸残基。

（2）肽段 G-3b 刚好包含甲硫氨酸，在指纹印迹过程中，甲硫氨酸残基会一定程度地被氧化为亚砜，由此产生的 "G-3b ox" 可以通过层析法进行分离与鉴定。肽段 A-3 和 A-9（以及 A-25）也存在着同样的氧化现象。

血红蛋白 X 也被消化并进行指纹印迹分析[11]，从实验结果（图 2 和图 3）可以很明显地看出血红蛋白 X 含有异常肽段，其特征与血红蛋白 C 和血红蛋白 G 的异常肽段相似。其肽段 A-4（β 链）被血红蛋白 C 的两个肽段 C-4a 和 C-4b 所取代，肽段 A-3 和 A-9（α 链）被血红蛋白 G 的三个肽段 G-3a、G-9a 和 G-3b 所取代。因此，血红蛋白 X 是自然产生的 α 链和 β 链均异常的成人血红蛋白。板野和鲁滨逊[15]在体外将人的单链异常血红蛋白分散再重组后，得到了这样的双链异常血红蛋白 X。

One can now write the constitution of D. K. P.'s four haemoglobins as follows:

$$\text{Haemoglobin A} = \alpha_2^A \beta_2^A$$
$$\text{Haemoglobin G} = \alpha_2^G \beta_2^A$$
$$\text{Haemoglobin C} = \alpha_2^A \beta_2^C$$
$$\text{Haemoglobin X} = \alpha_2^G \beta_2^C$$

We would deduce from these findings that D. K. P.'s genotype is α^A/α^G β^C/β^A (see also in refs. 16 and 17). It has also been suggested before by Atwater $et\ al.$[5] that D. K. P. is doubly heterozygous for the G and C abnormalities. Apparently, each gene causes the manufacture of the peptide chain characteristic of it; thus, the α^A gene produces α^A chains, the α^G gene produces α^G chains with lysine in place of asparagine in peptide A-3, and so on. We must next assume that each chain dimerizes as soon as it is formed; $2\alpha^A \rightarrow \alpha_2^A$ and $2\alpha^G \rightarrow \alpha_2^G$. This must be so, since no real hybrids, such as $\alpha^A \alpha^G \beta_2^A$, etc., are produced, although all these chains are present inside the same cells. Finally, the four types of subunit seem to assemble to complete molecules in a random fashion, leading to the four haemoglobins:

$$\alpha_2^A + \alpha_2^G + \beta_2^A + \beta_2^C \rightarrow \alpha_2^A \beta_2^A + \alpha_2^G \beta_2^A + \alpha_2^A \beta_2^C + A_2^G \beta_2^C.$$

It is important to realize the random nature of this final step in the assembly of the molecule, in contrast to the previous dimerization step, which is specific. The fact that the four haemoglobins occur in roughly equal amounts, with A predominating, implies that at least in this patient the rate of manufacture of the different chains is only slightly different. As in the other abnormal human haemoglobins, the phenotype corresponding to each gene is expressed and the product of all of them is to be found within the same cell.

Rather similar chemical results are being obtained by Dr. Park Gerald, Boston, working with an apparently unrelated patient who possesses four haemoglobins called at present A, D, C and X. However, the four components are not nearly in equal amounts, haemoglobins C and X greatly predominating. It will be interesting to see whether these haemoglobins D and X are identical chemically with D. K. P.'s haemoglobins G and X and why the proportions are so different. Very recently, Raper $et\ al.$[20] reported briefly on the occurrence of A, G, C, X in one patient, with X abnormal in both chains. The relationship between their haemoglobins and ours remains to be determined.

Full details of our experiments have been submitted for publication.

We wish to thank Miss Jean Atwater and Dr. L. M. Tocantins for their help and interest. We would also like to acknowledge the help of Miss Marianne Schick, who performed

现在 D. K. P. 的四种血红蛋白的构成可以描述如下：

$$血红蛋白\ A = \alpha_2^A\beta_2^A$$
$$血红蛋白\ G = \alpha_2^G\beta_2^A$$
$$血红蛋白\ C = \alpha_2^A\beta_2^C$$
$$血红蛋白\ X = \alpha_2^G\beta_2^C$$

根据这些结果我们可以推断出 D. K. P. 血红蛋白的基因型是 $\alpha^A/\alpha^G\ \beta^C/\beta^A$（也见参考文献 16、17）。阿特沃特等人[5] 之前也认为 D. K. P. 的异常血红蛋白 G 和血红蛋白 C 是双杂合子。肽链的特征显然取决于编码它的基因；因此，基因 α^A 决定 α^A 链，基因 α^G 决定 α^G 链以及肽段 A-3 上的赖氨酸残基取代天冬酰胺残基等。又由于同一细胞内虽然同时含有四种肽链但并不存在真正的杂合子（如 $\alpha^A\alpha^G\beta_2^A$ 等），我们可以推测出每条肽链一旦合成就会二聚体化，即 $2\alpha^A \rightarrow \alpha_2^A$ 和 $2\alpha^G \rightarrow \alpha_2^G$。最终这四种亚基似乎能够随机组装成这四种血红蛋白：

$$\alpha_2^A+\alpha_2^G+\beta_2^A+\beta_2^C \rightarrow \alpha_2^A\beta_2^A+\alpha_2^G\beta_2^A+\alpha_2^A\beta_2^C+A_2^G\beta_2^C$$

与先前独特的二聚体化步骤相比，能认识到血红蛋白分子组装过程中最后一步的随机性是重要的。事实上，四种血红蛋白中除了 A 组分的含量高一些以外，其他三种出现的概率是大致相等的，这表明人体内各种肽链的含量相差甚微，至少在该患者体内如此。与其他人的异常血红蛋白一样，对应每个基因的表型都会表达出来，并且它们所有的产物都能在同类细胞中检测到。

波士顿的帕克·杰拉尔德医生获得了非常相似的化学实验结果，他的研究对象是一名无血缘关系的患者，该患者具有四种血红蛋白，目前分别被称为血红蛋白 A、D、C 和 X。但这四种血红蛋白的含量不尽相同，其中血红蛋白 C 和 X 要远多于 A 和 D。值得关注的是，该患者的血红蛋白 D 和 X 在化学性质方面是否等同于 D. K. P. 的血红蛋白 G 和 X？它们各自的含量为什么会存在这样的差距？最近雷珀等人[20] 简要报道了一名患者体内同时存在血红蛋白 A、G、C 和 X，其中 X 为双链异常血红蛋白。这些报道中的血红蛋白和我们所研究的四种血红蛋白之间到底是什么样的关系，这一点还有待进一步研究。

有关我们研究的全部细节已经提交并等待发表。

我们向琼·阿特沃特小姐和托坎廷斯医生对本研究的帮助与关注表示感谢，并感谢玛丽安娜·希克小姐为我们进行了定量氨基酸分析。同时，这项工作还得到了

the quantitative amino-acid analysis. This work has been supported by a grant from the Institute for Arthritis and Metabolic Diseases, U.S. Public Health Service.

(**189**, 465-467; 1961)

C. Baglioni and V. M. Ingram: Division of Biochemistry, Department of Biology, Massachusetts Institute of Technology.

References:

1. Rhinesmith, H. W., Schroeder, W. A., and Pauling, L., *J. Amer. Chem. Soc.*, **79**, 4682 (1957). Rhinesmith, H. W., Schroeder, W. A., and Martin, N., *ibid*, **80**, 3358 (1958).

2. Hunt, J. A., and Ingram, V. M., CIBA Found. Symp. Biochem. Human Genetics (Naples), 114 (1959).

3. Hill, R. L., and Schwartz, H. C., *Nature*, **184**, 642 (1959).

4. Hunt, J. A., and Ingram, V. M., *Nature*, **181**, 1062 (1958).

5. Atwater, J., Schwartz, I. R., and Tocantins, L. M., *Blood*, **15**, 901 (1960).

6. Edington, G. M., and Lehmann, H., *Lancet*, **267**, 173 (1954).

7. Shooter, E. M., Skinner, E. R., Garlick, J. P., and Barnicot, N. A., *Brit. J. Haematol.*, **6**, 140 (1960).

8. Schwartz, H. C., Spaet, J. H., Zuelzer, W. W., Neel, J. V., Robinson, A. R., and Kaufman, S. F., *Blood*, **12**, 238 (1957).

9. Smithies, O., *Biochem. J.*, **61**, 629 (1955).

10. Kunkel, H. G., and Wallenius, G., *Science*, **122**, 788 (1955).

11. Ingram, V. M., *Biochim. Biophys. Acta*, **28**, 539 (1958).

12. Sanger, F., and Tuppy, H., *Biochem. J.*, **49**, 465 (1951).

13. Sjøquist, J., *Arkiv Kemi*, **14**, 291, 323 (1959).

14. Naughton, M. A., and Sanger, F., *Biochem. J.*, **70**, 4p (1958).

15. Itano, H. A., and Robinson, E., *Nature*, **183**, 1799 (1959).

16. Ingram, V. M., and Stretton, A. O. W., *Nature*, **184**, 1903 (1959).

17. Ingram, V. M., "Haemoglobin and Its Abnormalities", C. C. Thomas, 1961 (in the press).

18. Wittmann, H. G. (personal communication).

19. Hunt, J. A., and Ingram, V. M., *Nature*, **184**, 640 (1959).

20. Raper, A. B., Gammack, D. B., Huchns, E. R., and Shooter, E. M., *Biochem. J.*, 77. 10P (1960).

美国公共卫生署关节炎和代谢疾病研究所的资助。

（高如丽 翻译；金侠 审稿）

Gene Action in the *X*-chromosome of the Mouse (*Mus musculus* L.)

M. F. Lyon

Editor's Note

In 1960, geneticists Susumu Ohno and T. S. Hauschka studied a variety of female mouse cells, and reported that one of the two *X*-chromosomes always appeared different. Here English geneticist Mary F. Lyon suggests these condensed *X*-chromosomes can be of maternal or paternal origin in different cells of the same animal, and that they are genetically inactive. This could, she explained, account for the mottled appearance of female mice heterozygous for coat colour genes. The Lyon hypothesis, now widely accepted, states that in female mammals, one copy of the *X*-chromosome becomes inactivated early in development. This prevents the female from having twice as many *X*-chromosome gene products as the male.

OHNO and Hauschka[1] showed that in female mice one chromosome of mammary carcinoma cells and of normal diploid cells of the ovary, mammary gland and liver was heteropyknotic. They interpreted this chromosome as an *X*-chromosome and suggested that the so-called sex chromatin was composed of one heteropyknotic *X*-chromosome. They left open the question whether the heteropyknosis was shown by the paternal *X*-chromosome only, or the chromosome from either parent indifferently.

The present communication suggests that the evidence of mouse genetics indicates: (1) that the heteropyknotic *X*-chromosome can be either paternal or maternal in origin, in different cells of the same animal; (2) that it is genetically inactivated.

The evidence has two main parts. First, the normal phenotype of *XO* females in the mouse[2] shows that only one active *X*-chromosome is necessary for normal development, including sexual development. The second piece of evidence concerns the mosaic phenotype of female mice heterozygous for some sex-linked mutants. All sex-linked mutants so far known affecting coat colour cause a "mottled" or "dappled" phenotype, with patches of normal and mutant colour, in females heterozygous for them. At least six mutations to genes of this type have been reported, under the names mottled[3,4], brindled[3], tortoiseshell[5], dappled[6], and 26*K*[2]. They have been thought to be allelic with one another, but since no fertile males can be obtained from any except, in rare cases, brindled, direct tests of allelism have usually not been possible. In addition, a similar phenotype, described as "variegated", is seen in females heterozygous for coat colour mutants translocated on to the *X*-chromosome[7,8].

小鼠（小家鼠）X染色体上的基因作用

莱昂

编者按

1960 年，遗传学家大野干和豪施卡研究了各种各样的雌鼠细胞，并报道称其两条 X 染色体中的一条总是显得不一样。在本文中，英国的遗传学家玛丽·莱昂指出，在同一种动物的不同细胞中这些稠密 X 染色体来自母本或父本，它们缺乏遗传活性。她解释道，这可能就是毛色基因杂合体雌鼠呈现斑驳毛色的原因。莱昂假说现在已被人们广泛接受，它阐述了在雌性哺乳动物发育早期一套 X 染色体会失活，这防止了雌性产生比雄性多一倍的 X 染色体基因产物。

大野和豪施卡[1]指出，雌鼠的乳腺癌细胞以及卵巢、乳腺和肝脏的正常二倍体细胞中都有一条染色体是异固缩的。他们认为这条染色体是一条 X 染色体，并提出所谓的性染色质是由一条异固缩的 X 染色体组成的。他们实际上留下了一个问题，即到底是只有父本的 X 染色体表现为异固缩形态，还是来源于双亲中任何一方的 X 染色体都可以表现为异固缩形态。

当下学术界的讨论认为来自小鼠遗传学的证据可以表明：(1) 同一种动物不同细胞中异固缩的 X 染色体既可能来自于父本，也可能来自母本；(2) 异固缩染色体缺乏遗传活性。

这些证据主要包括两部分。首先是存在表型正常的 XO 型雌鼠[2]，这表明正常发育（包括性发育）只需要一条有活性的 X 染色体。第二个证据是和杂合的性连锁突变的雌鼠具有嵌合体表型有关。至今已知所有能影响毛色的性连锁突变都能使其相应的杂合雌鼠出现"斑驳"或者"斑纹"状表型，即正常颜色和突变颜色的皮毛斑块混杂在一起。现在至少已经报道了 6 种这一类型的基因突变，被称为斑驳[3,4]、斑点[3]、龟甲纹[5]、斑纹[6]和 26K[2]。人们认为它们彼此间互为等位基因，但是，除少量斑点突变外，没能够得到其他突变类型的有生育能力的雄鼠，因此通常不可能直接测定基因的等位性。此外，类似的表型（被描述为"杂色"）也可以在毛色突变能够易位到 X 染色体上[7,8]的杂合雌鼠中看到。

It is here suggested that this mosaic phenotype is due to the inactivation of one or other X-chromosome early in embryonic development. If this is true, pigment cells descended from cells in which the chromosome carrying the mutant gene was inactivated will give rise to a normal-coloured patch and those in which the chromosome carrying the normal gene was inactivated will give rise to a mutant-coloured patch. There may be patches of intermediate colour due to cell-mingling in development. The stripes of the coat of female mice heterozygous for the gene tabby, *Ta*, which affects hair structure, would have a similar type of origin. Falconer[9] reported that the black regions of the coat of heterozygotes had a hair structure resembling that of the *Ta* hemizygotes and homozygotes, while the agouti regions had a normal structure.

Thus this hypothesis predicts that for all sex-linked genes of the mouse in which the phenotype is due to localized gene action the heterozygote will have a mosaic appearance, and that there will be a similar effect when autosomal genes are translocated to the X-chromosome. When the phenotype is not due to localized gene action various types of result are possible. Unless the gene action is restricted to the descendants of a very small number of cells at the time of inactivation, these original cells will, except in very rare instances, include both types. Therefore, the phenotype may be intermediate between the normal and hemizygote types, or the presence of any normal cells may be enough to ensure a normal phenotype, or the observed expression may vary as the proportion of normal and mutant cells varies, leading to incomplete penetrance in heterozygotes. The gene bent-tail, *Bn*[10], may fit into this category, having 95 percent penetrance and variable expression in heterozygotes. Jimpy, *jp*, is recessive, suggesting that the presence of some normal cells is enough to ensure a normal phenotype, but Phillips[11] reported one anomalous female which showed the jimpy phenotype. Since it showed the heterozygous phenotype for *Ta* this animal cannot be interpreted as an *XO* female; it is possible that it represents an example of the rare instance when by chance all the cells responsible for the jimpy phenotype had the normal gene inactivated.

The genetic evidence does not indicate at what stage of embryonic development the inactivation of one X-chromosome occurs. In embryos of the cat, monkey and man sex-chromatin is first found in nuclei of the late blastocyst stage[12,13]. Inactivation of one *X* at a similar stage of the mouse embryo would be compatible with the observations. Since an *XO* female is normally fertile it is not necessary to postulate that both X-chromosomes remain functional until the formation of the gonads.

The sex-chromatin is thought to be formed from one X-chromosome also in the rat, *Rattus norvegicus*[14], and in the opossum, *Didelphis virginiana*[15]. If this should prove to be the case in all mammals, then all female mammals heterozygous for sex-linked mutant genes would be expected to show the same phenomena as those in the mouse. The coat of the tortoiseshell cat, being a mosaic of the black and yellow colours of the two homozygous types, fulfils this expectation.

(**190**, 372-373; 1961)

　　在此我们认为，这种嵌合体表型是胚胎发育早期一条或另一条X染色体失活的结果。如果确实如此，由那些携带突变基因但缺失遗传活性的染色体细胞衍生出来的色素细胞就会产生正常颜色的斑块，而由那些携带正常基因但缺乏遗传活性的染色体细胞衍生出来的色素细胞就会产生突变颜色的斑块。而发育时细胞的混合则可能产生中间颜色的斑块。斑纹基因（Ta，一种能够影响毛发结构的基因）杂合的雌鼠可能具有和其亲本相似类型的皮毛条纹。福尔克纳[9]曾报道称杂合子皮毛黑色区域的毛发结构类似于Ta半合子和纯合子黑色区域的毛发结构，而灰色区域的毛发则具有正常的结构。

　　因此，这一假说预示着所有性连锁基因小鼠的表型是由局部基因作用决定的，相应的杂合子都呈现嵌合体表型；并且当常染色体基因易位到X染色体上时，将会产生类似的效应；当表型不是由局部基因作用决定时，有可能产生各种类型的结果。除非在失活时这种基因作用被限制在具有非常少量细胞的后代中，否则除了极个别的特例外，这些原始细胞将包括两种类型。所以个体表型可能会介于正常型和半合子型之间，或者只要存在任何正常细胞就能够表现出正常的表型，或者可能由于正常细胞和突变细胞的比例不同从而该基因在不同细胞中表达水平不同而导致杂合子表现出不完全的外显率。弯尾基因Bn[10]可能就属于这种类型，其杂合子有95%的外显率和各种水平的基因表达。吉皮基因（jp）是一种隐性基因，表明只要存在一些正常细胞就能够表现出正常的表型，但是菲利普斯[11]曾报道有一例异常的雌性却显示出吉皮表型。由于该雌性显示出Ta杂合表型，因而这种动物就不可能是XO型的雌性个体；这有可能是一个极个别的特例，恰好所有决定吉皮表型的细胞中正常基因失活。

　　这样的遗传学证据并不能表明一条X染色体的失活具体发生在胚胎发育的哪个阶段。在猫、猴子和人类的胚胎中，最早能在囊胚晚期[12,13]的细胞核中发现性染色质。小鼠胚胎中一条X染色体在类似阶段失活，这与观察结果相吻合。既然XO型的雌性个体通常是有生育能力的，那就没必要假定直至生殖腺形成时，两条X染色体还都具有功能。

　　褐家鼠[14]和维几尼亚负鼠[15]中的性染色质也被认为是由一条X染色体形成的。如果在所有哺乳动物中都是如此，那么在所有性连锁突变基因杂合的雌性哺乳动物中就都有可能出现像小鼠中一样的现象了。龟纹猫的皮毛颜色就是黑色和黄色这两种纯合子皮毛颜色的嵌合，这正好是符合这一预期的。

（荆玉祥 翻译；陈新文 陈继征 审稿）

Mary F. Lyon: Medical Research Council, Radiobiological Research Unit, Harwell, Didcot.

References:

1. Ohno, S., and Hauschka, T. S., *Cancer Res.*, **20**, 541 (1960).

2. Welshons, W. J., and Russell, L. B., *Proc. U. S. Nat. Acad. Sci.*, **45**, 560 (1959).

3. Fraser, A. S., Sobey, S., and Spicer, C. C., *J. Genet.*, **51**, 217 (1953).

4. Lyon, M. F., *J. Hered.*, **51**, 116 (1960).

5. Dickie, M. M., *J. Hered.*, **45**, 158 (1954).

6. Phillips, R. J. S., *Genet. Res.* (in the press).

7. Russell, L. B., and Bangham, J. W., *Genetics*, **44**, 532 (1959).

8. Russell, L. B., and Bangham, J. W., *Genetics*, **45**, 1008 (1960).

9. Falconer, D. S., *Z. indukt. Abstamm. u, Vererblehre*, **85**, 210 (1953).

10. Garber, E. D., *Proc. U.S. Nat. Acad. Sci.*, **38**, 876 (1952).

11. Phillips, R. J. S., *Z. indukt. Abstamm. u. Vererblehre*, **86**, 322 (1954).

12. Austin, C. R., and Amoroso, E. C., *Exp. Cell Res.*, **13**, 419 (1957).

13. Park, W. W., *J. Anat.*, **91**, 369 (1957).

14. Ohno, S., Kaplan, W. D., and Kinosita, R., *Exp. Cell Res.*, **18**, 415 (1959).

15. Ohno, S., Kaplan, W. D., and Kinosita, R., *Exp. Cell Res.*, **19**, 417 (1960).

General Nature of the Genetic Code for Proteins

F. H. C. Crick *et al.*

Editor's Note

Early in the development of molecular biology, there came an urgent need to discover how the sequence of nucleotides in a molecule of DNA specifies the sequence of amino acids in a protein molecule. Francis Crick, the co-discoverer of the structure of DNA, took the lead in this endeavour. By 1965, the code had been completed: some groups of three nucleotides coded for specific amino acids, others had the property of bringing the elongation of a protein chain to a halt.

THERE is now a mass of indirect evidence which suggests that the amino-acid sequence along the polypeptide chain of a protein is determined by the sequence of the bases along some particular part of the nucleic acid of the genetic material. Since there are twenty common amino-acids found throughout Nature, but only four common bases, it has often been surmised that the sequence of the four bases is in some way a code for the sequence of the amino-acids. In this article we report genetic experiments which, together with the work of others, suggest that the genetic code is of the following general type:

(*a*) A group of three bases (or, less likely, a multiple of three bases) codes one amino-acid.

(*b*) The code is not of the overlapping type (see Fig. 1).

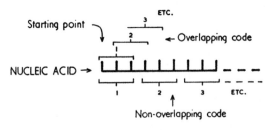

Fig. 1. To show the difference between an overlapping code and a non-overlapping code. The short vertical lines represent the bases of the nucleic acid. The case illustrated is for a triplet code.

(*c*) The sequence of the bases is read from a fixed starting point. This determines how the long sequences of bases are to be correctly read off as triplets. There are no special "commas" to show how to select the right triplets. If the starting point is displaced by one base, then the reading into triplets is displaced, and thus becomes incorrect.

蛋白质遗传密码的普遍特征

克里克等

编者按

在分子生物学发展的早期，就出现了一个亟待解决的问题：一种 DNA 分子中的核苷酸序列如何特异决定一种蛋白质分子中的氨基酸序列。DNA 结构的发现者之一弗朗西斯·克里克成为了这一攻坚领域的领头人。到 1965 年，蛋白质密码的破译工作已经全部完成：一些三核苷酸组合能编码特定的氨基酸，另外一些则具有将蛋白质链的延伸过程终止的功能。

现在有大量的间接证据表明，一种蛋白质多肽链的氨基酸序列是由遗传物质核酸的某一特定部分的碱基序列所决定的。虽然自然界普遍存在 20 种常见氨基酸，但只有 4 种常见碱基，因此人们通常推测 4 种碱基的序列以某种方式编码氨基酸序列。本文中，我们将结合其他人的研究成果，报道一些遗传学实验结果，研究表明遗传密码具备以下普遍特征：

(a) 三个碱基（或者三的倍数个碱基，这种可能性较小）为一组编码一种氨基酸。

(b) 遗传密码是不重叠的（见图 1）。

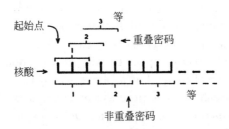

图 1. 显示重叠密码与非重叠密码的差别。短的竖线代表核酸序列中的碱基。本图是基于三联体密码概念而绘制的。

(c) 碱基序列是从一个固定的起始点开始被读取的。这决定着一段很长的碱基序列是如何被正确地读取为三联体的。其中并没有特别的"逗号"来提示如何选择正确的三联体。如果起始点出现了一个碱基的错位，那么读出的所有三联体密码就都将发生错位，从而导致整体读码错误。

(d) The code is probably "degenerate"; that is, in general, one particular amino-acid can be coded by one of several triplets of bases.

The Reading of the Code

The evidence that the genetic code is not overlapping (see Fig. 1) does not come from our work, but from that of Wittmann[1] and of Tsugita and Fraenkel-Conrat[2] on the mutants of tobacco mosaic virus produced by nitrous acid. In an overlapping triplet code, an alteration to one base will in general change three adjacent amino-acids in the polypeptide chain. Their work on the alterations produced in the protein of the virus show that usually only one amino-acid at a time is changed as a result of treating the ribonucleic acid (RNA) of the virus with nitrous acid. In the rarer cases where two amino-acids are altered (owing presumably to two separate deaminations by the nitrous acid on one piece of RNA), the altered amino-acids are not in adjacent positions in the polypeptide chain.

Brenner[3] had previously shown that, if the code were universal (that is, the same throughout Nature), then all overlapping triplet codes were impossible. Moreover, all the abnormal human haemoglobins studied in detail[4] show only single amino-acid changes. The newer experimental results essentially rule out all simple codes of the overlapping type.

If the code is not overlapping, then there must be some arrangement to show how to select the correct triplets (or quadruplets, or whatever it may be) along the continuous sequence of bases. One obvious suggestion is that, say, every fourth base is a "comma". Another idea is that certain triplets make "sense", whereas others make "nonsense", as in the comma-free codes of Crick, Griffith and Orgel[5]. Alternatively, the correct choice may be made by starting as a fixed point and working along the sequence of bases three (or four, or whatever) at a time. It is this possibility which we now favour.

Experimental Results

Our genetic experiments have been carried out on the B cistron of the r_{II} region of the bacteriophage $T4$, which attacks strains of *Escherichia coli*. This is the system so brilliantly exploited by Benzer[6,7]. The r_{II} region consists of two adjacent genes, or "cistrons", called cistron A and cistron B. The wild-type phage will grow on both *E. coli B* (here called B) and on *E. coli K*12(λ) (here called K), but a phage which has lost the function of either gene will not grow on K. Such a phage produces an r plaque on B. Many point mutations of the genes are known which behave in this way. Deletions of part of the region are also found. Other mutations, known as "leaky", show partial function; that is, they will grow on K but their plaque-type on B is not truly wild. We report here our work on the mutant P 13 (now re-named *FC* 0) in the *B*1 segment of the B cistron. This mutant was originally produced by the action of proflavin[8].

We[9] have previously argued that acridines such as proflavin act as mutagens because they

(d) 密码可能具有"简并性"，也就是说，通常几种三联体碱基能编码同一种特定氨基酸。

读　码

遗传密码不重叠（见图 1）的证据不是来自我们的工作，而是来自维特曼[1]、次田皓和弗伦克尔－康拉特[2]对亚硝酸诱发的烟草花叶病毒突变体的研究工作。在一个重叠三联体密码中，一个碱基的改变通常将引起多肽链上三个邻近的氨基酸的改变。他们就病毒蛋白质发生改变所开展的工作表明，用亚硝酸处理病毒核糖核酸（RNA）导致的结果是，通常一次只有一种氨基酸被改变。在少数的情况下出现过两种氨基酸被改变的结果（这可能是由于亚硝酸使同一段 RNA 分子中发生了两处独立的脱氨基作用），但被改变的氨基酸并不位于多肽链上的相邻位置。

布伦纳曾指出[3]，如果这样的密码是通用的（即整个自然界都使用同一套密码），那么所有重叠三联体密码是不可能存在的。此外，所有被详细研究过的异常人血红蛋白都只显示出单个氨基酸的变化[4]。这些新近的实验结果基本上排除了重叠密码存在的可能性。

如果密码是不重叠的，那么必然存在某种规则以指示如何沿着连续的碱基序列选择出正确的三联体（或四联体，或其他任何可能的类型）。一种显而易见的设想是，每组第四个碱基都是一个"逗号"；另一种想法是某些三联体是"有义"的，而其他的则是"无义"的，就像克里克、格里菲思和奥格尔[5]提出的无逗号密码那样。或者从某个固定的位点起始沿着碱基序列以每次读取三个（或四个，或任何其他可能的数量）的方式来做出正确选择，我们现在更倾向于这种可能性。

实 验 结 果

我们的遗传实验是在能侵染大肠杆菌菌株的 T4 噬菌体 r_{II} 区域的顺反子 B 上进行的。这是由本则尔[6,7]巧妙开发的一个的体系。r_{II} 区域由两个相邻的基因或"顺反子"组成，即顺反子 A 和顺反子 B。野生型噬菌体既可以在大肠杆菌 B（以下简称 B）上生长，也可以在大肠杆菌 K12（λ）（以下简称 K）上生长。噬菌体一旦丧失其中任何一个基因的功能就不能在 K 菌体上生长了，但这种噬菌体能在 B 菌体上产生一个 r 型噬菌斑。已知这两个基因的多个点突变都表现出这样的特性。也发现了这一区域的部分缺失。其他一些被称为"渗漏性的"突变则只表现出部分功能，也就是说，它们能在 K 菌体上生长，但它们在 B 菌体上所产生的噬菌斑类型并不同于真正的野生型。这里将报道我们利用顺反子 B 中的 B1 片段中的 P13 突变体（现被重命名为 FC 0）所做的研究工作。这一突变体最初是在二氨基吖啶的作用下产生的[8]。

我们[9]曾指出，二氨基吖啶等吖啶类物质之所以能作为诱变剂，是因为它们能

add or delete a base or bases. The most striking evidence in favour of this is that mutants produced by acridines are seldom "leaky"; they are almost always completely lacking in the function of the gene. Since our note was published, experimental data from two sources have been added to our previous evidence: (1) we have examined a set of 126 r_{II} mutants made with acridine yellow; of these only 6 are leaky (typically about half the mutants made with base analogues are leaky); (2) Streisinger[10] has found that whereas mutants of the lysozyme of phage $T4$ produced by base-analogues are usually leaky, all lysozyme mutants produced by proflavin are negative, that is, the function is completely lacking.

If an acridine mutant is produced by, say, adding a base, it should revert to "wild-type" by deleting a base. Our work on revertants of FC 0 shows that is usually reverts not by reversing the original mutation but by producing a second mutation at a nearby point on the genetic map. That is, by a "suppressor" in the same gene. In one case (or possibly two cases) it may have reverted back to true wild, but in at least 18 other cases the "wild type" produced was really a double mutant with a "wild" phenotype. Other workers[11] have found a similar phenomenon with r_{II} mutants, and Jinks[12] has made a detailed analysis of suppressors in the h_{III} gene.

The genetic map of these 18 suppressors of FC 0 is shown in Fig. 2, line a. It will be seen

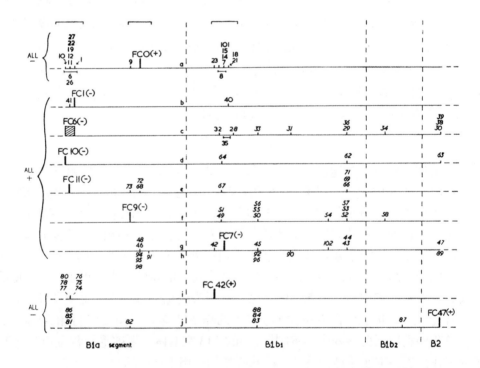

Fig. 2. A tentative map—only very roughly to scale—of the left-hand end of the B cistron, showing the position of the FC family of mutants. The order of sites within the regions covered by brackets (at the top of the figure) is not known. Mutants in italics have only been located approximately. Each line represents the suppressors picked up from one mutant, namely, that marked on the line in bold figures.

增添或缺失一个或多个碱基。支持这个观点的最有力的证据是，吖啶类物质诱导的突变体很少是"渗漏性的"，它们几乎总是导致基因功能的完全丧失。在我们的实验结果发表以后，又有两项实验数据支持我们先前的证据：（1）我们检测了一组由吖啶黄诱导的 126 种 r_{II} 突变体，其中只有 6 种是渗漏性的（一般来说，由碱基类似物诱导得到的突变体约有一半是渗漏性的）；（2）史屈辛格 [10] 发现，尽管由碱基类似物诱导的 T4 噬菌体溶菌酶的突变通常都是渗漏性的，但所有由二氨基吖啶诱导产生的溶菌酶突变却都是阴性的，也就是其功能完全丧失了。

如果吖啶诱发的突变体，比如说，是通过增添一个碱基而产生的话，那么缺失一个碱基就应该使其回复为"野生型"。我们关于 FC 0 回复突变体的研究工作表明，这通常不是通过回复最初的突变来促成的，而是通过在遗传图谱上原突变位点附近的位置引入第二个突变，换句话说，是通过在同一基因中的一个"抑制子"而起作用的。实验中有一例（或者可能是两例）突变的确回复到了真正的野生型，但至少有 18 例的回复突变产生的"野生型"实际上是表现出"野生型"性状的双突变体。其他研究者 [11] 在 r_{II} 突变体的研究中也发现了类似的现象，金克斯 [12] 还对 h_{II} 基因的抑制子做了详细的分析。

这 18 个 FC 0 抑制子的遗传图谱如图 2 中的 a 行所示。从图中可以看出，虽然

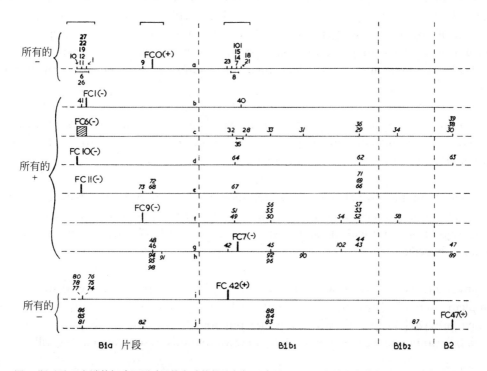

图 2. 顺反子 B 左端的初步图谱（只是大致的按比例），图中显示了 FC 家族突变的位置。括弧（在图顶部）括起来的区域内的位点的顺序目前并不清楚。斜体表示的突变只是大致的定位。每一行代表一种突变体（用加粗的线做了标记）的一系列抑制子。

that they all fall in the $B1$ segment of the gene, though not all of them are very close to FC 0. They scatter over a region about, say, one-tenth the size of the B cistron. Not all are at different sites. We have found eight sites in all, but most of them fall into or near two close clusters of sites.

In all cases the suppressor was a non-leaky r. That is, it gave an r plaque on B and would not grown on K. This is the phenotype shown by a complete deletion of the gene, and shows that the function is lacking. The only possible exception was one case where the suppressor appeared to back-mutate so fast that we could not study it.

Each suppressor, as we have said, fails to grown on K. Reversion of each can therefore be studied by the same procedure used for FC 0. In a few cases these mutants apparently revert to the original wild-type, but usually they revert by forming a double mutant. Fig. 2, lines b-g, shows the mutants produced as suppressor of these suppressors. Again all these new suppressors are non-leaky r mutants, and all map within the $B1$ segment for one site in the $B2$ segment.

Once again we have repeated the process on two of the new suppressors, with the same general results, as shown in Fig. 2, lines i and j.

All these mutants, except the original FC 0, occurred spontaneously. We have, however, produced one set (as suppressors of FC 7) using acridine yellow as a mutagen. The spectrum of suppressors we get (see Fig. 2, line h) is crudely similar to the spontaneous spectrum, and all the mutants are non-leaky r's. We have also tested a (small) selection of all our mutants and shown that their reversion-rates are increased by acridine yellow.

Thus in all we have about eighty independent r mutants, all suppressors of FC 0, or suppressors of suppressors, or suppressors of suppressors of suppressors. They all fall within a limited region of the gene and they are all non-leaky r mutants.

The double mutants (which contain a mutation plus its suppressor) which plate on K have a variety of plaque types on B. Some are indistinguishable from wild, some can be distinguished from wild with difficulty, while others are easily distinguishable and produce plaques rather like r.

We have checked in a few cases that the phenomenon is quite distinct from "complementation", since the two mutants which separately are phenotypically r, and together are wild or pseudo-wild, must be put together in the same piece of genetic material. A simultaneous infection of K by the two mutants in separate viruses will not do.

The Explanation in Outline

Our explanation of all these facts is based on the theory set out at the beginning of this

不是所有的抑制子都非常靠近 $FC\ 0$，但它们都处在该基因的 $B1$ 片段上。它们散布在大约为顺反子 B 的 1/10 大小的范围内，但并不是所有的都位于不同的位点。我们总共发现了 8 个位点，但大部分突变都处于或紧靠两个相近的位点簇。

在所有这些突变体中，抑制子都是非渗漏性的 r 型。这就是说，它们在 B 菌体上产生一个 r 型噬菌斑，且不能在 K 菌体上生长。这是基因完全缺失时的表型，表明相应的功能完全缺失了。不过，唯一可能的例外是，有一个抑制子似乎发生了快得难以对其进行研究的回复突变。

如前所述，每个抑制子都不能在 K 菌体上生长。因此可以用研究 $FC0$ 突变体相同的步骤来对这里的抑制子的回复突变进行研究。在少数情况下，这些突变体表面上看是回复到了最初的野生型，但通常它们是通过形成双突变体来回复的。图 2 中的 $b\sim g$ 行显示的突变体是作为这些抑制子的抑制子。所有这些新的抑制子都是非渗漏性的 r 型突变体，并且除了一个突变位点位于 $B2$ 片段上之外，所有突变位点都位于 $B1$ 片段上。

我们又在两个新的抑制子上重复了上述实验，得到了同样的普遍性结果，如图 2 中的 i 行和 j 行所示。

除了最初的 $FC\ 0$ 突变体外，其余所有的突变都是自发产生的。不过，我们用吖啶黄作为诱变剂得到了一组突变体（作为 $FC\ 7$ 突变体的抑制子）。我们获得的抑制子中突变位点的分布（见图 2，h 行）与自发产生的抑制子中突变位点的分布大致相似，同时所有的突变体也都是非渗漏性的 r 型。我们从所有突变体中选出一（小）部分并对其进行了检测，发现在吖啶黄作用下它们的回复突变率提高了。

综上所述，我们总计获得了大约 80 种独立的 r 型突变体，它们都是 $FC\ 0$ 的抑制子，或者是抑制子的抑制子，或者是抑制子的抑制子的抑制子。所有的突变位点都分布在基因的某个有限区域内，而且都是非渗漏性的 r 型突变体。

在 K 菌体上生长的各种双突变体（包含一种突变加上其抑制子）在 B 菌体上会产生不同类型的噬菌斑。有的与野生型之间完全无法区分，有的与野生型可以区分但具有一定的难度，而其他一些则很容易区分，并长出非常类似于 r 型的噬菌斑。

我们检测了其中的几种回复情况，发现该现象与"互补"现象很不一样，因为两个突变体单独存在时都是 r 噬菌斑表型，但合在一起时却表现出野生型或假野生型表型，这里所说的合在一起是指必须被放置在遗传物质的同一片段上。而 K 菌株同时被位于不同病毒上的两种突变体感染却是无效的（即不能产生噬菌斑）。

解 释 概 要

我们对于这些现象的解释都基于本文开始时所提出的理论。虽然我们没有直接

article. Although we have no direct evidence that the *B* cistron produces a polypeptide chain (probably through an RNA intermediate), in what follows we shall assume this to be so. To fix ideas, we imagine that the string of nucleotide bases is read, triplet by triplet, from a starting point on the left of the *B* cistron. We now suppose that, for example, the mutant *FC* 0 was produced by the insertion of an additional base in the wild-type sequence. Then this addition of a base at the *FC* 0 site will mean that the reading of all the triplets to the right of *FC* 0 will be shifted along one base, and will therefore be incorrect. Thus the amino-acid sequence of the protein which the *B* cistron is presumed to produce will be completely altered from that point onwards. This explains why the function of the gene is lacking. To simplify the explanation, we now postulate that a suppressor *FC* 0 (for example, *FC* 1) is formed by deleting a base. Thus when the *FC* 1 mutation is present by itself, all triplets to the right of *FC* 1 will be read incorrectly and thus the function will be absent. However, when both mutations are present in the same piece of DNA, as in the pseudo-wild double mutant *FC* (0+1), then although the reading of triplets between *FC* 0 and *FC* 1 will be altered, the original reading will be restored to the rest of the gene. This could explain why such double mutants do not always have a true wild phenotype but are often pseudo-wild, since on our theory a small length of their amino-acid sequence is different from that of the wild-type.

For convenience we have designated our original mutant *FC* 0 by the symbol + (this choice is a pure convention at this stage) which we have so far considered as the addition of a single base. The suppressors of *FC* 0 have therefore been designated − . The suppressors of these suppressors have in the same way been labelled as +, and the suppressors of these last sets have again been labelled − (see Fig. 2).

Double Mutants

We can now ask: What is the character of any double mutant we like to form by putting together in the same gene any pair of mutants from our set of about eighty? Obviously, in some cases we already know the answer, since some combinations of a + with a − were formed in order to isolate the mutants. But, by definition, no pair consisting of one + with another + has been obtained in this way, and there are many combinations of + with − not so far tested.

Now our theory clearly predicts that all combinations of the type + with + (or − with −) should give an *r* phenotype and not plate on *K*. We have put together 14 such pairs of mutants in the cases listed in Table 1 and found this prediction confirmed.

Table 1. Double Mutants having the *r* Phenotype

− With −	+ With +	
FC (1+21)	FC (0+58)	FC (40+57)
FC (23+21)	FC (0+38)	FC (40+58)
FC (1+23)	FC (0+40)	FC (40+55)
FC (1+9)	FC (0+55)	FC (40+54)
	FC (0+54)	FC (40+38)

证据表明顺反子 B 编码一条多肽链（可能通过一种 RNA 中间体），但下文中我们将假定事实就是如此。为了确定我们的想法，我们是这样设想的，核苷酸碱基串是从位于顺反子 B 左侧的一个起始点开始三个三个地被阅读的。比如，现在我们可以这样推测，突变体 $FC\,0$ 是通过在野生型序列中插入一个额外的碱基而得到的。那么这个在 $FC\,0$ 位点插入的额外碱基就会使得所有在 $FC\,0$ 突变位点右侧的三联体的读取错位一个碱基，从而导致读码错误。因此，原本由顺反子 B 编码的蛋白质的氨基酸序列将从这个位点开始被完全改变。这就解释了为什么这个基因突变会导致功能缺失。为了简化说明，我们现在假定 $FC\,0$ 的抑制子（比如 $FC\,1$）是通过缺失一个碱基而产生的。那么当 $FC\,1$ 的突变单独出现时，所有位于 $FC\,1$ 突变位点右侧的三联体都将发生读码错误，从而导致基因功能丧失。然而，就像假野生型双突变体 $FC\,(0+1)$ 中的情形那样，当这两种突变存在于同一 DNA 片段上时，尽管 $FC\,0$ 位点和 $FC\,1$ 位点之间的三联体的读取被改变了，但是这个基因的其余部分的读码将会恢复到最初的正常状态。这就解释了为什么这样的双突变体通常并不是真正的野生型，而是假野生型，因为根据我们的理论，它们的氨基酸序列中有一小段与野生型是不同的。

为了方便起见，我们将最初的突变体 $FC\,0$ 以 " $+$ "（此标记纯属习惯）来表示，我们一直认为该突变是由增添的单个碱基造成的。因此，$FC\,0$ 的抑制子用 " $-$ " 来表示。同样地，这些抑制子的抑制子标记为 " $+$ "，其抑制子继而又标记为 " $-$ "（见图 2）。

双 突 变 体

现在我们可以提出这样的问题：将我们获得的约 80 种突变中的任意一对突变同时置放在同一个基因上所得到的双突变体会有什么样的特性呢？很显然我们已经知道了某些情况下的答案，为了分离得到这些突变体，我们已经得到了一些由一个 " $+$ " 突变和一个 " $-$ " 突变组合而成的突变。但是，根据定义，用这种方法并没有得到任何由两个 " $+$ " 突变组成的双突变对，另外还有许多 " $+$ " 突变和 " $-$ " 突变的组合突变尚未被检测。

现在我们的理论清楚地预测到，所有 " $+$ " 突变和 " $+$ " 突变（或 " $-$ " 突变和 " $-$ " 突变）的组合都应该出现 r 噬菌斑表型，并且不能在 K 菌体上生长。我们将表 1 列出的实例中的 14 对这样的突变放在一起，结果证实了我们的预测。

表 1. 具有 r 噬菌斑表型的双突变体

" $-$ " 和 " $-$ " 的组合	" $+$ " 和 " $+$ " 的组合	
$FC\,(1+21)$	$FC\,(0+58)$	$FC\,(40+57)$
$FC\,(23+21)$	$FC\,(0+38)$	$FC\,(40+58)$
$FC\,(1+23)$	$FC\,(0+40)$	$FC\,(40+55)$
$FC\,(1+9)$	$FC\,(0+55)$	$FC\,(40+54)$
	$FC\,(0+54)$	$FC\,(40+38)$

At first sight one would expect that all combinations of the type (+ with −) would be wild or pseudo-wild, but the situation is a little more intricate than that, and must be considered more closely. This springs from the obvious fact that if the code is made of triplets, any long sequence of bases can be read correctly in one way, but incorrectly (by starting at the wrong point) in two different ways, depending whether the "reading frame" is shifted one place to the right or one place to the left.

If we symbolize a shift, by one place, of the reading frame in one direction by → and in the opposite direction by ←, then we can establish the convention that our + is always at the head of the arrow, and our − at the tail. This is illustrated in Fig. 3.

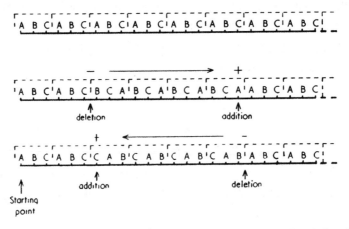

Fig. 3. To show that our convention for arrows is consistent. The letters *A*, *B* and *C* each represent a different base of the nucleic acid. For simplicity a repeating sequence of bases, *ABC*, is shown. (This would code for a polypeptide for which every amino-acid was the same.) A triplet code is assumed. The dotted lines represent the imaginary "reading frame" implying that the sequence is read in sets of three starting on the left.

We must now ask: Why do our suppressors not extend over the whole of the gene? The simplest postulate to make is that the shift of the reading frame produces some triplets the reading of which is "unacceptable"; for example, they may be "nonsense", or stand for "end the chain", or be unacceptable in some other way due to the complications of protein structure. This means that a suppressor of, say, *FC* 0 must be within a region such that no "unacceptable" triplet is produced by the shift in the reading frame between *FC* 0 and its suppressor. But, clearly, since for any sequence there are *two* possible misreadings, we might expect that the "unacceptable" triplets produced by a → shift would occur in different places on the map from those produced by a ← shift.

Examination of the spectra of suppressors (in each case putting in the arrows → or ←) suggests that while the → shift is acceptable anywhere within our region (though not outside it) the shift ←, starting from points near *FC* 0, is acceptable over only a more limited stretch. This is shown in Fig. 4. Somewhere in the left part of our region, between

　　初看起来，人们可能会预期所有的"＋"突变和"－"突变的组合都会表现为野生型或假野生型，但实际情况要比这个复杂一些，我们必须进行更加周密的考虑。这源于一个明显的事实，那就是，如果密码是三联体形式的，那么任意长度的一段碱基序列被正确读取的方式都只有一种，但被错误读取（于一个错误的起始点开始）的方式有两种，其"阅读框"或向右移动了一个碱基的位置，或向左移动了一个碱基的位置。

　　如果我们用"→"来表示阅读框在某处向一个方向移动了，而用"←"来表示阅读框向相反方向移动了，那么我们可以确立这样的约定，"＋"总是在箭头的头端，而"－"总是在尾端，如图 3 所示。

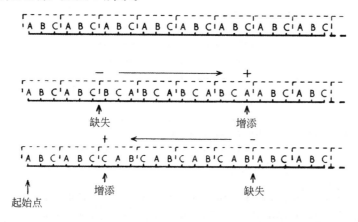

图 3. 我们对于箭头意义的约定始终是一致的。字母 A、B 和 C 分别代表一种不同的核酸碱基。为了说明的方便，我们仅仅显示了一种 ABC 不断重复的碱基序列（这将编码一条只由一种氨基酸组成的多肽链）。我们假设密码是三联体形式的。虚线代表假想的从序列左端开始每三个一组进行阅读的"阅读框"。

　　现在我们肯定会提出这样的问题：为什么我们得到的抑制子的突变位点不是遍布在整个基因上？最简单的假设就是，阅读框的移动产生了一些读取时"不能被接受的"三联体密码。比如，它们可能是"无义"的，或者它们代表"链的结束"，再或者由于某些和蛋白质结构复杂性有关的原因而不能被接受。这意味着，一个抑制子，如 FC 0 抑制子，发生突变的位点必然位于一定的区域内，使得 FC 0 和它的抑制子之间的阅读框的移动不至于产生"不能被接受的"三联体。但很明显的是，既然任意序列都存在**两种**可能的错读，因此我们可以预期的是，由"→"方向的移动和"←"方向的移动产生的"不能被接受的"三联体将会发生在遗传图谱的不同位置上。

　　对抑制子（每一个实例中都用箭头"→"或"←"标示）分布情况的分析表明，尽管"→"方向的移动在我们关心的基因区域内的任意位置都是可以接受的（尽管在此之外是不行的），那么从 FC 0 位点附近开始的"←"方向的移动只会在一个更为有限的片段内可以被接受，如图 4 所示。当出现"←"方向的移动时，在我们关

FC 0 or FC 9 and the FC 1 group, there must be one or more unacceptable triplets when a ← shift is made; similarly for the region to the right of the FC 21 cluster. Thus we predict that a combination of a + with a − will be wild or pseudo-wild if it involves a → shift, but that such pairs involving a ← shift will be phenotypically r if the arrow crosses one or more of the forbidden places, since then an unacceptable triplet will be produced.

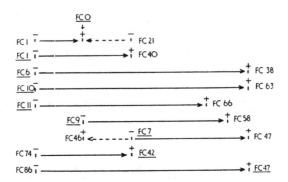

Fig. 4. A simplified version of the genetic map of Fig. 2. Each line corresponds to the suppressor from one mutant, here underlined. The arrows show the range over which suppressors have so far been found, the extreme mutants being named on the map. Arrows to the right are shown solid, arrows to the left dotted.

We have tested this prediction in the 28 cases shown in Table 2. We expected 19 or these to be wild, or pseudo-wild, and 9 of them to have the r phenotype. In all cases our prediction was correct. We regard this as a striking confirmation of our theory. It may be of interest that the theory was constructed before these particular experimental results were obtained.

Table 2. Double Mutants of the Type (+ with −)

+ / −	FC 41	FC 0	FC 40	FC 42	FC 58*	FC 63	FC 38
FC 1	W	W	W		W		W
FC 86		W	W	W	W	W	
FC 9	r	W	W	W	W		W
FC 82	r		W	W	W	W	
FC 21	r	W			W		W
FC 88	r	r			W	W	
FC 87	r	r	r	r			W

W, wild or pseudo-wild phenotype; W, wild or pseudo-wild combination used to isolate the suppressor; r, r phenotype.
* Double mutants formed with FC 58 (or with FC 34) give sharp plaques on K.

心区域左侧的某些位置，也就是在 *FC* 0 或 *FC* 9 位点与 *FC* 1 位点群之间，肯定会出现一个或更多个不能被接受的三联体。*FC* 21 位点簇右侧的区域也存在类似的情况。因此我们预计，如果一对"＋"突变和"－"突变的组合只涉及"→"方向的移动，就会是野生型或假野生型；而如果这种突变的组合只涉及一个"←"方向的移动并且箭头会跨过一个或更多个禁止区域的话，组合突变的结果将是 *r* 噬菌斑表型，因为那样会产生一个不能被接受的三联体。

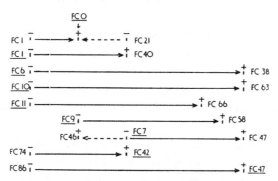

图 4. 图 2 所示遗传图谱的简化版。每一行对应由一种突变体（用下划线标记）产生的抑制子。箭头显示了到目前为止发现的抑制子的跨度范围，两端的突变体的名称在图中都有标注。向右的箭头用实线表示，向左的箭头用虚线表示。

我们在表 2 列出的 28 种情况下对这个预测进行了检验。我们预期其中 19 个会是野生型或假野生型，9 个会是 *r* 表型。结果表明，所有情况都与我们的预测完全一致。我们认为，这样的结果是对我们理论的有力验证。也许会让很多人感兴趣的是，我们的理论在这些实验结果获得之前就提出了。

表 2. "＋"和"–"组合类型的双突变体

+ –	*FC* 41	*FC* 0	*FC* 40	*FC* 42	*FC* 58*	*FC* 63	*FC* 38
FC 1	*W*	*W*	*W*		W		W
FC 86		W	W	W	W	W	
FC 9	*r*	*W*	W	W	*W*		W
FC 82	*r*		W	W	W	W	
FC 21	*r*	*W*			W		W
FC 88	*r*	*r*			W	W	
FC 87	*r*	*r*	*r*	*r*			W

W 代表野生型或假野生型表型；*W* 代表用于分离抑制子的野生型或假野生型组合；*r* 代表 *r* 表型。
* 由 *FC* 58（或 *FC* 34）产生的双突变在 *K* 菌体上会产生明显的噬菌斑。

Rigorous Statement of the Theory

So far we have spoken as if the evidence supported a triplet code, but this was simply for illustration. Exactly the same results would be obtained if the code operated with groups of, say, 5 bases. Moreover, our symbols + and − must not be taken to mean literally the addition or subtraction of a single base.

It is easy to see that our symbolism is more exactly as follows:

$$+ \text{ represents } + m, \text{ modulo } n$$
$$- \text{ represents } - m, \text{ modulo } n$$

where n (a positive integer) is the coding ratio (that is, the number of bases with code one amino-acid) and m is any integral number of bases, positive or negative.

It can also be seen that our choice of reading direction is arbitrary, and that the same results (to a first approximation) would be obtained in whichever direction the genetic material was read, that is, whether the starting point is on the right or the left of the gene, as conventionally drawn.

Triple Mutants and the Coding Ratio

The somewhat abstract description given above is necessary for generality, but fortunately we have convincing evidence that the coding ratio is in fact 3 or a multiple of 3.

This we have obtained by constructing triple mutants of the form (+ with + with +) or (− with − with −). One must be careful not to make shifts across the "unacceptable" regions for the ← shifts, but these we can avoid by a proper choice of mutants.

We have so far examined the six cases listed in Table 3 and in all cases the triples are wild or pseudo-wild.

Table 3. Triple Mutants having a Wild or Pseudo-wild Phenotype

FC (0 + 40 +38)

FC (0 + 40 +58)

FC (0 + 40 +57)

FC (0 + 40 +54)

FC (0 + 40 +55)

FC (1 + 21 +23)

The rather striking nature of this result can be seen by considering one of them, for example, the triple (*FC* 0 with *FC* 40 with *FC* 38). These three mutants are, by themselves, all of like type (+). We can say this not merely from the way in which they were obtained, but because each of them, when combined with our mutant *FC* 9(−),

218

理论的严格表述

到目前为止，我们好像是在实验证据支持三联体密码的前提下描述的，但这实际上仅仅是为了论述的方便。如果密码是其他形式的，比如说 5 个碱基一组，那么我们也将得到完全一样的结果。另外，我们的符号"＋"和"－"不应该从字面上理解为一个单一碱基的增加或减少。

显然，如下表述才能更准确地表示我们的符号系统的意义：

$$\text{"＋"代表} + m, \text{以} n \text{为模}$$
$$\text{"－"代表} - m, \text{以} n \text{为模}$$

其中，n（一个正整数）是编码比率（即编码一个氨基酸的碱基数目），而 m 代表任意整数个碱基，可正可负。

另外也可以看出，我们对阅读方向的选择带有随意性，遗传物质无论从哪一个方向被阅读，也就是说无论起始点是在基因的右边还是左边（即按照习惯所绘出的图示那样），我们得到的结果将都是相同的（大致近似的）。

三突变体和编码比率

上述略显抽象的描述对于讨论问题的普遍性是必要的，但幸运的是，我们有令人信服的证据表明编码比率确实是 3 或 3 的倍数。

我们是通过构建三个"＋"组合或三个"－"组合的三突变体得到这样的结论的。实验中必须小心地避让"←"方向的移位越过"不能被接受的"区域，但是我们可以通过对突变体进行适当选择来避免这类情况。

到目前为止，我们已经检测了表 3 列出的 6 种情况，所有这些三突变体都是野生型或假野生型。

表 3. 三突变体都有野生型或假野生型表型

FC (0 + 40 +38)
FC (0 + 40 +58)
FC (0 + 40 +57)
FC (0 + 40 +54)
FC (0 + 40 +55)
FC (1 + 21 +23)

我们得到的结果十分令人振奋，下面就以其中一个三突变体（FC 0 和 FC 40 和 FC 38）为例来说明。这个三突变体本身都属于相似的类型（＋）。我们这么说不只是基于这些突变体被获得的方式，而且是因为这三种突变中的任何一个和 FC 9（－）组合在一起时都表现为野生型或假野生型。然而，这三种突变不论是单独存在还是

gives the wild, or pseudo-wild phenotype. However, either singly or together in pairs they have an *r* phenotype, and will not grow on *K*. That is, the function of the gene is absent. Nevertheless, the combination of all three in the same gene partly restores the function and produces a pseudo-wild phage which grows on *K*.

This is exactly what one would expect, in favourable cases, if the coding ratio were 3 or a multiple of 3.

Our ability to find the coding ratio thus depends on the fact that, in at least one of our composite mutants which are "wild", at least one amino-acid must have been added to or deleted from the polypeptide chain without disturbing the function of the gene-product too greatly.

This is a very fortunate situation. The fact that we can make these changes and can study so large a region probably comes about because this part of the protein is not essential for its function. That this is so has already been suggested by Champe and Benzer[18] in their work on complementation in the r_{II} region. By a special test (combined infection on *K*, followed by plating on *B*) it is possible to examine the function of the *A* cistron and the *B* cistron separately. A particular deletion, 1589 (see Fig. 5) covers the right-hand end of the *A* cistron and part of the left-hand end of the *B* cistron. Although 1589 abolished the *A* function, they showed that it allows the *B* function to be expressed to a considerable extent. The region of the *B* cistron deleted by 1589 is that into which all our *FC* mutants fall.

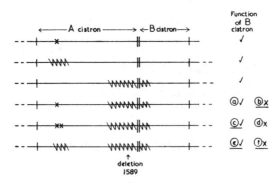

Fig. 5. Summary of the results with deletion 1589. The first two lines show that without 1589 a mutation or a deletion in the *A* cistron does not prevent the *B* cistron from functioning. Deletion 1589 (line 3) also allows the *B* cistron to function. The other cases, in some of which an alteration in the *A* cistron prevents the function of the *B* cistron (when 1589 is also present), are discussed in the text. They have been labelled (*a*), (*b*), etc., for convenience of reference, although cases (*a*) and (*d*) are not discussed in the paper. √ implies function; × implies no function.

Joining two Genes Together

We have used this deletion to reinforce our idea that the sequence is read in groups from a fixed starting point. Normally, an alteration confined to the *A* cistron (be it a deletion, an acridine mutant, or any other mutant) does not prevent the expression of the *B* cistron.

成对存在时都表现出 r 表型，并且不能在 K 菌体上生长，这就是说，基因的功能缺失了。然而，当这三种突变同时出现在同一个基因中时却能够部分地恢复该基因的功能，从而产生能在 K 菌体上生长的假野生型噬菌体。

这正是编码比率为 3 或 3 的倍数时，我们预期会出现的结果。

因此我们揭示编码比率的能力有赖于如下事实：至少在我们获得的一种复合"野生型"突变体中，在多肽链中增添或缺失至少一个氨基酸不会很大程度地干扰基因产物的功能。

这是一种非常幸运的情形。我们之所以能制备出这一系列的突变并研究如此大的一个基因区域，可能是因为该蛋白质的这一部分对于其功能而言是非必需的。钱普和本则尔[18] 在研究 r_{II} 区域内的互补现象时就曾提出过这种观点。通过一种特别的检测方法（联合侵染 K 菌株后再涂布到 B 菌株的菌体上），顺反子 A 和顺反子 B 的功能就能够被分别检测。一种特殊的缺失突变体 1589（见图 5），它覆盖了顺反子 A 的右端以及顺反子 B 左端的部分序列。尽管 1589 突变彻底破坏了顺反子 A 的功能，但它却允许顺反子 B 的功能在一定程度上得以表达。顺反子 B 中因为 1589 突变而缺失的区域处于我们所获得的所有 FC 突变分布的位置上。

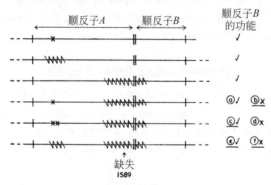

图 5. 1589 缺失实验结果的汇总。前两行显示，在没有发生 1589 缺失时顺反子 A 上的突变或缺失并不阻止顺反子 B 的功能。1589 缺失（第 3 行）发生时，顺反子 B 的功能仍能发挥。其他一些案例在正文中进行了讨论，在部分的这些案例中，顺反子 A 的变化（当 1589 缺失也存在的情况下）会阻止顺反子 B 的功能。为了论述中引用的方便，这些案例被标记为 (a)、(b) 等，虽然案例 (a) 和 (d) 并没有在本文中进行讨论。√代表功能正常，×代表功能丧失。

将两个基因连接在一起

我们使用的这一缺失突变强化了我们提出的序列是从一个固定的起始点按碱基组被读取的概念。正常情况下，一个局限于顺反子 A 区域内的碱基变化（可以是一个碱基的缺失、一种吖啶引起的突变或者是其他任何突变）都不会阻止顺反子 B 功

Conversely, no alteration within the *B* cistron prevents the function of the *A* cistron. This implies that there may be a region between the two cistrons which separates them and allows their functions to be expressed individually.

We argued that the deletion 1589 will have lost this separating region and that therefore the two (partly damaged) cistrons should have been joined together. Experiments show this to be the case, for now an alteration to the left-hand end of the *A* cistron, if combined with deletion 1589, can prevent the *B* function from appearing. This is shown in Fig. 5. Either the mutant *P*43 or *X*142 (both of which revert strongly with acridines) will prevent the *B* function when the two cistrons are joined, although both of these mutants are in the *A* cistron. This is also true of *X*142 *S*1, a suppressor of *X*142 (Fig. 5, case *b*). However, the double mutant (*X*142 with *X*142 *S*1), of the type (+ with −), which by itself is pseudo-wild, still has the *B* function when combined with 1589 (Fig. 5, case *c*). We have also tested in this way the 10 deletions listed by Benzer[7], which fall wholly to the left of 1589. Of these, three (386, 168 and 221) prevent the *B* function (Fig. 5, case *f*), whereas the other seven show it (Fig. 5, case *e*). We surmise that each of these seven has lost a number of bases which is a multiple of 3. There are theoretical reasons for expecting that deletions may not be random in length, but will more often have lost a number of bases equal to an integral multiple of the coding ratio.

It would not surprise us if it were eventually shown that deletion 1589 produces a protein which consists of part of the protein from the *A* cistron and part of that from the *B* cistron, joined together in the same polypeptide chain, and having to some extent the function of the undamaged *B* protein.

Is the Coding Ratio 3 or 6?

It remains to show that the coding ratio is probably 3, rather than a multiple of 3. Previous rather rough estimates[10,14] of the coding ratio (which are admittedly very unreliable) might suggest that the coding ratio is not far from 6. This would imply, on our theory, that the alteration in *FC* 0 was not to one base, but to two bases (or, more correctly, to an even number of bases).

We have some additional evidence which suggests that this is unlikely. First, in our set of 126 mutants produced by acridine yellow (referred to earlier) we have four independent mutants which fall at or close to the *FC* 9 site. By a suitable choice of partners, we have been able to show that two are + and two are −. Secondly, we have two mutants (*X*146 and *X*225), produced by hydrazine[15], which fall on or near the site *FC* 30. These we have been able to show are both of type −.

Thus unless both acridines and hydrazine usually delete (or add) an even number of bases, this evidence supports a coding ratio of 3. However, as the action of these mutagens is not understood in detail, we cannot be certain that the coding ratio is not 6, although 3 seems more likely.

222

能的发挥。相反，在顺反子 B 区域内没有任何碱基变化会阻止顺反子 A 发挥其功能。这意味着在这两个顺反子之间可能存在一个区域，它将两个顺反子各自分开，并使两个顺反子的功能相互独立地发挥。

我们认为，1589 缺失突变会导致这个分隔区域的丢失，因此两个顺反子（部分残缺的）就应该被连在一起。实验结果表明情况确实如此，因为如果 1589 缺失和顺反子 A 左端的碱基变化一起发生的话，那么能够阻止顺反子 B 功能的出现，这一现象如图 5 所示。当两个顺反子连在一起时，无论是 $P43$ 突变还是 $X142$ 突变（二者都能被吖啶强烈地诱导发生回复突变）都会阻止顺反子 B 发挥其功能，尽管这两种突变都是发生在顺反子 A 区域内。同样的情况也出现在 $X142$ 突变体的抑制子 $X142$ $S1$ 中（见图 5 中的案例 b）。但是由一个 "+" 和 "−" 组合而成的双突变体（如 $X142$ 和 $X142$ $S1$）本身是假野生型，与 1589 缺失一起发生时仍然能表现出顺反子 B 的功能（见图 5 中的案例 c）。我们用同样的方式也检测了本则尔列出的 10 种缺失突变体 [7]，这些突变都发生在 1589 缺失的左侧。在这些突变体中，有 3 种突变（386、168 和 221）会阻止顺反子 B 的功能（见图 5 中的案例 f），而另外 7 种突变中仍然能表现出顺反子 B 的功能（见图 5 中的案例 e）。我们猜测这 7 种突变体中每一种丢失的碱基数都是 3 的倍数。之所以预期缺失的碱基数可能不是随机的，而更经常的是丢失的碱基数等于编码比率的整数倍，是有理论依据的。

如果最终的结果表明 1589 缺失会产生一种蛋白质，其中的一部分由顺反子 A 编码，另一部分由顺反子 B 编码，它们连接在同一条多肽链上而且在某种程度上还具有未被损坏的 B 蛋白的功能，我们是不会吃惊的。

编码比率是 3 还是 6？

我们还需要说明为什么编码比率可能是 3，而不是 3 的倍数。以前对于编码比率的粗略估计 [10,14]（被认为是非常不可靠的）提示编码比率有可能接近 6。这就意味着，就我们的理论而言，FC 0 中发生变化的可能不是一个碱基，而是两个碱基（或者更确切地说是偶数个碱基）。

我们有更多的证据表明这是不太可能的。首先，在我们获得的一组由吖啶黄诱导产生的 126 个突变中，有 4 个是相互独立的突变，发生在 FC 9 位点或其附近的位点。通过配对检测，我们发现其中两个属于 "+" 类，两个属于 "−" 类。其次，我们还获得了两种由肼诱导产生的突变（$X146$ 和 $X225$）[15]，这两种突变发生在 FC 30 位点或其附近位点。我们发现它们都属于 "−" 类型。

因此，除非吖啶和肼通常都缺失（或增添）偶数个碱基，否则这个证据就支持编码比率为 3 结论。不过，因为我们对这些诱变剂的具体作用机理还不太清楚，因此我们还不能肯定地说编码比率不是 6，虽然看起来更可能是 3。

We have preliminary results which show that other acridine mutants often revert by means of close suppressors, but it is too sketchy to report here. A tentative map of some suppressors of $P\,83$, a mutant at the other end of the B cistron, in segment $B\,9a$, is shown in Fig. 6. They occur within a shorter region than the suppressors of $FC\,0$, covering a distance of about one-twentieth of the B cistron. The double mutant WT (2+5) has the r phenotype, as expected.

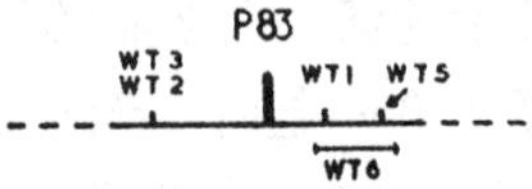

Fig. 6. Genetic map of $P\,83$ and its suppressors, WT 1, etc. The region falls within segment $B\,9a$ near the right-hand end of the B cistron. It is not yet known which way round the map is in relation to the other figures.

Is the Code Degenerate?

If the code is a triplet code, there are 64 (4×4×4) possible triplets. Our results suggest that it is unlikely that only 20 of these represent the 20 amino-acids and that the remaining 44 are nonsense. If this were the case, the region over which suppressors of the $FC\,0$ family occur (perhaps a quarter of the B cistron) should be very much smaller than we observe, since a shift of frame should then, by chance, produce a nonsense reading at a much closer distance. This argument depends on the size of the protein which we have assumed the B cistron to produce. We do not know this, but the length of the cistron suggests that the protein may contain about 200 amino-acids. Thus the code is probably "degenerate", that is, in general more than one triplet codes for each amino-acid. It is well known that if this were so, one could also account for the major dilemma of the coding problem, namely, that while the base composition of the DNA can be very different in different micro-organisms, the amino-acid composition of their proteins only changes by a moderate amount[16]. However, exactly how many triplets code amino-acids and how many have other functions we are unable to say.

Future Developments

Our theory leads to one very clear prediction. Suppose one could examine the amino-acid sequence of the "pseudo-wild" protein produced by one of our double mutants of the (+ with −) type. Conventional theory suggests that since the gene is only altered in two places, only two amino-acids would be changed. Our theory, on the other hand, predicts that a string of amino-acids would be altered, covering the region of the polypeptide chain corresponding to the region on the gene between the two mutants. A good protein on which to test this hypothesis is the lysozyme of the phage, at present being studied chemically be Dreyer[17] and genetically by Streisinger[10].

At the recent Biochemical Congress at Moscow, the audience of Symposium I was startled by the announcement of Nirenberg that he and Matthaei[18] had produced polyphenylalanine (that is, a polypeptide all the residues of which are phenylalanine) by adding polyuridylic acid (that is, an RNA the bases of which are all uracil) to a cell-free

我们有一些初步的结果表明，其他由吖啶诱发的突变经常是通过邻近位点的抑制子回复的，但是这些结果还太粗略不适合在这里报道。图 6 显示了 P 83 突变（这一突变发生在顺反子 B 的另一端，B 9a 片段内）的一些抑制子分布的初步图谱。可以看出，它们分布在比 FC 0 的抑制子更窄的区域内，大约占顺反子 B 的 1/20。正如预期的那样，双突变体 WT（2+5）呈现出 r 表型。

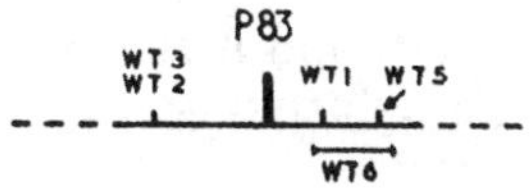

图 6. P 83 突变及其抑制子（WT 1 等）的遗传图谱。此区域位于顺反子 B 右端附近的 B 9a 片段中。此图谱与文中的其他各图的位置关系目前还不清楚。

密码是简并的吗?

如果遗传密码为三联体密码的话，那么就会有 64（4×4×4）种可能的三联体。我们的结果表明，不太可能只是其中 20 种三联体密码编码 20 种氨基酸，而剩余的 44 种三联体都是无义的。如果情况真是如此，那么 FC 0 家族的抑制子分布的区域（可能为顺反子 B 的 1/4）应该比我们观察到的要小得多，因为阅读框的移动可能会在更近的距离内产生一个无义读取。这个争论与我们假设的顺反子 B 产生的蛋白质的大小有关。但我们并不知道所产生的蛋白质的具体长度，不过顺反子 B 的长度表明这种蛋白质可能包含有大约 200 个氨基酸。因此，密码子可能是"简并的"，换句话说，通常而言，一种氨基酸能被不止一种的三联体所编码。众所周知，如果情况真是这样，我们也就能解释编码问题中的另一个主要难题了，即为什么不同微生物的 DNA 碱基组成差异很大，而它们的蛋白质的氨基酸组成却只有中等程度的差异 [16]。然而，具体有多少个三联体密码是编码氨基酸的，多少个具有其他功能，我们现在还无法说清楚。

未来的发展

我们的理论引出了一个非常清晰的预测。假设我们可以检测由"+"和"−"组合的双突变体中的一种产生的"假野生型"蛋白质的氨基酸序列。根据传统理论，既然该基因中只在两个位置发生了改变，那么应该只有两个氨基酸会发生改变。另一方面，我们的理论预测，两个突变位点之间的基因区域对应的多肽链区域上的一连串氨基酸都会发生改变。噬菌体的溶菌酶是验证这个假设的一种很好的蛋白质，目前德雷尔正从化学角度 [17]、史屈辛格正从遗传学角度 [10] 对此展开研究。

在最近一次于莫斯科召开的国际生物化学大会上，论坛 I 的听众都被尼伦伯格的言论所震惊，他说他和马太 [18] 已经通过向一个能合成蛋白质的无细胞系统注入多聚尿苷酸（也就是所有碱基都是尿嘧啶的 RNA）而制造出了多聚苯丙氨酸（就是说

system which can synthesize protein. This implies that a sequence of uracils codes for phenylalanine, and our work suggests that it is probably a triplet of uracils.

It is possible by various devices, either chemical or enzymatic, to synthesize polyribonucleotides with defined or partly defined sequences. If these, too, will produce specific polypeptides, the coding problem is wide open for experimental attack, and in fact many laboratories, including our own, are already working on the problem. If the coding ratio is indeed 3, as our results suggest, and if the code is the same throughout Nature, then the genetic code may well be solved within a year.

We thank Dr. Alice Orgel for certain mutants and for the use of data from her thesis, Dr. Leslie Orgel for many useful discussions, and Dr. Seymour Benzer for supplying us with certain deletions. We are particularly grateful to Prof. C. F. A. Pantin for allowing us to use a room in the Zoological Museum, Cambridge, in which the bulk of this work was done.

(**192**, 1227-1232; 1961)

F. H. C. Crick, F.R.S., Leslie Barnett, S. Brenner and R. J. Watts-Tobin: Medical Research Council Unit for Molecular Biology, Cavendish Laboratory, Cambridge.

References:
1. Wittman, H. G., Symp. 1, Fifth Intern. Cong. Biochem., 1961, for refs. (in the press).
2. Tsugita, A., and Fraenkel-Conrat, H., *Proc. U.S. Nat. Acad. Sci.*, **46**, 636 (1960); *J. Mol. Biol.* (in the press).
3. Brenner, S., *Proc. U.S. Nat. Acad. Sci.*, **43**, 687 (1957).
4. For refs. See Watson, H. C., and Kendrew, J. C., *Nature*, **190**, 670 (1961).
5. Crick, F. H. C., Griffith, J. S., and Orgel, L. E., *Proc. U.S. Nat. Acad. Sci.*, **43**, 416 (1957).
6. Benzer, S., *Proc. U.S. Nat. Acad. Sci.*, **45**, 1607 (1959), for refs. to earlier papers.
7. Benzer, S., *Proc. U.S. Nat. Acad. Sci.*, **47**, 403 (1961); see his Fig. 3.
8. Brenner, S., Benzer, S., and Barnett, L., *Nature*, **182**, 983 (1958).
9. Brenner, S., Barnett, L., Crick, F. H. C., and Orgel, A., *J. Mol. Biol.*, **3**, 121 (1961).
10. Streisinger, G. (personal communication and in the press).
11. Feynman, R. P.; Benzer, S.; Freese, E. (all personal communications).
12. Jinks, J. L., *Heredity*, **16**, 153, 241 (1961).
13. Champe, S., and Benzer, S. (personal communication and in preparation).
14. Jacob, F., and Wollman, E. L., *Sexuality and the Genetics of Bacteria* (Academic Press, New York, 1961). Levinthal, C. (personal communication).
15. Orgel, A., and Brenner, S. (in preparation).
16. Sueoka, N., *Cold Spring Harb. Symp. Quant. Biol.* (in the press).
17. Dreyer, W. J., Symp. 1, Fifth Intern. Cong. Biochem., 1961 (in the press).
18. Nirenberg, M. W., and Matthaei, J. H., *Proc. U.S. Nat. Acad. Sci.*, **47**, 1588 (1961).

组成这种多肽链的氨基酸残基都是苯丙氨酸）。这意味着，一串尿苷酸序列编码一个苯丙氨酸，而我们的工作表明很可能是三个尿苷酸编码一个苯丙氨酸。

通过各种化学或酶学的方法，人们已经能够合成序列完全确定或部分确定的多聚核糖核苷酸。如果这些也能用于产生具有特定序列的多肽，那么编码问题就可能可以通过实验攻克了，事实上包括我们自己在内的很多实验室都已经开始研究这个问题了。如果像我们结果所表明的那样，编码比率确实是 3，并且在整个自然界这一套密码都相同的话，那么遗传密码问题可能在一年内就会得到解决。

我们向艾丽斯·奥格尔博士为我们提供了某些突变体，并让我们使用她学位论文中的数据表示感谢，感谢莱斯利·奥格尔博士与我们进行了很有益的讨论，感谢西莫尔·本则尔博士为我们提供某些缺失突变。此外我们还要特别感谢潘廷教授允许我们使用位于剑桥的动物博物馆的房间，本研究的大部分工作都是在那里完成的。

（杜丽 翻译；昌增益 审稿）

Chemical Difference between Normal Human Haemoglobin and Haemoglobin-I

M. Murayama

Editor's Note

Haemoglobin's structure was, by this time, well established—the adult red-blood-cell protein contains two identical α polypeptide chains and two identical β chains. And amino-acid substitutions in the polypeptide chains had been shown to produce disease-causing variants. In particular, biologist Vernon M. Ingram had shown that that the abnormal haemoglobin I molecule differs from its normal adult counterpart because it contains the amino acid tryptophan at a particular site on its α chain. Here Makio Murayama from the National Institutes of Health, Bethesda, Maryland, refines the difference further, spelling out the chemical difference between normal human haemoglobin and haemoglobin I.

INGRAM and I have reported that a specific colour reaction on a "fingerprint" indicated that the tryptic peptide 23 of haemoglobin-I (Hb-I) contains tryptophan whereas the corresponding one in the normal haemoglobin (Hb-A) does not. Ingram reported that the peptide is in the α-chain of the molecule[2]. Hill and Konigsberg[3] found that there is only one tryptophan residue in the α-chain of Hb-A; it is in a pentapeptide with valyl as the NH_2-terminal and lysyl as the COOH-terminal residues. I wish to report that it is the lysyl residue of the pentapeptide which is exchanged with aspartyl in the genetic alteration of Hb-I. It is now known that the pentapeptide is the third tryptic peptide from the NH_2-terminal end of the α-chain[4,5] and that the interchange of lysyl for aspartyl takes place at the 16th from the NH_2-terminus.

The soluble peptides were prepared from chromatographically purified Hb-A and Hb-I as previously described[1]. One-dimensional electrophoresis was adequate for the present investigation for the examination and isolation of soluble peptides. The tryptophan-containing peptides were easily located by fluorescence under an ultra-violet lamp.

Results indicate that there were three tryptophan-positive peptides in both haemoglobins as shown in Fig. 1. The tryptophan-positive peptide which migrated the fastest at pH 6.4 towards the cathode in Hb-A vanished in Hb-I. (In our previous publication, Fig. 2, 4 tryptophan-positive peptides were shown for Hb-I; this is now known to be due to a trace of Hb-A present as a contaminant.) As it was previously reported the peptide 23 of Hb-I moved more rapidly towards the anode than the normal peptide. The peptide 23 of Hb-I contained tryptophan and an aspartyl residue (the latter accounted for its greater mobility towards the anode), whereas the corresponding peptide 23 of Hb-A did not.

正常的人血红蛋白和血红蛋白I之间的化学差异

村山

编者按

目前，血红蛋白的结构已经确立——成人红细胞血红蛋白包含两条相同的α多肽链和两条相同的β多肽链。已知多肽链中的氨基酸替换会产生致病的变异体。特别是生物学家弗农·英格拉姆曾指出，异常的血红蛋白I分子不同于正常成人的血红蛋白，是由于在其α链的一个特定位点上包含色氨酸。在本文中，来自马里兰州贝塞斯达国立卫生研究院的村山氏进一步细化了两者的不同，阐释了正常的人血红蛋白和血红蛋白I之间的化学差异。

英格拉姆和我已经报道过在"指纹图谱"上的一个特异性显色反应，该结果表明：血红蛋白I经胰蛋白酶酶解后产生的第23条肽段上包含色氨酸，而在正常血红蛋白（血红蛋白A）的相应肽段上却没有。英格拉姆曾报道这条肽段来自血红蛋白分子的α链[2]。希尔和柯尼希斯贝格[3]发现，在血红蛋白A分子的α链上只有一个色氨酸残基。并且这一色氨酸残基位于一条氨基末端为缬氨酸残基而羧基端为赖氨酸残基的五肽上。这里我要报道的是，血红蛋白I的遗传改变正是由这条五肽中的赖氨酸残基被天冬氨酸残基替换引起的。现在已经知道这条五肽是血红蛋白分子被胰蛋白酶酶解后从α链[4,5]氨基末端算起的第3条肽段，而赖氨酸残基被天冬氨酸残基替换发生在从氨基端算起的第16位上。

按之前所述的方法[1]，采用层析法纯化的血红蛋白A和血红蛋白I来制备可溶性肽段。本研究中检测和分离各种可溶性肽段采用的是单向电泳的方法。采用紫外灯下观察荧光的方法可以很容易地对含色氨酸的肽段进行定位。

结果如图1所示，在两种血红蛋白中都发现了3条含色氨酸残基的肽段。在pH值为6.4时向负极迁移最快的含色氨酸的肽段只存在于血红蛋白A中，而在血红蛋白I中这条肽段消失了。（在我们以前发表的文章中，如图2所示，在血红蛋白I中有4条含色氨酸的肽段，现在知道这是由于样品中存在血红蛋白A的污染而造成的。）正如以前所报道的，血红蛋白I经酶解后的第23条肽段向正极的移动比正常肽段更快。血红蛋白I的第23条肽段含有色氨酸残基和一个天冬氨酸残基（后者的存在使血红蛋白I能更快地向正极迁移），而在血红蛋白A相应的肽段中并不存在这两种

Because the tryptic digest of the α-chain of Hb-A contained but only one tryptophan residue the peptide with which it is associated vanished in Hb-I (Fig. 1). Thus the specific chemical difference between Hb-A and Hb-I must be located in the tryptophan-containing pentapeptide.

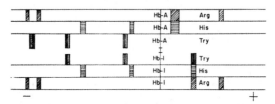

Fig. 1. Comparison by one-dimensional electrophoresis of tryptic digests of the normal and I haemoglobins at pH 6.4. Three peptides of both haemoglobins developed colour for tryptophan. Note that the peptide which moves the fastest toward cathode showing positive reaction for tryptophan in Hb-A is absent in Hb-I. It appears on the anode side in haemoglobin-I.

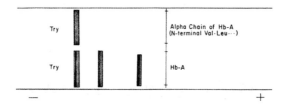

Fig. 2. Tracing of one-dimensional paper electrophoresis of tryptic digest of the normal haemoglobin (Hb-A) and the α-chain isolated from it. The α-chain contains only one tryptophan-positive peptide. It moves the fastest towards the cathode; it is this peptide which vanishes in Hb-I. The amino-acid interchange of this genetic alteration takes place in the tryptophan-positive pentapeptide of the α-chain.

The specific amino-acid composition of the tryptophan-positive pentapeptide of the α-chain of Hb-A was determined by the column chromatography method of Spackman, Stein and Moore[6] and found to consist of $(Ala)_2(Try)(Gly)(Lys)$; where Ala = alanyl, Try = tryptophyl, Gly = glycyl, and Lys = lysyl. In order to preserve tryptophan, leucine-aminopeptidase was used to hydrolyse the peptide which contained it. The NH_2-terminal residue of the pentapeptide was alanyl by the method of Sanger[7]. The amino-acid sequence of the pentapeptide was deduced in the following manner: the peptide was split with α-chymotrypsin and then the peptides were separated by one-dimensional paper electrophoresis; the tryptophan-positive segment behaved essentially neutral; the NH_2-terminal residue was alanyl and the amino-acid composition was $(Ala)_2(Try)$; so the sequence of the tripeptide was Ala-Ala-Try. The dipeptide contained glycine and lysine. The amino-acid sequence of the pentapeptide was deduced to be Ala-Ala-Try-Gly-Lys.

The NH_2-terminal residue of peptide 23 of Hb-I was alanyl, but it was valyl for peptide 23 of Hb-A. The amino-acid composition of Hb-I peptide 23 was identical with the sum of the tryptophan-positive pentapeptide and peptide 23 of Hb-A, however, with one exception—instead of lysyl, aspartyl residue was found. Since the pentapeptide is the third from NH_2-terminus of the α-chain, the tryptic peptide 23 must be the fourth; they are

230

残基。由于血红蛋白 A 的 α 链经胰蛋白酶消化后的产物中只有一个色氨酸残基，所以这一残基一定和血红蛋白 I 中消失的那条肽段（图 1）有关联。因此，血红蛋白 A 和血红蛋白 I 之间特定的化学差异就一定是位于这条含色氨酸的五肽上。

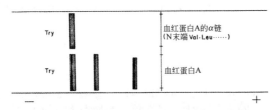

图 1. 在 pH 值为 6.4 时正常的血红蛋白 A 和血红蛋白 I 的胰蛋白酶消化产物单向电泳比较结果。通过对色氨酸显色发现这两种血红蛋白都有 3 条含色氨酸残基的肽段。需要注意的是，在血红蛋白 A 中向负极移动最快的、能呈现出色氨酸阳性反应的肽段并没有出现在血红蛋白 I 中，但血红蛋白 I 在正极一侧出现了一条含色氨酸残基的肽段。

图 2. 对正常血红蛋白 A 和从中分离出来的 α 链的胰蛋白酶消化产物进行单向纸电泳后的印迹。α 链只有一条含色氨酸的肽段，它向负极移动得最快，也正是血红蛋白 I 中消失的那条肽段。因此，造成血红蛋白 A 与血红蛋白 I 的遗传改变的氨基酸替换发生在 α 链中含色氨酸的五肽上。

采用斯帕克曼、斯坦和穆尔[6]的柱层析方法对血红蛋白 A 的 α 链中含色氨酸的五肽的特定氨基酸组成进行测定，结果发现其氨基酸组成是 (Ala)₂(Try)(Gly)(Lys)。其中，Ala 代表丙氨酸残基，Try 代表色氨酸残基，Gly 代表甘氨酸残基，Lys 代表赖氨酸残基。为保护色氨酸，我们选用亮氨酸氨肽酶水解含色氨酸的肽段。采用桑格[7]的方法测定出该五肽的氨基端是丙氨酸残基。通过下面的方法来推断该五肽的氨基酸序列：用 α–胰凝乳蛋白酶使该五肽裂解，然后用单向纸电泳分离酶切产物，结果发现含色氨酸的片段基本呈电中性，另外也知道了这一片段的氨基酸组成是 (Ala)₂(Try)，而氨基末端是丙氨酸残基，所以三肽序列是 Ala-Ala-Try。剩下的二肽则是由甘氨酸和赖氨酸组成的，因此可以推断出该五肽的氨基酸序列是 Ala-Ala-Try-Gly-Lys。

血红蛋白 I 的第 23 条肽段的氨基末端是丙氨酸残基，而血红蛋白 A 的第 23 条肽段的氨基末端是缬氨酸残基。就氨基酸组成来说，血红蛋白 I 的第 23 条肽段与血红蛋白 A 的那条含色氨酸的五肽再加上其第 23 条肽段是基本一致的，但有一处不同，即血红蛋白 A 中的赖氨酸残基在血红蛋白 I 中变成了天冬氨酸残基。由于这条五肽是从 α 链氨基末端算起的第 3 条肽段，那么经胰蛋白酶酶解后的第 23 条肽段就一定是从 α

simply linked together in Hb-I through aspartyl residue.

Attempts made in this laboratory to find out whether or not other amino-acids in peptide 23 might be involved in the genetic change have not been successful; the amino-acid sequence appears to be identical with those reported by Braunitzer et al.[4] and by Hill and Konigsberg[5].

Schroeder et al.[8] reported that human haemoglobin is composed of four polypeptide chains of two different kinds, referred to as α- and β-chains. In accordance with recommendations for the nomenclature of haemoglobins[9] $\alpha_2^A\beta_2^A$ designates Hb-A and $\alpha_2^{16Asp}\beta_2^A$ could be used to define Hb-I.

I thank Dr. Robert J. Hill and Dr. William Konigsberg, of the Rockefeller Institute, who provided me with the α-chain of Hb-A; they also collaborated with me on the amino-acid analyses of the pentapeptide. I also thank T. Viswanatha and Mr. H. B. Marsh for assistance.

(**196**, 226-227; 1961)

Makio Murayama: National Institute of Arthritis and Metabolic Diseases, National Institutes of Health, Public Health Service, U.S. Department of Health, Education and Welfare, Bethesda, Maryland.

References:

1. Murayama, M., and Ingram, V. M., *Nature*, **183**, 1798 (1959).

2. Ingram, V. M., *Nature*, **183**, 1795 (1959).

3. Hill, R. J., and Konigsberg, W., *J. Biol. Chem.*, **235**, PC, 21 (1960).

4. Braunitzer, G., Gehring-Müller, R., Hilschmann, N., Hilse, K., Hobom, G., Rudloff, V., and Wittmann-Liebold, B., *Z. physiol. Chem. Hoppe-Seyler's*, **325**, 283 (1961).

5. Hill, R. J., and Konigsberg, W., *J. Biol. Chem.*, **236**, PC, 7 (1961).

6. Spackman, D. H., Stein, W. H., and Moore, S., *Anal. Chem.*, **30**, 1190 (1958).

7. Fraenkel-Conrat, H., Harris, J. I., and Levy, A. L., *Methods of Biochemical Analysis*, **2**, 359 (Interscience Pub., Inc., New York, 1955).

8. Rhinesmith, H. S., Schroeder, W. A., and Martin, N., *J. Amer. Chem. Soc.*, **80**, 3358 (1958).

9. Gerald, P. S., and Ingram, V. M., *J. Biol. Chem.*, **236**, 2155 (1961).

链氨基末端算起的第 4 条肽段，在血红蛋白 I 分子中，这两肽段是通过天冬氨酸残基连接在一起的。

本实验室曾试图寻找第 23 条肽段中是否有其他氨基酸残基可能发生了遗传改变，但没有找到。我们得到的氨基酸序列与布劳尼策尔等 [4] 以及希尔和柯尼希斯贝格 [5] 报道的序列完全一样。

施罗德等 [8] 曾报道，人血红蛋白由 4 条多肽链组成，这 4 条多肽链分属 α 链和 β 链两种类型。为了与推荐采用的人血红蛋白命名法 [9] 保持一致，可以用 $\alpha_2^A\beta_2^A$ 表示血红蛋白 A，用 $\alpha_2^{16Asp}\beta_2^A$ 表示血红蛋白 I。

我要感谢洛克菲勒研究所的罗伯特·希尔博士和威廉·柯尼希斯贝格博士，他们给我提供了血红蛋白 A 的 α 链样品，并且也和我一起对五肽的氨基酸组成进行了分析。我还要感谢维斯瓦那斯和马什先生对我的帮助。

（荆玉祥 翻译；袁峥 审稿）

Fossil Hand Bones from Olduvai Gorge

J. Napier

Editor's Note

Here British palaeontologist John Napier studies the mechanics of hand bones recovered from the same floor at Olduvai Gorge from which Louis Leakey had recently reported a new hominid skull. Napier analyses the capacity of the fossil hand for exerting the two forms of grip on external objects that modern human beings are capable of. He concludes that the creatures concerned would certainly have been capable of the power grip in which the hand is used as a clamp but that the evidence did not prove that the precision grip used by the tips of the fingers and the thumb would have been possible.

IN *Nature* of December 17, 1960, Dr. L. S. B. Leakey reported on the discovery of a number of fossil bones of the hand and the foot on a living floor some 20 ft. below the uppermost limit of Bed I, Olduvai. Later (*Nature* of February 25, 1961), Dr. Leakey reported the discovery of a mandible and two parietal fragments of a juvenile from the same site and associated with a well-defined living floor of an Oldowan culture.

Fifteen of the hand bones pertaining to at least two individuals, an adult and a juvenile, have been identified and examined, and are described here. Their allocation is given in Table 1.

Table 1

Juvenile		Uncertain Age		Adult	
4 Middle phalanges	*I.U.*	1 Trapezium	*I.U.*	2 Proximal phalanges	*C.D.*
		1 Scaphoid	*I.U.*		
2 Terminal phalanges (fingers)	*C.U.*	1 Capitate	*C.D.*		
		1 Base of 2nd metacarpal	*I.D.*		
1 Terminal phalanx (thumb)	*C.U.*	2 Fragments of middle phalanges *I.D.*			

C, complete; I, Incomplete; D, damaged; U, undamaged.

The middle phalanges (Fig. 1, second row) from the juvenile hand, lacking only their epiphyses, constitute a series II–V from the right hand. They are robust bones, rather more so than phalanges of comparable length of juvenile *Gorilla* and adult *Homo sapiens*. They are strongly curved and, palmad, bear well-defined grooves which are situated in the distal half of the bone for the insertion of flexor digitorum superficialis.

234

来自奥杜威峡谷的手骨化石

内皮尔

编者按

本文中英国古生物学家约翰·内皮尔研究了在奥杜威峡谷同一生活面中发现的手骨的力学特征，最近路易斯·利基报道在那里发现了一件新的人科头骨。内皮尔分析了化石手骨是否像现代人类一样能够对外部物体施加两种形式的握力。他推断这种生物的手就像夹钳一样具有力量抓握，但不能证明化石手的拇指与其他手指的指尖能够进行精确抓握。

根据 1960 年 12 月 17 日的《自然》杂志上利基博士的报道，在奥杜威地层的第 I 层上界之下约 20 英尺的生活面中发现了大量的手、足骨化石。稍后（1961 年 2 月 25 日的《自然》），利基博士报道在同一遗址中的一个很清晰的奥杜威文化生活面中发现了一件未成年个体的下颌骨与两件顶骨破片。

经过鉴别与认定，15 块手骨至少属于两个个体，一个成年个体与一个未成年个体。这些材料的分类见表 1。

表1

未成年个体		年龄未定的个体		成年个体	
4 块中节指骨	I.U.	1 块大多角骨 1 块手舟骨	I.U. I.U.	2 块近节指骨	C.D.
2 块远节指骨（手指）	C.U.	1 块头状骨 1 块第二掌骨的基端	C.D. I.D.		
1 块远节指骨（拇指）	C.U.	2 块中节指骨破片	I.D.		

C，完整；*I*，不完整；*D*，有损；*U*，无损。

未成年个体的中节指骨（图 1，第 2 排）没有骨骺，构成了右手 II~V 指系列。与未成年大猩猩和成年智人的同等长度的指骨相比，这些骨显得更粗壮。它们强烈弯曲，掌型，骨末端一半处有很清晰的槽，以便指浅屈肌附着。

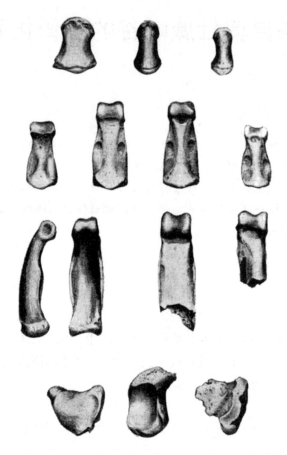

Fig. 1. Hand bone assemblage (drawings by Audrey Besterman). Top row. Juvenile. L. to R., terminal phalanx (thumb); 2 terminal phalanges (fingers). Second row. Juvenile. Middle phalanges, II–V. Third row. L. to R., proximal phalanx (adult), lateral and A.-P. views. 2 proximal phalanges (juvenile). Fourth row. Indeterminate age. L. to R. L. trapezium, L. scaphoid, L. capitate.

The adult proximal phalanges (Fig. 1, third row) are also more robust than bones of comparable length in modern man; they are strongly curved both longitudinally and transversely, fusiform in shape and deeply hollowed out on the palmar aspect; sharply defined fibrous flexor sheath ridges extend from the base of the bones to their necks.

The terminal phalanges (Fig. 1, top row) which are juvenile, having incompletely fused epiphyses, are characteristically *sapiens* in form. The terminal phalanx of the thumb is of particular interest; it is stout and broad and bears a deep impression for the insertion of flexor pollicis longus.

The carpal bones (Fig. 1, 4th row) are all damaged, but sufficient of their original form remains to determine their structural and functional affinities. The lunate surface of the scaphoid has a rectangular outline; and the tubercle, which is broken off at its root, was probably somewhat elongated, though not as long as in the anthropoid apes. The

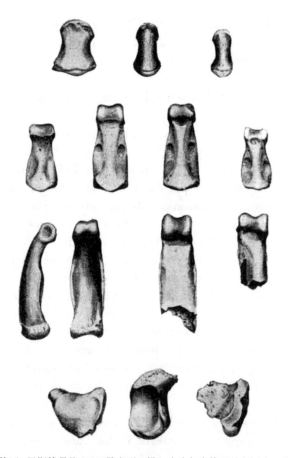

图 1. 手骨组合（奥德丽·贝斯特曼绘画）。最上面一排：未成年个体，从左到右，远节指骨（拇指）、两块远节指骨（手指）；第 2 排：未成年个体，中节指骨，II~V；第 3 排：从左到右，近节指骨（成年个体），侧面与掌侧视图、两块近节指骨（未成年个体）。第 4 排：年龄未确定，从左到右，大多角骨、手舟骨和头状骨。

　　成年个体的近节指骨（图 1，第 3 排）与现代人类的同等长度的骨相比也粗壮很多；纵横两个方向都强烈弯曲，呈纺锤形，掌面深陷；纤维状屈肌腱鞘脊非常明显，从骨基部延伸到颈部。

　　未成年个体的远节指骨（图 1，第 1 排）的骨骺融合不完全，形状上有智人的特征，拇指的远节指骨尤其重要，它宽大粗壮，因拇长屈肌附着而有很深的印痕。

　　腕骨（图 1，第 4 排）全部损坏，但凭借腕骨化石原始形状足以确定结构和功能上的亲缘关系。手舟骨的月状面轮廓呈矩形；结节根部折断，可能略被拉长，但其长度不如类人猿的结节。大多角骨表面呈清晰的马鞍状，但其他关节面提供的证

237

trapezium has a well-defined saddle-surface but the evidence provided by the other articular surfaces indicates that its "set" in the carpus was unlike that found in modern man and similar to the condition in *Gorilla*.

The capitate, though badly eroded, is generally more *sapiens* than ape-like.

Morphologically, the Olduvai hand bones cannot be closely matched with any known hominoid species living today. They bear, however, a greater similarity to juvenile *Gorilla* and adult *H.s. sapiens* than to adult *Gorilla*, *Pan* or *Pongo*. This is due largely to the absence in the fossil bones of any features peculiarly characteristic of brachiators. The adult gorilla hand has a number of specializations that are presumably related to its secondarily terrestrial mode of life and its great body-weight; these features, again, are absent from the fossil bone. The juvenile gorilla hand lacks the secondary specializations of the adult, and it is possibly for this reason alone that its bones have affinities with those of the fossil. The fossil bones differ from those of modern man in a number of features: (1) robustness; (2) dorsal curvature of the shafts of the phalanges; (3) distal insertion of the flexor digitorum superficialis; (4) strength of fibro–tendinous markings; (5) "set" of trapezium in the carpus; (6) the form of the scaphoid; (7) the depth of the carpal tunnel. The fossil bones resemble modern man in the following features: (1) presence of broad, stout terminal phalanges on fingers and thumb; (2) form of the distal articular surface of the capitate; (3) ellipsoidal form of metacarpo-phalangeal joint surfaces.

There seems little doubt that this assemblage of fossil hand bones all belong to the same species. The difference between the juvenile and adult hand bones are no greater than the differences between the bones of an adult and a juvenile gorilla. In view of this conclusion that the two individuals on the *F.L.K. NN* I site are co-specific, all the hand bones are taken into account in the discussion of the functional and systematic implications of the hand as a whole.

While morphologically the precise affinities of the Olduvai hand are indistinct, functionally there seems little reason to doubt that the hand is that of a hominid.

The hand of modern man is capable of two basic prehensile movements that have been termed precision grip and power grip[1]. The power grip is used by man when a secure and strong grip is required for performing an activity in which the elements of delicacy and precision are of secondary importance. In the power grip the object is held as in a clamp between the flexed fingers and the palm, reinforcement and counter-pressure being supported by the adducted thumb. The precision grip, nevertheless, is used by man where a delicate touch and a precise control of movement is required and is achieved by means of a grip between the palmar aspect of the terminal phalanx of the fully opposed thumb and the terminal phalanges of the fingers. The essential osteological correlates of the precision grip are: (1) a fully opposable thumb with a broad spatulate terminal phalanx; (2) broad terminal phalanges on the other digits; (3) a proportion in length between thumb and digits that would permit a full pulp-to-pulp contact between them when they are

据显示它在腕骨上的"接合方式"与现代人类的不一样，而类似于大猩猩的情形。

头状骨虽然侵蚀很严重，但大体能看出更像是智人的而非类似猿类的。

奥杜威手骨在形态上不与任何现存的已知人猿超科动物相一致。但它们与未成年大猩猩及成年智人的相似程度要大于它们与成年大猩猩、黑猩猩或猩猩的相似程度。这主要是由于骨化石没有呈现出任何臂行动物的独特特征。成年大猩猩的手发生了许多特化，这大概与其次生的地面生活方式及大体重有关，奥杜威手骨化石上也没有这些特征。未成年大猩猩的手没有成年个体的次生特征，原因可能是它的骨与奥杜威峡谷化石存在亲缘关系。奥杜威手骨化石与现代人类的骨相比有许多不同特点：(1) 粗壮程度；(2) 指骨骨干的背侧弯曲；(3) 指浅屈肌在远节附着；(4) 纤维–肌腱纹理的强度；(5) 腕骨上大多角骨的"接合方式"；(6) 手舟骨形状；(7) 腕管深度。手骨化石与现代人类的手骨存在下列相似特征：(1) 手指与拇指的远节指骨宽大粗壮；(2) 头状骨末端关节面的形状；(3) 掌骨与指骨间的关节面呈椭圆体形状。

毋庸置疑，这组手骨化石全都属于同一种类。未成年与成年个体手骨之间的差异不比未成年与成年大猩猩手骨之间的差异大，因此在 *F.L.K. NN* I 遗址的两个个体应该属于同一物种。在讨论手的功能与系统意义时，所有的手骨都应作为整体来考虑。

尽管奥杜威手骨在形态上还不能确定其准确的亲缘关系，但从功能上没有理由怀疑它们不是人科的手骨。

现代人类的手能够完成抓握的两种基本动作，即精确抓握和力量抓握[1]。当人类完成一个动作，对抓握的优美性与准确性要求不高，而对抓握的稳固性和力度有要求时，人类就运用力量抓握。进行力量抓握时，通过拇指的内屈反向增压，物体在弯曲的手指与手掌之间像被一把钳子夹住。然而，在要求有技巧性触摸和精确控制时，人类会采用精确抓握；这一动作依靠与其他手指完全相对的拇指的远节指骨掌面和手指的远节指骨抓握来完成。精确抓握在骨骼学上的表现特征是：(1) 与其他手指完全相对的拇指具有宽大的匙形远节指骨；(2) 其他手指的远节指骨宽大；(3) 拇指与其他手指间的长度比例可使它们在互相靠近时能完全紧密接触。虽然，毋庸置疑，奥杜威手骨化石具有精确抓握的前两个特征，但拇指掌骨的缺失使我们无法确定它是否具有最后一个特征。虽然大多角骨的马鞍状表面使人毫不怀疑拇指

approximated. While there appears to be little or no doubt that the Olduvai hand fulfils the first two requirements of the precision grip, there is no way to be certain about the last in the absence of thumb metacarpal. While the saddle-surface of the trapezium leaves no doubt that the thumb could be rotated medially about its own axis to face towards the other digits, there is no reason for supposing that the proportions between the thumb and index finger of the fossil form are exactly as in modern man; indeed, the "set" of the trapezium in the carpus suggests that this is not so. The question therefore is whether the thumb, having undergone rotation, is capable of pulp-to-pulp contact with the remaining digits. In the anthropoid apes while the thumb is opposable *per se*, pulp-to-pulp contact cannot be made owing to the marked disproportion of the length of fingers and thumb. It would seem, therefore, that the Olduvai hand was capable of power grip equal in performance, but, in view of the evidence of the attachment of the flexor tendons, comparatively greater in strength than in modern man. There is less certainty with regard to precision grip, which, while undoubtedly possible, may not have been as effective as in modern man. The overall picture presented by this assemblage is of a short powerful hand with strong, curved digits, surmounted by broad, flat nails and held in marked flexion. The thumb is strong and opposable, though possibly rather short.

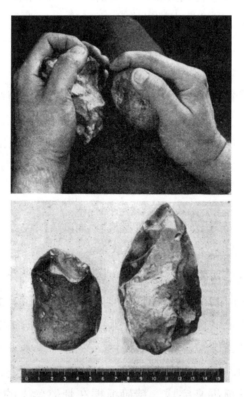

Fig. 2. Top, stone-on-stone technique of hand-axe construction using a power grip only. The flint core is being supported on the knee; bottom left, "Oldowan" pebble-tool; right, "Chellean" hand-axe made by me using the above technique.

At a recent conference at Burg Wartenstein, Austria, organized by the Wenner–Gren

能绕其自身的轴向内朝其他手指的方向转动，但还没有理由推测，化石形态的拇指和食指之间的比例与现代人类的相同；腕部大多角骨的"接合方式"也表明并非如此。接下来的问题是，能经受转动的拇指能否与其余的手指紧密接触？类人猿的拇指虽然可与其他手指相对，但由于拇指与其他手指的长度明显不成比例而不能紧密接触。因此，奥杜威手骨似乎在力量抓握方式上与现代人一致，但屈肌腱附着的证据表明它的抓握力量可能比现代人类相对大些。至于精确抓握就难以确定了，虽然可能性毋庸置疑，但抓握可能不如现代人类的有效。这一组手骨呈现出来的总体特征是手短而有力，手指粗壮、抓握时明显弯曲、手指上的指甲扁而宽。拇指虽然可能相当短，但是强壮，并可与其他手指相对。

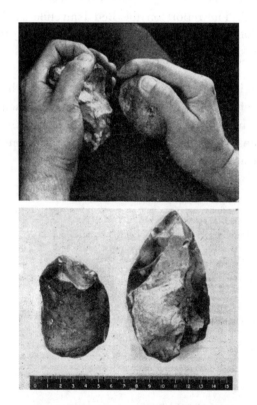

图 2. 上：仅用力量抓握石头碰石头地制造手斧，用手将燧石石核固定在膝盖上；下左："奥杜威文化"的卵石工具；下右：笔者用上述方法制造的"阿布维利文化"的手斧。

温纳－格伦人类学研究基金会最近在奥地利瓦尔特施泰因堡组织了一次大会，

Foundation for Anthropological Research, an attempt was made to produce a diagnosis for the genus *Homo*. It was agreed that such a diagnosis could not be made unless certain characters and character complexes were present in combination. Included in these characters, which are referable to the skull, the brain and the post-cranial skeleton, was the criterion that "the hand is capable of making tools of a recognizable culture". If one assumes that the artefacts of an early Oldowan culture found on the living floor were the work of the species the remains of which are found there, then this criterion is fulfilled and, in addition, an interesting conclusion is possible: that toolmaking was established in the human lineage long before the hand had assumed its modern human form.

If, on the other hand, one bears the possibility in mind that some other, more advanced form was the toolmaker and the known incumbents of the floor were its victims, then it is in the functional morphology of the hand itself that one must look for evidence of toolmaking. On anatomical grounds there is no doubt that the Olduvai hand was sufficiently advanced in terms of the basic power and precision grips to have used naturally occurring objects as tools to good advantage. There is less certainty about toolmaking, which involves not only a peripheral but also a central intellectual factor as Oakley[2] has long insisted. The report on the Bed I juvenile skull fragments, soon to be published, may throw some light on this question by indicating the approximate cranial capacity of the juvenile skull.

Given the intellectual ability, the construction of the crude, rather small pebble-tools of the type found on the living floor, is well within the physical capacity of the Olduvai hand. Precision grip, which is imperfectly evolved in the fossil hand, is not an essential requisite at this level of craftsmanship as personal experiments in the construction of "Oldowan" pebble-tools and "Chellean" hand-axes have shown (Fig. 2).

(**196**, 409-411; 1962)

John Napier: Primatology Unit, Royal Free Hospital School of Medicine, London.

References:

1. Napier, J. R., *J. Bone and Joint Surg.*, **38**, B, 902 (1956).

2. Oakley, K. P., *Man the Toolmaker*, Fourth ed. (Brit. Mus. (N. H.) 1949).

试图确定人属的鉴定标准。大会一致认为，除非某些特征和综合特征在一起出现，否则这样的鉴定标准就无法建立。涉及头骨、脑及颅后骨骼的这些特征也是"手具有制造可识别文化特征工具的能力"的判断标准。如果假设发现在生活面上的早期奥杜威文化的器物是由在此发现遗骸的物种所创造的，那么，这个判断标准是成立的；此外，可能会得出一个有趣的结论：在人类演化的谱系上制作工具的出现远远早于现代人手雏形的出现。

从另一方面说，如果有人认为某些更高等的物种才是工具制造者，而在生活面上发现的动物只是其牺牲品，那么就必须从手的功能形态学上寻找手本身制造工具的证据。就解剖学而言，在基本的有力抓握与精确抓握方面，奥杜威手骨无疑十分高级，能够利用现成的物体作为工具来获取利益。不太确定是否能制造工具，因为这不仅涉及外部环境因素，还涉及奥克利一直坚持的大脑的智力因素[2]。地层中第 I 层的未成年个体头骨破片的报道即将发表，通过指出未成年个体头骨的近似颅容量可能会阐明这个问题。

倘若奥杜威人类有此智力能力，那么在生活面上发现的天然的、相当小的卵石工具就肯定在其手的制造能力范围之内。奥杜威手骨化石还没有进化形成精确抓握的能力，但这并不是技术水平高低的一个先决条件，重造"奥杜威"卵石工具和"阿布维利文化"手斧的实验都已经说明了这一点（图 2）。

（田晓阳 翻译；冯兴无 审稿）

Molecular Biology, Eugenics and Euphenics[*]

J. Lederberg

Editor's Note

Eugenics, the deliberate improvement of the human stock by selective breeding, had been popular among scientists in Britain and the United States in the early 1930s but, with the end of the Second World War in Europe in 1945, had fallen from favour largely because of the German government's widespread use of sterilization. By the early 1960s, however, and with the prospect that the new science of molecular biology might provide better ways of improving on nature, eugenics made occasional reappearances on the scientific agenda. Joshua Lederberg was at Stanford University in California when he contributed this article to *Nature*; he had already been awarded a Nobel Prize for his work on bacteria and afterwards became the director of the Rockefeller Institute in New York.

THE risks of scientific prophecy are well known. But foresight about our scientific culture is as important for the culture to gather as it is difficult for the scientist to expound. His credentials to speak on the impact of science on human welfare are scarcely unique, but he does have a responsibility which stems from his technical judgment of the plausibility and especially the time-scale of scientific advance, which by furthering human power must impinge on policy.

Recent years have seen breath-taking advances in the molecular foundations of biology, at a pace that reminds us that the gross effort in science in one year now matches the total accumulation to the beginning of this century; as much scientific effort has been invested since 1950 as was in all previous history. These actuarial calculations cannot, of course, measure the intellectual value of the return, nor do they take account either of instrumentation multipliers or of the overloading of the communications net. The details of these advances are well told elsewhere. What must be noted here is the solution to the fundamental problems of genetics: the encoding of genetic information in the structure of DNA, and the enzymatic mechanism by which the nucleotide sequence is replicated. Intertwined with these developments have been the unification of terrestrial biology within a single biochemical genetic scheme, and the now rapid unravelling of the cellular mechanism of protein synthesis whereby the genetic information is translated into the working machinery of life.

Eugenics, the conscious betterment of man's genetic quality, has fascinated many idealistic

---

[*] Substance of an address to the Symposium on "The Future of Man" held at the Ciba Foundation, London, November 26-30, 1962. The full proceedings of the Symposium will be published by Messrs. J. and A. Churchill, Ltd.

分子生物学、优生学和人种改良学[*]

莱德伯格

编者按

20 世纪 30 年代初期，优生学曾得到英国和美国科学家的广泛关注，它是通过选育的方法对人类血统进行有意改良的一门学科。但是随着 1945 年第二次世界大战在欧洲的结束，由于德国政府大范围地推行绝育政策，使这门学科受到极大冷落。直到 20 世纪 60 年代初期，由于未来前景显示分子生物学这门新兴科学有可能为改良人类血统提供更好的途径，在此背景下优生学重新出现在科学的议事日程中。当约书亚·莱德伯格将这篇文章投到《自然》杂志的时候，他就职于加利福尼亚州斯坦福大学。此前他已经凭借在细菌研究方面的工作获得了诺贝尔奖，后来他成为了纽约洛克菲勒研究所的所长。

众所周知，提出科学预言需要承担风险。但是为我们的科学文化所做的预见对于文化的汇集非常重要，而这很难让科学家去阐述。科学家并非是唯一有资格在科学对人类的幸福所产生的影响方面发表言论的人，但他确有此责任，这种责任来源于他在专业技术上对可能性所具有的判断力，特别是对科学进展的时间尺度的判断力。而通过推进人类的力量，科学进展势必会对政策产生影响。

近年来，基础分子生物学已取得惊人的进展，其发展速度提醒我们：现在一年所取得的科学成就的分量相当于 20 世纪初研究成果的总和；同样，20 世纪 50 年代之后所投入的科研精力也相当于 50 年代之前所投入精力的总和。当然，即使不将研究设备的扩充或通信网络的超载考虑在内，这些统计估算也不能衡量这些科研投入的智力价值回报。这些研究进展的细节可以从其他渠道获知。此处值得一提的是解决遗传学基本问题的研究发现：DNA 结构中遗传信息的编码机制和核苷酸序列复制的酶学机制。以上研究进展揭示了陆上生物在生物化学水平上统一的遗传模式，以及遗传信息翻译成为生物性状的过程，也就是细胞内蛋白质合成的机制。

对人类的遗传质量进行有意改良的优生学，已经受到许多唯心主义者的关注。

---

[*] 发表于 1962 年 11 月 26 日至 30 日在伦敦汽巴基金会举行的"人类的未来"讨论会。完整的会议记录将由丘吉尔有限公司出版。

245

thinkers. Like other noble aims it has been perverted to justify unthinkable inhumanity; which does not help to assess its validity and feasibility by ethically proper means. The case for eugenics, ably presented by Huxley and by Muller, has one most trenchant argument against complacency: man's long pre-cultural evolution has given him a biological legacy which can be only fortuitously adapted to the physical power and technological complexity of the modern world. In a word, man, unless he grows less "human", may destroy himself.

Eugenic progress creeps within the joint constraints of our limited knowledge of human genetics and customary wisdom concerning its implementation. Even so, the eugenicists argue, some beginning must be made, to offset exigent counter-eugenic influences, perhaps to assure that some eugenic wisdom survives until the species can or must act.

The new biology is relevant here—ultimately it could diagnose, then specify, the actual DNA composition of ideal man. But clearly, this will not happen for some time ("if ever", most of my colleagues will reassure themselves, while they concentrate on more penetrating assaults on these secrets).

Having shared this view, I may record how easy and tempting it is to postpone consideration of the probable impact of biological knowledge on human affairs. It is difficult enough to make a fragmentary contribution to such knowledge, much more to be usefully concerned with its total consequences.

The emphasis on eugenics as the point of application of molecular biology overlooks the most immediate prospects for the understanding and then control of human development. To dramatize the antinomy, I propose the term "euphenics" as the counterpart of "eugenics", in the same sense that "phenotype" is opposed to "genotype".

Development is the translation of the genetic instructions of the egg, embodied in its DNA, to direct the unfolding of its substance to form the adult organism in all its aspects, which comprise its phenotype. The crucial problem of embryology is the regulation and execution of protein synthesis, how some DNA segments are made to call out their instructions, others suppressed, which underlies the orderly differentiation of cell types.

Until now, the major problems of human development—not only embryology, but also the phenomena of learning (in its neurobiological aspects), immunity (with its bearing on transplantation), neoplasia and senescence—could be approached at only the most superficial level. They are about to be transformed in the sense that genetics has been, as epiphenomena of protein and nucleic acid synthesis. The present intensity of effort suggests a span of from five to no more than twenty years for an analogous systematization.

On these premises it would be incredible if we did not have the basis of developmental engineering technique, for example, to regulate the size of the human brain by prenatal or early postnatal intervention.

246

像其他崇高的目标一样，优生学曾被误认为是不人道的做法，也无法通过合乎伦理的方法来评定它的有效性和可行性。赫胥黎和马勒巧妙地引用一个优生学的例子深刻驳斥了这种自满情绪：人类从长期史前进化中获得的生物学遗产很幸运地刚好能满足现代人的体能需求以及适应现代社会技术的复杂性。简单来说，如果人类长得像"现代人"，他就是自寻毁灭。

优生学受到有限的人类遗传学知识及关于其实施的传统观念的共同制约，缓慢地发展着。尽管如此，优生学家认为，为了抵消反优生学的影响，一些初步研究势在必行，这也许能够保证一些优生学的思想延续下去，直至人类能够或是必须采取行动的时候。

新兴分子生物学的价值在于，它最终可以诊断并详细说明理想人类的实际 DNA 的组成。但是，很显然，短期之内这一点是做不到的（"如果可以做到的话"，我的大部分同事们将恢复他们对研究的信心，专注于对这些未知的问题进行更深入的研究）。

分享过这些观点后，我要说暂时不考虑生物学知识对人类可能产生的影响，是多么轻松而吸引人的选择。对这些知识做一些零散的贡献已经非常困难了，更别说是对其整体的结果能够有所帮助。

对优生学是作为分子生物学一项应用的强调忽视了直接理解和调控人类的发育过程的可能。为了显示出这一矛盾，我建议将"人种改良学"与"优生学"对应起来，像"表现型"与"基因型"相对应一样。

发育是指卵细胞中 DNA 所携带的遗传信息经过翻译后，指导其自身物质演变形成完整的成熟有机体（包含它的表现型）的过程。胚胎学的关键问题是蛋白质合成的调控和执行，如何使得一些 DNA 片段表达而另外一些 DNA 片段受到抑制，这种调控机制为不同种类细胞的有序分化奠定了基础。

迄今为止，有关人类发育的主要问题不仅包括胚胎学，也包括学习现象（神经生物学方面）、免疫（移植耐受性）、肿瘤形成以及衰老，对这些问题的研究都只停留在最浅显的水平上。这些研究即将像遗传学一样产生巨变，例如遗传学不过是蛋白质和核酸合成过程的附带现象。以目前的研究力度来看，我们还需要 5 年～20 年的时间将该类问题系统化。

基于这些推测，如果我们没有发育的工程技术作为基础，例如通过产前或产后初期干预来调控人类脑的大小将令人难以置信。

The basic concept of molecular biology is the chain of information from DNA to RNA to protein. We are just now beginning to ask questions of mental mechanism from this point of view. The simplest and one of the oldest of speculations about memory is the modification of neuronal interconnexion through control of synthesis and deposition of durable proteins at the interfaces. A plausible link between electrical impulses and protein synthesis might be the accompanying shifts of potassium and sodium concentrations; these ions being also important cofactors for several enzymes involved in protein synthesis. Thus, cation balance could control the assembly of chosen polypeptide chains into a complex protein, the selective reactions of glutamine —$CONH_2$ in protein, or the imperfect specifications of degenerate RNA codes. Such speculations merely illustrate the relationship of mental science to molecular biology.

In another field of developmental engineering Medawar has already exhibited a *tour de force*, the abolition of immunity to transplants introduced in early life, which has clarified the biology of immunity and points to the solution of the transplantation problem. At present, human individuality is the bar to spare-part medicine: the organism rejects grafts from other individuals even of the same species, the alien tissue a life-extending kidney or heart notwithstanding. The solution to the homograft problem now partly resolved must be imminent, under intensive attack as an aspect of the cell biology of immunity, and of the molecular structure and cyto-synthesis of antibodies and tissue antigens. The management of the problems and opportunities it raises should be a prototype for the exercise of responsible power in biological engineering. There is no evident forethought of them, perhaps just because of their cataclysmic impact on medicine.

What if surgical finesse were now the only criterion of transplantability? The direct replacement of defective, diseased or worn-out organs could pre-empt all available surgical talent for years to come. Then, many potent régimes, once restrained by the side-effects on other organs, are now available to internal medicine. These tools, like present-day drugs, will also have an indispensable role in the treatment of healthy individuals.

The most nightmarish prospects arise from indifference to technological and procedural requirements with respect to the sources of indispensable, scarce life-saving organs. The orderly evolution of transplantation technique might be facilitated if organ transplants in man (with evident trivial exceptions) were already subject to formal registration as vital statistics.

Many social problems arise from technological imbalance, or at least have possible technological antidotes which can then be properly discussed here. For example, the political stability of the world might be enhanced if the present technology of the detection matched that of the power output of nuclear explosives; likewise for the moderation of human prolificity concurrently with infant mortality. In the present case intolerable stresses arising from the economics of human organs could be averted by further advances beyond the first stage of successful homotransplantation. These might include a eugenic programme on other species to facilitate their use as sources of organs.

分子生物学的基本概念是指从DNA到RNA再到蛋白质的信息链传递。现在我们刚开始根据这种观点来提出一些关于脑机制的问题。关于记忆的一个古老而简单的推测认为记忆是神经元连接处发生修饰作用而产生的，这种作用是通过对连接处的持久性蛋白（缝隙连接蛋白）的合成与沉积进行调控来完成的。电脉冲和蛋白质合成之间的联系可能是通过钾离子和钠离子浓度的转换来实现的；这些离子也是参与蛋白质合成的几个酶的重要辅助因子。所以，阳离子平衡可能能够控制：特定多肽链如何组装成复杂的蛋白质，蛋白质中谷氨酰胺的—$CONH_2$基因的选择性反应，或RNA简并密码的部分特异性。以上推测只能解释脑科学和分子生物学之间的关系。

在发育工程的另一领域中，梅达沃已经展示了一个**杰作**，那就是在生命早期不会对移植体产生免疫反应，并由此阐明了免疫生物学的机理并提出了解决移植问题的方法。目前，人类个体自身对器官移植产生阻挡：生物体会排斥其他个体甚至是相同物种个体上移植而来的器官，即使是可以延续生命的外源组织的肾或心脏。通过加强对免疫细胞生物学、分子结构细胞生物学以及抗体和组织抗原细胞合成的研究，目前已经得到部分解决的同种移植问题应该很快就能被彻底攻克。对器官移植所带来的问题和机遇的处理方式，应该可以作为我们解决生物工程问题的示范。或许正是因为它们对医学的巨大冲击，我们对这些研究并没有事先预料到。

如果外科手术是目前衡量可移植性的唯一标准，那又会如何？在未来的几年，直接替换缺陷的、患病的或穿孔的器官将优先占用外科所有可用人力资源。然后，由于对其他器官有过副作用而受到限制的药物组合将应用于内科医学。这些工具就像现在的药物一样，也将在治疗正常人群中发挥不可替代的作用。

如果对那些至关重要的、来源不充足的器官的需求在技术上和程序上没有得到足够的重视，那么前景将会是非常糟糕的。如果人类器官移植（除了少数明显的例外）已经作为人口统计正式注册，那么可能会促进移植技术的有序发展。

许多社会问题是由技术不均衡引起的，尽管有少数通过技术恢复均衡的方法，随后我们会对后者进行适当的讨论。例如，如果现有的核探测技术与核爆炸技术相匹配，那么世界政治将更加稳定；同样的平衡也存在于人类适度的多育与婴儿的死亡率之间。以目前的情况看，同种移植技术已经达到初步阶段，如果在此基础上继续发展的话，那么由人类器官紧缺所引起的巨大压力将得到改善。其中一个可能的优生学方案是将其他物种作为器官来源。如果这些器官的来源在遗传上是一致的，

The more difficult problem of heterotransplantation, from other species, would be mitigated if these sources were genetically uniform and could be specifically selected for immunological and functional suitability. At present the only animals which begin to fit these criteria are inbred mice. The industrial manufacture of specific proteins (either by chemical- or cyto-synthesis) would be an invaluable adjunct. Such precious proteins, for example, hormones or enzymes, are sometimes the functional purpose of a transplant. As antibodies or tissue antigens they would play a specific part in neutralizing the homograft rejection mechanism. As structural proteins they would be valuable for the manufacture of compatible parts and connexions.

The heart probably poses the most perplexing problems of supply and allocation. Yet, of all the vital organs, this should be the first to be simulated by a mechanical analogue—machines are already available for short-term use during surgery. Should the engineering effort be accelerated to produce a practical substitute for this efficient pump? These proposals stress engineering development, partly to illustrate a prevalent gap between academic science and its useful implementation in this aspect of human welfare. There are many equally insistent candidates for the succession the military uses of industrial technology.

Man's control of his own development, "euphenics", transmutes the means, and also the ends of eugenics, as have all the precedent cultural revolutions that have shaped the species: language, agriculture, political organization, the physical technologies. Eugenics is aimed at the design of a reaction system (a DNA sequence) that, in a given context, will develop to somehow defined goal. Few insights would be worth more than the design of human value—but will culture stand still merely to validate the eugenic criteria of a past generation? For a given end, the means will have shifted: the best inborn pattern for normal development will not always react best to euphenic control.

Within the framework of formal eugenics the disruptive effects of recombination may need further investigation. Most genes segregate independently of sex, but must then work in concert with the bio-cultural dimorphism of sex. This must impede stringent selection; or conversely, does rapid eugenics not imply the convergence of the sexes to a common goal? At a considerable cost in its rate the evolutionary process might be confined to sex-limited, -linked, or -irrelevant mutations, if any, which still affect personality. Euphenics can switch the entire programme to match the sexual or other role-defining polymorphisms. Education—the whole cultural apparatus—does this now.

Euphenics will, of course, open the way for a more comprehensive eugenics, if only through the systematization of knowledge of gene action. Even now the outlook for eugenic improvement of intelligence would be improved by a biochemical assay for it.

In our inquiry on his future, the aims of human existence are inseparable from the power and responsibility for human nature. It becomes more perplexing as biological technology dissolves the barriers around individual man and intrudes on his secret, germinal

并且能通过免疫和功能适应性的特定筛选，那么从其他物种进行异种移植这个较为困难的问题就能得以缓解了。目前，动物中只有近交系小鼠符合这些标准。通过工业生产特异蛋白质（化学合成或细胞合成）将是一个非常重要的辅助手段。这些珍贵的蛋白质，例如激素或酶，有时正是能否实现器官移植的功效的关键所在。比如抗体或组织抗原在消除同种移植排异反应中起特殊作用，又如结构蛋白在兼容和连接方面起重要作用。

在器官移植中，心脏的供应和分配问题可能是最复杂的。然而，在所有重要的器官中，心脏应当是首个被机械模拟的——人工心脏已经可以在外科手术中短期使用了。我们是否应该在工程方面投入更多的精力以生产一个合适的心脏替代品呢？这项提议强调了工程技术发展的重要性，在一定程度上说明了科学理论距离为人类创造福利还有一定差距。此外，军用技术也能为工业技术的发展提供很多好的候选技术。

人类调控其自身的发育，也就是所谓的"人种改良学"，改变了优生学的方法和目的，正如先前所有的文化革命一样，它塑造了人类的语言、农业、政治组织和物理技术。优生学的目标是设计出一个在特定的环境中通过某种方式达到已定目标的反应体系（DNA序列）。没有什么比设计人类的价值更有意义——那么，难道说文化存在的意义只是为了验证前人的优生学标准吗？因为正常发育的先天模式不会总是与人种改良的目标相适应，所以为了实现特定的改良目标，调控的具体方法要有所改变。

在正式的优生学框架内，可能需要对基因重组的副作用做进一步的研究。绝大多数基因分离与性别无关，但一定与生物文明的性别二态性有关。这一定阻止了严格的筛选；或者反过来说，快速优生学难道不正是意味着两性向一个共同目标汇聚吗？这种严格的选择和淘汰需要付出相当大的代价，进化过程可能会受到限性、伴性或无性突变的限制，如果有的话，还会对个性产生影响。人种改良学可以控制将整个方案转向为性别或其他角色分类的多态性。目前的教育体系，也就是所有的文化机构，正在发挥这样的作用。

当然，只要对基因的作用有系统化的认识，人种改良学就会开启一条通往更全面的优生学的道路，尽管现在通过生化检验已经能够进行优生学上的智力改良。

在我们对人种改良学前景的调查中发现，人类生存的目的与人的权利和责任是分不开的。以生物技术为手段的人种改良已经变得更复杂了，它可以消除围绕在个

continuity. The humanist premise of individual value must face the issue of a definition of man, taking full account of his psychosocial progeny. We now recognize genetic continuity in mechanistic terms as a nucleotide sequence—in due course this will itself be subordinate to the psychosocial machinery. While man perfects the knowledge of his own mechanism, he also vitalizes machines on to a convergent evolutionary pathway. Genetics is rapidly becoming a corollary of information theory. As he thus evolves from substance to concept, Is it the bond of genetics or of communication that qualifies "man" for the aspirations of humanistic fulfilment, apart from the other robots born of human thought?

(**198**, 428-429; 1963)

Joshua Lederberg: Kennedy Laboratories for Molecular Medicine, Department of Genetics, School of Medicine, Stanford University, Palo Alto, California.

体周围的壁垒，也可以探知个体的秘密——种质连续性。人道主义者提出，以个体价值为前提，就必须面临人的定义这个议题，也要充分考虑到他的后代的心理。现在我们认识到遗传连续性用生物学术语来说就是核苷酸序列，在适当的时候它本身将会附属于心理机制。人类在完善对自身机制认识的同时，也促使进化途径趋于一致。这些过程中，信息技术的理论为遗传学的迅速发展提供了条件。因此遗传学得以从客观存在发展到理论水平。暂且不提人类思想下诞生的其他自动机械，或许是生物学上的遗传或文化上的传承才使得"人类"得以迎来人道主义的实现？

（郭娟 翻译；金侠 审稿）

A New Species of the Genus *Homo* from Olduvai Gorge

L. S. B. Leakey *et al.*

Editor's Note

In 1960, not long after Leakey found *Zinjanthropus*, fragments of an altogether different kind of hominid started turning up at Olduvai. More years of searching accumulated enough material for Leakey to be sure that this new creature was not an "ape-man", but a member of our own genus, *Homo*. The small brain of this creature necessitated a bold revision of our own genus—extending what it means, in technical terms, to be "human". The new fossils also seemed to have been associated with primitive chipped-pebble tools. Partly because of this, Leakey and colleagues called the new hominid *Homo habilis*—"handy man"—thus binding humanity forever in the public mind with technology. Leakey's message was clear: humanity began at Olduvai.

THE recent discoveries of fossil hominid remains at Olduvai Gorge have strengthened the conclusions—which each of us had reached independently through our respective investigations—that the fossil hominid remains found in 1960 at site *F.L.K.N.N.* I, Olduvai, did not represent a creature belonging to the sub-family Australopithecinae[*].

We were preparing to publish the evidence for this conclusion and to give a scientific name to this new species of the genus *Homo*, when the new discoveries, which are described by L. S. B. and M. D. Leakey in the preceding article, were made.

An examination of these finds has enabled us to broaden the basis of our diagnosis of the proposed new species and has fully confirmed the presence of the genus *Homo* in the lower part of the Olduvai geological sequence, earlier than, contemporary with, as well as later than, the *Zinjanthropus* skull, which is certainly an australopithecine.

For the purpose of our description here, we have accepted the diagnosis of the family Hominidae, as it was proposed by Sir Wilfrid Le Gros Clark in his book *The Fossil Evidence for Human Evolution* (110; 1955). Within this family we accept the genus *Australopithecus* with, for the moment, three sub-genera (*Australopithecus*, *Paranthropus* and *Zinjanthropus*) and the genus *Homo*. We regard *Pithecanthropus* and possibly also *Atlanthropus* (if it is indeed distinct) as species of the genus *Homo*, although one of us (L. S. B. L.) would be prepared to accept sub-generic rank.

It has long been recognized that as more and more discoveries were made, it would

---

[*] See also *Nature* of March 7, pp. 967, 969, and preceding articles in this issue.

在奥杜威峡谷发现的一个人属新种

利基等

编者按

1960年，在利基发现东非人后不久，他在奥杜威又发掘出一种完全不同的人科动物化石破片。多年来收集的材料足以让利基相信这个新发现的物种不是"猿人"，而是我们人类的一个新成员，一个人属新种。这个新种的脑比较小，这使得我们必须对自己的属做大胆的修正——在专业术语范畴内拓宽"人"的含义。此外，这些新化石似乎也与原始的砾石打击工具有关。一定程度上由于这个原因，利基和同事们称这种新的人科为能人——"灵巧的人"，从而在公众的脑海中将人类和技术永远地捆绑在了一起。利基传递的信息非常明确：人类始于奥杜威。

最近在奥杜威峡谷发现的人科化石使我们更加相信，1960年在奥杜威 *F.L.K.N.N. I* 地点发现的人科化石并不属于南方古猿亚科，这是我们每个人通过自己独立的调查研究得到的一致结论*。

当我们有了这一新发现时（正如路易斯·利基及其夫人玛丽·利基在先前文章中所描述的），就准备发表关于这一结论的证据，并且给予这个属于人属的新标本一个科学的名字。

这些发现的检验结果使我们拓宽了判断新种的基础，并充分证实了人属存在于奥杜威地层序列的下部，比确定属于南方古猿亚科的东非人头骨或者更早，或者同时代，或者更晚。

为了描述方便，本文中我们采用了威尔弗里德·勒格罗-克拉克爵士在《人类进化的化石证据》（第110页，1955年）一书中提出的对于人科的判断标准。在这个科中我们暂时接受南方古猿属和人属，包括南方古猿属目前具有的三个亚属（南方古猿亚属、傍人亚属和东非人亚属）。尽管我们中的路易斯·利基准备接受亚属的分类标准，但我们还是把爪哇猿人，可能还有阿特拉猿人（如果确实有明显区别）作为人属中的种。

人们早已认识到，随着越来越多的发现，有必要对人属的判断标准进行修改。

---

* 参见3月7日《自然》第967、969页以及本期前面的文章。

become necessary to revise the diagnosis of the genus *Homo*. In particular, it has become clear that it is impossible to rely on only one or two characters, such as the cranial capacity or an erect posture, as the necessary criteria for membership of the genus. Instead, the total picture presented by the material available for investigation must be taken into account.

We have come to the conclusion that, apart from *Australopithecus* (*Zinjanthropus*), the specimens we are dealing with from Bed I and the lower part of Bed II at Olduvai represent a single species of the genus *Homo* and not an australopithecine. The species is, moreover, clearly distinct from the previously recognized species of the genus. But if we are to include the new material in the genus *Homo* (rather than set up a distinct genus for it, which we believe to be unwise), it becomes necessary to revise the diagnosis of this genus. Until now, the definition of *Homo* has usually centred about a "cerebral Rubicon" variably set at 700 c.c. (Weidenreich), 750 c.c. (Keith) and 800 c.c. (Vallois). This proposed new definition follows:

Family	hominidae	(as defined by Le Gros Clark, 1955)
Genus	*Homo*	Linnaeus

Revised diagnosis of the genus Homo. A genus of the Hominidae with the following characters: the structure of the pelvic girdle and of the hind-limb skeleton is adapted to habitual erect posture and bipedal gait; the fore-limb is shorter than the hind-limb; the pollex is well developed and fully opposable and the hand is capable not only of a power grip but of, at the least, a simple and usually well developed precision grip*; the cranial capacity is very variable but is, on the average, larger than the range of capacities of members of the genus *Australopithecus*, although the lower part of the range of capacities in the genus *Homo* overlaps with the upper part of the range in *Australopithecus*; the capacity is (on the average) large relative to body-size and ranges from about 600 c.c. in earlier forms to more than 1,600 c.c.; the muscular ridges on the cranium range from very strongly marked to virtually imperceptible, but the temporal crests or lines never reach the midline; the frontal region of the cranium is without undue post-orbital constriction (such as is common in members of the genus *Australopithecus*); the supra-orbital region of the frontal bone is very variable, ranging from a massive and very salient supra-orbital torus to a complete lack of any supra-orbital projection and a smooth brow region; the facial skeleton varies from moderately prognathous to orthognathous, but it is not concave (or dished) as is common in members of the Australopithecinae; the anterior symphyseal contour varies from a marked retreat to a forward slope, while the bony chin may be entirely lacking, or may vary from a slight to a very strongly developed mental trigone; the dental arcade is evenly rounded with no diastema in most members of the genus; the first lower premolar is clearly bicuspid with a variably developed lingual cusp; the molar teeth are variable in size, but in general are small relative to the size of these teeth in the genus *Australopithecus*; the size of the last upper molar is highly variable, but it is generally smaller than the second

---

* For the definition of "power grip" and "precision grip", see Napier, J. R., *J. Bone and Joint Surg.*, 38, B, 902 (1956).

尤其是现在已经清楚了不能只依赖于一个或两个特征，诸如颅容量或直立的姿势，作为是否为属中成员的必要判别条件。事实上，应该将调查中可获得材料的全部特征纳入考虑范围之中。

我们得到的结论是：除南方古猿（东非人）外，我们所讨论的在奥杜威第 I 层以及第 II 层下部发现的标本代表人属内一个单独的种，而不是南方古猿亚科的种。而且这个种明显不同于此前确认的人属内的种。但是如果我们把这个新种归入人属（而非设立一个不同的属——我们认为这样做是不明智的），则必须修正人属的判断标准。直到现在，人属的定义通常还是以"脑量的界限"为中心，而这个界限设为 700 毫升（魏登瑞）、750 毫升（基思）以及 800 毫升（瓦卢瓦），并不确定。这里建议新的定义如下：

科	人科	（1955 年由勒格罗 – 克拉克定义）
属	人属	林奈

人属的修正标准　人属具有如下的特征：骨盆带和下肢骨骼结构适应于惯常的直立姿势和两足行走的步态；上肢比下肢短；拇指发育良好，并与其他指的方向完全相反，手不仅具备有力抓握的能力，而且还至少具有简单并且通常发育良好的精确抓握的本领 *；颅容量变化较大，但平均而言大于南方古猿属中成员们的颅容量范围，尽管人属颅容量范围的下限同南方古猿属颅容量范围的上限有重叠；颅容量（平均而言）相对身体尺寸要大，其范围在早期形态的 600 毫升到多于 1,600 毫升之间；颅骨肌肉附着处的粗壮程度不等，从非常强健到难以发觉均可出现，但颞脊或颞线从未达到正中线；颅骨前额部没有显著的眶后狭缩（这在南方古猿属的成员中是常见的）；额骨的眶上区变化很大，从厚大而突出的眶上圆枕到完全缺乏任何眶上突出部分和平滑的眉弓区间变化；面骨骼形态在中度的突颌到直颌间变化，但没有向内凹入（或中凹），这在南方古猿亚科的成员中也是常见的；前端联合面轮廓外形可以从显著向后倾斜到向前倾斜之间变化，而颏隆起可能完全缺乏，也可能从纤细的颏三角变化到非常发达的颏三角；属中大多数成员的齿弓是均匀的圆形并且没有齿隙；第一下前白齿有明显的双齿尖，其中舌侧尖发育程度变异较大；臼齿的大小多变，但与南方古猿属牙齿的大小相比通常较小；最后一颗上臼齿的大小变化比较大，不过一般比第二上臼齿小，通常也比第一上臼齿小；第三下臼齿有时稍微大于第二下臼齿；从人猿超科总体上看：人属的犬齿较小，在磨耗初期后很少或没有重叠，但与南方古猿属的成员进行比较时，门齿、犬齿相对于臼齿和前白齿不是很小；

---

* 对于"力量抓握"和"精确抓握"的定义参见内皮尔，《骨与关节外科杂志》，第 38 卷，B，第 902 页（1956 年）。

upper molar and commonly also smaller than the first upper molar; the lower third molar is sometimes appreciably larger than the second; in relation to the position seen in the Hominoidea as a whole, the canines are small, with little or no overlapping after the initial stages of wear, but when compared with those of members of the genus *Australopithecus*, the incisors and canines are not very small relative to the molars and premolars; the teeth in general, and particularly the molars and premolars, are not enlarged bucco-lingually as they are in the genus *Australopithecus*; the first deciduous lower molar shows a variable degree of molarization.

Genus	*Homo*	Linnaeus
Species	*habilis*	sp. nov.

(*Note*: The specific name is taken from the Latin, meaning "able, handy, mentally skilful, vigorous". We are indebted to Prof. Raymond Dart for the suggestion that *habilis* would be a suitable name for the new species.)

A species of the genus *Homo* characterized by the following features:

A mean cranial capacity greater than that of members of the genus *Australopithecus*, but smaller than that of *Homo erectus*; muscular ridges on the cranium ranging from slight to strongly marked; chin region retreating, with slight or no development of the mental trigone; maxillae and mandibles smaller than those of *Australopithecus* and within the range for *Homo erectus* and *Homo sapiens*; dentition characterized by incisors which are relatively large in comparison with those of both *Australopithecus* and *Homo erectus*; canines which are proportionately large relative to the premolars; premolars which are narrower (in bucco-lingual breadth) than those of *Australopithecus*, but which fall within the range for *Homo erectus*; molars in which the absolute dimensions range between the lower part of the range in *Australopithecus* and the upper part of the range in *Homo erectus*; a marked tendency towards bucco-lingual narrowing and mesiodistal elongation of all the teeth, which is especially evident in the lower premolars (where it expresses itself as a marked elongation of the talonid) and in the lower molars (where it is accompanied by a rearrangement of the distal cusps); the sagittal curvature of the parietal bone varies from slight (within the hominine range) to moderate (within the australopithecine range); the external sagittal curvature of the occipital bone is slighter than in *Australopithecus* or in *Homo erectus*, and lies within the range of *Homo sapiens*; in curvature as well as in some other morphological traits, the clavicle resembles, but is not identical to, that of *Homo sapiens sapiens*; the hand bones differ from those of *Homo sapiens sapiens* in robustness, in the dorsal curvature of the shafts of the phalanges, in the distal attachment of *flexor digitorum superficialis*, in the strength of fibro–tendinous markings, in the orientation of trapezium in the carpus, in the form of the scaphoid and in the marked depth of the carpal tunnel; however, the hand bones resemble those of *Homo sapiens sapiens* in the presence of broad, stout, terminal phalanges on fingers and thumb, in the form of the distal articular surface of the capitate and the ellipsoidal form of the metacarpo-phalangeal joint surfaces; in many of their characters

一般而言，牙齿，特别是臼齿和前臼齿，并不像在南方古猿属中一样在颊舌向放大；第一下乳臼齿显示出不同程度的臼化。

属	人属	林奈
种	能人	新种

（**注释**：种名来源于拉丁文，含义是"有能力的、手灵巧的、有思想的、有技术的、精力充沛的人"。我们感谢雷蒙德·达特教授的提议，对于新种而言，能人是个非常适宜的名字。）

人属新种表现出如下的特点：

平均颅容量比南方古猿属成员的颅容量大，但是比直立人的要小一些；颅部肌肉附着处粗壮程度表现不一，在微小到非常明显的范围之间；下巴后缩，颏三角轻度发育或未形成；上颌骨和下颌骨比南方古猿的小，处于直立人和智人的范围之中；就齿系特点来说，门齿比南方古猿和直立人的相对要大；犬齿相对于前臼齿成比例增大；前臼齿比南方古猿的狭小（颊舌径），但是仍处于直立人的范围之内；臼齿的绝对尺寸处于南方古猿范围下限与直立人范围上限之间；所有牙齿都具有颊舌向变窄和近中远中径向延长的显著趋势，这一趋势在下前臼齿和下臼齿尤其明显（在下前臼齿跟座显著延长，在下臼齿远侧齿尖重新排列）；顶骨矢状曲度呈现从轻微（在人亚科的变化范围之内）到中等（在南方古猿亚科的范围之内）的变化；枕骨外矢状曲度比南方古猿和直立人的轻微，在智人的范围内；在曲度及其他一些形态学特性上，锁骨类似于但不完全等同于现代智人；手骨在粗壮程度、指骨干的背脊弯曲、指浅屈肌的远节附着、纤维肌腱的强度、腕骨大多角骨的方位、手舟骨的形态和腕管显著的深度等方面不同于现代智人；但手骨也有类似于现代智人的方面，表现在手指和拇指上存在宽而短的远节指骨、头状骨末端关节面的形状和掌-指骨关节表面的椭圆体形状；足骨的很多特征在现代智人的变化范围内；大趾是短小的、内收的并且是脚掌着地行走类型；存在非常显著的纵向弧和横向弧；另一方面，第三跖骨比现代人粗壮得多，距骨滑车的中、侧剖面的曲度没有显著的差异。

the foot bones lie within the range of variation of *Homo sapiens sapiens*; the hallux is stout, adducted and plantigrade; there are well-marked longitudinal and transverse arches; on the other hand, the 3rd metatarsal is relatively more robust than it is in modern man, and there is no marked difference in the radii of curvature of the medial and lateral profiles of the trochlea of the talus.

Geological horizon. Upper Villafranchian and Lower Middle Pleistocene.

Type. The mandible with dentition and the associated upper molar, parietals and hand bones, of a single juvenile individual from site *F.L.K.N.N.* I, Olduvai, Bed I.

This is catalogued as Olduvai Hominid 7.

Paratypes. (*a*) An incomplete cranium, comprising fragments of the frontal, parts of both parietals, the greater part of the occipital, and parts of both temporals, together with an associated mandible with canines, premolars and molars complete on either side but with the crowns of the incisors damaged, parts of both maxillae, having all the cheek teeth except the upper left fourth premolar. The condition of the teeth suggests an adolescent. This specimen, from site *M.N.K.* II, Olduvai, Bed II, is catalogued as Olduvai Hominid 13.

(*b*) The associated hand bones, foot bones and probably the clavicle, of an adult individual from site *F.L.K.N.N.* I, Olduvai, Bed I. This is catalogued as Olduvai Hominid 8.

(*c*) A lower premolar, an upper molar and cranial fragments from site *F.L.K.* I, Olduvai, Bed I (the site that yielded also the *Australopithecus* (*Zinjanthropus*) skull). This is catalogued as Olduvai Hominid 6. It is possible that the tibia and fibula found at this site belong with *Homo habilis* rather than with *Australopithecus* (*Zinjanthropus*). These limb bones have been reported on by Dr. P. R. Davis (*Nature*, March 7, 1964, p. 967).

(*d*) A mandibular fragment with a molar in position and associated with a few fragments of other teeth from site *M.K.* I, Olduvai, Bed I, This specimen is catalogued as Olduvai Hominid 4.

Description of the type. Preliminary descriptions of the specimens which have now been designated the type of *Homo habilis*, for example, the parts of the juvenile found at size *F.L.K.N.N.* I in 1960, have already been published in *Nature* by one of us (**189**, 649; **191**, 417; 1961). A further detailed description and report on the parietals, the mandible and the teeth are in active preparation by one of us (P. V. T.), while his report on the cranial capacity (preceding article) as well as a preliminary note on the hand by another of us (*Nature*, **196**, 409; 1962) have been published. We do not propose, therefore, of give a more detailed description of the type here.

Description of the paratypes. A preliminary note on the clavicle and on the foot of the adult,

地质层位 维拉方期上部和中更新世下部。

正型标本 从奥杜威 *F.L.K.N.N.* I 地点第 I 层中出土的幼年个体带有齿系的下颌骨以及相关的上臼齿、顶骨和手骨。

这件标本在奥杜威人科标本中编号为 7。

副型标本 (*a*) 一个不完整的颅骨，包含额骨破片和两侧顶骨部分、大部分枕骨和两侧颞骨部分，还有与之相应的下颌骨，其两侧的犬齿、前臼齿和臼齿都是完整的，但门齿齿冠破损；此外，还有部分双侧上颌骨，上附有除第四左上前臼齿外所有的颊齿。牙齿显示这件标本为青少年个体。这件出土于奥杜威 *M.N.K.* II 地点第 II 层的标本在奥杜威人科标本中编号为 13。

(*b*) 从奥杜威 *F.L.K.N.N.* I 地点第 I 层中出土的一成年个体，包括手骨、足骨以及可能的锁骨，其在奥杜威人科标本中编号为 8。

(*c*) 从奥杜威 *F.L.K.* I 地点第 I 层中（在这个位置上曾出土了南方古猿（东非人）的头骨）出土了一颗下前臼齿、一颗上臼齿和头骨破片。其在奥杜威人科标本中编号为 6。在这个地点发现的胫骨和腓骨可能属于能人而不是南方古猿（东非人）。这些肢骨由戴维斯报道在 1964 年 3 月 7 日出版的《自然》第 967 页上。

(*d*) 从奥杜威 *M.K.* I 地点的第 I 层出土了一件下颌骨破片和一颗臼齿，在原位上还有一些其他与之相联系的牙齿破片。其在奥杜威人科标本中编号为 4。

正型标本描述 关于标本（现在已定名为能人）的初步描述，例如 1960 年在奥杜威 *F.L.K.N.N.* I 地点发现的部分幼年个体的初步描述已经发表在《自然》（189 卷，第 649 页；191 卷，第 417 页，1961 年）上。我们之中的托拜厄斯在积极地准备着关于顶骨、下颌骨和牙齿进一步的详细描述和报道。托拜厄斯有关颅容量（先前的文章中）的研究和我们中另一位对于手的初步研究已经发表了（《自然》，196 卷，409 页，1962 年）。因此，我们在这里不能提供更多关于正型标本的描述。

副型标本描述 对副型标本（*b*）成年个体的锁骨和足骨的初步研究已经刊登

which represents paratype (*b*), was published in *Nature* (**188**, 1050; 1961), and a further report on the foot by Dr. M. H. Day and Dr. J. R. Napier was published in *Nature* of March 7, 1964, p. 969.

The following additional preliminary notes on the other paratypes have been prepared by one of us (P. V. T.).

Description of Paratypes

(*a*) *Olduvai Hominid* 13 *from* M.N.K. *II.* An adolescent represented by a nearly complete mandible with complete, fully-erupted lower dentition, a right maxillary fragment including palate and all teeth from P^3 to M^3, the latter in process of erupting; the corresponding left maxillary fragment with M^1 to M^3, the latter likewise erupting, the isolated left P^3; parts of the vault of a small, adult cranium, comprising much of the occipital, including part of the posterior margin of *foramen magnum*, parts of both parietals, right and left temporosphenoid fragments, each including the mandibular fossa and foramen ovale. The distal half of a humeral shaft (excluding the distal extremity) may also belong to Olduvai Hominid 13. The *corpus mandibulae* is very small, both the height and thickness at M_1 falling below the australopithecine range and within the hominine range. All the teeth are small compared with those of Australopithecinae, most of the dimensions falling at or below the lower extreme of the australopithecine ranges. On the other hand, practically all the dental dimensions can be accommodated within the range of fossil Homininae. The Olduvai Hominid 13 teeth show the characteristic mesiodistal elongation and labiolingual narrowing, in some teeth the L/B index exceeding even those of the type Olduvai Hominid 7, and paratype Olduvai Hominid 6. The occipital bone has a relatively slight sagittal curvature, the Occipital Sagittal Index being outside the range for australopithecines and for *Homo erectus pekinensis* and within the range for *Homo sapiens*. On the other hand, the parietal sagittal curvature is more marked than in all but one australopithecine and in all the Pekin fossils, the index falling at the top of the range of population means for modern man. Both parietal and occipital bones are very small in size, being exceeded in some dimensions by one or two australopithecine crania and falling short in all dimensions of the range for *Homo erectus pekinensis*. The form of the parietal—anteroposteriorly elongated and bilaterally narrow, with a fairly abrupt lateral descent in the plane of the parietal boss—reproduces closely these features in the somewhat larger parietal of the type specimen (Olduvai Hominid 7 from *F.L.K.N.N.* I).

(*b*) *Olduvai Hominid* 6 *from* F.L.K. *I.* An unworn lower left premolar, identified as P_3, an unworn, practically complete crown and partly developed roots of an upper molar, either M^1 or M^2, as well as a number of fragments of cranial vault. These remains were found at the *Zinjanthropus* site and level, some *in situ* and some on the surface. Both teeth are small for an australopithecine, especially in buccolingual breadth, but large for *Homo erectus*. The marked tendency to elongation and narrowing imparts to both teeth an L/B index outside the range for all known australopithecine homologues and even beyond the range for *Homo erectus pekinensis*. The elongating–narrowing tendency is more marked in this molar than in

在《自然》（188 卷，第 1050 页，1961 年）上，关于足骨的进一步研究由戴博士和内皮尔博士发表在 1964 年 3 月 7 日的《自然》第 969 页上。

以下这些关于其他副型标本的初步研究是由我们之中的托拜厄斯提供的。

副型标本的描述

(a) **奥杜威 M.N.K. II 地点发现的编号为 13 的人科化石**　标本代表一个年轻个体，包括一件近乎完整的下颌骨，附有全部萌出的下齿系、保存有腭和第三上前臼齿到第三上臼齿所有牙齿在内的右侧上颌骨破片，其中第三上臼齿还在萌出中。相对应的左侧上颌骨破片附着有第一上臼齿到第三上臼齿以及脱落的第三上前臼齿，其中第三上臼齿同样处于萌出状态；一个较小的成年个体颅骨包含大部分枕骨，包括**枕骨大孔**后缘部分，部分顶骨、左右颞蝶骨破片（每个都带有下颌窝和卵圆孔）。肱骨骨干的末端半部分（排除末端部分）可能也属于奥杜威第 13 号人科化石。**下颌体**非常小，第一下臼齿在高度和厚度上小于南方古猿亚科，同时它又在人亚科变化范围之中。和南方古猿亚科相比所有牙齿均较小，大多尺寸低于南方古猿亚科变化范围的下界。一方面，几乎所有牙齿的尺寸都能够纳入人亚科化石的变化范围之内。奥杜威第 13 号人科化石的牙齿表现出近中远中向延长和颊舌向缩小的特点。在有些牙齿中，齿冠指数（长/宽）甚至超过了奥杜威 7 号正型标本和 6 号副型标本。枕骨矢状曲度轻微，枕骨矢状指数处于南方古猿亚科和北京直立人的变化范围之外、智人变化范围之内。另一方面，顶骨矢状曲度要比所有南方古猿亚科标本（除一例以外）和所有北京直立人标本的都要大，指数下降到现代人群变化范围的上限。顶骨和枕骨尺寸很小，某些尺寸小于一两例南方古猿亚科标本，且所有尺寸达不到北京直立人的变化范围。顶骨前后延长、两侧变窄，顶骨凸面陡然侧斜——顶骨的形态再次展现了与正型标本（从奥杜威 F.L.K.N.N. I 点发现的第 7 个人科）稍大的顶骨的形态接近。

(b) **奥杜威 F.L.K. I 地点发现的编号为 6 的人科化石**　该化石包括一颗尚未磨损的左下前臼齿，被鉴定为第三下前臼齿，一个没有磨损的上臼齿，不是第一上臼齿就是第二上臼齿，齿冠几乎完整，齿根部分发育。此外，还有一些头骨穹隆破片。这些化石是在发现东非人的地点和底层发现的，有些出土于原位，有些采集于地表。两颗牙齿都比南方古猿亚科的小，特别是颊舌径，但比直立人的大。两颗牙齿延长和缩小的显著趋势使得其齿冠指数超出了所有已知的南方古猿亚科的同族体的变化范围，甚至超出北京直立人的变化范围。臼齿的延长－缩小趋势比正型标本（从奥

the upper molar belonging to the type specimen (Olduvai Hominid 7) from *F.L.K.N.N.* I.

(*c*) *Olduvai Hominid 8 from* F.L.K.N.N. *I.* Remains of an adult individual found on the same horizon as the type specimen, and represented by two complete proximal phalanges, a fragment of a rather heavily worn tooth (premolar or molar), and a set of foot-bones possessing most of the specializations associated with the plantigrade propulsive feet of modern man. Probably the clavicle found at this site belongs to this adult rather than to the juvenile type-specimen; it is characterized by clear overall similarities to the clavicle of *Homo sapiens sapiens*.

(*d*) *Olduvai Hominid 4 from* M.K. *I.* A fragment of the posterior part of the left *corpus mandibulae*, containing a well-preserved, fully-erupted molar, either M_2 or M_3. The width of the mandible is 19.2 m level with the mesial half of the molar, but the maximum width must have been somewhat greater, The molar is 15.1 mm in mesiodistal length and 13.0 mm in buccolingual breadth; it is thus a small and narrow tooth by australopithecine standards, but large in comparison with *Homo erectus* molars. There are several other isolated dental fragments, including a moderately worn molar fragment. These are stratigraphically the oldest hominid remains yet discovered at Olduvai.

Referred Material

Olduvai Hominid 14 from M.W.K. *II.* (1) A juvenile represented by a fragment of the right parietal with clear, unfused sutural margins; two smaller vault fragments with sutural margins; a left and a right temporal fragment, each including the mandibular fossa.

(2) A fragmentary skull with parts of the upper and lower dentition of a young adult from site *F.L.K.* II, Maiko Gully, Olduvai, Bed II, is also provisionally referred to *Homo habilis*. This specimen is catalogued as Olduvai Hominid 16. It is represented by the complete upper right dentition, as well as some of the left maxillary teeth, together with some of the mandibular teeth. The skull fragments include parts of the frontal, with both the external orbital angles preserved, as well as the supraorbital region, except for the glabella; parts of both parietals and the occipital are also represented.

Implications for Hominid Phylogeny

In preparing our diagnosis of *Homo habilis*, we have not overlooked the fact that there are several other African (and perhaps Asian) fossil hominids whose status may now require re-examination in the light of the new discoveries and of the setting up of this new species. The specimens originally described by Broom and Robinson as *Telanthropus capensis* and which were later transferred by Robinson to *Homo erectus* may well prove, on closer comparative investigation, to belong to *Homo habilis*. The Kanam mandibular fragment, discovered by the expedition in 1932 by one of us (L. S. B. L.), and which has been shown to possess archaic features (Tobias, *Nature*, **185**, 946; 1960), may well justify further investigation along these lines. The Lake Chad craniofacial fragment, provisionally described by M. Yves Coppens in 1962, as an australopithecine, is not, we are convinced,

杜威 *F.L.K.N.N.* I 地点发现的编号为 7 的人科化石）的上臼齿更加明显。

(*c*) **奥杜威 *F.L.K.N.N.* I 地点发现的编号为 8 的人科化石**　发现的一个成年个体的化石与正型标本处于同层位中，包括两个完整的近节指骨，一件磨损得相当严重的牙齿破片（前臼齿或臼齿）和一组足骨，这组足骨有很多同现代人直立行走的脚相关的特化。在这个点发现的锁骨属于成年的可能性胜过属于幼年正型标本的可能性，并与现代智人很接近。

(*d*) **奥杜威 *M.K.* I 地点发现的编号为 4 的人科化石**　该化石为一左下颌体后部破片，其上附有一颗保存很好且完全萌出的臼齿，不是第二下臼齿就是第三下臼齿。以该臼齿的近中部分为基准，测得的下颌宽度为 19.2 米（译者注：此处原文有误，应为厘米），但实际下颌的最大宽度要更大一些。臼齿近中远中径为 15.1 毫米，颊舌径为 13.0 毫米；因此按南方古猿亚科的标准，它是一颗又小又窄的牙齿，但和直立人的臼齿相比是较大的。同时还出土了一些其他单个的牙齿破片，包括一个中度磨损的臼齿破片。这些是至今在奥杜威发现的地层时代最古老的人科动物遗骸。

参 考 材 料

奥杜威 *M.N.K.* II 地点发现的编号为 14 的人科化石　（1）该化石属于一个幼年个体，包括一个右侧顶骨破片，有清晰且尚未愈合缝边缘；两个小的颅顶破片，有骨缝边缘；一个左侧颞骨破片和一个右侧颞骨破片，每个都带有下颌窝。

（2）从奥杜威 *F.L.K.* II 地点第 II 层中出土的年轻成年个体的头骨破片上具有部分上、下颌齿系，也临时划归能人。这件标本被编号为奥杜威人科 16，包括完整的上颌右侧齿系、部分左侧上颌骨牙齿和一些下颌骨牙齿。头骨破片包括部分额骨（保存下了两个外眶角），以及眶上区（除了眉间部分）；此外还有部分顶骨和枕骨。

人科系统演化的意义

在能人的判断标准中，我们注意到，随着许多新化石的出土和能人这一新种的建立，许多非洲（甚至是亚洲）的人科化石的地位需要重新考虑。通过进一步的比较研究，可以充分证明最初由布鲁姆和鲁滨逊记述为开普远人、后来由鲁滨逊转而归入直立人的标本属于能人。利基在 1932 年发现的卡纳姆下颌骨破片具有许多原始的特性（托拜厄斯，《自然》，第 185 卷，第 946 页，1960 年），这使得沿着这一线索进一步调查研究很必要。乍得湖发现的颅面破片由伊夫·柯庞在 1962 年进行过临时的描述，定为南方古猿亚科的类型，而我们确信它不属于这一亚科。我们认为发

a member of this sub-family. We understand that the discoverer himself, following his investigation of the australopithecine originals from South Africa and Tanganyika, now shares our view in this respect. We believe that it is very probably a northern representative of *Homo habilis*.

Outside Africa, the possibility will have to be considered that the teeth and cranial fragments found at Ubeidiyah on the Jordan River in Israel may also belong to *Homo habilis* rather than to *Australopithecus*.

Cultural Association

When the skull of *Australopithecus* (*Zinjanthropus*) *boisei* was found on a living floor at *F.L.K.* I, no remains of any other type of hominid were known from the early part of the Olduvai sequence. It seemed reasonable, therefore, to assume that this skull represented the makers of the Oldowan culture. The subsequent discovery of remains of *Homo habilis* in association with the Oldowan culture at three other sites has considerably altered the position. While it is possible that *Zinjanthropus* and *Homo habilis* both made stone tools, it is probable that the latter was the more advanced tool maker and that the *Zinjanthropus* skull represents an intruder (or a victim) on a *Homo habilis* living site.

The recent discovery of a rough circle of loosely piled stones on the living floor at site *D.K.* I, in the lower part of Bed I, is noteworthy. This site is geologically contemporary with *M.K.* I, less than one mile distant, where remains of *Homo habilis* have been found. It seems that the early hominids of this period were capable of making rough shelters or windbreaks and it is likely that *Homo habilis* may have been responsible.

Relationship to *Australopithecus* (*Zinjanthropus*)

The fossil human remains representing the new species *Homo habilis* have been found in Bed I and in the lower and middle part of Bed II. Two of the sites, *M.K.* I and *F.L.K. N.N.* I, are geologically older than that which yielded the skull of the australopithecine *Zinjanthropus*. One site, *F.L.K.* I, has yielded both *Australopithecus* (*Zinjanthropus*) and remains of *Homo habilis*, while two sites are later, namely *M.N.K.* II and *F.L.K.* II Maiko gully. The new mandible of *Australopithecus* (*Zinjanthropus*) type from Lake Natron, reported in the preceding article by Dr. and Mrs. Leakey, was associated with a fauna of Bed II affinities.

It thus seems clear that two different branches of the Hominidae were evolving side by side in the Olduvai region during the Upper Villafranchian and the lower part of the Middle Pleistocene.

(**202**, 7-9; 1964)

L. S. B. Leakey: Coryndon Museum, Centre for Prehistory and Palaeontology.
P. V. Tobias: University of Witwatersrand, Johannesburg.
J. R. Napier: Unit of Primatology and Human Evolution, Royal Free Hospital Medical School, University of London.

现者本人在南非和坦噶尼喀进行关于南方古猿亚科起源的调查后，如今他在这方面和我们的观点一致。我们认为它很有可能是能人的北方代表。

非洲之外，在以色列约旦河的乌贝迪亚发现的牙齿和头骨破片也可能属于能人而不是属于南方古猿。

文化联系

在 *F.L.K.* I 的生活面中发现鲍氏南方古猿（鲍氏东非人）头骨的时候，在奥杜威早期层位中没有任何其他类型已知的人科动物。因此假定这个头骨代表了奥杜威文化的创造者看来是合理的。然而，在随后的三个地点发现能人遗骸与奥杜威文化并存的现象极大地改变了这一观点。东非人和能人可能都会制造石器，但是后者可能是更先进石器的制造者，这一东非人头骨代表的可能是能人生活地域的一个闯入者（或者牺牲品）。

最近在 *D.K.* I 地点第 I 层的下部生活层发现了由石头松散堆成的类似圆圈形状，这是值得我们注意的。在地质学上这个点和 *M.K.* I 地点是同一时代，就在距离它不及 1 英里的地方发现了能人的遗骸。看来，这一时期的早期人科具有制造粗制的掩蔽物或防风墙的能力，而这一人类很可能是能人。

与南方古猿（东非人）的关系

在第 I 层和第 II 层的中、下部发现了代表新种能人的人类化石，*M.K.* I 和 *F.L.K.N.N.* I 两个地点的地质学年代要比出土南方古猿亚科东非人头骨的更远。其中 *F.L.K.* I 地点出土了南方古猿（东非人）和能人的遗骸，而另外两个地点——*M.N.K.* II 和 *F.L.K.* II 马伊科河年代稍近。路易斯·利基博士和夫人玛丽·利基在前文中报道过从纳特龙湖产出的新的南方古猿（东非人）下颌骨同第 II 层中的动物群是有密切联系的。

因此，我们可以清楚地了解到：在维拉方期上部和中更新世下部，人科的两个不同分支在奥杜威区域平行地演化着。

（张玉光 翻译；刘武 邢松 审稿）

Haemoglobin G_{Accra}

H. Lehmann *et al.*

Editor's Note

Here Hermann Lehmann and D. Beale from Cambridge University identify the specific amino-acid substitution responsible for haemoglobin G$_{Accra}$, a rare and abnormal variant of the iron-containing red-blood-cell protein. The blood sample, procured by F. S. Boi-Doku of Accra's Central Clinical Laboratory in Ghana, came from an apparently healthy individual in his sixties who had inherited two copies of the abnormal gene. All other homozygous abnormal haemoglobin variants known at that time produced disease, making haemoglobin G$_{Accra}$ all the more enticing to study and raising the possibility that other, similarly harmless, rare haemoglobin variants might exist.

IN 1954 the major haemoglobin types, A, M, S, C, D, E and F, had been described, and the observation of a new variant, haemoglobin G (ref. 1), was the first of a series of reports on rarer human haemoglobins which is still continuing. The haemoglobin then described—and differentiated from other haemoglobins with identical or similar electrophoretic properties as haemoglobin G$_{Accra}$—has never been found again, except in the family in which it was originally observed. Haemoglobin G$_{Accra}$ is abnormal in the β-chain[2]. The haemoglobins with identical electrophoretic and chromatographic properties hitherto examined were α-chain abnormal pigments. There are three: (1) haemoglobin G$_{Philadelphia}$ with which numerous other haemoglobins observed in West Africa, or in persons of West African origin, are identical[3]; (2) haemoglobin G$_{Chinese}$, which has only been found in persons of Chinese extraction[4]; (3) a haemoglobin found in a British family, haemoglobin G$_{Norfolk}$[5]. There is one other haemoglobin named G which also carries its substituted amino-acid residue in the β-chain-haemoglobin G$_{San\ José}$[6], but this pigment differs from haemoglobin G$_{Accra}$ in its electrophoretic properties, and should perhaps have been described under a different letter.

Haemoglobin G$_{Accra}$ is remarkable for the fact that it is the only rare haemoglobin for which an individual has been described the homozygous state of which could be established on the basis of family investigation and of laboratory findings[7]. This homozygote was found to be healthy without an enlarged spleen, without anaemia, and his red cells were morphologically indistinguishable from those containing only normal haemoglobin. This was, and still is, in contrast with what is seen in the haemoglobin S, C, D and E diseases, the only other homozygous conditions for adult haemoglobin variants so far described.

There was, therefore, a special interest in establishing the precise nature of the amino-acid substitution in haemoglobin G$_{Accra}$. Whatever its nature, it does not have a similar effect

血红蛋白G_{阿克拉}

莱曼等

编者按

在本文中，剑桥大学的赫尔曼·莱曼和比尔鉴定出了血红蛋白 G 阿克拉中特有的氨基酸替换。血红蛋白 G 阿克拉是含铁红细胞蛋白质的一种稀有异常变异体。加纳阿克拉中心临床实验室的博伊–多库从一位六十多岁、看起来健康的人身上取得血液样本，此人遗传了两个拷贝的异常基因。在那个时候，已知所有其他的纯合异常血红蛋白变异体都会引起疾病，这使得血红蛋白 G 阿克拉更值得研究，而且增加了其他稀有的非致病血红蛋白变异体存在的可能性。

到 1954 年人们已经发现了血红蛋白的主要类型，包括血红蛋白 A、M、S、C、D、E 和 F。对一种新的变异体——血红蛋白 G（参考文献 1）的观察研究是关于人稀有血红蛋白系列报道的第一个，这样的报道仍在继续。当时所描述的这种血红蛋白和其他与血红蛋白 G 阿克拉具有相同或类似的电泳性质的血红蛋白不同，除了在最初的家族观察到以外再也没有被发现过。血红蛋白 G 阿克拉在 β 链上发生异常 [2]，而迄今所检测到的电泳和色谱性质相同的血红蛋白都是在 α 链上发生异常。这样的血红蛋白有 3 种：（1）血红蛋白 G 费城，与在西非或西非血统的人中发现的许多其他血红蛋白都是相同的 [3]；（2）血红蛋白 G 中国人，只在有中国血统的人中发现过 [4]；（3）血红蛋白 G 诺福克 [5]，在一个英国家族中发现。还有另一种命名为 G 的血红蛋白，其 β 链中携带被替换的氨基酸残基，即血红蛋白 G 圣何塞 [6]，但是这种血红蛋白与血红蛋白 G 阿克拉在电泳性质上不同，也许应该用另一个字母来表示。

血红蛋白 G 阿克拉突出的地方在于它是唯一一个基于家族调查和实验室研究所确证的、在个体中以纯合状态存在的稀有血红蛋白 [7]。这个纯合子个体是健康的，脾没有肿大，没有贫血症，并且他的红细胞在形态上与只含有正常血红蛋白的红细胞没有区别。直到今天，这都与在血红蛋白 S、C、D 和 E 疾病中所见到的情况完全不同，是迄今为止已描述的成人血红蛋白变异体中唯一一例不同的情况。

因此，科学家对血红蛋白 G 阿克拉中的氨基酸替换的具体情况特别感兴趣。无论如何，这种替换与血红蛋白 S、C、D 和 E 中已知的氨基酸替换对红细胞的形态也

on the morphology and, presumably, the life span of the red cell which is produced by the known substitutions in haemoglobins S, C, D and E.

*Amino-acid Substitution in Haemoglobin G*_{Accra}. Haemoglobin G was purified from the blood of the homozygote and a peptide-chromatogram (fingerprint) was prepared according to Ingram[8] and Baglioni[9]. It was noted that peptide 6 according to the old numbering system, or tryptic peptide β*Tp*IX according to the new nomenclature[10], was missing and another new peptide was seen instead which had moved nearer towards the cathode (Fig. 1). The new peptide gave a positive colour reaction for histidine[11], as does the β*Tp*IX of haemoglobin A (β^A*Tp*IX). This peptide contains in haemoglobin A 16 amino-acid residues[12,13] (Table 1).

Table 1. Amino-acid Sequence of Tryptic Peptide IX of the β-Chain of Haemoglobin A

Val–leu–gly–ala–phe–ser–asp–gly–leu–ala–his–leu–asp–asn–leu–lys
67 68 69 70 71 72 73 74 75 76 77 78 79 80 81 82

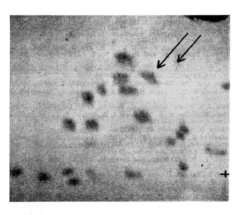

Fig. 1. Fingerprint of the tryptic digest of haemoglobin G_{Accra} . The right arrow points to the area where β^A*Tp*IX is missing, and the left arrow to the peptide which has taken its place—β^G*Tp*IX. +, Point of application.

The NH₂-terminal group[14], was valyl in both peptides, and leucine amino peptidase[4] which splits off amino-acids from the NH₂-terminal part of a peptide chain released valine and leucine in both. As the peptides had arisen from tryptic digestion which implies lysyl or arginyl as the COOH-terminal, and as the colour reaction and amino-acid analysis failed to show arginine[15], lysyl had to be the COOH-terminal group.

Table 2 shows a comparison of the results of quantitative amino-acid analysis on an EEL automatic analyser[16], after the acid hydrolysis of the two peptides from haemoglobins A and G. It will be seen that the results were the same in both, although the electrophoretic mobility of the G-peptide had indicated that it had a greater positive charge than the corresponding A-peptide. The only explanation was that one of the two aspartic acid residues of β^A*Tp*IX had been replaced by an asparaginyl in β^G*Tp*IX, and that this had been converted into aspartic acid on acid hydrolysis.

270

可能是对红细胞的寿命产生的影响不同。

血红蛋白G阿克拉**中的氨基酸替换**　从纯合子的血液中纯化出血红蛋白G，按照英格拉姆[8]和巴廖尼[9]的方法制作肽色谱图（指纹图谱）。人们注意到6号肽段（根据旧的计数系统）或胰蛋白酶消化肽段 $\beta TpIX$（根据新的命名法则）[10]消失了，而在更接近阴极的地方出现了另一个新的肽段（图1）。这个新肽段与组氨酸可以发生阳性的颜色反应[11]，如同血红蛋白A的 $\beta TpIX$（$\beta^A TpIX$）一样。在血红蛋白A中，该肽段有16个氨基酸残基[12,13]（表1）。

表1. 血红蛋白A的β链胰蛋白酶消化的肽段IX的氨基酸序列

Val–leu–gly–ala–phe–ser–asp–gly–leu–ala–his–leu–asp–asn–leu–lys
67　68　69　70　71　72　73　74　75　76　77　78　79　80　81　82

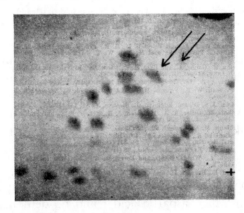

图1. 血红蛋白G阿克拉胰蛋白酶消化物的指纹图谱。右侧箭头指出了 $\beta^A TpIX$ 缺失的区域，左侧箭头指出了新出现的肽段 $\beta^G TpIX$。+表示加样点。

两个肽段的氨基末端[14]都是缬氨酰，亮氨酸氨肽酶[4]（可以从肽链的氨基末端切下氨基酸）处理两个肽段后释放出缬氨酸和亮氨酸。由于该肽段是由胰蛋白酶消化产生，这就意味着其羧基末端只能是赖氨酰基或精氨酰基。又通过颜色反应和氨基酸分析证明其羧基末端不是精氨酸[15]，所以只能是赖氨酰基。

表2显示了来自血红蛋白A和G的两个肽段经酸水解后在EEL自动分析仪上进行氨基酸定量分析的结果比较[16]。可以看到两者的结果相同，虽然G肽段的电泳迁移率表明它比A肽段相应的多了一个正电荷。唯一的解释就是在 $\beta^G TpIX$ 中，原 $\beta^A TpIX$ 的两个天冬氨酸残基之一被天冬酰胺酰所取代，而这一残基在酸水解时转变成了天冬氨酸。

Table 2. Molar Ratio of Amino-acid Residues after Acid Hydrolysis of Tryptic

Peptides IX from Haemoglobin A and Haemoglobin G$_{Accra}$

Residues	β^{A}TpIX		β^{G}TpIX	
Asp	2.8	2.7	2.6	2.6
Thr	trace	trace	trace	trace
Ser	1.0	1.2	1.1	1.3
Glu	trace	trace	trace	trace
Pro	trace	trace	trace	trace
Gly	2.1	2.0	1.9	2.2
Ala	2.5	2.4	2.3	2.1
Val	1.3	1.1	1.0	0.9
Leu	3.6	3.7	3.7	3.7
Phe	1.0	1.0	1.1	1.2
Lys	0.9	0.9	1.1	1.1
His	0.9	1.1	0.9	1.2

Although on electrophoresis at pH 6.4 the mobility of the A peptide is 0.0 and that of the G peptide + 0.17

(lysine = + 1.0) there seems to be no difference in amino-acid composition after acid hydrolysis.

To investigate this possibility, both peptides βTpIX were eluted with 10 percent (v/v) pyridine to prevent hydrolysis of asparaginyl, and then digested with pronase for 4 h using the same arrangement as had been used for trypsin, and the resulting smaller peptides and amino-acids were separated by high voltage electrophoresis at pH 6.4. Pronase attacks virtually all peptide linkages, although it is relatively inactive towards those which involve aspartic acid residues. Table 3 shows the electrophoretic mobility of these breakdown products, and their amino-acid composition found after acid hydrolysis. This latter method of course converts asparaginyl to aspartic acid. It will be seen that the number of acidic breakdown products in which aspartic acid could be demonstrated differed in the two peptides—four from haemoglobin A, and three from haemoglobin G. Three seemed to be the same in both haemoglobins. Free aspartic acid was obtained from both, and there were two peptides which had nearly the same electrophoretic mobility, and the same amino-acid composition in both. One peptide with the mobilities of −0.77 and −0.78 respectively was hydrolysed into aspartic acid and glycine with aspartyl as the NH$_2$-terminal residue. The other peptide with the mobilities of −0.64 and −0.62 respectively showed on hydrolysis only aspartic acid, glycine, and leucine with NH$_2$-terminal aspartyl and COOH-terminal leucyl[14,17]. The sequence 73–75 was therefore the same in both β^A and β^G, namely, asp-gly-leu, and the substitution of aspartyl in haemoglobin G had to be in position 79 of the β chain.

From β^{A}TpIX one other peptide yielding aspartic acid was obtained. It had a mobility of −0.68 but produced aspartic acid only. However, aspartic acid or an asp-asp peptide would have had a mobility of −1.00, hence this peptide had to consist of aspartyl and asparaginyl, and it represented the 79–80 sequence of the β^A-chain, namely, asp-asn.

This peptide was not found in β^{G}TpIX. However, free asparagine could be demonstrated in the neutral band before hydrolysis. No free asparagine was found in the neutral band of

表 2. 血红蛋白 A 和血红蛋白 G_{阿克拉}的胰蛋白酶消化的肽段 IX 酸水解后氨基酸残

基的摩尔比

氨基酸残基	$\beta^A TpIX$		$\beta^G TpIX$	
天冬氨酸	2.8	2.7	2.6	2.6
苏氨酸	痕量	痕量	痕量	痕量
丝氨酸	1.0	1.2	1.1	1.3
谷氨酸	痕量	痕量	痕量	痕量
脯氨酸	痕量	痕量	痕量	痕量
甘氨酸	2.1	2.0	1.9	2.2
丙氨酸	2.5	2.4	2.3	2.1
缬氨酸	1.3	1.1	1.0	0.9
亮氨酸	3.6	3.7	3.7	3.7
苯丙氨酸	1.0	1.0	1.1	1.2
赖氨酸	0.9	0.9	1.1	1.1
组氨酸	0.9	1.1	0.9	1.2

虽然在 pH 为 6.4 条件下电泳时，A 肽的迁移率是 0.0，G 肽的迁移率是 +0.17（赖氨酸 =+1.0），但是在酸水解后的氨基酸组成上看不出什么差别。

为了研究这种可能性，两种 $\beta TpIX$ 肽段都用 10%(v/v) 吡啶洗脱，以防止天冬酰胺酰水解，然后在与胰蛋白酶消化条件一样的情况下，用链霉蛋白酶消化 4 小时。在 pH 为 6.4 的条件下，通过高压电泳将得到的较小的肽段和氨基酸分离开。虽然链霉蛋白酶作用于含有天冬氨酸残基的肽键的活性较弱，但它几乎可以解开所有的肽键。表 3 显示了这些分解产物的电泳迁移率及其酸水解后的氨基酸组成。酸水解无疑可将天冬酰胺酰转换为天冬氨酸。可以看到在酸性水解产物中的天冬氨酸的数量在 2 个肽段中有差别——血红蛋白 A 中是 4 个，血红蛋白 G 中是 3 个。其中的 3 个在两种血红蛋白中似乎是一样的。从两种血红蛋白中都可以得到游离的天冬氨酸。还有两个肽段具有几乎相同的电泳迁移率和相同的氨基酸组成。迁移率分别是 –0.77 和 –0.78 的一种肽段可以水解成天冬氨酸和甘氨酸，氨基末端残基是天冬酰胺。而迁移率分别是 –0.64 和 –0.62 的另一种肽段水解后只生成天冬氨酸、甘氨酸和亮氨酸，其氨基末端是天冬酰胺，羧基末端是亮氨酰[14,17]。因此在 β^A 和 β^G 中，73~75 位的序列一样，即为 asp-gly-leu，在血红蛋白 G 中天冬酰胺的替换必定发生在 β 链的第 79 位。

从 $\beta^A TpIX$ 中还获得了另一个产生天冬氨酸的肽，其迁移率是 –0.68，但只产生天冬氨酸。然而天冬氨酸或 asp-asp 肽的迁移率是 –1.00。因此，该肽一定是由天冬氨酸和天冬酰胺酰组成，它是 β^A 链的第 79~80 位序列，即为 asp-asn。

在 $\beta^G TpIX$ 中没有发现这种肽。但是在水解前的中性带中却发现了游离的天冬酰胺。而在 $\beta^A TpIX$ 的中性带中并没有发现游离的天冬酰胺。由此得出结论：与

$\beta^A T p$IX. From this it could be concluded that the residue next to leucyl in position 79 of the β^G-chain was asparaginyl and haemoglobin G$_{Accra}$ could be defined as $\alpha_2^A \beta_2^{79\ Asn}$.

Table 3. Products of Pronase Hydrolysis of $\beta^A T p$IX and $\beta^G T p$IX

	Mobility*	Amino acids on acid hydrolysis	NH^{2-} terminal DNP	COOH-terminal
	−1.00	Asp		
	−0.77	Asp, gly	Asp	
	−0.68	Asp		
	−0.64	Asp, gly, leu	Asp	Leu
$\beta^A T p$IX	−0.52			
	0.00	Val, gly, leu, asp Ala, phe, ser		
	+0.53	His		
	+0.67	Leu, lys	Leu	
	+1.00	Lys		
	−1.00	Asp		
	−0.78	Asp, gly	Asp	
	−0.62	Asp, gly, leu	Asp	Leu
	−0.54			
$\beta^G T p$IX	0.00	Val, gly, leu, asp Ala, ser, phe		
	+0.51	His		
	+0.67	Leu, lys	Leu	
	+1.00	Lys		

* Electrophoretic mobility is measured at pH6.4 using as markers aspartic acid (−1.00) and lysine (+1.00).

The sequence asn-asn-leu in haemoglobin G would explain why pronase digestion produced the peptide asp-asn from $\beta^A T p$IX preferably to the free asparagine as obtained from $\beta^G T p$IX. In $\beta^G T p$IX due to the aspartyl → asparaginyl replacement the 79–80 linkage was more easily broken because of the absence of an acidic aspartyl residue.

Homozygote. The homozygote is now sixty-four years old, or even a little older, and is physically fit with the exception of a mild hypertension of 160/100 mm mercury. From 1923 until his retirement in 1963 he worked continually as a printer in the Accra Government Printing Department. Of his twenty children by two wives, all but one is alive and well. The age of the oldest offspring is thirty-six years and the youngest is eleven years old.

His blood shows the following values:

haemoglobin	12.6 g/100 ml
red cells	4,500,000/mm^3
packed cell volume	40 percent

β^G 链第 79 位亮氨酰相邻的残基是天冬酰胺酰，因此血红蛋白 G阿克拉还可以表示为 $\alpha_2^A\beta_2^{79\ Asn}$。

表 3. $\beta^A TpIX$ 和 $\beta^G TpIX$ 的链霉蛋白酶水解产物

	迁移率*	酸水解后得到的氨基酸	氨基末端的 DNP	羧基端
$\beta^A TpIX$	−1.00	Asp		
	−0.77	Asp, gly	Asp	
	−0.68	Asp		
	−0.64	Asp, gly, leu	Asp	Leu
	−0.52			
	0.00	Val, gly, leu, asp Ala, phe, ser		
	+0.53	His		
	+0.67	Leu, lys	Leu	
	+1.00	Lys		
$\beta^G TpIX$	−1.00	Asp		
	−0.78	Asp, gly	Asp	
	−0.62	Asp, gly, leu	Asp	Leu
	−0.54			
	0.00	Val, gly, leu, asp Ala, ser, phe		
	+0.51	His		
	+0.67	Leu, lys	Leu	
	+1.00	Lys		

* pH 6.4 时用天冬氨酸 (−1.00) 和赖氨酸 (+1.00) 作标记测定电泳迁移率。

血红蛋白 G 中的序列 asn-asn-leu 可以解释为什么 $\beta^A TpIX$ 经链霉蛋白酶消化产生了肽 asp-asn，而不是像 $\beta^G TpIX$ 那样产生游离的天冬酰胺。在 $\beta^G TpIX$ 中，由于天冬氨酰被天冬酰胺酰所替代，第 79~80 位间的肽键因为缺少了一个酸性的天冬氨酰残基而更容易断裂。

纯合子 纯合子来源于今年 64 岁的老人，也许更老一些，身体很健康，只是有些轻度的高血压（160/100 毫米汞柱）。他是一个印刷工人，从 1923 年开始一直在阿克拉政府的印刷部门工作，直到 1963 年退休。他的 20 个孩子（由两位妻子所生）除了一个之外其他的都还在世而且身体健康。这些孩子中最大的 36 岁，最小的只有 11 岁。

他的血液化验值如下：

血红蛋白	12.6 克 /100 毫升
红细胞	4,500,000/ 毫米^3
红细胞压积	40%

The mean corpuscular values are:

cell volume	$89\,\mu^3$
haemoglobin	$28\,\gamma\gamma$
haemoglobin concentration	32 percent

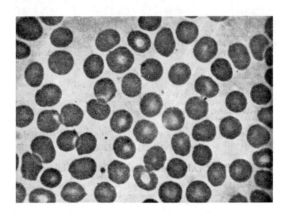

Fig. 2. Stained blood smear of the haemoglobin-G homozygote. The cells are not hypochromic and the target cells and other abnormal forms associated with the blood picture of haemoglobin S, C, D and E homozygotes are absent.

The red cell fragility showed an almost normal distribution: initial haemolysis at 0.5 percent sodium chloride, 50 percent haemolysis at 0.415 and 90 percent haemolysis at 0.3 percent sodium chloride. The only unusual feature was that 10 percent of the cells were still not haemolysed at 0.3 percent sodium chloride, and that complete haemolysis was only achieved in distilled water. There was no enlarged spleen, the red cells were normal in appearance and there were no target cells. Haemoglobin A$_2$ was within normal limits, and on haemoglobin F was demonstrable. These findings contrast with those in haemoglobinopathies of the homozygotes for the genes controlling haemoglobins S, C, D and E. Haemoglobin $\alpha_2^A \beta_2^{79\ Asn}$ seems to cause no haemoglobinopathy and the substitution of an asparaginyl residue for an aspartyl in position 79 of the β-chain seems to have no deleterious effect.

Differences in the Morbidity of Haemoglobin Variants. Of the rarer haemoglobins only haemoglobin G$_{Accra}$ has so far been seen in the homozygous state. It is probable that many of the other rare haemoglobins may be similarly harmless and that there is a relation between the effect a haemoglobin has on the red cell morphology in the homozygote and the advantage it confers on the heterozygote in natural selection which causes its high frequency in a population.

In the heterozygote for the genes for haemoglobin A, and for haemoglobins S, C, D and E respectively, the proportion of the abnormal variant is less than 50 percent. This has usually been interpreted as the outcome of a difference in the rate of production of normal and abnormal polypeptide chains within the cell[18]. An alternative explanation has recently been proposed by Levere, Lichtman and Levine[19]. From iron incorporation

平均红细胞体积如下：

细胞体积	89 微米 ^{3}
血红蛋白	28 $\gamma\gamma$
血红蛋白浓度	32%

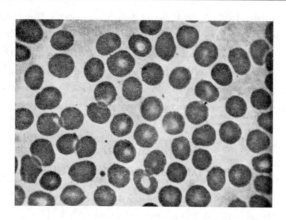

图 2. 血红蛋白 G 纯合子染色后血液涂片。细胞被充分染色，图中没有靶细胞以及其他与血红蛋白 S、C、D 和 E 纯合子血象相关联的异常形式。

红细胞的脆性实验显示出几乎正常的分布：在 0.5% 的氯化钠溶液中红细胞开始溶解，在 0.415% 的氯化钠溶液中有 50% 红细胞溶解，在 0.3% 的氯化钠溶液中有 90% 的红细胞溶解。唯一不正常的是在 0.3% 的氯化钠中仍有 10% 的红细胞没有溶解，只有在蒸馏水中才能完全溶解。脾没有肿大，红细胞外观正常，没有发现靶细胞。血红蛋白 A_2 的数量在正常范围之内，没有发现血红蛋白 F。这些结果与控制血红蛋白 S、C、D 和 E 的基因的纯合子引起的血红蛋白病的结果完全不同。血红蛋白 $\alpha_2^A\beta_2^{79\,Asn}$ 似乎不引起血红蛋白病，在 β 链 79 位上天冬酰胺酰代替了天冬氨酰似乎没有有害的作用。

血红蛋白变异体致病性的差异　在这些较稀有的血红蛋白中，迄今只发现血红蛋白 G 阿克拉 为纯合状态。很有可能还存在许多其他同样非致病的稀有血红蛋白。纯合子中血红蛋白对红细胞形态的作用可能与它给杂合子赋予的在自然选择过程中的优势之间存在着某种关系，这种优势使其在种群中保持有较高的频率。

在血红蛋白 A 以及血红蛋白 S、C、D 和 E 的基因杂合子中，异常变异体的比例分别小于 50%。这通常被解释为细胞内产生正常多肽链和异常多肽链的速率不同所致 [18]。近来莱维尔、利奇特曼和莱文提出了另一种解释 [19]。对具有贫血镰刀状细胞特征的（A + S）携带者进行铁掺入研究，从中他们得出结论：至少在血红蛋

studies in anaemic sickle-cell trait (A + S) carriers they concluded that, in the case of haemoglobin A + S at least, cells with a wide range of A : S proportions are produced—some with A < S and some with A > S. The first are rapidly eliminated and as a result the final over-all proportion of A : S becomes greater than 1.

In the case of haemoglobin A + G_{Accra} heterozygotes, the proportion of the two haemoglobins is exactly 1. We have established this by a number of investigations because by ordinary paper electrophoresis at alkaline pH the leading (A) band contains about 53 percent of the total haemoglobin. This proportion is also obtained when an artificial mixture of 50 percent A and 50 percent G is submitted to electrophoresis at alkaline pH. Furthermore, when open boundary electrophoresis of haemoglobin from the A + G heterozygote was performed at pH 6.5 (ref. 7) and the G fraction was the leading component, the G fraction then amounted to from 50 to 55 percent of the total pigment.

This equal proportion of A and G in the heterozygote would indicate that there is no elimination of cells with a higher proportion of G than A, and that the whole range of cells survives equally, those with A < G, and those with A > G.

One might predict that other haemoglobins which show an equal proportion of A and abnormal variants in the haemolysates from heterozygotes are similarly harmless as haemoglobin G_{Accra}. There would, therefore, be a series of haemoglobin types: (1) Harmless in homozygotes and heterozygotes (heterozygotes show 50 percent of the abnormal variant): example, haemoglobin G_{Accra}. (2) Harmful in homozygotes, harmless in heterozygotes (abnormal variant < 50 > 20 percent). Haemoglobins S, C, D and E would belong to this group. (3) Harmful in heterozygotes (abnormal variant < 20 percent). To this would belong the haemoglobins which are associated with haemolysis even in the heterozygote, such as haemoglobin Köln[20].

The correlation between the morbidity of a haemoglobin and its molecular structure might throw an important light on the function of haemoglobin as a whole.

We thank Mr. D. Irvine for valuable assistance, also Dr. G. Braunitzer and his colleagues of the Max Planck Institute für Biochemie, Munich, Germany, for advice and discussion.

(**203**, 363-365; 1964)

H. Lehmann and D. Beale: Medical Research Council Abnormal Haemoglobin Research Unit, University Department of Biochemistry, Cambridge.

F. S. Boi-Doku: Central Clinical Laboratory, Accra.

A + S 的情况下，机体可以产生 A∶S 比例范围很广的细胞——有些是 A < S，有些是 A > S。前一种细胞被迅速清除，结果 A 与 S 的最终比例就大于 1。

在血红蛋白 A + G_{阿克拉}杂合子这种情况下，两种血红蛋白的比例恰好是 1。我们通过多次研究证实了这一点，因为在碱性 pH 条件下的常规纸电泳中，前带（A）大约占总血红蛋白的 53%。当 50% 的 A 和 50% 的 G 人为混合后，在碱性 pH 条件下进行电泳也得到了同样的比例。另外，当对 A + G 杂合子的血红蛋白在 pH 6.5 条件下进行开放界面电泳时 (参考文献 7)，G 部分是前沿组分，G 部分占血红蛋白总量的 50%~55%。

在杂合子中，A 和 G 的比例相等可能表明：G 的比例比 A 高的细胞并没有被清除，所有细胞，不管是 A < G 还是 A > G，都同等地存活下来。

可以预测，在杂合子的溶血产物中 A 组分和异常变异体具有相同比例的其他血红蛋白，就像血红蛋白 G_{阿克拉}一样是非致病的。因此，一共可能有以下三类血红蛋白：（1）纯合子和杂合子（杂合子有 50% 的异常变异体）均是非致病的：如血红蛋白 G_{阿克拉}。（2）纯合子是致病的而杂合子（20% < 异常变异体 < 50%）是非致病的。血红蛋白 S、C、D 和 E 属于这种类型。（3）杂合子（异常变异体 < 20%）是致病的。即使在杂合子中与溶血相关的血红蛋白也属于这一类，如科隆血红蛋白 [20]。

对于血红蛋白的致病性及其分子结构的相互关系的阐述也许会对血红蛋白整体功能的研究具有重要的启示。

我们感谢欧文先生的大力协助，也感谢布劳尼策尔博士和他所在的德国慕尼黑马克斯·普朗克生物化学研究所的同事们的建议和讨论。

（荆玉祥 翻译；杨茂君 审稿）

References:

1. Edington, G. M., and Lehmann, H., *Lancet*, ii, 173 (1954).

2. Gammack, D. B., Huehns, E. R., Lehmann, H., and Shooter, E. M., *Acta Genet. Stalist. Meà.*, **11**, 1 (1961).

3. Schwartz, I. R., Atwater, J., and Tocantins, L. M., *Blood*, **15**, 901 (1960).

4. Swenson, R. T., Hill, R. L., Lehmann, H., and Jim, R. T. S., *J. Biol. Chem.*, **237**, 1517 (1962).

5. Huntsman, R. G., Hall, M., Lehmann, H., and Sukumaran, P. K., *Brit. Med. J.*, **1**, 720 (1963).

6. Hill, R. L., Swenson, R. T., and Schwartz, H. C., *J. Biol. Chem.*, **235**, 3182 (1960).

7. Edington, G. M., Lehmann, H., and Schneider, R. G., *Nature*, **175**, 850 (1955).

8. Ingram, V. M., *Biochim. Biophys. Acta*, **28**, 539 (1958).

9. Baglioni, C., *Biochim. Biophys. Acta*, **48**, 392 (1961).

10. Gerald, P. S., and Ingram, V. M., *J. Biol. Chem.*, **236**, 2155 (1961).

11. Whitehead, J. K., and Beale, D., *Clin. Chim. Acta*, **4**, 710 (1959).

12. Braunitzer, G., and Rudloff, V., *Deut. Med. Wochschr.*, **87**, 959 (1962).

13. Goldstein, J., Konigsberg, W., and Hill, R. J., *J. Biol. Chem.*, **238**, 2016 (1963).

14. Sanger, F., *Biochem. J.*, **39**, 507 (1945).

15. Jepson, J. B., and Smith, I., *Nature*, **172**, 1100 (1953).

16. Spackman, D. H., Stein, W. H., and Moore, S., *Anal. Chem.*, **30**, 1190 (1958).

17. Ambler, R. P., *Biochem. J.*, **89**, 349 (1963).

18. Itano, H. A., in *Abnormal Haemoglobins*, edit. by Jonxis, J. H. P., and Delatresnaye, J. F., 1 (Blackwell Sci. Publ., Oxford, 1959).

19. Levere, R. D., Lichtman, H. C., and Levine, J., *Nature*, **202**, 499 (1964).

20. Pribilla, W., in *Haemoglobin Colloquium*, edit. by Lehmann, H., and Betke, K., 73 (Georg Thieme Verlag, Stuttgart, 1962).

Homo *"habilis"* and the Australopithecines

J. T. Robinson

Editor's Note

This trenchant critique of *Homo habilis*, the new Olduvai hominid species proposed by Leakey and colleagues in 1964, exposes many problems with Leakey's conception of "handy man" that still remain unresolved. As well as finding problems with Leakey's casual use of nomenclature— a highly technical pursuit—the author John T. Robinson suggests that *Homo habilis* from Bed I at Olduvai looked much more like *Australopithecus africanus* than *Homo*—and that the stratigraphically later finds from Bed II might be better subsumed within *Homo erectus*. The problem of inadequate definition has dogged the study of *Homo habilis* to this day. The material Leakey assigned to this species is indeed highly heterogeneous, some closely resembling australopithecine material, other specimens more akin to *Homo*.

LEAKEY, Tobias and Napier have given a preliminary account of specimens which in their opinion belong to a hitherto unknown species of the genus *Homo*[1]. They re-defined the latter genus and proposed the species name *H. habilis*, with definition, for the new taxon which includes specimens discovered between late 1960 and late 1963.

Diagnosis of the New Taxon

Two critically important functions of the original description of a new taxon are: (*a*) to demonstrate that the population sample under discussion represents a taxon which is different from all recognized taxa and is therefore indeed new and in need of a name; (*b*) to place on record the new name selected for the taxon.

In order that the first of these functions be satisfied a differential diagnosis is necessary. This brings out the points of difference between the specimens comprising the new taxon and all closely related specimens which have been referred to already existing taxa. From this point of view the original description of *H. "habilis"* by Leakey *et al.* is unsatisfactory. Doubtless the authors intend to present further evidence elsewhere at a later date, but clearly the appropriate place is in the original description (in the technical taxonomic sense) so that others may judge the validity of the proposed taxon.

The proposed new definition of the genus *Homo* depends in part on the validity of the new species proposed for it. The new species is defined briefly as though proof had previously been presented that it is a new taxon—but in fact there is no discussion of "evidence" which clearly establishes that this is the case. The definition does include a few very broad diagnostic comparisons, for example, "... premolars which are narrower (in bucco-lingual breadth) than those of *Australopithecus* ...", but no evidence is presented to support them.

282

"能人"和南方古猿

鲁滨逊

编者按

1964年，利基教授及其同事们提出了一种新的奥杜威原始人种——能人。对能人的尖锐批判揭露出利基的"灵巧的人"概念中存在许多问题，这些问题仍未得到解决。此外还发现利基的临时命名在更高的学术层面上存在问题，本文作者约翰·鲁滨逊认为来自奥杜威第I层的能人看起来较之于人属更像南方古猿非洲种——而将后来从第II层中发现的标本归为直立人则更合适。定义不恰当的问题至今仍伴随在对能人的研究中。被利基归为这一物种的标本确实存在不同的种类，一些标本非常类似于南方古猿亚科，其他一些标本则更类似于人属。

利基、托拜厄斯及内皮尔已经对标本进行了初步描述，依照他们的观点，这些标本属于一种迄今为止还未知的人属种类[1]。他们重新定义了这个属，并且建议将其种名命名为能人，根据定义这个新种包括了1960年末至1963年末所发现的标本。

新分类单元的鉴别分析

一个新分类单元的原始描述有两个关键的重要作用：(a) 论证所研究的种群样本代表的分类单元不同于所有已知的分类单元，因而确实是新的并且需要一个名称；(b) 将这个分类单元选用的新名称记录在案。

为了实现其中第一个作用，需要进行鉴别分析。通过鉴别分析总结出属于新分类单元的标本与所有密切相关的属于已存在的分类单元的标本之间的差异点。从这个观点来看，利基等人所给出的"能人"的原始描述并不令人满意。作者无疑打算以后在其他地方给出更进一步的证据，但是显然这个工作最应该体现在原始描述中（在分类学的专业意义上），这样别人可以判断其所提出的分类单元的有效性。

所提出的人属的新定义在某种程度上取决于引出该定义的这一新种的有效性。这一新种定义得很简略，就好像之前已经有证据表明它是一种新的分类单元似的——但是事实上根本没有讨论出能够明确证实情况确实如此的"证据"。定义确实包括了少量而又非常宽泛的鉴别比较，例如，"……前臼齿要比南方古猿狭小（颊舌径）……"，但是没有给出证据来支持这些观点。

Furthermore, although reference is made to some previously known specimens with which the authors think their new species may be conspecific, no comparisons with the new material are actually included in their original description. In fact the following very curious statement is made: "The specimens originally described by Broom and Robinson as *Telanthropus capensis* and which were later transferred by Robinson to *Homo erectus* may well prove, on closer comparative investigation, to belong to *Homo habilis*." This statement clearly implies that the authors described a new species even though they believe that proper comparative investigation may well show that it is conspecific with specimens of which they are aware, the originals of which two of the authors have seen more than once and of which detailed descriptions exist in the literature. It is of interest also to note that the authors think that if the conspecificity be established in this case, then the material which was described more than a decade ago would belong to their new species rather than the other way about—as required by the *International Code of Zoological Nomenclature*.

One may thus conclude that: (*a*) the authors have not demonstrated that their proposed new species is in fact new; (*b*) the authors themselves are in doubt whether it is new; (*c*) since the validity of the new species has not been established, that of the new definition of *Homo* has not been established either. (In this general connexion see also the recent comment by Oakley and Campbell[2].)

Validity of the Proposed New Species

Although the authors do not present evidence in support of the few suggested differentially diagnostic features separating the proposed new species from the australopithecines, it is worth enquiring into the validity of these features.

Evidently a character which has impressed the authors as significant[1,3] is "a marked tendency toward the buccolingual narrowing and mesiodistal elongation of all the teeth, which is specially evident in the lower premolars ... and in the lower molars ... ". In particular it is held that the lower premolars are narrower than those of *Australopithecus* but fall within the range of *Homo erectus*. Through the courtesy of Dr. and Mrs. Leakey I was able to make a fairly detailed investigation, in 1961, of the mandible which has now been designated the holotype of *H. "habilis"*. My measurements of P_3, made in the same manner as those on the South African australopithecines used for comparison, gave the following mesiodistal lengths and buccolingual breadths in millimetres: left, 9.6×9.4 and right, 9.6×9.9. These figures yield length/breadth indices $(L \times 100)/B$ of 102.2 and 98.0 respectively. Leakey[3] lists these dimensions as 11.0×9.5 in both cases and the index as 115.8. However, as comparison is being made with measurements made by me on the South African australopithecine material, my measurements on the Olduvai specimen should yield a better comparison since all the measurements involved were made by the same person using the same technique.

The table compares the mean length/breadth index and the observed range of the index for P_3 M_1 and M_2 of *Paranthropus*, *Australopithecus* and Pekin Man (*Homo erectus*) with the corresponding values for the type mandible of *H. "habilis"*. P_3 of the latter falls within the

而且，尽管作者参考了一些之前已知的标本并由此认为他们的新种可能都属于同一物种，但是实际上在他们的原始描述中没有包括与新材料的对比。事实上他们得出了以下十分古怪的论点："通过进一步的比较研究，可以充分证明最初由布鲁姆与鲁滨逊记述为开普远人、后来由鲁滨逊转而归入直立人的标本属于能人。"这个陈述明显暗示着作者描述了一个新种，尽管他们意识到适当的比较研究能够充分显示出它与他们知道的标本属于同一物种，而这些原始标本两位作者见过不止一次，并且在文献中也有详细的描述。同样有意思的是我们注意到作者认为，如果在此情况下确定了其同种性，那么十多年前描述的材料应该归属于他们的新种而非其他情形——遵照《国际动物命名法规》的要求。

因而可能得出下列结论：(a) 作者没有证明他们提出的新种确实是全新的；(b) 连作者他们自己都怀疑它是否是全新的；(c) 既然新种的有效性还不确定，人属新定义的有效性也就尚未确定。(有关这方面还可以参见奥克利与坎贝尔 [2] 的最近评论。)

建立新种的有效性

尽管作者没有给出证据来支持他们所提出的少量能区分新种与南方古猿亚科的不同的鉴别特征，但是这些特征的有效性仍值得探究。

很显然，给作者留下深刻印象的一个特征 [1,3] 是"所有牙齿都具有颊舌向变窄和近中远中向延长的显著趋势，这一趋势在下前臼齿……和下臼齿……尤其明显"。作者特别坚持其下前臼齿比南方古猿的下前臼齿窄，但是却认为其落在直立人的范围之内。承蒙利基博士与利基夫人的允许，1961 年我得以对现在被认为是"能人"正型标本的下颌骨进行了相当细致的观察研究。我测量了第三下前臼齿，测量方法与测量南非南方古猿亚科的方法一致以便用于比较，得出近中远中径与颊舌径如下（单位：毫米）：左，9.6×9.4；右，9.6×9.9。由这些数据得出：齿冠指数 $(L \times 100)/B$ 分别为 102.2 与 98.0。利基 [3] 列出的测量数据两边一样，均为 11.0×9.5，而系数为 115.8。然而，由于他在比较时用了我的南非南方古猿亚科材料测量数据，因此我对奥杜威标本测量数据应该得到更好的比较结果，因为全部测量是由同一个人用同样的方法进行的。

表 1 将傍人、南方古猿和北京人 (直立人) 的第三下前臼齿、第一下臼齿及第二下臼齿的齿冠指数的平均值及系数的观测范围与"能人"正型下颌骨的对应值进

observed range of that for *Paranthropus* and not far outside that for Pekin Man. The fit is slightly less close with *Australopithecus*; but the sample for this form is made up of only four specimens representing three individuals. The highest value listed for *Paranthropus* belongs to a tooth which had not yet begun to erupt and the final dimensions would possibly have been slightly higher than at present, although the length dimension already is very close to the upper limit for the sample of 12 specimens. However, since we are here concerned with proportions it is legitimate to include the value for this specimen as the crown is intact and undamaged and it is highly improbable that any slight increase in the crown dimensions that may have occurred before eruption would alter the shape so disproportionately as to change the shape index significantly.

Table 1. Length/Breadth indices

	P_3			M_1			M_2		
	N	Mean	Range	N	Mean	Range	N	Mean	Range
Australopithecus	4	83.0	77.0–88.5	8	107.8	100.8–117.0	9	107.7	98.0–111.3
Paranthropus	14	85.2	76.0–112.3	19	106.2	100.7–110.8	12	108.4	101.4–117.3
H. "habilis" type	2	100.1	98.0–102.2	2	117.2	117.2	2	114.3	114.3
Pekin Man	13	86.3	75.0–96.8	11	106.1	96.8–112.0	7	104.4	98.3–115.4

In the case of M_1, the *H. "habilis"* value falls slightly outside the ranges for *Paranthropus* and Pekin Man, but barely outside that for *Australopithecus*. Leakey[3] mistakenly quotes 114.6 as the highest observed and published value for *Australopithecus* (*sensu stricto*) and concludes that his value for the type mandible (117.0) falls outside that for *Australopithecus*. In the case of M_2 the *H. "habilis"* value falls within the observed ranges for both *Paranthropus* and Pekin Man and very slightly outside that for *Australopithecus*. In both cases the *Australopithecus* sample size is small.

In one section of the original description Tobias refers to a lower molar, of Hominid 4, which he believes to be either M_2 or M_3, and writes: "The molar is 15.1 mm in mesiodistal length and 13.0 mm in buccolingual breadth; it is thus a small and narrow tooth by australopithecinae standards ...". The length/breadth index of this tooth is 116.2; the observed range for *Paranthropus* is 101.4–117.3 for M_2 and 106.6–124.1 for M_3—indeed, the "mean" index for the sample of 13 specimens for the latter tooth is 116.3. The ranges for this index for M_2 and M_3 of *Australopithecus* are 98.0–11.3 and 100.0–116.0 ($N = 10$) respectively.

The size can best be compared by using the module so that single values can be compared instead of pairs. The module for the aforementioned Olduvai tooth is 14.05. This certainly is smaller than are either M_2 or M_3 of *Paranthropus*. However, the lowest modules for the small collection of these two tooth of *Australopithecus* are, respectively, 13.75 and 13.20. These figures thus show that the Olduvai hominid 4 tooth is neither exceptionally small nor exceptionally narrow by australopithecine standards.

行比较。后者的第三下前臼齿落入傍人第三下前臼齿的观测范围，并且在北京人第三下前臼齿的范围外不远。尺寸与南方古猿不是非常吻合；但其样本仅由代表 3 个个体的 4 个标本构成。列出的傍人的最高值是还没有萌出的牙齿，其最终尺寸可能将比现在的值稍高一点，尽管其长度尺寸已经十分接近 12 个标本的上限。然而，既然我们在此涉及比例，那么把这个标本的值包括进来是合理的，因为牙冠完整无损，并且齿冠尺寸任何微小的增加（这在萌出之前可能已经发生）肯定不可能如此不成比例地改变其形状以至于显著地改变形状系数。

表1. 齿冠指数

	第三下前臼齿			第一下前臼齿			第二下前臼齿		
	标本数	平均值	范围	标本数	平均值	范围	标本数	平均值	范围
南方古猿	4	83.0	77.0～88.5	8	107.8	100.8～117.0	9	107.7	98.0～111.3
傍人	14	85.2	76.0~112.8	19	106.2	100.7~110.8	12	108.4	101.4~117.3
典型"能人"	2	100.1	98.0~102.2	2	117.2	117.2	2	114.3	114.3
北京人	13	86.3	75.0~96.8	11	106.1	96.8~112.0	7	104.4	98.3~115.4

就第一下臼齿而言，"能人"的值落在傍人与北京人范围外不远处，但是勉强落在南方古猿范围之外。利基[3]错误地引用 114.6 作为南方古猿（从狭义说）所公布的最高观测值，并得出结论认为其正型下颌骨的值 (117.0) 落在南方古猿之外。就第二下臼齿而言，"能人"的值落在傍人与北京人二者的观测范围之内，并且落在南方古猿第二下臼齿的值的范围外不远处。在这两种情形中，南方古猿的样本都比较小。

在一部分原始描述中，托拜厄斯提到人科 4 的一颗下臼齿，他认为它不是第二下臼齿就是第三下臼齿，他写道："臼齿近中远中径为 15.1 毫米，颊舌径为 13.0 毫米；因此按南方古猿亚科的标准，它是一颗又小又窄的牙齿……"这颗牙齿的齿冠指数为 116.2；傍人第二下臼齿观测值的范围是 101.4~117.3，第三下臼齿为 106.6~124.1——确实，对后者的牙齿，13 个标本样本的系数"平均值"是 116.3。对南方古猿的第二下臼齿与第三下臼齿而言，这个参数的范围分别是 98.0~113.3 与 100.0~116.0 (N =10)。

最好用模数比较大小，以便能用单个值代替成对值来进行比较。前述奥杜威牙齿的模数为 14.05。这个值一定比傍人的第二下臼齿或第三下臼齿小。然而，收集数量很少的南方古猿的这两颗牙齿的最低模数分别是 13.75 与 13.20。因此，按南方古猿亚科的标准，这些数字表明奥杜威人科 4 的牙齿既不是特别小也不是特别窄。

This evidence, along with that from the type specimen, therefore does not support the contention that the shape characteristics of the *H. "habilis"* mandibular teeth are recognizably different from those of the australopithecines. The *H. "habilis"* specimens here considered have values which fall mostly toward the upper end of the ranges of variation at present available for the australopithecines. But it is very clear that in no case are adequate samples available: the ranges of variation are certainly smaller than they should be and insufficient is known about *H. "habilis"* to know how representative the few known specimens are of the populations from which they came.

However, a far more trenchant criticism of the use made of the dental length/breadth index by Leakey *et al.* is that analysis of the index shows that it and the features on which it is based have extremely low phyletic valence so far as hominids are concerned. This is readily apparent from the extremely wide overlap of the ranges for this index for *Australopithecus*, *Paranthropus*, *Homo erectus* and *Homo sapiens*. It is not possible to distinguish taxonomically between these groups by means of this index if anything like adequate sample sizes are used. As may be seen from Table 1 the means for *Paranthropus* and Pekin Man are 85.2 and 86.3 respectively for P_3 and (not shown in Table 1) values for samples of modern Bantu and aboriginal Australians differ from that of Pekin Man by a few tenths of a unit. Thus, populations covering so great a span of the hominids have means for this index which are very closely similar. On the other hand, various local populations of modern man have mean values for this index which actually show greater variation than the foregoing. That is to say, the intra-species variation in the mean in modern man is actually greater than the intergeneric differences in the mean for three populations of *Australopithecus*, *Paranthropus* and *Homo erectus*. Furthermore, these differences between the means are small compared with the range of variation observed within any one of the populations concerned.

The endocranial capacity of *H. "habilis"* appears, on the scanty and indirect evidence available[8,1], to have differed little from that of the australopithecines with a range overlapping that of the latter substantially.

The conclusion that the foot of the new form has a fairly advanced and *Homo*-like structure while the hands appear to have been relatively more primitive does not help a great deal at present since not much evidence of these parts is available for either australopithecines or *H. erectus*. However, neither conclusion should occasion astonishment. If the australopithecines were erectly bipedal in posture and locomotor habit, as much evidence suggests, then the foot was being used in an essentially human fashion and is likely to have achieved a relatively advanced structure soon after the new locomotor habit was achieved. On the other hand, neither the australopithecines, as apparent tool-users, nor *H. "habilis"* as an apparent primitive tool-maker, were culturally advanced and therefore one might expect that the moulding of the hand under the influence of manipulative activity of the human sort was not far advanced. Thus one might expect the foot to have had a more modern-looking structure than the hand in both the australopithecine stage and the early hominine stage. The apparently fairly advanced foot *H. "habilis"* does not

因而，这个证据（连同从正型标本得到的证据一起）不支持以下论点：即"能人"与南方古猿的下颌齿的形状特征存在可被识别的差异。这里所研究的"能人"标本的值大部分落在现在获得的南方古猿亚科观测值变化幅度的上限。但是很清楚的是，那绝不是充足有效的样本：其变化幅度肯定小于真实值，并且对于"能人"的认识还远远不够，仅仅凭少数几个已知标本并不可知在多大程度上可以代表其来自的群体。

然而就人科而言，对利基等人使用牙齿齿冠指数的做法的更尖锐的批判认为，系数分析显示出系数及由此而来的特点的分类价值极低。从南方古猿、傍人、直立人以及智人的这一指数范围有很大重叠来看，这是显而易见的。即使用到的样本足够大，靠系数平均值不可能在分类学上对这些种类加以区别。正如从表 1 所示，傍人与北京人的第三下前臼齿的平均值分别是 85.2 与 86.3，现代班图人与澳大利亚土著居民样本的值（表 1 中没有列出）与北京人只相差一个单位的十分之几。可见，涵盖如此大范围的人科种群的系数平均值十分接近。另一方面，现代人类不同地区种群的系数平均值实际上表现出比前述更大的变化。也就是说，现代人类种内平均值的差异实际上比南方古猿、傍人以及直立人三个种群平均值的属间差异大。而且，与任何一个种群内观测到的变化幅度相比，平均值间的差异是很小的。

根据可用的少量间接证据[8,1]，"能人"的颅容量似乎与南方古猿的颅容量差异很小，其范围与后者有很大重叠。

这种新型人科的足具有类似人属的结构，进化程度颇高，然而手似乎相对更加原始。这个结论现在没有起到太大帮助，因为对于南方古猿亚科或直立人，这部分都没有太多的证据可用。然而，这个结论也不应引起如此震惊。大量证据表明，如果南方古猿亚科在体态与运动习惯上为两足直立行走，那么它本质上以人类的方式使用足，并且有可能在养成新的运动习惯之后不久获得了进化程度相对较高的结构。另一方面，南方古猿亚科（明显的工具使用者）和"能人"（明显的原始工具制造者）在文明程度上都不够先进，因此，人们可能认为在诸如这种人类操作活动的影响下手的结构的进化程度并不很高。于是有可能认为，在南方古猿亚科阶段与早期人科阶段，足已经具有比手更加类似于现代人的结构。因此"能人"明显相当进化的足不一定表示他与人属而非南方古猿属有亲缘关系——确实，大步行走的步态似乎与

therefore necessarily indicate affinity with *Homo* rather than *Australopithecus*—indeed, the fact that a striding gait appears to be inconsistent with the morphology of the *H. "habilis"* foot[9] suggests that it is unlikely to have been significantly more advanced than that of *Australopithecus*.

In view of the foregoing it seems to me that Leakey *et al.*, in their original description of *H. "habilis"*, have by no means provided a reasonable case for establishing a new species of *Homo* to accept the recently discovered Olduvai specimens. Furthermore, some of the distinguishing criteria used do not appear to be valid for the purpose in the light of available knowledge of early hominids.

Affinities of *H. "habilis"*

In assessing the material attributed to *H. "habilis"* it must be remembered that two groups of specimens are involved: one from Bed I and the other from Bed II. On the available dating evidence these two groups are separated by a significant time gap.

The morphological characteristics of the two groups are not the same. This is well shown, for example, by comparing the type mandible from *FLKNN* I with the mandible from *MNK* II (see Leakey and Leakey[5], Fig. 3). I have elsewhere[6,7] discussed the shape of the internal mandibular contour, seen in occlusal view, and its narrow V shape in both types of australopithecine but its relatively wide U shape in "Telanthropus"and all other forms of *Homo*. The type *H. "habilis"* mandible is damaged near the symphyseal region so that the partial right side has been displaced toward the left. However, the midline can be determined within very narrow limits and the internal contour of the left half is intact and undisturbed from very near the symphysis to a point behind M_2. It is therefore simple to reflect this contour on to the right side in a graphic reconstruction in order to determine the correct original position of the displaced right half. Carrying out this procedure shows that the type mandible had a typically australopithecine internal mandibular contour with the corpus thickness in the premolar region greater than the distance between the two halves of the mandible in that region. This is actually readily apparent on visual inspection since the corpus mandibulae is relatively thick compared with the breadth of the crowns of the teeth and much of this breadth is mesial to the teeth in the premolar region. In contrast the mandible from *MNK* II does not have this narrow V-shaped contour but has the wide U-shaped contour and relatively thin corpus of the sort normal for *Homo*. The more recent, Bed II mandible thus agrees with *Homo erectus*, including "Telanthropus", in this feature as well as its generally greater gracility, while the older Bed I mandible falls within the observed range of australopithecines in both respects. It is therefore by no means clear that the Bed I and Bed II groups of specimens necessarily belong to the same species.

The teeth of the type mandible show that the australopithecine affinities of this mandible are very clearly with *Australopithecus* and not with *Paranthropus*. The latter is characterized among other things, as witnessed by the South African, East African and Indonesian evidence, by a small canine as compared with the size of the premolars, especially in the

"能人"的足的形态学不一致 [9]，这个事实表明，他的足不太可能比南方古猿属的进化程度显著更高。

根据前述在我看来，似乎利基等人在对"能人"的原始描述中，对于最近发现的奥杜威标本的归属，并没有为建立一个人属新种提供一种合理的情形。而且，根据早期人科可用的知识，其采用的一些区别标准似乎并非有效。

"能人"的亲缘关系

在评价归为"能人"的材料时，一定要记得涉及两组标本：一组出自第 I 层，另一组出自第 II 层。根据现有的测年证据,这两组标本之间跨越一个很大的时间间隔。

两组标本的形态学特征是不一样的。举例来说，通过比较来自 *FLKNN* I 的正型标本下颌骨与来自 *MNK* II 的下颌骨 (参见路易斯·利基与玛丽·利基 [5]，图 3)，可以充分显示这一点。我已经在其他地方 [6,7] 讨论过在上下齿咬合面视图中所见下颌骨内部轮廓的形状，以及它在两类南方古猿亚科中狭窄的 V 字形状和在"远人"及所有其他人属类型中相对较宽的 U 字形状。"能人"下颌骨靠近联合部的位置被损坏，所以右边的一部分错位至左边。然而，我们能在很窄的范围内确定正中线，从十分靠近下颌联合处到第二下臼齿后面的尖端左半部分的内部轮廓都完整无缺并且没有变形。因此，通过绘图重构可以很容易地反映出右边的这个轮廓，从而确定错位的右半部分正确的原始位置。这个步骤的实现显示出正型标本下颌骨具有典型的南方古猿亚科下颌骨内部轮廓，其前白齿区的骨体厚度大于那个区下颌骨两半部分之间的距离。实际上通过目测都能发现这是相当明显的，因为与齿冠宽度相比下颌骨体相对较厚，对于前白齿区的牙齿，这个宽度的大部分在中央。相反，来自 *MNK* II 的下颌骨不具有这个窄 V 形轮廓，而具有宽 U 形轮廓，以及诸如人属正常种类的相对较薄的骨体。于是进一步而言，在这一特征及其通常更为纤细方面，第 II 层的下颌骨与包括"远人"在内的直立人的下颌骨一致，而较老的第 I 层的下颌骨在两方面都落在南方古猿亚科的观测范围之内。因此完全不能确定第 I 层与第 II 层的两组标本一定属于同一种类。

正型下颌骨的牙齿十分清楚地表明这个下颌骨与南方古猿而非傍人有亲缘关系。就如南非、东非以及印度尼西亚的证据所证实的那样，傍人的特点是犬齿与前白齿（特别是在下颌骨中）相比，尺寸较小。确实，这个类型的下犬齿如此小，以至于它

mandible. Indeed, the lower canines of this form are so small that they fall within the observed size range for modern living man while the premolars are the largest known among fossil and modern hominids. *Australopithecus*, on the other hand, has relatively large lower canines, *Sts* 3 from Sterkfontein being one of the largest hominid mandibular canines known, but the premolars are smaller than those of *Paranthropus*. Thus the relative size of canine to P_3 in the mandible is very different in the two australopithecines and any individual mandible can be assigned to the correct genus without hesitation on visual inspection if these teeth are present. The *H. "habilis"* canine to P_3 ratio is like that of *Australopithecus* and quite unlike that of *Paranthropus*. This fact is further supported by greater morphological resemblance in general between the teeth of the Olduvai specimen and those of *Australopithecus* as compared with *Paranthropus*.

On the other hand, the Bed II mandible shows much greater resemblance to the "Telanthropus" mandibles, but the latter can easily be distinguished from the Bed I type mandible.

In terms of the available evidence it would seem that there is more reason for associating the Bed I group of specimens with *Australopithecus* and the Bed II group with *Homo erectus* than there is for associating the Bed I and II groups with each other. This would therefore seem a perfectly reasonable course to adopt: placing the Bed I material as advanced representatives of *Australopithecus africanus* and the Bed II group as somewhat early members of *Homo erectus*.

However, the Bed I and II groups of specimens occurred in the same geographical area, both appear to have been tool-makers and there seems no obvious morphological reasons why the earlier group could not have been ancestral to the later group. Furthermore, it seems unlikely in terms of ecology and behaviour that two morphologically similar groups, both adapting at least to a significant extent by cultural means, would develop in the same general geographic area. Clearly at least two hominid lines did in fact exist simultaneously in this region: the material under discussion and *Paranthropus* (= "Zinjanthropus"). But in this case the *Paranthropus* line consisted of forms which differed markedly in morphology and evidently also in ecology and behaviour not only from the *H. "habilis"* material but also all other known hominids. If the Bed I and Bed II groups represented two different lines, this would indicate not only that they would be adaptively similar but also that three different hominid lineages existed in the same area. For these reasons it seems probable that they actually do represent the same lineage at two different time levels. If this is so, then it is reasonable to place them in the same species. On the other hand, as already seen, morphological considerations favour their being placed in two different taxa which already exist, in which case they belong to two different genera. The evidence in favour of the latter course is actually the stronger.

The two interpretations do not have to be mutually exclusive: the Bed I material may represent an advanced form of *Australopithecus* and Bed II specimens an early *H. erectus* and at the same time the latter may be a lineal descendant of the former. This seems to

292

们落在现生人类的尺寸观测范围之内，而在化石人科与现代人科中前臼齿是已知最大的。另一方面，南方古猿的下犬齿相当大，出自斯泰克方丹的 *Sts 3* 是已知最大的人科下颌骨犬齿之一，但是其前臼齿小于傍人的前臼齿。因此，两个南方古猿的犬齿与下颌骨上的第三下前臼齿相比大小十分不同，如果展示这些牙齿，通过目测就能毫不犹豫地给任一单个下颌骨指定出正确的属。"能人"的犬齿对第三下前臼齿的比率与南方古猿的类似，而与傍人的很不相似。与傍人相比，总体来说奥杜威标本的牙齿与南方古猿标本的牙齿之间在形态学上更为相似，这也进一步支持了这个事实。

另一方面，第 II 层的下颌骨表现出与"远人"下颌骨有更多类同之处，而后者可以很轻易地与第 I 层的正型标本下颌骨区别开。

根据可用的证据，似乎更有理由把第 I 层的标本组与南方古猿联系起来，把地层 II 的标本组与直立人联系起来，而不是把第 I 层与第 II 层两组互相联系起来。因此采用以下做法似乎是完全合理的：把第 I 层的材料作为南方古猿非洲种的高等代表，第 II 层的组作为直立人稍微早期的成员。

然而，第 I 层与第 II 层的两组标本出现在相同的地理区域，两者似乎都已经能够制造工具，而且似乎没有明显的形态学的原因能够解释为什么较早的类型不能是其后类型的祖先。而且，形态学上相似的两个类型至少都通过文化群落进行了很大程度的适应性改变，根据生态学与行为学，它们似乎不可能会在同一个普通地理区域发展。事实上很明显地，在这个区域同时存在至少两种人科类型：即所讨论的材料和傍人（＝"东非人"）。 但是在这种情形下，傍人种类包括与"能人"及所有其他已知的人科在形态学上明显不同，在生态学与行为学上也明显不同。如果第 I 层与第 II 层的两个组代表两个不同的种类，这将不仅表明它们有相似的适应性，而且表明三个不同的人科谱系存在于同一地区。基于这些原因，实际上它们有可能代表着不同时间层次的同一个谱系。如果真是这样，那么有理由把它们放入同一种类。另一方面，正如已经看到的，形态学上的因素有利于将它们归为已经存在的两个不同的分类单元，在这种情形下它们属于两个不同的属。实际上支持后一种做法的证据更加有力。

两种解释并不一定互斥：第 I 层的材料可能代表南方古猿的一种高等类型，而第 II 层的标本可能代表一种早期直立人，同时后者可能是前者的一直系后代。到目前为止，对我来说这似乎是最可能的解释。按照这个假说，第 I 层的标本代表了南

me to be by far the most probable interpretation. According to this hypothesis the Bed I specimens represent a transitional stage between *Australopithecus* and *H. erectus* just at that stage where the essentially tool-using stage of the former was giving way to the primarily tool-making condition of the latter. The widely held belief of recent years that *Paranthropus* at Olduvai was responsible for the stone implements found associated with it never did seem probable to me[7] in the light of the available evidence from the Sterkfontein Valley. However, it seems far more probable that the *H. "habilis"* material from Bed I represents the remains of the maker of the stone industry from that level.

If the interpretation suggested here is correct, then clearly no new species name is needed; the situation is simply one common in palaeontology when specimens are found which link two already existing taxa. Creating a new taxon here is no solution; the two taxa between which the new one falls are already so similar that insufficient morphological distance exists between them to justify the insertion of another species. As is well recognized, conventional Linnean taxonomy is not suited to dealing with a problem such as this and, if the hypothesis is correct, whatever solution is adopted must be a compromise of some sort. The more conservative approach would be simply to place the Bed I material in the specie *Australopithecus africanus* and the Bed II specimens in *H. erectus*. However, if it is a fact that the Bed I specimens were already primitive tool-makers and were ancestors of the Bed II material, the implication that the ends of a transitional sequence, which involved relatively little morphological and ecological change and did not occupy a geologically long period of time, should be in different genera seems very unsatisfactory, especially as the transition was gradual and did not involve a threshold followed by rapid re-adaptation.

A reasonable way of overcoming this difficulty would be to extend the genus *Homo* to include not only the new Olduvai material but also that at present in the genus *Australopithecus* (as distinct from *Paranthropus*). This genus would then include the whole sequence from the point where a shift to an omnivorous diet (by the inclusion of a substantial degree of carnivorousness) caused a new set of selection pressures to come into play favouring the whole complex of culture as a means of adaptation and thus caused the emergence of culture-bearing man. This suggestion comes very close to one made a long time ago by Mayr[10], but differs from it in not including *Paranthropus* since there is good evidence to indicate that the basic adaptation of the latter was quite different from that of the *Homo* lineage as defined here, a lineage which was separate from that of *Paranthropus* from at least early Pleistocene times as the evidence now clearly indicates. This difference is most conveniently indicated by a generic distinction. The known hominids would thus fall into the two genera *Paranthropus* and *Homo*.

If this were done, then it would seem to be useful to modify the species division at the same time so that the re-defined genus *Homo* includes only two species. Since these would belong to the same lineage they obviously could not be sharply defined. The first could be *H. transvaalensis* and would include the tool-using phase of the lineage involving small-brained forms which were primarily tool-using, had relatively poor communication and comparatively simple social structure. The second could be *H. sapiens*, including larger-

方古猿与直立人之间的过渡阶段，正是在那个阶段，由前者原本的使用工具阶段转变至后者最初的制造工具阶段。最近几年人们普遍认为奥杜威的傍人制造了同它一起被发现的石质工具，根据来自斯泰克方丹峡谷的可用证据，在我看来，这似乎从未可能[7]。然而，来自第 I 层的"能人"的材料似乎更有可能代表着当时的石器工业制造者的遗骸。

如果这里提出的解释是正确的，那么显然并不需要新的种名：发现某标本将两个已有分类单元联系在一起的情况在古生物学上很常见。在此创立一个新分类单元不是解决办法；新种介于其间的两个已有种已经是如此相似，以至于它们之间存在的形态学差异不足以在其间插入另一种类。正如大家所普遍认可的，传统的林奈分类学不适用于处理像这样的问题，如果假设正确，无论采用哪一种解决方案，都一定是某种情况的折中。更为保守的方法是简单地将第 I 层的材料归为南方古猿非洲种，而把第 II 层标本归为直立人。然而，如果第 I 层的标本确实已经是原始工具制造者并且是第 II 层标本的祖先，那么过渡期两端的类群，其形态学与生态学的改变相对较小，且占地质时代时间不长，若放在不同的属似乎并不令人满意，尤其当过渡是渐变的且并不涉及需要快速重新适应的开端时。

要克服这个困难，一个合理的方法可能是扩展人属的范围，不仅把新的奥杜威材料包括进来，而且把南方古猿属（性质明显与傍人属不同）的现有材料包括进来。那么这个属将包括从转变到杂食 (包括一定量的食肉) 那一阶段之后的整个部分。这种转变使得一组新的选择压力开始起作用，并促进形成作为一种适应手段的整个文化群落综合体，于是才使得具有文明意义的人类得以出现。这个建议十分接近于很久以前迈尔[10]提出的建议，但是不同点在于它不包括傍人在内，因为有充分的证据显示后者的基本适应变化完全不同于这里所定义的人属谱系的基本适应变化，正如现在证据清楚显示的那样，人属谱系与傍人属谱系至少从更新世早期就分开了。用属的区别来表示这种差异是最为方便的。于是，已知人科将落入傍人属与人属这两个属之内。

如果做到了这些，那么对种类划分的修改将是有益的，同时重新定义的人属就仅包括两个种。既然这些将属于同一个谱系，那么很明显不能将它们定义得过于清晰。第一种可能是德兰士瓦人，包括处于工具使用阶段、脑较小的类型，它们是原始工具使用者，在信息交流方面相对较差且社会结构相对较简单。第二种可能是智人，包括脑较大的工具制造者，它们拥有的信息交流手段有了极大的提高，社会结构相

brained tool-makers who possessed greatly improved means of communication and comparatively complex social structure. These two species would clearly intergrade but by the very nature of the situation in a single lineage it is not possible to have satisfactory division points for taxonomic purposes, once enough material is available, since there is genetic continuity between successive time levels throughout the sequence. In this event the whole lineage in which culture is a very important adaptive mechanism is included in a single genus and the two species defined in terms of two major stages of cultural development.

(**205**, 121-124; 1965)

J. T. Robinson: Departments of Anthropology and Zoology, University of Wisconsin, Madison 6, Wisconsin.

References:

1. Leakey, L. S. B., Tobias, P. V., and Napier, J. R., *Nature*, **202**, 7 (1964).

2. Oakley, K. P., and Campbell, B. G., *Nature*, **202**, 732 (1964).

3. Leakey, L. S. B., *Nature*, **191**, 417 (1961).

4. Robinson, J. T., *Transvaal Mus. Mem.*, No. 9 (Pretoria, 1956).

5. Leakey, L. S. B., and Leakey, M. D., *Nature*, **202**, 5 (1964).

6. Robinson, J. T., *Amer. J. Phys. Anthrop.*, **11**, 445 (1953).

7. Robinson, J. T., *S. Afr. J. Sci.*, **57**, 3 (1961).

8. Tobias, P. V., *Nature*, **202**, 3 (1964).

9. Day, M. H., and Napier, J. R., *Nature*, **201**, 969 (1964).

10. Mayr, E., *Cold Spring Harbor Symp. Quant. Biol.*, **15**, 109 (1950).

对复杂。很明显，一旦获得足够的材料，这两个种类将逐渐合一，但是在单一谱系情况下，不可能存在符合分类学目的的令人满意的划分点，因为贯穿进化过程的连续时间层次之间具有遗传上的连续性。在这个事件中，整个谱系（其中文化是一个十分重要的适应机制）被归为一个单一属和两个种内，这两个种是根据文明演变的两个主要阶段来定义的。

（田晓阳 张立召 翻译；赵凌霞 审稿）

Dimensions and Probability of Life

H. F. Blum

Editor's Note

Harold Blum, a biologist at Princeton University, was unusual in being prepared to bring general thermodynamic considerations to bear on the seemingly intractable question of how life began. Erwin Schrödinger had already pointed out in the 1940s that life, a highly ordered system, seemed to involve the production of negative entropy or "negentropy". Here Blum attempts to explore what thermodynamics has to say about the possible course of evolution, warning that the latter may have many more "degrees of freedom" than can be easily judged from the fragmentary evolutionary record. This could lead to an underestimation of the "probability of life", tending to compel an overly teleological view of life's origin.

LIFE constitutes a very thin layer of matter on the Earth's surface, the biosphere: continuous cyclic change provides for maintenance, replication and other activities of this system; carbon dioxide is reduced by photosynthesis with concurrent formation of high-energy compounds that are then oxidized in the metabolism of living organisms, carbon dioxide being returned to the atmosphere. At present a balance is maintained with virtually no change in biosphere and atmosphere of the total amount of matter taking part in this cycle. The energy for photosynthesis is provided by continuous inflow of radiant energy from the Sun; an equal amount of energy being re-radiated from the Earth, but in the form of a larger number of quanta of lower energy than those received. The net increase in number of quanta may be regarded as a decrease in order with attendant increase in probability and entropy of a Sun–Earth system of which the biosphere system is a part[1]. Thus, increase in order may occur in the biosphere, without contradiction of the second law of thermodynamics, so long as this average increase is less than the average decrease of order in the Sun–Earth system.

If one were to think of the biosphere as an open system continuously replicated in exact quantity and pattern—the same total amount of information being maintained—one would assume that no over-all change in order occurred in this system, the only change in entropy being in the Sun-Earth system. But evolution of the biosphere has been going on for millenia, and this has certainly involved increase in orderly arrangement and decrease in entropy, which has, of course, been compensated by corresponding increase in entropy within the including Sun-Earth system.

We may represent the changes pertaining to the Sun-Earth system by:

$$\Delta S = N\,\mathrm{k}\,\ln\frac{W_2}{W_1} = N\,\mathrm{k}\,\ln\frac{P_2}{P_1} \tag{1}$$

298

生命的维数与可能性

布卢姆

编者按

普林斯顿大学的生物学家哈罗德·布卢姆以不同寻常的方式准备将普通热力学与生命的起源这个看来难以解决的问题联系起来。埃尔温·薛定谔在 20 世纪 40 年代就已经指出，生命这个高度有序的系统似乎包含着产生"负熵"的过程。本文中布卢姆试图去探索热力学对可能的进化过程提供了哪些解释，并且提出进化过程中可能有更多的"自由度"，而并不是仅通过零碎的进化的记载就能够轻易判断的。这可能导致对"生命的可能性"过低估价，进而对生命的起源强加一个过于目的论式的认识。

生命构成了地球表面极薄的一层，即生物圈，持续周期性的交换供给着生物圈的维持、复制和其他活动；光合作用消耗二氧化碳，进而形成高能量的化合物，这些化合物又在生物体的新陈代谢中被氧化，转化为二氧化碳释放到大气中。目前生物圈和大气圈中参加循环的物质总量事实上没有变化，使得平衡得以维持。光合作用的能量是由持续进入的太阳辐射能提供的，等量的能量再从地球辐射出去，但与接收到的相比，辐射出去的能量含有更多能量较低的量子。量子数量的净增长可以被看作有序性的下降，并伴随着日地系统的可能性及熵的增加，而生物圈正是日地系统的一部分 [1]。因此，生物圈中可以出现有序性的增加，只要该平均增长小于日地系统中有序性的平均下降，就不会与热力学第二定律相矛盾。

如果把生物圈看作一个按照精确的数量及类型进行连续复制的开放系统，即该系统的信息总量保持不变，那么就会假设该系统中的有序性没有发生根本的变化，而只是日地系统中的熵发生变化。但是，生物圈的进化已经进行了千百万年，这无疑包括有序性的增加和熵的减少，而熵的减少当然被日地系统中熵的增加所相应补偿。

我们可以用下式来表示日地系统中的变化：

$$\Delta S = N\,k\,\ln\frac{W_2}{W_1} = N\,k\,\ln\frac{P_2}{P_1}$$

(1)

where S is the entropy; N is the number of molecules involved; k is the Boltzmann constant; ln indicates the natural logarithm; W is the number of arrangements of microscopic properties of the system; and P is the probability of that state. The greater the number of arrangements of microscopic properties—energy-levels of electrons, vibrations of atoms in molecules, velocities of molecules, etc.—the less the order, and the greater the probability. Thus, in accord with the second law of thermodynamics $W_2 > W_1$, and $P_2 > P_1$. We cannot hope to measure W or P, but by various means we may assess the relative change in these values for a given system in terms of ΔS, provided the system can be put into dimensions definable in terms of N and k.

In attempting to relate the increase of order in the evolving biosphere to the inverse change in the Sun-Earth system, let us first have analogy to a digital computer, to which successive questions are presented. The computer "answers" each question put to it by choosing one of two possible alternatives. If each question is postulated on the previous one, the selection of answers represents an increase in order with corresponding decrease in probability. We may describe this relationship by:

$$-\Delta s = \log_2 \frac{b_2}{b_1} \tag{2}$$

where $-s$, the negentropy, is used to measure the increase of information; $\log_2$ indicates the logarithm to the base 2; b is the number of questions answered; and $b_2 > b_1$. The number of answers is not concerned with meaningful content—it is not a measure of the knowledge gained from the computer operation. In such an operation interest is usually focused on only the accepted answers, the rejected alternatives being disregarded.

Evolution by natural selection may be compared to computer operation by assuming that each mutation which better adapts the species to the environment is chosen from among other mutations, "unsuccessful" mutations being lost, as a rule, from the record. In cultural evolution—man being a part of the biosphere—we may think of innovations as being chosen in analogous manner, remembering, however, that the mechanisms for biological and cultural evolution are widely different[2]. In attempting to include these cases within a common mathematical framework we may write:

$$-\Delta s = q \ln \frac{f_2}{f_1} = q \ln \frac{p_1}{p_2} \tag{3}$$

where q is a proportionality constant, f is the number of facets of orderly pattern, and p is the probability of that state: $f_2 > f_1$, $p_2 < p_1$. The term "facet" is used here in a very general sense to describe a unit of pattern which is related to the pattern as a whole; increase in number of facets indicates increase in order with corresponding decrease in probability. Nothing is specified as to the kind or properties of the facets, which might have many forms but in all cases represent choices that have been made between alternatives—successful mutations, and cultural innovations are examples. It is not meant to specify in equation (3) that each added facet of pattern depends on a choice between one of only two alternatives, as in equation (2) for the accumulation of choices.

S 为熵，N 为涉及的分子数量；k 为玻尔兹曼常数；ln 表示自然对数；W 是该系统的微观性质排列的数量；而 P 是该状态下的可能性。微观性质，即分子中电子的能量水平、分子中原子的振动和分子的速度等排列的数量越大，有序性越低，可能性越高。因此，根据热力学第二定律，$W_2>W_1$，$P_2>P_1$，我们不能去度量 W 或 P，但是在一个给定的系统中，只要该系统可表示成可以用 N 和 k 定义的维数，我们就可以用各种方法估计这些变量的相对变化，即 ΔS。

为了试着将生物圈中有序性的增加与日地系统中的相反变化联系起来，我们首先模拟一台数字计算机，并对它不断提出问题。该计算机通过从两个选项中选择一个来"回答"对它提出的每个问题。如果每个问题都以前一个问题为条件，则对答案的选择代表了有序性的增加和可能性的相应下降。我们可以用下式描述这种关系：

$$-\Delta s = \log_2 \frac{b_2}{b_1} \tag{2}$$

这里 $-s$ 表示负熵，用来度量信息的增加，$\log_2$ 表示底数为 2 的对数，b 是已回答问题的数量，且 $b_2>b_1$。回答问题的数量被认为没有实质的意义，因为它不是计算机在运算中获得知识的量度。在这种操作中，通常只关心已被选择的选项，而忽视了未被选择的选项。

自然选择下的进化可以与计算机的工作相比拟，只要假定运算规则为：每一个使物种更好地适应环境的突变均是从其他突变中选择出来的，而"不成功"的突变则从记录中去除。关于文化的进化——人类作为生物圈的一部分——我们可以类似地认为创新被选择下来，不过要记住，生物学进化与文化进化是非常不同的 [2]。如果试着把这些情况囊括到一个普遍的数学框架中，我们可以写成：

$$-\Delta s = q \ln \frac{f_2}{f_1} = q \ln \frac{p_1}{p_2} \tag{3}$$

这里 q 是一个比例常数，f 是有序类型的组合的数量，p 是该状态的可能性：$f_2>f_1$，$p_2<p_1$。这里所用的概念"组合"，是一般情况中描述类型的单位，作为整体与类型相关联；组合的数量增加，表明有序性增加而可能性相应下降。对组合的特性和种类没有特指，它可能有许多形式，但在所有情况下均代表在选项中被选择的选项——例如成功的突变和文化创新。但这并不意味着，公式（3）中每个新增的类型的组合都依赖于对仅有两个选项的选择，就像公式（2）中选择的累加那样。

If we let equation (1) describe the changes of order in the Sun-Earth system, and equation (3) describe those in the included biosphere system during the same period of time, we know that $-\Delta s < \Delta S$, if we could measure these in the same dimensions. But more quantitative comparison would demand that it be possible to measure q as an N times a k, and this presents difficulties that appear insurmountable since we lack a co-ordinate frame of reference[3]. The evolutionary record retains, for the most part, only evidence of selected facets, for example, successful mutations or cultural innovations, disregarding the rejected ones that would also have to be included in the assignment of common dimensions required for drawing up a thermodynamic balance sheet. Thus, when we deal with the evolution of the biosphere, and aspects of living systems which have been determined by that evolution, we may err gravely if we view these systems too rigidly in terms of physical dimensions. Seeing only a part of the total change, we may conclude that evolution has followed a much more "direct" course than it has, losing sight of the many alternative paths which might have been possible but were not followed. In so doing we are likely to accept into our thinking an excess of determinism and may slip inadvertently into a mechanistic finalism and teleology perhaps as misleading in some aspects as vitalistic finalism and teleology.

Although number of facets, like number of choices, does not measure meaningful content, we tacitly assume that the latter runs roughly parallel in both cases—the greater the number of choices made by the computer, the greater knowledge increase is to be expected; the greater the number of successful mutations, the greater the expectation of close adaptation to the environment—but no strict quantitative relationship can be predicted. It may be noted that, similarly, number of arrangements of microscopic properties measures order, but not meaningful content—free energy change and heat of reaction are properties of specific substances not measured in terms of entropy change, although quantitatively related thereto in systems of rigidly defined dimensions. We lack knowledge of any close quantitative relationship between order and meaningful content in the evolutionary changes going on in the biosphere, so in estimating change in number of facets we must rely on subjective judgment based on what seems to us to be the meaningful content. This is done whenever numerical values are assigned in preparing curves to describe evolutionary processes[4], which thus contain an unavoidable element of uncertainty.

Similar uncertainties enter into estimations of the probability of life—here and elsewhere in the universe. Such estimates should take into account the evolution of the biosphere, which may for convenience be divided into three phases, while recognizing that the divisions must be arbitrary since we deal with a continuous process: (1) a phase of chemical evolution during which the organic components of living systems and the mechanism of replication evolved; (2) a phase of biological evolution by natural selection; (3) a phase of man's cultural evolution—from our human view this is the culmination of the evolutionary process.

About a million species of living organisms are recognized today, and we may

如果我们用公式（1）来描述日地系统中有序性的变化，用公式（3）来描述同一时间段生物圈系统中所包含的有序性的变化，则我们知道如果能在同样的维数下度量它们的话，$-\Delta s < \Delta S$。但更多定量的比较则要求能够用 N 乘以 k 来度量 q，这就出现了难以逾越的困难，因为我们缺少一个参考坐标系 [3]，在绝大多数情况下，进化记录只保留被选择组合的证据，例如成功的突变或文化创新，而忽视被弃的选择，但为了保持热力学平衡，后者也应当被包括在共同维数的组成中。因此，当我们在研究生物圈的进化以及进化所选择的生命系统的状况时，如果认为这些系统在物理学维度方面极为严格，则会严重误入歧途。如果只看到整个变化的一部分，我们可能会认定进化所采取的过程比实际上"直接"得多，而看不到许多可能存在的其他途径，这些途径曾经是可能的但并未被采纳。这样做，我们可能会在头脑中接受过多的决定论，而毫不察觉地滑进机械的结局论和目的论，在某些情况下，这可能与灵活的结局论和目的论同样具有误导性。

尽管组合的数量与选择的数量一样不具有实质意义，我们假定后者在两种情况下是大致平行的——计算机所作选择的数量越多，知识增长的就越多；成功的突变越多，对环境的适应就越好——但并不能预期他们有严格的数量关系。类似地，可以注意到，微观性质的排列的数量可以度量有序性，但并无实质意义——自由能交换和反应热是特定物质的特征，且不以熵的变化来度量，尽管它们在具有严格确定维数的系统中是定量相关的。关于生物圈内进行的有意义的进化改变和有序性之间的数量关系，我们并不了解。因此在估计组合数量的变化时，我们必须依赖主观判断，而后者是基于那些对我们来说似乎是实质内容而做出的。在指定一些数值用于制作描述进化过程的曲线时 [4]，我们就会这样做，因而不可避免地包含不确定因素。

在估计地球上或宇宙中其他地方的生命可能性时，也会出现类似的不确定因素。这种估计应当考虑生物圈的进化，为了方便，可以将其划分为三个阶段，但应当认识到由于我们研究的是一个连续过程，因而这种划分是主观的：（1）化学进化阶段，包含生命系统中的有机化合物和相关复制机制；（2）自然选择下的生物进化阶段；（3）人类文化进化阶段，从我们人类的观点看这是进化过程的顶点。

目前，人们已经识别了大约一百万个物种。我们可以保守地估计，每个物种至

conservatively estimate that at least one thousand successful mutations were concerned in each one. Assuming the change in probability in the course of this process to be inversely proportional to the total number of choices of successful mutations, we estimate, in terms of equation (3) that the probability of biological evolution having arrived at its present state would be 10^{-9}. It seems again conservative to assume that a million innovations have been introduced in the course of cultural evolution, and multiplying the probability of this number of choices by the corresponding figure for biological evolution we decrease the probability of man's combined biological and cultural evolution to 10^{-15}. Presumably there were points in chemical evolution where choice of pathway also occurred, for example, where the rate of one reaction caused it to predominate over a thermodynamically more probable one[5]; but it would seem even more difficult to estimate the number. Assuming one thousand such choices, however, we reach a total figure of 10^{-18} for the probability of man having reached his present state. Conservative as this estimate seems, it approaches inversely some of those for the number of "habitable" planets in the universe that have been invoked to indicate the likelihood that man-like creatures may exist elsewhere.

Numerous uncertainties in the above estimate of man's probability must be obvious from the foregoing discussion, in which it was tacitly assumed that evolutionary choices are purely matters of chance, thus neglecting any change in meaningfulness resulting from these choices. For example, properties differ for chemical species; some of them are better fitted to be components of living systems and their environment than are others, as implied in Henderson's cogent phrase, "fitness of the environment"[6]; and natural selection depends on the characteristics of the environment and of the phenotype which impinge. Such things introduce "deterministic" factors which were neglected in the strictly "probabilistic" argument used in the foregoing estimates, which might be either too low or too high on this account. In any event, such estimates—open as they are to great uncertainty—may give our imagination pause in peopling the universe with living things, particularly with "intelligent" life approaching closely to the characteristics of man[7].

In general, examination of the uncertainties involved in putting living systems into the dimensions used so effectively in the study of physical systems may give us better perspective on explanations of life and of human activity, on extrapolations into space, and into past and future time.

I am greatly indebted to Dr. Roger S. Pinkham for kindly criticism throughout the development of these thoughts.

(206, 131-132; 1965)

H. F. Blum: National Cancer Institute, National Institutes of Health, Department of Health, Education and Welfare, Bethesda, Maryland, and Department of Biology, Princeton University.

少包含一千个成功的突变。假定该过程中可能发生的变化与选择的成功突变的总数成反比，根据公式（3），我们估计生物进化达到当前水平的可能性是 10^{-9}。假定在文化进化的过程中引入了一百万个创新（这看来也是保守的），把该选择数目的可能性乘以生物进化的相应数字，则人类生物学与文化进化的总可能性就降到了 10^{-15}。假设在化学进化的过程中也存在着一些路径选择的节点，例如，其中一个反应的比率导致它优于在热力学上更可能的另一个反应 [5]；不过看来估计出这些数字更加困难。然而，假设有 1,000 个这样的选择，我们就会得出人类能够达到今天这样的进化程度的总可能性为 10^{-18}。该估计看上去是保守的，不过可以反映出宇宙中其他"适宜"星球上存在类人生物的可能性。

显而易见，上述关于进化到人类的可能性估计的讨论中存在许多不确定性，由于这里默认假定进化选择纯粹是机会性的，因而忽视了来自这些选择的任何有意义的变化。例如，化学物质的特性是不同的，其中的一些是对生物系统及其环境来说更加适合的成分，正如亨德森那个恰如其分的词汇——"环境适应性"[6]。自然选择依赖于环境和生物的显性表型，这就引入了那些"决定性"因素，而它们在之前估计中作为严格的"可能性"论据时被忽略了，这使估计也许过高或过低。任何情况下这种估计都是开放的，因为有很大的不确定性，这种估计也许会中止我们关于宇宙中存在生命的想象，特别是关于存在与人类特征非常接近的"智慧"生物的想象 [7]。

总的来说，探究将生命系统转变成维数时包含的不确定性在物理学的研究中应用得非常有效，也许会为我们解释生命和人类活动以及推知空间、过去和未来提供更好的视角。

非常感谢罗杰·平卡姆博士在我进行以上思考时提出的中肯的批评意见。

（周志华 翻译；江丕栋 审稿）

References:

1. Blum, H. F., *Science*, **86**, 285 (1937).

2. Blum, H. F., *Amer. Scientist*, **51**, 32 (1963).

3. Blum, H. F., *Synthese*, **15**, 115 (1963); also in *Form and Strategy in Science* (Dordrecht, D. Reidel, 1964).

4. Cailleux, A., *C. R. Soc. Geol. France*, **20**, 222 (1950). *Bull. Soc. Prehist.*, France, **48**, 62 (1951). Hart, H., *The Technique of Social Progress* (New York, Henry Holt, 1931). *Symp. Sociological Theory*, edit. by Gross, L. (Evanston, Ill., Row Peterson, 1959). Meyer, F., *Problematique de l'Evolution* (Paris, Presses Universitaires de France, 1954). Price, J. deS., *Science Since Babylon* (New Haven, Yale Univ. Press, 1961).

5. Blum, H. F., *Amer. Scientist*, **49**, 474 (1961).

6. Henderson, L. J., *The Fitness of the Environment* (New York, Macmillan, 1917). And see Blum, H. F., *Time's Arrow and Evolution* (New York, Harpers, 1962).

7. Simpson, G. G., *Science*, 143, 769 (1964). Blum, H. F., *Science*, **144**, 613 (1964).

Genetic Code: the "Nonsense" Triplets for Chain Termination and Their Suppression

S. Brenner *et al.*

Editor's Note

By the mid-1960s, molecular geneticists had uncovered many of the detailed processes involved in the production of RNA molecules. These transfer information from the nuclear DNA to the organelles in cells where protein synthesis takes place. The general principle is that triplets of nucleotides (codons) in DNA are represented by corresponding triplets in RNA, and that each of these codons in general specifies a particular amino acid in the respective protein. This paper by Sydney Brenner and his colleagues at Cambridge discussed the nature of the genetic code in the light of experiments carried out with the bacterium *Escherichi coli*. Brenner was awarded a Nobel Prize in 2002.

THE nucleotide sequence of messenger RNA is a code determining the amino-acid sequences of proteins. Although the biochemical apparatus which translates the code is elaborate, it is likely that the code itself is simple and consists of non-overlapping nucleotide triplets. In general, each amino-acid has more than one triplet corresponding to it, but it is not known how many of the sixty-four triplets are used to code for the twenty amino-acids. Triplets which do not correspond to amino-acids have been loosely referred to as "nonsense" triplets, but it is not known whether these triplets have an information content which is strictly null or whether they serve some special function in information transfer.

The evidence that there are nonsense triplets is mainly genetic and will be reviewed later in this article. A remarkable property of nonsense mutants in bacteria and bacteriophages is that they are suppressible; wild-type function can be restored to such mutants by certain strains of bacteria carrying suppressor genes. It was realized early that this implies an ambiguity in the genetic code in the sense that a codon which is nonsense in one strain can be recognized as sense in another. The problem of nonsense triplets has become inextricably connected with the problem of suppression and, in particular, it has proved difficult to construct a theory of suppression without knowing the function of the nonsense triplet.

In this article we report experiments which allow us to deduce the structure of two nonsense codons as UAG and UAA. We suggest that these codons are the normal recognition signals in messenger RNA for chain termination and, on this basis, propose a theory of their suppression.

遗传密码：终止翻译的"无义"三联体及其抑制机制

布伦纳等

编者按

在 20 世纪 60 年代中期，分子遗传学家已经揭示了 RNA 分子产生的具体过程。这些 RNA 分子将核 DNA 信息转运到细胞器中，在细胞器中合成蛋白质。基本原理是 DNA 中的核苷酸三联体（密码子）用 RNA 中相应的三联体表示，一般而言在各个蛋白质中，每一个密码子确定一个特定的氨基酸。本文中，剑桥大学的辛迪·布伦纳及其同事们以大肠杆菌为材料，通过实验探索了遗传密码的本质。布伦纳于 2002 年获得了诺贝尔奖。

信使 RNA 中的核苷酸序列是编码蛋白质氨基酸序列的密码。尽管翻译该密码的生化结构很复杂，但是很可能密码本身很简单并且是由不重叠的核苷酸三联体构成的。一般而言，每个氨基酸都对应着不止一种三联体，但是目前尚不清楚 64 个三联体中有多少用来编码 20 种氨基酸。"无义"三联体泛指不编码氨基酸的核苷酸三联体。但目前尚不清楚这些三联体含有的遗传信息是完全无意义的，还是在信息传递过程中发挥着某些特定的功能。

无义三联体存在的证据主要来自遗传学方面，文章在后面会进行论述。细菌和噬菌体的无义突变体的显著特性是它们是可抑制的；某些携带抑制基因的细菌菌株可以恢复这些突变体的野生型功能。这暗示着遗传密码是否有意义并不能确切的界定，这一点人们早就已经意识到。因为在一种菌株中是无义的密码子在另一菌株中可能被识别是有义的。无义三联体的问题与抑制问题是紧密联系的，特别是在不知道无义三联体功能的情况下，事实证明难以建立抑制作用的学说。

本文中，我们讲述了能让我们推出 UAG 和 UAA 这两种无义密码子结构的实验。我们认为这两种无义密码子是信使 RNA 终止翻译时的正常识别信号。并以此为基础提出了它们的抑制机制。

Nonsense mutants and their suppressors. One class of suppressible nonsense mutants which has been widely examined includes the subset I ambivalent *r*II mutants[1], the suppressible mutants of alkaline phosphatase[2], the *hd* or *sus* mutants of phage λ[3], and many of the *amber* mutants of bacteriophage T4 (ref. 4). These mutants have been isolated in various ways and the permissive (*su*+) and non-permissive (*su*−) strains used have been different. When isogenic bacterial strains, differing only in the *su* locus, are constructed, it can be shown that all these mutants respond to the same set of suppressors. They are therefore of the same class and we propose that all these mutants should be called *amber* mutants.

We may now consider the evidence that these mutants contain nonsense codons. Garen and Siddiqi[2] originally noted that *amber* mutants of the alkaline phosphatase of *E. coli* contained no protein related immunologically to the enzyme. Benzer and Champe[1] also showed that the mutants exert drastic effects and suggested that *amber* mutants of the *r*II gene interrupt the reading of the genetic message. In the *r*II genes, a deletion, *r*1589, joins part of the A cistron to part of the B cistron; complementation tests show that this mutant still possesses B activity although it lacks the A function. It may therefore be used to test the effects of mutants in the A cistron on the activity of the B. Double mutants, composed of an A *amber* mutant together with *r*1589 did not have B activity on the *su*− strain; this effect is suppressed by the *su*+ strain which restores B activity. This result is explained by our finding that each *amber* mutant of the head protein produces a characteristic fragment of the polypeptide chain in *su*− bacteria[5]. More recently we have shown that the polarity of the fragment is such that it can only be produced by the termination of the growing polypeptide chain at the site of the mutation[6]. In *su*+ strains, both the fragment and a completed chain are produced, and the efficiency of propagation depends on which of the suppressors is carried by the strain. In *su*+I, the efficiency of propagation is about 65 percent[7], and the completed chain contains a serine at a position occupied by a glutamine in the wild type[6] as shown in Fig. 1.

Wild Type	**Ala.Gly.(Val,Phe)Asp.Phe.Gln.Asp.Pro.Ile.Asp.Ile.Arg** ...
*H*36 on **su**−	**Ala.Gly.Val.Phe.Asp.Phe.**
*H*36 on **su**+I	**Ala.Gly.Val.Phe.Asp.Phe.**
	and
	Ala.Gly.Val.Phe.Asp.Phe.Ser.Asp.Pro.Ile.Asp.Ile.Arg ...

Fig. 1. Amino-acid sequences of the relevant region of the head protein in wild-type *T4D* and in the *amber* mutant *H*36 on *su*− and *su*+I strains.

In addition to the *amber* mutants, there are other mutants, called *ochre* mutants, which are suppressed by a different set of suppressors[8]. *Ochre* mutants of the A cistron of *r*II abolish the B activity of *r*1589. This effect is not suppressed by *amber* suppressors, which shows that the *ochre* mutants are intrinsically different from the *amber* mutants.

310

无义突变体及其抑制基因 经大量实验证明，目前有一类可抑制的无义突变体，这类突变体包括子集 I 矛盾突变体 rII[1]、碱性磷酸酶可抑制突变体 [2]、λ 噬菌体 hd 或 sus 突变体 [3] 及 $T4$ 噬菌体的许多琥珀突变体（参考文献 4）。将这些突变体用不同的方法进行分离并且在分离过程中用不同的受纳菌株（su^+）和非受纳菌株（su^-）。当构建了仅在 su 位点不同的等基因细菌菌株时，发现所有这些突变体都对同一套抑制基因做出应答。由此推测它们应该属于同一类突变体，我们提议将所有这些突变体统称为琥珀突变体。

现在我们来寻找这些突变体含有无义密码子的证据。加伦和西迪基 [2] 首先发现了大肠杆菌的碱性磷酸酶琥珀突变体中没有碱性磷酸酶相关的免疫蛋白。本则尔和钱普 [1] 也表明突变体有着明显的表型并且提出 rII 基因的琥珀突变体中断遗传信息的读取。rII 基因中 r1589 缺失使部分 A 顺反子与部分 B 顺反子连接在了一起，互补测验表明该突变体仍然具有 B 活性但缺少了 A 功能。因此该测验方法可用于检测 A 顺反子的突变体对 B 活性的影响。su^- 菌株中由 A 琥珀突变体和 r1589 组成的双突变体不具有 B 活性，而在 su^+ 菌株中，由于该菌株可以恢复 B 活性，所以突变的作用被抑制。我们的发现可以解释这一结果：su^- 菌株中头部蛋白的每一个琥珀突变体都产生一种特征性的多肽链片段 [5]。最近，我们发现该片段的极性只在正生成的多肽链终止于突变位点时才可以产生 [6]。而 su^+ 菌株中，既可以产生这种片段，也可以产生完整的多肽链，增殖效率依赖于该菌株携带了哪一个抑制基因。su^+_I 菌株的增殖效率大概为 65%[7]，完整的多肽链是将野生型中谷氨酰胺变成了丝氨酸 [6]，如图 1 所示。

野生型	Ala.Gly.(Val,Phe)Asp.Phe.Gln.Asp.Pro.Ile.Asp.Ile.Arg······
$\underline{su^-}$ 菌株的 $H36$	Ala.Gly.Val.Phe.Asp.Phe.
$\underline{su^+}_I$ 菌株的 $H36$	Ala.Gly.Val.Phe.Asp.Phe.
	和
	Ala.Gly.Val.Phe.Asp.Phe.Ser.Asp.Pro.Ile.Asp.Ile.Arg······

图 1. 野生型 $T4D$ 及 su^- 菌株和 su^+_I 菌株的琥珀突变体 $H36$ 的头部蛋白中对应片段的氨基酸序列。

除了琥珀突变体，还有一种赭石突变体，赭石突变体是受不同于琥珀突变体的另一套抑制基因所抑制的 [8]。rII 的 A 顺反子的赭石突变体能够使 r1589 的 B 活性失活。而琥珀抑制基因则不能抑制这个作用，这表明赭石突变体与琥珀突变体本质上是不同的。

Thus, nonsense mutants may be divided into two types, *amber* and *ochre* mutants, depending on their pattern of suppression. In Table 1, which is abstracted from a larger set of results[8], it can be seen that strains, carrying the *amber* suppressors su^+_I, su^+_{II}, su^+_{III} and su^+_{IV}, suppress different but overlapping sets of *amber* mutants, but do not suppress any *ochre* mutants. This is the feature which distinguished the two classes of mutants from each other. Table 1 also shows that *ochre* mutants are suppressed by one or more of the strains carrying the *ochre* suppressors su^+_B, su^+_C, su^+_D and su^+_E. These suppressors are also active on various *amber* mutants, and we have not yet been able to isolate suppressors specific for *ochre* mutants. Table 1 also shows that the suppressor strains can be differentiated by the set of mutants they suppress.

Table 1. Suppression of *r*II Mutants by su^+ Strains of *E. coli Hfr H*(λ)

		Amber suppressors				Ochre suppressors			
		su^+_I	su^+_{II}	su^+_{III}	su^+_{IV}	su^+_B	su^+_C	su^+_D	su^+_E
Amber mutants									
*r*IIA	HD120	+	+	+	+	+	0	+	0
	N97	+	poor	+	0	poor	poor	+	0
	S116	0	0	+	0	0	+	+	0
	N19	+	poor	+	0	poor	0	+	0
	N34	+	+	+	+	+	+	+	0
*r*IIB	HE122	+	+	+	0	poor	0	+	0
	HB74	+	+	+	+	+	+	+	+
	X237	+	poor	+	0	0	+	+	0
	HB232	+	+	+	+	+	0	+	+
	X417	+	+	+	+	0	poor	+	0
Ochre mutants									
*r*IIA	HD147	0	0	0	0	+	+	+	+
	N55	0	0	0	0	+	+	+	0
	X20	0	0	0	0	+	0	+	0
	N21	0	0	0	0	+	0	0	+
*r*IIB	UV375	0	0	0	0	+	+	+	+
	360	0	0	0	0	+	0	+	+
	375	0	0	0	0	+	+	+	0
	HF208	0	0	0	0	+	0	0	+
	N29	0	0	0	0	0	0	+	0

Unlike some of the *amber* suppressors, all the *ochre* suppressors are weak[7,8]. This has made the isolation of *ochre* mutants of the head protein impossible. We therefore do not know the molecular consequences of *ochre* mutants, but we shall assume that, like the *amber* mutants, they too result in chain termination.

If we accept that both types of nonsense mutants result in termination of the polypeptide chain, we have to ask: at which level of information transfer is this effect exerted? We have recently shown that it is likely that chain termination occurs as part of protein synthesis,

因此，可以根据无义突变体的抑制模式将其分为两种类型：琥珀突变体和赭石突变体。表 1 是从大量实验结果中总结出来的 [8]，可以看出，携带琥珀抑制基因 su^+_I、su^+_{II}、su^+_{III} 和 su^+_{IV} 的菌株分别能抑制一个不同的琥珀突变体组，但各组之间相互重叠，但不抑制任何赭石突变体。这一特征将这两类突变体彼此区分开来。表 1 也表明，携带赭石抑制基因 su^+_B、su^+_C、su^+_D 和 su^+_E 的一个或多个菌株均可抑制赭石突变体。而这些抑制基因在各种琥珀突变体中也有活性，我们目前还无法分离到对赭石突变体特异的抑制基因。表 1 还表明携带抑制基因的菌株可以根据它们抑制的突变体类型而区分开来。

表 1. 大肠杆菌 $Hfr\ H\ (\lambda)$ 的 su^+ 菌株对 rII 突变体的抑制作用

		琥珀抑制基因				赭石抑制基因			
		su^+_I	su^+_{II}	su^+_{III}	su^+_{IV}	su^+_B	su^+_C	su^+_D	su^+_E
琥珀突变体									
$rIIA$	HD120	+	+	+	+	+	0	+	0
	N97	+	微弱	+	0	微弱	微弱	+	0
	S116	0	0	+	0	0	+	+	0
	N19	+	微弱	+	0	微弱	0	+	0
	N34	+	+	+	+	+	+	+	0
$rIIB$	HE122	+	+	+	0	微弱	0	+	0
	HB74	+	+	+	+	+	+	+	+
	X237	+	微弱	+	0	0	+	+	0
	HB232	+	+	+	+	+	0	+	+
	X417	+	+	+	+	0	微弱	+	0
赭石突变体									
$rIIA$	HD147	0	0	0	0	+	+	+	+
	N55	0	0	0	0	+	+	+	0
	X20	0	0	0	0	+	0	+	0
	N21	0	0	0	0	+	0	0	+
$rIIB$	UV375	0	0	0	0	+	+	+	+
	360	0	0	0	0	+	0	+	+
	375	0	0	0	0	+	+	+	0
	HF208	0	0	0	0	+	0	0	+
	N29	0	0	0	0	0	0	+	0

与某些琥珀抑制基因不同，所有赭石抑制基因的抑制作用都比较弱 [7,8]。这令我们无法分离到头部蛋白的赭石突变体。因此我们无法知道赭石突变体的分子机理，但是我们可以假定它们与琥珀突变体一样也导致了多肽链终止。

如果这两种类型的无义突变体都导致了多肽链终止的话，那么我们不得不问这个作用是发生在信息传递的哪个阶段？最近我们发现链终止可能是在蛋白质合成的某些阶段发生的，因为当遗传信息的读取阶段发生变化时，两种类型的突变体都消

since both types of mutants vanish when the phase of reading of the genetic message is altered[9]. This leads us to conclude that *amber* and *ochre* mutants produce different triplets which have to be read in the correct phase and which are recognized as signals for chain termination.

Decoding of amber and ochre triplets. We have done two types of experiments which allow us to deduce the structures of the *amber* and *ochre* triplets. First, we studied the production and reversion of *r*II *amber* and *ochre* mutants using chemical mutagens. We show that the two triplets are connected to each other and that we can define their possible nucleotide compositions. Next, we investigated head protein *amber* mutants, to define the amino-acids connected to the *amber* triplet. Comparison of these results with known amino-acid codons allows us to deduce the structures of the *amber* and *ochre* triplets.

The experiments with *r*II mutants depend on the specificity of the mutagenic agent, hydroxylamine. This reacts only with cytosine in DNA, and although the exact structure of the product has not yet been defined, the altered base (called U′) appears to act like T with high efficiency, producing base-pair transitions of the G–C→A–T type[10]. Hence, response of any particular site of the DNA to hydroxylamine is evidence for the existence of a G–C pair at that site. Usually, phage particles are treated with hydroxylamine and, since these contain double-stranded DNA, the alteration of C occurs on only one of the strands. In any given gene, only one of the strands is transcribed into messenger RNA (ref.11). This is the *sense* strand; it carries the genetic information proper and contains a nucleotide sequence which is the inverse complement of the sequence of the messenger RNA. The other strand, the *antisense* strand, has the same sequence as the messenger. In a phage, treated with hydroxylamine, the altered base, U′, could be on either the sense or the antisense strand of the DNA. Since the *r*II genes express their functions before the onset of DNA replication[12], only sense strand changes will register a phenotypic effect in the first cycle of growth; changes on the antisense strand, while still yielding altered DNA progeny, will go unexpressed (Fig. 2).

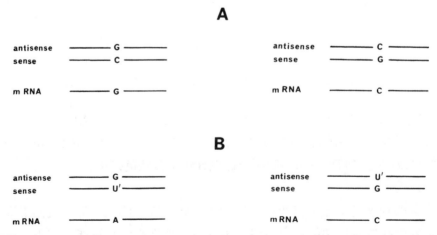

Fig. 2. Diagram illustrating the expression of the two types of G–C pairs (*A*) before and (*B*) after treatment with hydroxylamine.

失了 [9]。由此我们得出以下结论：琥珀突变体和赭石突变体产生了不同的三联体，这些三联体在正确的阶段必须得被读取并且它们被识别为链终止的信号。

琥珀三联体和赭石三联体的译码　为了推导琥珀三联体和赭石三联体的构成，我们进行了以下两项实验：首先，用化学诱变剂诱导产生了 rII 琥珀突变体和赭石突变体及它们的回复突变菌株。结果表明这两个三联体互相联系，我们可以确定出它们可能的核苷酸构成。其次，我们研究了琥珀突变体的头部蛋白来确定与琥珀三联体有关的氨基酸组成。将这些结果与已知氨基酸密码子进行比较后，我们便可以推导出琥珀三联体和赭石三联体的构成。

rII 突变体的实验依赖于诱变剂羟胺的特异性。羟胺只与 DNA 中的胞嘧啶作用，尽管反应产物的确切结构尚未确定，但改变后的碱基（称为 U'）表现出与碱基 T 高度相似的特点，即羟胺能够产生 G–C→A–T 类型的碱基对转换 [10]。因此，DNA 中任何特定位点对羟胺的应答都可以作为该位点存在 G–C 碱基对的证据。通常，噬菌体颗粒都是用羟胺来处理的，由于噬菌体颗粒含有双链 DNA，所以碱基 C 只在其中的一条链上发生改变。任何给定的基因中，都只有一条链可以转录成信使 RNA（参考文献 11）。该链称为有义链；有义链携带有正确的遗传信息，其核苷酸序列与信使 RNA 的序列是反向互补的。另一条链称为反义链，其序列与信使 RNA 相同。经羟胺处理过的噬菌体中，改变后的碱基 U' 既可能位于 DNA 的有义链上，也可能位于反义链上。由于 rII 基因在 DNA 复制开始之前表达并发挥功能 [12]，所以如果有义链发生变化，表型效应在生长的第一个周期就显现出来了，而反义链上发生变化虽然也可以产生出 DNA 变异的后代，但是却不会表达，如图 2 所示。

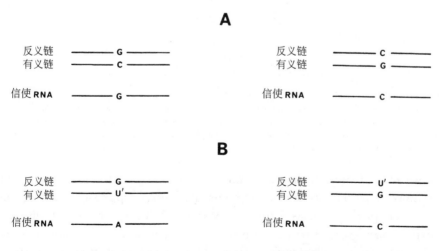

图 2. 羟胺处理前（A）和处理后（B）G–C 碱基对的两种转录形式

We can now examine the reversion properties of *amber* and *ochre* mutants. Champe and Benzer[12] studied reversion of a large number of *r*II mutants using different mutagens. They noted that no *amber* mutant was induced to revert by hydroxylamine. Some of the mutants they studied can now be identified as *ochre* mutants, and their results show that *ochre* mutants are equally insensitive to hydroxylamine. However, in their experiments, the treated mutant phages were plated directly on the bacterial strain which restricts the growth of *r*II mutants. They could therefore detect *sense* strand changes only, and any mutation on the antisense strand would not have been expressed. Strictly speaking, then, their results tell us that neither the *amber* nor the *ochre* triplet contains a C on the sense strand of the DNA, or, if any one does, it is connected by a C→U change to another nonsense codon. To extend this result and to recover all possible mutational changes, we grew the mutagenized phages in *E. coli B*, in which the *r*II functions are unnecessary, and then plated the progeny on strain *K* to measure reversion frequency. Table 2 shows that *amber* and *ochre* mutants are not induced to revert by hydroxylamine, and we conclude that in neither mutant, does the triplet in the DNA contain G–C pairs, or, if a G–C pair is present, that triplet is connected by a G–C→A–T transition to another nonsense codon. In other words, subject to the last important reservation, we can conclude that the codons on the messenger RNA contain neither G nor C.

Table 2. Reversion of *Amber* and *Ochre* Mutants after Allowing DNA Replication

		Reversion index $\times 10^{-7}$	
		Control	NH$_2$OH
amber mutants	S116	0.04	0.03
	HD26	0.1	0.2
	S24	0.4	0.9
	S99	0.1	0.1
	N19	0.15	0.13
	HD59	0.0	0.05
	HB232	0.1	0.4
ochre mutants	UV375	0.8	2.0
	360	0.6	0.8
	X27	0.3	0.5
	375	0.8	0.9
	X511	0.2	0.3
	UV256	1.0	280

Phages were incubated in a solution of M NH$_2$OH in 2 M NaCl and 0.05 M sodium phosphate (*p*H 7.5) for 2 h at 37°. The reaction was terminated by dilution into acetone broth. About 10^8 phage particles were used to infect a culture of *E. coli B* which was grown to lysis, and the progeny assayed on *E. coli B* and *E. coli K*12(λ). The reversion index is the *K/B* ratio. The control was treated in the same way except that hydroxylamine was omitted. The mutant UV256, which is not an *ochre* or an *amber* mutant, was used to check the efficacy of the mutagenic treatment.

However, we next discovered that *ochre* mutants can be converted into *amber* mutants by mutation. Since *ochre* mutants are not suppressed by *amber* suppressors, plating on strains carrying such suppressors selects for *amber* revertants. Wild-type revertants also grow, but the two can be distinguished by testing revertant plaques on the *su*⁻ strain. Twenty-six *r*II *ochre* mutants have been studied and, of these, 25 have been converted into *amber* mutants. A sample of the results is given in Table 3, which shows that the mutation is strongly

现在我们来探究琥珀突变体和赭石突变体回复突变的特点。钱普和本则尔[12]使用不同的诱变剂研究了很多 rII 突变体的回复突变。他们指出，琥珀突变体被羟胺诱导没有发生回复突变。同样赭石突变体也对羟胺不敏感，他们研究的琥珀突变体有些现在被鉴定为赭石突变体。但是在他们的实验中，处理过的突变噬菌体是直接涂在限制 rII 突变体的生长的细菌菌株上。因此他们只能检测到有义链发生的变化，而反义链发生的所有变化都没有表达出来。因此严格地说，他们的实验结果告诉我们，琥珀突变体和赭石突变体的 DNA 有义链都不含 C 碱基，或者是通过 C→U 变化而成为了另一个无义密码子。为了继续探究这一结果和回复所有可能的突变变化，我们让诱变处理过的噬菌体在大肠杆菌 B 菌株中生长，该条件下 rII 功能不是必需的，然后将其后代涂在 K 菌株上来测量回复突变的频率。表 2 表明羟胺不能诱导琥珀突变体和赭石突变体发生回复突变，我们推论这两种突变体 DNA 中的三联体都不包含 G–C 碱基对，或者，即使 G–C 碱基对存在的话，三联体也是通过 G–C→A–T 转换而变成了另一个无义密码子。换句话说，可以保守的估计，信使 RNA 上的密码子既不包含 G 也不包含 C。

表 2. 允许 DNA 复制后琥珀突变体和赭石突变体的回复情况

		回复指数（$\times 10^{-7}$）	
		对照	NH$_2$OH
琥珀突变体	S116	0.04	0.03
	HD26	0.1	0.2
	S24	0.4	0.9
	S99	0.1	0.1
	N19	0.15	0.13
	HD59	0.0	0.05
	HB232	0.1	0.4
赭石突变体	UV375	0.8	2.0
	360	0.6	0.8
	X27	0.3	0.5
	375	0.8	0.9
	X511	0.2	0.3
	UV256	1.0	280

噬菌体在 1 摩尔/升 NH$_2$OH 溶液（2 摩尔/升 NaCl，0.05 摩尔/升磷酸钠，pH 7.5）中于 37℃孵育。2 小时后在丙酮液体培养基中稀释来终止反应。实验中用大约 10^8 个噬菌体颗粒侵染大肠杆菌 B 菌株培养物，溶菌后分别对大肠杆菌 B 菌株和大肠杆菌 K12（λ）菌株上的噬菌体进行检测。回复指数用 K/B 比值来表示。对照组除了没有用羟胺处理外其余步骤均与实验组相同。选取突变体 UV256 作为阳性对照来检测诱变处理是否有效，该突变体不是赭石突变体或琥珀突变体。

然而，我们又发现赭石突变体可以突变为琥珀突变体。由于赭石突变体不受琥珀抑制基因所抑制，所以可以通过将突变噬菌体涂在携带琥珀抑制基因的菌株上来选择琥珀型回复突变体。野生型回复突变体也会生长，但是可以通过检测 su^- 菌株上的回复突变体噬菌斑来区分二者。我们研究了 26 个 rII 赭石突变体，其中 25 个被转变成了琥珀突变体。表 3 给出了结果中的一个样本，表明该突变被 2- 氨基嘌呤强

induced by 2-aminopurine, as strongly as the reversion of the *ochre* mutant to wild type. Other experiments, not reported here, show that the mutations of *ochre* mutants both to the *amber* and to the wild type are also induced by 5-bromouracil, but the induction is weaker than with 2-aminopurine. These results prove that the *amber* and *ochre* triplets differ from each other by only one nucleotide base, and must have the other two bases in common. 2-Aminopurine is a base analogue mutagen inducing the transition $A–T \rightleftharpoons G–C$ in both directions[13]. This tells us that one of the triplets has a G–C pair in the DNA. The experiment reported in Table 4 shows that *ochre* mutants cannot be induced to mutate to *amber* mutants with hydroxylamine, even after the treated phages have been grown in *E. coli B*. This shows that it is the *amber* triplet which has the G–C pair and the *ochre* which contains the A–T pair.

Table 3. Mutation of *ochre* Mutants to *amber* Mutants

	Reversion index $\times 10^{-7}$		
	Spontaneous	2-Aminopurine	
*r*IIA cistron	(wild type + *ambers*)	wild type	*amber*
N55	0.5	830	280
X20	0.05	100	2,100
X372	0.1	300	2,100
X352	0.06	340	1,500
HD147	0.3	370	50
*r*IIB cistron			
X511	0.2	710	65
N17	0.2	610	80
SD160	0.6	380	390
N29	1.0	3,900	330
AP53	0.7	350	15

Cultures of *E. coli B* in minimal medium with and without 2-aminopurine (600 μg/ml.) were inoculated with about 100 phages and grown to lysis. These were plated on *E. coli B* and on *E. coli K*12(λ) *su*⁺I. About 50 induced revertants were tested on *E. coli K*12(λ) *su*⁻ to measure the relative frequencies of *amber* revertants.

Table 4. Induction of the *ochre* → *amber* Mutation

		Reversion index $\times 10^{-7}$	
		r+	*amber*
360	Control	0.6	0.2
	Hydroxylamine	0.8	0.3
	2AP	200	1,200
UV375	Control	0.8	0.4
	Hydroxylamine	2.0	1.0
	2AP	660	140
X27	Control	0.3	<0.1
	Hydroxylamine	1.0	0.5
	2AP	7.0	73
375	Control	0.8	0.4
	Hydroxylamine	2.0	1.0
	2AP	1,400	1,700

Hydroxylamine treatment and growth of the mutagenized phages, and 2-aminopurine induction, were carried out as described in Tables 2 and 3.

烈诱导而成，其突变频率与赭石突变体回复为野生型的频率相当。其他实验（本文未报道）表明 5–溴尿嘧啶也可以诱导赭石突变体转变为琥珀突变体或者回复为野生型，但是突变频率比 2–氨基嘌呤要低。这些结果表明琥珀三联体和赭石三联体只有一个核苷酸碱基的差异，另外两个碱基相同。2–氨基嘌呤是一种碱基类似物诱变剂，$A\text{–}T \rightleftharpoons G\text{–}C$ 两个方向的转换都可以由它诱导发生[13]。这告诉我们 DNA 中的一个三联体有一对 G–C 碱基对。表 4 报道的实验结果表明羟胺不能诱导赭石突变体突变为琥珀突变体，甚至羟胺处理的噬菌体已在大肠杆菌 B 中生长。这表明具有 G–C 碱基对的是琥珀三联体，而赭石三联体含有的是 A–T 碱基对。

表 3. 赭石突变体转变为琥珀突变体的频率

	回复指数（$\times 10^{-7}$）		
	自发突变	2–氨基嘌呤诱导突变	
rIIA 顺反子	野生型和琥珀突变体	野生型	琥珀突变体
N55	0.5	830	280
X20	0.05	100	2,100
X372	0.1	300	2,100
X352	0.06	340	1,500
HD147	0.3	370	50
rIIB 顺反子			
X511	0.2	710	65
N17	0.2	610	80
SD160	0.6	380	390
N29	1.0	3,900	330
AP53	0.7	350	15

将大肠杆菌 B 菌株培养在有或无 2–氨基嘌呤（600 微克／毫升）的基本培养基中，与约 100 个噬菌体一起孵育直至发生溶菌。溶菌后将噬菌体后代涂在大肠杆菌 B 菌株和大肠杆菌 K12（λ）su^+_1 菌株上。将以得到的大约 50 个诱导回复突变体涂在大肠杆菌 K12（λ）su^- 菌株上以检测琥珀回复突变体所占的比例。

表 4. 赭石突变体到琥珀突变体的诱导突变

		回复指数（$\times 10^{-7}$）	
		r+	琥珀突变
360	对照	0.6	0.2
	羟胺	0.8	0.3
	2–氨基嘌呤	200	1,200
UV375	对照	0.8	0.4
	羟胺	2.0	1.0
	2–氨基嘌呤	660	140
X27	对照	0.3	<0.1
	羟胺	1.0	0.5
	2–氨基嘌呤	7.0	73
375	对照	0.8	0.4
	羟胺	2.0	1.0
	2–氨基嘌呤	1,400	1,700

上表中羟胺及 2–氨基嘌呤诱变处理的条件和诱变噬菌体的培养条件与表 2 和表 3 的相同。

Although the insensitivity of the mutants to reversion induction by hydroxylamine might suggest that they contain A–T base pairs only, the conversion of *ochre* mutants to *amber* mutants shows that, in one position, the *amber* mutant contains a G–C pair. The other two bases must be common to both triplets, but we cannot conclude that both are A–T pairs. In fact, both could be G–C pairs and the triplets may be connected to other nonsense triplets by G–C→A–T changes. However, we know that *amber* and *ochre* mutants can be induced by hydroxylamine from wild type. This proves that both triplets have at least one common A–T pair.

We now present an experiment which shows that the *amber* triplet has two A–T base pairs, and which also establishes the orientation of the pairs with respect to the two strands of DNA. Let us suppose that the *amber* triplet in the messenger RNA contains a U. This corresponds to an A in the sense strand of DNA of the *amber* mutant, implying that the wild-type DNA contains a G in this strand and a C in the antisense strand. When the wild-type DNA is treated with hydroxylamine to alter this C the change is not effective and normal messenger is still made (Fig. 2, right). On the other hand, if the *amber* triplet in the messenger contains an A, the mutant will be induced by the action of hydroxylamine on a C in the sense strand of the wild-type phage DNA, and provided that the U′ produced acts identically to U in messenger synthesis, mutant messenger will be made. This argument has been tested by the following experiment. Wild-type $T4r^+$ phages were treated with hydroxylamine to induce r mutants to a frequency of 1 percent. In set B, the phages were then grown on *E. coli B*, in which the rII functions are not required, to recover all mutants. In set K, the phages were grown through *E. coli* $K12(\lambda)$ su^-, to eliminate from the population all phages with an immediate mutant expression. *Amber* and *ochre* mutants were then selected and mapped. Table 5 summarizes the results. About the same number of rI mutants were recorded in each set, and since these mutants show no difference in growth on the two bacterial strains, this shows that the results may be compared directly. It will be seen that *amber* mutants at the sites, N97, S116, S24, N34, X237 and HB232, recur many times in set B, but are absent or rarely found in set K. At other *amber* sites, such as HB118, HB129, EM84 and AP164, mutants occur with approximately equal frequency in both sets. The first class fulfils the expectation for a C→U change on the sense strand, while the second class must arise by C→U changes on the antisense strand. This shows that the *amber* triplet in the messenger contains both an A and a U. The same should be true of the *ochre* mutants. However, as shown in Table 5, *ochre* mutants are not as strongly induced by hydroxylamine as are *amber* mutants, and we cannot separate the two classes with the same degree of confidence. Nevertheless, since we have already shown that the mutants are connected, it follows that the *ochre* triplet must also contain an A and a U. We conclude that the *amber* and *ochre* triplets are, respectively, either (UAG) and (UAA), or (UAC) and (UAU). If we had a strain which suppressed *ochre* mutants only, we could specify the third base by studying the induction of the *amber*→*ochre* change with hydroxylamine.

尽管羟胺诱导的突变体回复突变灵敏度低显示出三联体中可能只含有A-T碱基对，但从赭石突变体转变成琥珀突变体的数据来看，琥珀突变体的一个位点上应该含有G-C碱基对。而另外两个碱基一定是两个三联体共有的，但是我们无法推断他们都是A-T碱基对。事实上，它们有可能都是G-C碱基对，并且可能通过G-C→A-T变化而变成了其他的无义三联体。但是我们知道野生型可由羟胺诱导成为琥珀突变体和赭石突变体，这证明两个三联体至少共有一对A-T碱基对。

我们现在通过实验证明了琥珀三联体有两对 A-T 碱基对，并且也对 DNA 双链上的碱基对进行了定位。我们假设信使 RNA 中的琥珀三联体包含一个碱基 U，则相应地，琥珀突变体的 DNA 有义链上就是一个碱基 A，这意味着野生型 DNA 该链中含有的是 G，反义链对应为 C。用羟胺处理野生型 DNA 以改变这个 C 时并没有影响，仍可转录出正常的信使 RNA，如图 2 右所示。另一方面，如果信使 RNA 中的琥珀三联体含有一个 A 的话，那么羟胺就会对野生型噬菌体 DNA 有义链上的 C 发生作用而诱导产生突变体，假设信使 RNA 合成时 U' 产生的作用与 U 相同，就会产生突变的信使 RNA。这个假设已经通过如下的实验得到验证，用羟胺处理野生型 $T4r^+$ 噬菌体以诱导其产生频率为 1% 的 r 突变体。在 B 组中，让噬菌体在不需要 rII 功能的大肠杆菌 B 菌株上生长，以得到所有的突变体。在 K 组中，噬菌体通过侵染大肠杆菌 $K12$（λ）su^- 菌株而生长，以消除群体中所有具有即时突变体表达的噬菌体，然后选出琥珀突变体和赭石突变体并作图。表 5 对该结果进行了概括。每组中得到了大概同样数量的 rI 突变体，由于这些突变体在两种细菌菌株上的生长并无差别，表明或许可以直接比较这些结果。可以看到在 B 组中琥珀突变体在位点 N97、S116、S24、N34、X237 和 HB232 上出现许多次，但在 K 组中却完全没有或很少发现。其他的琥珀位点，例如 HB118、HB129、EM84 和 AP164，出现突变体的频率在两组中大概相同。第一类实现了在有义链上诱导 C→U 变化的期望，而第二类就必须在反义链上引起 C→U 变化。这表明信使 RNA 中的琥珀三联体既含有 A 碱基，也含有 U 碱基。赭石突变体应该也是这样。但是，正如表 5 所示，羟胺对赭石突变体的诱导效果不如对琥珀突变体的强，所以我们并不能以同样的置信度分离得到这两类突变体。但是由于我们已经表明突变体间是相关联的，因此可以认为赭石三联体也必定含有一个 A 和一个 U。从以上结果我们推论：琥珀三联体和赭石三联体分别是 UAG 和 UAA 或 UAC 和 UAU。如果我们有一种只抑制赭石突变体的菌株，那么就能够通过研究羟胺将琥珀突变体诱导成为赭石突变体这一变化而确定第三个碱基是什么。

Table 5. Hydroxylamine Induction of *amber* and *ochre* Mutants

A. No. of mutants isolated

	Set B	Set K
*r*I	2,010	1,823
Leaky or high reverting *r*II	720	508
non-suppressible *r*II	1,144	433
amber	319	121
ochre	83	82
Total	4,276	2,967

B. Recurrences found at different sites

Amber mutants			Ochre mutants		
No. found at each site			No. found at each site		
Site	Set B	Set K	Site	Set B	Set K
A cistron			A cistron		
*HB*118	27	15	*HD*147	2	0
*C*204	1	0	*HF*220	1	0
*N*97	44	1	*HF*240	1	0
*S*116	31	2	*N*55	19	19
*N*11	3	3	*X*20	9	8
*S*172	9	5	*HF*219	1	0
*S*24	44	3	*HF*245	1	0
*HB*129	14	25	*N*31	3	5
*S*99	12	16	*HM*127	0	1
*N*19	15	9	*N*21	2	2
*N*34	8	0			
B cistron			B cistron		
*HE*122	1	0	360	11	10
*EM*84	29	21	*UV*375	2	0
*HB*74	16	5	*N*24	6	4
*X*237	14	2	375	2	5
*AP*164	28	12	*N*17	5	3
*HB*232	21	1	*HF*208	1	2
*X*417	1	0	*N*7	4	12
*HD*231	1	1	*N*12	5	7
			*X*234	0	2
			*X*191	1	0
			*HE*267	5	0
			*AP*53	2	2

*T*4Br$^+$ was treated with M hydroxylamine (see Table 2) for 2 h at 37°C. Survival was 50 percent, and the frequency of *r* and mottled plaques, 1 percent. 1.2×10^7 phage particles were adsorbed to 10^9 cells of *E. coli B* (set B) and to *E. coli* K12(λ) *su⁻* (set K). After 8 min, the infected bacteria were diluted a thousand-fold into 2 litres of broth, incubated for 35' and lysed with CHCl₃. The burst sizes in both sets were 60. *r* mutants were isolated from each set using less than 2 ml. to ensure that the mutants selected had mostly arisen from independent events. These were picked and stabbed into B and K, and *r*I mutants and leaky mutants discarded. The *r*II mutants were then screened on *su⁺*III and *su⁺*B to select for *amber* and *ochre* mutants which were then located by genetic mapping.

表 5. 羟胺诱变分离琥珀突变体和赭石突变体

A. 分离到的突变体数		
	B 组	K 组
rI 突变体	2,010	1,823
渗漏或高回复突变的 rII 突变体	720	508
不能被抑制的 rII 突变体	1,144	433
琥珀突变体	319	121
赭石突变体	83	82
总数	4,276	2,967

B. 在不同位点重复出现的突变体数					
琥珀突变体			赭石突变体		
每个位点出现的次数			每个位点出现的次数		
位点	B 组	K 组	位点	B 组	K 组
A 顺反子			A 顺反子		
HB118	27	15	HD147	2	0
C204	1	0	HF220	1	0
N97	44	1	HF240	1	0
S116	31	2	N55	19	19
N11	3	3	X20	9	8
S172	9	5	HF219	1	0
S24	44	3	HF245	1	0
HB129	14	25	N31	3	5
S99	12	16	HM127	0	1
N19	15	9	N21	2	2
N34	8	0			
B 顺反子			B 顺反子		
HE122	1	0	360	11	10
EM84	29	21	UV375	2	0
HB74	16	5	N24	6	4
X237	14	2	375	2	5
AP164	28	12	N17	5	3
HB232	21	1	HF208	1	2
X417	1	0	N7	4	12
HD231	1	1	N12	5	7
			X234	0	2
			X191	1	0
			HE267	5	0
			AP53	2	2

将 T4Br+ 用 1M 羟胺于 37℃ 处理 2 小时（见表 2）后存活率为 50%，r 和斑点状噬菌斑的频率为 1%。用 1.2×10^7 个噬菌体颗粒去侵染 10^9 个大肠杆菌 B 菌株（B 组）和大肠杆菌 K12（λ）su⁻ 菌株（K 组）。8 分钟后，用 2 升的培养基将上述培养物稀释 1,000 倍，再孵育 35 分钟后用三氯甲烷将其溶解。两组中的噬菌体裂解量都是 60。每组中的 r 突变体用少于 2 毫升来分离，保证各组中筛选的突变是彼此独立的事件。剔除 rI 突变体和渗漏突变体之后，将挑选出来的这些突变体导入 B 株和 K 菌株。然后在菌株 su⁺III 和 su⁺B 上筛选 rII 突变体作为选择出来的琥珀突变体和赭石突变体，进一步通过遗传图谱定位。

Fortunately, we can resolve the ambiguity by determining the amino-acids to which the *amber* triplet is connected by mutation. In particular, we note that it should be connected to two and only two amino-acid codons by transitions, corresponding, in fact, to the two types of origin of the mutants described here. The third codon to which it is connected by a transition is the *ochre* triplet. As mentioned earlier, the head *amber* mutant *H*36 has arisen from glutamine (Fig. 1). This mutant was induced with hydroxylamine. We have evidence that two other mutants, *E*161 and *B*278 induced by 2-aminopurine and 5-bromouracil respectively, have arisen from tryptophan. In a recent study of two *amber* mutants of the alkaline phosphatase, Garen and Weigert[14] found one mutant to arise from glutamine and the other from tryptophan; and Notani *et al.*[15] have found an *amber* mutant to arise from glutamine in the RNA phage *f*2. In addition, we have examined 2-aminopurine induced revertants of 10 different head *amber* mutants. Ten to 12 independently induced revertants of each of the mutants have been screened for tryptophan containing peptides by examining the ^{14}C-tryptophan labelled protein. Among a total of 115 revertants, 62 are to tryptophan. Determination of glutamine involves sequence analysis and takes more time. So far, among the remaining 53 revertants, glutamine has been identified in one revertant of *H*36. These results suggest that the two amino-acids connected to the *amber* triplet are glutamine and tryptophan. If the *amber* triplet is (UAC), then one of these must be (CAC) and the other (UGC); if it is (UAG), then the corresponding codons are (CAG) and (UGG). Nirenberg *et al.*[16] have shown that poly AC does not code for tryptophan, but does for glutamine. However, they find that the triplet for glutamine clearly has the composition (CAA) and is definitely not (CAC). Since this latter triplet corresponds neither to glutamine nor to tryptophan we can eliminate the first alternative. We note with satisfaction that (UGG) is the composition of a codon assigned to tryptophan[16,17] , and this assignment of (UAG) to the *amber* triplet suggests that glutamine is (CAG).

We can also make a reasonable assignment of the order of the bases in the triplet. Our original argument was based on deductions from a few known triplets and from amino-acid replacement data; it will not be given here. The order of the bases follows directly from a recent demonstration by Nirenberg *et al.*[18] that the triplet CAG does, in fact, correspond to glutamine. The *amber* triplet is therefore UAG and the *ochre* triplet UAA. This assignment is supported by the following additional evidence. We have found a tyrosine replacement in 21 independent spontaneous mutants of the head *amber* mutant, *H*36 (ref. 19). This change must be due to a transversion because we have already accounted for all the transitions of the *amber* triplet. In support of this, we find that the change is not induced by 2–aminopurine. There are six possible transversions of the *amber* triplet, namely, AAG, GAG, UUG, UCG, UAU and UAC. It has recently been shown that both UAU and UAC correspond to tyrosine[20] which confirms the order. The spontaneous revertants of the *amber* mutants to leucine, serine and glutamic acid found by Weigert and Garen[21] are further evidence for the assignment. UUG, a transversion of the *amber* triplet, does in fact code for leucine,[22] and reasonable allocations for serine and glutamic acid are UCG and GAG, respectively. Weigert and Garen[21] also find revertants of an *amber* mutant to either lysine or arginine. This may be the final transversion expected since AAG is a codeword for lysine[20].

幸运的是，我们可以通过突变来确定琥珀三联体与何种氨基酸相对应，从而解决这个问题。特别是我们注意到它应该与通过转换形成的两种（仅此两种）氨基酸密码子有关，实际上这两种密码子对应本文描述的两种突变体。通过转换与第三个密码子相关的就是赭石三联体。正如早期提到的，头部蛋白琥珀突变体 H36 是由羟胺诱导谷氨酰胺突变产生的（图1）。另外，我们有证据证明，另外两个突变体 E161 和 B278 分别是由 2-氨基嘌呤和 5-溴尿嘧啶诱导色氨酸突变而产生。最近对碱性磷酸酶的两个琥珀突变体的研究中，加伦和魏格特 [14] 发现一个突变体由谷氨酰胺突变产生，另一个则由色氨酸突变产生，野谷等 [15] 则发现 RNA 噬菌体 f2 的一个琥珀突变体是由谷氨酰胺突变产生的。此外我们研究了 2-氨基嘌呤诱导的 10 种不同的头部蛋白琥珀突变体的回复突变体。通过检测 [14]C- 色氨酸标记蛋白从每一种突变体中可筛选到 10~12 个独立诱导的包含色氨酸肽段的回复突变体。在总共的 115 个回复突变体中，有 62 个是色氨酸突变产生的回复突变体。而谷氨酰胺突变产生的回复突变体是通过序列分析进行的，所以需要花费的时间更久一些。目前为止，在其余的 53 种回复突变体中，在 H36 一个回复突变体已检测出是谷氨酰胺突变。以上这些结果提示我们，与琥珀三联体相关的两个氨基酸是谷氨酰胺和色氨酸。如果琥珀三联体是 UAC，则这两者之一一定是 CAC，另一个是 UGC，如果琥珀三联体是 UAG，则对应的密码子就是 CAG 和 UGG。尼伦伯格等 [16] 已经表明多聚 AC 不编码色氨酸，但是编码谷氨酰胺。并且他们发现谷氨酰胺三联体肯定含有 CAA 这种组合，而一定不含 CAC。由于 CAC 既不编码谷氨酰胺也不编码色氨酸，所以可以排除第一种情况。可喜的是，我们注意到 UGG 是编码色氨酸的一个密码子组合 [16,17]，将 UAG 确定为琥珀三联体提示编码谷氨酰胺的应该是 CAG。

现在我们也可以对三联体的碱基顺序进行合理地分配了。最初，我们的论据是根据少数已知三联体以及氨基酸替代的信息推导出来的，这里就不表述了。关于碱基顺序的确定，其直接根据是尼伦伯格等人最近的实证 [18]，即证明三联体 CAG 确实编码的是谷氨酰胺。因此琥珀三联体是 UAG，而赭石三联体是 UAA。后来的其他证据也支持这个分配顺序。我们已经在头部蛋白琥珀突变体 H36 的 21 个独立自发突变体中发现了酪氨酸替代的情况（参考文献 19）。这种情况一定是由颠换突变引起的，因为我们已经推算了琥珀三联体的所有转换情况。为支持这一观点，我们还发现这一变化不是由 2-氨基嘌呤诱导产生的。琥珀三联体具有六种可能的颠换，即 AAG、GAG、UUG、UCG、UAU 和 UAC。最近又有研究表明 UAU 和 UAC 都编码酪氨酸 [20]，这使得碱基顺序得以确定。魏格特和加伦 [21] 发现琥珀突变体能够自发回复突变为亮氨酸、丝氨酸和谷氨酸，这为碱基分配顺序提供了进一步的支持证据。UUG 是琥珀三联体发生一个颠换产生的，实际上确实编码的是亮氨酸 [22]，而丝氨酸和谷氨酸应分别对应 UCG 和 GAG。另外，魏格特和加伦 [21] 还发现琥珀突变体既可以回复突变成赖氨酸，也可以回复突变为精氨酸。由于 AAG 是赖氨酸的密码子 [20]，所以上述情况可能是颠换突变的最终形式。

It should be noted that in the foregoing discussion it has been tacitly assumed that the *amber* and *ochre* signals are triplets. Examination of revertants of *amber* mutants has supported this assumption, since in 41 independent revertants of *H*36 the amino-acid replaced is always at the site of mutation, and never in adjacent positions. The 21 revertants that Weigert and Garen[21] isolated reinforce this conclusion.

Function of amber and ochre triplets and the mechanism of suppression. According to present-day ideas of protein synthesis, it is expected that the termination of the growth of the polypeptide chain should involve a special mechanism. Since the terminal carboxyl group of the growing peptide chain is esterified to an *s*RNA (ref. 23), chain termination must involve not only the cessation of growth, but also the cleavage of this bond. Since the *amber* mutants have been shown to result in efficient termination of polypeptide chain synthesis, it is reasonable to suppose that this special mechanism may be provided by the *amber* and *ochre* triplets.

We postulate that the chain-terminating triplets UAA and UAG are recognized by specific *s*RNAs, just like other codons. These *s*RNAs do not carry amino-acids but a special compound which results in termination of the growing polypeptide chain. There are many possible ways of formulating the mechanism in detail, but all are speculative and will not be considered here. The essential feature of this hypothesis is to make the process of chain termination exactly congruent with that of chain extension.

In suppressing strains, a mechanism is provided for competing with chain termination; it is easy to visualize this process as being due to two ways of recognizing the nonsense codon—one by the chain-terminating *s*RNA, and the other by an *s*RNA carrying an amino-acid. Mechanisms of suppression can be classified according to which *s*RNA carries the amino-acid to the nonsense codon.

Alteration in the recognition of the chain-terminating *s*RNA might allow the attachment of an amino-acid to this *s*RNA. This could be brought about either by modifying normal activating enzymes so as to widen their specificity, or by changing the chain terminating *s*RNAs to allow them to be recognized by activating enzymes.

Another possibility is that the region of an amino-acyl *s*RNA used for triplet recognition is modified so that it can recognize the nonsense triplet. Clearly, this alteration must not affect the normal recognition of its own codon by the amino acyl *s*RNA because such a change would be lethal. Either there must be more than one gene for the given *s*RNA, or else the change must produce an ambiguity in the recognition site so that it can read both its own codon and the nonsense codon. Such ambiguity could result not only from mutation in the *s*RNA gene but also by enzymatic modification of one of the bases in the recognition site. The ambiguity, however, must be narrowly restricted to prevent the suppression from affecting codons other than the *amber* and *ochre* triplets. Moreover, the amino-acids which are inserted by the *amber* suppressors must be those the codons of which are connected to UAG. It should be noted that this condition is fulfilled by *su*[+]I

326

应该注意到，在前面的讨论中已经默认了琥珀密码子和赭石密码子是三联体。琥珀突变体的回复突变体检验实验支持该假设，因为 H36 的 41 个独立的回复突变体中，被取代的氨基酸总是处于突变位点，而从来没有在相邻位置出现过。魏格特和加伦 [21] 分离到的 21 个回复突变体也进一步支持了该结论。

琥珀三联体和赭石三联体的功能及其抑制机制　根据目前的蛋白质合成的观点，可以预期多肽链生长的终止应该涉及一种特定机制。sRNA 正确识别后使延伸的肽链的羧基末端发生酯化（参考文献 23），所以链终止不仅使翻译停止了，也使这一化学键发生了断裂。因为已经表明琥珀突变体导致了多肽链合成的有效终止，所以可以合理地认为这一特定机制是通过识别琥珀三联体和赭石三联体而实现的。

我们假定链终止三联体 UAA 和 UAG 就像其他密码子一样，由特定的 sRNA 识别。这些 sRNAs 并不携带氨基酸，而是携带一种特殊的复合物，这种复合物可以导致正在生长的多肽链的终止。关于该机制，我们可以用许多可能的方式来详细叙述，但是它们都只是推导出来的，所以在这里不赘述了。这一假说的本质特征是要使链终止的过程与链延伸的过程完全一致。

在抑制菌株中，存在着一种可与链终止竞争的机制。该过程很容易被视为识别无义密码子的两种方式，一种是通过链终止 sRNA，另一种是通过携带一个氨基酸的 sRNA。可以根据哪个 sRNA 将氨基酸携带至无义密码子而对抑制机制进行分类。

链终止 sRNA 识别的不确定性可能允许将一种氨基酸与该 sRNA 联系起来。而这个过程可以通过修饰正常的激活酶进而扩大酶的专一性来实现，也可以通过改变链终止 sRNA 以使它可以被激活酶识别来实现。

还有一种可能性即氨基酰 sRNA 上用来识别三联体的区域被修饰以使其能够识别无义三联体，很明显，这种修饰一定不能影响氨基酰 sRNA 对其自身密码子的正常识别，因为这种变化是致命的。要么该 sRNA 一定有多于一个基因来编码，否则这一变化一定会在识别位点产生不确定性，导致其既可以读自身的密码子也可以读无义密码子。这种不确定性可能是由于 sRNA 基因存在突变引起的，也可能是由于识别位点的某一碱基被酶修饰造成的。然而，这一不确定性肯定被严密限制以阻止对琥珀三联体和赭石三联体以外的其他密码子产生抑制影响。此外，被琥珀抑制基因插入突变的氨基酸的密码肯定是与 UAG 相关的。被 su_1^+ 插入突变的丝氨酸密码就是这种情况，因为已经研究发现丝氨酸是琥珀突变体的一个回复突变体。这一学说

which inserts serine, since serine has been found as a reversion of an *amber* mutant. This theory does not easily explain the *ochre* suppressors. Since these recognize both *amber* and *ochre* mutants the *s*RNA must possess this ambiguity as well.

Another quite different possibility for suppression that has been considered is that the suppressors alter a component of the ribosomes to permit errors to occur in the reading of the messenger RNA (ref. 24). This is probably the explanation of streptomycin suppression[25], but suppression of *amber* and *ochre* mutants cannot be readily explained by this theory. It is scarcely likely that such a mechanism could be specific for only one or two triplets, and for this reason it might be expected to give us suppression of mutants which are not nonsense, but missense, and this has not been found[1,8]. Moreover, the efficiency of *amber* suppression argues strongly against such a mechanism. It is unlikely that a generalized error in reading nucleotides could produce the 60 percent efficiency of suppression found for su^+_1 without seriously affecting the viability of the cell.

It is a consequence of our theory that normal chain termination could also be suppressed in these strains. Since the *amber* suppressors are efficient we have to introduce the *ad hoc* hypothesis that the UAG codon is rarely used for chain termination in *Escherichia coli* and bacteriophage *T*4 and that UAA is the common codon. This is supported by the fact that all *ochre* suppressors thus far isolated are weak[7,8]. Another possibility is that neither is the common chain terminating triplet. We cannot exclude the existence of other chain terminating triplets which are not suppressible.

To summarize: we show that the triplets of the *amber* and *ochre* mutants are UAG and UAA, respectively. We suggest that the "nonsense" codons should be more properly considered to be the codons for chain termination. In essence, this means that the number of elements to be coded for is not 20 but more likely 21. We propose that the recognition of the chain-terminating codons is carried out by two special *s*RNAs.

We thank our colleagues for their advice, and Dr. M. Nirenberg for allowing us to quote his unpublished results.

(**206**, 994-998; 1965)

S. Brenner, A. O. W. Stretton and S. Kaplan: Medical Research Council, Laboratory of Molecular Biology, Cambridge, England.

References:

1. Benzer, S., and Champe, S. P., *Proc. U.S. Nat. Acad. Sci.*, **47**, 1025 (1961); **48**, 1114 (1962).

2. Garen, A., and Siddiqi, O., *Proc. U.S. Nat. Acad. Sci.*, **48**, 1121 (1962).

3. Campbell, A., *Virology*, **14**, 22 (1961).

4. Epstein, R. H., Bolle, A., Steinberg, C. M., Kellenberger, E., Boy de la Tour, E., Chevalley, R., Edgar, R. S., Susman, M., Denhardt, G. H., and Lielausis, A., *Cold Spring Harbour Symp. Quant. Biol.*, **28**, 375 (1963).

很难解释赭石抑制基因。因为它们既可以识别琥珀突变体，也能识别赭石突变体，该 sRNA 也一定具有上述不确定性。

关于抑制机制另一种相当不同的可能性是：抑制基因改变了核糖体组分以允许核糖体在读取信使 RNA 时出现错误（参考文献 24）。这可能是对链霉素抑制 [25] 的解释，但是该学说不能对琥珀突变体和赭石突变体的抑制给出很好的解释。因为这种机制不可能仅特异地针对一两种三联体，由于这个原因，所以可以预料突变体的抑制不是无义的，而是错义的，但是这种情况目前还没有报道 [1,8]。此外，琥珀抑制的效率与该机制一点都不吻合。如果在读取核苷酸的过程中出现的一个普遍性错误，能在 su^+_1 菌株中产生 60% 的抑制效率却对该细胞的生活力没有严重影响，这是不可能的。

我们的理论得到的结论是正常的链终止也可以在这些菌株中被抑制。由于琥珀抑制基因很有效，所以我们在这里提出一个特别的假说，即 UAG 密码子很少被用来引发大肠杆菌和 T4 噬菌体中的链终止，UAA 才是通用的终止密码子。至今分离到的所有的赭石抑制子的抑制作用都很弱这一事实可以用来支持该假说 [7,8]。另一种可能性是它们都不是通用的链终止三联体。我们不能排除存在其他的不可抑制的链终止三联体的可能性。

综上所述，我们的研究表明琥珀和赭石突变体的三联体分别是 UAG 和 UAA。我们认为"无义"密码子被当作终止密码子会更为恰当。从本质上说，这意味着 64 种三联体编码的氨基酸数目不是 20 而更可能是 21。我们提出链终止密码子的识别是由两种特殊的 sRNA 来实现的。

感谢我们的同事们提出的宝贵建议以及尼伦伯格博士允许我们引用他尚未发表的结果。

（刘皓芳 翻译；刘京国 审稿）

5. Sarabhai, A., Stretton, A. O. W., Brenner, S., and Bolle, A., *Nature*, **201**, 13(1964).

6. Stretton, A. O. W., and Brenner, S., *J. Mol. Biol.* (in the press).

7. Kaplan, S., Stretton A. O. W., and Brenner, S. (in preparation).

8. Brenner, S., and Beekwith, J. R. (in preparation).

9. Brenner, S., and Stretton, A. O. W. (in preparation).

10. Brown, D. M., and Schell, P., *J. Mol. Biol.*, **3**, 709 (1961). Freese, E., Bautz-Freese, E., and Bautz, E., *J. Mol. Biol.*, **3**, 133(1961). Schuster, H., *J. Mol. Biol.*, **3**, 447 (1961). Freese, E., Bautz, E., and Freese, E. B., *Proc. U.S. Nat. Acad. Sci.*, **47**, 845 (1961).

11. Tocchini-Valentini, G. P., Stodolsky, M., Aurisicchio, A., Sarnat, M., Graziosi, F., Weiss, S. B., and Geiduschek, E. P., *Proc. U.S. Nat. Acad. Sci.*, **50**, 935 (1963). Hayashi, M., Hayashi, M. N., and Spiegelman, S., *Proc. U.S. Nat. Acad. Sci.*, **50**, 664 (1963). Marmur, J., Greenspan, C. M., Palacek, E., Kahan, F. M., Levene, J., and Mandel, M., *Cold Spring Harbor Symp. Quant. Biol.*, **28**, 191 (1963). Bautz, E. K. F., *Cold Spring Harbor Symp. Quant. Biol.*, **28**, 205 (1963). Hall, B. D., Green, M., Nygaard, A. P., and Boezi, J., *Cold Spring Harbor Symp. Quant. Biol.*, **28**, 201 (1963).

12. Champe, S. P., and Benzer, S., *Proc. U.S. Nat. Acad. Sci.*, **48**, 532 (1962). Tessman, I., Poddar, R. K., and Kumar, S., *J. Mol. Biol.*, **9**, 352 (1964).

13. Freese, E., *J. Mol. Biol.*, **1**, 87 (1959). Freese, E., *Proc. U.S. Nat. Acad. Sci.*, **45**, 622 (1959). Howard, B. D., and Tessman, I., *J. Mol. Biol.*, **9**, 372 (1964).

14. Garen, A., and Weigert, M. G., *J. Mol. Biol.* (in the press).

15. Notani, G. W., Engelhardt, D. L., Konigsberg, W., and Zinder, N., *J. Mol. Biol.* (in the press).

16. Nirenberg, M., Jones. O. W., Leder, P., Clark, B. F. C., Sly, W. S., and Pestka, S., *Cold Spring Harbor Symp. Quant., Biol.*, **28**, 549 (1963).

17. Speyer, J. F., Lengyel, P., Basilio, C., Wahba, A. J., Gardner, R. S., and Ochoa, S., *Cold Spring Harbor Symp. Quant., Biol.*, **28**, 559 (1963).

18. Nirenberg, M., Leder, P., Bernfield, M., Brimacombe, R., Trupin, J., and Rottman, F., *Proc. U.S. Nat. Acad. Sci.* (in the press).

19. Stretton, A. O. W., and Brenner, S. (in preparation).

20. Trupin, J., Rottman, F., Brimacombe, R., Leder, P., Bernfield, M., and Nirenberg, M., *Proc. U.S. Nat. Acad. Sci.* (in the press). Clark, B. F. C., presented before the French Biochemical Society, February, 1965.

21. Weigert, M. G., and Garen, A., *Nature* (preceding communication).

22. Leder, P., and Nirenberg, M. W., *Proc. U.S. Nat. Acad. Sci.*, **52**, 1521 (1964).

23. Gilbert, W., *J. Mol. Biol.*, **6**, 389 (1963). Bretscher, M. S., *J. Mol. Biol.*, **7**, 446 (1963).

24. Davies, J., Gilbert, W., and Gorini, I., *Proc. U.S. Nat. Acad. Sci.*, **51**, 883 (1964).

25. Gorini, L., and Kataja, E., *Proc. U.S. Nat. Acad. Sci.*, **51**, 487 (1964).

Abnormal Haemoglobins and the Genetic Code

D. Beale and H. Lehmann

Editor's Note

Here biologists D. Beale and Hermann Lehman from Cambridge University describe how the degenerate genetic code influences different types of human haemoglobin, and settle an ongoing debate between themselves and Nobel laureate Francis Crick. Crick, best known for his work deciphering the structure of DNA, had pointed out that haemoglobin I, an abnormal type of the oxygen-binding protein, cannot be the outcome of a single mutation. But Beale and Lehman show that it can, identifying a single amino-acid substitution on the alpha polypeptide chain.

DR. F. H. C. CRICK has directed our attention to the latest work of Nirenberg *et al.*[1] which has considerably advanced the definition of the code whereby messenger RNA determines which amino-acid is incorporated into a protein. The code is "degenerate" and more than one arrangement of three bases or a codon can spell the same amino-acid. The first two bases are, however, somewhat more specific than the third. For example, for glutamic acid (Glu) they are guanine adenine (GA), but the third can be either A or G. However, the first two bases in the codon spelling aspartic acid (Asp) are also GA, but the third may be cytosine or uracil (C or U). Thus the coding for Glu is **GA**A or **GA**G. Some amino-acids are coded by any of the four possible bases in the third place of the codon. For example, the coding for threonine (Thr) is **AC**A or G or C or U. For a few amino-acids only, alternatives have also been found for the first two bases of the codon: leucine (Leu), for example, may be coded **UU**A or G or as **CU**G or C or U. From the considerable evidence for A being an alternative third part of the codon when G is one, one might write the second coding for leucine as **CU***A* or G or C or U. Table 1 gives the codons for amino-acids involved in substitutions of human haemoglobin; those surmised because A and G, and C and U, respectively, are alternatives in the third position, are in italics. Table 2 summarizes the amino-acid substitutions which have been reported in human haemoglobin. They include the abnormal haemoglobins fully described and those for which an amino-acid substitution is known but has not yet been finally allocated to a definite part in one of the polypeptide chains. Furthermore, as the β- and δ-polypeptide chains differ in only a very few amino-acid residues, likely to be point mutations, differences between these two chains have also been listed.

332

异常的血红蛋白及其遗传密码

比尔，莱曼

编者按

本文中，剑桥大学的生物学家比尔和赫尔曼·莱曼阐述了简并的遗传密码是如何影响不同类型的人血红蛋白的，并由此结束了他们与诺贝尔奖获得者弗朗西斯·克里克之间的不休争论。以提出 DNA 双螺旋结构而闻名的克里克曾经指出，血红蛋白 I（一种氧结合蛋白质的异常类型）不可能是单点突变的结果。但是比尔和莱曼指出那是可以的，并鉴定了发生在 α 多肽链上的单个氨基酸替换。

克里克博士已使我们的注意力集中到尼伦伯格等人最近的工作上[1]，他们的工作极大地促进了遗传密码的确定，而信使 RNA 正是通过遗传密码来决定用哪种氨基酸合成蛋白质的。遗传密码具有"简并性"，不止一组三碱基序列或者说不止一个密码子能够编码相同的氨基酸。然而，前两个碱基的特异性比第三个碱基稍强。例如，对于谷氨酸（Glu）来说，前两个碱基是鸟嘌呤（G）和腺嘌呤（A），但是第三个碱基可以是 A 或 G。不过编码天冬氨酸（Asp）的密码子前两个碱基也是 GA，但是第三个碱基可以是胞嘧啶或尿嘧啶（C 或 U）。因此，编码 Glu 的密码子是 **GA**A 或 **GA**G。有些氨基酸密码子的第三个碱基可以是四种碱基中的任何一个。比如，编码苏氨酸（Thr）的密码子可以是 **AC**A、**AC**G、**AC**C 或 **AC**U。只有少数一些氨基酸，编码它的密码子的前两个碱基也具有可替换性：如亮氨酸（Leu），可以被 UUA 或 UUG 编码，也可以被 **CU**G、**CU**C 或 **CU**U 编码。大量的证据表明当密码子的第三个碱基是 G 时就可以用 A 替换，于是编码亮氨酸的第二类密码子可以记为 **CU**A、**CU**G、**CU**C 或 **CU**U。表 1 给出了编码人血红蛋白中可能发生替换的氨基酸的密码子；由于密码子的第三个碱基可能分别被 A、G、C 或 U 中的一种碱基替换，所以这些猜测的碱基用斜体标注。表 2 总结了已经报道的人血红蛋白中发生的氨基酸替换。其中包括详细记述的异常血红蛋白，以及一些已知发生氨基酸替换但还未最终定位在某条多肽链的特定位置上的情况。另外，由于 β 多肽链和 δ 多肽链只有很少的几个氨基酸残基不一致，很可能是点突变，我们也列出了这两条链之间的差异。

Table 1. Messenger Ribonucleic-acid Codons for Amino-acids involved in
Substitutions in Human Haemoglobin

Lysine	**AA**	A	or	*G*					Proline	**CC**	A	or	*G*	or	C	or	U
Asparagine	**AA**				C	or	U		Leucine	**CU** *A* or *G* or C or U **UU** A or G							
Threonine	**AC**	A	or	G	or	C	or	U	Glutamic acid	**GA**	A	or	G				
Arginine	**AG** A or *G* **CG** A or *G* or C or *U*								Aspartic acid	**GA**				C	or	U	
Serine	**AG** C or U **UC** A or G or C or U								Alanine	**GC**	*A*	or	G	or	*C*	or	U
Methionine	**AU**	*A*	or	G					Glycine	**GG**	*A*	or	G	or	*C*	or	U
Glutamine	**CA**	A	or	G					Valine	**GU**	*A*	or	G	or	C	or	U
Histidine	**CA**				C	or	U		Tyrosine	**UA**					C	or	U

(Surmised bases in italics.)

A, Adenine; C, cytosine; G, guanine; U, uracil

Table 2. Amino-acid Substitutions in Human Haemoglobins

Substitution	Examples	Substitution	Examples
Lys → Asn	β61, Hikari[2]	Val → Glu	β67, M Milwaukee[23]
Lys → Asp	α16, I[3]	Gly → Asp	α15, J Oxford[24] (also described as "I" Interlaken[25]); α22 J Medellin[26]; α57 Norfolk[27]; β16, Baltimore[28]
Lys → Glu	β95, N[4]	Asp → Gly	α47, L Ferrara[29]
Thr → Lys	β87, D Ibadan[5]	Asp → Asn	β79, G Accra[30]
Asn → Lys	α68, G Philadelphia[6]	His → Tyr	α58, M Boston[23], α87 M Iwate[31], β2, Tokuchi[32]; β63 M Saskatoon[33]
Glu → Lys	α116, O Indonesia[7]; β6, C[8]; β7, Siriraj[9]; β26, E[10]; β121 O Arab[7]; γ6, F Galveston[11]	His → Asp	β143, Kenwood (or His-Glu)[32]
Glu → Gln	α30, G Chinese[12]	His → Asn	δ117, A$_2$[19,20]
Gln → Glu	α54, Mexico[13]	His → Arg	β63, Zurich[34]; δ117, A$_2$[19,20]
Gln → Arg	α54, Shimonoseki[14]	Leu → Arg	δ14, A'$_2$ (or B$_2$)[35]
Glu → Gly	β7, San Jose[15]	Ser → Thr	δ9, A$_2$[19,20]
Gly → Glu	β46 or β56, K Ibadan[16]	Thr → Asn	δ12, A$_2$[19,20]
Glu → Ala	β22 or β26, G Coushatta[17]; β43, G Galveston[18]; δ220, A$_2$[19,20]	Thr → Ser	δ50, A$_2$[19,20]
Ala → Glu	β70 or β76, Seattle[21]	Pro → Gln	δ126, A$_2$[19,20]
Glu → Val	β6, S[22]		

Fig. 1 illustrates the possible one-step mutations for the amino-acids listed in Table 2. This scheme covers all substitutions with one exception. As Dr. Crick has pointed out to us, haemoglobin I α16 Lys → Asp cannot have been the outcome of a single mutation but requires two mutations within one codon. As this is a point of considerable theoretical interest we have re-investigated haemoglobin I and have found that, in fact, the substitution is α16 Lys → Glu. Thus, at present, all amino-acid substitutions in human haemoglobin can be the outcome of single mutations. Degeneracy of the code must be assumed because the last base in the codon for glycine cannot be the same for the mutations Glu → Gly[15] and Gly → Asp[24-28]; similarly, it cannot be the same for arginine in His → Arg[34] and Gln → Arg[14]. The exclusion of the step Lys → Asp or Asp → Lys as a single mutation strongly supports the formula involving α116 Glu → Lys for haemoglobin O Indonesia[7]. The electrophoretic mobility suggests a double charge change from an acidic to a basic residue. The peptide involved (α*Tp*XII) contains both, one Glu (α116) and one Asp (α126), but as Glu → Arg, Asp → Arg, and Asp → Lys would require a double

表 1. 编码人血红蛋白可能发生替换的氨基酸的信使核糖核酸密码子

赖氨酸 (Lys)	**AA** A 或 *G*	脯氨酸 (Pro)	**CC** A 或 *G* 或 C 或 U
天冬酰胺 (Asn)	**AA** C 或 U	亮氨酸 (Leu)	**CU** *A* 或 G 或 C 或 U **UU** A 或 G
苏氨酸 (Thr)	**AC** A 或 G 或 C 或 U	谷氨酸 (Glu)	**GA** A 或 G
精氨酸 (Arg)	**AG** A 或 *G* **CG** A 或 *G* 或 C 或 *U*	天冬氨酸 (Asp)	**GA** C 或 U
丝氨酸 (Ser)	**AG** C 或 U **UC** A 或 G 或 C 或 U	丙氨酸 (Ala)	**GC** *A* 或 G 或 *C* 或 U
甲硫氨酸 (Met)	**AU** *A* 或 G	甘氨酸 (Gly)	**GG** *A* 或 G 或 *C* 或 U
谷氨酰胺 (Gln)	**CA** A 或 G	缬氨酸 (Val)	**GU** *A* 或 G 或 *C* 或 U
组氨酸 (His)	**CA** C 或 U	酪氨酸 (Tyr)	**UA** C 或 U

（猜测的碱基用斜体标注。）

A，腺嘌呤；C，胞嘧啶；G，鸟嘌呤；U，尿嘧啶

表 2. 人血红蛋白中的氨基酸替换

替换	例子	替换	例子
Lys → Asn	β61，光市 [2]	Val → Glu	β67，M 密尔沃基 [23]
Lys → Asp	α16，I [3]	Gly → Asp	α15，J 牛津 [24]（也可以表示为 "I" 因特拉肯 [25]）；α22 J 麦德林 [26]；α57 诺福克郡 [27]；β16，巴尔的摩 [28]
Lys → Glu	β95，N [4]	Asp → Gly	α47，L 费拉拉 [29]
Thr → Lys	β87，D 伊巴丹 [5]	Asp → Asn	β79，G 阿克拉 [30]
Asn → Lys	α68，G 费城 [6]	His → Tyr	α58，M 波士顿 [23]，α87，M 岩手 [31]，β2，德地 [32]；β63，M 萨斯卡通 [33]
Glu → Lys	α116，O 印尼 [7]；β6，C [8]；β7，诗里拉吉 [9]；β26，E [10]；β121，O 阿拉伯 [7]；γ6，F 加尔维斯顿 [11]	His → Asp	β143，肯伍德（或 His-Glu）[32]
Glu → Gln	α30，G 中国人 [12]	His → Asn	δ117，A₂ [19,20]
Gln → Glu	α54，墨西哥 [13]	His → Arg	β63，苏黎世 [34]；δ117，A₂ [19,20]
Gln → Arg	α54，下关市 [14]	Leu → Arg	δ14，A'₂（或 B₂）[35]
Glu → Gly	β7，圣何塞 [15]	Ser → Thr	δ9，A₂ [19,20]
Gly → Glu	β46 或 β56，K 伊巴丹 [16]	Thr → Asn	δ12，A₂ [19,20]
Glu → Ala	β22 或 β26，G 考沙塔 [17]；β43，G 加尔维斯顿 [18]；δ220，A₂ [19, 20]	Thr → Ser	δ50，A₂ [19,20]
Ala → Glu	β70 或 β76，西雅图 [21]	Pro → Gln	δ126，A₂ [19,20]
Glu → Val	β6，S [22]		

    图 1 说明了表 2 列出的氨基酸中可能发生的一步突变。除了一个例外，这个图涵盖了所有可能发生的替换。正如克里克博士指出，血红蛋白 I α16 处 Lys → Asp 不是单点突变的结果，而是需要某一密码子内双突变。鉴于这一点具有相当重要的理论意义，我们再一次对血红蛋白 I 展开研究，发现实际上 α16 处的替换是 Lys → Glu。因此，目前发现的人血红蛋白中所有的氨基酸替换都是单点突变的结果。我们必须假设密码子具有简并性，因为编码甘氨酸的密码子的最后一个碱基与突变 Glu → Gly[15] 和 Gly → Asp[24-28] 不可能相同；类似地，编码精氨酸的密码子的最后一个碱基与突变 His → Arg[34] 和 Gln → Arg[14] 也不可能相同。排除 Lys → Asp 或 Asp → Lys 为单点突变，这强有力地支持了印尼氧合血红蛋白中 α116 Glu → Lys 突变 [7] 的规则。电泳迁移率显示残基从酸性转变为碱性，发生了两次电荷改变。涉及的多肽（α*Tp*XII）有两处存在这种情况：一个 Glu（α116）和一个 Asp（α126），但是由于 Glu → Arg、

mutation the likely formula is $\alpha_2^{116Glu \rightarrow Lys}\beta_2$. By similar reasoning one can assume that haemoglobin Kenwood[32], for which two alternative formulae involving $\beta 143$ His $\rightarrow$ Glu and $\beta 143$ His $\rightarrow$ Asp have been proposed, is likely to be $\alpha_2\beta_2^{143His \rightarrow Asp}$.

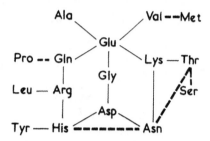

Fig. 1. Possible one-step mutations for amino-acids involved in substitutions in human haemoglobin. The dotted lines indicate differences between β- and δ-chains.

So far as human haemoglobin is concerned, there are four possible charge changes detectable by electrophoresis at pH 8.6 (that is, a pH at which charge changes do not include the influence of histidine) (Fig. 2). A loss of a negative charge can arise from the replacement of Glu or Asp by a neutral residue, or from replacing a neutral residue by Lys or Arg. The variant of haemoglobin A will then move in the position of haemoglobins S or G. A positive charge can be lost by substitution of a Lys or Arg by a neutral residue or of a neutral residue by Glu or Asp. The resulting variants will then have the mobility of haemoglobins J or K. A double charge change involving either the replacement of a negative by a positive or of a positive by a negative charge can arise by a single mutation only when Glu is substituted by Lys, or Lys by Glu. Thus the haemoglobins C, E, O are all Glu $\rightarrow$ Lys mutations, and the haemoglobins of the I and N type must be expected to involve a change of Lys $\rightarrow$ Glu.

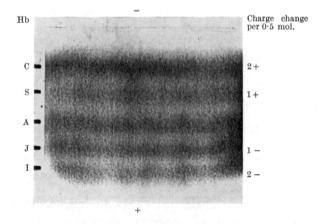

Fig. 2. Paper electrophoresis at pH 8.6 of haemoglobin A, and four variants showing respectively one and two additional positive or negative charges per half molecule.

It seems remarkable that of eleven glutamic acid residues, nine and possibly ten have been

336

Asp → Arg 及 Asp → Lys 可能需要双重突变，可能的形式是 $\alpha_2^{116Glu \to Lys}\beta_2$。根据类似的原因可以推测，肯伍德血红蛋白[32]中提出的包括 β143 His → Glu 和 β143 His → Asp 的两种可变形式可能是 $\alpha_2\beta_2^{143His \to Asp}$。

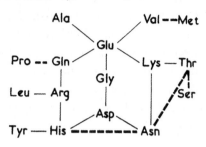

图 1. 人血红蛋白中发生替换的氨基酸可能的一步突变。虚线代表 β 链和 δ 链间的差异。

就人血红蛋白而言，在 pH 8.6 的条件下（也就是说，在这个 pH 值下不包括组氨酸影响的电荷变化）进行电泳，可检测到 4 种可能的电荷变化（图 2）。Glu 或 Asp 被一个中性残基替换，或者一个中性残基被 Lys 或 Arg 替换，会导致损失一个负电荷。然后变异的血红蛋白 A 将移动到血红蛋白 S 或血红蛋白 G 的位置。Lys 或 Arg 被一个中性残基替换，或者一个中性残基被 Glu 或 Asp 替换，会导致损失一个正电荷。产生的变异体将具有血红蛋白 J 或血红蛋白 K 的迁移率。只有 Glu 被 Lys 替换或者 Lys 被 Glu 替换时，一个单点突变才能引起双电荷改变（包括负电荷被正电荷替换或者正电荷被负电荷替换）。因此，血红蛋白 C、E、O 都是 Glu → Lys 突变，预计 I 型和 N 型的血红蛋白必然涉及 Lys → Glu 的突变。

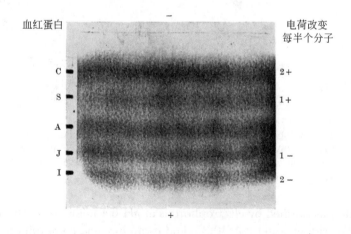

图 2. 在 pH 8.6 的条件下进行纸电泳，血红蛋白 A 及 4 种变异体分别显示出每半个分子增加 1~2 个正电荷或负电荷。

似乎比较显著的是，11 个谷氨酸残基中有 9 个（也可能是 10 个）发生了突变，

involved in mutations, three of them twice, yet of seventeen aspartic acid residues only one[30], or possibly two[29], have been subject to substitutions. On the other hand, changes to Asp and Glu occur with similar frequency, six and five times respectively. The genetic code would permit Glu and Asp to mutate equally well to Val, Ala, or Gly. In addition, whereas Glu can mutate to Lys and Gln, Asp can mutate to Asn, His or Tyr. All such mutations would involve charge changes in the molecule which would be detected by electrophoresis. The frequency of mutations from Glu to other residues and the rarity of replacements of Asp could suggest that Asp may be more important for the structure and function of the haemoglobin A molecule ($\alpha_2\beta_2$). This is not necessarily so for the haemoglobins A_2 ($\alpha_2\delta_2$) and F ($\alpha_2\gamma_2$) because only two Asp and two Glu respectively are common to all four polypeptide chains, α, β, γ and δ.

Analysis of haemoglobin I. Haemoglobin I[36] was isolated from a specimen kindly given to us by Miss J. Atwater, Philadelphia. The purified haemoglobin was "fingerprinted" by the usual methods (all relevant methods have recently been summarized[5]) and the fingerprints showed that αTpIII and αTpIV were missing and a new peptide moving towards the anode was observed (Fig. 3). It will be seen that the electrophoretic (horizontal) mobility of the new peptide is the same as that of βTpV and βTpV (oxidized). βTpV yields on hydrolysis three aspartic acid residues. Prolonged chromatography is required to remove traces of βTpV and particularly of βTpV (oxidized) from the abnormal peptide in haemoglobin I.

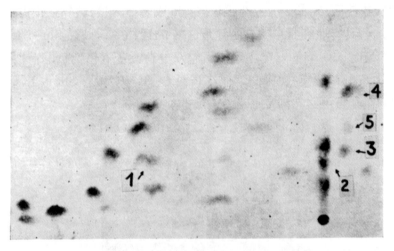

Fig. 3. The "fingerprint" of haemoglobin I. 1, $\alpha^t Tp$III missing; 2, $\alpha^t Tp$IV missing; 3, new peptide ($\alpha^t Tp$IIIIV); 4, βTpV; 5, βTpV (oxidized). Note that the βTpV peptides have the same electrophoretic mobility (horizontal) as the abnormal peptides from haemoglobin I, but they are well separated by chromatography (vertical).

The new peptide was purified by electrophoresis at pH 6.4 followed by chromatography in isoamyl alcohol:pyridine:water (30:30:35) and its amino-acid composition determined on an automatic analyser. This analysis (Table 3), when compared with a similar analysis for αTpIII and αTpIV, showed that a lysine residue, presumably at position 16 of the α-chain, had been replaced by one of glutamic acid or of glutamine (glutamine would have been converted to glutamic acid during hydrolysis).

338

其中的 3 个是 2 次突变，而 17 个天冬氨酸残基中只有 1 个 [30] 或 2 个 [29] 发生了替换。另一方面，突变为 Asp 和 Glu 的频率相似，分别为 6 次和 5 次。遗传密码可能允许 Glu 和 Asp 机会均等地突变为 Val、Ala 或 Gly。另外，Glu 可以突变为 Lys 和 Gln，Asp 可以突变为 Asn、His 或 Tyr。所有的这些突变都可能涉及分子中电荷的改变，因而可以通过电泳进行检测。Glu 突变为其他残基的突变频率和 Asp 很少发生替换都表明，Asp 对于血红蛋白 A 分子（$\alpha_2\beta_2$）的结构和功能可能具有更重要的作用。对于血红蛋白 A_2（$\alpha_2\delta_2$）和 F（$\alpha_2\gamma_2$）来说则并非如此，因为对于所有的 4 条多肽链 α、β、γ 和 δ，各自仅有 2 个 Asp 和 2 个 Glu 是共有的。

血红蛋白 I 的分析　血红蛋白 I 是从费城的阿特沃特小姐惠赠给我们的样品中分离出来的 [36]。采用通常的方法（最近已经对所有相关的方法进行了总结 [5]）对纯化的血红蛋白进行"指纹测定"，指纹图谱显示 αTpIII 和 αTpIV 缺失，并且观察到一个新的多肽朝正极移动（图 3）。由此可知新肽链的电泳（水平方向上）迁移率与 βTpV 和 βTpV（被氧化的）相同。βTpV 水解产生 3 个天冬氨酸残基。需用长效色谱技术从血红蛋白 I 的异常多肽中分离出微量的 βTpV，特别是氧化的 βTpV。

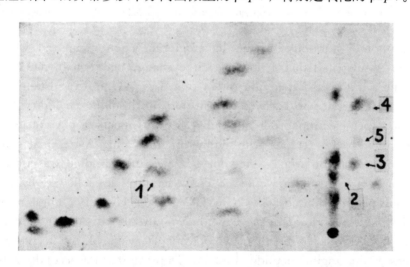

图 3. 血红蛋白 I 的"指纹测定"。1, α^TpIII 缺失；2, α^TpIV 缺失；3, 新的多肽（α'TpIII-IV）；4, βTpV；5, βTpV（被氧化的）。注意 βTpV 多肽与来自血红蛋白 I 的异常多肽的电泳迁移率（水平方向上）相同，但色谱法（垂直方向上）可以将它们很好地分离。

在 pH 6.4 的条件下进行电泳，随后用流动相为异戊醇∶吡啶∶水（30∶30∶35）的色谱进行分离，从而使这个新的多肽得到纯化，并用氨基酸自动分析仪分析其氨基酸组成。与 αTpIII 和 αTpIV 的类似分析相比较，这一分析结果（表 3）显示可能在 α 链的第 16 个氨基酸的位置上，一个赖氨酸残基被一个谷氨酸或者谷氨酰胺取代（在水解过程中谷氨酰胺可以转变为谷氨酸）。

Table 3. Amino-acid Analysis of Abnormal Peptide from Hb-I and the Peptides
α*Tp*III and α*Tp*IV from Hb-A

Amino-acid	α^I*Tp*III-IV		α^A*Tp*III		α^A*Tp*IV	
	μ Mole	Molar ratio	μ Mole	Molar ratio	μ Mole	Molar ratio
Asp	0.005	0.07	0.003	0.05	0.004	0.06
Thr	0.003	0.04	0.005	0.08	0.002	0.03
Ser	0.003	0.04	0.002	0.03	0.003	0.04
Glu	0.285	4.07	0.006	0.10	0.222	3.09
Pro	0.002	0.03	0.003	0.05	0.002	0.03
Gly	0.283	4.04	0.062	1.09	0.208	2.90
Ala	0.401	5.73	0.110	1.93	0.273	3.79
Val	0.082	1.17	0.004	0.07	0.075	1.04
Met	0.001	0.01	0.001	0.02	0.002	0.03
Leu	0.078	1.11	0.008	0.05	0.069	0.96
Tyr	0.068	0.97	0.001	0.02	0.062	0.86
Phe	0.004	0.06	0.003	0.05	0.002	0.03
Try*	+	(1.0)	+	(1.0)	–	(0.0)
Lys	0.004	0.06	0.057	1.00	0.005	0.07
His	0.065	0.93	0.005	0.08	0.058	0.80
Arg	0.070	1.00	0.001	0.02	0.072	1.00

* Detected in peptide by the Ehrlich reagent. A positive reaction was assumed to be due to one residue of tryptophan.

The electrophoretic mobility of the haemoglobin when compared with that of haemoglobin A suggested that a double charge change per α-chain had taken place, thus favouring a mutation Lys → Glu rather than Lys → Gln. To substantiate this, the new peptide was digested with chymotrypsin and the resulting chymotryptic peptides separated by paper electrophoresis at *p*H 6.4. By this means a chymotryptic peptide giving a transient yellow colour with ninhydrin (indicating *N*-terminal glycine) and giving a positive reaction when tested for tyrosine and histidine was isolated. Acid hydrolysis (by which any Gln would be converted to Glu) followed by amino-acid analysis of this peptide showed 3Gly, 2Glu, 2Ala, 1Val, 1His, 1Tyr. These results would correspond with a peptide, α15–24: Gly-Glu-Val-Gly-Ala-His-Ala-Gly-Glu-Tyr which would have been produced by the chymotryptic cleavage of the tryptophanyl and tyrosyl bonds of the original peptide (Table 4). To prove that the second residue of the chymotryptic peptide was the expected glutamyl rather than glutaminyl, the peptide was treated with leucine amino peptidase whereby glycine, glutamic acid and valine, but no glutamine, were quickly released.

This biochemical finding arising from a suggestion made by Dr. F. H. C. Crick is of interest because it is probably the first time that the genetic code has been used as an aid in the analysis of a protein, rather than that the analysis of a protein has been used for the unravelling of the genetic code.

表 3. 来自血红蛋白 I 的异常多肽与来自血红蛋白 A 的 α*Tp*III 和 α*Tp*IV 多肽的氨基酸分析

氨基酸	α^I*Tp*III-IV		α^A*Tp*III		α^A*Tp*IV	
	微摩尔	摩尔比	微摩尔	摩尔比	微摩尔	摩尔比
Asp	0.005	0.07	0.003	0.05	0.004	0.06
Thr	0.003	0.04	0.005	0.08	0.002	0.03
Ser	0.003	0.04	0.002	0.03	0.003	0.04
Glu	0.285	4.07	0.006	0.10	0.222	3.09
Pro	0.002	0.03	0.003	0.05	0.002	0.03
Gly	0.283	4.04	0.062	1.09	0.208	2.90
Ala	0.401	5.73	0.110	1.93	0.273	3.79
Val	0.082	1.17	0.004	0.07	0.075	1.04
Met	0.001	0.01	0.001	0.02	0.002	0.03
Leu	0.078	1.11	0.008	0.05	0.069	0.96
Tyr	0.068	0.97	0.001	0.02	0.062	0.86
Phe	0.004	0.06	0.003	0.05	0.002	0.03
Try*	+	(1.0)	+	(1.0)	−	(0.0)
Lys	0.004	0.06	0.057	1.00	0.005	0.07
His	0.065	0.93	0.005	0.08	0.058	0.80
Arg	0.070	1.00	0.001	0.02	0.072	1.00

* 借助埃利希试剂进行多肽的测定。推测阳性反应可能是由于存在一个色氨酸残基。

与血红蛋白 A 相比，血红蛋白的电泳迁移率显示每条 α 链都发生了双电荷改变，从而支持是 Lys → Glu 而不是 Lys → Gln。为了证实这一点，用胰凝乳蛋白酶消化新肽，将酶解后的多肽片段在 pH 6.4 的条件下通过纸电泳分离。用这种方法分离出胰蛋白酶酶解后的多肽：茚三酮染色显示出短暂的黄色（表明氮末端为甘氨酸），游离酪氨酸和组氨酸检测呈阳性。对这个肽链进行酸解（这样 Gln 可以完全转化为 Glu），随后对多肽氨基酸分析，结果显示为 3Gly、2Glu、2Ala、1Val、1His、1Tyr。这些结果可能对应多肽 α15~24：Gly-Glu-Val-Gly-Ala-His-Ala-Gly-Glu-Tyr，它应由胰蛋白酶切断原来多肽的色氨酰与酪氨酰之间的肽键而得到（表 4）。为了证明胰蛋白酶酶解肽段的第二个残基是预期的谷氨酰而非谷氨酰胺酰，将这个多肽用亮氨酸氨肽酶进行处理，结果快速释放出甘氨酸、谷氨酸和缬氨酸，而不释放谷氨酰胺。

由克里克博士提出的建议而引出的这一生化发现是很吸引人的，因为它可能是第一次将遗传密码用于辅助蛋白质分析，而不是像之前那样通过分析蛋白质来揭示遗传密码。

Table 4. The Chymotryptic Digestion of the Abnormal Tryptic Peptide in
Haemoglobin I

Note that peptide 4 just separated from 5 under the conditions of this experiment

Proposed peptide α^ATpIII-IV based on the known amino-acid sequence of α^ATpIII and α^ATpIV		α 12 13 14 15 16 17 18 19 20 21 22 23 24 25 26 27 28 29 30 31 Ala-Ala-Try-Gly-Glu-Val-Gly-Ala-His-Ala-Gly-Glu-Tyr-Gly-Ala-Glu-Ala-Leu-Glu-Arg ↑ ↑ ↑
Chymotryptic* peptides from the abnormal peptide in Hb I	Amino-acids found on hydrolysis	
1	Gly, Ala Glu, Leu	Gly Ala Glu Ala Leu
2	Gly, Glu, Val Ala, His, Tyr Leu	Gly Glu Val Gly Ala His Ala Gly Glu Tyr Gly Ala Glu Ala Leu
3	3 Gly, 2 Glu, 1 Val 2 Ala, 1 His, 1 Tyr	Gly Glu Val Gly Ala His Ala Gly Glu Tyr
4	Glu, Arg	Glu Arg
5	Ala, Try	Ala Ala Try

* The composition was found in all cases by paper electrophoresis, but in peptide 3 the amino-acids were also determined on the amino-acid analyzer; molar ratio (±0.2) is, therefore, also given.

We thank Dr. A. J. Munro for advice concerning the genetic code.

(**207**, 259-261; 1965)

D. Beale and H. Lehmann: Medical Research Council Abnormal Haemoglobin Research Unit, Department of Biochemistry, University of Cambridge.

References:

1. Nirenberg, M., Leder, P., Bernfield, M., Brimacombe, R., Trupin, J., Rottman, F., and O'Neal, C., *Proc. U.S. Nat. Acad. Sci.*, **53**, 1161 (1965).

2. Shibata, S., Miyaji, T., Iuchi, I., Ueda, S., and Takeda, I., *Clin. Chim. Acta*, **10**, 101 (1964).

3. Murayama, M., *Nature*, **196**, 276 (1962).

4. Kraus, L. (personal communication).

5. Watson-Williams, E. J., Beale, D., Irvine, D., and Lehmann, H., *Nature*, **205**, 1273 (1965).

6. Baglioni, C., and Ingram, V. M., *Nature*, **189**, 465 (1961).

7. Baglioni, C., and Lehmann, H., *Nature*, **196**, 229 (1962).

8. Hunt, J. A., and Ingram, V. M., *Biochim. Biophys. Acta*, **42**, 409 (1960).

9. Tuchinda, S., Beale, D., and Lehmann, H., *Brit. Med. J.*, **1**, 1583 (1965).

10. Hunt, J. A., and Ingram, V. M., *Biochim. Biophys. Acta*, **49**, 520 (1961).

11. Schneider, R. G., and Jones, R. T., *Science*, **148**, 240 (1965).

12. Swenson, R. T., Hill, R. L., Lehmann, H., and Jim, R. T. S., *J. Biol. Chem.*, **237**, 1517 (1962).

13. Jones, R. T., Koler, R. D., and Lister, R., *Clin. Res.*, **11**, 105 (1963).

14. Miyaji, T., Iuchi, I., Takeda, I., and Shibata, S., *Acta Haemat. Jap.*, **26**, 531 (1963).

15. Hill, R. L., Swenson, R. T., and Schwartz, H. C., *J. Biol. Chem.*, **235**, 3182 (1960).

16. Allan, N., Beale, D., Irvine, D., and Lehmann, H. (in preparation).

17. Schneider, R. G., Haggard, M. E., McNutt, C. W., Johnson, J. E., Bowman, B. H., and Barnett, D. R., *Science*, **143**, 687 (1964).

18. Bowman, B. H., Oliver, C. P., Barnett, D. R., Cunningham, J. E., and Schneider, R. G., *Blood*, **23**, 193 (1964).

19. Ingram, V. M., and Stretton, A. O. W., *Biochim. Biophys. Acta*, **62**, 456 (1962).

20. Ingram, V. M., and Stretton, A. O. W., *Biochim. Biophys. Acta*, **63**, 20 (1962).

21. Huehns, E. R., and Shooter, E. M., *J. Med. Genet.*, **2**, 48 (1965).

表 4. 对血红蛋白 I 中的异常胰蛋白酶多肽进行胰凝乳蛋白酶酶解后的酶解产物

注意肽段 4 仅能在此实验条件下从肽段 5 中分离出来

根据已知的 α^*Tp*III 和 α^*Tp*IV 氨基酸序列提出的 α^*Tp*III-IV 多肽序列		α 12 13 14 15 16 17 18 19 20 21 22 23 24 25 26 27 28 29 30 31 Ala-Ala-Try-Gly-Glu-Val-Gly-Ala-His-Ala-Gly-Glu-Tyr-Gly-Ala-Glu-Ala-Leu-Glu-Arg ↑ ↑ ↑
血红蛋白 I 中异常多肽的胰凝乳蛋白酶酶解肽段 *	水解得到的氨基酸	
1	Gly, Ala Glu, Leu	Gly Ala Glu Ala Leu
2	Gly, Glu, Val Ala, His, Tyr Leu	Gly Glu Val Gly Ala His Ala Gly Glu Tyr Gly Ala Glu Ala Leu
3	3Gly, 2Glu, 1Val 2Ala, 1His, 1Tyr	Gly Glu Val Gly Ala His Ala Gly Glu Tyr
4	Glu, Arg	Glu Arg
5	Ala, Try	Ala Ala Try

* 通过纸电泳检测所有情况的构成组分，但是在氨基酸分析仪中也对肽段 3 的氨基酸进行了分析；因此也给出了摩尔比（±0.2）。

我们对芒罗博士在遗传密码方面给予的建议表示感谢。

（郑建全 翻译；崔巍 审稿）

343

22. Ingram, V. M., *Biochim. Biophys. Acta*, **36**, 402 (1959).

23. Gerald, P. S., and Efron, M. L., *Proc. U.S. Nat. Acad. Sci.*, **47**, 1758 (1961).

24. Liddell, J., Brown, D., Beale, D., Lehamnn, H., and Huntsman, R. G., *Nature*, **204**, 269 (1964).

25. Marti, H. R., Pik, C., and Mosimann, P., *Acta Haemat.*, **32**, 9 (1964).

26. Gottlieb, A. J., Restrepa, A., and Itano, H. A., *Fed. Proc.*, **23**, 172 (1964).

27. Baglioni, C., *J. Biol. Chem.*, **237**, 69 (1962).

28. Baglioni, C., and Weatherall, D. J., *Biochim. Biophys. Acta*, **78**, 637 (1963).

29. Baglioni, C. (personal communication).

30. Lehamann, H., Beale, D., and Boi Doku, F. S., *Nature*, **203**, 363 (1964).

31. Miyaji, T., Iuchi, I., Shibata, S., Takeda, I., and Tamura, A., *Acta Haemat. Jap.*, **26**, 538 (1963).

32. Shibata, S., Iuchi, I., Miyaji, T., and Takeda, I., *Bull. Yamaguchi Med. Sch.*, **10**, 1 (1963).

33. Bayrakel, C., Josephson, A., Singer, L., Heller, P., and Coleman, R. D., *Proc. Soc. Haemat. Tenth Congr., Stockholm L*, **6** (1964).

34. Muller, C. J., and Kingma, S., *Biochim. Biophys. Acta*, **50**, 595 (1961).

35. Stretton, A. O. W. (personal communication).

36. Schwartz, I. R., Atwater, J., Repplinger, E., and Tocantins, L. M., *Fed. Proc.*, **16**, 115 (1957).

Haemoglobin J and E in a Thai Kindred

R. Q. Blackwell *et al.*

Editor's Note

By 1965, many abnormal variants of haemoglobin had been described, including haemoglobin J, which was originally discovered in an "American Negro woman" and subsequently identified in several other ethnic groups. Here R. Quentin Blackwell and colleagues from the U.S. Naval Medical Research Unit in Taiwan, and Chamras Thephusdin from the Royal Thai Army's Institute of Pathology in Bangkok, report the discovery of haemoglobin J in a Thai family. It was the first time this variant had been found in people from Thailand, and was surprising because there it is often associated with another abnormal haemoglobin variant, haemoglobin E.

SINCE its discovery in 1956 by Thorup *et al.*[1] in an American Negro woman, haemoglobin J has been identified in other ethnic groups including Gujerati Indians[2,3], Indonesians[4,5], French-Canadian[6] and Swedish-American[7] Caucasians, Algerians[8], Chinese in Singapore[9], Hakkanese Chinese in Taiwan[10], and Hawaiian-Chinese-Caucasians[11]. This communication concerns the occurrence of haemoglobin J in a Thai family; the family is of special interest because it has haemoglobin E in addition to J.

The presence of an electrophoretically fast-moving haemoglobin in addition to normal haemoglobin A was detected in one subject among 676 presumably healthy Thai soldiers stationed at Nakhornratchsima (Korat), Thailand. Results of the haptoglobin distribution[12,13] and haemoglobin E distribution[13] in that group of subjects were reported previously. Subsequent studies indicated that the electrophoretic mobilities of the haemoglobin under various pH conditions appeared to be identical to those of a fast haemoglobin found in members of a Hakkanese Chinese family in Taiwan[10]. Furthermore, both the Thai and the Chinese fast-moving haemoglobins appeared to have mobilities identical with that of a sample of haemoglobin J provided by Dr. Oscar Thorup from his original patient[1]. In addition, a sample of the Hakkanese Chinese haemoglobin was studied electrophoretically by Dr. H. Lehmann, who identified it as J. Pending completion of structure studies on the Thai haemoglobin it can be identified as J_{Korat}.

Subsequent investigations on blood samples from 16 other members of the Thai family by Smithies's vertical starch-gel electrophoresis procedure[14] at pH 9.0, with the *tris*-EDTA-borate buffer system[15] as used by Goldberg[16], disclosed the family to comprise a mixture of individuals possessing E, A+J, A+E and E+J type haemoglobin combinations. The distribution of the haemoglobin phenotypes in the kindred is shown in Fig. 1 and

一个泰国家族中的血红蛋白 J 和 E

布莱克韦尔等

编者按

截止到 1965 年，已经描述了许多异常的血红蛋白变异体，其中包括血红蛋白 J，这一蛋白最初是在一名"美国黑人妇女"体内发现的，随后在其他一些种族中得到鉴定。在本文中，来自位于中国台湾的美国海军第二医学研究所的昆廷·布莱克韦尔及其同事们和来自曼谷泰国皇家军队病理学研究所的占叻·泰普丁报道了在一个泰国家族中发现了血红蛋白 J。这是第一次在泰国人中发现该变异体，并且令人吃惊的是，此处的血红蛋白 J 常和另一个异常血红蛋白的变异体——血红蛋白 E 相关联。

自从 1956 年索普等人在一名美国黑人妇女体内发现了血红蛋白 J 以来 [1]，人们已在其他种族中发现了这种血红蛋白，包括古吉拉特印度人 [2,3]、印度尼西亚人 [4,5]、法裔加拿大 [6] 和瑞典裔美国 [7] 白种人、阿尔及利亚人 [8]、新加坡华人 [9]、中国台湾的客家人 [10] 和有华人血统的夏威夷白种人 [11]。本篇通讯报道了在一个泰国家族中发现的血红蛋白 J，该家族有特殊的意义，因为它除了含有血红蛋白 J 外，还有血红蛋白 E。

对 676 名驻扎在泰国那空叻差是玛（呵叻）的可能健康的泰籍士兵进行研究，发现在其中一名士兵的血样中除了有正常的血红蛋白 A，还能检测到一种电泳时快速移动的血红蛋白。此前曾报道过在这一群体中结合珠蛋白的分布 [12,13] 和血红蛋白 E[13] 的分布。随后的研究表明，该血红蛋白在多种不同的 pH 条件下的电泳迁移率与在一中国台湾客家人家族中发现的一种快速移动的血红蛋白相同 [10]。此外，在泰国人以及中国人中发现的快速移动的血红蛋白似乎与奥斯卡·索普博士提供的来自其最初病人 [1] 的一份血红蛋白 J 的样品具有相同的电泳迁移率。另外，莱曼博士还对一份中国客家人的血红蛋白样品进行了电泳研究，鉴定它为血红蛋白 J。在弄清楚这种泰国人中发现的血红蛋白的结构之前，我们可将其称为血红蛋白 J 呵叻以作区别。

随后我们用史密西斯提出的垂直淀粉凝胶电泳方法 [14] 研究了该泰国家族中其他 16 名成员的血液样品，实验在 pH 9.0 下进行，并选用戈德堡 [16] 采用的三羟甲基氨基甲烷–乙二胺四乙酸–硼酸盐缓冲系统 [15]，研究发现该家族的成员中存在 E、A+J、A+E 和 E+J 几种类型的血红蛋白组合。该家族血红蛋白表型的分布如图 1 所

representative electrophoretic patterns of the haemoglobins are illustrated in Fig. 2. The E type haemoglobin components in the present subjects were identified by comparison of their electrophoretic mobilities with those of numerous other haemoglobin E and A+E samples found in the earlier studies on the 676 Thai blood samples. In those investigations an authentic Thai blood sample with haemoglobins A+E, provided by Dr. Prawase Wasi, Siriraj Hospital, Bangkok, was used for comparison. Numerous studies have reported the presence of haemoglobin E in South-east Asia since its occurrence in Thais[17] was initially recognized. Our work[13] among the 676 Thai subjects, for example, disclosed that 39, or 5.8 percent, of the individuals were homozygous for the haemoglobin E gene and that 246, 36.4 percent, had haemoglobins A and E. The combined incidence in the Thai sample of 42.2 percent for the E and A+E genotypes, which is higher than those reported previously, has been confirmed in recent studies among North-eastern Thais by Wasi and NaNakorn[18].

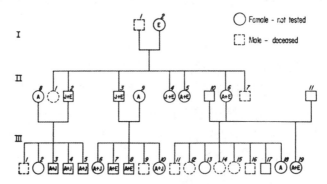

Fig. 1. Distributions of haemoglobin types E, A+E, A+J, and J+E among 3 generations of Thai kindred. Subject III-3 is the propositus.

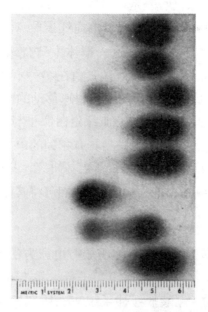

Fig. 2. Results of starch-gel electrophoresis[14-16] of haemoglobins from Thai kindred. From top downward the subjects and haemoglobin types are: 1, normal control, A; 2, II-8, A; 3, II-2, J+E; 4, III-4, A+J; 5, III-5, A+J; 6, I-2, E; 7, II-6, A+E; 8, III-3, propositus, A+J.

示，图 2 显示了有代表性的血红蛋白的电泳图谱。通过与此前对 676 名泰国人血样研究时发现的众多其他血红蛋白 E 和 A+E 样品的电泳迁移率相比较，我们在目前的研究对象中鉴定出了 E 型血红蛋白组分。在这些研究中，由曼谷诗里拉吉医院的普拉瓦设·瓦西博士提供的一份确定含有血红蛋白 A+E 的泰国人血液样品用于对比参照。自从在泰国人中发现了血红蛋白 E[17] 以来，已经有大量的研究报道称其存在于东南亚。例如，在我们对 676 名泰国士兵进行的研究中 [13]，发现有 39 名士兵(占 5.8%)是血红蛋白 E 基因的纯合子，有 246 名士兵（占 36.4%）同时有血红蛋白 A 和 E。这样，我们研究的泰国人血样中血红蛋白 E 和 A+E 基因型的总量就达到了 42.2%，这一数值高于此前报道的数据，而最近瓦西和纳那空 [18] 关于泰国东北部人群的研究已经证实了我们的这一结果。

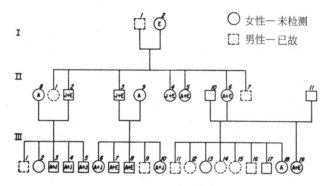

图 1. 某泰国家族的三代人中 E 型、A+E 型、A+J 型和 J+E 型血红蛋白的分布情况。编号为 III-3 的研究对象是在该家族中发现的先证者。

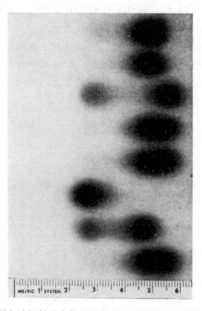

图 2. 该泰国家族成员的血红蛋白淀粉凝胶电泳 [14-16] 结果。从上到下，研究对象和血红蛋白类型依次是：1. 正常对照，A; 2. II-8，A; 3. II-2，J+E; 4. III-4，A+J; 5. III-5，A+J; 6. I-2，E; 7. II-6，A+E; 8. III-3，先证者，A+J。

Subject III-3 was the original case with type A+J haemoglobins found in our study of Thai subjects; his 2 brothers also had the same combination. Their mother, II-8, possessed only normal A-type haemoglobin; their father, II-2, possessed genes for both J and E haemoglobins, and the J haemoglobin genes in the sons were apparently inherited from him. Subject II-3, the brother of subject II-2, also had types J and E haemoglobins; his wife, subject II-9, had normal haemoglobin A. Among the children of subjects II-3 and II-9, one daughter, III-6, had types A and J haemoglobins and 2 sons, III-7 and III-8, had types A and E.

All 5 of the living siblings in the second generation of the kindred had E haemoglobins in combination either with A or J haemoglobin. Their mother, I-2, was homozygous for the E haemoglobin gene as illustrated in Fig. 2. Their father is presumed to have had A and J haemoglobins.

The present family provides the first reported examples of haemoglobin J in Thais and also of an association within a kindred of haemoglobins E and J. Due to the high density of the gene for haemoglobin E in the Thai race as described here, it is obvious that any other abnormal haemoglobin occurring in Thais would occur rather frequently in the same kindred with haemoglobin E. Among those members of the family studied thus far, subjects II-2, II-3 and II-4 are the only examples of the E+J haemoglobin combination in one individual. In future work this family will be examined clinically and haematologically.

We thank Dr. H. Lehmann for his help in identifying one of our haemoglobin J samples, and Drs. Prawase Wasi and Oscar Thorup for their generous contributions of abnormal haemoglobin samples. We also thank the medical officers and technicians of the Royal Thai Army Medical Corps for their aid in procuring blood specimens from the Thai family. Likewise the support of Capt. R. A. Phillips, Commanding Officer of NAMRU-2, and Col. James L. Hansen, director of the U.S. Army Component of the SEATO Medical Research Laboratory, Bangkok, is acknowledged. This work was supported in part by U.S. Public Law 480, Section 104(*c*).

(**207**, 767-768; 1965)

R. Quentin Blackwell, Boon-Nam Blackwell and Jeanette Tung-Hsiang Huang: Department of Biochemistry, U.S. Naval Medical Research Unit No. 2, Taipei, Taiwan, China.

Chamras Thephusdin: Royal Thai Army, Institute of Pathology, Bangkok, Thailand.

References:
1. Thorup, O. A., Itano, H. A., Wheby, M., and Leavell, B. S., *Science*, **123**, 889 (1956).
2. Raper, A. B., *Brit. Med. J.*, i, 1285 (1957).
3. Sanghvi, L. D., Sukumaran, P. K., and Lehmann, H., *Brit. Med. J.*, ii, 828 (1958).
4. Huisman, T. H. J., Noordhoek, K., and Da Costa, G. J., *Nature*, **179**, 322 (1957).

在我们这项对泰国人的研究中，编号为 III-3 的研究对象是我们最先发现的 A+J 型血红蛋白的案例；他的两个兄弟也有同样的组合。他们的母亲 II-8 只有正常的 A 型血红蛋白；他们的父亲 II-2 则同时有血红蛋白 J 和 E 的基因，因此其儿子的血红蛋白 J 基因显然是遗传自父亲。II-2 的兄弟 II-3 也有血红蛋白 J 和 E，II-3 的妻子 II-9 则只有正常的血红蛋白 A，在 II-3 和 II-9 的孩子中，一份女儿 III-6 有血红蛋白 A 和 J，两个儿子 III-7 和 III-8 有血红蛋白 A 和 E。

这个家族的第二代中健在的 5 人都有血红蛋白 E，它或与血红蛋白 A 结合，或与血红蛋白 J 组合。如图 2 所示，他们的母亲 I-2 是血红蛋白 E 基因的纯合子。因此可以推断他们的父亲同时带有血红蛋白 A 和 J。

所示家族是首次报道的泰国人中含有血红蛋白 J 的案例，也是在一个家族内同时存在血红蛋白 E 和 J 的案例。鉴于泰国人中血红蛋白 E 基因的密度较高（如本文所述），很明显，在泰国含有血红蛋白 E 的同一家族中，任何其他异常血红蛋白出现的频率会相当高。在已研究的该家族成员中，同一个体中存在血红蛋白 E+J 组合的例子只有编号为 II-2、II-3 和 II-4 的研究对象。今后我们还将对该家族进行临床学和血液学的研究。

我们感谢莱曼博士帮助我们鉴定了其中一份血红蛋白 J 样品，也感谢普拉瓦设·瓦西博士和奥斯卡·索普博士慷慨地给我们提供异常血红蛋白样品。我们还要感谢泰国皇家军队医疗团的医疗官员和技术人员帮助我们取得了该泰国家庭的血样。同时，我们也要感谢美国海军第二医学研究所指挥官菲利普斯上尉和泰国曼谷东南亚条约组织医疗研究实验室美国军队组主任詹姆斯·汉森上校对我们工作的支持。我们的部分研究工作得到美国 480 号公法第 104（*c*）节的支持。

（周志华 翻译；崔巍 审稿）

5. Lie-Injo, L. E., *Acta Haemat.*, **19**, 126 (1958).

6. McCabe, M. E., Lange, R. D., and Crosby, W. H., *Amer. J. Med.*, **23**, 329 (1957).

7. Wasi, P., Githens, J., and Hathaway, W., *Blood*, **16**, 1795 (1960).

8. Boulard, C., Cabannes, R., Duzer, A., and Scotto, J. -Cl., *Bull. Mem. Soc. Med. Hopit.* (Paris), **76**, 41 (1960).

9. Ager, J. A. M., Lehmann, H., and Vella, F., *Brit. Med. J.*, ii, 539 (1958).

10. Blackwell, R. Q., and Huang, J. T. -H. (in preparation).

11. Jim, R. T. S., and Yarbro, M. T., *Blood*, **15**, 285 (1960).

12. Blackwell, R. Q., and Thephusdin, C., *Nature*, **197**, 503 (1963).

13. Thephusdin, C., Blackwell, B. -N., Huang, J. T. -H., and Blackwell, R. Q., *Proc. Third Congr. Asian Pac. Soc. Haem. Jerusalem*, August 17-23, 1964 (in the press).

14. Smithies, O., *Biochem. J.*, 71, 585 (1959).

15. Aronsson, T., and Grönwall, A., *Scand. J. Clin. Lab. Invest.*, **9**, 338 (1957).

16. Goldberg, C. A. J., *Clin. Chem.*, 4, 484 (1958).

17. Chernoff, A. I., Minnich, V., and Chongchareonsuk, S., *Science*, **120**, 605 (1954).

18. Wasi, P., and Na-Nakorn, S. (personal communication).

Haemoglobin E in Vietnamese

R. Q. Blackwell *et al.*

Editor's Note

In his continuing search for abnormal haemoglobin variants in different ethnic groups, R. Quentin Blackwell and colleagues from the US Naval Medical Research Unit in Taiwan, China, screened 482 Vietnamese hospital inpatients for the presence of haemoglobin E. Here they report the presence of haemoglobin E in 17 patients, always expressed alongside haemoglobin A, the most common and normal adult version of the protein. Their results add to the catalogue of haemoglobin E-expressing ethnic groups, which most notably included groups from the Middle East and Asia.

SINCE its simultaneous discovery a decade ago in an American family of mixed Spanish, Guatemalan, and Indian ancestry[1] and in Thais[2], haemoglobin E has been found in a variety of ethnic groups a majority of which are located in the Middle East and Asia. It is particularly common among the peoples of south-east Asia where it had been found in relatively high incidence in Burmese[3], Thais[4-7] and Cambodians[8], and in lower incidences in Vietnamese[9], Malayans[10,11], Indonesians[12], Chinese[13,14] and Filipinos[15]. The results in this communication confirm the occurrence of haemoglobin E among Vietnamese.

Subjects for this investigation were residents of Saigon and environs. At the time of the work, during February and March 1964, they were patients in the Cho Quan Hospital, Saigon, where they had been admitted with diarrhoea during the cholera epidemic and were under treatment for dehydration. Red cells were obtained from residual heparinized blood samples collected for other diagnostic purposes and were preserved with merthiolate and by refrigeration prior to their transfer by air to the Biochemistry Department Laboratory of this Unit for analysis.

On receipt in the laboratory the red cells were washed with saline, mixed with equal volumes of distilled water, and frozen until used for electrophoretic analysis. After thawing, the samples were centrifuged and the supernatant haemolysates examined by Smithies's vertical starch-gel electrophoresis procedure[16]. The gel buffer employed was the *tris*-EDTA-borate buffer, pH 9.0, at the concentrations recommended by Goldberg[17].

Results of the study indicate that 17, or 3.53 percent, of the 482 subjects had A+E haemoglobins. No other abnormal haemoglobins were detected. Although no quantitative determinations were made of the relative amounts of haemoglobins A and E in the samples, haemoglobin E was always the minor component. By inspection the E component

354

越南人中的血红蛋白 E

布莱克韦尔等

编者按

来自位于中国台湾的美国海军第二医学研究所的昆廷·布莱克韦尔及其同事们，一直在寻找不同种族人群中的异常血红蛋白变异体。他们从越南医院 482 名住院患者中筛选含有血红蛋白 E 的患者。在本文中，布莱克韦尔等人报道了 17 名患者存在血红蛋白 E 并且总是伴随着血红蛋白 A 的表达，血红蛋白 A 是最普遍和常见的成人血红蛋白。他们的研究结果增加了表达血红蛋白 E 的种族人群数，其中来自中东和亚洲地区的人群血红蛋白 E 表达最显著。

自从十年前在一个具有西班牙、危地马拉和印度混合血统的美国家庭 [1] 及泰国人 [2] 中同时发现血红蛋白 E 后，人们又在许多种族中发现了血红蛋白 E，这些种族大多分布在中东和亚洲地区。血红蛋白 E 在东南亚人群中特别常见，其中缅甸人 [3]、泰国人 [4-7] 和柬埔寨人 [8] 呈现相对较高的发生率，而在越南人 [9]、马来人 [10,11]、印度尼西亚人 [12]、中国人 [13,14] 和菲律宾人 [15] 中发生率相对较低。本篇文章的研究结果证实了血红蛋白 E 存在越南人中。

本研究的对象是西贡市及近郊的居民。在我们进行研究时，即 1964 年 2 月至 3 月间，他们是霍乱流行期因患痢疾而在西贡赵泉医院住院的患者，当时正在接受脱水症状的治疗。从剩余的肝素化血液样本（原本供其他诊断用）中获得红细胞，用硫柳汞保存红细胞，冷冻后通过航空转运到美国海军第二医学研究所的生物化学实验室进行分析。

当实验室收到样品后，首先用生理盐水洗涤红细胞，然后将红细胞与等体积的蒸馏水混合并冷冻保存以供电泳分析。实验时，先将样品解冻，然后离心，按照史密西斯的垂直淀粉凝胶电泳方法 [16] 对上层溶血液进行检测。凝胶缓冲液是参照戈德堡推荐的浓度配制的 pH 为 9.0 的三羟甲基氨基甲烷–乙二胺四乙酸–硼酸盐缓冲液 [17]。

研究结果表明，在 482 名研究对象中，17 人（占 3.53%）有血红蛋白 A+E。实验中没有检测到其他的异常血红蛋白。尽管没有对样品中血红蛋白 A 和 E 的相对含量进行定量测定，但是血红蛋白 E 总是相对较少的成分。经检验，血红蛋白 E 大概占血红蛋白总量的 1/5 到 1/3。在所有样本中，血红蛋白 E 的含量都超过了血红蛋白 A_2

was estimated to comprise from one-fifth to one-third of the total; in all cases it exceeded the levels to be expected for A_2 haemoglobin, which has the same electrophoretic mobility as E in the buffer employed.

Three cases of haemoglobin E, 2.7 percent, were reported by Albahary et al.[9] among their 113 Vietnamese subjects. Based on the results of both investigations the incidence of the gene for haemoglobin E in the Vietnamese appears to be lower than those in Thais, Cambodians, Burmese and Malayans, similar to those of some of the Indonesians, and higher than those of Chinese and Filipinos.

Although our results with respect to haemoglobin E in Vietnamese agree with those of Albahary et al.[9], we were unable to confirm their finding of a slow-moving component which they called haemoglobin "Sud-Vietnam". No slow component other than E or A_2 was found among our 482 subjects.

We thank James W. Fresh, Charles Neave, Donald L. Pankratz and other members of the Cholera Treatment Team of this Research Unit, for collection and transport of the blood samples. This work was supported in part by a fund provided under U.S. Public Law 480, Section 104 (c).

(**207**, 768; 1965)

R.Quentin Blackwell, Jeanette Tung-Hsiang Huang, Li-Chin Chien: Department of Biochemistry, U.S. Naval Medical Research Unit No. 2, Taipei, Taiwan, China.

References:

1. Itano, H. A., Bergren, W. R., and Sturgeon, P., *J. Amer. Chem. Soc.*, **76**, 2278 (1954).

2. Chernoff, A. I., Minnich, V., and Chongchareonsuk, S., *Science*, **120**, 605 (1954).

3. Lehmann, H., Story, P., and Thein, H., *Brit. Med. J.*, i, 544 (1956).

4. Na-Nakorn, S., Minnich, V., and Chernoff, A. I., *J. Lab. Clin. Med.*, **47**, 490 (1956).

5. Wasi, P., and Na-Nakorn, S. (personal communication).

6. Flatz, G., Pik, C., and Sundharagiati, B., *Lancet*, ii, 385 (1964).

7. Thephusdin, C., Blackwell, B.-N., Huang, J. T.-H., and Blackwell, R. Q., *Proc. Third Congr. Asian Pac. Soc. Haem.* (In the press).

8. Brumpt, L., Brumpt, V., Coquelet, M. L., and De Traverse, D. M., *Rev. Hemat.* (Paris), **13**, 21 (1958).

9. Albahary, C., Dreyfus, J. C., Labie, D., Schapira, G., and Tram, L., *Rev. Hemat.* (Paris), **13**, 163 (1958).

10. Lehmann, H., and Singh, R. B., *Nature*, **178**, 695 (1956).

11. Vella, F., *Lancet*, i, 268 (1963).

12. Lie-Injo, L. E., and Oey, H. G., *Lancet*, i, 20 (1957).

13. Vella, F., *Acta Haemat.*, **23**, 393 (1960).

14. Blackwell, R. Q., Huang, J. T. -H., and Chien, L. -C., *Proc. Third Congr. Asian Pac. Soc. Haem. Jerusalem*, August 17-23, 1964 (in the press).

15. De La Fuente, V., Florentino, R. F., Alejo, L. G., Huang, J. T. -H., Chien, L. -G., and Blackwell, R. Q., *J. Philipp. Med. Assoc.* (in the press).

16. Smithies, O., *Biochem. J.*, **71**, 585 (1959).

17. Goldberg, C. A. J., *Clin. Chem.*, **4**, 484 (1958).

的预期水平，在我们所用的缓冲液中血红蛋白 A_2 与 E 具有相同的电泳迁移率。

在此之前,阿尔巴阿里等人[9]曾报道,他们研究的 113 个越南人中,3 人(占 2.7%)含有血红蛋白 E。基于这两项研究结果可以看出，血红蛋白 E 基因在越南人中的出现频率似乎比在泰国人、柬埔寨人、缅甸人和马来人中出现的频率低，与印度尼西亚人类似，比在中国人和菲律宾人中出现的频率高。

尽管我们关于越南人中血红蛋白 E 的研究结果与阿尔巴阿里等人的结果[9]一致，但是我们未能证实此前他们发现的一种迁移较慢的被称为血红蛋白"南越"的组分。在所有 482 名研究对象中，没有发现比血红蛋白 E 或 A_2 迁移更慢的组分。

我们感谢詹姆斯·弗雷什、查尔斯·尼夫、唐纳德·潘克拉茨和本研究所霍乱医疗组的其他成员在血样采集及转运中所做的工作。这项工作部分由美国 480 号公法第 104（c）节设立的一项基金资助。

（郑建全 翻译；崔巍 审稿）

New Model for the Tropocollagen Macromolecule and Its Mode of Aggregation

R. A. Grant *et al.*

Editor's Note

Collagen is the main component of connective tissue, present in ligament and tendon, blood vessels, skin and bone. This makes it one of the key structural proteins, which act as materials rather than enzymes. Like many biomaterials, it has a hierarchical structure: the protein strands are woven into triple helices and then bundled together in fibrils. Early electron-microscope studies showed the fibrils have periodic light and dark bands about 64 nanometres apart. Here researchers at the UK's Agricultural Research Council suggest an explanation, based on the notion that the protein strands have regularly spaced regions that can and cannot form crosslinks. The idea was on the right lines: the bands in fact come from a staggered arrangement of polypeptide chains.

THE molecular architecture of collagen has been intensively studied by a variety of physical and chemical techniques. According to current views collagen fibres are composed of fibrils which are themselves built up from a basic structural unit termed the tropocollagen macromolecule. Each tropocollagen unit is believed to have a three-stranded coiled-coil structure in which the most probable arrangement of the polypeptide chains would appear to be the collagen II model proposed by Rich and Crick[1]. The tropocollagen unit has been described as a rigid rod-like structure for which Boedtker and Doty[2], using physico-chemical methods, originally found a diameter of 13.6 Å, an average length of 3,000 Å and an average molecular weight of 345,000. Recently, Rice, Casassa, Kerwin and Maser[3] have suggested a length of 2,800 Å and a molecular weight not greater than 310,000.

One of the earliest observations made on biological material with the electron microscope was that native collagen fibres exhibited a conspicuous transverse banding with a repeat distance of approximately 640 Å (Schmitt, Hall and Jakus[4]; Wolpers[5]). There have since been many reports on the structure of collagen using shadowing, thin sections and positive staining. These have been reviewed in detail by Harrington and Von Hippel[6], Gross[7], Fitton Jackson[8] and Veis[9]. Much research has been devoted to clarifying the relationship of the basic tropocollagen unit to the native collagen fibre.

An important aspect of the problem is to explain how the observed periodicity of 640 Å of native collagen results from the combination of tropocollagen units of length 2,800 Å.

It is well known that native-type collagen fibres may be reconstituted from acetic acid

原胶原大分子及其聚集模式的新模型

格兰特等

编者按

胶原蛋白是结缔组织的主要组成部分，分布在韧带、肌腱、血管、皮肤和骨骼中。这使得胶原蛋白成为一种重要的结构蛋白，它主要充当组成材料，而不是酶。和许多生物材料一样，它具有多级结构：蛋白质链被编织成三股螺旋后捆绑在一起形成纤维。早期的电子显微镜研究表明蛋白纤维含有以 64 nm 为间隔周期的明暗带。本文中，英国农业研究委员会的研究者基于以下这一概念给出了一种解释，即蛋白质链有规则的间隔区，这些区域之间可以形成或不可以形成交联。这一想法大体正确：事实上，这种结合来自多肽链之间的交错排列。

人们通过多种物理和化学技术已经深入地研究了胶原蛋白的分子结构。根据现在的观点，胶原蛋白纤维是由以原胶原大分子为基本结构单元构成的纤维组成的。通常认为每个原胶原单位都有一个三链卷曲螺旋结构，这种结构中多肽链最可能的排列方式可能符合里奇和克里克[1]所提出的胶原蛋白 II 模型。原胶原单位被描述成一个刚性的棒状结构，起初，博德克和多蒂[2]利用物理化学方法发现其直径为 13.6 Å，平均长度为 3,000 Å，平均分子量为 345,000。最近，赖斯、卡萨萨、克尔温和马泽尔[3]认为其长度为 2,800 Å，分子量不超过 310,000。

利用电子显微镜对生物材料最早的观察结果之一表明，天然胶原纤维显示出一个明显的重复距离约为 640 Å 的横向带（施米特，霍尔和亚库斯[4]；沃尔佩斯[5]）。自此以后，有了许多使用投影、超薄切片和正染技术研究胶原蛋白结构的报道。哈林顿和冯·希佩尔[6]、格罗斯[7]、菲顿·杰克逊[8]和维斯[9]对此进行了详细的论述。许多研究都致力于阐明基本原胶原单位与天然胶原蛋白纤维之间的关系。

这一问题的一个重要的方面是解释长度为 2,800 Å 的原胶原单位是怎样形成了天然胶原蛋白中所观察到的 640 Å 周期性的。

众所周知，通过调节胶原蛋白乙酸溶液的 pH 和离子强度可以使天然胶原蛋白

solutions of collagen by adjustment of pH and ionic strength. It is also possible to reconstitute collagen in a number of forms differing characteristically from the native fibre. A particularly interesting form is the segment long spacing (SLS) crystallites (Gross, Highberger and Schmitt[10]) produced by the addition of adenosine triphosphate (ATP) to an acetic acid solution of collagen. The SLS crystallites appear to consist of tropocollagen units aggregated side by side with the ends of the molecules in register. A large number of bands, arranged at right angles to the long axes of the tropocollagen units, may be seen with the electron microscope using either positive- or negative-staining techniques. The fact that these bands are distributed asymmetrically is strong evidence that the structure of the tropocollagen macromolecule is polarized.

Another type of collagen fibril, produced by dialysing an acetic acid solution of collagen (tropocollagen) containing acid glycoprotein or chondroitin sulphate (Highberger, Gross and Schmitt[11]; Schmitt, Gross and Highberger[12]) has been termed the fibrous long spacing (FLS) form. The repeat distance of this form has been found to vary with the method of preparation and the different varieties produced have been further designated FLS— I, II, etc., depending on the spacing observed. In the case of the FLS form of collagen a maximum repeat distance of about 2,500 Å has been found. It has been suggested (Schmitt and Hodge[13]) that in the FLS form of fibre the tropocollagen units are aggregated end to end. The direction of polarization of the macromolecules was considered to be alternated in view of the symmetrical pattern obtained in the electron microscope after positive staining with phosphotungstic acid.

Application of the negative staining technique (Brenner and Horne[14]) to high-resolution studies of collagen with the electron microscope showed that it was possible to visualize, directly, tropocollagen macromolecules occurring in native collagen fibrils (Tromans et al.[15]; Olsen[16,17]). In their negatively stained preparations of collagen fibrils the most striking feature seen was the arrangement of slender, somewhat crooked filaments running roughly parallel to each other along the long axis of the fibril. These filaments, 15–20 Å across, were interpreted as being the tropocollagen macromolecules.

In a detailed study of negatively stained collagen specimens, Tromans[18] suggested that the light and dark bands resulted from differences in the density of the protein and in the penetration of the negative stain between the macromolecules. It was not found possible to demonstrate the precise positions of the ends of the macromolecules within the fibrils, nor was there any evidence in favour of a "quarter staggered" arrangement as suggested by the model of Hodge and Schmitt[19]. The recent descriptions of native and FLS-type collagen by Olsen[16,17] and Kuhn and Zimmer[20] have indicated that the macromolecules are aggregated by end overlap. This view is not in agreement with the model proposed by Schmitt and Hodge[13], in which it is suggested that end-to-end aggregation takes place by a coiling around each other of polypeptide chains projecting from each end of the tropocollagen macromolecule (Hodge and Petruska[21]).

This report is based on an electron microscope investigation on native, SLS and FLS

纤维重组。也可以使天然纤维重组为具有大量不同形态特征的胶原蛋白。通过向胶原蛋白的乙酸溶液中加入三磷酸腺苷（ATP）可以产生片段长间距微晶（SLS）（格罗斯，海贝格和施米特[10]），这是一个非常有趣的形态。片段长间距微晶可能是由原胶原分子末端并排聚集而成的。使用正染或负染技术，在电子显微镜下可能看到与原胶原单位长轴呈直角排列起来的大量的条带。这些条带的不对称分布可以有力地证明原胶原大分子结构是极性的。

通过透析含有酸性糖蛋白或硫酸软骨素的胶原蛋白（原胶原）的乙酸溶液而产生的另一种形态的胶原蛋白纤维（海贝格，格罗斯和施米特[11]；施米特，格罗斯和海贝格[12]）已经被定义为纤维长间距型（FLS）。人们发现这种形态的重复距离随着制备的方法不同而改变，根据观察到的间隔，将这些不同的产物种类进一步定义为纤维长间距 I 型、II 型等。在胶原蛋白的纤维长间距形态中，已发现的最大的重复距离约为 2,500 Å。有人认为（施米特和霍奇[13]）纤维长间距型纤维是由原胶原单位首尾相连聚集而成的。根据磷钨酸正染后在电子显微镜下所得到的对称构型可以认为，大分子的极化方向是交替变化的。

应用负染技术（布伦纳和霍恩[14]）在电子显微镜下对胶原蛋白的高分辨率研究表明，直接观察天然胶原纤维中的原胶原大分子是有可能的（特罗芒等[15]；奥尔森[16,17]）。在经过负染处理的胶原蛋白纤维制品中可以看到的最明显的特征是，稍弯曲的细丝大致平行地排列在纤维的长轴上。这些跨度大约为 15 Å~20 Å 的细丝被认为是原胶原大分子。

在对经过负染的胶原蛋白样本的详细研究中，特罗芒[18]认为明暗带是蛋白质密度和大分子间负染渗透不同导致的。人们发现要证明纤维中大分子末端的准确位置是不可能的，也没有任何证据可以支持霍奇和施米特[19]的模型所提出的"四分之一交错"排列。最近在对天然胶原蛋白和纤维长间距型胶原蛋白的描述中，奥尔森[16,17]、库恩和齐默[20]指出，大分子是通过末端重叠聚集形成的。这种观点与施米特和霍奇[13]所提出的模型不一致，后者认为原胶原大分子首尾相连聚集的发生是通过其各末端伸展出的多肽链彼此缠绕实现的（霍奇和彼特鲁斯卡[21]）。

这篇报道是以电子显微镜下观察天然胶原蛋白、片段长间距型胶原蛋白和纤维

forms of collagen, and on these same forms modified by treatment with various cross-linking agents. Our experiments have also included observations on the mode of aggregation of tropocollagen to the above forms. The effect of collagenase on native fibres was also investigated, in experiments which will be reported in greater detail elsewhere. Acetic acid solutions of rat-tail tendon and guinea-pig skin collagen were used as sources of tropocollagen. From these investigations, a model has been developed which explains, in a relatively simple manner, the production of the majority of collagen forms. Their characteristic spacings can be considered to be determined by a basically random process of aggregation of tropocollagen units.

It was considered important to elucidate the mechanism whereby the light and dark bands, seen in negatively stained preparations of collagen, are produced. Treatment with a cross-linking agent, such as glutaraldehyde, resulted in an increase in the size of the light bands (*A* bands), Fig. 1, *b* and *c*; a well-defined light band also appeared in the dark or *B* bands. It was obvious from marked changes in the chemical and physical properties of the treated fibres (for example, no longer being soluble when autoclaved with water) that a number of strong cross-links had been introduced. Since the introduction of intermolecular cross-links by mild chemical means resulted in an increase in the size of the light or electron-transparent bands we may infer that the light bands, seen in native unaltered collagen, also

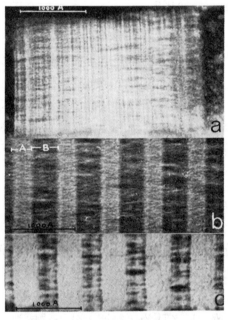

Fig. 1. *a*, Electron micrograph of negatively stained SLS crystallite. The length of the crystallite spans five *A* bands of the native fibril, shown in *b* at the same magnification. *b*, Negatively stained collagen fibril showing regions of light *A* bands and dark *B* bands. The rather flexible tropocollagen macromolecules are seen arranged roughly parallel to the fibril long axis. *c*, Negatively stained collagen fibril following treatment with glutaraldehyde. The increase in both size and density of the *A* bands is demonstrated. A well-defined light band within the *B* zone is also apparent.

362

长间距型胶原蛋白以及具有上述三种相同形态的改性胶原蛋白为基础的，其中改性胶原蛋白是用各种交联剂处理获得的。同时我们的实验也包括观察上述形态中原胶原的聚集模式。同时也对胶原酶对天然纤维的作用进行了研究并将在其他地方给出实验较详细的报道。大鼠尾部肌腱和豚鼠皮肤胶原蛋白的乙酸溶液被用来作为原胶原的来源。我们通过这些研究建立了一个模型，以一个相对简单的方式解释了多数胶原蛋白形态的形成。它们的特征间距被认为是由原胶原单位聚集的基本随机过程决定的。

人们认为阐明经过负染处理的胶原质制品的明暗带的产生机制是非常重要的。利用交联剂（如戊二醛）进行处理导致了明带（A 带）大小的增加（图1，b 和 c）；一个很清晰的明带也出现在暗带或 B 带中。从处理过的纤维的物理化学性质的显著变化（比如，当用蒸汽高压处理时不再可溶）明显可以看出纤维中产生了大量强有力的交联键。由于通过温和的化学方法产生的分子间交联键导致了明带或是电子透过带大小的增加，因此我们可以推断在天然的未作改变的胶原蛋白中所看到的明带也是由相邻大分子的结合区形成的。在天然胶原蛋白中这种分子间的结合可能是由

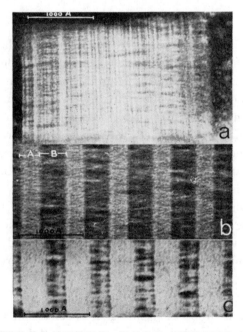

图 1. a，负染的片段长间距微晶的电子显微图片。微晶的长度横跨相同放大倍数下，示于图 b 的天然纤维的 5 条 A 带。b，负染的胶原蛋白纤维表现出 A 带明区和 B 带暗区。可以看到颇为柔韧的原胶原大分子沿着纤维长轴大致平行排列。c，经过戊二醛处理后的负染的胶原蛋白纤维。A 带增宽且密度增大。在 B 区也出现了一条清晰的明带。

result from regions of bonding between adjacent macromolecules. In native collagen this intermolecular bonding probably consists of polar and hydrogen bonds arising from polar amino acid side-chains arranged in a structurally complementary manner. It thus appears that the tropocollagen units, forming the fibril, are not laterally bonded uniformly throughout their length. Such a concept is in conformity with the finding that the primary structure of collagen consists of inhomogeneous amino-acid sequences (Grassman *et al.*[22]).

In order to deduce a simple rational explanation of the relationship of the length of the tropocollagen macromolecule (*c.* 2,800 Å) to the normal repeat distance (*D*) of 640 Å, use may be made of the observation that the length of an SLS crystallite spans 5 *A* bands of a native fibril when electron micrographs at the same magnification are compared (see Fig. 1, *a* and *b*). On average, the length of an *A* band (light) is about 0.4 *D*; hence the length of the tropocollagen unit equals the total distance spanned by 5 *A* bands plus the intervening *B* bands (=4.4 *D*). Careful examination of the electron micrographs gives no evidence of end-to-end junctions, all the macromolecules appearing to be aggregated together by lateral association with a good deal of cross-over in both *A* and *B* bands. The precise positions of the ends of tropocollagen macromolecules within the fibrils were difficult to determine. However, on careful inspection of the electron micrographs instances were noted of tropocollagen filaments appearing to end at the edge of an *A* band after passing through the band.

We suggest, on the basis of this evidence, that the tropocollagen macromolecule consists of alternating bonding and non-bonding regions, the bonding regions containing polar amino-acids which are arranged in a structurally specific manner. When two such bonding regions approach one another closely in proper alignment a strong lateral attraction results from the formation of many intermolecular polar and hydrogen bonds. All the 5 bonding regions may not be exactly equivalent and there is some evidence that the terminal bonding zones differ from the intermediate ones. Examination of the SLS form, by positive or negative staining, indicates that the tropocollagen macromolecule is polarized, and this is supported by the fact that in native fibres the band structure is asymmetrical. Tromans *et al.*[15] have suggested that the light bands be regarded as regions of crystallographic disorder. In view of the evidence from the cross-linking experiments we feel that this interpretation may be abandoned in favour of the more definitive explanation that the light bands correspond to regions of bonding between adjacent macromolecules.

The proposed model for tropocollagen (Fig. 2, 1) consists of a somewhat flexible filament of length *c.* 2,800 Å, divided into 9 zones consisting of 5 bonding zones of length about 0.4 *D* (265 Å) separated by 4 non-bonding zones of length about 0.6 *D* (375 Å). It should be appreciated that the measurements quoted in this report represent approximate values drawn from our own experiments and other results cited in the literature.

以结构互补方式排列的极性氨基酸侧链所形成的极性键和氢键所组成的。因此，组成纤维的原胶原单位似乎并不是自始至终均一地横向结合。这种观念与胶原蛋白的一级结构是由不均一氨基酸序列组成的这一发现（格拉斯曼等 [22]）是一致的。

对于原胶原大分子的长度（c 为 2,800 Å）和 640 Å 的正常重复距离（D）之间的关系，我们为了给出一个简单合理的解释，对相同放大倍数下的电子显微图片进行了比较，发现片段长间距微晶的长度跨越了天然纤维的 5 条 A 带（图 1，a 和 b）。平均来说，1 条 A 带（明带）的长度约为 0.4 D，因此原胶原单位的长度就等于 5 条 A 带的跨度加上其间的 B 带跨度（=4.4 D）。通过对电子显微图片的仔细观察并没找到有关首尾相连的证据，所有大分子似乎都是以横向结合的方式与 A 带和 B 带中的大量横向线条聚集而成。很难确定纤维中原胶原分子末端的准确位置。然而通过对电子显微图片的仔细观察发现，原胶原细丝经过整条带后似乎停止在 A 带边缘。

在这一证据的基础上，我们认为原胶原大分子是由结合区域和非结合区域交替组成的，结合区域包括以结构特异性方式排列的极性氨基酸。当两个结合区域适当地以排列整齐的方式彼此紧密靠近时，许多分子间极性键和氢键的形成导致一个强有力的横向吸引力的产生。所有的 5 个结合区都不是严格相同的，有证据表明结合区的末端与中间部分是有区别的。利用正染或负染技术对片段长间距型胶原蛋白的检测表明原胶原大分子具有极性，天然纤维中带结构的不对称性也支持这一观点。特罗芒等人 [15] 认为可以把明带看作微晶的无序区域。然而根据交联实验得到的证据，我们感觉这种解释可以被否定，而明带对应的是相邻大分子间的结合区域这一明确的解释更具有说服力。

提出的原胶原模型（图 2，1）是由长度（c）为 2,800 Å，有一定韧性的细丝组成的，分为 9 个区域，包括 5 个长度约为 0.4 D（265 Å）的结合区域以及将它们分开的 4 个长度约为 0.6 D（375 Å）的非结合区域。值得一提的是，本文提供的测量结果代表着我们自己从实验得到的近似值和从文献中所引用的其他结果。

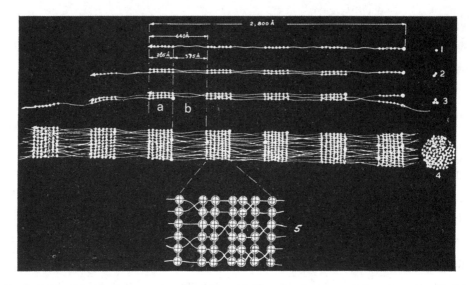

Fig. 2. 1, The diagram illustrates the proposed model for tropocollagen, consisting of a rather flexible filament divided into 9 zones. Five of these are bonding zones and four are considered as non-bonding zones. 1, 2, 3; Association of tropocollagen by random aggregation through bonding zones is illustrated. The overlap between two adjacent macromolecules can cover one to five bonding zones. The direction of the macromolecules, indicated by arrows, determines the polarization of the collagen fibril. 4, The establishment of a repeat pattern of 640 Å results in the formation of light (*a*) bands as bonding zones and dark (*b*) bands as non-bonding zones. 5, For diagrammatic purposes, the linking with an (*a*) band is shown to illustrate the possible flexibility of the macromolecules.

It need not necessarily be assumed that the non-bonding regions are completely devoid of polar amino-acids. Indeed it appears from the cross-linking experiments that there is some lysine in these parts of the macromolecule. Nevertheless, it seems that in the native fibre only a few intermolecular bonds are formed between the non-bonding zones.

When tropocollagen units, conforming to the above model, are aggregated together so that there is an initial random choice as to which bonding region on one molecule cross-links in a structurally complementary manner with a bonding region on a different molecule, a fibre with 640 Å periodicity results (Fig. 2). A model of a collagen fibril constructed in this way shows a close resemblance to the structure of native collagen revealed by negative staining. Moreover, aggregation of tropocollagen macromolecules in an apparently random manner to form collagen fibrils has been observed by us in the electron microscope. There is no need to postulate a two-stage process in which tropocollagen macromolecules are first polymerized by end-to-end linkage through the interaction of terminal peptide chains to form protofibrils, which are then displaced relatively to one another by 0.25 of the molecular length (Schmitt and Hodge[13]; Hodge and Petruska[21]). Nor is it necessary to suggest an initial polymerization of tropocollagen units by overlapping which is limited to the ends of the molecules, to be followed by "quarter staggering" of the protofibrils (Olsen[17]).

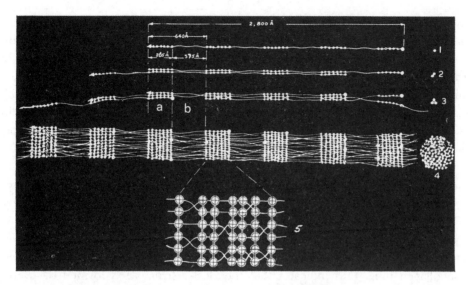

图 2.1，图示说明提出的原胶原模型，由 1 条被分成 9 个区的颇为柔韧的细丝组成。其中 5 个是结合区，另外 4 个被认为是非结合区。1、2、3 表示通过结合区随机聚合在一起的原胶原联合体。相邻两个大分子间的重叠可以覆盖 1 至 5 个结合区。大分子的方向（箭头所示）决定了胶原蛋白纤维的极性。4，一个 640 Å 重复模式的建立导致结合区明带（a）和非结合区暗带（b）的形成。5，图中（a）带之间的耦联说明了大分子可能的柔韧性。

没有必要去假设在非结合区间完全不存在极性氨基酸。实际上从交联实验可以知道，在大分子的非结合区有一些赖氨酸的存在。然而在天然纤维中，只有少数分子间的键在非结合区间形成。

当原胶原单位按上述模型聚集在一起时，就为一个分子的结合区以结构互补的方式与另一不同分子的结合区交联提供了最初的随机选择，产生了一条以 640 Å 为周期性的纤维（图 2）。按这种方式建立起来的胶原纤维模型与通过负染所表现出的天然胶原蛋白的结构非常相似。而且，在电子显微镜下我们已经观察到原胶原大分子是以一种非常明显的随机方式聚集起来从而形成胶原纤维的。没必要去假设以下包含两个阶段的过程：首先原胶原大分子通过末端肽链的相互作用首尾相连聚合在一起形成原纤维，然后由原纤维互相取代掉其分子长度的 0.25（施米特和霍奇[13]；霍奇和彼特鲁斯卡[21]）。也没有必要去考虑原胶原单位仅限于分子末端重叠的最初聚合以及随后的原纤维的"四分之一交错"（奥尔森[17]）。

It also follows from the model for native collagen that all the tropocollagen units in one fibril will face in the same direction. The direction of polarization of a fibril, as determined from the band pattern, will depend on the direction of polarization of the first tropocollagen macromolecule. If, therefore, the tropocollagen macromolecules are not initially all pointing in the same direction, the resulting collagen fibrils will also be polarized in different directions. This is of interest in view of the recent finding of Braun-Falco and Rupec[24] that in human dermis the collagen fibrils have an "antiparallel arrangement".

More direct evidence that the tropocollagen macromolecule consists of nine segments comes from experiments on the FLS form. In Fig 3a is shown an electron micrograph of FLS Type 1, made by dialysing an acetic acid solution of collagen containing serum glycoprotein; the repeat distance is about 2,500 Å while the distance measured between the outer edges of the light bands corresponds to the length of an SLS crystallite (c. 2,800 Å) or tropocollagen unit. This indicates that the FLS-1 structure is formed by overlap of the terminal bonding zones of tropocollagen as suggested by Olsen. However, the reason for the formation of this type of fibre may be that adsorption of negatively charged glycoprotein on to the tropocollagen units prevents the establishment of intermolecular bonds by the intermediate bonding zones, thus producing the wide spacing (Fig. 4; 1, 2). Moreover, when such an FLS-1 preparation was treated with glutaraldehyde the 640 Å repeat period was re-established (Fig. 3b). Presumably the glutaraldehyde was able to bridge between amino-groups in the blocked bonding zones to re-establish the native period. These experiments appear to provide clear evidence of the existence of 5 similar zones in the tropocollagen molecule separated by regions having different properties. The three intermediate bonding zones in tropocollagen can thus be suppressed with glycoprotein but revealed by subsequent treatment with glutaraldehyde (Fig. 4; 2, 3).

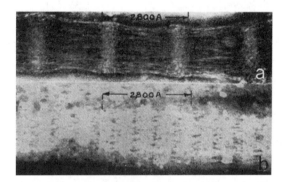

Fig. 3. a, FLS type 1, showing repeat period of about 2,500 Å. b, FLS Type 1, following treatment with glutaraldehyde. The re-establishment of the 640-Å period is clearly visible. Both preparations were negatively stained.

In the case of the SLS crystallite, negative staining has revealed a marked asymmetry of the band structure. On the other hand, although Olsen[17] has attempted to deduce, by photographic methods, a relationship between SLS and the native fibril, a number

根据这一模型，天然胶原蛋白中所有的原胶原单位都将朝同一方向排列。一条由带的类型决定的纤维的极性方向，将依赖于第一个原胶原大分子的极性方向。因此，如果原胶原大分子最初并不全都指向同一方向，就会导致胶原纤维也会在不同方向产生极性。有意思的是，最近布朗－法尔科和鲁佩克[24]发现人类皮肤的胶原纤维中存在"反向平行排列"。

更多支持原胶原大分子是由9个部分组成的直接证据来源于纤维长间距型胶原质实验。图3a显示的是纤维长间距I型胶原蛋白的电子显微图片，它是通过透析含有血清糖蛋白的胶原蛋白乙酸溶液得到的；重复距离约为2,500 Å，而明带外边缘间的实测距离则与一个片段长间距微晶（c为2,800 Å）或原胶原单位的长度一致。正如奥尔森所提出的，这表明纤维长间距I型结构是通过原胶原末端结合区重叠而形成的。然而，之所以形成这种类型的纤维可能是由于原胶原单位吸附了带负电荷的糖蛋白，阻止了中间结合区分子间键的形成，从而产生了宽间距（图4，1、2）。而且当这种纤维长间距I型胶原蛋白经过戊二醛处理后，640 Å的重复周期又会重新出现（图3b）。据推测戊二醛可以使在被阻碍的结合区内的氨基群之间关联，并重建天然周期。这些实验似乎能够提供明显的证据来证明原胶原分子中存在被不同性质区域分开的5个相似区域。原胶原中间的3个结合区能够被糖蛋白抑制，也可以通过随后的戊二醛处理而暴露出来（图4，2、3）。

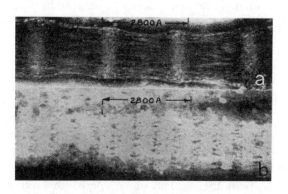

图3. a，纤维长间距I型胶原蛋白，显示了约为2,500 Å的重复周期。b，戊二醛处理后的纤维长间距I型胶原蛋白。640 Å的周期重新出现，清晰可见。制品经过负染处理。

在片段长间距微晶中，负染样品表明带结构具有明显的不对称性。另一方面，尽管奥尔森[17]曾尝试利用照相的方法来推断片段长间距和天然纤维之间的关系，但还是有很多地方是模糊不清的。在片段长间距型胶原蛋白中观察到的带状图案可能

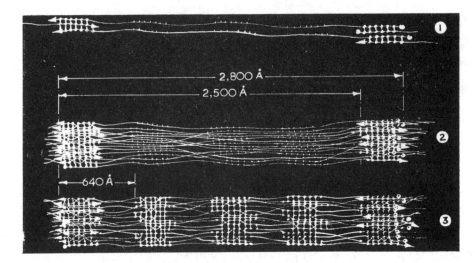

Fig. 4. 1 and 2, The establishment of a repeat period of 2,500 Å by tropocollagen macromolecules aggregated in the presence of glycoprotein. Linking between intermediate bonding zones is suppressed by glycoprotein, resulting in the observed long spacing. 2 and 3, FLS 1, showing re-establishment of 640- Å period by treatment with glutaraldehyde.

of points remain obscure. The band pattern observed in the SLS form may in part be due to the presence of negatively charged ATP molecules interacting with positively charged groups in tropocollagen, outside as well as within the bonding regions. If this interpretation is correct, the SLS pattern could be regarded, in part at least, as an artefact with a band structure (by negative staining) not directly related to the native fibre. Furthermore, in the case of positively stained SLS form, the staining with phosphotungstic acid or uranyl acetate undoubtedly reveals the position of basic and acidic groups in the macromolecule. Positive staining, however, does not seem to indicate whether or not these groups are involved in the formation of bonds between adjacent macromolecules.

The fundamental difference between the model proposed here and other interpretations lies in the division of the tropocollagen filament into five more or less equal bonding zones separated by four regions apparently capable of forming few intermolecular links. It should be emphasized that the bonding zones mentioned in this report have no connexion with the subunits of Petruska and Hodge[23]. Collagen fibres have also been observed with a repeat period of only 210–220 Å (Gross[7]). By assuming that, under the conditions of pH and ionic strength necessary to precipitate this form, only a portion of each bonding zone is able to interact with another, it is possible to derive this repeat period from the new model (Fig. 5). It may be noted that the short period of 210–220 Å is less than the length of a bonding zone (A band) and also one-third of the native period as would be expected if this model is correct.

This investigation has been restricted to the longitudinal and transverse structure of collagen fibres in a two-dimensional sense. Work is at present in progress to include three-dimensional (cross-sectional) aspects of the problem and a three-dimensional model

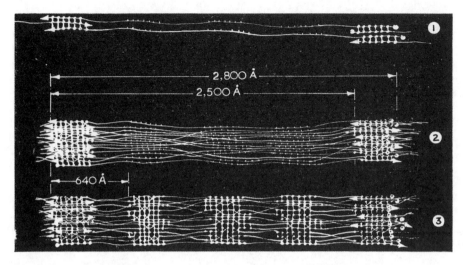

图 4. 1 和 2，在存在糖蛋白的情况下，原胶原大分子聚集产生 2,500 Å 的重复周期。中间结合区的连接受糖蛋白的抑制，导致长间距的产生。2 和 3，纤维长间距 I 型胶原蛋白经戊二醛处理后，640 Å 的周期重新出现。

在一定程度上是由于在结合区内外存在与原胶原中正电基团相互作用的带负电荷的三磷酸腺苷分子。如果这种解释是正确的，那么至少在一定程度上片段长间距模式可以被认为是与天然纤维没有直接关系的具有带结构（通过负染反应得到的）的一种人为产物。而且，在正染的片段长间距型胶原蛋白中，利用磷钨酸或醋酸双氧铀染色确实显示出了大分子中碱性基团和酸性基团的位置，然而正染似乎并不能说明这些基团是否参与了相邻大分子间键的形成。

本文所提出的模型与其他说法之间的根本区别在于，认为原胶原细丝被 4 个有可能形成少数分子间连接的区域分为 5 个几乎相同的结合区。需要强调的是，这篇报道中所提到的结合区与彼特鲁斯卡和霍奇 [23] 所说的亚单位无关。人们还观察到了仅有 210 Å~220 Å 重复周期的胶原蛋白纤维（格罗斯 [7]）。在能够沉淀这一胶原蛋白的酸碱度和离子强度的条件下，如果假设每个结合区只有一部分可能与其他区域发生相互作用，那么就有可能从新模型中导出这种重复周期（图 5）。值得注意的是，如果这个模型正确的话，210 Å~220 Å 的短周期要比一个结合区（A 带）的长度短，并且是预期的天然周期的三分之一。

这项研究局限于由胶原蛋白纤维的纵向和横向结构所构成的二维空间上。当前正在进行的工作包括将这一问题扩展到三维空间（截面）上并建立一个三维模型。

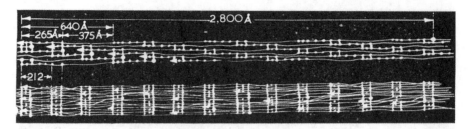

Fig. 5. The establishment of the observed short (210–220 Å) period is illustrated in the diagram. This can be considered to occur under conditions where each macromolecule is capable of interacting only with neighbours over a portion of the bonding zone.

is being developed. Notwithstanding that this is only a two-dimensional model, since aggregation takes place in an essentially random fashion the criteria of Smith[25] appear to be satisfied.

We thank the Wellcome Trust for a grant for the purchase of an electron microscope and ancillary apparatus.

(**207**, 822-826; 1965)

R. A. Grant, R. W. Horne and R. W. Cox: Agricultural Research Council, Institute of Animal Physiology, Babraham, Cambridge.

References:

1. Rich, A., and Crick, F. H. C., *J. Mol. Biol.*, **3**, 483 (1961).

2. Boedtker, H., and Doty, P., *J. Amer. Chem. Soc.*, **78**, 4267 (1956).

3. Rice, R. V., Casassa , E. F., Kerwin, R. E., and Maser, M. D., *Arch. Biochem. Biophys.*, **105**, 409 (1964).

4. Schmitt, F. O., Hall, C. E., and Jakus, M. A., *J. Cell. Comp. Physiol.*, **20**, 11 (1942).

5. Wolpers, C., *Klin. Wschr.*, **22**, 624 (1943).

6. Harrington, W. F., and Von Hippel, P. H., *Adv. Protein Chem.*, **16**, 1 (1961).

7. Gross, J., in *Comparative Biochemistry*, edit. by Florkin, M., and Mason, H. S., **5**, 307 (Academic Press, New York and London, 1963).

8. Fitton, Jackson, S., in *The Cell*, edit. by Brachet, J., and Mirsky, A. P., **6**, 387 (Academic Press, New York and London, 1964).

9. Veis, A., *The Macromolecular Chemistry of Gelatin* (Academic Press, New York and London, 1964).

10. Gross, J., Highberger, J. H., and Schmitt, F. O., *Proc. U.S. Nat. Acad. Sci.*, **40**, 679 (1954).

11. Highberger, J. H., Gross, J., and Schmitt, F. O., *Proc. U.S. Nat. Acad. Sci.*, **37**, 286 (1951).

12. Schmitt, F. O., Gross, J., and Highberger, J. H., *Symposia Soc. Exp. Biol.*, **9**, 148 (1955).

13. Schmitt, F. O., and Hodge, A. J., *J. Soc. Leather Trades Chem.*, **44**, 217 (1960).

14. Brenner, S., and Horne, R. W., *Biochim. Biophys. Acta.*, **34**, 103 (1959).

15. Tromans, W. J., Horne, R. W., Gresham, G. A., and Bailey, A. J., *Z. Zellforsch.*, **58**, 798 (1963).

16. Olsen, B. R., *Z. Zellforsch.*, **59**, 184 (1963).

17. Olsen, B. R., *Z. Zellforsch.*, **59**, 199 (1963).

18. Tromans, W. J., Ph. D. thesis, Cambridge University (1963).

19. Hodge, A. J., and Schmitt, F. O., *Proc. U.S. Nat. Acad. Sci.*, **46**, 186 (1960).

20. Kuhn, K., and Zimmer, E., *Naturwiss.*, **48**, 219 (1961).

21. Hodge, A. J., and Petruska, J. A., *Fifth Intern. Cong. Electron Microscopy, Philadelphia*, **2**, QQ-1 (Academic Press, New York and London, 1962).

22. Grassman, W., Hannig, K., and Nordwig A., *Z. Physiol. Chem.*, **333**,154 (1963).

23. Petruska, J. A., and Hodge, A. J., *Proc. U.S. Nat. Acad. Sci.*, **51**, 871 (1964).

24. Braun-Falco, O., and Rupec, M., *J. Invest. Dermatol.*, **42**, 15 (1964).

25. Smith, J. W., *Nature*, **205**, 356 (1965).

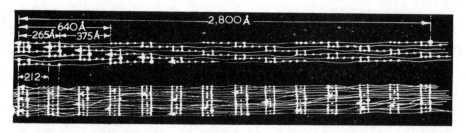

图 5. 图中所示的是观测到的短周期（210 Å~220 Å）的形成。只有当每个大分子都仅能与相邻大分子的部分结合区相互作用的情况下，才可能产生这样的条带。

尽管这只是一个二维空间模型，但聚集是以本质上符合史密斯 [25] 标准的随机方式发生的。

我们感谢维康基金会为购置电子显微镜和辅助设备提供资金。

（郭娟 翻译；周筠梅 审稿）

An Improved Method for the Characterization of Human Haemoglobin Mutants: Identification of $\alpha_2\beta_2^{95\text{GLU}}$, Haemoglobin N (Baltimore)

J. B. Clegg *et al.*

Editor's Note

In this paper, researchers from the Johns Hopkins University School of Medicine, Baltimore, describe an improved method for analysing abnormal haemoglobin proteins. Previous methods failed to fully digest these complex proteins and struggled to effectively separate α and β polypeptide chains from the small quantities of starting material. The new technique, devised by John Clegg, Mike Naughton and David Weatherall, relied on high-resolution chromatography to separate the chains, and a chemical step to convert them to a fully digestible form. The method was quick, sensitive and enabled the team to separate normal and abnormal chains, as demonstrated by their analysis of the abnormal haemoglobin N (Baltimore) protein.

ONE of the problems during investigation of human haemoglobin mutants has been the difficulty of separating the α and β chains in a pure form from small quantities of starting material. While chain separation is not a pre-requisite for all chemical investigations, such work can be carried out more easily on the isolated α and β chains. The method for the separation of the peptide chains of human haemoglobin described in this communication is relatively easy to use and has an advantage over those previously reported in that it will not only give separation of the α and β chains, but is also capable of resolving a charged mutant chain from the normal, thus enabling the globin of a heterozygous charged mutant to be fractionated into all its component chains.

A second difficulty usually encountered is the familiar one of incomplete digestion of proteins by proteolytic enzymes. Digestion of human α and β haemoglobin chains with trypsin normally leaves a trypsin-resistant "core" with the result that a region consisting of about 30 percent of the molecule cannot easily be examined by fingerprinting. However, conversion of the cysteine residues of the α and β chains to aminoethyl-cysteine by reaction with ethyleneimine[1] has been shown by Jones[2] to give derivatives which no longer contain the trypsin-resistant region. Thus, fingerprints of aminoethylated α and β chains show all the expected trypsin peptides plus additional ones due to splits at aminoethyl-cysteine.

The combination of a high-resolution chromatographic separation of the α and β chains and the conversion of the separated chains into aminoethyl derivatives susceptible to trypsin[2] affords a procedure considerably more rapid and more sensitive than any previously available for studying haemoglobin mutants. With the technique described

一种改进的研究人类血红蛋白突变体的方法：$\alpha_2\beta_2^{95GLU}$血红蛋白N（巴尔的摩）的鉴定

克莱格等

编者按

在这篇论文中，来自巴尔的摩约翰斯·霍普金斯大学医学院的研究者们描述了一种改进的分析异常血红蛋白的方法。此前的方法不能使这些复杂的蛋白质完全酶解，且很难高效地从少量原始材料中分离出 α 多肽链和 β 多肽链。由约翰·克莱格、迈克·诺顿以及戴维·韦瑟罗尔设计的新技术采用高分辨色谱法分离多肽链，通过化学步骤将它们转化为可以完全酶解的形式。这种方法快速、灵敏并且可以使研究团队将正常多肽链和异常多肽链分离，就像他们在对异常血红蛋白 N(巴尔的摩) 的分析中所展示的那样。

在研究人类血红蛋白突变体的过程中，长久以来困扰人们的一个难题是从少量的原始材料中分离出纯的 α 链和 β 链。虽然链的分离并不是所有化学研究的必要条件，但分离的 α 链和 β 链可使这类研究实施起来更容易。本文中描述的人类血红蛋白多肽链的分离方法比较容易操作，与此前报道的那些方法相比，它具有一个优点，因为它不仅可以分离 α 链和 β 链，而且也能从正常的链中分离出带电突变链，进而使得杂合的带电突变体的珠蛋白的全部组成链都得以分离。

经常遇到的第二类难题是，较为常见的蛋白水解酶不完全消化蛋白质。用胰蛋白酶消化人类血红蛋白的 α 链和 β 链，通常会留下一个胰蛋白酶耐受的"核心"，结果导致指纹图谱不能容易地检测这个约占整个分子 30% 的区域。然而琼斯[2] 指出，通过与吖丙啶 [1] 反应将 α 链和 β 链的半胱氨酸残基转化为氨乙基－半胱氨酸，得到的衍生物中就不包含胰蛋白酶耐受区域。因而，氨基乙基化的 α 链和 β 链的指纹图谱可以显示所有预期的胰蛋白酶多肽段以及在氨乙基－半胱氨酸处进行切割而得到的肽段。

α 链和 β 链的高分辨色谱分离产物与将分离链转化为对胰蛋白酶敏感[2] 的氨乙基衍生物相结合，为研究血红蛋白突变体提供了一个比以前更为快速和更为灵敏的方法。通过本文描述的技术，从少至 10 毫克的原珠蛋白中可回收到相当数量的分离

here, recoveries of the separated α and β chains are quantitative and sufficient material for aminoethylation and subsequent fingerprinting can be obtained from as little as 10 mg of starting globin. In addition, quantitative amino-acid analyses of peptides eluted from fingerprints of 2 mg of the digested chains have been routinely achieved.

Isolation and characterization of the β chain of haemoglobin N (Baltimore). An 8-month-old Negro female was found to have two haemoglobins by routine electrophoretic screening. Apart from a mild iron deficiency there were no significant haematological abnormalities either in the child or in the mother, who also carried the abnormal variant. The relevant haematological values and the proportions of normal and abnormal haemoglobins are summarized in Table 1. On starch-gel electrophoresis at pH 8.6 (Fig. 1) the abnormal haemoglobin migrated faster than haemoglobin J (Baltimore)[3] but slower than haemoglobin H. These electrophoretic characteristics are similar to those previously described for haemoglobin N[4] and the variant has therefore been designated as haemoglobin N (Baltimore). Comparative gel-electrophoresis experiments with other known haemoglobin mutants suggested that at pH 8.6 haemoglobin N (Baltimore) possesses four more negative charges per 68,000 molecular weight than does haemoglobin A. Hybridization with canine haemoglobin indicated that the abnormality in haemoglobin N (Baltimore) was in the β chain. Fingerprints of tryptic digests of whole haemoglobin N (Baltimore) or of isolated $\beta^{N(Baltimore)}$ chain showed no detectable differences from those of haemoglobin A or normal β chain.

Table 1. Haematological and Electrophoretic Data

Family member		Propositus	Mother
Age		8 months	24 years
Haemoglobin (g/100 ml.)		10.2	13.8
Red cell count (millions/mm³)		4.06	4.13
Reticulocytes (%)		5.1	1.1
MCV (μ^3)		81	102
MCH ($\mu\mu$g)		28	33
MCHC (%)		31	33
Haemoglobin constitution		A+N	A+N
Haemoglobin fractionation	HbA(%)	46.3	48.6
	HbN(%)	44.3	49.6
	HbA₂(%)	1.0	1.4
Alkali-resistant haemoglobin (%)		8.4	0.4

20 mg of the child's globin, prepared by 2 percent acid-acetone precipitation of a whole red-cell lysate, was dissolved in 2 ml. of a buffer consisting of 8 M urea, 0.05 M 2-mercaptoethanol, and 0.005 M Na₂HPO₄ , adjusted to pH 6.7 with phosphoric acid. The solution was dialysed at room temperature against three changes of a 50-fold excess of the same buffer for a total of 2.5 h and then applied to a 1 cm × 10 cm column of carboxymethyl-cellulose (0.7 m.equiv.g) equilibrated against the same buffer. After the column had been washed to remove any unretarded material the peptide chains were

的 α 链和 β 链，足以进行氨基乙基化和随后的指纹图谱分析。另外，对 2 毫克酶解后的肽链进行指纹图谱，洗脱下的肽段通常能进行定量氨基酸分析。

血红蛋白 N（巴尔的摩）β 链的分离与鉴定 通过常规的电泳分析，发现在一个 8 月龄的黑人女性体内存在两种血红蛋白。除了轻微的缺铁，无论孩子还是母亲体内（她也携带异常变异体）都没有发现明显的血液学上的异常。表 1 列出了相关的血液学值、正常的和异常的血红蛋白的比例。在 pH 8.6 的条件下进行淀粉凝胶电泳（图 1），发现异常的血红蛋白的迁移速率大于血红蛋白 J（巴尔的摩）[3]，但是小于血红蛋白 H。这些电泳特征与之前描述的血红蛋白 N[4] 相似，因此这种变异体被确定为血红蛋白 N（巴尔的摩）。将凝胶电泳的实验结果与其他已知的血红蛋白突变体进行比较发现，在 pH 8.6 的条件下每 68,000 分子量的血红蛋白 N（巴尔的摩）比血红蛋白 A 多 4 个负电荷。与犬科血红蛋白杂交结果显示，血红蛋白 N（巴尔的摩）的 β 链存在异常。胰蛋白酶消化整个血红蛋白 N（巴尔的摩）所得产物的指纹图谱，或者分离出的 $\beta^{N（巴尔的摩）}$ 链的指纹图谱表明，与血红蛋白 A 或者正常的 β 链相比没有可检测的差别。

表 1. 血液学和电泳数据

家庭成员		先证者	母亲
年龄		8 个月	24 岁
血红蛋白（克 /100 毫升）		10.2	13.8
红细胞量（百万 / 毫米³）		4.06	4.13
网状细胞（%）		5.1	1.1
红细胞平均容量（微米³）		81	102
红细胞平均血红蛋白量（皮克）		28	33
红细胞平均血红蛋白浓度（%）		31	33
血红蛋白组成		A+N	A+N
血红蛋白组分	HbA（%）	46.3	48.6
	HbN（%）	44.3	49.6
	HbA₂（%）	1.0	1.4
耐碱血红蛋白（%）		8.4	0.4

用 2% 的丙酮酸沉淀一个完整红细胞的溶出物，得到 20 毫克儿童的珠蛋白，将其溶解于 2 毫升缓冲液中（缓冲液成分：8 摩尔 / 升尿素，0.05 摩尔 / 升 2-巯基乙醇和 0.005 摩尔 / 升 Na_2HPO_4，用磷酸调 pH 为 6.7）。在室温下透析溶液，将透析袋放入 50 倍过量的相同缓冲液中，其间更换缓冲液 3 次，共透析 2.5 小时，然后将透析后的溶液上样到用同样的缓冲液平衡过的 1 厘米 ×10 厘米羧甲基纤维素柱里（0.7 m.equiv.g），洗脱去除所有未挂柱的物质后，通过线性钠离子梯度的方法

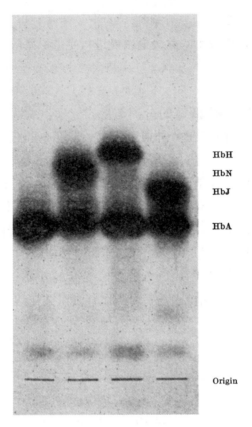

Fig. 1. Starch-gel electrophoresis (*tris*-EDTA-borate, pH 8.6) of haemolysates containing (left to right): haemoglobin A, haemoglobins A and N, haemoglobins A and H, and haemoglobins A and J.

eluted at a flow rate of 1 ml./min by means of a linear Na^+-ion gradient made by mixing 100 ml. of starting buffer with 100 ml. of a buffer consisting of 8 M urea, 0.05 M 2-mercaptoethanol, and 0.03 M Na_2HPO_4, adjusted to pH 6.7 with phosphoric acid. The column effluent was monitored continuously at 280 mμ and the resulting chromatogram is shown in Fig. 2. Of the three peaks obtained, two were found at the expected elution volumes of normal α and β chains, while the third peak emerged much earlier than normal β chain, indicating the presence of a more acidic β chain. This confirmed the findings of the hybridization experiments.

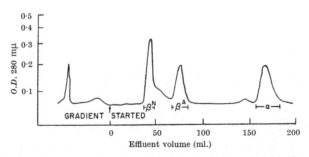

Fig. 2. Gradient elution chromatography on carboxymethyl-cellulose of globin from haemoglobin N heterozygote.

378

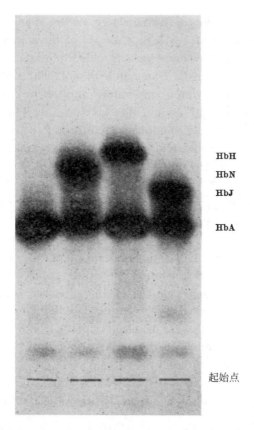

图1. 红细胞溶解物的淀粉凝胶电泳（三羟甲基氨基甲烷–乙二胺四乙酸–硼酸盐缓冲系统，pH 8.6），从左向右依次为血红蛋白A、血红蛋白A和N、血红蛋白A和H、血红蛋白A和J。

（将100毫升起始缓冲液与100毫升组分为8摩尔/升尿素、0.05摩尔/升2-巯基乙醇和0.03 摩尔/升 Na_2HPO_4 且用磷酸调 pH 为 6.7 的缓冲液进行混合而制得），以1毫升/分钟的速率洗脱多肽链。在280纳米下连续监测柱子洗脱液，得到的色谱图见图 2。在获得的三个峰中，两个出现在预期的正常 α 链和 β 链的洗脱体积处，然而第三个峰出现的时间早于正常的 β 链，表明存在一个酸性更强的 β 链。这证实了杂交实验的结果。

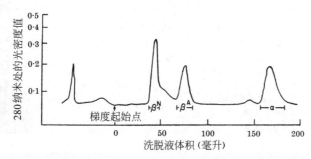

图 2. 血红蛋白 N 杂合体的珠蛋白在羧甲基纤维素层析柱上梯度洗脱的色谱图

379

The fractions corresponding to the two β chains were collected and solid *tris* added to each solution to give a concentration of 1 M. After the pH of the solutions had been adjusted to 9.2 with concentrated HCl, ethyleneimine was added to a final concentration of 0.5 M (that is, a 10-fold molar excess over the 2-mercaptoethanol). The aminoethylation reactions were then allowed to proceed at room temperature until no free sulphydryl groups were detectable by the nitroprusside test[5] (usually after about 2.5 h). The pH was then adjusted to 3 with concentrated HCl and the solutions passed through a "Sephadex *G*-25" column equilibrated with 0.5 percent formic acid, in order to remove urea, salts, etc. Finally, the recovered protein fractions were freeze dried.

Tryptic digests of amounts of protein greater than 10 mg were carried out at pH 9.0 for 2 h at room temperature in a radiometer pH-stat. Digestion was terminated by adjusting the pH to 4.7 and freeze-drying the solution. It was preferable to digest smaller amounts of protein in 1 percent NH_4HCO_3, pH 8.2 (the amounts of salt introduced by the automatic procedure being sufficient to cause smearing of peptide spots when it was necessary to apply all the digest to a single fingerprint). After digestion was complete, the NH_4HCO_3 was removed by repeated freeze-drying. For all the digestions the trypsin/protein weight ratio was 1/100. The dried digests were dissolved in pH 4.7 buffer (1.25 percent pyridine, 1.25 percent acetic acid) and aliquots corresponding to 2 mg of the original proteins were subjected to electrophoresis in this buffer on 108 cm × 57 cm sheets of Whatman No. 3 MM paper for 3.25 h at 33 V/cm in a "Varsol"-cooled tank. After drying, the papers were chromatographed overnight in *n*-butanol, acetic acid, water, pyridine (15 : 3 : 12 : 10)

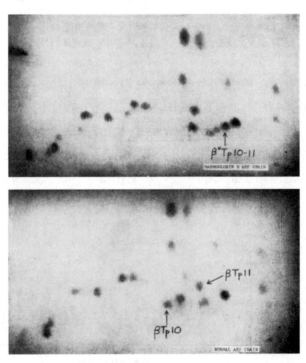

Fig. 3. Fingerprints of tryptic digests of AEβ[N (Baltimore)] (above) and normal AEβ chains (below)

收集两个 β 链对应的分离组分，在每一收集液中加入固体三羟甲基氨基甲烷使浓度为 1 摩尔 / 升。用浓盐酸将收集液的 pH 调到 9.2 后，加入吖丙啶使其终浓度为 0.5 摩尔 / 升（即摩尔数为 2– 巯基乙醇的 10 倍）。在室温条件下进行氨基乙基化反应，直至用硝普盐试验检测不到游离的巯基为止 [5]（通常为 2.5 小时后）。然后用浓盐酸将 pH 调为 3，再使溶液通过一个用 0.5% 的甲酸平衡过的"葡聚糖凝胶 G-25"柱，以便除掉尿素、盐等。最后将收集到的蛋白质组分冷冻干燥保存。

在雷迪美特 pH 恒定仪中将 pH 调为 9，用胰蛋白酶消化总量大于 10 毫克的蛋白质，室温条件下反应 2 个小时。然后将 pH 调为 4.7 以终止反应，并且冷冻干燥反应液。在 1% 的 NH_4HCO_3、pH 为 8.2 的条件下消化少量蛋白质效果会更好（当必须对所有消化物进行单一指纹图谱时，自动化程序引入的盐的量足以引起肽点的拖尾效应）。消化完成后，再通过一次冷冻干燥将 NH_4HCO_3 除去。对于所有消化物，胰蛋白酶与蛋白质的重量比是 1/100。消化物干粉溶于 pH 为 4.7 的缓冲液(1.25% 嘧啶，1.25% 醋酸）中，剩余的 2 毫克原始蛋白质在相同缓冲液中进行电泳，选用规格为 108 厘米 ×57 厘米的沃特曼 No.3 MM 纸，在"瓦索尔"冷却系统中于 33 伏特 / 厘米条件下电泳 3.25 小时。烘干后，在正丁醇∶醋酸∶水∶吡啶（15∶3∶12∶10）的条件下，

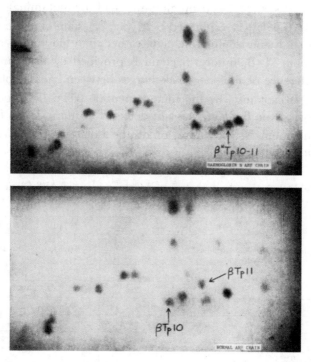

图 3. 用胰蛋白酶消化 $AE\beta^{N（巴尔的摩）}$（上图）和正常的 $AE\beta$ 链（下图）的指纹图谱

(ref. 6). The resulting fingerprints were stained by dipping in 0.02 percent ninhydrin-acetone and developing at $60°$. Photographs of the two fingerprints are shown in Fig. 3 and the significant differences between the mutant and the normal patterns are indicated by arrows. The other slight differences are due to the varying amounts of partially split peptides.

The fingerprint of the mutant β chain is characterized by the absence of peptides βTp10 and βTp11 and by the appearance of a new peptide which has a distinctive yellow colour. Since βTp10 also gives a yellow colour with ninhydrin, these observations suggest that the mutant peptide is a combination of βTp10 and βTp11. (No tryptic cleavage at the aminoethyl-cysteine residue in βTp10 has been found, presumably because aspartic acid is the adjacent amino-acid, the sequence in this region being AECys (aminoethyl-cysteine)^{93}-Asp^{94}-Lys^{95} (ref. 7).This absence of a split at AECys^{93} was previously reported by Jones[2].) The peptides to be analysed were cut from the fingerprints and eluted with 6 N HCl into 100 µl. capillary disposable pipettes by the method described by Sanger and Tuppy[8]. Collection of 30-40 µl. was sufficient to ensure complete removal of the peptide from the paper. The tubes were then sealed, and heated at $105°$ for 18 h. After hydrolysis, the contents of the tubes were washed into 0.5 ml. of distilled water and applied directly to a "Technicon" 5-column amino-acid analyser. This made it unnecessary to dry down the hydrolysates, a procedure which has been shown to reduce the yield of amino-acids, possibly because of further reaction with ninhydrin[9]. Table 2 gives the amino-acid analysis and composition of the mutant peptide. From this it is apparent that the lysine residue (95) of βTp10 has been replaced by glutamic acid (or glutamine), with the result that a conjugate peptide, β^NTp10-11, is formed by tryptic digestion of the mutant $\beta^{N(Baltimore)}$ chain. The evidence for a lysine to glutamic acid (rather than glutamine) change at position 95 in the haemoglobin N (Baltimore) β chain is provided by the starch-gel electrophoresis findings of a difference of four negative charges between haemoglobin N (Baltimore) and haemoglobin A. It explains why the mutant peptide β^NTp10-11, occurring as it does in the "core" region of the β chain, was not detected on the original fingerprints of tryptic digests of haemoglobin N (Baltimore) or of the isolated $\beta^{N(Baltimore)}$ chain which had not been reacted with ethyleneimine.

Table 2. Amino-Acid Analysis and Compositions of AEβ Chain Peptides

Amino-acid	AE$\beta^{N(Baltimore)}$			AEβ^A (ref. 7)		
	$\beta^{N(Baltimore)}$ Tp10-11			β^ATp10	β^ATp11	β^ATp10+β^ATp11
	Found (µmoles)	Residues	Nearest whole residue	Expected	Expected	Expected
Lys	-	-	-	1	-	1
His	0.043	1.6	2	1	1	2
Arg	0.028	1.0	1	-	1	1
Asp	0.078	2.8	3	1	2	3
Thr	0.051	1.8	2	2	-	2
Ser	0.033	1.2	1	1	-	1
Glu	0.085	3.0	3	1	1	2

连夜进行色谱分离（参考文献 6）。产生的指纹图谱浸泡在 0.02% 的水合茚三酮溶液中染色并在 60℃ 下显影。图 3 列出了两种指纹图谱的照片，用箭头标注突变体和正常样品之间的显著差异。其他微小的差异是由于不完全酶切多肽量的不同造成的。

突变 β 链的指纹图谱的特征是缺失肽段 βTp10 和 βTp11 并且出现了一条呈特殊黄色的新肽段。加入茚三酮后，βTp10 也显黄色，因此这些结果表明突变多肽是 βTp10 和 βTp11 的结合。（在 βTp10 中，未发现胰蛋白酶在氨乙基 – 半胱氨酸残基处的酶切位点，据推测因为邻近的氨基酸是天冬氨酸，这段区域的序列是 AECys（氨乙基 – 半胱氨酸）^{93}-Asp（天冬氨酸）^{94}-Lys（赖氨酸）^{95}（参考文献 7）。之前琼斯也曾报道过在 AECys^{93} 处酶切位点的缺失 [2]。）利用桑格和塔皮 [8] 描述的方法将多肽从指纹图谱上切下来，在 100 微升一次性毛细管中用 6 摩尔 / 升的盐酸洗脱，对得到的多肽进行分析。收集 30 微升 ~ 40 微升就能确保从纸上分离出全部多肽。将管密封，在 105℃ 下加热 18 小时。水解后，在 0.5 毫升蒸馏水中洗涤管内物质，然后直接利用"泰克尼康"5– 柱氨基酸分析仪进行分析。这样就不必干燥水解产物，已证明这个过程可以减少氨基酸的生成，可能是因为与水合茚三酮进一步反应 [9]。表 2 给出了氨基酸的分析结果和突变多肽的组成。从这里可以明显看出，βTp10 的赖氨酸残基（95）被谷氨酸（或者谷氨酰胺）代替，结果形成一个结合肽 β^{N}Tp10-11，它是通过胰蛋白酶消化突变的 β$^{N(巴尔的摩)}$ 多肽链形成的。通过淀粉凝胶电泳发现血红蛋白N（巴尔的摩）和血红蛋白 A 之间相差 4 个负电荷，因此为血红蛋白 N（巴尔的摩）β 链的 95 位上的赖氨酸突变为谷氨酸（而非谷氨酰胺）提供了证据。这解释了为什么突变多肽 β^{N}Tp10-11 在血红蛋白 N（巴尔的摩）的胰蛋白酶消化产物的原始指纹图谱或在未与吖丙啶发生反应的分离的 β$^{N(巴尔的摩)}$ 链的原始指纹图谱中都没有检测到，看起来它确实存在于 β 链的"核心"区域。

表 2. 氨基酸分析结果和 AEβ 链多肽的组成

氨基酸	AEβ$^{N(巴尔的摩)}$			AEβ^A（参考文献 7）		
	β$^{N(巴尔的摩)}$ Tp10-11			β^{A}Tp10	β^{A}Tp11	β^{A}Tp10 + β^{A}Tp11
	结果（微摩尔）	残基	最接近的残基整数	预期	预期	预期
Lys	-	-	-	1	-	1
His	0.043	1.6	2	1	1	2
Arg	0.028	1.0	1	-	1	1
Asp	0.078	2.8	3	1	2	3
Thr	0.051	1.8	2	2	-	2
Ser	0.033	1.2	1	1	-	1
Glu	0.085	3.0	3	1	1	2

Continued

Amino-acid	AEβN(Baltimore)			AEβA (ref. 7)		
	βN(Baltimore) Tp10-11			βATp10	βATp11	βATp10+βATp11
	Found (μmoles)	Residues	Nearest whole residue	Expected	Expected	Expected
Pro	0.026	0.9	1	-	1	1
Gly	0.026	0.9	1	1	-	1
Ala	0.030	1.1	1	1	-	1
Val	0.028	1.0	1	-	1	1
AE-Cys	0.032	1.1	1	1	-	1
Met	-	-	-	-	-	-
Lieu	-	-	-	-	-	-
Leu	0.076	2.7	3	2	1	3
Tyr	-	-	-	-	-	-
Phe	0.047	1.7	2	1	1	2
Tryp	-	-	-	-	-	-

Overall recovery of AEβN(Baltimore) Tp10-11 was 24 percent of amount originally applied to the fingerprint.

In accordance with the recommended nomenclature[10], haemoglobin N (Baltimore) is designated as $\alpha_2\beta_2{}^{95Glu}$.

The procedure outlined in this article has the following advantages over those previously described:

(1) The chromatographic system enables globin from a heterozygous charged mutant to be resolved quantitatively into all its component chains, thus simplifying subsequent chemical investigations. In addition, the high resolution achieved during chromatography eliminates the need for hybridization experiments to determine in which chain an amino-acid substitution is located. (It should be mentioned that the pH of the chromatographic system is not critical and can be varied from at least pH 6.5 to 7.2, the actual choice depending largely on the nature of the mutant globin to be fractionated and the degree of resolution required. As an example of this flexibility, Fig. 4 shows a separation of 11 mg of sickletrait globin into β^A, β^S and α components at pH 7.2.)

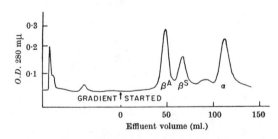

Fig. 4. Chromatography at pH 7.2 of globin from haemoglobin S heterozygote

(2) All the tryptic peptides of the amino-ethylated α and β chains can be resolved by fingerprinting.

续表

| 氨基酸 | AEβ$^{N(巴尔的摩)}$ | | | AEβ^A（参考文献 7） | | |
| | β$^{N(巴尔的摩)}$ Tp10-11 | | | β^{A}Tp10 | β^{A}Tp11 | β^{A}Tp10 + β^{A}Tp11 |
	结果（微摩尔）	残基	最接近的残基整数	预期	预期	预期
Pro	0.026	0.9	1	-	1	1
Gly	0.026	0.9	1	1	-	1
Ala	0.030	1.1	1	1	-	1
Val	0.028	1.0	1	-	1	1
AE-Cys	0.032	1.1	1	1	-	1
Met	-	-	-	-	-	-
Lieu	-	-	-	-	-	-
Leu	0.076	2.7	3	2	1	3
Tyr	-	-	-	-	-	-
Phe	0.047	1.7	2	1	1	2
Tryp	-	-	-	-	-	-

全部回收的AEβ$^{N(巴尔的摩)}$ Tp10-11占最初应用于指纹图谱量的24%。

根据有人提议的命名法 [10]，血红蛋白 N（巴尔的摩）被确定为 $\alpha_2\beta_2^{95Glu}$。

本文中描述的流程与之前提到的那些相比具有以下优点：

（1）色谱系统能定量地分离来自杂合带电突变体的珠蛋白的所有肽链组分，因而简化了后续的化学研究。另外，色谱实现的高分辨率使通过杂交实验检测氨基酸替换位于哪条链上变得不必要。（值得一提的是色谱系统的 pH 不是至关重要的，至少可以在 pH 6.5~7.2 之间变化，实际选择的 pH 值主要取决于需要分离的突变珠蛋白的性质和需要的分辨率。作为实际应用的一个例子，图 4 显示了在 pH 7.2 的条件下将 11 毫克镰刀型珠蛋白分离为 β^A、β^S 和 α 三个组分的结果。）

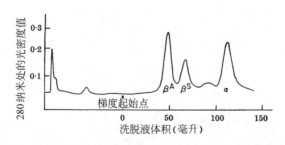

图 4. 血红蛋白 S 杂合体的珠蛋白在 pH 7.2 的条件下的色谱图

（2）可以通过指纹图谱分离氨乙基化的 α 链和 β 链的所有胰蛋白酶分解形成的片段。

(3) Quantitative amino-acid analysis of theses peptides can be determined on samples obtained by elution from a ninhydrin-stained fingerprint of as little as 2 mg of digested chain.

One of us (J. B. C.) thanks the Wellcome Foundation for a travel grant. This work was supported by U.S. National Institutes of Health grant *AM*-06006-03, U.S. National Science Foundation grant *CB*-2630, a continuation of *G*-24214, and U.S. Public Health Service grants *HE*-02799 and *TI-AM*-5260.

(**207**, 945-947; 1965)

J. B. Clegg and M. A. Naughton: Department of Biophysics.

D. J. Weatherall: Department of Medicine, Johns Hopkins University School of Medicine, Baltimore 5, Maryland.

References:

1. Raftery, M. A., and Cole, R. D., *Biochem. Biophys. Res. Comm.*, **10**, 467 (1963).

2. Jones, R. T., *Cold Spring Harbour Symp.*, **29**, 297 (1964).

3. Weatherall, D. J., *Bull. Johns Hopkins Hosp.*, **114**, 1 (1964).

4. Ager, J. A. M., and Lekmann, H., *Brit. Med. J.*, i, 929 (1958).

5. Katchalski, E., Benjamin, G. S., and Gross, V., *J. Amer. Chem. Soc.*, **79**, 4096 (1957).

6. Hill, R. L., Swenson, R. T., and Schwartz, H. C., *Blood*, **19**, 573 (1962).

7. Hill, R. J., Konigsberg, W., Guidotti, G., and Craig, L. C., *J. Biol. Chem.*, **237**, 1549 (1962).

8. Sanger, F., and Tuppy, H., *Biochem. J.*, **49**, 463 (1951).

9. Reider, R., and Naughton, M. A., *Bull. Johns Hopkins Hosp.*, **116**, 17 (1965).

10. "Recommendations of the International Society of Haematology on the Nomenclature of Abnormal Haemoglobins" (*Brit. J. Haemat.*, **11**, 121; 1965).

（3）仅需用从低至 2 毫克的酶解肽链的茚三酮染色指纹图谱洗脱获得的样本，便可对这些多肽进行定量的氨基酸分析。

我们中的一位研究者（克莱格）对威康基金会的旅费补助表示感谢。本研究由美国国立卫生研究院（基金 AM-06006-03）、美国国家科学基金会（基金 CB-2630 以及后续的 G-24214）和美国公共卫生署（基金 HE-02799 和 TI-AM-5260）资助。

（郑建全 翻译；刘京国 审稿）

A Biological Retrospect[*]

P. Medawar

Editor's Note

Peter Medawar had already won a Nobel Prize for his work on the limits to the compatibility of tissues taken from different individuals when he delivered this address on the state of biological research to the Zoological Section of the British Association for the Advancement of Science. The address deals only peripherally with molecular genetics, which had by 1965 transformed the pattern of biological research in Britain. Medawar suffered a severe stroke in 1969 when he was 54. Although he was physically handicapped, he remained mentally active until he died in 1987.

THE title of my presidential address, you will have discerned, is "A Biological Retrospect", and on the whole it has not been well received. "Why a biological *retrospect*?", I have been asked; would it not be more in keeping with the spirit of the occasion if I were to speak of the future of biology rather than of its past? It would indeed be, if only it were possible, but unfortunately it is not. What we want to know about the science of the future is the content and character of new scientific theories and ideas. Unfortunately, it is impossible to predict new ideas—the ideas people are going to have in ten years' or in ten minutes' time—and we are caught in a logical paradox the moment we try to do so. For to predict an idea is to have an idea, and if we have an idea it can no longer be the subject of a prediction. Try completing the sentence "I predict that at the next meeting of the British Association someone will propound the following new theory of the relationships of elementary particles, *namely* ...". If I complete the sentence, the theory will not be new next year; if I fail, then I am not making a prediction.

Most people feel more confident in denying that certain things will come to pass than in declaring that they can or will do so. Many a golden opportunity to remain silent has been squandered by anti-prophets who do not realize that the grounds for declaring something impossible or inconceivable may be undermined by new ideas which cannot be foreseen. Here is an instructive passage from the philosophic writings of a great British physiologist, J. S. Haldane (father of J. B. S.). It comes from *The Philosophy of a Biologist* of 1931, and its subject is the nature of memory in a very general sense that includes "genetic memory", for example, the faculty or endowment which ensures that a frog's egg develops into a frog and, indeed, into a particular kind of frog.

---

[*] Presidential address delivered to Section D (Zoology) on September 2, 1965, at the Cambridge Meeting of the British Association for the Advancement of Science.

388

生物学回顾*

梅达沃

编者按

彼得·梅达沃向英国科学促进会动物学分会发表该篇关于生物学研究状况的演讲之时，就已经因对不同个体之间组织不相容性的研究工作而获诺贝尔奖了。这篇演讲只是粗浅地论及了分子遗传学，自 1965 年以来分子遗传学已经改变了英国的生物学研究模式。1969 年，54 岁的梅达沃得了严重的中风，尽管他身体残疾了，但思维依然活跃，直至 1987 年去世。

大家可能已经注意到了，我演讲的题目是"生物学回顾"，但总的来说这个题目不是很受欢迎。有人曾问我，"为什么要**回顾**生物学？"讲生物学的发展趋势而不是生物学回顾，岂不是更适合这一场合？如果可以预测未来的话，这的确可以，但遗憾的是不能。对于未来的科学，我们想知道的是新科学理论和新科学思想的内容和特征。但遗憾的是，十年后甚至十分钟后人们的新思想都是无法预知的，人们一旦试图去预测未来，就会让自己陷入自相矛盾的境地。因为预测一种思想也就等于接受了这种思想，如果我们接受了这种思想，它就不再是预测的产物。让我们试图将下面这句话补充完整："我推测，下一届英国科学促进会会议上将有人提出基本粒子间相互关系的新理论，**叫做**……"如果我现在将这句话补充完整的话，那么下次会议时这个理论就不是新理论，而如果我无法说出这是一个什么样的理论，那么我就没有完成这次预测。

与推测将来可能或肯定会发生某事相比，大部分人在否定将来会发生某些事上更有把握。伪预言者们已经浪费了无数次保持缄默的宝贵机会，他们没有意识到不可预见的新理论可能会推翻用于宣称某事不可能或难以想象的依据。在这个问题上，英国伟大的生理学家约翰·斯科特·霍尔丹（约翰·伯登·桑德森·霍尔丹之父）1931 年在其哲学著作《生物学家的哲学》中的一段话非常具有启发性，这段话的主旨是在广义上说（包括"遗传记忆"在内）记忆的本质，例如先天因素，使一枚蛙卵发育成一只青蛙，确切地说是一只特定种类的青蛙。

---

* 本文系作者 1965 年 9 月 2 日在英国科学促进会剑桥会议上对 D 分部（动物学）发表的就职演说。

389

Haldane is very critical of the theories of memory propounded by Ewald Hering and Richard Semon, who "assume that memory in general is dependent on protoplasmic 'engrams', and that germ-cells are furnished with a system of engrams, functioning as guide-posts to all the normal stages of development". ("Engrams", I should explain, are more or less permanent physical memory traces or memory imprints that act as directive agencies in development[1].) "This theory", Haldane goes on to say, "has quite evidently all the defects of other attempts at mechanistic explanations of development. How such an amazingly complicated system of signposts could function by any physico-chemical process or reproduce itself indefinitely often is inconceivable[2]."

What Haldane found himself unable even to conceive is today a commonplace. Only twelve years after the publication of the passage I quote, Avery and his colleagues had determined the class of chemical compound to which genetic engrams belong. In the meantime our entire conception of "the gene" has undergone a revolution. Genes are not, as at one time or another people have thought them, samples or models; they are not enzymes or hormones or prosthetic groups or catalysts or, in the ordinary sense, agents of any kind. Genes are messages. I think Kalmus[3] was the first to use this form of words, but the idea that a chromosome is a molecular code script containing a specification of development is Schrödinger's[4].

My purpose in this address is to identify some of the great conceptual advances that have taken place during the past twenty-five or thirty years on four different planes of biological analysis. As I have pointed out elsewhere[5], working biologists tend nowadays to classify themselves less by "subjects" than by the analytical levels at which they work—a horizontal classification where the older was vertical. So we have molecular biologists, whose ambition is to interpret biological performances explicitly in terms of molecular structure; we have cellular biologists, biologists who work at the level of whole organisms (the domain of classical physiology), and biologists who study communities or societies of organisms. We can discern each of these four strata within each "subject" of the traditional, that is, the vertical, classification. There are molecular and cellular geneticists, geneticists in Mendel's sense, and population geneticists. So also in endocrinology or immunology: each is now studied at the molecular and cellular level as well as at the level of whole organisms. They abut into the population level, too: we study the effects of crowding and fighting on the adrenals and so indirectly on reproductive performance, and we study the epidemiological consequences of natural or artificial immunization and the evolutionary consequences of epidemics. I have noticed that a biologist's interests and understanding, and also, in a curious way, his loyalties, tend to spread horizontally, along strata; rather than up and down. Our instinct is to try to master what belongs to our chosen plane of analysis and to leave to others the research that belongs above that level or below. An ecologist in the modern style, a man working to understand the agencies that govern the structure of natural populations in space and time, needs much more than a knowledge of natural history and a map. He must have a good understanding of population genetics and population dynamics generally, and certainly of animal behaviour; more than that, he must grasp climatic physiology and have a feeling for whatever may concern him among

埃瓦尔德·赫林和理查德·西蒙"认为总体来说，记忆源于原生质的'印痕'，生殖细胞内存储着一个印痕系统，用以指挥细胞各个时期的正常发育"。（需要指出的是，"印痕"在一定程度上是指永恒的物理记忆痕迹或者用以指挥发育的记忆印记[1]。）霍尔丹对他们提出的记忆理论很不赞成，他接着说："该理论很明显具有其他发育机理假说的所有缺点。这个如此复杂的痕迹系统竟然能通过任何物理–化学过程起作用或者无限复制自我，这简直难以置信[2]。"

上述令霍尔丹难以相信的关于记忆痕迹的说法在今天已经是一个共识了。事实上，在我引用的这段文字发表仅 12 年后，埃弗里和他的同事们就确定了遗传痕迹物质所归属的化合物种类。同时，"基因"的整个概念也发生了一次变革。基因不是我们一度认为或其他人现在所认为的某种样本或者模型，也不再是普通意义上的酶、激素、辅基或者催化剂之类的成分。基因是信息。我认为这个概念是由卡尔马斯[3]首次提出的，而关于染色体是包含发育信息的分子编码图的说法则是由薛定谔[4]首次提出的。

我这次演讲的目的是阐述在过去 25 年～30 年内，生物学研究四个不同水平内所发生的一些重要概念演变。就像我在其他地方[5]指出过的，如今生物学研究人员趋向于根据分析水平的不同，而不是根据"学科"来自我分类，这是一种水平分类法，传统的则是垂直分类。按照这种分类，生物学家被划分成了四类，产生了致力于从分子结构的角度来解释生物学现象的分子生物学家，细胞生物学家，以整个生物体（传统生理学领域）为研究对象的生物学家以及以群体或生物体群落为研究对象的生物学家。我们可以在传统的，即垂直分类的各个"学科"中识别这 4 个层次，例如遗传家包括分子遗传学家、细胞遗传学家、孟德尔意义上的遗传学家以及种群遗传学家。这在内分泌学或免疫学中也同样如此：现在这些学科既在分子和细胞水平上研究，又在整个生物体水平上研究。此外，还有在群体水平上研究。我们研究集群行为和打斗行为对肾上腺素分泌的影响，以及随后对生殖行为的间接影响；我们研究自然免疫或人工免疫的流行病学结果以及流行病的进化结果。我注意到这样一个现象，在求知方面而言，生物学家对一门学科的兴趣和思维倾向往往在同一学科的不同水平之间横向发展，而不会在不同学科之间纵向移动。我们凭直觉试着去区分哪些是属于我们研究领域的内容，并把这个研究领域之外的那些内容留给其他研究者研究。一名研究生物种群时空结构的现代生态学家所应当掌握的不只是博物学知识和一张地图而已，他还应该精通种群遗传学、种群动态学、动物行为学，更为重要的是，他应该掌握气候生理学以及其他许许多多的相关生物学知识（包括我所提

the other conventional disciplines in biology (I have already mentioned immunology and endocrinology). There is no compelling reason why he should be able to talk with relaxed fluency about messenger-RNA, and it is not essential that he should ever have heard of it—though an unreasonable feeling that he "ought" to know something about it is more likely to be found in a good ecologist than in an indifferent one.

I shall now take one example from each of these four planes of biological analysis and try to show how our ideas have changed since the last Cambridge meeting of the British Association in 1938—a period that corresponds roughly with my own professional lifetime.

Population Genetics and Evolution Theory

Biologists of my generation were still brought up in what I call the "dynastic" concept of evolution. The course of evolution was unfolded to us in the form of pedigrees or family trees, and we used the old language of universals in speaking of the evolution of *the* dogfish, *the* horse, *the* elephant and, needless to say, of Man.

The dynastic conception coloured our thoughts long after the revival of Darwinism had made it altogether inappropriate. By the "revival of Darwinism" I mean the reformulation of Darwinism in the language of Mendelian genetics—the work, as we all so very well know, of Fisher, J. B. S. Haldane, Wright, Norton and, in a rather qualified sense, of Lotka and Volterra. The subject of evolutionary change, we now learned, was a population, not a lineage or pedigree: evolution was a systematic secular change in the genetical structure of a population, and natural selection was overwhelmingly its most important agent. But to those brought up in the dynastic style of thinking about evolution it seemed only natural to suppose that the outcome of an evolutionary episode was the devising of a new genotype—of that new genetical formula which conferred the greatest degree of adaptedness in the prevailing circumstances. This improved genetic formula, a new solution of the problem of remaining alive in a hostile environment, would be shared by the great majority of the members of the population, and would be stable except in so far as it might be modified by further evolution. The members of the population were predominantly uniform and homozygous in genetic makeup, and, to whatever degree they were so, would necessarily breed true. Genetic diversity was maintained by an undercurrent of mutation, but most mutants upset the hard-won formula for adaptedness and natural selection forced them into recessive expression, where they could do little harm. When evolution was not in progress natural selection made on the whole for uniformity. Polymorphism, the occurrence of a stable pattern of genetic inequality, was recognized as an interesting but somewhat unusual phenomenon, each example of which required a special explanation, that is, an explanation peculiar to itself.

These ideas have now been superseded, mainly through the empirical discovery that natural populations are highly diverse. Chemical polymorphism (allotypy[6]) is found wherever it is looked for intently enough by methods competent to reveal it. The molecular

到过的免疫学和内分泌学）。但是，我们不必强求他能够流利地阐述信使 RNA 的相关知识，甚至于他有没有听说过信使 RNA 都并不重要——尽管我们可能会不合理地将"应该"知道某些知识作为区分生态学家优劣的指标。

自 1938 年英国科学促进会剑桥会议召开以来（这一时期与我的研究生涯基本一致），生物学的理论发生了很大变化，下面，我将在四个生物学研究水平内各举一个例子来说明这种巨大变化。

种群遗传学和进化理论

我同时代的生物学家还是在进化的"王朝"概念下成长的。进化理论是通过家系或家谱的形式展开的，我们用传统的民间用语来描述角鲨、马、大象的进化，不用说，还有我们人类的进化。

在达尔文主义复兴后的很长一段时间内，王朝这一概念丰富了我们的思想，但它已经完全不合适了。我所说的"达尔文主义复兴"是指通过孟德尔遗传学的方式再现达尔文主义，如我们非常熟知的费希尔、霍尔丹、赖特、诺顿所做的工作以及一定程度上洛特卡和沃尔泰拉所做的工作。现在我们知道进化是发生在种群水平上的，而不是沿着世系或家系发生的：进化是一个种群在遗传结构上所发生的长期而系统的变化，其中自然选择起着至关重要的作用。但是，对于以王朝形式考虑进化的人来说，进化的结果是产生了一个新的基因型。新基因型的产生为种群在恶劣环境下的生存提供了一条解决途径，并且能够在种群内绝大多数个体之间共享，还能在未来的进化中保持稳定。一个种群内的绝大多数个体在遗传结构上大体是一致的或者是同源的，无论这种一致或同源性是何种程度，都能使种群的性状稳定地遗传。遗传多样性通过隐性突变的方式保留下来，但是大部分突变会破坏来之不易的适应性，因此它们在自然选择中被迫隐性表达，这样便能减小发生突变所引起的危害。在进化没有发生时，自然选择就保持一个群体上的同一性。多样性是稳定型遗传不均等的结果，它是一种有趣却不常见的现象，每一种多样性都需要有一个独特的解释，即自身特有的解释。

如今，研究结果发现自然种群是高度多样化的，从而取代了上述观点。只要留意并使用恰当的方法，处处可以发现化学多样性（同种异型性 [6]）。仅人类血液中已

variants known in human blood alone provide combinations that far outnumber the human race—variants of haemoglobin, non-haemoglobin proteins, and red-cell enzymes; of red-cell antigens and white-cell antigens; and of haptoglobins, transferrins and gamma globulins. Today it is no longer possible to think of the evolutionary process as the formulation of a new genotype or the inauguration of a new type of organism enjoying the possession of that formula. The "product" of evolution is itself a population—a population with a certain newly devised and well adapted pattern of genetic *in*equality. This pattern of genetic differentiation is determined and actively maintained by selective forces: it is the population as a whole that breeds true, not its individual members. We can no longer draw a distinction between an active process of evolution and a more or less stationary end-product: evolution is constantly in progress, and the genetical structure of a population is actively, that is dynamically, sustained.

These newer ideas have important practical consequences. The older outlook was embodied in that older, almost immemorial ambition of the livestock breeder, to produce by artificial selection a true breeding stock with uniform, and uniformly desirable characteristics; and this was also the ambition—sometimes kindly, but always mistaken— of old-fashioned "positive" eugenics. It now seems doubtful if, with free-living and naturally out-breeding organisms, such a goal can ever be achieved. Modern stockbreeders tend to adopt a very nicely calculated regimen of cross-breeding which, abandoning the goal of a single self-perpetuating stock, achieves a uniform marketable product of hybrid composition. The genetical theory underlying this scheme of breeding embodies, and was indeed partly responsible for, the newer ideas of population structure I have just outlined.

I cannot predict what new ideas will illuminate the theory of evolution in future, but it is not difficult to guess the contexts of thought in which they are likely to appear. The main weakness of modern evolutionary theory is its lack of a fully worked out theory of variation, that is, of *candidature* for evolution, of the forms in which genetic variants are proffered for selection. We have therefore no convincing account of evolutionary progress—of the otherwise inexplicable tendency of organisms to adopt ever more complicated solutions of the problems of remaining alive. This is a "molecular" problem, in the newer biological usage of that word, because its working out depends on a deeper understanding of how the physicochemical properties and behaviour of chromosomes and nucleoproteins generally qualify them to enrich the candidature for evolution; and this reflection is my cue to turn to conceptual advances in biology at the molecular level.

Physical Basis of Life

In the early 1930's no one knew what to make of the nucleic acids. Bawden and Pirie had not yet shown that nucleic acid was an integral part of the structure of tobacco mosaic virus, and we were still a decade from the astonishing discovery by Avery and his colleagues, in the Rockefeller Institute, that the agent responsible for pneumococcal transformations was a deoxyribonucleic acid.

知的分子变异体的组合，就远远超过人类种族的数量，如血红蛋白变异体、非血红蛋白变异体和红细胞酶系变异体；红细胞抗原变异体和白细胞抗原变异体；结合珠蛋白变异体、转铁蛋白变异体以及伽马球蛋白变异体。如今，我们不再认为进化只是产生了一个新基因型或者含有这种新基因的新生物体。种群进化的"产物"就是种群本身——该种群具有一种新的、适应性良好的遗传**不**均等性模式。选择力决定和维持这种遗传分化的模式，这是一种保持遗传稳定性的群体行为，而不是个体的行为。我们不能再区分进化的动态过程和一个多少算作稳定的终产物：因为进化是一个持续的过程，并且种群遗传结构处在连续的、动态的变化之中。

这些新理论具有重要的现实意义。以前的家畜饲养者根据旧的进化理论去人工筛选纯种的优良牲畜，可是结果往往不能如愿，这被认为是传统的"积极"优生学。如今看来，在野外放养和自然地异系繁殖条件下，家畜饲养者的育种目标是很难实现的。现代饲养者们抛弃了传统的筛选方法，采用杂交手段获得了适应市场需求的杂种牲畜。遗传理论奠定了这种育种方法的基础，上面我所提到的关于种群结构的新理论也在一定程度上促进了这种新型育种方法的产生。

我无法预测到以后的进化论将会有什么样的新思想，不过不难猜测它可能包含的内容。现代进化论的一大局限就是缺少完整的变异理论，即进化的**备选物**，也就是为选择提供遗传突变的形式。因此，对于进化过程（另一令人费解的趋势是面对继续生存的难题，生物体趋于采用更复杂的解决方案），我们没有令人信服的解释。用现代生物学术语来说，变异是"分子"水平上的问题，因为它的解决依赖于更深地理解个体的生化特性、染色体的行为以及核蛋白是如何使它们有资格丰富进化的备选物的，我的思路是将这一思考转向生物学在分子水平上的概念演变。

生命的物质基础

在 20 世纪 30 年代早期，没有人知道核酸是由什么构成的。鲍登和皮里尚未发现核酸是烟草花叶病毒的组成部分，距离埃弗里及其同事们在洛克菲勒研究所的惊人发现——脱氧核糖核酸是肺炎球菌的转化因子——也还有十年之久。

Since there was nothing very much to say about nucleic acids you may well wonder what everybody *did* talk about. One topic of conversation was the crystallization of enzymes. Sumner had crystallized urease in 1926 and Northrop pepsin in 1930; soon Stanley would crystallize tobacco mosaic virus, at that time still thought to be a pure protein. But the most exciting and, as it seemed to us, portentous discoveries were those of W. T. Astbury, whose X-ray diffraction pictures of silk fibroin and hair and feather keratins had revealed an essentially crystalline orderliness in ordinary biological structures. For some purposes, however, X-ray analysis was too powerful. The occasion called for resolving powers between those of the optical microscope and the X-ray tube, and this need was fulfilled by electron microscopy. I saw my first electron-photomicrograph in *Nature* in 1933; its resolving power was then one micron.

Electron microscopy has shown that cells contain sheets, tubes, bags and, indeed, micro-organs—real anatomical structures in the sense that they have firm and definite shapes and look as if only their size prevented our picking them up and handling them. Moreover, there is no dividing line between structures in the molecular and in the anatomical sense: macromolecules have structures in a sense intelligible to the anatomist and small anatomical structures are molecular in a sense intelligible to the chemist. (Intelligible *now*, I should add: as Pirie[7] has told us, the idea that molecules have literally, that is, spatially, a structure was resisted by orthodox chemists, and the credentials of molecules with weights above 5,000 were long in doubt.) In short, the orderliness of cells is a structural or crystalline orderliness—a "solid" orderliness, indeed, for "the so-called amorphous solids are either not really amorphous or not really solid"[4].

This newer conception represents a genuine upheaval of biological thought, and it marks the disappearance of what may be called the *colloidal* conception of vital organization, itself a sophisticated variant of the older doctrine of "protoplasm". The idea of protoplasm as a fragile colloidal slime, a sort of biological ether permeating otherwise inanimate structures, was already obsolete in the 'thirties; even then no one could profess to be studying "protoplasm" without being thought facetious or slightly mad. But we still clung to the colloidal conception in its more sophisticated versions, which allowed for heterogeneity and for the existence of liquid crystalline states, and it was still possible to applaud Hopkins's famous aphorism from the British Association meeting of 1913, that the life of the cell is "the expression of a particular dynamic equilibrium in a polyphasic system". For inadequate though the colloidal conception was seen to be, there was nothing to take its place. Peters's idea of the existence of a "cytoskeleton" to account for the orderly unfolding of cellular metabolism in time and place now seems wonderfully prescient, but there was precious little direct evidence for the existence of anything of the kind, and much that seemed incompatible with it.

The substitution of the structural for the colloidal conception of "the physical basis of life" was one of the great revolutions of modern biology; but it was a quiet revolution, for no one opposed it, and for that reason, I suppose, no one thought to read a funeral oration over protoplasm itself.

因为人们对核酸尚无甚了解，所以无话可说，你可能想知道大家**究竟**都在谈论些什么呢？话题之一就是酶的晶体结构。萨姆纳在 1926 年结晶了脲酶，诺斯罗普在 1930 年结晶了胃蛋白酶；随后不久，斯坦利获得了烟草花叶病毒的结晶，但当时还认为该病毒是单纯的蛋白质。但是最令人兴奋和具有预见性的是阿斯特伯里的发现，他通过 X 射线衍射的方法分析蚕丝蛋白、头发和羽毛的角蛋白后在一般生物结构中发现了重要的晶体秩序。然而，对于某些实验分析而言，X 射线分析技术显得过于强大了，它的分辨率远远超过了普通的光学显微镜，因而需要用分辨率介于两者之间的电子显微镜来填补分辨率中间的空白。我见到的第一幅电子显微图谱出现在 1933 年的《自然》上，那时的分辨率是 1 微米。

用电子显微镜观察发现，细胞的结构组成包括片层结构、管状结构、袋状结构以及微小的"细胞器"——从解剖学角度看，它们拥有固定而精确的形状，似乎只是因为个体太小，我们才无法将它们拿在手上进行操作。另外，分子结构和解剖结构之间没有明确的界限：大分子在解剖学家眼中是有结构的，而小的解剖学结构在化学家眼中是分子。（我需要补充的是，**现在**可以理解，正如同皮里 [7] 告诉我们的，传统化学家们不认为分子真的具有所谓的空间结构，并且长期以来怀疑分子量在 5,000 以上的大分子是否存在。）总而言之，细胞秩序是一种结构或晶体秩序，即一个"固体"秩序，而实际上，"这种所谓无定形固体既不是真的没有固定形状也不是纯粹的固体"[4]。

这个新概念是生物学史上的一次巨大变革，它标志着旧有的认为生命机体是**胶质**的说法的消失，胶质学说是一个由古老的"原生质体"学说衍生而来的复杂概念。原生质学说认为原生质是一种易碎的胶状物，是一种生物醚渗透结构或者是无生命的结构，这早在 30 年代就已经过时，那时如果有人声称自己在研究"原生质"，那么他肯定会被嘲笑或被认为是疯子。然而我们对较高深的胶质学说仍然深信不疑，此理论认为细胞内存在异质性和液态晶体秩序。同时，我们也赞成霍普金斯于 1913 年在英国科学促进会会议上所发表的著名论断，即细胞的生命是"一个多相系统内某种独特的动态平衡现象"。胶质理论虽不够完善，但是我们还没有找到一种新的理论来取代它。彼得斯提出了"细胞骨架"学说，解释了细胞内代谢发生的时空有序性，这个学说在今天看来是非常有预见性的，不过至今尚未找到很直接的证据支持该假说，相反，倒是有很多证据看起来与这个学说相矛盾。

"生命的物质基础"的胶质概念被结构概念替代，是现代生物学最伟大的变革之一；但是这个变革却是在潜移默化中发生的，因为没有人对它提出反对意见，我猜测正是由于这个原因，没有人为原生质体学说致悼词。

Cellular Differentiation in Embryonic Development

Embryology is in some ways a model science. It has always been distinguished by the exactitude, even punctilio, of its anatomical descriptions. An experiment by one of the grand masters of embryology could be made the text of a discourse on scientific method. But something is wrong; or has been wrong. There is no *theory* of development, in the sense in which Mendelism is a theory that accounts for the results of breeding experiments. There has therefore been little sense of progression or timeliness about embryological research. Of many papers delivered at embryological meetings, however good they may be in themselves—in themselves they are sometimes marvels of analysis, and complete and satisfying within their own limits—one too often feels that they might have been delivered five years beforehand without making anyone much the wiser, or deferred for five years without making anyone conscious of great loss.

It was not always so. In the 1930's experimental embryology had much the same appeal as molecular biology has today: students felt it to be the most rapidly advancing front of biological research. This was partly due to the work of Vogt, who had shown that the mobilization and deployment of cellular envelopes, tubes and sheets was the fundamental stratagem of early vertebrate development (thus relaying the foundations of comparative vertebrate embryology); but it was mainly due to the "organizer theory" of Hans Spemann, the theory that differentiation in development is the outcome of an orderly sequence of specific inductive stimuli. The underlying assumption of the theory (though not then so expressed) was that we should look to the chemical properties of the inductive agent to find out why the amino-acid sequence of one enzyme or organ-specific protein should differ from the amino-acid sequence of another. The reactive capabilities of the responding tissue were emphasized repeatedly, but only at a theoretical level, for "competence" did not lend itself to experimental analysis, and the centre of gravity of actual research lay in the chemical definition of inductive agents.

Wise after the event, we can now see that embryology simply did not have, and could not have created, the background of genetical reasoning which would have made it possible to formulate a theory of development. It is not now generally believed that a stimulus external to the system on which it acts can specify the primary structure of a protein, that is, convey instructions that amino-acids shall be assembled in a given order. The "instructive" stimulus has gone the way of the philosopher's stone, an agent dimly akin to it in certain ways. Embryonic development at the level of molecular differentiation must therefore be an unfolding of pre-existing capabilities, an acting-out of genetically encoded instructions; the inductive stimulus is the agent that selects or activates one set of instructions rather than another. It is just possible to see how something of the kind happens in the induction of adaptive enzymes in bacteria—a phenomenon of which the older description, the "training" of bacteria, reminds us that it too, at one time, was thought to be "instructive" in nature. All this applies only to biological order at the level of the amino-acid sequences of proteins or the nucleotide sequences of nucleic acids. Nothing is yet known about the genetic specification of order at levels above the molecular level.

398

胚胎发育中的细胞分化

从某种程度上讲，胚胎学是一门模式学科，其特征是对解剖结构的描述极为精确，甚至有些过分拘泥细节。一位胚胎学大师所做的实验可能只是将一系列科学方法的叙述做成文本。但是胚胎学在某处错了，或者一直存在错误。没有发育的**理论**，在这个意义上说，孟德尔遗传学说是一个理论，它解释了杂交实验的结果。因此关于胚胎学研究的进展或时效性没有意义。然而，在胚胎学会议上提交的许多文章无论它们本身多么好——许多文章是分析史上的杰作，在本领域内可称完美——总令人觉得这些文章如果提前五年发表，不会让人更明智；如果延迟五年发表，不会让人感到巨大的损失。

当然，事实也不尽如此。20 世纪 30 年代的实验胚胎学就像今天的分子生物学一样别具吸引力：那时，研究者们认为胚胎学是发展最快速的生物学研究的前沿。这部分要归功于沃格特的工作，他指出细胞膜、管状结构以及片层结构的流动性和展开是脊椎动物胚胎早期发育的主要策略（这为脊椎动物比较胚胎学的建立提供了基础），不过他的观点主要是基于汉斯·施佩曼的"组织者理论"，该理论认为发育分化是胚胎受到有序特定诱导刺激的结果。这一理论的言外之意（尽管当时没有这么说）是我们应该去研究诱导物的化学成分，以探知为什么不同酶或器官特异性的蛋白质的氨基酸序列会各不相同。虽然反复强调反应组织的反应能力，但只是在理论层面上，因为"全能性"尚未得到实验的验证，实际研究工作的重心应是确定诱导物的化学组分。

事后诸葛亮，现在我们知道仅仅胚胎学没有，也不可能形成遗传学理论的基础，而后者本来会使前者可能发展出一套发育理论。现在，人们一般不相信作用于系统的外来的诱导物能够指定蛋白质的一级结构，即传递氨基酸以特定顺序聚集起来的指令。具有"指导"作用的诱导物已经重蹈点金石的覆辙，在某些方面依稀与点金石类似。因此在分子水平的胚胎发育的分化是先前存在能力的展现和遗传编码指令的行为，诱导物是一种起选择或激活作用的物质，而不是具有其他作用的物质。这只有在诱导细菌适应酶（以前称为细菌"驯化"的现象，同时，也提醒我们它在自然界中被认为有"指导"作用）时，才能看到类似事件如何发生。这一切只适用于在蛋白质氨基酸序列水平或核酸的核苷酸序列水平上的生物顺序，而对于上述分子水平之上的顺序的遗传特性我们尚且知之甚少。

The function performed by the hierarchy of inductive stimuli as it occurs in vertebrate development is to determine the specificities of time and place: it is an inductive stimulus which determines that a lens shall form just here and just now—not elsewhere, and at no other time. As I see it, it is the inductive process that allows vertebrate eggs and embryos before gastrulation to indulge in the prodigious range of adaptive radiation to be seen in germs as disparate as a dogfish's egg and a human being's—a case I have argued elsewhere and need not go over here again[8].

Biology of the Organism: Animal Behaviour

If experimental embryology was the subject that seemed most exciting to students of the 'thirties, that most nearly on the threshold of a grand revelation, the study of animal behaviour (in the sense in which we now tend to use the word "ethology"), seemed just as clearly the most frustrating and unrewarding. Twenty years later it was the other way about: embryology had lost much of its fascination and many of the ablest students were recruited into research on behaviour instead. What had happened in the meantime?

In the early 1930's we had one new behavioural concept to ponder on: the idea that an animal might in some way apprehend a sensory pattern or a behavioural situation as a whole and not by a piecing together of its sensory or motor parts. That was the lesson of *Gestalt* theory. We had also learnt finally, and I hope for ever, the methodological lesson of behaviourism: that statements about what an animal feels or is conscious of, and what its motives are, belong to an altogether different class from statements about what it does and how it acts. I say the "methodological" lesson of behaviourism, because that word also stands for a certain psychological theory, namely, that the phenomenology of behaviour is the whole of behaviour—a theory of which I shall only say that, in my opinion, it is not nearly as silly as it sounds. Even the methodology of behaviourism seemed cruelly austere to a generation not yet weaned from the doctrine of privileged insight through introspection. But what comparable revolution of thought ushered in the study of animal behaviour in the style of Lorenz and Tinbergen and led to the foundation of flourishing schools of behaviour in Oxford, here in Cambridge, and throughout the world?

I believe the following extremely simple answer to be the right one. In the 'thirties it did not seem to us that there *was* any way of studying behaviour "scientifically" except through some kind of experimental intervention—except by confronting the subject of our observations with a "situation" or with a nicely contrived stimulus and then recording what the animal did. The situation would then be varied in some way that seemed appropriate, whereupon the animal's behaviour would also vary. Even poking an animal would surely be better than just looking at it: that would lead to anecdotalism: that was what bird-watchers did.

Yet it was also what the pioneers of ethology did. They studied natural behaviour instead of contrived behaviour, and were thus able for the first time to discern natural behaviour structures or episodes—a style of analysis helped very greatly by the comparative

400

在脊椎动物发育过程中，通过诱导刺激层次发挥的作用决定分化的时空特异性：诱导刺激决定晶状体现在就在这里形成——而不是别的地点和时间。在我看来，正是诱导过程允许脊椎动物的卵细胞和胚胎在原肠胚形成前进入适应辐射的惊人范围中，就如同在细菌与狗鲨和人类卵细胞发育过程中看到的区别一样——这点我曾在其他地方讨论过 [8]，不再在此重复。

个体生物学：动物行为

如果实验胚胎学看起来是最令 30 年代研究者兴奋的学科和最接近重大启示起点的学科，那么显然对动物行为的研究（在这个意义上，现在我们倾向于使用"动物行为学"这个词），似乎是最令人沮丧和不值得做的。20 年后，相反地，胚胎学失去了大部分魅力，许多最有能力的学者转而研究动物行为。当时发生了些什么事情呢？

在 20 世纪 30 年代初期，我们有一个新的行为学概念要仔细考虑：动物可能以某种方式从总体上领会感官模式或行为状况，而不是靠动物感官或运动部分的拼凑。这是**格式塔**理论的内容。我们后来也了解了行为主义方法论，并且我希望能一直如此：它陈述的是动物的感受是什么或能意识到什么以及它的动机是什么，它与动物做了什么和怎样做的陈述属于完全不同的一类。我称它为行为主义"方法论"是因为，这个词代表着特定的心理学理论，即行为的现象学是行为的全部——以我看来，这个理论远不像它的名字一样听起来无聊。尽管行为主义方法学看起来过于严谨以至于产生了尚未丢弃，通过内省获得的特殊洞察力的教条，但是洛伦兹和丁伯根研究动物行为的方式近乎引领了一场思维上的革命，并使牛津、剑桥和遍布世界各地的多家学校纷纷建立起动物行为学学院，又有哪个学科能与之相提并论呢？

我认为下面这个非常简单回复就是正确答案。在 30 年代，除了通过某种实验干预（即通过一定的方法刺激动物或者将动物置于一定的"情境"之下，然后观察并记录动物的行为外），对我们来说没有"科学地"研究行为的方法。当情境以某种看起来恰当的方式变化时，动物的行为也会发生变化。当然戳一下动物的反应效果肯定会强于单纯的观察，后者称得上是件趣闻轶事，但是只有观鸟者才会这么做。

不过，动物行为学的先行者们也做了这样的事情。他们研究自然状态下的行为而非人为的行为，因此他们能够首次识别自然行为结构或部分——一种通过比较法

approach, for the occurrence of the same or similar behavioural sequences in members of related species reinforced the idea that there was a certain natural connectedness between its various terms, as if they represented the playing out of a certain instinctual programme. Then, and only then, was it possible to start to obtain significant information from the study of contrived behaviour—from the application or withholding of stimuli— for it is not informative to study variations of behaviour unless we know beforehand the norm from which the variants depart.

The form of address I chose—to trace the recent growth and transformation of ideas in four "subjects" belonging to four levels of biological analysis—gave me no opportunity to mention some of the greatest innovations of modern scientific thought: the dynamical state of bodily constituents, for example, the perpetual flux of the material ingredients of the body. Nor have I said anything except by implication of the greatest discoveries in modern science, those which revealed the genetical functions of the nucleic acids. Yet I feel I have said enough of the growth of biology in the recent past to draw some morals, however trite. The history of animal behaviour—in particular the sterility of the older experimental approach—illustrates the danger of doing experiments in the Baconian style; that is to say, the danger of contriving "experiences" intended merely to enlarge our general store of empirical knowledge rather than to sustain or confute a specific hypothesis or pre-supposition. The history of embryology shows the dangers of an imagined self-sufficiency, for embryology is an inviable fragment of knowledge without genetics. (I often wonder what academics mean when they say of a certain subject that it is a "discipline in its own right"; for what science is entire of itself?) You may think our recent history entitles us to feel pretty pleased with ourselves. Perhaps: but then we felt pretty pleased with ourselves twenty-five years ago, and in twenty-five years time people will look back on us and wonder at our obtuseness. However, if complacency is to be deplored, so also is humility. Humility is not a state of mind conducive to the advancement of learning. No one formula will satisfy that purpose, for there is no one kind of scientist; but a certain mixture of confidence and restless dissatisfaction will be an ingredient of most formulae. Confident we may surely be, for the next twenty-five years will throw up several new ideas as profound and astonishing as any I have yet described, namely ... but I have no space left to tell you what they are.

(**207**, 1327-1330; 1965)

Peter. Medawar: F.R.S., Director, National Institute for Medical Research.

References:

1. See *The Mneme* by R. Semon, London, 1921, particularly pp. 24, 113, 180, 211; and Hering's paper "On Memory, a Universal Attribute of organized Matter" in *Alm. Akad. Wiss. Wien.*, **20**, 253 (1870).

2. *The Philosophy of a Biologist*, 162 (London, 1931).

3. "A Cybernetical Aspect of Genetics", *J. Hered.*, **41**, 19 (1950).

4. Schrödinger, E., *What is Life?*, especially pp. 19-20, 61-62, 68 (Cambridge, 1944). For Weismann's far-sighted views on the matter, see *The Architecture of Matter* by Toulmin, S., and Goodfield, J. (London, 1962).

非常便于分析的类型，因为相关物种中相同或相似行为序列的发生强化了在其多种行为方式间存在某种特定的自然联系的观点，可能它们代表着特定本能程序的发挥。接下来，也只有在那时，才可能开始从人为行为的研究中获得重要信息——使用或不使用刺激物，因为除非我们事先知道变体偏离的标准，否则研究的行为变异是无益的。

为了追溯历属生物分析四个水平的四个"主题"目前增加的观念和转变的观念，我选择的称谓形式没有机会提及一些现代科学思想中的最伟大的革新，即生物体组分的动态，如生物体组成成分是不停流动的，除了暗示，我也没谈及现代科学中最伟大的发现，即关于核酸的遗传功能的揭示。但我觉得我已经讲了许多近年来生物学的发展，这些足以使我们吸取一些教训，虽然是平庸的。动物行为的历史——尤其是成效甚微的传统实验方法——例证了以培根的方式做实验的危险性，也就是说，人为"实验"的危险性仅意味着扩大了经验知识的一般储存而不是支持或驳斥某个特定的假说或预想。胚胎学的历史表明了想象的自我满足思想的危险性，因为离开了遗传学，胚胎学是不独存的知识片段。（我经常在想，当有些人提起某个学科时常说这个学科独树一帜，可事实上哪门科学又能自成一体呢？）你可能对近来的历史感到非常满意。可是我们在25年前自我感觉极好，但是，25年后人们也会回过头来看我们并惊讶于我们的愚钝。然而，如果自满是悲叹的，谦逊也会这样。谦逊不是传导学术进步的心理状态，没有一个准则符合那个目的，因为没有这类的科学家，但是信心和永不自满的心态的组合将是绝大多数准则中的一个因素。当然，我们是有信心的，在将来的25年里将提出几个如同我曾描述过的那样深刻和令人吃惊的新观念，那就是……不过这次我没时间告诉你们具体究竟是什么了。

（高如丽 翻译；金侠 审稿）

5. In my Tizard Memorial Lecture, *Encounter* (August 1965).

6. A term coined by J. Oudin to describe γ-globulin variants: it might well be generalized to include all molecular polymorphism.

7. Pirie, N. W., "Patterns of Assumption about Large Molecules", *Arch. Biochem. Biophys.*, Supp. **1**, 21 (1962).

8. *J. Embryol. Exp. Morph.*, **2**, 172 (1954).

The International Biological Programme[*]

E. B. Worthington

Editor's Note

The context of the International Biological Programme which ran during the 1960s and 1970s, as explained here by its scientific director E. Barton Worthington, looks astonishingly prescient from today's perspective. The programme was essentially concerned with environmental biology and conservation, but it adopted the approach, now very much in vogue again, of regarding nature as a resource on which human welfare depends. It developed the notion of the biosphere, or as Worthington puts it, "the living layer around the Earth". And in considering such issues as human impact on fish stocks, human adaptability to environmental change and the robustness of ecosystems, the IBP anticipated some of the major themes in environmental research today.

THE International Biological Programme is now well into its first year—the first of seven or eight. It is concerned with "the biological basis of productivity and human welfare". It is divided into two phases: Phase I, which consists mainly of design and feasibility studies, to occupy most of 1965 and 1966; Phase II, the operational programme, to commence in 1967 and carry on for about five years. Beyond these simple facts I doubt if there are many people who know much about the Programme, for we have been so busy planning it that little thought has been given to advertising it.

The International Biological Programme is a programme which, because of the effects of various factors, particularly the climate, must extend over a number of years, since it is obvious that an investigation of growth in a summer such as we have had this year would be quite different from a similar study undertaken last year. It has, however, drawn some of its inspiration from the International Geophysical Year. Indeed, if one looks at the development of the sciences during the past few decades one is struck by the high peaks of achievement reached in international collaboration and also through international competition in the physical sciences. Research on the ionosphere, the atmosphere, and the geosphere has been greatly stimulated by organized activities—the International Polar Years, the International Geophysical Year and the International Years of the Quiet Sun which have developed one from the other. Programmes are going forward strongly, too, in connexion with the hydrosphere, as arranged by the Intergovernmental Oceanographic Commission and the International Hydrological Decade. But in such investigations the biosphere has been almost completely neglected. Yet on the natural resources of this Earth—especially the renewable ones which come from biological productivity—the future

---

[*] Substance of an Evening Discourse delivered on September 3 at the meeting in Cambridge of the British Association for the Advancement of Science.

406

国际生物学计划*

编者按

站在今天的角度看，国际生物学计划是一个非常具有前瞻性的计划，本文中该计划的科学主任巴顿·沃辛顿指出它是在 20 世纪六七十年代期间实施的。虽然该计划的重点是关于环境生物学和环境保护，但它采用了将自然看作人类福利所依赖的资源的方法，现在这种方法又流行了起来。该计划还提出了生物圈或沃辛顿所说的"围绕地球表面的生命层"的概念。国际生物学计划预见到了现在环境研究中的一些重大议题，如人类对鱼类资源的影响，人类对环境变化的适应性以及生态系统的稳定性。

现在是国际生物学计划实施的第一年——该项目计划实施七至八年，是关于"生产力和人类福利的生物学基础"方面的研究。它被分成两个阶段：阶段 I 主要是在 1965 年和 1966 年的大部分时间里设计计划并对其可行性进行分析；阶段 II 是方案执行阶段，开始于 1967 年并将持续大约 5 年。除了这些简单的事实外，我估计没有多少人对这个计划有更多的了解，因为我们一直忙于制订计划以至于几乎没有考虑去宣传它。

由于受到各种因素的影响，特别是气候影响，国际生物学计划必须持续若干年，因为很显然我们今年夏天对生长情况的调查结果可能显著不同于去年进行的类似的调查。然而，从国际地球物理年中我们能得到一些鼓励。实际上，如果人类回顾过去数十年间的科学发展，那么一定会对物理科学中通过国际合作和国际竞争所获得的辉煌成就而感到吃惊。通过有组织的活动大大促进了对电离层、大气层和地圈的研究，这些活动包括相继举办的国际极地年、国际地球物理年和国际宁静太阳年。由政府间海洋学委员会和国际水文十年计划举办的与水圈相关的计划也将得到迅速的发展。但是在诸如此类的研究中，生物圈已经几乎完全被忽略了。然而人类的将来必须依赖地球上的自然资源，特别是来源于生物生产力区域的可再生资源。相比于登陆月球或抓拍火星的特写镜头而言，这些对人类要重要得多。有人甚至认为，对人类来说保证北海鱼类的稳定产量要比探测与开采可能存在于北海下的石油更重要。

---

* 在 9 月 3 日剑桥的英国科学促进会会议上发表的晚间演说的内容。

of mankind depends. They are much more important to most human beings than getting to the Moon or taking close-up photographs of Mars. One could even argue that it is more important to humankind to ensure a sustained yield of fish from the North Sea than to discover and extract the oil that may lie beneath it.

What do we mean by biology within the context of the International Biological Programme? To some, biology means the life sciences as opposed to the earth sciences; to some it stimulates thoughts of the inner workings of the cell; and to some perhaps, merely the facts of human reproduction. For the purposes of an International Biological Programme, I think the best definition is the biosphere, that is the living layer around the Earth, on land, in the waters and in the air.

This word biosphere, by the way, seems to present a problem in semantics. I have heard it explained by a television astronomer as a concentric region of the solar system, comprising the orbits of Venus, the Earth and Mars, where there is some reason to believe that life as we know it exists. I use it, however, as the layer which supports life in, on and around the Earth, for the International Biological Programme is likely to have little time or opportunity to extend its activities to other planets.

Some biologists are inclined to think that the International Biological Programme is rather a waste of effort and money. They argue that opportunities for the exchange of knowledge and ideas are not lacking, but are almost overburdening, for international unions and associations exist in every branch of biology. From spring to autumn scarcely a week passes without some international biological congress, often several running contemporaneously, even in the same country. Is this not enough?

The counter argument is that we scientists concerned with environmental biology have to take but one look at our colleagues concerned with the physical environment to appreciate what a degree of organization can do. Few would deny that studies of this kind of biology, concerned with the organism in relation to its environment, which had such a splendid start in the past century, have been sadly overtaken in this century by studies in the physical sciences. In the long run, I believe that the International Biological Programme will be looked on as a major effort—and a successful one—to redress this balance. In doing so it will provide a part of the fundamental knowledge which today is so much needed so that biological science can be applied more effectively to the provision of human needs.

History

Before examining the content of the International Biological Programme we should consider how the idea came into being. British scientists have taken a prominent part in this. Sir Rudolph Peters some years ago, when president of the International Council of Scientific Unions (ICSU), took a major part in the conception of the International Biological Programme. Prof. H. W. Thompson of Oxford, the current president of ICSU, did a good job as midwife during the first International Biological Programme Assembly

在实施国际生物学计划的大背景下，我们所说的生物学具体指的是什么呢？对某些人来说，生物学指的是生命科学而不是地球科学；对有些人来说，它促使其思考细胞内部的运行方式；或许还有些人认为，它仅指人类繁殖这一事实。在国际生物学计划里，我认为最好的定义是生物圈，也就是围绕地球表面的，存在于陆地上、水中还有空中的生命层。

顺便说一下，生物圈这个词似乎在语义上存在一个问题。我曾听过一个电视天文学家将其解释为太阳系的同心区域，包括金星、地球和火星的轨道，也就是那些我们有理由相信存在生命的地方。然而，我指的是用以维持生命的地球的下层、表层和周围，就国际生物学计划而言，它可能没有时间或机会将其活动扩展到其他的行星上。

一些生物学家倾向于认为国际生物学计划是相当浪费精力和金钱的。他们认为交流知识和想法的机会并不缺乏而是几乎过盛，因为国际联盟和协会包含生物学的各个分支。从春季到秋季几乎每个星期都会有一些国际生物学大会，常常几个会议同时举行甚至是在一个国家举行。难道这还不够吗？

与之背道而驰的观点是我们环境生物学科学家只需稍微关注研究自然环境的同事们，就可以领悟组织能够做到何种程度。很少有人会不认可对这类涉及生物与环境之间联系的生物学的研究，它在上个世纪有着相当好的开端，但在本世纪却不幸地被自然科学研究超越了。从长远来看，我相信国际生物学计划在成功恢复这种平衡的过程中，将发挥主要作用。通过这样做，它将提供一些我们现在迫切需要的基本知识，以便更有效地应用生物科学来满足人类的需要。

历　史

在详阅国际生物学计划的内容之前，我们应该知道这个想法是如何产生的。英国的科学家在这里扮演了非常重要的角色。多年前，时任国际科学联合会理事会（ICSU）主席的鲁道夫·彼得斯爵士在国际生物学计划的构思中发挥了重要作用。去年，来自牛津大学的现任国际科学联合会理事会主席汤普森教授在巴黎举行的第一届国际生物学计划大会中扮演了一个非常好的助产士的角色。而作为国际生物科

held in Paris last year. Prof. C. H. Waddington, as president of IUBS, has been a good doctor in the pre- and post-natal clinics; while the president, officers and secretariat of the Royal Society, by priming and holding the feeding-bottle, have enormously helped the infant's early growth.

How the ideas leading up to the International Biological Programme have fluctuated and developed has been related by Prof. Montalenti, the Italian geneticist who led much of the preparatory work. Initially, it was thought that the programme should concentrate on the human organism, and especially population dynamics and genetics. Later there was a body of opinion which thought it could involve itself best on the preservation and conservation of biological systems in different parts of the world, especially those which are most endangered by economic progress. Finally, these and other approaches have been brought together in such a way as to be of interest to a rather broad spectrum of biological scientists. The objective, as defined in 1964, is to ensure a world-wide study of: (*a*) organic production on the land, in fresh waters, and in the seas, and the potentialities and uses of new as well as of existing natural resources; (*b*) human adaptability to changing conditions. The programme should not range through the entire field of biology but should be limited to basic biological studies, related to productivity and human welfare, which will benefit from international collaboration, and are urgent because of the rapid rate of the changes taking place in all environments throughout the world.

Thus, of two major problems now facing the human species on Earth—of controlling his rate of reproduction and of increasing the biological product on which he subsists—the International Biological Programme is concerned with the second. The advances in techniques to control reproduction which have been produced in the past decade have reached a point at which there is more need for sociologists to apply them than for biologists to invent new ones. Yet the attributes and potentialities of man are by no means left out of the programme, as I will explain.

Principles

The programme is based on a number of principles. First, it is a directional and a selected programme. It is not a free-for-all, nor is all research bearing on productivity and human adaptability automatically incorporated into it. The initials IBP can be regarded, if you like, as a kind of status symbol, and to ensure that this is not misused we have a sifting arrangement for projects at the national level and later at international level before admission into the programme. The tag of IBP does not, of course, mean that the project is necessarily good science, nor that other work on biological productivity conducted outside IBP is not good science. It merely implies that those immediately concerned with IBP agree that the project fits in and is likely to advance materially the objects of the programme.

Secondly, the programme is concerned throughout with the fundamental approach to research. It starts from the premise that the applied sciences of biological production, such as agriculture, forestry, fisheries, have in recent decades gone striding ahead more quickly

410

学联合会主席的沃丁顿教授，就像是产前产后诊所里一个称职的医生；同时英国皇家学会的主席、工作人员还有秘书兢兢业业的工作为国际生物学计划这个婴儿的早期成长做出了较大的贡献。

蒙塔伦蒂教授讲述了这些想法是如何促成国际生物学计划在曲折中发展的，这个意大利遗传学家为这项计划做了大量的筹备工作。首先，有人认为这个计划应该关注人体组织，特别是种群动力学和遗传学。之后，大量的主张认为该计划应致力于保存和保护世界不同地区的生物系统，特别是那些因为经济发展而濒临灭绝的生物系统。最后，这些观点和其他想法以这样一种方式整合起来以引起众多生物学家的注意。1964 年确定出其目标旨在保证世界范围内的以下研究：(a) 陆地上、淡水中以及海水中的有机质生产，新能源以及现有自然资源的潜力和使用；(b) 人类对环境变化的适应性。这项计划并没有囊括生物学的所有领域，而是仅限于与人类生产和人类福利相关的基础生物学研究，这些项目将在国际合作中受益，同时也是迫在眉睫的，因为全世界的环境都在发生着快速的变化。

因此，地球上的人类现在面临两个主要的问题——控制人类的繁殖速率和增加其赖以生存的生物产品——国际生物学计划关注的是第二点。过去十年内，人类已经拥有了控制繁殖率的技术，现在最需要的是社会学家去应用它而不是生物学家发明新的技术。然而这项计划并不忽视人类的作用和潜力，我将对此作出解释。

原　则

这个计划建立在许多原则的基础上。首先，它是一个有着明确目标的精心筛选出的计划。它不是对任何人开放的，也不是所有涉及人类生产力和人类适应性的研究就能自动加入这个计划。如果你愿意的话，可以将国际生物学计划看作一种社会地位的象征，在获准进入这个计划之前，为了确保其不被滥用，我们会对项目进行国家级水平的筛选，然后是国际级水平的筛选。当然，附有国际生物学计划这个标签的项目并不一定是好的科学，也不意味着在国际生物学计划管理范围之外的和生物生产力相关的项目就不是好的科学。它仅仅意味着那些与国际生物学计划直接相关的项目很快就会被认同并且可能在本质上促进项目的实施。

第二，这个计划普遍关注的是研究的基础方法。这个计划以生物生产的应用科学这一前提为出发点，例如农业、林业和渔业在近几十年里的发展速度远超过我们

than the fundamental understanding of the causes of productivity. Similarly the sciences of man, medicine and some of the applied social sciences, have in some ways outpaced fundamental understanding of mankind and the differences between one man and another. The International Biological Programme is an opportunity to restore the balance and to provide new organized knowledge as a springboard for the technologies to take further plunges into progress.

Thirdly, the International Biological Programme is urgent. This is not only because of the steady growth of the pressure of human populations on renewable resources, but also because many of the situations in the world, both biological and human, are changing so rapidly that they will soon cease to exist. The International Biological Programme is not a preservationist body, for our approach is dynamic not static, conservation rather than preservation; but the perpetuation of sample biological systems for future biologists to study is definitely among its objectives.

Fourthly, and this could perhaps almost go without saying, the International Biological Programme is limited to research which could benefit from international co-operation. There are some scientists, and great ones at that, who do not readily co-operate with others and whose special attributes lead them to penetrate deeply on a narrow front the unknown, behind locked doors so to speak, rather than, by a sharing of knowledge, to advance with others along a broad one. The International Biological Programme is clearly not designed for the isolationist researcher, although his findings might well be picked up and developed as a Programme project.

Divisions of the Programme

In order to get to grips with what inevitably is a very large content, the International Biological Programme is divided into seven sections. One of the first and most obvious divisions was into the biological communities of the land, of fresh waters, and of the sea, because terrestrial ecology, limnology, and oceanography have each developed their own discipline of research, although biological energy often flows from one to another. These three sections are known by their initials; PT for Productivity of Terrestrial Communities; PF for Productivity of Freshwater Communities; PM for Productivity of Marine Communities.

All three have their problems of conservation, including the definition, description and management of prescribed areas or samples of living communities. The problems of conservation on the land are more extensive and generally more complex than those of the waters. They include such concepts as national parks, nature reserves, and sites of special scientific interest. Therefore, there is a special section of IBP labelled CT for Conservation of Terrestrial Communities. Each of the three, moreover, presents different facets of the processes of production—photosynthesis (utilization of solar energy) and the nitrogen cycle—this section is called Production Processes (PP). Each of the three may reveal new resources of value to mankind or new methods applicable to the use and conservation of resources, which require fundamental study; this will be carried out

对生产力提高原因的基本理解。同样，人类科学、医学和一些应用社会科学在某些方面已经超过了对人类以及人与人之间差异的基本理解。国际生物学计划为重新建立这种平衡，以及提供可作为技术进步的跳板的系统化知识提供了契机。

第三，国际生物学计划是势在必行的。这不仅仅是因为人口对可再生资源需求的压力不断增长，也是因为许多世界形势（包括生物的和人类的）正在发生着如此快速的变化以至于他们在不久将会灭绝。由于我们的方法是动态的而不是静态的，是保护而不是保存，因此国际生物学计划不是一个保存机构；但是延续样本生物系统以便将来生物学家进行研究的目标是非常明确的。

第四，国际生物学计划仅限于通过国际合作能够获益的研究，这一点可能几乎是众所周知的。有一些科学家特别是一些优秀的科学家不愿意与其他人合作，可谓闭门造车，由于其特殊的性格导致他们深陷狭窄的未知前沿而不是通过知识的共享与其他人在宽广的前沿上共同进步。很明显国际生物学计划不是为孤立主义研究者制定的，尽管他的发现可能被重视并采纳为国际生物学计划中的一个项目。

计划的分工

这项计划将会很庞大，为能掌握要领，国际生物学计划分为七个部分。第一个也是最明显的部分是陆地、淡水和海洋生物群落，因为尽管生物能源常从一处流向另一处，但是陆地生态学、湖沼学和海洋学都已各自发展成为研究学科。这三个部分由于它们的缩写而为人们所熟知：PT 代表陆地生物群落生产力；PF 代表淡水生物群落生产力；PM 代表海洋生物群落生产力。

这三个部分都面临着保护方面的问题，包括对生物群落指定区域或现存标本的定义、描述和管理。与水域生物群落相比，陆地生物群落的保护问题更为广泛且通常更为复杂。这些问题包括国家公园、自然保护区及具有特别科学价值的地方等概念。因此，国际生物学计划中有一个特殊的陆地生物群落保护（CT）的部分。此外，这三个部分中的每一个都代表着生产过程的不同方面——光合作用（太阳能的利用）和氮循环——这个部分被称为生产过程（PP）。三个部分中的每一个都可能向人类展示出有价值的新能源或适用于使用和保护能源的新方法，这就需要进行基础研究；这将在被称为生物能源使用和管理（UM）的部分内进行。最后，人类问题产生了

within the section named UM for Use and Management of biological resources. Finally, the human problems, which bring in another series of disciplines, notably physiology and anthropology, will be studied within the Section of HA, Human Adaptability. This makes seven: PT, PP, CT, PF, PM, HA and UM.

Each of these sections is headed by an international convener and sectional committee, and is busy in producing a five-year programme of its own. Two of the conveners are British—Mr. Max Nicholson of CT and Prof. J. S. Weiner of HA. The seven sections are, of course, closely related to each other, but also to other programmes going on during the same period, for example PF to the International Hydrological Decade.

IBP consists of the sum of its parts, not only in the sense of containing seven different programmes, but also in the sense that the bulk of the research will be undertaken and financed nationally, not internationally. Many countries are at present busy preparing their national programmes, and of these one of the most advanced is that for the United Kingdom, now published by the Royal Society.

Examples of Research

Let us take a few examples of the kind of work which IBP will contain, and think especially of the British contribution. Although a great deal has yet to be learned about our own biological communities and our own people, some of the thinking has been on the lines that we should use the knowledge and abilities of a good many scientists overseas, where ignorance about biological productivity is still very great. Thus the U.K. programme in its PT section aims to establish two bases for IBP work in the warmer regions overseas, in co-operation, of course, with the countries concerned. One of these is likely to be in savannah country, at the Cambridge Nuffield Unit of Tropical Animal Ecology, situated in the Queen Elizabeth Park in Uganda; another at a site yet to be determined in an area of tropical rain forest. At these bases, in co-operation with local scientists, it is hoped to undertake investigations of productivity in depth—primary production of the trees, shrubs, grasses and herbs, secondary by herbivorous mammals and insects, tertiary by predators and parasites.

It is hoped to quantify as well as qualify each link in the food chain so as to obtain data to compare production in the wild with production from the tame, for example, where domestic animals and plants have been established on land used for agriculture.

Passing to another section, namely CT, it is pleasing to report that something has been actually achieved. In the spring of 1964 and again in 1965, expeditions went from Great Britain to the desert lands of Jordan, and made a survey especially of the Azraq Oasis and its environs, to the east of Amman. This has resulted in several things: a useful IBP booklet entitled *An Approach to the Rapid Description and Mapping of Biological Habitats*, and more important, in detailed recommendations about the creation of national parks in Jordan. Further preparatory work has been based on the Azraq Oasis where the park

另外一系列的学科特别是生理学和人类学，将在人类适应性（HA）的部分内得到研究。这个计划共分为七个部分：PT、PP、CT、PF、PM、HA 和 UM。

每个部分由一名国际会议召集人和分科委员会负责，他们正忙着制定各自的 5 年计划。会议召集人中有两个英国人——负责陆地生物群落保护部分的马克斯·尼克尔森先生和负责人类适应性部分的韦纳教授。当然，这七个部分不仅彼此之间紧密联系，并且与同一时期进行的其他计划也是紧密联系的，例如淡水生物群落生产力部分和国际水文十年计划之间存在联系。

国际生物学计划由它的各个部分共同组成，其意义不仅仅在于包含七个不同的部分，而且在于大部分研究的开展和资助都是在国家层面而非国际层面上进行的。目前许多国家正忙着准备他们自己的国家级项目，其中一个最先进的项目是英国的，目前由皇家学会公布。

研 究 例 子

让我们来举些例子来说明国际生物学计划所要从事的这种工作，尤其是要看看英国所做的贡献。虽然我们对国内生物群落和人群了解的还不多，但一些流行的想法是我们应当利用海外优秀科学家的知识和能力，那里对生物生产力仍然知之甚少。因而英国陆地生物群落生产力部分的目标是在海外较温暖地区建立两个基地，当然，这要与相关国家进行合作。其中一个基地可能在草原地区，坐落于乌干达的伊丽莎白女王公园的剑桥纳菲尔德热带动物生态学研究组；另一个设在热带雨林地区，场地仍未确定。我们希望与当地科学家们一起在基地中能够承担较深入的生产力方面的研究——树木、灌木、禾本科植物和草本植物的初级生产力，食草性哺乳动物和昆虫的二级生产力，食肉动物和寄生虫的三级生产力。

我们希望能定量并定性地分析食物链中的每个环节，以获取能够对野生生物与驯养生物（例如，陆地上已用于农业生产的家养动物和植物）的生产力进行比较的数据。

下面谈谈另一个被称为陆地生物群落保护的部分，令人高兴是这部分已经有了实质性的进展。在 1964 年和 1965 年春天，分别从英国直至约旦的沙漠地区组织探险，并且沿途作了调查，特别是对安曼东部的阿兹拉克绿洲及其附近地区进行了调查。这个调查导致几件事情的发生：它促成了一本名为"生物栖息地快速描述和绘制方法"的实用的国际生物学计划小册子的产生，更重要的是，它详细地介绍了约旦国家公园的建造。以阿兹拉克绿洲为基础，进一步的筹备工作正在进行着，在美国人

headquarters and warden are now established with American aid. Within a week or two, two experienced members of the Nature Conservancy's staff are going there to prepare a management plan, applying the experience in management of wild areas gained in this country. Meanwhile, in August H.M. The King of Jordan publicly announced that he will set up the Azraq National Park, and also that Jordan will participate in the IBP by establishing at the Park's centre an institute for biological and human investigations of desert and oasis conditions. This indeed will make a useful counterpart to the proposed IBP bases in savannah and rain forest to which I have already referred.

From the PF section I will draw two examples of British initiative: one is an intensive study of the biological productivity of a reach of the Thames, organized from the University of Reading. This is dealing with primary and secondary production and will include studies on the feeding and population dynamics of a considerable number of fish species. It will fill a major gap in our knowledge of British fresh waters, for most intensive work has been devoted to lakes and very little to rivers. The other example relates to research in tropical lakes. In terms of food supplies through fisheries, and water supplies for irrigated agriculture, they are of much greater importance to the world than the little lakes in our own country. So the programme includes the establishment, based on Britain, of a team of limnologists for work in the tropics, especially Africa. This takes advantage of the fact that we happen to have in Britain a rather particular expertise in tropical lakes.

During the present planning phase each of the seven sections has already held, or has arranged, a number of symposia consisting of selected specialists from all over the world to discuss methods of research in particular spheres and to prepare handbooks. The object here is not to "straight-jacket" the methods in use in IBP, for field biology is at present advancing very rapidly through the development of old methods and the devising of new ones; but it is to agree on those proved methods which can be advised for general use throughout the world and can be relied on to produce comparable results. A good example was a symposium held in Aberdeen and Cambridge in September concerning research on large herbivorous mammals. In Australia, studies of kangaroos and introduced sheep suggest that the former may be more effective converters of grass into meat than the latter; if only kangaroos produced wool instead of hair, they might be the basis of Australian prosperity. In Africa there are indications that the broad spectrum of indigenous African mammals, such as giraffe, antelopes, buffalo and wart hog, may be better agents of secondary production of meat than the exotic cattle, sheep and goats which have replaced them in many areas and are apt to depress the yield of the habitat. In Scotland we have the Island of Rhum, where management of red deer by the Nature Conservancy has resulted in more venison coming off the island than mutton under the former agricultural management and at the same time it seems as though the level of primary production of the vegetation has been raised. This symposium brought experiences of this sort together and advised on methods for future work under IBP from the three points of view—the ecological, the physiological and the pathological.

Another example I want to give is one with which I am personally involved just at present,

的帮助下公园的总部和管理区现在都已经完工。在一两个星期内，大自然保护协会两个经验丰富的工作人员将到那里制定管理计划，在这个国家获取有关野生保护区管理的经验。与此同时，八月约旦国王陛下公开宣布他将建立阿兹拉克国家公园；通过在公园中心建立研究所以便在沙漠和绿洲环境下进行生物和人类调查，约旦也将参与到国际生物学计划中。这确实与我之前所提到的在草原和雨林地区建立国际生物学计划基地不谋而合。

在淡水生物群落生产力部分，我将举两例英国人发起的项目：一个是由雷丁大学组织的对泰晤士河某河段生物生产力进行的一次深入研究。这涉及初级生产力和次级生产力，也将包括对相当数量的鱼类种群摄食及种群动态进行研究。由于大部分集中性的研究一直针对湖泊，而几乎没有河流，因此这一研究将填补我们在英国淡水知识中的一大空白。另一个例子与热带湖泊研究有关。从渔业供应食物、水灌溉农田方面来讲，它们对世界的重要性要比小湖对我们国家的重要性大得多。因此，这个计划包括成立一个在英国的湖泊生物学家团队，其工作致力于热带地区，特别是非洲。这利用了英国碰巧拥有关于热带湖泊的独特经验这一事实。

在目前的规划阶段里，七个部分中的每一个或已经开展起来或正在筹备中，由来自全世界的经过挑选的专家组成专题讨论会，商讨特殊领域的研究方法并制定手册。这里的目的不是"限制"用于国际生物学计划的方法，因为通过旧方法的发展和新方法的发明，目前生物学领域发展非常迅速；而是它将认同那些被证明了的、被建议在全世界范围内普遍使用的方法，并通过这些方法获得对比结果。一个很好的例子是九月在阿伯丁和剑桥举行的关于大型食草哺乳动物的研讨会。在澳大利亚对袋鼠和引进羊种进行的研究表明，前者比后者能更有效地将草转化为肉；如果只有袋鼠生产羊毛而不是毛发，那么它们将是澳大利亚繁荣的基石。在非洲有迹象表明大量的非洲本土哺乳动物诸如长颈鹿、羚羊、水牛和疣猪，相比于多年前就已经在许多地区取代它们的、使其栖息地产量趋于减少的外来牛、绵羊和山羊来说，是更好的肉类的次级生产者。在苏格兰拉姆岛，我们通过大自然保护协会对马鹿进行管理，从岛上获取的鹿肉比先前在农业管理下产出的要更多，与此同时，植物的初级生产水平似乎比以前提高了。本次研讨会将这些经验整理在一起并从三个方面对国际生物学计划未来的工作方法提出了建议，这三个方向分别是：生态学、生理学和病理学。

我要举的另一个例子是目前我亲自参与的研究，即大型人工湖的研究工作。建

namely research on large man-made lakes. The basic reason for creating these is generally for hydroelectric power: for example, at Kariba, Volta and Kainji in Africa, Brokopondo in Surinam, and a number of large impoundments in North America. Sometimes water conservation for irrigation is equally or more important, as in the Aswan High Dam in Egypt. But the influence of these great sheets of water will extend not only to producing power and water but also to many other human needs, and has a profound bearing on the biology of large regions. Anyone familiar with Africa, for example, will realize the fundamental influence of the great natural lakes on climate, vegetation, fisheries, water supply for man and stock, transport, sites for urban development, and the rest. Since the period of IBP happens to coincide with the period of constructing several new great impoundments and of the biological changes consequent on several of those already completed, their study will take quite a significant place in the programme. Here again, it is good to note that Britain has taken an initiative, and as I speak a group of eight scientists drawn from Liverpool and other universities in Britain and Nigeria, and supported financially by the Ministry of Overseas Development, are undertaking a preliminary investigation of the part of the River Niger where Mungo Park met his end in 1806, the reaches which, in a year or more, will form a five-hundred square mile lake impounded by the dam at Kainji. The group has now been joined by several American and African sociologists who are extending the aquatic study to the 50,000 or so people now living below the future water-level. Under the leadership of local people, assisted by scientists from outside, and with funds which we hope will be provided by the special fund of the United Nations, it is hoped to establish a co-ordinated programme of study on these great impoundments.

I have been speaking mainly about the purely biological aspects of IBP, but the section on human adaptability has so far been quite one of the most active. It brings in a lot of scientists in the spheres of medicine and of physical anthropology, and its programme is designed rather differently from some of the other sections. It is wide and comprehensive, divided into a large number of headings and subheadings. Research topics include environmental physiology, including tolerance of different human groups to cold, heat, and high altitude; fitness, growth and physique, which, incidentally, will include a survey of highly fit athletes; the genetics of populations, with international data centres to be developed on existing national ones such as the blood-group centre of the Medical Research Council in London; and lastly health, nutritional and epidemiological topics. Clearly a lot of this work will need to be done by groups of research workers traveling with their equipment to study communities in their natural environments. This involves mounting a good many quite complex expeditions to remote places. British workers under IBP/HA expect, for example, to work this autumn on the Hazda tribe around Lake Eyassi in Tanzania and the people of Bhutan in the very high Himalayas.

Organization

There is a school of scientific thought that resents all forms of direction or organization. Science is free: give us the funds and let us go unfettered in any direction where research leads, and we will show results. Good scientists should not become administrators; any

418

造这些大型人工湖最根本的原因通常是为了水力发电：例如，非洲的卡巴里、沃尔塔和卡因吉、苏里南的布罗科蓬多以及北美许多大型水库。有时对灌溉用水的保护同等重要或更加重要，就像埃及的阿斯旺高坝一样。而这些贮存的大量的水不仅用于发电和灌溉，而且能满足人类的许多其他需求，并且也对较大区域里的生物产生了深远的影响。例如，熟悉非洲的任何人都能认识到大型自然湖泊对天气、植被、渔业和人类用水以及水资源的贮存、运输、城市发展的地理位置和其他方面造成的重要影响。国际生物学计划执行的时期碰巧与在建的几座新的大型水库及已完工的几座水库引起的生物变化所处的时期相同，对它们的研究将在这项计划中占据重要地位。在这里，我们再次表扬英国采取的主动行动，在我演讲的这一刻，由 8 名分别来自利物浦以及英国和尼日利亚的其他一些大学的科学家组成的团队正承担着对尼日尔河部分河段的初步研究工作，该项目由海外发展部给予资金支持。1806 年芒戈·帕克曾来过这些河段，这些河段将在一年内或更长的时间里被卡因吉大坝拦截从而形成一个 500 平方英里的蓄水湖。几位来自美国和非洲的社会学家现在也加入了这个团队，将研究从水生生物扩展到约 5 万人，这些人现在居住的位置在将来的水平面以下。在当地人的领导下，通过外来科学家的帮助以及我们所期待的联合国提供的专项资金，有望制定关于这些大型水库的统筹方案。

上面我主要从纯生物方面谈了国际生物学计划，但是到目前为止，有关人类适应性的部分是相当热门的。它引进了许多来自医学和体质人类学领域的科学家，并且制定的计划显著不同于其他部分。它广泛而全面，分为非常多的课题和副课题。研究课题包括环境生理学（包括不同人群对冷、热和高海拔的耐受性）、适应性、生长发育和体质（顺便说一下，这包括调查非常健康的运动员）、人口遗传学（由现有的国家数据中心发展起来的国际数据中心，如伦敦医学研究理事会的血型中心）、最后是与健康、营养和流行病有关的课题。很显然，这些工作需要大量的研究人员带着他们的设备外出研究自然环境中的群落。这包括组织相当综合的探险队去偏远的地方。例如，隶属于国际生物学计划中人类适应性部分的英国工作者们期望在这个秋天研究坦桑尼亚埃亚西湖周围的哈次达族和海拔非常高的喜马拉雅山上的不丹人。

组　织

有一支科学思想学派反对所有形式的指导或组织机构。科学是自由的：给我们经费并允许我们在感兴趣的任何方向上进行不受限制的研究，我们将给出成果。好

who do so are so much loss to science. This is an attitude which some of us associate with the older universities, but it is quite widespread, within Government scientific service as well as outside it. I met it strongly expressed when discussing IBP recently in Uganda. Now this libertine approach to science does not face the realities of how the funds and facilities for research are to be produced, how the results are to be discussed and published, or how those results which have an element of application in them might be turned to practical use. Nevertheless, it is an attitude with which I personally, and many others who are concerned with IBP, have a good deal of sympathy. We have been influenced by it in thinking up the organization necessary for IBP, at least to the extent that we have devised this as lightly as possible. We are trying to build a simple structure from the foundations and to make the roof no heavier than is necessary to protect those inside from the weather.

Starting from the top there is ICSU which has set up the Special Committee for the International Biological Programme (SCIBP), comparable with its scientific committees for Oceanic Research (SCOR) and Antarctic Research (SCAR). SCIBP is a democratic organization which holds a general assembly about every two years. It has a Swiss president, Prof. Jean Baer, vice-presidents from Italy, Poland, the United States and Great Britain, and a number of other members, some representing the main international unions in the biological sciences and some drawn from the different bio-geographical regions of the world. It includes also the seven international conveners of the sections of IBP. Unesco, FAO, WHO, SCOR, and SCAR have representatives on it. Each of the conveners has also his sectional committee of specialist members drawn from all over the world.

The central office for IBP is at 7 Marylebone Road, Regent's Park, London, in accommodation generously provided and furnished by the Government and the Royal Society. Each of the seven international conveners has his own office, although only four of them as yet have whole-time staff.

Publications of IBP are so far very few: a journal called *IBP News* was started in October 1964, and so far three numbers have been published. We are also starting a series of handbooks, and the first of these, being a general guide to the activities of HA section, has just gone to press.

The central funds, for running the offices and holding meetings and for publications, come from national dues paid in by each participating country. These are on a rather modest scale and in fact are at present insufficient. I am glad to say they are augmented by grants and loans from ICSU and Unesco, and a particularly generous grant recently made by the U.S. National Academy of Sciences. It is hoped that other countries will follow this excellent example.

The finance I have spoken of is concerned only with the organizational framework and, by international standards, is quite small. The cost of the actual research undertaken under

的科学家不应该成为管理者；无论谁这样做对科学来说都是严重损失。我们中的一些人常将这种态度与历史悠久的大学相关联，其实这在政府科学服务部门的内部和外部也都相当普遍。最近在乌干达国际生物学计划的讨论中，我对这个观点深有感受。而这种自由主义的科学态度没有面对现实，即研究经费和设备从何而来，研究结果将怎样被讨论和发表，或者研究结果的应用价值怎样被实际应用。即便如此，我和许多其他关注国际生物学计划的人对这种态度都深有同感。我们在筹建国际生物学计划的必要组织时受其影响，至少在一定程度上我们尽可能精简组织结构。我们一直尝试着在地基上建立一个简单的组织结构，使屋顶轻到只够保护其内部免受外界环境的影响。

最高级的部门是国际科学联合会理事会，它设立了国际生物学计划特别委员会（SCIBP）、海洋研究科学委员会（SCOR）和南极研究科学委员会（SCAR）。国际生物学计划特别委员会是一个民主的组织，通常每两年组织一次会员大会。该组织的主席是来自瑞士的琼·贝尔教授，副主席分别来自意大利、波兰、美国和英国，还有许多会员，其中一些是生物科学领域内主要国际联盟的代表，还有一些来自世界不同的生物地理区域。这也包括国际生物学计划中七个部门的国际会议召集人。联合国教科文组织、联合国粮农组织、世界卫生组织、海洋研究科学委员会和南极研究科学委员会都有代表出席这个会议。每个会议召集人也属于他们各自部分的专家成员委员会，这些专家成员来自世界各地。

国际生物学计划的中央办公室位于伦敦的马里波恩路 7 号丽晶公园，由政府和皇家学会慷慨提供并装修。七个国际会议召集人都有属于他们自己的办公室，尽管至今他们中只有四人有全职的工作人员。

到目前为止，由国际生物学计划发行的出版物还很少：一本名为《国际生物学计划消息》的杂志创办于 1964 年 10 月，到目前为止已经出版了三期。我们也开始制定一系列手册，其中作为人类适应性部分通用指南的第一本小手册刚刚出版。

办公室日常运作、会议举行和出版物发行所需的经费主要由每个参会国提供。但他们提供的经费数量是有限的，事实上目前并不足以维持工作的正常开展。令人高兴的是国际科学联合会理事会和联合国教科文组织为我们提供了贷款和捐助，以及最近美国国家科学院提供了特殊慷慨的捐助，希望其他国家能效仿这样好的例子。

我所谈及的经费仅与组织框架有关，与国际标准相比是非常少的。如果国际生物学计划的各部分都体现价值，那么由其主持的对世界各处进行实际研究所需的费

the auspices of IBP in all parts of the world will naturally be quite large if the programme is to be at all worth while. As yet it is impossible to say how much it may cost, because very few national programmes have yet been prepared in detail although a number are in active preparation. If we take the British programme as a model, and if some twenty major nations have comparable programmes and twenty or thirty others prepare smaller programmes, the total is bound to run to several million pounds. The finance for national programmes will of course be provided nationally: in this first year of design and feasibility studies, the British Treasury has provided some £80,000 specially for IBP; added to the normal votes for the Royal Society and the Nature Conservancy. This will enable some appointments to be made and some research of a trial nature to be started. A number of other countries have acted somewhat similarly. Czechoslovakia, for example, in addition to financing its own planning phase, has provided from its own funds the office and staff for one of the international conveners, Prof. Malek, who is based in Prague.

In order to plan and carry out national programmes, many countries are setting up their own national committees, often with sub-committees equivalent to the international sectional committees of SCIBP. Thus the United Kingdom, through its committee structure which operates under the patronage of the Royal Society, brings about 150 British scientists into the active planning stage. The United States now has a national committee under Dr. Roger Revelle. Poland one under Prof. Petrusewicz; Japan not only has a national committee but has already submitted a programme. I believe that before long a major problem of SCIBP and of its Central Office will be, not so much to get IBP in motion which has been a preoccupation up until now, but to prevent it running away from itself by becoming too large and unmanageable.

Conclusion

The gestation period of the Programme has been long, but recent progress is at a much greater rate than we had dared to hope a couple of years ago, for the world needs an IBP. What, then, is the philosophy behind it? What is the driving force that causes many busy people about the world to devote their time and energy to helping it on? I suggest it is a spirit of service to our fellow men and to all the other animals and plants that make up the very varied living communities of the Earth. Even in these days of rather selfish approach to living, apparent in the political, commercial and even scientific spheres, I suppose there is some feeling of service in all of us, even if it is only an instinct towards conserving the race, after conserving the individual. Here in IBP is the opportunity for very many biologists, whose efforts may have been not perhaps of very high value to others, to bring their researches together, to focus them and to make them more useful. Here in IBP is the opportunity of showing the world the value of field biology as well as laboratory biology. New resources will undoubtedly be discovered, and so also will be the aesthetic as well as the scientific and economic value of plants and animals in their natural and also in their man-made environments.

An important part of the service IBP hopes to render will be to ensure that at least

用自然是相当大的。迄今为止，还不能估算出到底要花费多少钱，因为很少的国家项目已经有了详细的计划，尽管有一些正在积极地准备着。如果我们以英国的计划为例，若二十个左右的主要国家能有类似的计划并且还有二三十个其他准备好的小计划，那么执行这些计划所需的总费用一定会达到几百万英镑。当然，国家计划的经费将由该国提供：在制订计划和可行性研究的第一年，英国财政部专门为国际生物学计划拨款了 80,000 英镑；为英国皇家学会和大自然保护协会增加合法投票权。这有益于进行委任并开始开展一些野外实验研究。很多其他国家也有些许类似的做法。例如，捷克斯洛伐克除了支付规划阶段的资金外，还为其中一个国际召集人（在布拉格基地的马利克教授）提供办公室日常运转和员工费用。

为了制定和执行国家计划，许多国家正在成立他们自己的全国委员会，通常该委员会设置小组委员会，这类似于国际生物学计划特别委员会的国际分科委员会。因而英国在皇家学会的赞助下来运作委员会机构，并引入了约 150 名英国科学家积极筹备计划。现在美国有一个由罗杰·雷维尔博士负责的国家委员会，波兰的国家委员会由彼得鲁塞维奇教授负责；日本不仅有一个国家委员会而且已经提交了计划。我认为不久以后国际生物学计划特别委员会及其中央办公室的一个主要的问题不是获得较多的现在已经引起关注的国际生物学计划议题，而是防止它由于发展太大造成难以管理而最后脱离了自己的轨道。

结　论

这个计划经历了一个较长的酝酿期，但其目前的发展速率比我们两年前所希望的还要快，因为世界需要国际生物学计划。那么这一计划背后蕴含了什么样的哲学思想？能吸引世界上许多忙碌的人们把他们的时间和精力花费在该计划上的驱动力是什么？我认为，对于我们的同胞和其他组成地球上多样生物群落的动植物来说，它是一种服务的精神。即使是在生活方式（尤其在政治、商业甚至是科学领域中）相当自私的现在，我认为我们所有人都是有一些服务理念的，尽管保护好个体之后再保护种族仅是一种本能。对于那些曾经努力但没有给他人带来很大价值的生物学家们来说，国际生物学计划是一个机遇，因为它可以将这些研究者聚集起来，关注他们并使他们发挥自己的价值。国际生物学计划也为野外生物学和实验室生物学向全世界展示其价值提供了机会。毫无疑问新型能源将被发现，同样地，生活在自然和人工环境中的动植物的审美价值、科学价值及经济价值也将被发现。

国际生物学计划希望提供的服务中，一个重要的部分是确保至少这些令人惊奇

samples of these wonderful creations of evolution, so different and yet so similar in the several bio-geographical zones, will continue and flourish for subsequent generations of humans to appreciate and use. Nor should we forget the many and diverse races of man himself, with their separate and particular attributes. Unless we understand these matters and use our understanding to plan the future, they will go down before the axe and fire, the bulldozer or the hypodermic needle. This marvellous biological differentiation on the Earth would then tend towards dull uniformity and even obliteration.

(**208**, 223-226; 1965)

Dr. E. B. Worthington: Scientific Director (Designate).

的进化演变样本（它们在多个生物地理区域如此不同却又如此相似）将继续存在并生长良好，使得人类的后代将有机会欣赏和使用到。我们也不应该忘记人类自身种族的多样性，及其所具有的分散性和独特性。除非我们已经理解了这些问题并能根据我们的理解去规划未来，否则他们将在斧头、火、推土机或注射器针头面前衰退。地球上非凡的生物分化将趋于单一甚至消失。

（尹金 翻译；金侠 审稿）

Double Chromosome Reduction in a Tetraploid *Solanum*

G. E. Marks *et al.*

Editor's Note

Meiosis, the process that produces sex cells with a single set of chromosomes, occurs via a process of reductional division that halves chromosome number. Unusual plant hybrids, thought to be produced by double chromosome reduction, had been reported before but definitive proof of their origin was lacking. Here British botanist Jeffrey B. Harborne and colleagues present biochemical and morphological data confirming the production of an unusual *Solanum* hybrid by this method. The potato plant was more like its female parent, suggesting that the double reduction occurred in the male plant, which contained four sets of chromosomes.

REPEATED attempts to cross *Solanum demissum* ($2n = 6x = 72$) with *S. stoloniferum* ($2n = 4x = 48$) gave one seeded berry containing one seed. This yielded a plant (*M.* 271) which had $2n = 4x = 48$, instead of the expected $2n = 5x = 60$. The cytological, morphological and biochemical studies of *M.* 271 reported here provide good evidence that it is a true hybrid and therefore the product of double chromosome reduction.

The plants used in this investigation were as follows: *Parents. Solanum demissum* (C.P.C. 2249) (*dms* 2249). This clone, collected by Dr. J. G. Hawkes in 1949 in Hidalgo State, Real del Monte, Mexico, is morphologically typical of the species. It is both male and female fertile, producing 94.6 (±1.64) percent stainable pollen of a regular diameter and a mean of 96 seeds per berry on self-pollination. *Solanum stoloniferum* (C.P.C. 2282.2) (*sto* 2282.2). This clone was grown from seed collected by Dr. J. G. Hawkes in 1949 in Puebla State, Mexico. It is a form of the species characterized by dark green leaves, many interjected leaflets, a high leaflet index and blue flowers. It sets abundantly seeded berries by self-pollination.

*Hybrid M.*271. This one seed obtained in the cross *dms* 2249 (♀) × *sto* 2282.2 (♂) germinated and grew very slowly into a small plant with extremely short internodes and a tufted, almost cushiony habit; it produced only a single flower. By contrast, when grafted on tomato, its growth was upright and vigorous, it produced long internodes and flowering was profuse and continuous. Comparison of the leaf and flower characters of *M.* 271 with those of its parents (Table 1) shows that in most ways it resembles the *demissum* parent more closely than it does *sto* 2282.2. A predominance of *demissum* would also be expected in a pentaploid hybrid ($2n = 60$—3 *dms* +2 *sto* genomes). Morphologically, *M.* 271 resembles polyhaploid *S. demissum* ($2n = 3x = 36$)[1-3] and could be taken for one were it not for its chromosome number. An analysis of meiosis gave the results shown in Table 2.

426

四倍体茄属植物中染色体的双减数

马克斯等

编者按

减数分裂是产生只含有一套染色体的生殖细胞的过程，在这个过程中，染色体数目减半。曾有报道认为一些异常杂种植株可能是由染色体双减数引起的，但是缺乏关于其起源的确切证据。本文中，英国植物学家杰弗里·哈本及其同事们提供的生物化学和形态学方面的数据证实了茄属中的异常杂种植株是通过这种方式产生的。该马铃薯植株的性状更接近其母本，表明杂种产生是由于含有四套染色体的父本发生了染色体双减数。

反复尝试将 Solanum demissum（$2n = 6x = 72$）和 S. stoloniferum（$2n = 4x = 48$）杂交，使其最终获得含有一粒种子的结实浆果。这粒种子萌发长成的植株（M. 271）体细胞染色体数目为 $2n = 4x = 48$，而不是预期的 $2n = 5x = 60$。本文对 M. 271 进行了细胞学、形态学和生物化学方面的研究，均提供了确凿的证据证明 M. 271 确实是一个由于发生染色体双减数而产生的真杂种。

本研究所用的实验材料如下：**亲本**。Solanum demissum (C.P.C.2249) (dms 2249) 是由霍克斯博士于 1949 年在墨西哥雷亚尔–德尔蒙特的伊达尔戈州收集到的克隆株，其具有该物种典型的形态特征：雄性雌性均可育，花粉直径正常，可染花粉粒占 94.6($\pm$1.64)%，自花授粉时每个浆果平均可结 96 粒种子。Solanum stoloniferum (C.P.C. 2282.2) (sto 2282.2) 是霍克斯博士于 1949 年在墨西哥普埃布拉州所采集种子的后代，它是该物种的一种变型，其特点是长有深绿色叶子和许多嵌入小叶，小叶指数较高，花朵呈现为蓝色。它可以通过自花授粉产生大量的结实浆果。

杂种 M. 271 这是由 dms 2249（♀）× sto 2282.2（♂）杂交得到的种子，该种子萌发后逐渐长为节间极短的簇生盘状的小植株，仅生长出一朵花。相比之下，若将其嫁接在番茄上，则生长直立且旺盛，节间长，花多且持续产生。对 M. 271 与其亲本的叶片和花器官性状进行比较（表 1）表明，它在许多方面更接近亲本 demissum 而不是 sto 2282.2。五倍体杂种（$2n = 60$）应该也更偏向于 demissum，因为其基因组组成为 3 dms+2 sto。如果不考虑染色体数量，M. 271 应该在形态上更像多元单倍体 S. demissum（$2n = 3x = 36$）[1-3]。表 2 是减数分裂的分析结果。

Table 1. Leaf Characters of Parents and Hybrid

	dms 2249 Mean* (range)	*M.* 271 Mean* (range)	*sto* 2282.2 Mean* (range)
Pairs of lateral leaflets	2.90 (2–3)	2.00 (0)	3.40 (3–4)
Number of interjected leaflets	2.55 (1–4)	0.90 (0–1)	6.40 (3–10)
Terminal leaflet index $\left(\dfrac{\text{Breadth}}{\text{Length}}\%\right)$	58 (52–68)	61 (50–72)	66 (60–73)

* Based on ten measurements.

Table 2. Meiotic and Fertility Data

Plant	Meiosis in pollen mother cells							Percent ± *S.E.* stainable pollen
	No. of cells	Mean ± *S.E.* per cell					Term. coeff.	
		IV	III	II	I	Xta		
sto 2282.2	25	0	0	24.00	0	32.64 ± 0.51	0.56	60.4 ± 4.02
M. 271	25	0.08	1.48 ± 0.25	12.64 ± 0.36	17.96 ± 0.40	22.4 ± 0.73	0.72	0

M. 271 has no stainable pollen and attempts to backcross it by both parents as males gave one berry when *sto* 2282.2 was used. This contained two seeds, but neither germinated.

Possible Origins

There is no definite morphological evidence to suggest, let alone prove, that *M.* 271 is a true hybrid of *S. demissum* × *S. stoloniferum*. Before giving the biochemical evidence on this point, we may consider three other possible origins, as follows:

(1) *contamination*. It is possible that *M.* 271 is a hybrid between *dms* 2249 and a diploid *Solanum*, since three such species, *S. chomatophyllum*, *S. santolallae* and *S. verrucosum*, were growing in the same glasshouse as *S. demissum* and were being used in the same crossing programme. However, all attempts to cross *demissum* with these diploids failed, and, in addition, *M.* 271 shows no evidence of the characteristic morphology or phenolic pattern (see later) of any of these species. Hybrids between *demissum* and *verrucosum* have been recorded[4,5]: they differ both morphologically and cytologically from *M.* 271.

(2) *Female parthenogenesis*. Polyhaploid plants of *S. demissum* are occasionally produced as a result of female parthenogenesis in selfed progenies of *S. demissum*[3] or in progenies obtained from crosses between *demissum* and diploid species[2,6,7]. All had $2n = 3x = 36$ except for two plants: one which Dodds[2] found to have $2n = 3x + 3 = 39$ and another which Beamish[7] found to have $2n = 2x + 2 = 26$.

A tetraploid plant could arise parthenogenetically from a hexaploid only by an extremely abnormal maternal meiosis. Meiosis and embryo-sac development were therefore studied in ovules of *dms* 2249, using sections stained with Heidenhain's haematoxylin. Meiosis was regular and embryo-sac development was of the type normal for *S. demissum*[8] . Thus the

428

表 1. 亲本与杂种的叶片性状

	dms 2249 均值 * (范围)	*M.* 271 均值 * (范围)	*sto* 2282.2 均值 * (范围)
侧生小叶的对数	2.90(2~3)	2.00(0)	3.40(3~4)
嵌入小叶的数量	2.55(1~4)	0.90(0~1)	6.40(3~10)
顶生小叶指数 ($\frac{宽度}{长度}\%$)	58(52~68)	61(50~72)	66(60~73)

* 基于 10 次测量得到。

表 2. 减数分裂和育性分析

植株	花粉母细胞减数分裂							可染花粉百分比 ± 标准误差
	细胞数量	每个细胞的均值 ± 标准误差					顶端系数	
		IV	III	II	I	Xta		
sto 2282.2	25	0	0	24.00	0	32.64 ± 0.51	0.56	60.4 ± 4.02
M. 271	25	0.08	1.48 ± 0.25	12.64 ± 0.36	17.96 ± 0.40	22.4 ± 0.73	0.72	0

M. 271 不能产生可染花粉，因此尝试将其两亲本作为父本与其回交，当以 *sto* 2282.2 为父本时结出了一个浆果，它含有两粒种子，但均不能发芽。

可能的起源

迄今尚无确切的形态学证据表明 *M.* 271 是 *S. demissum* × *S. stoloniferum* 的真杂种，更不用说证明了。在获得生物化学证据之前，我们考虑了以下三种其他可能的起源：

（1）**污染** *M.* 271 有可能是一个由 *dms* 2249 与一种二倍体茄属植株产生的杂种，因为 *S. chomatophyllum*、*S. santolallae* 和 *S. verrucosum* 这三个物种的植株与 *S. demissum* 生长在同一个温室中，而且被用于同样的杂交项目。然而，所有试图将 *demissum* 与这些二倍体杂交的努力都以失败告终，另外 *M.* 271 也没有表现出任何上述物种的特征性的形态学或酚型特征（见后）。以前报道的 *demissum* 和 *verrucosum* 的杂种 [4,5] 的形态学和细胞学特征均与 *M.* 271 不同。

（2）**孤雌生殖** *S. demissum*[3] 自交后代或其与二倍体物种 [2,6,7] 杂交后代的孤雌生殖有时会产生多元单倍体植株。这些多元单倍体植株染色体组成均为 $2n = 3x = 36$，只有以下两个植株例外：一个是多兹 [2] 发现的植株 $(2n = 3x + 3 = 39)$，另一个是比米什 [7] 发现的植株 $(2n = 2x + 2 = 26)$。

仅通过六倍体母本植株的极度异常减数分裂而导致的孤雌生殖就可能产生四倍体植株。本研究利用海登海因苏木素切片染色法，观察了 *dms* 2249 胚珠中的减数分裂和胚囊发育的情况，结果发现其减数分裂正常，其胚囊发育是正常的 *S. demissum*

occurrence of a parthenogenetic tetraploid seedling of *demissum* is extremely unlikely.

(3) *Male parthenogenesis*. *M.* 271 could have conceivably arisen by male parthenogenesis, in which case it would be genetically equivalent to a segregate in a progeny raised by selfing *sto* 2282.2. Such a progeny was grown and all the individuals were found to resemble closely the parent plant; furthermore, all were fertile and set berries by selfpollination. Thus, this parental clone is highly homozygous, and it is therefore inconceivable that a male parthenogenetic seedling should resemble *demissum* rather than its *stoloniferum* parent.

Biochemical Evidence

Except for the anthocyanin pigments of flowers which exhibit dominant-recessive relationships in their inheritance, the genetic control of most chemical constituents is more or less quantitative and hybrids contain most or all of the substances present in the parents[9]. Thus chemical constituents characteristic of presumed parents can give good evidence of the origin of a putative hybrid. On this assumption, the flavonol glycosides of the flowers and the leaf alkaloids of *dms*, *sto* and *M.* 271 were examined, since these substances were known to show significant variation among the wild tuber-bearing *Solanum* species.

Phenolic patterns in flowers. Ten flavonol glycosides are known to occur (and vary) in potato flowers[10]. Their distributions in *M.* 271, in its putative parents and in three related diploid species were therefore examined (Table 3). The results eliminate the possibility that *M.*271 could have arisen by a chance hybridization between *demissum* and a related diploid species. The following points about the phenolic patterns of the flowers may be noted.

Table 3. Phenolic Constituents of Flowers

	dms 2249	*M.* 271	*sto* 2282.2	*chm** 2708	*san* 2519	*ver* 2644
Flavonol glycosides						
Kaempferol 3-glucoside	−	−	−	−	+	−
Kaempferol 3-rutinoside	−	−	−	−	−	+
Kaempferol 3-sophoroside	−	−	−	+	+	−
Kaempferol 3-(2^G-glucosylrutinoside)	−	−	−	+	−	+
Quercetin 3-glucoside	−	−	−	−	+	−
Quercetin 3-rutinoside	+	+	+	−	−	+
Quercetin 3-sophoroside	−	−	−	−	+	−
Quercetin 3-(2^G-glucosylrutinoside)	−	−	−	+	−	+
Myricetin 3-rutinoside	+	+	−	+	−	−
Luteolin 7-glucoside	−	−	+	−	−	−
Other phenolics						
Caffcic-glucose-quinic complex	+	+	−	−	−	+
Unknown colourless→mauve	−	−	−	+	−	−

* Also contains two other unidentified flavonol glycosides.

Key: *dms*—*demissum*; *sto*—*stoloniferum*; *chm*—*chomatophyllum*; *san*—*santollalae*; *ver*—*verrucosum*.

430

植株的类型 [8]。由此可见，通过 *demissum* 孤雌生殖产生四倍体是完全不可能的。

（3）**孤雄生殖** *M*. 271 可能是通过孤雄生殖产生的，在这种情况下，它与 *sto* 2282.2 自交后代的某个分离个体在遗传组成上将完全相同。种植这样的后代后，其所有个体均表现出与亲本植株相似的性状；而且所有个体均可育，并可通过自花授粉产生浆果。因此这个亲本克隆株是高度纯合的，另外，我们也很难相信一个孤雄生殖产生的幼苗会与 *demissum* 相似，而不是与其亲本 *stoloniferum* 相似。

生物化学证据

除了花朵的花青素在遗传上表现为显隐性关系外，绝大多数化学成分的遗传控制或多或少都是数量性的，而且杂种含有存在于亲本植株内的大多数或所有化学物质 [9]。因此，研究可能亲本的化学成分特征可以为某些假定的杂种的起源提供有力的证据。根据这一假设，本研究检测了材料 *dms*、*sto* 和 *M*. 271 花中的黄酮醇糖苷含量以及叶片中的生物碱含量，因为人们知道这些物质的含量在野生茄属块茎物种中存在显著差异。

花朵中酚的类型的分布特征 已知在马铃薯的花中含有 10 种黄酮醇糖苷，而且其含量存在差异 [10]。因此本研究检测了这些物质在 *M*. 271、其可能的亲本以及相关的 3 个二倍体物种中的分布特征（表 3）。研究结果表明，*M*. 271 不可能来自 *demissum* 和任何一个相关的二倍体物种间的偶然杂交。关于花中酚的类型的分布特征，以下几点值得关注。

表 3. 花中酚类成分分析

	dms 2249	*M*. 271	*sto* 2282.2	*chm* * 2708	*san* 2519	*ver* 2644
黄酮醇糖苷						
山奈酚 3– 葡糖苷	–	–	–	–	+	–
山奈酚 3– 芸香糖苷	–	–	–	–	–	+
山奈酚 3– 槐糖苷	–	–	–	+	+	–
山奈酚 3–(2^G– 葡糖基芸香糖苷)	–	–	–	+	–	+
槲皮素 3– 葡糖苷	–	–	–	–	+	–
槲皮素 3– 芸香糖苷	+	+	+	–	–	+
槲皮素 3– 槐糖苷	–	–	–	–	–	–
槲皮素 3–(2^G– 葡糖基芸香糖苷)	–	–	–	+	–	–
杨梅酮 3– 芸香糖苷	+	+	–	+	–	–
木樨草素 7– 葡糖苷	–	–	+	–	–	–
其他酚						
咖啡 – 葡萄糖 – 奎宁复合物	+	+	–	–	–	+
未知无色→紫红色	–	–	–	+	–	–

* 还包含其他两种未鉴定的黄酮醇糖苷。

缩写：*dms—demissum*; *sto—stoloniferum*; *chm—chomatophyllum*; *san—santollalae*; *ver—verrucosum*.

(1) The pattern of *M*. 271 is identical with that of *demissum*, agreeing with the morphological evidence (see earlier) showing that *demissum* is the predominant parent.

(2) The pattern of *M*. 271 does not disagree with the idea that *stoloniferum* is the other parent. One constituent of *stoloniferum*, it is true, is not present in the hybrid. This substance, luteolin 7-glucoside, though characteristic of the species, is not always present: it was lacking in four out of 38 clones examined[10]. Its absence from *M*. 271 is therefore not critical.

(3) The pattern of *M*. 271 is quite different from those of the three diploid species (Table 2), which makes it very unlikely that any of these species is involved in its parentage. Thus, all three not only contain kaempferol as well as quercetin glycosides but, in addition, the glycosidic patterns (sophoroside, glucoside and (2^G-glucosylrutinoside)) are quite different from *M*. 271 or its parents (which have only the 3-rutinoside). Production of (2^G-glucosylrutinoside) in flowers of *S. stoloniferum* and *S. chacoense* is known[10] to be controlled by a dominant gene *Gl* (3-rutinoside being the recessive character), and it is unlikely that this glycosidic form would not appear in hybrids involving *chomatophyllum* or *verrucosum*.

Alkaloids of the leaves. Steroidal alkaloids occur, combined with sugars (branched tri- or tetra-saccharides), quite characteristically in the genus *Solanum*, and their distribution has been studied intensively in recent years[11]. While glycosides of solanidine are widely distributed in the tuberous solanums, those based on demissidine are found only in *S. demissum*. Thus it should be possible to characterize hybrids of *demissum* with other species on the basis of their alkaloid content and this has, in fact, been done with *M*. 271.

The alkaloids of both parents were examined, with the results shown in Table 4. The clone of *S. demissum* contained the expected demissine[12] and *S. stoloniferum* contained solanine and α-chaconine, the two commonest potato alkaloids. *M*. 271 contained three glycosides of which two corresponded in R_F value with glycosides from the different parents, that is with demissine and α-chaconine; the third appeared to be a new demissidine glycoside. On acid hydrolysis, these glycosides together yielded demissidine, with traces of solanidine, and four sugars: glucose, galactose, xylose and rhamnose. Thus, *M*. 271 has the aglycones of both parents and also the constituent sugars of both, since the bound xylose is derived from *S. demissum* and the bound rhamnose can only have come from *S. stoloniferum*.

Since major intraspecific variation in alkaloid content is unknown in the potatoes, this result is effective proof that *M*. 271 is a true hybrid between *S. demissum* and *S. stoloniferum*.

(1) 杂种 M. 271 的酚型与 *demissum* 的相同，这与形态学证据是吻合的（见上文），表明 *demissum* 是其主要亲本。

(2) 杂种 M. 271 的酚型并不能否定 *stoloniferum* 是其另外一个亲本。的确，在杂种中没有检测到 *stoloniferum* 的一种成分，但也有研究表明虽然木樨草素 7- 葡糖苷是该物种的特征性物质，但并非总能被检测到：在所有检测过的 38 个克隆体中，4 个克隆体中未检测到木樨草素 7- 葡糖苷 [10]，因此该物质在 M.271 中不存在就不能证明 *stoloniferum* 不是该杂种的亲本之一。

(3) 杂种 M. 271 的酚型与本研究所分析的 3 个二倍体物种中的酚型存在显著差异（表 2），由此推测这些二倍体物种不可能参与该杂种的形成。因为这 3 种二倍体不仅含有山奈酚和槲皮素苷，还含有槐糖苷、葡糖苷和 2^G- 葡糖基芸香糖苷等糖苷，这与 M. 271 或其亲本（仅含 3- 芸香糖苷）显著不同。已知 *S. stoloniferum* 和 *S. chacoense* 的花朵产生的 2^G- 葡糖芸香糖苷 [10] 受一个显性基因 *Gl* 控制（3- 芸香糖苷是隐性性状），这种类型的糖苷应该在与 *chomatophyllum* 或 *verrucosum* 有关的杂种中出现。

叶片中的生物碱 茄属植物的重要特征是存在甾体生物碱与糖类（带分支的三糖或四糖），近年来已对它们的分布情况进行了深入研究 [11]。糖苷茄啶在有块茎的茄属中广泛分布，而糖苷垂茄碱仅在 *S. demissum* 中存在。因此，可以根据生物碱的含量区分来自 *demissum* 和其他物种的杂种。本研究正是对 M.271 进行了这样的分析。

检测了两个亲本的生物碱组成情况，结果见表 4。*S. demissum* 的克隆体含有预期的垂茄碱 [12]，而 *S. stoloniferum* 则含有两种最常见的马铃薯生物碱：茄碱和 α- 卡茄碱。杂种 M. 271 含有 3 种糖苷，其中两种的 R_F 值分别与来自不同亲本的糖苷相对应，即对应于垂茄碱和 α- 卡茄碱，第三种则是一种新的糖苷垂茄碱。这些糖苷碱在酸水解时共同产生垂茄碱、痕量的茄啶和 4 种糖：葡萄糖、半乳糖、木糖和鼠李糖。因此，M. 271 含有两个亲本的糖苷配基和糖类，其木糖来自 *S. demissum*，而鼠李糖则只能来自 *S. stoloniferum*。

因为在马铃薯中尚未发现其生物碱含量在种内有较大的差异，所以本研究结果有效地证明了 M.271 是 *S. demissum* 和 *S. stoloniferum* 间的真杂种。

Table 4. Alkaloids in the Leaves

Plant	Alkaloid	R^*a-s	R_F†	Products of acid hydrolysis	
				Aglycone(s)	Sugars
dms 2249	Demissine	0.94	0.29	Demissidine	Xylose, glucose and galactose
sto 2282.2	Solanine	1.0	0.28	Solanidine	Rhamnose, glucose and galactose
	α-chaconine	1.35	0.46		
M. 271	Demissine (?)	0.88	0.30	Demissidine + traces of solanidine§	Rhamnose, xylose, glucose and galactose
	New demissidine glycoside α-chaconine (?)	‡1.37	0.48		

* R_F relative to α-solanine: chromatograms run on citrate-buffered whatman No.1 paper in butanol-citric acid-water.

† Chromatograms run on silica-gel plates in acetic acid-ethanol (1 : 3).

‡ Identified as a mixture of two components, since this spot gave a strong Dragendorf reaction for alkalold but only a weak reaction with Clarke's reagent (used for detecting solanidine glycosides in the presence of demissidine glycosides, which do not react).

§ Solanidine was detected, in the presence of demissidine, by the use of Sarett's reagent (compare ref. 13). The crude mixture of alkaloid aglycones failed to respond to the nitroso test, indicating that tomadine was absent.

Origin of M. 271

The morphology and biochemistry of *M.* 271 indicate that it is far nearer its female parent, *demissum*, than it is to *stoloniferum*, suggesting that double chromosome reduction occurred in the latter, the male parent.

Meiosis was normal in the sample of pollen mother cells examined from *sto* 2282.2 (Table 1). However, pollen grain measurements from ten different anthers showed a range of size within each anther. Using within-anther variance as a measure of this variation, Bartlett's test showed a highly significant heterogeneity between anthers ($\chi^2[9]487.4$). Furthermore, the mean variance within anthers of *sto* 2282.2 was above that for other *stoloniferum* clones. Pollen stainability was also extremely variable in *sto* 2282.2, ranging from 12 to 75 percent per anther. Of a sample of 1,000 "good" pollen grains 5.6 percent had diameters equal to or below the modal diameter of pollen grains measured in a diploid species. Since, within species, the size of a pollen grain is often correlated with its chromosome number, it is quite likely that some of the small pollen grains found in *sto* 2282.2 were double-reduced, with 12 chromosomes.

Matsubayashi[14] found that certain clones of *stoloniferum* produced occasional tripolar spindles at *M*I of meiosis which resulted in hexads containing six nuclei of almost equal size. Such a mechanism could well be the source of double-reduced pollen grains. Unfortunately a re-investigation of meiosis in *sto* 2282.2 is impossible since the clone was lost through virus infection. There is therefore no information about the cytological mechanism of the presumed double reduction.

Discussion

Double reduction has been invoked on a number of occasions to account for the appearance of unusual hybrids. Thus Steere[15] obtained diploid progeny which resembled

表 4. 叶片中的生物碱

植物	生物碱	R^*a-s	R_F†	酸水解产物	
				糖苷配基	糖类
dms 2249	垂茄碱	0.94	0.29	垂茄碱	木糖、葡萄糖和半乳糖
sto 2282.2	茄碱	1.0	0.28	茄啶	鼠李糖、葡萄糖和半乳糖
	α–卡茄碱	1.35	0.46		
M. 271	垂茄碱 (?)	0.88	0.30	垂茄碱 + 痕量茄啶§	鼠李糖、木糖、葡萄糖和半乳糖
	新的糖苷垂茄碱 α–卡茄碱 (?)	‡1.37	0.48		

* R_F 与 α–茄碱的相对量：在丁醇–柠檬酸–水中，在浸有柠檬酸盐缓冲液的沃特曼 1 号纸张上跑色谱。
† 在乙酸和乙醇 (1:3) 的混合液中，在硅酸凝胶平板上跑色谱。
‡ 被鉴定为两种组分的混合物，因为在该点有强烈的生物碱德拉根多夫反应，但与克拉克试剂只有微弱反应（该试剂用于在不发生反应的垂茄碱糖苷存在时检测茄啶糖苷）。
§ 使用萨雷特试剂（见参考文献 13）在垂茄碱存在时检测到了茄啶。生物碱糖苷配基的粗混合物未对亚硝基检测发生反应，表明番茄碱缺失。

M. 271 的来源

M. 271 的形态学和生物化学特征表明，它更接近于其母本 *demissum*，而不是父本 *stoloniferum*，说明其父本染色体发生了双减数。

检查发现来自 *sto* 2282.2 的花粉母细胞样品的减数分裂正常（表 1）。不过，对 10 个不同花药的花粉粒大小的测定表明，每个花药中花粉粒大小存在差异。把同一花药内的数据方差作为这种变化的量度，巴特利特的测验结果表明在花药之间存在着明显的异质性（χ^2[9]487.4）。此外，*sto* 2282.2 花药内的均方差高于其他的 *stoloniferum* 克隆株。*sto* 2282.2 每个花药中花粉的可染能力也存在很大差异，从 12% 至 75% 不等。选取 1,000 个"好"的花粉粒样品进行测量，其中 5.6% 的直径等于或低于在二倍体物种中测量获得的花粉粒的标准直径。由于在同一物种内，花粉粒的大小通常与其染色体的数量相关，很可能在 *sto* 2282.2 中发现的某些较小的花粉粒发生了染色体双减数，只有 12 条染色体。

松林 [14] 发现 *stoloniferum* 的某些克隆株有时会在减数第一次分裂中期产生三极纺锤体，导致形成包含 6 个几乎同等大小核子的六核体。这种机制非常有可能是产生发生染色体双减数的花粉粒的原因。遗憾的是，由于病毒感染使克隆体丢失，未能对 *sto* 2282.2 的减数分裂进行再次研究。因此，尚缺乏双减数假说的细胞学证据。

讨　论

双减数常被用来解释许多异常杂种的产生。例如，斯蒂尔 [15] 通过将腋花矮牵牛 ($2n = 2x = 14$) 和碧冬茄 ($2n = 4x = 28$) 杂交，获得了与其两亲本均相似的二倍体后代。

both parents, from the cross *Petunia axillaris* $(2n = 2x = 14) \times P.\ hybrida$ $(2n = 4x = 28)$. Nishiyama[16] made similar observations on *Avena* hybrids, Karprechenko[17] on *Brassica* × *Raphanus* crosses and Ewart and Walker[18] on *Poinsettia*. In the cherry, three diploid cultivars are reputed to have arisen from crosses between diploid *Prunus avium* and a tetraploid cultivar, "May Duke"[19]; Darlington[20] thought that the chromosome number of the parents disproved the supposed origin of these cultivars but, in retrospect, it now seems that the original interpretation is worth reconsideration.

In none of these examples is there definite proof of double chromosome reduction: for the potato hybrid, although direct cytological evidence is lacking, the biochemical data appear to be decisive. The fact that Matsubayashi[14] independently obtained a tetraploid plant from the same cross that produced *M.* 271 suggests that double chromosome reduction may be less uncommon than we think. But it seems unlikely that it is an important general mechanism in Nature for decreasing chromosome numbers.

From a biochemical point of view, *M.* 271 is particularly interesting since it contains a new hybrid substance, a glycoside of demissidine not present in either parent. The identification of the sugar linkages in this glycoside will clearly be of interest, as it should throw some light on the control of alkaloid glycoside synthesis, about which very little is at present known.

(**208**, 359-361; 1965)

G. E. Marks, R. K. Mckee and J. B. Harborne: John Innes Institute, Bayfordbury, Hertford.

References:

1. Howard, H. W., and Swaminathan, M. S., *Genetica*, **26**, 381 (1953).

2. Dodds, K. S., *Nature*, **166**, 795 (1950).

3. Marks, G. E., *J. Genet.*, **53**, 262 (1955).

4. Propach, H., *Z. Indukt. Abstamn. Vererb.*, **74**, 376 (1937).

5. Kawakami, K., and Matsubayashi, M., *Sci. Rep. Myogo Univ. Agric.*, 3, 17 (1957).

6. Bains, G. S., and Howard, H. W., *Nature*, **166**, 795 (1950).

7. Beamish, K., *Amer. J. Bot.*, **43**, 297 (1955).

8. Walker, R. I., *Bull. Torrey Bot. Club*, **82**, 87 (1955).

9. Alston, R. E., and Turner, B. L., *Biochemical Systematics*, 295 (Prentice-Hall, 1963).

10. Harborne, J. B., *Biochem. J.*, **84**, 100 (1962).

11. Schreiber, K., *Die Kulturpflanze*, **11**, 422 (1963).

12. Schreiber, K., and Aurich, O., *Z. Naturforschg.*, 18*b*, 471 (1963).

13. Schreiber, K., Aurich, O., and Osske, G., *J. Chromatog.*, **12**, 63 (1963).

14. Matsubayashi, M., *Chromosome Information Service*, No. 3, 7 (1962) and personal communication.

15. Steere, W. C., *Amer. J. Bot.*, **19**, 340 (1932).

16. Nishiyama, I., *Cytologia*, 5, 146 (1933).

17. Karprechenko, G. D., in *Recent Advances in Cytology*, edit. by Darlington, C. D., 193 (Churchill, 1937).

18. Ewart, L. C., and Walker, D. E., *J. Heredity*, **50**, 203 (1960).

19. Knight, T. A., quoted by Darlington, C. D., *J. Genet.*, **22**, 19 (1928).

20. Darlingron, C. D., *J. Genet.*, **28**, 327 (1934).

西山 [16] 在燕麦属杂种研究中，卡尔佩琴科 [17] 在芸薹属植物和萝卜属植物杂交实验中，尤尔特和沃克 [18] 在对一品红的研究中都观察到了类似的现象。在对樱桃的研究中，研究者认为 3 个二倍体樱桃栽培品种是由二倍体物种欧洲甜樱桃和四倍体栽培品种"五月爵士樱桃"杂交产生的 [19]。达林顿 [20] 认为亲本中染色体的数量证明了这些栽培品种的假定起源是不正确的，但是现在看来最初的解释值得重新考虑。

以上这些研究例子中均没有给出染色体双减数的确切证据：在关于马铃薯杂种的研究中，尽管缺乏直接的细胞学证据，但提供了确切的生物化学数据。松林 [14] 从得到 M. 271 的同一杂交过程中还获得了一个四倍体植株，这一事实表明染色体双减数也许不像我们想象的那样少见。但是在自然界中这种机制可能并不是染色体数量减少的一个重要的常规机制。

在生化水平上，尤其令人感兴趣的是，M. 271 包含了一种新的糖苷垂茄碱，这种新的杂交物质在任何一个亲本中都不存在。有必要鉴别该糖苷中的糖链连接，以便帮助人们更好地了解糖苷生物碱的合成，因为目前我们对此还所知甚少。

（周志华 翻译；王秀娥 审稿）

"Pink Spot" in the Urine of Schizophrenics

R. E. Bourdillon *et al.*

Editor's Note

Here Raymond Bourdillon and colleagues add fuel to the "pink spot" controversy. Three years earlier, psychiatrist Arnold Friedhoff and biochemist Elnora Van Winkle used paper chromatography to identify a "pink spot" in the urine of patients with schizophrenia. Subsequent studies, often done non-blind and with small subject numbers, yielded conflicting results. So Bourdillon and colleagues set up a blind, controlled study of 808 samples, concluding that there is a strong link between the pink spot and schizophrenia. Diet and drugs were, they thought, unlikely to bias the results. But the theory was later abandoned when it was found that the pink spot was a probable metabolite of both tea and the anti-psychotic chlorpromazine.

SINCE Friedhoff and van Winkle[1] reported that they had isolated the compound 3,4-dimethoxyphenyl-ethylamine (DMPE)—the "pink spot"—from the urine of 15 out of 19 schizophrenics whereas they found it was absent in 14 mentally normal people, other workers have carried out similar investigations. Of these, Sen and McGeer[2], Kuehl *et al.*[3] and Horwitt[4] were confirmatory, Perry *et al.*[5] and Faurbye and Pind[6] were not, and Takesada *et al.*[7] found that the compound was also present in nearly half their controls. In all these series the number of individuals tested was small, and moreover it appears that in none of them were the tests done in ignorance of the diagnosis. In the past many metabolic abnormalities have been described in schizophrenia, but none has since been shown to have any fundamental bearing on the disease. It therefore seemed important to us to investigate the pink spot further, for if it could be shown to be causally related to schizophrenia this would provide important evidence favouring the abnormal methylation hypothesis as put forward by Osmond and Smythies[8] and Harley-Mason[9].

We have tested for the pink spot in the urine of 808 individuals to try to find out how often it is found: (*a*) in the different forms of schizophrenia; (*b*) in different types of mental disease; (*c*) in close relatives of schizophrenics; (*d*) in mentally normal people. To try to answer these problems, four experiments were planned:

Exps. I and II: Here two independent and "blind" surveys were carried out in which separate workers ascertained the incidence of the pink spot using different methods for its detection and also different criteria for the assessment of schizophrenia.

Exp. III: A small series of family studies in which the close relatives of pink spot positive schizophrenic propositi were investigated.

438

精神分裂症患者尿中的"粉红色斑点"

鲍迪伦等

编者按

本文中，雷蒙德·鲍迪伦及其同事们的研究使得关于"粉红色斑点"的争议更加激烈。3 年前，精神病专家阿诺尔德·弗里德霍夫和生物化学家埃尔诺拉·范温克尔采用纸层析法在精神分裂症患者的尿液中发现了"粉红色斑点"。随后进行的研究（经常是针对小部分患者进行的非盲法研究）产生了相互矛盾的结果。因此鲍迪伦及其同事们对 808 份样品作了一个盲点对照研究，结果显示粉红色斑点与精神分裂症密切相关。他们认为日常饮食和药物不会影响这种结果。但是后来，当发现粉红色斑点可能是茶和抗精神病药氯丙嗪的代谢物时，这种观点被抛弃了。

自从弗里德霍夫和范温克尔[1]报道他们从 19 名精神分裂症患者中的 15 个人的尿液中分离出化合物 3,4-二甲氧基苯基乙胺（DMPE）——"粉红色斑点"——而在 14 名精神正常的人中没有发现这一物质之后，其他工作者也开始进行类似的研究。其中，森和麦克吉尔[2]、屈尔等人[3]以及霍威特[4]的研究证实了这一结果，而佩里等人[5]以及福尔拜和平德[6]的研究没有得到这种结果，武贞等人[7]发现在他们近一半的对照样品中也存在该化合物。在这一系列实验中，被测个体的数目少，而且他们中没有一个是在忽视诊断的前提下进行检测的。过去，在精神分裂症患者中发现过很多代谢异常现象，但之后没有一种被证明与该病有根本上的关系。因此，进一步研究粉红色斑点很重要，因为如果这种斑点与精神分裂症密切相关，就可以作为支持异常甲基化假说的重要证据，这一假说由奥斯蒙德和斯迈西斯[8]以及哈利-梅森[9]提出。

我们已经测试了 808 个人尿液中的粉红色斑点，试图找出该斑点在以下几种情况中存在的概率：(a) 在不同类型的精神分裂症中；(b) 在不同类型的精神病中；(c) 在精神分裂症患者的近亲属中；(d) 在精神正常的人中。为了回答这些问题，我们设计了 4 组实验：

实验 I 和实验 II：这是两个独立的"盲法"实验，研究中每个工作人员都单独使用不同的检测方法和不同的评价标准来鉴定精神分裂症患者中粉红色斑点的发生率。

实验 III：这是一个小系列的家庭研究，对家族中粉红色斑点阳性的精神分裂症患者的近亲属进行调查研究。

439

Exp. IV: A survey of mentally normal individuals some of whom were apparently healthy in every way and some of whom were in-patients in the wards of two general hospitals.

Exp. I. The urines of 101 mental hospital in-patients, all "possible schizophrenics", were tested with the primary object of finding the incidence of pink spot in the disease. The patients, who were drawn from three mental hospitals, consisted of acute and chronic cases and some had been off drug therapy for up to two months. All had at one time been given a provisional diagnosis of schizophrenia, and this and the drug therapy were known to the investigator (R. E. B.) who was testing for the pink spot. After he had assessed the presence or absence of this, each patient was interviewed and re-assessed by a psychiatrist (S. A. L.) who was ignorant of the pink spot finding. It was she who made the final assessment, and in about 17 percent of the patients she thought the original diagnosis of schizophrenia was substantially in doubt; when this was so they were classed as non-schizophrenic. Where the diagnosis of schizophrenia was definitely upheld the patients were grouped into those who at any time had exhibited one or more of Schneider's first-rank symptoms[10] and those who had not (these are referred to as Schneider positive and Schneider negative respectively). Afterwards the patients were also sub-divided into classical schizophrenics and those who showed only paranoid features without formal thought disorder or flattening of affect. The former for convenience are referred to as "non-paranoid" and the latter as "paranoid". In five patients the information necessary for diagnosis was unobtainable either because the patients were unco-operative or because of language difficulties, and these five are classed as "impossible to assess".

The urine sample was, whenever possible, collected over 20h, though in some cases shorter periods of collection proved more practicable. The extraction and the paper chromatography were carried out as described by Friedhoff and van Winkle[1] and this is referred to here as Method 1. Where there was marked interference by drugs the presence or absence of the pink spot sometimes had to be classed as "impossible to assess".

Because a much more detailed investigation would be necessary to ascertain that 3,4-DMPE was in fact being demonstrated we have, throughout this article, preferred to use the term "pink stop" rather than "presence of DMPE" in reporting the results.

These are given in Table 1 and the statistical analysis in Table 2. This consists of various comparisons using the χ^2 test or Fisher's two-tailed test for exact probability, whichever seemed the more appropriate. The control group were 149 mentally normal individuals (tested by Method 1) none of whom had the pink spot. Given also (Table 3) are the percentages of individuals with the pink spot within sub-classes.

实验 IV：对精神正常的个体进行调查研究，其中一部分人从各方面看都明显健康，而另一部分是在两个综合医院住院的患者。

实验 I 为了调查粉红色斑点的可能发生率，对 101 例在精神病院住院的患者的尿样进行了检测。这些患者来自 3 个精神病院并且"可能都是精神分裂症患者"。他们中有急性和慢性患者，其中一些人已经停止药物治疗达 2 个月之久了。在同一时间对所有患者进行精神分裂症的临时诊断，负责检测粉红色斑点的研究者雷蒙德·鲍迪伦熟悉这种诊断和药物治疗。当他检测完斑点是否存在后，每个患者还要与不了解粉红色斑点的精神病学家雪莉·莱斯利会面并进行再次检测。据其做出的最后评估，她对约 17% 精神分裂症患者的最初诊断结果持怀疑态度；鉴于此，他们被归为非精神分裂症患者。已经确诊的精神分裂症患者被分成两组，一组是在任何时候都会表现出一种或多种施奈德一级症状的患者[10]，另一组是从来不表现症状的患者（这分别被称为施奈德阳性和施奈德阴性精神分裂症）。然后患者也被细分为典型的精神分裂症患者和只显示有妄想症特性而没有思维形式障碍或是情感偏激的患者。为了方便起见，前者被称为"非妄想症者"，后者称为"妄想症者"。由于患者不合作或语言障碍的原因，我们在 5 名患者中没有获得诊断所需的必要信息，因此将他们划分为"无法评估"类。

要尽可能收集超过 20 小时的尿样，尽管已经证实对某些患者而言短时间收集更有效。尿样的萃取和纸层析按弗里德霍夫和范温克尔[1]所描述的方法进行，本文中将其定义为方法 1。当被药物显著干扰时，粉红色斑点时有时无，我们有时不得不将其划分为"无法评估"类。

因为还有必要作更详细的研究以明确事实上正被我们所论证的 3,4-二甲氧基苯基乙胺，所以整篇文章中我们更倾向于使用术语"粉红色斑点"而不是"存在 3,4-二甲氧基苯基乙胺"来报道实验结果。

表 1 和表 2 分别表示研究的结果和统计分析。其中包括了对确切的发生概率用卡方检验或费希尔双边检验而得到的多项比较结果，任何一种方法似乎都更加合适。对照组是 149 个精神正常的个体，没有检测到粉红色斑点（通过方法 1 测定）。表 3 同样给出在亚群中粉红色斑点个体的出现比例。

Table 1. Experiment I

(*a*) Pink spot in relation to Schneider's first-rank symptoms

Schneider rating	Pink spot rating			
	Positive	Negative	Impossible to assess	Totals
Positive	35	11	6	52
Negative	7	16	4	27
Impossible to assess	4	0	1	5
Non-schizophrenic	0	16	1	17
Totals	46	43	12	101

(*b*) Pink spot in relation to non-paranoid and paranoid classification

Psychiatric assessment		Pink spot rating		
		Positive	Negative	Totals
Schneider positive	Non-paranoid	35	7	42
	Paranoid	0	4	4
Schneider negative	Non-paranoid	5	5	10
	Paranoid	2	11	13
Totals		42	27	69

Table 2. Experiment I. Comparison of Incidence of Pink Spot between Classes

Comparison	χ^2	d.f.	Probability
Schneider positive with Schneider negative	11.57	1	< 0.001
Schneider positive with non-schizophrenic	24.95	1	< 0.001
Schneider positive with control			$= 2.5 \times 10^{-39}$
Schneider negative with non-schizophrenic			$= 0.029$
Schneider negative with control			$= 3.1 \times 10^{-7}$
Non-paranoid with paranoid	20.18	1	< 0.001
Non-paranoid with non-schizophrenic	26.80	1	< 0.001
Non-paranoid with control			$= 8.06 \times 10^{-32}$
Paranoid with non-schizophrenic			N.S.
Paranoid with control			0.0099
Schneider positive with Schneider negative Within non-paranoid			0.0388
Schneider positive with Schneider negative within paranoid			N.S.
Paranoid with non-paranoid within Schneider positive			$= 0.0020$
Paranoid with non-paranoid within Schneider negative			N.S.

Table 3. Experiment I. Percentage of Individuals with Pink Spot within Sub-classes

Sub-class	Percentage with pink spot
Schneider positive, non-paranoid	83.3 ± 5.75
Schneider positive, paranoid	0.0
Schneider negative, non-paranoid	50 ± 15.8
Schneider negative, paranoid	15.4 ± 10.0

表 1. 实验 I

（a）与施奈德一级症状相关的粉红色斑点

施奈德评价	粉红色斑点评价			
	阳性	阴性	无法评估	合计
阳性	35	11	6	52
阴性	7	16	4	27
无法评估	4	0	1	5
非精神分裂症	0	16	1	17
合计	46	43	12	101

（b）与非妄想症和妄想症相关的粉红色斑点

精神病评估		粉红色斑点评价		
		阳性	阴性	合计
施奈德阳性	非妄想症	35	7	42
	妄想症	0	4	4
施奈德阴性	非妄想症	5	5	10
	妄想症	2	11	13
合计		42	27	69

表 2. 实验 I. 不同类群间粉红色斑点发生率的比较

比较	χ^2	自由度	概率
施奈德阴性和施奈德阳性	11.57	1	< 0.001
施奈德阳性和非精神分裂症	24.95	1	< 0.001
施奈德阳性和对照			$= 2.5 \times 10^{-39}$
施奈德阴性和非精神分裂症			= 0.029
施奈德阴性和对照			$= 3.1 \times 10^{-7}$
非妄想症和妄想症	20.18	1	< 0.001
非妄想症和非精神分裂症	26.80	1	< 0.001
非妄想症和对照			$= 8.06 \times 10^{-32}$
妄想症和非精神分裂症			无显著性差异
妄想症和对照			0.0099
施奈德阳性和非妄想性施奈德阴性			0.0388
施奈德阴性和妄想性施奈德阴性			无显著性差异
妄想症和施奈德阳性非妄想症			= 0.0020
妄想症和施奈德阴性非妄想症			无显著性差异

表 3. 实验 I. 亚群中粉红色斑点个体的出现比例

亚群	粉红色斑点比例
施奈德阳性，非妄想症	83.3 ± 5.75
施奈德阳性，妄想症	0.0
施奈德阴性，非妄想症	50 ± 15.8
施奈德阴性，妄想症	15.4 ± 10.0

Exp. II. In the second experiment a different psychiatrist (P. H.) and biochemist (A. P. R.) worked together. The urines of 296 psychiatric in-patients drawn from two hospitals were investigated for the pink spot. The patients had often been in hospital for many years and included a larger proportion of chronic cases than in the first experiment. The samples of urine were analysed in ignorance both of the diagnosis and of the nature and quantity of drugs being received, so that a different part of the work was "blind" from that in Exp. I. The psychiatrist selected the patients so as to include a variety of mental diseases, and he accepted the diagnosis given in the case-sheets and did not interview the patients personally as did the psychiatrist in Exp. I. He divided the cases into four groups: (*a*) "non-paranoid" schizophrenics; (*b*) "paranoid" schizophrenics; (*c*) schizophreniform syndromes (where although some features of schizophrenia were present it was uncertain whether or not the primary diagnosis was schizophrenia); (*d*) non-schizophrenics, which included such conditions as manic depressive psychosis, organic dementia and mental defect.

Since a different psychiatrist was scoring the cases and the diagnosis was assessed from the case sheets, the classes are probably not strictly comparable with those of Exp. I.

The chromatographic procedures were the same as those described by Friedhoff and van Winkle[1] (our "Method 1"), but a different extraction procedure was used and the technique is referred to by us as "Method 2". A volume of 300–500 ml. of a 16-h overnight sample of urine was adjusted to pH 9.0 with 2 N sodium hydroxide. This was extracted three times with 100-ml. portions of 1,2-dichloroethane and the extracts dried over anhydrous sodium sulphate. The first extract was examined by ultra-violet absorption at 279 mμ. A peak at this value is a useful indicator of DMPE in the absence of interfering substances such as drug metabolites. The ultra-violet absorption spectrum helps particularly to distinguish DMPE from any closely related compounds which could not satisfactorily be resolved from it by the chromatographic procedures used. The three extracts were then combined and evaporated to dryness on a rotary evaporator. The residue was redissolved in dichloroethane for application to the chromatography paper. Authentic 3,4-dimethoxyphenyl-ethylamino (K and K Laboratories) and mixed spots of urine extracts and authentic DMPE were applied alongside the urine extracts on the chromatography papers and the separation carried out.

The results of Exp. II are given in Table 4 and the statistical analysis in Table 5. The controls are 310 mentally normal individuals (tested by Method 2) one of whom had the pink spot. The results for the schizophrenform syndrome class have not been analysed because the group was probably very heterogeneous.

实验 II　第二个实验由另一位精神病学家哈珀和生物化学家保利娜·布里奇斯共同完成。他们检测了来自两个医院的 296 例住院精神病患者尿样中的粉红色斑点。这些患者经常长期住院，而且其中慢性病例的比例比第一次实验中的大。为了与实验 I 相区别，部分工作采用的是"盲法"实验，即尿样的分析是在不了解诊断情况和所使用药品的性能和数量的情况下进行的。精神病学家挑选了多种类型的精神病患者，同时他接受病历中的诊断结果，并没有像实验 I 中的精神病学家那样亲自会见患者。他把患者分为 4 组：(a)"非妄想症"精神分裂症；(b)"妄想症"精神分裂症；(c) 精神分裂症样综合征（尽管有一些精神分裂症的特征，但最初诊断不能确定是否是精神分裂症）；(d) 非精神分裂征，包括躁狂抑郁症、器官性痴呆和心理缺陷。

因为病例是由不同的精神病专家根据病历中的诊断结果进行评判的，所以这些分类可能没有实验 I 那样严格。

纸层析过程也是按照弗里德霍夫和范温克尔[1] 所述的方法（我们称为"方法 1"）进行的，但使用的萃取过程是不同的，是按照我们定义的方法 2 进行的。使用 2 摩尔/升的氢氧化钠溶液将过夜 16 小时的体积为 300 毫升～500 毫升尿样的 pH 值调至 9.0。然后用 100 毫升 1,2–二氯乙烷萃取 3 次后再用无水硫酸钠干燥萃取物。通过紫外分光光度计在波长为 279 纳米下检测第一次萃取物。在没有干扰物质（比如药物代谢物）的情况下，所检测到的峰值是鉴定 3,4–二甲氧基苯基乙胺的有用指标。紫外分光光度法特别有助于将 3,4–二甲氧基苯基乙胺从其他相近的混合物中区分出来，而层析法则不能达到这个水平。然后将 3 种萃取物混合在一起并在旋转蒸发器中脱水蒸干。残余物用二氯乙烷再次溶解后进行纸层析。高纯度的 3,4–二甲基苯乙胺（K&K 实验室）及其与尿液萃取物斑点的混合物和尿液提取物一起进行纸层析和分离实验。

表 4 和表 5 分别是实验 II 的研究结果和统计分析。以 310 例精神正常个体为对照，通过方法 2 对其进行检测发现只有一例呈现粉红色斑点。因为精神分裂症样综合征患者的种类非常多，所以没有分析这类患者的结果。

Table 4. Experiment II. Pink Spot Assessment in 296 Psychiatric Patients

Psychiatric assessment	Pink spot rating			
	Present	Absent	Impossible to assess	Totals
Non-paranoid schizophrenics	20	30	19	69
Paranoid schizophrenics	2	54	6	62
"Schizophreniform syndromes"	5	58	25	88
Non-schizophrenics	1	68	8	77
Totals	28	210	58	296

Table 5. Experiment II. Comparison of Incidence of Pink Spot between Classes

Comparison	χ^2	d.f.	Probability
Non-paranoid with paranoid	19.16	1	< 0.001
Non-paranoid with non-schizophrenic	27.05	1	< 0.001
Non-paranoid with control			$= 2.81 \times 10^{-18}$
Paranoid with non-schizophrenic			N.S.
Paranoid with control			N.S.

Exp. III. The pink spot was investigated in 20 close relatives of three schizophrenic propositi, all three of whom were found (by one or both of the two methods) to excrete the pink spot. None of the relatives was schizophrenic though some were reported as having other mental abnormalities.

The pink spot was not found in the urine of any of the close relatives, and these therefore are not significantly different from the control group of mentally normal individuals.

Exp. IV: Investigation of mentally normal individuals. These consisted of two groups. The first was made up of 265 healthy individuals, mostly undergraduates or university staff. The second was composed of 126 mentally normal in-patients, comprising 54 pre-operative and 20 post-operative cases, 10 patients with liver disease, 20 with chronic neurological states and 22 with general medical conditions. The pink spot was assessed by one or both of the two methods previously described and the series has acted as the control for the various comparisons.

All the controls were negative for the pink spot with the exception of one healthy individual, a woman aged 54. She suffered from migraine but was mentally normal and so was her family.

Comparison between Method 1 (Friedhoff and van Winkle) and Method 2 (ultra-violet + chromatography) in the investigation of the pink spot. In 133 people both methods for assessing the pink spot were used (Table 6).

表 4. 实验 II. 在 296 例精神分裂症患者中检测粉红色斑点

精神病评估	粉红色斑点评价			
	存在	不存在	无法评估	合计
非妄想性精神分裂症	20	30	19	69
妄想性精神分裂症	2	54	6	62
"精神分裂样综合征"	5	58	25	88
非精神分裂症	1	68	8	77
合计	28	210	58	296

表 5. 实验 II. 各种类型之间粉红色斑点发生率的比较

比较	χ^2	自由度	可能性
非妄想症和妄想症	19.16	1	< 0.001
非妄想症和非精神分裂症	27.05	1	< 0.001
非妄想症和对照			$= 2.81 \times 10^{-18}$
妄想症和非精神分裂症			无显著性差异
妄想症和对照			无显著性差异

实验 III 在 3 例精神分裂症患者的 20 个近亲属中检测粉红色斑点。使用一种或同时使用这两种方法均能在这 3 个患者中检测出粉红色斑点,而他们的近亲属中没有一人患有精神分裂症,尽管有些人被报道有其他精神异常症状。

在所有近亲属的尿液中没有检测到粉红色斑点,因此他们与作为对照的精神正常的个体相比无显著性差异。

实验 IV:对精神正常的个体进行的研究 他们被分为两组:第 1 组由 265 例健康个体组成,他们大部分是大学生或大学职工。第 2 组是由 126 例精神正常的住院患者组成,包括 54 例手术前患者和 20 例手术后患者、10 例有肝病的患者、20 例有慢性神经功能缺损的患者和 22 例有常见疾病的患者。使用前面提到的一种方法或同时使用两种方法检测粉红色斑点,同时该系列已作为各种比较的对照。

除了一个 54 岁的健康女性外,所有对照的红色斑点检测均为阴性。她患有偏头痛但精神正常,她的家人也是这样。

使用方法 1(弗里德霍夫和范温克尔)和方法 2(紫外分光光度法和层析法)研究粉红色斑点进行的比较 同时使用这两种方法对 133 例个体进行粉红色斑点检测(表 6)。

Table 6. Comparison of Method 1 and Method 2 in Investigation of the Pink Spot

Friedhoff and van Winkle (Method 1)	Ultra-violet + chrom. (Method II)		
	Pink spot positive	Pink spot negative	Totals
Pink spot positive	12	7	19
Pink spot negative	2	112	114
Totals	14	119	133

Though there is a significant correlation, $P = 1.06 \times 10^{-10}$, between the results given by the two methods, there are discrepancies.

There are several reasons which might account for this non-concordance: (1) Method 2 may be less sensitive in detecting the pink spot (as opposed to DMPE); (2) slightly different substances may be being estimated; (3) one method may be more likely to produce undetectable false readings because of the presence of drugs; (4) there may be observer discrepancies. There possibilities are being investigated and will be the subject of a further article.

The association previously found between the pink spot and schizophrenia has been confirmed using large numbers. The results of Exp. I give compelling evidence that its presence is particularly associated with Schneider-positive individuals. It is much rarer in those who are Schneider negative and rarer still (if present at all) in non-schizophrenics.

In Exp. I the same patients have also been classified as to whether or not they only have paranoid features. The presence of the pink spot is particularly associated with the non-paranoid group but is significantly higher among the paranoid group than among the mentally normal controls.

When the individuals are classified as being Schneider positive and non-paranoid this sub-class has a significantly higher incidence of the pink spot than does any other group (83.3 ± 5.75 percent). Thus to predict the presence of the pink spot the double classification is the most efficient.

Five individuals were so seriously disturbed that they could not be interviewed or assessed on the Schneider rating. The fact that four of them showed the pink spot suggests that these were true Schneider-positive non-paranoid cases as judged by our other results.

The second experiment using a different method for detecting the pink spot again shows the very strong association between the presence of the pink spot and "non-paranoid" schizophrenia. The lower frequency of the pink spot may be due to any of the four reasons mentioned above and/or to the cases being loss florid than in Exp. I. However, not enough have been scored by the Schneider rating to show whether the association is particularly with the Schneider-positive "non-paranoid" group although the data show a strong tendency in this direction.

表 6. 使用方法 1 和方法 2 研究粉红色斑点的比较

弗里德霍夫和范温克尔（方法 1）	紫外分光光度计法和层析法（方法 2）		
	阳性粉红色斑点	阴性粉红色斑点	合计
阳性粉红色斑点	12	7	19
阴性粉红色斑点	2	112	114
合计	14	119	133

虽然这两种方法的研究结果显著相关（P 值为 1.06×10^{-10}），但仍存在差异。

结果出现差异的原因可能有如下几点：（1）在检测粉红色斑点时，方法 2 的敏感性可能更低（3,4-二甲氧基苯基乙胺与之相反）；（2）可能检测到了一些差异较小的物质；（3）由于药物的存在，其中的一种方法更有可能产生觉察不到的错误读数；（4）可能存在测者差异。这些可能的原因正在被研究并作为将来更深层研究文章的主题。

通过检测大量样品，证实了先前提出的粉红色斑点和精神分裂症之间存在联系的观点。实验 I 的研究结果为粉红色斑点与施奈德阳性个体间存在密切联系提供了有力的证据。在施奈德阴性个体中这是很少见的，在非精神分裂症（如果存在的话）个体中出现的概率更小。

实验 I 中相同的患者也按照是否只有妄想症特征来分类。粉红色斑点与非妄想症组存在特别密切的关系，但在妄想症组中出现的概率明显高于精神正常的对照组。

当个体被划分为施奈德阳性和非妄想症类型时，其粉红色斑点发生率明显高于其他组（$83.3\% \pm 5.75\%$）。因此对于预测粉红色斑点的出现，这种双重分类法最有效。

由于 5 个人受到严重干扰，以至于不能与他们进行面谈或按照施奈德的方法对其进行分类。实际上，在其中的 4 个人中检测到了粉红色斑点，就像我们得到的其他结果一样，他们都是施奈德阳性非妄想症患者。

第二次实验使用不同的方法检测粉红色斑点，结果再次表明斑点的出现和"非妄想症"精神分裂症有紧密的联系。粉红色斑点出现的频率较低可能是由于上述提到的 4 个原因中的任何一个，或由于患者不如实验 I 中的健康。然而按照施奈德的方法，没有足够的证据显示施奈德阳性"非妄想症"组与粉红色斑点之间有明确关系，尽管数据强烈地倾向于这一方向。

The evidence from the two experiments suggests that the pink spot is highly associated with schizophrenia and is not often present in other forms of mental disorder or in close relatives or in controls. Moreover, from our survey it seems highly improbable that diet and institution life play any part in producing the pink spot. An important problem is whether the abnormal metabolite precedes or is a result of the disease. The family data point to the latter conclusion, but it is possible that the pink spot only appears shortly before the disorder manifests itself, and that more refined techniques might demonstrate it in sibs. It would be interesting to follow pink-spot-positive patients with florid schizophrenia during remissions since these data might indicate whether the acuteness of the disease was associated with the quantity of pink spot material. If it did, this might explain its absence in sibs. The possibility has to be considered that drugs might be the cause of the pink spot. This is refuted by our data. Thus the relationship between schizophrenia and pink spot is at least as marked in those who had been off drugs for two weeks or more as in those who had not. Moreover, the non-schizophrenics on drugs showed no elevation of the frequency of the pink spot over those not on drugs. However, a complicating factor in using the pink spot for diagnosis is that drugs interfere both with the chromatography and with the ultra-violet absorption and it is difficult to obtain patients (particularly acute cases) who are not on drug therapy.

We thank Dr. B. Finkleman, Prof. F. J. Fish, Dr. A. V. de P. Kelly, Dr. I. Leveson, and Dr. B. Ward for allowing us to examine and test their patients, and the staffs of the hospitals concerned for help in the collection of samples. We also thank the Hon. Miriam Rothschild for obtaining the co-operation of certain of the individuals investigated.

(**208**, 453-455; 1965)

R. E. Bourdillon, C. A. Clarke, A. Pauline Ridges and P. M. Sheppard: Nuffield Unit of Medical Genetics, Department of Medicine, University of Liverpool.
P. Harper: Department of Psychological Medicine, University of Liverpool.
Shirley A. Leslie: Psychiatric Registrar, Royal Liverpool Children's Hospital, Liverpool.

References:

1. Friedhoff, A. J., and van Winkle, E., *Nature*, **194**, 897 (1962).

2. Sen, N. P., and McGeer, P. H., *Biochem. Biophys. Res. Comm.*, **14**, 227 (1964).

3. Kuehl, F. A., Hichens, M., Ormond, R. E., Meisinger, M. A. P., Gale, P. H., Cirillo, V. J., and Brink, N. G., *Nature*, **203**, 154 (1964).

4. Horwitt, M. K., contribution to discussion at *Symp. Amine Metabolism in Schizophrenia*, Atlantic City, April 1965 (in the press).

5. Perry, T. L., Hansen, S., and Macintyre, L., *Nature*, **202**, 519 (1964).

6. Faurbye, A., and Pind, K., *Acta Psychiat. Scand.*, **40**, 540 (1964).

7. Takesada, M., Kakimoto, Y., Sano, I., and Kaneko, Z., *Nature*, **199**, 203 (1963).

8. Osmond, H., and Smythies, J., *J. Ment. Sci.*, **98**, 309 (1952).

9. Harley-Mason, J., *J. Ment. Sci.*, **98**, 313 (1952).

10. Schneider, K., *Fortschr. Neurol. Psych.*, **25**, 487 (1957).

两个实验给出的证据表明粉红色斑点与精神分裂症高度相关，而在其他的精神障碍类型、近亲属或对照组中并不经常出现。此外，根据我们的调查，日常饮食和日常生活不大可能对粉红色斑点的产生起到非常重要的作用。异常代谢物是先于疾病产生还是由疾病引起，这是一个重要的问题。家族数据指向后者，但粉红色斑点可能只在精神障碍发生前短暂地显现，而且许多精细技术可以验证出近亲属中的斑点。密切注意处于康复期的粉红色斑点呈阳性的妄想症型精神分裂症患者将很有趣，因为这些数据可能暗示疾病的严重程度是否与粉红色斑点的数量有关。如果真是这样的话，将有助于解释在近亲属中不出现粉红色斑点的原因。必须考虑药物可能导致出现粉红色斑点这种可能性。我们的研究数据否定了这一点。因此精神分裂症与粉红色斑点之间的关系至少应标记在那些已经停药两周以上或者没有服药的人身上。此外，服药的非精神分裂症患者与那些没有服药的患者相比，粉红色斑点出现的概率并没有提高。然而，在使用粉红色斑点作诊断时，一个复杂的因素是药物会干扰层析法和紫外吸收光谱法的检测，并且很难找到那些没有进行药物治疗的患者（尤其是急性患者）。

我们感谢芬克莱曼医生、菲什教授、凯利医生、莱韦森医生和沃德医生允许我们检测他们的患者，也感谢关心和帮助我们收集样品的医院的全体员工。我们还要感谢米丽娅姆·罗思柴尔德阁下在个体研究时所给予的合作。

（郭娟 翻译；刘京国 审稿）

The Juvenile Hormone[*]

V. Wigglesworth

Editor's Note

Here British entomologist Vincent Wigglesworth extols the virtues of "one of the most outstanding morphogenetic hormones known"—the juvenile hormone, which he himself had recently discovered. The hormone, which is released from a pair of endocrine glands behind the brain, controls metamorphosis by maintaining larval features in many different kinds of insects. Decapitating kissing bug larvae, Wigglesworth had shown, triggers premature metamorphosis. In this paper, his opening address to the third conference of European Comparative Endocrinologists, Wigglesworth sums up what is known about the juvenile hormone, including its role in metamorphosis and reproduction, and its chemical nature. He also believes that the hormone brings about "gene switching" and so may affect more complex behaviours, such as sociability.

IN choosing this topic for the opening address to the third conference of European Comparative Endocrinologists, I had in mind the consideration of the juvenile hormone in general terms—as an agent with properties that must be of interest to all endocrinologists. I shall not therefore attempt a complete or specialized survey of this very complex subject, but will select only a limited number of points for discussion.

A Morphogenetic Hormone

The juvenile hormone is a morphogenetic hormone—one of the most outstanding morphogenetic hormones known. The existence of a hormone responsible for maintaining larval characters was first revealed by the fact that decapitation in young *Rhodnius* larvae causes premature metamorphosis[1]. Parabiosis showed that this hormone was freely circulating in the blood; and by removing progressively more and more of the anterior parts of the head from the insects used in the parabiosis experiments, it was shown that the hormone came not from the brain but from the corpus allatum—an endocrine organ that buds off from the ectoderm at the base of the mouthparts and comes to lie just behind the brain[1]. This conclusion was confirmed by the implantation of corpora allata from young (3rd- or 4th-stage) larvae into 5th-stage larvae. These 5th-stage larvae then moulted not into adults but into giant 6th-stage larvae[2].

The juvenile hormone acts directly on the epidermal cells responsible for laying down the cuticle at moulting; restricted local application of the hormone results in a restricted

---

[*] Substance of an opening address to the third conference of European Comparative Endocrinologists held in Copenhagen during August 1965.

保幼激素[*]

威格尔斯沃思

编者按

本文中，英国昆虫学家文森特·威格尔斯沃思盛赞了由他自己在不久前发现的一种重要的形态发生激素——保幼激素的优点。保幼激素是由位于脑后的一对内分泌腺分泌的，在许多不同种类的昆虫体内，该激素通过保持幼虫特征来控制变态。威格尔斯沃思的实验表明，切除接吻虫幼虫的头部会使其提前变态。这篇文章是威格尔斯沃思在第三届欧洲比较内分泌学家大会上致的开幕词。在这里，他总结了已知的保幼激素的知识，包括它在变态和繁殖中的作用以及它的化学性质。他也相信该激素可以引起"基因转换"，并可能因此而影响到更为复杂的行为，例如社会行为。

选择保幼激素作为第三届欧洲比较内分泌学家大会开幕演讲的题目，是因为我认为一般来说保幼激素这种物质的属性一定会引起所有内分泌学家的兴趣。但是我并没有因此而试图对这个复杂的题目进行一个完整或专业的纵览，而只是选取了有限的几个点以供讨论。

形态发生激素

保幼激素是一种形态发生激素——是已知最重要的形态发生激素之一。这种负责保持幼虫特性的激素的发现源于切除红猎蝽幼小的幼虫头部引起了过早的变态发育[1]。异种共生实验显示这种激素在血液中自由循环；通过逐渐延长异种共生实验中昆虫头部前段的切除范围，发现这种激素不是来源于大脑，而是来源于咽侧体——外胚层的一种内分泌器官，位于口器的底部，正好在脑后[1]。将幼小的幼虫（第三龄或第四龄）的咽侧体移植到第五龄的幼虫后，发现这些第五龄的幼虫并没有蜕变为成虫，而是成为巨型第六龄幼虫[2]，此发现进一步证实了这个结论。

保幼激素直接作用于负责在蜕皮阶段产生外皮的表皮细胞；限制激素的局部应用会使受限部分的幼虫表皮变为成虫表皮[2]。例如，有可能产生一只带有幼虫翅膀

---

[*] 这篇文章是 1965 年 8 月于哥本哈根举行的第三届欧洲比较内分泌学家大会的开幕词。

local patch of larval cuticle in an otherwise adult insect[2]. It is possible, for example, to produce adult insects with one larval wing[3]. The self-same cell has the capacity for laying down larval cuticle or adult cuticle. That is most clearly shown in the sensory bristles or hairs: the trichogen cells which form the bristles persist from one stage to the next; in the presence of a large amount of juvenile hormone these cells lay down larval-type bristles; in the absence of the hormone they lay down adult-type bristles, and with intermediate amounts of juvenile hormone, bristles of intermediate type are developed[1]. Under natural conditions intermediate forms do not commonly occur. There is a strong tendency for the insect to develop either larval or adult characters.

Experiments by many authors have shown that the juvenile hormone controls metamorphosis in insects of all kinds. In holometabolous insects, which have a pupal stage between the larva and the adult, the juvenile hormone again controls the morphogenetic change; a large amount of the hormone ensures retention or re-development of larval characters; absence of the hormone results in metamorphosis to the adult; the presence of a very small amount of juvenile hormone leads to the appearance of the pupal form[4]. This result not only confirms the importance of the juvenile hormone in controlling morphogenesis, but it illustrates in a most striking way a point often made by C. M. Child: that the same inductor substance can evoke totally different results depending on its concentration or the timing of its action.

Here again intermediate forms rarely occur in Nature. But they can be induced experimentally: if the corpus allatum is removed from the last-stage larva of the honeybee[5] or of the giant silkmoth *Hyalophora*[6] the supply of juvenile hormone falls below the level necessary to produce the pupa, and monstrous forms intermediate between pupa and adult develop. Some caterpillars pass through a regular succession of morphological stages in successive instars. These forms also seem to be regulated by levels of juvenile hormone secretion—but the detailed evidence for this has not yet been fully worked out[7].

The Nature of Hormone-controlled Metamorphosis

It has long been recognized that the effect of the juvenile hormone is to control the realization or suppression of inborn potencies. In other words, it brings about "gene switching". In this regard it resembles the inductor substances which control morphogenesis in different parts of the body during differentiation; and the factors which lead to the differences in form if different individuals in environmentally induced polymorphism[8]. Indeed, there are a number of polymorphic changes in which there is evidence that the level of juvenile hormone activity may itself be involved, notably (i) the change over from the "solitary" to the "gregarious" form in locusts[9]; (ii) the switch from apterous to alate forms among parthenogenesis aphids[10]; (iii) the production of the soldier caste in termites[11,12].

Just how and where the juvenile hormone is acting is not known. Twelve years ago I wrote "it is a matter for discussion whether the simultaneous inheritance of the dual potentialities for larval and adult differentiation within these societies of cells is by way of the nucleus

454

的成虫 [3]。同一种细胞既能产生幼虫表皮，也能产生成虫表皮。这种现象在感觉器官的刚毛或绒毛上体现得最为明显：产生刚毛的毛原细胞持续存在于昆虫发育的过程中；当存在大量的保幼激素时，这些细胞产生幼虫型刚毛；当缺乏这种激素时，它们产生成虫型刚毛，当保幼激素处于中等水平时，就会产生处于中间状态的刚毛 [1]。在自然情况下中间状态并不普遍存在。昆虫的发育有很强的倾向，不是发育为幼虫就是发育为成虫。

很多科学家的实验结果显示保幼激素调控着各种昆虫的变态。对于完全变态的昆虫来说，在幼虫阶段和成虫阶段之间存在蛹的阶段，保幼激素在这里也调控着昆虫的形态变化；大量激素的存在会确保幼虫特性的维持或再发育；激素缺乏的结果是变态为成虫，只存在极小量保幼激素时幼虫会化成蛹 [4]。这个结果不仅肯定了保幼激素在调控变态发育中的重要性，而且用最惊人的方式阐明了蔡尔德经常提及的观点：相同的感应物质可能产生截然不同的作用，这取决于它的浓度和作用时间。

这里再次指出昆虫发育的中间状态在自然界中很少出现。但是它们可以在实验中被诱导产生：如果摘除蜜蜂 [5] 或巨蚕蛾 [6] 的最后阶段的幼虫的咽侧体，使供给的保幼激素降低到化蛹所必需的含量以下，这时就会产生介于蛹和成虫之间的巨型虫。一些幼虫会以连续中间状态的形式经过常规的连续形态学阶段。这些形态变化似乎也由保幼激素的分泌水平进行调节——但是有关这一点的详细证据仍不充分 [7]。

受激素调节的变态的本质

我们很久以前就已经认识到，保幼激素的作用是对某些先天潜能的实现或抑制进行调节。换句话说，它导致"基因转换"。从这方面考虑，它类似于可以在分化过程中调节身体不同部位形态发育的感应物质，也与由环境导致的昆虫多态性中导致个体间形态差异的因子 [8] 类似。实际上，有证据显示昆虫的多态性变化都涉及保幼激素的分泌水平，比较显著的是（i）蝗虫由"独居型"转变为"群居型" [9]；（ii）孤雌生殖的蚜虫由无翅转变为有翅 [10]；（iii）白蚁中兵蚁等级的产生 [11,12]。

保幼激素如何发挥作用以及在哪里发挥作用仍不得而知。12 年前我曾经写过"这些细胞群既能分化为幼虫又能分化为成虫的双重潜能，是取决于细胞核遗传还是细

or cytoplasm or both"[13]. Eight years ago I submitted that the juvenile hormone "controls the manifestation of alternative genetically controlled forms" and I suggested that "it is possible to conceive it as being concerned in the regulation of permeability relations within the cells—in such a way that the gene-controlled enzyme system responsible for larval characters is brought increasingly into action when the juvenile hormone is present"[14].

The position is unchanged today—except that tissue-specific puffing patterns in the chromosomes of *Drosophila*[15], on one hand, and the theories of enzyme induction in bacteria as developed by Jacob and Monod[16] *et al.*, on the other, render us much more prepared to accept the idea of a primary action of the hormone at the level of the gene. Experimental evidence has yet to come.

Reversal of Metamorphosis

The question was early raised whether the genetic system responsible for the production of larval characters was still capable of re-activation by the juvenile hormone in the adult insect. In other words, whether metamorphosis can be reversed. Of course, this can only be tested by making the adult moult again by exposing it artificially to the moulting hormone.

In general it can be said that in most adult tissues it is not possible to induce any reversal of metamorphosis. But there are certain undoubted examples of such reversal: in the abdominal cuticle of *Rhodnius*[17] and of *Oncopeltus*[18], in the thoracic cuticle of the earwig *Auisolabis*, where the "ecdysial line" characteristic of the larva can be re-induced in a moulting adult[19], and in the integument of Lepidoptera where larval cuticle can be re-induced in fragments of pupal and imaginal integument[20,21].

Reversal of metamorphosis is an abnormal phenomenon. There has been no selection for its occurrence and it is not surprising that it occurs with difficulty. Likewise the production of stages intermediate between the normal stages in metamorphosis is an abnormal phenomenon and selection will have acted against it.

Time of Action of the Juvenile Hormone

The most effective moment for exposing the tissues to the juvenile hormone seems to be just before they begin the synthetic activities characteristic of the larval stage. Hormone administered too early is less effective—presumably because it is broken down in metabolism before the time for its action has arrived[22,23].

But there are certain effects which are induced in the cells long before they become manifest. A very small dose of juvenile hormone administered soon after feeding in the 5th-stage larva of *Rhodnius* has no effect on the type of cuticle laid down: normal adult cuticle is produced over the abdomen. But it does have the effect of ensuring the survival of the trichogen cells (almost all of which normally break down and disappear from the

456

胞质遗传还是两者兼有，这是一个值得讨论的问题"[13]。8 年前我提出保幼激素"调节可供选择的遗传调控方式的表现"，并且我认为"它在细胞内很可能与渗透性关系的调节有关——通过这种方式，当存在保幼激素时，负责幼虫特性的由基因控制的酶系统能够更多地发挥作用"[14]。

此立场至今没有改变——只不过果蝇染色体中的组织特异性膨胀方式 [15] 以及雅各布和莫诺等人提出的细菌中酶诱导的理论 [16] 让我们更加容易接受激素在基因水平发挥初级作用这一观点。目前还没有获得具体的实验证据。

变态发育的逆转

人们很早就提出一个问题：负责产生幼虫特征的遗传系统在成虫中是否仍然有可能被保幼激素重新激活。换句话说，变态发育是否可以逆转。当然，这只能通过让成虫暴露于人工合成的蜕皮激素而使它再次蜕皮来进行检验。

一般来说，在大多数成虫组织中不可能引发任何变态发育逆转，但是也存在一些确定无疑发生了这种逆转的例子：对于红猎蝽 [17] 和突角长蝽 [18] 的腹部表皮以及矮蠼螋的胸部表皮来说，在正在蜕皮的成虫体内能诱导出幼虫的特征"蜕皮线"[19]，对于鳞翅目昆虫的表皮来说，能再次诱导蛹的片段和成虫的表皮成为幼虫表皮 [20,21]。

变态发育的逆转是一种异常现象。它的发生不被选择，这种情况很难发生也就不足为奇了。同样地，在变态反应中介于正常阶段之间的中间状态的产生也是异常现象，自然选择会抵制这种情况的发生。

保幼激素作用的时间

组织暴露于保幼激素下的最有效时刻似乎恰好是在它们开始显示幼虫阶段的合成活性特征之前。激素施用得太早，效率会降低——大概是因为在它起作用之前就在代谢过程中分解了 [22,23]。

但是，早在激素效应出现之前，在细胞中就已诱导出一些效应。对刚喂完食的第五龄的红猎蝽幼虫施用很小剂量的保幼激素，不会对表皮类型产生影响：腹部产生了正常的成虫表皮。但是它确保毛原细胞(在变态过程中毛原细胞几乎会全部脱落，并从腹部的背面消失) 全部继续存在，因此在成虫的腹部出现了多余的绒毛 [23]。

dorsum of the abdomen during metamorphosis), so that an excessive number of hairs appears on the abdomen of the adult[23].

The juvenile hormone has a comparable effect on the survival of the thoracic gland. The thoracic gland (the source of the moulting hormone) normally undergoes autolysis in *Rhodnius* within 24 h after moulting to the adult. This is a response to some hormonal factor in the newly moulted adult. But if the gland has been exposed to juvenile hormone during the pre-moulting period, it does not respond in this way and fails to undergo autolysis[24]. (The breakdown of certain muscles after metamorphosis in the silkmoth *Hyalophora cecropia* has recently been shown to be regulated in much the same way[25].)

Influence of the Juvenile Hormone on Behaviour

In certain caterpillars, behaviour is different in the final stage before pupation than it is before moulting in the earlier larval stages. The wax moth *Galleria* spins a tough cocoon before pupation, a flimsy web before a larval moult[26]. The sphingid *Mimas tiliae* crawls down the tree-trunk to the soil before pupation, but rests on the foliage before each larval moult[27]. These differences result from the presence or absence of juvenile hormone and are attributed to a direct effect on the nervous system. But the possibility remains that they could result from a nervous feed-back effect from other organs, for example from the distended silk glands.

Gonadotrophic Effects of the Juvenile Hormone

Secretion of the juvenile hormone ceases in *Rhodnius* in the 5th-stage larva before metamorphosis, but begins again in the adult. The hormone is then necessary for yolk formation in the female and for the full activity of the accessory glands in the male, which serve to produce the spermatophores to enclose the sperm[2]. In *Hyalophora cecropia*, in which the eggs are developed during the pupal stage, very little juvenile hormone is secreted in the adult female, but large amounts are produced in the adult male[28], which secretes a succession of spermatophores.

The precise nature of the gonadotrophic effect of the juvenile hormone is uncertain. The position is complicated by the fact that in some insects the secretion from the neurosecretory cells in the brain seems to be more important than the juvenile hormone in ensuring yolk production. In the case of *Rhodnius*, G. C. Coles[29] concluded that the juvenile hormone acts on the fat body cells and serves to activate those components of the gene system which lead to the synthesis of the specific proteins that are discharged into the blood and are taken up by the acolytes and added to the yolk. In the male locust, *Schistocerca*, T. R. Odhiambo[30] considers that the juvenile hormone activates the many systems concerned in protein synthesis in the nuclei of the accessory glands.

Metabolic Action of the Juvenile Hormone

The juvenile hormone is often said to be a "metabolic hormone"—by which is meant that it sets going metabolic processes, either of synthesis or of combustion, whether or not

保幼激素对于胸腺的继续存在也具有类似的作用。正常情况下，在红猎蝽蜕变为成虫后的 24 小时内胸腺（蜕皮激素的来源）自溶。这个过程是新蜕变的成虫中的某些激素因子作用的结果。但是如果在蜕变前这一时期，胸腺暴露在保幼激素中，就不会发生这样的反应，腺体也不会自溶 [24]。（最近发现惜古比天蚕蛾变态后特定肌肉的分解也通过与之非常相似的方式被调节 [25]。）

保幼激素对行为的影响

在某些毛虫中，即将化蛹前的最后阶段的行为与早先幼虫阶段蜕皮前的行为不同。蜡螟在化蛹前会织一个坚韧的茧，但幼虫每次蜕皮前只是结一个轻薄的网 [26]。椴天蛾化蛹前会爬下树干到土壤中，但幼虫每次蜕皮前都栖息在树叶上 [27]。这些差异是由保幼激素的存在或缺乏造成的，并且是对神经系统直接作用的结果。但也有可能是由来自其他器官（比如，来自膨大的吐丝腺）的神经反馈调节导致的。

保幼激素的促性腺作用

变态前，在第五龄的红猎蝽幼虫中保幼激素的分泌停止，但在成虫时期又重新开始分泌保幼激素。这种激素对于雌性的卵黄形成和雄性副腺（负责产生包裹精子的精子包囊）的完全活力是必不可少的 [2]。对于惜古比天蚕蛾，卵在化蛹阶段发育，雌性成虫几乎不分泌保幼激素，但连续不断地分泌精子包囊的雄性成虫却产生大量保幼激素 [28]。

保幼激素促性腺作用的精确机制仍不确定。在卵黄形成的过程中，一些昆虫脑部神经分泌细胞的分泌物似乎比保幼激素更重要，这个事实使情况更加复杂。以红猎蝽为例，科勒斯 [29] 得出如下结论：保幼激素作用于脂肪体细胞并激活了那些基因系统组分，引发特定蛋白质的合成。蛋白质被释放到血液中，之后被卵母细胞吸收，加入卵黄。对于雄性沙漠蝗来说，奥德西亚姆波 [30] 认为保幼激素激活了许多与副腺细胞核中蛋白质合成相关的系统。

保幼激素的代谢活动

保幼激素经常被称为"代谢激素"——这意味着它调节合成代谢过程或者氧化代谢过程，无论怎样，这两个过程都是正常运转的身体所必需的。从这个意义上讲，

these are required for the working body. I am sceptical about the existence of metabolic hormones in this sense; I think they are unphysiological. I fancy that in most cases of this kind the hormone is setting in motion some physiological process, and the observed changes in metabolism are feed-back effects. (But, of course, abnormally large doses of hormones may have an abnormal pharmacological effect on metabolism which has little relation to their normal influence in the body.)

A well-known example of a metabolic effect of the juvenile hormone is the accumulation of reserves of fat and glycogen in the fat body of the grasshopper *Melanoplus* when the corpora allata are removed[31]. But Odhiambo[32] has shown that in the locust *Schistocerca* the juvenile hormone acts on the central nervous system and causes continuous activity. After allatectomy the insect continues to feed normally but it becomes inactive; consequently reserves pile up. The accumulation of reserves is not a feed-back effect from the gonads; the same effects are seen in males with or without their accessory glands.

In the blowfly *Calliphora*[33] and in the cockroach *Leucophaea*[34] the corpora allata are necessary to maintain the normal high level of oxygen consumption. But here again the juvenile hormone may be initiating some physiological activity (perhaps the synthesis of ovarial proteins) which demands increased oxygen consumption. In these cases it is not known whether nervous or muscular activity is changed. In the bug *Pyrrhocoris* the effect of the corpus allatum on metabolism is seen only if the gonads are present: it appears to be a feed-back effect from the activated ovaries which are demanding nutrients[35]. Likewise in *Rhodnius*, the accelerated rate of digestion in the presence of the corpus allatum is a feed-back effect from the developing ovaries[36].

In *Leptinotarsa* the adult female goes into diapause when the juvenile hormone is absent or the corpora allata are removed. The rate of metabolism falls to a very low level and egg development ceases. In this state the thoracic muscles and their mitochondria degenerate almost completely. When juvenile hormone is supplied everything is restored: the beetles become active, the ovaries develop, muscles and sarcosomes are fully regenerated[37]. The metabolic effects are profound—but just where the juvenile hormone is acting is not known.

It is, of course, self-evident that hormones can influence the body only by bringing about chemical changes in the cells. In this sense they are always "metabolic hormones". Within a few hours the moulting hormone restores nucleoprotein synthesis in dormant epidermal cells of *Rhodnius*[14]; the juvenile hormone seems specifically to induce the synthesis of yolk proteins in the fat body of *Rhodnius*[29]; as Gilbert[38] has recently shown, the juvenile hormone will induce the fat body of the cockroach *Leucophaea* to synthesize ovarian lipids. But these, and many similar effects, are elements in a pattern of development evoked by the hormone. They are not simply quantitative changes in metabolism unrelated to growth requirements—as is commonly implied by the expression "metabolic hormone".

我怀疑代谢激素的存在；我认为它们不符合生理规律。我想在大多数情况下这种激素在动态中调节一些生理活动，并且在代谢中观察到的变化是具有反馈效应的。（但毫无疑问的是，异常大剂量的激素可能会对代谢产生一种异常的药理效应影响，这在体内与它们的正常影响几乎没有关系。）

一个著名的关于保幼激素代谢效应的例子是，当切除黑蝗的咽侧体后，其脂肪体内的脂肪和糖原储备发生积聚[31]。但是奥德西亚姆波[32]指出沙漠蝗的保幼激素作用于中枢神经系统并产生持续活性。咽侧体切除后，昆虫能继续正常进食，但是它变得不活跃；因此储备发生积聚。这种储备的积聚并不是来自性腺的反馈效应；同样的效应也发生在带有或切除了副腺的雄性昆虫中。

对于丽蝇[33]和蜚蠊[34]，咽侧体是保持正常高水平的氧气消耗所必需的。但是这里再次指出保幼激素也可能启动某些要求增加氧气消耗的生理活性（也许是卵蛋白的合成）。在这些例子中还不清楚神经或肌肉活性是否改变。只有当性腺存在时红蝽咽侧体在代谢中的作用才能看得出来：这看上去是由于正需要营养的被激活的卵巢所产生的反馈效应[35]。类似地，对于红猎蝽，当咽侧体存在时消化速率的提高是发育中的卵巢的反馈效应[36]。

对于瘦跗叶甲属昆虫，当保幼激素缺乏或咽侧体被切除时雌性成虫出现滞育。代谢率降到一个非常低的水平，卵的发育也停止了。在这种情况下胸部肌肉及其线粒体几乎全部退化了。一旦提供保幼激素，一切又恢复了：甲虫又变得活跃，卵巢开始发育，肌肉和肌细胞线粒体充分再生[37]。保幼激素对代谢的影响是深远的——但是激素作用位点仍不清楚。

当然，很明显激素只是通过引起细胞内的化学变化来影响整体。在这个意义上，它们永远是"代谢激素"。蜕皮激素在几个小时内就恢复了红猎蝽休眠表皮细胞的核蛋白的合成[14]；保幼激素似乎能特异性地诱导红猎蝽脂肪体中卵黄蛋白的合成[29]；如吉尔伯特[38]最近发表的结果所示，保幼激素能诱导蜚蠊的脂肪体合成卵巢的脂质。但是这些以及很多相似的效应只是激素引发的一种发育模式中的部分。它们不是与生长需求无关的代谢过程中的简单量变——如同通常"代谢激素"所表达的那样。

Chemical Nature of the Juvenile Hormone

The large accumulation of juvenile hormone in the adult moth of *Hyalophora cecropia* provided a source of active extract. Ether extracts from the abdomen of the male moth gave an orange-coloured oil rich in juvenile hormone activity[28]. This material was utilized to develop methods of assay[3,22] and methods for concentrating the active principle by countercurrent separation[22]. These procedures were applied to extracts from many sources and have shown that material with roughly the same partition properties, and with juvenile hormone activity, is widely spread throughout the animal and plant kingdoms: in the tissues of invertebrates and vertebrates, in higher plants, some bacteria and yeasts.

The material extracted from two of these sources, namely from the excrement of the mealworm, *Tenebrio*, and from yeast, was examined chemically by Schmialek[39]; the active principle was isolated and shown to be a mixture of *trans-trans*-farnesol and its aldehyde farnesal. The question arises whether the natural juvenile hormone in the insect has any relation to farnesol or is of a totally different nature.

At the present time the answer to this question is not known. Farnesol will certainly reproduce all the morphogenetic and gonadotrophic effects of the juvenile hormone in *Rhodnius*[40]. It is particularly effective if its stability is increased by blocking the alcohol end of the molecule in the form of an ether (for example, farnesyl methyl ether) or as the farnesyl acetone[41].

In the form of the methyl ether, a 0.06 percent solution of the active *trans-trans* isomer is far more effective than the natural extract from *cecropia*. 1.2 μg will cause a partial retention of larval characters in *Rhodnius*—a dose of 5.2 μg/g of body-weight[23]. In *Antheraea*, 20,000 μg of the natural extract from *cecropia* will cause a partial retention of pupal characters—a does of 4,000 μg/g of body-weight[42].

By repeated partition with methyl alcohol, chromatographic separation on silicic acid columns, followed by the crystallization of impurities, a "non-crystalline fraction" has been isolated which will produce this same effect in *Antheraea* at dose of 5 μg, that is, 1 μg/g of body-weight[42]. Accepting these results at their face value this product is about five times as effective as farnesyl methyl ether in *Rhodnius*.

Further attempted purification by gas-liquid chromatography leads to heat destruction of the natural substance, but one fraction isolated in this way had an activity twelve times that of the "non-crystalline fraction"[42]. Meyer, Schneiderman and Gilbert[43] report similar results, showing pyrolytic breakdown of the material in the gas-liquid chromatography column, but with some highly active fractions. Röller, Bjerke and McShan[44] have isolated a well-defined active substance which is certainly not farnesol.

The significance of these results is uncertain. Assay methods for material that is exerting its action over a period of many days of development, and which is continuously being

462

保幼激素的化学属性

保幼激素在惜古比天蚕蛾成虫中的大量积聚为含激素活性的提取物提供了来源。雄性飞蛾腹部的乙醚提取物是橙色油状物，富有保幼激素活性[28]。这种物质可以用于发展化验方法[3,22]，也可以用于发展通过逆流分离浓缩活性物质的方法[22]。很多原始材料的提取物适用这些过程，并且显示出在动植物中广泛存在具有大致相同分离特性和保幼激素活性的材料：如在无脊椎动物和脊椎动物的组织中，在高等植物、部分细菌和部分酵母中。

施麦阿来克对从其中两种原始材料（即拟步行虫幼虫的排泄物和酵母）提取出的物质进行了化学检验[39]；分离出活性物质，结果显示是反–反–法呢醇及其乙醛法呢醛混合物。这就提出了一个问题，即昆虫中的天然保幼激素是否与法呢醇有关或是具有完全不同的天然性质。

目前这个问题的答案仍然未知。法呢醇确实可以在红猎蝽身上重现保幼激素的所有形态发生效应和促性腺效应[40]。如果通过阻断醇基从而提高它的稳定性将会特别有效，包括醚（比如法呢甲醚）形式分子或法呢丙酮形式分子[41]。

在甲醚形式中，0.06% 的活性反–反异构体溶液远比惜古比天蚕蛾的天然提取物有效。1.2 微克就会使长红锥蝽一定程度地保留幼虫特性——剂量为 5.2 微克 / 克体重[23]。对于柞蚕，20,000 微克来自惜古比天蚕蛾的天然提取物能使蛹的特性得到一定程度的保留——剂量为 4,000 微克 / 克体重[42]。

通过甲醇多次分离、硅酸柱色谱分离以及随后的杂质结晶，分离出来一种"非晶体馏分"，它可以在柞蚕中产生同样的作用，剂量为 5 微克，即 1 微克 / 克体重[42]。如果在它们的表面数值上接受这些结果，那么这个产物在红猎蝽中的效率大约是法呢甲醚的 5 倍。

利用气液相色谱纯化会导致这种天然物质热分解，但是用这种方法分离出的一种馏分的活性是"非晶体馏分"的 12 倍[42]。迈耶、施奈德曼和吉尔伯特[43]报道了相似的结果，显示出材料在气液相色谱柱中发生热分解，但是伴有更高活性的馏分。罗勒、毕尔克和麦克沙恩[44]已经分离出了一种明确的活性物质，确定其不是法呢醇。

这些结果的重要性还不能确定。材料能在很多天内不断发挥作用，并且在代谢过程中持续分解，其化验方法还没有非常精确的定义。（通过稍微地修改施用材料的

broken down in metabolism, have no very precise meaning. (By modifying slightly the means of administering the material I have recently repeated the results on *Rhodnius* as described here with half the earlier dose, that is, with about 2.5 μg of the active isomer of farnesyl methyl ether per gram of body-weight.) At the present time farnesol derivatives are the only compounds of known chemical composition with juvenile hormone activity. They were identified as the substance with juvenile hormone activity in the extracts of non-insect material. They have been found in the extracts of *cecropia* and other silkmoths, Until other known compounds have been isolated, I am inclined to hold to the provisional hypothesis that the active group in the juvenile hormone is indeed the triple isoprene unit of farnesol, and that this exists in the natural hormone in some form that has not yet been defined.

(**208**, 522-524; 1965)

V. Wigglesworth: C. B. E., F. R. S., Department of Zoology, University of Cambridge.

References:

1. Wigglesworth, V. B., *Quart. J. Micro. Sci.*, 77, 191 (1934).

2. Wigglesworth, V. B., *Quart. J. Micro. Sci.*, 79, 91 (1936).

3. Wigglesworth, V. B., *J. Insect Physiol.*, 2, 73 (1958).

4. Piepho, H., *Verh. Dtsch. Zoo. Ges.* (Wilhelmshaven), 62 (1951).

5. Schaller, F., *Bull. Soc. Zoo Fr.*, 77, 195 (1952).

6. Williams, C. M., *Biol. Bull., Woods Hole*, 121, 572 (1961).

7. Staal, G. B., *Entom. Bericht.*, 25, 34 (1965).

8. Wigglesworth, V. B., *The Physiology of Insect Metamorphosis* (Cambridge University Press, 1954).

9. Joly, P., *Insectes Sociaux*, 3, 17 (1956).

10. Lees, A. D., *Symp. Roy. Entomol. Soc. Lond.*, 1 (Insect Polymorphism), 68 (1961).

11. Kaiser, P., *Naturwiss.*, 42, 303 (1955).

12. Lüscher, M., *Naturwiss.*, 45, 69 (1958).

13. Wigglesworth, V. B., *J. Embryol. Exp. Morph.*, 1, 269 (1958).

14. Wigglesworth, V. B., *Symp. Soc. Exp. Biol.*, 11, 204 (1957).

15. Beermann, W., *Chromosoma*, 5, 138 (1952).

16. Jacob, J., and Monod, J., *Cold Spring Harbor Symp. Quant. Biol.*, 26, 193 (1961).

17. Wigglesworth, V. B., *Naturwiss.*, 27, 301 (1939).

18. Lawrence, P., thesis, Univ. Camb. (1965).

19. Ozeki, K., *Sci. Papers Coll. Gen. Educ. Univ. Tokyo*, 11, 102 (1961).

20. Piepho, H., *Naturwiss.*, 27, 301 (1939).

21. Piepho, H., and Meyer, H., *Biol. Zbl.*, 70, 252 (1951).

22. Gilbert, L. I., and Schneiderman, H. A., *Trans. Amer. Micro. Soc.*, 79, 38 (1960).

23. Wigglesworth, V. B., *J. Insect Physiol.*, 9, 105 (1963).

24. Wigglesworth, V. B., *J. Exp. Biol.*, 32, 485 (1955).

25. Lockshin, R. A., and Williams, C. M., *J. Insect Physiol.*, 10, 642 (1965).

26. Piepho, H., *Z. Tierpsychol.*, 7, 424 (1950).

27. Piepho, H., Böden, E., and Holz, I., *Z. Tierpsychol.*, 17, 261 (1960).

28. Williams, C. M., *Nature*, 178, 212 (1956).

29. Coles, G. C., *Nature*, 203, 323 (1964).

30. Odhiambo, T. R., thesis, Univ. Cambridge (1965).

方法，我最近重复了上面描述的红猎蝽的实验结果，所用剂量是早期的一半，也就是大约每克体重用 2.5 微克法呢甲醚的活性异构体。）目前法呢醇衍生物是具有保幼激素活性的化合物中唯一已知的化学组分。它们在非昆虫原料的提取物中作为具有保幼激素活性的物质被鉴别出来。在惜古比天蚕蛾和其他蚕蛾的提取物中也发现了它们。在分离出其他已知化合物之前，我倾向于坚持这个临时性的假说，即保幼激素中的活性基团实际上是法呢醇的三异戊二烯单元，并且以某种还没有被定义的形式存在于天然激素中。

（李响 翻译；崔巍 审稿）

31. Pfeiffer, L. W., *J. Exp. Zool.*, **99**, 183 (1945).

32. Odhiambo, T. R., *Nature.*, **207**, 1314 (1965).

33. Thomsen, E., *J. Exp. Biol.*, **26**, 137 (1949).

34. Sägesser, H., *J. Insect Physiol.*, **5**, 204 (1960).

35. Novák, V. J. A., Slama, K., and Wenig, K., *The Ontogeny of Insects*, Symposium in Prague 1959, 147 (1961).

36. Wigglesworth, V. B., *J. Exp. Biol.*, **25**, 1 (1948).

37. Stegwee, D., *J. Insect Physiol.*, **10**, 97 (1964).

38. Gilbert, L. I. (unpublished results).

39. Schmialek, P., *Z. Naturf.*, **16***b*, 461 (1961).

40. Wigglesworth, V. B., *J. Insect Physiol.*, 7, 73 (1961).

41. Schmialek, P., *Z. Naturf.*, **18**b, 516 (1963).

42. Williams, C. M., and Law, J. H., *J. Insect Physiol.*, **11**, 569 (1965).

43. Meyer, A. S., Schneiderman, H. A., and Gilbert, L. I., *Nature.*, **206**, 272 (1965).

44. Röller, H., Bjerke, J. S., and McShan, W. H., *J. Insect Physiol.*, **11**, 1185 (1965).

Propagation and Properties of Hepatitis Virus

Z. F. Ch. Kachani

Editor's Note

Here German researcher Zarah. F. Kachani argues that the hepatitis virus is made of DNA, rather than RNA. She manages to isolate the virus from patients with acute and chronic forms of the illness, normal donors, and from sewage and faeces. It is now known that several different hepatitis viruses can trigger the illness, and that they store their genetic information in different forms. Hepatitis C, for example, uses RNA, whereas hepatitis B uses DNA. It is also recognised that transmission routes vary: hepatitis B can be transferred via blood, while hepatitis A can be transferred via contaminated sewage and faeces.

TESTING a newly developed chemical reaction, we found the virus of hepatitis to consist of pure deoxyribonucleic acid without proteins, contrary to other known forms. This was investigated and confirmed as follows: by treatment with alcohol the virus can be precipitated like the nucleic acids and can be dissolved afterwards without loss of activity. With other virus such treatment led to total inactivation[1,2]. This property of the hepatitis virus enables one to prepare highly concentrated and pure samples, which do not contain proteins from tissues or other virus.

The hepatitis virus itself cannot permeate into the cell, the protein hull of virus being, as is well known, responsible for the affinity to the host cells. However, if the virus is brought into the cell by special treatment, then propagation and formation of plaques occur as with other viruses. The fact that, in such cases, the virus can be propagated in many tissue cultures confirms its character as a pure nucleic acid[3].

Enzymes that cleave nucleic acids degrade the virus, but trypsin, papain, Na-deoxycholate and ether are without any effect.

In vitro, the virus can be fixed by actinomycin like other deoxyribonucleic acids.

The following factors also indicate that the nucleic acid of the hepatitis virus is deoxyribonucleic acid (DNA): (1) Total inactivation by DNase, but not RNase; (2) 5-iodo-2′-deoxyuridine yielding an inhibition of propagation *in vivo*, while substances influencing the metabolism of RNA, such as 8-azaguanine, 2-thiouracil and 5-fluorouracil, have no influence; (3) actinomycin inactivates the virus *in vivo* as well as *in vitro*; it is known that actinomycin blocks DNA, but not RNA, *in vitro*; (4) estimation by the method of Schmidt and Thannhauser[5] confirms the DNA character of the virus.

肝炎病毒的繁殖及其特性

卡切尼

编者按

在这篇文章中，德国研究人员扎拉·卡切尼认为肝炎病毒是由 DNA 而不是由 RNA 组成的。她分别从患有急性疾病和慢性疾病的患者体内、正常人体内以及污水和粪便中分离出该病毒。人们现在已经知道几种不同的肝炎病毒都能够引发疾病，并且它们以不同的形式存储自身的遗传信息。例如，C 型肝炎病毒使用 RNA 而 B 型肝炎病毒使用 DNA。现在人们也已经认识到它们的传播途径不同：B 型肝炎通过血液传播，而 A 型肝炎通过受污染的污水及粪便传播。

在验证一个新发现的化学反应时，我们发现肝炎病毒仅由脱氧核糖核酸组成而不包含蛋白质，这和其他已知肝炎病毒的组成形式不同。通过如下实验对这种肝炎病毒进行了检验和证实：经过乙醇处理后的肝炎病毒可以像核酸一样发生沉淀，将沉淀重新溶解后病毒的活性依然存在。而其他病毒在经过这样的处理后会完全丧失活性 [1,2]。利用肝炎病毒的这种特性，可以制备出高浓度的不含任何组织蛋白质或其他病毒蛋白质的纯肝炎病毒。

众所周知，病毒一般通过衣壳蛋白黏附到宿主细胞上，因此，肝炎病毒本身不能侵入细胞。然而，如果经过特殊的处理将肝炎病毒引入到细胞中，它会像其他病毒一样繁殖并形成噬菌斑。在这种情况下，肝炎病毒能够在多种组织培养物中繁殖这一事实证实了其仅由核酸构成 [3]。

核酸剪切酶能够降解肝炎病毒，但是胰蛋白酶、木瓜蛋白酶、脱氧胆酸钠和乙醚对它则没有任何影响。

在体外实验中，肝炎病毒同其他脱氧核糖核酸一样能被放线菌素固定。

以下几个方面也表明肝炎病毒中的核酸是脱氧核糖核酸（DNA）：（1）肝炎病毒能被 DNA 酶完全灭活，却不被 RNA 酶灭活；（2）在体内实验中，5-碘-2′-脱氧尿苷能抑制肝炎病毒的繁殖，然而影响 RNA 代谢的物质，如 8-氮鸟嘌呤、2-硫尿嘧啶和 5-氟尿嘧啶对其繁殖则没有影响；（3）在体内实验和体外实验中，放线菌素都能使肝炎病毒失活，而一般认为放线菌素在体外实验中抑制了 DNA 而不是 RNA；（4）采用施密特和汤恩豪瑟 [5] 的方法进行的推测证实了肝炎病毒是 DNA 型病毒。

The hepatitis virus can be propagated on the chicken embryo by the following technique: (1) Treating the chorio-allantoic membrane with hyaluronidase just before inoculation; this loosens the tissue and thereafter the virus can passively permeate into the tissue; (2) the cytotoxic effect, especially from sera, can be removed by dilution, otherwise it prohibits multiplication of the virus; (3) the effect of DNA in cleaving enzymes such as DNase I and DNase of tissue origin will be greatly reduced by adding 0.01 moles Na-citrate and shifting the pH up to 8.5.

In tissue cultures the cells were washed before inoculation with 1 M NaCl, pH 8.2. The inoculation fluid contained hyaluronidase. Further propagation for many passages in tissue cultures is possible only if the pH of the medium does not go below 7, since pH 6.5 strongly decreases the active virus particles. The optimum pH for preservation of the hepatitis virus is between 8.5 and 9.5.

Plaque formation on the chorio-allantoic membrane appeared on the third day, and in liver and spleen on the fourth day after inoculation, the foci being similar to those of other virus, both morphologically and histologically[6]. In tissue cultures the beginning of the cytopathogenic effect was seen on days 3–4 after inoculation.

Electron microscopy shows the virus to have a polyhedral shape (Fig. 1). Negative staining with 1 percent uranyl acetate makes it possible to measure the diameter, which is between 40 and 150 mμ. To determine whether the smaller particles are parts of larger ones, or are the virus itself, we separated the particles by ultracentrifugation. These were then tested for infectivity and diameters measured. The results showed that the smallest particles represent the virus itself.

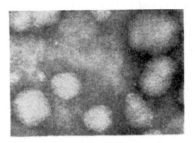

Fig. 1. Human hepatitis virus stained with uranyl acetate (× 100,000)

The virus was propagated from 144 cases of acute and chronic hepatitis epidemica, from serum hepatitis and from a series of normal blood donors. The virus was found in sewage as well as in blood and in faeces. All 144 virus stems tested were identical in electron micrographs and by immunological tests.

For preparation of immune serum from rabbits, the injection of the pure virus suspension with Freund's adjuvant is recommended together with simultaneous application of anabolic steroids. With immune sera prepared in this way 144 stems were tested by

通过如下方法可以在鸡胚中成功繁殖肝炎病毒：（1）接种前用透明质酸酶处理绒毛尿囊膜，这样可以使组织松弛，之后病毒被动进入组织中；（2）通过稀释能够消除细胞毒素特别是来自血清的毒素的影响，否则病毒的繁殖会受到抑制；（3）通过加入 0.01 摩尔的柠檬酸钠并将 pH 值调至 8.5，可以明显减少组织中的 DNA 酶 I 和 DNA 酶等剪切酶对 DNA 的降解作用。

接种前，用 pH 值为 8.2 的 1 摩尔 / 升 NaCl 溶液洗涤组织培养物中的细胞，接种液中要加入透明质酸酶。由于当 pH 值为 6.5 时病毒颗粒的活性会显著降低，因此只有当培养基的 pH 值不低于 7 时，在组织培养物中病毒才可能发生更多代繁殖。保存肝炎病毒的最适 pH 值介于 8.5~9.5 间。

接种后第 3 天，绒毛尿囊膜上形成噬菌斑，接种后第 4 天，在肝脏和脾脏上形成噬菌斑。从形态学和组织学来看 [6]，这些斑点与其他病毒的噬菌斑十分相似。接种 3~4 天后，组织培养物中就开始出现细胞病变了。

电子显微镜下观察发现肝炎病毒呈多面体形（图 1）。采用 1% 的醋酸铀负染法可以测算肝炎病毒的直径，其直径大小约在 40 纳米至 150 纳米之间。为了鉴别较小颗粒是较大的肝炎病毒的碎片还是肝炎病毒本身，我们用超速离心法分离不同尺寸的颗粒，然后检测了颗粒的感染性并测算了其直径，结果表明最小的颗粒就是病毒本身。

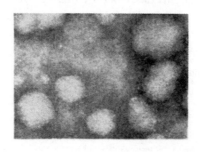

图 1. 醋酸铀染色的人肝炎病毒（放大 100,000 倍）

实验所用的病毒是由 144 个急性和慢性流行性肝炎病毒繁殖而来的，它们来自含肝炎病毒的血清以及一些正常的供血者。在污水、血液和粪便中也发现了肝炎病毒。电子显微镜观察和免疫学检测表明这 144 个肝炎病毒株属于同一家系。

在制备兔源免疫血清时，建议在注射含弗氏佐剂的纯病毒悬液的同时注射合成类固醇。用这种方法制备的免疫血清被用来进行豚鼠红细胞血凝反应和病毒中和反

neutralization and haemagglutination with erythrocytes of guinea-pig, and a complement fixation test. As has been stated, they all gave the same reaction.

Neutralizing and haemagglutination-inhibiting antibodies were found in patients after complete healing, while complement-fixing antibody was present as well as the virus only in chronically active cases. An efficient disinfectant is "Havisol" (Schulke and Mayr G.m.b.H., Hamburg), based on phenol. The 6 percent solution inactivates massive, protein-poor virus suspensions in 2 min. With more proteins present the virus was inactivated by "Havisol" and "Parmetol" in 2 percent solution after 15 min.

As expected, the virus was much more resistant to heat than other viruses, and heating to 75°C for 1 h had no effect at all. Heating above 170°C was necessary to inactivate pure virus suspension with 10 P. F. U., within 1 h. The same suspension was inactivated in the autoclave after 30 min at 2 atmospheres at 134°C. With protein and cations, present heating for 1 h at 195°C, or at two atmospheres, was required to kill the virus. The addition of 2 percent sodium carbonate led to the total loss of activity within 20 min at 100°C.

The hepatitis virus was more resistant to ultra-violet radiation. Irradiation of 0.5 ml. solutions in Petri dishes at a distance of 10 cm, using a mercury vapour lamp giving 2,540 Å emission, produced an area of inactivation of 8.5 cm diameter after 6 min.

Since it was not possible to inoculate man with the virus, we performed the classical experiment of Neefe et al.[7] on human volunteers, as with the propagated virus in chicken embryo. The results were exactly as described by these authors.

(**208**, 605-606; 1965)

Z. F. Ch. Kachani: Department of Colloid Chemistry, University of Kiel.

References:

1. Chargaff, E., and Davidson, J. N., in *The Nucleic Acids*, 1, 382 (Academic Press, New York, 1955).

2. Maassah, H. F., *J. Immunol.*, **90**, 265 (1963).

3. Herriot, R. M., *Science*, **134**, 256 (1961).

4. Tatum, E. L., *Proc. U.S. Nat. Acad. Sci.*, **48**, 1238 (1962).

5. Schmidt, G., and Thannhauser, S. J., *J. Biol. Chem.*, **161**, 83 (1945).

6. Kachani, Z. F. Ch., *Arch. Hyg. Bakt.*, **147**, 546 (1963).

7. Neefe, J. R., Stokes, J., Baty, J. B., and Reinhold, J. G., *J. Amer. Med. Assoc.*, **128**, 1076 (1945).

应，及补体结合实验以检测这 144 个病毒株。结果如前所述，它们的反应一致。

在完全康复的患者中发现了中和抗体和血凝抑制抗体，而补体结合抗体只和病毒共存于慢性肝炎患者体内。以苯酚为主要成分的"Havisol"（许尔克和马尔有限公司，汉堡）能有效地灭活病毒。用 6% 的"Havisol"溶液处理病毒悬液，2 分钟内大部分不含蛋白质的病毒都会失活。如果病毒中蛋白质含量较高，可以用 2% 的"Havisol"和"Parmetol"溶液作用 15 分钟使病毒失活。

同预期的一样，肝炎病毒比其他病毒更耐热，75℃ 加热 1 小时对它根本没有影响。要灭活 10 个蚀斑形成单位的纯病毒悬液，需要在高于 170℃ 的条件下加热 1 小时。该病毒悬液在 2 个大气压下 134℃ 加热 30 分钟后也能被灭活。如果病毒中含有蛋白质和阳离子，则要在 195℃ 加热 1 小时或者在 2 个大气压下加热才能灭活病毒。如果在病毒悬液中加入 2% 的碳酸钠，那么在 100℃ 的条件下，加热不到 20 分钟就可以使病毒完全失活了。

肝炎病毒抗紫外线辐射的能力也很强。用汞蒸气灯发射 2,540 埃的紫外线，照射距其 10 厘米的含有 0.5 毫升病毒溶液的培养皿，6 分钟后产生了一个直径为 8.5 厘米的失活区。

由于不能在人身上接种肝炎病毒，我们对志愿者采用了尼夫等 [7] 提出的经典实验方法，如同对鸡胚接种繁殖的肝炎病毒一样。我们得到的实验结果同这些作者在他们的论文中给出的结果完全一致。

（彭丽霞 翻译；金侠 审稿）

Three Haemoglobins K: Woolwich, an Abnormal, Cameroon and Ibadan, Two Unusual Variants of Human Haemoglobin A

N. Allan *et al.*

Editor's Note

Here, biologist Hermann Lehmann from Cambridge University and colleagues, continue their catalogue of haemoglobin variants, describing and characterising three types of haemoglobin K. Two, denoted Cameroon and Ibadan, are unusual variants of the more common haemoglobin A. The third, an abnormal haemoglobin denoted Woolwich, is of particular interest because it contains an amino-acid substitution at a position that had been thought immutable to change. The lysine group normally present in this position was thought to stabilize the molecule, and the team conclude that its replacement in haemoglobin K Woolwich may interfere with the molecule's "status quo".

HAEMOGLOBIN K was first described by Cabannes and Buhr[1] and is a "fast" variant of normal adult haemoglobin (haemoglobin A). It moves further towards the anode on electrophoresis at alkaline pH than haemoglobin A, but only just separates from it.

A haemoglobin K with its abnormality in the β-chain has been reported to occur in combination with haemoglobin S (ref. 2). The proportion of S : K was 3 : 2, and there was a mild haemoglobinopathy. The family in which this haemoglobin K was observed came from the West Indies and was of African ancestry. This haemoglobin K was first noted at Woolwich, England, and it will be denoted as K β Woolwich.

We have now found a second instance of the combination of a haemoglobin K with haemoglobin S. By hybridization with canine haemoglobin[3] this haemoglobin K could be shown to be a β-chain variant. Unlike haemoglobin K Woolwich, the proportion of this haemoglobin K to haemoglobin S was 3 : 1, similar to that of haemoglobin A in the sickle-cell trait (A + S). There was no haemoglobinopathy. Indeed, the carrier of this K + S combination was found in the course of routine screening of healthy blood donors at University College Hospital, Ibadan, Nigeria. The blood donor, a woman, came from Cameroon, and neither she herself nor her relatives could be examined further. This new haemoglobin K will be denoted as haemoglobin K β Cameroon (Figs. 1 and 2).

三种血红蛋白K：一种异常的伍力奇血红蛋白、两种人血红蛋白A的稀有变体——喀麦隆和伊巴丹血红蛋白

艾伦等

编者按

在本文中，来自剑桥大学的生物学家赫尔曼·莱曼及其同事们继续他们关于血红蛋白变体目录的研究，他们描述并定义了三种血红蛋白K。其中两种蛋白是常见的血红蛋白A的稀有变体，分别被命名为喀麦隆和伊巴丹。第三种是异常血红蛋白，被命名为伍力奇，它的分子中有一个原本被认为不可变更的氨基酸被取代，因此非常有趣。通常在这一位置上的赖氨酸残基被认为能够使蛋白分子稳定。研究组认为，血红蛋白K伍力奇中这一氨基酸的取代可能会影响蛋白分子的"现状"。

卡巴纳和布尔[1]首先对血红蛋白K进行了描述，它是正常成人血红蛋白（血红蛋白A）的一种"快速"变体。在碱性pH条件下，血红蛋白K在电泳中向正极移动的距离稍远于血红蛋白A，两者恰好能够区分开。

早先已经报道过β链异常的血红蛋白K与血红蛋白S组合存在的情况（参考文献2）。S与K的比例是3:2，伴有轻度血红蛋白病。携带此血红蛋白K的家庭来自西印度群岛，其祖先来自非洲。这种血红蛋白K首先在英国的伍力奇被发现，因此以Kβ伍力奇来表示。

现在我们发现了血红蛋白K与血红蛋白S组合存在的另一个实例。与犬血红蛋白的杂交表明，该血红蛋白K的β链发生了变异[3]。与血红蛋白K伍力奇不同，这种血红蛋白K与血红蛋白S的比例是3:1，这类似于镰刀形红细胞性状（A+S）中的血红蛋白A与S的比例，但不伴有血红蛋白病。实际上，尼日利亚伊巴丹大学学院附属医院在常规筛选健康供血者的过程中发现了这种血红蛋白K+S组合的携带者。供血者是一位来自喀麦隆的妇女，但是未能对她及其亲戚作进一步的检查。这种新的血红蛋白K以血红蛋白Kβ喀麦隆来表示（图1和图2）。

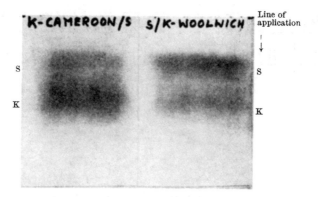

Fig. 1. Paper electrophoresis at pH 8.6 of haemoglobins K Cameroon + S (left) and S + K Woolwich (right).
The proportion of K Cameroon to S is that found for A in sickle-cell trait (see Fig. 2).

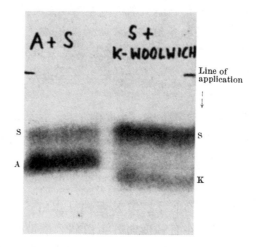

Fig. 2. Paper electrophoresis at pH 8.9 of haemoglobins from sickle-cell trait (A + S) and of haemoglobins
S + K Woolwich. The combination of haemoglobin S and K Woolwich shows a greater proportion of S
and results in a mild haemoglobinopathy.

A third sample of haemoglobin K, this time in combination with haemoglobin A, was
found in a Yoruba adolescent in Ibadan, in the course of an anthropological survey.
Hybridization with canine haemoglobin showed the variant to have its amino-acid
substitution in the β-chain. The proportion of A : K was 1 : 1, and there was no anaemia.
The body was repeatedly examined; but in the members of the family available for
examination no further instances of haemoglobin K were found. This haemoglobin K will
be denoted as haemoglobin K β Ibadan.

It has previously been discussed that some haemoglobin variants may be recognized as
"abnormal"when in combination with haemoglobin A they are found to form less than
half the total haemoglobin[4]. Others may be considered as merely unusual, and they can

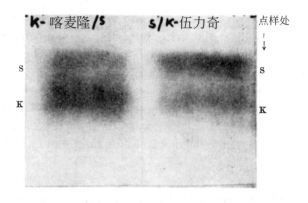

图 1. pH 8.6 条件下的血红蛋白 K 喀麦隆 + S(左) 和 S+K 伍力奇 (右) 的纸电泳。K 喀麦隆与 S 的比例与镰刀形红细胞性状血红蛋白 A 与 S 的比例一致（均为 3∶1）。(见图 2)

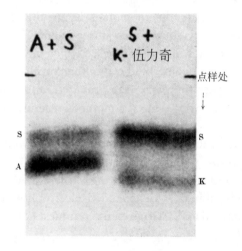

图 2. 来自镰刀形红细胞性状（A+S）的血红蛋白和来自血红蛋白 S+K 伍力奇的血红蛋白在 pH 8.9 条件下的纸电泳。血红蛋白 S 和血红蛋白 K 伍力奇的组合显示 S 所占比例较大，结果伴有轻度血红蛋白病。

　　第 3 例血红蛋白 K 是在伊巴丹进行人类学普查过程中在一个约鲁巴青年的血液中发现的，其血红蛋白 K 与血红蛋白 A 相结合。与犬血红蛋白的杂交表明，该变体的 β 链中发生了氨基酸替换。A 与 K 的比例是 1∶1，不伴有贫血症。反复检查了该患者，但在经检查的家庭成员中再没有发现血红蛋白 K 的例子。这种血红蛋白 K 被定名为血红蛋白 K β 伊巴丹。

　　以前曾经讨论过，当一些血红蛋白变体与血红蛋白 A 组合时，若血红蛋白变体占血红蛋白总量的一半以下时，即可被当作"异常"变体 [4]；但若它们与血红蛋白

be recognized by being found in equal proportion to haemoglobin A. The combination A + S (A > S) is an example of the first, abnormal type, and that for A + G Accra (A = G) represents the second unusual type. On two occasions a combination of unusual haemoglobins with haemoglobin S was observed. Haemoglobin S was present as the lesser fraction, as it is found in sickle-cell trait carriers with haemoglobin A (A > S). Haemoglobin J Baltimore[5] and haemoglobin D Ibadan[6] were the two unusual haemoglobins which in combination with S formed the greater part of the total pigment. In both instances the carriers of the haemoglobin mixtures had no sickle-cell disease but were sickle-cell trait carriers.

Of our three haemoglobins K, the Woolwich variant can be described as abnormal, and the combination haemoglobin K Woolwich and haemoglobin S (K < S) causes a mild haemoglobinopathy. Haemoglobin K Cameroon together with haemoglobin S results in the sickle-cell trait (K > S), and haemoglobins K Ibadan and A are found in equal proportions (K = A). The two latter haemoglobins K are therefore unusual rather than abnormal. It was of interest to investigate these three haemoglobins which are similar in their electrophoretic properties, and if possible to see which amino-acid substitution in the β-chain changed the physiological value of the resulting variant, and which did not. The haemoglobins were isolated, peptide maps were prepared, and when possible the variant peptides were separated and analysed using the methods recently listed in detail[6].

Haemoglobin K Woolwich. The peptide map of haemoglobin K Woolwich is shown in Fig. 3. It will be noted that the tyrosine peptide representing β^{A}TpXIII and the histidine peptide β^{A}TpXIV are missing, as well as a small spot near to β^{A}TpVI that is usually present in peptide maps of haemoglobin A. An extended electrophoretogram (Fig. 5) showed these differences more clearly. βTpXIII represents residues 121–132 of the β-chain and βTpXIV residues 133–144.

Peptide β^{A}TpXIII

121 122 123 124 125 126 127 128 129 130 131 132
Glu-Phe-Thr-Pro-Pro-Val-Gln-Ala-Ala-Tyr-Gln-Lys

↑

chymotryptic splitting

Peptide β^{A}TpXIV

133 134 135 136 137 138 139 140 141 142 143 144
Val-Val-Ala-Gly-Val-Ala-Asn-Ala-Leu-Ala-His-Lys

A 含量相等，则只被认为是稀有变体。A+S（A>S）组合是第一种异常类型的例子。而 A+G 阿克拉（A=G）代表第二种稀有类型。还观察到以下两种情况是稀有血红蛋白与血红蛋白 S 的组合。血红蛋白 S 所占比例较少，就像在镰刀形红细胞性状携带者的血红蛋白 A（A>S）组合中所发现的情况那样。血红蛋白 J 巴尔的摩 [5] 和血红蛋白 D 伊巴丹 [6] 是两种稀有血红蛋白，它们与血红蛋白 S 结合时构成总色素的大部分。在这两个例子中，含有组合血红蛋白的携带者没有镰刀形红细胞疾病，但他们是镰刀形红细胞性状的携带者。

在我们的 3 种血红蛋白 K 中，伍力奇变体可被描述为异常蛋白，血红蛋白 K 伍力奇和血红蛋白 S（K<S）的组合引起轻度血红蛋白病。血红蛋白 K 喀麦隆与血红蛋白 S 在一起就产生镰刀形红细胞性状（K>S），并发现血红蛋白 K 伊巴丹和血红蛋白 A 等量存在（K=A）。因此后两种血红蛋白 K 是稀有血红蛋白，而不是异常的血红蛋白。研究这 3 种血红蛋白是有趣的，因为它们的电泳性质相似。如果有可能，还应观察在 β 链中，哪个氨基酸的取代改变了所得变体的生理学功能，哪个取代不引起改变。于是我们分离了血红蛋白，进行了肽谱分析，并在可能的情况下对变异的肽链进行了分离和分析，采用的是最近详细阐述过的方法 [6]。

血红蛋白 K 伍力奇 图 3 显示了血红蛋白 K 伍力奇的肽谱，可以看到，代表 $\beta^ATpXIII$ 的酪氨酸肽和组氨酸肽 β^ATpXIV 缺失，在血红蛋白 A 肽谱中通常存在的靠近 β^ATpVI 的小点也消失。延长的电泳图（图 5）更加清楚地显示了这些差别。βTpXIII 代表了 β 链第 121~132 位残基，βTpXIV 代表了第 133~144 位残基。

$\beta^ATpXIII$ 肽

121 122 123 124 125 126 127 128 129 130 131 132
Glu-Phe-Thr-Pro-Pro-Val-Gln-Ala-Ala-Tyr-Gln-Lys

↑
胰凝乳蛋白酶裂解处

β^ATpXIV 肽

133 134 135 136 137 138 139 140 141 142 143 144
Val-Val-Ala-Gly-Val-Ala-Asn-Ala-Leu-Ala-His-Lys

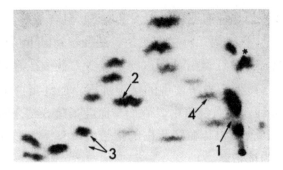

Fig. 3. Peptide map of haemoglobin K Woolwich. Note that at 1 β^ATpXIII (positive for tyrosine), and at 2 β^ATpXIV (positive for histidine) are missing. The incidental chymotryptic peptide β131-132 derived from β^ATpXIII is also missing at 3 (lower arrow) just below β^ATpVI 3 (top arrow). A new peptide which was found to represent β^ATpXIII-XIV can be seen at 4. The starred peptide is β^ATpV in its usual position—compare with Fig. 6.

A mutation affecting the mobility of both these tryptic peptides is likely to involve the lysine residue at position 132 because a change in the nature of this residue, apart from Lys→Arg, would result in a peptide bond at β132 which could not be broken by tryptic hydrolysis. In this case βTpXIII and βTpXIV would form one new peptide containing both tyrosine and histidine. Such a new peptide can be seen in Fig. 3, and it will be noted that it has moved somewhat towards the cathode, thus indicating a slight positive charge. This would eliminate the possibility of a Lys→Glu mutation which would have given rise to a negatively charged peptide. This mutation was already unlikely on the basis of the electrophoretic mobility of haemoglobin K Woolwich. The mutation Lys→Glu results in the acquisition of two negative charges per half molecule of haemoglobin and this would produce a variant with an electrophoretic mobility resembling that of haemoglobin I or haemoglobin N rather than that of haemoglobin K. The mutation Lys→Arg has already been excluded. It would not give rise to a charge change and would not resist the tryptic separation of the βXIII and βXIV peptides. The remaining single mutations permitted by the genetic code[7] are Lys→Gln, Lys→Asn, Lys→Thr and Lys→Met. The amino-acid analysis of the new peptide is given in Table 1. It will be seen that all the residues of βTpXIII and βTpXIV are present except for one lysine residue, and that one additional residue of glutamic acid was found. Glutamine is hydrolysed to glutamic acid when peptides are hydrolysed into their amino-acid constituents and the additional glutamic acid had to come from a glutamine, as the observed electrophoretic mobility of the new peptide demands. The lysine residue in position β132 of the β-chain had thus been substituted by one of glutamine. Haemoglobin K Woolwich may be described as $\alpha_2\beta_2^{132\text{Lys}\to\text{Gln}}$. This represents the first observation of a Lys→Gln mutation in human haemoglobin. The mutation also explains the absence of the small spot usually found near βTpVI. This spot represents a dipeptide β131-132 (Gln-Lys) resulting from an incidental chymotryptic splitting of the β130 tyrosyl peptide bond during tryptic hydrolysis. No such dipeptide could, of course, arise when β132 is Gln instead of Lys because the residues β131-132 would remain attached to βTpXIV.

480

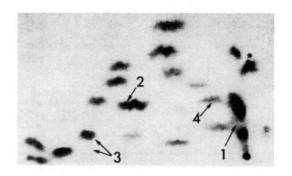

图 3. 血红蛋白 K 伍力奇的肽谱。注意 1 处的 β^ATpXIII（存在酪氨酸）和 2 处的 β^ATpXIV（存在组氨酸）缺失；来自 β^ATpXIII 的附带的胰凝乳蛋白酶裂解肽 β131-132 在 3 处（下箭头）也消失，其位置恰好在 β^ATpVI 3（上箭头）的下面。在 4 处可看到代表 β^ATpXIII-XIV 的一个新肽。带星号的肽是 β^ATpV，处于其正常位置，可与图 6 相比较。

　　影响这两个胰蛋白酶裂解肽迁移率的突变可能与第 132 位的赖氨酸残基有关，因为除去 Lys（赖氨酸）转变为 Arg（精氨酸）的情况外，该残基性质的变化都可能导致 β132 位形成的肽键不能被胰蛋白酶水解断裂。在这种情况下，βTpXIII 和 βTpXIV 可能形成含有酪氨酸和组氨酸的一个新肽。在图 3 中可以看到这个新肽，还可以注意到它有些向负极偏移，因而显示有轻微的正电荷变化。这就排除了产生 Lys → Glu（谷氨酸）突变的可能性，因为这种突变会产生一个带负电荷的肽。基于血红蛋白 K 伍力奇的电泳迁移率，这种突变本来就不可能发生。Lys → Glu 突变的结果使每半个血红蛋白分子得到 2 个负电荷，这会产生一个变体，其电泳迁移率类似于血红蛋白 I 或血红蛋白 N，而非血红蛋白 K。已经排除了 Lys → Arg 的突变，因为它不会产生电荷的变化，也不会抑制胰蛋白酶对 βXIII 和 βXIV 的裂解作用。遗传密码[7] 容许的其他点突变还有 Lys → Gln（谷氨酰胺）、Lys → Asn（天冬酰胺）、Lys → Thr（苏氨酸）和 Lys → Met（甲硫氨酸）。表 1 列出了新肽的氨基酸分析的结果，可以看出 βTpXIII 和 βTpXIV 中所有氨基酸残基都存在，只有一个赖氨酸除外，而且还发现了一个额外的谷氨酸残基。当肽被水解成其氨基酸组分时，谷氨酰胺被水解成谷氨酸。根据所观察到的新肽的电泳迁移率，额外的谷氨酸一定来自谷氨酰胺。因此，在 β 链 β132 位的赖氨酸残基必然被一个谷氨酰胺所取代。血红蛋白 K 伍力奇可以表示成 $\alpha_2\beta_2^{132Lys \to Gln}$。这是在人血红蛋白中首次观察到 Lys → Gln 突变。这种突变也可以解释通常在 βTpVI 附近存在的小点的缺失。这个点代表了一个二肽 β131-132（Gln-Lys），它是在胰蛋白酶水解作用期间 β130 酪氨酰肽键被胰凝乳蛋白酶附带裂解产生的。当 β132 是 Gln 而不是 Lys 时，由于 β131-132 残基可能仍然保持连接于 βTpXIV，那么就不会产生这种二肽。

Table 1. Haemoglobin K Woolwich

Amino-acid analysis (molar ratio) of the variant peptide $\beta^K Tp$ (XIII-XIV)

	Known values for haemoglobin A			haemoglobin K
	βTpXIII	βTpXIV	βTpXIII +βTpXIV	βTp(XIII-XIV)
Asp		1	1	0.9
Thr	1		1	1.2
Ser				0.3
Glu	3		3	3.8
Pro	2		2	2.1
Gly		1	1	1.1
Ala	2	4	6	5.8
Val	1	3	4	3.5
Leu		1	1	1.0
Tyr	1		1	0.7
Phe	1		1	0.8
Lys	1	1	2	1.0
His		1		0.9

Low recovery of valine is probably due to slight resistance of Val-Val bond to acid hydrolysis.

Haemoglobin K Cameroon. Unfortunately there was insufficient material available to make a complete structural investigation, and in spite of all efforts the donor could not be contacted again. The peptide map of this haemoglobin is shown in Fig. 4. It will be seen that the small spot usually found near βTpVI is missing and it is known that this spot represents the dipeptide β131-132 (Gln-Lys) arising from an incidental chymotryptic splitting of βTpXIII during tryptic digestion. A very small shift of βTpXIII towards the anode is also noticeable. This alteration in the electrophoretic mobility of βTpXIII was confirmed by an extended electrophoretogram (Fig. 5). The increase in negative charge of βTpXIII is compatible with a haemoglobin K which differs from haemoglobin A by an additional negative charge per half-molecule. This, in βTpXIII, could only be the substitution of a neutral amino-acid residue by Glu or Asp. The absence of the incidental chymotryptic peptide from βTpXIII suggests that there was an inhibition of the chymotryptic splitting of the β130 tyrosyl bond. Such an inhibition could be caused by a substitution of the neutral β129 Ala residue by an acidic residue. Thus it seems likely that in haemoglobin K Cameroon a Glu or an Asp has replaced the Ala at β130. Clearly these observations need substantiating when the opportunity arises.

482

表 1. 血红蛋白 K 伍力奇

变体肽 $\beta^K Tp(XIII\text{-}XIV)$ 的氨基酸分析（摩尔比）

	已知值的血红蛋白 A			血红蛋白 K
	βTpXIII	βTpXIV	βTpXIII +βTpXIV	βTp(XIII-XIV)
Asp		1	1	0.9
Thr	1		1	1.2
Ser				0.3
Glu	3		3	3.8
Pro	2		2	2.1
Gly		1	1	1.1
Ala	2	4	6	5.8
Val	1	3	4	3.5
Leu		1	1	1.0
Tyr	1		1	0.7
Phe	1		1	0.8
Lys	1	1	2	1.0
His		1		0.9

缬氨酸的回收率低可能是由于 Val（缬氨酸）–Val 肽键对酸水解有一定抗性。

血红蛋白 K 喀麦隆　遗憾的是没有足够的样品来进行完整的结构研究，尽管进行了不懈努力，但是仍然不能再联系到供血人。这种血红蛋白的肽谱见图 4。可以看到通常靠近 βTpVI 的小点缺失了。已知这个小点代表二肽 β131-132（Gln-Lys），它是胰蛋白酶消化时，胰凝乳蛋白酶附带裂解 βTpXIII 产生的。还可以注意到 βTpXIII 向正极有很小的偏移。延长的电泳图确认了 βTpXIII 电泳迁移率的这一变化（图 5）。βTpXIII 负电荷的增加与血红蛋白 K 一致，因为血红蛋白 K 与血红蛋白 A 的不同之处在于血红蛋白 K 每半个分子增加了一个负电荷。在 βTpXIII 中，这只可能是一个中性氨基酸残基被 Glu 或 Asp（天冬氨酸）所取代。来自 βTpXIII 的附带的胰凝乳蛋白酶裂解肽缺失，提示 β130 酪氨酰肽键被附带胰凝乳蛋白酶裂解的作用受到抑制。这种抑制作用的产生，可能是因为 β129 的中性残基 Ala（丙氨酸）被一个酸性氨基酸残基所取代。因此，很有可能是血红蛋白 K 喀麦隆的 β130 位上的 Ala 被 Glu 或 Asp 所取代。显然，如果有机会的话，这些观察结果还需要进一步确证。

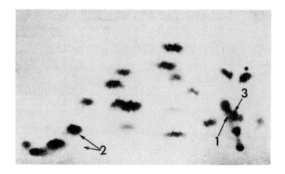

Fig. 4. Peptide map of haemoglobin K Cameroon. Note that at 1 β^ATpXIII (positive for tyrosine) is missing. The chymotryptic dipeptide β131-132 derived from β^ATpXIII is also missing at 2 (lower arrow), just below β^ATpVI 2 (top arrow). 3 indicates the new peptide β^KTpXIII. The starred peptide is β^ATpV in its usual position—compare with Fig. 6.

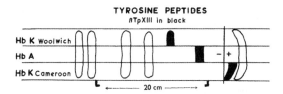

Fig. 5. Paper electrophoresis (90 min at pH 6.4) of tyrosine containing peptides from haemoglobin A compared with those from haemoglobins K Woolwich and K Cameroon. βTpXIII from haemoglobin K Woolwich is combined with βTpXIV and remains positively charged. βTpXIII from haemoglobin K Cameroon is negatively charged.

Haemoglobin K Ibadan. The peptide map of haemoglobin K Ibadan is shown in Fig. 6. It will be seen that the methionine peptide β^ATpV is missing, and a new methionine peptide has appeared moving further towards the anode with an electrophoretic mobility similar to that of βTpIII. This alteration in electrophoretic mobility of βTpV could be confirmed in an extended electrophoretogram (Fig. 7). βTpV represents residues 41-59 of the β-chain.

β^ATpV

41 42 43 44 45 46 47 48 49 50 51 52 53 54 55 56 57 58 59

Phe-Phe-Glu-Ser-Phe-Gly-Asp-Leu-Ser-Thr-Pro-Asp-Ala-Val-Met-Gly-Asn-Pro-Lys

↑

Splitting with cyanogen bromide

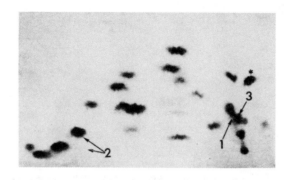

图 4. 血红蛋白 K 喀麦隆的肽谱。注意在 1 处的 β^ATpXIII(存在酪氨酸) 缺失。在 2 处（下箭头）来自 β^ATpXIII 的胰凝乳蛋白酶二肽 β131-132 也缺失。它正好位于 β^ATpVI 2 处（上箭头）的下方。3 表示新肽 β^KTpXIII。加星号的肽是 β^ATpV，处于它的正常位置，可与图 6 相比较。

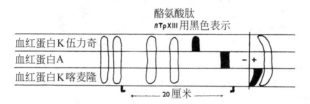

图 5. 来自血红蛋白 A 的含酪氨酸肽的纸电泳（pH 6.4，90 分钟），与来自血红蛋白 K 伍力奇和血红蛋白 K 喀麦隆的含酪氨酸纸电泳相比较。来自血红蛋白 K 伍力奇的 βTpXIII 与 βTpXIV 相组合，并保持带正电荷。来自血红蛋白 K 喀麦隆的 βTpXIII 带负电荷。

血红蛋白 K 伊巴丹　血红蛋白 K 伊巴丹的肽谱如图 6 所示，可以看到甲硫氨酸肽 β^ATpV 缺失，而一个新的甲硫氨酸肽出现，并进一步向正极移动，其电泳迁移率类似于 βTpIII。βTpV 电泳迁移率的这种变化可以在延长的电泳图中得到证实(图 7)。βTpV 代表着 β 链的 41~59 残基。

$$\beta^{A}TpV$$

41　42　43　44　45　46　47　48　49　50　51　52　53　54　55　56　57　58　59
Phe-Phe-Glu-Ser-Phe-Gly-Asp-Leu-Ser-Thr-Pro-Asp-Ala-Val-Met-Gly-Asn-Pro-Lys
↑
用溴化氰裂解

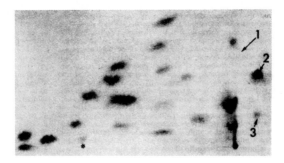

Fig. 6. Peptide map of haemoglobin K Ibadan. β^ATpV is missing at 1, and a new peptide can be seen at 2, with the electrophoretic mobility of β^ATpIII(3). The starred peptide is the incidental chymotryptic β131-132 from β^ATpXIII which is missing in Figs.3 and 4.

Fig. 7. Paper electrophoresis (90 min at pH 6.4) of methionine containing peptides from haemoglobin A and haemoglobin K Ibadan. Note the increase in mobility of the negatively charged βTpV from haemoglobin K Ibadan.

The amino-acid analysis of the new peptide, from haemoglobin A, is shown in Table 2. It will be seen that the new peptide contains one glycine residue less and one glutamic acid residue more than the haemoglobin A peptide. A mutation Gly→Glu would explain the increase in the negative charge of β^KTpV and would be compatible with the new variant being a haemoglobin K which has one negative charge per half-molecule more than haemoglobin A. Consistently low recoveries were obtained for phenylalanine. This can be seen also in a previous analysis of βTpV[8]. The Phe-Phe bond may be difficult to hydrolyse because of steric effects. Increasing the time of hydrolysis from 20 h to 40 h at 108° did not improve recovery.

Table 2. Haemoglobin A and Haemoglobin K Ibadan

Amino-acid analysis (molar ratio) of peptide βTpV

	β^ATpV	β^KTpV
Asp	3.0	2.9
Thr	1.0	1.0
Ser	2.0	1.8
Glu	1.0	2.0
Pro	2.0	1.9
Gly	1.8	1.0
Ala	1.2	1.0
Val	1.1	0.8
Met	0.8	Present*
Leu	1.1	1.1
Phe	2.4	2.5
Lys	1.2	1.1

* Methionine and sulphone together amounted to about 1 residue.

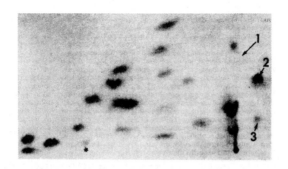

图 6. 血红蛋白 K 伊巴丹的肽谱。在 1 处的 β^ATpV 缺失，在 2 处可以看到一个新肽，它具有 β^ATpIII（3）的电泳迁移率。带星号的肽是来自 β^ATpIII 附带的胰凝乳蛋白酶解的二肽 β131-132，β^ATpXIII 在图 3 和图 4 中消失。

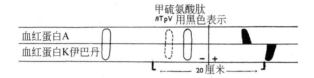

图 7. 来自血红蛋白 A 和血红蛋白 K 伊巴丹的含甲硫氨酸肽的纸电泳 (pH 6.4，90 分钟)。注意来自血红蛋白 K 伊巴丹的带负电荷的 βTpV 的迁移率增加。

    表2显示了来自血红蛋白A的新肽的氨基酸分析，可以看到，与血红蛋白A肽相比，这个新肽减少了一个甘氨酸残基，增加了一个谷氨酸残基。Gly（甘氨酸）→ Glu 的突变可以解释 β^KTpV 所带负电荷的增加，也可以与这种新变体是血红蛋白 K 的新变体相一致，因为与血红蛋白 A 相比，血红蛋白 K 每半个分子多含有一个负电荷。苯丙氨酸的回收率一直很低。这在以前对 βTpV 的分析 [8] 中也可以看到。Phe（苯丙氨酸）-Phe 肽键可能由于立体效应而难于水解。在 108°将水解时间从 20 小时增加到 40 小时也没能改善回收率。

表 2. 血红蛋白 A 和血红蛋白 K 伊巴丹

βTpV 肽的氨基酸分析（摩尔比）

	β^ATpV	β^KTpV
Asp	3.0	2.9
Thr	1.0	1.0
Ser	2.0	1.8
Glu	1.0	2.0
Pro	2.0	1.9
Gly	1.8	1.0
Ala	1.2	1.0
Val	1.1	0.8
Met	0.8	存在 *
Leu	1.1	1.1
Phe	2.4	2.5
Lys	1.2	1.1

* 甲硫氨酸和砜共约等同于 1 个残基。

487

$\beta^A TpV$ contains two glycine residues: $\beta46$ and $\beta56$. In order to determine which of the two possible glycine residues had been replaced by glutamyl, the methionyl bond at $\beta55$ was split with cyanogens bromide[9]. The resulting peptides were separated by paper electrophoresis at pH 6.4. $\beta^A TpV$ isolated from haemoglobin A was treated similarly. From both $\beta^K TpV$ and $\beta^A TpV$ two fragments each were obtained, one charged positively, and one charged negatively. The positively charged peptides obtained from $\beta^K TpV$ and $\beta^A TpV$ had the same electrophoretic mobility, whereas of the two negatively charged peptides, that from haemoglobin K had a greater mobility. If the amino-acid substitution Gly→Glu in haemoglobin K Ibadan had been at $\beta56$ no positively charged peptide would have resulted from the cleavage of the methionyl bond at $\beta55$. Both the positively charged peptides from A and K, presumably representing residues $\beta56$-59, stained transiently yellow with ninhydrin, indicating the N-terminal glycyl. The increased mobility of the negatively charged peptide from haemoglobin K Ibadan indicated that the mutation had occurred at position $\beta46$. The results of the amino-acid analysis of the four fragments are shown in Table 3. The basic amino-acids were not determined as this was not relevant to the position of the Gly→Glu mutation. It will be seen that the positively charged peptides (56–59) from $\beta^K TpV$ and $\beta^A TpV$ have an identical amino-acid composition. However, the negatively charged peptide representing residues 41–55 from $\beta^K TpV$ has one glutamic acid residue more and one glycine residue loss than the corresponding negatively charged peptide from $\beta^A TpV$. This indicates that of the two glycine residues 46 and 56 the first has been substituted by glutamyl and that the formula of haemoglobin K Ibadan is $\alpha_2\beta_2^{46Gly \rightarrow Glu}$. This is the first observation of a Gly→Glu mutation in human haemoglobin although a mutation Glu→Gly has been observed[10].

Table 3. Haemoglobin A and Haemoglobin K Ibadan

Amino-acid analysis (molar ratio) of peptides arising from splitting βTpV (β41-59) at β55
methionyl with cyanogen bromide

	Negatively charged peptides(β41-55)		Positively charged peptides (β56-59)	
	Hb A	Hb K	Hb A	Hb K
Electrophoretic mobility at pH 6.4 (Glu, −1.0, Lys, + 1.0)	−0.42	−0.59	+0.57	+0.57
Asp	2.0	2.2	1.0	1.0
Thr	1.0	1.0		
Ser	2.0	1.8		
Homoserine	Present	Present		
Glu	1.2	1.9		
Pro	1.0	1.1	1.0	0.9
Gly	1.0	0.3	0.9	1.0
Ala	1.1	1.0		
Val	1.1	1.0		
Leu	1.0	1.0		
Phe	2.4	2.3		

It has been suggested here that haemoglobin K Cameroon (mutation of β130 Ala to

β^ATpV 有 2 个甘氨酸残基：β46 和 β56。为了测定这两个可能的甘氨酸残基中哪一个被谷氨酰残基取代，将 β55 的甲硫氨酰键用溴化氰裂解 [9]。在 pH 值为 6.4 的条件下将所得的肽用纸电泳分离。从血红蛋白 A 分离得到的 β^ATpV 也用类似的方法处理。从 β^KTpV 和 β^ATpV 中各获得两个分别带正电荷和负电荷的片段。从 β^KTpV 和 β^ATpV 得到的正电荷肽段有相同的电泳迁移率，而在两条带负电荷的肽段中，来自血红蛋白 K 的肽段有较大的迁移率。如果血红蛋白 K 伊巴丹中的 Gly → Glu 取代发生在 β56 位，那么 β55 位的甲硫氨酰键的裂解就不会产生带正电荷的肽。血红蛋白 A 和 K 产生的 2 个带正电荷的肽，假定是 β56~59 的残基，茚三酮染色瞬间变黄，表明 N 端是甘氨酰。来自血红蛋白 K 伊巴丹的带负电荷的肽的迁移率有提高，这表明在 β46 位发生了突变作用。表 3 列出了这四个肽的氨基酸分析结果。由于与 Gly → Glu 的突变位置不相关，没有测定其碱性氨基酸。可以看到来自 β^KTpV 和 β^ATpV 的带正电荷的肽（56~59）有相同的氨基酸组分。但是代表 β^KTpV 的 41~55 位残基的带负电荷的肽比从 β^ATpV 中得到的相应的带负电荷的肽多一个谷氨酸残基，少一个甘氨酸残基。这就表明在 46 位和 56 位的两个甘氨酸残基中，前一个已被谷氨酰取代，因而血红蛋白 K 伊巴丹可表示为 $\alpha_2\beta_2^{46Gly \to Glu}$。这是第一次在人血红蛋白中观察到 Gly → Glu 的突变，虽然以前已经见到过 Glu → Gly 突变的发生 [10]。

表 3. 血红蛋白 A 和血红蛋白 K 伊巴丹

用溴化氰裂解 βTpV（β41~59）β55 甲硫酰胺处得到的肽的氨基酸分析（摩尔比）

	带负电荷的肽（β41~55）		带正电荷的肽（β56~59）	
	血红蛋白 A	血红蛋白 K	血红蛋白 A	血红蛋白 K
pH 6.4 电泳迁移率 (Glu, –1.0, Lys, +1.0)	–0.42	–0.59	+0.57	+0.57
Asp	2.0	2.2	1.0	1.0
Thr	1.0	1.0		
Ser	2.0	1.8		
Homoserine（高丝氨酸）	存在	存在		
Glu	1.2	1.9		
Pro	1.0	1.1	1.0	0.9
Gly	1.0	0.3	0.9	1.0
Ala	1.1	1.0		
Val	1.1	1.0		
Leu	1.0	1.0		
Phe	2.4	2.3		

这表明血红蛋白 K 喀麦隆（β130 Ala → Asp 或 Glu?）和血红蛋白 K 伊巴丹

Asp or Glu?) and haemoglobin K Ibadan ($\alpha_2\beta_2^{46\text{Glu}}$) are unusual haemoglobins whereas haemoglobin K Woolwich ($\alpha_2\beta_2^{132\text{Gln}}$) is abnormal. It is remarkable that the $\beta132$ lysine residue, which is substituted by one of glutamine in haemoglobin K Woolwich, is one of the most immutable residues in the evolution of haemoglobin. It is also found in myoglobin, and indeed no haemoglobin has as yet been described in which this $\beta132$ lysyl is not present. The lysyl in this position must obviously serve some purpose in stabilizing the molecule and its replacement by an amino-acid with a neutral side-chain may thus be expected to interfere with the *status quo* within the molecule.

(**208**, 658-661; 1965)

N. Allan: Subdepartment of Haematology, University College Hospital, Ibadan, Nigeria.

D. Beale, D. Irvine and H. Lehmann: Medical Research Council Abnormal Haemoglobin Research Unit, University Department of Biochemistry, Cambridge.

References:

1. Cabannes, R., and Buhr, L., *Pédiatrie*, **10**, 888 (1955).

2. O'Gorman, P., Allsopp, K. M., Lehmann, H., and Sukumaran, P. K., *Brit. Med. J.*, ii, 1381 (1963).

3. Huehns, E. R., Shooter, E. M., and Beaven, G. H., *J. Mol. Biol.*, **4**, 323 (1962).

4. Lehmann, H., Beale, D., and Boi Doku, F. S., *Nature*, **203**, 363 (1964).

5. Charache, S., and Conley, C. L., *Blood*, **24**, 25 (1964).

6. Watson-Williams, E. J., Beale, D., Irvine, D., and Lehmann, H., *Nature*, **205**, 1273 (1965).

7. Nirenberg, M., Leder, P., Bernfield, M., Brimacombe, R., Trupin, J., Rottman, F., and O'Neal, C., *Proc. U.S. Nat. Acad. Sci.*, **53**, 1161 (1965).

8. Bowman, B. H., Oliver, C. P., Barnett, D. R., Cunningham, J. E., and Schneider, R., *Blood*, **23**, 193 (1964).

9. Gross, E., and Witkop, B., *J. Biol. Chem.*, **237**, 1856 (1962).

10. Hill, R. L., Swenson, R. T., and Schwartz, H. O., *J. Biol. Chem.*, **235**, 3182 (1960).

($\alpha_2\beta_2{}^{46Glu}$）是稀有血红蛋白，而血红蛋白 K 伍力奇 ($\alpha_2\beta_2{}^{132Gln}$) 是异常血红蛋白。值得注意的是在血红蛋白 K 伍力奇中，被谷氨酰胺取代的 β132 赖氨酸残基是在血红蛋白演化过程中最稳定的残基之一。在肌红蛋白中也发现了这种情况，但在此之前却没有关于血红蛋白中不存在这个 β132 赖氨酰的记载。显然，这个位置上的赖氨酰必然对稳定分子起到某种作用。因此，当它被含有中性侧链的氨基酸取代后，可以预期蛋白质分子内部的"现状"会受到影响。

（荆玉祥 翻译；顾孝诚 审稿）

Drug-dependence

W. R. Brain*

Editor's Note

In this Inaugural Address to the School of Pharmacy, neurologist Walter Russell Brain discusses the issues surrounding drug-dependence and abuse. The increasing prescription of sedatives, stimulants and tranquilizers may, he argues, be cause for concern, but the beneficial effects cannot be ignored. Tranquilizers, for example, can help schizophrenics reintegrate with society and lessen the need for shock therapy. He argues for research into prescription and consumption patterns, with the hope that an integrated approach will influence future drug design. In the meantime, he adds, "there is still much to be said for the ideal of fighting our own battles if we can without the aid of pharmacy, but with the support of some philosophy which gives a meaning to life."

WHEN I chose the present subject, I hoped that before I presented it the new report of the Interdepartmental Committee on Drug Addiction would have been published. However, it is still in the printer's hands, so I can say nothing about that now. When the report appears, however, as I hope it will soon, I believe it will be self-explanatory and call for no comments from me. The report of the earlier Committee, of which also I was chairman, raised a general question which seems to me important enough to call for further discussion. I can best explain this by quoting from the report. We said: "Most of our witnesses affirmed that, today, drugs acting on the central nervous system are being used excessively, but they were unable to furnish records in support of this contention". We obtained some information about the quantity of barbiturates prescribed annually in England and Wales over the previous two years and said: "It is obvious that usage has expanded both progressively and substantially so that, in 1959, it was almost twice what it was in 1951". We also stated that analysis of National Health prescriptions showed that "barbiturates, other sedatives and hypnotics, together with analgesics and antipyretics (excluding dangerous drugs), account for no less than about 19 percent of all the prescriptions issued". There was also reason to believe that the prescription of tranquillizers had also increased, and we noted that the amount spent on one particular tranquillizer by nine selected mental hospitals had increased ten-fold over a course of five years. After mentioning the dangers of addiction to amphetamine and phenmetrazine, we noted that an analysis of some 214 million National Health Service prescriptions in 1959 indicated that some 5,600,000 or approximately 2.5 percent were for preparations of the amphetamines and phenmetrazines.

---

* Substance of the Address at the Inaugural Ceremony of the School of Pharmacy, University of London, delivered on October 13.

药物依赖

布雷恩 *

编者按

本文是神经学家沃尔特·罗素·布雷恩在药学院的就职演说，他围绕药物依赖和滥用问题进行了讨论。他认为越来越多的含有镇静剂、兴奋剂和安定药的处方应当引起关注，但是其益处不能被忽视。拿安定药来讲，它确实可以让精神分裂症患者重新回归社会，减少对休克疗法的需求。他主张对药物处方和消耗模式进行一些研究，希望拿出一个可以影响未来药物设计的综合方案。同时，他补充道："还有许多要说的是，我们的理想状态是在没有药物帮助的情况下，只是借助某种赋予生命一定意义的人生观，依靠身体自身与疾病做斗争。"

在我选择这个题目时，我希望在我演讲之前药物成瘾跨部门委员会的新报告就已经公布了。但是现在仍然在印刷中，因此关于那份报告我现在不能说任何事情。在这份报告公布后（正如我所希望的很快就会），我相信它会是不言自明的，不需要我进行评论。先前委员会（我也担任这个委员会的主席）的报告提出了一个普遍问题，这个问题对我来说很重要，需要进一步的讨论。引用报告的原文能更好地说明这个问题，即："我们大多数的见证人都承认，如今作用于中枢神经系统的药物正在被过量地使用，但是他们不能提供支持这一论点的证据。"我们获得了一些关于前两年英格兰和威尔士每年使用的巴比妥类药物数量的信息，并写道："很明显，药物用量已经逐步大幅增加，以至于 1959 年的用量几乎是 1951 年的两倍。"我们还陈述了对全民健康处方的分析，表明"巴比妥类药物、其他镇静剂和催眠剂以及止痛药和退烧药（不包括危险药物）在所有开出的处方药中不少于 19%"。还有理由相信开安定药的处方也增加了，我们注意到在调查的 9 所精神疾病医院消耗的某种安定剂的量在 5 年的时间里增加到原来的 10 倍。提到安非他明和苯甲吗啉易上瘾的危险，我们注意到对 1959 年全民健康服务的约有 2.14 亿份处方药的分析显示，大约有 560 万份或近似 2.5% 准备用安非他明和苯甲吗啉制剂。

---

* 本文是在伦敦大学药学院的就职仪式上的演讲内容，演讲致词是在 10 月 13 日。

Reviewing these facts, we made the following observations "To explain this trend in medication directed at the central nervous system, we have found no single answer. In part it must be due to the vigorous advertising of these drugs and their preparations by the pharmaceutical industry, both to the medical profession and to the public. To some extent the accelerated tempo and heightened anxieties of modern life have been held to blame; but this is an assertion based on assumption more than fact. Thirdly, and possibly of considerable consequence, there is the materialistic attitude adopted nowadays to therapeutics in general. This is one feature of an age which owes so much to science. For every deviation from health, great or small, a specific, chemical corrective is sought and, if possible, applied, and it is also widely believed that health may be positively enhanced by the use of drugs. When dealing with mental disease, psychotherapy may still be invoked. Often, however, a prescription is given for a drug when the patient's real need is a discussion of his psychological difficulties with the doctor. An obvious danger arises when the drugs so employed, far from being placebos, are undeniably potent, frequently toxic, and sometimes habit-forming as well. On the other hand, the newer drugs are proving of great value in psychiatry where they are to some extent replacing other methods of treatment. This increasing use of sedatives, stimulants and tranquillizers raises issues on which we do not as a Committee feel confident to pronounce. In particular, it is not for us to decide whether their occasional or even regular use is justified if it enables that person to lead a happier and more useful life. In any case, if resort to potentially habit-forming drugs is sometimes to be regarded as a symptom of psychological maladjustment, it should be treated as a symptom, and its cause sought, perhaps as much in social conditions as in the mind of the individual. These are questions which should be considered not only by doctors, but by all concerned with social welfare."

These are certainly difficult questions, and I do not propose to try to answer them all now. What I shall try to do instead is to clarify them, to look at their implications, and to suggest ways in which we may hope to contribute towards answering them. Basic to the whole problem are the meaning and implications of drug-dependence.

This term was introduced by the World Health Organization Expert Committee on Addiction-producing Drugs in their thirteenth report, published in 1964. In the fourth section, with the significant title "Terminology in Regard to Drug Abuse", it directs attention to some confusion which had arisen in the use of the term "drug addiction" and "drug habituation". Because, it said, "the list of drugs abused increased in number and diversity" it sought a term which could be applied to drug abuse generally, and since "the component in common appears to be dependence, whether psychic or physical or both", it adopted the term "drug dependence", which it defined as "a state arising from repeated administration of a drug, often on a periodic or continuous basis. ... Its characteristics will vary with the agent involved and this must be made clear by designating the particular type of drug-dependence in each specific case—for example, drug-dependence of morphine-type, of cocaine-type, of cannabis-type, of barbiturate-type, of amphetamine-type, etc.". In using the term "drug dependence", and relating it to the abuse of drugs, it may seem that the World Health Organization Committee was only emphasizing

鉴于以上这些事实，我们得到的结论如下："要解释针对中枢神经系统进行药物治疗这一趋势，我们发现答案并不是唯一的。部分原因一定是制药企业向从医人员和公众有力地宣传推广了这些药物及其制剂。在某种程度上也归咎于现代生活的快节奏和高度焦虑，但这个论断更多的是根据假设提出的，而不是根据事实。第三（这一点可能带来相当严重的后果），就是目前在治疗上人们总体上都持唯物主义的态度，这是科学时代的一个特征。对于每一次出现的健康问题，不管大小，如果可能的话，人们都会去寻找并服用某种特定的化学药物，人们也普遍相信药物能够让人变得更健康。对于精神疾病，也许仍然会采用心理治疗方法。但是当一名患者的真正需求是和医生讨论其心理问题时医生经常会给他开药。这样用药远比安慰剂有更明显的危险，毫无疑问用药是有效的，但通常具有一定毒性并且有时会成瘾。另一方面，较新的药物被证明在治疗精神疾病方面具有重要的价值，在某种程度上会取代其他的治疗方法。我们委员会不能够自信地宣布因镇静剂、兴奋剂和安定药的用量增加会产生问题，尤其当这些药物能够使得某些人过上更幸福并且更有益的生活时，不是我们来决定他们偶尔或者甚至定期地使用这些药物是否合理。不管怎样，如果依赖于有潜在成瘾性的药物有时被视为心理失调的症状，那么就应该视它为一种症状来治疗，而这种失调的原因可能与社会环境有关，也可能与个人心态有关。这些问题不仅仅需要由医生来考虑，还应当由所有关心社会福利的人来考虑。"

这些当然都是很难的问题，我现在不打算尝试回答所有的问题。反而我应该尽力去做的是讲清楚它们、认清它们的影响以及提出有助于我们回答这些问题的方法。所有问题的基础是药物依赖的含义及其影响。

这个术语是在 1964 年世界卫生组织药物依赖专家委员会公布的第 13 份报告中提出的。这份报告第四部分的标题很显眼，就是"与药物滥用相关的术语"，它将注意力转移到了在使用术语"药物成瘾"与"服药习惯化"所产生的混淆上。因为文中提到："滥用药物的数量和种类都有所增加"，我们需要寻找一个适用于通常的药物滥用的术语，而且由于"相同点似乎都是依赖性，无论是心理或生理上的或二者兼而有之"，因此它采用"药物依赖"这个术语，其定义是"重复使用一种药物，通常是周期性或者持续性地使用而产生的一种状态……其特征根据涉及药剂的不同而不同，而且必须指明每种情况下特定的药物依赖类型，比如吗啡类的药物依赖、可卡因类的药物依赖、大麻类的药物依赖、巴比妥类的药物依赖和安非他明类的药物依赖等等"。在"药物依赖"术语的使用和它与药物滥用的关系上，似乎世界卫生组织委员会只是强调了一些显而易见的问题。但我想说的是"药物滥用"这个术语本

the obvious; but my object is to suggest that the term "drug abuse" itself raises some important questions which are not only unanswered but which we have not at present the knowledge to answer.

At least one valuable thing has emerged from the philosophy of the present century—its emphasis on the importance of words. So let us begin by asking what the word "drug" means, and why it means what it does. *Webster's Dictionary* defines "drug" as "any substance used as a medicine, or in the composition of medicine, for internal or external use", and then it goes on, "whether or not a given substance should be included under the term drug depends upon the purpose for which it is sold (as regards the seller) or used (as regards the purchaser) ". So it would seem that the same substance may sometimes be a drug and sometimes not. When we get to the verb "to drug" a new meaning creeps in, for Webster gives three definitions: "(1) to affect or season with drugs or ingredients; especially to stupefy by a narcotic drug; (2) to tincture with something offensive or injurious; (3) to dose to excess with or as with drugs". So a sinister note has already appeared: though we begin with drugs as any ingredients of a medicine, the average person hearing that someone has been "drugged" would not imagine that he had been given penicillin. This is sometimes reflected in what patients will say to a doctor, for example: "I don't mind taking medicine, but I hope you won't put me on any drugs, doctor!" Or: "I hope that this is not a habit-forming drug". So far as the public Press is concerned, if you see the word "drugs" in a headline you may be fairly sure that what follows refers to some sinister aspect of pharmacy, either "drug-addiction" or the supposed danger of taking some particular drug or group of drugs. In the annual report of the Chief Medical Officer of the Ministry of Health, though there is a chapter called "Therapeutic Agents: Control and Toxicology", when reference is made to the possible dangers of therapeutic agents the word "drugs" appears, and the relevant committee is called not a Committee on the Safety of Therapeutic Agents, but a Committee on the Safety of Drugs.

This pejorative meaning of the word "drug" is relevant to this article, for the subject of drug-dependence is a highly emotional one. "Drug addiction" hits the headlines frequently in the Press, and is the subject of plays and broadcasts. As a social problem, it excites an interest out of all proportion to its magnitude. In Great Britain, the total number of addicts to heroin, cocaine and morphine is well short of a thousand, and we are fortunate compared with Canada, where there are at least four thousand, and the United State where there are a great many more. But the number of men and women who will die of lung cancer this year in Britain is 25,000. Most of these are men in middle life. Their sufferings, the distress of their families and the loss to the community are collectively enormously greater than the effects of what is normally called "drug addiction", yet these deaths excite no similar emotional interest, although they, too, are the indirect result of a form of drug addiction, addiction to cigarettes.

I am not a psychologist, and perhaps the psychologists can tell us why drug addicts to morphine, heroin and cocaine arouse such a disproportionate amount of interest; even without going very deeply into the matter, however, one may surmise that it is partly

身带来了一些不仅没有被解答的重要问题，并且以我们目前所知也没有办法解答。

本世纪的哲学至少带来了一件有价值的事情——就是强调词的重要性。因此让我们先看看"药物"这个词的含义，以及为什么它会有这种含义。《韦氏字典》对"药物"的定义是"作为药品或者药品成分的任何物质，内服或者外用"，接着它继续写道"一种特定的物质能否被包含在药物术语的范畴内取决于它售卖（与销售者有关）或者使用（与购买者有关）的目的"。因此有可能同一种物质有时候是药物而有时候不是。当我们开始使用其动词形式"给药"时，一个新的含义产生了，《韦氏字典》给出了三个定义："（1）用药物或者其成分影响或者调节，尤其是用麻醉药麻醉；（2）用刺激性或者损伤性的东西使略受影响；（3）服用过量的或等量的药物"。因此一个不好的注解已经出现了：尽管我们说药物是药品的任何成分，但是普通人在听到某人被"给药"后不会认为他是服用了青霉素。这有时候反映在病人对医生所说的话中，比如："医生，我不介意服药，但是请不要让我使用任何的药物！"或者："我希望这不是容易成瘾的药物"。就公共媒体而言，如果你在标题中见到了药物这个词，你可能会很确定之后的内容将是药物的一些不光彩的方面，或者是"药物成瘾"，或者是服用某个或某类药物可能具有的危险性。在卫生部首席医疗官的年度报告中尽管有一章称为"治疗药剂：管理和毒理学"，但每当提到治疗药剂可能带来危害时"药物"这个词就会出现，而且相关的委员会的名称也不是治疗药剂安全委员会，而是药物安全委员会。

"药物"这个词的贬义在本文中被提及，因为药物依赖这个话题总能调动起人们的情绪。"药物成瘾"经常见诸媒体的头条，并且是戏剧和广播节目的常见题材。作为一个社会问题，它能引起最广泛的关注。在英国，对海洛因、可卡因和吗啡成瘾的人的总数可能不到 1,000，加拿大至少有 4,000 人成瘾，而美国则拥有更多数量的瘾君子，相对来讲我们很幸运。但英国今年死于肺癌的总人数是 25,000，且大部分都是中年男性。他们遭受的痛苦、家人的悲伤和给社会带来的损失总体上要远远大于"药物成瘾"所带来的影响，但是这些死亡并没有引起类似的情绪上的关注。尽管肺癌也可以看作一种药物成瘾方式（香烟成瘾）的间接结果。

我不是心理学家，可能心理学家能够告诉我们为什么吗啡、海洛因和可卡因成瘾能够激发如此不成比例的关注。甚至在没有对这个问题进行非常深入的调查，我

because these drugs are potent, mysterious and potentially dangerous: and one of their most mysterious characteristics is their power of temporarily or permanently changing the personality. Add to that the association of "drug addiction" with crime and violence and we reach the conception of "potions ... drunk of Siren tears, distill'd from limbecks foul as hell within".

This is no exaggeration. The drug addict in this narrower sense is indeed a pathetic figure and a potentially dangerous one, too, if he or she becomes a source of infection to others. But, as the World Health Organization Committee recognizes, these drug addicts, though creating special problems, are only part of the broader spectrum of drug dependence.

It is obvious that there is nothing wrong with drug dependence as such. If a diabetic requires regular dosage of a drug to maintain his blood sugar at a normal level, or if a patient suffering from a collagen disease requires regular amounts of steroids, such patients are drug-dependent as already defined, their dependence being a physical one. But no one would describe this kind of drug-dependence as drug abuse: exactly the contrary. So it would seem that something more than mere drug-dependence is necessary to create drug-abuse, and if we look at the list of such drugs of both abuse and dependence which appears in the report of the World Health Organization Committee, I think the basis of the distinction becomes clear, for they are all drugs which are taken for their psychological effects. This brings us to the next stage of our enquiry, which is the crucial one. Is dependence on a drug which is taken for its psychological effects necessarily an abuse of that drug? Few people would, I suppose, quarrel with the view that this is true of morphine, heroin and cocaine, though someone recently wrote to the Press and complained that, since he found it necessary to take heroin, he did not see why his freedom to do so should be interfered with. But there is rather less unanimity about cannabis, and where do alcohol and tobacco stand? (I use the term tobacco here comprehensively to include whatever its pharmacologically potent constituents may be.)

Tobacco and alcohol may both be drugs of dependence. Their action is pharmacological, and in extreme cases the addict attempting to give them up suffers deprivation symptoms, which may be so severe that he is unable to do without them. The fact that they are both drugs which you prescribe and obtain for yourself, instead of through a doctor, and that there are social aspects of their use, is irrelevant to their pharmacological action. As in the case of other drugs of dependence, people vary very greatly in their liability to become addicted, and serious dependence is much more common in the case of tobacco than alcohol. I know of no evidence that tobacco-dependence has any bad psychological effects; but it undoubtedly may have bad physical effects, and it is here that the dependence becomes important, because it may make it extremely difficult for the addict to give up smoking. Alcohol-dependence, in its extreme forms, is recognized to be a manifestation of a psychological illness, and calls for treatment accordingly, but as in Britain that cannot be carried out without the willing co-operation of the patient, which is usually difficult to obtain, the serious alcohol addict is a difficult medical and social problem.

们就可以推测部分是因为这些药物都是药效强的、神秘的并且有潜在危害的。它们最神秘的特点之一是其能够暂时或者永久地改变人格的能力。再加上"药物成瘾"和暴力与犯罪的结合，我们就能理解这些描述："魔药——如喝了鲛人的泪珠，从心中地狱般的锅里蒸出来"。

这并非夸大其词。如果狭义上的药物成瘾者成为影响其他人的根源，他或她确实是一个悲情的角色，也是一个潜在的危险人物。但是世界卫生组织委员会承认这些药物成瘾者尽管带来了特殊的问题，但也仅仅是药物依赖这个大群体的一部分。

显然，药物依赖本身同样没有什么错。如果一个糖尿病患者需要常规剂量的药物来维持其血糖在正常的水平上，或者一位胶原病患者需要定期服用类固醇，这些患者属于已被定义过的药物依赖，他们的依赖是身体上的。但是没有人会将这种药物依赖描绘成药物滥用，事实上恰恰相反。因此似乎要成为药物滥用，那么仅仅有药物依赖是不够的。如果我们查看世界卫生组织委员会的报告中既是滥用药物又是依赖性药物的清单，我想它们的基本区别就显而易见了，即它们都是因为会产生心理作用才被服用的药物。这样就进入到我们调查的下一个很关键的阶段。因为心理作用而服用的药物的依赖一定是药物滥用吗？我想几乎没有人会否认在吗啡、海洛因和可卡因中的确是这样。尽管最近有人写信给出版社并抱怨他不明白为什么自己服用海洛因的自由应该受到干涉，因为他发现服用海洛因是必要的。但是关于大麻却很少有一致性意见，更别说酒精和烟草了。（我这里使用烟草这个词，无论其药理上的有效成分是什么。）

烟草和酒精可能都是依赖性的药物。它们具有药理学作用，并且在极个别情况下，想要戒掉的成瘾者都会受戒断症状的折磨，这些症状是如此严重以至于患者不得不继续服药。事实上一方面它们都是无须经过医生，自己就可以给自己开的药物；另一方面它们又具有社交方面的用途，而这与它们的药理学作用无关。和其他依赖性的药物一样，成瘾倾向的个体差别非常大，对烟草具有严重依赖的情况比酒精更常见。我知道没有证据表明对烟草的依赖有负面的心理影响，但是毫无疑问它可能对身体有不好的影响。这里的依赖性就很重要，因为这使得成瘾者极难戒烟。酒精依赖在极端情况下被认为是心理疾病的表现，并且需要相应的治疗，但是在英国这种治疗在没有病人的主动配合下是无法完成的，而这种配合常常很难获得，因此严重的酒精成瘾者对于医疗和社会都是一个很头疼的问题。

I have mentioned both tobacco and alcohol addiction not only because of their intrinsic importance, though that is considerable, but in order to bring out a point of more general relevance. The World Health Organization Committee points out that the characteristic feature of drug-dependence is the strong desire or need to continue taking the drug. Leaving on one side the chronic alcoholic, using that term to describe people whose consumption of alcohol leads to such psychological or physical ill-effects that they are in need of medical treatment, there are many others who habitually take alcohol in more moderate amounts and claim that they feel and are the better for it, and miss it, if for some reason they are unable to get it. They would agree with a patient of mine who once said: "Alcohol has been a very good friend to me". This is a point to which I shall return.

What are we to think of this steadily rising consumption of barbiturates and tranquillizers? Before we can answer this question, which is a very complex one, we need a great deal of information which is not at present available. As the first Drug Addiction Committee said in its report: "It is clear that there is scope and a need for operation research into the prescribing pattern in this country with particular reference to habit-forming drugs". We need to know first how many patients are receiving prescriptions for barbiturates, and what is the average amount prescribed for these patients per annum. Then we must know what proportion of them are receiving barbiturates for treatment of epilepsy. An epileptic patient on barbiturates is drug-dependent because he cannot give up the drugs without serious ill-effects, but this type of drug-dependence is a use and not an abuse of the drug. Having eliminated those prescriptions, the next thing we need to know is what proportion of the tablets prescribed the patient actually consumes. Doctors, I am sure, would over-estimate their patients' faith in their prescriptions if they supposed that the tablets they prescribe are regularly and faithfully consumed by every patient. So we need to know from an appropriate sample of patients what proportion of the barbiturates prescribed for them they actually take; and if they do not take them all, what do they do with the rest? This is a question with several practical implications. Are the superfluous tablets locked up, or thrown away, or left about, where perhaps a child can find them? Moreover, if the patient puts them safely away, does he himself know what they are when he turns them out again in a year's time? It has sometimes been suggested that the prescription of excessive amounts of barbiturates by some doctors may be a source of such drugs for the black market. In any event, I think it is clear that prescription figures are likely to be a very misleading guide to actual consumption.

Nevertheless, let us assume—and experience shows that this is a reasonable assumption—that there are patients who are drug-dependent on barbiturates, not in the sense that they are in a state of chronic barbiturate intoxication, but that they are more or less regular consumers of barbiturates, either as hypnotics or as general sedatives to blunt the sense of being stretched on "the rack of this tough life". If this is drug-dependence, is it a use or an abuse of the drug? In the passage I quoted from the first report of the Drug Addiction Committee, it will be remembered, occur the words "often, however, a prescription is given for a drug when the patient's real need is a discussion of his psychological difficulties with the doctor". No doubt, this is sometimes true, but we have to remember that the average

我提到的烟草和酒精成瘾不仅因为它们固有的重要性（虽然是值得考虑的），也是为了提出更具有普遍意义的一点。世界卫生组织委员会指出，药物依赖的典型特征是继续服药的强烈欲望或需求。暂且不说慢性嗜酒者，即那些由于消耗大量的酒精而导致心理或身体上的副作用以至于需要医学治疗的人，还有许多习惯于饮用大于适量酒精的人声称因为饮酒他们感觉良好，如果因为某种原因他们无法获得酒精的话也会非常想念。他们会同意我的一位患者曾经说过的："酒精已经是我的一个非常好的朋友。"这是我要说的一点。

我们对巴比妥类药物和安定药消耗量的稳定增长怎么看呢？在回答这个非常复杂的问题之前，我们需要很多现在尚未获得的信息。正如第一届药物成瘾委员会在报告里说的："很清楚的是，现在有对国内的开药模式尤其是针对成瘾性药物进行研究的机会和需求"。我们首先需要知道有多少患者正在使用巴比妥类药物，以及每年为这些患者开药的平均量是多少。然后我们必须知道有多大比例的人使用巴比妥类药物是为了治疗癫痫。癫痫患者对巴比妥类药物是具有依赖性的，因为他们一停药就会有严重的后果，但是这种形式的药物依赖是正常的使用而不是滥用。在排除了这些处方之后，下一步我们需要知道的就是在开出的药物中有多大的比例真正被患者用掉了。如果医生们认为每一位患者都会定期如实地服用开出的药物，我可以肯定地说他们高估了患者对他们所开处方的信心。因此我们需要在合适的患者样本中调查开出的巴比妥类药物中到底有多大比例真正被患者使用，以及如果他们没有全部按照处方用掉，剩下的如何处置？这是一个具有许多实际意义的问题。这些剩余的药物是被储存起来、丢弃还是放在儿童可及的地方？此外，如果患者将其安全地储存起来，那一年之后再翻出来时他自己是否还能记得这是什么药？有时候人们就认为一些医生开出的过量巴比妥类药物可能是黑市中这类药物的来源。不管怎样，我想显然开药的数目误导了实际消耗的药量。

不过，让我们假设一下——经验告诉我们这是一个合理的假设——那些对巴比妥类药物有依赖的患者（并非处于慢性巴比妥类药物中毒的状态中，仅仅是巴比妥类药物的或多或少的定期服用者），用来作为催眠剂或者通用镇静剂以麻木他们"在艰苦生活拷问台上"被拉长的感觉。如果这也是药物依赖，那么到底是正常使用还是药物滥用？在我引用的药物成瘾委员会第一份报告的一段中，大家会记起的一句话就是"然而，当一名患者的真正需求是和医生讨论其心理问题时，医生经常会给他开药"。毫无疑问有些时候确实是这样的，但是我们要记住，当今普通的医生太忙

doctor today is much too busy to be able to spare the time for that kind of psychotherapy except on rare occasions, and though it would undoubtedly sometimes be helpful, many people would doubt whether psychotherapy, even if it were available, is the most suitable form of treatment for many such patients. Underlying a critical attitude to the regular prescription of barbiturates or other sedatives for the purpose I have mentioned, I think I can detect a hidden assumption. Let us look for a moment at another instance, namely, the administration of tranquillizers, which amounts to drug-dependence, but is nevertheless universally regarded as a use and not an abuse: I refer to the modern pharmacological treatment of schizophrenia. The regular administration of chlorpromazine and similar drugs to schizophrenics has enabled many of them to leave a mental hospital and live in the community so long as they take the drug regularly. Comparable beneficial results have been achieved by the treatment of patients suffering from depression by the antidepressant drugs. The authors of a recent monograph on the subject summarized the results of these forms of treatment as follows: "In general, these psychopharmacological agents have facilitated the control of mood and behavioural disturbances. In the mental hospital setting, they have induced a more quiet and more orderly atmosphere. They have enabled the discharge of an increasing number of patients and the return of many, previously resistant to treatment, to their appropriate places in society. In addition, they have facilitated ambulatory and outpatient treatment and have lessened the need for shock therapy"[1]. Here, then, is a beneficent form of drug-dependence in the case of patients suffering from major psychological disorders. Why, then, or in what circumstances, should we question the use of similar drugs in the doctor's consulting room for the treatment of patients suffering from minor psychological disorders? Is it perhaps because we are inclined to draw a line between gross psychiatric illnesses, for which we feel the patient cannot in any way be held responsible, and which are due to the operation of unknown physical factors, which may turn out to be biochemical, and the minor psychological disorders for which we do not postulate any physical cause but tend to attribute to psychological causes, and to regard as reactions to stresses and strains to which, perhaps we think, the fortunate majority of us are too tough to succumb. We recognize that it is of no use to admonish the schizophrenic as a general rule, but would not many of our neurotic patients, we wonder, be better as a result of some effort on their own part? This, of course, raises a difficult philosophical question involving the relationship between the mind and the brain. I shall not discuss this further now, but I doubt whether it is philosophically sound to make any such distinction between the major and the minor psychological disorders. We cannot interpret schizophrenia or depression purely in psychological terms, nor, for that matter, can we at present give a physiological explanation for them. It often appears easier to give a psychological explanation of the minor neurotic disorders from which many people suffer; but that does not exclude the probability that they, too have physiological explanations which we shall one day discover. The mind and body constitute a complex unity, and when we are dealing with psychological disorders, we always need to take account of both psychological and physiological interpretations so far as we can achieve them. It may well be that for the kind of reasons I have already mentioned some patients receive prescriptions for sedatives indiscriminately; but that is a long way from saying their continuing use involves a form of drug-dependence which is an abuse. I quote once more

以至于不能抽出时间进行这种心理治疗，除了极少数的情况。尽管毫无疑问心理治疗有时候是有帮助的，但是很多人会怀疑这些心理治疗（即使可以提供的话）是不是治疗多数的这种患者的最合适的方式。在我上文提到的目的而定期开巴比妥类药物或者其他镇静剂的批判态度之下，我想我又发现了一种隐藏的假设。让我们来看另一个例子，即镇静剂的使用达到了依赖性的程度但是仍然被普遍认为是正常使用而不是滥用，我指的是精神分裂症的现代药物治疗。定期地服用氯丙嗪和类似的药物治疗精神分裂症已经使得患者离开了精神病医院并重新回到社会，只要他们能够定期地服用这些药物。用抗抑郁药来治疗受抑郁症折磨的患者也取得了类似的好结果。最近在有关这个主题的专著中作者们总结了这些方式的治疗效果："总体来说，这些治疗心理疾病的药物有助于控制情绪和行为紊乱。在精神病医院，它们创造了更加安静、更加有序的环境。它们使得出院患者的数量越来越多，并使许多先前对治疗很抵抗的患者回归到了社会中合适的位置。此外，它们也促进了不卧床和门诊患者的治疗并减少了对休克治疗的需求。"[1] 这就是药物依赖对患有重度心理紊乱的患者的有益方面。那么为什么或者我们在何种情况下才应该质疑医生在诊察室里开出类似的药物来治疗有轻度的心理疾病的患者呢？可能是因为我们倾向于对所有的心理疾病进行分类（我们无论如何也不应该让患者为此负责），哪些是由于未知的身体因素产生的（证明可能是生化方面的），对于轻度的心理紊乱我们不会假定任何身体因素而将其归因于心理原因，可以看作对压力和紧张的反应，可能我们认为很幸运我们大部分人足够坚强从而未被压垮。我们认识到一般来说劝诫精神分裂症患者是没有用处的，但是我们想知道神经官能症患者在自己的一些努力下能否好转？当然这提出了一个涉及思维和脑之间关系的困难的哲学问题。我现在不再深入谈这个问题，但是我怀疑在重度和轻度的心理紊乱之间所做的任何区分在哲学上是否合理。我们不能单纯用心理学术语去解释精神分裂症或者抑郁症，而且我们目前也不能够给出生理学方面的解释。对很多人患有的轻度神经官能症给出心理学解释常常比较简单，但是这不能排除某一天我们可能会发现它们也有生理学解释的可能性。思维和躯体形成一个复杂的整体，当我们处理心理紊乱时，我们常常需要尽可能地把目前已知的心理和生理两方面的解释考虑进来。恰恰是由于我已经提到的这类原因，患者会不加选择地接受镇静剂的处方。但是还不能说他们持续服用这些药物是药物滥用。我再一次引用第一届药物成瘾委员会的报告："当药物能够让他们过上更幸福更有益的生活时，不是我们来决定他们偶尔或者定期地服用这些药物是否合理"。如果你是把酒精当成朋友的人，你当然认同在适宜的病例中定期服用镇静剂。

from the report of the first Drug Addiction Committee: "it is not for us to decide whether their occasional or even regular use is justified if it enables a person to lead a happier and more useful life". If you are among those who regard alcohol as a friend, you may also think the same of the regular use of sedatives in suitable cases.

But, in my view, the real limitation to the best possible use of such drugs springs from our ignorance. Let me end by looking into the future. I foresee a day when we shall understand much more than we do now about the relationship between the brain and the mind. On one hand we shall have methods of making accurate psychological assessments of the personality, very possibly in terms of factors which find no place in our present psychological vocabulary. We shall then be able to interpret psychological disorders in terms of these functions, their mutual interplay, and their reactions to our experiences. Parallel with these developments, psychopharmacology will have been placed on a rational instead of an empirical basis. We shall think in terms of the normal biochemistry of nerve cells and synapses, their groupings and interactions, their disorders and the effect of drugs on them. As a result, correlating in this way psychology, physiology, biochemistry and pharmacology, we shall have not only a more comprehensive armamentarium of drugs, but much more precision and individualization in their use. All we shall need then will be enough doctors with time to use them. But it would be wrong to leave the matter there. If mind and body are a unity we must not concentrate on the physical treatment of stress to the exclusion of its psychological aspects. Here, however, we need to look beyond individual psychotherapy and take a broader view. As the first Drug Addiction Committee said in its report, the cause of psychological maladjustment should be sought perhaps as much in social conditions as in the mind of the individual. So far as we know, life has always been stressful, and looking back over history and prehistory, it becomes clear that, although we have our own peculiar stresses, we have eliminated a great many which our ancestors had to put up with. Indeed, stress seems inherent in the evolutionary principle. There will always be some who find life too much for them, and they will continue to need help from pharmacology and supportive psychotherapy. We are all aware of the challenge to international and social organization which some of our major current stresses present. Beyond that, there is still much to be said for the ideal of fighting our own battles if we can without the aid of pharmacy, but with the support of some philosophy which gives a meaning to life, and from which we can draw strength and support.

(**208**, 825-827; 1965)

Reference:

1. Benson, W. M., and Schicle, B. C., *Tranquillising and Antidepressive Drugs* (C. C. Thomas, Springfield, Illinois, 1962).

　　但是，在我看来，对这些药物可能的最佳用途的真正限制源自我们的无知。我想通过展望未来来结束这次演讲。我预知有一天我们能够比现在更加了解大脑和思维之间的关系。一方面我们会有能够对人格做出准确的心理学评估的方法，很有可能用一些在我们现在的心理学词汇中还没有的因素。然后我们能够根据这些功能、它们之间的相互作用和它们对于我们体验的反应来解释心理紊乱。与这些进展同步的是，精神药理学将建立在理性基础上而不是经验主义基础上。我们应该从神经细胞和突触的正常生物化学、它们的分类和联系、它们的紊乱和药物对它们的作用方面进行考虑。结果，以这种方式将心理学、生理学、生物化学和药理学结合起来，我们不仅能够有更加充足的药物，而且在使用这些药物时可以更加精确和个性化。那时我们所需要的就是足够的医生有时间来使用它们。但是仅仅做到这些是不对的。如果思维和躯体是一个整体，我们一定不能仅仅将压力的治疗局限于身体层面上，而忽略了心理方面。然而，此时我们需要一个更广阔的视角来看问题，而不只局限于个体的心理治疗。正如第一届药物成瘾委员会在其报告中所说，精神失调的原因可能与社会环境有关，也可能与个人心态有关。就我们目前所知，生活总是充满压力的。回望历史以及史前时期，显而易见，虽然我们有我们自己特有的压力，但我们已经免去了太多先辈们不得不承受的压力。的确在进化原则中压力可能是与生俱来的。有些人经常觉得生活让他们难以承受，他们需要持续的药物和支持性心理治疗的帮助。我们都意识到自身巨大的压力给国际和社会组织所带来的挑战。此外，还有很多要说的是，我们的理想状态是在没有药物帮助的情况下，只是借助某种赋予生命一定意义的人生观，依靠身体自身与疾病做斗争，我们能从这种人生观中获得力量和支撑。

（毛晨晖 翻译；张旭 审稿）

A New Model for Virus Ribonucleic Acid Replication

F. Brown and S. J. Martin

Editor's Note

How do viruses replicate? Many have genes encoded not in DNA but in RNA, and in 1963 Luc Montagnier and F. K. Sanders proposed that single-stranded viral RNA could become a replicating double-stranded form inside host cells. Here researchers at the UK's Animal Virus Research Institute propose a more complicated replication strategy that involves circular forms of single- and double-stranded RNA, inspired by the way some bacterial viruses (phage) replicate their DNA. It is now known that RNA viruses employ several different strategies for replication—some make their own replication enzymes, others use the RNA to imprint their genes in the DNA of the host.

SUBSTANTIAL evidence now exists to support the initial observations of Montagnier and Sanders[1] that the multiplication of virus RNA proceeds via a replicating form. This form is readily differentiated from the RNA which is present in the complete virus particles by its considerably slower rate of sedimentation in sucrose, lower buoyant density in caesium sulphate and resistance to degradation by ribonuclease. It has been postulated that this replicating form permits the preferential synthesis of new virus RNA (corresponding to the plus strand) on a primer (minus strand) which is initially coded for by the ingoing virus RNA. This replicating form is envisaged as containing a double-stranded region as well as a number of single-stranded tails, corresponding to partially formed plus strands (Fig. 1a)[2-4]. Such a molecule would possess a lower buoyant density than single-stranded RNA and it has been assumed that ribonuclease will remove the single-stranded tails, leaving only a double-stranded molecule (Fig. 1b).

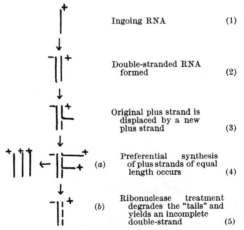

Ingoing RNA (1)

Double-stranded RNA formed (2)

Original plus strand is displaced by a new plus strand (3)

Preferential synthesis of plus strands of equal length occurs (4)

Ribonuclease treatment degrades the "tails" and yields an incomplete double-strand (5)

Fig. 1. Diagrammatic representation of the replicating mechanism of virus RNA which is at present accepted.

病毒RNA复制的新模式

布朗，马丁

编者按

病毒如何复制？很多病毒的基因并不是编码在 DNA 上而是编码在 RNA 上，1963 年吕克·蒙塔尼耶和桑德斯提出，在宿主细胞中的单链病毒 RNA 能够变为正在复制中的双链形式。本文中，来自英国动物病毒研究所的研究人员在一些细菌病毒 (噬菌体) 的 DNA 复制方式的启发下，提出了一种更为复杂的、涉及单链及双链 RNA 环状结构的复制策略。现在我们已经知道，RNA 病毒使用多种不同的复制策略——有些病毒利用自己的复制酶来进行复制，其他病毒则是利用 RNA 把它们的基因编码在宿主的 DNA 上。

现在存在实证支持蒙塔尼耶和桑德斯 [1] 最初观察到的病毒 RNA 繁殖经历的一种复制模式。通过其在蔗糖中相对较慢的沉降速度、在硫酸铯中较低的浮力密度以及对核糖核酸酶降解作用的耐受，可以很容易将这种形式的 RNA 和存在于完整病毒颗粒中的 RNA 区分开来。据推测，在这种复制模式中，侵入宿主的病毒首先以其 RNA 为模板编码产生引物链 (负链)，然后再根据引物链选择合成新的病毒 RNA (与正链一致)。估计这一复制模式中会出现一段双链区域，以及很多相当于正链片段的单链尾巴 (图 1a) [2-4]。这种 RNA 复合体分子的浮力密度比单链 RNA 分子的浮力密度更低，目前认为核糖核酸酶会降解掉单链尾巴，仅留下一个双链分子 (图 1b)。

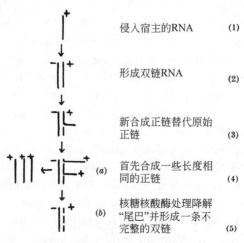

侵入宿主的RNA	(1)
形成双链RNA	(2)
新合成正链替代原始正链	(3)
(a) 首先合成一些长度相同的正链	(4)
(b) 核糖核酸酶处理降解"尾巴"并形成一条不完整的双链	(5)

图 1. 目前公认的病毒 RNA 复制机制的示意图

Certain features of this model, however, do not account for all the experimental evidence available. In the first place, it has been shown that the ribonuclease-resistant RNAs present in cells infected with EMC[1], foot-and-mouth disease[5] and poliomyelitis[6] viruses possess low but significant infectivity. Unless the minus strand is infective, which is extremely unlikely, the infectivity must be contained in the plus strand. In the model which is at present favoured, the plus strand of the ribonuclease-resistant RNA would not be an intact virus RNA strand but a collection of RNA molecules, each shorter in length than the virus RNA. All the evidence so far accumulated in a number of laboratories has shown that the intact virus RNA molecule is required for infectivity[7,8]. Secondly, the ribonuclease-resistant RNA in the model shown in Fig. 1b is a double-stranded molecule which has gaps between adjoining segments of the residual plus strands. It is questionable whether such a molecule would be stable to relatively high concentrations of ribonuclease, for example, 50 µg/ml.[9].

A model for the replicating form which accounts for these experimental findings should contain a complete double strand as the ribonuclease-resistant part of the structure. Additional experimental evidence which we have derived from an examination of the RNA molecules synthesized in baby hamster kidney cells infected with foot-and-mouth disease virus supports this supposition and has led us to propose a new hypothesis for virus RNA replication. Our hypothesis is based on a cyclic mechanism similar to that which is thought to be involved in the replication of DNA phages.

In a previous report from this laboratory, Brown and Cartwright[5] demonstrated that three peaks of ^{32}P-labelled RNA were formed in the presence of actinomycin D during foot-and-mouth disease virus replication in baby hamster kidney cells. These fractions were separated by sucrose-gradient sedimentation and have approximate sedimentation coefficients of 37S, 20S and 16S (peaks A, B and C, Fig. 2A).

This work has now been extended to a more detailed investigation of the individual fractions obtained from sucrose gradients. For the determination of the base composition, aliquots of each fraction were precipitated with 2 volumes of ethanol at $-20°$ following addition of purified yeast RNA. The precipitates were then hydrolysed in 0.3 N potassium hydroxide and the distribution of radioactivity in the 2'-3' -nucleotides was determined. Aliquots of individual fractions were precipitated with ethanol in the presence of unlabelled baby hamster kidney cell RNA, which was added to serve as an internal marker, and then recentrifuged in a second sucrose gradient. The results in Table 1 show that the RNA sedimenting in fractions 1–15 of virus-induced RNA has the same base composition as RNA extracted from purified virus[10,11]. Recycling of the individual fractions showed, however, that peak A was not homogeneous but consisted of a spectrum of fractions which differed in their sedimentation coefficients (Fig. 2B, C, D). In contrast, different fractions of the RNA from purified radioactive virus obtained in a similar way by sedimentation in a sucrose gradient had the same sedimentation coefficient (Fig. 3). The spectrum of sedimentation values present in virus-induced RNA has therefore been interpreted as being due to molecules of different chain-lengths. It is considered unlikely that these differences in sedimentation values, which are not found in RNA from

但是这种复制模式的某些特征不能解释所有已获得的实验证据。首先，研究发现在感染脑心肌炎病毒[1]、口蹄疫病毒[5]和脊髓灰质炎病毒[6]的细胞中，耐核糖核酸酶的 RNA 具有低的但是显著的感染性。负链具有感染性的可能性极小，因此这一感染性应该来源于正链。在目前较为认可的复制模式中，耐核糖核酸酶的 RNA 中的正链并不是完整的病毒 RNA 链，而是一组 RNA 分子，其每个分子的长度都比病毒 RNA 的长度短。到目前为止，许多实验室积累的证据表明，完整的病毒 RNA 分子对于感染是必需的[7,8]。其次，如图 1b 所示的模型中的耐核糖核酸酶 RNA 是一个双链的分子，其邻接的正链片段之间都有缺口。这样的分子在较高浓度的核糖核酸酶下（比如 50 微克 / 毫升[9]）能否保持稳定是值得怀疑的。

要想解释这些实验结果，病毒 RNA 复制形式的模型应该包含一个类似于耐核糖核酸酶的 RNA 部分结构的完整双链。在检测感染了口蹄疫病毒的幼仓鼠肾细胞中 RNA 分子合成情况时，我们得到的实验证据支持这一假设，并提出病毒 RNA 复制的一种新假说。这一假说基于一个环状机制，它类似于噬菌体 DNA 复制过程中所涉及的环状机制。

在本实验室先前的报道中，布朗和卡特赖特[5]证明，在存在放线菌素 D 的情况下，幼仓鼠肾细胞中的口蹄疫病毒复制时会形成 3 个 ^{32}P 标记的 RNA 高峰。这 3 个组分可以用蔗糖梯度沉降法进行分离，它们的沉降系数大约分别为 $37S$、$20S$ 和 $16S$（图 $2A$ 中的峰 A、峰 B 和峰 C）。

现在这项工作已经延伸到对通过蔗糖梯度得到的各种组分进行更加详细的研究。为了测定碱基组成，我们向等份的各组分中加入纯化的酵母 RNA，再加入 2 倍体积的乙醇后在 –20℃下沉淀。接着沉淀物在 0.3 摩尔 / 升的氢氧化钾溶液中进行水解，然后测定 $2', 3'$– 核苷酸的放射性分布情况。向等份的各组分中加入未标记的幼仓鼠肾细胞 RNA，作为内参，也用乙醇进行沉淀，然后在另一蔗糖梯度中再次离心。结果如表 1 所示，病毒感染组沉降分离后的第 1~15 组分沉淀得到的 RNA，在碱基组成上和提取的纯病毒的 RNA 是相同的[10,11]。然而对各组分进行二次沉降分析的结果表明，峰 A 并不是由单一成分组成的，而是包含一系列沉降系数互不相同的组分（如图 2 中的 B、C、D）。相比之下，对于来自纯化的具有放射性的病毒 RNA，用同样的蔗糖梯度沉降法分离得到的不同组分则具有相同的沉降系数（图 3）。因此，病毒感染得到的 RNA 具有多个不同的沉降系数值，这一现象应该解释为其中含有链长度不同的多种 RNA 分子。由于用相同步骤分离纯化的病毒中的 RNA 时并没有出现沉降系数的多样化，因此沉降系数多样化不太可能是由于存在多种稳定的构型造成的。

purified virus isolated by the same procedure, are due to the presence of different stable configurations. The results in Figs. 2 and 3 also reveal that peak *A* from virus-infected cells sediments more quickly than the RNA isolated from purified virus. This suggests that the RNA isolated from virus-infected cells contains some molecules which are larger than those which are incorporated into the virus particles. These molecules, as shown in Table 1, have the same base composition as virus RNA. Molecules of a length greater than that of virus RNA could not be formed on the basis of the displacement mechanism shown in Fig. 1. The maximum length that could be obtained by this mechanism would be the same as that of the primer RNA.

Table 1. Base Composition of Foot-and-Mouth Disease Virus RNA

Fraction No.	Base composition (counts/100 counts)			
	A	*U*	*G*	*C*
1	26.6	21.8	23.3	28.3
2	26.7	21.7	23.4	28.2
3	25.9	22.0	24.2	27.9
4	26.9	21.3	23.0	28.8
5	26.7	21.7	23.7	27.9
6	26.2	21.9	23.9	28.0
7	26.2	22.2	24.3	27.3
8	26.9	21.7	23.2	28.2
9	26.2	21.9	23.7	28.2
10	25.8	21.4	23.6	29.2
11	26.0	21.9	23.3	28.8
12	25.5	21.7	23.6	29.2
13	24.8	22.1	24.5	28.6
14	24.5	23.1	24.3	28.1
15	25.4	22.8	23.7	28.1
Purified virus RNA	26.0	21.8	24.1	27.8
RNase-resistant RNA	23.8	24.2	25.6	26.4
Calculated duplex	24.0	24.0	26.0	26.0

Individual fractions from gradients were precipitated with 1 mg yeast RNA any hydrolysed in 0.3 N potassium hydroxide for 18 h at 37°. The 2′-3′-nucleotides were separated by paper electrophoresis at pH 3.5 in sodium citrate buffer. The distribution of phosphorus-32 in the ultra-violet absorbing regions was determined after elution.

We have also examined the structure of the ribonuclease-resistant RNA obtained by two successive centrifugations in sucrose gradients of virus-induced RNA which had been treated with 0.01 μg RNase/ml. This has a base composition which is in agreement with its being a duplex structure consisting of a plus and a minus strand (Table 1). Heating this molecule at various temperatures up to 110°C for 5 min, followed by rapid cooling, has failed to produce any molecules sedimenting at the same position as virus RNA (peak *A*, Fig. 3). A typical result is shown in Fig. 4, which indicates that both heated and unheated RNA sediment in approximately the same position. Nevertheless, after heating at 110°C the ribonuclease-resistant RNA is degraded by the enzyme. Single-stranded RNA isolated from purified virus also sedimented to the same position as the ribonuclease-resistant RNA after heating under these conditions (Fig. 4*C*). Apparently, heating single-stranded virus RNA in this manner produces structural or configurational changes in the molecule which

510

图 2 和图 3 的结果也表明，被病毒感染细胞的峰 A 比从纯病毒中分离得到的 RNA 沉降得更快。这说明从病毒感染细胞中分离得到的 RNA 中包含一些分子，它们比组成病毒颗粒的 RNA 分子更大。正如表 1 所示，这些分子和病毒 RNA 具有相同的碱基组成。而通过图 1 所示的置换机制是无法形成比病毒 RNA 更长的分子的，因为通过置换机制得到的最长的分子也只能同 RNA 引物链一样长。

表1. 口蹄疫病毒RNA的碱基组成

管号	碱基组成（计数/100计数）			
	A	U	G	C
1	26.6	21.8	23.3	28.3
2	26.7	21.7	23.4	28.2
3	25.9	22.0	24.2	27.9
4	26.9	21.3	23.0	28.8
5	26.7	21.7	23.7	27.9
6	26.2	21.9	23.9	28.0
7	26.2	22.2	24.3	27.3
8	26.9	21.7	23.2	28.2
9	26.2	21.9	23.7	28.2
10	25.8	21.4	23.6	29.2
11	26.0	21.9	23.3	28.8
12	25.5	21.7	23.6	29.2
13	24.8	22.1	24.5	28.6
14	24.5	23.1	24.3	28.1
15	25.4	22.8	23.7	28.1
纯化病毒的RNA	26.0	21.8	24.1	27.8
耐核糖核酸酶的RNA	23.8	24.2	25.6	26.4
理论上的双链	24.0	24.0	26.0	26.0

分别向密度梯度分离得到的各组分中加入 1 毫克酵母 RNA 并沉淀，然后在 37℃、0.3 摩尔 / 升的氢氧化钾溶液中水解 18 小时。在 pH 为 3.5 的柠檬酸钠缓冲液中通过纸电泳对 2′,3′- 核苷酸进行分离。洗脱后测定 ^{32}P 在紫外吸收区的分布

我们也分析了病毒感染产生的 RNA 经 0.01 微克 / 毫升核糖核酸酶处理并进行连续两次蔗糖梯度沉降后得到的耐核糖核酸酶的 RNA 的结构。其碱基组成表明它是由一条正链和一条负链组成的双链结构（表 1）。在包括高达 110℃ 的各种温度下将该分子加热 5 分钟，迅速冷却后，并不能得到任何与病毒 RNA（图 3 中的峰 A）具有相同沉降位置的分子。图 4 显示了一个典型的结果，可以看出，不管是经过加热处理还是没有经过加热处理，这种 RNA 总是在几乎相同的位置发生沉降。不过，经过 110℃ 加热处理后，这种耐核糖核酸酶的 RNA 能够被酶降解。在此条件加热处理后，从纯化的病毒中分离得到的单链 RNA 的沉降位置也与耐核糖核酸酶的 RNA 的沉降位置相同（图 4C）。很明显，用这种方法加热单链病毒 RNA 可以改变其原来的分子

are sufficient to greatly reduce its sedimentation coefficient. The fact that heating the ribonuclease-resistant RNA at 110°C did not produce any molecules with sedimentation coefficients smaller than that of heated virus RNA has been taken as evidence for the absence of a double-stranded structure in which the plus strand is composed of segments (cf. Fig. 1*b*).

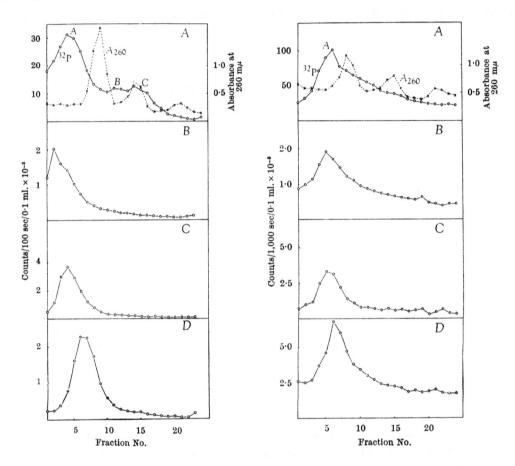

Fig. 2. Sedimentation in sucrose gradients (5–25 percent sucrose in 0.1 M acetate, *p*H 5.0) of "Sephadex G-200" filtered RNA from actinomycin *D* treated *BHK* cells, 4 h after infection with *FMD* virus. (*A*), Total RNA; (*B*), fraction 2 from first sedimentation recycled in a second gradient; (*C*), recycled fraction 4; (*D*), recycled fraction 6. The RNA was centrifuged for 16 h at 20,000 r. p. m. in an *SW* 25 rotor in a Spinco ultracentrifuge at 4°. Unlabelled *BHK* cell RNA was added to each sample before centrifuging to serve as an internal marker.

Fig. 3. Sedimentation in sucrose gradients of RNA from purified ^{32}P-labelled *FMD* virus. (*A*), Total RNA; (*B*), fraction 3 from first sedimentation recycled in second gradient; (*C*), recycled fraction 5; (*D*), recycled fraction 6. Unlabelled *BHK* cell RNA was added to each sample before centrifuging to serve as an internal marker.

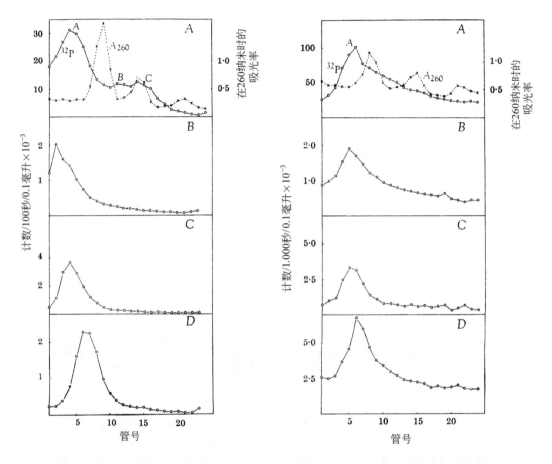

结构或者构型，并足以降低其沉降系数。耐核糖核酸酶的 RNA 在 110℃下加热处理后并不能得到沉降系数比经同样加热处理的病毒 RNA 的沉降系数更小的任何分子，这一事实已经作为并不存在那种正链是由一些小片段组成的双链结构的证据（参见图 1b）。

图 2. 幼仓鼠肾细胞用口蹄疫病毒感染 4 小时，再经放线菌素 D 处理后通过"葡聚糖 G-200"进行凝胶过滤，得到的 RNA 进行蔗糖梯度（0.1 摩尔/升乙酸盐中含 5% ~ 25% 的蔗糖，pH 5.0）沉降的情况。（A）总 RNA；（B）第一次沉降后的第 2 组分进行二次沉降的结果；（C）第 4 组分二次沉降的结果；（D）第 6 组分二次沉降的结果。在 4℃下，将 RNA 在斯宾诺超速离心机的 SW 25 转子上以每分钟 20,000 转离心 16 小时。以离心前加入未标记的幼仓鼠肾细胞 RNA 的样品作为内参。

图 3. 纯化的 [32]P 标记的口蹄疫病毒中的 RNA 进行蔗糖梯度沉降的情况。（A）总 RNA；（B）第一次沉降后的第 3 组分进行二次沉降的结果；（C）第 5 组分二次沉降的结果；（D）第 6 组分进行二次沉降的结果。以离心前加入未标记的幼仓鼠肾细胞 RNA 的样品作为内参。

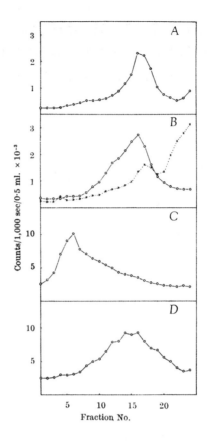

Fig. 4. Effect of heating at 110° in 10^{-2} M EDTA on the sedimentation in sucrose gradients of RNase-resistant RNA and virus RNA. (*A*), Unheated RNase-resistant RNA; (*B*), heated RNase-resistant RNA; ○, without RNase; ×, with 0.01 µg RNase/ml.; (*C*), unheated virus RNA; (*D*), heated virus RNA.

All the evidence at present available leads us to propose a replicating form which will release an intact double strand on ribonuclease treatment and also permit the synthesis of molecules with different chain-lengths. Neither of these requirements can be met with the displacement mechanism at present held. It is postulated that the ingoing virus RNA (plus strand) codes for a complementary minus strand (Fig. 5). This minus strand takes up the configuration of a circle, or is synthesized as a cyclic structure. The RNA polymerase which produces the new plus strands can rotate around this cyclic primer producing a long chain of plus strand, which is repetitive in sequence. This long chain is then broken down by an enzyme into segments of approximately the correct length for viral RNA. Molecules of shorter length may function as specific messenger RNAs for the synthesis of virus coat protein. This may be the function of the molecules present in peak *B* of the virus-induced RNA. Only those molecules of the correct length are finally incorporated into virus.

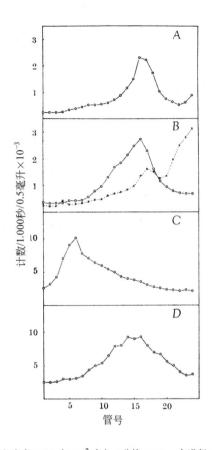

图 4. 耐核糖核酸酶的 RNA 和病毒 RNA 在 10^{-2} 摩尔／升的 EDTA 中进行 110℃ 加热处理对其蔗糖密度梯度沉降情况的影响。(*A*) 未加热的耐核糖核酸酶的 RNA；(*B*) 加热的耐核糖核酸酶的 RNA；○代表没有加核糖核酸酶的实验组，× 代表每毫升溶液中加入 0.01 微克核糖核酸酶的实验组；(*C*) 未加热的病毒 RNA；(*D*) 加热的病毒 RNA。

基于目前获得的所有证据，我们提出了一种复制模式，在这种复制模式中经核糖核酸酶处理后，会释放完整的双链，也允许合成链长度不同的多种分子。这两点完全不符合目前支持的置换机制。我们猜测，侵入宿主的病毒 RNA（正链）能够编码一条互补的负链（图 5），这条负链形成环状构型，或者其本来就被合成为环形结构。合成新正链的 RNA 聚合酶能够绕着这种环状的引物链合成一条在序列上重复的很长的正链。之后，在酶的剪切作用下，这条长链被切割成与病毒 RNA 长度相当的片段。那些长度较短的 RNA 片段可能充当合成病毒外壳蛋白的特异信使 RNA。病毒感染产生的 RNA 的峰 *B* 中的分子可能就具有这种功能。只有那些长度相当的 RNA 分子最后才会组装到病毒中。

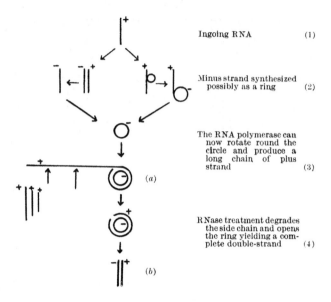

Fig. 5. Diagrammatic representation of the proposed model for virus RNA replication

This cyclic replicating form is consistent with the evidence available at present. On ribonuclease treatment the side chain will be removed and the ring probably opened at the growing point to yield a complete double strand. The hypothesis also explains the presence of virus-like RNA in virus-infected cells that has longer chain-lengths than the RNA extracted from purified virus.

(**208**, 861-863; 1965)

F. Brown and S. J. Martin: Animal Virus Research Institute, Pirbright, Woking, Surrey.

References:

1. Montagnier, L., and Sanders, F. K., *Nature*, **199**, 664 (1963).

2. Weissmann, C., Borst, P., Burdon, R. H., Billeter, M. A., and Ochoa, S., *Proc. U.S. Nat. Acad. Sci.*, **51**, 682 (1964).

3. Fenwick, M. L., Erikson, R. L., and Franklin, R. M., *Science*, **146**, 527 (1964).

4. Hausen, P., *Virology*, **25**, 523 (1965).

5. Brown, F., and Cartwright, B., *Nature*, **204**, 855 (1964).

6. Pons, M., *Virology*, **24**, 467 (1964).

7. Gierer, A., *Nature*, **179**, 1297 (1957).

8. Ginoya, W., *Nature*, **181**, 958 (1958).

9. Weissmann, C., Billeter, M. A., Schneider, M. C., Knight, C. A., and Ochoa, S., *Proc. U.S. Nat. Acad. Sci.*, **53**, 653 (1965).

10. Brown, F., and Cartwright, B., *Nature*, **199**, 1168 (1963).

11. Bachrach, H. L., Trautman, R., and Breese, S. S., *Amer. J. Vet. Res.*, **25**, 333 (1964).

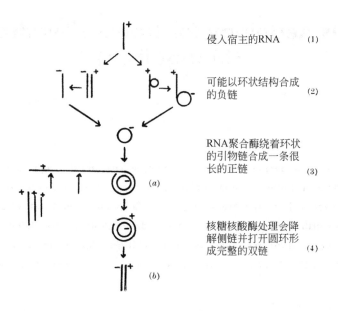

侵入宿主的RNA　　　　(1)

可能以环状结构合成的负链　　　　(2)

RNA聚合酶绕着环状的引物链合成一条很长的正链　　　　(3)

核糖核酸酶处理会降解侧链并打开圆环形成完整的双链　　　　(4)

图 5. 我们提出的病毒 RNA 复制模型的示意图

　　这种环状复制模式和目前获得的证据是相吻合的。经核糖核酸酶处理后，侧链会被移除，环可能会在生长点处被打开从而形成一条完整的双链。这种假设也能够解释在病毒感染细胞中存在着比从纯化病毒中提取的 RNA 更长的病毒状 RNA。

（彭丽霞 翻译；王晓晨 审稿）

Virus Aetiology for Down's Syndrome (Mongolism)

A. Stoller and R. D. Collmann

Editor's Note

Today it is known that people with Down's syndrome have an extra copy of chromosome 21, and that the condition is more common in children born to older mothers. In this paper, Australian mental health researchers Alan Stoller and R. D. Collmann suggest instead a viral cause. They plot 12 years worth of Australian data on graph, demonstrating a correlation between maternal hepatitis infection and the occurrence of "mongol" births nine months later. The theory did not survive, and today the ultimate cause of Down's syndrome still remains uncertain.

FOR many years now, we have been working on the epidemiology of Down's syndrome (mongolism) in the State of Victoria, Australia, and have charted its occurrences during 1942–64. Peaks of incidence, of 2-year duration have been recorded at 5–7 year intervals from 1942 until 1957 and, as a result of this, a further peak of occurrence for this congenital anomaly was forecast for 1962–63. This was the first time ever that such a forecast had been able to be made and, in fact, this eventuated[1,2]. On the basis of our original findings[1,2], we had postulated a hypothesis of an infective virus, of long incubation, affecting mostly, but not exclusively, the ovum of the ageing mother, either directly or through some immunity pattern. Our reasons for this were not only the perception by one of us (A. S.) of a possible clinical relationship between the exposure of the mother to infective hepatitis prior to conception, but also the epidemiological findings that cases of mongol births clustered significantly in time and place, that urban peaks of annual incidence were in every case greater than rural peaks (higher contact rates), and rural peaks followed on one year after urban peaks (suggesting a slow spread of infection out of the high-contact urban areas into the rural areas). Our latest investigations have consisted of an attempt to match the annual occurrence of mongolism in the Melbourne area with the annual occurrence of notifiable infectious diseases during 1952–64, 1952 being the year that infective hepatitis first became notifiable in the State of Victoria. Of all infections diseases, the incidence of infective hepatitis, charted nine months prior to that of mongolism, has alone shown concordance (Fig. 1): another link in the chain of evidence we had forged relating mongolism to the virus of infective hepatitis, or some process associated with the infection, affecting the ovum prior to, or about the time of, conception. The correlation coefficient between the incidences of mongols, per 100,000 live births, and those of infective hepatitis, per million of population, was 0.81, significant at the level of $P<0.01$.

唐氏综合征（先天愚型）的病毒病因学

斯托勒，科尔曼

编者按

现在人们已经知道唐氏综合征患者多一条 21 号染色体，而高龄孕妇所生的孩子中患病的情况更为常见。在本文中，澳大利亚心理健康研究者艾伦·斯托勒和科尔曼认为该病是由病毒引起的。他们将 12 年间澳大利亚的数据进行绘图分析，指出孕妇肝炎感染和 9 个月后产出"先天愚型"患儿这两者之间存在相关性。这个理论没能站住脚跟。迄今为止，唐氏综合征的根本原因仍然不确定。

多年来我们致力于研究澳大利亚维多利亚州唐氏综合征（先天愚型）的流行病学，并对 1942 年～1964 年间的发病情况进行了作图分析。从 1942 年到 1957 年，每隔 5~7 年就会出现一个持续 2 年的发病高峰，根据这一结果我们预测在 1962 年～1963 年间将会出现这种先天性异常疾病发病的下一个高峰。这是首次对该病的流行性进行预测，事实上也的确发生了 [1,2]。根据我们最初的发现 [1,2]，我们提出一个假说，即可能有一种潜伏期很长的传染性病毒，主要但非专一地感染高龄孕妇的卵子，这种病毒或直接感染或经由某种免疫机制间接感染。我们之所以提出这一假说，不仅由于我们研究团队中的一员斯托勒发现先天愚型患儿的出生和产妇在受孕前感染肝炎可能在临床上有联系；而且流行病学的研究发现先天愚型的发病在时间和地点上具有明显的群发性，即每年城市发病高峰的病例都比农村发病高峰的病例多（接触率更高），在城市发病高峰一年后，农村才出现发病高峰（表明传染病从高接触率的城市地区慢慢传播到农村地区）。在最近的一项调查中，我们试图将 1952 年～1964 年间墨尔本地区每年先天愚型的发病情况和每年呈报的传染性疾病的发病情况联系起来。1952 年维多利亚州开始呈现传染性肝炎。在所有的传染性疾病中，只有传染性肝炎与先天愚型的发病情况具有一致性（图 1），如图中所示，传染性肝炎的发病时间比先天愚型早 9 个月。这种一致性是我们把先天愚型和肝炎病毒感染，或者说与受孕前或受孕期间影响卵子的病毒感染过程联系起来的又一例证。每 100,000 个新生儿中先天愚型的患儿数与每 1,000,000 人中感染肝炎的人数之间的相关系数为 0.81，在 $P<0.01$ 的水平上两者具有显著的相关性。

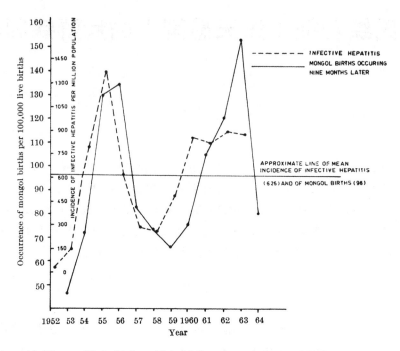

Fig. 1. Annual incidence of Infective hepatitis in Melbourne, and of mongol births 9 months later for the period 1952 to 1964.

We would, in fact, postulate that such a process of virus human interaction could well be the basis of other genetic anomalies. We have already noted some degree of concordance between peaks of hydrocephaly and mongolism[3], even though, in the former, no visible chromosomal abnormality is apparent. A prospective clinical investigation is being put into operation to test the association between infective hepatitis and the occurrence of mongolism; and immunity patterns for this virus, as soon as feasible, will need to be tested on a large group of mothers "at risk". Meanwhile, also, examination of the effects of viruses on human cells in culture, and of ovarian tissue in particular, might well open up an undreamed of prospect of obtaining insight into the production, and thence prevention, of many gene disturbances.

(**208**, 903-904; 1965)

Alan. Stoller and R. D. Collmann: Mental Health Research Institute, Royal Park, N.2., Victoria, Australia.

References:

1. Collmann, R. D., and Stoller, A., *Amer. J. Publ. Hlth.*, 52, 813 (1962).

2. Stoller, A., and Collmann, R. D., *Med. J. Austral.*, 1, 1 (1965).

3. Collmann, R. D., and Stoller, A., *J. Ment. Defic. Res.*, 76, 22 (1962).

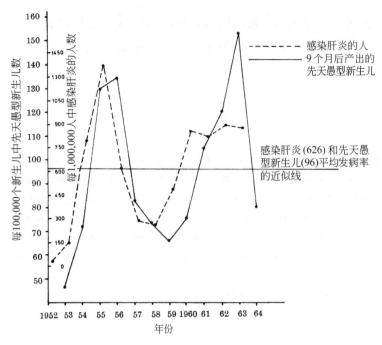

图 1. 1952 年～1964 年间墨尔本每年感染肝炎的人数以及 9 个月后产出的先天愚型新生儿数

事实上，我们推测病毒与人体相互作用的这一过程很可能是其他遗传异常发生的基础。我们已经发现脑积水发病高峰和先天愚型具有一定的相关性[3]，尽管在脑积水病例中没有观察到明显的染色体异常现象。正在进行一项具有前瞻性的临床调查以检测肝炎感染和先天愚型发病之间的联系；一旦可行，将要在大量"高危"的母亲中检测该病毒的免疫机制。同时，检测病毒对体外培养的人类细胞尤其是卵巢组织的影响也许会揭示出许多令人意想不到的基因异常发生的原因，进而预防许多基因异常的发生。

（彭丽霞 翻译；金侠 审稿）

521

Biological and Mental Evolution:
an Exercise in Analogy*

Editor's Note

Here the Hungarian writer and philosopher Arthur Koestler contributes a paper to *Nature* based on an address commemorating the mineralogist James Smithson, who bequeathed the funds for the Smithsonian Institution in Washington DC. Koestler draws an analogy between biological and mental evolution. He suggests that the former can be considered a "history of escapes from over-specialization", and that "breakthroughs" in art and science can be thought of this way too, for example in reference to Thomas Kuhn's model of scientific "revolutions" occurring via paradigm shifts. Koestler's argument looks somewhat dated now, but his suggestion that evolution has "structural constraints" that limit what random mutation may create remains relevant.

ALLOW me to take you on a ride on the treacherous wings of analogy, starting with an excursion into genetics. Creativity is a concept notoriously difficult to define; and it is sometimes useful to approach a difficult subject by way of contrast. The opposite of the creative individual is the pedant, the slave of habit, whose thinking and behaviour move in rigid grooves. His biological equivalent is the over-specialized animal. Take, for example, that charming and pathetic creature, the koala bear, which specializes in feeding on the leaves of a particular variety of eucalyptus tree and on nothing else; and which, in lieu of fingers, has hook-like claws, ideally suited for clinging to the bark of the tree—and for nothing else. Some of our departments of higher learning seem expressly designed for breeding koala bears.

Sir Julian Huxley has described over-specialization as the principal cause why evolution in all branches of the animal kingdom—except man's—seems to have ended either in stagnation or in extinction. But, having made his point, he drew a conclusion which you may find less convincing. "Evolution," be concluded, "is thus seen as an enormous number of blind alleys with a very occasional path to progress. It is like a maze in which almost all turnings are wrong turnings[1]." With due respect, I think this metaphor is suspiciously close to the old-fashioned behaviourist's views of the rat in the maze as a paradigm of human learning. In both cases the explicit or tacit assumption is that progress results from a kind of blind man's buff—random mutations preserved by natural selection, or random tries preserved by reinforcement—and that that is all there is to it. However, it is possible to dissent from this view without invoking a *deus ex machina*, or a Socratic *daimon*, by making

---

* Substance of an address delivered on September 18 at the Bicentennial Celebration commemorating the birth of James Smithson, held in Washington during September 16-18 (see *Nature*, **208**, 320; 1965).

生物进化与心理演进：类推法运用*

凯斯特勒

编者按

本文是匈牙利作家、哲学家阿瑟·凯斯特勒基于纪念矿物学家詹姆斯·史密森（其遗赠资金建立了位于华盛顿的史密森学会）的致词而发表在《自然》上的一篇论文。凯斯特勒运用类推法来研究生物进化和心理演进。他认为前者可以被看作一段"逃离高度特化的历史"，这也同样可以用来理解艺术和科学中的"突破"，比如托马斯·库恩建立的通过典范式转移而引发"革命"的科学模型。现在看来凯斯特勒的论点有些过时了，但是他提出的进化中存在着能够限制随机突变发生的"结构限制"这一论断仍然适用。

让我们从遗传学的角度出发去领略一下类推法的奥妙。创造是一个很难界定的概念，而有时候通过对照或许可以有效地解决这个难题。创造的对立面是教条、惯性思维，即思考和行动严格地按照常理来办。从生物学的角度来看，这等价于高度特化的动物。例如，树袋熊是一种迷人而又让人怜爱的动物，它专门以一种特定桉树叶为食，绝不以其他食物为食；其指被钩状的爪代替，非常适合紧紧握住树皮——除此以外，便也没有其他用处了。高校的一些院系就好像是为了"培养树袋熊"而设置的一样。

朱利安·赫胥黎爵士认为高度特化是动物界所有进化分支（除了人类之外）最终停滞或消失的主要原因。不过，为了表达他的观点，他做出这样一个也许你会认为并不那么可信的结论：将"进化"总结为"就好比是众多死胡同中偶然向前伸展出来的一条小道，这就像迷宫一样，无论向哪个方向转弯都是错误的[1]。"恕我直言，我认为这一比喻可能类似于传统行为学家利用老鼠迷宫研究人类学习模式的观点。两者都直接或默认地做出了这样的假设，即发展是源于一种类似捉迷藏的过程而获得的——源于经过自然选择或强化后所保存下来的随机突变——并且就是这样简单。然而，这种观点可能不被人接受，因为仅仅通过简单的假设，我们并不能排除这是

---

* 于 9 月 18 日发表在纪念詹姆斯·史密森诞辰 200 周年庆典上的演讲，该庆典于 9 月 16 日~18 日在华盛顿举行（见《自然》第 208 卷，第 320 页；1965 年）。

the simple assumption that, while random events no doubt play an important part in the picture, that is not all there is to it.

One line of escape from the maze is indicated by a phenomenon known to students of evolution by the ugly name of paedomorphism, a term coined by Garstang[2] some forty years ago. The existence of the phenomenon is well established; but there is little mention of it in the text-books, perhaps because it runs against the *Zeitgeist*. It indicates that in certain circumstances evolution can re-trace its steps, as it were, along the path which led to the dead-end and make a fresh start in a more promising direction. To put it simply, paedomorphism means the appearance of some evolutionary novelty in the larval or embryonic stage of the ancestral animal, a novelty which may disappear before the adult stage is reached, but which reappears in the adult descendant. This bit of evolutionary magic is made possible by the well-known mechanism of neoteny, that is to say, the gradual retardation of bodily development beyond the age of sexual maturity, with the result that breeding takes place while the animal still displays larval or juvenile features. Hardy[3], de Beer[4] and others have pointed out that if this tendency toward "prolonged childhood" were accompanied by a corresponding squeezing out of the later adult stages of ontogeny, the result would be a rejuvenation and despecialization of the race which would thus regain some of its lost adaptive plasticity. But of even greater importance than this re-winding of the biological clock is the fact that in the paedomorphic type of evolution selective pressure operates on the early, malleable stages of ontogeny. In contrast to this, gerontomorphism—the appearance of novel characters in the late-adult stages— can only modify structures which are already highly specialized. One is accordingly led to expect that the major evolutionary advances were due to paedomorphism and not to gerontomorphism—to changes in the larval or embryonic, and not in the adult, stage.

Let me give an example, which will make clearer what I am driving at. There is now strong evidence in favour of the theory, proposed by Garstang[2] in 1922, that the chordates, and thus we, the vertebrates, descended from the larval state of some primitive echinoderm, perhaps rather like the sea-urchin or sea-cucumber. Now an adult sea-cucumber would not be a very inspiring ancestor—it is a sluggish creature which looks like an ill-stuffed sausage, lying on the sea-bottom. But its free-floating larva is a much more promising proposition: unlike the adult, it has bilateral symmetry, a ciliary band presumed to be the forerunner of the neural fold, and other sophisticated features not found in the adult animal. We must assume that the sedentary adult residing on the sea-bottom had to rely on mobile larvae to spread the species far and wide in the ocean, as plants scatter their seeds in the wind; and that the larvae, which had to fend for themselves, exposed to much stronger selective pressures than the adults, gradually became more fish-like; and lastly because sexually mature while still in the free-swimming; larval state—thus giving rise to a new type of animal which never settled on the bottom at all and altogether eliminated the senile, sessile cucumber stage from its life-history.

It seems that the same re-tracing of steps to escape the dead-ends of the maze was repeated at each decisive evolutionary turning-point—the last time, so far as we know,

一种任意的巧合或者是苏格拉底式的迷局，毫无疑问随机事件在全局中发挥着重要作用，但这并不是全部。

逃离迷宫的一条路线可以用进化学习者所熟知的一种现象来说明，这一现象有一个难听的名字叫做幼稚形态，是加斯唐 [2] 在大约四十年前提出的一个概念。已经明确证实该现象是存在的；但是在教科书中却很少被提及，这也许是因为它违背了**时代精神**。该概念指出，在特定环境下进化可以重演，犹如沿着一条通往死胡同的道路，转入一个更有希望的方向重新开始一样。简单说来，幼稚形态就是指动物祖先的幼体或胚胎期出现的一些进化的新表型可能在成熟期到来前就消失了，但却又会重新出现在成体的后代当中。这种令人不可思议的进化可能是由人们熟知的幼态持续机制导致的，也就是说，越过了正常的性成熟年龄之后，身体发育逐步延迟，以至于在它开始出现繁殖行为时仍表现出幼体的形态特征。哈迪 [3]、德比尔 [4] 及其他学者均指出：如果这种"幼稚期延长"的趋势伴随着之后的成体发育阶段相应地减少，结果便产生了返老还童的特征和种族的去高度特化性，这种性质能够使其重新获得一些曾经失去的适应可塑性。但是，较之生物钟重绕，更为重要的是以下事实，即幼体形态类型的进化选择压力发生在个体发育可延展性的早期。相反地，老龄形态——在成熟阶段晚期出现的新表型——则只能调控已经高度特化的结构。可以认为进化过程主要是由幼稚形态而非老龄形态所致，即在幼体期或胚胎期而非在成熟期发生改变。

举个可以令我的讲述更为清楚的例子。1922 年，加斯唐 [2] 提出的一个现在已经得到强有力的证据支持的理论，即脊索动物及脊椎动物（包括人类）来源于一些原始棘皮动物的幼期形态，这种形态可能与海胆或者海参非常相似，但这却不是一种令人鼓舞的祖先形态，它行动迟缓，深躺海底，犹如一根未塞满的腊肠。然而，它的自由漂浮生活的幼虫是一个更有研究前景的命题：幼虫与成虫不同，它具有双面对称结构，一个被推测认为可能是神经褶前体的纤毛环，和其他一些成虫不具有的复杂特征。我们推测：驻留在海底生活的成虫依赖于可运动的幼虫在海里向更远处及更大范围散播其物种，就像植物通过风来散播种子一样；而且必须努力谋生的幼虫较成虫暴露在更大的选择压力下，便逐渐地变得与鱼类更为相似，最后在浮游期就发生了性成熟，从而产生出一种新的动物类型，这种类型的动物从来不驻留在海底生活，其衰老时的驻留生活状态在其生活史中完全消失。

仿佛避开迷宫的死胡同一样，在每一个决定进化方向的转折时刻都重复发生着重演的步骤——目前我们所知的最后一次转折是人类从一些原始的灵长类动物中分

when the line which bore our own species branched off from some ancestral primate. It is now generally recognized that the human adult resembles more the embryo of an ape than an adult ape. In both, the ratio of brain-weight to body-weight is disproportionately high; in both, the closing of the sutures of the skull is retarded to allow for further brain growth. The back to front axis through man's head—the direction of his line of sight—forms an angle of ninety degrees with his spinal column; a condition which, in apes and other mammals, is only found in the embryonic stage. The same applies to the angle between the uro-genital canal and the backbone, which accounts for the singularity of the human way of mating. Other embryonic—or, to use Bolk's[5] term, foetalized—features are the absence of brow-ridges, scantness of body-hair, retarded development of the teeth, and so on. As Haldane[6] has said: "If human evolution is to continue along the same lines as in the past, it will probably involve a still greater prolongation of childhood and retardation of maturity. Some of the characters distinguishing adult man will be lost." But there is a reverse to the medal, which Aldous Huxley gleefully showed us in *After Many A Summer*: artificial prolongation of the absolute life-span of man might provide an opportunity for features of the adult ape to re-appear in Methuselah. But this only by the way.

The essence of the process which I have described is a retreat from highly specialized adult forms of bodily structure and behaviour to an earlier, more plastic and less committed stage—followed by a sudden advance in a new direction. It is as if the stream of life had momentarily reversed its course, flowing uphill for a while, then opened up a new stream-bed—leaving the koala bear stranded on its tree like a discarded hypothesis. We have now reached the crucial point in our excursion, because it seems to me that this process of *reculer pour mieux sauter*—of drawing back to leap, of undoing and re-doing—is a basic feature of all significant progress, both in biological and mental evolution.

It can be shown, I think, that these two types of progress—the emergence of biological novelties and the creation of mental novelties—are analogous processes on different levels of the developmental hierarchy. But to demonstrate the connexion we must proceed stepwise from lower to higher organisms. One of the fundamental properties of living organisms is their power of self-repair, and the most dramatic manifestations of this power are the phenomena of regeneration (which Needham[7] called "one of the more spectacular pieces of magic in the repertoire of living organisms"). Primitive creatures, like flatworms, when cut into slices, can regenerate a whole animal from a tiny fragment; Amphibia can regenerate limbs and organs; and once more the "magic" is performed by *reculer pour mieux sauter*—the regression of specialized tissues to a genetically less committed, quasi-embryonic stage, a de-differentiation or de-specialization followed by a re-differentiation.

Now the replacement of a lost limb or lost eye is a phenomenon of a quite different order from the adaptive processes in a normal environment. Regeneration could be called a meta-adaptation to traumatizing challenges. The power to perform such meta-adaptations manifests itself only when the challenge exceeds a critical limit and can only be met by having recourse to the genetic plasticity of the embryonic stage. We have just seen that

支出来的时候。现在一般认为，相比成熟猿类的形态，成人与猿类的胚胎期更为相似。两者的脑重与体重比都高得不成比例；两者的大脑都因头骨缝闭合时间推迟而得以进一步发育。人体头部的后前轴（参照头部视线方向）与脊柱成九十度角；而这种情形对于猿类和其他哺乳动物来说只出现在胚胎时期。在泌尿生殖器道与脊柱间也呈现同样的九十度角，这是与人类所特有的交配方式相适应的。其他的胚胎的特征或似胚胎状（运用博尔克 [5] 的术语）的特征有：没有眉梁、缺少体毛、牙齿发育迟缓等等。正如霍尔丹 [6] 所说："如果人类进化是在原有基础上的继续发展，其幼体期也将很可能会进一步延长，而成熟期将被延后。一些区别于成人的特征也将消失。"但是，相反地，如奥尔德斯·赫胥黎在《多少个夏天之后》中欣慰地向我们展示一样：人为延长人类的绝对寿命可能会使成年猿类所具有的特征在高龄的个体中再次出现。但这只是顺便提一下而已。

我所描绘的这个过程的本质是从身体结构形式和行为方式高度特化的成年期回到一个更早、更具有可塑性而定向性较低的时期，该时期过后紧随着会出现一个朝着新方向的突发性的提高。这就像一条生命的溪流突然改变其进程，向上流动一会儿，之后就打开了一个新的河床，就像一个被抛弃的假说所描述的那样将树袋熊留在树上使其进退两难。现在，我们已经到达了进化中的关键时刻，因为似乎对我来讲，在生物进化和心理演进中，以退为进、撤销与重建是所有重要过程的一种基本特征。

我认为可以将生物学新事物的出现与新思维的产生这两种过程看作在不同的发展等级水平上的两个相互类似的过程。然而，为了论证两者的关系，我们应该遵循逐步由低等生物体到高等生物体的步骤。生命有机体的一个基本特征是具有自我修复功能，而再生现象则是这种功能的最生动的体现（尼达姆 [7] 曾称这种再生现象是"生命有机体中令人惊叹的魔法中的一个片断"）。当将原始生物体（如扁形虫）切成几个片断时，每个小片断都可以再生出一个完整的个体；两栖动物可以再生出肢体和器官；这种"魔法"再一次由以退为进来完成，即特化组织经过去分化或紧随着再分化的去特化退行到定向性较低的准胚胎期。

此时，丢失肢体或丢失眼的替换过程完全不同于在正常环境中的适应过程的顺序。再生可以被称为是挑战中受创伤后的一种超适应性，这种超适应性只有当其受到的挑战超过一定的临界点才会表现出来，且只有具备胚胎期遗传可塑性的生命有机体才会表现出来。我们刚才已经了解到系统发生的主要变化是在成体退回到胚胎

the major phylogenetic changes were brought about by a similar retreat from adult to embryonic forms. Indeed, the main line of development which led up to our species could be described as a series of operations of phylogenetic self-repair: of escapes from blind alleys by the undoing and re-moulding of maladapted structures.

Evidently, self-repair by the individual produces no evolutionary novelty, it merely restores the *status quo ante*. But that is all the individual needs in order to regain its normal adaptive balance in a static environment (assuming that the traumatizing disturbance was only a momentary one). Phylogenetic "self-repair", on the other hand, implies changes in the genotype to restore the adaptive balance in a changing environment.

As we move toward the higher animals, the power of regenerating physical structures is superseded by the equally remarkable power of the nervous system to reorganize its mode of function. (Ultimately, of course, these reorganizations must also involve structural changes of a fine-grained nature in terms of circuitry, molecular chemistry or both, and so we are still moving along a continuous line.) Lashley[8] taught his rats certain visual discrimination skills; when he removed their optical cortex, the learning was gone, as one would expect; but, contrary to what one would expect, the mutilated rats were able to learn the same tasks again. Some other brain area, not normally specializing in visual learning, must have taken over this function, deputizing for the lost area.

Similar feats of meta-adaptation have been reported in insects, birds, chimpanzees and so on. But let us get on to man, and to those lofty forms of self-repair which we call self-realization, and which include creativity in its broadest sense. Psycho-therapy, ancient and modern, from shamanism down to contemporary forms of abreaction therapy, has always relied on what Ernst Kris[9] has called "regression in the service of the ego". The neurotic with his compulsions, phobias and elaborate defence-mechanisms is a victim of maladaptive specialization—a koala bear hanging on for dear life to a barren telegraph pole. The therapist's aim is to regress the patient to an infantile or primitive condition; to make him retrace his steps to the point where they went wrong, and to come up again, metamorphosed, re-born. Goethe's *Stirb und Werde*, the inexhaustible variations of the archetype of death and resurrection, dark night and spiritual rebirth, all revolve around this basic paradigm—Joseph in the well, Jesus in the tomb, Buddha in the desert, Jonah in the belly of the whale.

There is no sharp dividing line between self-repair and self-realization. All creative activity is a kind of do-it-yourself therapy, an attempt to come to terms with traumatizing experiences. In the scientist's case the trauma is some apparent paradox of Nature, some anomaly in the motion of the planets, the sting of data which contradict each other, disrupt an established theory, and make nonsense of his cherished beliefs. In the artist's case, challenge and response are manifested in his tantalizing struggle to express the inexpressible, to conquer the resistance of his medium, to escape from the distortions and restraints imposed by the conventional styles and techniques of his time.

形式时产生的。事实上，发展到我们人类物种的主要线路可以被形容为是一系列系统发生的自我修复的结果，通过对适应不良的机体结构进行淘汰和重塑来避开进化过程中的死胡同。

显然，个体的自我修复并没有产生进化上的新事物，它只是恢复了原状。然而，这却是所有个体在静态环境中重获正常适应性平衡所必须经历的过程（假定损伤干扰只是暂时的）。另外，系统发生的"自我修复"也暗示着在变化的环境中基因型的改变能够修复适应性平衡。

当我们向着高等动物发展时，身体结构的再生功能则被功能同样显著的神经系统所替代，以便重组其功能模式。（当然，最后这些重组也必然涉及天然精细结构的改变，如循环结构、化学分子物质或两者兼有，所以我们仍朝着原有的线路进行演进。）拉什利 [8] 教授训练他的大鼠某些视觉辨别技巧，而正如我们可以预计的一样，当他将大鼠的视觉皮层切除后，这种辨别能力便消失了。然而，与我们预计的结果不同的是：这些被切除视觉皮层的大鼠可以通过再学习获得这种技巧，这说明大脑内其他非视觉学习区域必定接替了这种任务，从而取代了被切除的区域。

相似的超适应能力在昆虫、鸟类、黑猩猩等中也有过报道。但让我们回到人类，回到那些我们称之为自我实现的在广义上具有创造性的自我修复的高级形式。无论是古代还是现代的精神疗法，从萨满教到当代的情感发泄疗法，都依赖于恩斯特·克里斯 [9] 所提出的"自我机能的退行"。具有强迫症、恐惧症和精细防御机制的神经病患者是不适应特化的牺牲品，如同把树袋熊挂在单调乏味的电线杆上虚度宝贵生命。治疗师的目的在于引导病人恢复到婴幼儿或原来的状态，折回到错位点上，并再次向前发展、改变，进而获得重生。在歌德的"死去重生"中，死亡与复活、黑暗之夜和精神复活原型的无穷无尽的变化都围绕着这一基本模式——井里的约瑟，坟墓里的耶稣，沙漠里的佛陀，鲸鱼肚里的约拿。

自我修复与自我实现之间没有一个确切的分界线。所有创造性的活动都是一种自我完成的疗法，都是一种恢复精神创伤体验的尝试。对于科学家来讲，损伤是自然界的一种明显的自我矛盾形式，是行星运行的异常形式，是相互矛盾的棘手的数据，它使原有理论陷于混乱，使人们珍视的信念变得一文不值。对于艺术家来说，挑战与响应可以在以下情形中得到证明，当他们急切努力去表达那些难以表达的事物时，当他们去征服其媒介的阻力时，当他们努力挣脱其所处时代的传统的方法与形式强加给他们的扭曲与抑制时。

In other words, the so-called revolutions in the history of both science and art are successful escapes from blind alleys. The evolution of science is neither continuous nor strictly cumulative except for those periods of consolidation and elaboration which follow immediately after a major breakthrough. Sooner or later, however, the process of consolidation leads to increasing rigidity and orthodoxy, and so into the dead-end of over-specialization. The proliferation of esoteric jargons which seems to characterize this phase reminds one sometimes of the monstrous antlers of the Irish elk, and sometimes of the neurotic's elaborate defence-mechanisms against the threats of reality. Eventually, the process leads to a crisis, and thus to a new revolutionary break-through—followed by another period of consolidation, a new orthodoxy, and so the cycle starts again.

In the history of art, this cyclic process is even more obvious: periods of cumulative progress within a given school and technique end inevitably in stagnation, mannerism or decadence, until the crisis is resolved by a revolutionary shift in sensibility, emphasis, style.

Every revolution has a destructive and a constructive aspect. In science the destruction is wrought by jettisoning previously unassailable doctrines, including some seemingly self-evident axioms of thought. In art, it involves an equally agonizing re-appraisal of accepted values, criteria of relevance, frames of perception. When we discuss the evolution of art and science from the historian's detached point of view, this un-doing and re-doing process appears as a normal and inevitable part of the whole story. But when we focus our attention on any concrete individual who initiated a revolutionary change, we are immediately made to realize the immense intellectual and emotional obstacles he had to overcome. I mean not only the inertial forces of society; the primary locus of resistance against heretical novelty is inside the skull of the individual who conceives of it. It reverberates in Kepler's agonized cry when he discovered that the planets move in elliptical pathways: "who am I, Johannes Kepler, to destroy the divine symmetry of the circular orbits!". On a more down-to-earth level the same agony is reflected in Jerome Bruner's[10] experimental subjects who, when shown for a split second a playing card with a black queen of hearts, saw it as red, as it should be; and when the card was shown again, reacted with nausea at such a perversion of the laws of Nature. To unlearn is more difficult than to learn; and it seems that the task of breaking up rigid cognitive structures and reassembling them into a new synthesis cannot, as a rule, be performed in the full daylight of the conscious, rational mind. It can only be done by reverting to those more fluid, less committed and specialized forms of ideation which normally operate in the twilight below the level of focal awareness. Such intervention of unconscious processes in the creative act is now generally, if sometimes reluctantly, accepted even by behaviourists with a strong positivist bias. Allow me, therefore, to take it for granted that in the period of incubation—to use Graham Wallas's[11] term—the creative individual experiences a temporary regression to patterns of thinking which are normally inhibited in the rational adult.

But it would be a gross over-simplification to identify—as is sometimes done—these patterns with Freud's so-called "Primary Process". The primary process is supposedly

换句话讲，科学和艺术的历史上所谓的革命就是成功逃离死胡同后的结果。科学进展除了那些在产生重大突破后进一步巩固强化和精心加工的阶段之外都不是持续的，也并非严格地累积起来的。然而，巩固过程迟早将进入到越发刻板与保守的阶段，从而进入到高度特化的尽头。增殖是一个能表征这个阶段的深奥的专业术语，它时而让人想起爱尔兰麋鹿的巨大鹿角，时而让人想起神经病患者面对现实威胁的精细防御机制。最后，这一过程导致了危机的产生，并因此爆发一场新的革命，之后又到达了一个新的巩固时期，产生了新的正统观念，进而循环又开始了。

在艺术的历史上，这种循环过程甚至更为明显，即原有学派和技巧经过不断发展的累积期之后，不可避免地进入到发展停滞、特殊习惯成形的颓废期，直到在感性、重点、风格上出现的又一次革新，危机才得以解决。

每一次革命都具有破坏性和建设性两个方面。在科学上，破坏就等于抛弃以前的教条思想，包括一些看起来好像不证自明的公理性思想。在艺术上，破坏涉及对已经被人们接受的价值、中肯标准和理念框架进行公平而痛苦的再评估。当我们从历史学家各种超然的观点来探讨艺术和科学的发展时，毁灭和重建过程在整个历史中是正常的不可缺少的部分。但是，当我们将注意力聚焦于任何一个具体的具有革命导火线性质的事物时，便立刻认识到了那些不得不克服的巨大的心理障碍和情感障碍。这里，我指的不仅是社会的惯常势力，还有存在于头脑中的反对异端新事物的原有思想。当开普勒发现行星的运行轨道是椭圆形时，他极度痛苦地喊出："我是谁，约翰内斯·开普勒，怎么就毁灭了神圣的对称性的圆形轨道论了呢？"更为实际的是，杰罗姆·布鲁纳[10] 实验中的被试者们也曾表达过同样的痛苦，当在一瞬间亮出一张黑心皇后牌时，他们看到的却是红色，就像它本来就应该是红色一样，而当牌再次亮出时，被试者们便对这种颠倒自然规律的现象感到反胃。忘却比学习更加困难；在白天意识清醒时和处于理性思维状态时，要打破严格的认知结构并将其重新组合成一个新的系统似乎是行不通的，这已经成为一种规律。只有当大脑回复到更为流畅的、较少教条化和特化的构思状态时，这种任务才得以完成，而这种构思状态通常在黎明时分运作，因为此时集中意识水平较低。现在，无意识过程对创造性行为的干涉作用甚至被带有较强实证主义偏见的行为主义者所普遍接受，虽然他们有时并不十分情愿。因此，请允许我理所当然地认为，在潜伏期（用格雷厄姆·沃拉斯[11] 的术语来说），创造性个体经历了思维模式暂时性的退行，这种思维模式通常在理性的成年期被抑制。

但是，有时用弗洛伊德的所谓的"初级过程"来鉴别这些模式总显得高度单纯化。一般说来初级过程是缺乏逻辑性的，是被快乐原则所支配的，是易于混淆感知与幻

devoid of logic, governed by the pleasure principle, apt to confuse perception and hallucination, expressed in spontaneous action, and accompanied by massive affective discharge. I believe that between this very primary process, and the so-called secondary process governed by the reality principle, we must interpolate a whole hierarchy of cognitive structures which are not simply mixtures of primary and secondary processes, but are autonomous systems in their own right, each governed by a distinct set of rules. The paranoid delusion, the dream, the daydream, free association, the mentality of children at various ages and of primitives at various stages, should not be lumped together, for each has its own logic or rules of the game. But while clearly different in many respects, all these forms of ideation have certain features in common, since they are ontogenetically, and perhaps phylogenetically, older than those of the civilized adult. I have elsewhere[12] called them "games of the underground", because if not kept under restraint they would play havoc with the routines of disciplined thinking. But under exceptional conditions, when disciplined thinking is at the end of its tether, a temporary indulgence in these underground games may suddenly produce a solution which was beyond the reach of the conscious, rational mind—that new synthesis which Poincare[13] called the happy combination of ideas, and which I like to call "bisociation" (as distinct from associative routine). I have discussed this process in some detail in a recent book[12] and shall not dwell on its intricate details. The point I want to make here is that the creation of novelty in mental evolution follows the same pattern of *reculer pour mieux sauter*, of a temporary regression to a naive or juvenile level, followed by a forward leap, which we have found in biological evolution. We can carry the analogy further and interpret the Aha reaction, or "Eureka!" cry, as the signal of a happy escape from a blind alley—an act of mental self-repair, achieved by the de-differentiation of cognitive structures to a more plastic state, and the resulting liberation of creative potentials—the equivalent of the release of genetic growth-potentials in regenerating tissues.

It is a truism to say that in mental evolution social inheritance replaces genetic inheritance. But there is a less trivial parallel between phylogenesis and the evolution of ideas: neither of them proceeds along a continuous curve in a strictly cumulative manner. Newton said that if he saw farther than others it was because he stood on the shoulders of giants. But did he really stand on their shoulders or some other part of their anatomy? He adopted Galileo's laws of free fall, but rejected Galileo's astronomy. He adopted Kepler's planetary laws, but demolished the rest of the Keplerian edifice. He did not take as his point of departure their completed "adult" theories, but retraced their development to the point where it had gone wrong. Nor was the Keplerian edifice built on top of the Copernican structure. That ramshackle structure of epicycles he tore down and kept only its foundations. Nor did Copernicus continue to build where Ptolemy had left off. He went back two thousand years to Aristarchus. The great revolutionary turns in the evolution of ideas have a decidedly paedomorphic character. The new paradigm, to use Thomas Kuhn's[14] term, which emerges from the revolution is not derived from a previous adult paradigm; not from the aged sea-urchin but from its mobile larva, floating in the currents of the ocean. Only in the relatively peaceful periods of consolidation and elaboration do we find gerontomorphism—small improvements to a fully mature body of knowledge. In

觉的，是由自发行为所表达出来的，并且伴随着大量的情感释放。我认为，在这种初级过程与所谓的由现实原则所支配的次级过程之间应当插入一个完整的认知结构等级，这一等级并不是初级过程与次级过程的简单融合，而是在其自身范围的一个自治系统，且受一套不同的规律所支配。妄想症患者的错觉、梦、白日梦、自由联想，儿童不同年龄的心理状态以及不同阶段的原始心理都有着各自的逻辑和规则，因而不能归并在一起考虑。但是，尽管所有这些构思过程在许多方面都存在明显差异，它们依然具有一定的共性特征，因为它们在个体发生上，或者可能在系统发生上都早于那些正常的成人。我在以前的著作[12]中称它们为"地下游戏"，这是因为如果不对其进行克制，他们将严重破坏那些因训练而形成的常规思维。但是，在异常条件下，当常规思维的使用已经达到极限时，短暂地放任这些地下游戏可能会突然产生出一个解决方案，这一解决方案是意识和理性头脑所不能到达的，它是一个新的合成，普安卡雷[13]称其为思想的愉悦合并，而我则喜欢称其为"异缘联想"以区别于联合程序。我已经在最近的一本书[12]里详细地讨论了这一过程，这里就不再赘述。在本文中我想说明的一点是：在心理演进中新事物的产生遵循着与以退为进相似的模式，即经过短暂退行至幼稚水平后，又跳跃式地向前演进，这在生物进化中同样被发现过。我们仍然可以通过运用类推法来解释"啊哈"的反应或"找到了！"的惊呼，它们可以被看作避开死胡同后的一种幸福信号，即一种心理自我修复行为，它经过认知结构去分化后变得更具可塑性，并且同时释放创造性潜能，而这种创造性潜能相当于再生组织中释放的遗传性生长潜能。

众所周知，在心理演进中社会遗传取代了基因遗传。但是，在系统发生与思想进化之间存在一些微小的相似，即两者都不符合连续曲线那样严格的累积方式。牛顿曾说过，如果说他比别人看得更远，那是因为他站在了巨人的肩膀上。但是，他真的就站在巨人的肩膀上或是他们身体的其他部位上吗？他吸收了伽利略的自由落体定律，但却反对其在天文学上的观点。他接受了开普勒的行星定律，但却推翻其所有其他建树。他没有直接以他们完整而"成熟"的理论作为基础继续前进，而是折回到他们理论发展中的错误点。而开普勒的理论大厦也不是在哥白尼的理论上建成的，他将摇摇欲坠的本轮理论推翻，而只保留其基础部分。同样，哥白尼也不是在托勒密的理论上继续前进，而是回溯到两千年前的阿利斯塔克的理论。思想演进中那些伟大的革命性的转折点具有明显的"幼稚形态"特点。用托马斯·库恩[14]的话来讲，从革命中产生的新的范式并不是来源于先前的成熟的范式；不是来源于成年的海胆，而是来源于那些可游动的、在海浪中漂浮的幼虫。只有在相对平静的巩固强化和精心加工的时期，我们才会发现老龄形态，即对完全成熟的知识体系微

the history of art the process is again all too obvious; there is no need to elaborate on it.

I began with a wistful remark about the treacherous wings of analogy, aware of the fact that those who trust these waxen wings usually share the fate of Icarus. But it is one thing to argue from analogy, and quite another to point to an apparent similarity which has perhaps not been paid sufficient attention, and then to ask whether that similarity has some significance or whether it is trivial and deceptive. I believe that the parallel between certain processes underlying biological and mental evolution has some significance. Biological evolution could be described as a history of escapes from over-specialization, the evolution of ideas as a series of escapes from the bondage of mental habit; and the escape-mechanism in both cases is based on the same principles. We get an inkling of them through the phenomena of regeneration—the remoulding of structures and reorganization of functions—which only enter into action when the challenge exceeds a critical limit. They point to the existence of unsuspected "meta-adaptive" potentials which are inhibited or dormant in the normal routines of existence, and, when revealed, make us sometimes feel that we move like sleepwalkers in a world of untapped resources and unexplored possibilities.

It could be objected that I have presented a reductionist view; that it is sacrilegious to call the creation of a Brahms symphony or of Maxwell's field equations an act of self-repair, and to compare it with the mutation of a sea-squirt larva, the regeneration of a newt-tail, the relearning process in the rat or the rehabilitation of patients by psycho-therapy. But I think that such a view is the opposite of sacrilegious. It points, however tentatively, at a common denominator, a factor of purposiveness, without invoking a *deus ex machina*. It does not deny that trial and error are inherent in all progressive development. But there is a world of difference between the random tries of the monkey at the typewriter, and the process which I called, for lack of a better name, *reculer pour mieux sauter*. The first means reeling off all possible responses in the organism's repertory until the correct one is hit on by chance and stamped in by reinforcement. The second may still be called trial and error, but of a purposive kind, using more complex, sophisticated methods: a groping and searching, retreating and advancing towards a goal. "Purpose," to quote Herbert J. Muller[15], "is not imported into Nature and need not be puzzled over as a strange or divine something. ... It is simply implicit in the fact or organisation." This directiveness of vital processes is present all along the line, from conscious behaviour down to what Needham[7] called "the striving of the blastula to grow into a chicken". How tenacious and resourceful that striving is has been demonstrated by experimental embryology, from Speeman to Paul Weiss—though its lessons have not yet been fully digested.

Thus to talk of goal-directedness or purpose in ontogeny has become respectable again. In phylogeny the monkey still seems to be hammering away at the typewriter, perhaps because the crude alternatives that had been offered—amorphous entelechies, or the Lysenko brand of Lamarckism—were even more repellent to the scientific mind. On the other hand, some evolutionary geneticists are beginning to discover that the typewriter is structured and organized in such a way as to defeat the monkey, because it will print

小的提高。在艺术历史上，这个过程依然是十分明显的，这里也不必再赘述。

关于类推法的奥妙，我想以一个引人深思的评论作为开始，我意识到这样一个事实，那就是那些信任蜡质双翼的人往往要面对伊卡洛斯般的命运。但是，这只是类推法争论中的一个方面，另一个方面指的是或许还没有被我们足够重视的外观上的相似性，接着则要思考这种相似性是否具有一定的意义，抑或是微不足道的、具有欺骗性的。我相信，某些潜藏于生物进化过程与心理演进过程之间的相似性具有一定的意义。生物进化可以被描述为是一个逃离高度特化的历史，而心理演进则是逃离心理习惯束缚的一系列事件；两者的逃离机制都是建立在同样的原理上。通过再生现象（即结构重组和功能重塑）我们对它们产生了一个粗浅的认识，再生现象只有当面临的挑战超过临界限度时才发挥作用。它们都指出"超适应"潜能毫无疑问是存在的，它通常以压抑或潜伏的形式存在，而当我们揭开它时，则会使我们感到自己就像梦游者一样，在一个充满着待开发资源和待探索的可能事物中遨游。

人们可能会反对我提出的缩影式观点；将一曲勃拉姆斯交响乐或麦克斯韦电磁方程组的创作称为自我修复行为，并将其与真海鞘幼虫的突变、蝾螈尾部的再生、大鼠的再学习过程或心理治疗后患者的复原一起相提并论，这是亵渎神明的。但是，我的看法却刚好相反，因为它试探性地指出了目的形成因素的共同特性，而没有当成任意的巧合。不能否定尝试与错误在所有发展进程中是固有存在的。但是，猴子在打字机上的随机尝试与我所称之为以退为进的过程完全不是一回事。前者指的是在有机体所有可能的反应库中随意抽取，直到偶然地做出了一次正确的输入为止，之后通过强化而铭刻于心；而后者可能仍然被称为尝试和错误，但是却是有目的性地通过利用更为复杂的方法，如探索和搜寻，向着目标迂回前进。引用赫伯特·马勒 [15] 的话来讲，"目的不是被引入到自然界的，是不应该像陌生的或神圣的事情一样让人迷惑的。……它只是隐含在事实或组织之中"。重要过程的发展趋向性是确定的，沿着从有意识的行为到尼达姆 [7] 所称的"囊胚发育成为一个幼仔的努力"的线路进行。这种努力是不屈不挠的，是足智多谋的，这已经被实验胚胎学研究（从施佩曼到保罗·韦斯）所证实——尽管这些研究成果尚未被人们完全接受。

因此，关于个体发生的目的性或目标的讨论又重新变得很有价值。在系统发生的研究中，猴子似乎依然在打字机上敲打，或许这是因为那些已经被提出的粗略的备选方案——模糊的生命原理或拉马克学说下的李森科主义——被科学思想更多地排斥。另一方面，一些进化遗传学家开始发现打字机的这种结构和组织形式对猴子

only meaningful words and sentences. In recent years the rigid, atomistic concepts of Mendelian genetics have undergone a softening process and have been supplemented by a whole series of new terms with an almost holistic ring. Thus we learn that the genetic system represents a "micro-hierarchy" which exercises its selective and regulative control on the molecular, chromosomal and cellular level; that development is "canalized", stabilized by "developmental homeostasis" or "evolutionary homeostasis"[16] so that mutations affect not a single unit character but a "whole organ in a harmonious way"[17], and, finally, that these various forms of "internal selection" create a restricted "mutation spectrum"[18] or may even have a "direct, moulding influence guiding evolutionary change along certain avenues"[19]—and all this happens long before external, Darwinian selection gets to work. But if this is the case, then the part played by a lucky chance mutation is reduced to that of the trigger which releases the co-ordinated action of the system; and to maintain that evolution is the product of blind chance means to confuse the simple action of the trigger, governed by the laws of statistics, with the complex, purposive processes which it sets off. Their purposiveness is manifested in different ways on different levels of the hierarchy, from the self-regulating properties of the genetic system through internal and external selection, culminating perhaps in the phenomena of phylogenetic self-repair: escapes from blind alleys and departures in new directions. On each level there is trial and error, but on each level it takes a more sophisticated form. Some twenty years ago, Tolman and Krechevsky[20] created a stir by proclaiming that the rat learns to run a maze by forming hypotheses; soon it may be permissible to extend the metaphor and to say that evolution progresses by making and discarding hypotheses.

Any directive process, whether you call it selective, adaptive or expectative, implies a reference to the future. The equifinality of developmental processes, the striving of the blastula to grow into an embryo, regardless of the obstacles and hazards to which it is exposed, might lead the unprejudiced observer to the conclusion that the pull of the future is as real and sometimes more important than the pressure of the past. The pressure may be compared to the action of a compressed spring, the pull to that of an extended spring, threaded on the axis of time. Neither of them is more or less mechanistic than the other. If the future is completely determined in the Laplacian sense, then there is nothing to choose between the actions of the two springs. If it is indeterminate in the Heisenbergian sense, then indeterminacy works in both directions, and the distant past is as blurred and unknowable as the future; and if there is something like a free choice operating within the air-bubbles in the stream of causality, then it must be directed towards the future and oriented by feed-back from the past.

(**208**, 1033-1036; 1965)

References:

1. Huxley, J., *Man in the Modern World*, 13 (New York, 1948).

2. Garstang, W., *J. Linnean Soc. Lond. (Zoology)*, **35**, 81 (1922).

3. Hardy, A. C., in *Evolution as a Process* (New York, 1954).

也是不利的，因为它打印出来的只是一些有意义的词和句子。近些年来，孟德尔遗传学严格的原子论概念已经经历了一个软化过程，并且一整套的新术语已经被补充进去，这些术语几乎可以自成一个完整的体系。因此，我们了解到遗传系统表现出了一种可以在分子、染色体和细胞水平发生选择性调控的"微层次"；我们了解到发育是被"发育动态平衡"或者"进化动态平衡"[16]所"定向的"，从而突变不仅仅是改变一个特征，而是"以和谐的方式影响到所有的器官"[17]，最后，这些不同形式的"内在选择"创建了一个有限的"突变谱"[18]，或者甚至是"直接引导着进化向着某一个方向演进"[19]，并且这些不同形式的内在选择都发生在达尔文自然选择（外在选择）发挥作用之前。但是，如果情况确实如此的话，那么由碰巧产生的随机突变引起的变化部分则是引发系统协同变化的触发器，而要维持进化是偶然变化的产物这一主张，就意味着用已经启动的复杂的有目的的过程去扰乱那个遵循统计学规律的简单的触发行为。从遗传系统在内外部选择下的自我调控，到系统发生的自我修复这一可能的终点，即避开死胡同之后再开启新的方向，它们的目的性在不同的等级水平上表现方式显然不尽相同。在任一水平上都存在尝试和错误，但在任一水平上都以一种更为复杂的方式进行。大约二十多年前，托尔曼和克雷契夫斯基[20]宣布的关于大鼠学会走迷宫的形成假说在科学界引起了轰动；不久，假说在隐喻意义上得到延伸，并又有了关于进化发展的制造和抛弃假说。

任何定向过程，不管你将其称之为选择、适应或是预期，都暗示着未来的可能情形。不考虑所面临的那些障碍和危险时，发育过程（囊胚长成胚胎的努力过程）最终得到同样的结果，这可能会使公正的观测者得出这样的结论：未来的引力是真实存在的，并且有时它比过去的压力更加重要。压力好比是被压缩的弹簧在时间轴上所表现出来的行为，引力则好比是被拉伸的弹簧在时间轴上的行为，两者的力量大小相当。如果未来完全由拉普拉斯的理论所决定，那么不用在这两种弹簧的行为之间做出选择。如果在海森堡的理论中它是不确定的，那么在两个方向上都有不确定的功，则遥远的过去就如未来一样模糊而不可知；而且如果有什么事物（比如自由选择）像一个在众多因果关系的洪流之中的气泡一样，则它必定会在过去的反馈信息的引导下向着未来发展。

（曾菊平 翻译；刘京国 审稿）

4. de Beer, G. R., *Embryos and Ancestors* (Oxford, 1940).

5. Bolk, L., *Das Problem der Menschwerdung* (Jena, 1926).

6. Haldane, J. B. S., *The Causes of Evolution*, 150 (London, 1932).

7. Needham, A. E., *New Scientist*, London, November 2, 1961.

8. Lashley, K. S., *Brain Mechanisms and Intelligence* (Chicago, 1929).

9. Kris, E., *Psychoanalytic Explorations in Art* (New York, 1952).

10. Bruner, J. S., and Postman, L., *J. of Personality*, XVIII (1949).

11. Wallas, G., *The Art of Thought* (London, 1954).

12. Koestler, A., *The Act of Creation* (New York, 1964).

13. Poincare, H., in *The Creative Process* (Berkeley, Calif., 1952).

14. Kuhn, T. H., *The Structure of Scientific Revolutions* (Chicago, 1962).

15. Muller, H. J., *Science and Criticism* (New Haven, Conn., 1943).

16. Cannon, H. G., *The Evolution of Living Things* (Manchester, 1958).

17. Waddington, C. H., *The Listener*, London (Nov. 13, 1952).

18. Spurway, H., in *Supplemento. La Ricerca Scientifica* (Pallanza Symp.), 18 (Cons. Naz. delle Richerche, Rome, 1949).

19. For a survey of literature in this field see Whyte, L. L., *Internal Factors in Evolution* (London, 1965).

20. Krechevsky, I., *Psychol. Rev.*, **39** (1932).

Biological Systems at the Molecular Level

B. Askonas *et al.*

Editor's Note

This report of a meeting of the International Union of Pure and Applied Biophysics in Naples offers a snapshot of what biophysics meant in the 1960s. By this time the key characteristics of the molecular structures of proteins were known, although a seemingly inordinately large part of the discussion here was taken up with issue of how the various amino acids along a protein chain are rotationally ordered. There is little sense yet that it is the more global three-dimensional arrangement that determines a protein enzyme's function. The possibility of cloning antibody-forming immune cells to make specific types of antibody, mentioned here, does however presage the immensely valuable production of "monoclonal antibodies" in the mid-1970s.

A SYMPOSIUM was organized by the Commission of Molecular Biophysics of the International Union of Pure and Applied Biophysics during September 8-11, under the auspices of the International Laboratory of Genetics and Biophysics, Naples—the local arrangements being made by Prof. A. Buzatti-Traverso.

The first day's discussions, under the chairmanship of Prof. H. A. Scheraga (Cornell), were concerned fundamentally with the problem of predicting the conformations of proteins from their primary structure.

In the first paper Dr. S. Lifson (Rehovoth) gave an account of his recent statistical-mechanical investigations of conformational changes in polypeptides. His method consisted in defining sequences, sequence partition functions and sequence-generating functions and in using them to derive an equation for the contribution of each chain element to the partition function of the whole molecule. The results make possible, for example, an assessment of the relative importance of hydrogen bonding and hydrophobic interactions in determining the conformation of poly-L-alanine at different temperatures.

The remaining papers were more closely concerned with the investigation of specific conformations and the limitations imposed on them by interactions between non-bonded atoms. Given that the conformation of the peptide group is generally *trans* and planar, the most important variables determining the conformation of a polypeptide chain are the angles ψ and φ for rotation about the $N-C_\alpha$ and $C_\alpha-C^1$ bonds respectively. Prof. G. N. Ramachandran (Madras) reviewed the earlier work in which he and his colleagues had studied the values of ψ and φ that are allowed when fixed minimum distances of approach are set for the atoms of a dipeptide unit comprising the two peptide groups and the β-carbon joined to an α-carbon atom. This had shown that only certain regions in a

分子水平的生物系统

阿斯康纳斯等

编者按

这篇是在那不勒斯召开的国际纯粹与应用生物物理学联合会的某次会议的报告。它为20世纪60年代生物物理学的含义提供了一个简要的全景概括。尽管看起来这里所讨论的绝大部分内容都是关于各种氨基酸是如何在多肽链中有序旋转的，但那时候人们已经知道蛋白质分子结构中的关键特点。那时尚未了解到是更为全局的三维排布决定了蛋白酶的功能。这里提到了通过克隆抗体的免疫细胞来产生特异性抗体的可能性，不管怎样，这确实预言了20世纪70年代中期具有重大意义的"单克隆抗体"的生产。

9月8日至11日，由国际纯粹与应用生物物理学联合会的分子生物物理学委员会组织的一次研讨会，在位于那不勒斯的国际遗传和生物物理实验室的支持下召开，本次会议的当地筹办工作由布扎蒂-特拉韦尔索教授负责。

第一天的讨论由施拉格教授（康奈尔）担任主席并主持会议，从根本上涉及了由蛋白质一级结构来预测其构象的问题。

在第一个报告中，里夫森博士（雷霍沃特）给出了他最近对多肽构象变化的统计力学研究结果。他的方法包括了定义出序列，序列分区函数以及序列生成函数，以及运用它们得出等式，此等式关乎构成整个分子的分区函数的每一个肽链元素的贡献。例如，这些结果使得我们有可能去评估氢键和疏水相互作用在决定多聚-L-丙氨酸在不同温度下的构象中的相对重要性。

其余报告与特定构象及其被非键合原子间相互作用所强加的局限性的研究密切相关。既然肽基团的构象大体上都是反式和平面化的，那么决定多肽链构象的最重要变量分别是 $N-C_\alpha$ 旋转时的键角 ψ 和 $C_\alpha-C^1$ 旋转时的键角 φ。拉马钱德兰（马德拉斯）教授回顾了早先的工作，在这项工作中，他和他的同事们对组成两个肽基团的二肽单元的原子以及连接 α-碳原子的 β-碳原子间的最小距离确定时的 ψ 和 φ 的允许值进行了研究。他们的研究结果表明，在 ψ、φ 图中，只有特定区域是空间允许的，多肽中已经提出的大多数构象，或在相关结构分析中观测到的大多数构象都分布或

ψ, φ-diagram are sterically allowed and that these regions enclose, or nearly enclose, most of the conformations that have been proposed for polypeptides or observed in analyses of relevant structures. He then described an investigation of the effect of allowing the angle $NC_\alpha C^1$ to vary by $\pm 5°$ from the tetrahedral value (while the deviations from $110°$ of the other angles at C_α were minimized), which showed that the allowed regions in the ψ, φ-diagram are increased, by the introduction of this flexibility, to embrace most of the observed conformations that were previously just forbidden.

Prof. Ramachandran went on to consider the properties of helical polypeptide chains, in which the angles ψ and φ are kept constant at each α-carbon, with particular reference to the main chain—NH $\cdots$ O hydrogen bonding. He showed that families of helices of the right- and left-handed 3.6_{13} or α-type and of the 3.0_{10} type (with hydrogen bonds from peptides 1 to 4, and 1 to 3, respectively) are sterically allowed, with hydrogen bonds of acceptable length that depart from colinearity by less than $30°$, provided that the bond angle at C_α is close to $110°$. On the other hand, right- and left-handed π-helices (1-5 hydrogen bonded) are allowed only when the angle at C_α is $115°$, and the 2.2_7 ribbon structure (1-2 bonded) is permitted only with hydrogen-bond-angles greater than $20°$. The triple helix structure for collagen with two inter-chain hydrogen bonds (angles of $27°$ and $30°$) for every three residues, as proposed in Madras, falls within a fully allowed region in the ψ, φ-diagram, and a new suggestion was put forward that this structure also includes additional CH $\cdots$ O hydrogen bonds between neighbouring chains. When collagen has the sequence gly-pro-hyp in all three chains, however, it appears that only one hydrogen bond can occur for every three residues and the preferred structural parameters closely follow those given by Rich and Crick.

Finally, Prof. Ramachandran discussed the conformations of amino-acid side-chains that have been observed in crystal structures, showing (in keeping with earlier reviews) that they usually adopt fully staggered conformations, but that any one of the possible variants may be adopted in response to a particular environment.

Developments of the limiting-contact approach to the analysis of polypeptide conformations were also described by Dr. G. Némethy. In these studies the influence of various amino-acid side-chains on the conformation of the dipeptide unit was investigated, the results being presented again in terms of the allowed regions in the ψ, φ-diagram. It was shown that the addition of a γ-carbon atom reduces the percentage of allowed conformations form 16 percent for C_β alone to 14 percent, even though there are three rotational positions giving staggered conformations of the side-chain from which to choose. The addition of atoms beyond C_γ in an unbranched side-chain does not limit the range of allowed conformations any more, but the presence of more than one δ-atom, as in leu, reduces the allowed conformations to 11 percent of the possible, and branching at C_β, as in val, ile or thr, reduces this total to only 4.5 percent.

From all these analyses by the limiting-contact method it is clear, therefore, that local steric restrictions, quite apart from the interaction of groups widely separated in the primary

者基本分布在这些特定区域内。他接下来描述了一份当允许角 $NC_{\alpha}C^1$ 发生偏离四面体值 $\pm 5°$（当 C_{α} 的其他角偏离 $110°$ 的值最小时）以内时所产生的影响的研究报告。这表明通过引入这种灵活性，ψ、φ 图中的允许区域增加了，以接受先前禁止的多数观察到的构象。

拉马钱德兰教授接下来考虑螺旋多肽链的特征，其每个 α 碳原子的键角 ψ 和 φ 都保持一致（特别参考了主链中的 $NH \cdots O$ 氢键）。他指出右手螺旋家族和左手螺旋家族的 3.6_{13} 或 α 型，以及 3.0_{10} 型（其肽 1 与肽 4，以及肽 1 与肽 3 分别形成氢键）在空间上都是允许的，倘若 C_{α} 的键角接近 $110°$，其氢键长度位于可接受的范围并且距离共线性最大不超过 $30°$。另一方面，只有在 C_{α} 的键角为 $115°$ 时才允许存在右手和左手 π 螺旋（1–5 氢键连接）；只有在氢键–氢键夹角大于 $20°$ 时才允许存在 2.2_7 带状结构（靠 1–2 氢键连接）。正如马德拉斯所提出的那样，在胶原蛋白的三重螺旋结构中，每三个残基内就有两个链间氢键（键角分别为 $27°$ 和 $30°$），落在 ψ，φ 图中的完全允许范围内，并且提出了一个新建议，即这一结构也包括相邻链间额外的 $CH \cdots O$ 氢键。胶原蛋白在全部三条链中均含有 gly（甘氨酸）-pro（脯氨酸）-hyp（羟脯氨酸）序列，但看起来每三个残基中只能产生一个氢键，并且这种优势结构的参数接近里奇和克里克给出的数据。

最后，拉马钱德兰教授讨论了在晶体结构中已经观测到的氨基酸侧链的构象，显示（与此前综述一致）它们经常采用完全交错的构象，但是，在特殊的环境中，任一可能变体都有可能被采用。

内梅蒂博士也描述了用于分析多肽构象的限制接触法的发展。在这些研究中，调查了各种氨基酸侧链对二肽单元构象的影响，这些结果同样用 ψ，φ 图的允许区域来表示。尽管存在三种旋转位置能够形成侧链的交错构象以供选择，但结果显示，每增加一个 γ 碳原子，所允许构象的百分比会从 C_{β} 单独存在时的 16% 减少为 14%。在无分支侧链中，C_{γ} 之外原子的加入一点也不限制所允许构象的范围，但是一个以上 δ 原子的存在，例如在 leu（亮氨酸）中，会使可能的允许构象减少至 11%，并且 C_{β} 处所产生的分支，例如 val（缬氨酸），ile（异亮氨酸）或 thr（苏氨酸），会使这一总值减少至只有 4.5%。

因此，从所有这些利用限制接触法分析得出的结论很明确：除了一级结构中距离很远的基团间的相互作用之外，局部空间位阻大大限制了多肽链能够形成的构象

structure, severely restrict the number of conformations accessible to a polypeptide chain and go some way (though not yet far enough) towards making practicable the calculation of conformation from amino-acid sequence. A general method for handling such calculations has been pioneered by Prof. A. M. Liquori (Naples), and his colleagues, who described his preliminary attempts to calculate the helical conformations of minimum potential energy, taking into account the forces between non-covalently-bonded atoms. These calculations have led to diagrams of the potential energy as functions of ψ and φ which indicate, for example, that the right-handed α-helix with the original Pauling and Corey parameters is a very stable conformation even without consideration of the hydrogen-bonding that stabilizes it still further. There are four other potential-energy minima, including one corresponding to the left-handed α-helix which is shown, encouragingly, to be less stable than its right-handed counterpart.

Analysis of the structure of myoglobin has shown that these stable helical values of ψ and φ also occur frequently in non-helical regions of the molecule. Prof. Liquori suggested, therefore, that it might be interesting to investigate an idealized structure for myoglobin in which all the dihedral angles were constrained to take the values at the closest of these favoured pairs. Here again, of course, the aim was to reduce the number of conformations that has to be considered to manageable proportions.

Unfortunately there are very few reliable data from which the conformational potential energy of a polypeptide can be calculated. Prof. Liquori presented some evidence to suggest that the general features of the ψ, φ potential energy diagram for helices do not depend very sensitively on the exact shape of the interaction potential curves used in the calculations. He emphasized his belief that while precise energy values are not yet available, proper use of the relatively reliable van der Waals's radii may enable progress to be made towards the prediction of conformations.

In a final short paper in this session, Dr. D. C. Phillips (London) described the structure of lysozyme, recently determined at the Royal Institution, remarking particularly on the occurrence in it of some residues forming an anti-parallel pleated sheet and of some others in the conformation of the 3.0_{10} helix with corresponding hydrogen bonding. Nearly all the helices in the molecule appear to be distorted to some extent from the standard α-structure and there is evidence that the hydrogen bonds in them sometimes depart from linearity by $20°$ or more. He described how comparison of the main features of the structure with the varying hydrophobicity of the amino-acid residues in the primary structure had suggested that the polypeptide chain folds itself from the terminal amino-end, forming first a compact unit with a hydrophobic core, then an extended arm of hydrophilic residues, partly in the β-conformation, and finally a coil that nearly closes the gap between the two parts, leaving a cleft that appears to be the active region, before winding itself around the terminal amino-end. Following ideas developed in collaboration with Dr. P. Dunnill, he noted that analysis of the distribution of hydrophobicity along a polypeptide chain might be useful in predicting conformations if enough guiding principles could be established.

544

的数量，并且在某种程度上（尽管还不够成熟）使根据氨基酸序列来计算构象变得可行。利阔里教授（那不勒斯）和他的同事们首先提出一种处理这种计算的普遍方法，利阔里描述了在计算具有最小势能的螺旋构象的同时将由非共价键连接的原子间的作用力考虑进去的初步尝试。这些计算推导出作为 ψ、φ 的函数的势能图，例如，它表明带有最初的鲍林和科里参数的右手 α 螺旋即使在不考虑使它更加稳定的氢键时也是一种非常稳定的构象。还有另外四个势能最小值，包括已经展示过的一种相当于左手 α 螺旋的结构，令人鼓舞的是，这种左手 α 螺旋结构不如与之对应的右手 α 螺旋结构稳定。

对肌红蛋白结构的分析表明，这些稳定螺旋的 ψ、φ 值也常常出现在分子的非螺旋区域。因此利阔里教授认为对肌红蛋白的一种理想结构进行研究可能是有趣的，这种结构中所有二面角都受到限制，取最接近这些有利于形成配对的数值。当然，这里进行这样研究的目的是为了将不得不考虑的构象个数减少至容易管理的程度。

不幸的是，几乎没有可靠的数据能够用于计算多肽构象的势能。利阔里教授给出的一些证据显示，螺旋所具备的 ψ、φ 势能图的总体特征并不十分敏感地依赖于计算中所使用的相互作用潜在曲线的准确形状。他强调他所相信的，即在仍未获得准确能量值的情况下，合理运用相对可靠的范德华半径也许可以完成预测构象的过程。

在这部分的最后一个小报告中，菲利普斯博士（伦敦）描述了最近由皇家研究院确定的溶菌酶的结构，特别引人注意的是在蛋白内部的一些氨基酸残基形成了一种反平行的折叠片，其他一些残基形成了具有相应氢键的 3.0_{10} 螺旋构象。分子中几乎所有螺旋看上去都发生了相对于标准 α 螺旋结构的不同程度的扭曲，有证据表明，它们中的氢键偏离线性达 20°或更多。他描述了此结构的主要特征与一级结构中氨基酸残基的不同疏水性间的比较，结果认为多肽链从氨基末端开始自我折叠，首先形成一个具有疏水核心的紧密单元，然后形成由亲水残基组成的部分为 β 构象的延伸臂，最终，一个卷曲结构几乎封闭了这两部分间的间隔，在尚未围绕氨基末端完成自我缠绕时，留下了一个看似是活性区域的裂缝。随后的想法是在与丹尼尔博士的合作中发展而来的，他注意到如果建立起足够的指导原则，那么对多肽链的疏水性分布的分析将可能有助于预测构象。

A preliminary calculation of the dihedral angles ψ, φ in lysozyme had shown them to be nearly all in allowed regions of a potential energy diagram calculated for a peptide unit by the use of Lennard–Jones type interaction potentials. This diagram, which was similar to that calculated by Brant and Flory, differed from the original limiting-contact diagram mainly in allowing a greater range of conformations near those appropriate to the 3.0_{10} and the left-handed α-helices. This was presumably because in these regions the occurrence of marginally short contacts, which are forbidden in the limiting-contact analysis, are outweighed, in the calculation of potential energies, by the presence of a large number of favourable contacts. The effect on the ψ, φ-diagram is very similar to that reported by Prof. Ramachandran to result from allowing the $NC_{\alpha}C^1$ angle to depart from $110°$.

These papers provoked a lively discussion in the course of which Dr. F. H. C. Crick (Cambridge) remarked that the study of allowed helical conformations has really progressed very little beyond the results obtained by Donohue from the careful measurement of models. He urged strongly that rigorous attempts should be made to distinguish the important conformational variables, by comparison of the relative energies involved, from among covalent bond-lengths and bond angles, hydrogen bond lengths and angles, van der Waals's contact distances and rotations about bonds, suggesting that calculations based on simple crystal structures might provide necessary criteria for the establishment of valid energy parameters. Dr. J. C. Kendrew (Cambridge) described the helices found in myoglobin, noting that they are less regular than was first supposed and that some residues, particularly at the carboxyl ends of α-helices, are in the 3.0_{10} conformation. There was general agreement that models based only on α-helices and "random coils" must be considered inadequate. In view of the present very large number of investigations in which peptide conformations are described in terms of ψ, φ-diagrams, a plea was made for the general adoption of the standard conventions for labelling rotations that have been drawn up by Dr. G. Némethy. These are shortly to be published in the leading journals.

The second day, under the chairmanship of Dr. G. M. Edelman (New York), was devoted to the structural basis of the immune response. Dr. A. Nisonoff (Urbana) gave an extensive review of work done on structure of immunoglobulins over many years, leading up to the present-day concept of a divalent multi-chain structure consisting of two heavy chains, and two light chains with two antigen-binding sites per molecule. Even specific antibody directed to a single antigen is highly heterogeneous. Four classes of immunoglobulins are known, which share light chains but differ in heavy chains. Although the main antigen-binding site is on the heavy chain, a consensus of opinion is that both light and heavy chains contribute to the configuration of the antigen-binding site. Recent studies suggested that the two heavy chains are held together by only one S—S bond. In closing, Dr. Nisonoff discussed recent work by Drs. Hilschmann and Craig and Dr. Putnam and his colleagues, who have determined a partial amino-acid sequence of three Bence Jones proteins of type I (equivalent or similar to light chains of type I immunoglobulins). The C-terminal half of the molecule is constant in the three Bence Jones proteins, while the

初步计算溶菌酶中的二面角 ψ、φ，结果表明它们几乎都落在了运用伦纳德–琼斯型相互作用势计算得到的多肽单元势能图中的允许区域。此图与布兰特和弗洛里的计算结果相似，而与最初利用限制接触法获得的图不同，主要区别在于它允许更大范围的接近于 3.0_{10} 和左手 α 螺旋的构象。这可能是因为在这些区域中发生了少量短接触，这在限制接触法分析中是不允许的，在大量有利接触存在的情况下，它们在势能的计算中被高估了。这对 ψ、φ 图的影响与拉马钱德兰教授所报道的在允许 $NC_{\alpha}C^{1}$ 角偏离 110° 时所得到的结论非常相似。

这些报告引发了活跃的讨论，在此期间，克里克博士（剑桥）评论道，除了多诺霍从对模型的仔细测量中所获得的结果外，对于被允许的螺旋构象的研究几乎没有取得进展。克里克博士极力主张通过对所涉及的能量进行比较，来进行缜密的尝试以区分出由共价键长和键角、氢键长和键角、范德华接触距离及围绕键的旋转所引起的重要的构象变量，由此指出，基于简单晶体结构的计算也许能为有效能量参数的建立提供必要的标准。肯德鲁博士（剑桥）描述了肌红蛋白中发现的螺旋，它们并不像最初想象的那样规则，一些残基特别是在 α 螺旋的羧基末端的残基形成了 3.0_{10} 的构象。仅基于 α 螺旋和"无规卷曲"的模型肯定会被认为是不完善的，这一观点大家基本达成一致。鉴于目前有非常大量的调查以 ψ、φ 图的形式描述肽段构象，由此恳请大家普遍采用由内梅蒂博士所绘图中所示旋转的标准协议。这些将很快发表于前沿期刊中。

第二天，在埃德尔曼博士（纽约）主持下，主要致力于讨论免疫应答的结构基础。尼索诺夫博士（乌尔瓦纳）全面综述了多年来对免疫球蛋白结构所做的工作，引入当前的这一概念，即二价多链结构由两条重链和两条轻链组成，每个分子带有两个抗原结合位点。甚至指向单个抗原的特异性抗体也是高度异质性的，已知存在四类免疫球蛋白，它们轻链相同，但重链不同。虽然主要抗原结合位点位于重链，但是一致认为轻链和重链都对抗原结合位点的构型有贡献。近期的研究表明两条重链只通过一个 S–S 键结合在一起。最后，尼索诺夫博士讨论了希尔施曼博士、克雷格博士和帕特南博士及其同事们近期的工作，他们测定了三种 I 型本周蛋白的部分氨基酸序列（等于或类似于 I 型免疫球蛋白的轻链）。三种本周蛋白分子中靠近 C 末端的一半保持一致，但靠近 N 末端的另一半则表现出很大的差异性。米尔斯坦博士（剑桥）给出了几种抗原蛋白结构的进一步的数据，表明结果与变异源于单点

N-terminal half shows wide variation. Dr. Milstein (Cambridge) gave further data on the structures of several antigen proteins and suggested that the results were incompatible with the concept that the variations are due to single point mutations. We all await further sequence studies to throw light on the constant and variable regions of the antibody molecules.

Next, Prof. Jerne (Pittsburg) discussed the cellular kinetics of the antibody response when antigen is injected into the whole animal. Studying haemolytic antibody formation by individual spleen cells, using his plaque assay technique, he observed the events following the injection of sheep red blood cells. There is a rapid rise and fall in the number of antibody-forming cells. Subsequent to the injection of sheep red blood cells one is dealing with a cell population multiplying for a few days, after which time the cells reach an end-stage and to not divide into further antibody-forming cells. Dosage of sheep red blood cells affects the slope of the increase in antibody-forming cells as well as the time (in days) taken to reach a peak number of antibody-forming cells. At low doses of sheep red blood cells only two-thirds of the animals tend to respond and the slope of the increase of antibody-forming cells is much flatter. This is not easily explained by a population of reactive cells multiplying at a certain rate. Prof. Jerne discussed various possibilities. He suggested that perhaps the most likely explanation would be that lymph gland has several compartments, and that contact of antigen with reactive cells can only occur in certain parts of the lymph gland, perhaps the germinal follicles.

In the third paper in this section, Dr. M. Cohn (La Jolla) gave a lucid discussion of the potentiality of single cells. Since it has not been possible to clone antibody-forming cells in tissue culture, three approaches have been used. Different investigators have studied: (1) antibodies formed by single cells; (2) immunoglobulins formed by clones of transplantable murine plasma cell tumours; (3) the use of fluorescent staining techniques to detect different types of immunoglobulins in individual cells. The various studies agreed in finding that at least 90–95 percent of the cells make only one antibody, one class of immunoglobulin and one type of light or heavy chain. In an investigation testing for genetic markers on heavy chains less than 5 percent of the cells had a potential to express both alleles. Therefore, only one structural gene appears to be expressed by the majority of cells. The significance of the low percentage of apparently multi-potential cells is not clear at present. No two myeloma proteins have been found to be alike. Therefore, the number of possible light and heavy chains must be very large.

In closing, Dr. Cohn discussed theories of antibody formation, germ-line versus soma. Soma would rely on mutation during the life of the cell and be less useful than the germ-line, which would follow Mendelian genetics.

Dr. B. Askonas (London) discussed the processing of antigens and the role of information at macromolecules in the immune response. She stated that the problem of control of antibody synthesis by antigen runs far behind the other problems. The fate of antigen was discussed at the cellular and biochemical level. Radioactive antigen is taken up by

548

突变这一概念不符。我们都期待进一步的序列研究能弄清楚抗体分子中的恒定区和变化区。

接下来，杰尼教授（匹兹堡）讨论了将抗原注入动物时的抗体应答的细胞动力学。通过采用他的斑块测定技术，研究了由单个脾细胞形成溶血抗体的过程，他观测了在注射了绵羊的红细胞后发生的现象。抗体生成细胞的数量出现快速增长和减少。在注入绵羊的红细胞后，随后几天内，被处理的细胞数量倍增，之后细胞达到末期，不再进一步分裂产生抗体生成细胞。绵羊的红细胞的注入剂量影响着抗体生成细胞的增长斜率，同时也影响着抗体生成细胞的数量达到峰值所需的时间（天数）。当注入低剂量的绵羊红细胞时，只有三分之二的动物倾向于产生反应，而且抗体生成细胞增长曲线的斜率非常平缓。用反应细胞的数量以某一特定速率倍增来解释这些现象并不容易。杰尼教授讨论了多种可能性。他认为也许最有可能的解释是淋巴腺具有很多隔间，而抗原与反应细胞的接触只能发生在淋巴腺的某些部分，有可能是新发生的滤泡。

在这一部分的第三个报告中，科恩博士（拉荷亚）就单个细胞的潜能作了浅显易懂的讨论。由于还不可能在组织培养中克隆抗体生成细胞，那么只好采用以下三种方法。不同的研究者已经对此进行了研究：（1）由单一细胞系生成抗体；（2）由可移植的鼠类血浆细胞瘤的克隆生成免疫球蛋白；（3）用荧光染色技术来检测单个细胞系中的不同类型的免疫球蛋白。这些不同的研究都发现，至少90%～95%的细胞只能产生一种抗体、一类免疫球蛋白，以及一种类型的轻链或重链。在一项检测重链遗传标记的研究中，不到5%的细胞具有同时表达两个等位基因的潜力。因此，看上去大多数细胞只表达一个结构基因。目前只有少量明显具有多潜能的细胞的意义尚不清楚。目前没有发现任何两种骨髓瘤蛋白是相似的，因此，可能存在的轻链和重链的数量必然非常庞大。

最后，科恩博士讨论了抗体形成的理论，比较了种系细胞形成的方式和体细胞形成的方式。体细胞形成的方式依赖细胞周期中发生的突变并且不如种系细胞中发生的突变有效，后者遵循孟德尔遗传学规律。

阿斯康纳斯博士（伦敦）讨论了抗原的处理以及在免疫应答过程中在大分子层面上信息所扮演的角色。她指出通过抗原对抗体合成进行控制这一问题的开展远远落后于其他问题。她在细胞和生化水平上讨论了抗原的命运。放射性抗原被遍布于

phagocytic cells throughout the lymph gland; in the secondary response it is particularly concentrated in the dendritic cells in the germinal centre. Although a major part of the antigen taken up is degraded very rapidly, the remaining antigen persists for weeks and is lost only gradually from the cells. The failure to detect radioactively labelled antigen in antibody-forming cells by G. J. V. Nossall and J. H. Humphrey and their collaborators has shown that there can be only very few antigenic determinants present in the antibody-forming cell.

The question of how antigen stimulates the potential antibody-forming cell is still a vital problem. Whether it does so by direct interaction with the reactive cell or has to go through an intermediary cell needs further clarification. Since the phagocytic cells, the macrophages, take up the antigen they have been implicated as possible intermediary cells. RNA preparations containing antigen can be extracted from macrophages and they are highly active in inducing antibody, but whether this is an essential step in the induction of antibody is not clear. Suggestions that antigen-free informational molecules are transferred from macrophages to the reactive cells have also been made, but this has not been shown convincingly. Further experimentation is required to throw light on this problem.

On the third day a discussion on allosteric enzymes was held with Dr. F. Jacob (Paris) as chairman. The fact that a combination of one molecule of ligand with a macromolecule can influence the combination of another, the same or different, has been known for a long time. The term "allosteric" proteins was introduced by Monod and Jacob to describe proteins in which such interactions occur. Although it is to be expected that many proteins are in some degree allosteric, the introduction of this term has been especially useful in directing attention to a particularly important class of phenomena involving enzymes which, potentially at least, provide an explanation of the regulation of metabolic processes in the organism.

The original observations of such interactions were on haemoglobin which may be considered as a type case of an allosteric protein. This was the subject of Dr. Wyman's talk. Dr. Wyman discussed mainly the haem-haem interaction, considered as a model of interaction between sites for the same ligand.

The second speaker, Dr. J. Monod (Paris), described a model which aims at explaining both the interactions between similar and between different ligands, in terms of quaternary structures of proteins. In this model an allosteric protein is considered to be a polymer with an axis of symmetry, which can exist under at least two different states which are assumed to differ between them by the degree of association between the sub-units. The two states are supposed to differ in their affinity for the ligands which the protein can bind so that the presence of a given ligand can push the equilibrium towards a given state.

The third speaker was Dr. H. K. Schachman (Berkeley, California), who gave a physico-chemical description of the enzyme aspartyl transcarbamylase of *E. coli*. Dr. Schachman showed that the enzyme is made of different sub-units, some of which possess a site

淋巴腺内的吞噬细胞吸收；在二级应答中，放射性抗原特别集中在树突状细胞的生发中心。尽管被吸收抗原的主要部分很快就被降解了，但是抗原的剩余部分仍能维持几周，并且只能从细胞中逐渐消失。诺萨尔、汉弗莱和他们的同事们未能在抗体生成细胞中检测到放射性标记的抗原，这表明在抗体生成细胞中只会有极少量抗原决定簇出现。

抗原如何刺激潜在的抗体生成细胞仍是一个关键问题。它是直接与反应细胞作用还是必须通过中间细胞，这一点尚需进一步明确。因为吞噬细胞，如巨噬细胞吸收抗原，所以它们已经被认为可能是中间细胞。从巨噬细胞中可以提取出包含抗原的 RNA 制剂，它们在诱导抗体的过程中有很高的活性，但是这是否是抗体诱导中的必需步骤仍不清楚。有人也已经建议将无抗原信息的分子从巨噬细胞内转移到应答细胞中，但这尚未给出令人信服的结果，仍需进一步实验来阐明这个问题。

第三天，雅各布博士（巴黎）作为主席主持了一场关于变构酶的讨论。一个大分子与一个配体分子的结合可以影响它与另外一个相同或不同的配体分子的结合，人们很早之前就已经了解到这一事实。莫诺和雅各布引入"变构"蛋白这一术语，用来描述蛋白质中发生的这种相互作用。尽管希望许多蛋白都存在一定程度的变构，但此术语的引入在引起人们对包括酶在内的特别重要的一类现象的关注方面仍然是非常有用的，这些酶至少能为生物体内代谢过程的调节提供一种解释。

对这种相互作用的最初观察是在血红蛋白中进行的，血红蛋白可能被认为是某一类型的变构蛋白，这是怀曼博士的演讲主题。怀曼博士主要讨论了血红素间的相互作用，这被认为是相同配体的位点间相互作用的模型。

第二个报告人莫诺博士（巴黎）描述了一个模型，旨在根据蛋白质四级结构来解释相似配体间和不同配体间的相互作用。在这个模型中，变构蛋白被认为是拥有一个对称轴的多聚体，它至少以两种不同形态存在，假定用亚基间的结合程度来区别这几种不同形态。这两种形态的蛋白和能够与之结合的配体间的亲和力应当不同，因此已知配体的存在可以推动化学平衡向已知状态发展。

第三个报告人是沙赫曼博士（伯克利，加利福尼亚），他从物理化学的角度描述了大肠杆菌的天冬氨酸转氨甲酰酶。沙赫曼博士指出该酶由不同亚基组成，其中一些

specific for one of the substrates, aspartate, and others a site specific for CTP which inhibits the reaction catalysed by the enzyme. Isolated sub-units still exhibit affinity for their respective ligand but without co-operative effects, those being restricted to the complex polymer.

In the discussion, many other enzymes were discussed which exhibit similar behaviour. This is the case, for example, of the enzyme *d*CMP amino hydrolase which has been extensively investigated by Dr. Scarano. In contrast, other complex enzymes appear to operate on a different scheme. This is the case, for example, of the enzyme glutamine synthetase of *E. coli*, investigated by Dr. E. Stadtman, an enzyme the activity of which is susceptible to partial inhibition by eight different compounds of widely different structures.

In the final session on "Molecular Aspects of Differentiation", with Prof. J. Brachet (Brussels) as chairman, the main topic discussed was the synthesis of nucleic acids and proteins during early development. Prof. Monroy (Palermo) described in detail the significance of the events following the process of fertilization in sea-urchin eggs. Ribosomes from unfertilized eggs are not capable of incorporating amino-acids, while those obtained from fertilized eggs are capable of doing so. Prof. Monroy presented the following simple experiment. When RNA from unfertilized eggs was added to liver ribosomes the latter incorporated amino-acids. But no incorporation was detected when RNA from unfertilized eggs was added to ribosomes from unfertilized eggs. He has therefore suggested that some inhibitor is present on the ribosomes which prevents them from synthesizing protein. Though the unfertilized egg has a store of messenger RNA, the ribosomes become active only after fertilization. Prof. Monroy further indicated that inhibition on the inactive ribosomes could be lifted considerably by treating them with trypsin and removing the trypsin by washing through a sucrose layer.

Dr. D. Brown (Baltimore) examined the synthesis of ribosomal, soluble and DNA-like RNA during development of *Xenopus*. The kind of RNA synthesized varies conspicuously as development proceeds. Dr. Brown has been able to show that DNA-like RNA and soluble RNA are synthesized during late cleavage phase and the synthesis continues after gastrulation. The synthesis of ribosomal RNA starts only at the onset of gastrulation and increases as development proceeds.

The final paper was one on cell interactions and carcinogenesis by Dr. L. Sachs (Rehovoth), who discussed *in vitro* studies on the mechanism of carcinogenesis by polyoma virus and by carcinogenic hydrocarbons. In the experiments with polyoma it was shown that virus infection can induce the synthesis of cellular DNA after normal cell DNA synthesis has been repressed by contact inhibition or by X-irradiation, that each cell is induced to synthesize about double its DNA content, and that this induction is not dependent on the replication of viral DNA, but is a function of the viral genome. It was suggested that all the known experimental findings on cell-virus interactions with the small DNA tumour viruses can be explained by the synthesis of a messenger RNA early after virus infection that mediates the induction of cellular enzymes required for DNA synthesis by way of

552

亚基拥有一个底物（天冬氨酸）的特异结合位点，其他亚基则具有结合三磷酸胞苷的特异位点，与三磷酸胞苷的结合可以抑制该酶的催化反应。分离的亚基仍然对它们各自的配体具有亲和力，但是没有表现出那些限于发生在复杂多聚体中的协同效应。

在讨论中，也探讨了其他许多表现出相似行为的酶。例如已经被斯卡拉诺博士广泛研究的脱氧胞苷一磷酸氨基水解酶。相反，其他复杂的酶表现为不同的方式，例如，斯塔特曼博士研究的大肠杆菌谷氨酰胺合成酶，这种酶的部分活性容易被8种结构差异极大的化合物所抑制。

布拉谢教授（布鲁塞尔）作为主席主持了最后一部分关于"分化的分子方面"的会议，讨论的主题是早期发育中核酸与蛋白质的合成。蒙罗伊教授（巴勒莫）详细地描述了海胆卵受精后所发生的一系列事件的重要性。从未受精的卵中获得的核糖体不能结合氨基酸，而那些从受精卵中获得的核糖体则可以结合氨基酸。蒙罗伊教授介绍了下面的简单实验。将从未受精的卵中提取出的 RNA 加入肝脏的核糖体中，该核糖体能够结合氨基酸。但是，将从未受精的卵中提取的 RNA 加入到从未受精的卵中提取的核糖体中，却没有检测到这种结合。因此他认为在核糖体上存在可以阻止其合成蛋白质的一些抑制因子。虽然未受精的卵中存有很多信使 RNA，但是核糖体只在受精后才变得有活性。蒙罗伊教授进一步指出，先用胰蛋白酶处理非活性的核糖体，然后用蔗糖梯级溶液洗去胰蛋白酶，可以显著降低非活性核糖体的抑制作用。

布朗博士（巴尔的摩）检查了非洲爪蟾发育过程中核糖体的、可溶性的和 DNA 状的 RNA 的合成。在发育过程中，各种 RNA 的合成差异显著。布朗博士已能够表明 DNA 状 RNA 和可溶性 RNA 在卵裂后期合成并一直持续到原肠胚形成后。核糖体 RNA 的合成只起步于原肠胚形成的开始期，并且随着发育进行而增加。

最后是萨克斯博士（雷霍沃特）的关于细胞相互作用和癌症发生的一个报告，他讨论了对多瘤病毒和致癌烃类致癌机理的体外研究。多瘤病毒的实验显示出在正常细胞的 DNA 合成已经被接触抑制或被 X 射线抑制后，病毒感染能够引发细胞 DNA 的合成，被引发的每一个细胞的合成大约能使 DNA 的含量加倍，而且这种引发并不依赖于病毒 DNA 的复制，而是病毒基因组的功能。有人提出，所有已知关于细胞病毒与小 DNA 肿瘤病毒间的相互作用的实验发现都可以这样解释，即在病毒感染后的早期，信使 RNA 的合成通过改变细胞表面来调节 DNA 合成所需细胞酶的诱导。致癌烃的实验表明，在体外，这些化学物质能够直接且迅速地引发正常细

alteration of the cell surface. In the experiments with carcinogenic hydrocarbons it was shown that these chemicals can directly and rapidly induce *in vitro* a high frequency of transformation of normal cells to tumour cells. Such *in vitro* investigations provide evidence on the similarities between the two types of carcinogenesis.

(**208**, 1048-1050; 1965)

胞向肿瘤细胞的高频转变。这种体外研究为这两类致癌作用间的相似性提供了证据。

（孙玉诚 翻译；崔巍 审稿）

Globin Synthesis in Thalassaemia:
an *in vitro* Study

D. J. Weatherall *et al.*

Editor's Note

Thalassaemia is an inherited disease of the blood which takes several forma, each corresponding to an abnormal configuration of the haemoglobin molecule. David Weatherall, a British physician and academic scientist, was the first to recognize the medical importance of these diseases, which are particularly common in the Middle East. This study describes the proofs that he and his colleagues obtained, that haemoglobin in thalassaemia is indeed wrongly synthesized, because the genes concerned have been mistakenly reproduced.

THE inherited haemoglobinopathies are of two main types. There are those, like sickle-cell disease, in which a structural change in one of the peptide chains of the globin fraction can be demonstrated, and there are others in which, despite evidence of an inherited defect in globin synthesis, no structural change in the globin moiety can be found. Disorders of haemoglobin synthesis of the second type are often found in patients with the clinical and haematological findings of thalassaemia and are, therefore, designated the "thalassaemia syndromes"[1].

Human adult haemoglobin has a major and a minor component called haemoglobins A and A_2 respectively[2]. Haemoglobin A has two α- and two β-peptide chains ($\alpha_2\beta_2$) and haemoglobin A_2 has two α- and two δ-chains ($\alpha_2\delta_2$) (ref. 3). Foetal haemoglobin, which has usually disappeared by the age of one year, has two α- and two γ-chains ($\alpha_2\gamma_2$) (ref. 4). There is good evidence that the structures of the α-, β-, γ-, and δ-chains are determined by separate pairs of genes[3].

In some patients with the clinical picture of thalassaemia, foetal haemoglobin synthesis persists beyond the first year of life and haemoglobin A_2-levels exceed the normal range of 1.5–3.5 percent of the total haemoglobin[2]. Furthermore, the genetic determinant for this type of thalassaemia behaves as though it were allelic or closely linked to the β-chain locus and also interacts with the sickle-cell (β^s-chain) gene[5]. These observations suggest that this form of thalassaemia results from an inherited defect in β-chain synthesis, the increased amounts of haemoglobins F and A_2 reflecting an attempt at compensation for the deficit in β-chains. The disorder is therefore designated β-thalassaemia.

Some patients with the clinical picture of thalassaemia do not have increased levels of haemoglobin A_2 and F, however, and the genetic determinant segregates separately from the β-chain gene[6]. Some, but not all, individuals in this group carry variable quantities

地中海贫血中珠蛋白的合成：一项体外研究

韦瑟罗尔等

编者按

地中海贫血是一种血液遗传性疾病，包括几种类型，每种类型对应于一种血红蛋白分子的异常构型。英国医生兼学术科学家戴维·韦瑟罗尔首先认识到了这些在中东地区尤为常见的疾病的医学重要性。这个研究描述了他和同事们所获得的实验证据，即在地中海贫血中血红蛋白的合成确实出现了错误，因为相关基因的复制出现了错误。

遗传性血红蛋白病有两种主要类型，一些能被证明的血红蛋白病（如镰状细胞贫血病）在其珠蛋白部分的一条肽链上发生了结构改变，而另一些血红蛋白病，尽管有证据表明，在珠蛋白合成时发生了遗传性缺陷，但珠蛋白部分未发现结构改变。在具有地中海贫血的临床和血液学表现的患者中，经常可以见到第二种类型的血红蛋白合成异常，因此被称为"地中海贫血综合征"[1]。

成年人血红蛋白有一种主要组分和一种次要组分，分别被称为血红蛋白 A 和血红蛋白 A_2[2]。血红蛋白 A 有两条 α 肽链和两条 β 肽链（$\alpha_2\beta_2$），血红蛋白 A_2 有两条 α 肽链和两条 δ 肽链（$\alpha_2\delta_2$）（参考文献 3）。通常到一岁时消失的胎儿血红蛋白，有两条 α 肽链和两条 γ 肽链（$\alpha_2\gamma_2$）（参考文献 4）。有确凿的证据表明，α 链、β 链、γ 链和 δ 链的结构是由单独成对的基因决定的 [3]。

在某些有地中海贫血临床表现的患者中，胎儿血红蛋白的合成可持续到一周岁以后，且血红蛋白 A_2 的水平超出占总血红蛋白 1.5%~3.5% 的正常范围 [2]。此外，该型地中海贫血的遗传定子似乎与 β 链基因座是等位基因或紧密连锁，并与镰状细胞（β^s 链）基因相互作用 [5]。这些观测结果表明该型地中海贫血症是由 β 链合成期间遗传缺陷以及血红蛋白 F 和血红蛋白 A_2 的量增加（反映出试图补偿 β 链的缺陷）引起的。因此，这种疾病被称为 β–地中海贫血。

然而，某些有地中海贫血临床表现的患者并不出现血红蛋白 A_2 和血红蛋白 F 水平的升高，且遗传定子与 β 链基因是单独分离的 [6]。其中，本组中部分个体并不是全部携带有不同数量的血红蛋白 H 和巴特血红蛋白。血红蛋白 H 是正常 β 链的四聚

of haemoglobins H and Bart's. Haemoglobin H is a tetramer of normal β-chains (β_4), while haemoglobin Bart's is composed of four normal γ-chains (γ_4) (refs. 7, 8). It has been suggested, therefore, that this form of thalassaemia results from an inherited defect in α-chain synthesis, α-thalassaemia[9], the resulting excess of β-chains or γ-chains aggregating to form haemoglobins H or Bart's respectively. This concept of the genetic basis of the thalassaemias, while serving as a useful model for their further investigation, is incomplete since there is increasing evidence for the existence of several types of β-thalassaemia and, probably, of α-thalassemia[1].

Little is known about the mechanisms involved in the control of normal α- and β-chain synthesis in man, or about the factors which maintain δ-chain synthesis at about 1/40 of the level of β-chain synthesis. Furthermore, although there is evidence that the reticulocyte ribosomes of thalassaemic individuals have a reduced capacity for globin synthesis[10], there has been no direct evidence for defective α- and β-chain synthesis in these disorders. The chemical structure of these chains appears to be normal[11] and the nature of the apparent defect in synthesis remains quite obscure. The experiments to be described here were carried out in an attempt to clarify some of these problems by comparing the *in vitro* incorporation of radioactive amino-acids into α-, β-, and δ-chains in the reticulocytes of thalassaemic and non-thalassaemic persons.

Reticulocyte-rich blood was obtained from 19 persons with the conditions listed in Tables 1, 2 and 3. 2–8 ml. of washed red cells was suspended in an amino-acid mixture[12] containing 25–100 μc. of uniformly labelled ^{14}C-leucine, and incubated in a Dubnoff metabolic shaker at 37°C, aliquots being removed at various times and immediately frozen. Haemoglobin fractions were purified by dialysis, column chromatography, and starch-block electrophoresis as previously described[13].

Table 1. Relative Specific Activities of α- and β-Chains of Haemoglobin A prepared from Red Cells of Patients with a Variety of Haematological States. ^{14}C-Leucine was used in Each Case

Clinical disorder	Incubation time (min)	Specific activity of α-chain (c.p.m./mg)	Specific activity of β-chain (c.p.m./mg)
Hereditary spherocytosis	4	15	28
	15	88	77
	45	138	163
	240	422	439
Hereditary spherocytosis	2	39	93
	4	105	210
	200	1,617	1,652
Hereditary spherocytosis	2	189	616
	4	691	1,470
	300	8,960	9,296
Pyruvate kinase deficiency	4	418	595
	15	2,240	2,030
	60	3,780	4,116
	240	6,930	6,370

体（β₄），而巴特血红蛋白是由 4 条正常的 γ 链组成（γ₄）（参考文献 7，8）。因此，可以认为该型地中海贫血是由于 α 链合成期间的遗传缺陷引起的，即 α–地中海贫血[9]，由此产生的过量 β 链或 γ 链聚集，分别形成了血红蛋白 H 或巴特血红蛋白。关于地中海贫血遗传学基础的这种概念是不完整的，尽管在对其进一步研究中是个有用的模型，因为越来越多的证据表明存在多种类型的 β–地中海贫血，可能也存在多种类型的 α–地中海贫血[1]。

关于人体内正常 α 链与 β 链合成控制的机制，或者是什么因素让 δ 链的合成保持在 β 链合成水平的 1/40 左右，目前所知甚少。此外，尽管有证据表明，在地中海贫血个体中网织红细胞核糖体合成珠蛋白的能力下降[10]，但缺乏这些疾病 α 链、β 链合成缺陷的直接证据。这些链的化学结构看上去是正常的[11]，但在合成期间明显缺陷的本质仍然相当模糊。本文将要描述的实验是通过比较在体外将放射性氨基酸掺入地中海贫血患者和非地中海贫血患者的网织红细胞的 α 链、β 链和 δ 链中实现的，进而试图阐明其中的一些问题。

富含网织红细胞的血液来自 19 个受试者，其基本状况见表 1、表 2 和表 3。将 2 毫升 ~8 毫升洗过的红细胞悬浮于含有 25 微居里 ~100 微居里标记均匀的 ¹⁴C–亮氨酸的氨基酸混合液中[12]，在 37℃ 下，在杜布诺夫代谢摇床中温育，在不同时间取出等份样品并立即冷冻。按以前所述的透析、柱层析和淀粉阻滞电泳纯化血红蛋白组分[13]。

表 1. 由各种血液病患者红细胞制备的血红蛋白 A 的 α 链和 β 链的相对比活。
各例中均使用 ¹⁴C–亮氨酸。

临床疾病	温育时间 （分钟）	α 链的比活 （每分钟计数 / 毫克）	β 链的比活 （每分钟计数 / 毫克）
遗传性球形红细胞增多症	4	15	28
	15	88	77
	45	138	163
	240	422	439
遗传性球形红细胞增多症	2	39	93
	4	105	210
	200	1,617	1,652
遗传性球形红细胞增多症	2	189	616
	4	691	1,470
	300	8,960	9,296
丙酮酸激酶缺陷	4	418	595
	15	2,240	2,030
	60	3,780	4,116
	240	6,930	6,370

Table 2. Relative Specific Activities of α- and β-Chains of Haemoglobin A prepared from the Red Cells of Patients with β-Thalassaemia and Haemoglobin H Disease (α-Thalassaemia). ^{14}C-Leucine was used in Each Case

Clinical disorder	Incubation time (min)	Specific activity of α-chain (c.p.m./mg)	Specific activity of β-chain (c.p.m./mg)
Thalassaemia major	30	494	266
Thalassaemia major	30	840	470
Thalassaemia major	30	1,638	1,043
Thalassaemia major	30	242	61
Thalassaemia major	30	89	24
Thalassaemia major	60	126	180
Thalassaemia major	30	494	266
	180	3,064	1,904
	12 h	4,718	3,924
Thalassaemia major	3	627	373
	10	945	441
	30	1,638	1,043
	180	4,382	2,618
Haemoglobin H disease	30	1,660	277
Haemoglobin H disease	30	260	65
Haemoglobin H disease	30	80	10
Haemoglobin H disease	30	267	19
Haemoglobin H disease	240	67	13
	20 h	100	43
Haemoglobin H disease	10	35	2
	30	127	11
	180	189	32

Table 3. Distribution of Radioactivity in the Peptide Chains of Haemoglobins A and A$_2$ after 60-min Incubation of the Red Cells with ^{14}C-Leucine

Haemoglobin type	Specific activity of α-chain (c.p.m./mg)	Specific activity of β-chain (c.p.m./mg)	Specific activity of δ-chain (c.p.m./mg)
A	59.3	52.7	—
A$_2$	32.1	—	6.9

In some experiments ribosomes were prepared[14] before fractionation of the haemoglobins. Globin was prepared[15] from either purified haemoglobin fractions or, in some cases, directly from the washed whole-cell lysates without further purification. The α- and β-chains were separated by gradient elution chromatography on carboxymethylcellulose columns in urea/mercaptoethanol buffers as recently described[16]. In some experiments, chain separation was also achieved by hybridization with canine haemoglobin or by counter-current distribution[17]. The purity of each chain was checked by fingerprinting[16] and the radioactivity measured, after plating out about 0.5 mg of protein on aluminium planchets, using a low-background gas-flow counter[13]. In order to determine the distribution of radioactivity in finished or partly finished chains on the ribosomes, 5–10 mg of unlabelled carrier globin was added to the ribosomal pellets and the α- and β-chains of the mixture separated by carboxymethylcellulose chromatography[16].

Non-thalassaemic Reticulocytes

The specific activities of the separated α- and β-chains of purified haemoglobin A from

表 2. 由 β–地中海贫血患者和血红蛋白 H 病（α–地中海贫血）患者红细胞制备的血红蛋白 A 的 α 链和 β 链的相对比活。各例中均使用 ^{14}C–亮氨酸。

临床疾病	温育时间（分钟）	α 链的比活（每分钟计数 / 毫克）	β 链的比活（每分钟计数 / 毫克）
重型地中海贫血病	30	494	266
重型地中海贫血病	30	840	470
重型地中海贫血病	30	1,638	1,043
重型地中海贫血病	30	242	61
重型地中海贫血病	30	89	24
重型地中海贫血病	60	126	180
重型地中海贫血病	30	494	266
	180	3,064	1,904
	12 小时	4,718	3,924
重型地中海贫血病	3	627	373
	10	945	441
	30	1,638	1,043
	180	4,382	2,618
血红蛋白 H 病	30	1,660	277
血红蛋白 H 病	30	260	65
血红蛋白 H 病	30	80	10
血红蛋白 H 病	30	267	19
血红蛋白 H 病	240	67	13
	20 小时	100	43
血红蛋白 H 病	10	35	2
	30	127	11
	180	189	32

表 3. 红细胞在 ^{14}C–亮氨酸中温育 60 分钟后，血红蛋白 A 和 A_2 肽链中放射性分布情况。

血红蛋白类型	α 链的比活（每分钟计数 / 毫克）	β 链的比活（每分钟计数 / 毫克）	δ 链的比活（每分钟计数 / 毫克）
A	59.3	52.7	–
A_2	32.1	–	6.9

在一些实验中，核糖体是在血红蛋白分离前制备的 [14]。珠蛋白或是由纯化的血红蛋白组分制备，或是在某些情况下由洗过的全细胞裂解物（未经进一步纯化）直接制备 [15]。α 链和 β 链的分离采用最近描述的方法 [16]，即在羧甲基纤维素柱上使用尿素 / 巯基乙醇缓冲液进行的梯度洗脱层析法。在一些实验中，通过与犬血红蛋白杂交或逆流分布法也能实现链的分离 [17]。采用指纹图谱法检测每种链的纯度 [16]，并将约 0.5 毫克蛋白铺在铝板上，用低背景气流计数仪测量放射性强度 [13]。为确定核糖体上已合成或部分合成的肽链的放射性分布情况，将 5 毫克 ~ 10 毫克未标记的载体珠蛋白加到核糖体沉淀物中，并用羧甲基纤维素层析法分离混合物中的 α 链和 β 链 [16]。

非地中海贫血的网织红细胞

表 1 总结了从各种非地中海贫血病症纯化的血红蛋白 A 的单独的 α 链和 β 链的

a variety of non-thalassaemic conditions are summarized in Table 1 and the distribution of radioactivity in a typical chain separation, after 30 min incubation, is shown in Fig. 1. Under all conditions investigated the specific activities of the α- and β-chains were very similar at incubation times of 5 min to 5 h. At shorter times, however, the specific activity of the β-chain was always significantly greater than that of the α-chain. The pattern of radioactivity in separated α- and β-chains obtained from addition of carrier globin to ribosomal pellets was always the same in this group. Thus, the radioactivity under the α-chain peak always exceeded that under the β-chain peak, the ratio α-chain/β-chain ranging from 1.5 to 2.1/1. Clear peaks of radioactivity were seen for both chains in each case, the radioactivity peak usually being slightly displaced from the protein peak. These results indicate that since the number of leucine residues in α- and β-chains is the same, α- and β-chain synthesis is normally synchronous. The fact that the specific activity of the β-chain is greater than that of the α-chain at short times of incubation, and the presence of an excess of radioactivity under the α-chain peak from the ribosomes after long periods of incubation, implies that a small pool of α-chain exists on the ribosomes. The existence of only a small peak of radioactivity associated with the β-chain from the ribosomes indicates that release of β-chains from the ribosomes is much more rapid than that of α-chains, but is not instantaneous. These observations suggest that finished β-chains are probably necessary for the removal of α-chains from the ribosomes. A similar situation has been described in the rabbit reticulocyte[18] where the discrepancy in chromatographic behaviour between radioactivity (newly made chain) and carrier chain was also noted. Whether this means that the newly made chain, while still on the ribosome, is chemically different from that found in finished haemoglobin is not clear, but this would seem very likely.

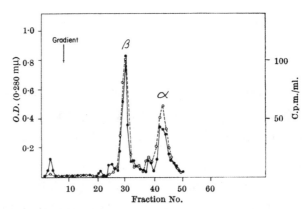

Fig. 1. Incorporation of radioactivity into the α- and β-chains of haemoglobin A prepared from the red cells of a patient with hereditary spherocytosis. Incubation period 45 min. ●—●, O.D. (0.280 mμ); ○---○, radioactivity.

In order to follow the incorporation of radioactive amino-acids into the α- and δ-chains of haemoglobin A_2, reticulocytes from an individual with hereditary spherocytosis were incubated with ^{14}C-leucine for 60 min and haemoglobins A and A_2 isolated[13]. The α- and β-chains of haemoglobin A and the α- and δ-chains of haemoglobin A_2 were then

比活，图 1 则给出了一种典型的分离链在温育 30 分钟后放射性的分布情况。在研究的所有病症中，当温育时间为 5 分钟 ~ 5 小时，α 链和 β 链的比活非常相似。但是，当温育时间较短时，β 链的比活总是显著大于 α 链的比活。在这一组，向核糖体沉淀物中加入载体珠蛋白，所得到的单独的 α 链和 β 链的放射性模式总是相同的。这样，α 链峰的放射性总是超过 β 链峰的放射性，α 链和 β 链的比值在 1.5∶1~2.1∶1 的范围内。在各例中都可见到两条链的清晰的放射性峰，放射性峰通常略偏离蛋白质峰。这些结果表明，由于 α 链和 β 链中亮氨酸残基的数目相同，所以 α 链和 β 链的合成通常是同步的。在较短的温育时间下，β 链的比活大于 α 链的比活，而长时间温育后，核糖体中 α 链峰出现过量的放射性这一事实表明核糖体上有一个小的 α 链库。只有一个小的与核糖体 β 链相关的放射性峰，这表明从核糖体中释放的 β 链要比 α 链快得多，但并不是瞬时的。这些观测结果表明，已合成的 β 链也许对核糖体中 α 链的释放是必要的。在兔网织红细胞中已描述了类似的情况 [18]，人们也发现了带放射性的新合成链和载体链的色谱行为有区别。我们还不清楚这是否意味着仍位于核糖体上的新合成链与在已合成的血红蛋白中发现的化学结构不同，但很可能是这样的。

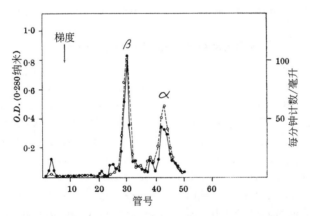

图 1. 掺入到由一位遗传性球形红细胞增多症患者的红细胞制备的血红蛋白 A 中 α 链和 β 链的放射性。温育时间为 45 分钟。●—●，$O.D.$ (0.280 纳米)；○ --- ○，放射性。

为了跟踪放射性氨基酸掺入到血红蛋白 A_2 的 α 链和 δ 链的情况，将一位遗传性球形红细胞增多症个体的网织红细胞与 ^{14}C–亮氨酸温育 60 分钟，并将血红蛋白 A 和 A_2 分离 [13]。然后将血红蛋白 A 的 α 链和 β 链及血红蛋白 A_2 的 α 链和 δ 链分开，

separated and the specific activities of the whole haemoglobin fractions and separated chains determined (Table 3). As in previous experiments[13] the specific activity of haemoglobin A was more than twice that of haemoglobin A_2. The specific activities of the α- and β-chains of haemoglobin A were very similar. The specific activity of the α-chain of haemoglobin A_2 was less than that of haemoglobin A—this finding being compatible with the finding[13] that, in the *in vitro* system, haemoglobin A_2 synthesis in reticulocytes is retarded before that of haemoglobin A. This observation cannot, however, explain the ratio of specific activity of α-chain to δ-chain of 5/1 in haemoglobin A_2. It has recently been suggested that one reason for the relatively slow rate of δ-chain synthesis might lie in the presence of one or more "slow points" during assembly of the δ-chain[20]. In such an event, δ-chain clearance from the ribosomes would be slow relative to α-chain clearance. Thus, in investigation of radioactive incorporation, the time taken to achieve uniform δ-chain labelling would be longer than that required for uniform α-chain labeling, resulting in the marked difference in specific activity between α-chain and δ-chain observed in these experiments. These findings are thus compatible with the recent observation that uniform labelling of the δ-chain does in fact take a long period of incubation[20]. Another explanation would be the exchange of α-chains between newly made haemoglobin A and haemoglobin A_2, the synthetic rate of which falls off so rapidly in peripheral blood, but this mechanism seems very unlikely.

These results suggest, therefore, that there are "slow points" in δ-chain synthesis giving rise to a relatively slow rate of clearance from the ribosomes. It is unlikely that the slow rate of δ-chain synthesis is associated with a quantitative reduction in messenger RNA since, although this would result in the production of fewer δ-chains, their specific activity would be similar to that of α-chains, if the rates of assembly of the two chains were comparable.

β-Thalassaemia Reticulocytes

The findings in the experiments utilizing cells from persons with β-thalassaemia, which were quite different from those in non-thalassaemic samples, are summarized in Table 2. The specific activity of the α-chain of purified haemoglobin A always exceeded that of the β-chain, the ratios α/β ranging from 1.5/1 to 7/1. These values were obtained at incubation times of 30 min–5 h, the differences being less marked after longer periods of incubation (Fig. 2). These ratios were similar to those recently reported[19]. In order to rule out the possibility of variation of leucine pool size for the two chains in this disorder, experiments were also performed with ^{14}C-lysine and ^{14}C-valine, similar differences in specific activity being noted. Such differences between the specific activities of the α- and β-chains in finished haemoglobin A could occur if a large intracellular pool of β-chain existed at any given time. After introduction of the radioisotope, newly made (and therefore labelled) β-chain would be diluted out by pre-existing unlabelled β-chains present in the pool. To examine this possibility, two experiments were carried out in which a chain separation was performed on a washed, whole-cell lysate without prior purification of haemoglobin fractions (Fig. 3). The recovery of protein and radioactivity exceeded 90 percent in each case. Fingerprinting of each peak showed only α- and β-chain peptides, no new spots being observed. In both experiments the amount of radioactivity under

测定全血红蛋白组分和单独链的比活（表 3）。与之前的实验 [13] 相同，血红蛋白 A 的比活是血红蛋白 A_2 比活的两倍多。血红蛋白 A 的 α 链和 β 链的比活非常接近，血红蛋白 A_2 的 α 链比活低于血红蛋白 A 的 α 链比活，这一结果与体外系统中网织红细胞血红蛋白 A_2 的合成滞后于血红蛋白 A 的合成的结果 [13] 是一致的。不过，这一观测结果不能解释为什么在血红蛋白 A_2 中 α 链和 δ 链的比活的比值是 5∶1。近来有人认为 δ 链的合成速度较慢的一个原因可能是 δ 链组装时存在一个或多个"限速位点"[20]。在这种情况下，δ 链从核糖体中的释放将相对慢于 α 链。这样，在放射性掺入的研究中，δ 链均一标记所需的时间将会长于 α 链均一标记所需的时间，导致在这些实验中观察到的 α 链和 δ 链的比活显著不同。这些结果也与最近观察到的 δ 链均一标记实际上确实需要很长的温育时间相一致 [20]。另一解释是新生成的血红蛋白 A 和血红蛋白 A_2 间的 α 链相互交换，在外周血中，血红蛋白 A_2 的合成速率迅速下降，但这种机制似乎很不可能。

因而，这些结果表明，在 δ 链的合成中存在"限速位点"，导致其从核糖体中释放的速率相对慢。δ 链合成的速率降低不可能与信使 RNA 的量减少有关，因为尽管信使 RNA 量的减少会导致合成的 δ 链减少，但如果两条链组装的速度相似，那么它们的比活将与 α 链的比活相似。

β- 地中海贫血的网织红细胞

使用 β-地中海贫血患者的细胞的实验结果与使用非地中海贫血患者的样品得到的结果完全不同，表 2 总结了这些结果。纯化的血红蛋白 A 中 α 链的比活总是高于 β 链的比活，当温育时间为 30 分钟~5 小时，两者比值在 1.5∶1~7∶1 范围内，较长时间温育后，两者的差异将不明显（图 2）。这些比值与最近报道的比值相似 [19]。为了排除在这种疾病中两种链的亮氨酸库大小存在差异的可能性，我们还用 ^{14}C- 赖氨酸和 ^{14}C-缬氨酸进行了实验，也发现了类似的比活差异。如果在任何给定时间，细胞内都存在一个大的 β 链库，则已合成的血红蛋白 A 的 α 链和 β 链之间的比活就会出现这种差异，在导入放射性同位素之后，新合成的（即被标记的）β 链会被库中已存在的、未标记的 β 链稀释。为了检查这种可能性，进行了两次实验，没有预先纯化血红蛋白组分，而是用洗涤过的全细胞裂解物进行链分离（图 3）。在各例中蛋白质与放射性的回收率都超过 90%。每个峰的指纹图谱只显示出 α 链和 β 链肽段，未观察到新斑点。在两个实验中，β 链峰的放射性量都与纯化的血红蛋白 A（用等量的初始红细胞裂解物制备）的 β 链的放射性量相似。在一次实验中，在 β 链峰之

the β-chain peak was similar to that found in the β-chain from purified haemoglobin A, which had been prepared from an equivalent amount of the original red-cell lysate. In one experiment a large peak of protein and radioactivity was eluted before the β-chain peak, this being the γ-chain of haemoglobin F, while in another too little haemoglobin F was present for clear separation of β- and γ-chains. These results thus excluded the presence of a large intracellular pool of β-chain, with subsequent dilution of newly labelled chains, as a basis for the observed differences in specific activity in α- and β-chains.

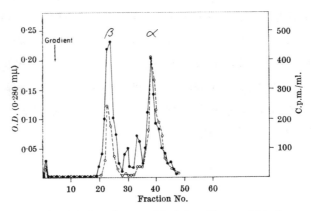

Fig. 2. Incorporation of radioactivity into the α- and β-chains of haemoglobin A prepared from the cells of a patient with β-thalassaemia major. Incubation time 60 min. ● — ●, *O.D.* (0.280 mμ); ○ --- ○, radioactivity.

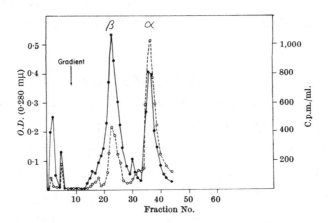

Fig. 3. Distribution of radioactivity compared with protein in a washed whole-cell lysate prepared from the red cells of a patient with β-thalassaemia major. Incubation time 60 min. ● — ●, *O. D.* (0.280 mμ); ○ --- ○, radioactivity.

Such differences in specific activity could occur, however, if there were a large block (that is, rate-limiting step) at some point during the assembly of the β-chain or its release from the ribosome, associated with the presence on the ribosomes at any given time of many completed or partially completed chains. As in the case of normal δ-chains already discussed, relatively long periods of time would then be required to achieve uniform labelling of newly synthesized β-chain. At long times, each newly synthesized chain would

566

前洗脱出一个大的蛋白质和放射性峰，这是血红蛋白 F 的 γ 链，而在另一实验中，只有很少的血红蛋白 F，难以明确分开 β 链和 γ 链。因此，这些结果排除了在细胞内存在大的 β 链库的可能性，也不存在随后新标记的链被稀释，以及由此可能导致的 α 链和 β 链比活的差异。

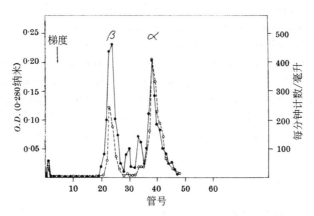

图 2. 掺入到由一位重型 β–地中海贫血患者细胞制备的血红蛋白 A 中 α 链和 β 链的放射性。温育时间为 60 分钟。●—●，O.D.(0.280 纳米)；○ --- ○，放射性。

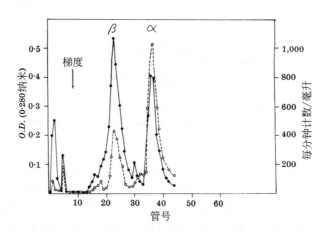

图 3. 在一位重型 β–地中海贫血患者红细胞制备的洗涤过的全细胞裂解物中蛋白质与放射性分布的比较情况。温育时间为 60 分钟。●—●，O.D.(0.280 纳米)；○ --- ○，放射性。

　　然而，如果在 β 链组装或从核糖体上释放的过程中的某个点存在大的阻塞（即限速步骤），与在任何给定的时间核糖体上存在大量已合成或部分合成的肽链有关，也会发生这种比活的差异。正如已经讨论过的关于正常 δ 链的情况一样，要使新合成的 β 链达到均一标记则需要较长的时间。当时间较长时，每条新合成的肽链就会

then have the same specific activity as an α-chain synthesized from the same amino-acid pool. The time required to achieve uniform labelling would thus depend on the severity of the block, that is, of the ease with which unlabelled β-chains present at the time of introduction of the radioactive amino-acid were cleared from the ribosomes.

One consequence of a rate-limiting step in synthesis, of the sort already discussed, would be the absence of radioactivity associated with β-chain on the ribosomes, in contrast to normal ribosomes where both labelled α- and β-chains can be shown to be present after a few minutes incubation (see earlier). Prolonged periods of incubation would be required to clear pre-existing unlabelled β-chains from the ribosomes and replace them with labelled chains. Ribosomes were therefore prepared from the red cells of two patients, homozygous for β-thalassaemia, after incubation of the cells with ¹⁴C-leucine for 45 min. After addition of carrier globin to the ribosome pellet, the chains were separated by column chromatography. No peak of radioactivity corresponding to the β-chain was observed, while a large peak was seen in the α-chain region in each case (Fig. 4).

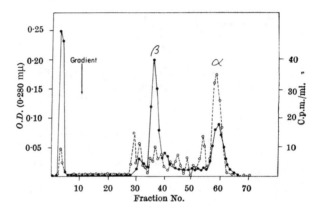

Fig. 4. Distribution of radioactivity after addition of 10 mg unlabelled carrier globin to the ribosomal pellet prepared from the cells of a patient with β-thalassaemia major, Incubation time 45min. ● — ●, O. D. (0.280 mμ); ○ --- ○, radioactivity.

These findings are thus compatible with the concept of a gross defect in β-chain synthesis at the ribosomal level in β-thalassaemia. Whether, since α-chains appear to require finished β-chains for their release from the ribosomes, this results in a secondary accumulation of finished α-chain on the ribosomes is uncertain, although this seems likely since the relative amount of radioactivity in the α-chain fraction was several times that observed in ribosomes obtained from non-thalassaemic reticulocytes. Further attempts to demonstrate the site of this proposed block in β-chain synthesis have been made by comparing the ratios of the specific activities of α- and β-chains of purified haemoglobin A prepared from four parallel incubations of aliquots of the same red cell sample using ¹⁴C-labelled leucine, arginine, histidine, and tyrosine. Since these amino-acid are not distributed evenly along the peptide chain, the specific activities of non-uniformly labelled β-chains might be expected to differ with each amino-acid. The specific activities of individual peptides[14] of thalassaemic β-chain were also measured. No definite site of delay in synthesis has yet been demonstrated, although the results have not excluded a "slow point" either at, or

568

与从同样的氨基酸库中合成的 α 链有相同的比活。这样，使链达到均一标记所需的时间取决于阻塞的强度，即在轻度阻塞时，那些在导入放射性氨基酸时存在的未标记的 β 链将从核糖体中释放出来。

这类合成过程（已讨论过的那种）中限速步骤的后果之一，是造成核糖体上与 β 链相关的放射性的缺失，这与正常核糖体上标记的 α 链和 β 链均可在温育几分钟后出现的情况相反（见前文）。释放核糖体中已存在的未标记 β 链并用标记的 β 链将其替换，需要很长的温育时间。将两位纯合 β–地中海贫血患者的红细胞与 ^{14}C–亮氨酸温育 45 分钟，然后制备核糖体。向核糖体沉淀物中加入载体珠蛋白后，用柱层析分离两种肽链。在各例中均未观察到对应于 β 链的放射性峰，而在 α 链区域则见到一个大的峰（图 4）。

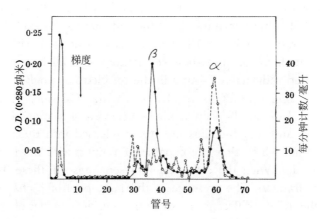

图 4. 将 10 毫克未标记的载体珠蛋白加到由一位重型 β–地中海贫血患者细胞制备的核糖体沉淀物中后的放射性分布情况。温育时间为 45 分钟。●—●，O.D.(0.280 纳米）；○---○，放射性。

因此，这些结果与 β–地中海贫血患者的 β 链在核糖体水平的合成存在严重缺陷这一观点是一致的。由于 α 链似乎需要已合成的 β 链才能从核糖体中释放，这是否会导致已合成的 α 链在核糖体上二次积累尚不清楚，尽管有可能如此，因为 α 链组分放射性的相对量是在非地中海贫血网织红细胞的核糖体中观察到的放射性相对量的数倍。为进一步确定在 β 链合成中这种提议的阻塞的位点，由同一红细胞样品的四种平行温育小样（分别用 ^{14}C–亮氨酸、^{14}C–精氨酸、^{14}C–组氨酸和 ^{14}C–酪氨酸温育）制备纯化血红蛋白 A，比较其 α 链和 β 链比活的比值。由于这些氨基酸在肽链中的分布不均匀，因此使用每种氨基酸非均一标记 β 链的比活预计会不同。同时还测量了地中海贫血每个 β 链的比活 [14]。尚未确定合成中确切的延迟位点，尽管这些结果尚未排除位于或者靠近 β 链的羧基末端存在一个"限速位点"。

near, the carboxyl-terminal end of the β-chain.

α-Thalassaemia (Haemoglobin H Disease) Reticulocytes

The results of six ^{14}C-leucine incubation experiments performed on the cells of persons with haemoglobin H thalassaemia are shown in Table 2. At incubation times of 30 min–5 h the specific activity of the α-chain of purified haemoglobin A was 1.5–15 times that of the β-chain (Fig. 5). This difference became less marked with more prolonged incubation of the red cells. Because of these findings, which were surprisingly like those in β-thalassaemia, the possibility of a large intracellular pool of β-chain diluting out newly made and, therefore, labelled β-chains again had to be examined. Washed whole-cell lysates from two persons with haemoglobin H disease were converted to globin without prior purification of the haemoglobins and the α- and β-chains separated (Fig. 6). The radioactivity in the β-chain fraction exceeded that in the α-chain fraction by a factor of 2.3/1 in one case and 3.0/1 in the second, while there was an associated increase in the optical density values for the β-chain peak. In a separate experiment using identical amounts of the same two samples of washed whole-cell lysates, the amounts of haemoglobin A and H were determined and the radioactivity in each fraction measured. The excess of both radioactivity and optical density associated with the β-chain (over that present in the α-chain) of the whole-cell lysate could all be accounted for by the amount of protein and radioactivity found in the purified haemoglobin H fraction in the second experiment. Similar results were observed with ^{14}C-lysine. These results suggest that in haemoglobin H disease β-chain synthesis occurs at a rate of about 2–3 times that of α-chain synthesis, and that β-chains are freely released into the red cell where they form a large pool from which they are capable of uniting with newly made α-chains as these become available. When the cell is haemolysed, most of these β-chains appear in the haemoglobin H fraction. The presence of this large pool of β-chains at any one time probably explains the striking difference between the specific activities of the α- and β-chains of haemoglobin A in haemoglobin H disease. Newly-made labelled α-chains will combine with unlabelled β-chains which were already present in this pool. Whether the excess β-chains exist as haemoglobin H (β$_4$) in the red cell is uncertain, but at least in an *in vitro* system haemoglobin H readily combines with α-chains to form haemoglobin A[21].

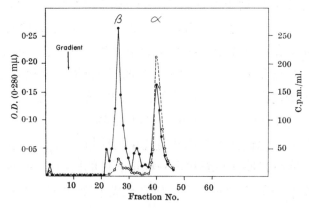

Fig. 5. Distribution of radioactivity in the α- and β-chains of haemoglobin A from the cells of a patient with haemoglobin H disease. Incubation time 45 min. ● — ● , *O.D.* (0.280 mμ); ○ --- ○ , radioactivity.

α– 地中海贫血（血红蛋白 H 病）的网织红细胞

表 2 给出了对血红蛋白 H 地中海贫血患者的细胞进行 6 次 ^{14}C–亮氨酸温育的实验结果。当温育时间在 30 分钟~5 小时之间时，纯化血红蛋白 A 中 α 链的比活是 β 链的 1.5~15 倍（图 5）。随着红细胞温育时间的延长，这一差异将变得不太明显。由于这些结果与 β–地中海贫血的结果惊人地相似，所以必须再次检测大的胞内 β 链库稀释了新合成的和标记的 β 链可能性。将来自两位血红蛋白 H 病患者洗涤过的全细胞裂解物转化成珠蛋白（未事先纯化血红蛋白），并将 α 链和 β 链分开（图 6）。β 链组分的放射性超过了 α 链组分的放射性，在一个患者中超过了 2.3 倍，在另一个患者中超过了 3.0 倍，同时 β 链峰的光密度值出现了相应增长。在一次独立实验中，使用相同的洗涤过的全细胞裂解物的两个等量样品测定了血红蛋白 A 和 H 的量及各组分的放射性。在第二次实验中，纯化后血红蛋白 H 组分中蛋白质和放射性的量可以解释为什么全细胞裂解物中与 β 链相关的放射性和光密度值过量(超出 α 链的量)。采用 ^{14}C–赖氨酸也观察到类似的结果。这些结果表明，在血红蛋白 H 病中，β 链合成的速率约为 α 链合成的 2~3 倍，β 链被自由释放到红细胞中，并在红细胞内形成一个大库，当有 α 链时，便可与新生成的 α 链结合。当细胞发生溶血时，这些 β 链中的绝大多数会出现在血红蛋白 H 的组分中。在任一时刻都存在这样一个大的 β 链库，这也许可以解释在血红蛋白 H 病中血红蛋白 A 的 α 链和 β 链比活的显著不同。新合成的标记的 α 链将与库中已经存在的未标记 β 链相结合。尚不能确定红细胞中过量的 β 链是否以血红蛋白 H（β_4）的形式存在，但至少在体外系统中，血红蛋白 H 可迅速与 α 链结合从而形成血红蛋白 A[21]。

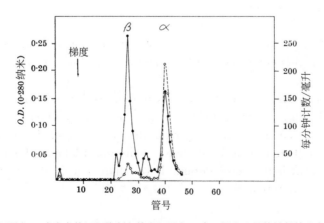

图 5. 由一位血红蛋白 H 病患者的红细胞制备的血红蛋白 A 中 α 链和 β 链的放射性分布情况。温育时间为 45 分钟。●—●，O.D.(0.280 纳米)；○ --- ○，放射性。

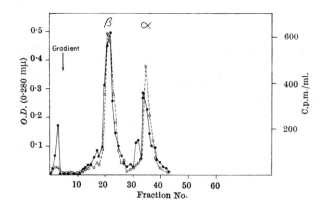

Fig. 6. Distribution of radioactivity compared with protein in a washed whole cell lysate prepared from the red cells of a patient with haemoglobin H disease. Incubation time 45 min. ● — ● , *O.D.* (0.280 mμ); ○ --- ○ , radioactivity.

These experimental results must, of course, be interpreted with caution since the behaviour of cells in an *in vitro* system may not fully reflect their *in vivo* properties. Furthermore, this type of experiment utilizing peripheral blood only measures the last vestiges of protein synthesis in reticulocytes. However, certain tentative conclusions can be drawn. It appears that in non-thalassaemic individuals, α- and β-chain synthesis is synchronous and that, as previously suggested[18], completed β-chains are required for the release of α-chains from the ribosomes. This probably results in a small number of finished α-chains being present on the ribosomes at any given time. Whether the peaks of radioactivity seen on ribosomal chain separations represent completed or partially completed chains is uncertain, although the high resolution of the chromatographic system used would have resulted in the separation of chains differing by as little as one charged residue[16].

From the results of experiments on the kinetics of α- and δ-chain synthesis of haemoglobin A_2, it seems likely that at least one mechanism whereby δ-chain synthesis occurs at a slower rate than α- and β-chain synthesis is the slow ribosomal release of δ-chains. This slow release may well be due to the presence of one or more "slow points" during assembly of the δ-chain.

In both forms of thalassaemia the specific activity of the α-chain of haemoglobin A exceeded that of the β-chain, even after long periods of cell incubation.

For α-thalassaemia, it has been clearly shown that α-chain synthesis occurs at about half the rate of β-chain synthesis, giving rise to a pool of free β-chain in the cell. Thus, although the rate of β-chain synthesis is greater than that of α-chain, the effect of the pre-existing pool of free β-chain is to dilute newly synthesized labelled β-chains. The specific activity of the β-chain is thus considerably lower than that of the α-chain even though synthesis of the latter is impaired.

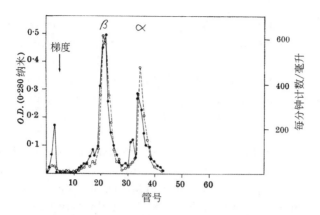

图 6. 由一位血红蛋白 H 病患者的红细胞制备的洗涤过的全细胞裂解物的蛋白质与放射性分布的比较。温育时间为 45 分钟。●—●，O.D. (0.280 纳米)；○ --- ○，放射性。

当然，必须慎重解释这些实验结果，因为体外系统中细胞的行为也许不能完全反映其在体内的特性。此外，这类使用外周血的实验仅测定了网织红细胞中蛋白质合成的最后痕迹。不过，我们仍然可以得出一些初步的结论。在非地中海贫血的个体内，α 链和 β 链的合成似乎是同步的，并且如前所述 [18]，从核糖体中释放 α 链需要已合成的 β 链，这也许会导致在任何给定的时间里，少量的已合成的 α 链会停留在核糖体上。还不能确定核糖体链分离的放射性峰是否代表已合成的或部分合成的肽链，尽管所用的色谱系统分辨率很高，能够分离只差一个带电残基的肽链 [16]。

根据血红蛋白 A₂ 中 α 链和 δ 链合成动力学实验的结果，看起来 δ 链合成比 α 链和 β 链合成缓慢，至少有一种机制，那就是 δ 链从核糖体中释放较慢。这种缓慢释放很可能是由于在 δ 链组装的过程中存在一个或多个"限速位点"。

在两种地中海贫血中，即使细胞经过长时间的温育，血红蛋白 A 中 α 链的比活也都是超过 β 链的比活的。

对 α–地中海贫血来说，很显然 α 链的合成速率大约是 β 链合成速率的一半，导致在细胞中形成大的游离的 β 链的库。因此，尽管 β 链的合成速率大于 α 链合成速率，但已存在的游离 β 链的库的作用是稀释新合成的已标记的 β 链，因此，尽管 α 链的合成被削弱了，但 β 链的比活明显低于 α 链的比活。

In β-thalassaemia, no such pool of β-chains could be demonstrated, nor was a large pool of free α-chain detected. However, excess α-chains do occur. Some are bound to γ-chain (as haemoglobin F), others appear to be attached to ribosomes, and some excess α-chain has been observed in chain separations of globin prepared from red cell stroma. These observations are in keeping with the suggestion[22] that excess α-chain may be insoluble and precipitates in the cell to form the inclusion bodies characteristic of β-thalassaemia.

From the date presented, the most probable explanation of the marked difference of specific activity between α- and β-chains in β-thalassaemia is a block or "slow point" during the assembly of the β-chain, resulting in an accumulation of completed or partially completed β-chains on the ribosomes. It seems likely that since α-chains require β-chains for their release, there is also a secondary accumulation of α-chains on the ribosomes. It is difficult to imagine a mechanism which could give rise to the differences in the specific activities of the α- and β-chains noted here, and yet at the same time leave the rate of assembly of the β-chain unaltered. If, for example, the defect were simply in the number of ribosomes active in β-chain synthesis, there would be a deficit in β-chains but those which were made would have the same specific activity as α-chains. Similarly, a net reduction of messenger RNA for β-chains would lead to a deficit in β-chains, but would not be expected to give rise to different specific activities of the α- and β-chains in completed haemoglobin A unless the rates of assembly of the two chains were also different. Experiments using multiple ^{14}C-labelled amino-acids have tentatively ruled out the N-terminal and central region of the chain as a site of a block in synthesis, but it could occur at or near the carboxyl-terminal end of the molecule. Whichever is true, it seems quite probable that in different instances of β-thalassaemia the site of the defect may not always be the same. The heterogeneity of β-thalassaemias already recognized[1] may be, in some part, a reflexion of this. The similarity of the kinetics of peptide chain synthesis in thalassaemic β-chains and the δ-chains of normal haemoglobin A_2 suggests that a rewarding approach to the β-thalassaemia problem might be in the search for rate-limiting steps in synthesis. A reduction in the rate of α-chain synthesis in haemoglobin H disease has been shown and a similar search for the presence of "slow points" in α-chain synthesis in this disorder is indicated.

In the absence of evidence indicating structural changes in the α- and β-chains of haemoglobin A in thalassaemia, it seems likely that explanations for these disorders may be found in changes in one or more of the factors which affect the process of assembly of the protein on the ribosomes. Mutations leading to changes in messenger RNA codewords, for example, while producing no change in the sequence of amino-acids, might, nevertheless, alter the rates of translation by requiring recognition of the altered codons by other sRNA anti-codons. It is conceivable that if synthesis of the α- or β-chains is dependent on a polycistronic messenger RNA, then changes in messenger RNA at a site remote from the α- or β-chain cistron might in some way influence the rate of assembly or the α- or β-peptide chain[23]. However, this mechanism seems less probable than one directly involving the α-or β-chain messenger RNA. Alternatively, it is possible that mutations leading to changes in specific sRNAs or ribosomal RNA could also influence the rate of assembly of one peptide

在 β–地中海贫血中，没能证实有这样的 β 链库，也未检测到大的游离 α 链库。但是的确出现了过量的 α 链。有些与 γ 链结合（如血红蛋白 F），其他的似乎附着在核糖体上，在由红细胞基质制备的珠蛋白的链分离中可以观察到过量的 α 链。这些观测结果均与以下建议 [22] 相一致，即过量的 α 链在细胞中可能不溶，并沉淀形成 β–地中海贫血的特征性包涵体。

根据已有的数据，在 β– 地中海贫血中 α 链和 β 链比活所出现的显著差异，最可能的解释是在 β 链组装的过程中出现了一个阻塞或"限速位点"，导致已合成或部分合成的 β 链在核糖体上积累。这很可能是由于 α 链的释放需要 β 链，所以在核糖体上也会存在 α 链的二次积累。很难设想一种引起在此观察到的 α 链和 β 链的比活存在差异的机制，而与此同时 β 链的组装速率不发生改变。举例来说，如果这种缺陷仅仅是由于 β 链合成时起作用的核糖体数量不足而引起的，那么 β 链的合成会不足，但所合成 β 链的比活应该与 α 链的相同。同样，β 链信使 RNA 的净减少会导致 β 链不足，但是也不会导致所合成的血红蛋白 A 的 α 链和 β 链的比活不同，除非两条链的组装速率也不同。用多种 ^{14}C 标记的氨基酸进行实验，初步排除了合成的阻塞位点位于链的 N 端或中间区域的可能性，但是可以位于或靠近分子的羧基末端。无论哪一种说法是对的，在不同的 β– 地中海贫血中缺陷的位点很可能并不总是相同的。已经认识到的 β– 地中海贫血的异质性 [1] 在某种程度上也许正好反映了这一点。地中海贫血 β 链和正常血红蛋白 A_2 中 δ 链合成动力学的相似性表明，解决 β– 地中海贫血这一问题的有效方法可能是寻找其合成过程中的限速步骤。已经证明血红蛋白 H 病中 α 链的合成速率降低，这意味着可以用同样的方法寻找该疾病中 α 链合成中"限速位点"的存在。

在没有证据表明地中海贫血中血红蛋白 A 的 α 链和 β 链结构改变的情况下，发现的影响核糖体上蛋白质组装过程的某一种或几种因子的改变似乎可以解释这些疾病。例如，导致信使 RNA 密码子改变（但氨基酸序列不变）的突变，由于需要 sRNA 反密码子识别改变的密码子，从而可能会改变翻译的速率。可以想象，如果 α 链或 β 链的合成依赖于一个多顺反子信使 RNA，那么远离 α 链或 β 链信使 RNA 顺反子的一个位点发生改变会在某种程度上影响组装速率或 α 肽链或 β 肽链 [23]。然而，这种机制的可能性似乎低于直接影响 α 链或 β 链信使 RNA 的可能性。另外，导致特异性 sRNA 或核糖体 RNA 改变的突变也可能会影响一条肽链相对于另外一条肽链的组装速率。例如，在兔血红蛋白分子中，一种亮氨酸特异性 sRNA 只编码约有三十

chain relative to another. An instance where a leucine-specific *s*RNA is involved in the coding of only one out of some 30 leucine residues in the rabbit haemoglobin molecule is already known[24]. The recent advances in the techniques for isolating and studying messenger RNA and *s*RNA from mammalian cells may eventually provide the necessary means of investigating the problem at this level.

We thank Dr. Marion Erlandson and Dr. Carl Smith for their help in obtaining blood samples for these investigations. We also thank Dr. S. H. Boyer, Dr. C. L. Conley, Dr. H. M. Dintzis and Dr. G. Von Ehrenstein for their advice.

This investigation was supported in part by grant *HE*-02799, U.S. Public Health Service, the National Heart Institute; grant *T*1-*AM*-5260, the National Institute of Arthritis and Metabolic Diseases; great *AM*-06006-03, National Institutes of Health; and grant *CB*-2630, National Science Foundation.

(**208**, 1061-1065; 1965)

D. J. Weatherall, J. B. Clegg and M. A. Naughton: Departments of Medicine and Biophysics, The John Hopkins University School of Medicine, Baltimore, Maryland.

References:

1. Weatherall, D. J., *The Thalassaemia Syndromes* (Blackwell Scientific Publications, Oxford, 1965).

2. Kunkel, H. G., Ceppellini, R., Muller-Eberhard, U., and Wolf, J., *J. Clin. Invest.*, **36**, 1615 (1957).

3. Baglioni, C., in *Molecular Biology*, edit. by Taylor, J. H., part I, 405 (Academic Press, New York, 1963).

4. Schroeder, W. A., and Matsuda, G., *J. Amer. Chem. Soc.*, **80**, 1521 (1958).

5. Ceppellini, R., in *Ciba Found. Symp. Biochem. Genet.*, edit. by Wolstenholme, G. E. W., and O'Connor, C. M., 133 (Churchill, London, 1959).

6. Cohen, F., Zuelzer, W. W., Neel, J. V., and Robinson, A. R., *Blood*, **14**, 816 (1959).

7. Jones, R. T., Schroeder, W. A., Balog, J. E., and Vinograd, J. R., *J. Amer. Chem. Soc.*, **81**, 3161 (1959).

8. Hunt, J. A., and Lehmann, H., *Nature*, **184**, 872 (1959).

9. Ingram, V. M., and Stretton, A. O. W., *Nature*, **184**, 1903 (1959).

10. Burka, E. R., and Marks, P. A., *Nature*, **199**, 706 (1963).

11. Guidotti, G., cited by Ingram, V. M., *Medicine*, **43**, 759 (1964).

12. Vinograd, J., and Hutchison, W. D., *Nature*, **187**, 216 (1960).

13. Rieder, R. F., and Weatherall, D. J., *J. Clin. Invest.*, **44**, 42 (1965).

14. Dintzis, H. M., *Proc. U.S. Nat. Acad. Sci.*, **47**, 247 (1961).

15. Anson, M. C., and Mirsky, A. E., *J. Gen. Physiol.*, **13**, 469 (1930).

16. Clegg, J. B., Naughton, M. A., and Weatherall, D. J., *Nature*, **207**, 945 (1965).

17. Ingram, V, M., and Stretton, A. O. W., *Biochim. Biophys. Acta*, **62**, 456 (1962).

18. Baglioni, C., and Colombo, B., *Cold Spring Harb. Symp. Quant. Biol.*, **29**, 347 (1964).

19. Heywood, J. D., Karon, M., and Weissman, S., *Science*, **146**, 530 (1964).

20. Ingram, V. M., *Medicine*, **43**, 759 (1964).

21. Huchns, E. R., Beaven, G. H., and Stevens, B. C., *Biochem. J.*, **92**, 444 (1964).

22. Fessas, P., and Loukopoulos, D., *Science*, **143**, 590 (1964).

23. Stent, G. S., *Science*, **144**, 816 (1964).

24. Weisblum, B., Gonano, F., Von Ehrenstein, G., and Benzer, S., *Proc, U.S. Nat. Acad. Sci.*, **53**, 328 (1965).

个亮氨酸残基中的一个亮氨酸残基 [24]。分离和研究哺乳动物的信使 RNA 和 *s*RNA 技术的最新进展为在这一水平上研究该问题提供了必要的方法。

我们感谢玛丽昂·厄兰森博士和卡尔·史密斯博士为本研究提供血液样品。也感谢博耶博士、康利博士、丹特齐斯博士和冯·埃伦施泰因博士给予本实验的建议。

本研究得到了：美国公共卫生署，国立心脏病研究所编号为 *HE*-02799 的研究经费的支持；美国国立关节炎和代谢病研究所编号为 *T*1-*AM*-5260 的研究经费的支持；美国国立卫生研究院编号为 *AM*-06006-03 的研究经费的支持；以及美国国家自然科学基金会编号为 *CB*-2630 的研究经费的支持。

（周志华 翻译；杨茂君 审稿）

Biochemistry and Mental Function[*]

S. S. Kety

Editor's Note

In 1951, American neuroscientist Seymour Kety became the first scientific director of the National Institute of Mental Health, Bethesda, and here he reflects on the Institute's work. Alongside Carl Schmidt, Kety had devised a method to measure circulation and metabolism in the brains of animals and people in states of normal and altered consciousness. These elegant studies showed that metabolism varied between affective states and brain regions. Deeply interested in the biological roots of mental illness, he also describes various neurochemical theories of schizophrenia, and emphasizes the need for a rigorous, heuristic approach to the study of mental health.

IF we are ever to understand and rationally to meliorate the disturbed processes which underlie mental illness, it will be by investigation of the clinical problems themselves and examination of the mental, social, neural and biological elements which comprise behaviour. It is all of these which the Americans call "psychiatric research", and in Great Britain, the Mental Health Research Fund has added much to its vigour for more than a decade.

Although they are not alone in their importance to psychiatry[1], the biological sciences have a significance which is not attenuated by community with the social and psychological sciences, nor is their power less real by having been only partially demonstrated. It is the area, tentative as yet, between biology and human psychology that I have chosen to dwell on, and I shall do so largely in terms of the work of my associates and myself in the Laboratory of Clinical Science at the National Institutes of Health, Bethesda, without any pretension that this will constitute an adequate review of a growing field.

The time is not yet at hand, if, in fact, it will ever be reached, when one can speak meaningfully of the biochemistry of mental state. There are, however, a few areas where one can see the beginnings of correlations and significant interrelationships and these include consciousness, intellectual function and affect.

My interest in consciousness goes back to the Science and Philosophy Club at the Central High School in Philadelphia, a club which bore the brave motto *"Felix qui potuit rerum cognoscere causas"*, and in which a great teacher, Edwin Landis, introduced us to Berkeley,

---

[*] Substance of the Third Mental Health Research Fund (38, Wigmore Street, London, W.1) Lecture delivered on February 26.

生物化学与心理功能[*]

凯蒂

编者按

1951 年，美国神经科学家西莫尔·凯蒂成为贝塞斯达市国家心理健康研究所的第一任科学主任，在本文中他对研究所的工作进行了仔细思考。凯蒂与卡尔·施密特一起设计出一种方法，用来测量在正常意识状态和意识改变状态下动物与人的脑循环及代谢。这些研究巧妙地展现出在不同情感状态下不同脑区域中代谢情况的差异。怀着对精神疾病的生物学病因的极大兴趣，他也提出了几种精神分裂症的神经化学理论，并且强调心理健康的研究需要采用一种严谨的、探索性的方法。

如果我们想要理解并合理地改善那些导致精神疾病的心理失常过程，就必定要研究它们本身的临床问题，并对构成其表现的心理、社会、神经和生物学要素进行检查。美国人将以上过程称为"精神病学研究"，而在英国，心理健康研究基金十多年来为这一研究增添了许多活力。

对于精神病学来说，虽然它们的重要性不是唯一的[1]，但生命科学的重要性不会因为社会科学与心理科学的参与而减弱，也不会因为只得到了部分阐明而降低其真实性。我选择目前还处于试验性阶段、介于生物学与人类心理学之间的领域来进行论述，这在很大程度上将依据我的同事们和我自己在贝塞斯达市国立卫生研究院临床科学实验室所做的工作，毫不夸张地说，这将构成这个正在发展的领域的全面性综述。

时间尚未来到，假如这一时刻确实会来临的话，也是在人们能够对心理状态的生物化学进行有意义的谈论时。不过，人们逐渐能够在包括意识、智力功能和情感等一些领域中看到某些相关性和显著的相互作用。

我对意识产生兴趣，要追溯到费城中心中学科学与哲学俱乐部，这个俱乐部支持那句勇气十足的格言，即"幸福属于能够理解事物起因的人"；在俱乐部里，一位了不起的教师埃德温·兰迪斯为我们介绍了伯克利、马赫和爱丁顿，还将我们引

---

[*] 发表于 2 月 26 日，第三届英国心理健康研究基金会（伦敦市威格莫尔街道 38 号）报告的内容。

Mach, Eddington, and the fathomless problem of the nature of consciousness. Many years later I was introduced to the cerebral circulation by Carl Schmidt, through his definitive work in the rhesus monkey, and began to feel how much might be learned from measurements in man, whose brain, with its subjective wealth, and whose diseases could not be replicated in animals. Making use of some fundamental principles of inert gas exchange, the Fick principle, and a little calculus, which there is no time to discuss in any detail, we were eventually able to make what still appear to be satisfactory measurements of blood flow, oxygen and glucose consumption in the conscious human brain under a variety of physiological and pathological conditions. In Table 1 are presented some normal values representing, in fact, the average of the first investigations of healthy volunteers of about twenty-five years of age[2]. Measurement of glucose consumption followed later, confirming the thesis that the major substrate for oxidative energy in the brain is glucose, the utilization of which represented an almost stoichiometric equivalent of the oxygen consumed. From these measurements it was possible to compute the rate of energy utilization by the human brain which turned out to be close to 20 W. In comparison with the enormous expenditure of energy which modern computers require, this represents a remarkable degree of efficiency and miniaturization.

Table 1. Over-all Blood Flow and Energy Metabolism of the Normal Human Brain

Blood flow ml./100 g/min	54
Oxygen consumption ml./100 g/min	3.3
Respiratory quotient (CO_2/O_2)	0.99
Glucose consumption mg/100 g/min	4.9

We were anxious to examine states markedly different from the normal in functional level and chose states of altered consciousness (Table 2). It was quite apparent that there was a rough correlation between level of consciousness and over-all oxygen and energy utilization by the brain[3]. In anaesthesia, for example, where the cerebral oxygen consumption was reduced by 40 percent, there appeared to be support *in vivo* to the earlier *in vitro* investigations of Quastel[4], who had shown that anaestheties interfere with the oxygen consumption of brain slices. But all these data merely tell us that the oxygen consumption and energy-level of the brain are reduced in states of depressed consciousness; they do not explain the coupling between function and metabolism which is one of the most interesting topics of present concern. One could argue that the primary effect in any of these conditions was on the metabolic "power supply" of the brain necessary for the maintenance of consciousness. An alternative hypothesis, however, would be that the primary site of interference was in the interaction between neurons at the synapses, which once inhibited, depressed both the functional activity and the energy requirements of the system. This interesting problem of the coupling between function and metabolism must await clarification by the work of those like McIlwain, Rodnight, Larrabee and Chance, among others.

入了关于意识本质这个难理解的问题中。许多年后，卡尔·施密特通过他在恒河猴身上所做的权威性的工作，为我介绍了脑循环。我开始感觉到，由于人类的脑具有它的主观性资源，所以人类的脑和脑的疾病在动物身上是不能重复的，我们对人类进行测量，从中可能学到很多。利用惰性气体交换的一些基本原理、菲克原理以及少量微积分（这里无暇进行细节的讨论），我们最终能够对处于清醒状态的人脑在多种生理和病理条件下的血流量以及氧与葡萄糖的消耗量进行似乎还算满意的测量。表 1 中列出的是一些正常值，实际上是对一批大约 25 岁的健康志愿者进行初次研究的平均值 [2]。之后对葡萄糖消耗量进行测量，结果证实了脑内氧化产能的主要底物是葡萄糖，对葡萄糖的利用代表了几乎化学计量相等的耗氧量。通过这些测量，有可能计算出接近 20 瓦的人脑能量利用率。与现代计算机所需的巨大能量消耗相比，这表现了非常高效和小型化的特点。

表 1. 正常人脑的总血流量和能量代谢

血流量 毫升 /100 克 / 分钟	54
耗氧量 毫升 /100 克 / 分钟	3.3
呼吸商（CO_2/O_2）	0.99
葡萄糖消耗量 毫克 /100 克 / 分钟	4.9

我们期望对那些在功能水平上与正常状态具有显著差别的状态进行检查，并选择了一些意识改变的状态（表 2）。非常明显的是，在意识水平和人脑的总氧气与能量利用之间存在着一种粗略的相关性 [3]。例如，在麻醉时，脑的耗氧量会减少 40%，体内实验结果似乎支持夸斯特尔早期进行的体外研究 [4]，他曾经指出，麻醉会影响脑切片的耗氧量。但是所有这些数据只能告诉我们，在意识处于受抑制状态时，脑的耗氧量与能量水平会下降；他们并没有解释功能与代谢之间的联结关系，这是目前关注的最有趣的话题之一。有人可能会认为在上述任何条件下，主要是对维持意识所必需的脑的代谢"能量供应器"产生影响。然而，另一种假设认为影响的主要部位是在突触处神经元之间的交互作用，这种交互作用一旦受到抑制，系统的功能活性和能量需求都会降低。这个关于功能与代谢之间的联结作用的有趣的问题还必须等待麦基尔韦恩、罗德奈特、拉腊比和钱斯以及其他人的研究工作来阐明。

Table 2. Cerebral Oxygen Consumption in States of Depressed Consciousness
(Expressed as Percentage of the Value in Healthy Young Men)

Senile psychosis	82
Diabetic acidosis	82
Insulin hypoglycaemia	79
Surgical anaesthesia	64
Insulin coma	58
Diabetic coma	52

There are states of altered consciousness, however, in which such a neat correlation with total cerebral metabolism and energy does not exist (Table 3). Normal sleep is one such state; the poetic description of the wakening brain by Sir Charles Sherrington in "Man on His Nature" is well known:

"Suppose we choose the hour of deep sleep. Then only in some sparse and out of the way places are nodes flashing and trains of light-points running ... the great knotted headpiece of the whole sleeping system lies for the most part dark. ... Should we continue to watch the scheme we should observe after a time an impressive change which suddenly accrues. In the great head end ... spring up myriads of twinkling stationary lights and myriads of trains of moving lights of many different directions. ... the great topmost sheet of the mass, that where hardly a light had twinkled or moved, becomes now a sparkling field of rhythmic flashing points with trains of travelling sparks hurrying hither and thither. The brain is waking and with it the mind is returning."

Table 3. Cerebral Oxygen Consumption in Various Mental States
(Expressed as Percentage of the Value in Normal Control States)

Normal sleep	97
Schizophrenia	100
LSD psychosis	101
Mental arithmetic	102

Not only did our results[5] force the rejection of a simple cerebral ischaemic theory for sleep which dated back to Alcmaeon; they challenged as well the generally accepted "Sherringtonian" notion which equated sleep with neuronal inactivity. More recent neurophysiological findings are more consonant with what we learned about the nature of sleep. Evarts[6], in very elegant investigations of the activity of individual neurons which he has observed through microelectrodes chronically implanted in the cortex of unanaesthetized cats, has found no net decrease in cortical neuronal activity during natural sleep. He has, on the other hand, demonstrated characteristic alterations in the activity of individual neurones or groups of neurones, some showing inhibition when the animal sleeps, but others coming into greater activity at that time.

The results in schizophrenia[7], during LSD psychosis[8] or in mental arithmetic[9] all reinforce the concept that the brain, unlike the heart or the liver or kidney, is an organ

表2. 意识处于受抑制状态时脑的耗氧量
（以相对于健康年轻人数值的百分比表示）

老年性精神病	82
糖尿病性酸中毒	82
胰岛素低血糖	79
外科麻醉	64
胰岛素昏迷	58
糖尿病昏迷	52

然而，有些改变的意识状态，总脑代谢与能量之间不存在这样的净相关性（表3）。正常睡眠就是这样一种状态。查尔斯·谢灵顿爵士在《关于人类的本性》一书中对于觉醒状态的大脑所做出的富于诗意的描述是广为人知的：

"假设我们选取深度睡眠的时刻。那么，只在某些零星、偏远的位置，才会有节点闪烁着，以及带有亮点的队列在延续着……整个睡眠系统中最复杂的中枢位于最黑暗的角落里。……继续观望这一图景，过一段时间后，我应该能观察到突然产生的令人印象深刻的变化。在那伟大的首端……无数闪烁着的、静止的亮点和无数向不同方向运动的亮点组成的队列陡然跃起。……在那总体的最高层，曾经罕有光亮闪烁或者移动的地方，现在变成了一片闪亮的场所，到处是有节奏的闪烁光点与匆忙移动的发光队列。大脑醒来了，意识恢复了。"

表3. 各种精神状态下脑的耗氧量
（以相对于正常对照状态时数值的百分比表示）

正常睡眠	97
精神分裂症	100
LSD 精神异常	101
心算	102

我们的结果[5]不仅使我们拒绝接受关于睡眠的简单的脑缺血性理论（这个理论要回溯至阿尔克迈翁），而且向已被普遍认可的"谢灵顿的"将睡眠等同于神经元处于静止状态的观点提出了挑战。最近神经生理学的研究结果与我们关于睡眠本质的认识相一致。埃瓦茨[6]对单个神经元的活动进行了非常巧妙的研究，他将微电极长期植入未经麻醉的猫的脑皮层进行观测，发现在自然睡眠期间，皮层的神经元活动没有净减少。另一方面，他还论证了单个神经元或神经元群活动的特征性改变，在动物睡眠时有些表现为抑制状态，而当时另外一些则更活跃。

精神分裂症[7]、LSD 精神异常[8]或者心算[9]时的结果都强化了这样一个概念，

for computation and communication. In such functions there is no necessary correlation between the energy utilized and the efficiency of the process or the quality of the output. To differentiate these alterations of consciousness in terms of the cerebral oxygen consumption would be like trying to correlate the nature of a radio programme with the power used.

Some of our more recent investigations have attempted to examine the energy utilization of many structures within the living brain. The first approach to measurement of oxygen consumption is in defining the local perfusion rates. Using basic principles similar to those of the nitrous oxide method, we have related the quantity of an inert diffusible substance taken up by a small tissue region to its perfusion[10]. If the tracer is radioactive, one can measure its uptake during a standard time interval in the various structures of the brain by autoradiography. In the autoradiogram, density is related to the concentration of tracer which in turn can be related to the blood flow during the physiological state just prior to the abrupt killing of the animal. Under most physiological conditions there is reason for believing that the blood flow is determined by the oxygen consumption, so that in a rough way the autoradiographic density gives information on the differential energy utilization in various structures of the brain. Such investigations have revealed a remarkable differentiation of cortical blood flow in the unanaesthetized brain with the primary sensory areas showing far greater activity[11]. This differentiation does not appear to be present in the brain of the foetus or the neonate. Anaesthesia obscures this differentiation, reducing the areas of greater cortical oxygen consumption to a relatively homogeneous average value, while there is evidence that sensory stimulation results in a recognizable increase in blood flow and, presumably, oxygen consumption along the appropriate sensory pathways[12].

The maturation of intellectual function and its maldevelopment depends on many processes in addition to oxygen consumption. In 1949, when Sokoloff first became associated with us, we undertook an investigation of cerebral blood flow and oxygen consumption in patients with hyperthyroidism. This resulted in the quite surprising finding that although the total oxygen consumption of such patients was markedly elevated, there was no significant increase in the oxygen consumption of the brain[13]. These results demonstrated that the effects of thyroid hormone were not uniformly applied to the metabolism of all cells in the body. A finding such as this requires an explanation and Sokoloff set about to find one. Using radioactive thyroxine he learned that the hormone crossed the blood-brain barrier and was available to the cells of the brain. He knew also that the brain was peculiar, in that its oxidative processes were almost entirely confined to a single substrate, glucose, and that, although Richter, Waelsch and others had demonstrated an active protein synthesis in the mature brain, that process was still considerably slower in brain than in liver and could scarcely account for a significant fraction of the cerebral oxygen consumption. These considerations led him to the hypothesis that thyroxine neither stimulated oxidative metabolism directly not uncoupled it from phosphorylation (which were the prevailing concepts) but acted on some specialized process such as protein synthesis.

584

即脑与心脏、肝脏或肾脏不同，它是一个负责计算和交流的器官。在实现这些功能时，能量的利用与过程的效率或输出质量之间不一定有相关性。要根据脑的耗氧量来区别这些意识改变，就如同试图在广播节目的性质与其所用的能量之间建立关联一样。

我们最近的一些研究尝试检查活脑内多种结构的能量利用。第一种测量耗氧量的方法是定义局部灌注率。应用与一氧化二氮方法类似的基本原理，我们把一小块组织区域吸收的某种惰性扩散物质的量与其灌注联系起来 [10]。如果示踪物具有放射性，就可以通过放射自显影法测量脑的不同结构在标准时间间隔内的吸收量。在放射自显影图中，密度与示踪物浓度有关，依次地，浓度与动物在被突然杀死前所处的生理状态期间的血流量有关。在绝大多数生理条件下，我们有理由相信血流量是由耗氧量决定的，因此放射自显影图像密度大致提供了脑的各种结构中能量利用差别的信息。这些研究已经揭示了未麻醉的脑中皮层血流量具有显著的差异，同时其初级感觉区的活性要大得多 [11]。这种差异似乎并不存在于胎儿或新生儿的脑中。麻醉通过降低具有较大皮层耗氧量的区域的耗氧量至相对均匀的平均值来淡化这种差异，不过有证据表明，感觉刺激会导致可识别的血流量增加，由此推测，相应感觉传导通路的耗氧量也增加 [12]。

除耗氧量外，智力功能的成熟与发育障碍还取决于很多过程。1949 年索科洛夫与我们首次合作，进行一项针对甲亢患者的脑血流量与耗氧量的研究。该研究得出了令人非常惊讶的发现：尽管这些患者总的耗氧量显著地升高了，但是脑的耗氧量没有显著增加 [13]。这些结果证明，甲状腺激素的作用并非对生物体内所有细胞的新陈代谢都一样适用。诸如这样的发现需要一个解释，索科洛夫便开始寻找原因。通过使用放射性甲状腺素，他认识到激素可以穿过血–脑屏障从而为脑细胞所利用。他还发现，脑是非常特殊的，因为脑中的氧化过程几乎只局限于葡萄糖这一种底物。尽管里克特、韦尔施和其他一些人曾证明过在成熟的脑中有一种活性蛋白质的合成，但是与在肝脏中相比，该过程在脑中还是要慢很多，而且它几乎不能对脑耗氧量的主要部分给出解释。基于上述考虑，他做出了如下假设，即甲状腺素既不直接刺激氧化代谢也不参与磷酸化作用的解耦联（这是普遍流行的看法），而是作用于一些特殊的过程，如蛋白质合成。

In 1954 he came to the National Institute of Mental Health and began a highly productive collaboration with Kaufman. In 1959 they were able to report that l-thyroxine, administered to normal animals *in vivo* or added directly to the incubation medium *in vitro*, stimulated the rate of amino-acid incorporation into protein in cell-free, rat liver homogenates. This stimulation of protein synthesis *in vitro* occurred in the absence of changes in oxygen consumption or oxidative phosphorylation. They further suggested that the characteristic effects of thyroxine on energy metabolism were secondary to the stimulation of reactions which required energy such as protein biosynthesis[14].

The mitochondria are clearly involved in this process since it is these structures alone rather than the microsomes or the cell sap which differentiate hyperthyroid rats from euthyroid controls in the ability to stimulate amino-acid incorporation into protein[15]. Thyroxine in the presence of mitochondria and an oxidizable substrate apparently produces a soluble factor that can be isolated and which will in itself stimulate protein synthesis. The evidence indicates that this factor stimulates the transfer of *s*RNA-bound amino-acids into microsomal protein[16,17]. Much remains to be done in identifying the factor and further defining its action on the microsome and the precise step in protein synthesis at which it occurs. The demonstration of a stimulation of protein synthesis by thyroxine, however, clearly defined as it is by *in vitro* investigations and confirmed *in vivo*, appears to be the fundamental mechanism of action of this hormone and explains much of its physiological effects.

Recently, Sokoloff, in collaboration with Klee[18], has reexamined the effects of this hormone on the brain. Although thyroxine does not stimulate protein synthesis in mature brain, explaining its failure to increase cerebral oxygen consumption in adult man, it does so significantly in neonatal brain, and again it is the mitochondria which differentiate the two.

These findings help to explain the well-known clinical effects of the thyroid hormone on the development of the brain and of intellectual function in infants compared with its relatively minor effects on these functions in the adult. They corroborate and offer a mechanism for Eayrs's finding of the requirement for thyroxine in the dendritic proliferation of immature cerebral cortex. Thus, this work forms a crucial link between the absence of thyroid hormone and the retarded cerebral development in cretinism.

In 1956, soon after the Laboratory of Clinical Science, National Institute of Mental Health, was organized, some of us became interested in the cluster of thought disorders which is called schizophrenia, and gave attention to the hypothesis which attempted to explain many of the mental symptoms of schizophrenia on the basis of an abnormal degradation of circulating epinephrine to abnormal oxidation products such as adrenochrome or adrenolutin[19]. That hypothesis seemed especially plausible because it took cognizance of the evidence for genetic factors as well as the importance of stressful life experiences in the pathogenesis of the mental disorder. The difficulty in testing the hypothesis lay in the lack of knowledge concerning the normal metabolism of epinephrine,

1954 年，他来到国家心理健康研究所，与考夫曼开始了一个卓有成效的合作。1959 年，他们报道称：将 1- 甲状腺素直接给予正常动物进行体内实验，或者体外直接添加至培养基内，它均能提高无细胞系的大鼠肝脏匀浆中氨基酸合成蛋白质的速率。这种刺激蛋白质体外合成的作用发生在耗氧量或氧化磷酸化未变化的情况下。他们进一步提出，甲状腺素对能量代谢的特征性作用次于对需要能量的反应的刺激作用，如蛋白质生物合成 [14]。

很明显，这一过程涉及线粒体，只有它能够将甲状腺机能亢进的大鼠与甲状腺机能正常的对照大鼠在刺激氨基酸合成蛋白质方面区分开 [15]，而不是微粒体或者细胞液。在线粒体和某种可氧化的底物存在时，甲状腺素明显地产生出一种可以分离出来的可溶性因子，这种因子本身刺激蛋白质合成。有证据表明，这种因子刺激 sRNA 结合的氨基酸转化为微粒体蛋白质 [16,17]。我们仍然需要做很多工作来鉴别该因子，并进一步阐明它对微粒体的作用以及它出现于蛋白质合成过程中的确切步骤。然而，甲状腺素对蛋白质合成的刺激作用似乎是这种激素的基本作用机制，并解释了该激素的多种生理作用，这一论证是通过体外实验明确阐释的并在体内实验中得到确认。

最近，索科洛夫与克莱 [18] 合作重新考察了这种激素对脑的影响。尽管甲状腺素不促进成熟脑中的蛋白质合成这一事实解释了它不能增加成人脑的耗氧量，但是它在新生儿大脑中有非常显著的影响。线粒体依然可以用来区分这两种情况。

这些发现有助于解释甲状腺激素对婴儿脑与智力功能的发育这一众所周知的临床作用，相比之下，它对成人的这些功能的作用相对较小。他们证实了埃尔斯关于甲状腺素是未成熟的脑皮层树突增殖所必需的这一发现，并提出了一种作用机制。因此，这一工作表明呆小症患者的甲状腺激素缺乏与脑发育迟滞之间存在重要的关联。

1956 年，国家心理健康研究所临床科学实验室组建后不久，我们中的一些人开始对思维障碍症候群即精神分裂症产生了兴趣，并注意了试图根据循环肾上腺素异常降解为异常氧化产物（如肾上腺素红或肾上腺黄素）来解释精神分裂症所具有的多种精神症状的假说 [19]。这种假说看起来似乎很有道理，因为它既认识到遗传因素的证据，又认识到充满压力的生活经历对于精神障碍发病机理的重要性。检验这一假说的困难在于我们缺乏关于肾上腺素正常代谢的知识，更不要说在精神分裂症中可能出现的异常情况。在 1956 年，人们能够对给予的约为 5% 的肾上腺素作出解释，

let alone its possible abnormality in schizophrenia. In 1956 one could account for some 5 percent of administered epinephrine which was excreted unchanged in the urine, while the remaining 95 percent was disposed of by unknown mechanisms.

Isotopic techniques, which had been so valuable in the tracing of other metabolic pathways, were not readily applied to this problem because the pharmacological potency of epinephrine prevented the administration of enough of the hormone labelled with carbon-14 which was then available to permit characterization of its products. It was apparent that to use isotopic techniques to advantage for studies of the metabolism of epinephrine, especially in man, would require an isotopically labelled epinephrine of unheard-of specific activity. We were finally successful in having a few millicuries of 7-^{3}H-epinephrine synthesized. The tritium label made possible the high specific activities required, while its position at C7 met our expectation that the label would be retained through the various possible metabolic degradations.

In 1957, Armstrong, McMillan and Shaw identified the first major metabolite of epinephrine (vanillylmandelic acid, VMA, or 3-methoxy-4-hydroxymandelic acid) in the urine of a patient with phaeochromocytoma and in normal urine[20].

A few years before, Julius Axelrod had joined the Laboratory, bringing with him great interest and competence in the catecholamines. Although the metabolism of adrenaline to VMA was generally regarded as involving first deamination by monoamine oxidase and then O-methylation, Axelrod, on the basis of pharmacological and biochemical evidence, postulated the existence of an alternative pathway with O-methylation as the first step followed by deamination. He then proceeded to demonstrate in the urine the existence of that hypothetical compound which he designated "O-methylepinephrine" or "metanephrine", a second major metabolite of epinephrine[21]. He described and characterized the enzyme responsible for this conversion (catechol-O-methyltrans-ferase) and the requirement of S-adenosylmethionine as the methyl donor[22]. He suggested that O-methylation rather than deamination was the principal enzymatic process involved in the inactivation of circulating epinephrine and later went on to show that norepinephrine was metabolized through completely analogous pathways by the same enzymes[23]. Fig. 1 shows the present state of knowledge of the metabolism of these two catecholamines with a number of additional minor metabolites which Axelrod *et al.* have identified. Together all these metabolic products account for some 98 percent of administered epinephrine or norepinephrine and, presumably, a similar accountability would hold for these substances when they are released into the circulation under physiological conditions.

这部分随尿液原样排出，而其余的 95% 通过未知机制发生代谢。

同位素技术在其他代谢途径的示踪研究中具有极大价值，但尚不适用于这一问题，因为肾上腺素的药理作用强度不允许给予足量的 C-14 标记的激素，而给予足量的 C-14 标记的激素使获得产物的特征成为可能。很显然，若要利用同位素技术研究肾上腺素代谢，尤其是在人体内，就需要用前所未闻比活度的同位素标记肾上腺素。我们最终成功地合成了几个毫居里的 7-^{3}H-肾上腺素。用氚标记使得所需的高比活度成为可能，而它所处的 C7 位置满足我们的期望，即在各种可能的代谢降解过程中标记将被保留下来。

1957 年，阿姆斯特朗、麦克米伦和肖在一位嗜铬细胞瘤患者的尿液和正常人尿液中首次鉴别出肾上腺素的主要代谢产物（香草扁桃酸、VMA 或 3-甲氧基-4-羟基扁桃酸）[20]。

几年前，尤利乌斯·阿克塞尔罗德加入了实验室，带来了其对儿茶酚胺浓厚的研究兴趣和很强的研究能力。尽管人们通常认为从肾上腺素到 VMA 的代谢过程首先涉及单胺氧化酶的脱氨基作用，随后进行氧位甲基化作用，但是根据药理学和生物化学方面的证据，阿克塞尔罗德认为可能存在另外一种途径，即氧位甲基化作用为第一步，随后才是脱氨基作用。接着，他开始证明在尿液中存在这种假设的化合物，他将其称之为"氧位甲基化肾上腺素"或"甲氧基肾上腺素"，它是肾上腺素的第二种主要代谢产物[21]。他描述了负责这一转化过程的酶及其特征（儿茶酚氧位甲基转移酶），而且需要 S-腺苷甲硫氨酸作为甲基供体[22]。他提出，在涉及循环肾上腺素失活的主要酶促过程中，涉及的是氧位甲基化作用而非脱氨基作用，后来他又进一步表明，去甲肾上腺素是借助同一种酶通过完全类似的途径产生代谢变化的[23]。图 1 显示出目前我们对这两种儿茶酚胺代谢过程的认识情况，其中一些是由阿克塞尔罗德等人鉴别的一些额外的次要代谢产物。所有的这些代谢产物可以解释给予的肾上腺素或去甲肾上腺素中约 98% 的部分，据推测，当这些物质在各种生理条件下释放到循环中，可以得到相似程度的解释。

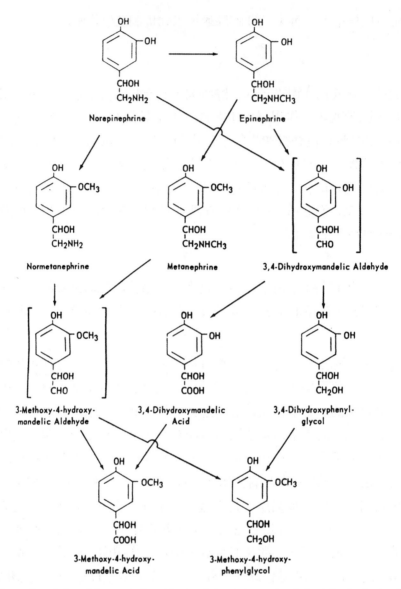

Fig. 1. Present knowledge of the metabolism of epinephrine and norepinephrine (after Axelrod and Kopin)

With the background of information on the normal degradation of the hormone which was thus provided, it was then possible to examine the metabolism of epinephrine in schizophrenic patients and normal volunteers using the tritium-labelled substance[24,25]. We were unable to find any evidence for a significant abnormality in the metabolism of intravenously administered epinephrine among the schizophrenics either qualitatively or quantitatively, the four normal metabolites and the unchanged hormone accounting for 98 percent of the tritium in the urine in both groups of subjects.

The synthesis of tritiated epinephrine which was stimulated by that hypothesis but, most important, the work of Axelrod *et al.* have, however, had important implications for

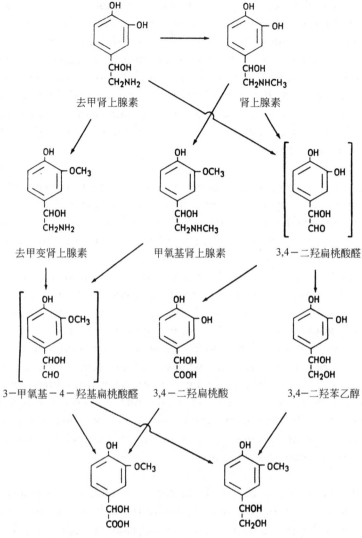

图 1. 目前认识的肾上腺素和去甲肾上腺素代谢过程（继阿克塞尔罗德和科平之后）

　　基于提供的关于激素正常降解的背景信息，我们就有可能利用氚–标记物检查肾上腺素在精神分裂症患者和正常志愿者体内的代谢过程 [24,25]。对于精神分裂症患者，我们未能发现静脉给予肾上腺素后代谢过程发生显著异常的任何定性或定量证据；在两组对象中，四种正常的代谢产物和未发生改变的激素可以解释尿液中氚含量为98％这一现象。

　　这个假设促进了氚标记的肾上腺素的合成，但是，最重要的是阿克塞尔罗德等人的工作对精神病学已经产生了重要的意义。肾上腺素代谢产物的鉴定以及在尿液

psychiatry. The identification of the metabolites of epinephrine and the development of methods for their estimation in urine make it possible to obtain information on the secretion of this hormone in a variety of physiological and pathological states and in response to drugs. Investigations by Axelrod and Kopin, among others, with norepinephrine were a logically related step, and in the past few years the storage and release of norepinephrine at the sympathetic nerve endings and the factors which control these processes have become one of the most exciting fields of pharmacology[26]. The insights which such investigations have given us into the possible actions of drugs which affect mood will be discussed later.

Thirteen years ago Osmond and Smythies, in conjunction with Harley-Mason[27], advanced the interesting hypothesis that there was an accumulation of an abnormal methylated compound with hallucinogenic properties in schizophrenia. They were led to this possibility by the fact that the potent psychotomimetic drug, mescaline, was almost identical with trimethylated dopamine. In the same communication Harley-Mason pointed out that the dimethyl derivative (3,4-dimethoxyphenyl-ethylamine), which had interesting behavioural effects, could possibly be formed by transmethylation *in vivo*. In 1961, Pollin, Cardon and I[28] tested this hypothesis by observing the mental effects of methionine given orally to a small number of chronic schizophrenic patients who had been maintained on a monoamine oxidase inhibitor. We reasoned that under those conditions it was conceivable that the levels of *S*-adenosylmethionine, which Cantoni[29] had shown to be an important methyl donor, could be increased and the biological transmethylation of amines facilitated. In some of the patients there was a temporary but quite obvious exacerbation of psychotic symptoms associated with methionine administration. These observations have now been confirmed by several other groups[30-32]; in addition, Brune and Himwich found similar effects with betaine, another methyl donor[30].

Further work by investigators in our laboratory has tended to support some of this reasoning. The absence of information regarding tissue levels of *S*-adenosylmethionine led Baldessarini and Kopin[33] to devise an ingenious assay of high specificity. By means of this they found a considerable elevation of *S*-adenosylmethionine in brain and liver of rats following methionine feeding. Axelrod[34] demonstrated the presence in normal mammalian tissue of an enzyme capable of methylating normal metabolites, that is, tryptamine and serotonin, to their dimethyl derivatives in the presence of *S*-adenosylmethionine. Dimethyltryptamine has been shown to be a potent psychotomimetic agent[35].

In 1962, Friedhoff and Van Winkle[36] detected a substance which behaved like 3,4-dimeth oxyphenylethylamine in the urine of a substantial fraction of schizophrenic patients and which appeared to be absent from normal urine. That finding, which has had substantial confirmation by Bourdillon and Ridges[37], lends further support to the hypothesis of Osmond, Smythies and Harley-Mason, especially since it was that compound to which Harley-Mason had directed attention ten years previously.

592

中其含量的测定方法的发展，使得获取肾上腺素在各种生理状态、病理状态以及在药物作用下的分泌信息成为可能。由阿克塞尔罗德、科平和其他一些人利用去甲肾上腺素展开的研究在逻辑上是有关联的步骤，并且在过去的几年中，去甲肾上腺素在交感神经末梢的储存和释放以及控制这些过程的因素已成为药理学中最令人兴奋的研究领域之一 [26]。这些研究使我们了解影响情绪的药物的可能效应，这些观点将在后面加以讨论。

13 年前，奥斯蒙德、斯迈西斯与哈利–梅森 [27] 合作提出了一个有趣的假说，即在精神分裂症患者体内存在具有致幻作用的异常甲基化化合物的积累。强效的拟精神病药墨斯卡灵与三甲基化的多巴胺几乎一样的事实使他们开始考虑这一可能性。在同一篇通讯中，哈利–梅森指出，具有有效的行为学作用的二甲基衍生物（3,4– 二甲氧基苯乙胺）有可能在体内通过转甲基作用形成。1961 年，波林、卡登和我 [28] 给少量使用单胺氧化酶抑制剂维持治疗的慢性精神分裂症患者口服甲硫氨酸，通过观察甲硫氨酸产生的心理效应对这一假说进行检验。我们推测，在那样的条件下，可以想象 S–腺苷甲硫氨酸（坎托尼 [29] 指出，它是一种重要的甲基供体）的水平会升高并且胺的生物转甲基作用得到增强。有些患者出现了短暂的但是非常明显的与甲硫氨酸使用相关的精神病症状的恶化。其他一些研究组已经证实了这些观测结果 [30-32]；此外，布龙和希姆威奇发现另外一种甲基供体甜菜碱具有类似的作用 [30]。

我们实验室的研究人员的进一步工作倾向于支持其中的一些推测。由于缺乏组织中有关 S–腺苷甲硫氨酸水平的信息，巴尔代萨里尼和科平 [33] 设计出一种精巧的具有高度特异性的检测方法。通过这种方法他们发现，在喂食了甲硫氨酸之后，大鼠的脑和肝脏中的 S–腺苷甲硫氨酸含量明显增加。阿克塞尔罗德 [34] 证明了在正常哺乳动物的组织中存在着一种酶，这种酶在有 S–腺苷甲硫氨酸存在时能够使正常的代谢产物（即色胺和 5–羟色胺）发生甲基化作用从而形成其二甲基衍生物。二甲基色胺已经被证明是一种强效的拟精神病药物 [35]。

1962 年，弗里德霍夫和范温克尔 [36] 在很大一部分精神分裂症患者的尿液中检测到一种表现类似于 3,4–二甲氧基苯乙胺的物质，该物质似乎不存在于正常人的尿液中。该发现得到了鲍迪伦和布里奇斯 [37] 的确认，这为奥斯蒙德、斯迈西斯和哈利–梅森的假说提供了进一步的支持，尤其是因为哈利–梅森早在 10 年前就特别关注该化合物了。

Our observations of the effect of methionine in schizophrenic patients, as well as the findings of Friedhoff and Van Winkle, are open to a number of alternative explanations which have not been ruled out. Nevertheless, the hypothesis that the accumulation of one or more methylated compounds plays a significant part in some forms of schizophrenia remains a plausible and parsimonious explanation of a number of different and independent observations and seems worthy of further evaluation.

The possible interrelationships between the biogenic amines and affective states have become a subject of lively interest and productive investigations in the relatively few years which have elapsed since the pioneering studies of Gaddum and Vogt in Great Britain, Erspamer in Italy, Rapport and Woolley in the United States. Interest centred at first on serotonin after remarkable demonstration by Shore et al.[38] of a depletion of that amine from the brain during reserpine-induced depression and its elevation following the antidepressant monoamine oxidase inhibitors. As evidence accumulated, however, it was learned that these drugs also affected noradrenaline- and dopamine-levels in the brain and that the catecholamine precursor, dopa, promptly and effectively reversed the depressant actions of reserpine in animals, suggesting to some an equally important role for catecholamines in the action of these drugs and possibly in affective states. It has been difficult to explain, however, the action of two effective antidepressant drugs in terms of the central biogenic amines. These agents, amphetamine and imipramine, are not especially active as monoamine oxidase inhibitors and have not been shown to elevate the levels of norepinephrine or serotonin in the brain.

Recently, Kopin was able to demonstrate, in the isolated, perfused heart, a differential metabolism of tritiated noradrenaline released under different circumstances[39]. When the catecholamine was liberated in a manner which did not provoke its characteristic effects on the heart, that is, by reserpine, it appeared largely as deaminated products in the perfusate. On the other hand, when its release was accompanied by cardiac stimulation as with stimulation of the cardiac sympathetic nerves or by tyramine, O-methylated products appeared in the perfusate. These observations have suggested the generalization that catechol-O-methyl transferase is the enzyme normally involved in the degradation of norepinephrine which is released physiologically and perhaps its O-methylated metabolites are indicative of adrenergic activation, at least in the periphery.

The release and metabolism of noradrenaline in the brain, however, remained quite a mystery, since the blood-brain barrier prevented the uptake by the brain of radioactive norepinephrine and the amount of label which could be applied through synthesis from tagged tyrosine was hardly enough for fractionation. In 1964, Glowinski, who had applied Feldberg's technique to injection into the lateral ventricles of rats, joined Axelrod and Kopin and succeeded in developing what appears to be a valid technique for labelling, at a high specific activity, the norepinephrine stores within the brain by injecting the tritiated form intraventricularly[40]. the label distributes itself quite rapidly in a pattern similar to that of endogenous norepinephrine and shows the same intracellular localization. Furthermore, it follows a curve of disappearance from the brain similar to that of ^{14}C-norepinephrine endogenously produced from ^{14}C-tyrosine.

594

我们观测到的甲硫氨酸在精神分裂症患者体内的作用结果，以及弗里德霍夫与范温克尔的发现都存在大量仍未被排除掉的其他解释的可能性。不过，一种或多种甲基化的化合物的积累在某些形式的精神分裂症中起着重要作用的假说，对于大量不同的、独立的观测结果来说，仍然是一种似有道理且又过于简单的解释，并且看来好像值得进一步评价。

由于英国的加德姆和沃格特、意大利的厄斯帕莫以及美国的拉波特和伍利的开创性研究，在随后的几年里，生物胺与情感状态之间可能的相互关系已成为一个引人注目的、成果颇丰的研究课题。在肖尔等人 [38] 对利血平诱导的抑郁期间脑内胺的耗竭以及在使用抗抑郁药单胺氧化酶抑制剂之后胺含量升高做出了出色的论证后，研究兴趣最初集中于 5-羟色胺。然而，随着证据的积累，我们认识到这些药物也会影响脑内去甲肾上腺素和多巴胺的水平，并且儿茶酚胺前体多巴会快速而有效地逆转利血平在动物体内的抑制作用，表明儿茶酚胺在这些药物的作用中和可能在情感状态中起着某些同等重要的作用。但是，始终难以从主要的生物胺的角度来解释两种有效的抗抑郁药物的作用机制。安非他明和丙咪嗪这些药剂作为单胺氧化酶抑制剂不是特别有效，也还未曾表现出能够提高脑内去甲肾上腺素或 5-羟色胺的水平。

最近，科平能够论证分离的、灌注后心脏在不同条件下释放的氚标记的去甲肾上腺素的代谢 [39]。当儿茶酚胺以不引发其对心脏的特征性作用的方式释放时，即通过利血平的方式，那么它作为脱氨基产物就会在灌注液中大量出现。另一方面，当儿茶酚胺的释放同时伴随着心刺激（如同心交感神经产生的刺激）或者酪胺，那么灌注液中就会出现氧位甲基化的产物。这些观测结果提示了一般性的结论，即去甲肾上腺素的降解过程一般都要涉及儿茶酚氧位甲基转移酶，它是生理性释放的，也许它的氧位甲基化的代谢产物标志着肾上腺素的活性，至少在外周如此。

然而，去甲肾上腺素在脑内的释放和代谢仍然是个谜，因为血-脑屏障阻止了脑对放射性去甲肾上腺素的吸收，并且通过由标记的酪氨酸合成的标记物的量几乎不能满足分馏法。1964 年，格洛温斯基（他曾经应用费尔德伯格的技术对大鼠的侧脑室进行注射）参与到阿克塞尔罗德与科平的合作中来，并成功地开发出一种看来很有效的、可获得高比活度的标记技术，通过心室内注射氚标记物使得去甲肾上腺素存储在脑内 [40]。标记物以一种与内源性去甲肾上腺素类似的形式非常快速地扩散，并且显示出相同的细胞内定位。此外，它遵循着与由 ^{14}C-酪氨酸内源性生成的 ^{14}C-去甲肾上腺素类似的从脑内消失的曲线。

With convincing evidence that they were studying the metabolism of endogenous norepinephrine in the brain, they examined the effects of a number of psychoactive drugs. Reserpine caused a rapid depletion and the predominant formation of deaminated products as it did in the heart. On the other hand, monoamine oxidase inhibitors, amphetamine and imipramine, all of which are antidepressant or euphoriant drugs, were followed by an increase in O-methylated norepinephrine products in the brain[41]. If one may generalize from Kopin's findings in the heart and infer physiological activity from an increase in norepinephrine O-methylation, these findings are compatible with the thesis that the drugs which induce depression or elevation of mood do so by depressing of facilitating the release of physiologically active norepinephrine in the brain or altering its availability at effector sites.

Such a hypothesis as well as the possibility that normal and abnormal changes in mood are dependent on alterations of catecholamines in the brain remain to be validated. The possibility of labelling norepinephrine in the brain has overcome a major obstacle in the way of elucidating its physiological role there.

Another recent development in this laboratory has some clear-cut implications, this time for cardiology. Despite the expectation that monoamine oxidase inhibitors should elevate the levels of the sympathetic neurotransmitter, these agents are found to have hypotensive and other sympatholytic effects which, though undesirable in psychiatry, have been found useful in the treatment of hypertensive disease and angina pectoris. An explanation of this paradoxical effect has been advanced by Kopin et al. [42], who presented evidence for the normal synthesis and accumulation of octopamine, the β-hydroxylated derivative of tyramine, in the region of the sympathetic nerve endings, its enhancement by monoamine oxidase inhibitors and its release by sympathetic stimulation. Their hypothesis that this relatively inactive amine may replace norepinephrine and act as a false neurochemical transmitter appears capable of explaining the partial sympathetic blockade observed after chronic inhibition of monoamine oxidase.

Much of what I have outlined is illustrative of an important generalization from the history of science, a principle which, though taken for granted by most scientists, nevertheless requires reinforcement today. The most practical way to attack a major medical problem or to bridge a great hiatus is not usually head-on, but by strengthening and extending the foundations on both sides and narrowing the gap which lies between. This is best accomplished when the scientists themselves choose their logical next steps, which each will do from his knowledge of the state of the field, the feasibility of an approach, the likelihood and significance of its being successful.

Nearly a hundred years ago in England, Thudichum, whom many regard as the father of modern neurochemistry, advanced a hypothesis that many forms of insanity were the result of toxic substances produced within the brain by faulty metabolism. But, more important, he went on to suggest that these processes, then quite obscure, would be obvious when we understood the biochemistry of the brain to its utmost detail[43]. It was

　　根据他们研究脑内源性去甲肾上腺素代谢令人信服的证据，他们研究了多种精神药物的作用。利血平导致快速耗竭，并主要形成就像其在心脏代谢一样的脱氨基产物。另一方面，单胺氧化酶抑制剂安非他明和丙咪嗪都是抗抑郁药物或称安乐药，会引起脑内氧位甲基化的去甲肾上腺素产物的增加[41]。如果人们能从科平在心脏中的发现归纳出结论，并从去甲肾上腺素的氧位甲基化作用增加推断其生理活性的话，那么这些发现就与下面的论点相吻合，即导致情绪抑郁或高涨的药物是通过抑制促进脑内有生理活性的去甲肾上腺素的释放，或者通过改变其在效应物部位的利用度来发挥效应。

　　情绪的正常和异常变化取决于脑内儿茶酚胺的变化这样一种假说和可能性尚有待于证实。对脑内去甲肾上腺素进行标记的实现，克服了在阐明它的生理作用方面的主要障碍。

　　该实验室另一项新近的进展在心脏病学领域具有某些明确的影响。尽管预期单胺氧化酶抑制剂会提高交感神经神经递质的水平，却发现这些药物具有降血压和其他一些抗交感神经的作用，尽管这些作用并不是精神病治疗法所需要的，但已经发现其对高血压和心绞痛的治疗确实是有用的。科平等人[42]对这种自相矛盾的作用提出了一个解释，他们提供了章鱼胺（酪氨 β–羟基化的衍生物）在交感神经末梢区域的正常合成与积累、单胺氧化酶抑制剂使其增加、交感神经刺激使其释放的证据。他们的假说认为，这种相对无活性的胺会取代去甲肾上腺素并作为假性神经化学递质，该假说似乎可以解释单胺氧化酶在长期抑制之后所观察到的局部交感神经阻滞。

　　我所描述的大部分内容都是对科学历史重要的概括性的说明，原理尽管已经为大多数科学家所承认，但在今天仍需要强化。解决一个主要医学难题或者桥接巨大裂缝时，最实用的方法并不是通常的正面冲锋，而是在两个方面加强和扩展基础并且缩短横亘其间的裂隙。当科学家自己选择他们逻辑的前进步骤，从他对所研究领域的认识、方法的可行性以及获得成功的可能性和重要性出发，就可以最完美地达到成功。

　　在大约 100 年前的英格兰，被很多人看作现代神经化学之父的图迪休姆提出了一个假说，认为很多种形式的精神错乱是由于代谢缺陷使得脑内产生有毒物质而造成。但他进一步提出，更为重要的是，当我们最大可能详细地理解脑内的生物化学反应时，这些过程（当时还所知甚少）将是显而易见的[43]。他在此后的 10 年中致

in the latter area that he spent the next ten years in what was to become the classical isolation, description and characterization of the chemical constituents of the brain.

It is not difficult to predict what would have resulted had Thudichum spent those years and the funds made available to him by the Privy Council in a premature search for the toxins of insanity. With the tools, techniques and knowledge available to him at that time, it is extremely unlikely that he would have found any of those hypothetical substances; it is equally unlikely that he would have made his fundamental contributions to our present knowledge of the biochemistry of the brain.

(**208**, 1252-1257; 1965)

Seymour S. Kety: Laboratory of Clinical Science, National Institute of Mental Health, National Institutes of Health, Bethesda, Maryland.

References:

1. Kety, S. S., *Science*, **132**, 1861 (1960).

2. Kety, S. S., and Schmidt, C. F., *J. Clin. Invest.*, **27**, 476 (1948).

3. Kety, S. S., in *Metabolism of the Nervous System*, edit. by Richter, D., 221 (Pergamon Press, London, 1957).

4. Quastel, J. H., *Anesthes. Analg.*, **31**, 151 (1952).

5. Mangold, R., Sokoloff, L., Conner, E., Kleinerman, J., Therman, P. G., and Kety, S. S., *J. Clin. Invest.*, **34**, 1092 (1955).

6. Evarts, E. V., Bental, E., Bihari, B., and Huttenlocher, P. R., *Science*, **135**, 726 (1962).

7. Kety, S. S., Woodford, R. B., Harmel, M. H., Freyhan, F. A., Appel, K. E., and Schmidt, C. F., *Amer. J. Psychiat.*, **104**, 765 (1948).

8. Sokoloff, L., Perlin, S., Kornetsky, C., and Kety, S. S., *Ann. N. Y. Acad. Sci.*, **66**, 468 (1957).

9. Sokoloff, L., Mangold, R., Wechsler, R. L., Kennedy, C., and Kety, S. S., *J. Clin. Invest.*, **34**, 1101 (1955).

10. Kety, S. S., in *Methods in Medical Research*, edit. by Bruner, H. D., **8**, 228 (Year Book Publishers, Chicago, 1960).

11. Landau, W. M., Freygang, W. H., Rowland, L. P., Sokoloff, L., and Kety, S. S., *Trans. Amer. Neurol. Assoc.*, **80**, 125 (1955).

12. Sokoloff, L., in *Regional Neurochemistry*, edit. by Kety, S. S., and Elkes, J., 107 (Pergamon Press, Oxford, 1961).

13. Sokoloff, L., Wechsler, R. L., Mangold, R., Balls, K., and Kety, S. S., *J. Clin. Invest.*, **32**, 202 (1953).

14. Sokoloff, L., and Kaufman, S., *Science*, **129**, 569 (1959).

15. Sokoloff, L., and Kaufman, S., *J. Biol. Chem.*, **236**, 795 (1961).

16. Sokoloff, L., Kaufman, S., Campbell, P. L., Francis, C. M., and Gelboin, H. V., *J. Biol. Chem.*, **238**, 1432 (1963).

17. Sokoloff, L.: The action of thyroid hormones on protein synthesis, as studied in isolated preparations and in the whole rat. *Proc. Second Intern. Congr. Endocrinol.*, London, Aug. 17, 1964 (Elsevier Publishing Co., Amsterdam; in the press).

18. Klee, C. B., and Sokoloff, L., *J. Neurochem.*, **11**, 709 (1964).

19. Hoffer, A., Osmond, H., and Smythies, J., *J. Ment. Sci.*, **100**, 29 (1954).

20. Armstrong, M. D., McMillan, A., and Shaw, K. N. F., *Biochim. Biophys. Acta*, **25**, 422 (1957).

21. Axelrod, J., *Science*, **126**, 400 (1957).

22. Axelrod, J., and Tomchick, R., *J. Biol. Chem.*, **233**, 702 (1958).

23. Axelrod, J., *Physiol. Rev.*, **39**, 751 (1959).

24. La Brosse, E. H., Axelrod, J., and Kety, S. S., *Science*, **128**, 593 (1958).

25. La Brosse, E. H., Mann, J. D., and Kety, S. S., *J. Psychiat. Res.*, **1**, 68 (1961).

26. Kopin, I. J., *Pharmacol. Rev.*, **16**, 179 (1964).

27. Osmond, H., and Smythies, J., *J. Ment. Sci.*, **98**, 309 (1952).

28. Pollin, W., Cardon, P. V., and Kety, S. S., *Science*, **133**, 104 (1961).

29. Cantonl, G. L., *J. Biol. Chem.*, **204**, 403 (1953).

30. Brune, G. G., and Himwich, H. E., in *Recent Advances in Biological Psychiatry*, **5**, 144 (Plenum Press, New York, 1963).

31. Alexander, F., Curtis, G. C., Sprince, H., and Crosley, A. P., *J. Nerv. Ment. Dis.*, **137**, 135 (1963).

32. Park, L. C., Baldessarini, R. J., and Kety, S. S., *Arch. Gen. Psychiat.*, **12**, 346 (1965).

力于后一问题，他的研究成为脑内化学组分经典的分离、描述和鉴定。

不难预料图迪休姆在那些年的努力所取得的成果以及他因为对精神错乱毒素的早期研究而获得枢密院的资助。采用当时所具备的设备、技术和知识，他发现那些假设的物质中的任何一种几乎都是不可能的；他同样不大可能对我们现代的脑生物化学知识做出根本性的贡献。

（王耀杨 翻译；李素霞 审稿）

33. Baldessarini, R. J., and Kopin, I. J., *Anal. Biochem.*, **6**, 289 (1963).

34. Axelrod, J., *Science*, **134**, 343 (1961).

35. Szara, S., *Fed. Proc.*, **20**, 855 (1961).

36. Friedhoff, A. J., and Van Winkle, E., *J. Nerv. Ment. Dis.*, **135**, 550 (1962).

37. Bourdillon, R. E., and Ridges, A. P., in *Amine Metabolism in Schizophrenia*, edit. by Himwich, H. E., Smythies, J. R., and Kety, S. S. (to be published).

38. Shore, P. A., Pletscher, A., Tomich, E. G., Carlsson, A., Kuntzman, R., and Brodie, B. B., *Ann. N. Y. Acad. Sci.*, **66**, 609 (1957).

39. Kopin, I. J., and Gordon, E., *J. Pharmacol.*, **140**, 207 (1963).

40. Glowinski, J., Kopin, I. J., and Axelrod, J., *J. Neurochem.*, **12**, 25 (1965).

41. Glowinski, J., and Axelrod, J., *J. Pharmacol.* (in the press).

42. Kopin, I. J., Fischer, J. E., Musacchio, J. M., Horst, W. D., and Weise, V. K., *J. Pharmacol.*, **147**, 186 (1965).

43. Thudichum, J. L. W., *A Treatise on the Chemical Constitution of the Brain*, 13 (Baillière, Tindall and Cox, London, 1884).

Malaria and the Opening-up of Central Africa

R.Weatherall

Editor's Note

In the nineteenth century, the British government was much occupied with the exploration of tropical Africa, partly because of the government's decision to do what it could to suppress the slave trade. This paper, based on a book by Michael Gelfand, who had taken up the post of professor at the University of Salisbury in Southern Rhodesia (now Harare in Zimbabwe), describes how those efforts were confounded by malaria until the development of powerful insecticides in the early twentieth century, such as DDT. But the author's conclusion that "in Central Africa the conquest of malaria is now almost complete" will be regarded now as premature.

IN his inaugural address to the University College of Rhodesia and Nyasaland, Prof. Michael Gelfand took the opportunity to review the part which malaria played in delaying the opening-up of Central Africa to European traders and settlers[1]. His original paper is illustrated with reproductions of some fine contemporary pictures which add to the interest of the text.

The earliest in the field were the Portuguese, who were setting up trading posts along the west coast of Africa as early as 1443. Somewhat later they were establishing footholds along the east coast too, but the "Angel of Death" effectively blocked their penetration into the interior. When quinine became available, the Portuguese took to it more readily than the British and put it to more effective use. Its introduction, although spread over a longer period of time, "had an impact on medicine similar to that of antibiotics today".

The legend that cinchona bark was introduced to Europe by the Countess of Chinchon, wife of the Viceroy of Peru, has been convincingly disproved. What is clear is that Jesuit missionaries were in contact with Indians who were aware of the medical properties of cinchona bark, at Loxa, in Brazil, and in Peru, about the beginning of the seventeenth century. Its first recorded importation into Europe was made by Barnabe de Cobo, after exploring parts of Mexico and Peru in 1632. News of it soon spread from Spain to Italy and the Netherlands, and it was introduced to England by James Thomson of Antwerp in 1650.

In many places its acceptance by doctors was very varied, in part perhaps on religious grounds, but in England, resort to its use was stimulated by some serious epidemics of ague, so that the weekly publication *Mercurius Politicus* could report that, by 1658, the bark was on sale by several London chemists. Its fame was heightened by Robert Talbot, who among many others treated King Charles II and the Dauphin Charles of France. He

疟疾和中部非洲的开放

韦瑟罗尔

编者按

19 世纪，英国政府忙于非洲热带地区的探险，这部分是由于政府废止奴隶贸易的决定。本文是根据南罗得西亚索尔兹伯里（现津巴布韦的哈拉雷）大学教授迈克尔·盖尔芬德的著作编写成的。它描述了 20 世纪初期在开发出强效杀虫剂（如 DTT）前，疟疾是如何给那些进行艰难尝试的人们带来困惑的。但是许多科学家认为，作者得出的"在中部非洲现在基本完成攻克疟疾"的结论还为时过早。

迈克尔·盖尔芬德教授在罗得西亚和尼亚萨兰大学的就职演说中，利用这个机会评述了在中部非洲向欧洲商人和移民开放问题上疟疾所起到的延缓作用 [1]。他在原始文章中插入了一些精美的、同时代的复制图片，这增加了人们对文章的兴趣。

最早来到中非的是葡萄牙人，他们早在 1443 年就在非洲西海岸建立了贸易站。随后又在东海岸有了立脚点，但"死神"有效地抑制了他们向内地的扩展。当药物奎宁出现后，葡萄牙人比英国人更容易获得它并将其更加有效地利用。尽管该药的广泛使用经历了较长时间，但它的引进"对医学的影响类似于今天的抗生素"。

已经令人信服地证实秘鲁的钦琼伯爵夫人将金鸡纳树皮引入欧洲这个传说是虚构的。现在清楚的是，大约 17 世纪初期在巴西的洛查和秘鲁的耶稣会传教士与了解金鸡纳树皮药学特性的印第安人已经有往来。1632 年巴纳比·德科沃勘探了墨西哥和秘鲁部分地区后，记录下了该药物首次进入欧洲。这则消息很快从西班牙传到意大利和荷兰，1650 年来自安特卫普的詹姆斯·汤姆森将该药引入英国。

不同地方的医生对该药物的接受程度是不一样的，部分可能是由于宗教原因，但是英国在经历几次严重的疟疾大流行后刺激了该药的使用，因此《英共和联邦政治信息公报》周刊报道，直至 1658 年该药已经在伦敦多家药房销售。罗伯特·塔尔博特用此药治好了查尔斯 II 世国王和法国查尔斯皇太子的病，提高了该药的声誉。

made a fortune out of its secret use. Sydenham was convinced of its efficacy as a medicine, and administered the powdered bark mixed with red wine. As early as 1659 Willis found that the bark relieved acute attacks of ague, although relapses were common; and it was not until 1768 that James Lind observed that, in cases of fever, the drug was most effective if given early in large doses.

In 1745, Claude Touissant de Cagarage attempted to produce an extract of quinine, as did Bernadino Antonio Gomes, of Lisbon, in 1810. Ten years later, Pierre Joseph Pelletier and Joseph Caventou were the first to isolate two of the four alkaloids in the substance. This discovery so stimulated the demand for cinchona bark that exploitation of the forests of Peru, Ecuador and Bolivia was carried to such an extent that fears arose about their exhaustion; the English and the Dutch took up its cultivation, and by 1862 the Dutch had almost established a monopoly in its supply, mainly based on Java.

In comparison with the Portuguese, the British were slow in making attempts to trade with or settle in Central Africa. One of the earliest schemes for settlement was made by William Bolts; it was sponsored by Maria Theresa of Austria. The expedition set out from Leghorn in 1776. It consisted of 152 Europeans, who set up stations along the Masoomo River, in Delagoa Bay, but malaria soon made its presence felt; local Africans rose against the enfeebled party, which was later attacked by the Portuguese, and within three years the entire scheme had collapsed.

These experiences might be taken as representative of a series of disasters which were to follow. About that time some people thought that the Niger joined up with the Nile, others that it had a confluence with the Gambia River. To explore the Niger, the British African Association sent out a Major Houghton, but he, after travelling through the kingdom of Bambouk, was robbed of his possessions and later died. His task was taken over by a Dr. Mungo Park, who left England with two servants in 1795. He reached the Niger in 1796, followed it for 300 miles, and then, emaciated with fever, returned home.

In 1803 Mungo Park led a fresh expedition. This time he asked for a mosquito net and two pairs of trousers for each man. The party left England in 1805; and, after having to contend with malaria, dysentery, incessant rain, swollen rivers and mud, they reached the Niger. There they built a schooner, and sailed down the Niger past Timbuktu, after which the boat capsized and all were drowned. Of the 44 Europeans in the expedition, 35 died of malaria.

An expedition financed by the Navy and led by Captain James Kingston Tuckey set out from England in 1816. The party entered the mouth of the Congo, and some members worked their way up the River as far as Soondy Nsanga, 280 miles from Cape Padron, where, stricken by disease, they had to abandon all hope of progressing farther. Of the 44 Europeans in the party, 18 died of malaria. Dr. McKerrow, a member of the team, gave a good description of the symptoms of malaria. He noticed that the men most seriously affected were those who had visited African villages or slept in the open. As a medicine

他因该药的用途尚不为人所知而大发其财。西德纳姆非常确信其药效并将树皮粉末和红酒混合进行使用。早在 1659 年威利斯就发现金鸡纳树皮能够缓解急性疟疾的发作，尽管常有复发；直到 1768 年詹姆斯·林德才发现发热患者早期大量服用该药效果最好。

1745 年，克劳德·图森·德卡加拉热尝试制备奎宁粗提物，1810 年里斯本的伯纳迪诺·安东尼奥·戈梅斯对此又进行了尝试。10 年后，皮埃尔·约瑟夫·佩尔蒂埃和约瑟夫·卡文图从该物质中首次分离出了 4 种生物碱中的 2 种。这个发现极大地刺激了对金鸡纳树皮的需求，导致秘鲁、厄瓜多尔和玻利维亚的森林被大量采伐，以致人们开始担心其会因如此采伐而灭绝。于是英国人和荷兰人开始种植金鸡纳树，到 1862 年荷兰人几乎已经垄断了其供应，主要在爪哇地区。

与葡萄牙人相比，英国人尝试在中部非洲进行贸易或移民要慢一些。最早的移民计划之一由威廉·伯尔茨提出，并由奥地利的玛丽亚·特蕾西亚赞助。探险是 1776 年从意大利里窝那开始的。此次探险共包含 152 名欧洲人，他们在德拉瓜湾沿着马苏莫河建立站点，但是疟疾很快就体现出了其危害性。当地的非洲人起来反抗这个乏力的团队，随后又受到葡萄牙人的袭击，但是 3 年内整个计划还是以失败而告终。

这些经历可能被认为是随后要发生的一系列灾难的代表。那时候有些人认为尼日尔河与尼罗河是连接的，另一些人认为它和冈比亚河相汇合。为了开发尼日尔河，英国非洲协会派出霍顿少校，但是他在穿过邦布王国后遭到抢劫，不久就死亡了。芒戈·帕克医生接管了他的工作，他于 1795 年带着两名仆人离开英国并于 1796 年到达尼日尔河，沿河走了 300 英里后开始发热，最后虚弱地返回了家乡。

1803 年芒戈·帕克开始了新的探险。这一次他为每个人准备了一个蚊帐和两条裤子。这个团队于 1805 年离开英国，在经过和疟疾、痢疾、连夜雨、暴涨的河流和淤泥斗争之后到达了尼日尔河。他们在那里建了一艘帆船并沿着尼日尔河向下经过廷巴克图，后来船倾覆了，所有人都被淹死了。在探险的 44 名欧洲人中有 35 人死于疟疾。

1816 年一支由海军资助并由詹姆斯·金斯顿·塔基船长带领的探险队从英国出发。这个团队进入刚果河口后，一些成员沿河向上游航行远至松迪恩桑加——一个离帕德隆角 280 英里的地方，他们在此地遭遇了疾病的袭击从而不得不放弃继续前进的所有希望。团队的 44 人中有 18 人死于疟疾。该团队的麦克罗医生详细地描述了疟疾的症状。他注意到最严重的患者是那些到访过非洲农村或露宿过野外的人。

he made some use of cinchona bark, but only as a last resort, and the results were unpromising.

Another expedition, sponsored by the Admiralty, left England in 1822, with the view of exploring the east coast of Africa, Madagascar and parts of Arabia. A small detachment attempted to make its way up the Zambesi, reaching as far as Senna. Of the three Europeans in the group, one had already died, and the other two died on the journey back to their base.

The failure of the Congo expedition, sent out in 1816, intensified the desire to solve the problem of the Niger. In 1822 Hugh Clapperton with two companions set out to cross the desert from Tripoli. They discovered Lake Chad; then Oudney, one of the group, died, leaving Clapperton to struggle as far as Sokoto by way of Kano, whence he was able to return alive.

Clapperton made another attempt in 1825, with five Europeans in the party. From the Bight of Benin, one of the group reached Yaourie before being murdered by his followers. The remainder went on to Jannah, where they were all ill with fever; but against formidable odds, Clapperton reached Katunga, crossed the Niger, and moved on to Kano. Weakened with dysentery and malaria he died at Sokoto, so that, of the Europeans, four died.

At about the same time, a further expedition set out to reach the Niger by way of Tripoli and Timbuktu. It was led by Major Alexander Gordon, who was the first known European to reach Timbuktu, but he was murdered by Arabs soon afterwards. In 1827, a Frenchman, René Caillié, starting from Freetown, reached the Niger and became the second European to see Timbuktu; he succeeded in returning to France alive.

The riddle of the Niger still remained unresolved. To solve it, Richard Lander, who had been a member of Clapperton's expedition, and his brother, John, offered their services to the British Government. They left Portsmouth in 1830, followed Clapperton's route as far as Bussa, and set sail down the Niger with four Negroes as a crew. They reached Eboe, near the Atlantic, but were then captured by Ibu traders and held by different people until finally released, when they found their way to the Atlantic along one of the subsidiary channels of the Niger delta. A boat picked them up and took them to Rio de Janeiro before they could return to England. They had been advised to take two to five grains of quinine every six hours.

Interest in the Niger rose still higher. A company at Liverpool financed an expedition which was led by Richard Lander and had the use of three small steamboats. Special attention was paid to the physique and fitness of the crews. But before the ships reached the Niger, fever began to take its toll. Moving up to Eboe, two of the boats were surrounded by dense vegetation and swamps on either side, and the men were exposed to relentless rain and "torrents of sandflies and mosquitoes".

606

他使用了一些金鸡纳树皮作为药物，但那只是最后一根救命稻草，而且效果并不理想。

1822 年另一支由海军部赞助的探险队离开英国，目的是开发非洲东海岸、马达加斯加和部分阿拉伯半岛。一部分人试图沿赞比西河而上，直达赛纳河。团队的 3 名欧洲人中，有 1 人先死了，其余 2 人死在返回基地的路上。

始于 1816 年的刚果探险的失败强化了人们解决尼日尔河问题的愿望。1822 年，休·克拉珀顿和两个同伴从的黎波里出发穿越沙漠。他们发现了乍得湖，后来一个队员奥德莱死了，剩下克拉珀顿及另一同伴艰难地经过卡诺到达索科托，并在那里活着返回了家乡。

1825 年，克拉珀顿进行了另一次尝试，团队中有 5 名欧洲人。从贝宁湾开始，到达亚乌雷后一名成员被同伴处死。其余的人继续前进并到了坚奈，在那里所有人都患上了发热的疾病，但是经过痛苦的磨难后，克拉珀顿到达了卡通加并穿过尼日尔河最后转移到卡诺。他由于罹患痢疾和疟疾而身体虚弱，在索科托去世，因此 5 名欧洲人中 4 人死了。

几乎是在同一时间，另一支探险队出发了，经的黎波里和廷巴克图到达尼日尔河。此次探险由亚历山大·戈登少校带领，他是第一个已知的到达廷巴克图的欧洲人，但不久之后就被阿拉伯人谋杀了。1827 年，法国人勒内·卡耶从弗里敦出发到达尼日尔河，并成为第二个看到廷巴克图的欧洲人，他后来成功地回到了法国。

尼日尔河之谜仍然未得到解决。为了揭开其神秘面纱，曾是克拉珀顿探险队成员的理查德·兰德及其弟弟约翰开始为英国政府服务。1830 年他们离开朴次茅斯，沿着克拉珀顿曾航行过的路线远至布萨，和 4 名黑人船员沿着尼日尔河向下航行。他们到达了邻近大西洋的埃博，但是被奴隶贩卖者俘获并在不同的人手中周转，直到他们发现了通向大西洋的尼日尔河三角洲的一条支流后才最终逃脱。他们搭乘一条船并被载至里约热内卢，直到返回英格兰。他们曾被建议每 6 小时服用 2~5 粒奎宁。

人们对尼日尔河更感兴趣了。利物浦的一个公司资助了由理查德·兰德带领并使用 3 艘小型蒸汽船的探险活动。这次探险特别注重了船员的体格和体质。但是船还没有到达尼日尔河，发热就已经开始了。向埃博行进过程中，其中的 2 艘船都有一侧陷入茂密的植被和沼泽中，船员们受到持续的暴雨、"无穷的白蛉和蚊子"的折磨。

Further up stream, one of the boats, the *Quorra*, ran aground and remained stuck fast from November until the following March. At the end of March, MacGregor Laird penetrated up stream as far as Fundah, but got no further, and what had been the supply ship, the *Alburkah*, reached Raba and then turned round. At the end of two years, of the total European complement of the three ships, numbering 82, 64 were dead.

By this time the British Navy was active along the west coast of Africa, attempting to suppress the slave trade. Some indication of the health risks to which the crews were exposed is conveyed by the fact that in 1834, of 792 men serving in seven British ships, 204 died. Nevertheless, in 1841 the British Government and members of the public jointly financed the Great Niger Expedition. Three special ships were built, with the *Wilberforce* joining them later. Each boat had a special system of ventilation: during the night as few men as possible were to remain on deck, and when up river, all the white crew had to sleep below. Special clothing was provided, and dry clothing was to be readily available. The men selected were all robust and in the prime of life.

The ships left Woolwich in 1841, and in August, entered the Niger. On September 4 a virulent attack of fever struck the crew of the *Albert*; soon after, the *Soudan* was sent down stream with all the sick. In October the plan to reach Raba was given up. Sickness and deaths continued so that the British Government decided to recall the expedition. Of a total of 145 Europeans, 42 died—almost all of them of malaria.

In his book on this expedition, M'Williams, surgeon of the *Albert*, reported that the practice of blood-letting, which had been the procedure of first choice for almost all the earlier expeditions, was of no value whatsoever, while quinine given at a late stage and in large doses was of some benefit; but he missed the point, which his own records would have shown, that quinine was more effective if given early. He gave details of eight autopsies in each of which he found the gall bladder distended with bile, "the colour and consistence of tar". The colon was generally empty except for "dark pultaceous matter viscid and tenacious".

By this time it was obvious that Africans were much less susceptible to malaria than Europeans, although these too became more resistant if they lived in the country for long periods. This realization, along with a better appreciation of the value of quinine, marked the turning-point in the opening-up of Central Africa. Quinine had been put to medicinal use in the Navy by Sir William Burnett, although it was not given a fair trial. But Alexander Bryson, who was later to become Director General of the Naval Medical Service, recommended that it should be used as a prophylactic and administered to all members of crews on going ashore and on their return, as well as to those who remained on board in swampy places.

With considerations like these in mind, a new vessel, the *Pleiad*, set sail from Plymouth in 1854, under the command of Mr. Beecroft—later succeeded by Dr. William Blaikie. With 12 Europeans and 53 coloured men on board, it entered the mouth of the Niger

再向上游行进中,其中一艘船"柯拉号"搁浅了,从 11 月到次年 3 月一直停留在那里。3 月末,麦格雷戈·莱尔德逆游而上直达丰达,但是没能走得更远,而供给船"阿尔布加号"到达了拉巴后就返回了。在 2 年探险结束时,3 艘船上 82 名欧洲船员中共有 64 人死亡。

到这个时候英国海军已活跃在非洲的西海岸,试图镇压那里的奴隶交易。1834 年,服务于 7 艘英国船只上的 792 人中有 204 人死亡,这一事实表明了船员所面临的健康风险。尽管这样,1841 年英国政府和市民共同资助了"伟大的尼日尔河探险"。人们建造了 3 艘专门的船,随后"威尔伯福斯"号也参与进来。每艘船上都装有特殊的通气系统,在夜间尽可能少地让人留在甲板上,向上游行驶时所有的白人船员都必须睡在船舱里面。同时船员配备了特别的衣物,而且可以随时得到干衣物。挑选的船员都非常强壮而且正值壮年。

1841 年这些船离开伍力奇,8 月到达尼日尔河。9 月 4 号"艾伯特号"上的船员遭受致命的发热袭击,不久,"苏丹号"载着所有患者沿下游返回。10 月,探险队放弃了抵达拉巴的计划。疾病和死亡还在继续,因此英国政府决定终止这次探险。在总共 145 名欧洲人中,42 人死亡,几乎全部死于疟疾。

"艾伯特号"上的外科医生威廉姆斯在其描述这次探险的书中写道:在几乎所有的早前探险活动中,放血疗法被认为是治疗此病的第一选择,但没有任何效果,而晚期大剂量服用奎宁则有一些效果。但是他忽略了他的记录中本来可以找到的关键点,即如果早期服用奎宁效果会更好。他详细描述了 8 例患者的尸检情况,其中每个患者的胆囊由于胆汁而膨胀,"胆汁类似于柏油的颜色和稠度";除了"黑色柔软的黏稠物质"外,结肠通常是空的。

到现在很明显的是欧洲人比非洲人更容易患疟疾,尽管有些欧洲人在当地居住长时间后也能变得更有耐受性。意识到这个问题以及正确评价奎宁的价值是中部非洲开放过程中的转折点。虽然奎宁还没有受到公正的评判,但威廉·伯内特先生已将其应用到海军的医疗服务中。后来成为海军医疗服务主管的亚历山大·布赖森建议将奎宁作为预防性药物,并给所有上岸和返回的船员以及沼泽地区的船员服用。

考虑到以上几点,一艘新船"昂星团号"于 1854 年从普利茅斯出发,船长是比克罗夫特先生,后期换成了威廉·布莱基医生。船上共有 12 名欧洲人和 53 名有色

and travelled as far as Tshomo; scurvy broke out among the crew, yet although the Europeans were subject to great fatigue and went ashore in unhealthy places, while some of them slept on deck, none of them died. They took three to four grains of quinine every morning, and sometimes in the afternoon.

At this stage David Livingstone came into the picture. It was through reading M'Williams's account of the Niger expedition that he devised his famous pill, consisting of quinine and purgatives. That was in 1850. With it he first treated an English party and members of his own family. His procedure was to give doses large and early; by means of it he was able to cross Africa from coast to coast. His confidence in the pill was so great that he severed connexion with the London Missionary Society, so that he could operate on a wider scale.

By then a national hero, Livingstone had no difficulty in persuading the Foreign Office to sponsor an expedition to ascend the Zambesi as far as Chobe and plant a mission somewhere near the Batoka plateau. When the party reached Africa the members took two grains of quinine every day. Livingstone doubted if that was enough, but he felt that he could stave off serious attacks of fever by extra doses. Altogether, at first things went fairly well in spite of personal dissensions, and in spite of the fact that with a good deal of malaria about, some individuals became more seriously affected. Livingstone also believed in the therapeutic value of physical exertion.

In April 1859, he discovered the Shire Highlands, and in September, Lake Nyasa. Soon afterwards he heard of the fate of the Helmore-Price expedition to Linyanti—an "unhealthy place". There was no medically trained person in the group, and out of nine Europeans, six lost their lives.

Further experience of malaria, to which Europeans made a varying response, induced Livingstone to abandon the prophylactic use of quinine for a time. He ran into difficulties over the U.M.C.A. Mission at Magomero, and a series of disasters followed, including the death of Bishop Mackenzie, who lost his supply of quinine when a boat capsized. Worse was to follow: Livingstone's wife died of malaria; the Mission at Magomero, which had been moved to Chibisa, had to be closed down, and he was recalled to England.

The two doctors, John Kirk and Charles Meller, now with extensive experience of malaria, realized that the disease was not so simple as Livingstone had imagined. They identified the dysenteric kind called "blackwater fever". Meller distinguished the asthenic and hepatic forms, and experience convinced him of the prophylactic value of quinine, with doses of up to five grains taken daily.

The next steps in the conquest of malaria are more widely known. In 1880, Laveran, a French Army doctor, discovered the cyst-like bodies of the protozoon in the red corpuscles of human blood. This observation was only slowly taken up, but it was confirmed by Marchiafava in 1884; and in 1889, the tertian, quartan and malignant types of the disease were distinguished. Following the discoveries of Manson and Theobald Smith, that insects

人种，船进入了尼日尔河口并一直航行到乔莫。船员中暴发了坏血病，但是尽管这些欧洲人极度疲劳而且在危险的地方登岸，尽管有些人睡在甲板上，但没有一个人死亡。他们每天早上服用 3~4 粒奎宁，有时候下午也服用。

这一时期戴维·利文斯通进入了人们的视线。通过阅读威廉姆斯尼日尔河探险的报道后，1850 年他发明了包含奎宁和泻药的著名药片。他首次用该药治疗了一个英国团队及其自己的家人。他的方法是早期大剂量给药，借助于这种方法他能够跨越非洲，从东海岸到达西海岸。他对此药非常有信心，以至于与伦敦传道会建立联系以便更大范围地推广该药。

利文斯通此时已是国家英雄，他成功说服外交部资助一次新的探险，这次航行沿着赞比西河直到乔贝，并在巴托卡高原附近留下了一个代表团。当这个团队到达非洲时每个成员每天服用两粒奎宁。利文斯通不能肯定这个剂量是否足够，但他觉得能用额外的剂量来抵御严重的发热袭击。尽管存在个人纷争，或有些人严重感染疟疾的事实，但事情起初进展得非常顺利。利文斯通也相信体能的治疗价值。

他在 1859 年 4 月和 9 月分别发现希雷高原和尼亚萨湖。不久他听说了赫尔莫–普赖斯到利尼扬蒂沼泽地区——一个"危险的地区"——探险的命运。他们的团队中没有经过医学训练的人，导致 9 名欧洲人中 6 人死亡。

欧洲人对疟疾的更多经历产生了不同的反应促使利文斯通一度放弃将奎宁作为预防性药物使用。他在麦高麦罗的"中非大学传教会"任务中遇到了困难，随后出现了一系列灾难，包括毕肖普·麦肯齐因一艘船倾覆得不到奎宁而去世，随后出现了更糟糕的事情：利文斯通的妻子死于疟疾。这样已经转移到奇比萨的麦高麦罗任务不得不终止，而利文斯通也被召回英国。

那时对疟疾有丰富经验的两名医生约翰·柯克和查尔斯·梅勒意识到，这个疾病并非如利文斯通想象的那样简单。他们发现了一种称为"黑尿热"的痢疾样疾病。梅勒将其区分为虚弱型和肝型，而且经验告诉他日均服用 5 粒奎宁预防效果最好。

攻克疟疾的下一阶段更广为人知。1880 年，法国军医拉韦朗在人类血液的红细胞中发现了疟原虫。这个发现逐渐被人接受并于 1884 年得到了马尔基亚法瓦的证实。1889 年相继发现了该疾病的 3 种类型：间日疟，三日疟和恶性疟。随后，曼森和西奥博尔德·史密斯发现昆虫能够作为疾病的载体。英国军医罗斯根据曼森的建议开

can act as vectors of disease, on Manson's suggestion, Ross, a British Army doctor, started work in India, and in 1897 found the oocyst in the outer wall of a mosquito's stomach. He worked out the life-cycle of the avian type of the plasmodium in the following year.

Up to 1914, quinine was the only drug known to be effective against *Plasmodium falciparum*, *P. vivax* and *P. malariae*, but not against the gametocytes. The outbreak of the First World War stimulated the search for other preventatives, particularly in Germany where supplies of quinine might be cut off. An observation by Guttmann and Ehrlich, in 1891, that methylene blue had some action against the plasmodium served as a starting-point for Schulemann, who after a series of trials synthesized plasmoquin in 1925, the first artificial anti-malarial. The discovery of atebrin soon followed. This proved to be a valuable drug for prophylactic use in the Second World War. Still more potent drugs, chloroquine and amodiaquine, were isolated shortly afterwards. These two are excellent, having a complete prophylactic action against almost all forms of malaria, while producing no side-effects. Two further anti-malarials to be discovered were proguanil and pyrimethamine ("Daraprim"), the latter being particularly long-lasting. It was discovered by George Hitching of the Burroughs Wellcome Laboratories.

The conquest of malaria has also been greatly assisted through the use of insecticides, such as pyrethrum, Paris green, benzene hexachloride (BHC) and DDT. In the case of some of these there was long delay between their first discovery and exploitation. BHC, for example, was isolated by Faraday in 1825, yet its insecticidal properties were only discovered in the United States in 1933; and a German chemist, Zeidler, synthesized DDT in 1874, yet its properties as an insecticide were first noticed by Paul Müller in the Geigy Laboratories, in Basle, in 1939. The Second World war did a great deal to stimulate the production and exploitation of these compounds. As an example of insecticidal potency, one might mention BHC, which six months after application on mud walls is capable of killing 80 percent of *Anopheles gambiae*.

Taken together, all these developments have completely transformed the situation in relation to the conquest of malaria. In Nyasaland, for example, as late as 1897, the death rate among European settlers averaged between 9 and 10 percent, mainly from fever, in a young population. In Northern Rhodesia, during 1907–08, the death rate from malaria and blackwater fever combined was 30.4 per 1,000; in 1925, even before the aforementioned developments, it had fallen to 2.8.

Such results can be compared with those associated with the construction of the Kariba Dam, which was started in 1956 and completed by 1960. Not once was work held up or even interrupted because of disease. This huge undertaking involved the importation of enormous numbers of African workers from many parts of the country and the employment of many Europeans who had not previously built up any immunity through exposure to malaria. For health and comfort, the living quarters were placed on high ground. All workers were informed about the dangers of heat stroke, and employers were encouraged to allow their workers a period for acclimatization. A survey of the locality

612

始在印度工作，于 1897 年在蚊子的胃外壁发现了卵囊。次年，他发现了禽疟原虫的生活周期。

直到 1914 年，奎宁一直是已知的有效治疗恶性疟原虫、间日疟原虫和三日疟原虫的唯一药物，但对配子体的治疗效果不好。第一次世界大战的爆发刺激了对其他预防性药物的开发，尤其是在奎宁供应被切断的德国。1891 年，古特曼和埃尔利希发现亚甲基蓝有抗疟原虫的作用，这成为舒勒曼实验的起点。1925 年，舒勒曼经过一系列试验合成了扑疟喹，该药是第一个人工合成的抗疟药。随后发现了阿的平。第二次世界大战证实该药是非常好的预防性药物。不久以后人们分离出了更强效的药物氯喹和阿莫地喹。这两种药效果非常好，对所有类型的疟原虫都有预防作用而且无副作用。宝来威康实验室的乔治·希青发现了另外两种抗疟药：氯胍和乙胺嘧啶（"达拉匹林"），后者药效尤其持久。

杀虫剂的使用为攻克疟疾做出了一定的贡献，比如除虫菊杀虫剂、巴黎绿、六氯环己烷（六六六）和 DDT。就这些杀虫剂而言，其首次发现和利用之间还有一段间隔。比如 1825 年法拉第分离出六六六，而 1933 年在美国才发现其杀虫特性。1874 年德国化学家蔡德勒合成了 DDT，而在 1939 年巴塞尔嘉基实验室的保罗·米勒首次发现其杀虫特性。第二次世界大战在很大程度上刺激了这些化合物的生产和开发。对于杀虫剂效价的实例，人们可能会想到六六六，该药喷洒到泥墙上 6 个月后能够杀死 80% 的冈比亚按蚊。

总体来说，所有的这些发展已经完全改变了人们攻克疟疾的形势。比如在尼亚萨兰，迟至 1897 年，欧洲青年群体移居者的平均死亡率在 9%~10% 之间，病因主要是发热。在北罗得西亚，1907 年～1908 年间由疟疾和黑尿热引起的人口死亡率是 30.4‰；但到 1925 年，即使在上述提到的发展出现之前，死亡率就降到了 2.8‰。

这样的结果可以与修建的卡里巴坝有关的结果相比较。该大坝始建于 1956 年，竣工于 1960 年。工程从没有因为疾病而延迟甚至受到干扰过。这个大型工程招募了大量来自非洲各个地区的工人以及许多之前未暴露于疟疾环境中而形成任何免疫力的欧洲人。为了健康和舒适，住房建于高地上。所有工人都被告知了发热疾病的危险，并且提倡雇主们允许工人们有一段环境适应期。对该地区的调查显示原住民和当地

showed a spleen rate of 80 percent, and a parasite rate of more than 30 percent among the original, local community. All the interior walls of dwellings were sprayed with BHC, and this procedure was repeated three times a year.

As a preliminary measure, an attack was made against mosquito breeding-places through spraying with "High Spread Malariol" (Shell). Survey counts showed that the operation was effective, so that there was no need to repeat it later. All workers were required to take the prophylactic drug with which they were provided—at first, 0.4 g of camoquin weekly for the Europeans, and 100 mg of mepacrin for the Africans. Later, daraprim was substituted at the rate of 25 mg each per week.

All windows of the European houses were screened, and this helped to keep away millions of other insects as well as mosquitoes. As a consequence, there was not one death from malaria among Europeans living on the site during the first two years. There were two deaths, however, of men living in temporary camps outside the recognized limits, and nearly all the European cases of fever gave a history of irregular prophylaxis, or fishing or hunting at night beyond the controlled area.

These results show that apart from human factors, in Central Africa the conquest of malaria is now almost complete. They also show that in Africa it is a mistake to think of medicine in terms of individual territories.

(**208**, 1267-1269; 1965)

Reference:
1. Gelfand, M., *Rivers of Death* (Supplement to *The Central African Journal of Medicine*, 11, No. 8; August, 1965).

人群中脾肿率是 80%，寄生虫感染率超过 30%。住宅的所有内墙壁都喷洒六六六，每年三次。

他们采取了初步的措施，在蚊子滋生的场所喷洒了"高度扩散性的疟蚊杀"（谢尔）。调查数据显示该措施效果非常好，因此以后不必重复。要求所有工人按照他们提供的药量服用预防性药物——最初是欧洲人每周服用 0.4 克卡莫喹，非洲人每周服用 100 毫克米帕林。后来，全部替换成每周服用 25 毫克乙胺嘧啶。

欧洲人住宅的所有窗户都被封闭，这有助于阻挡成数百万的其他昆虫和蚊子。结果在最初的两年内在他们所居住的地方，没有一个欧洲人因为疟疾而死亡。但是，居住于认可范围之外临时住所的人中，有两例死亡；而且几乎所有发热的欧洲人都曾不规律服用预防性药物或者在控制区域之外夜间垂钓或打猎。

这些结果显示除了人为因素，在中部非洲攻克疟疾现在基本完成。这也表明在非洲从单个区域角度考虑医学问题是错误的。

（毛晨晖 翻译；金侠 审稿）

Characterization of Glucose-6-phosphate Dehydrogenase among Chinese

P. W. K. Wong *et al.*

Editor's Note

Glucose-6-phosphate dehydrogenase (G-6-PD) deficiency is an inherited condition that can cause a particular type of anaemia. This common enzyme defect is found across the globe, and here Paul Wong and colleagues identify three variants of G-6-PD deficiency found in Chinese people. The variants differ in their pattern on an eletrophoretic gel, and in the way they cause anaemia—one patient only developed anaemia after ingesting toxic drugs. Other "triggers" of G-6-PD-related anaemia have since been identified, and it is now known that certain infections, medicines and foods can sometimes cause this anaemia in G-6-PD-deficient people.

VARIATIONS in the physico-chemical characteristics of glucose-6-phosphate dehydrogenase (G-6-PD) have been described both in individuals with normal[1-4] and deficient[5-11] enzyme activity. Recently, Kirkman, McCurdy and Naiman[12] have reported on the characteristics in three unrelated Chinese males with G-6-PD deficiency. They found that the G-6-PD migrated slightly faster than normal G-6-PD but the migration was not quite as fast as that of (A-) G-6-PD observed in Negroes. The enzyme had slightly more thermolability, a bimodal pH optimum curve similar to that of the Mediterranean variant, abnormally low Kms for G-6-P and TPN and relatively greater utilization of G-6-P analogues.

This communication reports on the character of G-6-PD in a group of normal and deficient Chinese subjects. For these investigations, blood was collected in ACD (formula A, U.S.P.) solution. G-6-PD activity was determined by the method of Glock and McHean as modified by Zinkham[13]. Purification of erythrocyte G-6-PD was performed according to the method of Kirkman *et al.*[8]. Starch-gel electrophoresis was performed as described by Shows *et al.*[10].

Starch-gel electrophoresis was performed on haemolysate of 38 normal males and 29 normal females of Chinese origin. Of this group, 36 were Cantonese, 21 were Fukianese, and one each came from Shanghai, Chieking, Shangtung, Hupei and Hunan. The remaining five subjects came from the Chinese mainland, but their place of origin is unknown. All 67 were found to show the single band normally seen in Caucasians.

Detailed investigations were carried out in four normal and four G-6-PD-deficient Chinese

中国人体内的葡萄糖-6-磷酸脱氢酶的特征

黄保罗（音译）等

编者按

葡萄糖-6-磷酸脱氢酶(G-6-PD) 缺陷是一种遗传疾病，能引起某种特定类型的贫血病。这种常见的酶缺陷在全球广泛存在，本文中黄保罗及其同事们鉴定了在中国人体内发现的 G-6-PD 缺陷的三种变异。这些变异的凝胶电泳模式和导致贫血的方式都不同——一个病人只有在摄入毒性药剂后才会发展为贫血。目前已经鉴别出其他一些能够"引发"与 G-6-PD 有关的贫血因素。并且现在已知某些感染、药品和食品有时能引发 G-6-PD 缺陷人群出现这种贫血。

在酶活性正常 [1-4] 的个体和缺乏 [5-11] 的个体中葡萄糖-6-磷酸脱氢酶（G-6-PD）的物理化学性质都存在差异。最近，柯克曼、麦柯迪和奈曼 [12] 报道了三位彼此无关且 G-6-PD 缺陷的中国男性的特征。发现他们的 G-6-PD 比正常的 G-6-PD 迁移得快一些，但是没有黑人体内观察到的 A 型 G-6-PD 迁移得那么快。该酶的热不稳定性略有提高，具有与地中海型变异类似的双峰状最适 pH 曲线，G-6-P（葡萄糖-6-磷酸）和 TPN（三磷酸吡啶核苷酸）的米氏常数都异常低，而对 G-6-P 类似物的利用率相对较高。

这篇通讯文章报道了一组酶正常和缺陷的中国人体内的 G-6-PD 的特征。在这些研究中，血液保存在酸性柠檬酸盐葡萄糖（根据《美国药典》中的 *A* 配方）溶液中。用格洛克和麦克希恩提出并经辛克罕 [13] 改进的方法测定 G-6-PD 的活性。依据柯克曼等人 [8] 提出的方法提纯红细胞中的 G-6-PD。用肖 [10] 等人描述的方法进行淀粉凝胶电泳。

对 38 位正常中国男性和 29 位正常中国女性的溶血液进行淀粉凝胶电泳。这些受试者中，36 个是广东人，21 个是福建人，另有 5 人来自上海、山东、湖北和湖南等地，其余 5 位受试者也来自中国大陆但籍贯未知。所有 67 名受试者的电泳结果都只出现了高加索人中常见的单一条带。

另外，对来自中国香港的 4 位正常的和 4 位 G-6-PD 缺陷的受试者进行了仔

subjects from Hong Kong. Patient 1 is a healthy 2-year-old male who has a haemoglobin level of 14–15g percent. The deficiency of G-6-PD was detected only following a haemolytic episode after the use of drugs. Patients 2, 3 and 4 have the typical clinical picture of congenital non-spherocytic haemolytic disease, with persistent reticulocytosis and a haemoglobin level of 6–8 g percent.

Erythrocyte G-6-PD (expressed as units/100 ml. red blood cell) was decreased in all four deficient subjects, ranging from 0 to 15 units as compared with an average value of 228 units in Chinese controls and 210 units in Caucasian controls. Leucocyte G-6-PD (expressed as μM TPN reduced/h/10^6 white blood cell) was also decreased. In patients 1 and 2 the values were 0.35 and 0.61 as compared with 1.05 in Chinese controls and 1.02 in Caucasian controls. Starch-gel electrophoresis performed on purified enzyme showed patients 1, 2 and 3 having a single band which migrated 100 percent (corresponding to that normally seen in Caucasians), while patient 4 showed a single band at 106 percent (corresponding to that usually seen in Pamaquin-sensitive Negro males (Fig. 1)). In all four patients the Km for G-6-P was reduced to half the normal values seen in both Chinese and Caucasian controls, but the Km for TPN was only slightly reduced in patient 4 and not reduced in the other three. Investigations of pH optima were carried out twice on purified enzyme from each patient. Patients 1, 2 and 3 showed a single peak at pH 8, while patient 4 showed a bimodal distribution (Fig. 2). Relative thermostability studies were carried out at 41°C and there was no significant difference between the patients and controls. Finally, substrate specificity was tested with G-6-P analogues. There was a slight but definite increase of activity using both galactose-6-phosphate and 2-desoxy-glucose-6-phosphate as substrates. The increase was more marked in patient 4.

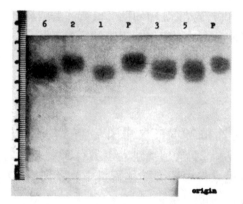

Fig. 1. Vertical starch-gel electrophoresis patterns of red blood cells G-6-PD. Slots 6, 3 and 5 contain preparations from Caucasians. Slots *P* contain preparations from pamaquin-sensitive Negroes. Slot 1 contains preparation from patient No. 3 and slot 2 contains preparation from patient No. 4.

细的研究。患者 1 是个健康的 2 岁男孩，其血红蛋白水平为 14 克 /100 毫升 ~15 克 /100 毫升，只是在对他用药后的溶血期才检测到 G–6–PD 缺乏。患者 2、3 和 4 都有先天性非球形红细胞溶血病的典型临床症状，他们患有持续的网状细胞增多症，其血红蛋白水平为 6 克 /100 毫升 ~8 克 /100 毫升。

在所有 4 名酶缺陷的受试者中，红细胞 G–6–PD（以每 100 毫升红细胞中的酶活力单位数表示）都减少了，其值在 0~15 个酶活力单位之间，与之相比，作为对照的中国人的平均含量为 228 单位，作为对照的高加索人的平均值为 210 单位。白细胞 G–6–PD（以每 10^6 个白细胞每小时能使 TPN 降低多少微摩尔来表示）也减少了。在患者 1 和患者 2 中，得到的值分别是 0.35 和 0.61，与之相比，作为对照的中国人的值是 1.05，作为对照的高加索人的值是 1.02。对提纯后的酶进行淀粉凝胶电泳，结果显示，来自患者 1、2 和 3 的酶只有迁移率 100% 的单一条带（和高加索人的正常可见结果一致），而来自患者 4 的酶则只显示出迁移率 106% 的单一条带（和对扑疟喹啉敏感的黑人男性中的常见结果一致）（图 1）。在全部 4 位患者中，G–6–P 的米氏常数都降低至对照中国人和对照高加索人中的正常值的一半，但是只在患者 4 中 TPN 的米氏常数略有降低，而在另外 3 位患者中并没有降低。对来自每一位患者的纯化出的酶进行两次最适 pH 的测定。来自患者 1、2 和 3 的纯化出的酶显示出位于 pH 8 单峰，而来自患者 4 的纯化出的酶则显示出双峰式分布（图 2）。相关热稳定性研究在 41℃ 条件下进行，患者组和对照组之间没有显著差异。最后，用 G–6–P 类似物检验底物特异性。在使用半乳糖–6–磷酸和 2–脱氧–葡萄糖–6–磷酸作为底物时，酶活性都有轻微但明确的升高。对于患者 4，升高得更为明显。

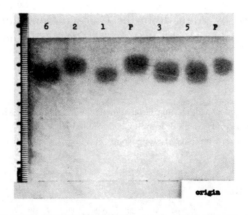

图 1. 红细胞中 G–6–PD 的垂直淀粉凝胶电泳图。6 号、3 号和 5 号上样孔中加入了来自高加索人种受试者的样品，上样孔 P 中加入了来自对扑疟喹啉敏感的黑人的样品。1 号上样孔中加入了来自患者 3 的样品，2 号上样孔中加入了来自患者 4 的样品。

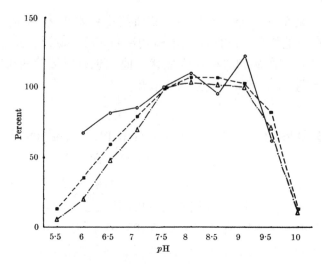

Fig. 2. *p*H Optimum curves for Chinese patients and a Caucasian control. Activity is expressed as percentage of that at *p*H 7.5. ■ , Caucasian control; △ , patient No. 3; ○ , patient No. 4.

These results indicate that several variants of G-6-PD exist among Chinese. One variant (patient 4) shows a persistent non-spherocytic haemolytic anaemia associated with G-6-PD deficiency. This variant is characterized by a fast band on electrophoresis, a decrease of *Km* for both G-6-P and TPN, and a bimodal *p*H optimum curve. This corresponds closely to the patients described by Kirkman, McCurdy and Naiman[12]. Another variant (patients 2 and 3) shows a non-spherocytic haemolytic anaemia with a normal electrophoretic pattern and a single *p*H optimum peak at *p*H 8.0. A third variant (patient 1) shows a G-6-PD deficiency without spontaneous haemolytic anaemia, where the haemolytic process can be induced by toxic drug ingestion. Other variants are likely to be found as individuals with normal and deficient enzyme activity are studied more extensively from various parts of China and South-East Asia.

This work was supported by grants from the Chicago Community Trust, the Kettering Foundation and the U.S. Public Health Service (*T*1-*HD*-00036 and *T*1-*AM*-05186).

(**208**, 1323-1324; 1965)

Paul. W. K. Wong, Ling-Yu Shih and David Yi-Yung Hsia: Genetic Clinic, Children's Memorial Hospital, Department of Paediatrics, Northwestern University Medical School, Chicago, Illinois.

Y. C. Tsao: Department of Paediatrics, University of Hong Kong.

References:

1. Boyer, S. H., Porter, I. H., and Weilbacher, R. G., *Proc. U.S. Nat. Acad. Sci.*, **48**, 1868 (1962).

2. Porter, I. H., *et al.*, *Lancet*, i, 895 (1964).

3. Nance, W. E., and Uchida, I., *Amer. J. Human Genet.*, **16**, 380 (1964).

4. Long, W. K., Kirkman, H. N., and Sutton, H. E., *J. Lab. Clin. Med.*, **65**, 81 (1965).

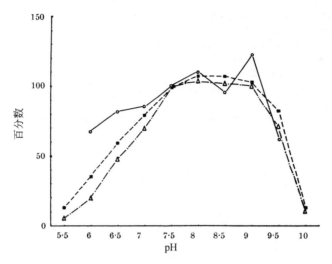

图 2. 来自中国患者和 1 个作为对照的高加索人的酶的最适 pH 曲线。酶活性用占 pH 7.5 时酶活性的百分数来表示。■ 代表作为对照的高加索人；△代表患者 3；○代表患者 4。

　　这些结果表明在中国人体内存在多种 G–6–PD 变异。其中一种 G–6–PD 变异（患者 4）表现出与 G–6–PD 缺陷相关的持续性非球形细胞溶血性贫血。这种变异的特征是在电泳中出现快速迁移的条带，G–6–P 和 TPN 的米氏常数都降低，而且最适 pH 曲线具有双峰。这与柯克曼、麦柯迪和奈曼 [12] 所描述的患者的情况颇为吻合。另一种变异（患者 2 和 3）也表现出非球形细胞溶血性贫血，具有正常的电泳模式和位于 pH 8.0 处的最适 pH 单峰。第三种变异（患者 1）则表现出 G–6–PD 缺陷，但并没出现自发性溶血性贫血，不过在摄入有毒药剂会诱发溶血过程。对中国和东南亚各地区酶活性正常和缺乏的个体进行广泛研究，有可能还会发现其他的变异。

　　这项工作受到了芝加哥社区信用社、凯特林基金会和美国公共卫生署（T1-HD-00036 和 T1-AM-05186）的资助。

（王耀杨 翻译；崔巍 审稿）

5. Kirkman, H. N., Riley, H. D., and Crowell, B. B., *Proc. U.S. Nat. Acad. Sci.*, **46**, 938 (1960).

6. Marks, P. A., Banks, J., and Gross, R. T., *Nature*, **194**, 454 (1962).

7. Kirkman, H. N., Schettini, F., and Pickard, B. M., *J. Lab. Clin. Med.*, **63**, 726 (1964).

8. Kirkman, H. N., *et al.*, *J. Lab. Clin. Med.*, **63**, 715 (1964).

9. Ramot, B., *et al.*, *J. Lab. Clin. Med.*, **64**, 895 (1964).

10. Shows, jun., T. B., Tashian, R. E., Brewer, G. J., and Deru, R. J., *Science*, **145**, 1056 (1964).

11. Grossman, A., *et al.*, *Paediatrics* (in the press).

12. Kirkman, H. N., McCurdy, P. R., and Naiman, J. L., *Cold Spring Harbor. Symp. Quant. Biol.*, **29**, 391 (1964).

13. Zinkham, W. H., *Paediatrics*, **23**, 18 (1959).

Thymus and the Production of Antibody-plaque-forming Cells

J. F. A. P. Miller *et al.*

Editor's Note

From the eighteenth century onwards, physicians recognised that a person infected by a bacterial or viral disease would thereafter be immune (or partially immune) to further infections of the same kind. By the beginning of the twentieth century, the cause of this immunity was found to be the production of antibodies against proteins carried by the bacteria or viruses responsible for the infection. The mechanism by which antibodies are produced was, however, obscure. Much interest centred on the role of the thymus gland, which is intimately connected with blood circulation, but its functions were not understood until the mid-1960s. One difficulty was that the removal of the thymus gland from experimental animals reduced their capacity to resist infection for a period of several months. This paper identified the cause of this delay. J. F. A. P. Miller, the first author of this paper, migrated to Australia soon after its publication and eventually became director of the Walter and Eliza Hall Institute in Melbourne, Victoria.

ALTHOUGH the thymus itself does not play an active part in immune responses[1], its presence is essential for the normal development of immunological faculties. Neonatal thymectomy in many species considerably impairs the capacity of an animal to produce some types of immune responses[2]. Thymectomy in adult life has no immediate effect but, after a period of 6–9 months, reduces the capacity to react to a newly encountered antigen[3]. The possible mechanisms by which the thymus exerts its influence on the immunological system have been discussed elsewhere[4], and it has been concluded that the immunological defects encountered after thymectomy are primary and not secondary to infection or autoimmunity[5]. Evidence is presented here to show that thymectomy practically inhibits the development of the capacity to produce antibody-plaque-forming cells following the injection of sheep erythrocytes. This impairment is evident in clinically healthy suckling baby mice thymectomized at birth.

Mice of the inbred strain *CBA*, F_1 hybrids from crosses between *T6* and *Ak* mice, and non-inbred Swiss (*SWS*) mice were thymectomized on the day of birth. A sham operation involving thoracotomy, but not removal of the thymus, was performed in litter-mates. The mice were given an intraperitoneal injection of 0.15 ml. of a 5 percent suspension of sheep erythrocytes in saline on the 10th day of life and killed at intervals of 2 days–3 weeks after injection. Cell suspensions were prepared from their spleens and assayed for the number of antibody-plaque-forming cells by plating out the suspension in agar gel containing sheep erythrocytes according to the technique described by Jerne *et al.*[6]. Routine histological investigations were performed on the thymus area in all thymectomized

624

胸腺与抗体空斑形成细胞的产生

米勒等

编者按

从 18 世纪开始，医生们就认识到，得过细菌性或者病毒性疾病的人此后再遭遇相同感染时会产生免疫（或部分免疫）。到了 20 世纪初期，人们发现这种免疫的原因是能够对引起感染的细菌或病毒携带的蛋白产生抗体。但是抗体产生的机制尚不清楚。大家的注意力集中在了与血液循环密切相关的胸腺的作用上，但直到 20 世纪 60 年代中期人们才认识到胸腺的功能。研究中遇到的一个难题是，实验动物在切除胸腺几个月后其抗感染能力才会下降。这篇文章确定了免疫力延迟的原因。本文的第一作者米勒在文章发表后不久便移民到澳大利亚，后来成为位于维多利亚州首府墨尔本的沃尔特–伊丽莎·霍尔研究所的主任。

尽管在免疫应答中，胸腺本身并不直接发挥作用[1]，但它的存在对免疫机能的正常发展是必要的。切除许多种系新生动物的胸腺，会极大程度地损害该动物产生某些类型的免疫应答能力[2]。在成年期切除胸腺并不立即产生影响，但是 6~9 个月后，对遭遇到的新抗原的免疫能力降低[3]。胸腺作用于免疫系统的可能机制在其他文章中已经讨论过了[4]，得出的结论是：胸腺切除后出现的免疫缺陷是原发性的，而不是由感染或自身免疫继发性的[5]。该证据表明，对切除胸腺的个体注射绵羊红细胞，实际上抑制了机体产生抗体空斑形成细胞的能力。对于刚出生就切除了胸腺的健康乳鼠，这种损伤是很明显的。

近交系 *CBA* 小鼠、*T6* 和 *Ak* 小鼠杂交的子一代（F_1）小鼠以及非近交系瑞士（*SWS*）小鼠在出生当日进行胸腺切除术。对同窝出生的小鼠进行开胸但不切除胸腺的假手术。向 10 日龄小鼠的腹腔内注射 0.15 毫升含 5% 绵羊红细胞的生理盐水，在注射后间隔 2 天 ~3 周处死这些小鼠。根据杰尼等人[6]描述的技术制备动物脾脏的细胞悬液，并将其涂布于含有绵羊红细胞的琼脂凝胶，用来检测抗体空斑形成细胞的数量。对所有胸腺切除小鼠的胸腺区域进行常规组织学检查，以确认胸腺切除是否完全。

mice to ensure that thymectomy had been complete. Mice with thymus remnants were discarded from the experiments.

The spleens of normal or operated mice aged between 1 week and 4 months and not challenged with sheep erythrocytes gave usually less than 1 antibody-plaque-forming cell per million spleen cells. The results obtained in immunized mice are shown in Figs. 1, 2 and 3 and analysed statistically in Table 1. In sham-thymectomized mice, the number of antibody-plaque-forming cells rose sharply from 2 to 3 days after immunization to reach a peak level at 4, 5 or 6 days, depending on the strain. Thereafter, the number fell rapidly to reach a low, yet significant, level at about 10 days after immunization. By contrast, neonatally thymectomized mice produced very few antibody-plaque-forming cells, the assays being made at intervals from 2 days to 3 weeks after challenge. The difference between the mean peak levels of the thymectomized and control group was highly statistically significant for the three strains of mice used (Table 1).

Table 1. Antibody Plaque Formation by Spleens and Thymuses of Normal,
Sham-operated and Thymectomized Mice

Strain	Treatment given	Cells plated	No. of mice	Age at immunization	Age at death	Antibody-plaque-forming cells per 10^6 cells	
						(Peak) level ± S.E.	P values
$(Ak \times T6)F_1$	Sham-thymectomy at birth	Spleen	8	10 days	14 days	204±27	< 0.001
	Thymectomy at birth	Spleen	10	10 days	14 days	5±2	
CBA	Sham-thymectomy at birth	Spleen	5	10 days	16 days	106±6	< 0.001
	Thymectomy at birth	Spleen	5	10 days	16 days	4±2	
SWS	Sham-thymectomy at birth	Spleen	6	10 days	15 days	276±53	< 0.001
	Thymectomy at birth	Spleen	8	10 days	15 days	15±2	
SWS	None	Spleen	6	56 days*	61 days*	945±82	Not significant
	Thymectomy at 42 days*	Spleen	6	56 days*	61 days*	901±65	
	None	Thymus	6	56 days*	61 days*	6±1	

* Average age.

有胸腺残留的小鼠从实验中剔除。

对于 1 周龄至 4 月龄之间的正常小鼠或切除胸腺的小鼠，如果没有用绵羊红细胞进行免疫处理，其脾脏中每百万个脾细胞通常只产生不足一个抗体空斑形成细胞。经过免疫处理的小鼠获得的结果见图 1、图 2 和图 3，对结果的统计分析见表 1。进行了假胸腺切除手术的小鼠，免疫后 2~3 天抗体空斑形成细胞的数量迅速增加，不同品系的小鼠分别于第 4 天、第 5 天或第 6 天达到峰值。之后，细胞数量急剧下降，大约在免疫后第 10 天降到一个比较低但仍然显著的水平。与此对比，在免疫后间隔 2 天到 3 周的时间内，对切除了胸腺的新生小鼠进行检测，其只产生了很少量的抗体空斑形成细胞。所使用的 3 种品系的小鼠，胸腺切除组与对照组平均峰值上的差别在统计学上极为显著（表 1）。

表1. 正常小鼠进行假胸腺切除术和胸腺切除术后小鼠脾脏与胸腺的抗体空斑形成

品系	给予的处理	涂布的细胞	小鼠数量	免疫时的年龄	死亡时年龄	抗体空斑形成细胞数量/10^6细胞	
						(峰位)均值±标准误差	P值
$(Ak \times T6)F_1$	出生日假切除胸腺	脾脏	8	10日	14日	204±27	< 0.001
	出生日切除胸腺	脾脏	10	10日	14日	5±2	
CBA	出生日假切除胸腺	脾脏	5	10日	16日	106±6	< 0.001
	出生日切除胸腺	脾脏	5	10日	16日	4±2	
SWS	出生日假切除胸腺	脾脏	6	10日	15日	276±53	< 0.001
	出生日切除胸腺	脾脏	8	10日	15日	15±2	
SWS	无	脾脏	6	56日*	61日*	945±82	无显著性
	第42日*切除胸腺	脾脏	6	56日*	61日*	901±65	
	无	胸腺	6	56日*	61日*	6±1	

* 平均年龄。

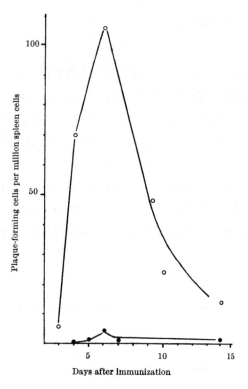

Fig. 1. Number of antibody-plaque-forming cells in spleens of *CBA* mice at various intervals after immunization with sheep erythrocytes given at 10 days of age. Each point represents the average value of assays on 2–5 mice. ○ , Mice sham-thymectomized at birth; ● , mice thymectomized at birth.

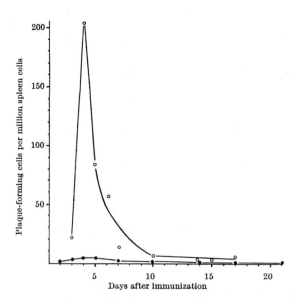

Fig. 2. Number of antibody-plaque-forming cells in spleens of (*Ak* × *T6*) *F*₁ mice at various intervals after immunization with sheep erythrocytes given at 10 days of age. Each point represents the average value of assays on 2–10 mice. ○ , Mice sham-thymectomized at birth; ● , mice thymectomized at birth.

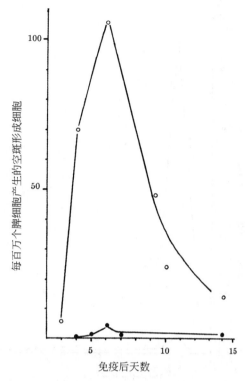

图 1. 10 日龄 *CBA* 小鼠用绵羊红细胞免疫后, 脾脏产生的抗体空斑形成细胞数量随时间变化图。每一点代表对 2~5 只小鼠测定的平均值。○代表在出生当日进行假胸腺切除手术的小鼠, ●代表在出生当日进行胸腺切除手术的小鼠。

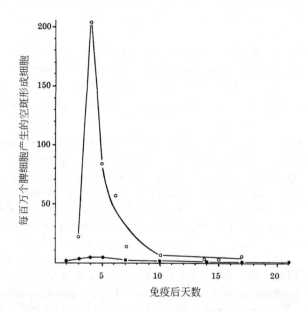

图 2. *Ak* 和 *T6* 小鼠杂交的子一代 10 日龄小鼠用绵羊红细胞免疫后, 脾脏产生的抗体空斑形成细胞数量随时间变化图。每一点代表对 2~10 只小鼠测定的平均值。○代表在出生当日进行假胸腺切除手术的小鼠, ●代表在出生当日进行胸腺切除手术的小鼠。

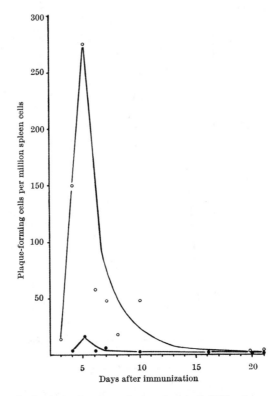

Fig. 3. Number of antibody-plaque-forming cells in spleens of *SWS* mice at various intervals after immunization with sheep erythrocytes given at 10 days of age. Each point represents the average value of assays on 2–8 mice. ○ , Mice sham-thymectomized at birth; ● , mice thymectomized at birth.

Thymectomy of 6-weeks-old *SWS* mice did not interfere with their capacity to produce antibody-plaque-forming cells when immunized at about 9 weeks of age (Table 1). In contrast to the spleens of normal *SWS* mice immunized with sheep erythrocytes, very few antibody-plaque-forming cells were present in the thymuses of the same mice (Table 1).

The results demonstrate that, in the absence of the thymus from birth, the capacity to produce antibody-plaque-forming cells in response to an injection of sheep erythrocytes has failed to develop. Very few antibody-plaque-forming cells appeared in the spleens during the first 3 weeks after immunization of neonatally thymectomized mice of two inbred strains and one non-inbred stock. This deficiency is presumably a primary effect of neonatal thymectomy and not secondary to infection or ill-health: it was evident in very early life in suckling baby mice the growth rate of which was identical to that of litter-mate controls. In a similar experiment it has been reported that colony-bred mice thymectomized at birth also failed to produce normal numbers of antibody-plaque-forming cells when challenged at 4 weeks of age[7]. Thymectomy in adult mice, by contrast, had no immediate effect on the response to sheep erythrocytes although a delayed effect was observed as reported elsewhere[3]. The amount of antibody produced per plaque-forming cells was judged to be within normal limits in the thymectomized mice as estimated by the size of the plaques in our experiments and in those of Takeya *et al.*[7].

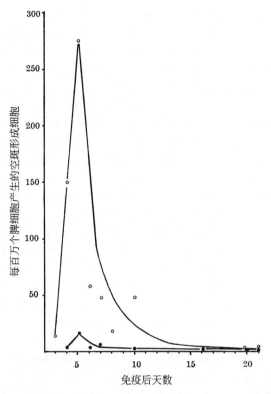

图 3. 10 日龄 *SWS* 小鼠用绵羊红细胞免疫后，脾脏产生的抗体空斑形成细胞数量随时间变化图。每一点代表对 2~8 只小鼠测定的平均值。○代表在出生当日进行假胸腺切除手术的小鼠，●代表在出生当日进行了胸腺切除手术的小鼠。

SWS 小鼠在 6 周龄时切除胸腺，在约 9 周龄时进行免疫，其产生抗体空斑形成细胞的能力并不会受到影响（表 1）。与用绵羊红细胞免疫的正常 *SWS* 小鼠的脾脏相比，相同小鼠的胸腺中只有极少量的抗体空斑形成细胞（表 1）。

研究结果表明，如果出生后就切除胸腺，则不能发展出应答绵羊红细胞而产生抗体空斑形成细胞的能力。切除胸腺的 2 种近交系新生小鼠和 1 种非近交系新生小鼠，在免疫后前 3 周其脾脏中极少有抗体空斑形成细胞产生。这种缺陷可能是新生动物胸腺切除的原发效应，而不是因感染或者健康状况欠佳的继发效应，这在乳鼠生命初期是很明显的，因为切除胸腺的小鼠的生长速度与同窝出生的对照小鼠是相同的。已有类似的实验报道显示：群体繁殖的小鼠若出生时即切除胸腺，在 4 周龄时进行免疫，它们也不能产生正常数量的抗体空斑形成细胞[7]。与此形成对照的是，小鼠成年后切除胸腺并不立即影响其对绵羊红细胞的应答，但另有报道称观察到了延迟效应[3]。在我们的实验以及竹谷等人[7]的实验中，对于胸腺被切除的小鼠，根据空斑大小估算的每个抗体空斑形成细胞产生抗体的量是在正常范围内的。

It can be seen that the thymus itself produced only very few antibody-plaque-forming cells (Table1) after immunization of adult mice. This observation has also been made in other strains of mice[8]. By contrast, specific antibody-plaque-forming cells have been demonstrated in the thymus of rabbits 5 days after a single systemic injection of 5 μg of somatic polysaccharide of *Salmonella enteritidis*[9]. Since significant numbers of plaque-forming cells have been detected in the peripheral blood in these rabbits[10], and since radical alterations in the structure of the thymus have been reported after administration of endotoxin[11], it is conceivable that circulating antibody-forming cells may have penetrated into the thymus of rabbits immunized with somatic polysaccharide.

In conclusion, it seems that the thymus itself fails to produce significant numbers of antibody-producing cells in response to an antigenic stimulus, but its presence from birth is essential to ensure that such cells will develop in the periphery. Whether the initial development of these cells takes place within the thymus and their final maturation occurs after emigration from the organ cannot be decided on the basis of present evidence. Experiments using thymus tissue enclosed in cell-impenetrable chambers in neonatally thymectomized mice have suggested that a humoral thymus factor plays a part in the maturation of potentially immunologically competent cells[12].

The investigations were supported by grants to the Chester Beatty Research Institute (Institute of Cancer Research, Royal Cancer Hospital) from the Medical Research Council and the British Empire Cancer Campaign for Research, from the Tobacco Manufacturers Standing Committee, and Public Health Service grant CA-03188-08 from the National Cancer Institute, U.S. Public Health Service.

Note added in proof. Since this paper was submitted, similar results obtained by Friedman have been published[13].

(**208**, 1332-1334; 1965)

J. F. A. P. Miller, P. M. de Burgh and G. A. Grant : Chester Beatty Research Institute, Institute of Cancer Research, Royal Cancer Hospital, Fulham Road, London, S. W. 3.

References:

1. Fagraeus, A., *Acta Med. Scand.*, **130**, Suppl. 204, 3 (1948). Bjorneboe, M., Gormsen, H., and Lundquist, T., *J. Immunol.*, **55**, 121 (1947). Harris, S., and Harris, T. N., *J. Exp. Med.*, **100**, 269 (1954).

2. Miller, J. F. A. P., *Lancet*, ii, 748 (1961); *Proc. Roy. Soc.*, B, **156**, 415 (1962). Archer, O. K., Pierce, J. C., Papermaster, B. W., and Good, R. A., *Nature*, **195**, 191 (1962). Jankovic, B. D., Waksman, B. H., and Arnason, B. G., *J. Exp. Med.*, **116**, 159 (1962). Sherman, J. D., Adner, M. M., and Dameshek, W., *Blood*, **23**, 375 (1964).

3. Taylor, R. B., *Nature* (following communication). Metcalf, D., *Nature* (following communication). Miller, J. F. A. P., *Nature* (following communication).

4. Miller, J. F. A. P., *Science*, **144**, 1544 (1964); *Brit. Med. Bull.*, **21**, 111 (1965).

5. McIntire, K. R., Sell, S., and Miller, J. F. A. P., *Nature*, **204**, 151 (1964). Miller, J. F. A. P., and Howard, J. G., *J. Reticuloendothelial Soc.*, **1**, 369 (1964).

6. Jerne, N. K., Nordin, A. A., and Henry, C., in *Cell-Bound Antibodies*, edit. by Amos, B., and Koprowski, H., 109 (Wistar Institute Press, Philadelphia, 1963).

7. Takeya, K., Mori, R., and Nomoto, K., *Proc. Jap. Acad.*, **40**, 572 (1964).

8. Friedman, H., *Proc. Soc. Exp. Biol.*, **117**, 526 (1964).

可以看到，成年小鼠在进行免疫后，胸腺本身只产生很少的抗体空斑形成细胞(表1)。在其他品系的小鼠身上也观察到了这种现象 [8]。与此不同的是，用5微克肠炎沙门氏菌的菌体多糖对兔子进行单次体内注射，5天后在兔子胸腺中会出现特异性的抗体空斑形成细胞 [9]。因为在这些兔子的外周血中检测到数量显著的抗体空斑形成细胞 [10]，且有报道称注射内毒素后胸腺的结构会发生根本改变 [11]，因此可以认为，在菌体多糖免疫后的兔子体内，循环的抗体空斑形成细胞可能是渗透到胸腺中的。

总之，看来胸腺本身并不能产生数量显著的抗体形成细胞以应答抗原刺激，但出生后胸腺的存在对于确保外周区域中产生抗体形成细胞是必需的。基于目前的证据，还不能确定这些细胞是否最初在胸腺发育但从胸腺迁移出去之后才最终成熟。有实验将胸腺组织包裹于细胞不能透过的小室中，然后将其植入切除胸腺的新生小鼠体内，结果说明，胸腺体液因子对具有潜在免疫活性的细胞的成熟有一定的作用 [12]。

这项研究由医学研究理事会和大英帝国癌症运动组织提供给切斯特·贝蒂研究所（皇家肿瘤医院肿瘤研究所）的基金支持，同时也受到来自烟草制造商常设委员会提供的基金以及美国公共卫生署国家癌症研究所的公共卫生署基金 CA-03188-08 的支持。

附加说明　自本文提交后，弗里德曼所得到的类似结果已经发表 [13]。

（吴彦 翻译；金侠 审稿）

9. Landy, M., Sanderson, R. P., Bernstein, M. T., and Lerner, E. M., *Science*, **147**, 1591 (1965).

10. Landy, M., Sanderson, R. P., Bernstein, M. T., and Jackson, A. L., *Nature*, **204**, 1320 (1964).

11. Rowlands, D. T., Claman, H. N., and Kind, P. D., *Amer. J. Path.*, **46**, 165 (1965).

12. Osoba, D., and Miller, J. F. A. P., *Nature*, **199**, 653 (1963); *J. Exp. Med.*, **119**, 177 (1964). Osoba, D., *J. Exp. Med.*, **122**, 633 (1965).

13. Friedman, H., *Proc. Soc. Exp. Biol.*, **118**, 1176 (1965).

Decay of Immunological Responsiveness after Thymectomy in Adult Life

R. B. Taylor

Editor's Note

When neonatal mice have their thymus glands removed, lymphoid tissues fail to develop properly and the immune system becomes severely compromised. But in the early 1960s the role of the adult thymus was debated. In the first of three back-to-back papers addressing this issue, R. B. Taylor of the UK Medical Research Council's centre at Mill Hill describes the effects of removing the adult thymus in mice. At first there is little change, but months later a rapid decline in immunological response is seen. Taylor concludes that the adult thymus continues to exert a stimulatory influence on the production of competent cells, but that a low rate of production may still occur after the thymus is removed.

THE remarkable effects of neonatal thymectomy are by now well documented[1]. Primary among these is a failure of the lymphoid tissues to develop properly, which is marked particularly by a deficiency of small lymphocytes. Other features, probably resulting from this, include a defective capacity to perform all kinds of immune responses, and a progressive wasting disease ending in early death.

By contrast, only minimal effects have been found to follow from thymectomy in adult life. In mice, the operation had no effect on growth rate, breeding behaviour, longevity or susceptibility to common infections[2], although blood lymphocyte counts and lymphoid organ weights were somewhat depressed[3]. A wasting disease has been reported in guinea-pigs thymectomized at 150–160 g body-weight[4]. No significant depression of antibody response has been detected after thymectomy of adult rabbits[5,6]. These results have led to the assumptions that once the lymphoid tissues have been formed, the thymus ceases to perform in its developmental role, and that the function of lymphopoiesis is then taken over by the other lymphoid tissues[7]. Yet no distinct morphological change has been described in the thymus on the attainment of immunological maturity, and it continues its high rate of cell production even in adult life. Indeed, it is still necessary for the recovery of immunological responsiveness after this has been depressed by irradiation[8], and thus it must be able to resume its lymphopoietic function, if only in response to a stimulus such as might be provided by destruction of lymphoid tissues. Even without, this stimulus, however, the thymus must play some part in normal turnover of lymphoid cells, since thymectomy in adult life largely prevents recovery from immunological paralysis[9,10]. It therefore seemed probable that the lymphoid tissues might not be able to maintain themselves indefinitely in the absence of the thymus, and that if mice were left for a sufficiently long time after thymectomy in adult life they should eventually show a decline in their primary immune responsiveness.

成年期胸腺切除后免疫应答能力的下降

泰勒

编者按

当新生小鼠的胸腺被移除后，淋巴组织不能正常发育，免疫系统受到严重损害。但是在 20 世纪 60 年代早期，成体胸腺的作用尚处于争论之中。在关于这个研究连续发表的三篇论文中的第一篇中，英国医学研究理事会中心（位于米尔山）的泰勒描述了小鼠成年期胸腺移除后的效应。起先，只有一点点变化，但是数月后，其免疫应答迅速下降。泰勒推断成体胸腺会持续地对免疫活性细胞的生成施加影响；但是当移除胸腺后，免疫活性细胞仍以低速率生成。

目前已经详细论述了新生动物胸腺切除的显著影响 [1]。其中最主要的影响是淋巴组织不能正常发育,其显著特征是缺乏小淋巴细胞。其他症状可能都是由此引发的,包括各种免疫应答能力的缺陷以及因慢性消耗性疾病而夭折。

与此不同的是，研究发现在成年期切除胸腺后所带来的影响微乎其微。对小鼠进行胸腺切除手术后，尽管血液中淋巴细胞数量和淋巴器官的重量有所降低 [3]，但生长速度、繁殖行为、寿命或对普通感染的易感性都没有受到影响 [2]。有报道称体重为 150 克 ~ 160 克的豚鼠在胸腺被切除后会得一种消耗性疾病 [4]。成年家兔在切除胸腺后检测到抗体反应没有明显降低 [5,6]。根据以上结果产生了这样一种猜想，即一旦淋巴组织形成，胸腺就停止执行它在发育过程中的功能，其他淋巴组织取代其生成淋巴细胞的功能 [7]。目前还没有关于胸腺达到免疫成熟后发生明显形态学变化的报道，而且甚至在成年期，胸腺仍然高速率地生成细胞。实际上，辐射引起免疫能力下降 [8] 后，胸腺对于免疫应答能力的恢复依然是必不可少的，因此只要对类似免疫组织破坏这样的刺激做出应答，那它必须恢复生成淋巴细胞的功能。然而，即使没有这样的刺激，胸腺对于淋巴细胞的正常周转必然也起着某些作用，因为成年期切除胸腺在很大程度上阻碍了机体从免疫麻痹状态的恢复 [9,10]。因此，看来在胸腺不存在的情况下淋巴组织也许不能无限期地自我维持，而且如果小鼠在成年期切除胸腺后存活足够长时间，它们最终将出现初次免疫应答能力下降。

In the first experiment the immune responsiveness of *CBA* mice was assessed by the ability of their lymphoid cells to cause a graft-versus-host (GVH) reaction after transfer to young (*C57BL×CBA*) F_1 hybrid mice. The strength of the GVH response was estimated by the degree of spleen enlargement in the hosts. The 4-point assay procedure outlined by Simonsen[11] and developed by Michie[12] was then used to relate the spleen enlargement response to the dose of lymphoid cells, so that a quantitative comparison could be made of the immune potency of cell suspensions made from thymectomized and control mice. *CBA* female mice were thymectomized (or sham-operated) at 2–3 months of age. Groups of 8–16 mice were tested by GVH assay at intervals up to one year after thymectomy. In each assay the pooled lymph node (cervical, axillary, brachial, inguinal and mesenteric) and spleen cells of one "test" mouse were compared for immune potency against those from one "control" mouse. At the intervals 4, 9, 25 and 29 weeks the thymectomized mice were compared with sham-operated controls of the same age. By 52 weeks the potency of these controls might be expected to have declined somewhat through age alone[13]; therefore both thymectomized and sham-operated mice were compared with untreated controls aged 4–6 months. To obtain a measure of the overall responsiveness of the whole "test" animal as a percentage of the control, each value for the relative potency of a cell suspension was multiplied by the fraction: total number of lymphoid cells obtained from the "test" mouse/total number of lymphoid cells obtained from the "control" mouse.

The results of the assays are shown in Fig. 1. The responsiveness of the thymectomized mice remained level with controls for about 25 weeks, then dropped fairly sharply, and thereafter remained about 15 percent of normal. This fall in responsiveness was due mainly to a fall in the immune potency of the cell suspensions, but also to a fall in the numbers of cells recovered. The responsiveness of sham-operated mice fell only slightly, and at 52 weeks after operation was still about 75 percent of the 4-6-month-old controls.

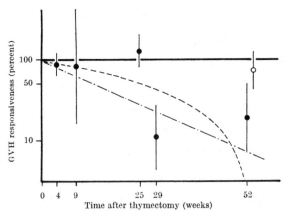

Fig. 1. Results of GVH assays. Each point represents the mean and standard deviation of the responsiveness of thymectomized relative to control mice (computed on log-transformed data). ●, Thymectomized versus sham-operated controls of the same age (4-29 weeks groups) or versus 4–6 month untreated mice (52-week group); ○, sham-operated (14 months old) versus 4–6 month untreated mice. *P*<0.01 for difference between 25 and 29 week groups; *P*<0.001 for difference between thymectomized and sham-operated groups at 52 weeks. Hypothetical linear (– – –) and exponential (– · –) decay curves are shown.

在第一个实验中，将 *CBA* 小鼠的淋巴细胞转移到年幼的（*C57BL×CBA*）F_1 杂交小鼠体内后，通过引起移植物抗宿主（GVH）反应的能力来评定 *CBA* 小鼠的免疫应答能力。GVH 反应的强度通过宿主脾脏增大的程度来衡量。由西蒙森 [11] 提出、并经米基 [12] 改进的 4 点测定法被用来分析脾脏增大与淋巴细胞量之间的关系，由此能够定量比较从胸腺切除小鼠与对照小鼠制备的细胞悬液的免疫效力。*CBA* 雌性小鼠在 2~3 月龄进行胸腺切除（或假手术操作）。8~16 只小鼠一组，在胸腺切除后 1 年内，每隔一段时间用 GVH 分析法进行检测。在每次分析中，比较"测试组"小鼠与"对照组"小鼠的淋巴结（颈部的、腋下的、臂的、腹股沟的和肠系膜的）和脾细胞的免疫效力。在 4 周、9 周、25 周和 29 周时将胸腺切除小鼠与同龄的假手术对照组小鼠进行比较。到 52 周时，对照组小鼠的免疫效力可能单纯因年龄因素会稍有下降 [13]，因此胸腺切除组小鼠和假手术对照组小鼠都与 4~6 月龄未经处理的对照组小鼠进行比较。为获得所有"测试"动物整体的应答能力相对于对照动物的百分比，每一组细胞悬液相对免疫效力的值都要乘以如下分数：从"测试组"小鼠获得的淋巴细胞总量 / 从"对照组"小鼠获得的淋巴细胞总量。

分析结果如图 1 所示，在大约 25 周内胸腺切除小鼠的免疫应答能力保持在与对照组小鼠一样的水平上，之后应答能力则急剧下降，然后会维持在对照组的 15% 左右。这一免疫应答能力的降低主要是由于细胞悬液免疫效力的降低，但也与恢复的细胞数量下降有关。假手术组小鼠的应答能力只是稍微降低，在手术后 52 周时应答能力仍然是 4~6 月龄对照组小鼠的 75% 左右。

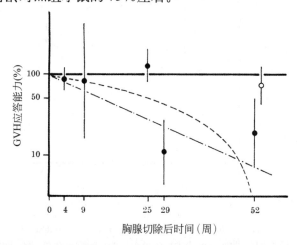

图 1. GVH 分析结果。每个点代表胸腺切除小鼠相对于对照小鼠的应答能力的均值和标准差（数据进行对数变换后再计算）。●代表胸腺切除组与同龄（4~29 周组实验）的假手术对照组或与 4~6 月龄未处理小鼠（52 周组实验）的比较；○代表假手术组（14 月龄）与 4~6 月龄未处理小鼠的比较。25 周组与 29 周组之间有差别 $P<0.01$；52 周时胸腺切除组与假手术组有差别 $P<0.001$。（– – –）显示假设的线性衰减曲线，（–·–）显示假设的指数衰减曲线。

In another experiment immune responsiveness was tested by the ability to produce circulating antibody to bovine serum albumin (BSA). The results have been collated from four experiments which were set up with other primary objects in mind. *CBA* mice of both sexes were thymectomized or sham-operated either at 4 weeks or at 10–14 weeks of age. They were challenged at intervals thereafter by subcutaneous injection of BSA in Freund's adjuvant, and the serum antibody titres estimated 20 and 40 days later by a modification[14] of Farr's antigen-binding-capacity method[15]. Since the absolute titres varied from one experiment to another, the titres from thymectomized mice have been presented as a percentage of the controls in each group (Fig. 2). The decay of responsiveness was less obvious than it was in the GVH experiment, but cell-transfer experiments indicate that a small difference in anti-BSA titre can reflect a larger difference in the number of competent cells[16]. However, the general pattern was similar in that the decline did not reach its most rapid phase for a considerable time after thymectomy. (The apparent initial fall which can be seen in the mice thymectomized at 4 weeks old reflects an increase in the responsiveness of controls, due to normal growth processes, rather than a decrease in the thymectomized mice.) This rapid phase can be seen at about 10 weeks after thymectomy in the mice thymectomized at 4 weeks old. In those thymectomized at 10–14 weeks old the time varied: in one experiment the responsiveness was still normal at 16 weeks, but in another it had already reached its lower level by 12 weeks after thymectomy (indicated by asterisk) and showed no further fall at 32 weeks. An effect of thymectomy was also seen in the general condition of the mice. While not amounting to a serious wasting disease, it appeared as a distinct ruffled condition of the fur, most obvious on the underside, and was associated with a slight loss of weight.

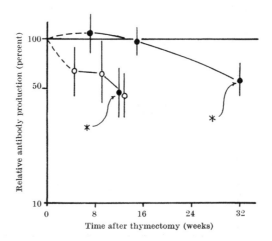

Fig. 2. Results of antibody titrations. Each point represents the mean antibody titre in thymectomized mice as a percentage of that in controls, with standard deviation (computed on log-transformed data). Results from 20 and 40 day bleedings have been combined. Symbols represent four separate experiments: ○, thymectomized at 4 weeks old; ●, thymectomized at 10–14 weeks old. * See text.

It may be concluded that even in the adult the thymus has a part to play in the maintenance of the numbers of competent cells in the body. If the effect of thymectomy

在另一个实验中，用牛血清白蛋白（BSA）产生循环抗体的能力来衡量免疫应答能力。结果由原计划用于其他目的的四个实验整理而来。对两种性别的 *CBA* 小鼠在 4 周龄或 10~14 周龄时切除胸腺或进行假手术处理。之后每隔一段时间皮下注射混悬于弗氏佐剂的 BSA，20 天和 40 天后用改良 [14] 的法尔氏抗原结合能力法 [15] 测定血清抗体的滴度。由于各组实验间绝对滴度的变化太大，因此我们把每组中胸腺切除小鼠的滴度表示成相对于对照组小鼠的百分比（图 2）。应答能力的下降不如在 GVH 实验中那么明显，但是细胞转移实验显示抗 BSA 滴度的微小变化能够反映出免疫活性细胞数量更大的变化 [16]。但是，基本的模式是类似的，在胸腺切除后相当长时间内应答能力的衰减并没有达到其最快速阶段。（4 周龄切除胸腺的小鼠初期应答能力明显降低是因为对照组小鼠由于在正常生长过程中应答能力的增加，而不是因为胸腺切除小鼠的应答能力降低。）4 周龄切除胸腺的小鼠会在胸腺切除约 10 周后进入快速衰减阶段。而那些 10~14 周龄切除胸腺小鼠的应答能力明显衰减的时间则是不一致的：在一次实验中，应答能力直到 16 周还是正常的，但在另一次实验中胸腺切除后 12 周应答能力就降到较低水平（星号所示），直到 32 周才没有再降低。从小鼠的总体情况也可以看到胸腺切除的影响。尽管小鼠没有发生严重的消耗性疾病，但皮肤明显褶皱，特别是下侧皮肤褶皱最明显，这与体重稍微降低有关。

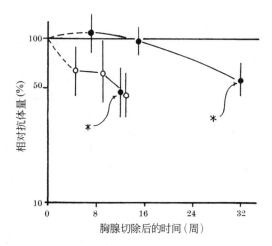

图 2. 抗体滴定的结果。每个点代表胸腺切除小鼠滴度的均值（以占对照组小鼠滴度的百分比来表示）和标准差（数据进行对数转换后再计算）。20 天和 40 天取血的结果已经合并。符号代表了 4 个独立的实验：○代表 4 周龄时切除胸腺；●代表 10~14 周龄时切除胸腺。* 见正文内容

我们可以得出以下结论：甚至在成体内，胸腺对于维持机体内活性细胞的数量也起着一定的作用。如果胸腺切除能够完全断绝新的活性细胞的供给，那

was to cut off entirely the supply of new competent cells, then the ensuing decay of responsiveness should reflect the rate of loss of the competent cells originally present. According to two simple hypotheses their numbers might be expected to decay in linear fashion of they had a definite life-span, or exponentially if their life depended on chance events—such as encounter with antigens. Support for the latter possibility may be drawn from investigations of irradiated human patients[17] in whom certain of the circulating lymphocytes were distinguishable by the presence of acentric chromosome fragments, which could be seen after stimulation of mitosis *in vitro* with phytohaemagglutinin. During division these fragments fail to become attached to the mitotic spindle, and thus are not transmitted to the daughter cells. The half-time between irradiation and either division or death of the cells was 366 days.

In the present experiments, however, the decay curve of immune responsiveness after thymectomy did not readily fit with either expectation (Fig. 1). In particular, although the responsiveness fell to a sub-normal level, it appeared to attain equilibrium by about 7 months after thymectomy, and showed no further decline towards the levels seen after neonatal thymectomy or irradiation. The same was shown by at least one of the BSA experiments (Fig. 2, asterisk). This suggests either that there is a distinct population of cells with a very long life, or that the production of competent cells can continue at a sub-normal rate in the absence of the thymus. In view of the much more severe and permanent depression of responsiveness which can be obtained after neonatal thymectomy, or adult thymectomy followed by irradiation, this latter process would presumable take place only in mature, unirradiated lymphoid tissue.

Although the early part of the curve approximates the linear, rather than the exponential mode of decay, it might still be consistent with an exponential loss of cells if these were being replaced by differentiation from some other cell type. This could be an intermediate form which had already undergone the thymus-dependent phase in its maturation. The eventual exhaustion of such cells could then account for the more rapid fall between 25 and 29 weeks. In conclusion, it is suggested that the thymus continues in adult life to exert a stimulatory influence on the production of competent cells, but that a low rate of production may still occur after thymectomy. It will be of interest to see if this residual responsiveness differs qualitatively from the major fraction, which decays more rapidly after thymectomy.

(**208**, 1334-1335; 1965)

R. B. Taylor: Division of Experimental Biology, National Institute for Medical Research, Mill Hill, London, N. W. 7.

References:

1. Good, R. A., and Gabrielsen, A. E., edit. by, *The Thymus in Immunobiology* (Harper and Row, New York, 1964).

2. Miller, J. F. A. P., *Adv. Cancer Res.*, **6**, 291 (1961).

3. Metcalf, D., *Brit. J. Haematol.*, **6**, 324 (1960).

么随后发生的应答能力下降应该反映了最初存在的活性细胞的损失速度。根据两种简单的假说，如果这些细胞有一定的寿命，那么可以预料它们的数量会以线性形式减少，而如果它们的寿命取决于类似遭遇抗原这样的偶发事件，那么其数量会以指数形式减少。对后一种可能性的支持来源于对受辐射的人类患者的研究 [17]，这些患者体内的某些循环淋巴细胞因存在无着丝点染色体碎片而容易辨识，在体外用植物凝血素刺激诱发有丝分裂后可以看到这些碎片。在分裂过程中，这些碎片不能附着在有丝分裂纺锤体上，因而不能转入子细胞。细胞从受到辐射到分裂或死亡的半衰期是 366 天。

然而，在该实验中，胸腺切除后免疫应答能力的衰减曲线并不能与任何一种预期很好地吻合（图 1）。特别是，尽管应答能力降到了低于正常的水平，但似乎在胸腺切除约 7 个月后又达到了平衡，并不继续下降到出生即切除胸腺或受辐射中出现的水平。BSA 实验中至少有一次也显示了相同情况（图 2 中星号所示）。这表明，要么存在一群非常长寿的细胞，要么在胸腺不存在的情况下活性细胞能以低于正常水平的速率继续产生。考虑到在出生即切除胸腺或成年切除胸腺后受辐射的情况中观察到更为严重和持久的应答能力下降，后一种情况大概只在成年期未受辐射的淋巴组织中发生。

尽管曲线的开始部分是近似线性而非指数模式的衰减，但如果这些细胞被某些其他细胞类型的分化替换它仍可能符合细胞指数下降。这可能是在其成熟过程中已经经历过胸腺依赖性阶段的一种中间态形式。这些细胞的最终耗竭可以解释在 25~29 周时更为急剧的下降。总之，可以认为成体胸腺会持续地对免疫活性细胞的生成施加影响，但是切除胸腺后，免疫活性细胞仍以低速率生成。比较一下这些残留的应答能力和胸腺切除后更加迅速下降的主要应答能力在性质上是否不同应该是很有趣的。

（吴彦 翻译；陈新文 陈继征 审稿）

4. Comsa, C., *Acta Endocr. (Kbh.)*, **26**, 261 (1957).

5. Harris, T. N., Rhoads, J., and Stokes, J., *J. Immunol.*, **58**, 27 (1948).

6. Maclean, L. D., Zak, S. J., Varco, R. L., and Good, R. A., *Transpl. Bull.*, **4**, 21 (1957).

7. Leading article, *Brit. Med. J.*, ii, 840 (1962).

8. Miller, J. F. A. P., *Nature*, **195**, 1318 (1962).

9. Claman, H. N., and Talmage, D. W., *Science*, **141**, 1193 (1963).

10. Taylor, R. B., *Immunology*, **7**, 595 (1964).

11. Simonsen, M., *Prog. Allergy*, **6**, 349 (1962).

12. Michie, D. (personal communication).

13. Makinodan, T., and Peterson, W. J., *Proc. U.S. Nat. Acad, Sci.*, **48**, 234 (1962).

14. Mitchison, N. A., *Proc. Roy. Soc.*, B, **161**, 275 (1964).

15. Farr, R. S., *J. Infect. Dis.*, **109**, 239 (1958).

16. Taylor, R. B. (unpublished observations).

17. Norman, A., Sasaki, M. S., Ottoman, R. E., and Fingerhut, A. H., *Science*, **147**, 745 (1965).

Delayed Effect of Thymectomy in Adult Life on Immunological Competence

D. Metcalf

Editor's Note

In the early 1960s a flurry of research papers attempted to elucidate the function of the adult thymus. It was well known that neonatal thymectomy (thymus removal) triggers rapid atrophy of the lymphoid organs, and Australian physiologist Donald Metcalf had previously shown that adult thymectomy essentially slowed this process. Here Metcalf shows that the effects of adult mouse thymectomy are apparent in around half the animals at about 18 months, suggesting that the adult thymus has a continuing influence on the immunological response. His is the second of three *Nature* papers, published back-to-back, to arrive at similar conclusions. Today it is accepted that the thymus plays a functional, albeit diminished role, in adulthood.

NEONATAL thymectomy leads to the rapid development of lymphoid organ atrophy and well-characterized immunological deficiencies[1-3], but thymectomy performed in adult life leads only to the slow development of a moderate degree of lymphoid atrophy[4] with no loss of immunological competence when the animals are tested immediately after operation[5,6]. However, adult thymectomy combined with whole-body irradiation does lead to immunological deficiencies of the same general nature as those following neonatal thymectomy[7]. Recent investigations have shown that continuous repopulation of normal haemopoietic organs occurs in normal life[8,9]. This suggests that whole-body irradiation by causing cell damage and repopulation in haemopoietic organs may merely accelerate a process which occurs continually throughout life, albeit at a much slower rate. These considerations prompted a re-examination of the long-term effects of adult thymectomy on immunological competence.

Six-week-old mice of the long-lived $(AKR \times C57BL)F_1$ strain were thymectomized or sham-operated and challenged 1 week, 11 months or 18 months after operation with intraperitoneal sheep red cells.

Haemagglutinin titres in mice challenged immediately after thymectomy did not differ from those in control mice (Table 1). However, in mice tested 11 months after thymectomy, some lower titres were found, particularly in the early ($19S$) phase of the response (Fig. 1). In mice tested 18 months after operation, approximately half the thymectomized mice produced no detectable haemagglutinins (Table 1), the remainder producing titres within the control range. The titres in the thymectomized mice challenged 18 months after operation corresponded almost exactly with those observed by Miller[7] in young adult mice subjected to thymectomy and 350-r. whole-body irradiation. The spleen is the major

成体胸腺切除对免疫功能的延迟性影响

梅特卡夫

编者按

在 20 世纪 60 年代早期，一系列研究论文试图阐明成体胸腺的功能。众所周知，新生动物胸腺切除（胸腺移除）能引起淋巴器官迅速萎缩，并且澳大利亚生理学家唐纳德·梅特卡夫先前曾指出，成体胸腺切除本质上会减慢淋巴器官萎缩的过程。在本文中，梅特卡夫指出：大约一半的成年小鼠在摘除胸腺后 18 个月出现了明显的淋巴器官萎缩现象，这意味着成体胸腺对免疫反应具有持续的影响。本文是连续在《自然》杂志上发表的能得出类似结论的三篇文章中的第二篇。现在，人们认为胸腺在成年期发挥功能性作用，尽管其作用减小。

新生动物胸腺切除将导致淋巴器官的快速萎缩和明显的免疫缺陷 [1-3]，但是当胸腺切除后立即检测动物则会发现成体胸腺切除只引起缓慢的淋巴样中度萎缩 [4]，而免疫功能并没有丧失 [5,6]。但是，将成体胸腺切除并进行全身辐射，那么成体也会出现像新生动物胸腺切除后那样的免疫缺陷 [7]。最近的研究显示，正常个体中常规的造血器官可以持续再生 [8,9]。这就提示我们，全身辐射造成造血器官的细胞损伤和再生，可能仅仅是加快了个体中持续发生着的某个过程，虽然是以很慢的速度进行的。这些考虑促使我们重新研究成体胸腺切除对免疫功能的长期影响。

将 6 周龄的长寿命的 $(AKR \times C57BL)F_1$ 系小鼠进行胸腺切除术或假手术。手术 1 周、11 个月或 18 个月后，用腹腔注射的绵羊红细胞进行刺激。

胸腺切除后立即进行刺激的小鼠的血凝素滴度与对照组小鼠的血凝素滴度没有区别（表 1）。但是，胸腺切除 11 个月后再对小鼠进行检测就会发现血凝素滴度有所降低，特别是在反应的早期（19S）（图 1）。而在手术 18 个月后进行检测时发现，胸腺切除的小鼠约有一半不能产生可检测的血凝反应（表 1），其余小鼠的血凝素滴度则在对照组范围之内。胸腺切除 18 个月后接受刺激的小鼠的血凝素滴度，几乎与米勒 [7] 在实验中发现的进行了胸腺切除并接受 350 伦琴全身辐射的年轻成年小鼠的

source of haemagglutinin-production (Metcalf, unpublished results) and in the immunized thymectomized mice challenged 18 months after operation, a general disorganization of the structure of the spleen lymphoid nodules was observed with an obvious decrease in the numbers of germinal centres and small lymphocytes and a diminished number of pyroninophilic mitotic cells.

Table 1. Primary Haemagglutinin Response of $(AKR \times C57BL)F_1$ Mice to Sheep Red Blood Cells

Type of mouse	No. of animals	Haemagglutinin titres ($\log_2$)											
		0	1	2	3	4	5	6	7	8	9	10	11
Normal 2 months old. No antigen	14	12		1		1							
Normal 18-20 months old. No antigen	18	18											
Thymectomy + antigen. 1 week after operation	13									5	7	1	
Sham-thymectomy+antigen. 1 week after operation	14									5	6	3	
Thymectomy + antigen. 11 months after operation	23			1		3	2	6	10	1			
Sham-thymectomy + antigen. 11 months after operation	31							1	13	13	3		1
Thymectomy + antigen. 18 months after operation	13	6	1				2			2	2		
Sham-thymectomy + antigen. 18 months after operation	48				4	4	4	7	11	12	3	2	1

Immunizing does: 0.2 ml. of 20 percent R.B.C. i.p. Sera titrated on day 10 following immunization.

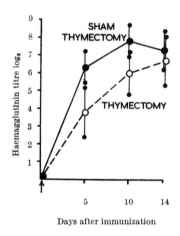

Fig. 1. Haemagglutinin titres in $(AKR \times C57BL)$ F_1 mice tested 11 months after thymectomy or sham-operation at six weeks of age. No. of mice: thymectomy 23, sham-thymectomy 31. Bars represent standard deviations.

The observations recorded here suggest that simple thymectomy performed on adult mice does lead to demonstrable immunological deficiencies in at least some mice, provided sufficient time is allowed to elapse before the animals are tested. The present results are only preliminary and need confirmation with other mouse strains and other antigens. The irregular nature of the loss of immunological competence in individual mice following thymectomy, suggests a deletion or loss of responsiveness in some mice of clones of cells

648

血凝素滴度完全符合。脾脏是产生血凝反应的主要来源（梅特卡夫未发表的实验结果），而在胸腺切除 18 个月后被刺激的免疫小鼠中，可以观察到脾脏淋巴小结结构普遍被破坏了，同时还伴随生发中心与小淋巴细胞数量的明显下降以及嗜派洛宁有丝分裂细胞的减少。

表 1. $(AKR \times C57BL)F_1$ 小鼠对绵羊红细胞的初次血凝反应

小鼠类型	动物数量	血凝素滴度（$\log_2$）											
		0	1	2	3	4	5	6	7	8	9	10	11
正常 2 月龄，无抗原	14	12		1			1						
正常 18~20 月龄，无抗原	18	18											
胸腺切除 + 手术 1 周后抗原刺激	13									5	7	1	
假胸腺切除 + 手术 1 周后抗原刺激	14									5	6	3	
胸腺切除 + 手术 11 个月后抗原刺激	23			1		3	2	6	10	1			
假胸腺切除 + 手术 11 个月后抗原刺激	31								1	13	13	3	1
胸腺切除 + 手术 18 个月后抗原刺激	13	6	1				2			2	2		
假胸腺切除 + 手术 18 个月后抗原刺激	48				4	4	4	7	11	12	3	2	1

免疫剂量：腹腔注射 0.2 毫升 20% 的红细胞。免疫 10 天后对血清进行滴定检测。

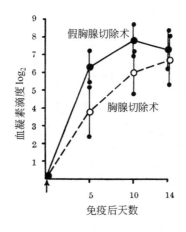

图 1.6 周龄 $(AKR \times C57BL)$ F_1 系小鼠进行胸腺切除术或假胸腺切除术 11 个月后检测的血凝素滴度。
受试小鼠数量：进行胸腺切除术的有 23 只，进行假胸腺切除术的有 31 只。图中竖线代表标准差。

本文中记录的观测结果表明，如果在经过足够长的时间后才进行检测，那么即便是只对成体小鼠胸腺进行切除，也至少有一些小鼠出现了明显的免疫缺陷。本文的结果只是初步的，还需要利用其他种系的小鼠和其他的抗原加以验证。这种胸腺切除后个体小鼠免疫功能的丧失是无规律的，这表明某些种系的小鼠中对所用抗原有应答能力的细胞克隆缺失或丧失了其应答功能。这种丧失明显是随机发生的，并

capable of responding to the antigen used, this loss apparently being on a random basis and time dependent.

The present result re-emphasize that the thymus has a continuing influence on immunological responsiveness throughout adult life, but do not indicate the processes involved. The thymus is known to produce the humoral factor L.S.F. throughout life[10] and a humoral influence of the thymus has been shown to be necessary for the full responsiveness of competent cells following antigenic stimulation[11,12]. Since there is continuous spleen and lymph node cell repopulation throughout life, the thymus could also serve as a production site for immunologically competent cells, continually reseeding the peripheral organs.

This work was supported by the Carden Fellowship Fund of the Anti-Cancer Council of Victoria.

(**208**, 1336; 1965)

Donald. Metcalf: Cancer Research Laboratory, Walter and Eliza Hall Institute, Post Office Royal Melbourne Hospital, Victoria, Australia.

References:

1. Miller, J. F. A. P., *Lancet*, ii, 748 (1961).

2. Miller, J. F. A. P., *Science*, **144**, 1544 (1964).

3. Martinez, C., Kersey, J., Papermaster, B. W., and Good, R. A., *Proc. Soc. Exp. Biol.*, **109**, 193 (1962).

4. Metcalf, D., *Brit. J. Haematol.*, **6**, 324 (1960).

5. Fichtelius, K. E., Laurell, G., and Philipsson, L., *Acta Pathol. Microbiol. Scand.*, **51**, 81 (1961).

6. MacLean, L. D., Zak, S. J., Varco, R. L., and Good, R. A., *Transpl. Bull.*, **4**, 21 (1957).

7. Miller, J. F. A. P., *Nature*, **195**, 1318 (1962). Miller, J. F. A. P., Doak, S. M. A., and Cross, A. M., *Proc. Soc. Exp. Biol.*, **112**, 785 (1963).

8. Harris, J. E., Barnes, D. W. H., Ford, C. E., and Evans, E. P., *Nature*, **201**, 884 (1964).

9. Metcalf, D., and Wakonig-Vaartaja, R., *Proc. Soc. Exp. Biol.*, **115**, 731 (1964). Metcalf, D., and Wakonig-Vaartaja, R., *Lancet*, i, 1012 (1964).

10. Metcalf, D., *Brit. J. Cancer*, **10**, 442 (1956).

11. Osoba, D., and Miller, J. F. A. P., *Nature*, **199**, 653 (1963).

12. Levey, R. H., Trainin, N., and Law, L. W., *J. Nat. Cancer Inst.*, **31**, 199 (1963).

且具有时间依赖性。

本文结果再次强调，胸腺对于整个成年期免疫应答的影响是持续性的，但是无法指出其中涉及的过程。人们已经知道，胸腺能终生性地产生一种叫做肝抑制因子的体液因子 [10]。而且对于抗原刺激后活性细胞完整应答来说，胸腺的体液影响是必需的 [11,12]。既然脾脏和淋巴结细胞在持续地、终生性地再生着，那么胸腺也可能就是担负着免疫活性细胞的生成，并发挥持续地供应外周器官的作用。

此项工作由维多利亚州抗癌协会卡登研究基金支持。

（李响 翻译；陈新文 陈继征 审稿）

Effect of Thymectomy in Adult Mice on Immunological Responsiveness

J. F. A. P. Miller

Editor's Note

Here the French-born biologist Jacques Miller provides further evidence for the continuing immunological function of the thymus in later life. His adult mouse thymectomy experiments point to a delayed immunological decline that is evident after 6 months, suggesting that the adult thymus helps maintain a pool of immunologically competent cells. Removing the thymus, he explains, has no immediate effect because the pool can buffer the blow. But when the pool itself becomes depleted, owing to the limited life span of its cells, the immunological defects become apparent. This is the last of three papers, published back-to-back, to arrive at the same conclusion. It is now known that the thymus builds up its stock of infection-fighting T-lymphocytes early in life, and has a less active role in adulthood.

THYMECTOMY in adult animals has been associated with a lowering of the population of lymphocytes in blood, thoracic duct lymph, lymph nodes and spleen[1,2]. However, no significant defects in the capacity for rejecting allogeneic skin homografts[2] or for producing serum antibodies[2,3] have been observed in animals thymectomized in adult life and challenged within 1–2 months after thymectomy. This is in marked contrast to the severe immunological defects which occur following neonatal thymectomy[4]. These observations have suggested that, during early life, the thymus is essential for the complete and normal development of some immunological faculties. Evidence that the function of the thymus in initiating immunogenesis is not necessarily restricted to early life has, however, been produced. Adult mice thymectomized and afterwards exposed to total body irradiation have shown severe deficiencies in immunological functions[2,5] indicating that the thymus in the adult is still essential to re-establish immune mechanisms when the immunological apparatus has been damaged or destroyed. It has been shown, furthermore, that thymectomy in adult immunologically tolerant mice prevented the reappearance of reactivity with respect to that antigen[6]. The thymus in the adult would thus appear to be essential for the correction of specific immunological defects. The work described here gives further evidence for a continuing immunological function of the thymus through adult life.

In one experiment inbred *CBA* mice and *(Ak×T6) F$_1$* mice were thymectomized at 2–3 months of age and challenged at various intervals up to 2 years after thymectomy with 0.3ml. of a 5 percent suspension of sheep erythrocytes in saline. 4–6 days after challenge, the mice were killed and their spleens removed to assay the number of antibody-plaque-forming cells per million spleen cells, according to the technique described by Jerne *et al.*[7].

成年小鼠胸腺切除对免疫应答的影响

米勒

编者按

在本文中法裔生物学家雅克·米勒对于胸腺在成熟期继续发挥免疫功能提供了进一步的证据。在他的成年小鼠胸腺切除实验中，小鼠在 6 个月后免疫能力明显下降，这表明成年胸腺帮助维持一些免疫活性细胞。他解释说，移除胸腺后没有立即产生效应，是因为这些免疫活性细胞能够减缓这一影响。但是由于免疫活性细胞寿命有限，当其耗尽后，免疫缺陷就变得明显。这是三篇连续发表的论文中的最后一篇，得出了与其他两篇相同的结论。我们现在知道胸腺在发育早期积聚抗感染 T 淋巴细胞，而在成年期其活性减弱。

成年动物胸腺切除与血液、胸导管淋巴、淋巴结和脾脏中的淋巴细胞数量的降低有关 [1,2]。然而，成体在切除胸腺后 1~2 个月内进行刺激，并没有观察到其排斥同种异源皮肤移植能力 [2] 或血清抗体产生能力 [2,3] 的明显缺陷。这与新生动物胸腺切除后出现的严重免疫缺陷形成了鲜明的对比 [4]。这些观察结果提示，在发育的早期，胸腺对于一些免疫功能完全地、正常地发育是必不可少的。然而，已有证据表明胸腺启动免疫发生的功能并不只局限于发育的早期。成年小鼠胸腺切除后接受全身辐射会发生严重的免疫功能缺失 [2,5]，这说明，当免疫器官被损伤或受到破坏时，成体中的胸腺对于免疫机能的重建仍然是必需的。此外，有实验表明，对免疫耐受的成年小鼠进行胸腺切除会使其无法再对相应抗原做出反应 [6]。所以，看起来成体的胸腺对于特异的免疫缺陷的修复是很重要的。本文所述的工作将进一步为胸腺在整个成体阶段具有持续的免疫功能提供证据。

在一组实验中，对 2~3 月龄的近交系 CBA 小鼠和（Ak×T6）F_1 小鼠进行了胸腺切除手术，在胸腺切除后 2 年内的不同时间，用 0.3 毫升溶于生理盐水的 5% 绵羊红细胞悬液刺激。刺激后 4~6 天，处死小鼠取出脾脏，根据杰尼等所述的方法 [7] 检测每百万个脾细胞中抗体空斑形成细胞的数量。还取出淋巴结、一小部分脾脏和胸腺

Lymph nodes, a small part of the spleen and the thymus area were also removed and kept for histological examination. The results of the plaque assays are shown in Table 1. There was no difference in the capacity of thymectomized and sham-thymectomized mice to produce antibody-plaque-forming cells when injected with sheep erythrocytes 2 months after the operation. However, thymectomized mice challenged 9 or more months after operation produced less antibody-plaque-forming cells than controls.

Table 1. Effect of Thymectomy in Adult Life on the Production of
Antibody-plaque-forming Cells in the Spleen

Strain	Treatment at 2–3 months	Age (months) at immunization	No. of mice	No. of mice* showing following No. of antibody-plaque-forming cells per 1,000,000 spleen cells:				
				0–100	101–300	301–500	501–800	801–1,200
$(Ak \times T6)F_1$	Sham-thymectomy	4	5					5
	Thymectomy	4	6					6
	Sham-thymectomy	9	6					6
	Thymectomy	9	8			6	2	
CBA	Sham-thymectomy	4	4					4
	Thymectomy	4	4					4
	Sham-thymectomy	18	5			1	4	
	Thymectomy	18	7	1	4	1	1	
	Sham-thymectomy	24	3			2	1	
	Thymectomy	24	3	3				

* Killed at time of peak titre after immunization.

In another experiment, *C3Hf/Bi* mice were thymectomized or sham-operated at 2–4 months of age and killed from 4 to 60 weeks later. Cell suspensions were prepared from the spleens and mesenteric lymph nodes and 5–6 million cells in 0.05 ml. buffered Ringer phosphate solution were injected intravenously into new-born $(C3Hf \times C57BL)F_1$ mice. Half of each litter (about 3–4 mice) received the cells and the other half received only the buffered solution. The capacity of the cell suspensions from a given donor to produce a graft-versus-host (GVH) reaction was determined by Simonsen's method of spleen assay[8]. The relative spleen weight (mg spleen/10 g bodyweight) of the injected baby mice was determined at 10 days of age. Spleen indices were calculated by dividing the average relative spleen weights of the cell-injected mice by the relative spleen weights of the litter-mate controls. The average spleen index of 36 control mice was 1.00 ± 0.13 and animals with indices $\geqslant 1.5$ were considered to show definite evidence of a GVH reaction. The results shown in Table 2 indicate that lymphoid cells from thymectomized mice are as effective as cells from control mice in producing a GVH reaction when collected up to 6 months after thymectomy. After a period of 6–9 months, cells from only 4 of 14 thymectomized mice were effective. Cells from 3 of 8 sham-thymectomized mice aged more than 1 year were ineffective in producing a GVH reaction possibly as a result of the decline of immunological capacity which occurs with advancing age[9]. In the thymectomized group, cells from only 2 of 16 mice killed 9–14 months after thymectomy produced signs of GVH reaction.

区以进行组织学检查。空斑的检测结果见表1。对于进行胸腺切除手术和假胸腺切除手术的小鼠，在术后2个月时注射绵羊红细胞后产生抗体空斑形成细胞的能力并没有区别。但是，在术后9个月或更长时间后接受刺激，那么胸腺切除小鼠比对照组小鼠产生的抗体空斑形成细胞的数量要少。

表1. 成体胸腺切除对脾脏中抗体空斑形成细胞数量的影响

种系	2~3 个月时的处理方式	免疫时的年龄（月）	小鼠数量	每 1,000,000 个脾细胞中产生下列数量的抗体空斑形成细胞的小鼠数量 *				
				0~100	101~300	301~500	501~800	801~1,200
$(Ak \times T6)F_1$	假胸腺切除	4	5					5
	胸腺切除	4	6					6
	假胸腺切除	9	6					6
	胸腺切除	9	8			6	2	
CBA	假胸腺切除	4	4					4
	胸腺切除	4	4					4
	假胸腺切除	18	5			1	4	
	胸腺切除	18	7	1	4	1	1	
	假胸腺切除	24	3			2	1	
	胸腺切除	24	3	3				

* 免疫后出现滴度峰值时处死。

在另一组实验中，对 2~4 月龄的 C3Hf/Bi 小鼠进行胸腺切除手术或假手术，在术后 4~60 周时处死。制备脾脏和肠系膜淋巴结细胞悬浮液，并将 0.05 毫升含 500 万~600 万个细胞的林格磷酸盐缓冲液通过静脉注射到新生的 $(C3Hf \times C57BL)$ F_1 小鼠中。对每窝小鼠中的一半（约 3~4 只）注射细胞悬液，另外一半只注射缓冲液。按照西蒙森的脾细胞分析方法 [8] 测定来自特定供体的细胞悬浮液产生移植物抗宿主（GVH）反应的能力。在接受注射的新生小鼠 10 日龄时,测量其脾脏的相对重量（每 10 克体重所对应的以毫克为单位的脾脏重量）。脾脏指数的计算方法是用注射了细胞悬液的小鼠的相对脾脏重量的平均值除以同窝对照组小鼠的相对脾脏重量。36 只对照组小鼠的脾脏指数的平均值是 1.00 ± 0.13，脾脏指数 $\geqslant 1.5$ 的动物被认为是显示 GVH 反应发生的确凿证据。表 2 中给出的结果表明，胸腺切除后 6 个月内收集的来自胸腺切除小鼠的淋巴细胞与来自对照组小鼠的淋巴细胞在产生 GVH 反应的效率上是没有差别的。手术后 6~9 个月，14 只胸腺切除小鼠中只有 4 只的细胞能有效地产生 GVH 反应。超过 1 岁龄的 8 只假胸腺切除的小鼠中有 3 只的细胞不能有效地产生 GVH 反应，这可能是由于随着年龄增长免疫功能发生下降造成的 [9]。在胸腺切除组中，胸腺切除后 9~14 个月时处死的 16 只小鼠中只有 2 只的细胞能产生 GVH 反应。

Table 2. Effect of Thymectomy in Adult Life on the Capacity of Lymphoid Cells
to produce Graft-versus-Host Reactions

Treatment at 2–4 months	No. of mice with lymphoid cells capable of producing GVH reactions* at following periods (weeks) after operation		
	4–25	26–40	41–60
Sham-thymectomy	8/8	7/8	5/8
Thymectomy	12/13	4/14	2/16

*Expressed as fraction of mice with cells producing, in new-born recipients, an average spleen index $\geq$ 1.5.

Histological examination of the spleen and lymph nodes of old thymectomized animals revealed a striking depletion of cells in the periarteriolar lymphocytic sheaths and primary lymphoid follicles.

It is now well established that some small lymphocytes are long-lived cells[10] capable of initiating graft-versus-host reactions and possibly other types of immunological reactions[11]. The results presented here demonstrate that thymectomy of adult mice is associated with a decline in immunological capacity which becomes significant after a period of 6–9 months. They suggest that the thymus continues to function during adult life in order to maintain an adequate pool of immunologically competent cells. Thymectomy in the adult, unlike thymectomy in the new-born, has no immediate effect on immunological capacity, presumably because an adequate pool of competent cells has already been constructed. Only when the pool becomes depleted, owing to the limited life span of its cells, do immunological defects with respect to newly encountered antigens become evident.

This work was supported by grants to the Chester Beatty Reasearch Institute (Institute of Cancer Research, Royal Cancer Hospital) from the Medical Research Council and the British Empire Cancer Campaign for Research, from the Tobacco Manufacturers Standing Committee, and from the National Cancer Institute, U.S. Public Health Service (grant *CA*-03188-08).

(**208**, 1337-1338; 1965)

J. F. A. P. Miller: Chester Beatty Research Institute, Institute of Cancer Research, Royal Cancer Hospital, Fulham Road, London, S.W.3.

References:

1. Metcalf, D., *Brit. J. Haematol.*, 6, 324 (1960). Bierring, F., *Ciba Found. Symp. on Haemopoicsis: Cell Production and its Regulation*, edit. by Wolstenholme, G. E. W., and O'Connor, M., 185 (Churchill, London, 1960).

2. Miller, J. F. A. P., Doak, S. M. A., and Cross, A. M., *Proc. Soc. Exp. Biol.*, 112, 785 (1963).

3. Harris, T. N., Rhoads, J., and Stokes, J. A., *J. Immunol.*, 58, 27 (1948). MacLean, L. D., Zak, S. J., Varco, R. L., and Good, R. A., *Transplant. Bull.*, 4, 21 (1957). Fichtelius, K. E., Laurell, G., and Philipsson, L., *Acta Path. Microbiol. Scand.*, 51, 81 (1961).

4. Miller, J. F. A. P., *Lancet*, ii, 748 (1961). *Proc. Roy. Soc.*, B, 156, 415 (1962). Good, R. A., Dalmasso, A. P., Martinez, C., Archer, O. K., Pierce, J. C., and Papermaster, B. W., *J. Exp. Med.*, 116, 773 (1962).

5. Miller, J. F. A. P., *Nature*, 195, 1318 (1962). Globerson, A., Fiore-Donati, L., and Feldman, M., *Exp. Cell. Res.*, 28, 455 (1962).

6. Claman, H. N., and Talmage, D. W., *Science*, 141, 1193 (1963). Taylor, R. B., *Immunology*, 7, 595 (1964).

表 2. 成体胸腺切除对淋巴细胞产生移植物抗宿主反应能力的影响

2~4 个月时的处理方式	术后不同阶段（周）淋巴细胞能够产生 GVH 反应 * 的小鼠数量		
	4~25	26~40	41~60
假胸腺切除	8/8	7/8	5/8
胸腺切除	12/13	4/14	2/16

* 用平均脾脏指数 ≥1.5 的小鼠与新生受体的比例来表示。

对老龄的胸腺切除小鼠的脾脏和淋巴结进行组织学检测，结果显示在动脉周围淋巴鞘和初级淋巴滤泡中的细胞有明显的衰竭。

现在可以肯定的是，一些小淋巴细胞是长寿细胞 [10]，它们可以引发移植物抗宿主反应，而且还可能引发其他类型的免疫反应 [11]。本文给出的结果表明，成年小鼠的胸腺切除与免疫能力的下降有关，这种免疫能力的下降在胸腺切除 6~9 个月后十分明显。这些结果表明胸腺在成年期会继续发挥作用，以维持有足够的免疫活性细胞。成体胸腺切除并不像新生动物胸腺切除那样很快地影响到免疫功能，这可能是因为成体中已经储备了充足的免疫活性细胞。只有当这些储备的细胞由于本身寿命的限制而消耗殆尽的时候，机体对新出现抗原的免疫缺陷才会显现出来。

此项工作受到提供给切斯特·贝蒂研究所（皇家肿瘤医院肿瘤研究所）的基金的支持，这些基金来自医学研究理事会和大英帝国癌症运动组织、烟草制造商常设委员会以及美国公共卫生署国家癌症研究所（基金 CA-03188-08）。

（李响 翻译；陈新文 陈继征 审稿）

7. Jerne, N. K., Nordin, A. A., and Henry, C., in *Cell-Bound Antibodies*, edit. by Amos, B., and Koprowski, H., 109 (Wistar Institute Press, Philadelphia, 1963).

8. Simonsen, M., *Prog. Allergy*, **6**, 349 (1962).

9. Makinodan, T., and Peterson, W. J., *Proc. U.S. Nat. Acad. Sci.*, **48**, 234 (1962).

10. Little, J. R., Brecher, G., Bradley, T. R., and Rose, S., *Blood*, **19**, 236 (1962).

11. Gowans, J. L., McGregor, D. D., Cowen, D. M., and Ford, C., *Nature*, **196**, 651 (1962).

UGA: A Third Nonsense Triplet in the Genetic Code

S. Brenner *et al.*

Editor's Note

Proteins are assembled by the ribosome using a template of RNA, on which each triplet of nucleotide bases (codon) in the template sequence encodes an amino acid in the protein. South African biologist Sidney Brenner, working with Francis Crick in Cambridge, played a key role in deducing the genetic code that related RNA codons to amino acids. By 1967 the role of almost all the 64 codons was known; here Brenner, Crick and their coworkers elucidate the function of the only codon still outstanding, designated UGA. They find that, like the codons UAA and UAG, it is a "nonsense" triplet, encoding no amino acid at all. It is now known to be a "stop" codon, signalling termination of protein synthesis.

MOST of the sixty-four triplets of the genetic code[1] have been allocated to one or other of the twenty amino-acids. The two known nonsense triplets (UAA, *ochre* and UAG, *amber*) are believed to signal the termination of the polypeptide chain. The only other triplet so far unallocated is UGA, for which binding experiments give uncertain or negative results.

In this article we show that UGA is "unacceptable" in our system (*Escherichia coli* infected with bacteriophage *T*4) and present suggestive evidence that it is nonsense; that is, that it does not stand for any amino-acid. Theoretical arguments make it likely that there is no transfer RNA (*t*RNA) to recognize it. The reason for this apparent absence of function is not yet known. Neither is it known whether UGA is nonsense in other organisms.

Evidence that UGA may be nonsense in *E. coli* has also been presented by Garen *et al.*[2]. They investigated the reversion of *amber* and *ochre* mutants in the alkaline phosphatase gene of *E. coli*. *Amber* mutants (UAG) reverted, as expected, to seven different amino-acids including tryptophan which is coded by UGG. *Ochre* mutants (UAA) reverted to six of these amino-acids, but not to tryptophan. This negative result makes it unlikely that UGA stands for tryptophan (see also Sarabhai and Brenner[3]) and suggests that it might be a nonsense codon.

Mutant X655 contains UGA. Much of our genetic work has been concerned with the left-hand end of the B cistron of the *r*II region of bacteriophage *T*4. We have made extensive and detailed investigations of this region which are being reported elsewhere[4]. The mutant *X*655 occurs in the middle of this region. In brief our proof that *X*655 contains the triplet UGA consists in converting it to an *ochre* (UAA), using mutagens the behaviour of which is already known.

UGA：遗传密码中的第三个无义三联密码子

布伦纳等

编者按

蛋白质是由核糖体按照 RNA 模板装配而成的，在模板序列的核苷酸碱基中，每个三联体（密码子）编码蛋白质中的一个氨基酸。与弗朗西斯·克里克一同在剑桥大学工作的南非生物学家悉尼·布伦纳在推测与氨基酸对应的 RNA 密码子的遗传密码方面作出了重大贡献。到 1967 年时，人们已经破译了几乎全部的 64 个密码子；在本文中，布伦纳、克里克及其同事们阐明了唯一一个尚有争议的密码子——UGA 的功能。他们发现：与密码子 UAA 和 UAG 一样，UGA 是一个"无义"三联密码子，不编码任何氨基酸。现在我们知道它是一个"终止"密码子，标志着蛋白质合成的终止。

在遗传密码 [1] 的 64 个三联密码子中，可以与 20 种氨基酸一一对应的占大多数。两种已知的无义密码子（赭石密码子 UAA 和琥珀密码子 UAG）被人们看作是多肽链终止的信号。到目前为止，唯一一个尚未确定的密码子就是 UGA，因为由结合实验给出的结果要么是不确定的，要么是阴性的。

在本文中我们要证明 UGA 在我们的系统(被噬菌体 T4 感染的大肠杆菌)中是"不可被接受"的。这可以作为它是无义密码子的证据，也就是说，它不编码任何氨基酸。理论上的说法是，有可能没有能识别它的转运 RNA（tRNA）。目前人们还不知道为什么这种功能会明显缺失，更不清楚 UGA 在其他生物体中是否也是无义的。

加伦等人也举出了能证明 UGA 在大肠杆菌中有可能是无义的证据 [2]。他们对大肠杆菌的碱性磷酸酶基因中琥珀突变体和赭石突变体的回复突变产物进行了研究。正如预期的那样，琥珀突变体（UAG）可以回复成 7 个不同的氨基酸，其中包括由 UGG 编码的色氨酸。而赭石突变体（UAA）则只能回复成除色氨酸以外的其他 6 个氨基酸。上述阴性结果说明 UGA 编码色氨酸的可能性不大（参见萨拉巴伊和布伦纳的文章 [3]），并且暗示着它有可能是一个无义密码子。

突变体 X655 含有 UGA。我们在基因研究方面的大部分工作涉及噬菌体 T4 rII 区域的 B 顺反子的左手末端。我们已对该区域进行了广泛而细致的研究，研究结果即将在其他出版物上发表 [4]。突变体 X655 出现在该区域的中间。简言之，我们认为 X655 含有三联密码子 UGA 的证据是可以利用性能已知的诱变剂使之转变成赭石密码子（UAA）。

X655 was induced from wild type by 2-aminopurine, and identical mutants are also found after treatment of wild type phage with hydroxylamine. This shows that it differs from an acceptable triplet by a G–C to A–T base pair change in the DNA. It is not suppressed by any *amber* or *ochre* suppressor (Table 1) and is therefore neither UAG nor UAA. The reversion properties of X655 are shown in Table 2. It is strongly induced to revert to r^+ by 2-aminopurine, as is expected, but there is no induction to r^+ by hydroxylamine. Thus the triplet in the DNA either contains no G–C pairs or, if it does contain one, it is connected to another unacceptable triplet by a G–C to A–T transition.

Table 1. Suppression Properties of X655 and Its Derivatives

Mutant	Triplet	su^-	*Amber* suppressors			*Ochre* suppressors		
			su_I^+	su_{II}^+	su_{III}^+	su_B^+	su_C^+	su_D^+
X655	UGA	0	0	0	0	0	0	0
X655 ochre	UAA	0	0	0	0	+	+	+
X655 amber	UAG	0	+	+	+	+	+	+

Phage stocks were plated on the following strains[9]: su^-, CA244; su_I^+, CA266; su_{II}^+, CA180; su_{III}^+, CA265; su_B^+, CA165; su_C^+, CA167; and su_D^+, CA248.

Table 2. Reversion of X655

	Reversion index (in units of 10^{-7})			
	Spontaneous	2-Aminopurine	Hydroxylamine direct	Hydroxylamine after growth
to r^+	4	312	5	6
to ochre	4	51	1,090	533

X655 was treated with 2-aminopurine and hydroxylamine as previously described[4,5]. Total phage was assayed on *E. coli* B and r^+ revertants on CA244 (su^-). *Ochre* revertants were selected on CA248 (su_D^+) and distinguished from r^+ revertants by picking and stabbing about 300 plaques into CA248 and CA244.

The triplet is in fact connected to UAA by a transition, because X655 can be converted to an *ochre* and this change is induced by 2-aminopurine (Table 2). The nature of the transition is more precisely specified by the finding that the conversion to an *ochre* is induced by hydroxylamine and that the *ochre* triplet produced does not require any replication for expression. Using a previous argument[5] this result suggests that the change arises from a G→A change in the messenger RNA. Because X655 is not an *amber*, this proves that it contains the triplet UGA. To confirm that an *amber* at the site of X655 would be suppressed by *amber* suppressors the X655 *ochre* has been converted to an *amber* by mutation and its properties tested (Table 1).

Other occurrences of UGA. In three cases we have been able to produce the triplet UGA by selected phase shifts in our region. When (+ −) phase shifts are made over the first part of the B cistron, the two phase shift mutants frequently do not suppress each other. We have shown[4] that these barriers to mutual suppression are due to the generation of unacceptable triplets in the shifted frame. One of these barriers, b_9, has been identified as an *amber* and two others,

X655 是用 2–氨基嘌呤诱导野生型产生的，在用羟胺处理野生型噬菌体后也能得到完全相同的突变体。这说明该密码子不同于 DNA 中由 G–C 到 A–T 的碱基对替换产生的可接受密码子，它不会被任何琥珀抑制基因或赭石抑制基因所抑制（表 1），因此不可能是 UAG 或者 UAA。X655 的回复性质示于表 2。正如所预计的，它很容易被 2–氨基嘌呤诱导从而回复成 r^+，但用羟胺诱导就不能使其回复成 r^+。因此，DNA 中的这个三联密码子或者不含 G–C 碱基对，或者只含有一个 G–C 碱基对，而且这个碱基对是通过 G–C 到 A–T 的转换连接到另一个不可接受的三联密码子上的。

表 1. X655 及其衍生物的抑制特性

突变体	三联密码子	su^-	琥珀抑制基因			赭石抑制基因		
			su_I^+	su_{II}^+	su_{III}^+	su_B^+	su_C^+	su_D^+
X655	UGA	0	0	0	0	0	0	0
X655 赭石型	UAA	0	0	0	0	+	+	+
X655 琥珀型	UAG	0	+	+	+	+	+	+

将噬菌体液加到下面这些菌株[9] 中：su^-，CA244；su_I^+，CA266；su_{II}^+，CA180；su_{III}^+，CA265；su_B^+，CA165；su_C^+，CA167；su_D^+，CA248。

表 2. X655 的回复性质

回复指数（单位是 10^{-7}）				
	自发	2–氨基嘌呤	羟胺直接	羟胺培养后
回复成 r^+	4	312	5	6
回复成赭石密码子	4	51	1,090	533

和以前描述过的方法一样，X655 都是用 2–氨基嘌呤和羟胺处理的[4,5]。总噬菌体在大肠杆菌 B 中进行实验，而 r^+ 回复子在 CA244（su^-）菌株中进行。赭石回复子在 CA248（su_D^+）菌株中进行选择，并通过在 CA248 和 CA244 菌株中挑选 300 个噬菌斑和 r^+ 回复子进行鉴别。

事实上，该三联密码子就是通过转换连接到 UAA 上的，因为 X655 能被 2–氨基嘌呤诱导转换成赭石密码子（表 2）。根据以下发现可以更加确切地判断转换的本质，即羟胺能够诱导其转换成赭石密码子，而且产生的密码子不需要进行复制就可以表达。由之前的讨论可知[5]：上述结果说明该转换是由信使 RNA 中发生了 G → A 的替换而引起。因为 X655 不是琥珀密码子，所以由此可以证明它含有 UGA 三联密码子。为了进一步证实 X655 位置上的琥珀密码子能够被琥珀抑制基因所抑制，我们通过突变将 X655 赭石密码子转换成了琥珀密码子，并检测了它的性质（表 1）。

出现 UGA 的其他情况。 在三次实验中，我们已经通过在一定区域内进行的特定相位移动产生了 UGA 三联密码子。当在 B 顺反子的第一部分进行（＋ －）相位移动时，这两个相位移动突变体很少会相互抑制。我们曾指出[4]：相互抑制受到阻碍的原因是在移动的读码框内产生了不可接受的密码子。其中一个障碍物 b_9 已被确定

b_3 and b_4, as *ochres*. Three barriers, b_2, b_5 and b_6, have now been identified as UGA by their base-analogue induced reversion to *ochres*. In each case the identification has been checked by converting the *ochre* to an *amber* at the same site.

Tryptophan is represented by the single codon UGG. It would therefore be expected to mutate by transitions to both UAG (*amber*) and UGA, and thus in such cases *amber* and UGA mutants should occur in close pairs. The *amber* mutant, *HB*74, which maps close to *X*655, is an example of this. Genetic crosses between it, *X*655, and the *ochre* and *amber* derived from *X*655, show that *HB*74 maps identically to the *amber* derived from *X*655, as expected (Table 3).

Table 3. Recombination between Various Mutants

	*X*655	*X*655 *ochre*	*X*655 *amber*	*HB*74	Triplet
*X*655	0				UGA
*X*655 *ochre*	0	0			UAA
*X*655 *amber*	+	0	0		UAG
*HB*74	+	0	0	0	UAG

The phages were crossed in *E. coli B* and the complexes irradiated with ultra-violet light to stimulate recombination (see ref. 4). In the Table, 0 means that r^+ recombinants were not significantly above the reversion rate, which was between 2 and 9×10^{-7}; in those experiments where positive results were obtained (+), the frequency was between 2 and 6×10^{-5}.

So far we have found the expected pairs consisting of UGA and an *amber* in two other cases. In the A cistron, a mutant *X*665[*] is found with the *amber* mutant *N*97, and in the B cistron, *N*65 is paired with the *amber* mutant *X*237. Both *N*97 and *X*237 are likely to have arisen from UGG (tryptophan) which is confirmed by the finding that they respond only poorly to the *amber* suppressor su_{II}^+ which inserts glutamine[5]. Both *X*665 and *N*65 have been converted into *ochre* mutants, showing that they contain the triplet UGA. These *ochres* have also been converted to *ambers* at the same site. Mapping investigations, analogous to those in Table 3, are consistent with these allocations.

Unacceptability of UGA. There is very good evidence that the amino-acid sequence coded by the first part of the B cistron is not critical for the function of the gene[4]. It can be replaced by varying lengths of the A cistron using deletions that join the two genes. Moreover, an extensive (− +) frame shift can be made without noticeable effect on the function. Of the fifteen known base-analogue mutants in the region, thirteen are either *ochres* or *ambers*; one, *HD*263, is temperature sensitive and *X*655 is UGA. The extreme bias towards *amber* and *ochre* chain-terminating mutants confirms the dispensability of the region[4]. These results make it unlikely that the unacceptability of UGA in *X*655 and the three barriers results from the insertion of an amino-acid, and strongly suggest that it is nonsense.

---

* This is not a misprint for X655.

是琥珀密码子，而另外两个，即 b_3 和 b_4，则被鉴定为赭石密码子。还有三个障碍物 b_2、b_5 和 b_6 暂时被看作是 UGA，因为它们能够在碱基类似物的诱导下回复成赭石密码子。在每次实验中，我们都会通过将同一位点的赭石密码子转变为琥珀密码子来验证密码鉴定的准确性。

编码色氨酸的唯一密码子是 UGG。因而它应该可以通过转换突变成 UAG（琥珀密码子）和 UGA，所以在这种情况下，琥珀密码子和 UGA 突变体应该会成对出现。与 $X655$ 配对的琥珀突变体 $HB74$ 就是一个琥珀密码子。将它与 $X655$ 以及 $X655$ 的赭石和琥珀突变体进行基因交配，结果发现：和预期一样，$HB74$ 的基因图谱与来自 $X655$ 的琥珀突变体完全相同（表3）。

表3. 不同突变体之间的重组

	$X655$	$X655$ 赭石型	$X655$ 琥珀型	$HB74$	三联密码子
$X655$	0				UGA
$X655$ 赭石型	0	0			UAA
$X655$ 琥珀型	+	0	0		UAG
$HB74$	+	0	0	0	UAG

使噬菌体在大肠杆菌 B 中发生基因交配，并用紫外光照射复合物以便刺激重组（见参考文献4）。在上表中，0 代表 r^+ 重组子并没有显著高于回复突变率 $2 \times 10^{-7} \sim 9 \times 10^{-7}$。在那些结果为阳性（+）的实验中，频率都介于 $2 \times 10^{-5} \sim 6 \times 10^{-5}$ 之间。

到目前为止，我们已经根据预期在另外两个实验中找到了由 UGA 和琥珀密码子组成的密码对。我们发现在 A 顺反子中，突变体 $X665^*$ 和琥珀突变体 $N97$ 是同时存在的，而在 B 顺反子中，$N65$ 和琥珀突变体 $X237$ 会配对。$N97$ 和 $X237$ 很有可能都来源于 UGG（色氨酸），以下发现可以证实这一点，即它们仅对插入谷氨酰胺的琥珀密码子抑制基因 su_{II}^+ 反应不敏感 [5]。$X665$ 和 $N65$ 都能被转变成赭石突变体，这说明它们含有 UGA 三联密码子。这些赭石密码子也都能在同一位点处被转变成琥珀密码子。这些定位结果与通过类似于表3的基因图谱研究所得到的结果相符。

UGA 的不可接受性。我们有足够充分的证据可以证明：由 B 顺反子第一部分编码的氨基酸序列不会对该基因的功能产生至关重要的影响 [4]。通过能够连接两个基因的缺失序列，B 顺反子第一部分编码的氨基酸序列可以被长度不同的 A 顺反子片段所替换。此外，还可以进行读码框的 (–+) 移动而不会显著地影响到基因的功能。在该区域 15 个已知的碱基类似突变体中，有 13 个不是赭石突变体就是琥珀突变体；另外一个，$HD263$，属于温度敏感型，而 $X655$ 是 UGA。这种对琥珀或者赭石链终止突变体的极端倾向性足以证明该区域是可有可无的 [4]。这些结果说明 $X655$ 中的 UGA 和三个障碍物的不可接受性不太可能源自一个氨基酸的插入，从而强烈地支持了 UGA 是无义密码子的观点。

---

* 此处并非 $X655$ 的误印。

In addition, the UGA mutant $X665$ in the A cistron has been combined with the deletion $r1589$ and has been found to remove the B activity of this phage. This is the test for nonsense originally used by Benzer and Champe[6].

In all these cases, however, it could be argued that UGA might code cysteine, especially as the two known triplets for cysteine are UGU and UGC. If the B protein already contained a cysteine essential for its function the effect of UGA elsewhere might be to produce an S–S bridge between the cysteine inserted by UGA and the (hypothetical) essential one, and thus inactivate the protein. Nevertheless we regard this as unlikely for two reasons, one genetic and one chemical.

The genetic evidence concerns the anomalous minutes produced by certain (++) combinations in the B cistron[4]. In some regions of the first part of the B cistron combinations of two (+) phase shift mutants are able to grow to some extent on the restrictive host, *E. coli* $K12$. The plaques produced are minute, however, showing that the wild type phenotype is very far from being completely restored. A detailed analysis of one set of these combinations showed that minutes are obtained only from pairs of (+) mutants which straddle barrier b_6. The presence of the barrier is obligatory because, if it is removed by mutation, the (++) doubles are unable to grow at all on *E. coli* $K12$. The minutes are clearly due to a phase error of one sort of another and the phase error is dependent on the barrier b_6 which we now know to be UGA. This result shows that UGA cannot be associated with any normal amino-acid reading and points strongly to the conclusion that it is nonsense.

The chemical reason for UGA not coding for cysteine comes from the work of Khorana *et al.*[7]. They have shown that poly $(UGA)_n$ when used as a messenger in a cell-free system derived from *E. coli* induces the production of poly methionine (corresponding to AUG) and also poly aspartic acid (corresponding to GAU). No other amino-acid appears to be incorporated. In particular, no poly cysteine was found. For various reasons this evidence is not completely decisive, but it at least makes it unlikely that UGA is cysteine.

Function of UGA. It might be thought that the sequence containing UGA was nonsense because it was the signal for the beginning or ending of a gene (or operon). In other words, that it produced its effect during the synthesis of the messenger RNA on the DNA template of the gene. This explanation is highly unlikely because the effects of UGA depend on it being read in phase. The phenotypic effect of $X655$ can be removed when the mutant is placed in a (− +) shifted frame[4], and the barriers b_2, b_5 and b_6 are of course produced by phase shifts. That is, the base sequence UGA actually occurs at these places in the wild type messenger RNA but in such a way that it is out of phase when the message is read correctly. Because we have no reason to suspect that RNA polymerase synthesizes messenger RNA in groups of three bases at a time these results imply that the phenotypic effects of UGA must occur during protein synthesis.

It thus seems unlikely that UGA codes for any amino-acid, and in particular it does not appear

666

此外，已发现 A 顺反子中的 UGA 突变体 *X*665 可以和缺失子 *r*1589 重组，我们还发现它能够去除该噬菌体的 B 活性。这就是本则尔和钱普起先用来检测无义密码子的方法 [6]。

但是针对所有这些实验，有人提出 UGA 或许可以编码半胱氨酸，尤其是因为人们已经知道半胱氨酸的两个三联密码子是 UGU 和 UGC。如果 B 蛋白已经含有一个对其功能有重要影响的半胱氨酸，那么 UGA 在其他位置处的功能也许就是在 UGA 插入的半胱氨酸和（假定存在的）基本半胱氨酸之间形成一个 S–S 键，以便使蛋白质失活。尽管如此，我们还是可以举出两方面的理由来驳斥这一点，一方面是基因上的，另一方面是化学上的。

在基因上的证据与由 B 顺反子中某些（++）重组产生的反常菌斑有关 [4]。在 B 顺反子第一部分的某些区域中，两个（+）相位移动突变体的重组子能够在限制性宿主大肠杆菌 *K*12 中生长到一定程度。但产生的菌斑非常小，这说明野生表型还远远没有得到完全恢复。在对其中一组重组子进行深入研究后发现：只有跨过障碍物 b_6 的（+）突变体对才能生成菌斑。这种障碍物的存在是必不可少的。因为，如果它被突变移除，那么（++）突变体对将完全不能在大肠杆菌 *K*12 中生长。菌斑的产生显然是由某种相位误差造成的，而相位误差又是由障碍物 b_6 决定的，我们现在知道障碍物 b_6 就是 UGA。这一结果说明 UGA 不可能与任何正常氨基酸的编码密码相关，并可以使我们明确地得到 UGA 是无义密码子的结论。

在化学上证明 UGA 不能编码半胱氨酸的证据源自霍拉纳等人的工作 [7]。他们发现：当多聚 $(UGA)_n$ 在大肠杆菌来源的无细胞系统中被用作信使时，它会诱导生成多聚甲硫氨酸（密码子是 AUG）以及多聚天冬氨酸（密码子是 GAU）。此外别无其他的氨基酸产生。尤其是没有发现多聚半胱氨酸。虽然可以列出多种原因来证明该结果并不具备完全确定性，但它至少可以用于否定 UGA 编码半胱氨酸的可能性。

UGA 的功能。有人可能会认为含有 UGA 的序列是无义的，因为它是一个基因（或者操纵子）开始或者终止的信号。换句话说，只有在基因的 DNA 模板上合成信使 RNA 时它才发挥作用。这个解释不可能成立，因为 UGA 的功能取决于它的同相读取。当突变体被置于（–+）移动的读码框中时，*X*655 的表型效应就会丧失 [4]，而且障碍物 b_2、b_5 和 b_6 显然是由相位移动产生的。也就是说，碱基序列 UGA 确实存在于野生型信使 RNA 的这些位点上，但是当信息被正确读取时，它们会因此而处在相位不符合的状态。因为我们没有理由怀疑 RNA 聚合酶是以三个碱基为单位同时合成信使 RNA 的，所以这些结果暗示着 UGA 的表型效应只会出现在蛋白质的合成过程之中。

因此，这样看起来 UGA 编码任何氨基酸都不太可能，尤其是不会编码半胱氨

to code for either cysteine (UGU and UGC) or tryptophan (UGG). The wobble theory of codon-anticodon interaction developed by one of us[8] makes the prediction that because of a wobble in the recognition mechanism at the third place of the codon no tRNA molecule can recognize XYA alone without at the same time recognizing either XYG or both XYU and XYC. Such theoretical arguments cannot be considered conclusive, but they certainly suggest that UGA is a triplet for which no tRNA exists. For this reason we think it unlikely that UGA produces the efficient termination of the polypeptide chain, but more direct evidence will be needed to establish this point.

Conclusion. We have thus established that in the phage-infected cell UGA is certainly "unacceptable" in the rII cistrons, although it remains to be seen whether this is true for other species. We have produced reasons why it is unlikely to code for any amino-acid. We are confident that there must be weighty reasons if even a single triplet is not used in the genetic code, because otherwise natural selection would have certainly allocated it to an amino-acid. At the moment we are inclined to believe that UGA may be necessary as a "space" to separate genes in a polycistronic message. It is possible to make a plausible theory for *E. coli* along these lines, but we prefer to leave the discussion of this until we have more experimental evidence to support it. This we are at present attempting to obtain.

We thank Drs. A. Garen, H. G. Khorana and A. Sarabhai for interesting discussions and for showing us their papers in advance of publication. One of the authors (E. R. K.) is a holder of a United States Churchill Foundation scholarship.

(**213**, 449-450; 1967)

S. Brenner, L. Barnett, E. R. Katz and F. H. C. Crick: M.R.C. Laboratory of Molecular Biology, Hills Road, Cambridge.

Received December 22, 1966.

References:

1. For the structure of the genetic code and the evidence for nonsense triplets see the papers in the Cold Spring Harbor Symposium XXXI on "The Genetic Code", 1966 (in the press).

2. Weigert, M. G., Lanka, E., and Garen, A., *J. Mol. Biol.*, **23**, 391 (1967).

3. Sarabhai, A., and Brenner, S., in preparation.

4. Barnett, L., Brenner, S., Crick, F. H. C., Shulman, R. G., and Watts-Tobin, R. J., *Phil. Trans. Roy. Soc.* (in the press).

5. Brenner, S., Stretton, A. O. W., and Kaplan, S., *Nature*, **206**, 994 (1965).

6. Benzer, S., and Champe, S. P., *Proc. U.S. Nat. Acad. Sci.*, **48**, 1114 (1962).

7. Morgan, A. R., Wells, R. D., and Khorana, H. G., *Proc. U.S. Nat. Acad. Sci.* (in the press).

8. Crick, F. H. C., *J. Mol. Biol.*, **19**, 548 (1966).

9. Brenner, S., and Beckwith, J. R., *J. Mol. Biol.*, **13**, 629 (1965).

酸（UGU 和 UGC）或者色氨酸（UGG）。我们中的一位作者提出过密码子－反密码子相互作用的摆动理论 [8]，该理论预言，由于密码子的第三位识别是可以发生摆动的，因而所有能够识别 XYA 的 tRNA 分子都能同时识别 XYG 或者 XYU 和 XYC。虽然这一理论依据不能被认为是绝对正确的，但确实可以说明 UGA 是一种不对应于任何 tRNA 的三联密码子。基于这个原因，我们认为 UGA 不太可能会有效地终止多肽链，不过这一点还需要更多直接的证据来证明。

结论。我们已经证实在噬菌体感染的细胞中，UGA 在 rII 区顺反子中确实是"不可被接受"的。然而，在其他物种中它是否也会如此还有待于进一步的研究。我们提出了几个它不可能编码任何氨基酸的理由。我们相信：即便只有一个三联密码子未被用于基因编码中，也一定有不同凡响的原因，否则的话自然选择必然会将其分配给某一个氨基酸。目前我们倾向于认为在多顺反子的序列中或许需要用 UGA 作为一个"空格"来分隔基因。根据这些结果我们完全可以提出一套似乎适用于大肠杆菌的理论，但我们宁愿暂时不去讨论它，直到我们有更多的实验证据来验证。这就是我们目前正在努力去获得的。

感谢加伦、霍拉纳和萨拉巴伊博士对我们的工作进行了饶有兴致的讨论，还要感谢他们能在自己的论文发表之前先拿给我们看。我们中的一位作者（卡茨）得到了由美国丘吉尔基金会授予的奖学金的资助。

（毛晨晖 翻译；陈新文 陈继征 审稿）

Specific Binding of the λ Phage Repressor to λ DNA

M. Ptashne

Editor's Note

In 1961, French biologists François Jacob and Jacques Monod described a negative form of gene regulation whereby so-called repressor proteins switch off target genes. But the underlying molecular mechanism was unclear. Here molecular biologist Mark Ptashne solves the problem by showing that a protein called the λ repressor protein binds directly to specific DNA sequences, suggesting that repressors exert their effects by blocking the transcription of DNA to RNA. Although the protein was originally isolated from a virus, Ptashne and others went on to show that the same mechanism occurs in yeast, plants, fruit flies and humans, and is a key form of gene regulation.

THERE are many examples, in bacteria and their phages, of a group of genes controlled by the product of another gene—a regulator gene. In the classical cases discussed by Jacob and Monod[1], the control is negative and the product of the regulator gene is called a repressor. Repressors act by switching off their target genes; in order to activate these genes the repressor itself must be inactivated. These facts were learned from genetic experiments which do not reveal how repressors work at the molecular level.

The isolation in recent months of two repressors[2,3] makes possible biochemical experiments exploring the mechanism of repression. Many of the models for this mechanism propose different sites for the action of the repressor. According to the simplest model, the repressor binds to a site on the DNA, directly preventing the transcription from DNA to RNA. According to other models the repressor interacts with *m*RNA or *s*RNA to block translation of the genetic message from RNA to protein. A prediction of the first model is that an isolated repressor will bind *in vitro* to DNA containing the receptor site for that particular repressor, but not to DNA lacking this site. The experiments reported here confirm this expectation.

The Genetic System

The protein made by the C_1 gene of phage λ, called the λ phage repressor, is used in these experiments. This repressor blocks the expression of the other phage genes, keeping the phage chromosome dormant within its host, *E. coli*. Only a very short segment of the phage genome is involved in this control[4,5], a region including the C_1 gene and the sites which determine the sensitivity of the phage to the repressor. Two phages which differ only in this segment are λ and λ*imm*[434] (see ref. 4). Phage λ*imm*[434], which contains almost all the other known genes of phage λ, makes and is sensitive to the 434 phage repressor only. Therefore, a critical test of specificity is that the isolated λ repressor should bind to λ DNA but not to λ*imm*[434] DNA.

λ噬菌体阻遏物与λ DNA的特异性结合

普塔什尼

编者按

1961年，法国生物学家弗朗索瓦·雅各布和雅克·莫诺描述了一种负向的基因调节作用，被称为阻遏物的蛋白可以借此关闭靶基因。但能够解释上述现象的分子作用机制尚不明晰。在本文中，分子生物学家马克·普塔什尼指出，用一种被称为λ阻遏物的蛋白直接与特殊DNA序列结合即可解释上述作用，他认为阻遏物发挥作用的方式是阻断由DNA向RNA的转录过程。尽管这种蛋白最初是从病毒中分离出来的，但普塔什尼和其他研究者随后发现在酵母、植物、果蝇和人类中都存在同样的机制，而且这种机制还是基因调节的主要形式。

在细菌和它们的噬菌体中有很多这样的例子：一组基因被另外某个基因——调节基因的产物所调控。在雅各布和莫诺[1]讨论过的那些经典例子中，调控都是负向的，所以调节基因的产物被称为阻遏物。阻遏物的作用是关闭靶基因；为了激活这些靶基因，就必须使阻遏物自身失活。上述结论可以从遗传学实验中得到，但遗传学实验并不能揭示在分子水平上阻遏物是如何起作用的。

在最近几个月中分离出了两种阻遏物[2,3]，这使得通过生物化学实验来研究阻遏机制成为可能。在为解释这一机制而提出的多种模型中，阻遏物的作用位点并不相同。根据最简单的模型，阻遏物结合于DNA上的某个位点并直接阻止了从DNA到RNA的转录过程。而根据其他一些模型，阻遏物通过与信使RNA（mRNA）或小的非编码RNA（sRNA）的相互作用阻止了遗传信息从RNA到蛋白质的翻译过程。由第一个模型可以预测：在体外条件下，一个分离出来的阻遏物能够结合含这一阻遏物特异结合位点的DNA，而不能结合缺乏这种位点的DNA。本文报道的实验证实了这一预测。

遗 传 系 统

由λ噬菌体C₁基因编码的蛋白质被称为λ噬菌体阻遏物，我们在下述实验中使用的就是这种蛋白质。该阻遏物能够阻断其他噬菌体基因的表达，使噬菌体的染色体在它的宿主大肠杆菌内只能处于休眠状态。在噬菌体的基因组中只有一个非常短的DNA片段参与了这一调控过程[4,5]，该区域包括C₁基因和若干能决定噬菌体对阻遏物敏感程度的位点。仅在这个片段上有所不同的两种噬菌体是λ噬菌体和λimm⁴³⁴噬菌体（见参考文献4）。λimm⁴³⁴噬菌体含有λ噬菌体中几乎所有的其他已知基因，唯一的不同就在于它只能合成434噬菌体阻遏物并只对434噬菌体阻遏物敏感。因此，对于阻遏物特异性来说，决定性的检验标准就是：分离出来的λ阻遏物应该与λ DNA结合，而不是与λimm⁴³⁴ DNA结合。

671

Characterization of the λ Phage Repressor

The isolation of the λ phage repressor was achieved by destroying the host DNA with ultra-violet light, thereby drastically decreasing cellular protein synthesis[3]. These irradiated cells were infected with many λ phages which, under the conditions of the experiment, synthesized little or no phage protein except repressor. The infected cells were fed radioactive amino-acids, and a single labelled protein was separated from the background label on a DEAE-cellulose column. This protein was identified as the product of the C_1 gene by two criteria: first, it was missing from cells infected with phages bearing amber mutations in the C_1 gene, and second, it was made in modified form by phages which produce temperature sensitive repressors as a result of mutation in the C_1 gene. Electrophoresis and sedimentation of the repressor indicate that it is an acidic protein with a sedimentation coefficient of about 2.8S, which corresponds to a molecular weight of approximately 30,000.

The binding experiments to be described were performed mainly with the repressor made by the mutant phage λind^-. The ind^- mutation renders the repressor insensitive *in vivo* to many conditions which inactivate the wild type repressor[6]. For example, a small dose of ultra-violet light delivered to a λ-lysogen will inactivate the wild type but not the ind^- repressor. The ind^- repressor can be isolated in the same way as the wild type and has approximately the same sedimentation coefficient, but it chromatographs separately from wild type on DEAE-cellulose columns (Fig. 1). The altered chromatographic behaviour may be due to a charge or a conformational change. However, the fact the this mutation in the C_1 gene, from wild type to ind^-, also changes the behaviour of the protein on DEAE provides further proof that this protein is coded for by the C_1 gene.

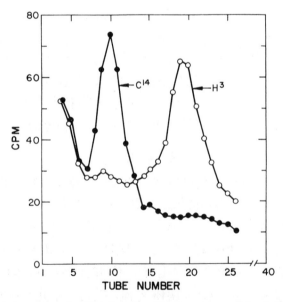

Fig. 1. DEAE-cellulose chromatography of λind^- and λ wild type repressors. Extracts of *E. coli* phage-infected cells containing ^{14}C-labelled λind^- repressor in one case and ^{3}H-labelled λ wild type repressor in the other were applied to a DEAE-cellulose column. Fractions from a salt gradient were collected and assayed as described previously[3]. The label used in this and the other experiments reported in this paper was ^{14}C or ^{3}H labelled reconstituted protein hydrolysate.

λ 噬菌体阻遏物的特性

对 λ 噬菌体阻遏物的分离可以通过用紫外光破坏宿主 DNA 来实现，因而大大减少了胞内蛋白质的合成 [3]。我们用大量 λ 噬菌体侵染这些被辐射过的细胞，在我们的实验条件下，λ 噬菌体几乎不能合成除阻遏物之外的其他噬菌体蛋白。在这些被感染的细胞中加入放射性标记的氨基酸，最后用二乙氨基乙基纤维素（DEAE-纤维素）柱层析就可以从背景标记中分离出一种单标蛋白。根据以下两点即可确定这种蛋白就是 C_1 基因编码的产物：第一，在用 C_1 基因中发生琥珀突变的噬菌体侵染过的细胞内并不存在这种蛋白；第二，在 C_1 基因中突变的作用下生成温度敏感型阻遏物的噬菌体将使这种蛋白以被修饰的形式存在。由阻遏物的电泳和沉降实验可知，该阻遏物是一种酸性蛋白，其沉降系数约为 2.8S，与此对应的分子量在 30,000 上下。

在下文要描述的阻遏物结合实验中，我们主要用到的是由 λ*ind*⁻ 突变型噬菌体合成的阻遏物。由 *ind*⁻ 突变体合成的阻遏物在体内对许多能使野生型阻遏物失活的条件均不敏感 [6]。例如，用小剂量紫外光照射 λ 溶原菌能使野生型阻遏物失活，但不能使 *ind*⁻ 突变型阻遏物失活。用分离野生型阻遏物的方法同样可以分离 *ind*⁻ 突变型阻遏物，两者的沉降系数也大致相同，但是利用 DEAE-纤维素柱层析即可将 *ind*⁻ 突变型阻遏物与野生型阻遏物分开（图 1）。这种在层析时特性发生改变的现象可能是由于电荷或者构象的变化造成的。不过，C_1 基因从野生型向 *ind*⁻ 型的突变也能改变蛋白质在 DEAE 柱层析时的行为，这一事实为证明这种蛋白是由 C_1 基因编码产生提供了进一步的证据。

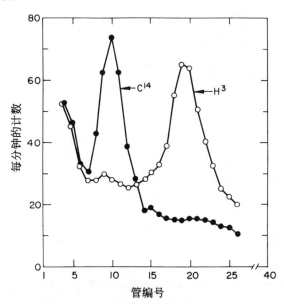

图 1. λ*ind*⁻ 型和 λ 野生型阻遏物通过 DEAE-纤维素柱的层析结果。两个用 DEAE-纤维素柱进行分析的样品都是大肠杆菌噬菌体感染过的细胞的提取物，其中一个含有 ¹⁴C 标记的 λ*ind*⁻ 型阻遏物，另一个含有 ³H 标记的 λ 野生型阻遏物。收集用盐梯度洗脱下来的组分并按以前描述过的方法进行检测 [3]。在这个实验和本文报道的其他实验中所用的标记物都是 ¹⁴C 或 ³H 标记的重组蛋白的水解产物。

Binding of the Repressor to DNA

The labelled λ*ind⁻* repressor was mixed with λ DNA and sedimented through a sucrose gradient. Fig. 2 shows that some of the label sedimented with the DNA, indicating that the λ*ind⁻* repressor binds to DNA. Fig. 2 also shows that this binding is specific: when the repressor was mixed with DNA from phage λ*imm*[434], no binding was observed. This experiment has been performed with several different preparations of repressor and phage DNA.

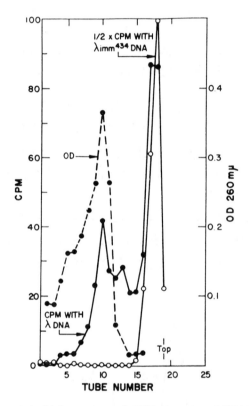

Fig. 2. Specific binding of the λ*ind⁻* repressor to λ DNA. A portion of ^{14}C-labelled λ*ind⁻* repressor, pooled and concentrated from the peak fractions of a DEAE column run, was mixed with 100 μg of λ or λ*imm*[434] DNA. The DNA was preheated at 70°C for 15 min. to minimize aggregation. The solutions were made 0.01 M EDTA, 0.1 M KCl, and 10⁻⁴ M Cleland's reagent, and 5 μg of commercial *s*RNA was added as an additional inhibitor of possible endonuclease activity. After 5 min. incubation at 37°C, the final volume of 0.7 ml. was layered on a 5–25 percent sucrose gradient containing 0.05 M KCl, 10⁻⁴ M Cleland's reagent, and 0.5 mg/ml. BSA as carrier. The gradients were spun at 41,000 r.p.m. for 5 h in an *SB*269 rotor in an IEC centrifuge. Fractions were collected and the DNA peak located by absorbance at 260 mμ. Each fraction was then precipitated on a Millipore filter with TCA and counted in a gas flow counter. The optical density profiles at 260 mμ from the tubes containing λ and λ*imm*[434] DNA are essentially identical. Phages λ and λ*imm*[434] were purified by several bandings in CsCl according to the method of Thomas and Abelson[9], and DNA was then extracted and purified from the two phage preparations using the same phenol and buffer solutions.

阻遏物与 DNA 的结合

　　将标记的 λ*ind⁻* 阻遏物与 λ DNA 混合并通过蔗糖密度梯度进行沉淀。由图 2 可知：有一部分带有标记的阻遏物会与 DNA 一起沉降，这说明 λ*ind⁻* 阻遏物结合了 DNA。图 2 还表明这种结合具有特异性：当阻遏物与 λ*imm*⁴³⁴ 噬菌体的 DNA 混合时是观察不到结合现象的。在这个实验中，我们采用了几种不同的阻遏物和噬菌体 DNA 样品。

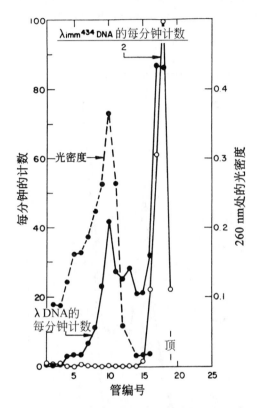

图 2. λ*ind⁻* 阻遏物与 λ DNA 的特异性结合。用 DEAE– 纤维素柱分离 ¹⁴C 标记的 λ*ind⁻* 阻遏物，先收集峰位洗脱部分并浓缩，然后与 100 μg 的 λ DNA 或 λ*imm*⁴³⁴ DNA 混合。将所用 DNA 在 70℃ 下预热 15 min 以减少 DNA 的聚集。溶液中包括 0.01 M EDTA（译者注：乙二胺四乙酸）、0.1 M KCl 和 10⁻⁴ M 克莱兰氏试剂，然后再加入 5 μg 商业 *s*RNA 作为一种抑制核酸内切酶活性的附加抑制剂。在 37℃ 下孵育 5 min 后，将终体积为 0.7 ml 的样品铺展到 5%~25% 的蔗糖密度梯度上，该蔗糖密度梯度中还含有 0.05 M KCl、10⁻⁴ M 克莱兰氏试剂以及作为载体的 0.5 mg/ml 牛血清白蛋白。将梯度样品置于一台 IEC 离心机的 *SB*269 转子中以每分钟 41,000 转的速度离心 5 h。收集分离出的组分，DNA 峰的位置是在光吸收为 260 nm 处。然后在密理博滤膜上对每一组分进行 TCA（译者注：三氯乙酸）沉淀，并用气流式计数器对沉淀物进行放射性计数。含 λ DNA 的组分管与含 λ*imm*⁴³⁴ DNA 的组分管在 260 nm 处的光密度分布图基本一致。按照托马斯和埃布尔森提出的方法 [9]，我们用几轮 CsCl 沉降来纯化 λ 噬菌体和 λ*imm*⁴³⁴ 噬菌体，然后用同样的苯酚和缓冲液从这两个噬菌体样品中提取并纯化 DNA。

The repressor used in the binding experiments was not isotopically pure. Depending on the fractions pooled from a DEAE column run, as much as 50 percent of the label might be present in impurities other than the repressor. In order to guarantee that the label sedimenting with the DNA was in the repressor and not in some contaminant, a double label binding experiment was performed. ^{14}C-labelled λ*ind*[−] repressor was isolated on a DEAE column from a mixture of extracts which included the ^{3}H-labelled products of cells infected with the phage λC[1]*sus*34. This phage bears in its C[1] gene the amber mutation *sus*34 which blocks production of the repressor[7]. Therefore the ^{3}H will have labelled all the proteins made except the repressor. Fig. 3 shows that about half the ^{14}C label but none of the ^{3}H label sedimented with the DNA.

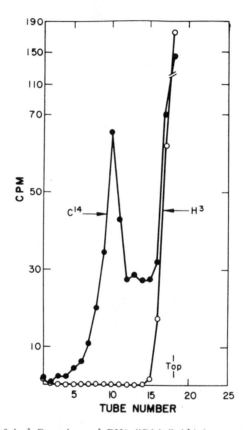

Fig. 3. Selective binding of the λ C[1] product to λ DNA. ^{14}C-labelled λ*ind*[−] repressor was isolated on a DEAE column from an extract which also contained the ^{3}H-labelled gene products of the phage λC[1]*sus*34. This mixture was then tested for binding to λ DNA as described in Fig. 2. The 260 mμ *OD* profile of the λ DNA is essentially identical to that shown in Fig. 2. The fractions were precipitated with TCA, dissolved in 1/2 ml. 0.1 M NaOH, and counted in 10 ml. of scintillation fluid containing toluene and Triton *X*-100 in the ratio 3:1 plus 0.4 percent PPO and 0.005 percent POPOP[10].

The repressor does not bind to denatured DNA. Fig. 4 shows that no counts were displaced from the top of the tube when a mixture of denatured DNA and labelled λ*ind*[−] repressor was

　　结合实验中所用的阻遏物并不是纯的同位素标记。由于在收集从 DEAE 柱流出的组分时存在误差，所以有时可能会有高达 50% 的标记出现在杂质中而不是阻遏物中。为了保证与 DNA 一起沉降的标记在阻遏物中而不是在杂质中，我们进行了一次双标结合实验。利用 DEAE 柱，我们从含有 λC₁sus34 噬菌体侵染细胞的 ³H 标记产物的混合提取物中分离出了 ¹⁴C 标记的 λind⁻ 阻遏物。λC₁sus34 噬菌体的 C₁ 基因发生了 sus34 琥珀突变，这一突变可以阻断该噬菌体中阻遏物的产生 [7]。因此，³H 将能够标记除阻遏物之外的所有蛋白。从图 3 中可以看出：与 DNA 一起沉降的有大约一半的 ¹⁴C 标记，但没有 ³H 标记。

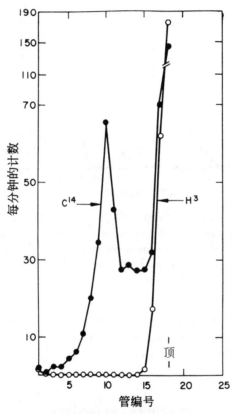

图 3. λ C₁ 产物与 λ DNA 的选择性结合。利用 DEAE 柱从同时含有 ³H 标记的 λ C₁sus34 噬菌体基因产物的抽提物中分离出 ¹⁴C 标记的 λind⁻ 阻遏物。然后按照图 2 中所描述的方法检测所得混合物与 λ DNA 的结合情况。λ DNA 在 260 nm 处的光密度分布图与图 2 中所示的结果基本一致。对收集的组分进行 TCA 沉淀，然后在 1/2 ml 0.1 M 的 NaOH 溶液中溶解沉淀，最后将其加入 10 ml 闪烁液中进行放射性计数，闪烁液的成分是：比例为 3:1 的甲苯和曲拉通 X-100（译者注：化学名称为聚乙二醇辛基苯基醚，是一种优异的表面活化剂）、0.4% 的 2,5– 二苯基恶唑和 0.005% 的 1,4–双 [2–(5–苯基) 恶唑基] 苯 [10]。

　　阻遏物并不结合变性的 DNA。图 4 表明：当变性 DNA 与标记的 λind⁻ 阻遏物的混合物在蔗糖梯度中沉淀时，管子顶部的放射性计数并没有转移到别的地方。上

677

sedimented through a sucrose gradient. The repressor preparation used in this experiment was found to bind efficiently to native DNA, hence its failure to bind here must have been due to the changed configuration of the DNA.

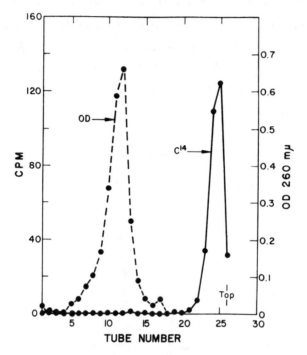

Fig. 4. Binding of λ*ind*⁻ repressor to denatured λ DNA. 100 μg of λ DNA was denatured in 0.1 M NaOH and then neutralized with HCl. The salt concentration was adjusted to 0.1 M KCl and a binding experiment was performed using ¹⁴C-labelled λ*ind*⁻ repressor as described in Fig. 2. The gradient was spun for only 2.5 h because of the increased sedimentation coefficient of denatured λ DNA in 0.05 M KCl[11].

In several experiments, the wild type repressor was tested for binding to DNA. Some binding was detected, but the results were not as striking as with the λ*ind*⁻ repressor. It is possible that the *ind*⁻ form is less susceptible to inactivation during the isolation procedure or that it binds more tightly to DNA.

Nature of the Binding

In order of binding to have been detected under the experimental conditions used, the repressor must bind very tightly to DNA. A close examination of Figs. 2 and 3 shows that some of the bound repressor washes off the DNA as it sediments. This suggests that the dissociation constant is of the same order of magnitude as the concentration of DNA binding sites (called operators) in the peak tubes. Assuming a small number of operators per phage genome (there are probably one or two), this value is roughly $10^{-9}-10^{-10}$ M. A repressor-operator affinity in this range *in vivo* is suggested by the magnitude of derepression observed with the *lac* operon. Since a 1,000-fold increase in β-galactosidase synthesis occurs on induction, the dissociation constant of the repressor–operator complex should be 1,000-fold less than the concentration

678

述实验所用的阻遏物曾被发现能够有效结合天然状态的 DNA，因此，该阻遏物在此不能结合变性 DNA 的原因一定是 DNA 的构型发生了改变。

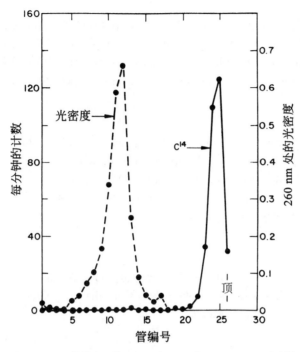

图 4. λ*ind*⁻ 阻遏物与变性 λ DNA 的结合。使 100 μg λ DNA 在 0.1 M NaOH 中变性，然后用 HCl 中和。将盐浓度调整到 0.1 M KCl，然后按照图 2 中所描述的方法用 ¹⁴C 标记的 λ*ind*⁻ 阻遏物进行结合实验。因为变性的 λ DNA 在 0.05 M KCl 中的沉降系数变大，所以将样品铺到梯度上后只离心 2.5 h 即可 [11]。

我们通过几次实验检测了野生型阻遏物与 DNA 的结合情况。有些情况下可以检测到这种结合，但是效果不及用 λ*ind*⁻ 阻遏物时那么明显。这可能是因为 *ind*⁻ 形式在分离过程中更不容易失活，或者它与 DNA 的结合更加紧密。

结合的特性

为了得到在所用实验条件下已经检测到的结合效果，阻遏物就必须与 DNA 结合得十分紧密。仔细观察图 2 和图 3 后发现，在沉降过程中一些本来与 DNA 结合的阻遏物脱离下来了。这表明：在与峰位对应的收集管中，解离常数与 DNA 结合位点（被称为操纵基因）的浓度具有相同的数量级。假设在每个噬菌体的基因组中只有少数几个操纵基因（可能是一个或者两个），那么该浓度值大致为 10^{-9} M 至 10^{-10} M。在浓度处于这一范围内的体内条件下，阻遏物与操纵基因之间的亲和力可以通过用乳糖操纵子进行实验得到的去阻遏作用幅度来推断。因为诱导后 β– 半乳糖苷酶的合成

of free repressor in the cell[8]. The concentration of free *lac* repressor has been estimated at 10^{-7} M (ref. 2), implying that the dissociation constant is of the order of 10^{-10} M.

The finding of the λ*ind*⁻ repressor to λ DNA was noticeably weaker when the complex was sedimented through a sucrose gradient containing 0.1 M KCl instead of the 0.05 M KCl used in the experiments described here. In a gradient containing 0.15 M KCl, no binding was detected. This observation suggests that the binding is partly electrostatic.

The finding that the λ repressor binds specifically and with high affinity to λ DNA strongly suggests that the simplest model for the mechanism of action of the repressor is correct—namely, that the repressor blocks transcription from DNA to RNA by directly binding to DNA. This conclusion is further supported by the recent observation of Dr. W. Gilbert that the *lac* repressor binds specifically to *lac* DNA and is removed by IPTG (W. Gilbert, to be published).

I thank Nancy Hopkins for technical assistance, and also Drs. W. Gilbert, J. D. Watson and S. E. Luria for help. The work was supported by grants from the U.S. National Science Foundation and the U.S. National Institutes of Health.

(**214**, 232-234; 1967)

Mark Ptashne: Department of Biology, Harvard University.

Received April 9, 1967.

References:

1. Jacob, F., and Monod, J., *J. Mol. Biol.*, **3**, 318 (1961).

2. Gilbert, W., and Mueller-Hill, B., *Proc. U.S. Nat. Acad. Sci.*, **56**, 1891 (1966).

3. Ptashne, M., *Proc. U.S. Nat. Acad. Sci.*, **57**, 306 (1967).

4. Kaiser, A. D., and Jacob, F., *Virology*, **4**, 509 (1957).

5. Isaacs, L. N., Echols, H., and Sly, W. S., *J. Mol. Biol.*, **13**, 963 (1965).

6. Jacob, F., and Campbell, A., *C.R. Acad. Sci., Paris*, **248**, 3219 (1959).

7. Jacob, F., Sussman, R., and Monod, J., *C.R. Acad. Sci., Paris*, **254**, 4214 (1962).

8. Sadler, J. R., and Novick, A., *J. Mol. Biol.*, **12**, 305 (1965).

9. Thomas, jun., C. A., and Abelson, J., *Procedures in Nucleic Acid Research*, edited by Cantoni, G. L., and Davies, D. R., 553 (Harper and Row, New York, 1966).

10. Modified according to E. Kennedy, pers. comm., from Patterson, M. S., and Greene, R. C., *Ann. Chem.*, **37**, 854 (1965).

11. Studier, F. W., *J. Mol. Biol.*, **11**, 373 (1965).

增加了 1,000 倍，所以阻遏物 – 操纵基因复合物的解离常数应该是胞内游离阻遏物浓度的 1/1,000[8]。游离乳糖操纵子阻遏物的浓度估计值是 10^{-7} M（参考文献 2），这意味着解离常数的数量级应为 10^{-10} M。

如果让复合物在含 0.1 M KCl 的蔗糖梯度中，而不是在本文所述实验中所用的含 0.05 M KCl 的蔗糖梯度中沉降，则会发现 λind^- 阻遏物与 λ DNA 的结合明显变弱。而如果用含 0.15 M KCl 的蔗糖梯度，则检测不到结合现象。上述结果表明这种结合有一部分是通过静电作用完成的。

λ 阻遏物与 λ DNA 的结合具有特异性和很高亲和力这一发现显然说明：解释阻遏物作用机制的最简单模型是正确的，即阻遏物通过直接与 DNA 结合而阻断了从 DNA 到 RNA 的转录过程。吉尔伯特博士最近的观察结果进一步证实了上述结论，他发现乳糖操纵子阻遏物能够特异地结合乳糖操纵子 DNA，并且这种结合作用可以被异丙基–β–D–硫代半乳糖苷消除（吉尔伯特即将发表的结果）。

感谢南希·霍普金斯在技术上给了我帮助，还要感谢吉尔伯特博士、沃森博士和卢里亚博士的帮助。此项研究受到了美国国家科学基金会和美国国立卫生研究院的资助。

（吕静 翻译；陈新文 陈继征 审稿）

Absorption Spectrum of Rhodopsin: 500 nm Absorption Band

R. Hubbard

Editor's Note

Ruth Hubbard at Harvard University was the wife of George Wald who, among others, was awarded a Nobel Prize for the understanding of the absorption of light by the pigments in the human eye. Her individual work is illustrated by this account of the absorption of rhodopsin, the principle absorber of light energy in the retina at the extreme red end of the visible spectrum. In the same issue of *Nature* she published a second paper dealing with the absorption at the blue end of the spectrum. Taken together, these papers amounted to the identification of the optically active chemical responsible for human vision.

RHODOPSIN, the photosensitive pigment of vertebrate rods and of a number of invertebrate photoreceptors, consists of a colourless protein (opsin), which carries the 11-*cis* isomer of retinal (retinaldehyde, vitamin A aldehyde) as chromophore. Bownds[1] showed that in cattle rhodopsin the chromophore is bound in Schiff base linkage to the ε–amino-group of an internal lysine residue in opsin.

Collins[2] and Morton, Pitt, and their co-workers[3] first demonstrated that retinaldehyde forms a Schiff base with an aliphatic amino-group on opsin[3]. This raises the question of how the absorption properties of rhodopsin arise from this type of attachment, for in aqueous solution λ_{max} of 11-*cis* retinal is at 380 nm, that of the Schiff bases at about 363 nm, and that of rhodopsin near 500 nm. (The precise position of λ_{max} in rhodopsins from various animals differs somewhat, presumably because of species differences in the opsins.)

To explain these absorption characteristics, Kropf and Hubbard[4] started with the observation first made by Ball *et al.*[5], that although the formation of a Schiff base of retinal with an aliphatic amine,

$$\overset{H}{C_{19}H_{27} \cdot C = O} + H_2N \cdot R \qquad \overset{H}{C_{19}H_{27} \cdot C = N \cdot R} + H_2O,$$

$$\text{retinal} \qquad\qquad\qquad \text{Schiff base}$$

视紫红质的吸收光谱：500 nm吸收带

哈伯德

编者按

哈佛大学的露丝·哈伯德是乔治·沃尔德的妻子，后者因为发现人类视网膜中色素的光吸收而与其他人一起共同获得了诺贝尔奖。这篇关于视紫红质吸收的报道是由哈伯德独立完成的，视紫红质是视网膜中主要吸收可见光谱红端光能的吸收剂。在同一期《自然》杂志上，她发表了另一篇与可见光谱蓝端的吸收有关的论文。总的来说，这两篇论文加深了人们对形成人类视觉的光学活性物质的认识。

视紫红质，即脊椎动物视杆细胞和很多无脊椎动物光感受细胞中的光敏色素，是由一种无色的蛋白质（视蛋白）构成的，视蛋白上含有作为生色团的视网膜醛（视黄醛，维生素A醛）的11-顺式异构体。鲍恩兹[1]指出：在牛的视紫红质中，生色团被限于与视蛋白内一个赖氨酸残基的 ε- 氨基相连的席夫碱中。

柯林斯[2]以及莫顿、皮特及其同事[3]首先证明，视黄醛与视蛋白上的一个脂肪族氨基形成了一种席夫碱[3]。这引出了下面的问题，即这种类型的连接如何使视紫红质具有吸收性质，因为在水溶液中11-顺式视网膜醛的 λ_{max} 位于380 nm处，席夫碱的 λ_{max} 位于363 nm附近，而视紫红质的 λ_{max} 则位于接近500 nm处。（不同动物视紫红质 λ_{max} 的准确位置会略有差异，也许是因为视蛋白的种类有所不同吧。）

为了解释这些吸收特性，克罗普夫与哈伯德[4]从由鲍尔等人[5]最先进行的观测入手，即尽管视网膜醛与脂肪胺可以生成一种席夫碱：

$$C_{19}H_{27} \cdot \overset{H}{C} = O + H_2N \cdot R \quad C_{19}H_{27} \cdot \overset{H}{C} = N \cdot R + H_2O,$$

视网膜醛   席夫碱

is accompanied by a shift in λ_{max} toward shorter wavelengths, the protonation of the imino group,

$$\overset{H}{C_{19}H_{27} \cdot C} = N \cdot R + H^+ \qquad \overset{H}{\underset{H}{C_{19}H_{27} \cdot C}} = \overset{+}{N} \cdot R.$$

shifts λ_{max} to approximately 440 nm.

The hypothesis assumes that the chromophores of the visual pigments are derived from the protonated Schiff base of 11-*cis* retinal by secondary interactions with reactive groups on opsin, which increase the mobility of electrons within the molecular orbitals of the chromophore and so shift λ_{max} to longer wavelengths (compare ref. 6). The triad of wavelengths that is therefore significant for this formulation are the absorption maximum of the Schiff base in its free and protonated forms and that of rhodopsin—approximately 363, 440 and 500 nm.

A direct experimental test of this hypothesis will have to wait until more is known about the amino-acid sequence and the three-dimensional structure of rhodopsin. Indirect support comes, however, from the observation that the absorption maxima of all the visual pigments that are known to have 11-*cis* retinal as chromophore—by now rather a large number—span more than 100 nm (approximately 440 to 565 nm). Yet they all lie at λ_{max} of the protonated Schiff base, or at longer wavelengths.

I report here observations concerning the denaturation of cattle rhodopsin which support the protonated Schiff base hypothesis.

All known chemical and physical protein denaturants denature opsin more readily than rhodopsin[7,8]. The denaturation of opsin is usually measured by its loss of the ability to regenerate rhodopsin; that of rhodopsin, by its bleaching in the dark to 11-*cis* retinal and denatured opsin[8].

Cattle rhodopsin is stable in 8 M urea, though this reagent denatures opsin in much lower concentrations (Fig. 1, also ref. 9). Guanidine hydrochloride (Gu-HCl), which is usually a more effective protein denaturant than urea, denatures cattle rhodopsin, but again in much higher concentrations than those required to denature opsin (Fig. 1). The logarithm of the rate at which rhodopsin is denatured increases proportionately with the logarithm of the Gu-HCl concentration (Fig. 2): more precisely, doubling the Gu-HCl concentration increases the rate constant (k) by a factor of ten.

684

其间会伴随着 λ_{max} 向短波方向的位移，但亚氨基的质子化反应：

$$\begin{array}{ccc}
H & & H \quad + \\
C_{19}H_{27} \cdot C = N \cdot R + H^+ & & C_{19}H_{27} \cdot C = N \cdot R \\
& & H
\end{array}$$

却使 λ_{max} 移至大约 440 nm 处。

该假说认为视色素中的生色团是由 11- 顺式视网膜醛的质子化席夫碱通过与视蛋白上的反应基团发生次级相互作用得到的，从而增加了电子在生色团分子轨道内部的活动性，因此使 λ_{max} 发生红移（对比参考文献 6 中的结果）。因而对该假说而言，三个重要的波长是游离态和质子化态席夫碱的最大吸收以及视紫红质的最大吸收——大致位于 363 nm、440 nm 和 500 nm。

我们将不得不等到对视紫红质中氨基酸序列和三维结构有更多了解的时候，再用实验直接验证这一假说。但是对所有视色素最大吸收值的观察可以提供间接的证据：目前已知有相当大数量的视色素含有 11- 顺式视网膜醛生色团，它们的 λ_{max} 跨度超过 100 nm（大约从 440 nm 到 565 nm）。不过它们都位于质子化席夫碱的 λ_{max} 附近或更长的波长处。

这里我报道的是对牛视紫红质变性的观测结果，由此可以证实质子化席夫碱的假说。

与使视紫红质发生变性相比，所有已知的化学蛋白质变性剂和物理蛋白质变性剂都更容易使视蛋白发生变性 [7,8]。视蛋白的变性通常是用丧失视紫红质再生能力的程度来衡量的；而视紫红质的变性则是由它在黑暗中褪色变成 11- 顺式视网膜醛和变性视蛋白而测定的 [8]。

牛视紫红质在 8 M 尿素溶液中很稳定；但在浓度大幅降低的情况下，尿素仍可使视蛋白发生变性（参见图 1，另见参考文献 9）。在通常情况下，盐酸胍（Gu-HCl）是一种比尿素更为有效的蛋白质变性剂，它能使牛视紫红质变性，但所需浓度同样要比使视蛋白变性的浓度高很多（图 1）。视紫红质变性速率的对数随 Gu-HCl 浓度对数的增加而成比例地增加（图 2）：更确切地说，Gu-HCl 浓度加倍会使速率常数 (k) 增大至 10 倍。

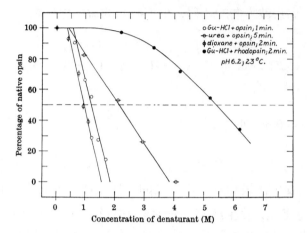

Fig. 1. Denaturation of opsin and rhodopsin as a function of the concentrations of several denaturants. The denaturation of opsin was measured as loss in its ability to form rhodopsin when incubated with excess 11-*cis* retinal in the dark. This assay took about 10 min, during which the denaturation continued. Estimates of native opsin therefore were too low and progressively more so as the concentration of denaturant increased. The lines therefore descend somewhat too steeply. Also the periods of incubation before the addition of 11-*cis* retinal differ for the three denaturants. The comparison therefore is only qualitative. The denaturation of rhodopsin was measured as irreversible bleaching in the dark. This involved only the measurement of the absorption spectrum in the presence of 0.07 M hydroxylamine, and entailed no systematic errors.

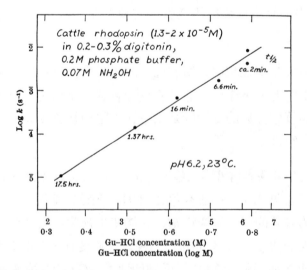

Fig. 2. The rate of denaturation of rhodopsin as a function of Gu-HCl concentration. Each point represents a separate experiment in which the unimolecular rate constant (k) for the denaturation was measured for a given Gu-HCl concentration. The half-times of denaturation ($t_{1/2}$) for each experiment are shown.

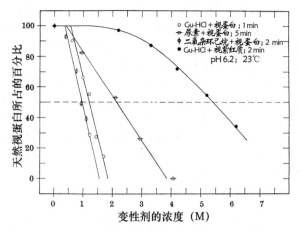

图 1. 视蛋白和视紫红质变性与几种变性剂浓度的关系曲线。视蛋白变性是由用过量 11- 顺式视网膜醛在黑暗中培养时形成视紫红质能力的丧失程度来衡量的。该实验过程大约需要 10 min，其间一直在发生变性反应。这样天然视蛋白所占百分比的估值就会偏低，而且当变性剂浓度增加时情况更是如此。因此曲线的下降有点过于陡峭。另外，三种变性剂在加入 11- 顺式视网膜醛之前的培养时间也各不相同。因此这只是定性的对比。视紫红质的变性由它在黑暗中的不可逆褪色来衡量。本结果只涉及存在 0.07 M 羟胺时测定的吸收光谱，因此不会产生系统误差。

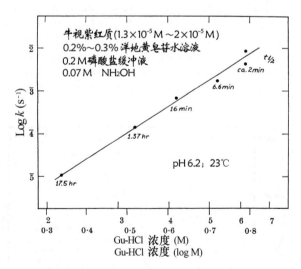

图 2. 视紫红质的变性速率与 Gu-HCl 浓度的关系曲线。每个点对应的是由一次单独实验得到的结果，在这些实验中测定了给定 Gu-HCl 浓度下变性反应的单分子速率常数（k）。由每次实验得到的变性反应半衰期（$t_{1/2}$）均示于图中。

A closer examination reveals that Gu-HCl exerts two effects on cattle rhodopsin, one rapid and reversible, the other slower and irreversible. The reversible interaction results in a shift in λ_{max} from 499 nm to shorter wave lengths, in the limit to 495.5 nm, without change in the half band width (Fig. 3), and is reversed rapidly by decreasing the concentration of Gu-HCl. I shall refer to the molecule with λ_{max} 495.5 nm as altered rhodopsin. It is not clear whether this shift in λ_{max} involves a series of intermediate compounds each with its own λ_{max} or a mixture of native rhodopsin (λ_{max} 499 nm) with increasing proportions of the altered pigment. Fig. 4 shows that the entire shift occurs within a rather narrow range of Gu-HCl concentrations, suggesting that what is being dealt with is probably one transition from the native to the altered state rather than smaller sequential steps.

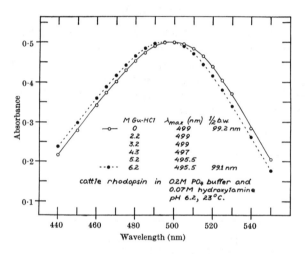

Fig. 3. The wavelength position of the absorption band of cattle rhodopsin as a function of Gu-HCl concentration. As the concentration of Gu-HCl increased, λ_{max} shifted to shorter wavelengths, with no change in half band width (1/2 bw). The shift in λ_{max} was reversed when the Gu-HCl concentration was lowered.

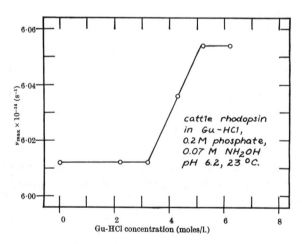

Fig. 4. Position of the absorption maximum (λ_{max}) of native and altered rhodopsin as a function of the Gu-HCl concentration.

688

　　进一步检验发现 Gu-HCl 可以对牛视紫红质产生两种效应：一种是快速而可逆的，另一种则较慢且不可逆。可逆的相互作用使 λ_{max} 从 499 nm 位移到较短波长处而不会改变半峰宽（图 3），极限情况下可以达到 495.5 nm，且随着 Gu-HCl 浓度的降低上述效应会快速逆转。我在下文中将说明这种 λ_{max} 为 495.5 nm 的分子就是发生可逆变性的视紫红质。现在仍不清楚 λ_{max} 从 499 nm 位移至 495.5 nm 到底是与一系列具有自身 λ_{max} 的中间化合物有关，还是与随着发生可逆变性色素比例的增加天然视紫红质（λ_{max} 为 499 nm）组成的变化有关。从图 4 中可以看出整个位移发生在一段相当窄的 Gu-HCl 浓度区间内，这意味着我们正在研究的可能是一种从天然状态到可逆变性状态的质变，而非一些较小的连续量变。

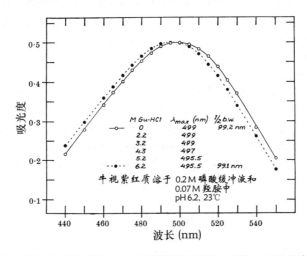

图 3. 牛视紫红质吸收带中波长的位置与 Gu-HCl 浓度的关系曲线。随着 Gu-HCl 浓度的增加，λ_{max} 位移到了较短波长处，而半峰宽（1/2 bw）并没有发生变化。当 Gu-HCl 浓度减少时，λ_{max} 又会向相反方向移动。

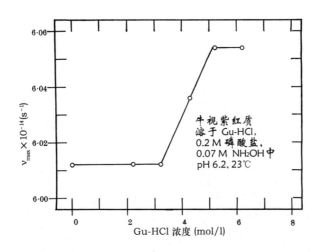

图 4. 天然和可逆变性视紫红质最大吸收的位置（λ_{max}）与 Gu-HCl 浓度的关系曲线。

Either interpretation presupposes that Gu-HCl shifts the absorption spectrum of rhodopsin by altering the conformation of opsin. Gu-HCl, however, carries a positive charge, localized primarily on the central carbon atom of the symmetrical guanidinium ion, $\overset{+}{C} \cdot (NH_2)_3$. At the high concentrations at which Gu-HCl affects the absorption spectrum of rhodopsin there could perhaps be a direct interaction between the charged guanidinium ion and the molecular orbitals of the retinyledene chromophore. But the abruptness with which λ_{max} of rhodopsin shifts with increasing Gu-HCl concentration (Fig. 4) argues against a non-specific solvent effect on the chromophore and in favour of a conformational attack on opsin.

The irreversible denaturation of rhodopsin by Gu-HCl is slower. Its first product is a pigment with λ_{max} 445 nm (Fig. 5, III), which is slowly converted to a pigment with λ_{max} 362.5 nm (Fig. 5, III and IV). There is no way to decide from there experiments whether the reversible alteration of rhodopsin to the pigment with λ_{max} 495.5 nm precedes the irreversible formation of the 445 nm pigment, or whether the two are parallel and independent processes. All concentrations of Gu-HCl which alter rhodopsin also denature it (compare Figs. 2 and 4). The 495.5 nm pigment is never stable therefore and its absorption spectrum must be measured while it disappears. This is done in the presence of hydroxylamine, which in sufficiently low concentrations does not decrease the stability of the 495.5 nm pigment, while converting the 445 nm pigment rapidly to 11-*cis* retinaldehyde oxime (λ_{max} 362.5 nm) and denatured opsin (Figs. 4 and 5). This is the only way to measure the absorption spectra of the 495.5 and 445 nm pigments, for they overlap too extensively to permit of more direct resolution (compare Fig. 5, I, II and III).

两种解释都需要预先假定 Gu-HCl 是通过改变视蛋白的构象而使视紫红质的吸收光谱发生位移的。但是，Gu-HCl 所带的一个正电荷主要定域在对称胍离子（$\overset{+}{C} \cdot (NH_2)_3$）的中心碳原子上。当高浓度的 Gu-HCl 影响到视紫红质的吸收光谱时，在带电胍离子与视黄基生色团分子轨道之间有可能会存在着一种直接的相互作用。但视紫红质 λ_{max} 随 Gu-HCl 浓度增加而发生突然性变化的现象（图 4）说明并非来自非特异性溶剂对生色团的影响，而倾向于视蛋白构象发生了变化。

视紫红质在 Gu-HCl 作用下发生不可逆变性的速度是比较慢的。它的第一个产物是 λ_{max} 为 445 nm 的色素（图 5，III），这种色素会慢慢地转化成另一种 λ_{max} 为 362.5 nm 的色素（图 5，III 和 IV）。从这些实验中无法判断视紫红质可逆转化成 λ_{max} 为 495.5 nm 色素的过程是否要先于不可逆生成 445 nm 色素的过程，也不能说明这两个过程是否相互平行且互不影响。所有会使视紫红质发生可逆变性的 Gu-HCl 浓度也会使其发生不可逆变性（对比图 2 和 4）。因此 λ_{max} 为 495.5 nm 的色素并不是稳定的，我们只能在它消失的瞬间测量它的吸收光谱。实验中需要用到羟胺，但羟胺的浓度很低，不会降低 495.5 nm 色素的稳定性，但 445 nm 色素会快速转化为 11- 顺式视黄醛肟（λ_{max} 为 362.5 nm），并使视蛋白变性（图 4 和 5）。以下是测定 495.5 nm 和 445 nm 色素吸收光谱的唯一方法，因为它们的重叠部分太多以至于不能用更为直接的方法进行分析（比较图 5，I、II 和 III）。

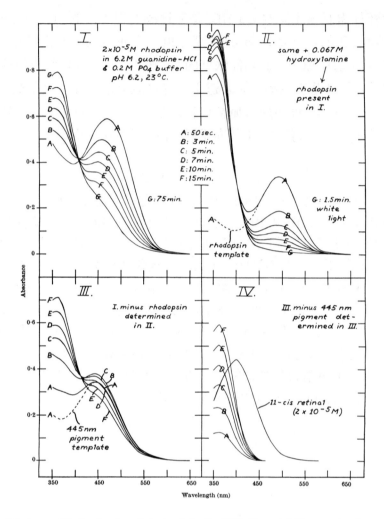

Fig. 5. Analysis of the irreversible denaturation of rhodopsin by Gu-HC1. Upper left (I), successive absorption spectra (*A* to *G*) recorded with a Cary model 14 spectrophotometer, after adding a small volume of a concentrated rhodopsin solution in 2 percent aqueous digitonin to buffered 6.2 M Gu-HC1. Upper right (II), the experiment was repeated with another aliquot of the same rhodopsin solution in the presence of 0.067 M hydroxylamine. This concentration of hydroxylamine does not labilize altered rhodopsin, so that the absorption spectra in (II) show the amounts of altered rhodopsin present in the experiment in (I). The dotted portion of curve II*A* is the absorption spectrum of this rhodopsin preparation below 450 nm, corrected to the appropriate absorbance at λ_{max}. Similar template spectra have been constructed for curves II*B*, *C* and so on. The relative absorbance and λ_{max} (362.5 nm) of the ultraviolet absorption band in spectra II*A* to *F* identify the product of the denaturation as 11-*cis* retinaldehyde oxime. Lower left (III), differences between successive absorption spectra in (I) and the corresponding template spectra in (II) (that is, curves I*A* minus II*A*, I*B* minus II*B* and so on). This yields the combined absorption spectra of the 445 and 362.5 nm pigments present in the experiment in (I). The dotted portion of curve III*A* was obtained by estimating the absorption of the 362.5 nm pigment present in the solidly drawn spectrum III*A* from the initial absorption of rhodopsin before denaturation (determined separately) and the amounts of the 362.5 and 445 nm pigments formed when a given amount of rhodopsin is denatured (that is, by comparing the differences between successive spectra in (II) and (III)). The absorption spectra of the 362.5 nm pigment shown at the lower right (IV) were obtained by subtracting successive 445 nm pigment template spectra obtained by scaling template III*A* to the successive absorbances at 445 nm, from the solidly drawn spectra in (III) (that is, curves III*A* minus template III*A*, III*B* minus template III*B*, etc.). (IV) Also shows the absorption spectrum of 11-*cis* retinal at the initial rhodopsin concentration in this solvent system. It has λ_{max} at 400 nm, a surprisingly long wavelength, and a specific absorbance at λ_{max} about 0.6 that of the Schiff base and of 11-*cis* retinaldehyde oxime, both of which have λ_{max} at 362.5 nm (compare ultraviolet absorption bands in (I) and (II)).

692

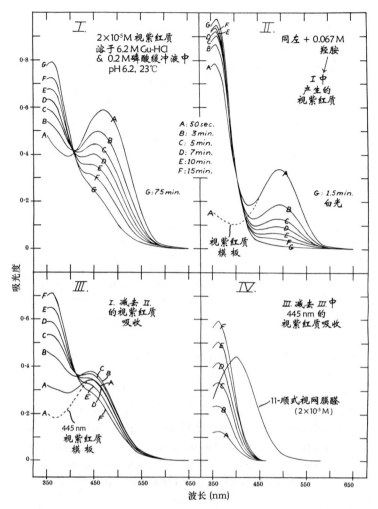

图 5. 视紫红质在 Gu-HCl 作用下发生不可逆变性的分析结果。左上图（I），将少量视紫红质浓缩液（2% 洋地黄皂苷水溶液）加入 6.2 M Gu-HCl 缓冲液中，再用卡里 14 型分光光度计记录连续吸收光谱（A 到 G）。右上图（II），用另一份同样的视紫红质溶液在含有 0.067 M 羟胺的情况下重复该实验。因为该浓度的羟胺不会降低可逆变性视紫红质部分的稳定性，所以（II）中的吸收光谱能体现出由（I）中实验所产生的可逆变性视紫红质的量。曲线 IIA 中的虚线部分是这种视紫红质制剂在 450 nm 以下的吸收光谱，并根据 λ_{max} 处的吸光度进行过修正。用类似方法建立了曲线 IIB、C 等的模板光谱。谱线 IIA ~ F 中紫外吸收带的相对吸光度和 λ_{max}（362.5 nm）说明，不可逆变性产物就是 11- 顺式视黄醛肟。左下图（III）：（I）中连续吸收光谱与（II）中所对应的模板光谱的差谱（即，曲线 IA 减去 IIA，IB 减去 IIB，依此类推）。这样得到的是（I）中实验里色素表现出来的 445 nm 和 362.5 nm 的混合吸收光谱。曲线 IIIA 中的虚线部分，是通过估计光谱 IIIA（译者注：原稿可能有误，此处应为 IIA）实线部分中 362.5 nm 色素吸收与视紫红质发生不可逆变性前初始吸收的差（单独测定），以及当给定量视紫红质发生变性时所形成的 362.5 nm 与 445 nm 色素的量而得到的（即，通过比较（II）与（III）中连续光谱的差异而得到）。显示于右下图（IV）中的 362.5 nm 色素的吸收光谱，是通过从（III）中光谱的实线部分里减去将色素模板 IIIA 与 445 nm 处连续吸收相对照而得到的连续 445 nm 色素模板而得到的（即，曲线 IIIA 减去模板 IIIA，IIIB 减去模板 IIIB，依此类推）。（IV）还显示出 11- 顺式视网膜醛在这一溶剂体系中于初始视紫红质浓度下的吸收光谱。它在波长非常长的 400 nm 处出现 λ_{max}，并且在 λ_{max} 处的比吸光度与席夫碱和 11- 顺式视黄醛肟相比约为 0.6，后两者在 362.5 nm 处出现 λ_{max}（对比（I）和（II）中的紫外吸收带）。

693

The absorption spectra of the two consecutive products of the irreversible denaturation (with λ_{max} 445 and 362.5 nm) strongly suggest that they are the protonated and free forms of a Schiff base of 11-*cis* retinal. The way in which they arise in the experiment shown in Fig. 5 further suggests that the retinal is still attached at the chromophoric site on opsin for the following reasons. At the *p*H of this experiment (6.2) essentially all the amino-groups on opsin are protonated, for even the most conservative estimates place the *p*Ks of the various classes of amino-groups in proteins well above *p*H 7. Retinal, however, forms Schiff bases only with unprotonated amino-groups[3] (see the first equation in this article). The very fact that the denatured product is a Schiff base therefore implies that retinal is attached at the same site as in native rhodopsin.

Furthermore, the *p*K of the Schiff bases of retinal (that is, of the imino-group) is considerably lower than that of the amino groups from which they arise[3], so that whereas at *p*H 6.2 all the amino-groups of opsin are protonated, the imino-group of the Schiff base of retinal should not be. The fact that at this *p*H the initial product of denaturation is protonated and gradually loses its proton as shown in Fig. 5, III, strongly suggests that the rhodopsin chromophore itself is protonated. Gu-HC1, so to speak, peels the opsin away from the chromophore and so reveals the unenhanced absorption spectrum of the protonated Schiff base.

This raises the question as to why the protonated Schiff base has not been seen as an intermediate in other types of denaturation of rhodopsin[7,8] except, of course, by strong acid[3]. The answer probably lies in the extraordinary stability of the unprotonated Schiff base in concentrated Gu-HC1. Fig. 5, I, II and IV, shows that λ_{max} of the unprotonated imine at 362.5 nm is unchanged throughout (that is, for 75 min at 23°C) whereas Schiff bases of retinal usually hydrolyse within about 15 min at this *p*H, as evidenced by a shift in λ_{max} to that of free retinaldehyde[3]. It must therefore be assumed that high concentrations of Gu-HC1 so lower the activity of water that a 6 M solution behaves like a non-aqueous system.

This research was supported by a grant from the US Public Health Service.

(**221**, 432-435; 1969)

Ruth Hubbard: Biological Laboratories, Harvard University, Cambridge, Massachusetts.

Received November 6, 1968.

References:

1. Bownds, D., *Nature*, **216**, 1178 (1967).

2. Collins, F. D., *Nature*, **171**, 469 (1953).

3. Pitt, G. A. J., Collins, F. D., Morton, R. A., and Stok, P., *Biochem. J.*, **59**, 122 (1955); Morton, R. A., and Pitt, G. A. J., *ibid.*, **59**, 128 (1955).

4. Kropf, A., and Hubbard, R., *Ann. NY Acad. Sci.*, **74**, 266 (1958).

5. Ball, S., Collins, F. D., Dalvi, P. D., and Morton, R. A., *Biochem. J.*, **45**, 304 (1949).

6. Hubbard, R., and Kropf, A., *J. Gen. Physiol.*, **49**, 381 (1965-66).

不可逆变性的两种连续产物的吸收光谱（λ_{max} 为 445 nm 和 362.5 nm）显然说明，它们分别是 11- 顺式视网膜醛的一种席夫碱的两种形态——质子化态和游离态。图 5 显示出它们出现于实验中的方式，这种方式进一步说明视网膜醛仍然连接在视蛋白的生色团位置上，原因有以下几点。在这次实验的 pH 值（6.2）下，视蛋白中几乎所有的氨基都会被质子化，因为即使按照最保守的估计，蛋白质中各种氨基基团的 pK 值也会远远高于 pH 值 7；但是，视网膜醛只与未质子化的氨基基团形成席夫碱 [3]（参见本文中第一个反应式）。因此，变性产物是一种席夫碱这一事实说明视网膜醛的连接位置与在天然视紫红质中相同。

此外，视网膜醛席夫碱（也就是亚氨基）的 pK 值比生成它们的氨基所具有的 pK 值要低很多 [3]，因此，尽管在 pH 值 6.2 时视蛋白的所有氨基都被质子化，但视网膜醛席夫碱的亚氨基却不会完全被质子化。在这个 pH 值下，初始的变性产物是质子化的，并会逐渐失去质子，如图 5，III 所示，这显然说明视紫红质生色团本身是质子化的。可以这样讲，Gu-HCl 的作用是将视蛋白与生色团剥离，因而表现为质子化席夫碱吸收光谱未出现增强的现象。

这样就引出了以下这个问题，即为什么在其他类型视紫红质的变性过程中从未看到过质子化席夫碱作为中间产物出现 [7,8]，当然，在强酸条件下 [3] 除外。答案可能就在于未质子化席夫碱在高浓度 Gu-HCl 中的超常稳定性。图 5 中的 I、II、IV 表明，未质子化亚胺在 362.5 nm 的 λ_{max} 始终没有发生变化（即在 23℃ 下长达 75 min 的时间内），但视网膜醛席夫碱在这个 pH 值下通常会在约 15 min 之内水解，证据是 λ_{max} 位移到了自由视黄醛的位置 [3]。因此我们必须假定，高浓度的 Gu-HCl 显著降低了水的活性，以至于浓度为 6 M 的溶液就像不含水的体系一样。

本研究得到了美国公共卫生署提供的一项基金的资助。

（王耀杨 翻译；杨志 审稿）

695

7. Radding, C. M., and Wald, G., *J. Gen. Physiol.*, **39**, 923 (1955-56).

8. Hubbard, R., *J. Gen. Physiol.*, **42**, 259 (1958-59).

9. Abrahamson, F. W., and Ostroy, S. E., *Prog. Biophys. and Mol. Biol.*, **17**, 181 (1967).

10. Hubbard, R., and St George, R. C. C., *J. Gen. Physiol.*, **41**, 501 (1957-58).

Early Stages of Fertilization *in vitro* of Human Oocytes Matured *in vitro*

R. G. Edwards *et al.*

Editor's Note

Robert Edwards worked at the University of Cambridge for most of his career. His goal was to develop means of producing human embryos by *in vitro* fertilization in cases where the union of sperm and ovum is difficult or impossible within the female body. Artificial fertilization is now widely practised throughout the world, and stems uniquely from Edwards' pioneering work with surgeon Patrick Steptoe. Here Edwards and his student Barry Bavister, working with Steptoe, report the first evidence of fertilization of a human egg *in vitro*.

THE technique of maturing human oocytes *in vitro* after their removal from follicles provides many eggs for studies on fertilization[1]. Their fertilization *in vitro* would yield a supply of embryos for research or clinical use, but in previous attempts the incidence of fertilization was too low to be useful[2].

A possible solution to the problem of obtaining "capacitated" spermatozoa has recently emerged from experiments on hamster eggs, where the addition of epididymal spermatozoa to eggs in tubal[3] or follicular[4] secretions can lead to a high incidence of fertilization. Study of the conditions leading to capacitation of hamster spermatozoa and fertilization *in vitro*[5] has led to the use of a medium based on Tyrode's solution, but with extra bicarbonate (final concentration 3 mg/ml.); also added were sodium pyruvate (9.0 μg/ml.), bovine serum albumin (2.5 mg/ml.), phenol red (20 μg/ml.) and penicillin (100 IU/ml.)[5] . After equilibration with 5 percent CO_2 in air, the *p*H of this medium was 7.6. Bicarbonate has been shown to stimulate the respiration and motility of rabbit spermatozoa *in vivo* and *in vitro*[6].

We have adapted the conditions found successful in the hamster for work on the human oocyte. The preliminary results reported in this article indicate that human eggs can be fertilized in conditions similar to those found most suitable for the fertilization of hamster eggs *in vitro*.

Maturation of Eggs and Insemination

Oocytes were released from Graafian and smaller follicles into a medium composed of equal amounts of Hank's solution containing heparin, and medium 199 (Microbiological Associates, Inc.) supplemented with 15 percent foetal calf serum (Microbiological Associates, Inc.) and buffered (*p*H 7.2) with phosphate buffer. Penicillin (100 IU/ml.) was added to these media. Some oocytes were transported from Oldham to Cambridge (a journey taking about 4 h), in medium 199 supplemented with 15 percent foetal calf serum.

698

体外成熟后的人类卵母细胞体外受精的早期阶段

爱德华兹等

编者按

罗伯特·爱德华兹的绝大部分职业生涯是在剑桥大学度过的。他的目标是为那些难以或不能在自身体内实现精子和卵子结合的女性患者找到通过体外受精产生人类胚胎的途径。目前，人工授精技术已经在全世界得到了广泛的应用，这类技术毫无例外地来源于爱德华兹和外科医生斯特普托的开创性研究。在本文中，爱德华兹和他的学生巴里·巴维斯特与斯特普托合作，第一次报道了可证明人类卵子能在体外受精的证据。

可使从卵泡中分离出来的人类卵母细胞在体外发育成熟的技术为受精研究提供了大量的卵子 [1]。它们的体外受精将会为研究或临床应用提供胚胎，但是在先前的实验中卵子的受精率非常低以至无法使用 [2]。

最近在对仓鼠卵子进行实验时发现，将仓鼠附睾的精子添加到输卵管 [3] 或卵泡 [4] 的分泌液中可使其中的卵子达到很高的受精率，这为解决如何获得"获能"精子这一难题提供了一种可能的解决途径。在研究使仓鼠精子在体外条件下获能并受精的条件时发现 [5]，应使用一种基于台罗德溶液的培养基，不过还需在培养基中另外添加碳酸氢盐（终浓度为 3 mg/ml）以及丙酮酸钠（9.0 μg/ml）、牛血清白蛋白（2.5 mg/ml）、酚红（20 μg/ml）和青霉素（100 IU/ml）[5]。用含 5% CO_2 的空气平衡后，这种培养基的 pH 值为 7.6。体内和体外的研究结果都表明，碳酸氢盐能够刺激兔精子的呼吸作用和运动能力 [6]。

我们已经把这种在仓鼠身上实验成功的条件应用到人类卵母细胞上。由本文中报道的初步结果可知，人的卵子在类似于最适合仓鼠卵子体外受精的条件下也可以成功受精。

卵子的成熟和人工授精

将卵母细胞从格拉夫卵泡和小卵泡中分离，并放入含有等量的以下两种成分的培养基中：含肝素的汉克氏溶液和加入 15% 胎牛血清（微生物联合有限公司）并用磷酸缓冲溶液缓冲（pH 值为 7.2）的 M199 培养基（微生物联合有限公司）。将青霉素（100 IU/ml）加入上述培养基中。有些卵母细胞是从奥尔德姆运送到剑桥的（路上大约需要 4 h），运送过程中卵母细胞一直被保存在添加了 15% 胎牛血清的 M199 培养基中。

The oocytes were cultured soon after liberation from their follicles, or after transport to Cambridge, in follicular fluid obtained from the same or different ovaries. Where follicular fluid was in short supply it was supplemented by Hank's, Brinster's[7] or Bavister's[5] medium. Oocytes were cultured under a gas phase of 5 percent CO_2 in air, in droplets of medium under liquid paraffin previously equilibrated with medium 199 containing 15 percent foetal calf serum, and with the same gas mixture. After 38 h in culture many of the oocytes had extruded their first polar body and reached metaphase of the second meiotic division (metaphase-II).

Ejaculated spermatozoa were washed once with Bavister's medium to remove seminal plasma, and were then re-suspended at a concentration of 10^6/ml. in more of the same medium. Human follicular fluid was added to some samples of spermatozoa before they were added to the oocytes. Oocytes were washed through one or two droplets of Bavister's medium and then pipetted into the sperm suspension. At different intervals after insemination, living eggs were examined by phase contrast microscopy, and then fixed in acetic-saline (20 percent acetic acid in normal saline solution) or acetic-alcohol (1:3), stained with aceto-orcein, and re-examined. In one-celled eggs, the occurrence of penetration was inferred from the presence of spermatozoa in the zona pellucida or in the perivitelline space; the presence in the vitellus of pronuclei, with mid-pieces or tails of spermatozoa, and the extrusion of the second polar body, were regarded as evidence of fertilization.

Fertilization

Fifty-six human eggs were inseminated. When examined for evidence of fertilization, twenty of them were found still to be in dictyotene, having failed to mature *in vitro*. In one of these eggs, a spermatozoon was seen in the perivitelline space. Two other eggs were degenerate.

The remaining thirty-four eggs had matured *in vitro*, as judged by the presence of chromosomes or polar bodies. Many of these eggs were so heavily coated with spermatozoa attached to the zona pellucida that it was difficult to discern internal details. Sixteen of them were in metaphase-II, and showed no evidence of penetration.

In eleven eggs spermatozoa had begun to move through the zona pellucida. One or more spermatozoa were deeply embedded in the zona pellucida in six eggs (Fig. 1). When fixed and stained, each of these eggs was found to be at metaphase-II. Spermatozoa were seen in the perivitelline space of the other five eggs (Fig. 2), and in four of them the spermatozoa were motile. When fixed and stained, four of these eggs were seen to be in metaphase-II, and one had a small nucleus.

在将卵母细胞从卵泡中分离出来后，或者在将其置于从相同或不同卵巢中获取的卵泡液中并运送到剑桥后，即对其进行培养。如果卵泡液短缺，可以补充汉克氏溶液、布林斯特[7]培养基或巴维斯特[5]培养基。将培养基微滴中的卵母细胞培养于含 5% CO_2 的混合空气中，微滴表面覆盖着液体石蜡并且之前曾用相同空气条件、含 15% 胎牛血清的 M199 培养基平衡过。培养 38 h 后，许多卵母细胞已排出第一极体，并且进入第二次减数分裂的中期（中期 –II）。

将射出的精子用巴维斯特培养基洗涤一次以除去精浆，然后用更多的巴维斯特培养基重悬精子，使其浓度达到 10^6/ml。在将一些精子样品加入卵母细胞中之前，要先向这些精子样品中加入人的卵泡液。用 1~2 滴巴维斯特培养基洗涤卵母细胞，再用微管将其吸取到精子悬液中。在人工授精后的不同时间点用相差显微镜观察存活的卵子，然后将其在醋酸 – 生理盐水（含 20% 醋酸的生理盐水）或醋酸 – 酒精混合液（1 : 3）中固定，经醋酸地衣红染色后重新观察。对于单细胞期的卵子，有无发生穿卵可以通过观察在透明带或卵周隙中是否存在精子来判断。在原核卵黄处存在可观察到中部或者尾部的精子以及第二极体的排出都被认为是受精发生的证据。

受　　精

采取人工授精处理的共有 56 枚人类卵子。我们在检测受精是否发生时发现：有 20 枚卵子仍处在核网期，它们未能完成体外成熟。我们在其中 1 枚卵子的卵周隙中观察到了 1 个精子。另外还有 2 枚卵子发生了退化。

其余 34 枚卵子均出现了染色体或者极体，可以据此认为它们已完成体外成熟。这些卵子大多被附着在透明带上的精子紧紧包裹，以至于很难分辨其内部结构。其中有 16 枚卵子处在中期 –II 阶段，而且没有出现精子穿卵的迹象。

在 11 枚卵子中，精子已经开始穿过透明带。一个或多个精子深深地植入到其中 6 枚卵子的透明带中（图 1）。对其进行固定和染色后发现：这 6 枚卵子都处在中期 –II 阶段。我们在另外 5 枚卵子的卵周隙中发现了精子（图 2），而且其中 4 枚卵子中的精子仍能移动。在对这 5 枚卵子进行固定和染色后发现：其中 4 枚卵子正处于中期 –II 阶段，另外 1 枚卵子有一个小细胞核。

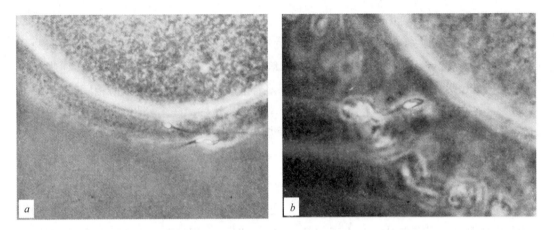

Fig. 1. Spermatozoa lying in the zona pellucida of living human eggs 24 h after insemination. (*a*, × 675; *b*, ×1,200.)

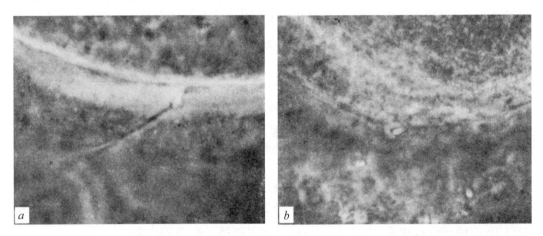

Fig. 2. Spermatozoa in the perivitelline space of human eggs. In *a*, the sperm head is lying in the vitellus, the mid-piece is still embedded in the zona pellucida, and the main piece lies outside the zona. This egg was examined 13 h after insemination; the spermatozoon was still active. (× 1,650.) In *b*, the whole of the spermatozoon was in the perivitelline space, although only the sperm head can be seen in the illustration. This spermatozoon was still active. (× 750.)

Seven eggs had well formed pronuclei. Two of them possessed two pronuclei each (Fig. 3), four had three pronuclei and one had five. Spermatozoa were seen in the zona pellucida or perivitelline space of three of these eggs.

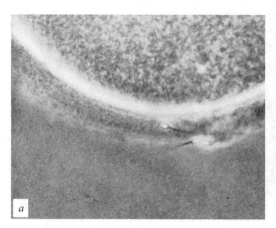

图 1. 人工授精 24 h 后处于存活人卵透明带中的精子。(*a*，放大 675 倍；*b*，放大 1,200 倍)

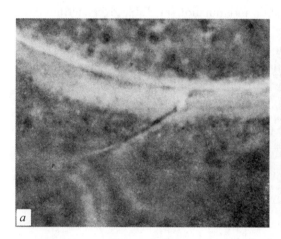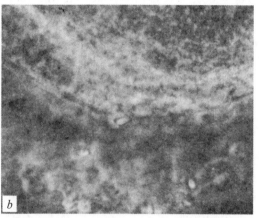

图 2. 处于人卵卵周隙中的精子。*a*，精子的头部位于卵黄中，中部仍嵌在透明带里，而主要部分则在透明带之外。这是在人工授精 13 h 后观察到的卵子的情况；其中的精子仍可移动（放大 1,650 倍）。*b*，整个精子都位于卵周隙中，不过我们只能在图里看到精子的头部。这个精子仍能移动（放大 750 倍）。

　　有 7 枚卵子已经很好地形成了原核。其中 2 枚卵子各有 2 个原核（图 3），4 枚卵子各有 3 个原核，另外 1 枚卵子有 5 个原核。其中 3 枚卵子的透明带或卵周隙中有精子存在。

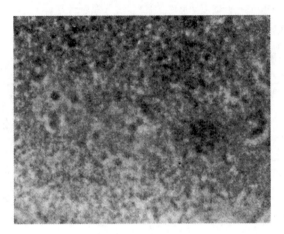

Fig. 3. Two pronuclei in a living human egg examined 22 h after insemination. Polar bodies were seen in this egg, but a sperm mid-piece could not be unequivocally identified. ($\times$ c. 675.)

These observations are related to the time after insemination in Table 1. Penetration of spermatozoa into the perivitelline space was first seen in eggs examined 7–7.25 h after insemination, and pronuclei at 11.5 h. Spermatozoa were found in the zona pellucida or perivitelline space of eggs in metaphase–II as late as 27 h after insemination. Many of these spermatozoa were immotile, and it is doubtful whether they would have penetrated further.

Table 1. Details of Human Oocytes Examined at Various Intervals after Insemination *in vitro*

| Time after insemination (h) | 1. Failed to mature *in vitro* | | 2. Matured *in vitro* | | | |
	Germinal vesicle	Vacuolated or degenerate	Unpenetrated	Spermatozoa in zona pellucida	Spermatozoa in perivitelline space	Pronucleate
6–6.5	1		3			
7–7.5	1				3	
8–9	1	1	2	1	1†	
9.5–10.75	3		3			
11.5	5		1	1		1
12.5–13.5	6*				1	1
22–31	3	1	7	4		5
	20	2	16	6	5	7

* In one of these eggs, a spermatozoon was present in the perivitelline space.

† This egg may have been in metaphase of the first meiotic division.

Mid-pieces and tails of spermatozoa were not seen with certainty in any of the living eggs, but after fixation structures of a size similar to mid-pieces (as illustrated by other workers[8]) were recognizable in three eggs (Fig. 4). In two eggs, each with two pronuclei, the first and second polar bodies could be identified. In one of these eggs, the two pronuclei were lying close to each other (Fig. 3); on staining, a mass of coarser chromatin was observed in one pronucleus in

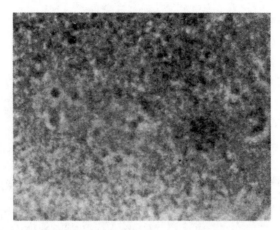

图 3. 人工授精 22 h 后在一存活人卵中观察到的 2 个原核。在这个卵子中可以看到极体，但不能明确地分辨出精子的中部（放大约 675 倍）。

上述观察结果与人工授精后所经过的时间之间的关系示于表 1。最早发现精子穿过卵周隙的时间是在人工授精 7 h~7.25 h 之后，而看到原核则是在 11.5 h 之后。人工授精 27 h 后可在处于中期 –II 阶段卵子的透明带或卵周隙中发现精子。这些精子中有不少已不能移动，我们很怀疑它们是否还会进一步穿卵。

表 1. 体外人工授精后在不同时间点观察到的人类卵母细胞的详细结果

人工授精后经过的时间（h）	1. 未能完成体外成熟的卵子		2. 完成体外成熟的卵子			
	生发泡	空泡化或退化	未穿卵	在透明带中的精子	在卵周隙中的精子	形成原核
6 ~ 6.5	1		3			
7 ~ 7.5	1				3	
8 ~ 9	1	1	2	1	1†	
9.5 ~ 10.75	3		3			
11.5	5		1	1		1
12.5 ~ 13.5	6*				1	1
22 ~ 31	3	1	7	4		5
	20	2	16	6	5	7

* 在其中 1 枚卵子的卵周隙中发现了 1 个精子。
† 该卵子可能已处于第一次减数分裂的中期。

在任何存活的卵子中都没有明确观察到精子的中部和尾部，但是经过固定后，我们在 3 枚卵子（图 4）中发现了大小类似于精子中部的结构（正如其他研究者所描述过的那样 [8]）。在 2 枚各有 2 个原核的卵子中可以分辨出第一极体和第二极体。在其中 1 枚卵子中，两个原核还相互靠得很近（图 3）；染色后，在两个原核相互靠

a region where the two pronuclei were apposed.

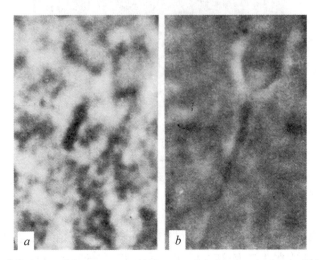

Fig. 4. In *a*, a small body resembling a sperm mid-piece was found in a stained egg which contained three pronuclei and at least one polar body. (× 3,300.) In *b*, a spermatozoon present in the perivitelline space of another egg has been photographed in a position illustrating the size of the mid-piece, for comparison with *a*. (× 3,300.)

As controls, seventeen eggs were cultured without the addition of spermatozoa. Eight of them failed to mature and displayed germinal vesicles. Of the remainder, one had a small mass of chromatin instead of metaphase chromosomes, seven had metaphase chromosomes, and in one the contents were obscured by overlying cumulus cells. Pronuclei were not seen in any of these eggs.

Of the seventy-three inseminated and control eggs, thirty-one failed to mature or degenerated. Thirteen of these failures occurred in one set of experiments involving a total of fifteen eggs. If this group of eggs is excluded, about 70 percent of the oocytes matured in culture. After insemination, eighteen of the thirty-four oocytes that had matured *in vitro* had spermatozoa in the zona pellucida, spermatozoa in the perivitelline space, or pronuclei. Judged on the criteria of fertilization given, it is highly probable that most of these eighteen eggs were undergoing fertilization.

Capacitation and Embryonic Development

The possible function of follicular fluid in the capacitation of spermatozoa and the fertilization of human eggs *in vitro* requires further examination. All eggs in the present series were matured in follicular fluid obtained from the same or a different ovary, and then washed in order to reduce any deleterious effect of follicular fluid on spermatozoa. We have found that spermatozoa of some animal species can be immobilized or agglutinated by follicular fluid, and heat treatment of the fluid does not necessarily abolish these effects. In some of the present experiments, however, we pre-incubated spermatozoa in follicular fluid before adding them to

近的区域中可以看到一个原核内有大量较粗大的染色质块。

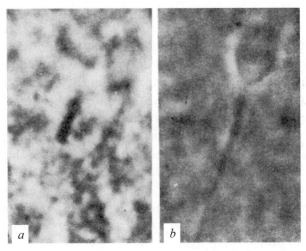

图 4. *a*，在 1 枚染色卵子中发现的 1 个类似于精子中部的小体，这枚被染色的卵子含有 3 个原核和至少
1 个极体（放大 3,300 倍）。*b*，为了与图 *a* 进行比较，我们在能说明精子中部尺寸的位置上拍摄了位于
另外 1 枚卵子卵周隙中的 1 个精子（放大 3,300 倍）。

作为对照，我们在没有加入精子的条件下培养了 17 枚卵子。其中 8 枚卵子未能
完成体外成熟，还呈生发泡样形态。在其余的卵子中，1 枚出现了一小团染色质而
非中期染色体，7 枚出现了中期染色体，另外 1 枚由于卵丘细胞覆盖于其上而无法
看清内部结构。在所有这些作为对照的卵子中都没有发现原核。

在 73 枚人工授精的和作为对照的卵子中，有 31 枚未能完成体外成熟或发生了
退化。其中 13 例失败发生在一组总共包括 15 枚卵子的实验中。如果将这组卵子排除，
则约有 70% 的卵母细胞能在体外培养时成熟。人工授精后，在 34 个完成体外成熟
的卵母细胞中有 18 个被发现在透明带、卵周隙里存在精子，或出现了原核。根据前
面给出的受精判断标准，这 18 枚卵子中的大多数很可能正在经历受精过程。

精子获能和胚胎发育

卵泡液在精子获能和人卵体外受精过程中可能发挥的作用还有待于进一步研究。
目前系列实验中所用的卵子都是在来自相同或不同卵巢的卵泡液中发育成熟的，随
后进行洗涤以减少卵泡液对精子的有害作用。我们曾发现卵泡液会使某些动物的精
子不能移动或者发生凝集，而对卵泡液进行加热处理不一定能起到消除这种有害效
应的作用。但是，在现在的一些实验中，我们是用卵泡液预孵育精子然后将其加入

the eggs. Our impression is that this pre-incubation led to the attachment of more spermatozoa to the zona pellucida, and to a higher incidence of penetrated and pronucleate eggs.

The failure of spermatozoa to pass completely through the zona pellucida or into the vitellus may reflect the existence of different layers in the zona with different requirements for penetration, or it may signify that the final movement of spermatozoa through the zona pellucida depends to a large extent on sperm/egg association. Complete penetration might be achieved by conferring greater activity on spermatozoa; mouse eggs can be fertilized *in vitro* by uterine spermatozoa[9] in a medium richer in pyruvate and albumin[10] than our media. Delayed fertilization may well have occurred in the conditions of our culture, and led to anomalies in the eggs. Thus the egg with a spermatozoon in the perivitelline space, polar bodies and a single small nucleus might have been activated parthenogenetically. The presence of several pronuclei in some eggs may have arisen through polyspermy, but is more probably due to the abnormal movement of chromosomes along the spindle or to fragmentation of the female pronucleus; multipronucleate eggs are common after delayed fertilization *in vivo* of eggs of various mammalian species[11].

Problems of embryonic development are likely to accompany the use of human oocytes matured and fertilized *in vitro*. When oocytes of the rabbit and other species were matured *in vitro* and fertilized *in vivo*, the pronuclear stages appeared normal but many of the resulting embryos had sub-nuclei in their blastomeres, and almost all of them died during the early cleavage stages (ref. 12 and unpublished work of R. G. E.). Abnormal development might have been the result of incomplete RNA synthesis when oocytes are removed from follicles for maturation *in vitro*; small amounts of RNA are synthesized by *Xenopus* oocytes in response to luteinizing hormone (LH) during the final period of maturation[13]. When maturation of rabbit oocytes was started *in vivo* by injecting gonadotrophins into the mother, and completed in the oviduct or *in vitro*, full term rabbit foetuses were obtained (ref. 12 and unpublished work of R. G. E.). Another developmental problem could be that some oocytes blocked during maturation at anaphase of the first meiotic division[1,14] could yield polyploid or heteroploid embryos at fertilization. Fortunately human oocytes in the present and earlier[1] work were rarely blocked at this stage.

Clinical Use of Human Embryos

These potential difficulties with human embryos may be solved when the conditions necessary for maturation *in vitro* are better understood, or, as in rabbits, by initiating maturation *in vivo* by administering gonadotrophins or clomiphene to the mother. When women were injected with gonadotrophins, maturing oocytes were recovered from excised pieces of ovary[15]. The timing of the stages of oocyte maturation in this work[15] was very similar to that exhibited by oocytes matured *in vitro*[1], and should indicate the appropriate moment to remove oocytes in preparation for fertilization *in vitro*.

Fertilized human eggs could be useful in treating some forms of infertility, and many infertile patients will probably be older women. If the "production line" of eggs in the ovary, inferred

到卵子中去的。我们感觉这种预孵育处理能使更多的精子黏附到卵子的透明带上，从而提高了穿卵和卵子原核形成的几率。

精子不能完全穿过透明带或到达卵黄也许意味着透明带中存在着不同的层，每一层都对精子穿卵有不同的要求，也可能意味着精子穿过透明带的最后过程在很大程度上取决于精子／卵子的相关性。通过使精子达到更高的活性或许可以实现完全穿卵；在比我们的培养基更富含丙酮酸盐和白蛋白的培养基[10] 中，用从子宫内冲取的精子[9] 能使小鼠卵子在体外完成受精。另外，在我们的培养条件下也很可能发生了延迟受精，从而使卵子出现异常。这样，那些已有一个精子进入卵周隙、出现极体以及单个的小细胞核的卵子可能已经被以孤雌生殖方式激活。某些卵子中存在好几个原核的原因可能与多精入卵有关，但可能性更大的原因是染色体沿着纺锤体的不规则移动，或者是源自雌性原核的碎片化；许多哺乳动物的卵子在体内发生延迟受精之后通常会出现多原核卵子[11]。

胚胎发育中出现的问题很可能与实验时使用了在体外成熟并受精的人类卵母细胞有关。当兔子和其他动物的卵母细胞在体外成熟而在体内受精时，它们看起来都能正常地发育到原核阶段，但是许多由此得到的胚胎在形成卵裂球时出现了亚核，并且几乎所有这样的胚胎在分裂早期就会死亡（参考文献 12 和爱德华兹尚未发表的研究结果）。胚胎发育异常或许是因为当把将要在体外培养成熟的卵母细胞从卵泡中取出时，RNA 的合成还没有全部完成；处于成熟最后阶段的爪蟾卵母细胞会在促黄体生成激素（LH）的作用下合成一小部分 RNA[13]。如果我们通过给雌兔注射促性腺激素来促发兔卵母细胞在体内的成熟过程，那么不管后续的成熟过程是在输卵管中完成还是在体外完成，都能够得到足月的胎兔（参考文献 12 和爱德华兹尚未发表的研究结果）。胚胎发育中的另一个问题是：一些在成熟过程中受阻于第一次减数分裂后期的卵母细胞[1,14] 可能在受精时会形成多倍体或异倍体胚胎。幸运的是，人类卵母细胞在目前和早先[1] 的研究工作中极少在这个阶段发生阻滞。

人类胚胎的临床应用

如果我们能对卵母细胞体外成熟所需的条件了解得更为透彻，那么在人类胚胎发育过程中可能出现的困难或许就会迎刃而解。或者像在兔子中的情形一样，采用给供卵雌兔注射促性腺激素或克罗米酚从而在体内促发卵母细胞成熟的方法来解决上述问题。给妇女注射促性腺激素后，从卵巢被切除部分中可以分离出正在成熟的卵母细胞[15]。这样得到的卵母细胞在成熟过程的时间进度上[15] 非常类似于体外成熟的卵母细胞[1]，这使我们可以推算出何时取出卵母细胞才能适合于对其进行体外受精。

受精的人类卵子或许能有助于治疗某些不孕症，而且许多不孕患者可能会是年龄较大的妇女。如果从研究小鼠卵母细胞[16] 中推断出来的卵巢内卵子的"生产线"

from studies on mouse oocytes[16], also occurs in humans, the eggs of older women will have more anomalies of bivalent association than those of younger women. A higher incidence of trisomic and polysomic embryos, and hence of mongols and abortions, would thus be expected in these patients.

The clinical use of human embryos will require the development of operative techniques for the recovery of follicular oocytes and for the transfer of eggs into human oviducts and uteri. Preliminary work using laparoscopy has shown that oocytes can be recovered from ovaries by puncturing ripening follicles *in vivo*, and that a few of the eggs transferred into the oviducts can be recovered following salpingectomy (unpublished work of P. C. S. and R. G. E.). Improvements in equipment and techniques may give better results and avoid resorting to laparotomy.

We thank especially Professor C. R. Austin for his encouragement and advice, and Drs. C. Abberley, G. Garrett and L. Davies for their help. One of us (R. G. E.) is indebted to the Ford Foundation and another (B. D. B.) to the Medical Research Council for financial assistance. We thank Professor N. Morris and Drs. M. Rose, J. Bottomley and S. Markham for ovarian tissue.

(**221**, 632-635; 1969)

R. G. Edwards and B. D. Bavister: Physiological Laboratory, University of Cambridge.
P. C. Steptoe: Oldham General Hospital, Oldham.

Received December 13, 1968.

References:

1. Edwards, R. G., *Lancet*, ii, 926 (1965).

2. Edwards, R. G., Donahue, R. P., Baramki, T. A., and Jones, H. W., *Amer. J. Obstet. Gynec.*, **96**, 192 (1966).

3. Yanagimachi, R., and Chang, M. C., *J. Exp. Zool.*, **156**, 361 (1964).

4. Barros, C., and Austin, C. R., *J. Exp. Zool.*, **166**, 317 (1967).

5. Bavister, B. D., *J. Reprod. Fertil.* (in the press).

6. Williams, W. L., Weinman, D. E., and Hamner, C. E., *Fifth Intern. Cong. Anim. Reprod. A. I., Trento*, 7, 288 (1964).

7. Brinster, R. L., *Exp. Cell Res.*, **32**, 205 (1963).

8. Dickmann, Z., Clewe, T. H., Bonney, W. A., and Noyes, R. W., *Anat. Rec.*, **152**, 293 (1965).

9. Whittingham, D. G., *Nature*, **220**, 592 (1968).

10. Whitten, W. K., and Biggers, J. D., *J. Reprod. Fertil.*, **17**, 399 (1968).

11. Austin, C. R., *The Mammalian Egg* (Blackwell, Oxford, 1961).

12. Chang, M. C., *J. Exp. Zool.*, **128**, 379 (1955).

13. Davidson, E., Allfrey, V. G., and Mirsky, A. E., *Proc. US Nat. Acad. Sci.*, **52**, 501 (1964).

14. Edwards, R. G., *Nature*, **208**, 349 (1965).

15. Jagiello, G., Karnicki, J., and Ryan, R. J., *Lancet*, i, 178 (1968).

16. Henderson, S. A., and Edwards, R. G., *Nature*, **218**, 22 (1968).

也适用于人类，那么和年龄较小的妇女相比，年龄较大的妇女的卵子会更容易出现与染色体二倍性相关的异常。由此可以推测，在年龄较大的妇女中出现三倍体和多倍体胚胎从而产生先天愚型后代和流产胎儿的可能性也会更大。

由体外受精得到的人类胚胎要在临床上得到应用尚需要以下两项相关技术的进一步发展，即获取卵泡卵母细胞的技术以及将受精卵重新植入人体内输卵管和子宫的技术。利用腹腔镜技术进行的初步研究表明，通过对体内正在成熟的卵泡进行穿刺可以从卵巢中获取卵母细胞，也可以通过输卵管切除术取出一些迁移到输卵管中的卵子（斯特普托和爱德华兹尚未发表的研究结果）。医疗设备和技术的进步或许会带来更好的结果，还能避免采用剖腹手术。

在此特别要感谢奥斯汀教授给我们的鼓励和建议，以及阿伯利博士、加勒特博士和戴维斯博士的帮助。我们中的一位作者（爱德华兹）受到了福特基金会的资助，另一位（巴维斯特）受到了医学研究理事会的资助。还要感谢莫里斯教授、罗斯博士、博顿利博士和马卡姆博士给我们提供了卵巢组织。

（彭丽霞 翻译；王敏康 审稿）

Remains of Hominidae from Pliocene/ Pleistocene Formations in the Lower Omo Basin, Ethiopia

F. C. Howell

Editor's Note

The search for fossil hominids moved from Olduvai Gorge in the Rift Valley of East Africa to the Omo valley in Ethiopia—further north, and back in time. This is Clark Howell's field report on work by a team from the University of Chicago which, over two seasons, looked at sediments laid down between four and two million years ago, antedating Olduvai and with the potential to plumb further back in human evolution. Baldly, the report is a long list of teeth, tentatively assigned to one known form of australopithecine or another. There was little sign, as yet, of the palaeoanthropological revolution that finds from Ethiopia would spark in the 1970s and 1980s, with "Lucy" (*Australopithecus afarensis*) and still more ancient, primitive forms at humanity's root.

DURING two seasons of geological and palaeontological study in the lower Omo Basin, Ethiopia, a series of remains of Hominidae were recovered from deposits of Pliocene/Pleistocene age. These deposits are the Omo Beds (including the Shungura formation and the Usno formation) which are now known to range in age from >4.0 m.y. to <2.0 m.y. These discoveries have been made by an international expedition with a contingent from the University of Chicago under my leadership, and a contingent from the Museum National d'Histoire Naturel, Paris, under the leadership of C. Arambourg and Y. Coppens. The hominids are from a series of horizons rich in fossil vertebrates, especially mammals, and are of particular interest as they antedate those recovered from Olduvai Gorge, Tanzania. Hence the fossil record of Hominidae in eastern Africa is extended farther back through the earlier Pleistocene into the terminal Pliocene. This preliminary article records the specimens recovered by the Chicago contingent of the expedition.

Hominid remains have now been recovered from twelve localities in the lower Omo Valley. The temporal relationships of the hominid localities are set out in Table 1 and their positions keyed to a succession of volcanic tuffs which are useful and important marker horizons within the >500 m thick Omo Beds series. Radiometric (K/Ar) age determinations are available for: (*a*) three tuffs (*B, D, I*) in the main area of Omo Beds exposures[1]; (*b*) from a basalt underlying the fossiliferous sediments at two localities (Brown Sands and White Sands) north-east of that area (F. H. Brown, personal communication); and (*c*) from a basalt overlying fossiliferous sediments (seemingly without Hominidae) at the Yellow Sands locality situated at the south-western foot of Nkalabong Mountain at the northern end of the basin (F. H. Brown; and R. E. Leakey, K. W. Butzer, F. J. Fitch and J. A. Miller, personal communications).

在埃塞俄比亚奥莫下游盆地的上新世/更新世地层中发现的人科化石

豪厄尔

编者按

对原始人类化石的搜寻从位于东非大裂谷的奥杜威峡谷转移到了奥莫河谷——更靠北，年代也更古老。这篇由克拉克·豪厄尔撰写的调查报告记录了芝加哥大学探险队在两个季度的时间里对奥莫河谷400万年前到200万前的沉积物的考察结果，这些沉积物比奥杜威的年代更早并且有望由此进一步追溯人类的进化过程。坦率地说，这篇报告列举了大量的牙齿，作者暂时把它们分别归到了南方古猿的两个已知种类。到目前为止，来自埃塞俄比亚的发现还没有标引出远古人类的进化线，也再没有发现能够与20世纪七八十年代发现的"露西"（南方古猿阿法种）一样引人注目的更古老、更原始的人类祖先化石。

在对埃塞俄比亚奥莫下游盆地进行地质学和古生物学研究的两个季度的时间里，我们从上新世／更新世年代的堆积物中发掘出了一系列人科化石。这些堆积物就是现在已知距今400多万年前到不到200万年前的奥莫组（包括上古拉地层和乌斯诺地层）。发掘工作由我领导的芝加哥大学分队和阿朗堡、科庞领导的巴黎国立自然历史博物馆分队组成的国际探险队共同完成。这些原始人类是从一系列富含脊椎动物化石，尤其是哺乳动物化石的地层中发现的；因为它们比在坦桑尼亚的奥杜威峡谷发现的标本年代更久远，所以受到了特别的关注。由此，东非的人科化石记录可以向前提早到早更新世时期甚至延伸到上新世末期。这篇初步报告记录了由这支探险队的芝加哥分队发现的化石。

现在已经在奥莫下游河谷的12处地点发掘出了原始人类的化石。表1中标明了这些原始人类遗址的时间关系，它们的位置是揭开一系列火山凝灰岩之谜的钥匙；在500多米厚的奥莫层序列内，这些火山凝灰岩具有实用且重要的标志。我们将放射性（K/Ar）年代测定法应用于以下几种岩石：(*a*) 位于奥莫组暴露出来的主要区域的三层凝灰岩（*B*、*D*、*I*层）[1]；(*b*) 位于上述区域东北方向的两处地点（褐沙滩和白沙滩）、覆盖着含有化石的沉积物的一层玄武岩（布朗，个人交流）；(*c*) 位于盆地北端恩卡拉邦山脉西南山脚下的黄沙滩所在地、被含有化石的沉积物（看来似乎未见人科化石）所覆盖的一层玄武岩（布朗，利基、巴策、菲奇和米勒，个人交流）。

Table 1. Temporal Relations of the Pliocene/Pleistocene Hominidae Recovered from the Lower Omo Basin

Tuffs of main Omo Beds succession

Secondary	Primary	
	J	
	I	K/Ar age = 1·81–1·87 m.y. (I_2)
	H	
U		Hominid localities = *Locality* 7; *Locality* 74
	G	
T		Hominid locality = *Locality* 28
	F	
S		Hominid locality = *Locality* 26
	E	
R		Hominid localities = *Locality* 9; *Locality* 64
	D	K/Ar age = 2·37–2·56 m.y.
Q		Hominid localities = *Localities* 2, 45, 51, 54
P		White Sands and Brown Sands
	C	localities: K/Ar age = ≦3·1 m.y.
	B	K/Ar age = 3·75 m.y. (B_4)
		Yellow Sands locality: K/Ar age
	A	= >4·0 m.y.

Brown Sands and White Sands Localities

Eight hominid teeth or parts thereof were recovered from the White Sands locality (Fig. 1). These represent a minimum of two, probably three, individuals. The series comprises the following (length and breadth in mm, and length/breadth or shape index in percent, are given successively where appropriate in parentheses)

White Sands.	*W*-23.	Left P_4 (11.8×10.5; 112)
	W-508.	Right M_1 (13.25×12.2; 108)
	W-749.	Right M^2 (13.2×14.6; 90.4)
	W-752.	Right M_1 (14.1×13.0; 108.4)
	W-578.	Mesial quadrant of crown of left M_1
	W-750.	Fragment of P^3
	W-751.	Disto-lingual quadrant of P_4 crown
	W-753.	Right dm^1 (11.35×12.0; 94.5)

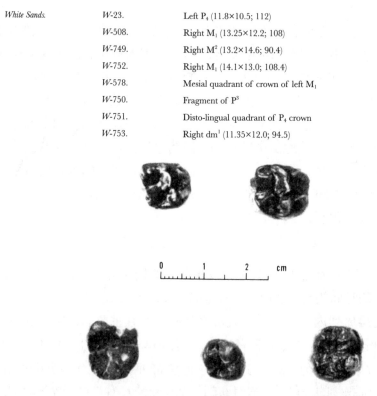

Fig. 1. Hominid teeth from White Sands locality. Above, right dm^1 (*W*-753), right M^2 (*W*-749); below, left P_4 (*W*-23), right M_1 (*W*-752), right M_1 (*W*-508).

714

表 1. 从奥莫下游盆地发掘出的上新世 / 更新世人科化石的时间关系

主奥莫层序列的凝灰岩

第二	第一	
	J	
	I	K/Ar 年代 = 181 万年 ～ 187 万年 (I₂)
	H	
U	G	原始人类遗址 = 第 7 号地点；第 74 号地点
T	F	原始人类遗址 = 第 28 号地点
S	E	原始人类遗址 = 第 26 号地点
R	D	原始人类遗址 = 第 9 号地点；第 64 号地点
Q		K/Ar 年代 = 237 万年 ～ 256 万年
P	C	原始人类遗址 = 第 2、45、51、54 号地点
		白沙滩和褐沙滩所在地：
	B	K/Ar 年代 = ≦310 万年
		K/Ar 年代 = 375 万年 (B₁)
	A	黄沙滩所在地：K/Ar 年代 = >400 万年

表 1. 时间关系（K/Ar 年代 = 181 万年 ～ 187 万年 (I₂)；原始人类遗址 = 第 7 号地点；第 74 号地点；原始人类遗址 = 第 28 号地点；原始人类遗址 = 第 26 号地点；原始人类遗址 = 第 9 号地点；第 64 号地点；K/Ar 年代 = 237 万年 ～ 256 万年；原始人类遗址 = 第 2、45、51、54 号地点；白沙滩和褐沙滩所在地：K/Ar 年代 = ≦310 万年；K/Ar 年代 = 375 万年 (B₁)；黄沙滩所在地：K/Ar 年代 = >400 万年)

褐沙滩和白沙滩所在地

从白沙滩所在地发掘出来了 8 枚原始人类的牙齿或其中的部分（图 1）。这些标本至少代表了 2 个或 3 个个体。该系列包括如下几部分（在对应位置的圆括号中依次给出用毫米表示的长度和宽度，以及用百分数表示的长宽比或称形状指数）：

白沙滩：	W-23.	左 P_4（11.8 × 10.5；112）
	W-508.	右 M_1（13.25 × 12.2；108）
	W-749.	右 M^2（13.2 × 14.6；90.4）
	W-752.	右 M_1（14.1 × 13.0；108.4）
	W-578.	左 M_1 近中侧的四分之一齿冠
	W-750.	P^3 碎片
	W-751.	P_4 远中舌侧的四分之一齿冠
	W-753.	右 dm^1（11.35 × 12.0；94.5）

0　　1　　2　　厘米

图 1. 在白沙滩所在地发现的原始人类牙齿。上排：右 dm^1（W-753），右 M^2（W-749）；下排：左 P_4（W-23），右 M_1（W-752），右 M_1（W-508）。

Eleven teeth or parts thereof were recovered from the Brown Sands locality (Fig. 2). These seem to represent a minimum of six individuals. The series comprises

Brown Sands.	*B*-27.	Right I^1
	B-14*a*.	Left P^4 (9.85×12.80; 75.4)
	B-14*b*.	Left P^4 (9.2×12.5; 73.6)
	B-4.	Left P^4 (9.0×12.5; 72. 0)
	B-23*a*.	Left P^4 (8.15×12.35; 65.9)
	B-23*b*.	Left M^1 (11.5×13.2; 87.1)
	B-39*a*.	Left P^3 (8.35×12.6; 66.2)
	B-39*b*.	Left P^4 (9.3×13.4; 60.9)
	B-39*c*.	Disto-buccal quadrant of M^1 crown
	B-20.	Lingual half of dm^2 crown
	B-28.	Distal half of dm_2 crown

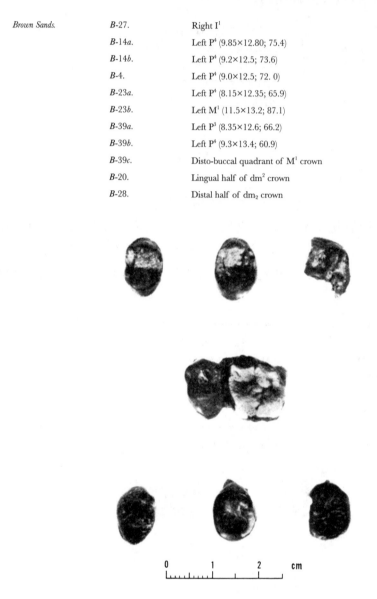

Fig. 2. Hominid teeth from Brown Sands locality. Above, left P^3, left P^4, inc. M^1 (*B*-39*a*, *b*, *c*); middle, left P^4, left M^1 (*B*-23*a*, *b*); below, left P^4 (*B*-4); left P^4 (*B*-14*b*); left P^4(?) (*B*-14*a*).

Except for several milk teeth and an upper incisor (from Brown Sands) the teeth from these two localities represent exclusively the permanent premolar and molar dentition. The complete dm^1 (*W*-753) is larger than the few known (and reported) specimens referred to either *Australopithecus africanus* (four specimens from two individuals) or to *A. robustus* (a single worn specimen). Its overall morphology and that of the incomplete dm^2 (*B*-20)

从褐沙滩所在地发掘出来了 11 枚牙齿或其中的部分（图 2）。这些标本似乎至少代表了 6 个个体。该系列包括：

褐沙滩：	B-27.	右 I^1
	B-14a.	左 P^4（9.85×12.80；75.4）
	B-14b.	左 P^4（9.2×12.5；73.6）
	B-4.	左 P^4（9.0×12.5；72.0）
	B-23a.	左 P^4（8.15×12.35；65.9）
	B-23b.	左 M^1（11.5×13.2；87.1）
	B-39a.	左 P^3（8.35×12.6；66.2）
	B-39b.	左 P^4（9.3×13.4；60.9）
	B-39c.	M^1 远中颊的四分之一齿冠
	B-20.	dm^2 齿冠的舌半侧
	B-28.	dm_2 齿冠的远舌侧

图 2. 在褐沙滩所在地发现的原始人类牙齿。上排：左 P^3、左 P^4、相关的 M^1（B-39a、b、c）；中央：左 P^4、左 M^1（B-23a、b）；下排：左 P^4（B-4），左 P^4（B-14b），左 P^4（?）（B-14a）。

除了几枚乳齿和一枚上门齿（来自褐沙滩）以外，从这两处地点得到的牙齿都属于恒前臼齿和恒臼齿。完整的乳齿 dm^1（W-753）比少数几个已知的（和已报道的）被认为是南方古猿非洲种（来自 2 个个体的 4 个标本）和南方古猿粗壮种（只有一个磨损的标本）的标本都要大。其整体形态和不完整的 dm^2（B-20）的形态都与南

717

resemble the australopithecines, while the incomplete dm$_2$ (*B*-28) has some morphological resemblance to samples from Swartkrans referred to *A. robustus*, as opposed to samples from Sterkfontein and Makapan Limeworks referred to *A. africanus*. The single upper incisor (*B*-27) has a mesiodistal length (9.2 mm) within the known australopithecine range.

The single P^3 (*B*-39*a*) is within the known size range of samples referred to *A. africanus*, but smaller than known homologues referred to *A. robustus* (or to *A. boisei*). Its shape index is within the lower range of samples from Sterkfontein and from Swartkrans. Two of the five complete P^{4}s from Brown Sands were associated with other teeth (see list earlier). All of these teeth fall within the known size range (based on eleven teeth) of samples referred to *A. africanus*, but are narrower than the known P^{4}s of samples referred to *A. robustus* (or to *A. boisei*). They are as short as, or shorter than, the smallest known specimens (of a sample of twenty teeth) of homologues referred to *A. robustus*. In the two associated specimens from Brown Sands, P^3 is smaller than P^4, not markedly so as in robust australopithecines, but more like the condition often found in *A. africanus*. The overall morphology of these teeth, particularly in the pattern of the primary fissure system and the lack of talon development, agrees most closely with specimens from Sterkfontein referred to *A. africanus*.

Only two complete upper molars are known, an M^1 (*B*-23*b*) from Brown Sands and an M^2 (*W*-749) from White Sands. The M^1 is within the known size range of homologues referred to *A. africanus*; it is shorter than known homologues referred to *A. robustus* (or to *A. boisei*), though barely within the known lower range in breadth and shape index. The M^2 is closer to the known size range of homologues referred to *A. africanus* than to those referred to *A. robustus* (or to *A. boisei*); its shape index falls within the ranges of both. The morphology of both teeth, and the incomplete M^1 (*B*-39*c*) as well, is broadly similar to that met within the Sterkfontein sample referred to *A. africanus*.

In the lower dentition only P$_4$ and M$_1$ are known. The White Sands P$_4$ (*W*-23) is unusually narrow. Its length and breadth dimensions and shape index diverge markedly from homologues referred to *A. robustus*, and also fall outside the known range of *A. africanus* (in which, however, only five specimens of P$_4$ are known). Its shape index even exceeds that of Olduvai hominid 7 (the type of "*Homo habilis*"). Both M$_1$s from White Sands (*W*-508, *W*-752) are within the range of variation of homologues referred to *A. africanus* for all dimensions. The former is smaller than the smallest known M$_1$ in the Swartkrans sample of *A. robustus*, while the latter is the same size as the smallest M$_1$ of that sample. The morphology of P$_4$ and of M$_1$ diverges from their homologues in the Swartkrans sample of *A. robustus* and approaches more closely the Sterkfontein sample of *A. africanus*.

In general, the teeth from these two localities suggest the presence of one hominid species which resembles the Sterkfontein sample referred to *A. africanus*. Most of the sample is therefore assigned to *A.* cf. *africanus*. An incomplete dm$_2$ (*B*-28) from Brown Sands and a complete dm^1 (*W*-753) from White Sands suggest that a second hominid, with resemblances in the deciduous dentition to known robust australopithecines, may also be represented.

方古猿很相像，而不完整的 dm$_2$（B-28）与来自斯瓦特克朗斯遗址的南方古猿粗壮种标本在形态上具有某种相似性，但与来自斯泰克方丹遗址和马卡潘石灰厂的南方古猿非洲种标本截然相反。唯一一枚上门齿（B-27）的近中远侧长度（9.2 毫米）处于已知的南方古猿的范围之内。

唯一一枚 P^3（B-39a）的尺寸处于南方古猿非洲种标本的已知尺寸范围之内，但要小于已知的南方古猿粗壮种（或南方古猿鲍氏种）标本。其形状指数处于斯泰克方丹和斯瓦特克朗斯标本的下限范围。在褐沙滩发现的 5 枚完整 P^4 中，有 2 枚与其余牙齿有关联（见前面的列表）。所有这些牙齿的尺寸都处于南方古猿非洲种标本的已知尺寸范围之内（由 11 枚牙齿得到的范围），但是比南方古猿粗壮种（或南方古猿鲍氏种）的已知 P^4 标本要窄一些。它们与南方古猿粗壮种的最小已知标本（20 枚牙齿中的一个标本）一样短，甚至更短。在从褐沙滩发掘出来的两个关联标本中，P^3 比 P^4 小，不如粗壮南方古猿中那么明显，而与南方古猿非洲种中常见的情况更相似。这些牙齿的整体形态，尤其是最初裂隙系统的模式以及爪部发育的欠缺，都与在斯泰克方丹发现的南方古猿非洲种标本非常吻合。

现在人们只知道两枚完整的上白齿，其中一枚是来自于褐沙滩的 M^1（B-23b），另一枚是来自于白沙滩的 M^2（W-749）。M^1 的长度落在南方古猿非洲种对应标本的已知尺寸范围之内，而比南方古猿粗壮种（或南方古猿鲍氏种）的已知标本要短一些，但是其宽度和形状指数仍落在已知的下限范围之内。与南方古猿粗壮种（或南方古猿鲍氏种）的同类标本相比，M^2 的大小更接近于南方古猿非洲种标本的已知尺寸范围；其形状指数落在二者的范围之内。这两枚牙齿的形态以及不完整的 M^1 都与属于南方古猿非洲种的斯泰克方丹标本非常相似。

在下齿系中，只有 P$_4$ 和 M$_1$ 已知。来自白沙滩的 P$_4$（W-23）异常狭窄。其长度、宽度和形状指数与南方古猿粗壮种的同类标本有很明显的差别，并且也落在了南方古猿非洲种（不过只有 5 个 P$_4$ 标本是已知的）的已知范围之外。其形状指数甚至超过了 7 号奥杜威原始人类（即"能人"类型）的对应值。来自白沙滩的两枚 M$_1$（W-508、W-752）的所有尺寸都处于南方古猿非洲种标本的变异范围之内。前者的尺寸小于斯瓦特克朗斯南方古猿粗壮种标本中已知的最小 M$_1$，而后者与该标本的最小 M$_1$ 大小相同。P$_4$ 和 M$_1$ 的形态与斯瓦特克朗斯南方古猿粗壮种的对应标本之间存在差异，而更接近于斯泰克方丹的南方古猿非洲种标本。

总之，在这两处地点发现的牙齿说明，存在着一种与斯泰克方丹南方古猿非洲种标本类似的原始人类。因此，大部分标本被归入南方古猿非洲种的名下。来自褐沙滩的一枚不完整的 dm$_2$（B-28）和来自白沙滩的一枚完整的 dm^1（W-753）表明：还可能存在着另外一种原始人类，它们具有与已知的粗壮南方古猿相似的乳齿齿系。

Localities below Omo Beds Tuff *D* and Tuff *C*

Arambourg and Coppens[2,3] have already reported a hominid mandible, without tooth crowns preserved, from a locality (their Omo 18) below Tuff *D*. Two K/Ar determinations on this tuff indicate an age of 2.37–2.56 m.y. (ref. 1). This fossiliferous zone thus seems to postdate the White Sands and Brown Sands localities, and it is certainly no older (this is also suggested by the faunal assemblages). A dozen isolated teeth have now been recovered by my contingent from one locality (2) below Tuff *C* and three localities below Tuff *D* (Fig. 3). These represent at least six or seven individuals. The series comprises

Omo Beds Locality 2.	2–79.	Incomplete $M_{2/3}$ (—; 13.8; —)
	2–89.	Left M_1 (13.7×11.5; 119.0)
Omo Beds Locality 45	45–2.	Right M_1 (12.9×12.0; 108.0)
Omo Beds Locality 54.	54–20.	Incomplete right dm_2
Omo Beds Locality 51.	†51–1.	Left M_2 (14.0×12.6; 111.0)
	*51–2.	Left M^1 (13.1×14.7; 89.0)
	†51–3.	Left M^1 (13.1×13.9; 94.2)
	‡51–4.	Left M^2 (or M^1?) (13.3×13.2; 100.7)
	*51–5.	Right I^1 (9×—; —)
	†51–6.	Incomplete right M^2
	*51–7.	Lingual half of right M (14.0×—; —)

* Younger individual.
† Older individual.
‡ Another younger individual, if 51–4 is an M^1.

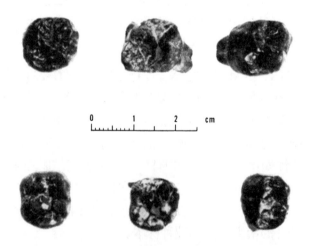

Fig. 3. Hominid teeth from localities (51, 45, 2) below Tuff *D*, main Omo Beds. Above, left (?) M^2 (51–4); left M^1 (51–3); left M^1 (51–2); below, left M_2 (51–1); right M_1 (45–2); left M_1 (2–89).

Only the trigonid area is preserved in the single deciduous lower molar (*L*54–20) and it seems to have australopithecine affinities. The anterior dentition is represented only by

奥莫组凝灰岩 D 层和凝灰岩 C 层之下的地点

阿朗堡和科庞[2,3]曾报道过一件原始人类的下颌骨，该下颌骨上没有保存下来任何齿冠，是在凝灰岩 D 层之下的一处地点（他们称之为奥莫第 18 号地点）发现的。对这种凝灰岩进行两次 K/Ar 测定的结果显示，其年龄在 237 万年～256 万年之间（参考文献 1）。因此这一含化石区的埋藏时间似乎晚于白沙滩和褐沙滩所在地的年代，并且其真实年代肯定不会比上述估计值更久远（动物区系的组成也暗示了这一点）。我的分队在 1 处位于凝灰岩 C 层之下的地点（第 2 号）以及 3 处位于凝灰岩 D 层之下的地点发现了 12 枚单独的牙齿（图 3）。这些牙齿代表了至少 6 个或 7 个个体。该系列包括：

奥莫第 2 号地点：	2–79.	不完整的 $M_{2/3}$（— ；13.8；—）
	2–89.	左 M_1（13.7×11.5；119.0）
奥莫第 45 号地点：	45–2.	右 M_1（12.9×12.0；108.0）
奥莫第 54 号地点：	54–20.	不完整的右 dm_2
奥莫第 51 号地点：	†51–1.	左 M_2（14.0×12.6；111.0）
	*51–2.	左 M^1（13.1×14.7；89.0）
	†51–3.	左 M^1（13.1×13.9；94.2）
	‡51–4.	左 M^2（或 M^1？）（13.3×13.2；100.7）
	*51–5.	右 I^1（9×— ；—）
	†51–6.	不完整的右 M^2
	*51–7.	右 M 的舌半侧（14.0×— ；—）

* 年轻个体。

† 老年个体。

‡ 如果 51–4 是 1 枚 M^1，那么它就代表另一个年轻个体。

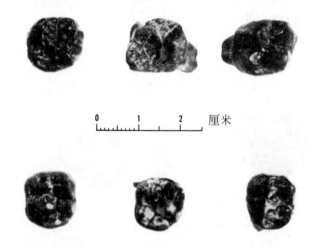

图 3. 在主奥莫组凝灰岩 D 层之下地点（第 51 号、45 号、2 号）发现的原始人类牙齿。上排：左（?）M^2（51–4），左 M^1（51–3），左 M^1（51–2）；下排：左 M_2（51–1），右 M_1（45–2），左 M_1（2–89）。

唯一的一枚乳下臼齿（$L54–20$）只保存下来了下三尖，看起来似乎与南方古猿具有亲缘关系。前面的齿系只有上中门齿作为代表（$L51–5$）。其大小与南方古猿粗

the upper medial incisor (*L51*–5). The size of this tooth is similar to both robust or gracile australopithecines; the morphology of its lingual surface, however, differs somewhat from both, and the morphology of the buccal surface is rather more like homologues referred to *A. africanus.*

The remaining teeth of this series represent upper molars (four) and lower molars (five). The first lower molars from Locality 2 (*L2*–89) and Locality 45 (*L45*–2) agree in size with homologues referred to *A. africanus*, and are as small as or smaller than the Swartkrans homologues referred to *A. robustus.* The second lower molar from Locality 51 (*L51*–1) is substantially smaller than its homologue in the Swartkrans sample of *A. robustus*, and slightly smaller than its homologue in the Sterkfontein and Makapan Limeworks samples referred to *A. africanus.* In respect to shape and particularly details of their fissure pattern and occlusal surface morphology these lower molars, and the incomplete specimens from Locality 2 (*L2*–79) and Locality 51 (*L51*–7) as well, have resemblances to lower molars from South African sites referred to *A. africanus.*

The first upper molars from Locality 51 are within the known size range of both *A. africanus* and *A. robustus.* The second upper molar (*L51*–4) from that locality is smaller than known homologues referred to *A. robustus;* its length falls within the known *A. africanus* range, though it is rather narrower. The morphology of these upper molars is clearly australopithecine, and there are some specific resemblances to the *A. africanus* sample from Sterkfontein. The sample from this horizon is therefore assigned to *A.* cf. *africanus.*

Localities below Omo Beds Tuff *E*

Hominid remains have been recovered from only two localities between Tuffs *D* and *E*. These represent only three isolated teeth (Fig. 4). They are

Omo Beds Locality 64.	64–2.	Left dm$_2$ (13.8×12.4; 111.2)
Omo Beds Locality 9.	9–11.	Incomplete right M$_3$ (>15.7×>14.0; >112.1)
	9–12.	Left M^3 (13.4×17.6; 76.1)

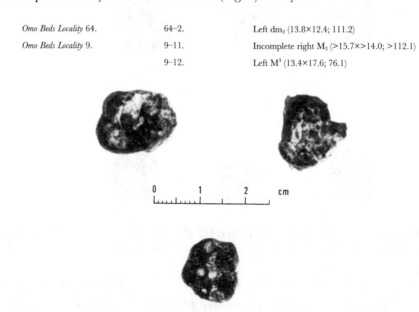

Fig. 4. Hominid teeth from localities (9, 64) below Tuff *E*, main Omo Beds. Above, left M^3 (9–12); right M$_3$ (9–11); below, left dm$_2$ (64–2).

壮种和纤细种都很相似；但其舌面形态与这两者稍有不同，颊面形态与南方古猿非洲种的同类标本更为相像。

该系列的其余牙齿还有上臼齿（4 枚）和下臼齿（5 枚）。在第 2 号地点（L2–89）和第 45 号地点（L45–2）发现的第一下臼齿的尺寸与南方古猿非洲种的同类标本一致，而与来自斯瓦特克朗斯的南方古猿粗壮种标本一样小，或者更小一些。来自第 51 号地点（L51–1）的第二下臼齿比来自斯瓦特克朗斯的南方古猿粗壮种标本小很多，而比来自斯泰克方丹和马卡潘石灰厂的南方古猿非洲种标本稍小一点。由形状，尤其是裂隙模式细微之处和咬合面形态，可以看出：这些下臼齿以及在第 2 号地点（L2–79）和第 51 号地点（L51–7）发现的不完整标本都与来自南非遗址的南方古猿非洲种的下臼齿具有相像之处。

在第 51 号地点发现的第一上臼齿的大小与南方古猿非洲种和南方古猿粗壮种的已知尺寸范围均相符。来自该地点的第二上臼齿（L51–4）比南方古猿粗壮种的已知标本小；其长度落在已知的南方古猿非洲种的范围之内，但是宽度要更小一些。这些上臼齿的形态显然对应于南方古猿，并与来自斯泰克方丹的南方古猿非洲种标本具有某些特定的相似性。因此从该地层得到的这一标本被归入南方古猿非洲种。

奥莫组凝灰岩 E 层之下的地点

仅在凝灰岩 D 层和 E 层之间的两处地点发现了原始人类的化石。这些标本只包括 3 枚单独的牙齿（图 4）。它们是：

奥莫第 64 号地点： 64–2. 左 dm$_2$（13.8 × 12.4；111.2）
奥莫第 9 号地点： 9–11. 不完整的右 M$_3$（>15.7 × >14.0；>112.1）
9–12. 左 M^3（13.4 × 17.6；76.1）

0　　1　　2　厘米

图 4. 在主奥莫组凝灰岩 E 层之下地点（第 9 号、64 号）发现的原始人类牙齿。上排：左 M^3（9–12），右 M$_3$（9–11）；下排：左 dm$_2$（64–2）。

723

The dm$_2$ is exceptionally large, whether comparison is made with other deciduous teeth from the Omo Beds series or with known australopithecines. It is as large as the largest dm$_2$s from the Swartkrans sample (thirteen teeth) referred to *A. robustus*, and larger than two specimens from Kromdraai. It is quite outside the known range of variation of samples referred to *A. africanus* (six teeth, two each from Sterkfontein, Makapan Limeworks, and Taungs, the last from a single individual). The tooth shows australopithecine affinities, with its general form like that of the Taungs child and most of its morphological details comparable with these described for the Swartkrans sample referred to *A. robustus*.

The upper and lower third molars from Locality 9 probably represent only one individual. Both teeth are large. The M^3 is in the range of samples referred to either *A. africanus* or *A. robustus*, though it is smaller than Olduvai hominid 5 referred to *A. boisei*. Its shape index and occlusal morphology agree best with specimens referred to *A. robustus* or *A. boisei*. The M$_3$, which lacks some enamel on its lingual and mesial margins, was also quite large. Its estimated dimensions are within the range for homologues referred to *A. robustus* and in the upper part of the range of specimens referred to *A. africanus*. Most of its morphological features can be seen in specimens of robust australopithecines from Swartkrans and from Peninj (Natron basin). The size of the deciduous molar and its morphology and the shape indices and morphological details of the two molars are similar to homologues which have been referred to the robust australopithecines *A. robustus* and *A. boisei*, but specific assignment of these specimens must wait pending results of studies by P. V. Tobias and others on the taxonomy of known robust australopithecines.

Locality below Omo Beds Tuff F

Two permanent teeth have been recovered from a locality (Locality 26) between Tuff *F* and Tuff *E* (Fig. 5), probably representing two different individuals. The specimens are

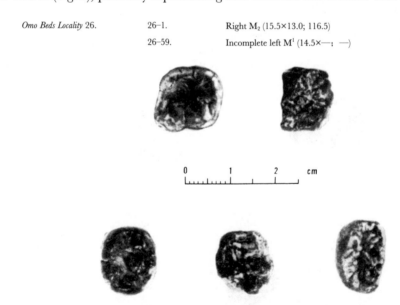

| *Omo Beds Locality* 26. | 26–1. | Right M$_2$ (15.5×13.0; 116.5) |
| | 26–59. | Incomplete left M^1 (14.5×—; —) |

Fig. 5. Hominid teeth from localities below Tuff *F* (26) and below Tuff *G* (28), main Omo Beds. Above, right M^2 (28–58); left M^1 (26–59); below, right M$_2$ (26–1); right M$_2$ (28–31); right M$_3$ (28–30).

724

　　无论是与来自奥莫层序列的其他乳齿相比，还是与已知的南方古猿相比，这个 dm_2 都格外大。它同斯瓦特克朗斯南方古猿粗壮种标本（13 枚牙齿）中最大的那些 dm_2 一样大，比来自克罗姆德拉伊的两个标本要大一些。其大小与南方古猿非洲种标本（6 枚牙齿，两两分别来自斯泰克方丹、马卡潘石灰厂和汤恩，来自汤恩的标本属于同一个个体）的已知变异范围之间有很大的差距。这枚牙齿显示出与南方古猿具有亲缘关系，总体形状与在汤恩发现的幼儿标本类似，且大部分形态细节与对斯瓦特克朗斯南方古猿粗壮种标本的描述相符。

　　在第 9 号地点发现的上、下第三臼齿可能只代表一个个体。这两枚牙齿都很大。尽管 M^3 比归属于南方古猿鲍氏种的 5 号奥杜威原始人类小，但尚处于南方古猿非洲种或南方古猿粗壮种标本的范围之内。其形状指数和咬合形态与南方古猿粗壮种或鲍氏种标本非常吻合。在舌缘和近中缘缺少珐琅质的 M_3 也很大。估计其大小落在南方古猿粗壮种对应标本的范围之内，位于南方古猿非洲种标本的上限。其大部分形态特征与来自斯瓦特克朗斯和佩宁伊（纳特龙盆地）的粗壮南方古猿标本一致。乳臼齿的大小和形态以及这两枚臼齿的形状指数和形态细节都与被归为粗壮南方古猿的南方古猿粗壮种和南方古猿鲍氏种相似，但是这些标本的具体分类要等到托拜厄斯等人对已知粗壮南方古猿的分类学研究有了结果之后才能定夺。

奥莫组凝灰岩 F 层之下的地点

　　在凝灰岩 F 层和凝灰岩 E 层之间的一处地点（第 26 号地点）发现的两枚臼齿（图 5），很可能分属于两个不同的个体。这两个标本是：

奥莫第 26 号地点：	26–1.	右 M_2（15.5×13.0；116.5）
	26–59.	不完整的左 M^1（14.5×—；—）

0　　　1　　　2　　厘米

图 5. 在主奥莫组凝灰岩 F 层之下地点（第26号）和凝灰岩 G 层之下地点（第28号）发现的原始人
上排：右 M^2（28–58），左 M^1（26–59）；下排：右 M_2（26–1），右 M_2（28–31），右 M_3（28–30）。

The complete M_2 crown is elongate, its length falling within the range of homologues referred to both *A. africanus* and *A. robustus*. It is slightly narrower, however, than homologues of either; its shape index is at the upper end of the *A. robustus* range, and is comparable with that of Olduvai hominid 7 (type of "*H. habilis*"). The morphology of this tooth resembles some specimens from the Sterkfontein sample referred to *A. africanus*. The incomplete M^1, which lacks small parts of the mesio-buccal and disto-lingual margins of the crown, is clearly australopithecine in its morphology. Both specimens are reasonably assigned to *A.* cf. *africanus*.

Locality below Omo Beds Tuff G

Three hominid teeth have been recovered from one locality (Locality 28) between Tuff *F* and Tuff *G* (Fig. 5). These seem to represent two individuals, one represented by an upper molar and the other by two lower molars. They are

Omo Beds Locality 28.	28–30.	Right M_3 (16.7×12.4; 135.0)
	28–31.	Right M_2 (ca. 15.0×13.0; ca. 115.0)
	28–58.	Worn right M^2 (15.5×14.0; 110.7)

The upper molar ($L28$–58), although very worn, is clearly australopithecine. This tooth has a very high length/breadth index, substantially above the range of teeth referred to *A. robustus* and even somewhat above the range of teeth referred to *A. africanus*. The M_2 ($L28$–31) has a shape index comparable with its homologue from Locality 26 ($L26$–1) and with Olduvai hominid 7. Its dimensions also accord well with the lower part of the known range of specimens referred to *A. africanus*. The M_3 ($L28$–30) is a long tooth, within the range of variation of specimens referred to both *A. africanus* and *A. robustus*. This tooth is very narrow, however, so that its shape (length/breadth) index is unusually high and outside the known range of specimens referred to either *A. africanus* or *A. robustus*. Olduvai hominid 7 (type of "*H. habilis*") lacks an M_3, but this tooth is preserved in a paratype specimen, Olduvai hominid 4 from site *MKI*. It is interesting to note that M_3 is shorter and slightly wider than the Locality 28 homologue, and its shape index is nearly twenty points lower. The morphology of both these lower molars from Locality 28 is similar to homologues from Sterkfontein referred to *A. africanus*. The specimens are therefore assigned to *A.* cf. *africanus*.

Localities above Omo Beds Tuff G

Hominid mandibles of australopithecine type have been recovered from two localities above Tuff *G* (and below Tuff *H*, although this tuff is hardly exposed in the northern sector of the main Omo Beds). Locality 7 (specimen No. *L7*–125) has yielded a complete mandibular corpus, lacking both ascending rami, with all permanent teeth except for three incisors. Locality 74 (specimen No. *L74*–21) has yielded a right mandibular corpus, without ascending ramus, preserving C, P_4 and the roots of P_3 and M_1. The second lower molar was evidently largely, or wholly, erupted to judge from the anterior part of its preserved socket.

这枚完整 M_2 的齿冠是细长的，其长度既落在南方古猿非洲种标本的范围之内，也落在南方古猿粗壮种标本的范围之内。不过其宽度比上述两者的同类标本略窄；其形状指数处于南方古猿粗壮种标本范围的上限，与 7 号奥杜威原始人类（"能人"类型）的相当。这枚牙齿的形态与来自斯泰克方丹的南方古猿非洲种标本中的某些标本很相像。另外一枚是缺少小部分近中颊侧和远中舌侧边缘齿冠的不完整 M^1，它在形态上显然应归属于南方古猿。有理由认为这两个标本都属于南方古猿非洲种。

奥莫组凝灰岩 G 层之下的地点

在凝灰岩 F 层和凝灰岩 G 层之间的一处地点（第 28 号地点）发现了 3 枚原始人类的牙齿（图 5）。这些牙看似代表两个个体，其中一枚上臼齿属于一个个体，另有两枚下臼齿属于另一个个体。它们是：

奥莫第 28 号地点：	28–30.	右 M_3（16.7×12.4；135.0）
	28–31.	右 M_2（约 15.0×13.0；约 115.0）
	28–58.	磨损的右 M^2（15.5×14.0；110.7）

虽然上臼齿（L28–58）磨损得很厉害，但是仍然可以清楚地判断出是南方古猿的。这枚牙齿的长宽比指数非常高，远远超出了南方古猿粗壮种的牙齿范围，甚至比南方古猿非洲种的牙齿范围还要略高一些。M_2（L28–31）的形状指数与在第 26 号地点发现的同类标本（L26–1）以及 7 号奥杜威原始人类的相当。它的尺寸也与南方古猿非洲种标本的已知范围的下半部分吻合得很好。M_3（L28–30）是一枚长牙，其大小既落在南方古猿非洲种标本的变异范围之内，也处于南方古猿粗壮种标本的变异范围之间。但是这枚牙齿宽度很小，所以它的形状指数（长宽比）特别高，落在了南方古猿非洲种或南方古猿粗壮种标本的已知范围之外。7 号奥杜威原始人类（"能人"类型）缺少一枚 M_3，但是在来自 MKI 遗址的 4 号奥杜威原始人类中，这枚牙齿以副模标本的形式被保存了下来。有趣的是：该 M_3 比在第 28 号地点发现的同类标本更短，并且还要略宽一些，其形状指数低了将近 20 个点。来自第 28 号地点的这两枚下臼齿的形态都与斯泰克方丹的南方古猿非洲种标本类似。因此这些标本被归入南方古猿非洲种。

奥莫组凝灰岩 G 层之上的地点

在凝灰岩 G 层之上（位于凝灰岩 H 层之下，但在主奥莫组北部，这种凝灰岩很少会暴露出来）的两处地点发现了属于南方古猿类型的原始人类下颌骨。在第 7 号地点（标本编号 L7–125）出土了一件完整的下颌体，但缺少两侧的上升支，除 3 枚门齿以外其余所有恒齿都在。在第 74 号地点（标本编号 L74–21）发现了一件右下颌体，也缺少上升支，但 C、P_4 以及 P_3 的牙根和 M_1 的牙根被保存了下来。从其留存牙槽的前面部分判断，其第二下臼齿显然有很大一部分，甚至是全部都已经萌生出来了。

The hemimandible (*L*74–21) has a very deep, robust body (Figs. 6 and 7). The height of the body exceeds that of all previously known robust australopithecines, except for one value for the big adult *SK*12 specimen from Swartkrans. The thickness of the body is only exceeded in one dimension by that specimen and the Peninj (Natron basin) specimen. The symphysial height is great, at the upper end (or in excess of) the known robust australopith range, and this is equally true for the thickness values of this region. Both C and P_4 are large-crowned teeth. The length (8.8 mm) and breadth (9.7 mm) of C exceeds known values for the Swartkrans sample referred to *A. robustus* but is within the range of the Sterkfontein and Makapan Limeworks samples referred to *A. africanus*. The morphology of this tooth resembles fairly closely, however, that of the robust australopithecines. Only the roots of P_3 are preserved, but its length (approx. 12.8 mm) can be estimated from the contact facets on C and P_4. This value is greater than the known range of the Swartkrans sample referred to *A. robustus*, and that of the Peninj specimen, and also outside the 95 percent confidence limits of the former sample. This extreme elongation and apparent molarization of P_3 are the most distinctive and hitherto unknown features in such early hominids. The size of P_4 is within the upper part of the known range of homologues referred to *A. robustus*. Its length (13.0 mm) exceeds that of specimens referred to *A. africanus*, and its breadth (13.75 mm) is barely within that range. It has a higher shape index (94.4) than the range recorded for such homologues. The crown is strongly molarized, with expanded talonid, its overall morphology being similar to that of the Peninj individual or some specimens (for example, *SK*9) from Swartkrans.

0 1 2 3 4 5 cm

Fig. 6. Superior (left) and inferior (right) views of hominid mandible (*L*74–21) from Locality 74, above Tuff *G*, main Omo Beds.

　　该半侧下颌骨（*L*74–21）具有一个很深且强壮的下颌体（图 6 和图 7）。该下颌体的高度超过了除斯瓦特克朗斯的大型成年 *SK*12 标本之外其他所有之前已知的粗壮南方古猿的对应值。该下颌体的厚度只在一个方向上没有 *SK*12 标本和佩宁伊标本（纳特龙盆地）的大。其联合部高度很大，处于（或超过）已知粗壮南方古猿对应范围的上限；这一区域的厚度值也是如此。C 和 P₄ 都是大齿冠牙齿。C 的长度（8.8 毫米）和宽度（9.7 毫米）均超过斯瓦特克朗斯南方古猿粗壮种标本的已知数值，但是落在了来自斯泰克方丹和马卡潘石灰厂的南方古猿非洲种标本的范围之内。然而，这枚牙齿的形态与粗壮南方古猿非常相像。虽然 P₃ 只有根部被保存了下来，但是可以通过其与 C 和 P₄ 的接触面来估计它的长度（约 12.8 毫米）。这个值高于斯瓦特克朗斯南方古猿粗壮种标本和佩宁伊标本的已知范围，并且还落在了前者的 95% 置信区间之外。P₃ 的极度伸长和明显的白齿化特征非常显著，这是迄今在如此早期的原始人类中还未见到过的。P₄ 的尺寸处于南方古猿粗壮种标本的已知范围的上半部分。其长度（13.0 毫米）超过了南方古猿非洲种标本，其宽度（13.75 毫米）也几乎不在南方古猿非洲种的范围之内。它的形状指数（94.4）高于同类标本范围的已有记录。齿冠白齿化严重，具有扩展的下白齿远中部，其整体形态与佩宁伊个体或来自斯瓦特克朗斯的一些标本（例如 *SK*9）相似。

0　1　2　3　4　5　厘米

图 6. 在主奥莫组凝灰岩 *G* 层之上第 74 号地点发现的原始人类下颌骨（*L*74–21）的嚼面视（左图）和腹面视（右图）。

Fig. 7. Lateral (above) and lingual (below) views of hominid mandible (L74–21) from Locality 74, above Tuff G, main Omo Beds.

The Locality 7 mandible (Figs. 8, 9 and 10) clearly represents a very robust australopithecine. Compared with seven specimens from Swartkrans and that from Peninj, the Locality 7 individual exceeds all in most dimensions of the mandibular body, including height, thickness and overall robusticity, and in the size and robusticity of the symphysial region. The body is exceptionally deep and the roots of the canines, premolars and molars are very long and robust. The incisors, however, appear to have been shorter-rooted and less robust. The superior and inferior transverse symphysial tori are very robust, the latter exceptionally so, and the genioglossal fossa is extensive and deep. Mental spines are preserved on the postero-inferior surface of the inferior transverse torus. The alveolar planum exceeds far back to the midlevel of the last premolar. The digastric fossae are extensive and situated on the inner surfaces (rather than the inferior margin) of the body. The dental arcade is slightly parabolic with the premolar–molar series diverging slightly posteriorly from the nearly transversely aligned incisors and canines. The lingual surfaces of the mandibular bodies converge anteriorly to produce a A-shape to the inner mandibular contour, a consequence of the excessive thickening of their upper portion (comparable with the alveolar prominence) beginning just distal of the level of the second molar and continuous anteriorly as the alveolar planum.

730

0 1 2 3 4 5 厘米

图 7. 在主奥莫组凝灰岩G层之上第74号地点发现的原始人类下颌骨（L74–21）的侧面视（上图）和舌面视（下图）。

第 7 号地点的下颌骨（图 8、图 9 和图 10）显然代表了一个非常粗壮的南方古猿。与斯瓦特克朗斯的 7 个标本以及佩宁伊的标本相比，第 7 号地点处个体下颌体的大部分尺寸，包括高度、厚度和整体粗壮性以及联合区的大小和粗壮性，都超过了所有其他个体。下颌体格外深，犬齿、前白齿和白齿的齿根很长、很粗壮。但是门齿的根部似乎比较短，粗壮性也稍差些。上、下横向联合圆枕都很粗壮，尤其是后者，颏舌肌窝又宽又深。颏棘存于下横圆枕的后下方面上。牙槽平面远远超过了最后一枚前白齿的中间位置。二腹肌窝很宽阔，位于下颌体的内表面（而不是下缘）之上。齿弓略呈抛物线形，前白齿－白齿系列比几乎横向排列的门齿和犬齿稍微偏后一点儿。下颌体的舌面向前集中从而与下颌内廓形成一个 A 字形，这是它们的上半部分（相当于牙槽突出）发生大范围变厚的结果，这里的上半部分仅仅是从第二白齿水平的远端开始，然后继续向前而成为牙槽平面。

Fig. 8. Right lateral view of hominid mandible (*L7–125*) from Locality 7, above Tuff *G*, main Omo Beds.

Fig. 9. Left lateral view of hominid mandible (*L7–125*) from Locality 7, above Tuff *G*, main Omo Beds.

0 1 2 3 4 5 厘米

图 8. 在主奥莫组凝灰岩 *G* 层之上第 7 号地点发现的原始人类下颌骨（*L*7–125）的右侧视。

0 1 2 3 4 5 厘米

图 9. 在主奥莫组凝灰岩 *G* 层之上第 7 号地点发现的原始人类下颌骨（*L*7–125）的左侧视。

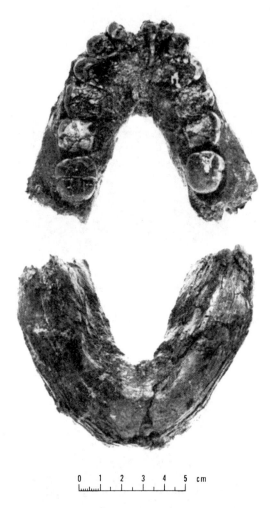

0 1 2 3 4 5 cm

Fig. 10. Superior (above) and inferior (below) views of hominid mandible (*L7–125*) from Locality 7, above Tuff *G*, main Omo Beds.

The incisors are diminutive, the canines small, and the premolars and molars extremely large. The preserved right I₂ is smaller than that of any robust australopithecine yet known. The length (7.8 mm) of the left C is within the known robust australopithecine range, while its breadth (9.6 mm) is slightly above that range. The lengths of P₃ (10.4 and 11.2 mm) and of P₄ (11.7 mm) are within the known robust australopithecine range, but both teeth are exceptionally wide (P₃=approximately 17.5 mm, P₄=18.9 mm) and outside the known range of such samples. Hence these teeth have low shape indices, a condition also found in Olduvai hominid 5, referred to *A. boisei*. M₁ is both wider (18.7 mm) and longer (approximately 16.8 mm) than known homologues of robust australopithecines, and hence has a lower shape (length/breadth) index (approximately 89.8). The length (approximately 16.2 mm) of M₂ is within the range of robust australopithecine homologues, but the width (18.0 mm) exceeds that range, and the shape index (approximately 90.0) is correspondingly lower. The length (approximately 18.2 mm),

0　1　2　3　4　5　厘米

图 10. 在主奥莫组凝灰岩 *G* 层之上第 7 号地点发现的原始人类下颌骨（*L7-125*）的嚼面视（上图）和腹面视（下图）。

　　门齿小得出奇，犬齿也小，而前臼齿和臼齿却非常大。保存下来的右 I_2 比目前已知的所有粗壮南方古猿标本都要小。左 C 的长度（7.8 毫米）处于已知的粗壮南方古猿的范围之内，而其宽度（9.6 毫米）则略微超过了粗壮南方古猿的范围。P_3 的长度（10.4 毫米和 11.2 毫米）和 P_4 的长度（11.7 毫米）都处于已知的粗壮南方古猿的范围之内，但是这两枚牙齿格外宽（P_3 = 约 17.5 毫米，P_4 =18.9 毫米），这一数值落在了此类标本的已知范围之外。因此这些牙齿的形状指数偏低，该现象也会在属于南方古猿鲍氏种的 5 号奥杜威原始人类中被观察到。M_1 比粗壮南方古猿的已知标本更宽（18.7 毫米），也更长（约 16.8 毫米），因此形状指数（长宽比）（约 89.8）较低。M_2 的长度（约 16.2 毫米）处于粗壮南方古猿标本的范围之内，但是宽度（18.0 毫米）却超过了它的范围，所以形状指数（约 90.0）相应较低。M_3 的长度（约 18.2

735

breadth (14.8 mm) and shape index (approximately 123.6) of M_3 fall within the known range of robust australopithecine homologues. The details of the pattern of the occlusal surface of M_3 are sufficiently well preserved to indicate a close resemblance to homologues referred to robust australopithecines.

The crowns of the premolar and molar teeth are substantially worn. Dentine is exposed across the full width of P_3, and on the buccal side of P_4. The occlusal surfaces of M_1 and of M_2 have extensive dentine exposure except for the marginal rims of enamel. The surface of M_2 is worn deeply concave, even more so than that of M_1. The wear tends to be greatest on the lingual portion of the occlusal surface, as is frequently (but not invariably) so in individuals considered to represent robust australopiths. The mandible and dentition of this individual indicate that it represents a very robust variety of australopith in many respects comparable with hominid 5 recovered from Olduvai Bed I, site *FLK* I, and referred to *A. boisei*.

The Omo Beds of the lower Omo basin are now known to span a substantial range of Pliocene/Pleistocene time. Remains of Hominidae from a range of time hitherto largely unknown have now been recovered from a succession of fossiliferous horizons within this series of beds. Hominids known reasonably certainly from this general range of late Cenozoic time have included up to now a distal left humerus from Kanapoi, south-east Turkana[4], and a right temporal bone from the Upper Fish Beds of the Chemeron Beds series, Baringo basin[5]. In each case, the faunal evidence suggests broad equivalence with the earliest range of time represented in the oldest sediments of the Omo Beds Series, that is, probably between 3 and 4 million years.

A robust form of australopithecine, tentatively referred to *A.* cf. *boisei*, is known from the massive mandible from Locality 7 with a geological age of approximately 2.0 m.y. Another hemimandible from this same time range also has robust australopithecine affinities; but it is evidently unique in the extreme enlargement and presumably attendant molarization of the anterior premolar. For the moment, it is best treated as *Australopithecus* sp. Some teeth from earlier horizons, including specimens from Localities 9 and 64, as well as other localities worked by the French contingent of the expedition (Y. Coppens, personal communication), also have robust australopithecine resemblances. This suggests that antecedents of robust australopithecines extend back at least another half million, and probably more, years. The series of isolated teeth from nine other localities, ranging in age between about 3 and 2 million years, generally diverge dimensionally and/or morphologically from their homologues usually attributed to robust australopithecines. Their strongest resemblances are with specimens customarily referred to the small australopithecine, *A. africanus.* On the basis of present evidence it is not unreasonable to refer these specimens to *A.* cf. *africanus.*

If these attributions are confirmed, then the hominid samples from the Omo Beds would indicate the coexistence of (at least) two australopithecine taxa through much of the range of Pliocene/Pleistocene time. And this would have been the case not only in the Omo

736

毫米）、宽度（14.8 毫米）和形状指数（约 123.6）都落在粗壮南方古猿对应标本的已知范围之内。M_3 咬合面形式的细部保存状况非常好，足以表明其与粗壮南方古猿的同类标本非常相似。

前臼齿和臼齿的齿冠磨损得很严重。在 P_3 的整个宽度上以及 P_4 的颊侧都暴露出了牙本质。除了珐琅质边缘上的一圈以外，M_1 和 M_2 咬合面的其余部分都出现了牙本质的大面积暴露。M_2 的表面被磨蚀成很深的凹坑，甚至比 M_1 还要严重。在咬合面的舌侧部分，磨蚀似乎是最大的，这和被认为代表了粗壮南方古猿的个体经常（但并非必然）发生的情况一样。从这一个体的下颌骨和齿系可以看出：它代表的是一种非常粗壮的南方古猿类型，在很多方面类似于来自奥杜威第 I 层 *FLK* I 遗址的 5 号原始人类，并且被认为属于南方古猿鲍氏种。

现在人们已经知道奥莫下游盆地的奥莫组跨越了上新世 / 更新世时期的一段相当长的时间。现在已经从这组地层里的一系列化石层中发现了迄今为止鲜为人知的一段时期的人科化石。到目前为止，有理由认为肯定来自新生代晚期这一大致范围的原始人类化石包括：来自图尔卡纳东南卡纳波伊的左肱骨远端 [4] 以及来自巴林戈盆地科莫龙地层序列的上菲什组的右颞骨 [5]。在每个例子中，来自动物群的证据都表明，其与奥莫层序列最古老沉积物所代表的最早的时间范围基本一致，也就是说，这些化石的年代可能介于 300 万年前到 400 万年前之间。

南方古猿中的一种粗壮类型，现在暂时被归为南方古猿鲍氏种，是根据在第 7 号地点发现的这件所处地质年代约为 200 万年前的巨型下颌骨得到的。另一件来自同一时间段的半侧颌骨也与粗壮南方古猿有亲缘关系；不过，第一前臼齿极度放大和可能与此相伴的臼齿化作用显然是它的独特之处。目前，最好的处理方法是把它归入南方古猿类。在早期地层中发现的牙齿，包括来自第 9 号地点和第 64 号地点的标本以及由本探险队法国分队（科庞，个人交流）从其作业的其他地点得到的标本，也都与粗壮南方古猿相近。这表明粗壮南方古猿祖先的出现年代至少还要再提前 50 万年，甚至还可能提前更长的时间。在其他 9 处地点发现的单独牙齿系列，年代介于约 300 万年前到 200 万年前之间，其大小和 / 或形态都与通常被归入粗壮南方古猿的同类标本迥然不同。与它们最相近的是习惯上被看作是小型南方古猿，即南方古猿非洲种的标本。从现有的证据来看，将这些标本暂时归入南方古猿非洲种并不是没有道理。

如果这些属性能够得以证实，那么在奥莫组发现的原始人类标本将表明：在上新世 / 更新世时代的大部分时间里，有（至少）两类南方古猿类群共存。不仅在奥

area, but even in broadly similar habitats, to judge from available palaeo-environmental data. The respective ecological niches of these creatures, however, remain essentially unknown.

The work of the Omo Research Expedition has been greatly helped by the full cooperation of the Governments of Ethiopia and Kenya. In particular, I thank Ato Kibbede Mikael, Minister of the Government of Ethiopia, for his interest and encouragement. In Nairobi, Dr. and Mrs. L. S. B. Leakey, and other members of the staff of the Centre for Prehistory and Palaeontology, have continually offered valuable assistance and hospitality. My work has been supported by grants-in-aid from the US National Science Foundation and from the Wenner–Gren Foundation for Anthropological Research.

(**223**, 1234-1239; 1969)

F. Clark Howell: Department of Anthropology, University of Chicago.

Received June 17, 1969.

References:
1. Howell, F. C., *Nature*, **219**, 567 (1968).

2. Arambourg, C., and Coppens, Y., *CR Acad. Sci. Paris*, **265**, 589 (1967).

3. Arambourg, C., and Coppens, Y., *S. Afr. J. Sci.*, **64**, 58 (1968).

4. Patterson, B., and Howells, W. W., *Science*, **156**, 64 (1967).

5. Martyn, J., and Tobias, P. V., *Nature*, **215**, 476 (1967).

莫地区如此，而且从可得到的古环境数据判断，甚至在与此大体相似的其他栖息地也是如此。然而，这些生物各自的生态位实质上还是未知的。

奥莫研究探险队的工作在埃塞俄比亚政府和肯尼亚政府的通力合作下得到了极大的帮助。特别是要感谢埃塞俄比亚政府大臣阿托·基比德·米卡埃尔对我们的关注和鼓励。在内罗毕，利基博士及其夫人以及史前考古学和古生物学中心的其他工作人员一直在为我们提供宝贵的支持和热情的款待。我的工作受到了来自美国国家科学基金会和来自温纳－格伦基金会人类学研究基金的资助。

（刘皓芳 翻译；冯兴无 审稿）

New Finds at the Swartkrans Australopithecine Site

C. K. Brain

Editor's Note

The site of Swartkrans, near Sterkfontein, had been opened up by Broom and Robinson, who had discovered abundant remains of a robust australopithecine as well as a more advanced hominid, at first called *Telanthropus*, later referred to *Homo erectus*. This report from the site by C. K. Brain contains descriptions of stone tools similar to those found at Olduvai Gorge in East Africa, as well as a dramatic and grisly discovery—an australopithecine skull pierced by the canine teeth of a leopard. Brain re-interpreted at least some of the Swartkrans site not as a hominid living space, but the den of leopards, containing the remains of some of their meals.

SWARTKRANS, one of the five South African australopithecine cave sites, has produced remains of more than sixty individuals of the large hominid, *Paranthropus robustus*, and is the only cave site where fossils of *Homo* have been found in direct association with those of australopithecines.

The new investigations described here have clarified a number of the important Swartkrans issues. A new series of hominid fossils has been discovered (including two vertebrae which will be described in detail by Dr. J. T. Robinson[1]), and it has been possible to associate indisputable stone culture with the hominid remains. The tools have been critically examined by Dr. Mary Leakey, who has compared them with an equivalent series from Olduvai Gorge in Tanzania and has found striking similarities with the assemblage from Bed II of that site[2]. In the absence of absolute dating for the Swartkrans fossils, cultural comparisons of this sort assume particular significance. Unexpected new information has come to light on the nature of the second hominid at Swartkrans[3], while a study of the bone accumulation as a whole has led me to conclude that the australopithecine remains, together with many of the other fossils, probably represent food remains of carnivores, especially of leopards.

Swartkrans is situated in the Sterkfontein valley, approximately 8 miles north of Krugersdorp, Transvaal. It was first scientifically investigated in 1948 by the late Dr. Robert Broom and By Dr. J. T. Robinson, who transferred the Transvaal Museum's operations to the cave from Sterkfontein, on the opposite side of the valley. Excavation of *in situ* breccia continued throughout 1949 and resulted in the recovery of about 3,000 fossils, including numerous *Paranthropus* specimens and remains of a second hominid, named *Telanthropus capensis*[4], but subsequently reclassified as *Homo erectus*[5].

在斯瓦特克朗斯南方古猿遗址的新发现

布雷恩

编者按

靠近斯泰克方丹的斯瓦特克朗斯遗址是由布鲁姆和鲁滨逊发掘的，他们在那里发现了一种粗壮型南方古猿的大量化石以及一种更为高等的原始人类，这种原始人类起初被命名为远人，后来被归入直立人。布雷恩在这篇关于该遗址的报告中不仅谈到了一个不可思议的发现——一件曾被豹子的犬齿刺穿的南方古猿头骨，还论证了这里的石器与在东非奥杜威峡谷发现的类似。布雷恩的最新诠释是：在斯瓦特克朗斯遗址中至少有一部分并非原始人类的居住之所，而是豹的巢穴，里面有豹留下的食余垃圾。

斯瓦特克朗斯是南非的 5 个南方古猿洞穴遗址之一，在此出土了 60 多件大型原始人科——傍人粗壮种个体的标本，这里是唯一一处发现人属化石与南方古猿化石相伴生的洞穴遗址。

本文描述的新发现澄清了许多关于斯瓦特克朗斯的重要争议。在这里发现了一系列新的原始人类化石（包括两块脊椎骨，鲁滨逊博士 [1] 将对此进行详细记述），因而使我们有可能将无可置疑的石器文化与人科联系起来。玛丽·利基博士曾对这些石器进行过审慎的检查，她将这些石器和坦桑尼亚奥杜威峡谷相当的石器进行了比较，发现它们与在奥杜威遗址第 II 层发现的石器有着惊人的相似性 [2]。由于不知道斯瓦特克朗斯化石的绝对年代，所以这种文化比较就具有特别的意义。我们已经公布了一些关于斯瓦特克朗斯第二种人科标本特性的出人意料的新信息 [3]，而对堆积的骨骼化石进行的整体研究令我得出如下结论：南方古猿化石以及许多其他化石很可能是食肉动物的，尤其是豹的，食余垃圾。

斯瓦特克朗斯位于斯泰克方丹河谷，距离南非德兰士瓦省克鲁格斯多普北部约有 8 英里。1948 年，已故的罗伯特·布鲁姆博士和鲁滨逊博士最先对此地进行了科学考察，他们将德兰士瓦博物馆对斯泰克方丹的研究工作转移到了河谷对面的洞穴。现场角砾岩的挖掘工作一直持续到 1949 年年底，最终发现了约 3,000 件化石，包括数量众多的傍人标本和另一种人科的标本，后者被命名为开普远人 [4]，但是后来又被归入到直立人 [5] 中。

The palaeontological work of 1949 revealed a substantial seam of pure travertine along the north wall of the excavation, and this was exploited commercially by a lime-miner in 1950 and 1951, who left the site in a chaotic state. Robinson excavated the site again in 1952, but the cave was then abandoned until the new investigations, reported here, were started in April 1965.

New Investigations

The new fieldwork at Swartkrans has had two important achievements: first, order has been restored after the chaotic mining episode and excavation of undisturbed breccia is now possible; second, a large and representative sample of 14,000 bone fragments has been recovered.

Our first task was to sort through a substantial dump on the hillside below the cave. Unexpectedly, we found very numerous blocks of fossiliferous breccia, obviously not derived from the 1950–51 mining operations. In fact, more than 200 tons of breccia were recovered from the dump and stockpiled for subsequent attention. Enquiries among old residents of the area revealed[6] that Swartkrans had, in fact, been mined for lime on a previous occasion between 1930 and 1935. The miners had later refilled their excavations with rubble, thereby confusing the situation and leading Robinson[7] to believe that, unlike Sterkfontein, Swartkrans had not been modified by mining before its palaeontological exploration.

After working through the hillside dump, rubble was removed from the excavation and hand-sorted. This work is now almost complete and, after a grid over the excavation area has been erected, selected blocks of breccia from different stratigraphic levels will be removed for detailed comparative study of their fossil contents. It is fortunate that a substantial bank of *in situ* breccia survived the first mining operations and remains today as an inclined surface along the breadth of the cave's south wall. A profile is thus preserved of the whole thickness of the original breccia deposit.

Form of Original Cave

Clearing of the site has clarified the extent of the deposit and form of the original cave, and there will have to be some modification of the original interpretation[8]. In this interpretation, the cave was reconstructed as having comprised two filled chambers, an outer and an inner, with pink unstratified breccia in the former and a brown, stratified deposit in the latter. The two deposits were separated by an unconformable contact, the brown breccia proving to be younger than the pink. After our new investigations, separation into inner and outer chambers remains valid, but study of the east end of the deposit strongly suggests that the outer-cave breccia is not the oldest in the Swartkrans cave and that a still older breccia occupies the eastern extension of the site. This is separated from the normal australopithecine-bearing deposit by a steeply inclined unconformity. Each of the three unconformable contacts between the Swartkrans breccias seems to have resulted from progressive slumping into an underground, lower cavern system. Neither the oldest of the breccias, nor the underlying cavern, has yet been investigated.

1949 年的古生物学工作揭示出了沿着发掘地点的北墙有一大层纯石灰华基质，在 1950 年和 1951 年，一名石灰矿矿主对其进行了商业开采，将此地槽蹋得一片狼藉。鲁滨逊曾于 1952 年再度挖掘此地，但是自此之后这个洞穴一直处于被遗弃的状态，直到 1965 年 4 月重新开始研究，在本文中将报道一些最新的考察结果。

最新的考察结果

在斯瓦特克朗斯新开展的野外工作已经取得了两项重要的进展：首先，清理出胡乱开采期间搞乱的地层顺序，从而使对原状的角砾岩进行发掘成为可能；其次，出土了近 14,000 件有代表性的骨骼碎片标本。

我们的第一项任务就是筛选在山洞之下的山腰上的大量堆积物。出乎意料的是，我们发现了大量含有化石的角砾岩块，显然它们在 1950 年～ 1951 年的采矿作业中没有被动过。实际上，我们在这些堆积中清理出了 200 多吨角砾岩，并将它们储存起来以备后续的研究。对该地区老住户的调查结果显示 [6]：斯瓦特克朗斯其实在 1930 年到 1935 年间就曾有过一次石灰的开采经历。后来，开采者又用砾石将他们挖掘过的地方填满，因此混淆了视听，从而令鲁滨逊 [7] 相信：与斯泰克方丹不同，斯瓦特克朗斯的地层在进行古生物学发掘之前并没有受到过采矿事件的破坏。

在整理完山腰上的堆积之后，将砾石从发掘地点移走并对其进行人工分拣。目前该项工作基本完成，在发掘区安装好方格之后，我们就可以从不同的地层取出选中的角砾岩块以便对它们所含的化石进行详细的比较研究了。幸运的是，大量原位角砾岩在第一次采矿作业中并未受到破坏，目前它们沿着山洞南墙的宽度形成了一个斜面。因而原始的角砾岩沉积物在整个厚度上的剖面都被保存了下来。

洞穴的原始形状

对该处遗址的清理结果澄清了沉积物的范围和洞穴的原始形状，这就需要对最初的解释 [8] 进行一些修正。最初的解释认为：洞穴被重建成由两个充满堆积的洞室构成，包括一个外室和一个内室，外室填充的是粉色的非成层角砾岩，而内室填充的则是褐色的成层沉积物。这两种堆积物被一个不整合的接触面分开，研究表明粉色角砾岩的年代比褐色角砾岩更久远。我们的最新调查结果显示，内外室之间的分离仍然有效，但是对沉积物东端的研究强有力地证明：外洞的角砾岩并非斯瓦特克朗斯洞中最古老的，在该遗址向东延伸的地方还存在着更为古老的角砾岩。一个陡倾的不整合面将其与正常的含有南方古猿的沉积物分开。在斯瓦特克朗斯角砾岩之间的这三个不整合面似乎都是由渐进滑动进入地下的低地大洞穴系统引起的。现在还没有人研究过那些最古老的角砾岩和下面的大洞穴。

In 1958 (ref. 8), the entrance to the outer cave was interpreted as an incline, opening obliquely into the hillside. The walls of the cave can now be seen, however, on all sides except the south-east, where extensive decalcification has occurred. On the evidence of the new exposures, it seems very likely that the original entrance was a precipitous, shaft-like one, perhaps no more than 5 feet in cross-section, linking the cavern with the surface above. Such shaft-like entrances to caverns are a common feature on the undulating dolomite hillsides in the area today. Soil, rock and other debris are funnelled down a shaft of this sort during rain storms, to build up a talus cone in the cavern below the discharge point of the shaft. The discharge point itself is marked by a concentration of rocks and coarse particles in the sediment, while the finer fractions tend to be carried into the further recesses of the cave. It is likely that a vertical shaft discharged in this area, from a hillside depression perhaps 50 feet above it (Fig. 1).

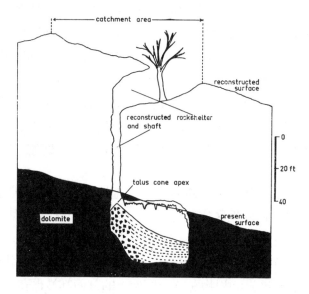

Fig. 1. Diagrammatic N–S section through the Swartkrans hillside. The upper reconstructed part has been removed by erosion since the accumulation of the fossiliferous deposit.

The Swartkrans cave system has developed at the intersection of two fault planes, clearly visible on an aerial photograph. They intersect at an angle of about 87 degrees and it is probable that a large rock-shelter may have existed on the surface in this area, with the shaft descending from it to the underground cavern. If this interpretation of cave form is correct, it is important to realize that the cavern containing the fossiliferous breccia would have been inaccessible to most animals, other than bats and owls. The deposit would, nevertheless, reflect events in the catchment area on the surface during the accumulation period. Relics left by animals or men in the upper rock shelter would ultimately find their way down to the subterranean fossilization site. Stratigraphy in a talus cone is difficult to follow because it is steeply inclined close to the apex, but becomes progressively flatter in those parts of the cave further from the shaft (see Fig. 1). Today, erosion has removed the overlying hillside with its shaft and rock shelter. The original form of these missing parts of the cave system is thus a matter for speculation, but it may be reconstructed on the

在 1958 年（参考文献 8）时，人们认为外洞的入口是一个斜坡，开口向山腰内部倾斜。但现在可以看到洞穴中除东南面以外的所有侧面。在东南面曾发生过大面积的脱钙作用。根据新清理出的露头判断，原来将大洞穴与其上表面相连接的入口很可能是一个陡峭的杆状通道，或许横截面还不足 5 英尺。目前，通往大洞穴的这种杆状入口在该地区连绵起伏的白云岩山腰区域已经成为一种很常见的现象。在下暴雨时，土壤、岩石和其他碎片从这种杆状通道漏下去，于是在大洞穴里杆状通道的排放点之下就形成了一个岩屑锥。排放点本身就是以沉积物中岩石和粗粒的丰度来确定的，较细的碎片倾向于被带到洞穴中更深的地方。很可能通过约 50 英尺以上的山腰凹陷处的一条垂直杆状通道可以排泄到这个区域（图 1）。

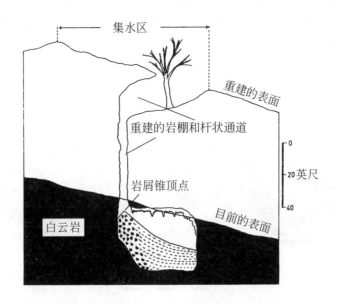

图 1. 穿过斯瓦特克朗斯山腰的南 – 北向剖面图。这里重建出来的上半部分已经在下部的含化石沉积物开始堆积时逐渐被剥蚀而不复存在了。

斯瓦特克朗斯洞穴系统形成于两个断层面的交汇处，这从航空照片上可以很清楚地看到。它们的交角约为 87 度，在该处的地表之上可能曾经有过一个巨大的岩棚，杆状通道就是从这里开始向下降至地面以下的大洞穴的。如果对洞穴形式的这种解释是正确的，那么意识到以下这一点就很重要，即除蝙蝠和猫头鹰以外的大部分动物很难进入到这个具有含化石角砾岩的大洞穴中。尽管如此，由沉积物也可以推断出岩屑积累时期地表集水区所发生的变迁。动物或人类在上方的岩棚上留下的遗骸最终会沿着伸入地下的通道下坠而堆积成化石。很难判断岩屑锥中的地层层序，因为在靠近顶点处它是陡峭倾斜的，而在洞穴中离杆状通道较远的部分它又逐渐变得平缓了（见图 1）。今天，侵蚀作用已经将覆盖于其上的山腰以及山腰上的杆状通道和岩棚一并剥蚀掉了。因此，该洞穴系统中这些缺失部分的原始形状就成为了我们

known geological conformation of the immediate area.

New Hominid Finds

The current fieldwork at Swartkrans has produced a little more than 11,000 bone fragments among which are seventeen recognizable hominid fossils. The hominids, which I shall describe briefly, are all presumed to represent *Paranthropus robustus*. It has been observed[6] that teeth in two of the mandible fragments (*SK* 1587 and *SK* 1588) were unusually small, with dimensions falling outside the observed range for *Paranthropus* from the site. At the time, it was suggested that these two mandibles may have represented a hominid other than *Paranthropus*, but this now seems unlikely—instead, the hominids probably extend the range of tooth dimensions for the Swartkrans *Paranthropus* sample. Measurements were taken on all teeth which had not suffered damage or undue wear. In the following descriptions mesio-distal length is quoted first, followed by bucco-lingual breadth, in millimetres.

Cranial and Vertebral Remains

SK 3978 Perfectly preserved and undistorted infant mandible lacking rami (see Fig. 2). Left dm_1 (10.2×7.8 mm) slightly worn, dm_2 (13.0×10.5 mm) unworn; right dm_1 (9.9×8.2 mm), dm_2 (12.9×10.7 mm). Sockets are present for deciduous incisors and canines while the unerupted left M_1 is visible in its crypt. Found August 8, 1967.

Fig. 2. An infant *Paranthropus* mandible (*SK* 3978) in superior (*a*), inferior (*b*) and lateral view (*c*).

SK 1585 A natural endocranial cast from a hominid, presumably *Paranthropus* (see Fig. 3). The skull was blasted out by lime miners in the early 1930s and was broken into several pieces before being dumped on the hillside. Two of the surrounding parts of the cranial vault have been recovered, but the face has not yet been found. It is clear that the skull originally came to rest in the cave on its right side and was filled with fine-grained sediment to the level of the top of the foramen magnum. The filling was subsequently hardened with calcium carbonate and is almost completely undistorted. The state of the sutures suggests that the individual was subadult. The endocast has recently been prepared and studied by Dr. R. Holloway, who estimates the capacity of the skull to have been 530 cc and whose detailed report will be published elsewhere. Found January 17, 1966.

需要推测的对象，不过根据邻近区域的已知地质构造还是有可能将其重建出来的。

新发现的人科化石

目前在斯瓦特克朗斯遗址出土的骨骼碎片略多于 11,000 件，其中包括 17 件可识别的人科化石。我在下文中简要记述的人科化石都权且属于傍人粗壮种。我曾经观察到 [6] 其中两件下颌骨碎片中的牙齿（SK 1587 和 SK 1588）小得异乎寻常，其尺寸落在来自该遗址的傍人的变异范围之外。当时，有人认为这两件下颌骨可能代表了一种不属于傍人属的人科化石，但是现在看来这似乎是不可能的——与此相反，其所代表的人科化石有可能扩大了斯瓦特克朗斯傍人标本的牙齿尺寸变异范围。我们测量了所有没有损坏或没有过度磨耗的牙齿。在下面的描述中，近中－远中长度在前，颊－舌宽度在后，单位都是 mm。

头骨和椎骨化石

SK 3978　保存完好且没有变形的幼年下颌骨，但下颌支缺损（见图 2）。左 dm_1（10.2 mm×7.8 mm）有轻微磨损，dm_2（13.0 mm×10.5 mm）未磨损；右 dm_1（9.9 mm×8.2 mm），dm_2（12.9 mm×10.7 mm）。乳门齿和犬齿的牙槽都保存下来了，可以看到未萌出的左 M_1 尚包埋于牙床中。发现于 1967 年 8 月 8 日。

图 2. 傍人幼年下颌骨（SK 3978）的嚼面视（a）、腹面视（b）及侧面视（c）。

SK 1585　人科天然颅内模标本，可能是傍人的（见图 3）。它的头骨是在 20 世纪 30 年代早期被石灰矿工爆破出来的，在被遗弃到山腰上之前就已经碎成了好几块。其中的两块头盖骨骨片已经找到，但还没有发现面部的骨片。显然该头骨最初在洞穴里埋藏时是向右侧侧卧的，而颅腔被细粒的沉积物填充至枕骨大孔顶部的水平。后来这些填充物被钙质胶结而硬化，几乎没有变形。骨缝愈合的状态表明该个体尚未成年。霍洛韦博士已于最近修理出了该颅内模并对其进行了研究，他估计这个头骨的颅容量可达 530 cc（译者注：即立方厘米），他的详细报告将另行发表。发现于 1966 年 1 月 17 日。

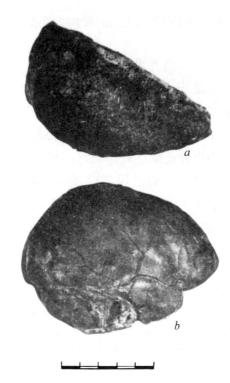

Fig. 3. A natural endocranial cast (*SK* 1585) from a Swartkrans hominid in superior (*a*) and right lateral view (*b*).

SK 3981   Two vertebrae, thought to belong to *Paranthropus*; (*a*) a last thoracic, (*b*) a last lumbar. These specimens have been described in detail by Dr. J. T. Robinson[1]. Found March 1968.

SK 1587   (*a*) Portion of the left corpus of a mandible with roots of P_2, complete P_4 (10.6×11.6 mm), M_1 (14.7×12.9 mm); (*b*) isolated right M_2 (15.2×12.9 mm) complete, from the same block and almost certainly from the same individual. Found July 19, 1966.

SK 1588   Part of a left corpus of a mandible with roots of P_3, complete P_4 (11.0×11.6 mm), M_1 (14.0×12.3 mm) and root of M_2. The specimen is of particular interest because the dimensions of M_1 are well below the hitherto observed range for *Paranthropus* from Swartkrans. Found October 6, 1966.

SK 1590   From an adult skull: two palatal fragments (*a*) and (*b*), with an incomplete right I^2, C, P^3 (9.25×13.3 mm), P^4 (10.2×14.5 mm) and M^1 (13.2×14.9 mm), all worn; crushed and distorted mandible (*c*) with parts of premolar and molar teeth. Associated were the head of a femur (*d*) and parts of an innominate bone in poor condition. Found April 13, 1965.

SK 1586   Almost complete adult mandible, severely shattered during fossilization, with right I_1, I_2, M_1 (broken), M_2 (15.0×14.9 mm) and M_2 (16.7×15.0 mm); also left I_1, M_1, M_2 (15.0×13.8 mm). One of the most robust mandibles from Swartkrans. Found September 1, 1965.

SK 1648   Anterior part of an adult mandible, severely distorted with worn right M_1 and M_2 (15.7×14.5 mm), left M_1 (incomplete) and roots of other teeth. Found November 4, 1966.

Isolated Teeth

SK 1593   Right P_3 (9.8×12.2 mm), well worn, with almost complete root system. Found March 24, 1966.

SK 3974   Right M_1 (14.9×13.4 mm), perfect crown, slightly worn, without roots. Found July 14, 1967.

SK 1594   Right M_1, moderately worn, most of the buccal half including the root. Found September 12. 1966.

SK 3976   Left M_2 (17.3×16.0 mm), complete and slightly worn. Found May 5, 1967.

SK 14001   Left P^3 (10.5×12.8 mm), complete crown but partly damaged root system. Found July 14, 1967.

SK 1589   Right P^4 (11.0×14.4 mm), complete crown of an unerupted tooth. Found October 14, 1966.

748

图 3. 斯瓦特克朗斯遗址出土的人科天然颅内模标本（*SK* 1585）的背侧视（*a*）及右侧视（*b*）。

SK 3981　被认为属于傍人的两块椎骨：（*a*）一块末尾的胸椎骨，（*b*）一块末尾的腰椎骨。鲁滨逊博士[1]已经描述过这些标本的细节。发现于 1968 年 3 月。

SK 1587　（*a*）带有 P_2、完整 P_4（10.6 mm × 11.6 mm）和 M_1（14.7 mm × 12.9 mm）牙根的左下颌体；（*b*）单独完整的右 M_2（15.2 mm × 12.9 mm），它们来自同一地点，并且几乎可以肯定属于同一个体。发现于 1966 年 7 月 19 日。

SK 1588　带有 P_3、完整 P_4（11.0 mm × 11.6 mm）和 M_1（14.0 × 12.3 mm）牙根以及 M_2 牙根的左下颌体。因为 M_1 的尺寸远远小于迄今为止观察到的斯瓦特克朗斯傍人的最小值，所以这件标本引起了人们特别的兴趣。发现于 1966 年 10 月 6 日。

SK 1590　来自一个成人的头骨：两块上腭碎片（*a*）和（*b*），含不完整的右 I^2、C、P^3（9.25 mm × 13.3 mm）、P^4（10.2 mm × 14.5 mm）和 M^1（13.2 mm × 14.9 mm），所有这些牙齿都有磨损；压碎变形的下颌骨（*c*），含部分前臼齿和臼齿。与此相连的还有一块股骨头（*d*）和一些保存状况不佳且无法鉴定的骨骼碎块。发现于 1965 年 4 月 13 日。

SK 1586　几乎完整的成人下颌骨，在化石化过程中破碎严重，含右 I_1、I_2、M_1（断裂）、M_2（15.0 mm × 14.9 mm）和 M_3（16.7 mm × 15.0 mm）；还有左 I_1、M_1 和 M_2（15.0 mm × 13.8 mm）。这是在斯瓦特克朗斯发现的最粗壮的下颌骨之一。发现于 1965 年 9 月 1 日。

SK 1648　一件成人下颌骨的前部，变形严重，含磨损的右 M_1 和 M_2（15.7 mm × 14.5 mm）、左 M_1（不完整）以及其他牙齿的牙根。发现于 1966 年 11 月 4 日。

单独的牙齿

SK 1593　右 P_3（9.8 mm × 12.2 mm），磨损严重，牙根系统几乎完整。发现于 1966 年 3 月 24 日。

SK 3974　右 M_1（14.9 mm × 13.4 mm），牙冠完整，有轻微磨损，没有牙根。发现于 1967 年 7 月 14 日。

SK 1594　右 M_1，中度磨损，颊半侧的大部分带有牙根。发现于 1966 年 9 月 12 日。

SK 3976　左 M_2（17.3 mm × 16.0 mm），牙齿完整，有轻微磨损。发现于 1967 年 5 月 5 日。

SK 14001　左 P^3（10.5 mm × 12.8 mm），牙冠完整，但牙根系统有部分损伤。发现于 1967 年 7 月 14 日。

SK 1589　右 P^4（11.0 mm × 14.4 mm），一颗未萌出的完整牙冠。发现于 1966 年 10 月 14 日。

SK 3977   Right M³ (15.5×17.9 mm), slightly worn. Found May 5, 1967.
SK 3975   Left M³ (14.6×16.7 mm), slightly worn. Found July 28, 1967.
SK 14000   Upper molar broken through middle, showing pulp cavity. Found November 11, 1967.

The following eight hominid fossils were recently found in blocks of breccia brought back to the Transvaal Museum during the Broom/Robinson operations at Swartkrans. They came to light when the breccia was processed in acetic acid. All are assigned to *Paranthropus robustus*.

SK 1596   Isolated upper canine, heavily worn, with almost complete root system.
SK 1591   Isolated left M¹ (13.0×14.4 mm), fairly worn with part of the root system.
SK 1524   Isolated left M³; most of the mesial portion of an unerupted and incompletely developed crown.
SK 1514   Mandibular fragment in very poor condition with roots of two molars and crown of one.
SK 1595   Fragment of a maxilla with slightly worn left dm² (11.5×12.4 mm) and pieces of erupting anterior teeth.
SK 1512   Palatal fragment with roots of several cheek teeth and part of a premolar.
SK 1592   Palatal fragment with broken right P⁴, M¹, M² and part of M³; also the root of one of the anterior teeth.
SK 14003   Part of a crushed skull with left M¹ (incomplete), part of right M¹, M² and complete M³ (15.0×16.2 mm).

Stone Culture

The presence of a stone culture at Swartkrans was first suspected in 1956 when several quartzite tools, encased in breccia, were found there[8]. Recent examination of breccia blocks, blasted from the outer cave by the lime miners, has led to further recovery of occasional pieces of rock, foreign to the immediate vicinity of the cave and, in some cases, artificially worked. To date, 195 foreign stones have been found, fifty-one of which show signs of artificial fracture. The whole collection has been studied in detail by Miss S. M. Johnston, who has very kindly made her results available to me (see Table 1).

Table 1

Rock type	No. of pieces	Percent of total foreign stones	Average wt (g)	No. of waterworn pieces	No. of worked pieces
Quartzite	148	75.9	507	87	33
Vein quartz	36	18.5	226	3	8
Diabase	6	3.1	450	1	5
Other	5	2.5	166	0	5
	195	100.0	337 (overall average)	91	51

The most numerous rock type present, quartzite, has its origin on the Witwatersrand escarpment a few miles to the south of Swartkrans; vein quartz occurs sporadically in the dolomite formation, while the closest outcrop of diabase is about 1 mile south-east of the cave, on the opposite side of the valley. Although fifty-one of the stones showed some signs of artificial fracture, only thirty of them could be termed artifacts and these have been studied and described in detail by Dr. Mary Leakey[2]. All the tools, with the exception of the two diabase bifaces and one quartzite chopper, have edges so sharp that they were clearly made close to the cave entrance shortly before their inclusion in the deposit. The

SK 3977　右 M³（15.5 mm×17.9 mm），有轻微磨损。发现于 1967 年 5 月 5 日。
SK 3975　左 M³（14.6 mm×16.7 mm），有轻微磨损。发现于 1967 年 7 月 28 日。
SK 14000　上白齿，中间裂穿，露出髓腔。发现于 1967 年 11 月 11 日。

下面的 8 件人科化石标本是我们刚刚从布鲁姆 / 鲁滨逊在开采斯瓦特克朗斯期间带回到德兰士瓦博物馆的角砾岩块中发现的。它们是在用醋酸浸洗角砾岩时出现的。我们把它们归入傍人粗壮种。

SK 1596　单独的上犬齿，磨损严重，牙根系统几乎完整。
SK 1591　单独的左 M¹（13.0 mm×14.4 mm），有相当程度的磨损，牙根系统不完整。
SK 1524　单独的左 M³；一颗未萌出且发育不完全的牙冠的大部分近中侧部分。
SK 1514　保存状况很差的下颌骨碎片，含两枚白齿的牙根和一枚白齿的牙冠。
SK 1595　上颌骨的碎片，含有轻微磨损的左 dm²（11.5 mm×12.4 mm）和正在萌出的几颗前齿碎片。
SK 1512　上腭的碎片，含几枚颊齿的牙根和一枚前白齿的一部分。
SK 1592　上腭的碎片，含断裂的右 P⁴、M¹、M² 和部分 M³；还有一枚前齿的牙根。
SK 14003　一件被压碎的头骨的一部分，含左 M¹（不完整）、部分右 M¹、M² 和完整的 M³（15.0 mm×16.2 mm）。

石 器 文 化

对斯瓦特克朗斯遗址曾存在过石器文化的揣测始于 1956 年，当时在该遗址出土了几个包裹在角砾岩中的石英质石器 [8]。最近对石灰开采者从外洞爆破出来的角砾岩块的研究使我们发现了更多的特殊石块，这些石块的岩性与洞穴中的及近邻地区的迥然相异，其中有些石块有人工痕迹。到目前为止，我们已经找到了 195 块外来石块，其中有 51 块具有人工痕迹。约翰斯顿小姐对全部采集物都进行过详细的研究，她非常友好地把她的研究成果提供给我（见表 1）。

表 1

岩石类型	块数	占全部外来石块的百分比	平均重量 (g)	水蚀石块的数量	加工过的石块的数量
石英岩	148	75.9	507	87	33
脉石英	36	18.5	226	3	8
辉绿岩	6	3.1	450	1	5
其他	5	2.5	166	0	5
合计	195	100.0	337（总平均值）	91	51

存在数量最多的岩石类型是石英岩，这种岩石的来源地是距离斯瓦特克朗斯南部几英里处的威特沃特斯兰德悬崖；脉石英零星出现于白云岩地层中，而最近的辉绿岩露头是在离洞穴东南约 1 英里处的河谷对岸。尽管有 51 块石头显示出人工痕迹，但是其中只有 30 块可以被称作是人工制品，玛丽·利基博士 [2] 曾对它们进行过详细的研究和论述。除了两件辉绿岩两面器和一件石英岩单刃砍砸器外，所有工具都具有很锋利的边缘，所以它们显然是在被埋进沉积物之前不久才在洞穴入口附近制

diabase tools both have weathered cortices, but it is known that diabase weathers while it is enclosed in a damp breccia deposit. The one quartzite chopper had been waterworn after manufacture, but before fossilization. It was originally assumed that all the foreign stones must have been brought to Swartkrans by human (or proto-human) agency; it now seems likely, however, that an old river gravel may have existed on the Swartkrans hillside when the cave deposit was accumulating and that some of the pebbles found their way naturally into the cavern, together with the soil and debris which were washed down into it. Tools that were made at the cave entrance were presumably fashioned on pebbles which happened to be in the immediate vicinity.

In view of the possibility that the Swartkrans foreign stones may have originated in a river gravel, a detailed study of contemporary gravels in the area has been made. The cave is situated 700 feet from the present course of the Blaaubank River, but this stream does not normally flow except after heavy rain. Sparse river gravels occur along the sides of the valley and, from three separate localities within 1.5 miles of the cave, Miss Johnston was able to collect a sample of 152 foreign stones for comparison with the Swartkrans specimens. The results of this comparison indicate that the Swartkrans foreign stones are very similar to those associated with the stream course in the vicinity of the cave. Waterworn quartzite pebbles proved to be slightly more abundant in the Swartkrans sample than in the gravel, but the percentage abundance of vein quartz pieces did not differ significantly between the two. Weights of the pieces in the two samples were found to be almost exactly comparable. The stones in the gravel sample also did not differ significantly in shape from those in the Swartkrans collection.

On the basis of this study, it seems that the foreign stones at Swartkrans were almost certainly derived from an old gravel associated with the stream course. Artificial transportation need not be invoked because, if the gravel lay within the catchment area of the cave entrance, it would have been washed into the underground cavern together with other surface debris. The stone tools are nevertheless of special significance, because they are indicative of sporadic hominid activity near the cave entrance during the Swartkrans accumulation period. Two different hominids are known at the site, *Homo erectus* and *Paranthropus robustus*, but it is not certain which was responsible for the stone tools. Three of the stones in the collection show evidence of heat-spalding, suggesting that they had been in a fire shortly before inclusion in the deposit. On this slender, but interesting, evidence it would be premature to suggest that the Swartkrans hominids had mastered the deliberate use of fire. Grass fires, started by lightning, are a recurrent feature of the Transvaal highveld and, when sweeping through a leaf-strewn rock shelter, can generate a good deal of heat. Nothing resembling a bone tool has yet been found at Swartkrans, in spite of examination of every fragment in the collection.

Interpretation of the Bone Accumulation

In the interpretation of any bone accumulation preserved in a cave deposit, the original form of the cave is of crucial significance because it is this that determines the kinds of animals likely to have made use of the site. As already described, Swartkrans seems to have

造出来的。两件辉绿岩工具的外表都被风化了，但是我们知道当辉绿岩被封在潮湿的角砾岩沉积物中时也会发生风化。有一件石英岩单刃砍砸器在制造出来之后和埋藏之前受到了水的冲蚀。最初假定所有的外来石块都一定是被人类（或早期原始人）带到斯瓦特克朗斯的；但是现在看来在斯瓦特克朗斯的山腰上似乎存在过一层古老的河相砾石，在洞穴沉积物的堆积过程中，其中一些鹅卵石便自然而然地顺道掉进了洞穴中，同时被冲入其中的还有土壤和岩屑。在洞穴入口处出土的工具很可能取材于这些就近的鹅卵石。

考虑到斯瓦特克朗斯的外来石块可能来源于河相砾石，因而我们对这一区域目前的砾石进行了详细的研究。该洞穴位于距离布拉奥班河现在的河道 700 英尺处，但是这条河通常不会有水流，除非是在暴雨之后。沿着河谷的两岸很少见到河相砾石，而约翰斯顿小姐从距离洞穴 1.5 英里以内的 3 处不同地点收集到了 152 块外来石块以便与斯瓦特克朗斯样本进行比较。比较结果显示：斯瓦特克朗斯的外来石块非常类似于在洞穴附近的那些与河道有关联的砾石。被水冲蚀的石英岩鹅卵石在斯瓦特克朗斯样本中确实比在砾石中略多一些，但是二者在脉石英岩块中的百分比丰度却没有明显的差异。两个样本中石块的重量也几乎完全相等。砾石样本中石块的形状也与在斯瓦特克朗斯收集到的石块的形状没有太大的差别。

基于此项研究，看来几乎可以肯定斯瓦特克朗斯的外来石块就是来自与河道有关的古砾石。这个过程并不需要人工运输，因为：如果砾石位于洞口的集水区之内，那么它们早就会和地面的其他碎屑一起被冲到地下的洞穴里。不过出土的石器还是具有很重要的意义的，因为它们说明在斯瓦特克朗斯的堆积时期靠近洞口的地方曾存在过零星的原始人类活动。已知在该遗址曾有两种不同的人科成员——直立人和傍人粗壮种，但还不确定石器的制造者是其中的哪种。在出土的物品中有三件石器显示出受过热的迹象，说明它们在埋藏进沉积物之前刚刚被火灼烧过。用这个羸弱但又很有趣的证据来证明斯瓦特克朗斯的原始人类已经掌握了用火的技术还为时尚早。由闪电引发的草地着火是德兰士瓦高地频频发生的一种现象，当火苗横扫遍布树叶的岩棚时，就可以产生大量的热。尽管已检查过出土物品中的每一块碎片，但是到现在为止在斯瓦特克朗斯还未发现任何类似于骨制品的化石。

对骨骼堆积物的解释

在解释存于洞穴沉积物中的任何骨骼堆积物时，洞穴的原始形态具有至关重要的意义，因为它决定了可能利用过该处遗址的动物类型。如上所述，斯瓦特克朗斯

consisted of a hillside rock-shelter, probably supporting several large trees, but enclosing a vertical shaft which descended to an underground cavern, in which the fossiliferous deposit accumulated (Fig. 1). The latter cavern was simply a receptacle into which surface-derived debris was funnelled, while the nature of the rock-shelter on the surface was presumably such that it is likely to have been used by a wide variety of animals, as well as primitive man. The sample of 14,000 fossil bone fragments currently being analysed comes from a thickness of deposit which probably took several thousand years to accumulate. At Swartkrans we are not dealing with an impressive concentration of bones, but rather with a more or less even scatter throughout the calcified soil which originally filled the cave. It is probably not necessary to account for the collection of more than about ten bones per year and these could have found their way into the cave in a variety of ways. Some of the possibilities will be considered.

It is, for instance, known that many animals, when injured or sick, will retreat into caves to die. If unmolested, complete skeletons or scattered undamaged bones can be expected to result. Nothing approaching a complete or undamaged skeleton has been found in the Swartkrans breccia and natural deaths are not thought to have been important. The same is true of artificial burials which are unlikely to have occurred during the australopithecine period.

A recent study of important factors in Southern African bone accumulations (my work, in preparation) has shown that porcupines are of the greatest significance as collectors of bones in caves. They are known to horde large numbers of bones in their lairs, but all their collections so far studied have been characterized by a high percentage of gnawed bones on which the marks of the rodents' incisors are unmistakable. The final figure for the percentage of porcupine-gnawed bones in the Swartkrans sample has yet to be determined, but it will probably be less than 5. It is therefore concluded that, although porcupines almost certainly contributed to the Swartkrans bone collection, their influence was small.

Second to porcupines, primitive men rated high in importance as collectors of bone fragments in caves during the South African Stone Age. A detailed comparative study of human bony food remains from various Southern African caves has led me to conclude that these accumulations are characterized by the extreme fragmentation of the bones (my work, in preparation). A high percentage abundance of long-bone flakes and unrecognizable fragments is indicative of human activity, in which the bones are broken with stone tools for the extraction of marrow. As already mentioned, fifty-one stone artefacts have been recovered from the Swartkrans breccia, many of them chopping tools suited to the breaking of bones for the extraction of marrow. Bone flakes and fragments do occur in the accumulation but they are not abundant, suggesting that human feeding activity occurred sporadically within the catchment area of the Swartkrans cave but that this was not of major importance in the building up of the bone collection.

似乎曾有过一个位于山腰的岩棚，可能还长有几棵大树，不过这几棵大树包围着一条通往地下洞穴的垂直杆状通道，而含有化石的沉积物正是在这个洞穴里堆积起来的（图 1）。这里所说的洞穴只是一个容纳地表岩屑的容器，而我们可以认为地面上岩棚的环境很可能不仅适合于原始人类，还适合于很多种动物的栖息。现在正在分析的标本包括 14,000 件骨头碎片化石，是从很厚的沉积物中出土的，这些沉积物的堆积可能经历了数千年。在斯瓦特克朗斯，我们研究的不仅是高度集中的化石，而且还有那些与化石同时堆积并填满洞穴的钙化土壤。也许无需解释每年约有 10 块以上的骨头会堆积到洞中，这些骨头可能通过很多种途径进入洞穴。对其中一些可能途径的探讨如下。

例如，我们都知道许多动物在受伤或生病之后会隐退至洞穴中苟延残喘至死。这些动物的尸体埋藏后如果未受干扰的话，完整的骨架或分散的未受损骨骼就有可能会被保存下来。在斯瓦特克朗斯的角砾岩中还从未发现过任何几近完整或未受损的骨架，因此它们是自然死亡的可能性不大。在南方古猿时代也不可能出现人为埋藏这些动物的现象。

一项对南非化石富集点的重要影响因素的最新研究结果（我的工作，完稿中）表明：豪猪对洞穴内骨骼的堆积贡献最大。我们知道它们会在自己巢穴中堆积大量的骨骼，但是从迄今为止的研究结果来看，它们的所有收集物都以高比例的啃咬过的骨头为特征，在这些骨头上的啮齿动物门齿痕迹清晰可辨。现在还未最终确定豪猪咬过的骨骼在斯瓦特克朗斯标本中所占的比例，不过这个值可能会小于 5%。因此可以得出如下结论：虽然几乎可以肯定豪猪对斯瓦特克朗斯的骨骼堆积有一定贡献，但是它们的影响并不大。

仅次于豪猪，在南非石器时代原始人类也是洞穴中骨骼碎片的主要堆积者。在对若干南非洞穴中人类骨质食余垃圾进行了详细的比较研究之后，我发现这些堆积物的特征是骨骼的破碎程度极高（我的工作，完稿中）。长骨的骨片和无法辨认的碎片所占比例很高说明有人类活动过，人类会用石器将骨骼打断以摄食其中的骨髓。正如前面已提到的那样，在斯瓦特克朗斯的角砾岩中曾发现了 51 件石器，其中有不少用于砍砸的工具适合用来剁碎骨头以吸取骨髓。骨片和骨渣在堆积物中的确存在，但是它们的数量并不是很多，这说明人类的摄食活动在斯瓦特克朗斯洞穴的集水区只是偶尔出现，因此人类活动在骨骼的堆积过程中也不起主要作用。

Carnivore Activity

The overall impression given by the Swartkrans bones is that they represent carnivore food remains; they come predominantly from antelope, baboons, australopithecines, hyraxes, and so forth, and are usually fragmentary, often showing damage apparently caused by carnivore teeth. Antelope are characteristically represented by a wide variety of damaged skeletal parts, but remains of many other animals are essentially cranial in origin, with very little else. The Swartkrans breccia has yielded an interesting series of carnivores, some of which were almost certainly involved in the collection of bones at the site. Forms identified to date are as follows (taxonomy according to Ewer[9-11]): hyaenids: *Crocuta crocuta venustula* (spotted hyaena), *Hyaena brunnea dispar* (brown hyaena), *Lycyaena silbergi nitidula*, *Leecyaena forfex*; sabre-toothed cats: *Therailurus* sp., *Megantereon* sp., *Nimravidae indet.*; felids: *Panthera pardus incurva* (leopard), *Panthera* sp. (lion); canids: *Canis mesomelas* (jackal), *Vulpes* sp. (fox); viverrids: *Cynictis penicillata* (mongoose).

Both the *Crocuta* and the *Hyaena* are regarded as early representatives of the extant spotted and brown hyaenas, while *Lycyaena* and *Leecyaena* represent extinct genera, the former showing characteristics which have been interpreted as "reflecting a secondary adaptation to a more predacious, less scavenging habit—a change presumably brought about under the selective pressure of competition with more efficient scavengers"[11]. *Leecyaena* is characterized by anterior cheek teeth resembling those of an advanced *Hyaena* and could possibly be ancestral to *H. bellax*, described from Kromdraai.

The three different sabre-toothed cats are represented by rather fragmentary material. Ewer has pointed out that, as a result of highly efficient carnassial shear, these carnivores were adapted to the slicing of meat and had premolars so reduced that they "must have been unable to deal with more than the very smallest bones"[11]. The conclusion that sabre-toothed cats left the skeletons of their prey almost intact has important implications, in that a niche was thus created for various species of hyaena, not necessarily capable of crushing the more resistant bones. Hyaenas like *Lycyaena* and *Leecyaena* may have owed their livelihood to the remains left by the abundant sabre-tooths of the day. As Ewer[11] has remarked: "The period of hyaena abundance in the Lower Pleistocene is thus also a period of abundant sabre-tooths and the disappearance of the latter during the Middle Pleistocene corresponds with the shrinkage of the hyaenid fauna to those few species which have succeeded in surviving today, in association with modern Felinae". Regrettably little is known about the behaviour of South African sabre-tooths and it is uncertain whether they were dominant to the associated hyaenas or whether they were forced to retreat with their prey into the seclusion of caves or trees. If so, it is likely that they would have contributed bones to the Swartkrans accumulation and that some of these may have been modified by the loss competent bone-crushing hyaenas, such as *Lycyaena* and *Leecyaena* before fossilization. (It is unlikely that the bones found today as fossils were ever thoroughly worked over by spotted or brown hyaenas; if they had, few would have remained to become fossilized.) Sabre-tooth involvement at Swartkrans can neither be demonstrated nor disproved, but it is very likely that the bone accumulation was added to by these cats and that the remains were modified by primitive hyaenas before being fossilized.

食肉动物的活动

斯瓦特克朗斯的骨骼化石给人留下的总体印象是：它们代表了食肉动物的食余垃圾；这些骨骼主要来源于羚羊、狒狒、南方古猿和蹄兔等，它们大多已成碎片，常常带有显然由食肉动物的牙齿所造成的损伤。羚羊的残骸主要是骨架上的不同部位，但是许多其他动物的残骸则主要是头骨，其他部位的几乎没有。在斯瓦特克朗斯角砾岩中已经出土了一系列有趣的食肉动物，几乎可以肯定其中的一些参与了该遗址内骨骼的堆积活动。迄今为止已经鉴定出来的动物种类如下（根据尤尔的分类方法 [9-11]）。鬣狗类：斑鬣狗、棕鬣狗、狼鬣狗、李氏鬣狗；剑齿虎类：假剑齿虎、巨剑齿虎、猎猫；猫类：豹、狮；犬类：豺、狐狸；灵猫类：黄猇。

化石斑鬣狗和棕鬣狗被认为是现生斑鬣狗和棕鬣狗的早期代表者，而狼鬣狗和李氏鬣狗则代表已灭绝的属。狼鬣狗的特征被认为是"反映了对更多偏向食肉、更少偏向食腐的习性的一种次生适应性——这种变化可能是由与更强大的食腐动物进行竞争而产生的选择压力引起的"[11]。李氏鬣狗的特征是前面的颊齿与一种进步鬣狗的很像，所以李氏鬣狗可能是来自克罗姆德拉伊的贝拉斯鬣狗的祖先。

三种不同的剑齿虎标本均破损严重。尤尔指出：由于食肉动物可以用裂齿进行有效的撕裂，所以它们习惯于将肉切成薄片，并且前臼齿退化以至于它们"只能应付极其小的骨头"[11]。认为剑齿虎会将其猎物的骨架几乎完整地保留下来很有意义，因为各类鬣狗将各取所需，这样剑齿虎也就无需具有粉碎更具韧性的骨头的能力了。狼鬣狗和李氏鬣狗等鬣狗类动物当时也许可以靠大量剑齿虎留下的骨架残骸来维持生计。正如尤尔 [11] 曾指出的："因而早更新世时期的鬣狗繁盛阶段也是剑齿虎繁盛的时期，而剑齿虎在中更新世时期的灭绝也对应于鬣狗种群的缩减，幸存到今天的只有极少的几个与现代猫亚科有关的物种"。遗憾的是，对于南非剑齿虎的行为我们知之甚少，因而不确定它们是否支配着与之相关的鬣狗的数量，也不知道它们是否要将猎物带到洞穴或者密林深处才能进食。如果是这样的话，那么它们就有可能会对斯瓦特克朗斯的骨骼堆积有一定贡献，或许其中一部分骨骼在化石化之前曾被骨骼撕咬能力稍差的鬣狗啃过，例如狼鬣狗和李氏鬣狗等。（现今发现的骨骼化石不太可能曾被斑鬣狗或棕鬣狗彻底啃过；如果真的被它们啃过，就不会再有什么可以留下来变成化石了。）剑齿虎在斯瓦特克朗斯的存在既没有被证实也没有被证伪，但是这些猫科动物很有可能参与了骨骼的堆积活动，并且在这些残骸变成化石之前又经过了原始鬣狗的啃咬。

The Role of Leopards

Leopards are well represented among the Swartkrans fossils, where remains of at least thirteen individuals have been found. As *Panthera pardus incurva*[10], they differ from the living leopard on the subspecific level only, with an indication that the fossil form may have been a little smaller than its contemporary counterpart. It has already been noted that the presence of sabre-toothed cats could have been important to the survival of primitive hyaenas; in the same way, the direct association of leopards and spotted hyaenas was probably a crucial factor in the building up of the Swartkrans bone accumulation.

Where leopards share their hunting area with spotted hyaenas, it is well known that, if they are to retain their prey, they are obliged to feed in places inaccessible to the scavengers. I recently studied fifteen leopard kills in the Kruger National Park. Spotted hyaenas were attracted to all these kills, but in twelve cases the leopards took their prey into trees sufficiently promptly to avoid interference. In three they were not quick enough and, as a result, lost their prey to the hyaenas.

In a woodland habitat, leopards have little difficulty in finding suitable trees for storage of their prey, but the situation in open country is rather different. Swartkrans is situated today on the open highveld, an area of undulating grassland, almost devoid of trees except along the watercourses. Study of the composition of the antelope fauna from the Swartkrans deposit suggests that, during the infilling of the cave, open grassland conditions prevailed as they do today, the fossil fauna being dominated by gregarious plains-living species such as springbuck, gazelles and alcelaphines of various sizes. Leopards hunting on the open highveld have to make use of suitable cliffs, caves and the few available trees for the protection of their prey. An interesting and highly significant correlation is found to exist between the occurrence of large trees and caves in the dolomitic areas of the Transvaal highveld. In this gently undulating countryside, caves are typically of the shaft or sinkhole variety, with depressions surrounding their entrances. Whereas the hillsides themselves are devoid of trees, the cave entrances characteristically support several large stinkwood (*Celtis kraussiana*) or wild fig (*Ficus* spp.) trees, which flourish there as a result of protection afforded to them from frost and fire. Leopards are thus inevitably attracted to the dolomitic caves as places of safe retreat, while the associated trees are invaluable for the storage of their prey. Food remains of leopards are consequently introduced into the catchment areas of the cave entrances and some will ultimately find their way down to the subterranean fossilization sites. Assuming that the Swartkrans deposit accumulated during a period of perhaps 20,000 years, it would be surprising if the cave and its associated trees had not been used on innumerable occasions by leopards as a feeding place and lair. The cave was inhabited throughout its life by owls whose regurgitated pellets have built up a concentration of microfaunal remains in the deposit, to give a remarkably complete picture of the animals hunted by the owls in the vicinity of the cave. In the same way, the larger faunal remains at Swartkrans are very likely to contain evidence of the range of animals hunted by leopards in this particular area.

Studies on living leopards have made it clear that the nature of their food remains is

豹 的 角 色

在斯瓦特克朗斯的化石中有很多是来自豹的，目前已经发现了至少 13 头豹的个体残骸。作为弯豹亚种 [10]，它们仅在亚种水平上与现生的非洲豹有所不同，化石种可能比现生种略小一点。如前所述，剑齿虎的存在可能对于原始鬣狗的生存来说至关重要；同样，豹和斑鬣狗之间的直接关联也有可能是斯瓦特克朗斯骨骼堆积过程中的一个关键因素。

当豹与斑鬣狗同在一个狩猎区的时候，大家都知道，如果豹想保住自己的猎物，就必须在食腐动物难以到达的地方进食。最近，我研究了克鲁格国家公园中的 15 起豹捕猎行为。所有这些捕猎行为都引起了斑鬣狗的注意，但是豹能足够及时地把猎物带到树上以躲避鬣狗抢食的情况有 12 次。在其余的 3 次中，豹的动作不够迅速，结果使自己的猎物被鬣狗夺走。

在林地中，豹想找到适合挂放猎物的树并不困难，但是在旷野就很难办到了。今天的斯瓦特克朗斯位于开阔的高原之上，是一片波浪起伏的草原，除了水道沿岸以外几乎没有树木。对斯瓦特克朗斯沉积物中羚羊动物群的研究显示：在该山洞被填充期间，开阔的草原环境和现在没有什么两样，动物群的主要成员是生活在草原上的群居物种，如各种大小的跳羚、瞪羚和狷羚等。在开阔的高原上进行狩猎的豹不得不利用合适的悬崖、山洞和少数便于利用的树来保护它们的猎物。在德兰士瓦高原的白云岩地区，我们发现在出现大树和出现山洞之间存在着一种有趣且极其重要的相关性。在这片起伏不大的乡间，洞穴的类型多为杆状或形状不一的坑状，在它们的入口周围有洼地。虽然山腰上缺少树木，但在山洞的入口处通常有几棵大臭木或野生无花果树；这些大树在那里繁茂地生长着，因为受到的保护可以使它们躲开霜冻和火灾。因此豹被吸引到白云岩山洞并将此处作为安全庇护之所也在意料之中，因为洞口的树木非常适于储存它们的猎物。因此，豹的食余垃圾会落到洞口的集水区，其中有一部分还会最终掉到地下的化石堆积区。假设斯瓦特克朗斯的沉积物堆积了大约 20,000 年，那么如果说该洞穴和与之相关的树木未曾被豹无数次地用于进食或栖息的话就太令人惊奇了。洞穴里一直有猫头鹰居住，它们的吐余构成了沉积物中小动物化石的主要部分，由此可以非常全面地了解猫头鹰在洞穴附近曾捕食过哪些动物。同样，斯瓦特克朗斯化石中的大型动物很可能可以为我们提供在这一特定区域被豹捕食的动物种类的证据。

对现生豹的研究使我们很清楚地了解到豹的食余垃圾的性质在很大程度上受制

considerably influenced by the presence or absence of hyaenas (my work, in preparation). In the Kruger Park study area, where the leopards are constantly harried by hyaenas, they were found to do considerable damage to the skeletons of their prey (largely impala). They would typically return to the carcass in the tree during a period of 3–4 days, by which time most of the body had been consumed. The head and lower leg segments would invariably be left, usually hanging on strips of skin, but the rest of the skeleton generally disappeared. As part of this investigation, leopards were also studied in South West Africa, in an area where no hyaenas or other significant scavengers occur. Here the pattern of leopard food remains was completely different. The leopards made no attempt to drag their prey into inaccessible places, except where the kill had been made close to human habitation. They typically ate where they killed and did comparatively little damage to the prey skeletons, eating only a small quantity of meat before moving on to make another kill a few days later. It is consequently impossible to define the characteristics of "typical" leopard food remains. Their nature will depend on the availability of food and the pressure applied to the leopards by dominant scavengers such as spotted hyaenas.

The greater part of the Swartkrans bone accumulation shows characteristics consistent with those of leopard food remains, and suggests a situation where the leopards had been regularly harried by hyaenas. There are, however, several features of the bone collection which require special explanation. These involve the disproportions in the skeletal parts preserved. In the case of bovid skeletons, it has already been established that certain parts are more resistant to destruction than others and it is possible to predict which parts are likely to survive any particular treatment[12]. At Swartkrans, for instance, bovid distal humeri are almost five times as common as proximal ones. Such a disproportion is to be expected and means simply that the carnivore involved was able to chew away the relatively fragile proximal end of the humerus, but had trouble with the resistant distal one. The difficulty is encountered when we compare the recognizable remains of primate skeletons (such as of baboons and australopithecines) with those of bovids. The latter are typically represented by a wide variety of damaged skeletal parts, while the former are largely cranial in origin with very little else. In fact, the Swartkrans *Paranthropus* sample consists of 190 separate pieces, representing a minimum of at least sixty individuals. Of these fossils, only eleven are post-cranial bones.

The probable reasons for this remarkable state of affairs have been suggested by a series of feeding experiments with cheetahs that I have recently carried out. A natural group of five wild cheetahs was caught on Valencia Ranch in South West Africa and maintained in a large semi-natural enclosure there, through the cooperation of Mr. A. F. Port. Within a short period the cheetahs tolerated human observers and fed readily on the animals provided. Previous observations had indicated that cheetah damage to bovid skeletons, from animals in the 50–150 pound range, was minimal. It was restricted to fragile parts such as ribs, vertebral processes and scapulae. These observations were confirmed in the feeding experiments and, when the cheetahs had fed on a springbuck for instance, almost the whole skeleton remained. When a baboon of equivalent weight was eaten, however, a very different result was obtained. The cheetahs were able to consume the whole of the vertebral column, the hands and feet, as well as to do appreciable damage to the ends of

于鬣狗的存在与否（我的工作，完稿中）。在克鲁格公园研究区，豹经常遭到鬣狗的掠夺，而鬣狗对猎物（主要是黑斑羚）骨架的破坏相当严重。在 3 天～4 天之内，它们一定会回到树丛中的尸体那儿去，大部分尸体在这段时间里会被吃掉。剩下的一般是头部和小腿部分，通常连着丝丝缕缕的毛皮，但是骨架的其余部分在大多数情况下会消失不见。研究非洲西南部的豹也是本项课题中的一部分，在那里没有鬣狗或其他重要食腐动物的出没。豹的食余垃圾的形式在此处完全是另一番景象。除非捕杀行为是发生在靠近人类住所之处，否则豹根本不会将它们的猎物拖到其他动物难以接近的地方。它们一般会在捕杀动物的地方就地进食，对猎物骨架的破坏也很小，因为一次捕猎只吃掉猎物的一小部分肉，几天之后便进行一次新的捕杀。因此很难定义"有代表性"的豹子食余垃圾的特征。这些食余垃圾的性质将取决于食物的可利用率以及斑鬣狗等主要食腐动物给豹造成的压力。

斯瓦特克朗斯的大部分化石堆积所表现出来的特征与豹的食余垃圾相符，并且由此可以说明豹经常遭遇被鬣狗掠夺的处境。然而，这些骨骼堆积还有其他一些特征需要另行解释。例如为什么保存下来的骨骼部位不成比例。在牛科动物骨架的例子中，我们已经证明其中某些部位将比另一些部位更不容易被破坏，因而可以预测出哪些部位可能会在特定的环境中幸存[12]。例如在斯瓦特克朗斯，牛科动物肱骨远端的出现频率是近端部分的近 5 倍。上面提到的不成比例应该是预料之中的事情，这只不过说明相关的食肉动物只能咀嚼相对脆弱的肱骨近端，而对于有韧性的远端则比较吃力。在比较可识别的灵长类动物骨架残骸（例如狒狒和南方古猿）与牛科动物骨架残骸时，我们遇到了一些困难。后者通常包括各种受损的骨骼部位，而前者则主要是头盖骨，其他部位很少。事实上，斯瓦特克朗斯的傍人样本由 190 个单独的碎片组成，代表了至少 60 个个体。在这些化石中，只有 11 件是颅后骨骼。

我最近对猎豹的摄食行为进行了一系列实验，由此得到了这些不寻常事件的可能原因。我们在非洲西南部的巴伦西亚牧场捕获了一个由 5 只野生猎豹组成的自然豹群，并且在波特先生的配合下把它们关进了一个半自然的大型围场之内。在短时间内，这些猎豹接受了人类观察者，并且顺从地吃下了人类提供的动物。之前的观察表明：猎豹对牛科动物的骨架破坏很小，这些牛科动物的体重在 50 磅～150 磅之间。损伤部位局限于脆弱的部分，例如肋骨、椎突及肩胛骨。上述观察结果在喂食实验中得到了证实，例如当猎豹被喂以跳羚时，几乎会剩下整副骨架。然而，当喂以一只同等重量的狒狒时，就会得到完全不同的结果。猎豹除了会严重破坏狒狒的肢骨末端以外，还能吃掉其全部脊柱、前脚和后脚。造成这种差异的原因显然在于

the limb bones. The reason for this difference clearly lies in the fact that the construction of a baboon skeleton is considerably less robust than that of a bovid of equivalent live weight. The baboon vertebral column is fragile enough to be consumed, while the hands and feet are found to be fleshy and palatable.

It is not suggested that cheetahs were involved in the Swartkrans situation: they are simply amenable experimental animals. Nevertheless, conclusions drawn from them on the comparative robusticity of prey skeletons will be equally applicable to leopards, which are known to be capable of doing far more extensive damage to the skeletons of their prey. A leopard is potentially capable of eating virtually a whole baboon, with the exception of the skull, which is invariably left. Conclusions drawn about baboons will be equally applicable to hominids, built on the generalized primate pattern, provided that their larger size is taken into account. One of the reasons for the fact that so few primate limb bones have been recognized in the Swartkrans accumulation may lie in the fact that the diagnostic ends of such bones are typically chewed off by leopards, with the result that damaged and undiagnostic shaft pieces remain. Many of the indeterminate bone fragments at Swartkrans could well be of australopithecine origin.

One of the australopithecine skull pieces form Swartkrans appears to bear direct evidence of leopard activity. The specimen, *SK* 54, consists of much of the calvaria of a subadult hominid thought to have been a *Paranthropus*. It was found at the site in 1949 and subsequently prepared by Dr. Robinson in Wisconsin, using the acetic acid technique. As shown in Fig. 4, the fossil consists of left and right parietal bones, part of the occipital and much of the frontal. The frontal has been distorted downwards, approximately across the width of the coronal suture, while the whole specimen has been slightly flattened by pressure in the deposit. From a comparison of the size of the parietal bones with those of other Swartkrans specimens, it is estimated that the endocranial capacity of this skull was approximately 500 cm^3. The two holes in the back of the skull, one in each parietal bone, close to lambda (see Fig. 4), are particularly interesting. Each is slightly elongated in a lateral direction and is approximately 6 mm in its shortest diameter. It is clear that the holes were made by two pointed objects, whose tips diverged slightly. On the internal surface of the parietals, flakes of bone have been lifted in a characteristic manner, suggesting that the bone was fresh and pliable when the injury was inflicted. If the two holes were made simultaneously, the points of the two sharp objects responsible must have been approximately 33 mm apart.

狒狒骨架的结构远远不如同等重量牛科动物的骨架坚固。狒狒的脊柱软到足以能够被吃掉，而它的前脚和后脚既肥美又可口。

这不能说明猎豹参与了斯瓦特克朗斯的骨骼收集：它们只是受人摆布的实验动物。然而，根据猎豹的摄食行为而得到的关于猎物骨架相对强度的结论也能适用于豹，我们知道豹对猎物骨架的破坏程度远远高于猎豹。豹具有吃掉几乎整个狒狒的潜在能力，不过头骨通常是要被剩下来的。从狒狒中得到的结论也同样适用于具有普通灵长类动物结构的原始人类，但要考虑到它们的体型更大。在斯瓦特克朗斯的堆积物中极少能发现灵长类肢骨化石的原因之一可能在于以下事实：这些骨头的可鉴定部位通常会被豹嚼碎，剩下的只有不完整的无法鉴定的杆状碎片。斯瓦特克朗斯的许多无法鉴定的骨骼碎片很可能来自南方古猿。

在斯瓦特克朗斯发现的南方古猿头骨碎片中有一块似乎能为我们提供豹活动的直接证据。该标本，即 SK 54，由一个接近成年的人科成员的大部分颅顶组成，这个人科成员被认为是傍人。该标本于 1949 年在斯瓦特克朗斯遗址被发现，随后威斯康星的鲁滨逊博士用醋酸浸洗法对其进行了处理。如图 4 所示，该化石包括左右顶骨、部分枕骨和大部分额骨。额骨已经向下扭曲，几乎贯穿了冠状缝的全部宽度，而整个标本由于在沉积物中受到压力而略微变平了。从比较该顶骨的大小和其他斯瓦特克朗斯标本顶骨大小的结果来看，这个头骨的颅腔容量约为 500 cm³。格外有意思的是：在头骨后面、靠近人字缝处有两个孔，每块顶骨上各有一个（见图 4）。每个孔在外侧方向上都有所拉长，最短直径约为 6 mm。这两个孔显然是由两个尖头物戳出来的，该尖头物的尖端略有一点分开。在顶骨的内表面，骨片以一种很有特色的方式被提高了，这说明顶骨在受到损伤时还是鲜嫩的。如果这两个孔是同时被戳出来的，那么两个尖头物的尖端之间必定相距 33 mm 左右。

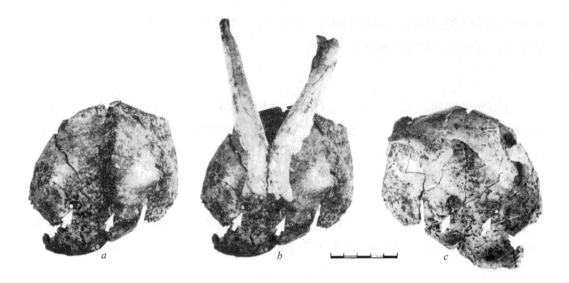

Fig. 4. Part of a hominid skull (*SK* 54) from Swartkrans showing holes thought to have been made by canine teeth. (*a*) Posterior view of left and right parietals and part of the occipital (arrows indicate position of holes on either side of lambda); (*b*) same view as in (*a*) to show how spacing of holes is matched by that of canine tips in the type mandible of the Swartkrans leopard; (*c*) internal view showing flakes of bone lifted when the parietals were pierced.

In a recent survey of the evidence for intrahuman killing in the Pleistocene, Roper[13] mentions this particular specimen and quotes Ardrey's argument[14] that the holes indicate that the australopithecine was struck twice on the back of the head with a pointed object. I believe that it is more likely that the two holes were made simultaneously by the canine teeth of a carnivore spaced about 33 mm apart. The question remains as to which species may have been involved and in what position the australopithecine's head could have been held when the damage was inflicted.

The Swartkrans fossils include remains of carnivores ranging in size from lion to mongoose. In many cases they represent forms closely related to living species so that, although the fossil material is fragmentary, modern skulls may be used to give an indication of typical canine spacings. Making use of the collection in the Transvaal Museum, distances between upper and lower canine tips have been measured on skulls of seven extant species (Table 2).

Table 2. Distances between Upper and Lower Canine Tips on Skulls of Seven Carnivores

Carnivore	*n*	Upper canine spacing, mm		Lower canine spacing, mm	
		Range	Mean	Range	Mean
Lion, *Panthera leo*	17	60–81	67	53–69	59
Brown hyaena, *H. brunnea*	17	44–57	52	43–57	52
Spotted hyaena, *Crocuta crocula*	8	49–57	54	42–54	46
Male leopard, *Panthera pardus*	10	33–47	39	31–41	36
Female leopard, *Panthera pardus*	14	33–42	38	30–38	33
Jackal, *Canis mesomelas*	20	21–28	25	21–26	23
Fox, *Vulpes chama*	11	14–19	16	11–16	13
Mongoose, *Cynictis penicillate*	25	10–13	11	9–13	10

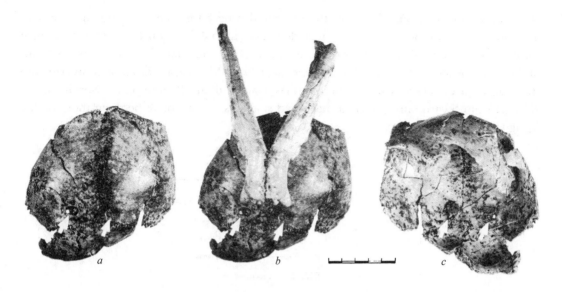

图 4. 一件斯瓦特克朗斯人科头骨（SK 54）的一部分，上面的孔被认为是犬齿造成的。(a) 左右顶骨以及部分枕骨的后面观（箭头指向的是人字缝两边的孔的位置）；(b) 与 (a) 相同的视图，目的是为了展示两个孔之间的间距与斯瓦特克朗斯豹典型下颌骨上的犬齿齿尖之间的间距相符；(c) 显示出当顶骨被刺穿时骨片抬高的内侧面观。

在最近的一项对更新世时期人类之间相互杀戮证据的调查中，罗珀 [13] 提到了这个特殊标本，并且引用了阿德里的论据 [14]，阿德里认为这两个孔表明南方古猿的头后部曾被一个尖头物击中了两次。我相信更可能的情况是这两个孔是由某种食肉动物相距约 33 mm 的两枚犬齿同时造成的。那么这两个孔究竟是哪个物种造成的以及当损伤发生时南方古猿的头部处于什么位置，这些仍然是疑问。

斯瓦特克朗斯化石包括的食肉动物化石从狮子到黄鼬大小不等。这些化石所代表的动物类型大多与现生物种有很密切的关系，所以尽管化石化材料非常破碎，但是我们可以用现代头骨来说明典型的犬齿间距。利用德兰士瓦博物馆的收藏品，我们对 7 件现存物种头骨的上下犬齿齿尖之间的距离进行了测量（表 2）。

表 2. 7 件食肉动物头骨上的上、下犬齿齿尖之间的距离

食肉动物	n	上犬齿之间的间距，mm		下犬齿之间的间距，mm	
		范围	平均值	范围	平均值
狮子	17	60~81	67	53~69	59
棕鬣狗	17	44~57	52	43~57	52
斑鬣狗	8	49~57	54	42~54	46
雄豹	10	33~47	39	31~41	36
雌豹	14	33~42	38	30~38	33
胡狼	20	21~28	25	21~26	23
南非狐	11	14~19	16	11~16	13
黄鼬	25	10~13	11	9~13	10

Results are plotted in Fig. 5, where the estimated ranges for some of the extinct species are also shown. On the basis of rather inadequate material, it is concluded that the known Swartkrans sabre-toothed cats had canine spacings intermediate between those of lions and spotted hyaenas. The type skull of *Leecyaena forfex* (*SK* 314) is comparable in size with the skull of a *Crocuta* and had upper canines spaced about 46 mm apart. Specimens of *Lycyaena* from Swartkrans appear to have had spacings intermediate between those of the living hyaenas and leopards.

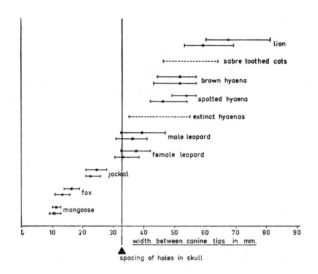

Fig. 5. Canine spacings in some South African carnivores. Observed ranges and means are plotted for upper and lower teeth while estimated ranges for Swartkrans sabre-tooths and extinct hyaenas are included.

It will be seen that the figure of 33 mm, representing the distance between the holes in the fossil skull, falls well within the leopard rang and coincides with the mean figure for lower canine spacings in the female skulls measured. It must be remembered, however, that the fossil leopards seem to have been a fraction smaller than their living counterparts. It can naturally not be proved that a leopard (or any other specific carnivore) was responsible for the damage to the fossil skull, but considering the various lines of evidence from Swartkrans, it is very probable that leopard canines did, in fact, produce the holes. Similar holes in the skulls of baboons eaten by leopards in the Suswa caves of Kenya have been reported by Simons[15]. A leopard typically kills its prey with a firm grip across the throat but, when subsequently dragging a dead animal to a protected feeding place, will often grip the head in its jaws. Fig. 6 shows how this may have been done in the case of the Swartkrans *Paranthropus* child. Under the weight of the child's body, the leopard's lower canines may have penetrated the rather thin parietal bones, while the upper canines were firmly embedded in the face.

图 5 中画出了所得的结果，某些已灭绝物种的估计范围也在图中标出。根据不太完整的材料推断：已知的斯瓦特克朗斯剑齿虎的犬齿间距介于狮子和斑鬣狗之间。李氏鬣狗的正型标本头骨（*SK* 314）在大小上与斑鬣狗的头骨相当，上犬齿之间的间距约为 46 mm。斯瓦特克朗斯狼鬣狗标本的犬齿间距似乎介于现生鬣狗和豹之间。

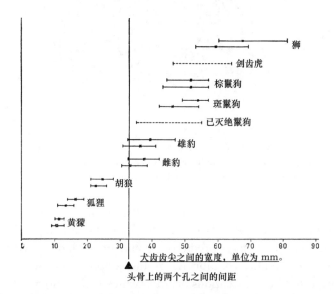

图 5. 几种南非食肉动物的犬齿间距。图中分别标出了上犬齿间距和下犬齿间距的范围和平均值，还包括斯瓦特克朗斯剑齿虎和已灭绝鬣狗的犬齿间距的估计范围。

从图中可以看到：头骨化石上两个孔之间的距离 33 mm 正好处于豹的范围之内，并与在雌豹头骨中测得的下犬齿间距的平均值相符合。但是，一定要记住，化石豹的个头似乎比现生豹小。虽然现在还无法证明是豹（或者某一种其他的食肉动物）对头骨化石造成了这一损伤，但是根据来自斯瓦特克朗斯的各种证据，事实上这两个孔是由豹的犬齿造成的可能性还是很大的。西蒙斯 [15] 曾报道过：在肯尼亚的苏苏瓦山洞穴，被豹吃掉的狒狒的头骨上也有类似的孔。一头豹通常会通过咬断猎物的喉咙来杀死它们，随后豹常常用自己的上下颌叼着死亡动物的头部以便把它拖到一个安全的进食场所。图 6 以斯瓦特克朗斯傍人小孩为例绘出了可能发生的过程。为了抵消小孩的体重，豹的下犬齿可能刺穿了非常薄的顶骨，而上犬齿则深深地嵌入面部。

Fig. 6. Reconstruction showing how the observed damage to the skull of a Swartkrans *Paranthropus* child could have been caused by a leopard. The leopard's lower canines are thought to have penetrated the parietals of the dead child while it was being dragged to a feeding place.

The evidence presented here suggests that *Paranthropus* individuals were preyed upon by leopards in the same way as were baboons, hyraxes and antelope. Leopard predation on humans still occurs today; such predation would presumably have been much more prevalent at a stage of human evolution when australopithecines were neither physically formidable nor protected by the weapons of an advanced technology.

The renewed work at Swartkrans would not have been possible without the support of the Wenner-Gren Foundation for Anthropological Research; in particular I wish to thank Mrs. L. Osmundsen for her interest in the project. I also thank the trustees of the Transvaal Museum and the South African CSIR for their support; the University of the Witwatersrand, present owners of Swartkrans, for help and permission to continue research there; Professor P. V. Tobias and Professor F. Clark Howell for helpful discussions; the South African National Parks Board for research facilities in the National Parks; Mr. and Mrs. A. F. Port for the opportunity to study living carnivores on Valencia Ranch, South West Africa; and my wife, Mr. A. C. Kemp and Mrs. E. A. Voigt for their help with the research.

(**225**, 1112-1119; 1970)

C. K. Brain: Transvaal Museum, Pretoria.

Received January 16, 1970.

References:
1. Robinson, J. T., *Nature* (in the press).

图 6. 示意斯瓦特克朗斯傍人小孩头骨上的损伤可能是由豹引起的情景复原图。在豹将死亡小孩拖到进食之处的过程中，它的下犬齿刺穿了死亡小孩的顶骨。

这里列出的证据说明：豹捕食傍人个体的方式与捕食狒狒、蹄兔和羚羊是相同的。豹子捕食人类的现象至今仍有发生；在人类进化的早期，这种捕食行为想必远比现在普遍得多，那时的南方古猿既没有威慑的外形，也没有技术先进的武器用以保护自己。

如果没有温纳－格伦基金会人类学研究基金的支持，就很难在斯瓦特克朗斯再次开展研究；我要特别感谢奥斯姆德森夫人对本项目的关注。还要感谢德兰士瓦博物馆和南非科学及工业研究委员会的管理人员对我的支持；感谢斯瓦特克朗斯现在的所有者——南非威特沃特斯兰德大学（又名金山大学）支持我并允许我继续在那里开展研究；感谢托拜厄斯教授和克拉克·豪厄尔教授与我进行了很有意义的讨论；感谢南非国家公园管理局为我提供国家公园中的研究设备；感谢波特先生及夫人给了我在非洲西南部巴伦西亚牧场研究现生食肉动物的机会；还要感谢我的妻子、肯普先生和沃伊特夫人对本研究给予的帮助。

（刘皓芳 翻译；董为 审稿）

2. Leakey, M., *Nature* (in the press).

3. Clarke, R. J., Howell, F. C., and Brain, C. K., *Nature* (in the press).

4. Broom, R., and Robinson, J. T., *Nature*, 164, 322 (1949).

5. Robinson, J. T., *S. Afric. J. Sci.*, 57, 3 (1961).

6. Brain, C. K., *S. Afric. J. Sci.*, 63, 378 (1967).

7. Robinson, J. T., *Ann. Transv. Mus.*, 22, 1 (1952).

8. Brain, C. K., *Transv. Mus. Memoir*, No. 11 (1958).

9. Ewer, R. F., *Proc. Zool. Soc. Lond.*, 124, 815 (1953); 124, 839 (1953); 125, 587 (1954).

10. Ewer, R. F., *Proc. Zool. Soc. Lond.*, 126, 83 (1956).

11. Ewer, R. F., in *Background to Evolution in Africa* (edit. by Bishop, W. W., and Clark, J. D.), 109 (University of Chicago Press, 1967).

12. Brain, C. K., *Scient. Pap. Namib Desert Res. Stn.*, 32, 1 (1967); 39, 13 (1969).

13. Roper, M. K., *Current Anthrop.*, 10, 427 (1969).

14. Ardrey, R., *African Genesis*, 300 (Collins, London, 1961).

15. Simons, J. W., *Bull. Cave Exploration Group E. Afric.*, 1, 51 (1966).

RNA-dependent DNA Polymerase in Virions of RNA Tumour Viruses

D. Baltimore

Editor's Note

David Baltimore was a newly graduated PhD when he published this remarkable paper in 1970. Until then, molecular biologists had taken it as an article of faith that genetic information flowed from DNA to RNA to protein: this was known as "Crick's central dogma", but there was then mounting evidence that the genetic code for RNA viruses could be incorporated in the genomes of cells that they infected. Baltimore went on to become the first director of the Whitehead Institute at the Massachusetts Institute of Technology, president of the Rockefeller Institute (later university) in New York and president of the California Institute of Technology. He won a Nobel Prize for this work in 1975.

DNA seems to have a critical role in the multiplication and transforming ability of RNA tumour viruses[1]. Infection and transformation by these viruses can be prevented by inhibitors of DNA synthesis added during the first 8–12 h after exposure of cells to the virus[1-4]. The necessary DNA synthesis seems to involve the production of DNA which is genetically specific for the infecting virus[5,6], although hybridization studies intended to demonstrate virus-specific DNA have been inconclusive[1]. Also, the formation of virions by the RNA tumour viruses is sensitive to actinomycin D and therefore seems to involve DNA-dependent RNA synthesis[1-4,7]. One model which explains these data postulates the transfer of the information of the infecting RNA to a DNA copy which then serves as template for the synthesis of viral RNA[1,2,7]. This model requires a unique enzyme, an RNA-dependent DNA polymerase.

No enzyme which synthesizes DNA from an RNA template has been found in any type of cell. Unless such an enzyme exists in uninfected cells, the RNA tumour viruses must either induce its synthesis soon after infection or carry the enzyme into the cell as part of the virion. Precedents exist for the occurrence of nucleotide polymerases in the virions of animal viruses. Vaccinia[8,9]—a DNA virus, Reo[10,11]—a double-stranded RNA virus, and vesicular stomatitis virus (VSV)[12]—a single-stranded RNA virus, have all been shown to contain RNA polymerases. This study demonstrates that an RNA-dependent DNA polymerase is present in the virions of two RNA tumour viruses: Rauscher mouse leukaemia virus (R-MLV) and Rous sarcoma virus. Temin[13] has also identified this activity in Rous sarcoma virus.

Incorporation of Radioactivity from ³H-TTP by R-MLV

A preparation of purified R-MLV was incubated in conditions of DNA polymerase assay. The preparation incorporated radioactivity from ³H-TTP into an acid-insoluble product

致癌RNA病毒粒子中的RNA依赖性DNA聚合酶

巴尔的摩

编者按

戴维·巴尔的摩于 1970 年发表这篇里程碑式的文章时，他还是个刚毕业的博士生。那时，遗传信息从 DNA 传递到 RNA 再传递到蛋白质已经被分子生物学家们奉为信条，这就是所谓的"克里克中心法则"。但此后越来越多的证据表明，RNA 病毒的遗传密码可以被整合到受其感染的细胞的基因组中。巴尔的摩后来成为麻省理工学院怀特黑德研究所的第一任所长、纽约洛克菲勒研究所（后来的洛克菲勒大学）的所（校）长以及加州理工学院的院长。他凭借本文所述的工作赢得了 1975 年的诺贝尔奖。

DNA 在致癌 RNA 病毒的复制和转化能力上似乎起到了关键作用[1]。在病毒接触到细胞后最初的 8 小时～12 小时内，加入 DNA 合成的抑制剂可以阻止由这些病毒引起的感染和转化[1-4]。虽然杂交研究的结果尚无法证明病毒特异性 DNA 的存在[1]，但在必需的 DNA 合成过程中似乎产生了对感染病毒来说具有遗传特异性的 DNA[5,6]。另外，致癌 RNA 病毒粒子的形成对放线菌素 D 很敏感，因此病毒粒子的形成似乎包括了 DNA 依赖性 RNA 的合成过程[1-4,7]。能够解释这些结果的一个模型是：遗传信息从感染 RNA 传递到一个 DNA 拷贝，然后以此 DNA 拷贝作为模板合成病毒 RNA[1,2,7]。这一模型需要一种特殊的酶，即 RNA 依赖性 DNA 聚合酶。

迄今为止，在任何类型的细胞中还从未发现过以 RNA 为模板合成 DNA 的酶。如果这种酶在未感染的细胞中不存在，那么它就一定是在致癌 RNA 病毒感染细胞后不久被诱导合成出来的，或者是作为病毒粒子的一部分被带入细胞的。有先例表明在动物病毒的病毒粒子中存在着核苷酸聚合酶。DNA 型牛痘病毒[8,9]、双链 RNA 型呼肠孤病毒（Reo）[10,11] 和单链 RNA 型水泡性口炎病毒（VSV）[12] 都已被人们发现含有 RNA 聚合酶。本文所述的研究表明：RNA 依赖性 DNA 聚合酶存在于两种致癌 RNA 病毒粒子中——劳舍尔小鼠白血病病毒（R–MLV）和劳斯肉瘤病毒。特明[13] 还鉴定过劳斯肉瘤病毒中这种酶的活性。

R–MLV 介导 ^{3}H– 胸苷三磷酸（^{3}H–TTP）的放射活性掺入

将纯化的 R–MLV 制剂在 DNA 聚合酶检测条件下进行孵育。在该制剂的制备过

(Table 1). The reaction required Mg^{2+}, although Mn^{2+} could partially substitute and each of the four deoxyribonucleoside triphosphates was necessary for activity. The reaction was stimulated strongly by dithiothreitol and weakly by NaCl (Table 1). The kinetics of incorporation of radioactivity from ^{3}H-TTP by R-MLV are shown in Fig. 1, curve 1. The reaction rate accelerates for about 1 h and then declines. This time-course may indicate the occurrence of a slow activation of the polymerase in the reaction mixture. The activity is approximately proportional to the amount of added virus.

Table 1. Properties of the Rauscher Mouse Leukaemia Virus DNA Polymerase

Reaction system	pmoles ^{3}H-TMP incorporated in 45 min
Complete	3.31
Without magnesium acetate	0.04
Without magnesium acetate + 6 mM $MnCl_2$	1.59
Without dithiothreitol	0.38
Without NaCl	2.18
Without dATP	<0.10
Without dCTP	0.12
Without dGTP	<0.10

A preparation of R-MLV was provided by the Viral Resources Program of the National Cancer Institute. The virus had been purified from the plasma of infected Swiss mice by differential centrifugation. The preparation had a titre of $10^{4.88}$ spleen enlarging doses (50 percent end point) per ml. Before use the preparation was centrifuged at 105,000g for 30 min and the pellet was suspended in 0.137 M NaCl–0.003 M KCl–0.01 M phosphate buffer (pH 7.4)–0.6 mM EDTA (PBS–EDTA) at 1/20 of the initial volume. The concentrated virus suspension contained 3.1 mg/ml. of protein. The assay mixture contained, in 0.1 ml., 5 μmoles Tris-HCl (pH 8.3) at 37°C, 0.6 μmole magnesium acetate, 6 μmoles NaCl, 2 μmoles dithiothreitol, 0.08 μmole each of dATP, dCTP and dGTP, 0.001 μmole [^{3}H-$methyl$]–TTP (708 c.p.m. per pmole) (New England Nuclear) and 15 μg viral protein. The reaction mixture was incubated for 45 min at 37°C. The acid-insoluble radioactivity in the sample was then determined by addition of sodium pyrophosphate, carrier yeast RNA and trichloroacetic acid followed by filtration through a membrane filter and counting in a scintillation spectrometer, all as previously described[12]. The radioactivity of an unincubated sample was subtracted from each value (less than 7 percent of the incorporation in the complete reaction mixture).

程中，³H–TTP 的放射活性被掺入到一种不溶于酸的产物中（表 1）。这一反应需要 Mg^{2+}，不过 Mn^{2+} 也可以部分替代 Mg^{2+}，此外 4 种脱氧核苷三磷酸都是必需的。二硫苏糖醇可以在很大程度上影响反应，而 NaCl 对反应也有微弱的促进作用（表 1）。图 1 中的曲线 1 代表 R–MLV 掺入 ³H–TTP 的放射活性的动力学曲线。反应速率在最初的 1 个小时左右不断加快，之后出现下降。这样的时间进程可能表明，在反应混合物中发生了缓慢的聚合酶激活。反应后产物的放射活性大致与加入的病毒量成正比。

表 1. 劳舍尔小鼠白血病病毒的 DNA 聚合酶的性质

反应系统	反应 45 分钟后掺入的 ³H– 胸苷酸 (³H–TMP) 的 pmol 数
完全的反应系统	3.31
无醋酸镁	0.04
无醋酸镁，但加入了 6 mM $MnCl_2$	1.59
无二硫苏糖醇	0.38
无 NaCl	2.18
无脱氧腺苷三磷酸（dATP）	<0.10
无脱氧胞苷三磷酸（dCTP）	0.12
无脱氧鸟苷三磷酸（dGTP）	<0.10

R–MLV 制剂由美国国家癌症研究所病毒资源组提供。通过差速离心法将病毒从受感染的瑞士小鼠血浆中纯化出来。每 ml 制剂的滴度为 $10^{4.88}$ 脾肿大剂量（终点值的 50%）。使用前，将制剂在 105,000 g 下离心 30 分钟，并将沉淀物悬浮于 0.137 M NaCl、0.003 M KCl、0.01 M 磷酸盐缓冲液（pH 值 7.4）、0.6 mM EDTA（PBS–EDTA）（译者注：PBS 为磷酸缓冲液，EDTA 为乙二胺四乙酸）中，使最终体积为初始体积的 1/20。浓缩的病毒悬浮液含有 3.1 mg/ml 蛋白。在 0.1 ml 反应混合物中含有 5 µmol Tris–HCl（pH 值 8.3，37℃）（译者注：Tris–HCl 为三羟甲基氨基甲烷盐酸盐）、0.6 µmol 醋酸镁、6 µmol NaCl、2 µmol 二硫苏糖醇以及 dATP、dCTP 和 dGTP 各 0.08 µmol，0.001 µmol [³H–甲基]–TTP（每 pmol 为 708 次放射性计数 / 分钟）（新英格兰核公司）和 15 µg 病毒蛋白。反应混合物在 37℃下孵育 45 分钟。加入焦磷酸钠、载体酵母 RNA 和三氯乙酸，接着用滤膜过滤后再在闪烁谱仪上计数，由此测定样品中不溶于酸的物质的放射性，上述所有操作在我们以前发表的文章中均已涉及 [12]。测定得到的每个数值都要减去未孵育样品的放射活性（小于完全反应混合物中掺入程度的 7%）。

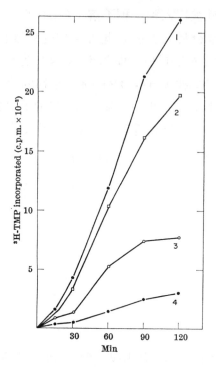

Fig. 1. Incorporation of radioactivity from ³H-TTP by the R-MLV DNA polymerase in the presence and absence of ribonuclease. A 1.5-fold standard reaction mixture was prepared with 30 μg of viral protein and ³H-TTP (specific activity 950 c.p.m. per pmole). At various times, 20 μl. aliquots were added to 0.5 ml. of non-radioactive 0.1 M sodium pyrophosphate and acid insoluble radioactivity was determined[12]. For the preincubated samples, 0.06 ml. of H_2O and 0.01 ml. of R-MLV (30 μg of protein) were incubated with or without 10 μg of pancreatic ribonuclease at 22°C for 20 min, chilled and brought to 0.15 ml. with a concentrated mixture of the components of the assay system. Curve 1, no treatment; curve 2, preincubated; curve 3, 10 μg ribonuclease added to the reaction mixture; curve 4, preincubated with 10 μg ribonuclease.

For other viruses which have nucleotide polymerases in their virions, there is little or no activity demonstrable unless the virions are activated by heat, proteolytic enzymes or detergents[8-12]. None of these treatments increased the activity of the R-MLV DNA polymerase. In fact, incubation at 50°C for 10 min totally inactivated the R-MLV enzyme as did inclusion of trypsin (50 μg/ml.) in the reaction mixture. Addition of as little as 0.01 percent "Triton N-101" (a non-ionic detergent) also markedly depressed activity.

Characterization of the Product

The nature of the reaction product was investigated by determining its sensitivity to various treatments. The product could be rendered acid-soluble by either pancreatic deoxyribonuclease or micrococcal nuclease but was unaffected by pancreatic ribonuclease or by alkaline hydrolysis (Table 2). The product therefore has the properties of DNA. If 50 μg/ml. of deoxyribonuclease was added to a reaction mixture there was no loss of

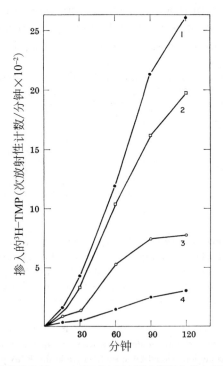

图 1. 在存在和不存在核糖核酸酶时，R–MLV 的 DNA 聚合酶催化掺入 ^{3}H–TTP 的放射活性。用 30 μg 病毒蛋白和 ^{3}H–TTP（比活度为每 pmol 950 次放射性计数 / 分钟）制备 1.5 倍的标准反应混合物。分别在不同时间点将 20 μl 的等分样品加入 0.5 ml 非放射性的 0.1 M 焦磷酸钠中，然后测定酸不溶性物质的放射性 [12]。预先孵育的样品指的是：将 0.06 ml 水和 0.01 ml R–MLV（30 μg 蛋白）分别在加入和不加入 10 μg 胰核糖核酸酶的条件下于 22℃ 下孵育 20 分钟，冷却后加入检测系统组分的浓缩混合物直至终体积达到 0.15 ml。曲线 1 代表没有进行过任何处理的实验组；曲线 2 代表预先孵育的实验组；曲线 3 代表在反应混合物中加入 10 μg 核糖核酸酶的实验组；曲线 4 代表用 10 μg 核糖核酸酶预先孵育的实验组。

对于病毒粒子中含核苷酸聚合酶的其他病毒，如果不提前用加热、加入蛋白水解酶或加入去垢剂来激活病毒粒子的话，就只能检测到很少的放射活性或根本检测不到放射活性 [8-12]。所有这些处理均不会增加 R–MLV 的 DNA 聚合酶的活性。事实上，在 50℃ 下孵育 10 分钟就可以使 R–MLV 的酶完全失活，和向反应混合物中加入胰蛋白酶（50 μg/ml）的效果一样。添加浓度仅为 0.01% 的"曲拉通 N–101"（一种非离子型去垢剂）同样会使活性显著降低。

反应产物的特性

可以根据反应产物对各种处理的敏感性来研究它的特性。用胰脱氧核糖核酸酶或微球菌核酸酶处理该产物可使其溶于酸，但用胰核糖核酸酶处理或加碱水解并不能使该产物受到任何影响（表 2）。因此，该产物具有 DNA 的性质。如果向反应混合物中加入 50 μg/ml 的脱氧核糖核酸酶，不溶于酸的产物将不会有任何损失。因此，

acid-insoluble product. The product is therefore protected from the enzyme, probably by the envelope of the virion, although merely diluting the reaction mixture into 10 mM $MgCl_2$ enables the product to be digested by deoxyribonuclease (Table 2).

Table 2. Characterization of the Polymerase Product

Expt.	Treatment	Acid-insoluble radioactivity	Percentage undigested product
1	Untreated	1,425	(100)
	20 μg deoxyribonuclease	125	9
	20 μg micrococcal nuclease	69	5
	20 μg ribonuclease	1,361	96
2	Untreated	1,644	(100)
	NaOH hydrolysed	1,684	100

For experiment 1, 93 μg of viral protein was incubated for 2 h in a reaction mixture twice the size of that described in Table 1, with ³H-TTP having a specific activity of 1,133 c.p.m. per pmole. A 50 μl. portion of the reaction mixture was diluted to 5 ml. with 10 mM $MgCl_2$ and 0.5 ml. aliquots were incubated for 1.5 h at 37°C with the indicated enzymes. (The sample with micrococcal nuclease also contained 5 mM $CaCl_2$.) The samples were then chilled, precipitated with trichloroacetic acid radioactivity was counted. For experiment 2, two standard reaction mixtures were incubated for 45 min at 37°C, then to one sample was added 0.1 ml. of 1 M NaOH and it was boiled for 5 min. It was then chilled and both samples were precipitated with trichloroacetic acid and counted. In a separate experiment (unpublished) it was shown that the alkaline hydrolysis conditions would completely degrade the RNA product of the VSV virion polymerase.

Localization of the Enzyme and Its Template

To investigate whether the DNA polymerase and its template were associated with the virions, a R-MLV suspension was centrifuged to equilibrium in a 15–50 percent sucrose gradient and fractions of the gradient were assayed for DNA polymerase activity. Most of the activity was found at the position of the visible band of virions (Fig. 2). The density at this band was 1.16 g/cm³, in agreement with the known density of the virions[14]. The polymerase and its template therefore seem to be constituents of the virion.

上述产物未受酶破坏的原因很可能是由于病毒粒子的包封，不过只要用 10 mM 的 $MgCl_2$ 稀释反应混合物即可使产物被脱氧核糖核酸酶降解掉（表 2）。

表 2. 聚合酶产物的特性

实验	处理方法	不溶于酸的产物的放射活性	未被降解产物所占的百分比
1	未处理	1,425	(100)
	20 μg 脱氧核糖核酸酶	125	9
	20 μg 微球菌核酸酶	69	5
	20 μg 核糖核酸酶	1,361	96
2	未处理	1,644	(100)
	NaOH 水解	1,684	100

对于实验 1，取 93 μg 病毒蛋白孵育 2 小时，孵育所用的反应混合物的体积是表 1 中所述的 2 倍，其中 ^3H–TTP 的放射性比活度为每 pmol 1,133 次放射性计数 / 分钟。用 10 mM $MgCl_2$ 将 50 μl 反应混合物稀释到 5 ml，然后等分成 0.5 ml 并分别加入表中所示的酶，再在 37℃ 下孵育 1.5 小时（加入微球菌核酸酶的样品中还含有 5 mM $CaCl_2$）。然后将样品冷却，用三氯乙酸沉淀后进行放射性计数。对于实验 2，取两份标准的反应混合物在 37℃ 下孵育 45 分钟，然后再向其中一份样品中加入 0.1 ml 浓度为 1 M 的 NaOH 并煮沸 5 分钟。随后使其冷却，再将两个样品都用三氯乙酸沉淀并进行放射性计数。有一个独立实验（尚未发表）的结果可以说明，用碱水解能完全降解掉在 VSV 病毒粒子聚合酶作用下产生的 RNA 产物。

酶的定位及其模板

为了研究 DNA 聚合酶及其模板是否与病毒粒子有关，将 R–MLV 的悬浮液在 15% ~ 50% 的蔗糖梯度中进行离心，达到平衡后测定该梯度下各组分的 DNA 聚合酶活性。结果发现，大部分活性落在了病毒粒子的可见带部分（图 2）。此带处的密度为 1.16 g/cm³，与该病毒粒子的已知密度相符 [14]。这样看来聚合酶及其模板可能就是病毒粒子的组成部分。

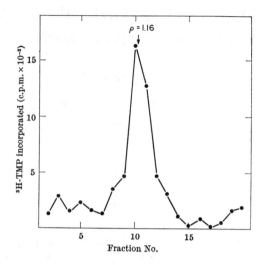

Fig. 2. Localization of DNA polymerase activity in R-MLV by isopycnic centrifugation. A preparation of R–MLV containing 150 µg of protein in 50 µl. was layered over a linear 5.2 ml. gradient of 15–50 percent sucrose in PBS–EDTA. After centrifugation for 2 h at 60,000 r.p.m. in the Spinco "SW65" rotor, 0.27 ml. fractions of the gradient were collected and 0.1 ml. portions of each fraction were incubated for 60 min in a standard reaction mixture. The acid-precipitable radioactivity was then collected and counted. The density of each fraction was determined from its refractive index. The arrow indicates the position of a sharp, visible band of light-scattering material which occurred at a density of 1.16.

The Template is RNA

Virions of the RNA tumour viruses contain RNA but no DNA[15,16]. The template for the virion DNA polymerase is therefore probably the viral RNA. To substantiate further that RNA is the template, the effect of ribonuclease on the reaction was investigated. When 50 µg/ml. of pancreatic ribonuclease was included in the reaction mixture, there was a 50 percent inhibition of activity during the first hour and more than 80 percent inhibition during the second hour of incubation (Fig. 1, curve 3). If the virions were preincubated with the enzyme in water at 22°C and the components of the reaction mixture were then added, an earlier and more extensive inhibition was evident (Fig. 1, curve 4). Preincubation in water without ribonuclease caused only a slight inactivation of the virion polymerase activity (Fig. 1, curve 2). Increasing the concentration of ribonuclease during preincubation could inhibit more than 95 percent of the DNA polymerase activity (Table 3). To ensure that the inhibition by ribonuclease was attributable to the enzymic activity of the added protein, two other basic proteins were preincubated with the virions. Only ribonuclease was able to inhibit the reaction (Table 3). These experiments substantiate the idea that RNA is the template for the reaction. Hybridization experiments are in progress to determine if the DNA is complementary in base sequence to the viral RNA.

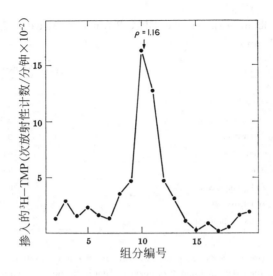

图 2. 用等密度梯度离心法对 R–MLV 中 DNA 聚合酶活性的定位。将 50 μl 含 150 μg 蛋白质的 R–MLV 样品铺在 5.2 ml 用 PBS–EDTA 配制的 15% ~ 50% 的蔗糖线性梯度上面。使用斯平科公司"SW65"型转头在 60,000 转 / 分钟的条件下离心 2 小时，然后按每份 0.27 ml 收集梯度组分，并从每个梯度组分中取出 0.1 ml 在标准反应混合物中孵育 60 分钟。随后收集酸不溶性物质并对其进行放射性计数。通过测定每个组分的折射率来确定其密度。箭头指示出了在密度为 1.16 处光散射物质的一个尖锐的可见条带。

模 板 是 RNA

致癌 RNA 病毒粒子中含有 RNA，但是没有 DNA[15,16]。因此，病毒粒子 DNA 聚合酶的模板很可能是病毒 RNA。为进一步证实 RNA 就是模板，研究了核糖核酸酶对反应的影响。当向反应混合物中加入 50 μg/ml 的胰核糖核酸酶时，在第 1 小时末有 50% 的活性被抑制，在第 2 小时末有超过 80% 的活性被抑制（图 1，曲线 3）。如果在 22℃ 的水中把病毒粒子与酶放在一起进行预孵育，然后再加入反应混合物的组分，那么抑制作用就会出现得更早而且更加严重（图 1，曲线 4）。如果是在没有核糖核酸酶的水中预孵育，那么病毒粒子的聚合酶活性只会发生轻微的失活（图 1，曲线 2）。在预孵育期间，增加核糖核酸酶的浓度能使 DNA 聚合酶的活性被抑制 95% 以上（表 3）。为了确认抑制作用源自所加入蛋白质中的核糖核酸酶的活性，用另外两种碱性蛋白与病毒粒子一起进行了预孵育。结果只有核糖核酸酶能够抑制反应（表 3）。这些实验证实了 RNA 就是反应模板的设想。目前正在进行杂交实验以确定产生的 DNA 是否与病毒 RNA 的碱基序列互补。

Table 3. Effect of Ribonuclease on the DNA Polymerase Activity of Rauscher Mouse Leukaemia Virus

Conditions	pmoles ^{3}H-TMP incorporation
No preincubation	2.50
Preincubated with no addition	2.20
Preincubated with 20 µg/ml. ribonuclease	0.69
Preincubated with 50 µg/ml. ribonuclease	0.31
Preincubated with 200 µg/ml. ribonuclease	0.08
Preincubated with no addition	3.69
Preincubated with 50 µg/ml. ribonuclease	0.52
Preincubated with 50 µg/ml. lysozyme	3.67
Preincubated with 50 µg/ml. cytochrome c	3.97

In experiment 1, for the preincubation, 15 µg of viral protein in 5 µl. of solution was added to 45 µl. of water at 4°C, containing the indicated amounts of enzyme. After incubation for 30 min at 22°C, the samples were chilled and 50 µl. of a 2-fold concentrated standard reaction mixture was added. The samples were incubated at 37°C for 45 min and acid-insoluble radio-activity was measured. In experiment 2, the same procedure was followed, except that the preincubation was for 20 min at 22°C and the 37°C incubation was for 60 min.

Ability of the Enzyme to Incorporate Ribonucleotides

The deoxyribonucleotide incorporation measured in these experiments could be the result of an RNA polymerase activity in the virion which can polymerize deoxyribonucleotides when they are provided in the reaction mixture. The VSV RNA polymerase and the R-MLV DNA polymerase were therefore compared. The VSV RNA polymerase incorporated only ribonucleotides. At its pH optimum of 7.3 (my unpublished observation), in the presence of the four common ribonucleoside triphosphates, the enzyme incorporated ^{3}H-GMP extensively[12]. At this pH, however, in the presence of the four deoxyribonucleoside triphosphates, no ^{3}H-TMP incorporation was demonstrable (Table 4). Furthermore, replacement of even a single ribonucleotide by its homologous deoxyribonucleotide led to no detectable synthesis (my unpublished observation). At pH 8.3, the optimum for the R-MLV DNA polymerase, the VSV polymerase catalysed much less ribonucleotide incorporation and no significant deoxyribonucleotide incorporation could be detected.

表 3. 核糖核酸酶对劳舍尔小鼠白血病病毒中 DNA 聚合酶活性的影响

条件	掺入到样品中的 ^{3}H– 胸苷酸（^{3}H–TMP）的 pmol 数
未进行预孵育	2.50
预孵育时不加任何物质	2.20
预孵育时加 20 μg/ml 核糖核酸酶	0.69
预孵育时加 50 μg/ml 核糖核酸酶	0.31
预孵育时加 200 μg/ml 核糖核酸酶	0.08
预孵育时不加任何物质	3.69
预孵育时加 50 μg/ml 核糖核酸酶	0.52
预孵育时加 50 μg/ml 溶菌酶	3.67
预孵育时加 50 μg/ml 细胞色素 c	3.97

在实验 1 的预孵育过程中，将 5 μl 含 15 μg 病毒蛋白的样品溶液加入 45 μl 含有如表中所示的一定量酶的 4℃水中。在 22℃下孵育 30 分钟后，将样品冷却并加入 50 μl 两倍浓缩的标准反应混合物。随后将此样品在 37℃下孵育 45 分钟，并测定酸不溶性物质的放射性。在实验 2 中，除了 22℃下的预孵育时间为 20 分钟和 37℃下的孵育时间为 60 分钟以外，其余步骤同实验 1。

酶掺入核糖核苷酸的能力

在上述实验中测到的脱氧核糖核苷酸被掺入的现象，可能是由病毒粒子中的一种 RNA 聚合酶的活性造成的，这种酶能够在反应混合物中存在脱氧核糖核苷酸时使脱氧核糖核苷酸发生聚合反应。因此，对 VSV 的 RNA 聚合酶与 R–MLV 的 DNA 聚合酶进行了比较。VSV 的 RNA 聚合酶只掺入核糖核苷酸。在溶液 pH 值达到该酶最适 pH 值 7.3（尚未发表的观察结果）且 4 种常见核苷三磷酸都存在的情况下，该酶能大量地掺入 ^{3}H– 鸟苷酸（^{3}H–GMP）[12]。但是，如果所用的是 4 种脱氧核苷三磷酸，那么在该 pH 值下是观察不到 ^{3}H–TMP 的掺入的（表 4）。此外，即使只将一种核苷三磷酸替换成与之对应的脱氧核苷三磷酸，也会导致检测不到合成作用（尚未发表的观察结果）。当 pH 值为 R–MLV 的 DNA 聚合酶的最适 pH 值——8.3 时，VSV 的聚合酶在催化核糖核苷酸掺入的能力上大为减弱，并且也检测不到脱氧核糖核酸被掺入的明显迹象。

Table 4. Comparison of Nucleotide Incorporation by Vesicular Stomatitis Virus and Rauscher Mouse Leukaemia Virus

Precursor	pH	Incorporation in 45 min (pmoles)	
		Vesicular stomatitis virus	Mouse leukaemia virus
³H-TTP	8.3	<0.01	2.3
³H-TTP (omit dATP)	8.3	N.D.	0.06
³H-TTP (omit dATP; plus ATP)	8.3	N.D.	0.08
³H-GTP	8.3	0.43	<0.03
³H-GTP	7.3	3.7	<0.03

When ³H-TTP was the precursor, standard reaction conditions were used (see Table 1). When ³H-GTP was the precursor, the reaction mixture contained, in 0.1 ml., 5 μmoles Tris-HCl (pH as indicated), 0.6 μmoles magnesium acetate, 0.3 μmoles mercaptoethanol, 9 μmoles NaCl, 0.08 μmole each of ATP, CTP, UTP; and 0.001 μmole ³H-GTP (1,040 c.p.m. per pmole). All VSV assays included 0.1 percent "Triton N-101" (ref. 12) and 2–5 μg of viral protein. The R-MLV assays contained 15 μg of viral protein.

The R-MLV polymerase incorporated only deoxyribonucleotides. At pH 8.3, ³H-TMP incorporation was readily demonstrable but replacement of dATP by ATP completely prevented synthesis (Table 4). Furthermore, no significant incorporation of ³H-GMP could be found in the presence of the four ribonucleotides. At pH 7.3, the R-MLV polymerase was also inactive with ribonucleotides. The polymerase in the R-MLV virions is therefore highly specific for deoxyribonucleotides.

DNA Polymerase in Rous Sarcoma Virus

A preparation of the Prague strain of Rous sarcoma virus was assayed for DNA polymerase activity (Table 5). Incorporation of radioactivity from ³H-TTP was demonstrable and the activity was severely reduced by omission of either Mg^{2+} or dATP from the reaction mixture. RNA-dependent DNA polymerase is therefore probably a constituent of all RNA tumour viruses.

Table 5. Properties of the Rous Sarcoma Virus DNA Polymerase

Reaction system	pmoles ³H-TMP incorporated in 120 min
Complete	2.06
Without magnesium acetate	0.12
Without dATP	0.19

A preparation of the Prague strain (sub-group C) of Rous sarcoma virus[16] having a titre of 5×10^7 focus forming units per ml. was provided by Dr. Peter Vogt. The virus was purified from tissue culture fluid by differential centrifugation. Before use the preparation was centrifuged and the pellet dissolved in 1/10 of the initial volume as described for the R-MLV preparation. For each assay 15 μl. of the concentrated Rous sarcoma virus preparation was assayed in standard reaction mixture by incubation for 2 h. An unincubated control sample had radioactivity corresponding to 0.14 pmole which was subtracted from the experimental values.

These experiments indicate that the virions of Rauscher mouse leukaemia virus and Rous

表 4. 水泡性口炎病毒和劳舍尔小鼠白血病病毒掺入核苷酸能力的对比

前体	pH 值	45 分钟的掺入量（pmol）	
		VSV	R–MLV
³H–TTP	8.3	<0.01	2.3
³H–TTP（未加 dATP）	8.3	未定义	0.06
³H–TTP（未加 dATP，加 ATP）	8.3	未定义	0.08
³H–GTP	8.3	0.43	<0.03
³H–GTP	7.3	3.7	<0.03

当所用前体为 ³H–TTP 时，用标准反应条件进行实验（见表 1）。当所用前体为 ³H– 鸟苷三磷酸（³H–GTP）时，0.1 ml 反应混合物中含有：5 μmol Tris–HCl（pH 值如表中所示），0.6 μmol 醋酸镁，0.3 μmol 巯基乙醇，9 μmol NaCl，腺苷三磷酸（ATP）、胞苷三磷酸（CTP）和尿苷三磷酸（UTP）各 0.08 μmol 以及 0.001 μmol ³H–GTP（每 pmol 1,040 次放射性计数 / 分钟）。在所有的 VSV 检测中均加入 0.1% 的 "曲拉通 N–101"（参考文献 12）及 2 μg ~ 5 μg 病毒蛋白。在检测 R–MLV 时加入 15 μg 病毒蛋白。

R–MLV 聚合酶只掺入脱氧核糖核苷酸。在 pH 值为 8.3 时很容易观察到 ³H–TMP 被掺入，但如果用 ATP 代替 dATP，合成过程就会完全被抑制（表 4）。而且，当存在 4 种核糖核苷酸时，几乎观察不到 ³H–GMP 被掺入的现象。在 pH 值为 7.3 时，R–MLV 聚合酶也没有掺入核糖核苷酸的活性。因此，R–MLV 病毒粒子中的聚合酶是对脱氧核糖核苷酸具有高度特异性的。

劳斯肉瘤病毒中的 DNA 聚合酶

用劳斯肉瘤病毒布拉格株制剂来检测 DNA 聚合酶的活性（表 5）。结果表明 ³H–TTP 的放射活性被掺入，而且当反应混合物中缺少 Mg²⁺ 或 dATP 时，酶活性都会大大降低。因此，RNA 依赖性 DNA 聚合酶很可能是所有致癌 RNA 病毒的一种组分。

表 5. 劳斯肉瘤病毒 DNA 聚合酶的性质

反应系统	反应 120 分钟后掺入到样品中的 ³H–TMP 的 pmol 数
完全的反应系统	2.06
无醋酸镁	0.12
无 dATP	0.19

由彼得 – 沃格特博士提供的劳斯肉瘤病毒布拉格株（C 亚株）制剂[16] 的滴度为每 ml 5×10⁷ 个病灶形成单位。用差速离心法从组织培养液中纯化病毒。使用前先对样品进行离心，然后将沉淀物溶解为上述 R–MLV 实验中初始体积的 1/10。对于每次检测，取 15 μl 浓缩的劳斯肉瘤病毒样品，在标准反应混合物中孵育 2 小时后进行测定。未孵育的对照样品的放射性相当于 0.14 pmol，这个值要从各组实验数据中减去。

这些实验表明：在劳舍尔小鼠白血病病毒和劳斯肉瘤病毒的病毒粒子中含有一

sarcoma virus contain a DNA polymerase. The inhibition of its activity by ribonuclease suggests that the enzyme is an RNA-dependent DNA polymerase. It seems probable that all RNA tumour viruses have such an activity. The existence of this enzyme strongly supports the earlier suggestions[1-7] that genetically specific DNA synthesis is an early event in the replication cycle of the RNA tumour viruses and that DNA is the template for viral RNA synthesis. Whether the viral DNA ("provirus")[2] is integrated into the host genome or remains as free template for RNA synthesis will require further study. It will also be necessary to determine whether the host DNA-dependent RNA polymerase or a virus-specific enzyme catalyses the synthesis of viral RNA from the DNA.

I thank Drs. G. Todaro, F. Rauscher and R. Holdenreid for their assistance in providing the mouse leukaemia virus. This work was supported by grants from the US Public Health Service and the American Cancer Society and was carried out during the tenure of an American Society Faculty Research Award.

(**226**, 1209-1211; 1970)

David Baltimore: Department of Biology, Massachusetts Institute of Technology, Cambridge, Massachusetts 02139.

Received June 2, 1970.

References:

1. Green, M., *Ann. Rev. Biochem.*, **39** (1970, in the press).

2. Temin, H. M., *Virology*, **23**, 486 (1964).

3. Bader, J. P., *Virology*, **22**, 462 (1964).

4. Vigier, P., and Golde, A., *Virology*, **23**, 511 (1964).

5. Duesberg, P. H., and Vogt, P. K., *Proc. US Nat. Acad. Sci.*, **64**, 939 (1969).

6. Temin, H. M., in *Biology of Large RNA Viruses* (edit. by Barry, R., and Mahy, B.) (Academic Press, London, 1970).

7. Temin, H. M., *Virology*, **20**, 577 (1963).

8. Kates, J. R., and McAuslan, B. R., *Proc. US Nat. Acad. Sci.*, **58**, 134 (1967).

9. Munyon, W., Paoletti, E., and Grace, J. T. J., *Proc. US Nat. Acad. Sci.*, **58**, 2280 (1967).

10. Shatkin, A. J., and Sipe, J. D., *Proc. US Nat. Acad. Sci.*, **61**, 1462 (1968).

11. Borsa, J., and Graham, A. F., *Biochem. Biophys. Res. Commun.*, **33**, 895 (1968).

12. Baltimore, D., Huang, A. S., and Stampfer, M., *Proc. US Nat. Acad. Sci.*, **66** (1970, in the press).

13. Temin, H. M., and Mizutani, S., *Nature*, **226**, 1211 (1970) (following article).

14. O'Conner, T. E., Rauscher, F. J., and Zeigel, R. F., *Science*, **144**, 1144 (1964).

15. Crawford, L. V., and Crawford, E. M., *Virology*, **13**, 227 (1961).

16. Duesberg, P., and Robinson, W. S., *Proc. US Nat. Acad. Sci.*, **55**, 219 (1966).

17. Duff, R. G., and Vogt, P. K., *Virology*, **39**, 18 (1969).

种 DNA 聚合酶。该酶的活性能被核糖核酸酶抑制，这暗示它是一种 RNA 依赖性 DNA 聚合酶。这样看来，很可能所有的致癌 RNA 病毒都具有这种活性。这种酶的存在强有力地支持了以前的一个假设 [1-7]：在致癌 RNA 病毒的复制循环中，遗传特异性 DNA 的合成是一个早期事件，而且 DNA 是病毒 RNA 的合成模板。到底病毒 DNA（"前病毒"）[2] 是整合到了宿主基因组中还是仍然作为 RNA 合成的自由模板？这个问题有待于进一步的研究。还有一个问题也有必要解决：到底是宿主中的 DNA 依赖性 RNA 聚合酶还是一种病毒特有的酶催化了由 DNA 合成病毒 RNA 的过程？

我要感谢托达罗博士、劳舍尔博士和霍尔登里德博士协助提供了小鼠白血病病毒。本工作得到了美国公共卫生署和美国癌症协会的资助，是我在获得一个美国学会的教师研究奖期间进行的。

（荆玉祥 翻译；孙军 审稿）

RNA-dependent DNA Polymerase in Virions of Rous Sarcoma Virus

H. M. Temin and S. Mizutani

Editor's Note

When Howard Temin proposed his "DNA provirus hypothesis" in 1964, it challenged dogma and was met with scepticism. Based on the results of genetic experiments and the effects of metabolic inhibitors, he proposed that viral RNA is copied to DNA in infected cells. Then in 1970, Temin and Satoshi Mizutani, and American virologist David Baltimore, working independently, reported the discovery of a viral enzyme, now known as reverse transcriptase, that makes DNA from an RNA template. The results offered clear-cut support for the DNA provirus hypothesis, and helped earn Temin and Baltimore a Nobel Prize. As predicted, the discovery has also led to major advances in our understanding of gene transcription, cancer and human retroviruses.

INFECTION of sensitive cells by RNA sarcoma viruses requires the synthesis of new DNA different from that synthesized in the S-phase of the cell cycle (refs. 1, 2 and unpublished results of D. Boettiger and H. M. T.); production of RNA tumour viruses is sensitive to actinomycin D[3,4]; and cells transformed by RNA tumour viruses have new DNA which hybridizes with viral RNA[5,6]. These are the basic observations essential to the DNA provirus hypothesis—replication of RNA tumour viruses takes place through a DNA intermediate, not through an RNA intermediate as does the replication of other RNA viruses[7].

Formation of the provirus is normal in stationary chicken cells exposed to Rous sarcoma virus (RSV), even in the presence of 0.5 µg/ml. cycloheximide (our unpublished results). This finding, together with the discovery of polymerases in virions of vaccinia virus and of reovirus[8-11], suggested that an enzyme that would synthesize DNA from an RNA template might be present in virions of RSV. We now report data supporting the existence of such an enzyme, and we learn that David Baltimore has independently discovered a similar enzyme in virions of Rauscher leukaemia virus[12].

The sources of virus and methods of concentration have been described[13]. All preparations were carried out in sterile conditions. Concentrated virus was placed on a layer of 15 percent sucrose and centrifuged at 25,000 r.p.m. for 1 h in the "SW 25.1" rotor of the Spinco ultracentrifuge on to a cushion of 60 percent sucrose. The virus band was collected from the interphase and further purified by equilibrium sucrose density gradient centrifugation[14]. Virus further purified by sucrose velocity density gradient centrifugation gave the same results.

劳斯肉瘤病毒粒子中的RNA依赖性DNA聚合酶

特明，水谷哲

编者按

当霍华德·特明在 1964 年提出 "DNA 前病毒假说" 时，该假说因为与经典理论冲突而遭到了人们的质疑。特明根据遗传学实验的结果和代谢抑制剂的作用效果提出，在受感染的细胞中病毒的 RNA 复制为 DNA。随后在 1970 年，特明和水谷哲以及美国病毒学家戴维·巴尔的摩分别独立地研究并报道了一项发现，即有一种病毒酶能从 RNA 模板合成 DNA，这种酶现在被称为逆转录酶。上述结果显然支持了 DNA 前病毒假说，这项成果使特明和巴尔的摩赢得了诺贝尔奖。正如所预测的那样，这一发现对我们理解基因转录、癌症和人类逆转录病毒也产生了重大的影响。

敏感细胞被 RNA 肉瘤病毒感染需要合成新的 DNA，这种新合成的 DNA 不同于在细胞周期的 S 期合成的 DNA（参考文献 1、2 以及伯蒂格和特明尚未发表的研究结果）。致癌 RNA 病毒的产生对放线菌素 D 敏感 [3,4]。被致癌 RNA 病毒感染的细胞中含有与病毒 RNA 杂交的新 DNA[5,6]。这些基本的观测结果都支持 DNA 前病毒假说，即致癌 RNA 病毒的复制是通过一个 DNA 中间体进行的，而不是像其他 RNA 病毒那样是通过一个 RNA 中间体进行的 [7]。

即使在 0.5 μg/ml 放线菌酮存在时，在暴露于劳斯肉瘤病毒（RSV）的稳定期鸡细胞中也能正常形成前病毒（我们尚未发表的研究结果）。这一研究结果以及在牛痘病毒和呼肠孤病毒的病毒粒子中聚合酶的发现 [8-11] 都说明：在 RSV 的病毒粒子中或许存在着一种能以 RNA 为模板合成 DNA 的酶。我们在本文中要报告能支持这种酶存在的证据，另外我们也获悉戴维·巴尔的摩已经独立地在劳舍尔小鼠白血病病毒的病毒粒子中发现了一种类似的酶 [12]。

病毒来源和浓缩方法如前所述 [13]。所有操作均在无菌条件下进行。将浓缩的病毒置于 15% 的蔗糖层上面，底部是 60% 的蔗糖，用斯平科超速离心机的 "SW 25.1" 型转头在 25,000 转 / 分钟的速度下离心 1 小时。从界面处收集病毒带，然后用平衡的蔗糖密度梯度离心法进一步纯化 [14]。也可以用蔗糖速度密度梯度离心法进一步纯化病毒，得到的效果是一样的。

The polymerase assay consisted of 0.125 µmoles each of dATP, dCTP, and dGTP (Calbiochem) (in 0.02 M Tris-HCl buffer at pH 8.0, containing 0.33 M EDTA and 1.7 mM 2-mercaptoethanol); 1.25 µmoles of $MgCl_2$ and 2.5 µmoles of KCl; 2.5 µg phosphoenolpyruvate (Calbiochem); 10 µg pyruvate kinase (Calbiochem); 2.5 µCi of ^{3}H-TTP (Schwarz) (12 Ci/mmole); and 0.025 ml. of enzyme (10^8 focus forming units of disrupted Schmidt–Ruppin virus, $A_{280 \, nm} = 0.30$) in a total volume of 0.125 ml. Incubation was at 40°C for 1 h. 0.025 ml. of the reaction mixture was withdrawn and assayed for acid-insoluble counts by the method of Furlong[15].

To observe full activity of the enzyme, it was necessary to treat the virions with a non-ionic detergent (Tables 1 and 4). If the treatment was at 40°C the presence of dithiothreitol (DTT) was necessary to recover activity. In most preparations of virions, however, there was some activity: 5–20 percent of the disrupted virions, in the absence of detergent treatment, which probably represents disrupted virions in the preparation. It is known that virions of RNA tumour viruses are easily disrupted[16,17], so that the activity is probably present in the nucleoid of the virion.

Table 1. Activation of Enzyme

System	^{3}H-TTP incorporated (d.p.m.)
No virions	0
Non-disrupted virions	255
Virions disrupted with "Nonidet"	
At 0° + DTT	6,730
At 0° – DTT	4,420
At 40° + DTT	5,000
At 40° – DTT	425

Purified virions untreated or incubated for 5 min at 0°C or 40°C with 0.25 percent "Nonidet P-40" (Shell Chemical Co.) with 0 or 1 percent dithiothreitol (DTT) (Sigma) were assayed in the standard polymerase assay.

The kinetics of incorporation with disrupted virions are shown in Fig. 1. Incorporation is rapid for 1 h. Other experiments show that incorporation continues at about the same rate for the second hour. Preheating disrupted virus at 80°C prevents any incorporation, and so does pretreatment of disrupted virus with crystalline trypsin.

聚合酶检测所用试剂的总体积为 0.125 ml，其中含有：脱氧腺苷三磷酸（dATP）、脱氧胞苷三磷酸（dCTP）和脱氧鸟苷三磷酸（dGTP）各 0.125 μmol（卡尔生物化学公司）（溶于 pH 值为 8.0 的 0.02 M Tris–HCl 缓冲液中，其中还含有 0.33 M EDTA 和 1.7 mM 2– 巯基乙醇）；1.25 μmol MgCl$_2$ 和 2.5 μmol KCl；2.5 μg 磷酸烯醇丙酮酸盐（卡尔生物化学公司）；10 μg 丙酮酸激酶（卡尔生物化学公司）；2.5 μCi ^{3}H– 胸苷三磷酸（^{3}H–TTP）（许瓦兹制药有限公司）（12 Ci/mmol）；0.025 ml 酶（通过破碎活性为 10^8 个病灶形成单位的施密特 – 鲁宾病毒得到，$A_{280\ nm}$ = 0.30）。在 40℃ 下孵育 1 小时。取 0.025 ml 反应混合液，用弗朗法 [15] 对酸不溶性物质进行放射性计数检测。

为了观察这种酶的完整活性，必须用非离子型去垢剂来处理病毒粒子（表 1 和表 4）。如果处理过程是在 40℃ 下进行的，那么就需要用二硫苏糖醇（DTT）来恢复酶活性。不过，大多数没有用去垢剂处理的病毒粒子样品也有一定的活性——是破碎病毒粒子活性的 5% ~ 20%，这很可能说明在制样过程中有一些病毒粒子破碎了。众所周知，致癌 RNA 病毒粒子是很容易发生破碎的 [16,17]，所以活性很可能存在于病毒粒子的拟核中。

表 1. 酶的激活

系统	被掺入的 ^{3}H–TTP（衰变次数 / 分钟）
无病毒粒子	0
未经破碎处理的病毒粒子	255
用 "诺乃洗涤剂" 破碎的病毒粒子	
在 0℃ 下，加入 DTT	6,730
在 0℃ 下，未加入 DTT	4,420
在 40℃ 下，加入 DTT	5,000
在 40℃ 下，未加入 DTT	425

用标准的聚合酶检测法测定：未经处理的纯化病毒粒子，或者用 0.25% 的 "诺乃洗涤剂 P–40"（壳牌化学公司）处理并在不加或加入 1% DTT（西格玛公司）的条件下于 0℃ 或 40℃ 孵育 5 分钟的纯化病毒粒子。

图 1 显示了破碎病毒粒子的掺入动力学曲线。掺入过程在 1 小时之内进行得很快。另有一些实验结果显示：掺入过程在第 2 个小时内仍能以大致相同的速率持续进行。在 80℃ 下预热破碎的病毒粒子会使掺入过程完全终止，而用结晶胰蛋白酶预处理破碎的病毒粒子也会有同样的现象出现。

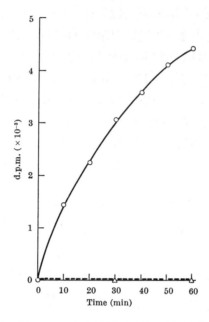

Fig. 1. Kinetics of incorporation. Virus treated with "Nonidet" and dithiothreitol at 0°C and incubated at 37°C (O—O) or 80°C (Δ- - -Δ) for 10 min was assayed in a standard polymerase assay. O, Unheated; Δ, heated.

Fig. 2 demonstrates that there is an absolute requirement for $MgCl_2$, 10 mM being the optimum concentration. The date in Table 2 show that $MnCl_2$ can substitute for $MgCl_2$ in the polymerase assay, but $CaCl_2$ cannot. Other experiments show that a monovalent cation is not required for activity, although 20 mM KCl causes a 15 percent stimulation. Higher concentrations of KCl are inhibitory: 60 percent inhibition was observed at 80 mM.

Table 2. Requirements for Enzyme Activity

System	^{3}H-TTP incorporated (d.p.m.)
Complete	5,675
Without $MgCl_2$	186
Without $MgCl_2$, with $MnCl_2$	5,570
Without $MgCl_2$, with $CaCl_2$	18
Without dATP	897
Without dCTP	1,780
Without dGTP	2,190

Virus treated with "Nonidet" and dithiothreitol at 0°C was incubated in the standard polymerase assay with the substitutions listed.

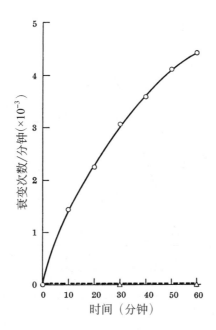

图 1. 掺入过程的动力学曲线。在 0℃ 下用"诺乃洗涤剂"和 DTT 处理病毒，并在 37℃ （O—O）或 80℃ （Δ - - - Δ）下孵育 10 分钟，然后用标准的聚合酶检测法进行测定。O 代表未加热的实验结果；Δ 代表加热的实验结果。

　　图 2 表明 MgCl$_2$ 是掺入过程所必需的，其最适浓度为 10 mM。由表 2 中的数据可知：在聚合酶检测实验中，MnCl$_2$ 可以替代 MgCl$_2$，但 CaCl$_2$ 不行。虽然 20 mM KCl 能使活性增加 15%，但另一些实验结果显示一价阳离子并不是活性所必需的。高浓度的 KCl 会起抑制作用：当 KCl 浓度为 80 mM 时，观察到的抑制程度可达 60%。

表 2. 酶活性检测所需条件

系统	被掺入的 ³H−TTP（衰变次数 / 分钟）
完全的反应系统	5,675
缺 MgCl$_2$	186
缺 MgCl$_2$，但有 MnCl$_2$	5,570
缺 MgCl$_2$，但有 CaCl$_2$	18
缺 dATP	897
缺 dCTP	1,780
缺 dGTP	2,190

在 0℃ 下用"诺乃洗涤剂"和 DTT 处理病毒样品，然后在标准的聚合酶检测实验条件下孵育，在每组实验中所用的替换物如表中所示。

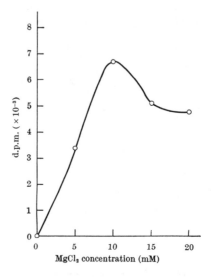

Fig. 2. MgCl₂ requirement. Virus treated with "Nonidet" and dithiothreitol at 0°C was incubated in the standard polymerase assay with different concentrations of MgCl₂.

When the amount of disrupted virions present in the polymerase assay was varied, the amount of incorporation varied with second-order kinetics. When incubation was carried out at different temperatures, at broad optimum between 40°C and 50°C was found. (The high temperature of this optimum may relate to the fact that the normal host of the virus is the chicken.) When incubation was carried out at different pHs, a broad optimum at pH 8–9.5 was found.

Table 2 demonstrates that all four deoxyribonucleotide triphosphates are required for full activity, but some activity was present when only three deoxyribonucleotide triphosphates were added and 10–20 percent of full activity was still present with only two deoxyribonucleotide triphosphates. The activity in the presence of three deoxyribonucleotide triphosphates is probably the result of the presence of deoxyribonucleotide triphosphates in the virion. Other host components are known to be incorporated in the virion of RNA tumour viruses[18,19].

The data in Table 3 demonstrate that incorporation of thymidine triphosphate was more than 99 percent abolished if the virions were pretreated at 0° with 1 mg ribonuclease per ml. Treatment with 50 µg/ml ribonuclease at 20°C did not prevent all incorporation of thymidine triphosphate, which suggests that the RNA of the virion may be masked by protein. (Lysozyme was added as a control for non-specific binding of ribonuclease to DNA.) Because the ribonuclease was heated for 10 min at 80°C or 100°C before use to destroy deoxyribonuclease it seems that intact RNA is necessary for incorporation of thymidine triphosphate.

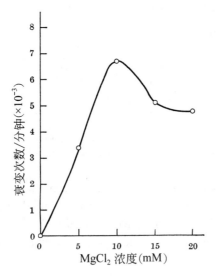

图 2. 对 MgCl₂ 的需求。在 0℃ 下用 "诺乃洗涤剂" 和 DTT 处理病毒样品，然后在 MgCl₂ 浓度不同的标准聚合酶检测体系中孵育病毒。

当聚合酶检测实验中所用的破碎病毒粒子的量发生变化时，掺入的量也随之变化，变化规律符合二级动力学。尝试了各种不同温度下的孵育之后，我们发现最适宜的温度范围很宽，在 40℃ ~ 50℃ 之间都可以。（最适温度范围处于较高的温度区可能与这种病毒的宿主通常是鸡有关。）尝试了在各种不同的 pH 值下进行孵育后，我们发现最适的 pH 值也有很宽的范围，在 8 ~ 9.5 之间。

表 2 说明 4 种脱氧核苷三磷酸对于完整活性来说都是必需的，但是，当仅仅加入 3 种脱氧核苷三磷酸时也会表现出一定的活性，甚至在只有 2 种脱氧核苷三磷酸时仍存在 10% ~ 20% 的活性。只加入 3 种脱氧核苷三磷酸时显示出的活性，很可能源自病毒粒子中存在脱氧核苷三磷酸。现在已经知道宿主的其他一些组分可以被掺入致癌 RNA 病毒粒子之中 [18,19]。

表 3 中的数据说明：如果在 0℃ 下用终浓度为 1 mg/ml 的核糖核酸酶预处理病毒粒子，那么胸苷三磷酸的掺入将被破坏 99% 以上。在 20℃ 下用 50 μg/ml 的核糖核酸酶处理并未完全阻止胸苷三磷酸的掺入，这说明病毒粒子的 RNA 有可能被蛋白质遮盖住了。（加入溶菌酶作为核糖核酸酶与 DNA 之间非特异性结合的对照。）因为使用前已将核糖核酸酶在 80℃ 或 100℃ 下加热了 10 分钟，这样就完全破坏了脱氧核糖核酸酶，所以完整的 RNA 对于胸苷三磷酸的掺入来说似乎是必需的。

Table 3. RNA Dependence of Polymerase Activity

Treatment	^{3}H-TTP incorporated (d.p.m.)
Non-treated disrupted virions	9,110
Disrupted virions preincubated with ribonuclease A (50 μg/ml.) at 20°C for 1 h	2,650
Disrupted virions preincubated with ribonuclease A (1 mg/ml.) at 0°C for 1 h	137
Disrupted virions preincubated with lysozyme (50 μg/ml.) at 0°C for 1 h	9,650

Disrupted virions were incubated with ribonuclease A (Worthington) which was heated at 80°C for 10 min, or with lysozyme at the indicated concentration in the specified conditions, and a standard polymerase assay was performed.

To determine whether the enzyme is present in supernatants of normal cells or in RNA leukaemia viruses, the experiment of Table 4 was performed. Normal cell supernatant did not contain activity even after treatment with "Nonidet". Virions of avian myeloblastosis virus (AMV) contained activity that was increased ten-fold by treatment with "Nonidet".

Table 4. Source of Polymerase

Source	^{3}H-TTP incorporated (d.p.m.)
Virions of SRV	1,410
Disrupted virions of SRV	5,675
Virions of AMV	1,875
Disrupted virions of AMV	12,850
Disrupted pellet from supernatant of uninfected cells	0

Virions of Schmidt–Ruppin virus (SRV) were prepared as before (experiment of Table 2). Virions of avian myeloblastosis virus (AMV) and a pellet from uninfected cells were prepared by differential centrifugation. All disrupted preparations were treated with "Nonidet" and dithiothreitol at 0°C and assayed in a standard polymerase assay. The material used per tube was originally from 45 ml. of culture fluid for SRV, 20 ml. for AMV, and 20 ml. for uninfected cells.

The nature of the product of the polymerase assay was investigated by treating portions with deoxyribonuclease, ribonuclease or KOH. About 80 percent of the product was made acid soluble by treatment with deoxyribonuclease, and the product was resistant to ribonuclease and KOH (Table 5).

Table 5. Nature of Product

Treatment	Residual acid-insoluble ^{3}H-TTP (d.p.m.)	
	Experiment A	Experiment B
Buffer	10,200	8,350
Deoxyribonuclease	697	1,520
Ribonuclease	10,900	7,200
KOH	—	8,250

A standard polymerase assay was performed with "Nonidet" treated virions. The product was incubated in buffer or 0.3 M KOH at 37°C for 20 h or with (A) 1 mg/ml. or (B) 50 μg/ml. of deoxyribonuclease I (Worthington), or with 1 mg/ml. of ribonuclease A (Worthington) for 1 h at 37°C, and portions were removed and tested for acid-insoluble counts.

表 3. 聚合酶活性的 RNA 依赖性

处理方法	被掺入的 ³H–TTP（衰变次数 / 分钟）
未经处理的破碎病毒粒子	9,110
在 20℃ 下用核糖核酸酶 A（50 μg/ml）预孵育 1 小时的破碎病毒粒子	2,650
在 0℃ 下用核糖核酸酶 A（1 mg/ml）预孵育 1 小时的破碎病毒粒子	137
在 0℃ 下用溶菌酶（50 μg/ml）预孵育 1 小时的破碎病毒粒子	9,650

向破碎的病毒粒子中加入在 80℃ 下加热了 10 分钟的核糖核酸酶 A（沃辛顿），或者如表中所示浓度的溶菌酶，然后在特定的条件下孵育，之后对病毒粒子进行标准的聚合酶检测实验。

为了确定这种酶到底是存在于正常细胞的上清液中还是存在于 RNA 白血病病毒中，我们又进行了表 4 中的实验。即使在使用"诺乃洗涤剂"处理后，正常细胞的上清液也不具有活性。而禽类成髓细胞瘤病毒（AMV）的病毒粒子具有活性且在用"诺乃洗涤剂"处理后，其活性会提高 10 倍。

表 4. 聚合酶的来源

来源	被掺入的 ³H–TTP（衰变次数 / 分钟）
SRV 病毒粒子	1,410
破碎的 SRV 病毒粒子	5,675
AMV 病毒粒子	1,875
破碎的 AMV 病毒粒子	12,850
未感染细胞上清液中的破碎沉淀物	0

用前文所述的方法（表 2 中的实验）制备施密特 - 鲁宾病毒（SRV）。由差速离心法得到禽类成髓细胞瘤病毒（AMV）的病毒粒子和未感染细胞的离心沉淀物。所有破碎的样品都在 0℃ 下用"诺乃洗涤剂"和 DTT 处理，并用标准的聚合酶检测实验进行分析。每个管中所用的样品分别来自 45 ml SRV 培养液、20 ml AMV 培养液和 20 ml 未感染细胞培养液。

为了研究聚合酶检测实验所得产物的性质，我们用脱氧核糖核酸酶、核糖核酸酶或 KOH 分别处理了部分样品。在用脱氧核糖核酸酶处理后有大约 80% 的产物能溶于酸，但产物能耐受核糖核酸酶处理或 KOH 处理（表 5）。

表 5. 产物的性质

处理方法	残留的酸不溶性 ³H–TTP（衰变次数 / 分钟）	
	实验 A	实验 B
缓冲液	10,200	8,350
脱氧核糖核酸酶	697	1,520
核糖核酸酶	10,900	7,200
KOH	—	8,250

用标准的聚合酶检测实验分析"诺乃洗涤剂"处理后的病毒粒子。将产物在下列条件下进行孵育：37℃ 下在缓冲液或 0.3 M KOH 中孵育 20 小时；或者加入（A）1 mg/ml 或（B）50 μg/ml 脱氧核糖核酸酶 I（沃辛顿）或 1 mg/ml 核糖核酸酶 A（沃辛顿）在 37℃ 下孵育 1 小时。孵育后除去上清液，并对酸不溶性物质进行放射性计数检测。

To determine if the polymerase might also make RNA, disrupted virions were incubated with the four ribonucleotide triphosphates, including ^{3}H-UTP (Schwarz, 3.2 Ci/mmole). With either $MgCl_2$ or $MnCl_2$ in the incubation mixture, no incorporation was detected. In a parallel incubation with deoxyribonucleotide triphosphates, 12,200 d.p.m. of ^{3}H-TTP was incorporated.

These results demonstrate that there is a new polymerase inside the virions of RNA tumour viruses. It is not present in supernatants of normal cells but is present in virions of avian sarcoma and leukaemia RNA tumour viruses. The polymerase seems to catalyse the incorporation of deoxyribonucleotide triphosphates into DNA from an RNA template. Work is being performed to characterize further the reaction and the product. If the present results and Baltimore's results[12] with Rauscher leukaemia virus are upheld, they will constitute strong evidence that the DNA provirus hypothesis is correct and that RNA tumour viruses have a DNA genome when they are in cells and an RNA genome when they are in virions. This result would have strong implications for theories of viral carcinogenesis and, possibly, for theories of information transfer in other biological systems[20].

This work was supported by a US Public Health Service research grant from the National Cancer Institute. H. M. T. holds a research career development award from the National Cancer Institute.

(**226**, 1211-1213; 1970)

Howard M. Temin and Satoshi Mizutani: McArdle Laboratory for Cancer Research, University of Wisconsin, Madison, Wisconsin 53706.

Received June 15, 1970.

References:
1. Temin, H. M., *Cancer Res.*, **28**, 1835 (1968).

2. Murray, R. K., and Temin, H. M., *Intern. J. Cancer* (in the press).

3. Temin, H. M., *Virology*, **20**, 577 (1963).

4. Baluda, M. B., and Nayak, D. P., *J. Virol.*, **4**, 554 (1969).

5. Temin, H. M., *Proc. US Nat. Acad. Sci.*, **52**, 323 (1964).

6. Baluda, M. B., and Nayak, D. P., in *Biology of Large RNA Viruses* (edit. by Barry, R., and Mahy, B.) (Academic Press, London, 1970).

7. Temin, H. M., *Nat. Cancer Inst. Monog.*, **17**, 557 (1964).

8. Kates, J. R., and McAuslan, B. R., *Proc. US Nat. Acad. Sci.*, **57**, 314 (1967).

9. Munyon, W., Paoletti, E., and Grace, J. T., *Proc. US Nat. Acad. Sci.*, **58**, 2280 (1967).

10. Borsa, J., and Graham, A. F., *Biochem. Biophys. Res. Commun.*, **33**, 895 (1968).

11. Shatkin, A. J., and Sipe, J. D., *Proc. US Nat. Acad. Sci.*, **61**, 1462 (1968).

12. Baltimore, D., *Nature*, **226**, 1209 (1970) (preceding article).

13. Altaner, C., and Temin, H. M., *Virology*, **40**, 118 (1970).

14. Robinson, W. S., Pitkanen, A., and Rubin, H., *Proc. US Nat. Acad. Sci.*, **54**, 137 (1965).

为了测定这种聚合酶是否也能合成 RNA，我们将破碎的病毒粒子与包括 ^{3}H– 尿苷三磷酸（^{3}H–UTP）（许瓦兹制药有限公司，3.2 Ci/mmol）在内的 4 种核苷三磷酸一起孵育。在孵育混合液中，无论加入 $MgCl_2$ 还是 $MnCl_2$，都未能检测到任何掺入。在用脱氧核苷三磷酸进行孵育的一个平行实验中，我们发现被掺入的 ^{3}H–TTP 达到 12,200 衰变次数 / 分钟。

这些结果表明：在致癌 RNA 病毒的病毒粒子中存在着一种新的聚合酶。这种酶并不存在于正常细胞的上清液中，而是存在于禽类肉瘤病毒和白血病病毒等致癌 RNA 病毒的病毒粒子中。这种聚合酶似乎能够催化将脱氧核苷三磷酸掺入到以 RNA 为模板的 DNA 中的反应。目前我们正在进一步表征这种反应及其产物。如果我们的结果和巴尔的摩用劳舍尔小鼠白血病病毒取得的结果 [12] 能够得到确认的话，这两项结果将有力地支持如下观点，即 DNA 前病毒假说是正确的，并且致癌 RNA 病毒在细胞中时具有一个 DNA 基因组，而在病毒粒子中时则有一个 RNA 基因组。这一结果会对病毒致癌作用的机理，以及还可能会对其他生物系统中的遗传信息传递理论产生重大的影响 [20]。

本工作受到了国家癌症研究所获得的一项美国公共卫生署研究基金的资助。霍华德 · 特明拥有一项由国家癌症研究所提供的研究事业发展基金。

（荆玉祥 翻译；孙军 审稿）

15. Furlong, N. B., *Meth. Cancer Res.*, **3**, 27 (1967).

16. Vogt, P. K., *Adv. Virus. Res.*, **11**, 293 (1965).

17. Bauer, H., and Schafer, W., *Virology*, **29**, 494 (1966).

18. Bauer, H., *Z. Naturforsch.*, **21**b, 453 (1969).

19. Erikson, R. L., *Virology*, **37**, 124 (1969).

20. Temin, H. M., *Persp. Biol. Med.* (in the press).

Fertilization and Cleavage *in vitro* of Preovulator Human Oocytes

R. G. Edwards *et al.*

Editor's Note

Although *in vitro* fertilization (IVF) had yielded full term pregnancies in certain animals, in 1970 human IVF success remained elusive. Robert Edwards realised the importance of harvesting eggs just before ovulation, and set up a collaboration with gynaecological surgeon Patrick Steptoe, who used laparoscopy to recover eggs from infertile women at the appropriate point in their cycle. With further help from assistant Jean Purdy, the team trialled various culture media, and here they report the production of eight- or sixteen-celled human embryos by this method. Eight years later, the team were delighted to announce the birth of the first "test tube" baby, Louise Brown. IVF went on to become a major type of infertility treatment.

HUMAN oocytes removed from patients shortly before ovulation should assist the study of early human development[1]. Previously, oocytes in dictyotene had been prepared for fertilization by maturing them in culture after their recovery from excised ovaries. In animals, fertilization of oocytes *in vitro* and *in vivo* after their maturation in culture results in embryos incapable of sustained growth; foetal development to full term is achieved by recovering oocytes just before ovulation, for example in metaphase of the first meiotic division, and completing their maturation *in vitro* before fertilization. We now present data on the fertilization and cleavage *in vitro* of human ova recovered by laparoscopy just before ovulation; preliminary observations have been given elsewhere[2,3].

Development beyond the pronuclear stage has proved difficult in several species, for example in mice. We have few human embryos for study, and have based our work on observations of animal embryos. Whittingham[4] obtained viable young by fertilizing mouse ova *in vitro* with uterine spermatozoa, culturing the eggs to the 2-cell stage, and then transferring them into a recipient female. Whitten has shown[5] that mouse ova recovered in their pronuclear stage can now be grown to blastocysts with complete success. The media used by Whittingham and Whitten were developed from earlier work[6,7]. In addition to these media, we have also used Ham's F10 medium[8] which has been used with rabbit embryos[9], and Waymouth's medium MB752/1[10] and medium 199[11] which are both widely used in tissue culture.

Oocytes and Spermatozoa

Patients were given injections of human menopausal gonadotrophin and chorionic gonadotrophin to induce follicular growth and maturation. Laparoscopy was performed 30–32 h after the injection of HCG, and each follicle was aspirated separately[1]. The

排卵前的人类卵母细胞在体外的受精和卵裂

爱德华兹等

编者按

虽然已经能够利用体外受精（IVF）在某些动物中实现满期妊娠，但在 1970 年时，人们仍未确定是否能将 IVF 成功应用于人类。罗伯特·爱德华兹意识到在即将排卵前收集卵子的重要性，并与妇科医生帕特里克·斯特普托建立了合作关系，斯特普托曾通过腹腔镜手术提取过处于月经周期适当时间的不孕妇女的卵子。在助手琼·珀迪的进一步协助下，该研究小组尝试了各种不同的培养基，他们在本文中报道了以这种方法培养产生的 8 细胞期或 16 细胞期的人类胚胎。8 年以后，该研究小组兴奋地向世人宣布：第一个"试管"婴儿——路易丝·布朗诞生了。随后 IVF 逐渐成为治疗不孕症的主要方式。

从即将排卵的患者体内取出的人类卵母细胞应该有助于我们研究人类的早期发育 [1]。以前的做法是：从切除的卵巢中提取卵母细胞，然后将处于核网期的卵母细胞培养成熟以用于受精。在培养基中成熟的动物卵母细胞，无论是通过体外受精还是体内受精，所产生的胚胎都不能生长。只能通过提取出即将排出的卵母细胞，例如处于第一次减数分裂中期的卵母细胞，并于受精前在体外培养到成熟阶段，这样胎儿才能发育到足月。现在我们要给出一些关于在即将排卵时用腹腔镜手术提取出的人类卵子在体外受精和卵裂的结果；最初的观察结果已经发表在了其他出版物上 [2,3]。

已经证明有几个物种（例如小鼠）的胚胎在体外发育时很难超越原核期。由于我们仅有极少数人类胚胎可供研究，所以我们的研究成果基本上来自对动物胚胎的观察。惠廷厄姆 [4] 曾用穿入子宫的精子使小鼠卵子在体外受精，将受精卵培养到 2 细胞期后再将其转移到受体母鼠中，通过这样的方法他得到了可存活的幼鼠。惠滕 [5] 曾指出：现在已经可以成功地把原核期提取的小鼠卵子培养到囊胚期。惠廷厄姆和惠滕所用的培养基是从前人的研究工作中 [6,7] 发展而来的。除这些培养基外，我们还使用了此前曾被用于兔胚胎培养 [9] 的哈姆氏 F10 培养基 [8]，以及被广泛用于组织培养的韦莫斯 MB752/1 培养基 [10] 和 M199 培养基 [11]。

卵母细胞和精子

为了诱导卵泡的生长和成熟，我们给患者注射了人绝经期促性腺激素和人绒毛膜促性腺激素。在给患者注射人绒毛膜促性腺激素 30 小时~32 小时后进行腹腔镜手

oocytes were suspended in droplets consisting of fluid from their own follicle (where available), and the medium being tested for fertilization. After incubation for 1–4 h at 37°C the oocytes were washed through two changes of the medium under test before being placed in the suspensions of spermatozoa. Preovulatory oocytes would be ready for fertilization, that is, in metaphase of the second meiotic division, by 3–4 h after collection. Many oocytes were obviously not preovulatory, and therefore unsuitable for fertilization, but all were placed in the fertilization droplets in order to simplify our procedure.

Ejaculated spermatozoa were supplied by the husband. The spermatozoa were washed twice by gentle centrifugation in the medium under test, and made up to a final concentration of between 8×10^5 and 2×10^6/ml. depending on the quality of the sample. The higher numbers were used with samples of poor quality containing many inactive spermatozoa, cellular inclusions, other debris, or viscous seminal fluid. The fertilization droplets were approximately 0.05 ml.

Oocytes classified as preovulatory were surrounded by layers of silvery-appearing corona and cumulus cells in a viscous matrix[1]. Oocytes classified as non-ovulatory were enclosed in a few layers of corona cells. Atretic oocytes had few or no cells. Initially, each oocyte was placed in its own fertilizing droplet, but later all the oocytes recovered from a patient were grouped together. The mass of cells surrounding preovulatory oocytes led inevitably to some dilution of the numbers of spermatozoa in the fertilization droplet.

Fertilization

Bavister's medium had been capable of sustaining fertilization in previous work[12-14], and was therefore used extensively. This medium was slightly modified during the work by reducing the sodium chloride to 0.75 g percent and increasing the KCl to 0.039 g percent, following the analysis of the amount of Na^+ and K^+ in human follicular fluid. Whittingham's medium was used unmodified, and both Waymouth's medium (Flow Laboratories) and Ham's F10 (Flow Laboratories) were modified by raising the pH to 7.5–7.6 with extra bicarbonate and supplementing with bovine serum albumin (0.36 g percent). Sodium pyruvate was also added to Waymouth's medium (1.1 mg percent). The gas phase was either 5 percent CO_2 in air, or 5 percent CO_2, 5 percent O_2 and 90 percent N_2.

The cumulus cells began to dissociate within 2–3 h of insemination, although corona cells remaining closely or loosely attached prevented the examination of some oocytes by low-power microscopy 15 h after insemination. The cells were left in place after attempts to remove them had led to damage of some oocytes. Criteria for judging fertilization were: (*i*) the observation by phase-contrast microscopy of two pronuclei, two polar bodies and, if possible, a sperm tail in the cytoplasm of the oocyte; (*ii*) identification of pronuclei in the ova in culture, using a stereoscopic low-power microscope (this method was inexact, and even impossible when cells persisted around the oocyte, but the eggs remained relatively

术，手术时分开单独吸取每个卵泡 [1]。将卵母细胞悬浮于由其自身卵泡液（如果能够得到的话）和用于受精试验的培养基组成的小液滴中。在 37℃下培养 1 小时 ～ 4 小时后，用待测培养基冲洗卵母细胞两次，然后将其置于精子的悬浮液中。排卵前的卵母细胞在被收集 3 小时～ 4 小时后就可以用于受精了，也就是说，这时正处于第二次减数分裂的中期。显然，其中有许多卵母细胞并不处在排卵前的状态，因此不适于进行受精，但是为了简化操作，我们还是把所有的卵母细胞都放到了受精小液滴中。

排出的精子由患者配偶提供。将精子在待测培养基中低速离心并冲洗两次，根据精子的质量进行调整使其终浓度介于 8×10^5/毫升 ～ 2×10^6/毫升之间。在质量较低的样品中含有许多无活力精子、细胞内含物、其他碎片或者黏稠的精液，对于这种样品就需要有更高数量的精子。受精小液滴的体积约为 0.05 毫升。

一般认为：排卵前的卵母细胞会被黏性基质中的数层银色冠细胞和卵丘细胞包裹 [1]；而尚未进入排卵状态的卵母细胞则被封闭在较少层的冠细胞之中。处于闭锁状态的卵母细胞只有极少的细胞包裹或者完全没有细胞包裹。一开始，每个卵母细胞都被置于其自身的受精小液滴中，但是随后所有从同一患者中提取的卵母细胞都被聚集到了一起。那些包围在排卵前卵母细胞周围的大量细胞必然会使受精小液滴中的精子数受到一定程度的稀释。

受 精 过 程

以前的研究工作曾证明巴维斯特培养基能够维持受精过程 [12-14]，因此该培养基被广泛应用。在分析了人类卵泡液中的 Na^+ 含量和 K^+ 含量之后，我们对此培养基进行了稍许调整：将 NaCl 的浓度降低到 0.75 克 /100 毫升，同时将 KCl 的浓度提高到 0.039 克 /100 毫升。对于惠廷厄姆培养基，我们在使用时未作任何调整；对于韦莫斯培养基（弗洛实验室）和哈姆氏 F10 培养基（弗洛实验室），我们都作了一些改良：加入额外的碳酸氢盐将 pH 值提高到 7.5 ~ 7.6，同时补充牛血清白蛋白（浓度为 0.36 克 /100 毫升）。在韦莫斯培养基中还加入了丙酮酸钠（浓度为 1.1 毫克 /100 毫升）。培养所用的气体环境为含 5% CO_2 的空气，或者为由 5% CO_2、5% O_2 和 90% N_2 组成的混合气体。

虽然在人工授精 15 小时后冠细胞仍然或紧或松地黏附着卵母细胞，这使得我们在低倍显微镜下无法仔细观察某些卵母细胞，但卵丘细胞在人工授精 2 小时～ 3 小时之内就开始脱离下来了。我们多次尝试将周围的这些细胞分离下来，结果导致一些卵母细胞被破坏，之后我们就保留了它们。判断是否受精的标准包括：(1) 通过相差显微镜能在卵母细胞的细胞质中观察到两个原核、两个极体，并且还有可能看到一条精子尾巴；(2) 用低倍体视显微镜能分辨出在培养的卵中有原核（这种方法并

undisturbed for further development); (*iii*) cleavage of the egg, preferably after the identification of pronuclei.

Data on the incidence of fertilization are given in Table 1. Only a few ova could be spared for examination by phase-contrast microscopy. Fertilization was observed in Bavister's medium, modified Waymouth's medium and in Whittingham's medium, but not in modified Ham's F10 on the sole occasion it was used. Low rates of fertilization sometimes seemed to be a consequence of the poor quality of the spermatozoa from some patients.

Table 1. Incidence of Fertilization *in vitro*

Medium	O₂ in gas phase	Total No. of oocytes	Non-ovulatory	Fertilization					
				Phase-contrast Microscopy			Low-power microscopy		
				Unfertilized	Fertilized	Others	1-celled	Pronucleate and cleaved	Pronuclei not seen, egg cleaved
Bavister's	20 percent	105	30	26	3+1*+1?	9	18	11†+2?‡	4
	5 percent	76	–	–	–	–	53§	6+2? ‡	15
	Combined	181	30	26	3+1*+1?	9	71§	17†+4?‡	19
Whittingham's	20 percent	19	3	5	–	3§	5§	2	1
Modified Waymouth's	20 percent	10	2	1	2	2	1	1	1?
Modified Ham's F10	5 percent	2	–	–	–	–	2	–	–

* Sperm in perivitelline space.

† Four of these eggs were transferred while pronucleate into the oviduct of a rabbit.

‡ Probably pronucleate.

§ One egg in each of these groups was found to possess two or more pronuclei between 48 and 72 h after insemination, and might have been undergoing rudimentary parthenogenetic development.

Cleavage

Between 12 and 15 h after insemination, the oocytes were gradually transferred from the medium used for fertilization into various other media for cleavage. At least five embryos were cultured in each medium, except for medium 199. All media were adjusted to a *p*H of approximately 7.3. A total of thirty-eight embryos cleaved in culture (Table 2). Two pronuclei had been observed in many of them. Almost all eggs cleaved twice, and a few completed their fourth cleavage.

不精确，当卵母细胞周围有许多细胞时，甚至无法进行观察，但卵的进一步发育没有受到太大的影响）；(3) 卵的分裂，尤其是在观察到原核之后。

我们在表 1 中给出了受精发生率的数据。用相差显微镜只能观察到少数几个卵子。我们在巴维斯特培养基、经过改良的韦莫斯培养基和惠廷厄姆培养基中都观察到了受精的发生，但是在唯一一次使用改良的哈姆氏 F10 培养基时却没有观察到受精。造成受精率较低的原因有时似乎与某些患者的精子质量欠佳有关。

表 1. 体外受精的发生率

培养基	气相中的氧气含量	卵母细胞总数	尚未进入排卵状态的	受精					
				相差显微镜			低倍显微镜		
				未受精的	受精的	其他	单个细胞	有原核，有卵裂	未见原核，有卵裂
巴维斯特培养基	20%	105	30	26	3+1*+1?	9	18	11†+2?‡	4
	5%	76	—	—	—	—	53§	6+2?‡	15
	混合	181	30	26	3+1*+1?	9	71§	17†+4?‡	19
惠廷厄姆培养基	20%	19	3	5	—	3§	5§	1	1
改良的韦莫斯培养基	20%	10	2	1	2	2	1	1	1?
改良的哈姆氏 F10 培养基	5%	2	—	—	—	—	2	—	—

* 精子处在卵周隙中。
† 在形成原核时，将其中 4 个卵移植到兔子的输卵管中。
‡ 可能形成了原核。
§ 每组实验中都有一个卵在人工授精后 48 小时～72 小时之内被发现存在两个或多个原核，并且这些卵可能一直在进行初期的孤雌发育。

卵　裂

在人工授精后 12 小时～15 小时之内，我们将卵母细胞从受精所用的培养基中逐渐地转移到用于卵裂的各种培养基中。除 M199 培养基外，在其他每种培养基中都至少培养了 5 个胚胎。所有培养基的 pH 值都被调至接近 7.3。在培养时共有 38 个胚胎发生了卵裂（表 2）。我们在其中的很多胚胎中发现了两个原核。几乎所有的卵都分裂了两次，有一些卵甚至完成了第 4 次分裂。

Table 2. Cleavage of the Embryos *in vitro*

Medium	Osmolality (mosm/kg)	O₂ in gas phase (percent)	Total No. embryos	Final No. of cells in the embryos							
				1	2	3	4	4–8	8	8–16	16 or more
Whittingham's	280–290	20	9	1*			1	2	4	1	
	280–290	5	2						2		
Whitten's	270	5	5	1*		1			1	2†	
	325	5	3						2†	1	
Waymouth's with 15 percent inactivated foetal calf serum	325	20	6	1‡				2	2	1	
199 with 20 percent inactivated foetal calf serum	292	20	1				1				
Ham's F10 with 20 percent inactivated foetal calf serum	287	5	2			2					
	300–305	5	10§					1¶	1**	4	3

* Reverted to 1-cell after initial cleavage.

† Both embryos had cells of dissimilar size, and cleared erratically.

‡ Probably reverted to 1-cell after cleavage, but vision obscured by corona cells.

§ One embryo became infected at 2-cell stage.

¶ Cleavage erratic.

** Removed from culture and photographed in 8-cell stage.

Many ova cleaved regularly and evenly in Whittingham's medium, and most reached the 8-celled stage. One embryo behaved anomalously, for it divided into two cells and then reverted to one cell.

Whitten's medium is similar to Whittingham's except for having a lower osmotic pressure. Development became distorted when embryos were cultured in this medium. Cleavage became irregular, a 2-celled egg reverted to one cell, and cytoplasmic division seemed to occur without nuclear division in three embryos. For example, one embryo cleaved very rapidly from 3 cells to an apparently normal 8-celled embryo, but when flattened and stained[15] was found to have only three nuclei. When the osmotic pressure of this medium was increased by adding more Na^+ and K^+ ions, cleavage became more regular although slower than in other media.

Waymouth's medium supplemented with 15 percent v/v foetal calf serum failed to support development from the 8-celled stage, perhaps because of the high osmotic pressure: ova placed in it showed retractions of the cell surface. One embryo cultured in medium 199 became arrested in the 4-cell stage.

表 2. 胚胎在体外的卵裂

培养基	摩尔渗透压浓度（毫渗摩/千克）	气相中的氧气含量（%）	胚胎总数	胚胎中最终的细胞数							
				1	2	3	4	4~8	8	8~16	≥16
惠廷厄姆培养基	280~290	20	9	1*			1	2	4	1	
	280~290	5	2						2		
惠滕培养基	270	5	5	1*		1			1	2†	
	325	5	3						2†	1	
添加 15% 灭活胎牛血清的韦莫斯培养基	325	20	6	1‡			2	2			
添加 20% 灭活胎牛血清的 M199 培养基	292	20	1				1				
添加 20% 灭活胎牛血清的哈姆氏 F10 培养基	287	5	2			2					
	300~305	5	10§					1¶	1**	4	3

* 第一次卵裂后又回到了单细胞状态。
† 两个胚胎都具有大小不等的细胞，而且明显不稳定。
‡ 卵裂后很可能回到了单细胞状态，但因被冠细胞包围而看起来模糊不清。
§ 一个胚胎在 2 细胞期被感染。
¶ 卵裂不稳定。
** 在 8 细胞期从培养基中取出并拍照。

有很多在惠廷厄姆培养基中培养的卵发生了正常且均匀的卵裂，而且大多数达到了 8 细胞期。只有一个胚胎表现异常，因为它在分裂成两个细胞之后又退回到了单细胞状态。

惠滕培养基与惠廷厄姆培养基很相似，只是渗透压低一些。但是在将胚胎放在惠滕培养基中进行培养时，它们的发育就会变得不正常。卵裂变得毫无规则，有一个 2 细胞期的卵又重新回到了单细胞状态，有 3 个胚胎似乎在细胞核没有发生分裂的情况下出现了细胞质的分裂。例如，其中有一个胚胎很快地从 3 个细胞卵裂到了看似正常的 8 细胞期，但是在将其压片染色 [15] 后我们却发现这个胚胎中只有 3 个细胞核。如果通过添加更多的 Na+ 离子和 K+ 离子增加惠滕培养基的渗透压，则虽然卵裂速度要比用其他培养基培养时慢，但卵裂过程会变得更为正常。

用添加 15% 体积比胎牛血清的韦莫斯培养基进行培养不能使卵细胞发育到 8 细胞期以后的状态，这可能是由于渗透压太高造成的：在这一培养基中培养的卵细胞出现了表面萎缩。在 M199 培养基中培养的一个胚胎发育到 4 细胞期就中止了发育。

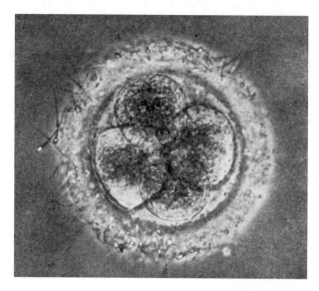

Fig. 1. A 4-celled egg grown in Waymouth's medium supplemented with calf serum. Development of this embryo had evidently been arrested.

Ham's F10 supplemented with 20 percent foetal calf serum was used at two osmotic pressures by adjusting the amount of water added; at 287 milliosmols/kg cleavage appeared abnormal in two embryos. At 300–305 milliosmols/kg seven embryos showed excellent cleavage to a stage approaching or beyond the 16-cell stage. Another embryo photographed in the 8-celled stage (Fig. 2) would probably have cleaved further.

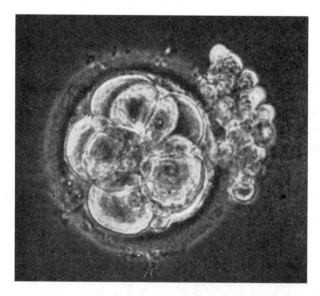

Fig. 2. An 8-celled egg grown in Ham's F10 supplemented with calf serum. It was removed from culture during cleavage, and could have been capable of further development.

The embryos were inspected at various times during culture. The first cleavage occurred

810

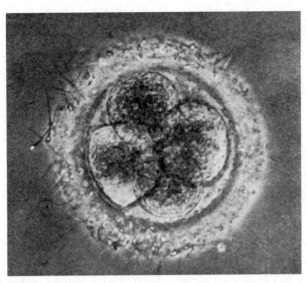

图 1. 一个在添加了牛血清的韦莫斯培养基中培养的 4 细胞期卵。这个胚胎显然已经停止了发育。

通过调节水的含量，我们制备了具有两种不同渗透压的含 20% 胎牛血清的哈姆氏 F10 培养基：在渗透压为 287 毫渗摩 / 千克的培养基中，有 2 个胚胎的卵裂看起来不太正常；在渗透压为 300 毫渗摩 / 千克～305 毫渗摩 / 千克的培养基中，有 7 个胚胎显示出完美的卵裂，接近或者超过了 16 细胞期。我们拍摄了另一个处于 8 细胞期的胚胎（图 2），这个胚胎很可能还会继续卵裂。

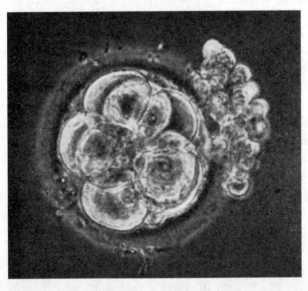

图 2. 一个在添加了牛血清的哈姆氏 F10 培养基中培养的 8 细胞期卵。这个胚胎是在卵裂期间被移出培养基的，要不然它也许还能继续发育。

我们在培养过程中的多个时间点对胚胎进行了检查。结果发现，第一次卵裂发

before 38 h post-insemination. The second cleavage occurred between 38 and $46\frac{1}{2}$h in seven embryos cultured in Whittingham's medium or Ham's F10, and the third cleavage between 51 and 62 h in four embryos. The fourth cleavage occurred before 85 h in embryos grown in Ham's F10. Timings of cleavage in other embryos did not differ from these estimates, but the intervals between recordings were too long for accurate estimates.

The embryos were left in culture until it was clear that development had ceased, that is, approximately 48 h after the previous recording of cell division. Many embryos were now displaying fragmentation of blastomeres or other forms of degeneration. The blastomeres of many embryos that had cleaved normally possessed a single nucleus as judged by phase-contrast microscopy. Whole mounts or flattened preparations of two 16-celled, three 8-celled, and two 4–5-celled embryos revealed the same number of nuclei as cells in each embryo. One of the 16-celled embryos possessed mitoses. One embryo classified while living as 16-celled possessed twenty-one nuclei, although some were smaller than others.

Comment

The amount of free follicular fluid added to the fertilization droplets must have been small, except for that in the viscous cumulus masses surrounding the oocyte. Various factors in follicular fluid are believed to be necessary for stimulating the spermatozoa and inducing changes in the acrosome[16], and steroids present in follicular fluid or synthesized by the granulosa cells could also be involved. Recent studies have indicated that the acrosome is a modified lysosome[17], and capacitation could involve agents known to destabilize lysosomal membranes. Among the most potent of these agents are progesterone and other progestogens[18]. Progesterone has been identified in human follicular fluid aspirated with the oocytes (unpublished work of K. Fotherby and of ourselves), and granulosa cells cultured *in vitro* can also synthesize progesterone and other progestogens[19]. Ultrastructural examination has revealed the acrosomal changes following capacitation are seen as the spermatozoa penetrate between the cumulus cells[20]. We have measured the levels of oestradiol-17β and of LH in many of these fluids[21], and are currently measuring progesterone and 17α-hydroxyprogesterone (unpublished work of K. Fotherby and of ourselves). These fluids could then be tested for their efficacy in inducing the acrosome change in spermatozoa in relation to the known levels of the different steroids. It would seem to be a wise precaution to remove the cells surrounding the oocytes before testing agents inducing capacitation and fertilization *in vitro*.

Cleavage occurred in all ova seen to possess pronuclei, and in some others that could not be examined because of the enveloping cells. Human embryos may have metabolic needs similar to mouse embryos, as judged by their cleavage in simple defined media. Firm conclusions about the value of different media obviously cannot be drawn when so few ova were available. On occasions, four or five embryos were obtained from one

生在人工授精后 38 小时之内。有 7 个在惠廷厄姆培养基或哈姆氏 F10 培养基中培养的胚胎发生第二次卵裂的时间是在人工授精后 38 小时 ~ 46.5 小时之间，有 4 个胚胎的第三次卵裂发生在 51 小时 ~ 62 小时之间。使用哈姆氏 F10 培养基培养的胚胎发生第四次卵裂的时间均在人工授精后 85 小时之内。其他胚胎的卵裂时间与上述估计值相差无几，但是由于各次记录之间的时间间隔太长，因而无法得到精确的结果。

胚胎一直在培养基中培养，直到我们可以确认发育已经停止，更确切地说是在距离之前记录到细胞分裂的大约 48 小时之后。这时，在很多胚胎中出现了卵裂球的碎片化或者其他形式的退化现象。由在相差显微镜下的观察结果可以断定，许多发生正常卵裂的胚胎中的卵裂球具有单个的细胞核。通过检查 2 个处于 16 细胞期的胚胎、3 个处于 8 细胞期的胚胎以及 2 个处于 4 细胞期 ~ 5 细胞期的胚胎的整体包埋封片或压片标本，我们发现每个胚胎中的细胞核数都等于细胞数。有一个 16 细胞期的胚胎正在经历有丝分裂。另外一个被认为正处于 16 细胞期的胚胎则含有 21 个细胞核，不过其中有一些细胞核要比其他细胞核小。

评　论

加入到受精小液滴中的游离卵泡液必须很少，除非卵母细胞被大量有黏性的卵丘细胞包围。一般认为，卵泡液中的各种因子对于增强精子活力和诱导顶体反应是必需的 [16]，而且存在于卵泡液中或者由卵泡颗粒细胞合成的类固醇也有可能会参与这一过程。最近的研究表明，顶体就是一个被修饰过的溶酶体 [17]，一些已知能够降低溶酶体膜稳定性的物质可能参与了精子的获能过程。在这些物质中效力最强的是孕酮和其他一些孕激素 [18]。我们在与卵母细胞一同吸出的人卵泡液中曾发现过有孕酮存在（福瑟比和我们尚未发表的研究结果），而且体外培养的卵泡颗粒细胞也可以合成孕酮和其他孕激素 [19]。对超微结构的检查结果显示，精子获能后出现的顶体反应发生在当精子从卵丘细胞之间穿透时 [20]。此前我们已经在多种卵泡液中测定过 17β- 雌二醇和促黄体生成激素（LH）的含量 [21]，目前我们正在测定孕酮和 17α- 羟孕酮的含量（福瑟比和我们尚未发表的研究结果）。然后，我们就可以测定这些卵泡液在诱导精子顶体反应上的效能，这种效能与各种类固醇的已知含量有关。在检验可在体外条件下诱导精子获能和受精的化学物质时，预先清除掉卵母细胞周围的细胞似乎是一种明智的做法。

卵裂发生在所有能看到有原核的卵以及其他一些因被细胞包裹而无法观察到原核的卵中。根据人类胚胎和小鼠胚胎在一些配方明确的简单培养基中所发生的卵裂情况，就可以推断出人类胚胎的代谢需要可能和小鼠胚胎的类似。由于可利用的人卵细胞数目非常少，所以要准确判断不同培养基的效能显然是不可能的。在有些情

patient, and comparisons could be made between different media. Results with defined media were inferior to those obtained from Ham's F10 supplemented with foetal calf serum and with an osmotic pressure of around 300 osmols. Among various differences, F10 contains salts of heavy metals and more pyruvate than the other media. Development beyond the 8 or 16-cell stage might demand the presence of human oviductal or uterine secretions, and these could be collected with a small chamber placed in the uterus for a few hours[22]. The premature recovery of preovulatory oocytes might be another cause of the arrested development, and they might have to be aspirated even closer to the time of ovulation.

Analysis of the chromosome complement of the embryos can be undertaken now that some cleavages have been timed, although perhaps better delayed until later stages of development are available. Identification of a Y chromosome in some embryos is now critically needed to furnish formal proof to confirm the morphological evidence of fertilization used currently. The quinacrine dyes might detect the Y chromosome in interphase nuclei and mitoses, as they do in somatic cells[23,24], although caution will be necessary, for they fail to stain the Y in most spermatogonia[25], Chromosomal analysis of the embryos should also reveal anomalies in early development, for example, non-disjunction and especially triploidy arising through polyspermy or failure of polar body formation. When more preovulatory oocytes are available their parthenogenetic activation might be used to produce haploid strains of cells, as in mice[26].

One or more embryos have been produced from twenty-nine of the forty-nine patients under treatment in this work. The normality of embryonic development and the efficiency of embryo transfer cannot yet be assessed, although conditions for implantation in the treated patients should be favourable[1].

We thank Mr. B. D. Bavister, Dr. D. G. Whittingham and Professor C. R. Austin for their help, Mr. C. Richardson for his assistance with photographs, and the Ford Foundation, the Manchester Regional Hospital Board and the Oldham and District Hospital Management Committee for financial support. We also thank Searle Scientific Services for their steroid assay.

(**227**, 1307-1309; 1970)

R. G. Edwards, P. C. Steptoe and J. M. Purdy: Physiological Laboratory, University of Cambridge, and Oldham General Hospital, Lancashire.

Received August 25, 1970.

References:

1. Steptoe, P. C., and Edwards, R. G., *Lancet*, i, 683 (1970).

2. Edwards, R. G., in *Symposium on Actual Problems in Fertility, Stockholm* 1970 (Plenum, New York, in the press).

814

况下，我们可以从一名患者身上得到 4 个 ~ 5 个胚胎，这样就可以对不同的培养基进行比较。用配方明确培养基得到的效果比不上用渗透压约为 300 渗摩、添加了胎牛血清的哈姆氏 F10 培养基。F10 培养基含有重金属盐且丙酮酸盐的含量较其他培养基高，这是 F10 培养基与其他培养基的不同之处之一。8 细胞期或 16 细胞期以后的胚胎发育有可能需要人输卵管或子宫的分泌物，可以通过将一个小的容器放在子宫中几小时来收集这些分泌物[22]。过早地取出排卵前卵母细胞可能是导致胚胎发育阻滞的另一个原因，也许吸出卵母细胞的时间应该更接近于排卵期。

既然我们已经确定了一些卵裂过程的时间，那么下一步就可以着手进行胚胎的染色体组分析了，不过也许推迟到能获得更靠后的胚胎发育阶段再进行这一分析会更好一些。为了提供正式的证据来证实目前所使用的受精的形态学证据，我们现在急需从一些胚胎中鉴定出 Y 染色体的存在。利用喹吖因类染料染色也许能检测出间期核中和有丝分裂期间的 Y 染色体，就像在体细胞中一样[23,24]，不过必须小心谨慎，因为喹吖因类染料不能使大多数精原细胞中的 Y 染色体着色[25]。对胚胎的染色体分析应该也能揭示出胚胎早期发育的异常，例如由多精受精或未能形成极体造成的不分离现象甚至出现三倍体。如果我们能得到更多的排卵前卵母细胞，也许就可以利用孤雌激活来产生单倍体细胞系，就像已经在小鼠中建立的细胞系那样[26]。

在这项研究中接受治疗的共有 49 位患者，我们从其中 29 位患者处获得了一个或多个胚胎。目前我们还无法评估正常发育胚胎的比率以及胚胎移植的成功率，尽管受试患者自身的胚胎移植条件应该是很有利的[1]。

感谢巴维斯特先生、惠廷厄姆博士和奥斯汀教授给予我们的帮助，感谢理查森先生帮助我们拍摄照片，感谢福特基金会、曼彻斯特地区医院董事会和奥尔德姆地区综合医院管理委员会在经费方面的资助。还要感谢瑟尔科学服务部帮我们作了类固醇的检测实验。

（吕静 翻译；王敏康 审稿）

3. Steptoe, P. C., in *Schering Symposium on Intrinsic and Extrinsic Factors in Early Mammalian Development* 1970 (edit. by Raspé, G.) (Advances in the Biosciences, 6) (Pergamon/Vieweg, in the press).

4. Whittingham, D. G., *Nature*, **220**, 592 (1968).

5. Whitten, W. K., in *Schering Symposium on Intrinsic and Extrinsic Factors in Early Mammalian Development* 1970 (edit. by Raspé, G.) (Advances in the Biosciences, 6) (Pergamon/Vieweg, in the press).

6. Brinster, R. L., *Exp. Cell Res.*, **32**, 205 (1963).

7. Whitten, W. K., and Biggers, R. L., *J. Reprod. Fert.*, **17**, 390 (1968).

8. Ham, R. G., *Exp. Cell Res.*, **29**, 515 (1963).

9. Daniel, J. C., and Olson, J. D., *J. Reprod. Fert.*, **15**, 453 (1968).

10. Waymouth, C., *J. Nat. Cancer Inst.*, **22**, 1003 (1959).

11. Morgan, J. F., Morton, H. J., and Parker, R. C., *Proc. Soc. Exp. Biol. Med.*, **73**, 1 (1950).

12. Bavister, B. D., *J. Reprod. Fert.*, **18**, 544 (1969).

13. Edwards, R. G., Bavister, B. D., and Steptoe, P. C., *Nature*, **221**, 632 (1969).

14. Bavister, B. D., Edwards, R. G., and Steptoe, P. C., *J. Reprod. Fert.*, **20**, 159 (1969).

15. Tarkowski, A. K., *Cytogenetics*, **5**, 394 (1966).

16. Yanagimachi, R., *J. Exp. Zool.*, **170**, 269 (1969).

17. Allison, A. C., and Hartree, E. F., *J. Reprod. Fert.*, **22**, 501 (1970).

18. Weissman, G., in *Lysosomes in Biology and Pathology* (edit. by Dingle, J. T., and Fell, Honor B.) (North Holland, Amsterdam, 1969).

19. Channing, C. P., and Grieves, J. A., *J. Endocrinol.*, **43**, 391 (1969).

20. Bedford, J. M., *J. Reprod. Fert.*, Suppl. 2, 35 (1967).

21. Abraham, G. E., O'Dell, W. D., Edwards, R. G., and Purdy, J. M., *Acta Endocrinol.*, Suppl. No. 147, 332 (1970).

22. Edwards, R. G., Talbert, L., Isralestam, D., Nino, H. V., and Johnson, M. H., *Amer. J. Obst. Gynec.*, **102**, 388 (1968).

23. Pearson, P. L., Bobrow, M., and Vosa, C. G., *Nature*, **226**, 78 (1970).

24. George, K. P., *Nature*, **226**, 80 (1970).

25. Pearson, P. L., and Bobrow, M., *J. Reprod. Fert.*, **22**, 177 (1970).

26. Fraham, C. F., *Nature*, **226**, 165 (1970).

Human Blastocysts Grown in Culture

P. C. Steptoe *et al.*

Editor's Note

Several months after they reported the successful production of eight- and sixteen-celled human embryos through *in vitro* fertilization (IVF), Patrick Steptoe, Robert Edwards and Jean Purdy report further advances. Improved tissue culture techniques meant the team were now able to keep their IVF embryos alive for longer. Two out of six IVF embryos developed into blastocysts— thin-walled hollow structures containing a cluster of cells called the inner cell mass from which the embryo can arise. Development to this key embryonic stage was seen as an important step along the pathway to producing the first human IVF birth, which took another seven years.

WE have already described the culture of cleaving human embryos to the sixteen celled stage[1], and we now wish to give details of a few embryos that have developed much further, including two that reached fully developed blastocysts. Methods were similar to those described before. Preovulatory oocytes recovered by laparoscopy[2] were fertilized in Bavister's medium[3], and transferred after 12–15 h into Ham's F 10 supplemented with human or foetal calf serum. Preliminary details of this work have been presented elsewhere[4].

Six embryos were cultured. The *p*H of the medium was 7.3, the osmotic pressure 300 mOsmol/kg, and the gas phase consisted of 5% oxygen, 5% carbon dioxide and 90% nitrogen. Two of the embryos developed to early morulae, one having more than twenty-three nuclei, and the other twenty-nine nuclei and one mitosis. After the sixteen cell stage, the blastomeres of these embryos became slightly mottled and the cell outlines were less distinct—an appearance similar to that found in many mouse morulae.

A blastocoelic cavity appeared in four embryos. In two of them, the cavity was small and eccentric and the inner cell mass and trophoblast failed to become fully differentiated. Staining revealed only sixteen and twenty nuclei in the two embryos. The small number of cells indicated that division had become out of step with the early development of the blastocoel. Two embryos which developed through typical morulae into fully expanded blastocysts will be described separately.

After 123 h culture in Ham's F 10 with 20% foetal calf serum, irregular light patches appeared throughout the tissue of one morula (blastocyst 1). Eleven hours later, a large blastocoelic cavity was seen, with thin cellular membranes traversing part of it. The cavity became increasingly clear as differentiation continued. After 147 h in culture, the blastocyst was greatly expanded, and apart from one small vesicle the blastocoelic cavity was completely clear. The zona pellucida was thin and stretched, and the trophoblast

在培养基中生长的人类囊胚

斯特普托等

编者按

就在帕特里克·斯特普托、罗伯特·爱德华兹和琼·珀迪报道怎样通过体外受精（IVF）成功获得 8 细胞和 16 细胞人类胚胎的几个月后，他们又报道了一些新的进展。对组织培养技术的改进意味着他们现在已经能够延长体外受精胚胎的存活期。在 6 个体外受精的胚胎中有 2 个发育成了囊胚——一种由被称为内细胞团的一组细胞组成的薄壁中空结构，由此可以产生胚胎。发育到这一关键胚胎发育阶段被认为是通往成功获得第一个人类试管婴儿的重要步骤，7 年之后世界上第一个试管婴儿诞生。

我们已经描述过怎样将卵裂中的人类胚胎培养到 16 细胞阶段 [1]，现在我们想详细描述一些生长到更靠后阶段的胚胎，其中有两个已经完全发育到了囊胚阶段。所用的方法和以前描述过的类似：将由腹腔镜手术获得的排卵前卵母细胞 [2] 放在巴维斯特培养基 [3] 中受精，在 12 小时 ~ 15 小时后转移到添加人血清或者胎牛血清的哈姆氏 F10 培养基中。我们已经把这项研究的初步结果发表在了其他出版物上 [4]。

一共培养了 6 个胚胎。培养基的 pH 值为 7.3，渗透压为 300 毫渗摩 / 千克，而气相由 5% O_2、5% CO_2 和 90% N_2 组成。有两个胚胎生长到了早期桑葚胚，其中一个的细胞核超过了 23 个，而另一个含有 29 个细胞核和一个有丝分裂相。经过 16 细胞阶段之后，这些胚胎的卵裂球变得轻度杂色且细胞轮廓变得不那么清晰了——外观与在很多小鼠桑葚胚中看到的情形类似。

在 4 个胚胎中出现了囊胚腔。其中有两个囊胚腔很小并且是偏心的，其内细胞团和滋养层未能发生完全的分化。染色结果显示，在这两个胚胎中分别只有 16 和 20 个细胞核。这么少的细胞数量说明细胞分裂已经和囊胚腔的早期发育变得不一致了。下面我们将分别描述那两个经过典型的桑葚胚发育至完全膨胀的囊胚的胚胎。

用含有 20% 胎牛血清的哈姆氏 F10 培养基培养 123 小时后，我们发现在其中一个桑葚胚（囊胚 1）的组织内各部分均出现了不规则的光斑。11 小时之后，可以看到一个巨大的囊胚腔，其中的一部分被细胞形成的薄膜所穿过。随着分化的进行，囊胚腔变得越来越清晰。培养 147 小时后，囊胚充分膨胀，除了一个小的囊泡以外，囊胚腔已完全清晰。透明带很薄并处于被拉伸的状态，紧贴在透明带下方的滋养层

formed a clear layer of cells immediately beneath the zona. The blastocoel occupied three-quarters or more of the embryo, and the inner cell mass was distinct—far more so than is usual in mouse blastocysts. Giant cell transformation was not seen, but the embryo was not examined sufficiently closely to ensure that it had not occurred. The blastocyst was cultured for a further 2 h in its original medium plus 5 μg/ml. of "Colcemid". During this period the blastocyst contracted away from the zona, making it unsuitable for photography.

The blastocyst was placed in 1% sodium citrate for 2 min and fixed in fresh acetic methanol[5]. After staining with aceto-orcein, it was found to possess 112 nuclei and at least sixteen mitoses (Fig. 1). Several of the mitoses overlapped in the region of the inner cell mass and were unscoreable. Others were discrete but insufficiently spread so that we could only estimate that they were probably diploid.

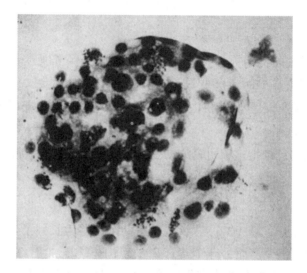

Fig. 1. Stained preparation of blastocyst 1. Many mitoses can be seen over the region of the inner cell mass (lower right) and a few elsewhere. The nuclei are even-sized.

The second blastocyst (blastocyst 2) developed approximately 24 h after the other. It was cultured in Ham's F 10 containing 25% human and 25% foetal calf serum. A blastocoelic cavity appeared after 159 h in culture. This cavity corresponded with that seen in the early stages of expansion in blastocyst 1, but the cell membranes traversing it were more pronounced and persisted for 12 h. Some of the trophoblastic cells had evidently failed to release their contents into the blastocoel. The thin zona pellucida and the shape of the inner cell mass were similar to those seen in blastocyst 1. Small extensions, somewhat resembling those seen in guinea-pig blastocysts[6], seemed to be passing through the zona pellucida (Fig. 2).

会形成一个清晰的细胞层。囊胚腔占据了胚胎体积的 3/4 甚至更多，而内细胞团很清晰——远远超过了通常在小鼠囊胚中所见到的清晰度。没有看到巨细胞转化，但因为我们对胚胎的观察不足够仔细，所以不能确定是否真的没有发生过。在原来的培养基中加入 5 微克 / 毫升"秋水仙酰胺"后再将囊胚继续培养 2 小时。在此期间，囊胚会因收缩而远离透明带，所以不适合拍照。

将囊胚在 1% 的柠檬酸钠中放置 2 分钟，并用新鲜的醋酸 – 甲醇固定 [5]。经醋酸 – 地衣红染色后，我们发现它含有 112 个细胞核和至少 16 个有丝分裂相（图 1）。有几个分裂相在内细胞团区域重叠，因此无法对其作出判断。另一些虽然是相互分离的，但没有充分展开，因此我们只能猜测它们有可能是二倍体。

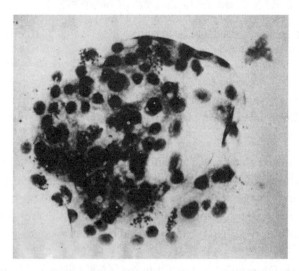

图 1. 染色的囊胚 1 标本。在内细胞团区域（右下方）以及其他一些区域可以看到许多有丝分裂相。细胞核的大小是均一的。

第二个囊胚（囊胚 2）的发育时间比第一个囊胚晚大约 24 小时。它是在含 25% 人血清和 25% 胎牛血清的哈姆氏 F 10 培养基中培养出来的。培养 159 小时后囊胚腔出现。这个囊胚腔与我们在囊胚 1 膨胀阶段早期所见到的囊胚腔相同，但是穿越它的细胞性膜更加明显并且持续了 12 小时之久。一些滋养层细胞显然未能将它们的内容物释放到囊胚腔内。薄薄的透明带和内细胞团的形态与在囊胚 1 中所见到的情况类似。一些小的且与在豚鼠囊胚中所见到的有些类似的延伸物 [6] 似乎穿过了透明带（图 2）。

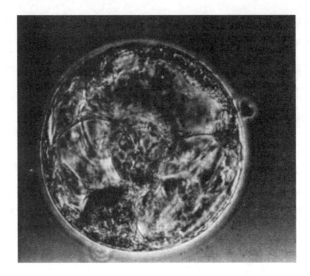

Fig. 2. Blastocyst 2 before it was fixed and stained. The thin zona pellucida and the underlying trophoblast can be seen; an extension is evidently emerging through the zona (at right). The inner cell mass (lower part of the embryo) is distinct. The cellular membranes persisting across the blastocoel can be seen.

The embryo was kept at room temperature for several hours before staining. It was found to have 110 nuclei. Bodies resembling sex chromatin were seen in a few nuclei.

We seem to have achieved this improved embryonic development through better handling of the cultures than previously. The preovulatory oocytes were kept at room temperature under reduced oxygen tension and 5% carbon dioxide until the spermatozoa were added. Small microdrops were used for culture, and enlarged when the embryos were eight celled. The embryos were left undisturbed for long periods after this stage.

We thank the Ford Foundation, the Oldham and District Hospitals Management Committee and the Manchester Regional Hospital Board for financial support, Mr. C. Richardson for photography and Professor C. R. Austin for encouragement.

(**229**, 132-133; 1971)

P. C. Steptoe, R. G. Edwards and J. M. Purdy: Oldham General Hospital, Lancashire, and Physiological Laboratory, University of Cambridge.

Received November 18, 1970.

References:

1. Edwards, R. G., Steptoe, P. C., and Purdy, J. M., *Nature*, 227, 1307 (1970).

2. Steptoe, P. C., and Edwards, R. G., *Lancet*, i, 683 (1970).

3. Bavister, B. D., *J. Reprod. Fert.*, 18, 544 (1969).

4. Edwards, R. G., *Harold C. Mack Symposium on the Biology of Fertilization and Implantation*, Detroit, October 1970 (in the press).

5. Tarkowski, A. K., *Cytogenetics*, 5, 394 (1966).

6. Amoroso, E. C., *Memoirs of the Society for Endocrinology*, 6, 50 (1959).

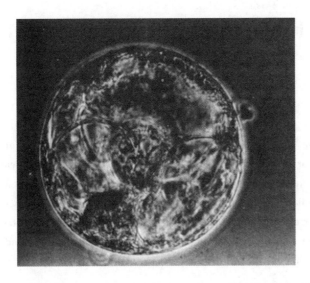

图 2. 在被固定和染色之前的囊胚 2。可以看到薄薄的透明带和位于其下的滋养层。显然有一条延伸物穿过了透明带(在右侧)。内细胞团(胚胎的下半部分)很清晰。可以看到一些由细胞形成的膜横穿过囊胚腔。

在染色前几小时,胚胎一直被保存在室温下。我们发现它含有 110 个细胞核。在若干个细胞核内可以见到类似于性染色质的小体。

看来,我们是通过较以往更好的处理方法达到了这一更靠后的胚胎发育阶段。在加入精子之前,排卵前的卵母细胞一直被保存在室温、较低氧分压及含 5% CO_2 的环境中。用小的微滴进行培养,并在胚胎达到 8 细胞期时扩大微滴的大小。在这一阶段之后很长时间内就不再干扰胚胎发育了。

感谢福特基金会、奥尔德姆地区综合医院管理委员会以及曼彻斯特地区医院理事会为我们提供了资金上的支持,还要感谢理查森先生对此进行拍照以及奥斯汀教授对我们的鼓励。

(毛晨晖 翻译;王敏康 审稿)

The DNA Replication Mystery

Editor's Note

Although by the early 1970s molecular biology was busily working out the functioning of important cellular components, the question of how DNA itself is replicated was unanswered. This News and Views article written by the staff of *Nature* is concerned with an enzyme called polymerase I, isolated from the bacterium *E. coli*, but, as the second paragraph of this article explains, the properties of the enzyme could not adequately account for the replication of DNA. Luckily, a second enzyme, polymerase II, was discovered by De Lucia and Cairns. Polymerase I was at the same time recognised to be an enzyme involved in the repair of mismatched sequences of DNA.

AS a group, molecular biologists have never fought shy of publicity and their propagandists have fed the popular image of the juggernaut subject advancing by enormous strides on all fronts. All fronts bar one, which is usually glossed over; for, although anybody reared on the current rash of molecular biology texts might be excused for thinking otherwise, the precise mechanism by which DNA is duplicated is almost as obscure today as ever.

To be sure, in the nineteen sixties Kornberg and his collaborators isolated from *Escherichia coli* an enzyme which can polymerize DNA and, for example, replicate infectious bacteriophage DNA *in vitro*. And until last year this enzyme, DNA polymerase I, was the only known enzyme with this capability. Small wonder therefore that many molecular biologists bent over backwards devising all sorts of ingenious schemes to explain how it might replicate DNA *in vivo*. But they laboured under one great disadvantage and probably in vain, for the properties of polymerase I (it makes DNA very slowly, it can only synthesize DNA in one, the 5′ to 3′ direction, and it is an effective exonuclease) are those expected of a DNA repair enzyme rather than a replicase. In other words, polymerase I has all the properties of an enzyme which edits DNA sequences, excising regions of mismatched base pairs and replacing them with the correct sequence, rather than replicates them.

The days of tortuous scheming are, however, over; the pressure is off polymerase I for two related reasons. First, at the end of 1969 De Lucia and Cairns isolated six mutant *E. coli* which lack or contain defective polymerase I but nonetheless survive and replicate their DNA normally. Second, several groups have detected and begun to isolate a second DNA polymerase, polymerase II, in these mutants. With at least two candidates at last in the lists, those molecular biologists concerned with defining the mechanism of DNA duplication are

DNA复制之谜

编者按

尽管在20世纪70年代初以前，分子生物学家们一直忙于研究重要细胞组分的功能，但人们对于DNA自身如何复制这一问题仍不清楚。这篇"新闻与观点"栏目中的文章是由《自然》杂志的编辑撰写的，文中提到一种从大肠杆菌中分离出来的被称为聚合酶I的酶。但是，正如本文第二段中所阐释的，该酶的性能不足以充分说明DNA的复制。幸运的是，德卢西亚和凯恩斯发现了第二种酶——聚合酶II。与此同时，人们认识到聚合酶I是一种参与DNA序列错配修复的酶。

作为一个群体，分子生物学家们从来不回避公众，他们的宣传员们在人们心目中塑造了分子生物学这个无往而不胜的主题正在各个前沿领域阔步前进的形象。但有一个前沿领域例外，而该领域常常被掩饰过去，因为任何一个受现有尚不成熟的分子生物学教科书所培养的人另有不同想法。虽然这是情有可原的，但人们对DNA复制精确机制的认识直至今天仍是一如既往地模糊不清。

诚然，在20世纪60年代，科恩伯格及其合作者们从大肠杆菌中分离出了一种酶，它能聚合DNA，譬如在体外复制出具有感染力的噬菌体DNA。而且直到去年，DNA聚合酶I还是已知的唯一一种具有这种能力的酶。因此，无怪乎许多分子生物学家回过头来想尽办法设计出各种别出心裁的方案以解释它怎样才可能在体内复制DNA。但是，他们的劳动处于一个相当大的不利因素之下，甚至很可能会徒劳无功：因为聚合酶I的特性（复制DNA的速度很慢，只能按一种方向——$5' \rightarrow 3'$方向合成DNA，并且是一种有效的核酸外切酶）表明：与其说它是一种DNA复制酶，倒不如说它是DNA修复酶。换言之，聚合酶I具有一个能编辑DNA序列的酶所具有的所有特征：可以剪切有错配碱基对的区域，并替换上正确的序列，而不是复制它们。

然而，这种费尽心机的设计时代已经结束，用以下两个相关联的原因就可以解释人们为什么得以从聚合酶I的窘境中解放出来。首先，1969年末，德卢西亚和凯恩斯分离出了6株缺少聚合酶I或含有缺陷型聚合酶I的大肠杆菌突变株，但菌体仍能存活，并能正常复制DNA。其次，有好几个研究小组已经在这些突变株中检测到并着手分离出了第二种DNA聚合酶——聚合酶II。在那些致力于阐明DNA复制机制的分子生物学家们现在的研究清单上终于有了至少两个候选对象，他们目前

currently far more intent on characterizing DNA polymerase II and searching for further species of DNA polymerase than on striving to accommodate polymerase I in hypothetical schemes of DNA replication. Few of them would now quarrel with the statement that the chief and probably sole function of polymerase I is DNA repair. If there are any waverers still sitting on the fence they should be finally decided by what Kelley and Whitfield have to say, on page 33 of this issue of *Nature*, about one of the six mutants isolated by De Lucia and Cairns.

De Lucia and Cairns, when they began their mutant hunt, argued that if polymerase I is a repair enzyme it should be possible to select mutant *E. coli* with a defective repair mechanism, because of a defective polymerase I, but still capable of normal DNA replication. They proved their point by isolating six such mutants, the first of which to be characterized proved to be an amber mutant. But although these strains lacked polymerase I activity and were unusually sensitive to ultraviolet light, there was no proof that the mutations were actually in the structural gene which specifies polymerase I. It might, for example, still be argued that the mutations were in a regulatory gene and result in a drastic reduction in the amount of polymerase I made. If that were the case, the handful of perfectly normal polymerase I molecules might all be sequestered immediately for DNA replication, which would proceed normally, leaving none available for DNA repair.

Because of such arguments it became a matter of considerable importance to find proof that any one of the six mutants De Lucia and Cairns isolated (all those mapped so far are in the same gene) contain a lesion in the structural gene directly specifying polymerase I. For, clearly, if the enzyme molecule itself contains a mutation which reduces its capacity to polymerize DNA, and cells containing such mutant enzyme replicate their DNA but lack a repair mechanism, the straightforward conclusion is that the polymerase I is involved in repair but not replication. Of course, anybody absolutely determined to keep all options open could say that all the mutations by chance occur in that part of the polymerase I molecule, admittedly a large protein comprising about 1,000 amino-acids, which is involved in repair, leaving the part responsible for DNA replication intact. But that is surely less plausible.

Fortunately as Kelley and Whitfield have now convincingly shown, *pol 6*, one of the half dozen mutants, turns out to be a temperature sensitive mutation. They purified polymerase I from this mutant and wild type *E. coli* and compared the properties *in vitro* of the two enzyme. The wild type enzyme has the same activity at 37°C and 52°C and its optimum temperature is 55°C. By contrast the mutant enzyme had less activity at 52°C than 37°C and an optimum temperature of about 45°C. Further, the two enzymes differ in their response to various synthetic DNAs and, although immunologically similar, sensitive complement fixation tests at various temperatures indicate that they differ in precise conformation. In short, the *pol 6* mutation is in the structural gene specifying the enzyme. Polymerase I is therefore certainly involved in DNA repair, and although it is premature to state categorically that that is the enzyme's sole function, the odds are stacked very high against the idea that polymerase I has a role in DNA replication.

826

(**230**, 11-12; 1971)

更迫切要做的事情是：表征 DNA 聚合酶 II 的性质以及寻找其他种类的 DNA 聚合酶，而非极力设法让聚合酶 I 适应假想的 DNA 复制方案。现在，绝大多数分子生物学家都不会对以下陈述提出争议，即聚合酶 I 的主要功能是 DNA 修复，并且这很可能是它唯一的功能。如果还有摇摆不定者对此心存疑虑，相信等他们看了凯利和惠特菲尔德在本期《自然》杂志第 33 页上就德卢西亚和凯恩斯分离的 6 株突变体之一所发表的不得已的论述之后，就应该能作出最后的定论了。

当德卢西亚和凯恩斯开始寻找突变体的时候，他们主张：如果聚合酶 I 是一种修复酶，那么就应当可以筛选出一种由有缺陷的聚合酶 I 所导致的修复机制缺陷型大肠杆菌突变株，但这种突变体仍能进行正常的 DNA 复制。他们分离出了 6 株这样的突变体，从而证实了自己的观点，其中第一个被确认性质的突变体是一个琥珀突变体。然而，尽管这些突变株缺少聚合酶 I 的活性并对紫外线异常敏感，但仍无证据表明突变就发生在编码聚合酶 I 的结构基因中。譬如，仍有可能争论，突变是发生在调节基因上，从而导致制备的聚合酶 I 数量急剧降低。如果事实果真如此，那么少数几个完全正常的聚合酶 I 分子就有可能全部被马上分开以便于 DNA 复制，而在复制正常进行的情况下，也就再没有酶可供用于 DNA 修复了。

由于存在以上争论，所以找出证据证明德卢西亚和凯恩斯分离出的所有 6 株突变体（目前它们的突变体图谱都定位在同一个基因上）中的任意一株在直接编码聚合酶 I 的结构基因上存在缺陷就显得尤为重要了。因为显然，如果酶分子自身发生了一个能降低其聚合 DNA 能力的突变，并且含有这种突变酶的细胞能够复制 DNA 但却缺乏修复机制，那么就可以直截了当地得到结论：聚合酶 I 参与修复，而不参与复制。当然，那些决心非要穷尽一切可能性的人会说，所有的突变会恰巧发生在聚合酶 I 分子的某一部分，该部分是由 1,000 个氨基酸组成的大蛋白，这个大蛋白参与修复，而负责 DNA 复制的那部分蛋白则保持完整。但是，这种假设显然不太可能成立。

幸运的是，凯利和惠特菲尔德现在已经令人信服地证明了 6 株突变体中的 1 株——聚合酶 6 属于温度敏感型突变。他们对来自该突变型和野生型大肠杆菌中的聚合酶 I 进行了纯化，并比较了这两种酶的体外性质。野生型酶的活性在 37℃ 下和 52℃ 下是一致的，其最适温度为 55℃。相比之下，突变型酶在 52℃ 下的活性要低于在 37℃ 下的活性，并且其最适温度约为 45℃。此外，这两种酶对于各种人工合成的 DNA 的反应不同。尽管两者有相似的免疫特性，但在不同温度下的敏感补体结合实验表明，两者的精细构象并不相同。简言之，聚合酶 6 突变发生在编码这个酶的结构基因上。因此，聚合酶 I 肯定参与了 DNA 修复过程。尽管断言聚合酶 I 的唯一功能是参与 DNA 的修复为时尚早，但聚合酶 I 能在 DNA 复制中起到某种作用的观点几乎没有成立的可能性。

（吕静 翻译；顾孝诚 审稿）

Polymorphism of Human Enzyme Proteins

W. H. P. Lewis

Editor's Note

The common notion that protein enzymes have, in any species, a well defined "primary structure"—their linear sequence of amino acids—has long been known to be a simplification. Here pathologist W. H. P. Lewis provides one of the early challenges to that notion, reporting a high degree of variability in the sequences of several human enzymes. Lewis uses the technique of electrophoresis, which separates molecules of different charge and shape, to identify several varieties or "polymorphisms" of common enzymes in different populations. These polymorphisms are heritable, and Lewis concludes that it is unlikely that any two individuals are identical in all their enzymes. We now know that even in individuals, specific enzymes can vary considerably in their activity too.

SINCE Landsteiner described the ABO blood group system[1], it has become increasingly obvious that human individuality is demonstrable at the molecular level. It seems probable that almost every protein, structural or enzymatic, can be made to reveal some degree of variability in human populations. The application of starch gel electrophoresis[2] during the past decade has been especially fruitful in revealing molecular variability of particular enzymes in man. Using this method it has been possible to show common variants in approximately one enzyme in three in one of the major population groups[3]. This finding is the more remarkable in view of the fact that starch gel electrophoresis can reveal differences only in charge or size. If, as seems to be the case, the type of structural variation revealed by this technique results from the substitution of a particular amino-acid for another of a different charge, then because a charge difference will result from only about one in three of all possible amino-acid substitutions, it seems likely that many more differences in primary enzyme structure occur.

Enzymes which have been studied so far have been chosen principally because the reactions which they catalyse can be coupled to a suitable chromogenic reaction, so that the site of enzyme activity can be revealed as a sharp zone after electrophoresis. There is no reason to suppose, therefore, that selection on this basis would lead to the study of enzymes which are more commonly variable than those which are not amenable to this type of study. Some of the enzymes which have been studied are listed in Table 1.

人类酶蛋白的多态性

卢亦思

编者按

通常认为，任何物种中的蛋白质酶都具有固定的"一级结构"（氨基酸的线性排列顺序），然而，大家早就知道这样的看法其实过于简单了。这篇文章是最早对这一看法提出异议的文章之一，在本文中病理学家卢亦思报道了几种人体酶的氨基酸系列的高度变异性。卢亦思利用能够根据电荷和大小差异而对分子进行分离的电泳技术鉴定出了存在于不同人群中的常见酶的几种不同形式，或者说"多态性"。这样的多态性是可以遗传的。卢亦思的结论是，任意两个个体其所有酶都相同这样的情形是不太可能的。现在我们已经知道：即使在一个个体中，特定酶的活性也可以发生着相当大的变化。

自兰德施泰纳描述了 ABO 血型分类系统以来 [1]，日益显而易见的是，人类的个体差异可以在分子水平上得到验证。似乎可能的是：几乎每一种蛋白质，无论是结构性蛋白还是酶蛋白，都能够用于揭示人群的一定程度的变异性。在过去的十年中，通过利用淀粉凝胶电泳法 [2]，人们相当有成效地揭示出了人体中特定酶分子的变异性。正是通过使用这种方法，人们才发现：在任意一个主要的人群中，大概每三种酶中就会有一种存在常见的变异体 [3]。考虑到淀粉凝胶电泳法只能揭示出酶蛋白所带电荷差异或大小差异这一事实，上述发现就更加值得我们思考。非常可能的情形是，利用这种技术检测出的结构差异是由一个特定的氨基酸被另一个带电不同的氨基酸所替代而引起的，而在所有可能的氨基酸替代中能导致电荷变化的只占约 1/3，所以发生在酶的一级结构水平的差异可能比检测到的还要多。

到目前为止，研究中选择酶的依据主要是看它们所催化的反应是否能和一个合适的显色反应偶联起来，因而在电泳后如果能产生显色条带就表明有酶的活性。没有理由认为根据这个原则选择酶会导致所研究的酶多为更容易变异的酶，而忽略了不适合用这种方法研究的酶。表 1 中列出了一些已被研究过的酶。

Table 1. Frequencies of the Commonest Allele in Different Population Groups of Some of the Enzymes which have been Studied by Starch Gel Electrophoresis

	Europeans	Negroes
Red cell acid[30] phosphatase	0.6–0.7	0.8
6-Phosphogluconate[31] dehydrogenase	0.96	0.72–0.87
Phosphoglucomutase		
PGM$_1$[16]	0.76	0.76–0.79
PGM$_2$[16]	1.0	0.95–0.99
PGM$_3$[22]	0.74*	0.66*
Peptidase A[5,19]	1.0	0.70–0.90
Peptidase D[18] (prolidase)	0.97	0.92
Adenosine deaminase[32]	0.90	0.90
Adenylate kinase[33]	0.95	0.99

*The commonest allele is not the same in the two populations studied for PGM$_3$.

The peptidases illustrate well the sort of information that can be obtained by the application of starch gel electrophoresis in association with a specific staining method. The occurrence of these enzymes in human tissues has been established for some time[4]; they catalyse the hydrolysis of the peptide bonds of small peptides consisting of two or three amino-acids. Their precise function is not known but they are probably involved in the terminal degradation of proteins both in intestinal absorption and protein catabolism. Like most other enzymes which have been studied, peptidases have been obtained from red blood cells because blood samples are readily obtained from families—an essential prerequisite for genetic studies.

After starch gel electrophoresis at pH 7.5, peptidase activity can be detected by the following reaction sequence[5],

$$\text{peptide} \xrightarrow{\text{peptidase}} \text{amino-acids}$$

$$\text{amino-acid} \xrightarrow{\text{amino-acid oxidase}} \text{keto-acid} + NH_4OH + H_2O_2$$

$$H_2O_2 + \text{dianisidine} \xrightarrow{\text{peroxidase}} \text{oxidized dianisidine} + H_2O$$

This results in a brown precipitate at the site of activity. Using this method and various dipeptides and tripeptides, it has been possible to characterize six distinct peptidases in human red blood cells[5,6] (Table 2). Their relative electrophoretic mobilities are shown in Fig. 1. Although some of these enzymes have overlapping substrate specificities there are wide differences in their patterns of substrate specificity. Moreover, three of these enzymes apparently contain at least one reactive sulphydryl group while the rest do not[5]. Further, gel filtration studies indicated that these enzymes also differ in molecular size[7].

表 1. 利用淀粉凝胶电泳法研究过的某些酶的最常见等位基因在不同人群中的出现频率

	欧洲人	黑人
红细胞酸性 [30] 磷酸酶	0.6~0.7	0.8
6–磷酸葡萄糖酸 [31] 脱氢酶	0.96	0.72~0.87
磷酸葡萄糖变位酶（PGM）		
PGM$_1$ [16]	0.76	0.76~0.79
PGM$_2$ [16]	1.0	0.95~0.99
PGM$_3$ [22]	0.74*	0.66*
肽酶 A [5,19]	1.0	0.70~0.90
肽酶 D [18]（脯肽酶）	0.97	0.92
腺苷脱氨酶 [32]	0.90	0.90
腺苷酸激酶 [33]	0.95	0.99

* 在研究两个人群中的 PGM$_3$ 时发现其最常见的等位基因并不相同。

这些肽酶的例子很好地表明了将淀粉凝胶电泳法与一种特异染色方法结合在一起利用时所能获得的那类信息。人们确认人体组织中的确存在这些肽酶已经有一段时间了 [4]；它们催化由两三个氨基酸组成的小肽的肽键的水解。虽然尚不知道它们的精确功能，但它们很有可能参与了蛋白质的最终降解，涉及在肠道中的吸收以及蛋白质的分解代谢。正如大多数其他被研究过的酶一样，肽酶也是从红细胞中获得的，因为从不同家庭成员获得血液样本是件比较容易做到的事，而这正是开展遗传学研究的必要前提。

在经过 pH 值为 7.5 条件下的淀粉凝胶电泳之后，可以通过如下反应顺序来检测肽酶的活性 [5]：

$$肽 \xrightarrow{肽酶} 氨基酸$$

$$氨基酸 \xrightarrow{氨基酸氧化酶} 酮酸 + NH_4OH + H_2O_2$$

$$H_2O_2 + 联茴香胺 \xrightarrow{过氧化物酶} 氧化的联茴香胺 + H_2O$$

这些反应将导致在具有活性的位点处产生棕色沉淀。利用上述方法以及不同的二肽和三肽，人们才得以确定人类红细胞中存在的六种不同肽酶的特征 [5,6]（表 2）。它们的相对电泳迁移率示于图 1。尽管其中一些酶的底物特异性有所重叠，但其底物特异性的模式却有很大的不同。此外，这些酶中有三种明显至少含有一个活性巯基，而其余的酶则不含有 [5]。另外，凝胶过滤研究结果显示这些酶在分子大小上也有所不同 [7]。

Table 2. Relative Activities of Human Red Blood Cell Peptidases against Various Amino-acid Peptides and Leucyl-β-naphthylamide (Leu-β-NA)

	A	B	C	D	E	F
Val-Leu	+++	−	−	−	−	−
Gly-Leu	+++	−	+	−	−	−
Leu-Gly	+++	−	+	−	−	−
Gly-Phe	++	−	+	−	−	−
Leu-Ala	++	−	++	−	−	−
Leu-Leu	++	−	++	−	+	−
Gly-Trp	++	−	++	−	−	−
Lys-Leu	+	+	+++	−	+	−
Lys-Tyr	+	+	+++	−	+	−
Pro-Phe	++	+	+++	−	−	−
Pro-Leu	++	+	+++	−	−	−
Phe-Leu	++	++	+++	−	+	−
Ala-Tyr	++	+++	+	−	±	−
Phe-Tyr	++	++	+++	−	+	−
Leu-Tyr	+++	++	+++	−	+	−
Leu-Gly-Gly	−	+++	−	−	±	−
Leu-Gly-Phe	−	+	−	−	±	−
Leu-Leu-Leu	−	++	−	−	+	+
Phe-Phe-Phe	−	++	−	−	++	±
Tyr-Tyr-Tyr	−	+	−	−	+	+
Phe-Gly-Phe-Gly	−	−	−	−	++	−
Leu-β-NA	−	−	−	−	+	−
Leu-Pro	−	−	−	++	−	−
Phe-Pro	−	−	−	++	−	−

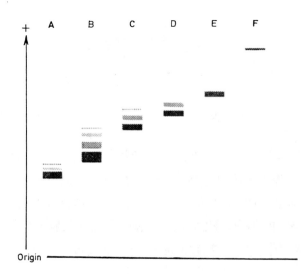

Fig. 1. Relative electrophoretic mobilities of human red cell peptidases at pH 7.5.

表 2. 人类红细胞肽酶对不同的氨基酸组成肽以及亮氨酰 – β – 萘酰胺（Leu– β –NA）的相对活性

	A	B	C	D	E	F
缬氨酰 – 亮氨酸	+++	–	–	–	–	–
甘氨酰 – 亮氨酸	+++	–	+	–	–	–
亮氨酰 – 甘氨酸	+++	–	+	–	–	–
甘氨酰 – 苯丙氨酸	++	–	+	–	–	–
亮氨酰 – 丙氨酸	++	–	++	–	–	–
亮氨酰 – 亮氨酸	++	–	++	–	+	–
甘氨酰 – 色氨酸	++	–	++	–	–	–
赖氨酰 – 亮氨酸	+	+	+++	–	+	–
赖氨酰 – 酪氨酸	+	+	+++	–	+	–
脯氨酰 – 苯丙氨酸	++	+	+++	–	–	–
脯氨酰 – 亮氨酸	++	+	+++	–	–	–
苯丙氨酰 – 亮氨酸	++	++	+++	–	+	–
丙氨酰 – 酪氨酸	++	+++	+	–	±	–
苯丙氨酰 – 酪氨酸	++	++	+++	–	+	–
亮氨酰 – 酪氨酸	+++	++	+++	–	+	–
亮氨酰 – 甘氨酰 – 甘氨酸	–	+++	–	–	±	–
亮氨酰 – 甘氨酰 – 苯丙氨酸	–	+	–	–	±	–
亮氨酰 – 亮氨酰 – 亮氨酸	–	++	–	–	+	+
苯丙氨酰 – 苯丙氨酰 – 苯丙氨酸	–	++	–	–	++	±
酪氨酰 – 酪氨酰 – 酪氨酸	–	+	–	–	+	+
苯丙氨酰 – 甘氨酰 – 苯丙氨酰 – 甘氨酸	–	–	–	–	++	–
亮氨酰 – β – 萘酰胺	–	–	–	–	+	–
亮氨酰 – 脯氨酸	–	–	–	++	–	–
苯丙氨酰 – 脯氨酸	–	–	–	++	–	–

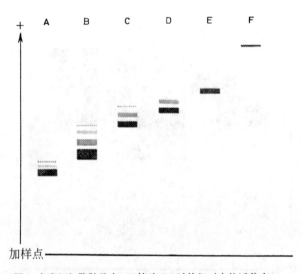

图 1. 人类红细胞肽酶在 pH 值为 7.5 时的相对电泳迁移率。

Biochemical evidence suggests that the six peptidases are distinct proteins, each with a characteristic primary structure and a characteristic specificity. It would be expected therefore that the primary structure of each peptidase would be coded for by a separate gene locus, so that a structural variant occurring in one peptidase would not be reflected in any of the others. This seems to be the case, and genetic studies have revealed further structural differences between these enzymes.

Peptidase A, a dipeptidase with a wide specificity (Table 2), has been studied in several different population groups. The variants shown in Fig. 2 are all rare except Pep A1, Pep A2–1 and Pep A2. Pep A1 is the commonest phenotype in all the population groups studied so far, while Pep A2–1 and Pep A2 occur in appreciable frequencies only in African populations, or in populations of African origin such as the West Indian group. Table 3 shows that the frequencies of these variants in different populations are quite similar, although Pep A2–1 and Pep A2 seem to occur most often in the Babinga pigmies[8], and are less common, as would be expected, in groups of mixed origin, such as the West Indians. A similar situation has been found with some other enzymes, for example glucose-6-phosphate dehydrogenase[9] and phosphoglucomutase, locus PGM_2[10].

Table 3. Frequencies of Peptidase A Variants in Various Population Groups

| | Peptidase A | | | | | | | | |
	Total	1	2–1	2	3–1	4–1	5–1	6–1	7–1
Europeans	3,129	3,123	—	—	1	3	1	1	1
British resident Negroes	636	554	78	3	1	—	—	—	—
"Bantu" (South Africa)	100	87	11	2	—	—	—	—	—
Yoruba (Nigeria)	155	127	25	3	—	—	—	—	—
Cape Coloured (South Africa)	177	169	8	—	—	—	—	—	—
Cape Malay (South Africa)	104	96	8	—	—	—	—	—	—
Asiatic Indians	615	613	—	—	1	—	1	—	—

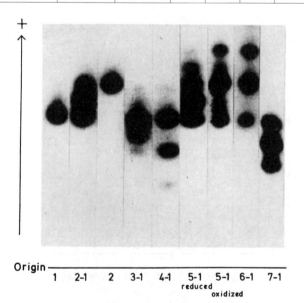

Fig. 2. Electrophoretic patterns of peptidase A variants.

生物化学证据表明：这六种肽酶是不同的蛋白质，每一种都有独一无二的一级级结构和底物特异性。可以假设性地认为每种肽酶的一级结构都是由一个独立基因位点编码的，因此，发生在一种肽酶中的结构变化就不会在任何其他肽酶中体现。事实似乎就是如此，而且遗传学研究揭示出了这些酶之间更多结构上的不同。

作为底物特异性很低的二肽酶（表2），肽酶A已经在几个不同的人群中被研究过。图2中所示肽酶A的变异体中，除肽酶A1、肽酶A2-1和肽酶A2以外，其他都非常罕见。在迄今为止研究过的所有人群中，肽酶A1是最常见的，而肽酶A2-1和肽酶A2仅在非洲人群或者源自非洲的人群如西印度人中以一定频率出现。表3中的数据说明这些变异体在不同人群中的出现频率是非常相近的，但肽酶A2-1和肽酶A2似乎只在巴宾加俾格米人中最为常见 [8]，而在混合起源的人群如西印度人中出现的频率较低，正如人们所预期的那样。对于其他一些酶，如6-磷酸葡萄糖脱氢酶 [9] 和磷酸葡萄糖变位酶（PGM$_2$ 编码基因） [10]，情形也类似。

表 3. 不同人群中肽酶 A 变异体的出现频率

	肽酶 A								
	总计	1	2–1	2	3–1	4–1	5–1	6–1	7–1
欧洲人	3,129	3,123	—	—	1	3	1	1	1
在英国居住的黑人	636	554	78	3	1	—	—	—	—
"班图人"（南非）	100	87	11	2	—	—	—	—	—
约鲁巴人（尼日利亚）	155	127	25	3	—	—	—	—	—
开普有色人种（南非）	177	169	8	—	—	—	—	—	—
开普马来人（南非）	104	96	8	—	—	—	—	—	—
移民美洲的印度人	615	613	—	—	1	—	1	—	—

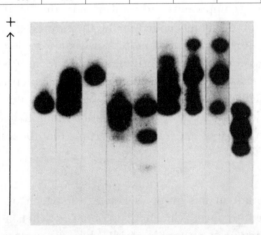

加样点 1 2-1 2 3-1 4-1 5-1 5-1 6-1 7-1
还原形式
氧化形式

图 2. 肽酶 A 变异体的电泳图谱。

Family studies involving quite a large number of individuals indicate that the Pep A2–1 pattern occurs in people who are heterozygous for a pair of alleles which occur at an autosomal locus and which in homozygotes give rise to the Pep A1 and Pep A2 types. These alleles have been designated *Pep A^1* and *Pep A^2*. This type of inheritance is sometimes referred to as co-dominant because both alleles are expressed in the heterozygote in contrast to the dominant-recessive inheritance in which only the dominant allele is apparently expressed in the heterozygote.

This type of study also makes possible inferences about the molecular structure of an enzyme. The electrophoretic pattern of the presumed heterozygote Pep A2–1 consists of three main zones of activity, a zone with a mobility equivalent to each of the presumed homozygotes, Pep A1 and Pep A2, and a hand with almost exactly intermediate mobility, which is the most active. This type of electrophoretic pattern is thought to occur when the enzyme in homozygotes consists of at least two identical subunits[11]. If the subunit coded for by the *Pep A^1* allele is represented by α^1 and that coded for by *Pep A^2* is represented by α^2, then if in the heterozygous individual both alleles are active and their corresponding subunits combine at random, there are three possible structures, $\alpha^1\alpha^1$, $\alpha^1\alpha^2$ and $\alpha^2\alpha^2$. Moreover, if the two subunits are synthesized at an equal rate and are equally active and stable, then the three zones would be expected to have activity approximately in the ratio of 1:2:1, and this seems to be the case.

Another interesting feature of this particular enzyme is the occurrence of a series of rare phenotypes (Fig. 2). All these types are very rare, with frequencies of 1 in 1,000 or less in most population groups studied so far. Family studies suggest that these phenotypes occur in individuals who are heterozygous for one of a series of rare alleles, *Pep A^3*, *Pep A^4*, *Pep A^5*, *Pep A^6* and *Pep A^7*, and the common allele *Pep A^1*. In red blood cells two of these rare types, Pep A3–1 and Pep A4–1, have an electrophoretic pattern in which the characteristic electrophoretically slower components are weaker than the band corresponding to Pep A1. This might result if the subunits coded for by the rare alleles *Pep A^3* and *Pep A^4* were catalytically less active, were synthesized more slowly or were less stable. It seems likely that the last possibility is correct, for when these phenotypes are seen after electrophoresis of a tissue extract, for example placenta, the pattern is symmetrical and the three main bands have activity roughly in the ratio of 1:2:1. It seems likely that the stability difference is revealed in the red blood cell because of the relatively long half life of this type of cell, and presumably protein synthesis has ceased by the time that the nucleus is lost.

Another interesting variant in this group is Pep A5–1, which occurs in two forms; the pattern which is termed "reduced" in Fig. 2 is obtained when a fresh sample from an individual of this phenotype is electrophoresed, while the pattern termed "oxidized" is obtained if the sample is stored for a few days at 4°C before electrophoresis. If the stored sample is treated with mercaptoethanol, the resultant electrophoretic pattern is similar to that of the fresh sample. On the other hand, when a fresh sample is treated with oxidized glutathione it has an electrophoretic pattern indistinguishable from that of the stored

涉及大量个体的家族性研究结果显示，肽酶 A2-1 出现在常染色体位点上一对等位基因的杂合子人群中，而该等位基因的纯合子形式将产生肽酶 A1 和肽酶 A2。这些等位基因被命名为肽酶编码基因 A^1 和肽酶编码基因 A^2。这种类型的遗传方式有时被称作共显性，因为两个等位基因都在杂合子中表达了，它与显－隐性遗传方式不同的是，后者似乎只有显性基因才能在杂合子中被表达。

这样的研究使得我们去推测酶的分子结构成为可能。假想的肽酶 A2-1 杂合子在电泳图谱上应该包括三条主要的活性带，其中两条带的迁移率分别与假定的纯合子肽酶 A1 和肽酶 A2 相同，还有一条带的迁移率几乎精确地介于两者的中间位置，这条带活性最强。这样的电泳图谱之所以出现，被认为是因为纯合子的酶包含了至少两个相同的亚基[11]。如果用 α^1 代表由肽酶编码基因 A^1 编码的亚基，用 α^2 代表由肽酶编码基因 A^2 编码的亚基，那么如果在杂合子个体中这两个等位基因都被表达，并且它们对应的亚基随机组合，则有可能会出现以下三种结构：$\alpha^1\alpha^1$、$\alpha^1\alpha^2$ 和 $\alpha^2\alpha^2$。此外，如果这两个亚基的合成速率相同并且活性和稳定性也相同的话，那么就可以推出这三条电泳带的活性之比约为 1:2:1，而实验结果似乎正是如此。

与这种特定肽酶有关的另一个有意思的现象是一系列罕见表型的发生（图 2）。所有这些表型都非常罕见，它们在迄今为止研究过的大多数人群中的出现频率等于或低于 1/1,000。家族研究结果表明，这些表型的出现是因为杂合子个体中的等位基因中的一个是如下等位基因：肽酶编码基因 A^3、肽酶编码基因 A^4、肽酶编码基因 A^5、肽酶编码基因 A^6 和肽酶编码基因 A^7；而另一个是常见的肽酶编码基因 A^1。在红细胞中，两种罕见肽酶，肽酶 A3-1 和肽酶 A4-1，具有这样的电泳图谱：其中特征性的、电泳速度慢的组分的活性比对应于肽酶 A1 的那条带要更弱。这可能是由于罕见等位基因肽酶编码基因 A^3 和肽酶编码基因 A^4 编码的亚基催化活性更低、合成速度更慢或者更不稳定而造成的结果。似乎最后一种可能性最大，因为在组织（如胎盘）提取物经过电泳后，可以观察到这些表型，这时电泳带的模式呈对称状态，并且三条主要染色带的活性比例约为 1:2:1。似乎可能的情形是，这种稳定性的差别可以在红细胞中显现出来，因为红细胞的半衰期相对较长，而且很可能在其细胞核消失的时候，蛋白质的合成过程就已经停止了。

在这一组中另一种有趣的变异体是肽酶 A5-1，它以两种形式发生：图 2 中标有"还原形式"字样的结果是利用从该表型的个体中获得的新鲜样品进行电泳后得到的，而标为"氧化形式"的结果是将样本在 4℃ 下储存几天后再进行电泳得到的。如果储存样品先用巯基乙醇处理的话，那么所得到的电泳图谱就和新鲜样本的电泳图谱类似。另一方面，当用氧化型谷胱甘肽处理新鲜样品时，其电泳图谱就会变得与储

sample. It seems likely that the enzyme coded for by the variant allele, *Pep A*[5], contains a reactive sulphydryl group, which is not present in other peptidase A subunits, and which undergoes an exchange reaction with oxidized glutathione so that a half molecule of oxidized glutathione forms a disulphide bond with the enzyme[12]. The addition of the acidic peptide glutathione to the enzyme molecule would explain the increase in electrophoretic mobility after storage of haemolysates. This variant reacts with various reagents known to react with sulphydryl groups and the electrophoretic changes produced by these reagents are consistent with the hypothesis I have outlined[13].

It seems probable that the genetic code is common to most species[14]. If so, it is possible to postulate the particular amino-acid substitution that has occurred in the subunit coded for by the *Pep A*[5] allele. The only amino-acid which has a reactive sulphydryl group which is found in proteins is cysteine. This amino-acid is neutral, and because there is an electrophoretic difference between the subunit coded for by *Pep A*[5] and that coded for by the common allele *Pep A*[1], in that they migrate further towards the anode so that there has been an increase in net negative charge, cysteine must have been substituted for an amino-acid which would be positively charged at *p*H 7.5. This could be either lysine or arginine. Examination of the genetic code shows that while the substitution of cysteine for arginine could result from a change in single base pair, the substitution of cysteine for lysine could not. It seems possible therefore that in this variant a particular arginine residue has been replaced by cysteine. A similar hypothesis was proposed for Pep A6–1, suggesting that in this case glutamic acid is substituted for a particular lysine residue[12].

A similar series of rare alleles has been described for several other enzymes, including peptidase B[5,15], phosphoglucomutase[16], and phosphohexose isomerase[17].

A polymorphism of peptidase D (prolidase) has also been described[18]. Variants occur in all the major population groups so far studied. Recently variants have also been described of peptidase C in African pigmies[19]. In each case a variant of a particular peptidase is not associated with any change in electrophoretic pattern of any of the other peptidases. This finding supports the biochemical evidence that these enzymes are distinct proteins, and suggests that they are coded for by separate gene loci. Moreover, studies of two families indicate that the loci controlling the structures of these pairs of enzymes are not closely linked, but are either well separated on the same chromosome or situated on different chromosomes.

Another interesting group of isozymes which illustrates this type of variation in human populations is phosphoglucomutase (PGM). These enzymes are very specific phosphotransferases which catalyse the reaction

$$\text{glucose-1-phosphate} \rightleftharpoons \text{glucose-6-phosphate}.$$

Using glucose-1-phosphate as the substrate it is possible to detect these isozymes after electrophoresis using glucose-6-phosphate dehydrogenase as a reagent, which oxidizes

存样品的电泳图谱难以相互区分。由变异体等位基因肽酶编码基因 A^5 编码的酶可能含有一个活性巯基，这样一个巯基在其他肽酶 A 亚基中并不存在，而且它能和氧化型谷胱甘肽发生交换反应，从而使半个分子的氧化型谷胱甘肽与该酶之间形成一个二硫键[12]。用酸性谷胱甘肽被加合到酶分子上就可以解释为什么血液裂解液经过储存后电泳迁移率会增加。这个变异体能与已知能与巯基发生反应的多种试剂发生反应，而由这些试剂导致的电泳图谱变化与我所提出的假设是互相吻合的[13]。

遗传密码似乎对大多数物种而言是相同的[14]。如果真是这样的话，那么我们就能推断出由等位基因肽酶编码基因 A^5 编码的亚基中发生了特定氨基酸的替换。蛋白质中唯一一个含有活性巯基的氨基酸是半胱氨酸。这种氨基酸是中性的，既然由肽酶编码基因 A^5 编码的亚基与由常见等位基因肽酶编码基因 A^1 编码的亚基在电泳图谱上有所不同，前者更靠近阳极，所以出现了净负电荷的增加，由此可推断，半胱氨酸一定是替换了一个 pH 值 7.5 时带正电的氨基酸。这种氨基酸可能是赖氨酸或者精氨酸。对遗传密码的分析结果表明：尽管半胱氨酸替换精氨酸可以由一个碱基对的变化而引起，但半胱氨酸替换赖氨酸却不能。因此，可能的情况是，在这个变异体中有一个特定位置的精氨酸残基被替换成了半胱氨酸。对于肽酶 A6–1 也有类似的假设提出，即谷氨酸替换了一个特定位置的赖氨酸残基[12]。

类似的罕见等位基因系列也在其他几种酶中被发现，包括肽酶 B[5,15]、磷酸葡萄糖变位酶[16] 和磷酸己糖异构酶[17]。

肽酶 D（脯肽酶）的多态性现象也被报道过[18]。在迄今为止研究过的所有主要人群中都有肽酶 D 的变异体出现。肽酶 C 的变异体最近在非洲俾格米人中被报道过[19]。在这些报道的例子中，每一种特定肽酶的变异体都与任意其他肽酶的电泳图谱的变化不相符。这一发现作为生物化学证据，也证明了这些肽酶确实是不同的蛋白质，并暗示它们是由独立的基因位点编码的。此外，人们对两个家族开展的研究所得结果表明，控制这些成对酶的结构的基因位点并不是紧密连锁的，它们或者在同一染色体上但完全独立，或者位于不同的染色体上。

能用于表征人群中出现这种变异性的另一组有趣的同工酶是磷酸葡萄糖变位酶（PGM）。这些酶都是特异性很强的磷酸转移酶，可以催化下述反应：

$$1\text{-磷酸葡萄糖} \rightleftharpoons 6\text{-磷酸葡萄糖}$$

用 1– 磷酸葡萄糖为底物可以在电泳后检测出这些同工酶，方法是：用 6– 磷酸葡萄糖脱氢酶为一种反应试剂，它能够氧化由磷酸葡萄糖变位酶催化产生的 6– 磷酸葡萄

the glucose-6-phosphate, formed by the action of the phosphoglucomutase, with the concomitant reduction of NADP to NADPH. The reduced coenzyme reacts with phenazine methosulphate and a tetrazolium salt resulting in a deposit of insoluble formazan blue at the site of enzyme activity. This method, used after electrophoresis at pH 7.4, has revealed a polymorphism of phosphoglucomutase in human populations[20]. The electrophoretic patterns of the three phenotypes, PGM1, PGM2−1 and PGM2, are shown in Fig. 3. This polymorphism occurs in most populations, although gene frequencies differ.

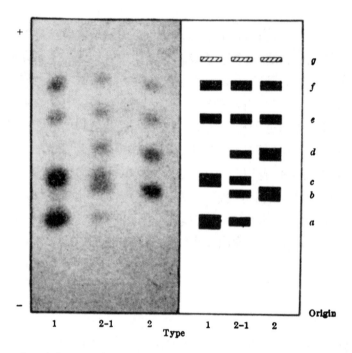

Fig. 3. Photograph and diagram of the electrophoretic patterns after electrophoresis of samples from individuals of the three common PGM types. (From ref. 20.)

Extensive family studies have shown that the different electrophoretic types result from the occurrence of a pair of autosomal alleles. Individuals who are PGM1 or PGM2 are apparently homozygous for one of these alleles. The electrophoretic pattern (Fig. 3) consists of two bands, which in PGM2 migrate further towards the anode than those given by PGM1. In PGM2−1 the electrophoretic pattern consists of four bands, corresponding in mobility to each of the isozymes seen in the homozygotes. The family studies indicate that individuals who are PGM2−1 are heterozygotes, and the products of each allele seem to be synthesized in this phenotype.

Electrophoresis of red cell lysates also reveals a second series of isozymes migrating further towards the anode which do not show any variation associated with that which occurs in the slower moving group (PGM$_1$). It was suggested that these faster moving isozymes (PGM$_2$) were controlled by a second gene locus, and later some variants were described which supported this hypothesis[10,16]. Moreover, the available evidence suggests that the

糖，并伴随着烟酰胺腺嘌呤二核苷酸磷酸酯（NADP）被还原成烟酰胺腺嘌呤二核苷酸磷酸（NADPH）。还原后的辅酶能与吩嗪 N– 甲硫酸盐和一种四唑盐反应，在具有酶活性的位置产生出不溶的蓝色甲䐶沉积物。通过在 pH 值 7.4 条件下电泳后再使用这种方法染色，人们揭示出了人群中磷酸葡萄糖变位酶的多态性[20]。三种表型 PGM1、PGM2–1 和 PGM2 的电泳图谱示于图 3。这种多态性发生于大多数人群中，尽管基因频率有所不同。

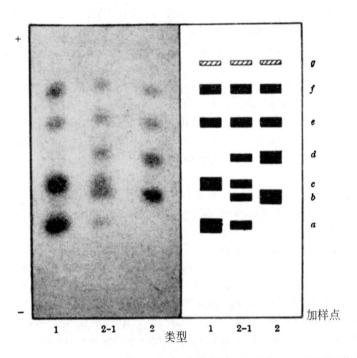

图 3. 对三种常见的 PGM 类型个体样品进行电泳检测后拍摄的电泳图谱照片和所作的图解。（来自参考文献 20）

　　大量家族研究的结果表明：这些不同的电泳类型是由一对常染色体等位基因的出现产生的。具有 PGM1 或者 PGM2 表型的个体从表观上看是其中一个等位基因的纯合子。其电泳图谱（图 3）显示为两条带（译者注：根据下文的描述，这里所说的两条带应该指的是靠近阴极的两条带），这两条带在 PGM2 中比在 PGM1 中更靠近阳极。PGM2–1 的电泳图谱包含四条带，分别对应于纯合子中每一种同工酶的迁移率。家族研究结果表明，PGM2–1 的个体是杂合子，在这种表型中似乎每个等位基因的产物都被合成了。

　　红细胞裂解物的电泳结果也表明：第二个系列的同工酶更加靠近阳极，而且它们没有显示出任何与移动更慢的 PGM₁ 中所发生的变化相关的变化。有人提出这些移动速度较快的同工酶（PGM₂）是受另一个基因位点控制的，随后人们发现了一些能支持上述假说的变异体[10,16]。此外，目前已有证据显示，控制 PGM₁ 和 PGM₂ 一

loci controlling the primary structures of PGM_1 and PGM_2 are not closely linked, but are either well separated on the same chromosome or on different chromosomes[21].

Examination of placental extracts by electrophoresis revealed a third series of isozymes, PGM_3, which are relatively less active and which migrate further towards the anode than either PGM_1 or PGM_2. This group of isozymes of phosphoglucomutase also have common variation, which is genetically determined[22], as studies of twin placentae have shown. This group of isozymes is apparently present in most tissues, although it has not been possible to demonstrate significant activity in red blood cells. Fibroblast cultures, however, have been used for the study of the segregation of the variants of PGM_3 in families, and the results agree well with those obtained from the study of twin placentae[21].

The significance of the occurrence of these three different forms of phosphoglucomutase in man is not clear. It is interesting, however, that the electrophoretic patterns of the three groups of isozymes are quite similar, and that they seem to have similar if not identical molecular weights[23,24]. It is possible that they have related primary structures which have evolved by gene duplication from a single locus.

The implications of this type of variation in the primary structure of proteins in human populations are quite wide ranging, both for theoretical human biology and in the practical aspects of the study of the human organism. It is possible that common variants of a particular enzyme provide a pool of variability within the species, such that the constraints of the environment for the species as a whole are minimized, and the species is less vulnerable to environmental changes. Moreover, available techniques indicate that at least one in three enzymes has this type of variation. The chance of any two unrelated individuals being identical in terms of protein structure therefore seems to be extremely small. So far there is no experimental evidence to support the hypothesis that variation in the primary structure of enzyme proteins is the basis of much of the observed variation in the gross appearance of individuals. It seems reasonable to suppose, however, that many of the characters which are inherited in an obscure manner, and which are apparently controlled by many genes, are the result of structural variation at a molecular level of several different proteins.

The effect of a particular variant may well differ from one tissue to another. For example, a variant enzyme with a shorter half life than the usual form of the enzyme may not have any appreciable effect in a tissue with a fairly rapid cellular turnover, such as intestinal mucosa. On the other hand, in the erythrocyte, which has a relatively long half life and in which protein synthesis presumably ceases at about the time that the nucleus is lost, the activity of a particular structural form of an enzyme may be lost quite soon after the cell is released into the circulation. Such an unstable enzyme may result in a more rapid destruction of the erythrocyte, as for example occurs with some rare variants of enzymes in the anaerobic glycolytic pathway, such as pyruvate kinase[25], or there may be no obvious effect until the individual is exposed to a toxic compound such as a drug or a particular type of food, as is seen with two different variants of glucose-6-phosphate

级结构的基因位点并非紧密连锁，它们或者在同一染色体上但完全独立，或者位于不同的染色体上 [21]。

在用电泳检查胎盘提取物时发现了第三个系列的同工酶——PGM$_3$，它比 PGM$_1$ 或 PGM$_2$ 的活性弱一些，并且电泳后比 PGM$_1$ 或 PGM$_2$ 更靠近阳极。这一组磷酸葡萄糖变位酶的同工酶也具有共同的变异，而且是由遗传决定的 [22]，这一点已通过对双胞胎的胎盘研究得到了证实。尽管现在还不能证实这一组同工酶在红细胞中有明显的活性，但它们似乎存在于大多数组织中。然而，通过体外培养成纤维细胞对家族中 PGM$_3$ 变异体进行的分离工作所获得的结果与通过双胞胎胎盘的研究所获得的结果吻合得很好 [21]。

现在还不清楚这三种不同形式的磷酸葡萄糖变位酶在人体中产生的重要性。但有趣的是：这三组同工酶的电泳图谱非常相似，而且它们的分子量也非常接近，甚至可以说完全一样 [23,24]。它们的一级结构有可能是相关的，由一个单一的位点通过基因加倍而产生。

人群中蛋白质一级结构的这类变异涉及多方面含义，包括理论人类生物学以及人作为生物而被研究的实践方面。一种特定酶的多个常见变异体的出现可能为一个物种内部的多样性提供了一种储存，这样环境对某个物种的整体所产生的约束作用就降到了最低程度，因而降低了物种对环境变化的脆弱性。此外，利用已有技术得到的结果表明，至少有 1/3 的酶具有这种类型的变异。因此，任意两个不相干的个体从蛋白质结构而言完全相同的几率似乎是极其微小的。目前还没有实验证据能支持以下假说：酶蛋白一级结构的变异是个体之间在总体外观上存在差异的主要基础。但似乎可以合理地认为：通过一种并非清晰的方式遗传、表观上看受到多个基因控制的许多个体特征实际上是几种不同蛋白质在分子水平上发生结构变异的结果。

一种特定酶蛋白的变异体在不同组织中产生的效应可能是不同的。例如，一种半衰期比普通酶更短的变异体酶在像肠黏膜这样细胞代谢速度较快的组织中可能不会产生什么明显的效应。另一方面，在半衰期相对较长且蛋白质合成大约在细胞核消失时就会停止的红细胞中，一种具有特定结构形式的酶的活性会在细胞被释放进入到血液循环系统后以比较快的速度丢失。这样一种不稳定的酶可能会加速红细胞的破坏，在无氧糖酵解过程中发挥作用的一些罕见变异体酶（如丙酮酸激酶）就是其中的一个例子 [25]。或者，在个体接触到有毒化合物（如药物或特定类型的食物）之前变异体酶并不表现出明显的效应，比如 6- 磷酸葡萄糖脱氢酶的两种不同的变异

dehydrogenase[26,27]. In one case, treatment with the antimalarial drug, primaquine, induces a haemolytic crisis and, in the other, ingestion of the fava bean has a similar effect.

Another consequence of the occurrence of structural variants of enzymes would be that heterozygotes would have a least two structural forms of the enzyme, and if the enzyme were a dimer there would be three isozymes, four for a trimer and five for a tetramer. The proportions in which these isozymes were present in any tissue would clearly depend on several factors, including the stability of the subunits and of different combinations of subunits. The apparent activity would also depend on the relative catalytic efficiency of particular combinations of subunits. It also seems possible that these isozymes are present in different proportions throughout the life of the cell. Whether or not this phenomenon has any significance for the whole organism is not known.

Another possible effect of a difference in half life, and therefore of activity, of an enzyme may occur when the enzyme affected is part of a metabolic scheme which is made up of different pathways. If we consider a hypothetical three-enzyme system such as that illustrated, and if in the normal situation with normal concentrations of α, and β, 25 percent of A is converted to D, then if α is replaced by α' which is less stable or catalytically less active, the rate of conversion of A to B would be reduced and more of A could be metabolized to D.

$$A \rightleftharpoons^{\alpha} B \rightleftharpoons^{\beta} C$$
$$\Updownarrow \gamma$$
$$D$$

Such a situation in itself might not, in most conditions, have any demonstrable effect on the individual. But if the individual is faced with a metabolic challenge in which the concentrations of B or C are critical, then the effect may well be quite dramatic, as in the case of primaquine sensitivity[26,27].

On the other hand, the accumulation of a high concentration of D may in itself be toxic, and this situation is found with some inborn errors of metabolism, such as galactosaemia[28,29], although in these cases the level of activity of the affected enzyme is usually very low indeed, and perhaps beyond the limits of detection. There are of course many possible variations within even the simple scheme I have outlined. It seems probable that these "enzymopathies" are part of the same general phenomenon as the rare variants, but in this case the affected individual can apparently synthesize only an unstable enzyme, as an inactive enzyme is perhaps completely incapable of synthesizing the particular enzyme protein.

If the enzyme concerned in a particular metabolic error is common to most tissues, the study of the enzyme in erythrocytes makes it possible to study the enzyme without recourse to biopsy procedures. The occurrence of a polymorphism can provide good evidence that the enzymes under investigation in different tissues are the products of the

体[26,27]。在一个案例中，用抗疟药伯氨喹进行治疗时导致了溶血的危险后果；而在另一个案例中，食入蚕豆也产生了类似的效应。

酶结构变异体发生的另一种情形是，杂合子个体将至少产生酶的两种结构形式，如果这种酶以二聚体形式存在的话，就会形成三种同工酶，如果是三聚体酶就会形成四种，如果是四聚体就会形成五种。这些同工酶在任何组织中存在的比例显然取决于几个方面的因素，包括亚基的稳定性以及亚基之间各种组合的稳定性。其表观活性也取决于特定亚基组合形式的相对催化效率。另一种似乎可能的情形是，这些同工酶在细胞整个生命过程中的不同阶段以不同的比例存在。目前还不清楚这一现象对生物个体的整体而言是否具有重要意义。

由酶的半衰期以及活性的差别而导致的另一个可能的效应也许会发生于当所影响的酶参与一个代谢网络的时候，这个代谢网络一般由多条途径组合而成。如果我们考查一个如下图所示的假想的三酶系统，而且如果在正常情况下两种酶 α 和 β 的浓度正常，并有 25% 的 A 转变成了 D，那么，如果 α 被更不稳定或者催化活性更弱的 α' 所替代的话，则 A 向 B 的转化率就会降低，更多的 A 就会被代谢转变成 D。

$$A \; \overset{\alpha}{\rightleftharpoons} \; B \; \overset{\beta}{\rightleftharpoons} \; C$$
$$\big\Updownarrow \gamma$$
$$D$$

在大多数情况下，这样一种情形本身也许不会对个体产生任何可被测量的影响。但如果个体面对的是一种 B 或者 C 的浓度处于导致代谢危机的临界值的情形，那么效应就很可能是相当剧烈的，就像伯氨喹敏感性导致的结果那样[26,27]。

另一方面，高浓度 D 的积聚本身也许就是有毒性的，这种情况出现在某些先天性代谢障碍中，比如半乳糖血症[28,29]，不过在这些例子中所影响的酶的活性水平实际上总是很低，而且可能已经超出了检测的底限。即便是在我上面描述过的这个简单系统中，也势必会出现很多种可能的变体。似乎可能的是，这些"酶缺陷症"与罕见变异体一样，都是同一种常见现象中的一部分。但在这种情况下受到影响的个体表观上看只能合成一种不稳定的酶，因为一种无活性的酶有可能完全不能合成特定的酶蛋白。

如果一种与特定代谢障碍相关的酶普遍存在于大多数组织中，那么对红细胞中的酶进行研究就可以使不借助活组织检查来研究酶成为可能。多态性的发生可以为以下观点提供有力的证据，即不同组织中被研究的酶都是同一个基因位点的产物，

same gene locus and are therefore probably structurally identical.

(**230**, 215-218; 1971)

W. H. P. Lewis: Department of Pathology, St Helier Hospital, Carshalton, Surrey.

References:

1. Landsteiner, K., *Wein. Klin. Wschr.*, **14**, 132 (1901).

2. Smithies, O., *Biochem. J.*, **61**, 629 (1955).

3. Harris, H., *Proc. Roy. Soc.*, B, **174**, 1 (1969).

4. Adams, E., McFadden, M., and Smith, E. L., *J. Biol. Chem.*, **198**, 663 (1952).

5. Lewis, W. H. P., and Harris, H., *Nature*, **215**, 351 (1967).

6. Lewis, W. H. P., *Symp. Zool. Soc.*, **26**, 93 (1970).

7. Lewis, W. H. P., and Harris, H., *Ann. Human Genet.*, **33**, 89 (1969).

8. Santachiara-Benerecetti, S. A., *Atti. Assoc. Genet. Ital.*, **14**, 145 (1969).

9. Boyer, S. H., Porter, I. H., and Weilbacher, R. G., *Proc. US Nat. Acad. Sci.*, **48**, 1868 (1962).

10. Hopkinson, D. A., and Harris, H., *Nature*, **208**, 410 (1965).

11. Shaw, C. R., *Science*, **149**, 936 (1965).

12. Lewis, W. H. P., Corney, G., and Harris, H., *Ann. Human Genet.*, **32**, 35 (1968).

13. Hopkinson, D. A., and Sinha, K. P., *Ann. Human Genet.*, **33**, 139 (1969).

14. Crick, F. H. C., *Proc. Roy. Soc.*, B, **167**, 331 (1967).

15. Blake, N. M., Kirk, R. L., Lewis, W. H. P., and Harris, H., *Ann. Human Genet.*, **33**, 301 (1969).

16. Hopkinson, D. A., and Harris, H., *Ann. Human Genet.*, **30**, 167 (1966).

17. Detter, J. C., Ways, P. O., Giblett, E. R., Baughan, M. A., Hopkinson, D. A., Povey, S., and Harris, H., *Ann. Human Genet.*, **31**, 329 (1968).

18. Lewis, W. H. P., and Harris, H., *Ann. Human Genet.*, **32**, 317 (1969).

19. Sanachiara-Benerecetti, S. A., *Amer. J. Human Genet.*, **22**, 232 (1970).

20. Spencer, N., Hopkinson, D. A., and Harris, H., *Nature*, **204**, 742 (1964).

21. Parrington, J. M., Cruikshank, G., Hopkinson, D. A., Robson, E. B., and Harris, H., *Ann. Human Genet.*, **32**, 27 (1968).

22. Hopkinson, D. A., and Harris, H., *Ann. Human Genet.*, **31**, 395 (1968).

23. McAlpine, P. J., Hopkinson, D. A., and Harris, H., *Ann. Human Genet.*, **34**, 177 (1970).

24. Monn, E., *Intern. J. Protein Res.*, **1**, 73 (1969).

25. Valentine, W. N., Tanaka, K. R., and Miwa, S., *Trans. Assoc. Amer. Physicians*, **74**, 100 (1961).

26. Carson, P. E., Flanagan, C. L., Ickes, C. E., and Alving, A. S., *Science*, **124**, 484 (1956).

27. Szeinberg, A., Sheba, C., Hirshom, N., and Bodongi, E., *Blood*, **12**, 603 (1957).

28. Harris, H., *Human Biochemical Genetics* (Cambridge University Press, 1959).

29. Crome, L., and Stern, J., *The Pathology of Mental Defect* (Churchill, London, 1967).

30. Hopkinson, D. A., Spencer, N., and Harris, H., *Nature*, **199**, 969 (1963).

31. Giblett, E. R., *Genetic Markers in Human Blood*, 491 (Blackwell, Oxford, 1969).

32. Spencer, N., Hopkinson, D. A., and Harris, H., *Ann. Human Genet.*, **32**, 9 (1968).

33. Fildes, R. A., and Harris, H., *Nature*, **209**, 261 (1966).

因而它们很可能具有相同的结构。

（毛晨晖 翻译；昌增益 审稿）

Experimentally Created Incipient Species of *Drosophila*

T. Dobzhansky and O. Pavlovsky

Editor's Note

One of the earliest complaints about Darwin's theory of evolution was that there was no evidence that new species can be created in laboratory conditions. The paper by Theodosius Dobzhansky and his colleague Olga Pavlovsky describes their successful creation of an incipient species of the fruit fly. Dobzhansky was one of the outstanding geneticists of the 1930s and played an important part in the theoretical underpinning of modern genetics.

EXPERIMENTAL creation of new biological species by means of allopolyploidy—doubling the chromosome sets in interspecific hybrids—has been known for several decades. One of the classics in this field is the work of Karpechenko[1], who obtained a tetraploid *Raphanobrassica* (radocabbage) from hybrids between *Raphanus* (radish) and *Brassica* (cabbage). *Raphanobrassica* is almost fully fertile *inter se*, but forms highly sterile hybrids with the parental species. Some allopolyploid species existing in nature have been experimentally resynthesized; for example, the mint, *Galeopsis tetrahit*[2], bread wheats[3] and others. Though widespread and important in some plant families, species formation by allopolyploidy is uncommon in the living world at large. A different mode of origin of species is prevalent among sexually reproducing and out-breeding organisms. This is accumulation of genetic differences between geographically separated (allopatric) populations or races, followed by a gradual emergence of reproductive isolation between them[4,5]. Species formation through genetic divergence and its fixation by reproductive isolating mechanisms is a slow process, generally requiring many generations. For this reason, this kind of speciation has not been observed or reproduced in experiments. In a sense, we are in a situation today similar to that experienced by Darwin more than a century ago: differentiation of species is inferred from copious indirect evidence, but has not actually been observed. The experiments described here are therefore very unusual: what may be regarded as the crucial stage of the speciation process has taken place, and in part deliberately induced in laboratory experiments.

Superspecies *Drosophila paulistorum*

Drosophila paulistorum, a member of the *willistoni* species group of *Drosophila*, occurs in the American tropics, from Guatemala and Trinidad to southern Brazil. *D. paulistorum* is a compound of six semispecies or incipient species—genetically too different to be regarded as races of the same species, yet not different enough to be rated as six fully differentiated species[6,7]. It is difficult to cross the semispecies because of a pronounced ethological (sexual) isolation. No hybridization has been detected in nature. In the laboratory, when females

通过实验创建的果蝇端始种

多布赞斯基，帕夫洛夫斯基

编者按

人们对达尔文进化论最早的不满之一在于没有新物种可以在实验室条件下被创建出来的证据。特奥多修斯·多布赞斯基及其同事奥尔加·帕夫洛夫斯基的这篇论文描述了他们对自己成功创建出的一个果蝇端始种的研究。多布赞斯基是20世纪30年代的杰出遗传学家之一，他在奠立现代遗传学的理论基础方面作出过重要贡献。

利用异源多倍性——使物种间杂种的染色体组加倍——通过实验创建新生物学物种在几十年前就已经为人们所知。卡尔佩琴科的研究堪称该领域中的经典之作之一 [1]；他从萝卜和甘蓝的杂交种获得了四倍体萝卜甘蓝。萝卜甘蓝的种内交配几乎是完全可育的，但与其亲本物种的杂交种是高度不育的。一些存在于自然界中的异源多倍体物种已经通过实验重新合成，例如薄荷、野芝麻 [2]、普通小麦 [3] 等等。虽然通过异源多倍性形成物种在某些植物科属中广泛存在，并且还十分重要，但在整个生物界中并不算常见。物种起源的另一种不同模式在有性繁殖和远系繁殖的生物中是普遍存在的，即随着遗传差异在地理上隔离的（异域的）种群或地理宗之间积累，这些地理种群或地理宗间会逐渐出现生殖隔离 [4,5]。遗传分化导致的物种形成以及生殖隔离机制对新物种的固定是一个缓慢的过程，通常需要经过很多世代。基于这个原因，这类物种形成从来没有在实验中被观察到过，或用实验重复出来过。从某种意义上讲，我们今天的处境与一个多世纪之前达尔文所面临的困难类似：物种分化是根据丰富的间接证据推断出来的，但是没有被真正观察到过。因此，本文中所描述的实验非同寻常：可以被认为是物种形成中最关键的阶段发生了，并且在一定程度上是在实验室条件下人为诱导出来的。

圣保罗果蝇超种

圣保罗果蝇是威氏果蝇中的一种，分布于美洲的热带地区，从危地马拉和特立尼达一直到巴西南部。圣保罗果蝇是 6 个半分化种或者说端始种的混合体，即它们在遗传学上非常不同以至于不能被看作是同一物种下的不同地理宗；但它们之间的差异也没有达到可以被看作是 6 个完全分化的物种的程度 [6,7]。半分化种之间由于存在显著的行为（性）隔离而难于进行杂交；在自然界中也从未发现过它们杂交。在

and males of two semispecies are placed together, most of the matings occur within a semispecies. The exceptions, females of one semispecies inseminated by males of another, give progenies of apparently normal and vigorous hybrid daughters and sons. The hybrid females are fertile when backcrossed to males of either parental semispecies. The hybrid males are completely sterile.

The combination of these two isolating mechanisms permits sympatric coexistence of two, or even three, semispecies. Thus, the Central American semispecies occurs alone from Guatemala to Costa Rica and western Panama, but in central Panama it meets the Amazonian and Orinocan semispecies, The distribution of the Amazonian semispecies extends from Panama to Guiana, Rio Negro and the estuary of the Amazon (Belem). The Orinocan semispecies has been found in northern Colombia, Venezuela, Trinidad and Guiana. The Interior semispecies occurs on upper Orinoco, Rio Negro, upper Amazon and their tributaries. The Andean semispecies occurs alone in central and southern Brazil, Bolivia, Peru and Ecuador, but it meets the Amazonian and Interior semispecies in central and southern Colombia, Trinidad, Venezuela, Guiana and along the Amazon and Rio Negro. Finally, the Transitional semispecies occurs on the Pacific Coast of Colombia, in northern Colombia and northern Venezuela[7]. The semispecies are indistinguishable morphologically but they can be distinguished cytologically[8,9]; the easiest in practice and usually unambiguous method of recognition is outcrossing a strain to be diagnosed to six tester strains, one from each semispecies. Fertile hybrids are produced with only one of the six testers, that which represents the semispecies to which belongs the strain being tested.

Career of the Llanos Strain

A laboratory strain was established from a female captured on March 19, 1958, south of Villavicencio, in the Llanos of Colombia. The strain was classified as belonging to the Orinocan semispecies, because it crossed easily to, and gave fertile hybrids with, other Orinocan strains then available in the laboratory. The behaviour of the Llanos strain changed, however, between 1958 and 1963; in 1963 and thereafter, the male hybrids obtained from crosses of Llanos to Orinocan strains proved to be completely sterile. The possibility that the change in the behaviour of the Llanos strain might have resulted from accidental contamination was ruled out on two grounds. First, a cytological study of the Llanos strain in 1959 showed two polymorphic chromosomal inversions which were, at that time, found in no other strain. Llanos still carries these chromosomal polymorphisms. Second, in 1963, Llanos was systematically outcrossed to all other strains of *D. paulistorum* then available in the laboratory, and gave fertile male hybrids with none. The reality of the change is not in doubt[10,11].

实验室条件下，当两个半分化种的雌性个体和雄性个体被放置在一起时，绝大多数交配行为发生在同一个半分化种内。少数情况下，当一个半分化种的雄性个体给另一个半分化种的雌性个体受精后，雌性个体会产下看来既正常又有活力的雌性和雄性杂种。雌性杂种与任一半分化种的雄性亲本回交时都是可育的；雄性杂种则是完全不育的。

这两种隔离机制的组合使得 2 种甚至 3 种半分化种可以同域共存。因此，中美洲半分化种只出现在从危地马拉到哥斯达黎加及巴拿马西部，但是在巴拿马中部，它遇到了亚马孙半分化种和奥里诺科半分化种。亚马孙半分化种的分布范围是从巴拿马延伸至圭亚那、里奥内格罗河区以及亚马孙河的河口地区（贝伦）。奥里诺科半分化种分布于哥伦比亚北部、委内瑞拉、特立尼达以及圭亚那。内陆半分化种出现在奥里诺科河上游、里奥内格罗河区、亚马孙河上游以及它们的支流流域。安第斯半分化种只出现在巴西中部和南部、玻利维亚、秘鲁和厄瓜多尔，但它与亚马孙半分化种和内陆半分化种在哥伦比亚中部和南部、特立尼达、委内瑞拉、圭亚那地区以及亚马孙河和里奥内格罗河沿岸区相遇。最后，过渡半分化种出现在哥伦比亚的太平洋沿岸、哥伦比亚北部和委内瑞拉的北部 [7]。这些半分化种从形态上无法分辨，但从细胞学上可以将它们区分开 [8,9]。操作上最为简便并且通常都能得到明确结果的分辨方法是：将待测品系与 6 个半分化种的各一个品系进行异型杂交，可育的杂交后代只会从 6 个检验组合中的一个获得，这说明待测品系所属的半分化种与对应的试验品系相同。

亚诺斯品系的来龙去脉

1958 年 3 月 19 日在哥伦比亚亚诺斯大草原上的比亚维森西奥市南部捕获到一只雌性果蝇，并从它建立了一个实验室品系。这一品系被归为奥里诺科半分化种，因为它容易与当时实验室中拥有的其他奥里诺科果蝇杂交，并产下可育的杂交种。然而，在 1958 年到 1963 年间，亚诺斯品系的行为发生了改变；1963 年及那以后，由亚诺斯品系果蝇和奥里诺科品系果蝇杂交得到的雄性杂交种被证明是完全不育的。基于如下两个依据，由于意外污染而导致亚诺斯品系果蝇在行为上发生变化的可能性被排除了。首先，1959 年对亚诺斯品系的细胞学研究发现了当时其他品系所没有的两种多态性染色体倒位。亚诺斯品系现在仍携带有这些染色体多态性。其次，1963 年曾将亚诺斯品系与实验室中的所有其他品系的圣保罗果蝇进行了系统性的异型杂交，没有得到任何雄性可育的杂种。因此，上述行为改变的真实性是毋庸置疑的 [10,11]。

In 1964, Professor H. L. Carson sent us a strain established from a female collected at Marco, on the upper Amazon, Brazil. This strain gave fertile hybrids only with Llanos. Later, many other strains of the Interior semispecies, to which Marco belongs, were found in equatorial South America[7,12]. Are we to conclude that the Llanos strain became transformed from a representative of the Orinocan semispecies to one of the Interior semispecies? Such a conclusion is not warranted, although we obviously do not know whether the Llanos strain would have given fertile male hybrids with the Interior semispecies in 1958 or 1959. What we do know is that although the Llanos strain has a pronounced sexual isolation from at least some Interior strains, it mates freely with the Orinocan strain. The isolation is fairly strong between Interior and Orinocan.

The lack of isolation between the Llanos and Orinocan strains was ascertained in 1964 with "male-choice" tests[10]; Virgin females of two strains, Llanos and an orange-eyed mutant found in an Orinocan strain from Georgetown, Guiana, were confined with males of one or the other of these strains. After about half the females had become inseminated, the females were dissected and the presence or absence of sperm in their seminal receptacles was determined under a microscope (see Table 1).

Table 1. Inseminated (+) and Uninseminated (–) Females of Two Strains of *Drosophila paulistorum*, Exposed to Males of One of These Strains

	Llanos ♀♀		Orinocan ♀♀	
	+	–	+	–
Llanos ♂♂	56	34	36	54
Orinocan ♂♂	44	47	44	48

In tests of this kind, the "choice" is actually exercised by the females who accept some males in preference to others. Table 1 shows that Llanos males are accepted by females of their own kind more easily than by Orinocan females; Orinocan males are accepted equally by both kinds of females. In 1969, the experiments were repeated, using an observation chamber[13]. Virgin but mature females and males were introduced, in equal numbers, into a saucer-shaped chamber with a glass top, and the matings that occurred were recorded with the aid of a hand lens. To make the Llanos individuals distinguishable from the Orinocan, one or the other kind had one of their wings slightly clipped. Table 2 summarizes the observations made by Lee Ehrman[13] on matings between the Llanos strain and an Orinocan strain derived from Georgetown, Guiana.

Table 2. Matings Observed between Llanos and Orinocan Strains

Llanos ♀ × Llanos ♂	Llanos ♀ × Orinocan ♂	Orinocan ♀ × Llanos ♂	Orinocan ♀ × Orinocan ♂	Isolation coefficient
34	22	23	23	+0.11 ± 0.10

1964 年，卡森教授给我们寄来了一个由从巴西亚马孙河上游的马科地区采集到的一只雌蝇所建立的品系。这个品系只跟亚诺斯品系产生可育的杂交种。随后，更多内陆半分化种品系（前述马科果蝇品系即属于该半分化种）在南美洲赤道地区被发现 [7,12]。我们是否可以由此得出结论——亚诺斯品系已经从典型的奥里诺科半分化种转变为内陆半分化种？这样一个结论是没有道理的，尽管我们显然并不知道在1958 年或 1959 年期间，亚诺斯品系的果蝇是否已经能跟内陆半分化种产生可育的雄性杂交种。我们所知道的是：虽然亚诺斯品系至少与一些内陆品系果蝇之间存在着明显的性隔离，但它却可以跟奥里诺科果蝇进行自由交配；而内陆品系和奥里诺科品系之间的性隔离是相当强的。

亚诺斯品系和奥里诺科品系果蝇之间不存在性隔离这一事实已经在 1964 年通过"雄性挑选"测验而得到证实 [10]。来自两个品系的未交配过的雌蝇，包括亚诺斯品系和在圭亚那乔治敦发现的奥里诺科品系的橙眼突变体，被限制只能与其中一个品系的雄蝇进行交配。在半数左右的雌性个体受孕之后，通过对雌蝇进行解剖，在显微镜下确认她们的受精囊内是否存在精子（见表 1）。

表 1. 圣保罗果蝇两个品系的雌蝇在与其中一个品系的雄蝇共处后受精（＋）和不受精（－）
　　　的个体数量

	亚诺斯品系♀♀		奥里诺科品系♀♀	
	＋	－	＋	－
亚诺斯品系♂♂	56	34	36	54
奥里诺科品系♂♂	44	47	44	48

在这类测验中，"挑选"实际上是通过雌性个体更倾向于接受某些雄性个体而实现的。表 1 的数据表明：亚诺斯品系的雄蝇更容易被它们自己品系的雌蝇所接受；而奥里诺科品系的雄蝇则均等地被所有两个品系的雌蝇所接受。1969 年，使用一个观察室重复了这个实验 [13]：将等量未交配过的成年雌蝇和雄蝇放入一个顶部带有玻璃盖的茶碟形观察室中，然后借助放大镜记录它们之间的交配情况。为了能区分亚诺斯品系个体和奥里诺科品系个体，其中一个品系的果蝇的翅膀被略微剪短。表 2总结了李·埃尔曼对亚诺斯品系果蝇与来自圭亚那乔治敦的奥里诺科果蝇之间交配情况的观察结果 [13]。

表 2. 亚诺斯品系和奥里诺科品系之间的交配情况

亚诺斯♀× 亚诺斯♂	亚诺斯♀× 奥里诺科♂	奥里诺科♀× 亚诺斯♂	奥里诺科♀× 奥里诺科♂	隔离系数
34	22	23	23	+0.11±0.10

The isolation coefficient is not significantly different from zero. This coefficient is +1 if the isolation is complete because only matings between likes take place, and −1 if matings only between unlikes are observed. Tests of the mating preferences of Llanos flies with other strains of the Orinocan and Interior semispecies, made by S. Perez-Salas[12], gave the isolation coefficients shown in Table 3.

Table 3. Isolation Coefficients of Crosses between Llanos and Other Strains

Llanos + Sarare, Orinocan	+0.33 ± 0.09
Llanos + Valparaiso, Interior	+0.18 ± 0.09
Llanos + Leticia, Interior	+0.12 ± 0.09
Llanos + Mitu, Interior	+0.24 ± 0.10
Llanos + Ocamo, Interior	+0.57 ± 0.08
Llanos + Ayacucho, Interior	+0.18 ± 0.12

Slight, but sometimes statistically significant, preferences for mating within a strain are often found in experiments with *D. paulistorum* in which two strains of the same semispecies are derived from ancestors collected in different localities. The Llanos strain crosses fairly easily with strains of both Orinocan and Interior semispecies, although the Interior semispecies are more strongly isolated ethologically from each other, and even more so from the other semispecies. We conclude that, having developed a sterility of male hybrids with the Orinocan semispecies, the Llanos strain did not acquire ethological isolation from the Orinocan. But can ethological isolation be superimposed on the hybrid sterility by means of artificial selection?

Selection for Isolation

Several workers[14-16] have induced or intensified ethological isolation between strains of the same or of different species. The techniques used in all these experiments are identical in principle. Two strains between which the isolation is to be developed are made homozygous for different recessive mutant genes. Females and males of the two strains are placed together and they are allowed to mate freely and to produce offspring. The part of that offspring coming from matings between the strains will be wild type in phenotype, whereas the progeny of matings of females and males of the same strain will show the recessive mutant traits. The wild type flies are discarded, and the mutants are allowed to become parents of the next generation. The progenies of the flies that mate within a strain are thus included, and those of the matings between the strains are excluded from parentage. This imposes a selective advantage on genetic constitutions which favour matings within and discriminate against matings between strains.

The variant of this technique that had to be used in experiments with *D. paulistorum* is rather laborious. The Llanos strain has produced a sex linked recessive mutant, rough eye.

隔离系数跟零没有显著差别。如果隔离是彻底的，那么这个系数就是 +1，因为交配只发生在同类个体之间；而如果只观察到异类个体之间发生交配，那么这个系数就是 –1。由佩雷斯-萨拉斯进行的亚诺斯品系果蝇与其他奥里诺科半分化种品系及内陆半分化种品系果蝇之间的交配偏好实验 [12] 得到了表 3 所示的隔离系数。

表 3. 由亚诺斯品系和其他品系杂交得到的隔离系数

亚诺斯 + 萨拉雷，奥里诺科	+0.33 ± 0.09
亚诺斯 + 瓦尔帕莱索，内陆	+0.18 ± 0.09
亚诺斯 + 莱蒂西亚，内陆	+0.12 ± 0.09
亚诺斯 + 米图，内陆	+0.24 ± 0.10
亚诺斯 + 奥卡莫，内陆	+0.57 ± 0.08
亚诺斯 + 阿亚库乔，内陆	+0.18 ± 0.12

在用从不同地点采集的祖先而衍生出的圣保罗果蝇同一半分化种的两个品系进行实验时，总能发现一些微小但有时统计上显著的交配偏好。亚诺斯品系果蝇与奥里诺科半分化种品系及内陆半分化种品系果蝇都可相当容易地进行杂交，不过内陆半分化种品系之间有更强的行为隔离，而且它们与其他半分化种间的隔离还会更严重一些。我们得到的结论是：虽然亚诺斯品系与奥里诺科半分化种品系间的雄性杂交种已经产生不育性，但亚诺斯品系并没有形成跟奥里诺科品系的行为隔离。然而，能否通过人工选择使其在杂种不育的基础上产生行为隔离呢？

隔离的选择

若干研究者 [14-16] 已经在同一物种的不同品系间或者不同物种间诱导出或强化了行为隔离。所有这些实验中所采用的技术在原理上都是一样的。要建立行为隔离的那两个品系分别是带有不同隐性突变基因的纯合体。把两个品系的雌性个体和雄性个体放在一起，并允许它们自由地交配和生育后代。由不同品系的个体交配而产生的那些后代在表现型上将是野生型，而由同一品系的雌雄个体交配产生的后代则会表现出隐性突变体性状。淘汰野生型果蝇，留下突变体作为下一代的亲本。因此，在家系中只保留由同一品系果蝇交配产生的后代，而那些由不同品系果蝇交配产生的后代就都被排除。这样就从家系的遗传构成上施加了祖护品系内交配、排斥品系间交配的选择压力。

在圣保罗果蝇实验中必须使用的这项技术经过改进后变得十分繁复。亚诺斯品系的果蝇已经产生了一个性连锁的隐性突变体——糙眼突变体。奥里诺科半分化种

An autosomal recessive mutant, orange eye, and sex-linked mutants, veinless wing and yellow body appeared in the Georgetown, Guiana, strain of the Orinocan semispecies. (Rough, orange and veinless were found by Mr. B. Spassky, and yellow by O. P.) In October 1966, pairs of experimental populations were started according to the following design. Between fifty and a hundred (usually seventy) virgin females of Llanos rough mutant are placed in a culture bottle with males of the same strain and of one of the Orinocan mutant strains. In another culture bottle, fifty to a hundred Orinocan mutant females are exposed to a mixture of the same Orinocan mutant males and of Llanos rough eyed males. Among the progeny, the hybrid females, coming from matings of unlike parents, are distinguishable by being wild type in phenotype. Non-hybrid females, coming from matings of like parents, show the traits of the respective mutants. The hybrid females are discarded, and the non-hybrid ones used as parents of the next generation. Because rough, yellow and veinless are sex linked, hybrid sons of the mutant mothers are not distinguishable from the non-hybrids. This is inconvenient but not fatal for the experiments, because the hybrid males are completely sterile. Because orange is autosomal it permits discrimination of hybrid and non-hybrid females as well as males. Six pairs of experimental populations were made:

1*A*: rough ♀♀ + rough ♂♂ + orange ♂♂

1*B*: orange ♀♀ + orange ♂♂ + rough ♂♂

2*A*: rough ♀♀ + rough ♂♂ + veinless ♂♂

2*B*: veinless ♀♀ + veinless ♂♂ + rough ♂♂

3*A*: rough ♀♀ + rough ♂♂ + yellow ♂♂

3*B*: yellow ♀♀ + yellow ♂♂ + rough ♂♂

Populations 1*A* and 1*B* are at present (December 1970) in the seventy-third generation of selection. Populations 2*A* and 2*B* were discontinued after sixty-five generations, and 3 *A* and 3 *B* after sixty generations of selection. The proportions of the like and the unlike males to which the females were exposed were adjusted to supply in different generations the strongest challenge to the discriminating abilities of the females. At the same time, enough mutant progeny must be produced to serve as parents of the next generation. In the early generations, the two kinds of males were usually equally numerous; later the like males were less numerous than the unlike. The total numbers of males were about equal to the numbers of females.

The selection for ethological isolation was successful in all six populations, although in none has anything like complete isolation been achieved. It can be stated that, as a rule, the proportions of hybrids among the progenies have decreased from early generations to the later ones. The progress of the selection seems, however, very uneven. For example, the percentages of hybrid females obtained in every tenth generation in population 1*A* were: F_1, 41.9; F_{10}, 17.3; F_{20}, 53.5; F_{30}, 31.8; F_{40}, 12.7; F_{50}, 13.9; F_{60}, 8.3; F_{70}, 31.8.

的圭亚那乔治敦品系中出现了一个常染色体的隐性突变体——橙眼，以及两种性连锁的突变体——无脉翅和黄体色（糙眼、橙眼和无脉翅是由斯帕斯基先生发现的；黄体色是由奥尔加·帕夫洛夫斯基发现的）。在 1966 年 10 月，按照以下方案开始了实验种群成对实验。在一个培养瓶中放置 50 只～100 只（通常是 70 只）未交配过的亚诺斯品系糙眼突变雌蝇，同时放入同一品系的雄蝇以及其中一种奥里诺科突变品系的雄蝇。在另一个培养瓶中，将 50 只～100 只奥里诺科突变雌蝇与同一突变品系的奥里诺科雄蝇及亚诺斯品系的糙眼突变雄蝇放在一起。在后代中，由不同品系亲本交配产生的杂种雌蝇可以通过表现型上的野生型性状而被区分出来。由相同品系亲本交配得到的非杂种雌蝇则表现出各自突变体的性状。淘汰杂种雌蝇，用非杂种雌蝇作为下一代的亲本。因为糙眼、黄体色和无脉翅是性连锁的，所以无法区分突变母本产生的雄性杂交子代和雄性非杂交子代。这虽然有所不便，但对本实验没有致命的影响，因为雄性杂交子代是完全不育的。由于橙眼是常染色体遗传的，因此它使我们既可以区分雄性杂交子代和非杂交子代也可以区分雌性杂交子代和非杂交子代。总共有如下 6 对实验种群：

1A：糙眼♀♀ + 糙眼♂♂ + 橙眼♂♂
1B：橙眼♀♀ + 橙眼♂♂ + 糙眼♂♂
2A：糙眼♀♀ + 糙眼♂♂ + 无脉翅♂♂
2B：无脉翅♀♀ + 无脉翅♂♂ + 糙眼♂♂
3A：糙眼♀♀ + 糙眼♂♂ + 黄体色♂♂
3B：黄体色♀♀ + 黄体色♂♂ + 糙眼♂♂

目前（1970 年 12 月）已经对种群 1A 和 1B 进行了 73 代的选择。种群 2A 和 2B 在选择了 65 代之后及种群 3A 和 3B 在选择了 60 代之后都没有再继续。雌蝇所面对的同品系和不同品系雄蝇的比例在不同的世代都被调整到足以对雌蝇的辨别能力构成最大的威胁。与此同时，必须保证能产生足够多的突变体后代以便用作下一代的亲本。在早期世代中，两类雄蝇的数目通常会一样多；后来，相同品系雄蝇的数目就会少于不同品系雄蝇的数目。雄蝇的总数大致与雌蝇的总数相当。

对行为隔离的选择在所有 6 个种群中都获得了成功，尽管没有哪一个达到了完全隔离。可以说，从早期世代到晚期世代，后代中杂交种的比例通常会有所下降。然而，选择的进展似乎很不均匀。例如：在种群 1A 中，每隔 10 个世代所得到的杂种雌蝇的百分比是：F_1，41.9；F_{10}，17.3；F_{20}，53.5；F_{30}，31.8；F_{40}，12.7；F_{50}，13.9；F_{60}，8.3；F_{70}，31.8。

The unevenness is largely an artefact. It is caused in part by the inability mentioned earlier to distinguish the hybrid and the non-hybrid males with the sex-linked mutant rough; the proportions of fertile individuals among the rough-eyed male progenies in the different generations are therefore not exactly known. Another disturbing factor is that the rough, veinless, and yellow mutant markers considerably reduce the viability of their carriers. Only orange has a satisfactory viability. The hybrid females, being wild type, have an advantage in survival.

Ethological Isolation Observed

The proportions of hybrid offspring obtained in our experimental populations do not measure reliably the degree of the ethological isolation between the strains used. We accordingly used the observation chamber technique previously mentioned. Twelve females and twelve males of each of two strains (forty-eight flies in all) were introduced in a chamber without etherization. The chambers were observed for about 3 h, and the matings that took place were recorded. A female could mate only once during this time interval, whereas males were free to mate repeatedly. Our results are summarized in Table 4.

Table 4. Observed Matings between Selected and Unselected Strains of *Drosophila paulistorum*

	Strains		Matings				Isolation
Date	A	B	A♀×A♂	A♀×B♂	B♀×A♂	B♀×B♂	coefficient
March 1970	Llanos U	Orange S	41	13	11	40	0.55±0.08
March 1970	Rough S	Orinocan U	47	17	6	40	0.59±0.08
October 1969	Rough S	Orange S	52	4	4	41	0.67±0.07
June 1970	Rough S	Orange S	45	7	3	54	0.82±0.05
March 1970	Llanos U	Veinless S	60	2	12	42	0.76±0.06
October 1969	Rough S	Veinless S	43	8	10	41	0.64±0.08
June 1970	Rough S	Veinless S	44	8	11	45	0.64±0.07
March 1970	Llanos U	Yellow S	48	8	13	43	0.62±0.07
October 1969	Rough S	Yellow S	50	6	13	32	0.62±0.08
June 1970	Rough S	Yellow S	27	16	4	66	0.65±0.07

S, Selected; U, unselected.

The designation Llanos unselected refers to the wild type Llanos strain which had not been exposed to challenges of hybridization with Orinocan strains; rough selected is the rough-eyed mutant which had been so exposed for fifty generations by October 1969, and fifty-eight generations by March 1970; Orinocan unselected is the wild type Orinocan strain from Georgetown, Guiana; orange selected, veinless selected and yellow selected are the three mutants which arose in the Georgetown strain, and were selected for as many generations as their rough selected counterparts.

All the tests reported in Table 4 show statistically highly significant isolation coefficients, ranging from 0.55±0.08 to 0.82±0.05. It should be recalled that unselected Llanos with unselected Orinocan shows little, if any, preferential mating (isolation coefficient +0.11± 0.10, Table 2). Without doubt, the selection has developed an ethological isolation where perhaps only a trace of preference for mating within a strain existed before selection. It is furthermore remarkable that the tests in which one of the strains had been selected for

这种不均匀性在很大程度上是人为造成的。前文曾提到我们无法区分携带性连锁的糙眼突变的杂种雄蝇和非杂种雄蝇就是其中一部分原因；因此，我们不能准确地知道各世代糙眼雄性后代中可育个体的比例。另一个干扰因素是：糙眼、无脉翅和黄体色等突变标记显著地降低了携带者的生存力。只有橙眼个体的生存力还算令人满意。杂种雌蝇，因为是野生型，所以在存活率方面具有优势。

观察到的行为隔离

我们在实验种群中得到的杂交后代的比例不能可靠地衡量所选用品系之间的行为隔离程度。因此我们采用了前面提到过的观察室技术。从两个品系中各抽取 12 只雌蝇和 12 只雄蝇（总共 48 只果蝇）关在一个观察室中（不进行醚麻醉）。观察大约 3 小时，并记录所发生的交配事件。在这个时间间隔内雌蝇只能交配一次，而雄蝇可以自由地重复交配。表 4 给出我们的实验结果。

表 4. 施加选择的和未施加选择的圣保罗果蝇品系之间的交配情况

时间	品系		交配情况				隔离系数
	A	B	$A♀×A♂$	$A♀×B♂$	$B♀×A♂$	$B♀×B♂$	
1970 年 3 月	亚诺斯 U	橙眼 S	41	13	11	40	0.55±0.08
1970 年 3 月	糙眼 S	奥里诺科 U	47	17	6	40	0.59±0.08
1969 年 10 月	糙眼 S	橙眼 S	52	4	4	41	0.67±0.07
1970 年 6 月	糙眼 S	橙眼 S	45	7	3	54	0.82±0.05
1970 年 3 月	亚诺斯 U	无脉翅 S	60	2	12	42	0.76±0.06
1969 年 10 月	糙眼 S	无脉翅 S	43	8	10	41	0.64±0.08
1970 年 6 月	糙眼 S	无脉翅 S	44	8	11	45	0.64±0.07
1970 年 3 月	亚诺斯 U	黄体色 S	48	8	13	43	0.62±0.07
1969 年 10 月	糙眼 S	黄体色 S	50	6	13	32	0.62±0.08
1970 年 6 月	糙眼 S	黄体色 S	27	16	4	66	0.65±0.07

S：施加选择的；U：未施加选择的。

表中的"亚诺斯 U"指的是从未面对过与奥里诺科品系进行杂交这类挑战的野生亚诺斯品系；"糙眼 S"指的是到1969年10月和1970年3月为止分别已经面对过50代和58代这样的挑战的糙眼突变体；"奥里诺科 U"指的是来自圭亚那乔治敦的野生奥里诺科品系；"橙眼 S"、"无脉翅 S"和"黄体色 S"是乔治敦品系果蝇的三种突变体，它们被选择了跟对等的"糙眼 S"果蝇一样多的世代。

表 4 报告的所有测试都有统计上极显著的隔离系数，数值范围从 0.55±0.08 到 0.82±0.05。别忘了在未施加选择的亚诺斯品系和未施加选择的奥里诺科品系之间近乎没有什么交配偏好（隔离系数为 +0.11±0.10，表 2）。毫无疑问，连续选择已经建立起行为隔离，而在施加选择之前品系内可能仅存在极微弱的交配偏好。更值得关注的是：在测试中只曾对两个品系中的一个施加过隔离选择而对另一个不施加选择，

isolation and the other was unselected showed about as much isolation as did the tests where both strains had been selected (Table 4).

Status of the Llanos Strain

In 1958 and 1959, the Llanos strain was giving fertile hybrids with strains of the Orinocan semispecies. From 1963 on, it gave sterile hybrid males with the same strains. The reason for this change is uncertain. It may be that the change is related to the geographic origin of the Llanos strain. The region where it was collected lies between the known distribution areas of the Interior and Orinocan semispecies[7]. These semispecies seem more closely related to each other than to the remaining four semispecies. Nowhere do they coexist sympatrically although each of them does so with other semispecies. The genetic instability of the Llanos strain may have resulted from its being a form intermediate between, or even a hybrid of, Orinocan and Interior semispecies[12]. Another possibility is that the change was brought about by an alteration in the population of mycoplasma-like intracellular symbionts, which seem to be different in the different semispecies[17,18].

At any rate, no appreciable reproductive isolation arose between the changed Llanos and Orinocan strains. Without ethological isolation, hybrids between them can be produced freely. The hybrids seem to be vigorous. Hybrid females are fertile, and so are the males in the offspring of the backcrosses. There is no barrier to gene flow. The gradual building up of ethological isolation changes the situation. The isolation coefficients listed in Table 4 are mostly below the mean, but well within the range of isolation coefficients encountered in experiments with semispecies found in nature—0.28 ± 0.10 to 0.94 ± 0.03 (ref. 12). There are good reasons to think that ethological isolation is probably stronger in nature than under laboratory conditions. This makes some of the semispecies able to coexist in the same territory, sympatrically, without mixing. Orinocan and Interior semispecies are an exception—they are not sympatric. Isolation coefficients between strains of these semispecies obtained in laboratory experiments range from 0.28 ± 0.10 to 0.65 ± 0.08 (ref. 12), lower than we achieved (Table 4).

We conclude that the selected Llanos strain is comparable with the naturally existing semispecies. It could not, however, maintain itself if it were sympatric with the Orinocan semispecies. To render artificial selection possible, the Llanos strain was made homozygous for the mutant gene rough eye. This mutant reduces the viability of its carriers, whereas the hybrids with Orinocan, which have the rough eye suppressed by its dominant normal allele, are more vigorous. Natural selection for ethological isolation would occur if the hybrids were, on the contrary, at a disadvantage. The last step needed to make the Llanos semispecies capable of sympatric coexistence with the Orinocan semispecies is to free it from the rough-eye mutant, without disturbing the rest of its genotype responsible for the ethological isolation. This may not be easy to achieve, but it can be attempted.

所得到的隔离效果与在测试中曾同时对两个品系都施加过选择的隔离效果是相同的（表4）。

亚诺斯品系的现况

在1958年和1959年，亚诺斯品系与奥里诺科半分化种的品系杂交会产生可育的杂交后代。从1963年起，这两种品系的果蝇杂交却产生了雄性不育的杂交种。目前还不完全清楚为什么会发生这样的变化。也许这种变化与亚诺斯品系的地理起源有关。亚诺斯品系果蝇的采集区域位于内陆半分化种和奥里诺科半分化种的已知分布区之间[7]。这3个半分化种相互之间的进化关系似乎比它们与其他4个半分化种之间的关系更近。没有任何地方这3个半分化种是共同存在的，尽管它们各自都与其他半分化种有一定的重叠分布。亚诺斯品系的遗传不稳定性有可能是因为它是一种介于奥里诺科半分化种和内陆半分化种之间的中间形式，甚或是这两者的杂交种[12]。另一种可能性是这种变化源于种群中细胞内类菌原体共生体的改变，这些共生体在不同的半分化种中似乎是不一样的[17,18]。

不管怎样，在改变后的亚诺斯品系和奥里诺科品系之间没有出现明显的生殖隔离。因为不存在行为隔离，所以它们之间的杂交种可以自由形成。这些杂交种看上去很强健。杂种雌蝇是可育的，回交后代中的雄蝇也是如此。因此不存在基因流障碍。但是逐渐形成的行为隔离改变了这一状况。表4中列出的隔离系数大多低于平均值，不过都稳稳地落在实验得到的、自然界中半分化种的隔离系数的范围之内——从0.28 ± 0.10到0.94 ± 0.03（参考文献12）。我们有合适的理由认为，自然界中的行为隔离很可能会强于在实验室条件下发生的行为隔离。这使得一些半分化种可以在同一个区域内共存，即分布区域重叠但并不发生混交。奥里诺科半分化种和内陆半分化种是个例外——它们并不同域。室内实验检测到这些半分化种的品系之间的隔离系数为$0.28 \pm 0.10 \sim 0.65 \pm 0.08$（参考文献12），低于我们得到的数值（表4）。

我们的结论是：经过选择的亚诺斯品系与自然条件下存在的半分化种之间具有可比性。然而，当它与奥里诺科半分化种同域共存时，就无法维持自身的品系了。为了使人工选择得以实现，我们选用了亚诺斯品系糙眼突变的纯合体。这种突变会降低基因携带者的生存力，但它与奥里诺科半分化种的杂交种却具有更强的生存力，因为显性的正常等位基因抑制了糙眼表征的表达。与之相反，当杂种处于劣势时，对行为隔离的自然选择就会发生。使亚诺斯半分化种能够与奥里诺科半分化种同域共存所需的最后一个步骤是：在不影响负责行为隔离的其他基因型的前提下，弃除糙眼突变体。这也许不太容易做到，但可以去尝试。

This work was supported by grants from the US Atomic Energy Commission and the National Science Foundation (International Biological Program).

(**230**, 289-292; 1971)

Theodosius Dobzhansky and Olga Pavlovsky: The Rockefeller University, New York City, New York 10021.

Received January 14, 1971.

References:

1. Karpechenko, G. D., *Z. Indukt. Abstamm.-u. Vererbungsl.*, **48**, 1 (1928).

2. Müntzing, A., *Hereditas*, **16**, 105 (1932).

3. McFadden, E. S., and Sears, E. R., *J. Hered.*, **37**, 81, 107 (1946).

4. Mayr, E., *Animal Species and Evolution* (Belknap, Cambridge, 1963).

5. Dobzhansky, Th., *Genetics of the Evolutionary Process* (Columbia University Press, New York and London, 1970).

6. Dobzhansky, Th., and Spassky, B., *Proc. US Nat. Acad. Sci.*, **45**, 419 (1959).

7. Spassky, B., Richmond, R. C., Perez-Salas, S., Pavlovsky, O., Mourão, C. A., Hunter, A. S., Hoenigsberg, H., Dobzhansky, Th., and Ayala, F. J., *Evolution* (in the press).

8. Kastritsis, C. D., *Chromosoma*, **23**, 180 (1967).

9. Kastritsis, C. D., *Evolution*, **23**, 663 (1969).

10. Dobzhansky, Th., and Pavlovsky, O., *Proc. US Nat. Acad. Sci.*, **55**, 727 (1966).

11. Dobzhansky, Th., and Pavlovsky, O., *Genetics*, **55**, 141 (1967).

12. Perez-Salas, S., Richmond, R. C., Pavlovsky, O., Kastritsis, C. D., Ehrman, L., and Dobzhansky, Th., *Evolution*, **24**, 519 (1970).

13. Ehrman, L., *Evolution*, **19**, 459 (1965).

14. Koopman, K. F., *Evolution*, **4**, 135 (1950).

15. Knight, G. R., Robertson, A., and Waddington, C. H., *Evolution*, **10**, 14 (1956).

16. Kessler, S., *Evolution*, 20, 634 (1966).

17. Williamson, D. L., and Ehrman, L., *Genetics*, **55**, 131 (1967).

18. Kernaghan, R. P., and Ehrman, L., *Chromosoma*, **29**, 291 (1970).

这项研究工作得到了来自美国原子能委员会和美国国家科学基金会（国际生物学计划）的经费支持。

（刘振明 翻译；张德兴 审稿）

Directed Genetic Change Model for X Chromosome Inactivation in Eutherian Mammals

D. W. Cooper

Editor's Note

The idea that one of the two X chromosomes inherited by female animals has to be inactivated was established by Mary Lyon in 1961. In humans the choice is random, but in female marsupials the X chromosome from the father is inactivated specifically. In this paper, Australian geneticist D. W. Cooper argues that this marsupial pattern is ancestral to that of placental animals (called eutherian, and including humans). It suggests that a controlling genetic element in the male-derived X in early female embryogenesis can hop at random to either X chromosome in a manner analogous to the chromosomal hopping of genetic factors in maize proposed by Barbara McClintock. The hypothesis has since been revised, but shows the wide impact of McClintock's Nobel-Prize-winning work.

IN 1961 Lyon put forward the single active X hypothesis to account for dose compensation of genes on the sex chromosomes of female eutherian mammals[1]. Under her hypothesis in the adult organism either the maternally or the paternally derived X is active in any one cell, but never both. The question of which X is to be inactivated is settled at a very early embryonic stage[2-4]. In female diploid individuals there is usually something near a 1:1 ratio of cells with active paternal to cells with active maternal X chromosomes[5]. The inactive X is detected cytologically as the heterochromatic Barr body or sex chromatin in interphase nuclei and because it synthesizes its DNA late, that is, it labels late[5]. Once established the inactivation is very stable. In abnormal individuals with more than two sex chromosomes all X chromosomes except one synthesize their DNA late and form Barr bodies[6]. Evidence from the phenotype of women with abnormal numbers of X chromosomes or X chromosomes with deletions suggests that the second X has some active genes[7]. With these exceptions, it can be said that cytological observations and studies on a number of sex linked genes have yielded considerable data supporting Lyon's hypothesis and none definitely against it[5,8,9].

Marsupial mammals are the closest extant group to eutherians[10]. Their system of sex determination is XX/XY[11,12] with the Y being male determining[12,13]. They have a late labelling X[11,14,15]. Sharman has obtained evidence that in kangaroos the late labelling X is always paternal in origin[14], this is, there is paternal rather than random X inactivation, a hypothesis which is supported by data on the inheritance of the enzymes glucose-6-phosphate dehydrogenase (G6PD)[16] and phosphoglycerate kinase (PGK)[17]. On the basis

真兽亚纲哺乳动物中X染色体失活的
定向遗传改变模型

库珀

编者按

1961年，玛丽·莱昂提出一种观点，认为雌性动物遗传继承的两条X染色体中的一条必然发生失活。在人类中，失活染色体的选择是随机的；但在雌性有袋动物中，却总是来自父本的那条X染色体发生特异性失活。澳大利亚遗传学家库珀在本文中提出理由证明，有袋动物的这种模式是有胎盘动物（即真兽亚纲动物，其中包括人类）的祖先模式。文中指出：在早期雌性动物的胚胎发育过程中，父源X染色体中的遗传控制元件会随机地跳到两条X染色体中的一条上，这种跳跃方式类似于芭芭拉·麦克林托克所发现的玉米中的遗传因子在染色体上的位置转移行为。虽然库珀的假说已经被后人修正，但从本文仍可以看出麦克林托克赖以获得诺贝尔奖的成果有多么大的影响力。

1961年，莱昂提出了单一活性X染色体假说，用来解释雌性真兽亚纲哺乳动物的性染色体基因的剂量补偿效应[1]。她在假说里提出：在成年生物体的每一个细胞中，不是来自母本就是来自父本的那条X染色体会具有活性，但永远不可能两者都具有活性。到底哪条X染色体会失活这一问题在胚胎的早期就已经决定了[2-4]。在雌性二倍体个体中，母源X染色体有活性的细胞与父源X染色体有活性的细胞的比例通常接近于1:1[5]。在间期细胞核中，细胞学检测显示失活的X染色体以异染色质巴氏小体或性染色质的形式存在，并且由于失活染色体上的DNA合成延迟，它被标记的时间也晚[5]。失活一旦建立，就能达到很稳定的状态。在具有多于两条性染色体的异常个体中，除其中一条外，其余所有X染色体都较晚地合成DNA并且形成巴氏小体[6]。从含有异常数目的X染色体或者X染色体存在缺失的妇女的表型中得出的证据显示，第二条X染色体（译者注：指失活的那条）上的某些基因仍是有活性的[7]。将这些例外情况考虑在内，可以这样说：对许多性连锁基因的细胞学观察和研究已经给出了大量能证实莱昂假说的数据资料，并且没有资料明确反对这一观点[5,8,9]。

有袋哺乳动物是现存的与真兽亚纲哺乳动物最接近的种群[10]。它们的性别决定系统是XX/XY[11,12]，其中Y染色体是雄性决定因素[12,13]。它们有一条延迟标记的X染色体[11,14,15]。沙曼发现袋鼠中延迟标记的X染色体总是源自父本[14]，即父源X染色体失活而不是X染色体随机失活，葡萄糖-6-磷酸脱氢酶（G6PD）[16]和磷酸甘油酸激酶（PGK）[17]的遗传数据也支持这一假说。基于这一发现，我希望能够提出

of this finding I wish to propose a model of X inactivation in eutherians. The model postulates that random X inactivation in eutherians has evolved from an ancestral paternal X inactivation which has been retained in marsupials. It proposes that during male meiosis in both groups the X undergoes a directed genetic change. A controlling element, analogous to the kind described by McClintock in maize[18], is introduced into the X, probably by the Y. Then at a certain point in the very early development of eutherian females this controlling element is excised and later reinserted at random into one of the two chromosomes, thus setting up random X inactivation. Once reinserted, it remains fixed in that chromosome. The purpose of this article is to examine the adequacy with which the model accounts for data at present available and to propose new investigations which should test it more critically.

Maize Controlling Elements

Controlling elements may be exemplified by the activator–dissociator (Ac–Ds) system of maize[18]. This system is in many respects like an operon system in which both the "regulator" element (Ac) and the "operator" element (Ds) have the episome-like property of occasionally being able to move from one part of the genome to another (although there is no evidence that either is capable of autonomous existence in the cytoplasm). Controlling element is the term used to describe entities like Ac and Ds, in contradistinction to gene which is reserved for conventional Mendelian units. Apart from being capable of this occasional transposition, controlling elements segregate at meiosis in the normal way. When the Ds element is present at a particular locus, genes adjacent to it on the same chromosome have their activity suppressed to some degree. Their expression is said to be unstable. Action of Ds depends on the existence of one or more Ac elements in the genome, not necessarily on the same chromosome or its homologue. Ac also affects the activity of genes adjacent to it in the chromosome. The occasional transposition of Ac and Ds probably occurs during the replication cycle of the chromosome. The genes adjacent to the old location usually return to normal activity while those adjacent to the new locus become unstable in their expression. Removal, however, is often associated with the appearance of a new stable allele of the adjacent gene. The Ac–Ds system has a number of other peculiarities, only one of which need concern us here. Ds may undergo mutation-like events which McClintock terms "changes of state". This term serves to describe the fact that the initial response of Ds to Ac may differ from subsequent ones. For example, when first isolated Ds responded to Ac by promoting the formation of dicentric chromatid bridges (a property from which Ds derives its name). In further responses, these were chiefly absent and only activity of adjacent genes was affected.

Controlling Element of Mouse X

Several workers have studied a region of the mouse X called the inactivation centre, which governs the degree to, or the frequency with, which the translocated autosomal material is inactivated when the associated X material is inactivated as a consequence of random X inactivation[19-21]. The inactivation of X-linked genes is likewise under control of this region of the X[22,23]. Cattanach calls the inactivation centre a controlling element, because there is

一个真兽亚纲动物 X 染色体失活的模型。这个模型假定真兽亚纲动物的 X 染色体随机失活是从其祖先的一条父源 X 染色体的失活演变而来的，在有袋动物中至今仍保留着这种父源 X 染色体失活的现象。假说认为：这两个类群的雄性亲本在减数分裂时都发生了 X 染色体的定向遗传学改变。一种控制元件（与麦克林托克所描述的在玉米中发现的那种 [18] 类似）被导入到 X 染色体中，它可能是从 Y 染色体导入的。然后在雌性真兽亚纲动物发展早期的某个特定时间点上，这个控制元件被切除，随后随机地插入到这两条染色体中的一条，从而形成了 X 染色体随机失活。一旦插入，它就会被整合到那条染色体中。本文旨在检验这个模型是否足以解释目前已有的数据，并提出应该采用哪些新的研究手段对其进行更准确的测试。

玉米控制元件

也许可以用玉米的激活因子 – 解离因子（Ac–Ds）系统来诠释控制元件的概念 [18]。这个系统在诸多方面类似于一个操纵子系统，其中"调控"元件（Ac）和"操纵"元件（Ds）具有游离基因那样的特性，偶尔能够从基因组的一个部位移动到另一个部位（尽管没有证据证明它们能够在细胞质中自主存在）。控制元件是用来描述像 Ac 和 Ds 这类元件的术语，与孟德尔定义的传统基因不同。除了能够偶尔发生转座外，控制元件在减数分裂时的分离是正常的。当 Ds 元件出现在某个特殊位点时，与它邻近的处于同一条染色体上的基因的活性会受到一定程度的抑制。人们认为这些基因的表达不稳定。Ds 的活动依赖于基因组中一个或多个 Ac 元件的存在，这些 Ac 元件不必出现在同一条染色体上或者其同源染色体上。Ac 也会影响与其邻近的位于同一条染色体上的基因的活性。在染色体的复制周期中，可能 Ac 和 Ds 偶尔会发生转座。转座后，与旧位点相邻的基因常常恢复正常的活性，而邻近新位点的基因则在表达上变得不稳定。然而，控制元件的剪切一般与邻近基因中出现一个新的稳定等位基因有关。Ac–Ds 系统有许多特性，在本文中我们仅关注其中一种。Ds 可能会发生与麦克林托克称之为"状态改变"类似的突变。这种描述可以用以说明，Ds 对 Ac 的初始应答可能与后续应答不同。例如，当第一次分离时，Ds 对 Ac 的应答是促进双着丝粒染色单体桥的形成（Ds 的命名就是源自这一性质）。在进一步的应答中，这些作用基本消失了，只有邻近基因的活性会受到影响。

小鼠 X 染色体的控制元件

几位学者研究了小鼠 X 染色体中被称为失活中心的一个区域，当 X 染色体的随机失活引起相关的 X 染色体物质的失活时，易位的常染色体物质也失活，其发生程度和频率受失活中心的调控 [19-21]。X 染色体相关基因的失活也同样受到 X 染色体失活中心这一区域的控制 [22,23]。卡塔纳克将这段失活中心称作控制元件，因为他在研

an indication from his work on the flecked (X^T, T (1X), Ct) X autosome translocation that it behaves like a maize controlling element[19,24]. This translocation involves the insertion of autosomal material from linkage group 1 into the X, making it about 20% larger[19,25]. The inserted material bears genes from albino (c) to ruby eye (ru-2). The locus of insertion is near jimpy (jp) (Fig. 1), but whether on the side distal or proximal to Gyro (Gy) is not yet known. Cattanach's early data indicated change from one generation to the next in the amount of inactivation of the autosomal material in the X^T chromosome which could not all be satisfactorily explained by meiotic crossing over between X^T chromosomes governing different levels of inactivation[24]. He proposed instead that the controlling element was undergoing "changes of state". In a more recent investigation, the results obtained could all be explained by meiotic crossing over, but whether these later results render the change of state interpretation of the earlier results invalid is doubtful[19]. (There is some discussion as to how many controlling elements exist in the inactive X of the mouse[8,19]. Since only one region of the chromosome has so far been clearly implicated as being a controlling element, I will assume for the purposes of this paper that there is only one.)

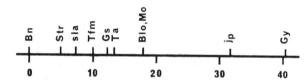

Fig. 1. Diagram of portion of the mouse X linkage group, showing position of sex linked genes discussed in the text. The diagram is from ref. 48. *Bn*, Bent tail; *Str*, striated; *sla*, sex linked anaemia; *Tfm*, testicular feminization; *Gs*, greasy; *Ta*, tabby; *Blo*, blotchy; *jp*, jimpy; *Gy*, gyro. The units are centimorgans.

Directed Change and Primary Paternal Inactivation

Cattanach and Perez have made an observation which suggests that the properties of the paternal X may differ from those of the maternal[26]. They have recently shown that female mice heterozygous for X^T chromosomes tend to have lower levels of variegation for albino when the rearranged X is inherited from the father rather than the mother. A maternal effect is not responsible for the differences between reciprocal crosses. This can be explained by postulating that a directed genetic change at male meiosis alters the properties of the X^T. As a consequence, the X^T inherited from the father associates less frequently with the controlling element.

If the primary step in establishing random X inactivation is paternal X inactivation, there should be mutant genes which convert the former to the latter. These genes may be defective controlling elements which cannot be transferred or genes governing enzymes mediating their transfer. There is some evidence which can be interpreted to mean that such genes do exist, the chief of which involves Searle's translocation. This, designated T16H, is a reciprocal X autosome translocation in the mouse[27-30]. With the exception of Cattanach's translocation, all mouse X autosome translocations render the male sterile. Hence all females heterozygous for Searle's translocation receive their rearranged

究 X 染色体 – 常染色体斑点易位 (X^T, T(1X), *Ct*) 时发现，失活中心的行为与玉米控制元件类似 [19,24]。发生这种易位时，连锁群 1 的常染色体组分插入到 X 染色体中，使 X 染色体增大了约 20% [19,25]。插入部分包含从白化基因 (*c*) 到红眼基因 (*ru-2*) 的片段。插入位点靠近吉皮基因 (*jp*)（图 1），但还不能确定插入位点是远离还是靠近陀螺基因 (*Gy*)。卡塔纳克早期的实验数据表明，X^T 染色体中失活的常染色体物质的量在每一代中都有所不同，仅仅用控制不同失活水平的 X^T 染色体在减数分裂中的交叉互换来解释这些变化并不能完全令人满意 [24]。对此，他提出这些控制元件发生了"状态改变"。在一项近期的研究中，用减数分裂交叉互换就可以解释所有获得的结果，但是这些最近的结果能否推翻基于状态改变对先前结果的解释还值得怀疑 [19]。（已经有一些关于小鼠失活的 X 染色体上到底存在多少控制元件的讨论 [8,19]。因为到目前为止，在染色体上只有一个区域被明确界定为控制元件，所以我认为要说明只存在一个控制元件应是这篇文章的目的。）

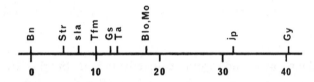

图 1. 小鼠 X 染色体连锁群的部分图解，图中标明了文中讨论过的性连锁基因的位置。本图来自参考文献 48。*Bn*, 弯尾基因；*Str*, 条纹基因；*sla*, 伴性贫血症基因；*Tfm*, 睾丸雌化基因；*Gs*, 多脂基因；*Ta*, 斑纹基因；*Blo*, 斑点基因；*jp*, 吉皮基因；*Gy*, 陀螺基因。以厘摩为单位。

定向改变和最初的父源性失活

卡塔纳克和佩雷斯观察到父源 X 染色体的特性可能与母源 X 染色体不同 [26]。他们最近发现：当重组的 X 染色体来自父本而不是母本时，X^T 染色体杂合的雌性小鼠的白化病变异（译者注：指个体中白化与非白化表型区域随机杂合呈现）水平降低。而母本效应无法解释在正反交实验间出现的结果差异。如果假定父本在减数分裂时发生的定向遗传改变更改了 X^T 染色体的特性，那么上述现象就可以得到解释。因此，父本遗传的 X^T 染色体不太经常与控制元件相关。

如果建立 X 染色体随机失活的最初步骤是父源 X 染色体的失活，那么就应该有使父源 X 染色体失活转变为 X 染色体随机失活的突变基因存在。这些突变基因可能是不能够发生转移的有缺陷的控制元件，或者是介导控制元件发生转移的一些酶的调控基因。有证据表明这些基因的确存在，其中最主要的是瑟尔易位。这个被称作 T16H 的基因是小鼠中 X 染色体 – 常染色体相互易位形成的 [27-30]。除卡塔纳克易位外，所有小鼠 X 染色体与常染色体之间的易位都导致雄性不育。因此所有瑟尔易位的雌性杂合体的重组染色体都来自母本，它们的正常 X 染色体和常染色体来自父本。由

chromosomes from their mother and their normal X and autosomes from their father. Since the normal X is almost always inactive in these heterozygotes, Searle's translocation exhibits paternal X inactivation. In this respect it is unlike all other mouse X autosome reciprocal translocations, which by contrast show random inactivation of either the normal or rearranged chromosome.

It is particularly significant that the break in Searle's translocation maps is the same region of the X as the controlling element. The break in the X is given as 0.85 centimorgans from *Ta* on the side distal from *Blo* (ref. 30 and Fig. 1). But the precise relationship of the breakpoint to genes in this part of the linkage group remains obscure because the translocation evidently leads to some cross-over suppression[31]. Cattanach *et al.* have concluded that their controlling element lies very close to *Ta*[19]. The data of Grahn *et al.* suggest that the controlling element may be 8.2 centimorgans from *Gs* on the side distal from *Ta*[21]. Problems of classification have not been completely resolved in their material and so this mapping is tentative. At present there seems to be nothing against the suggestion of Cattanach *et al.* that the break in Searle's translocation may be in the controlling element, thus impairing its function[19]. In my model, its transfer to the maternal X is prevented.

An alternative explanation of non-random inactivation in Searle's translocation is that the translocated X segment can only be switched off for the segment containing the controlling element. Switching off of the re-arranged X would lead to the breakdown of dosage compensation for genes on the segment lacking the controlling element. Selection of cells with correct dose compensation would then account for the observed preponderance of cells with inactive normal chromosomes[5]. The finding that at least one and probably two other mouse X autosome translocations with break points near *Ta* show random X inactivation makes this explanation unlikely[20].

There are other examples of apparent or possible paternal X inactivation in eutherians. Studies on the G6PD of mules indicate that in most tissues there is preferential inactivation of the paternal donkey chromosomes[32,33]. This is interpreted here to mean that the lack of homology between the two distantly related X chromosomes hinders the incorporation of the controlling element into the maternal X. An X autosome translocation heterozygote with preferential late labelling of the normal X has been described in the domestic cow, without any information on the parental origin of the arrangement[34]. The explanation given for the behaviour of Searle's translocation is, of course, also applicable to this.

Unsolved Problems

The model advanced explains how random X inactivation evolved, how the choice of chromosome to be inactivated is made, its stability once made, and a number of hitherto unexplained aspects of X inactivation, particularly the behaviour of Searle's and Cattanach's translocations. It cannot, however, represent the whole truth, because several difficult questions remain unanswered.

870

于正常的 X 染色体在杂合状态时几乎是没有活性的，因此瑟尔易位表现为父源 X 染色体失活。在这方面，它与所有其他的小鼠 X 染色体 – 常染色体相互易位不同，相比之下，其他的小鼠 X 染色体 – 常染色体易位表现为正常或者重组的染色体随机失活。

瑟尔易位图谱的切点与 X 染色体上作为控制元件的区域相同，这一点特别重要。X 染色体的切点位于 *Blo* 的末端，距离 *Ta* 0.85 厘摩处（参考文献 30 和图 1）。但是断开点与连锁群这部分基因之间的确切关系仍旧是模棱两可的，因为易位明显导致了一些交叉抑制[31]。卡塔纳克等人认为控制元件的位置与 *Ta* 很接近[19]。格兰等人的数据显示：在 *Ta* 末端，距离 *Gs* 8.2 厘摩处可能是控制元件的位置[21]。由于分类问题尚未完全解决，因此这只是一个暂定的基因图谱。卡塔纳克等人认为瑟尔易位的切点可能位于控制元件上，因而削弱了控制元件的功能[19]。到目前为止，还没有事实可以反驳这个观点。在我的模型里，控制元件转移插到母源 X 染色体的情况是不会发生的。

对于瑟尔易位中非随机失活的另一种解释是：易位的 X 染色体片段只能被含有控制元件的片段所关闭。重组 X 染色体的关闭将导致缺少控制元件片段的基因剂量补偿中止。因此，对具有正确剂量补偿的细胞的选择可以用于解释为什么会观察到正常染色体失活的细胞在数量上占优势[5]。在小鼠中发现，至少有一个且可能还有另外两个易位切点在 *Ta* 附近的 X 染色体 – 常染色体易位现象体现为 X 染色体的随机失活。这一发现使上述解释变得不可能[20]。

在真兽亚纲动物中还有其他一些明显的或者可能的父源 X 染色体失活的例子。对骡子 G6PD 的研究发现：在大多数组织中，父源驴染色体优先失活[32,33]。本文所作的解释是：两个关系较远的 X 染色体之间缺少同源性，这阻碍了控制元件插入到母源 X 染色体中。在驯养的牛中发现，X 染色体 – 常染色体的易位杂合体中大多会出现标记延迟的正常 X 染色体，但没有任何关于这种排列的父本起源的信息[34]。当然，针对瑟尔易位的行为所给出的解释也可用于解释这一现象。

尚未解决的问题

这个模型解释了 X 染色体随机失活是如何形成的、染色体选择性失活是如何做到的、一旦形成后的稳定性和许多至今尚未解决的关于 X 染色体失活的问题，尤其是解释了瑟尔易位和卡塔纳克易位的现象。然而，它并不能解释所有的现象，因为还有几个难题仍未解决。

How can XO individuals in man and mouse be viable when their X is from the father, as it often can be[35-37]? Obviously the controlling element must somehow be jettisoned or inactivated very early in development if the model is correct. But why should this happen to the paternal X in XO individuals but not in normal females?

Only one X is active in women and men with multiple X chromosomes[6]. Where do the extra controlling elements come from in cases where the extra chromosomes are from the mother[37]? Perhaps the controlling element is retained during female meiosis and is included in a polar body, except when non-disjunction occurs leading to a multiple X gamete. This will explain the human data except for situations where four X chromosomes come from the mother, as in some XXXXY and XXXX individuals[37]. For these female germ line polysomy must be invoked.

Retention of the controlling element cannot occur in the mouse, however; Searle's translocation gives normal segregation ratios for genes near the controlling element locus[31], which it would not do if the chromosome with the controlling element was preferentially included in a polar body. The only report of multiple X individuals in the mouse concerns XXY types, which were the result of fertilization of a normal ovum by an XY sperm[38]. It is thus of some importance to my model to discover if more than one X can be inherited from the maternal parent in a viable mouse.

Other problems are posed by the fact that in women heterozygous for a normal X and either Xqi (isochromosomes for the long arm of the X), Xpi (isochromosomes for the short arm), Xq- (lack of a long arm), Xp- (lack of a short arm) or Xr (a ring X) invariably have the abnormal X late labelling, irrespective of whether the abnormal X is paternal or maternal in origin[37,39,40]. The only exception to this rule seems to be one report of an XXr individual with a late labelling normal X[41]. How does the controlling element select the abnormal X? And if Xqi, Xpi, Xq-, and Xp- can all be inactivated, where is the locus of the controlling element? Must it be at or near the centromere in man?

Some explanations is needed to account for situations where the second X is apparently active. In marsupials these are the ovaries and uterus of the early pouch young kangaroo[17] and possibly the ovaries of bandicoots[11]. In placental mammals, oocytes[5] and the early embryo[5,42-44] lack sex chromatin. Possibly the controlling element is not fixed in either X at these stages. Or alternatively the controlling element could act in adult somatic tissue in an operator-like manner as the site of action of a repressor protein, which is absent at these early stages. In this connexion, one may note that Steele[45] has recently shown that human female foetuses and newborn female infants have more G6PD activity than males of either class, while hypoxanthine guanine phosphoribosyl transferase was essentially the same in both sexes at these stages. The structural loci for both enzymes are sex linked. In this instance induction of greater activity of the one active X seems more likely than switching on of the second X.

对于人和小鼠中的 XO 个体，如果 X 染色体与通常情况下一样来自父本，那么这些个体是怎么存活的 [35-37]？如果这个模型正确，显然控制元件在发展的早期就必然会以一定方式被抛弃或者失活。但是为什么这种现象发生在 XO 个体的父源 X 染色体上，而不会发生在正常的雌性个体中？

具有多条 X 染色体的女性和男性中只有一条 X 染色体是有活性的 [6]。如果多余的染色体来自母亲 [37]，那么多出的控制元件来自哪里？也许母本减数分裂时，控制元件被保留了下来，分布在一个极体里，但在染色体不分离导致多倍 X 染色体配体形成时除外。这可以解释人类的绝大多数现象，但不能解释在 XXXXY 和 XXXX 个体中 4 个 X 染色体都来自母本的情况 [37]。因为这种情况肯定会导致雌性生殖细胞系的多体性。

然而，在小鼠中控制元件不能被保留；在瑟尔易位中，靠近控制元件位点的基因的分离比例是正常的 [31]，如果含有控制元件的染色体优先分布到一个极体里，那么基因分离就不会发生。小鼠中报道的唯一一个 X 染色体多倍体的例子是 XXY 型，它是正常卵子与 XY 精子受精的结果 [38]。因此，在活体小鼠中发现不止一条 X 染色体来自母本这一点对我的模型来说很重要。

另一些难题是基于以下事实：女性杂合体总是由一条正常的 X 染色体和一条标记延迟的异常 X 染色体 Xqi（X 染色体长臂的等臂染色体）、Xpi（短臂的等臂染色体）、Xq-（缺少一个长臂）、Xp-（缺少一个短臂）或 Xr（环形 X 染色体）组成，不论异常的 X 染色体最初是来源于父本还是母本 [37,39,40]。这个规则的唯一例外似乎来自一篇关于 XXr 个体的报道，这一个体含有延迟标记的正常 X 染色体 [41]。控制元件是如何选择异常的 X 染色体的？如果 Xqi、Xpi、Xq- 和 Xp- 都能被失活，那么控制元件的位点在哪里？它一定会在男性着丝粒上或其附近吗？

对于第二条 X 染色体明显有活性的现象，我们需要给出一定的解释。在有袋动物中，卵巢和子宫是抚育小袋鼠的早期袋状结构 [17]，在袋狸中可能是卵巢 [11]。在有胎盘的哺乳动物中，卵母细胞 [5] 和早期的胚胎 [5,42-44] 缺少性染色质。可能在这些阶段，控制元件还没有整合到任何一条 X 染色体中。又或者控制元件在成年体细胞组织中能以类似操纵基因的形式作为阻遏蛋白的活性位点，这种活性位点在早期阶段是不存在的。关于这一点，我们注意到斯蒂尔 [45] 最近发现：人类女性胎儿和新出生的女婴与对应阶段的男性相比有更多的 G6PD 活性，而在这些阶段中次黄嘌呤－鸟嘌呤磷酸核糖转移酶在两性中并没有本质上的差别。这两种酶的结构位点都是性连锁的。在这个例子中，诱导有活性的那条 X 染色体产生更大的活性要比能开启第二条 X 染色体的活性的可能性更大。

Possible Experimental Tests

Although it cannot explain all data on X inactivation, the model explains sufficient to make worthwhile investigations to test its chief assumptions, namely, directed genetic change of the X by the Y being the primary step in setting up random X inactivation followed by transfer of the inactivation to the maternal X. If these are correct, two kinds of mutation should exist. There should be mutations in the gene on the Y responsible for the directed genetic change. Males carrying them would give rise only to males. If the mutation is found in the mouse, matings with XO mice will give both male and female offspring, the latter being all XO. There should also be mutations which affect the transfer of the controlling element, and which will often result in paternal X inactivation. Such genes could be either X-linked or autosomal. Hopefully both sexes carrying them will be fertile, so that reciprocal crosses can be made to establish their nature rigorously, something which cannot be done with Searle's translocation. It is also possible that genes converting random X inactivation to paternal X inactivation are normally present in some eutherian species. Conversely, it is possible but rather less likely that there are marsupials which, unlike kangaroos, possess random X inactivation. There is need for a large survey of both marsupial and eutherian mammals to detect more sex-linked enzyme polymorphisms. So far no exceptions have been discovered to Ohno's thesis[5,46] of homology between the sex chromosomes of all eutherian and marsupial X chromosomes. Hence the best tactic in search for sex-linked isoenzymes is to use those enzymes known to be sex-linked in man, for example, G6PD[46] and PGK[47], a procedure which is now being followed in this laboratory.

I thank Professor R. A. Brink, Drs. B. M. Cattanach and D. L. Hayman and Professor G. B. Sharman for helpful comments. My experimental work on X inactivation in marsupials is supported by grants from the Australian Research Grants Committee.

(**230**, 292-294; 1971)

D. W. Cooper: Department of Genetics and Human Variation, La Trobe University, Bundoora, Victoria 3083.

Received December 10, 1970; revised February 23, 1971.

References:

1. Lyon, M. F., *Nature*, **190**, 372 (1961).

2. Axelson, M., *Hereditas*, **60**, 347 (1968).

3. Austin, C. R., *The Sex Chromatin* (edit. by Moore, K. L.), 241 (Saunders, Philadelphia, 1966).

4. Issa, M., Blank, C. E., and Atherton, G. W., *Cytogenetics*, **8**, 219 (1969).

5. Ohno, S., *Sex Chromosomes and Sex-Linked Genes* (Springer, Berlin, Heidelberg, New York, 1967).

6. Barr, M. L., *The Sex Chromatin*, 129 (edit. by Moore, K. L.) (Saunders, Philadelphia, 1966).

7. Hamerton, J. L., *Nature*, **219**, 910 (1968).

8. Lyon, M. F., *Ann. Rev. Genet.*, **2**, 31 (1968).

9. Lyon, M. F., *Phil. Trans. Roy. Soc.*, B, **259**, 41 (1970).

10. Simpson, G. G., *Evolution*, **13**, 405 (1959).

可能的实验检测

尽管这个模型不能解释所有 X 染色体失活的现象，但它却充分解释了一些有价值的研究结果，这些研究则验证了它的主要观点，即：X 染色体的定向遗传改变最先一步是通过 Y 染色体的失活，然后失活被转移到了母源 X 染色体中。如果上述观点正确，那么就应该存在两种突变。一种是 Y 染色体上负责定向遗传改变的基因发生的突变。父本只能将它们传给雄性后代。如果这种突变发生在小鼠中，那么与 XO 小鼠交配产生的雄性和雌性后代中的后者全都是 XO 型。另一种是影响控制元件易位的基因发生的突变，这种突变经常会导致父源 X 染色体失活。这样的基因既可以位于 X 染色体上，也可以位于常染色体上。带有这些突变基因的两性个体都是有望可育的，因此可以用正反交实验来严格解释它们的本质，这是瑟尔易位无法做到的。使 X 染色体失活转变为父源 X 染色体失活的基因也有可能会正常存在于某些真兽亚纲物种中。相反，可能性更小的情况是：有些有袋动物与袋鼠不同，它们发生 X 色体随机失活。现在需要对有袋动物和真兽亚纲哺乳动物进行大量的调查以检测出更多的性连锁酶多态性。大野乾指出 [5,46]，所有真兽亚纲动物的性染色体和有袋动物的 X 染色体之间具有同源性。到目前为止还没有发现例外。因此，寻找性连锁的同工酶的最好策略是利用人类中已知的性连锁酶，比如 G6PD[46] 和 PGK[47]，这也是本实验室现阶段正在进行的工作。

感谢布林克教授、卡塔纳克博士、海曼博士和沙曼教授提出的宝贵意见。本人在有袋动物 X 染色体失活方面的实验研究承蒙澳大利亚科研拨款委员会的基金支持。

（郑建全 翻译；梁前进 审稿）

11. Hayman, D. L., and Martin, P. G., *Comparative Mammalian Cytogenetics*, 191 (edit. by Benirschke, K.) (Springer, New York, 1969).

12. Sharman, G. B., *Science*, **167**, 1221 (1970).

13. Sharman, G. B., Robinson, E. S., Walton, S. M., and Berger, P. J., *J. Reprod. Fertil.*, **21**, 57 (1970).

14. Sharman, G. B., *Nature*, **230**, 230 (1971).

15. Marshall-Graves, J. A., *Exp. Cell. Res.*, **46**, 37 (1967).

16. Richardson, B. J., Czuppon, A., and Sharman, G. B., *Nature New Biology*, **230**, 154 (1971).

17. Cooper, D. W., VandeBerg, J. L., Sharman, G. B., and Poole, W. E., *Nature New Biology*, **230**, 155 (1971).

18. McClintock, B., *Brookhaven Symp. Biol.*, No. 18, 162 (1965).

19. Cattanach, B. M., Perez, J. N., and Pollard, C. E., *Genet. Res.*, **15**, 183 (1970).

20. Russell, L. B., and Montgomery, C. S., *Genetics*, **64**, 281 (1970).

21. Grahn, D., Leu, R. A., and Hulesch, J., *Genetics*, **64**, 2 (2) s25 (abstr.) (1970).

22. Cattanach, B. M., *Genetics*, **60**, 168 (1968).

23. Cattanach, B. M., Pollard, C. E., and Perez, J. N., *Genet. Res.*, **14**, 223 (1969).

24. Cattanach, B. M., and Isaacson, J. H., *Genetics*, **57**, 331 (1967).

25. Cattanach, B. M., *Genet. Res.*, **8**, 253 (1966).

26. Cattanach, B. M., and Perez, J. N., *Genet. Res.*, **15**, 43 (1970).

27. Searle, A. G., *Heredity*, **17**, 297 (1962).

28. Ford, C. E., and Evans, E. P., *Cytogenetics*, **3**, 295 (1964).

29. Ohno, S., and Lyon, M. F., *Chromosoma*, **16**, 90 (1965).

30. Lyon, M. F., *Genet. Res.*, **7**, 130 (1965).

31. Lyon, M. F., Searle, A. G., Ford, C. E., and Ohno, S., *Cytogenetics*, **3**, 306 (1964).

32. Giannelli, F., Hamerton, J. L., Dickson, J., and Short, R. V., *Heredity*, **24**, 175 (1969).

33. Hook, E. B., and Brustman, L. D., *Genetics*, **64**, 2 (2), s30 (abstr.) (1970).

34. Gustavsson, I., Fraccaro, M., Tiepolo, L., and Lindsten, J., *Nature*, **218**, 183 (1968).

35. Cattanach, B. M., *Genet. Res.*, **12**, 125 (1968).

36. Morris, T., *Genet. Res.*, **12**, 125 (1968).

37. Race, R. R., and Sanger, R., *Brit. Med. Bull.*, **25**, 99 (1969).

38. Cattanach, B. M., *Genet. Res.*, **2**, 156 (1961).

39. Klinger, H. P., Lindsten, J., Fraccaro, M., Barrai, I., and Dolinar, Z. J., *Cytogenetics*, **4**, 96 (1965).

40. Rowley, J., Muldal, S., Lindsten, J., and Gilbert, C. W., *Proc. US Nat. Acad. Sci.*, **51**, 779 (1964).

41. Pfeiffer, R. A., and Buchner, T., *Nature*, **204**, 804 (1964).

42. Austin, C. R., in *The Sex Chromatin*, 241 (edit by Moore, K. L.) (Saunders, Philadelphia, 1966).

43. Kinsey, J. D., *Genetics*, **55**, 337 (1967).

44. Hill, R. N., and Yunis, J. J., *Science*, **155**, 1120 (1961).

45. Steele, M. W., *Nature*, **227**, 496 (1970).

46. Ohno, S., *Ann. Rev. Genet.*, **3**, 495 (1969).

47. Valentine, W. N., Hsieh, H. S., Paglia, D. E., Anderson, H. M., Baughan, M. A., Jaffe, E. R., and Garson, O. M., *Trans. Assoc. Amer. Phys.*, **81**, 49 (1968).

48. Hawkes, S. G., *Mouse News Letter*, **43**, 16 (1970).

Pneumococci Insensitive to Penicillin

D. Hansman *et al.*

Editor's Note

Since the introduction of penicillin to treat human infections three decades earlier, capsular pneumococci bacteria were thought highly sensitive to the antibiotic. And although penicillin-insensitive mutants had been selected *in vitro*, resistant wild strains were not recognized until 1967, when penicillin-resistant pneumococci were isolated from a patient with lung damage. In this paper David Hansman and colleagues report two further pneumococci strains, isolated in Australia and New Guinea, that are relatively insensitive to both penicillin and cephalosporin antibiotics. The team note that penicillin is much used in New Guinea, and that transmission probably occurred between villages, much as cross-infection with tetracycline-resistant pneumococci has occurred within hospitals.

PNEUMOCOCCI of all capsular types have been regarded as highly sensitive to penicillin[1] since its introduction for the treatment of human infections in 1940. Although penicillin-insensitive mutants can be selected *in vitro* by repeated subculture in the presence of sub-inhibitory concentrations of penicillin[2-6], resistant wild strains were not recognized until 1967, when penicillin-insensitive pneumococci (type 23) were isolated from a patient with hypogammaglobulinaemia and bronchiectasis, who had received much antibiotic therapy, including penicillin[7]. We now report further strains of pneumococci, isolated in Australia and New Guinea, relatively insensitive to both penicillin and cephalosporin antibiotics.

These penicillin-insensitive pneumococci were isolated from an aboriginal child at Ernabella, South Australia, in 1967, and from fifteen New Guineans at Anguganak in the West Sepik district during a fourteen week period in 1969. Pneumococci were typed by the specific capsular reaction, using typing sera from the Statens Seruminstitut, Copenhagen, and the Danish system of naming types has been followed. The pneumococcus from the aborigine, which was isolated from the upper respiratory tract, was identified as a type 6 strains. Most of the Sepik isolates were also from carriers, but two were isolated from subjects receiving penicillin for respiratory infections: all were identified as type 4.

The penicillin-insensitive strains were Gram-positive cocci with the morphology of pneumococci; cultural characteristics and α-haemolysis on horse blood agar were also typical; bile solubility and sensitivity to ethyl hydrocupreine hydrochloride ("Optochin") were demonstrated. Diffuse turbidity was produced in horse serum broth and long chains, typical of rough pneumococci, were not formed. Capsules were readily demonstrated by the specific capsular reaction. When inoculated by the intraperitoneal route, both type 4

对青霉素不敏感的肺炎双球菌

汉斯曼等

编者按

自从 30 多年前青霉素被用于治疗人类的感染性疾病以来，人们一直认为具有荚膜的肺炎双球菌对青霉素很敏感。虽然已在体外筛选出对青霉素不敏感的突变株，但直到 1967 年从一位肺部损伤的病人体内分离出耐青霉素的肺炎双球菌时，抗性野生株才被人们所认识。在这篇文章中，戴维·汉斯曼及其同事们报道了另外两种来自澳大利亚和新几内亚的肺炎双球菌株，它们对青霉素和头孢菌素类抗生素都相对不太敏感。该研究小组指出：在新几内亚，青霉素的使用量非常大，这种耐药性菌株很可能已经在村子与村子之间发生了传染，就像在医院内部发生的对四环素耐药的肺炎双球菌的交叉感染一样。

自从 1940 年青霉素被用于治疗人类的感染性疾病以来，人们一直认为所有具有荚膜的肺炎双球菌都对青霉素很敏感 [1]。尽管在体外通过青霉素亚抑菌浓度下的不断重复传代培养已经可以筛选出对青霉素不敏感的突变株 [2-6]，但直到 1967 年抗性野生株才被人们所认识，当时从一位患有低丙种球蛋白血症和支气管扩张的患者体内分离出了对青霉素不敏感的肺炎双球菌（23 型）。这位患者曾接受过多种抗生素的治疗，其中也包括青霉素 [7]。本文报道了另外两种来自澳大利亚和新几内亚的肺炎双球菌菌株，它们对青霉素和头孢菌素类抗生素都相对不太敏感。

这些对青霉素不敏感的肺炎双球菌是 1967 年从居住在南澳大利亚厄纳贝拉的一个土著小孩和 1969 年在安古干纳克的西塞皮克区逗留 14 周期间从 15 个新几内亚人体内分离得到的。用来自哥本哈根丹麦国立血清研究所的分型血清进行特异荚膜反应将肺炎双球菌分型，并按照丹麦命名系统对其命名。从土著小孩的上呼吸道中分离得到的肺炎双球菌菌株被确认为 6 型。大多数来自塞皮克区的肺炎双球菌菌株是从带菌者体内分离出来的，但有两个是从因呼吸道感染而接受过青霉素治疗的患者体内分离得到的，它们都被确认为 4 型。

青霉素不敏感菌株是具有肺炎双球菌形态的革兰氏阳性球菌；其培养特征和在马血琼脂中的 α 溶血反应也很典型；实验证实它可溶于胆汁并对盐酸乙基氢化叩卟啉（"奥普托欣"）敏感。在马血清培养液中可产生弥漫性混浊，但没有形成粗糙型肺炎双球菌所特有的长链。通过特异性的荚膜反应很容易鉴别出有荚膜。当腹腔接种时，4 型和 6 型菌株都对小白鼠有毒性，接种量为 100 个或以下活力单位时即可

879

and type 6 strains were virulent for white mice, and inoculum of 100 viable units or less producing a fatal infection.

Disk diffusion sensitivity tests showed resistance to penicillin but sensitivity to ampicillin, tetracycline, chloramphenicol, erythromycin, lincomycin and also to sulphadiazine and trimethoprim. When tested by the radial streak method, using as inoculum a standard 2 mm loopful of a 4 h culture in serum broth, growth occurred to within 2 mm of a disk containing 1 U (0.6 µg) of penicillin G, whereas sensitive pneumococci tested as controls showed a zone of inhibition of 6 to 7 mm. Quantitative sensitivity tests were carried out by the plate titration method with horse blood agar, using antibiotic solutions which had been freshly prepared. Nine strains, including the Ernabella strain, were tested. The results (Table 1) indicated that the Ernabella and Anguganak isolates were uniformly insensitive to penicillin G, methicillin and cephalosporin antibiotics. Penicillin sensitive pneumococci of 16 different types (including types 1, 2, 3, 4 and 6) tested as controls were all inhibited by 0.02 µg of penicillin G/ml.

Table 1. Quantitative Antibiotic Sensitivity Tests with Insensitive and Control Pneumococci

| | Minimal inhibitory concentration (µg/ml.) | | Resistance ratio |
	Test strains	Control strains	
Penicillin G	0.5	0.02	25
Methicillin	5 to 10	0.2	25 to 50
Ampicillin	0.1 to 0.2	0.05	2 to 4
Cephaloridine	0.5	0.05	10
Cephalothin	2.0	0.1	20

The minimal inhibitory concentrations of penicillin and methicillin were 25 to 50 times that required to inhibit the sensitive pneumococci used as controls. There was also a significant degree of resistance to cephaloridine and cephalothin. It may be noted that the penicillin-insensitive variants of pneumococci described by Gunnison et al.[6] were also insensitive to cephalothin. Quantitative tests by the plate titration method usually showed sharp end-points, indicating that populations of these isolates were largely homogeneous in their resistance to penicillins and cephalosporins.

Pneumococci resistant to sulphonamides were first encountered in 1940, 5 yr after the introduction of these drugs. Resistance appeared in many pneumococcal types and there was cross-resistance, so that pneumococci resistant to one sulphonamide were resistant to all. But resistance usually occurred only in strains isolated from subjects who had prolonged treatment with sulphonamides.

For many years it was assumed that pneumococci were invariably sensitive to those antibiotics which are effective in coccal infections. In 1963, however, strains resistant to tetracycline were recognized[8-11]. Such pneumococci are virulent for man, causing

产生致死性感染。

纸片扩散药敏试验的结果表明：它们对青霉素有抗性，而对氨苄西林、四环素、氯霉素、红霉素、林可霉素、磺胺嘧啶和甲氧苄氨嘧啶敏感。用径向条痕法进行试验时，接种物为一个在血清培养液中培养了 4 h 的 2 mm 标准菌环，供试菌种可在含有 1 U（0.6 µg）青霉素 G 的 2 mm 区域内生长，而作为对照的敏感型肺炎双球菌则出现了一个 6 mm ~ 7 mm 的受抑制区域。应用马血琼脂平板滴定法对包括厄纳贝拉菌株在内的 9 个菌株进行定量敏感性检测，所用抗生素溶液均为新鲜配制。实验结果（表 1）表明，来自厄纳贝拉和安古干纳克地区的分离菌株都表现出对青霉素 G、甲氧西林和头孢菌素类抗生素不敏感。在实验中用于对照的 16 种不同类型（其中包括 1、2、3、4 和 6 型）的青霉素敏感型肺炎双球菌都能被 0.02 µg/ml 的青霉素 G 所抑制。

表 1. 不敏感型肺炎双球菌和对照肺炎双球菌的定量抗生素敏感性检测

	最小抑菌浓度（µg/ml）		抗性比
	供试菌株	对照菌株	
青霉素 G	0.5	0.02	25
甲氧西林	5 ~ 10	0.2	25 ~ 50
氨苄西林	0.1 ~ 0.2	0.05	2 ~ 4
头孢噻啶	0.5	0.05	10
头孢噻吩	2.0	0.1	20

青霉素和甲氧西林的最小抑菌浓度是抑制敏感型肺炎双球菌对照菌株所需浓度的 25 到 50 倍。这些菌株对头孢噻啶和头孢噻吩也有明显的耐药性。应该注意的是，冈尼森等人 [6] 所描述的肺炎双球菌的青霉素抗性突变株对头孢噻吩也不敏感。用平板滴定法进行的定量敏感性检测通常会出现急剧变化的终点，说明这些被分离菌株的种群在对青霉素和头孢菌素的耐药性方面基本一致。

在磺胺类药物被应用了 5 年之后，1940 年人们首次发现肺炎双球菌对磺胺类药物有抗性。很多类型的肺炎双球菌已表现出耐药性，而且还存在交叉耐药性，以致对某一种磺胺类药物耐药的肺炎双球菌会对所有磺胺类药物都表现出耐药性。但这种耐药性通常只会出现在从长期接受磺胺治疗的患者体内分离得到的菌株中。

多年以来，人们一直认为肺炎双球菌对能够治疗球菌感染的抗生素总是很敏感。然而在 1963 年，科学家们发现了对四环素有耐药性的菌株 [8-11]。这类肺炎双球菌引

pneumonia and meningitis.

It may seem surprising that antibiotic-insensitive pneumococci should appear in a remote area such as the Sepik district, but much penicillin is used because respiratory and other infections are common in New Guinea. During the preceding 10 yr, the 507 inhabitants of the two villages at Anguganak had received 1,357 courses of procaine penicillin, which represents, on an individual basis, a course of penicillin every 3.7 yr. Moreover, during campaigns to control yaws, penicillin was administered to the entire population. Additional factors may be a high pneumococcal carrier rate, especially in children, and the heavy colonization of such carriers (ref. 12 and unpublished results). The isolation of the insensitive type 4 strain from fifteen individuals suggested that transmission had occurred, as these subjects lived in two nearby villages. This is analogous to cross-infection with tetracycline-resistant pneumococci which has occurred within hospitals[10,13].

In all other respects than antibiotic sensitivity, the penicillin-insensitive isolates have the characteristics of encapsulated pneumococci (results presented at the annual meeting of the Australian Society for Microbiology in 1970). Pneumococci of these types, 4 and 6, commonly cause infections in man, and epidemics of pneumonia due to type 4 pneumococci have been reported.

At present, the incidence of penicillin-insensitive pneumococci in Australia seems to be low: during 1965 to 1969 inclusive 1,242 smooth pneumococci were examined by a standard technique for sensitivity to penicillin G and other antibiotics (unpublished results of D. H.). Penicillin-insensitive strains from two subjects were detected during this period, and incidence of only 0.2%. The results of a preliminary examination of pneumococci from several districts in New Guinea suggest that the incidence of insensitive strains is greater than in Australia; studies of the distribution and frequency of such resistant strains are in progress.

We thank Dr. Phyllis Rountree of the Royal Prince Alfred Hospital, Sydney, for the pneumococcus isolated at Ernabella. These investigations were supported in part by a grant from the Wellcome Trust.

(**230**, 407; 1971)

David Hansman: Department of Bacteriology, The Adelaide Children's Hospital, North Adelaide 5006.

H. N. Glasgow, John Sturt, Lorraine Devitt and R. M. Douglas: Pneumonia Research Project, Anguganak and Port Moresby, New Guinea.

Received October 23, 1970.

起的肺炎和脑膜炎对人类来说是致命的。

　　虽然对抗生素不敏感的肺炎双球菌居然能出现在像塞皮克这样的偏远地区看似有点令人惊讶，但因为呼吸系统和其他感染在新几内亚很常见，所以青霉素在当地被大量使用。在过去的 10 年中，安古干纳克地区两个村子中的 507 位居民已经接受了 1,357 个疗程的普鲁卡因青霉素治疗。这意味着，对每个个体而言，平均每 3.7 年就要接受一个疗程的青霉素。除此之外，在开展防治雅司病（译者注：一种由雅司螺旋体引起的热带传染病）的活动时，全体居民都注射过青霉素。导致大量使用青霉素的其他一些因素可能还包括：肺炎双球菌的携带率很高，尤其是在儿童中；以及这些携带者的高度聚居（参考文献 12 及一些尚未发表的研究结果）。能从 15 个人体内分离出不敏感的 4 型菌株说明已经发生了传染，因为他们居住在两个邻近的村子里。这与在医院内部发生的对四环素耐药的肺炎双球菌的交叉感染类似[10,13]。

　　除对抗生素敏感程度不同以外，青霉素不敏感菌株的所有其他特征都与具有荚膜的肺炎双球菌相同（研究结果公布于 1970 年澳大利亚微生物学会学术年会）。4 型和 6 型肺炎双球菌通常会引起人类感染，而且已经有过由 4 型肺炎双球菌引起肺炎流行的报道。

　　目前，澳大利亚的青霉素不敏感肺炎双球菌感染发生率较低：从 1965 年到 1969 年，我们用标准方法检测了 1,242 个光滑型肺炎双球菌对青霉素 G 和其他抗生素的敏感性（戴维·汉斯曼尚未发表的研究结果）。在此期间检测到其中两位受试者体内出现了青霉素不敏感菌株，发生率仅为 0.2%。我们在新几内亚的一些地区对肺炎双球菌进行了初步的检测，结果表明，当地不敏感型菌株的发生率要高于澳大利亚；对此类抗药菌株的分布和发生率的研究还在进行中。

　　在此我们要感谢悉尼皇家阿尔弗雷德王子医院的菲莉丝·朗特里博士为厄纳贝拉地区肺炎双球菌的分离所做的工作。英国维康基金会为上述研究提供了一部分资金。

（董培智 翻译；王昕 审稿）

References:

1. Garrod, L. P., and O'Grady, F., *Antibiotic and Chemotherapy*, 57 (Livingstone, Edinburgh, 1968).

2. McKee, C. M., and Houck, C. L., *Proc. Soc. Exp. Biol. Med.*, **53**, 33 (1943).

3. Rake, G., McKee, C. M., Hamre, D. M., and Houck, C. L., *J. Immunol.*, **48**, 271 (1944).

4. Eriksen, K. R., *Acta Pathol. Microbiol. Scand.*, **22**, 398 (1945).

5. Eriksen, K. R., *Acta Pathol. Microbiol. Scand.*, **23**, 498 (1946).

6. Gunnison, J. B., Fraher, M. A., Pelcher, E. A., and Jawetz, E., *Appl. Microbiol.*, **16**, 311 (1968).

7. Hansman, D., and Bullen, M. M., *Lancet*, ii, 264 (1967).

8. Evans, W., and Hansman, D., *Lancet*, i, 451 (1963).

9. Richards, J. D. M., and Rycroft, J. A., *Lancet*, i, 553 (1963).

10. Turner, G. C., *Lancet*, ii, 1292 (1963).

11. Schaedler, R. W., Choppin, P. W., and Zabriskie, J. B., *New Engl. J. Med.*, **270**, 127 (1964).

12. Rountree, P. M., Beard, M., Arter, W., and Woolcock, A. J., *Med. J. Austral.*, **1**, 967 (1967).

13. Hansman, D., and Andrews, G., *Med. J. Austral.*, **1**, 498 (1967).

Mitochondrion as a Multifunctional Organelle

R. B. Flavell

Editor's Note

Mitochondria—membrane-enclosed organelles found in most eukaryotic cells—are commonly described as the powerhouses of the cell, since their primary function is the manufacture of the molecular energy source ATP (which, as Richard B. Flavell says in this paper, involves electron transport and oxidation). Flavell, a molecular biologist at Cambridge, was concerned to correct this over-simplification, pointing out that the mitochondrion has many other essential tasks too, which do not depend on the presence of mitochondrial DNA. In particular, he says, it is involved in protein synthesis; we would today add calcium-ion signalling as a key role. Flavell argues that this means one should not interpret the invisibility of mitochondria in microscopy of anaerobes as evidence of their absence.

THAT the mitochondrion is the site of the electron transport chain and oxidative phosphorylation is well known and, indeed, the mechanism of oxidative phosphorylation is an active area of biochemical research. Within the outer mitochondrial membrane, however, many other kinds of metabolic reactions also occur. Thus, mitochondria contain enzymes involved in the Krebs cycle, the β oxidation of fatty acids, fatty acid biosynthesis[1], synthesis of some amino-acids[2], lipid catabolism, the urea cycle and glutamate and aspartate biosynthesis[3] as well as several oxidative enzymes not directly involved with oxidative phosphorylation, such as rotenone-insensitive NADH, cytochrome c reductase, monoamine oxidase and kynurenine hydroxylase[1]. This list is not exhaustive and will no doubt be extended in the years ahead.

Many of these enzymes or sequences of enzymes are biosynthetic or catabolic and consequently have end products not directly involved in energy metabolism, so that the mitochondrion has a biosynthetic and catabolic as well as an energetic role, which would suggest that unless all mitochondrial end products—biosynthetic, catabolic and energetic—can be provided by alternative metabolic systems, mitochondria are essential for cell survival in all conditions. This conclusion is rarely discussed in general articles on mitochondria and infrequently recognized in research reports on mitochondria. Thus a recently published article[4] began: "Mitochondria are DNA-containing, self-replicating organelles the functions of which are dispensable to yeast cells grown in the presence of a fermentable energy source". The issue is especially significant in the question whether or not anaerobic yeast cells possess mitochondria.

There are conflicting reports of the presence of mitochondria in anaerobically grown yeast cells. Electron microscopical studies suggest that anaerobic yeast cells lack mitochondria[5,6], but several workers have reported that if lipid precursors are included in

线粒体是一个多功能的细胞器

弗拉维尔

编者按

线粒体是在大多数真核细胞中存在的、膜包被的细胞器，它们通常被描述为细胞的动力工厂，因为其主要功能是制造生物分子的能量来源——三磷酸腺苷（正如理查德·弗拉维尔在本文中所述，三磷酸腺苷参与了电子传递和氧化过程）。剑桥大学的分子生物学家弗拉维尔认为应该对这种过于简单的概括进行修正，他指出线粒体还有很多不依赖于线粒体 DNA 存在的其他重要功能。他特别提到了线粒体在蛋白质合成中的作用；现在我们当然还要将钙离子信号转导添加到它的关键功能里。弗拉维尔认为：这说明人们不应该把运用显微镜术在厌氧生物中看不见线粒体的现象当作线粒体不存在的证据。

众所周知，线粒体是电子传递链和氧化磷酸化的场所，而且氧化磷酸化机制确实也是生化研究中的一个活跃领域。然而，在线粒体外膜内，还发生着许多其他类型的代谢反应。因此，线粒体包含的酶包括参与克雷布斯循环、脂肪酸 β- 氧化作用、脂肪酸生物合成 [1]、某些氨基酸合成 [2]、脂分解代谢、尿素循环、谷氨酸盐和天冬氨酸盐生物合成 [3] 的酶以及几种不直接参与氧化磷酸化反应的氧化酶，诸如对鱼藤酮不敏感的还原型烟酰胺腺嘌呤二核苷酸（NADH）、细胞色素 c 还原酶、单胺氧化酶和犬尿氨酸羟化酶 [1]。这张单子并未穷尽所有，在今后的若干年中肯定还会有所扩展。

在这些酶或系列酶中，有许多是生物合成方面的或者分解代谢方面的，它们的最终产物并不直接参与能量代谢，因此线粒体不仅有能量方面的作用，还有生物合成和分解代谢方面的作用。这可能表明：除非线粒体的所有最终产物，包括生物合成的、分解代谢的和能量方面的，都可以由其他代谢系统来提供，否则在任何情况下线粒体对于细胞生存都是不可缺少的。这一结论在关于线粒体的一般文章中很少被论及，在与线粒体有关的研究报告中也难得被承认。无怪乎有一篇最近发表的文章 [4] 以这样的话作为开篇语："线粒体是含有 DNA 的、能够进行自我复制的细胞器，其功能对于生长在有可发酵能源环境中的酵母细胞来说是可有可无的"。这一观点在讨论厌氧酵母细胞是否拥有线粒体这一问题时尤其重要。

关于在无氧环境中生长的酵母细胞里到底有没有线粒体的问题，目前所报道的结果是相互矛盾的。电子显微镜的研究表明厌氧酵母细胞不具有线粒体 [5,6]；但也有几位

the defined growth medium, mitochondrial profiles can be seen in electron micrographs of anaerobically grown yeast cells, although some disagree[7].

Linnane *et al.* did not observe mitochondria in electron micrographs of anaerobic yeast cells[8], but they did recognize "reticular membranes" which appeared to differentiate into mitochondria in aerobic conditions (see also refs. 5, 7), and detailed biochemical studies[9,10] have confirmed their homology with mitochondria. Although the reticular membranes lack the enzymes and cytochromes for oxidative phosphorylation, they do possess mitochondrial DNA, mitochondrial ATPase and succinate dehydrogenase. These mitochondria have a similar—if not identical—"structural protein" fraction to those isolated from aerobic cells. These membrane structures in extracts from anaerobic cells possess similar percentages of extracted protein to those in extracts from aerobic cells[9], giving unequivocal biochemical evidence that mitochondria exist in anaerobic yeast cells, as would be expected from knowledge of their diverse functions.

In spite of this demonstration of the similarities between the membrane structures found in anaerobic yeast cells and the mitochondria in aerobic yeast cells, the membrane structures were called "promitochondria", not mitochondria, because they had neither the appearance of aerobic mitochondria in the electron microscope nor the ability to catalyse oxidative phosphorylation, but gained both these properties when the cells were exposed to oxygen[9,10]. I suggest that because the area of specialized cytoplasm enclosed by the outer mitochondrial membrane has diverse metabolic functions, the mitochondrion is incompletely defined by its appearance in the electron microscope and its oxidative phosphorylation activity. "Promitochondrion" is, therefore, inappropriate. It seems more appropriate to consider the units of specialized cytoplasm, organized within outer mitochondrial membranes around the mitochondrial DNA of anaerobic cells, as mitochondria lacking some of the components found in mitochondria of aerobic cells. Equally extreme variation in mitochondrial composition is found in some muscle cells, in which the synthesis of inner membrane and oxidative phosphorylation proteins is highly derepressed in response to large demands for ATP[11] and considerable variation in mitochondrial enzyme composition is expected from knowledge of the diversity of mitochondrial metabolism. Enzyme levels are frequently regulated by the concentration of their specific end product(s), so that in cells with different requirements for particular mitochondrial end products, mitochondria of differing enzyme composition are expected[12] and tissue specific differences in mitochondria from the same organism have been described[13].

Changes that occur in mitochondria after the transfer of yeast cells from an anaerobic to an aerobic environment, or from high glucose repression to low glucose repression, have been studied[14,15]. In these conditions a number of mitochondrial components, particularly those concerned with energy production, are preferentially synthesized in a process described as "mitochondrial biogenesis". This term has also been used in studies of the synthesis of mitochondrial components in animal cells[16-18]. It does not imply the synthesis of new mitochondria from pre-existing mitochondria as is almost certainly

888

研究者提出，如果在配方明确的培养基中含有脂质前体，那么从厌氧生长的酵母细胞的电子显微镜照片里就可以看到线粒体的轮廓，不过有些人并不同意这一观点 [7]。

虽然林纳内等人并没有在厌氧酵母细胞的电子显微镜照片中观察到线粒体 [8]，但是他们确实看到了"网状膜"，该网状膜在有氧条件下似乎可以分化成线粒体（参见参考文献 5 和 7），而且详细的生化研究 [9,10] 也证实了它们与线粒体的同源性。尽管这些网状膜不含有氧化磷酸化反应所需要的酶和细胞色素，但是它们确实含有线粒体 DNA、线粒体腺苷三磷酸酶和琥珀酸脱氢酶。这些线粒体的"结构蛋白"部分与从有氧细胞中分离出来的部分即使不完全相同，至少也比较相近。厌氧细胞提取物中的这些膜结构与有氧细胞提取物中的膜结构具有同样百分比的蛋白提取物 [9]，这一生物化学证据确凿无疑地证明，在厌氧酵母细胞中存在线粒体，此论点与我们根据对线粒体具有多种功能的认识所料想到的情况一致。

尽管上述证据已经表明，在厌氧酵母细胞中发现的膜结构与有氧酵母细胞中的线粒体具有相似性，但这种膜结构被人们称为"原线粒体"，而非线粒体，因为它们既无有氧线粒体在电子显微镜下的外观，也不能催化氧化磷酸化反应，然而一旦细胞被暴露于氧气之下，它们就会获得这两个特性 [9,10]。我认为：鉴于被线粒体外膜包被的特定细胞质区域具有多种不同的代谢功能，因而仅由其在电子显微镜下的外观及其氧化磷酸化活性来定义线粒体是片面的。因此，"原线粒体"这一名称并不恰当。似乎更恰当的说法是：把分布在厌氧细胞线粒体 DNA 周围、线粒体外膜内的特定细胞质单元看作是与有氧细胞中的线粒体相比缺少了某些成分的线粒体。人们在某些肌肉细胞中也发现了线粒体组分存在同等极端变异的情况，由于需要大量三磷酸腺苷，这些线粒体中的内膜合成及氧化磷酸化蛋白质的合成被高度去抑制 [11]，根据我们对线粒体代谢多样性的了解，可以料想到线粒体酶的组分也存在着相当大的变异。酶的水平通常受其特定终产物的浓度所调控，因此可以预料在对特定线粒体终产物有不同需求的细胞中会出现具有不同酶组分的线粒体 [12]，而且也有人曾描述过来自同一生物体的线粒体存在组织特异性上的差异 [13]。

有些人曾研究过将酵母细胞从无氧环境转移到有氧环境，或者从高葡萄糖阻遏作用环境转移到低葡萄糖阻遏作用环境之后线粒体内所发生的变化 [14,15]。在这些情况下，若干线粒体组分，尤其是那些与产生能量有关的组分，都在被称为"线粒体生物生成"的过程中优先合成。在研究动物细胞中的线粒体组分合成时，也用到了这一术语 [16-18]。这并不意味着从已经存在的线粒体合成新的线粒体，就像在

the case in all eukaryotes. Each mitochondrial component may be formed *de novo*, but formation of new organelles is by growth and division of pre-existing mitochondria. I suggest, therefore, that "mitochondrial differentiation" is more appropriate to describe the changes that occur when yeast cells are transferred from an anaerobic to an aerobic environment.

Descriptions[20,21] have been published of mitochondria isolated from "petite" strains of yeast which lack mitochondrial DNA, or the mitochondrial DNA of which has a severely restricted information content because it is composed almost entirely of adenine and thymine. Such mitochondria, which are of a similar size and buoyant density to those in the "grande" parent strain, lack the components of the electron transport chain but possess adenosine triphosphatase and other mitochondrial enzymes as well as an inner membrane[21]. Perlman and Mahler[21], therefore, raised the question of why mitochondria continue to be synthesized when their oxidative phosphorylation capacity is lost. Knowledge of the other essential functions of mitochondria provides the answer.

The presence of mitochondria in cells lacking mitochondrial DNA implies that in yeast the role of mitochondrial DNA in maintenance of mitochondria is limited, that is, the organelle is not genetically autonomous.

It has been suggested (refs. 22 and 23 and article in preparation by Flavell and Raven) that mitochondria have evolved from a prokaryote, able to catalyse oxidative phosphorylation, which sheltered in a primitive amoeboid-like cell and became stabilized there in a symbiotic relationship. At some time before the presumed symbiosis was established, the host cell must have been capable of living without the prokaryote, which would have been genetically autonomous. Since then considerable co-evolution of host cell and prokaryote has taken place.

The prokaryote has lost almost all its genetic information. Organization and compartmentalization have evolved in the eukaryotic cell such that the mitochondrion is now a fully integrated unit of the cytoplasm. Many biosynthetic and catabolic activities—as well as oxidative phosphorylation—must occur inside the outer mitochondrial membrane to be useful to the cell. This is well illustrated by auxotrophic strains of Neurospora which have metabolic deficiencies at the malate dehydrogenase step of the Krebs cycle, and other auxotrophic strains which have a metabolic block at the dehydroxy acid dehydratase step in isoleucine and valine biosynthesis. These strains, however, possess considerable malate dehydrogenase and dehydroxy acid dehydratase activity respectively, but possess metabolic blocks, because these enzymes are not localized in the mitochondria as in prototrophic strains, but remain outside the organelle in the cytoplasm[24,25].

Although multienzyme complexes can be isolated intact with the mitochondria, this physical association of related enzymes is probably the result of evolution rather than

所有真核生物中几乎一定会发生的过程那样。每一种线粒体组分都可以从头生成，但新细胞器是通过已存在线粒体的生长和分裂而形成的。因此，我认为用"线粒体分化"来描述那些发生在把酵母细胞从无氧环境转移到有氧环境时的变化会更贴切一些。

有些人曾在发表的文章中 [20,21] 描述过从酵母"小"菌株中分离得到的线粒体。这些酵母菌株或者不具有线粒体 DNA，或者虽然具有线粒体 DNA，但因其几乎全部由腺嘌呤和胸腺嘧啶构成，所以仅含有非常有限的信息量。这些线粒体的大小和浮力密度与"大"亲本菌株中的线粒体相似，它们缺少电子传递链组分，但除了内膜之外还含有腺苷三磷酸酶和其他线粒体酶 [21]。因此珀尔曼和马勒 [21] 提出了这样的疑问：为什么在线粒体缺失氧化磷酸化能力时仍然可以继续被合成。对线粒体其他重要功能的认识给出了该问题的答案。

在缺少线粒体 DNA 的细胞中仍存在线粒体意味着，酵母中的线粒体 DNA 在维持线粒体方面的作用有限，也就是说，该细胞器不具有遗传自主性。

有人提出（参考文献 22 和 23 以及弗拉维尔和雷文正在准备之中的文章）：线粒体是从能够催化氧化磷酸化反应的原核生物演化而来的，它藏身于一种原始的变形虫样的细胞中，并以共生关系在那里稳定下来。在这种假设共生关系被确立之前的某一段时期，宿主细胞肯定曾一度不依赖于原核生物而存活，这时原核生物是遗传自主的。随后宿主细胞和原核生物发生了明显的协同演化。

原核生物几乎失去了自己的全部遗传信息。组织结构和区室化已在真核细胞中演化发生，因此现在的线粒体已经充分整合成了细胞质的一个单元。包括氧化磷酸化在内的许多生物合成和分解代谢活动都必须在线粒体外膜之内发生，这样才能对细胞有所用处。用脉孢菌的营养缺陷型菌株和其他营养缺陷型菌株可以很好地证明这一点，前者在克雷布斯循环的苹果酸脱氢酶一步存在代谢缺陷，后者则在异亮氨酸和缬氨酸合成过程中的脱羟基酸脱水酶一步存在代谢障碍。然而，虽然这些菌株分别具有相当高的苹果酸脱氢酶活性和脱羟基酸脱水酶活性，但还是存在代谢障碍，因为与原养型菌株一样，这些酶并没有定位在线粒体上，而是仍位于细胞器之外的细胞质中 [24,25]。

尽管多酶复合物可以完整地与线粒体一道分离出来，但线粒体和相关酶的这种物理联系很可能是演化的结果，而并非早先自主性的残留。不能把现在的线粒体看

remains of a former autonomy. The present-day mitochondrion cannot be viewed as a symbiotic prokaryote or in any sense as self-sufficient, but is similar to other cell compartments or organelles such as Golgi bodies, glyoxysomes, microbodies and so on, which, although not having a prokaryotic origin, have evolved, as has the mitochondrion, to facilitate cytoplasmic organization, substrate and product channelling and metabolic pathway isolation.

Recently, great research emphasis has been placed on the electron transport system, oxidative phosphorylation, mitochondrial DNA and mitochondrial protein synthesis, which together with the endosymbiotic theory for the origin of mitochondria has, I think, propagated a view which fails to recognize the mitochondrion as a specialized area of cytoplasm of diverse functions and variable composition, the presence of which does not depend on mitochondrial DNA and is essential for eukaryotic life, even in anaerobic conditions. These features, however, cannot be ignored if we are to obtain a proper understanding of mitochondrial structure, function and replication.

(**230**, 504-506; 1971)

R. B. Flavell: Plant Breeding Institute, Trumpington, Cambridge CB2 2LQ.

Received February 24, 1971.

References:

1. Smoly, J. M., Kuylenstierna, B., and Ernster, L., *Proc. US Nat. Acad. Sci.*, **66**, 125 (1970).

2. Berquist, A., LaBrie, D. A., and Wagner, R. P., *Arch. Biochem. Biophys.*, **134**, 401 (1969).

3. Mahler, H. R., and Cordes, E. H., *Biological Chemistry*, 398 (Harper and Row, New York, 1966).

4. Goldring, E. S., Grossman, L. I., Krupruck, D., Cryer, D. R., and Marmur, J., *J. Mol. Biol.*, **52**, 323 (1970).

5. Wallace, P. G., and Linnane, A. W., *Nature*, **201**, 1191 (1964).

6. Morpurgo, G., Serlupi-Crescenzi, G., Tecce, G., Valente, F., and Venetacci, D., *Nature*, **201**, 897 (1964).

7. Polakis, E. S., Bartley, W., and Meek, G. A., *Biochem. J.*, **90**, 369 (1964).

8. Linnane, A. W., Vitols, E., and Nowland, P. G., *J. Cell. Biol.*, **13**, 345 (1962).

9. Criddle, R. S., and Schatz, G., *Biochemistry*, **8**, 322 (1969).

10. Plattner, H., Salpeter, M. M., Saltzgaber, J., and Schatz, G., *Proc. US Nat. Acad. Sci.*, **66**, 1252 (1970).

11. Lehninger, A., *The Mitochondrion*, 29 (Benjamin, New York, 1964).

12. Woodward, D. O., Edwards, D. L., and Flavell, R. B., *Symp. Soc. Exp. Biol.*, **24**, 55 (1970).

13. Kun, E., and Volfin, P., *Biochem. Biophys. Res. Commun.*, **23**, 696 (1966).

14. Clark-Walker, G. D., and Linnane, A. W., *J. Cell. Biol.*, **34**, 1 (1967).

15. Roodyn, D. B., and Wilkie, D., *The Biogenesis of Mitochondria* (Methuen, London, 1968).

16. Beattie, D. S., *J. Biol. Chem.*, **243**, 4027 (1968).

17. Kroon, A. M., and De Vries, H., *FEBS Lett.*, **3**, 208 (1969).

18. Haldar, D., Freeman, K., and Work, T. S., *Nature*, **211**, 9 (1966).

19. Luck, D. J. L., *J. Cell. Biol.*, **24**, 445 (1965).

20. Nagley, P., and Linnane, A. N., *Biochem. Biophys. Res. Commun.*, **5**, 989 (1970).

21. Perlman, P. S., and Mahler, H. R., *Bioenergetics*, **1**, 113 (1970).

22. Sagan, L., *J. Theoret. Biol.*, **14**, 225 (1967).

23. Nass, S., *Intern. Rev. Cytol.*, **25**, 55 (1969).

24. Munkres, K. D., Benveniste, K., Gorski, J., and Zuiches, C. A., *Proc. US Nat. Acad. Sci.*, 263 (1970).

25. Altmiller, D. H., and Wagner, R. P., *Biochem. Genet.*, **4**, 243 (1970).

作是一种共生的原核生物，或者从任何意义上来说认为它们可以像原核生物那样自给自足，而只能认为它们具有类似于其他细胞区室或细胞器（如高尔基体、乙醛酸循环体和微体等）的结构。尽管其他细胞区室或细胞器并不具有原核生物的起源，但它们和线粒体一样，都是朝着有利于细胞质组织架构、底物和产物疏导以及代谢路径分隔的方向而演化。

最近，人们把大部分的研究重点放在了电子传递系统、氧化磷酸化、线粒体DNA和线粒体蛋白质的合成上。我认为上述研究重点与线粒体起源的内共生理论都未能认识到：线粒体是具有多种功能和可变组分的细胞质中的一个特定区域，线粒体的存在并不依赖于线粒体DNA，而且即便在无氧条件下，线粒体对真核生物的生存来说也是必需的。如果我们要获得对线粒体结构、功能和复制方面的正确认识，就不能忽视这些特征。

（刘皓芳 翻译；顾孝诚 审稿）

Transmission of Two Subacute Spongiform Encephalopathies of Man (Kuru and Creutzfeldt–Jakob Disease) to New World Monkeys

D. C. Gajdusek and C. J. Gibbs, jun.

Editor's Note

By the late 1960s it was known that Creutzfeldt–Jakob Disease and its South Pacific counterpart, kuru, could be transmitted experimentally from man to chimpanzee, but the ape's large size and expensive lifestyle made it an impractical choice for laboratory study. Here Daniel Carleton Gajdusek and Clarence J. Gibbs demonstrate transmissibility of the two human spongiform encephalopathies to four species of New World monkey, providing, they claim, a more pragmatic laboratory model of the brain-wasting diseases. Gajdusek, who later received a Nobel Prize for his work on kuru, had lived amongst the kuru-stricken South Fore people of Papua New Guinea, where, he concluded, the disease was transmitted by the ritualistic eating of dead relatives' brains.

NOW that kuru and Creutzfeldt–Jakob disease have been transmitted to four species of New World monkeys, spider monkeys (*Ateles geoffroyi*), capuchin monkeys (*Cebus* sp.), squirrel monkeys (*Saimiri sciurens*), and woolly monkeys (*Lagothrix lagothica*), these more readily available animals may replace the chimpanzee in the laboratory study of these two subacute degenerative diseases of the human brain.

Both diseases have been transmitted to the squirrel monkey and to the spider monkey using 10% suspensions of human brain, inoculated intracerebrally, with incubation periods of about 2 yr. Human kuru brain suspension has caused the same disease in the capuchin monkey 45 months after intracerebral inoculation (Fig. 1). Furthermore, experimental kuru in the spider monkey has similarly been transmitted to the squirrel monkey with an incubation period of 22 months, and experimental Creutzfeldt–Jakob disease of the chimpanzee to the squirrel monkey with incubation periods of 9 and 19 months, and to the woolly monkey after 21 months of incubation (Fig. 2). Thus, these diseases of man and the experimental diseases in the chimpanzee have now been transmitted to smaller, less expensive, more readily available and more easily cared for hosts than the chimpanzee.

两种人类亚急性海绵状脑病（库鲁病和克雅氏病）传染给了新大陆猴

盖杜谢克，小吉布斯

编者按

到 20 世纪 60 年代晚期的时候，人们就已经知道克雅氏病以及与之相关的流行于南太平洋岛上的库鲁病可以通过实验从人类传染给黑猩猩，但因为类人猿体型庞大、饲养成本昂贵，所以将它用作实验研究是很不实用的。丹尼尔·卡尔顿·盖杜谢克和克拉伦斯·吉布斯在本文中证实这两种人类海绵状脑病可以传染给四种新大陆猴，他们认为上述发现为实验室研究这两类脑损耗疾病提供了更实际的解决方案。后来盖杜谢克因为在库鲁病方面的研究而获得了诺贝尔奖。他曾与库鲁病泛滥的巴布亚新几内亚南弗部落一起生活，他得出的结论是：这种病的传播源于当地人在仪式上经常食用死亡亲属的脑。

既然现在已经可以把库鲁病和克雅氏病传染给四种新大陆猴——蜘蛛猴、卷尾猴、松鼠猴和绒毛猴，那么在实验室中研究这两种人脑亚急性退化性疾病时就可以用这些更容易得到的动物来取代黑猩猩。

用 10% 的人脑悬液进行脑内接种，在潜伏了大约两年后，松鼠猴和蜘蛛猴都被传染上了这两种疾病。人类库鲁病脑悬液使脑内接种 45 个月后的卷尾猴染上了同样的疾病（图 1）。此外，蜘蛛猴的实验性库鲁病同样能够在潜伏 22 个月后传染给松鼠猴；黑猩猩的实验性克雅氏病也能在潜伏 9 个月和 19 个月后传染给松鼠猴，还能在潜伏 21 个月后传染给绒毛猴（图 2）。由此可见，现在我们已经能够把这两种人类疾病和黑猩猩中的实验性疾病传染给比黑猩猩更小、更便宜、更容易获得并且更易于饲养的寄主。

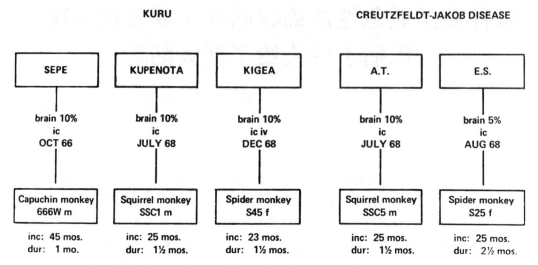

Fig. 1. Transmission of kuru and Creutzfeldt–Jakob disease to New World monkeys.

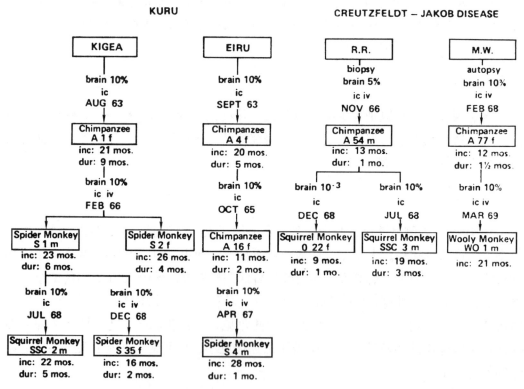

Fig. 2. Transmission of experimental kuru and Creutzfeldt–Jakob disease to New World monkeys.

Experimental kuru in the chimpanzee[1,2] was previously passed successfully to the spider monkey[3] with incubation periods of 23–28 months. It is now in third spider monkey to spider monkey passage and the incubation period on the second passage was 16 months (Fig. 2).

896

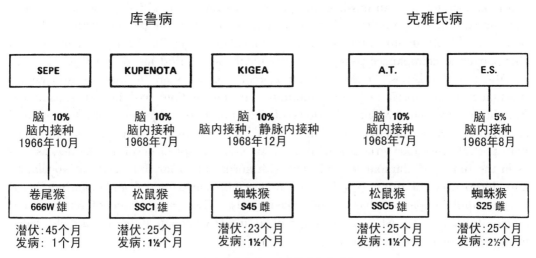

图 1. 库鲁病和克雅氏病传染给新大陆猴。

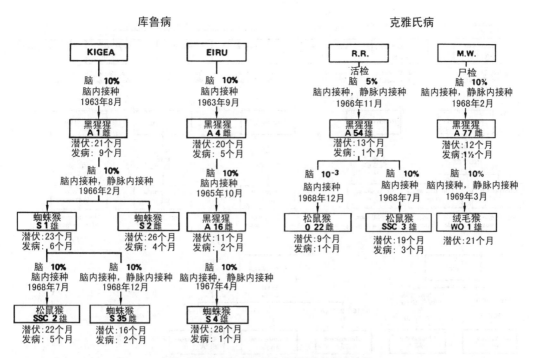

图 2. 实验性库鲁病和克雅氏病传染给新大陆猴。

之前我们曾把黑猩猩中的实验性库鲁病 [1,2] 成功地传染给了潜伏 23~28 个月后的蜘蛛猴 [3]。现在正处于第三个从蜘蛛猴到蜘蛛猴的传染过程之中，第二个传染过程的潜伏期为 16 个月（图 2）。

Kuru, known to be transmissible to the chimpanzee since 1965[1,2], has so far been transmitted from eleven different human patients to eighteen chimpanzees with incubation periods of 14–39 months after intracerebral inoculation (Fig. 3). We are now in the fifth chimpanzee to chimpanzee passage and the incubation period has dropped to 10 to 18 months. Infection is possible by either intracerebral or peripheral (combined intravenous, subcutaneous, intraperitoneal and intramuscular) inoculation, and by using bacterial-free filtrates through "Millipore" filters of 220 nm pore diameter. Both human and chimpanzee brains infected with kuru contain more than 10^6 chimpanzee infectious units per ml. Infectious virus is present in tissue pools of spleen, liver and kidney, as well as in the brains of chimpanzees affected with kuru, but so far intracerebral inoculation into chimpanzees of urine, blood, serum and cerebrospinal fluid from similarly affected humans and chimpanzees has not produced disease. Tissue suspensions of placenta and amnion from a patient who was delivered of an infant while late in her disease also failed to cause disease in the chimpanzee.

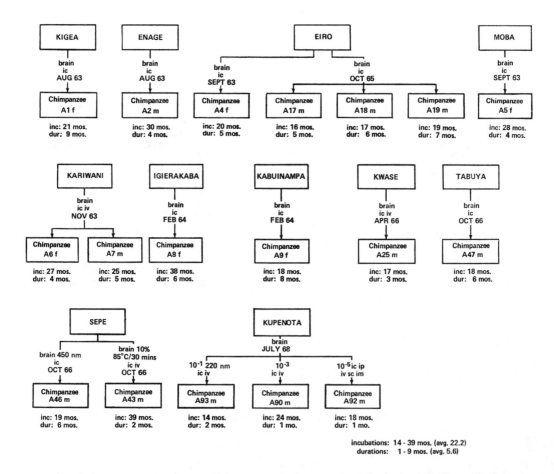

Fig. 3. Primary transmission of kuru from brain tissues of eleven human patients to eighteen chimpanzees.

　　自 1965 年起人们就已经知道库鲁病可以传染给黑猩猩[1,2]，到目前为止该病已经通过脑内接种的方式从 11 位人类患者传染到了 18 只黑猩猩中，潜伏期为 14~39 个月（图 3）。现在我们正处于第五个从黑猩猩到黑猩猩的传染过程之中，并且潜伏期已经降到了 10~18 个月。通过脑内或者外周（包括静脉内、皮下、腹腔内和肌肉内）接种经由孔径为 220 纳米的"密理博"滤膜过滤的无菌滤液就有可能导致感染。感染库鲁病的人脑和黑猩猩脑中的病毒含量都在 10^6 黑猩猩感染单位 / 毫升以上。传染性病毒除了存在于患库鲁病黑猩猩的脑内之外，还存在于脾、肝和肾的组织液中，但迄今为止通过给黑猩猩脑内接种同样感染库鲁病的人和黑猩猩的尿液、血液、血清和脑脊液均未使其染病。用一位在库鲁病晚期还分娩婴儿的女性患者的胎盘和羊膜组织悬液也不能使黑猩猩发病。

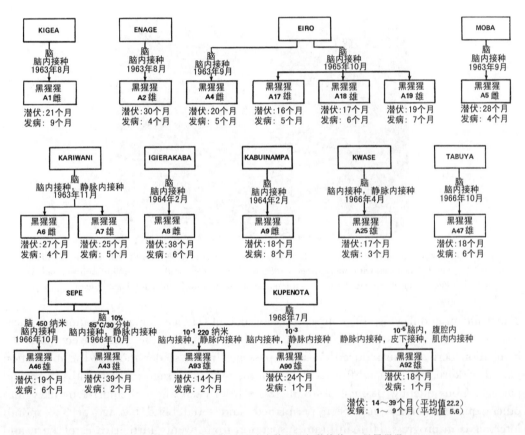

图 3. 库鲁病首次从 11 位人类患者的脑组织传染给 18 只黑猩猩。

Creutzfeldt–Jakob disease was first transmitted from man to the chimpanzee in 1968[4,5]; we have now transmitted the disease by cerebral inoculation brain tissue from ten human patients to eleven chimpanzees with incubation periods of 11–14 months. Furthermore, the brain tissue of one of these patients inoculated only peripherally (intravenously, intraperitoneally and intramuscularly) has caused the disease after 16 months of incubation. Serial passage of experimental Creutzfeldt–Jakob disease in the chimpanzee, with incubation periods of 10–14 months, has been possible using brain suspensions at dilutions of 10^{-1} and 10^{-3}.

A preliminary neuropathological survey by Dr. Kenneth Earle of the Armed Forces Institute of Pathology, and Dr. Peter Lampert of the University of California, San Diego School of Medicine, has revealed the presence of a gliosis, neuronal loss and status spongiosis in all the clinically positive New World monkeys, consistent with the diagnosis of experimental kuru or Creutzfeldt–Jakob disease (Fig. 4) and the detailed neuropathology of the individual animals is to be reported elsewhere (unpublished results of P. W. Lampert).

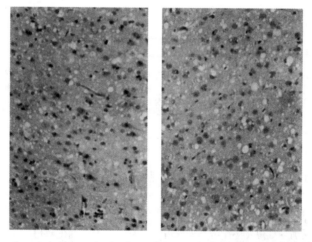

Fig. 4. Spongiform changes in the cerebral cortex of two squirrel monkeys: on the left (squirrel monkey SSC-1 in Fig. 1) inoculated with brain suspension from a kuru patient, and, on the right (squirrel monkey SCC-5 in Fig. 1), inoculated with brain suspension from a patient with Creutzfeldt–Jakob disease.

The squirrel monkey is, of course, the animal of choice for investigation of these diseases, should it be as sensitive an indicator of the viruses as the chimpanzee and if the incubation period were reduced on serial passage. In the spider monkey the incubation period of kuru was reduced from 23–28 months in first passage to 16 months in second passage. Consequently, passage of brain tissue from each affected squirrel monkey into other squirrel monkeys has been performed; and serial blind passage at 3–6 month intervals is in progress. Human brain suspension from twenty further cases of kuru and twenty of Creutzfeldt–Jakob disease has also been inoculated into New World monkeys. Titration of human and chimpanzee kuru brain suspensions known to be infectious at 10^{-5} dilution for the chimpanzee has been done in squirrel and capuchin monkeys.

克雅氏病首次从人传染给黑猩猩是在 1968 年 [4,5]，现在我们已经通过脑内接种 10 位人类患者的脑组织悬液将这种疾病传染到 11 只黑猩猩中，潜伏期为 11~14 个月。此外，其中一位患者的脑组织仅仅通过外周（静脉内、腹腔内和肌肉内）接种就已在 16 个月的潜伏期后使黑猩猩发病。使用稀释 10 倍和 10^3 倍的脑悬液已经可以在黑猩猩中以 10~14 个月的潜伏期连续传播实验性克雅氏病。

美国武装部队病理学研究所的肯尼思·厄尔博士和加州大学圣迭戈医学院的彼得·兰珀特博士经过初步的神经病理学研究发现：所有临床阳性的新大陆猴都出现了与实验性库鲁病或克雅氏病症状一致的神经胶质增生、神经元丢失和海绵状改变症状（图 4）。每个动物个体的详细神经病理学结果将在其他文章中报道（兰珀特尚未发表的结果）。

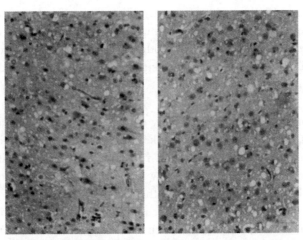

图 4. 两只松鼠猴大脑皮质中的海绵状改变：左图（图 1 中的松鼠猴 SSC-1），接种了一位库鲁病患者脑悬液之后的结果；右图（图 1 中的松鼠猴 SCC-5），接种了一位克雅氏病患者脑悬液之后的结果。

当然，松鼠猴是为研究这两种疾病而特殊挑选的动物，它作为病毒的指示物是否会与黑猩猩的敏感度一样高？在连续传播时潜伏期是否会减少？蜘蛛猴的库鲁病潜伏期在第一次传播时为 23~28 个月，到第二次传播时就减少至 16 个月。因此，我们用每只染病松鼠猴的脑组织感染其他松鼠猴；并且还在进行间隔期为 3~6 个月的盲目继代移植。另外 20 位库鲁病患者和 20 位克雅氏病患者的脑悬液也被接种到新大陆猴中。已知当稀释 10^5 倍时对黑猩猩有感染性的人和黑猩猩的库鲁病脑悬液已在松鼠猴和卷尾猴中进行滴定。

To achieve a degree of isolation which is unobtainable in any single primate holding facility we have used three widely separated facilities, in California, Louisiana, and Maryland. So far, dozens of additional chimpanzees and monkeys inoculated with brain and other organ suspensions from other subacute and chronic neurological disorders have lived in each colony in close contact with animals inoculated with material from kuru and Creutzfeldt–Jakob disease. Thus far, no neurological disease has appeared in any of these "control" animals, which further serves to indicate that experimental kuru and Creutzfeldt–Jakob disease are not communicable.

The squirrel, spider and capuchin monkeys and the chimpanzee are the only species out of twenty-three sub-human primates to develop kuru after inoculation with tissues from human kuru victims. Creutzfeldt–Jakob disease has been transmitted directly with human tissues to the squirrel and spider monkeys, as it was transmitted earlier to the chimpanzee. Furthermore, experimental Creutzfeldt–Jakob disease of the chimpanzee has been transmitted to the woolly monkey (Fig. 1). With the exception of the chimpanzee, none of the sixteen Old World species including the gibbon ape, and in addition the two other species of New World monkeys that were used*, has developed clinical disease after intracerebral inoculation with brain suspensions from patients suffering from kuru or Creutzfeldt–Jakob disease and from chimpanzees in which these diseases have been induced. The successful transmission of both the natural human and the experimental chimpanzee diseases into the squirrel monkey has therefore provided a new laboratory model, more convenient than the chimpanzee, for the study of these diseases.

We have no explanation for the susceptibility of the three species of New World monkeys to kuru, and the three species of these monkeys to Creutzfeldt–Jakob disease, whereas all the Old World monkeys studied have been resistant. There is still need for attempts to discover a broader host range for these slow virus infections in the hope that laboratory investigations may be further facilitated. In this regard it may be noted that our attempts to transmit both diseases to some thirty other mammalian and avian hosts by intracerebral inoculation of brain suspensions have been unsuccessful so far. But, in view of the finding of minimal kuru-like neuropathology in some inoculated animals dying without clinical neurological disease, we are involved in a further programme of blind passage in the species under suspicion.

Eckroade et al[6]. have successfully transmitted mink encephalopathy to the squirrel and stump-tailed monkey. This disease and scrapie are the two naturally occurring diseases

---

* Species used in transmission attempts were six New World monkeys: squirrel monkey (*Saimiri sciurens*), capuchin monkey (*Cebus* sp.), spider monkey (*Ateles geoffroyi*), woolly monkey (*Lagothrix lagothricha*), owl monkey (*Aotus trivirgatus*), and marmoset (*Callithrix* sp.); fifteen Old World monkeys: rhesus monkey (*Macaca mulata*), cynomolgous macaque (*M. irus*), stump-tailed macaque (*M.speciosa*), bonnet monkey (*M. radiata*), barbary ape (*M. sylvania*), pig-tailed macaque (*M. nemestrian*), African green monkey (*Cercopithecus aethiops* sp.), baboon (*Papio anubis*), langur (*Colobus* sp.), sooty mangabey (*Cercocebus atys*), patas (*Erythrocebus patas*), slow loris (*Nycticebus coucang*), talapoin (*Cercopithecus miopithecus*), tree shrew (*Ptilocercus lowii*), and bush baby (*Galago senegalensis*); two apes: chimpanzee (*Pan troglodytes troglodytes*) and gibbon ape (*Hylobatus lar*).

为了能达到在任何一个单独的灵长类动物圈养所中不可能达到的分离程度，我们使用了三个相隔很远的圈养所，分别位于加利福尼亚、路易斯安那和马里兰。迄今为止，另外几十只接种其他亚急性和慢性神经系统疾病患者脑悬液和其他器官悬液的黑猩猩和猴一直与每个聚居地内接种库鲁病和克雅氏病物质的动物在生活中密切接触。直到现在，在这些用于"对照"的动物中，没有一只染上神经系统疾病，这一结果进一步说明实验性库鲁病和克雅氏病是不会传染的。

松鼠猴、蜘蛛猴和卷尾猴以及黑猩猩是 23 种近似人类的灵长目动物中仅有的在接种人类库鲁病患者组织后能感染上库鲁病的物种。克雅氏病能直接由人类组织传染到松鼠猴和蜘蛛猴，就像它在较早时候能传染黑猩猩一样。此外，黑猩猩的实验性克雅氏病已能传染给绒毛猴（图 1）。除黑猩猩外，包括长臂猿在内的 16 种旧大陆猴以及另外 2 种新大陆猴*在脑内接种库鲁病或克雅氏病患者脑悬液和已感染这两种疾病的黑猩猩脑悬液后都没有产生临床疾病。因此，将人类的自然疾病和黑猩猩的实验性疾病成功地传染给松鼠猴就为研究这两种疾病提供了一类比黑猩猩更便利的新实验对象。

我们无法解释这 3 种新大陆猴对库鲁病的易感性，也不能解释新大陆猴中的 3 个物种对克雅氏病的易感性，而所有用于研究的旧大陆猴都具备抵抗力。我们仍需进一步尝试拓展这些缓慢病毒感染的寄主范围以期能使实验室研究更加便利。在这方面要说明的是：我们曾试图通过脑内接种脑悬液将这两种疾病传染给大约 30 只其他哺乳动物和鸟类寄主，但到目前为止尚未取得成功。然而，因为在一些无临床神经系统疾病死亡的接种动物中发现了极微小的库鲁病样神经病理学特征，所以我们还在继续进行可疑物种的盲目继代移植。

埃克罗德等人 [6] 已成功地将水貂脑病传染给了松鼠猴和短尾猴。这种病和羊瘙痒症是两类自然发生的疾病，我们将它们与库鲁病和克雅氏病归为一类并称之为亚

---

* 尝试用于传染的物种是 6 种新大陆猴：蜘蛛猴、卷尾猴、松鼠猴、绒毛猴、猫头鹰猴和绒猴；15 种旧大陆猴：恒河猴、食蟹猴、短尾猴、帽猴、地中海猕猴、豚尾猴、非洲绿猴、狒狒、叶猴、乌白眉猴、赤猴、懒猴、侏长尾猴、树鼩和灌丛婴猴；两种猿：黑猩猩和长臂猿。

which we have grouped with kuru and Creutzfeldt–Jakob disease and called subacute spongiform virus encephalopathies[7]. Scrapie virus has not yet produced clinical disease in a primate and, although sheep, goat and mouse strains of this virus have been inoculated into many species of monkeys and into the chimpanzee, all animals have remained well for periods of 3 to 5 yr. Reports of the transmission of scrapie to mink and of mink encephalopathy to goats and mice[8-10] reinforce the suspicion that the agents of these diseases are closely related. The crucial question of whether the viruses of kuru and Creutzfeldt–Jakob disease are related cannot be answered until serological identification of the viruses is possible and more is known of their properties. Whether one or both are related to the viruses of the two pathologically similar animal diseases remains to be determined.

(**230**, 588-591; 1971)

D. Carleton Gajdusek and Clarence J. Gibbs, jun.: National Institute of Neurological Diseases and Stroke, National Institutes of Health, Bethesda, Maryland.

Received December 10, 1970; revised February 9, 1971.

References:

1. Gajdusek, D. C., Gibbs, jun., C. J., and Alpers, M., *Nature*, **209**, 794 (1966).

2. Gajdusek, D. C., Gibbs, jun., C. J., and Alpers, M., *Science*, **155**, 212 (1967).

3. Gajdusek, D. C., Gibbs, jun., C. J., Asher, D. M., and David, E., *Science*, **162**, 693 (1968).

4. Gibbs, jun., C. J., Gajdusek, D. C., Asher, D. M., Alpers, M. P., Beck, E., Daniel, P. M., and Matthews, W. B., *Science*, **161**, 388 (1968).

5. Beck, E., Daniel, P. M., Matthews, W. B., Stevens, D. L., Alpers, M. P., Asher, D. M., Gajdusek, D. C., and Gibbs, jun., C. J., *Brain*, **92**, 699 (1969).

6. Eckroade, R. J., Zu Rhein, G. M., Marsh, R. F., and Hanson, R. P., *Science*, **169**, 1088 (1970).

7. Gibbs, jun., C. J., and Gajdusek, D. C., *Science*, **165**, 1023 (1969).

8. Marsh, R. F., thesis, Univ. Wisconsin (1968).

9. Zlotnik, I., and Barlow, R. M., *Vet. Rec.*, **81**, 55 (1967).

10. Barlow, R. M., and Rennie, J. C., *J. Comp. Pathol.*, **80**, 75 (1970).

急性海绵状脑病 [7]。迄今为止羊瘙痒症病毒还没有使灵长类动物产生临床疾病。尽管人们已经尝试过将这种病毒的绵羊、山羊和小鼠株接种到许多种猴子和黑猩猩中，但所有动物在 3 到 5 年内都很健康。关于羊瘙痒症传染给水貂以及水貂脑病传染给山羊和小鼠的报道 [8-10] 使我们更加怀疑在这些疾病的病原体之间是否存在着密切的关联。目前人们对库鲁病和克雅氏病的病毒是否相关这一重要问题还无法作答，除非可以通过血清学鉴定这些病毒并了解有关它们的更多特性。到底是其中一种还是两种都与这两种病理相似的动物疾病病毒相关，至今仍没有定论。

（李梅 翻译；袁峥 审稿）

Inhibition of Prostaglandin Synthesis as a Mechanism of Action for Aspirin-like Drugs

J. R. Vane

Editor's Note

Aspirin is one of the most widely used of synthetic drugs. It has been generally available since the beginning of the twentieth century, but for most of that time the basis for its effectiveness in treating pain, fever and other conditions has been unknown. In 1971, the physiologist John Vane published a mechanism for the action of aspirin and drugs like it which asserted that the drug functions by inhibiting the synthesis of the materials called prostaglandins—hormones produced in living cells which, in excess, can cause inflammation and of pain. Vane was awarded a Nobel Prize for his work in 1982 and was knighted in the same year.

THERE have been many attempts to link the anti-inflammatory actions of substances like aspirin with their ability to inhibit the activity of endogenous substances. Collier[1,2] calls aspirin an "anti-defensive" drug, and is largely responsible for studying its possible antagonism of the activity of endogenous substances such as kinins[3,4], slow reacting substance in anaphylaxis (SRS-A) (ref. 5), adenosine triphosphate[6], arachidonic acid[7,8] and prostaglandin $F_{2\alpha}$ (refs. 2 and 7).

A possible mechanism for some of the actions of anti-inflammatory acids was discovered by Piper and Vane[10] who found that lungs could release a previously undetected substance which, because of its action, they called "rabbit aorta contracting substance" or RCS. When isolated perfused lungs of sensitized guinea-pigs were challenged, RCS, along with histamine, SRS-A, prostaglandin E_2 (PGE_2) and prostaglandin $F_{2\alpha}$ ($PGF_{2\alpha}$) (ref. 11), were released. The release of RCS, which could also be provoked by bradykinin and SRS-A, was antagonized by aspirin-like drugs, as was the evoked bronchoconstriction. Because RCS has a short half-life (<5 min) it has not been isolated and its chemical nature is unknown. However, the finding that arachidonic acid, a prostaglandin precursor which induces bronchoconstriction, also releases RCS from perfused lungs[12] makes it possible that RCS is a prostaglandin or has a structure intermediate between arachidonic acid and PGE_2 or $PGF_{2\alpha}$. This release of RCS and the associated bronchoconstriction are also antagonized by aspirin-like drugs.

Prostaglandin release can often be equated with prostaglandin synthesis[13], for many tissues can be provoked to release more prostaglandin than they contain. The possibility arises, therefore, that anti-inflammatory substances such as aspirin inhibit the enzyme(s) which generate prostaglandins. The experiments described below were designed to test this possibility. Aspirin and indomethacin strongly inhibit prostaglandin synthesis; this may be the mechanism underlying some of their therapeutic actions.

抑制前列腺素合成是阿司匹林样药物的一种作用机制

文

编者按

阿司匹林是应用最广泛的合成药物之一。早在 20 世纪初，阿司匹林就已经得到了普遍的应用，但在那个时候的大部分时间里，人们并不了解阿司匹林治疗疼痛、发烧和其他一些病症的机理。1971 年，生理学家约翰·文提出了阿司匹林和阿司匹林样药物的一种作用机制，即认为这种药物的效用是通过抑制前列腺素的合成来实现的。前列腺素是活细胞产生的激素，过量时会引起炎症和疼痛。文因为这项工作在 1982 年获得了诺贝尔奖，并在同一年被授予英国爵士头衔。

为了把阿司匹林类物质的消炎作用与它们抑制内源性物质活性的能力联系起来，人们已经进行了很多次尝试。科利尔 [1,2] 称阿司匹林是一种"抗防御性"药物，他主要研究阿司匹林是否有可能拮抗激肽 [3,4]、过敏反应慢应物质（SRS-A）（参考文献 5）、三磷酸腺苷 [6]、花生四烯酸 [7,8] 和前列腺素 $F_{2\alpha}$（参考文献 2 和 7）等内源性物质的活性。

派珀和文 [10] 发现了消炎酸产生某些作用的一种可能机制。他们发现肺能释放出一种以前未检测到的物质。根据其生物活性，他们把这种物质称为"兔主动脉收缩物质"或简称为 RCS。当离体灌注的致敏豚鼠肺受到刺激时就会释放出 RCS，还有组胺、SRS-A、前列腺素 E_2（PGE_2）和前列腺素 $F_{2\alpha}$（$PGF_{2\alpha}$）（参考文献 11）。血管舒缓激肽和 SRS-A 也能激发 RCS 的释放，这种 RCS 的释放可以被阿司匹林样药物拮抗；阿司匹林样药物还可以拮抗激发产生的支气管收缩。因为 RCS 的半衰期很短（< 5 min），所以人们一直没有将它分离出来，也不知道它的化学性质。然而，人们发现有一种前列腺素的前体——花生四烯酸能诱导支气管收缩，并能激发灌注肺 [12] 释放 RCS，这说明 RCS 可能是一种前列腺素，或者具有介于花生四烯酸和 PGE_2/$PGF_{2\alpha}$ 之间的某种结构。由花生四烯酸诱导的 RCS 释放和与之相关联的支气管收缩都能被阿司匹林样药物所拮抗。

前列腺素的释放通常等同于前列腺素的合成 [13]，因为很多组织在受激时能够释放出的前列腺素多于本身所含有的量。因此，阿司匹林等消炎物质就有可能会抑制合成前列腺素的酶。下面的实验就是为验证这种可能性而设计的。阿司匹林和消炎痛强烈地抑制了前列腺素的合成，这或许就是它们发挥某些治疗作用的潜在机制。

Cell-free homogenates of guinea-pig lung synthesize prostaglandins E_2 and $F_{2\alpha}$ from arachidonic acid and the following is based on the procedure of Ånggård and Samuelsson[14]. Lungs from four adult guinea-pigs were excised rapidly and washed in ice-cold medium (a modified Bucher medium containing 20 mM KH_2PO_4, 72 mM K_2HPO_4, 27.6 mM nicotinamide, and 3.6 mM $MgCl_2$; pH 7.4). The lung tissue was homogenized in an MSE blade homogenizer at full speed for 1 min with a tissue: medium ratio of 1:4. The resultant suspension was transferred to a Potter–Elvehjem homogenizer and further homogenized by six up and down strokes of the "Teflon" pestle. The homogenate was then centrifuged at 900g for 15 min and the supernatant fluid was used. Fresh homogenates were made on the morning of each experiment.

Arachidonic acid was dissolved in ethanol (0.1 ml./mg) and diluted with a 0.2% (w/v) sodium carbonate solution (0.9 ml./mg), thus giving a solution of arachidonic acid of 1 mg/ml. This was further diluted to 200 μg/ml. with the modified Bucher medium.

Flasks containing 10 μg of arachidonic acid (0.05 ml.) and lung homogenate (1 ml.) were incubated aerobically at 37°C with gentle shaking for 30 min. A zero-time sample was taken. The reactions were stopped by heating the flasks in a boiling water bath until the protein in the sample coagulated (30–60 s) and then diluting five or ten-fold with 0.9% (w/v) saline. The samples were frozen if kept overnight, or kept on ice until assayed.

Prostaglandin-like activity was assayed[15] on isolated stomach strips[16] and colons[17] from the rat, superfused[18] in series with Krebs solution containing a mixture of antagonists[19] to make the assay more specific. Activity was assayed by bracketing the contractions induced by injections of diluted samples between smaller and larger contractions induced by the standards. In the dose range used (2–20 ng), PGE_2 contracted the stomach strips, but had no effect on the colons, whereas $PGF_{2\alpha}$ contracted the colons but had a much weaker effect than PGE_2 on the stomach strips. Activity was assayed on the rat colons in terms of $PGF_{2\alpha}$ and on the stomach strips in terms of PGE_2. But because $PGF_{2\alpha}$ had a small effect in the PGE_2 assay and the enzyme preparation also partially inactivated the PGE_2 generated[14], less emphasis has been placed on the PGE_2-like activity assayed in these experiments.

In some experiments, the reactions were terminated by acidifying to pH 3 with hydrochloric acid and extracting twice with ethyl acetate. The combined extracts were evaporated to dryness under reduced pressure. The residue was taken up in 0.2 ml. ethanol and chromatographed in the A I system[20] on thin-layer chromatography plates[21] with markers of 5 μg authentic prostaglandin E_2 and $F_{2\alpha}$. The strips on the developed chromatograms corresponding to the marker spots were separated, as was the area in between them. The rest of the chromatogram was divided into 1–3 cm strips. Each section was scraped into a test tube and shaken with 2 ml. Krebs solution. The supernatant was assayed on the rat colons and stomach strips.

豚鼠肺的无细胞匀浆能利用花生四烯酸合成前列腺素 E_2 和 $F_{2\alpha}$，以下是安加德和萨穆埃尔松 [14] 的实验步骤。先迅速剥离 4 只成年豚鼠的肺，然后用冰预冷的培养基（改良的布赫培养基，含 20 mM KH_2PO_4、72 mM K_2HPO_4、27.6 mM 烟酰胺、3.6 mM $MgCl_2$；pH 值 7.4）洗涤。将肺组织和布赫培养基按 1：4 的比例在 MSE 叶片匀浆器中全速研磨 1 min 以匀浆肺组织。然后把得到的悬浮液转移到波特 – 埃尔维耶姆匀浆器中，再用"特氟隆"研棒上下研磨 6 次以进一步匀浆。将匀浆物在 900 g 下离心 15 min，取上清液备用。在每次实验当天的清晨制备新鲜匀浆。

将花生四烯酸溶解于乙醇（0.1 ml/mg）中，然后用 0.2%（重量体积比）的碳酸钠溶液（0.9 ml/mg）稀释，得到 1 mg/ml 的花生四烯酸溶液。然后再用改良的布赫培养基将上述溶液稀释到 200 μg/ml。

将含有 10 μg 花生四烯酸（0.05 ml）和肺组织匀浆物（1 ml）的瓶子在 37℃ 下的空气浴中温和振荡 30 min。保留一份零时间点时的样品。终止该反应的方法是：将瓶子在沸水浴中加热，直到样品中的蛋白质发生凝固（30 s ~ 60 s），然后用 0.9%（重量体积比）的生理盐水稀释 5 倍或 10 倍。样品在分析之前需要一直置于冰上，过夜的样品需要冷冻起来。

可以用从大鼠中分离的胃条 [16] 和结肠 [17] 来测定前列腺素样活性 [15]，并用含有混合拮抗物 [19] 的克雷布斯溶液连续灌流 [18] 以使检测结果更加准确。检测活性的方法是：将由注入稀释样品所诱导的收缩与一系列标准样品所诱导的较小或较大的收缩进行比较。在所用的剂量范围（2 ng ~ 20 ng）内，PGE_2 可使胃条收缩，但对结肠没有作用；而 $PGF_{2\alpha}$ 可使结肠收缩，但对胃条的收缩作用明显不如 PGE_2 对胃条的收缩作用强烈。因此 $PGF_{2\alpha}$ 的活性用大鼠结肠检测，PGE_2 的活性用胃条检测。但因为 $PGF_{2\alpha}$ 对 PGE_2 检测有少许影响，而且酶的制备过程也会使生成的 PGE_2 部分失活 [14]，所以在这类实验中，人们不太强调 PGE_2 样活性的检测。

在某些实验中，终止反应的方法是用盐酸酸化至 pH 值为 3，并用乙酸乙酯进行两次抽提。混合抽提物在减压条件下蒸发至干。残留物用 0.2 ml 乙醇溶解后，以 5 μg 明确的前列腺素 E_2 和 $F_{2\alpha}$ 为标准物在 A I 系统 [20] 的薄层色谱板上进行层析分离 [21]。在所得的层析图上，对应于标准物位置的条带以及它们之间的区域都被分离出去。将层析图的其余部分分成 1 cm ~ 3 cm 的小条带。将每一部分分别刮到一个试管中，然后加入 2 ml 克雷布斯溶液并摇匀。用上清液进行大鼠结肠和胃条的分析。

The zero-time samples contained (per ml.) 60–150 ng of $PGF_{2\alpha}$-like activity and 120–750 ng of PGE_2-like activity. This activity varied between samples of the same enzyme preparation by less than 5% and did not increase when the enzyme was incubated without arachidonic acid for 30 min. Incubation with arachidonic acid for 30 min increased $PGF_{2\alpha}$-like activity by 220–520 ng/ml. and PGE_2-like activity by 100–500 ng/ml., according to the enzyme preparation. Variation between different control samples of the same preparation was less than 7%.

Tests for Inhibition

Results were expressed as the generation of $PGF_{2\alpha}$ or PGE_2-like activity (30 min sample activity minus zero-time sample activity). To test for inhibition of prostaglandin synthesis, varying amounts of indomethacin, sodium acetylsalicylate, sodium salicylate or other substances were added to the incubation flasks in volumes of 0.1 ml. or less; inhibition of generation by a drug was expressed as the percentage inhibition of the control generation.

Indomethacin, sodium aspirin and sodium salicylate all inhibited the generation of $PGF_{2\alpha}$-like activity. The degree of inhibition varied from one enzyme preparation to another, but with any batch there was a linear relationship between percentage inhibition and log concentration of indomethacin or aspirin. The media used to dissolve the anti-inflammatory substances did not influence prostaglandin synthesis. The results from all experiments are shown in Fig. 1. The ID_{50} for indomethacin was 0.27 μg/ml. (0.75 μM), whereas that for aspirin was 6.3 μg/ml. (35 μM). Thus, on a weight basis, indomethacin was twenty-three times more potent than aspirin as an inhibitor of synthesis of $PGF_{2\alpha}$, and on a molar basis forty-seven times more potent. Sodium salicylate was less potent than aspirin as an inhibitor of synthesis of $PGF_{2\alpha}$-like activity and there was much more variation in the results (Fig. 1). Similar results were obtained when the activity of the samples was assayed on stomach strips in terms of PGE_2.

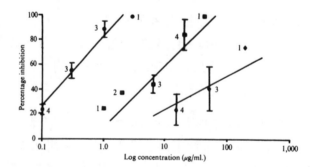

Fig. 1. Concentration (μg/ml.) of indomethacin (●), aspirin (■) and salicylate (♦) plotted on a log scale against the percentage inhibition of prostaglandin synthesis (assayed as $PGF_{2\alpha}$ on rat colons). The lines are those calculated for best fit. Numbers by the points indicate number of experiments. When three or more estimates were averaged, the standard error of the mean is shown.

零时间点的样品含有（每 ml）60 ng ~ 150 ng PGF$_{2\alpha}$ 样活性和 120 ng ~ 750 ng PGE$_2$ 样活性。同批次酶制剂不同样品之间的活性差异小于 5%；将酶在无花生四烯酸的条件下孵育 30 min 后，活性并没有增加。在有花生四烯酸的条件下孵育 30 min 后，根据酶制剂的不同情况 PGF$_{2\alpha}$ 样活性增加了 220 ng/ml ~ 520 ng/ml，PGE$_2$ 样活性增加了 100 ng/ml ~ 500 ng/ml。同批次酶制剂的不同对照样本之间的差异小于 7%。

抑制作用的检测

实验结果以产生的 PGF$_{2\alpha}$ 样活性或 PGE$_2$ 样活性（30 min 时间点的样品活性减去零时间点的样品活性）来表示。为了检测对前列腺素合成的抑制作用，我们将不同剂量的消炎痛、乙酰水杨酸钠、水杨酸钠或其他物质分别加入孵育瓶中，体积不超过 0.1 ml。用相对于对照物的抑制百分率来表示药物的抑制作用。

消炎痛、阿司匹林钠和水杨酸钠都能抑制 PGF$_{2\alpha}$ 样活性的产生。不同批次酶制剂被抑制的程度不一样，但是在每一批实验中，抑制百分率与消炎痛或阿司匹林浓度的对数都呈线性关系。用于溶解消炎物质的溶剂不会影响前列腺素的合成。全部实验结果如图 1 所示。消炎痛的 ID$_{50}$（译者注：即抑制率达 50% 所需的药液浓度）为 0.27 μg/ml（0.75 μM），而阿司匹林的 ID$_{50}$ 为 6.3 μg/ml（35 μM）。因此，在重量相同的条件下，消炎痛抑制 PGF$_{2\alpha}$ 合成的效果是阿司匹林的 23 倍；而在摩尔数相同的条件下，消炎痛的抑制作用则为阿司匹林的 47 倍。水杨酸钠对 PGF$_{2\alpha}$ 样产物合成的抑制作用比阿司匹林弱，而且实验结果之间的差异也要大很多（图 1）。各个样品的 PGE$_2$ 胃条分析也给出了类似的结果。

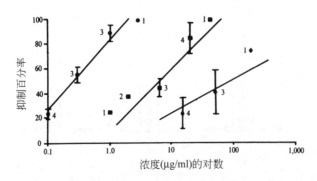

图 1. 消炎痛（●）、阿司匹林（■）和水杨酸盐（◆）的浓度（μg/ml）对数相对于前列腺素合成抑制百分率（基于用大鼠结肠对 PGF$_{2\alpha}$ 的活性分析）的曲线图。直线是通过计算得到的最佳拟合结果。标注的数字代表实验次数。当在三个或更多的估计值中取平均时，图中给出了平均标准误。

Hydrocortisone (50 µg/ml.) inhibited synthesis of $PGF_{2\alpha}$ and PGE_2-like activity by less than 20%; morphine (50 µg/ml.) or mepyramine (50 µg/ml.) had no effect. None of the drugs tested decreased the contractions of the assay tissues induced by PGE_2 or $PGF_{2\alpha}$; indeed there was sometimes a small potentiation of the responses.

Two samples (1.0 ml.) of enzyme were incubated without arachidonic acid: 1 µg $PGF_{2\alpha}$ was added to one and 1 µg PGE_2 to the other. After 30 min of incubation, the activity remaining was equivalent to 0.85 µg $PGF_{2\alpha}$ and 0.6 µg PGE_2.

In experiments with two different lung homogenates, samples containing arachidonic acid (10 µg/ml.) were incubated, extracted and the activity was separated by thin-layer chromatography in the A I system as described. In the first experiment, the zones corresponding to the prostaglandin markers showed substantial activity in the control 30 min incubation sample (160 ng $PGF_{2\alpha}$ and 50 ng PGE_2/ml. original sample), whereas the zero time sample showed much less (40 ng $PGF_{2\alpha}$ and 5 ng PGE_2/ml.). Similar samples incubated with indomethacin (5 µg/ml.) or aspirin (40 µg/ml.) showed little or no increase in $PGF_{2\alpha}$ or PGE_2-like activity over the zero-time sample.

In the second experiment, lower concentrations of indomethacin (1 µg/ml.) and aspirin (20 µg/ml.) were used. As Fig. 2 shows, indomethacin reduced the generation of activity in the $PGF_{2\alpha}$ zone to 25% of that in the control incubation but the activity in the PGE_2 zone was only reduced to 78%. With aspirin (20 µg/ml.) activity in the $PGF_{2\alpha}$ zone was reduced to about 56% and in the PGE_2 zone to 50%. Further identification of the PG-like activity in these experiments was considered unnecessary, for the enzyme system was the same as that used by Ånggård and Samuelsson[13], who identified PGE_2 and $PGF_{2\alpha}$ as the active products of the incubation.

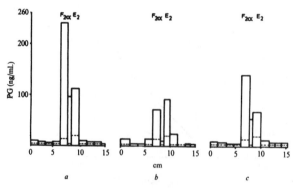

Fig. 2. Prostaglandin-like activity in samples of lung homogenate incubated for 30 min with arachidonic acid (10 µg/ml.), extracted with ethyl acetate and separated by thin-layer chromatography in the A I system. The chromatogram was divided into strips and the zone corresponding to the $PGF_{2\alpha}$ marker was assayed on rat colons in terms of $PGF_{2\alpha}$. The other zones were assayed on stomach strips in terms of PGE_2. The dotted lines represent the amount of PG-like activity in the sample at the start of the incubation. Indomethacin (1 µg/ml.) and aspirin (20 µg/ml.) reduced the generation of $PGF_{2\alpha}$ and PGE_2. a, Control; b, indomethacin; c, aspirin.

氢化可的松（50 μg/ml）对合成 $PGF_{2\alpha}$ 和 PGE_2 样产物的抑制仅能使其活性减少不到 20%；吗啡（50 μg/ml）或美吡拉敏（50 μg/ml）没有抑制作用。这些被检测的药物都不能降低由 PGE_2 和 $PGF_{2\alpha}$ 诱导的组织收缩，有时收缩反而会轻微增强。

两份酶样品（1.0 ml），一份加入 1 μg $PGF_{2\alpha}$，另一份加入 1 μg PGE_2，然后在无花生四烯酸的条件下孵育。孵育 30 min 后，$PGF_{2\alpha}$ 和 PGE_2 的剩余活性分别相当于 0.85 μg 的 $PGF_{2\alpha}$ 和 0.6 μg 的 PGE_2。

在两个不同的肺组织匀浆实验中，样品在有花生四烯酸（10 μg/ml）的条件下孵育，然后萃取，并用前面提到过的 A I 系统薄层层析法分离活性物质。在第一个实验中，孵育了 30 min 的对照样品在与前列腺素标准物相对应的区域中（原始样品：160 ng $PGF_{2\alpha}$/ml，50 ng PGE_2/ml）显示出很强的活性，而零时间点样品的活性则要小很多（40 ng $PGF_{2\alpha}$/ml，5 ng PGE_2/ml）。同样的样品与消炎痛（5 μg/ml）或阿司匹林（40 μg/ml）共孵育后，$PGF_{2\alpha}$ 和 PGE_2 样活性与零时间点样品的活性相比增加很少或者没有增加。

在第二个实验中，我们采用了更低浓度的消炎痛（1 μg/ml）和阿司匹林（20 μg/ml）。如图 2 所示，跟对照的孵育过程相比，消炎痛使 $PGF_{2\alpha}$ 区域的活性减少到 25%，但 PGE_2 区的活性却只下降至 78%。阿司匹林（20 μg/ml）使 $PGF_{2\alpha}$ 区的活性降低到约 56%，使 PGE_2 区的活性降低至 50%。没有必要对这些实验中的 PG 样活性进行进一步的鉴定，因为所用的酶系统与安加德和萨穆埃尔松 [13] 的酶系统没有区别，而安加德和萨穆埃尔松已经确认这种酶系统的孵育活性产物就是 PGE_2 和 $PGF_{2\alpha}$。

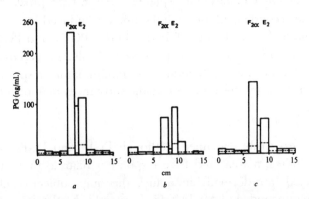

图 2. 经花生四烯酸（10 μg/ml）孵育 30 min、乙酸乙酯萃取和 A I 系统薄层层析分离的肺匀浆样品中的前列腺素样活性。层析图被划分为不同的条带，用大鼠结肠检测与 $PGF_{2\alpha}$ 标准物相对应的区域的 $PGF_{2\alpha}$ 活性。用胃条检测其他区域的 PGE_2 活性。虚线表示样品在孵育开始时的 PG 样活性。消炎痛（1 μg/ml）和阿司匹林（20 μg/ml）抑制了 $PGF_{2\alpha}$ 和 PGE_2 的合成。a，对照；b，消炎痛；c，阿司匹林。

The results show that the three anti-inflammatory acids tested inhibit the synthesis of prostaglandins. It is not yet known how the inhibition is brought about. If it is by competition with arachidonic acid for the active site of the enzyme, this might explain why all of these anti-inflammatory substances contain an acidic group. It would also explain why hydrocortisone, an anti-inflammatory substance of a different type, has little or no inhibitory action against the prostaglandin synthesizing enzyme(s).

Correlation with Therapeutic Actions

Anti-inflammatory acids have three principal actions; antipyretic, anti-inflammatory and analgesic. They also antagonize bronchoconstriction and some other smooth muscle contractions induced by substances such as bradykinin[1] and have a tendency to induce gastro-intestinal irritation. Can any of these actions be explained by a direct inhibition of prostaglandin synthesis?

First, we consider antipyretic action. When injected into the third ventricle of cats, PGE_1 induces fever[22,23]. It is the most potent substance yet found with this action, being much more active than 5-hydroxytryptamine, PGE_2 or $PGF_{2\alpha}$; $PGF_{2\alpha}$ has little pyretic activity. The antipyretic substance 4-acetamidophenol does not antagonize fever induced by PGE_1. These facts are compatible with the idea that the rise in temperature in a fever is induced by synthesis and release of RCS or of a known prostaglandin such as PGE_1, either in the temperature-regulating area of the hypothalamus or at a place from which it can reach this area. Anti-inflammatory substances might reduce temperature in a fever by preventing such prostaglandin synthesis. A necessary corollary of such a mechanism of action is that a prostaglandin whose synthesis is inhibited by these drugs does not regulate normal body temperature, which is unaffected by antipyretic drugs.

Anti-inflammatory action can also be explained. An E-type prostaglandin is found in exudate[24] during the secondary phase of inflammation induced by carageenin in the rat. Prostaglandins, identified as a mixture of PGE_1, PGE_2, $PGF_{1\alpha}$ and $PGF_{2\alpha}$, have also been isolated from fluid perfusing the skin of patients with allergic eczema[25]. In rat or man, PGE_1 or PGE_2 injected intradermally induces an inflammatory response[26]. Thus abolition of prostaglandin synthesis by this group of acidic substances may be a basis for their anti-inflammatory action.

Analgesic action is less easily explained. Although PGE_1 and PGE_2 and larger doses of $PGF_{1\alpha}$ and $PGF_{2\alpha}$ induce a weal and flare response similar to that caused by histamine release when injected intradermally in man[26], the only subjective effect reported was a sensation of warmth and a slight itching. It is unlikely, therefore, that any of these prostaglandins mediate skin pain. Thus, unless another untested prostaglandin or RCS is involved, there seems no link between a peripheral analgesic action of the anti-inflammatory acids and inhibition of prostaglandin synthesis.

Prostaglandin infusions, however, induce headache[27], so the relief by aspirin-like drugs

结果表明：这三种消炎酸都能抑制前列腺素的合成。但目前尚不清楚这种抑制作用是如何产生的。如果作用机制是通过与花生四烯酸竞争酶的活性位点，那么就有可能解释为什么这些消炎物质都含有一个酸性基团，并且还可以解释为什么另一类消炎物质——氢化可的松就几乎或者完全不具有竞争前列腺素合成酶的能力。

与治疗作用的关系

消炎酸有三种基本功能：退热、消炎和止痛。它们也可以抑制支气管收缩以及其他一些由血管舒缓激肽[1]等物质诱导的平滑肌收缩，并有可能刺激胃肠。这些效应是否都能用直接抑制前列腺素的合成来解释呢？

首先，我们来看退热作用。当 PGE_1 被注射到猫的第三脑室时，会引起发热[22,23]。在人们已知的所有物质中，PGE_1 是具有这种功效的最有效的物质，它比 5-羟色胺、PGE_2 和 $PGF_{2\alpha}$ 的活性高出很多；$PGF_{2\alpha}$ 基本上没有引起发热的功效。而可以退热的物质 4-乙酰氨基酚不能缓解由 PGE_1 所引起的发热。这些事实与以下观点一致，即认为在下丘脑的温度调节区域或者在其他可以通向该区的地方存在着 RCS 或某种已知前列腺素，如 PGE_1 等，的合成和释放过程，发烧时的体温升高就是由这种合成和释放过程引起的。消炎类物质通过抑制前列腺素的合成或许可以起到退烧的作用。由这种作用机制必然可以推出：其合成可被这些药物抑制的前列腺素不参与调节身体的正常体温，因为正常体温不受退热药物的影响。

消炎作用也能得到解释。在用角叉菜胶诱导大鼠发炎的第二阶段，我们在分泌物[24]中发现了一种 E 型前列腺素。分离过敏性湿疹患者皮肤中的化脓液体，我们得到了含有 PGE_1、PGE_2、$PGF_{1\alpha}$ 和 $PGF_{2\alpha}$ 的前列腺素混合物[25]。皮内注射 PGE_1 或 PGE_2 通常会引起大鼠或人类的炎症反应[26]。因此，这组酸性物质对前列腺素合成的抑制可能就是它们能起到抗炎作用的基础。

止痛作用不太容易得到解释。虽然由 PGE_1 和 PGE_2 以及较大剂量的 $PGF_{1\alpha}$ 和 $PGF_{2\alpha}$ 可以诱导类似于给人皮内注射组胺释放剂时由组胺释放引发的风团和潮红反应[26]，但被报道的唯一主观反应仅仅是感到温暖和轻微的瘙痒。由此可见，这些前列腺素能传递皮肤疼痛的可能性都不太大。所以除非还有另外一种未检测出来的前列腺素或 RCS，否则很难在消炎酸的外周止痛作用与前列腺素的合成被抑制之间建立联系。

注射前列腺素会诱发头痛[27]，因此阿司匹林样药物对此类疼痛（或因发炎引起

of such pain (or of pain induced by inflammation) may be explained by an inhibition of prostaglandin synthesis. Collier[28] develops further the possible link between prostaglandin production and pain.

The anti-bronchoconstrictor action may also be due to inhibition of the synthetic pathway for prostaglandins. We have already shown that challenge of sensitized guinea-pig lungs or injection of bradykinin or partially purified SRS-A into unsensitized lungs induces the release of RCS, PGE_2 and $PGF_{2\alpha}$ (refs. 10 and 11). Because the release of RCS was abolished by anti-inflammatory acids, as was the bronchoconstriction induced by bradykinin and SRS-A (ref. 1), we postulated that RCS may be the mediator of the bronchoconstrictor response. Vargaftig and Dao[12] have shown that arachidonic acid also releases RCS and that the release is inhibited by aspirin-like drugs and we have confirmed this (unpublished work). At the time of our first publication on RCS (ref. 10), we were unsure of the contribution that this substance made to the contractions of the tissues used to assay simultaneously the release of prostaglandins. We have assessed this contribution recently; whereas RCS contracts rat stomach strip, it has much less effect on chick rectum or rat colon (Piper and Vane, unpublished work). Re-examination of the tracings from these experiments shows that aspirin-like drugs not only prevented RCS release but also reduced activity on chick rectum and rat colon, indicative of a reduced output of PGE_2 and $PGF_{2\alpha}$ (see, for instance, Fig. 7 in Piper and Vane[10]).

Prostaglandin $F_{2\alpha}$ is bronchoconstrictor[5,29], so the antagonism of bronchoconstriction induced by bradykinin, SRS-A, arachidonic acid and so on may be due to antagonism of release (which probably means synthesis) of RCS or of synthesis and release of $PGF_{2\alpha}$ or both. The ratio of activities against bradykinin-induced bronchoconstriction (indomethacin 2; aspirin 1, salicylic acid 0.03; ref. 1) certainly fits with the relative lack of activity of sodium salicylate against the formation of $PGF_{2\alpha}$. Until RCS can be stabilized or generated in a pure form, however, its contribution to the process of anaphylaxis and to bronchoconstriction induced by anaphylactic mediators cannot be assessed.

Side Effects

The aspirin-like drugs all induce gastro-intestinal symptoms which may include peptic ulceration[30]. Prostaglandin synthesis and release can be provoked by many different forms of mechanical stimulation, including gentle massage[13]. Contractions of the gastro-intestinal tract churn the contents. It is possible, therefore, that the associated mechanical stimulation of the mucosa leads to synthesis intramurally of a prostaglandin which in some way protects the mucosa from damage. Inhibition of prostaglandin synthesis by aspirin-like drugs would remove this protective mechanism.

Whether inhibition of prostaglandin synthesis accounts for all the activities of the anti-inflammatory acids remains to be elucidated. Clearly, the blood concentrations[31] of indomethacin in man after an oral dose of 200 mg (7.5 µg/ml. at peak; 3.2 µg/ml. after 4 h), even when 90% binding[31] to plasma proteins is allowed for, are higher than concentrations needed to inhibit prostaglandin synthesis in these experiments.

916

的疼痛）的缓解也许可以用抑制前列腺素的合成来解释。科利尔[28]对前列腺素合成与疼痛之间的可能联系进行了进一步的阐述。

抗支气管收缩作用可能也是由前列腺素的合成通路被阻抑引起的。我们已经证明，刺激致敏豚鼠肺或者给脱敏的豚鼠肺注射血管舒缓激肽或部分纯化的 SRS-A 会诱发 RCS、PGE_2 和 $PGF_{2\alpha}$ 的释放(参考文献 10 和 11)。由于消炎酸能阻止 RCS 的释放，也能阻止由血管舒缓激肽和 SRS-A 诱导的支气管收缩（参考文献 1），因此我们推测 RCS 有可能就是支气管收缩反应的传递者。瓦尔加夫季格和达奥[12]曾指出：花生四烯酸也能释放出 RCS，而阿司匹林样药物会阻止 RCS 的释放。并且我们的实验也证实了这一点(尚未发表的研究结果)。在第一次发表有关 RCS 的论文(参考文献 10)时，我们还不能确定这种物质对能用于同步检测前列腺素释放量的组织的收缩有什么贡献。最近我们对这方面的贡献进行了分析，发现 RCS 虽然能使大鼠胃条收缩，但是对小鸡直肠或大鼠结肠的收缩作用非常有限（派珀和文，尚未发表的研究结果）。再次检查这些实验中的细节，我们发现阿司匹林样药物不仅阻止了 RCS 的释放，而且降低了小鸡直肠和大鼠结肠的活性，这说明 PGE_2 和 $PGF_{2\alpha}$ 的排出量下降了（比如，见派珀和文[10]文章中的图 7）。

前列腺素 $F_{2\alpha}$ 是支气管收缩剂[5,29]，因此阿司匹林样药物对由血管舒缓激肽、SRS-A 和花生四烯酸等诱导产生的支气管收缩的抑制可能是因为其抑制了 RCS 的释放（很可能就是合成），或者抑制了 $PGF_{2\alpha}$ 的合成和释放，或者兼而有之。几种药物对由血管舒缓激肽诱导产生的支气管收缩的抑制活性之比（消炎痛为 2；阿司匹林为 1；水杨酸为 0.03；参考文献 1）显然与水杨酸钠相对缺乏抑制 $PGF_{2\alpha}$ 合成的活性相吻合。除非 RCS 可以稳定存在或以纯净形式产生，否则很难评估它在由过敏性物质引起的过敏和支气管收缩过程中的作用。

副 作 用

所有阿司匹林样药物都会引起包括消化性溃疡在内的胃肠道症状[30]。有很多种不同形式的机械刺激，如轻轻地按摩[13]，可以激发前列腺素的合成和释放。胃肠道的收缩使其内容物受到搅动。因此，与黏膜相关联的机械刺激可能会导致壁内前列腺素的合成，从而在某种程度上保护了黏膜使其免受伤害。而阿司匹林样药物对前列腺素合成的抑制会使这种保护机制失效。

用抑制前列腺素的合成能否解释消炎酸的所有功效还有待于进一步阐明。显然，在成人口服 200 mg 消炎痛之后，即使将消炎痛与血浆蛋白的结合率[31]放宽至 90%，人体血液中的消炎痛浓度[31]（峰值为 7.5 μg/ml；4 h 后为 3.2 μg/ml）依然高于实验中抑制前列腺素合成所需的浓度。

One fact that needs explanation is the apparent lack of activity of sodium salicylate as an inhibitor of prostaglandin synthesis, for this substance has about the same potency as aspirin in antipyretic and anti-inflammatory tests[1]. One possibility is that salicylate is more potent as an inhibitor of synthesis of PGE_1 from dihomo-γ-linolenic acid than it is of the synthesis of PGE_2 from arachidonic acid. Certainly, with the probability of a series of iso-enzymes synthesizing from different substrates various prostaglandins (perhaps including RCS) with widely different pharmacological properties, the number of degrees of freedom is more than sufficient to allow explanation of the variations in potencies and properties within the whole group of aspirin-like substances.

There are several other implications of these results, some of which are listed below. First, inhibition of prostaglandin synthesis, perhaps using a more active enzyme preparation[32] than the one used here, together with different substrates, may provide a simple and rapid primary screen for anti-inflammatory drugs of the indomethacin type. Second, one of the few simple *in vitro* antagonisms shown by aspirin-like drugs is against arachidonic acid-induced contractions of some isolated smooth muscle preparations such as the guinea-pig ileum[7]. This suggests that some of the contractile actions of arachidonic acid may be brought about by an intramural synthesis of a prostaglandin. This problem deserves attention, as does the possibility that bioassay of prostaglandins may be improved by addition of a synthesis inhibitor, such as indomethacin, to the fluid bathing the assay tissues.

There is also the possibility that use of anti-inflammatory acids could be extended to ameliorate conditions thought to be brought about by prostaglandin release. For example, some evidence[33] suggests that release of $PGF_{2\alpha}$ occurs during labour. It may therefore be worthwhile testing an anti-inflammatory acid as an inhibitor of unwanted abortion or miscarriage. It would also be interesting to know whether these drugs reduce the efficacy of the intra-uterine device which may work as a contraceptive through prostaglandin release[34].

These results show that biologists now have a simple means of preventing prostaglandin synthesis and release and thereby assessing the functions of prostaglandins in individual cells or tissues, or in the body as a whole. The conclusions described here are supported by the results discussed in the next two articles[35,36].

I thank Mr. N. Pitman for technical help and the Wellcome Trust and Medical Research Council for grants.

(*Nature New Biology*, **231**, 232-235; 1971)

J. R. Vane: Department of Pharmacology, Institute of Basic Medical Sciences, Royal College of Surgeons of England, Lincoln's Inn Fields, London WC2A 3PN.

Received May 6, 1971.

需要解释的一个事实是：虽然水杨酸钠所具有的退热和消炎功效 [1] 与阿司匹林相差无几，但它对前列腺素合成的抑制作用非常有限。一种可能性是，水杨酸盐对合成二高 -γ- 亚麻酸来源的 PGE_1 的抑制作用要比对合成花生四烯酸来源的 PGE_2 的抑制作用更强烈。当然，由于一系列同工酶在不同底物上有可能会合成药理学性质迥然不同的前列腺素（可能包括 RCS），因此人们在解释所有阿司匹林样物质的不同药效和性能时有充分的自由度。

这些结果还隐含着其他几种可能，以下列出了其中一部分。首先，或许可以将对前列腺素合成的抑制作为一种初步筛选消炎痛类消炎药的简单而快捷的方法，这也许需要采用一种比现在所用的酶制剂活性更高的酶制剂 [32] 并配合不同的底物。第二，阿司匹林样药物在体外表现出的其中一种拮抗作用就是拮抗由花生四烯酸诱导的某些离体平滑肌样品的收缩，比如豚鼠回肠等 [7]。这表明一些花生四烯酸诱导的收缩反应可能源自内壁中某种前列腺素的合成过程。这一问题值得重视。另外，也许可以通过在待测组织的液体浴中加入一种合成抑制剂（例如消炎痛）来提高前列腺素的生物鉴定水平，这种可能性也值得引起注意。

另一种可能是：消炎酸还可以用来缓解由前列腺素释放引起的症状。例如，一些证据 [33] 表明在分娩过程中会释放出 $PGF_{2\alpha}$。因此，检测消炎酸对意外流产或早产的抑制作用或许是一件有意义的事情。而且了解这类药物是否会降低依靠前列腺素的释放 [34] 而起避孕作用的宫内节育器的有效性也很有意义。

这些研究结果表明：生物学家们现在已经掌握了一种能抑制前列腺素合成和释放的简单方法，因而能够分析前列腺素在单个细胞、组织或者整个机体中的功能。后面两篇文章 [35,36] 所讨论的结果支持了本文中的这些结论。

感谢皮特曼先生在技术上给予我的支持，以及维康基金会和医学研究理事会在研究经费上给予的资助。

（彭丽霞 翻译；孙军 莫楹 审稿）

References:

1. Collier, H. O. J., *Adv. Pharmacol. Chemother.*, 7, 333 (1969).

2. Collier, H. O. J., *Nature*, 223, 35 (1969).

3. Collier, H. O. J., and Shorley, P. G., *Brit. J. Pharmacol. Chemother.*, 15, 601 (1960).

4. Collier, H. O. J., and Shorley, P. G., *Brit. J. Pharmacol. Chemother.*, 20, 345 (1963).

5. Berry, P. A., and Collier, H. O. J., *Brit. J. Pharmacol. Chemother.*, 23, 201 (1964).

6. Collier, H. O. J., James, G. W. L., and Schneider, C., *Nature*, 212, 411 (1966).

7. Jacques, R., *Helv. Physiol. Pharmacol. Acta*, 23, 156 (1965).

8. Berry, P. A., thesis, Council for National Academic Awards, London, 91 (1966).

9. Collier, H. O. J., and Sweatman, J. F., *Nature*, 219, 864 (1968).

10. Piper, P. J., and Vane, J. R., *Nature*, 223, 29 (1969).

11. Piper, P. J., and Vane, J. R., in *Prostaglandins, Peptides and Amines* (edit. by Mantegazza, P., and Horton, E. W.), 15 (Academic Press, London, 1969).

12. Vargaftig, B. B., and Dao, N., *Pharmacology* (in the press).

13. Piper, P. J., and Vane, J. R., in *Prostaglandins* (edit. by Ramwell, P., and Shaw, J.) (NY Acad. Sci., in the press, 1971).

14. Ånggård, E., and Samuelsson, B., *J. Biol. Chem.*, 240, 3518 (1965).

15. Ferreira, S. H., and Vane, J. R., *Nature*, 216, 868 (1967).

16. Vane, J. R., *Brit. J. Pharmacol. Chemother.*, 12, 344 (1957).

17. Regoli, D., and Vane, J. R., *Brit. J. Pharmacol. Chemother.*, 23, 351 (1964).

18. Gaddum, J. H., *Brit. J. Pharmacol. Chemother.*, 8, 321 (1953).

19. Gilmore, N., Vane, J. R., and Wyllie, J. H., *Nature*, 218, 1135 (1968).

20. Gréen, K., and Samuelsson, B., *J. Lipid Res.*, 5, 117 (1964).

21. Willis, A. L., *Brit. J. Pharmacol.*, 40, 583P (1970).

22. Milton, A. S., and Wendlandt, S. J., *J. Physiol.*, 207, 76P (1970).

23. Feldberg, W., and Saxena, P. N., *J. Physiol.* (in the press, 1971).

24. Willis, A. L., in *Prostaglandins, Peptides and Amines* (edit. by Mantegazza, P., and Horton, E. W.), 31 (Academic Press, London, 1969).

25. Greaves, M. W., Søndergaard, J., and McDonald-Gibson, W., *Brit. Med. J.*, 2, 258 (1971).

26. Crunkhorn, P., and Willis, A. L., *Brit. J. Pharmacol.*, 41, 49 (1971).

27. Bergström, S., Carlson, L. A., and Weeks, J. R., *Pharmacol. Rev.*, 20, 1 (1968).

28. Collier, H. O. J., *Nature* (in the press).

29. James, G. W. L., *J. Pharm. Pharmacol.*, 21, 379 (1969).

30. Goodman, L. S., and Gilman, A., *The Pharmacological Basis of Therapeutics*, fourth ed. (Collier-Macmillan, London, 1970).

31. Hucker, H. B., Zacchei, A. G., Cox, S. V., Brodie, D. A., and Cantwell, N. H. R., *J. Pharmac. Exp. Therap.*, 153, 237 (1966).

32. Nugteren, D. H., Beerhuis, R. K., and Van Dorp, D. A., *Recl. Trav. Chim. Pays-Bas Belg.*, 85, 405 (1966).

33. Karim, S. M. M., *Brit. Med. J.*, 4, 618 (1968).

34. Chaudhuri, G., *Lancet*, i, 480 (1971).

35. Smith, J. B., and Willis, A. L., *Nature New Biology*, 231, 235 (1971).

36. Ferreira, S. H., Moncada, S., and Vane, J. R., *Nature New Biology*, 231, 237 (1971).

Aspirin Selectively Inhibits Prostaglandin Production in Human Platelets

J. B. Smith and A. L. Willis

Editor's Note

In this, one of a trio of papers published back-to-back, pharmacologists J. B. Smith and A. L. Willis suggest a mechanism of action for aspirin. They show that platelets in the blood of volunteers who have taken aspirin can no longer produce prostaglandins, complex chemicals thought to regulate inflammation. Similar effects are shown for two other anti-inflammatory drugs, sodium salicylate and indomethacin, leading them to suggest that the anti-inflammatory effects of all three drugs occur by inhibition of prostaglandin production.

ASPIRIN reduces the adhesiveness to glass of platelets in citrated plasma[1], reduces platelet aggregation by washed connective tissue fragments[2], and inhibits the second wave of aggregation induced by ADP, adrenaline and thrombin[3-5]. Aspirin also inhibits the release from washed pig or human platelets of permeability factors which differ from 5-hydroxytryptamine and histamine and cause contraction of the guinea-pig ileum[6]. One of these factors could be prostaglandin E_2 (PGE_2). This compound increases vascular permeability[7,8] and contracts guinea-pig ileum[9]. When washed human platelets are incubated with thrombin it is formed and appears extracellularly, together with prostaglandin $F_{2\alpha}$ ($PGF_{2\alpha}$)[10].

The following experiments, which were initiated independently of those described in the accompanying two articles[11,12], were designed to test whether aspirin and other anti-inflammatory drugs inhibit the production of prostaglandins, which may be important mediators of inflammation.

The effects of aspirin and other drugs were investigated on the production of prostaglandins and "the release reaction" induced by thrombin, that is, release of 5-hydroxytryptamine, adenine nucleotide and lysosomal enzymes[13]. We looked for the release of a lysosomal phospholipase A (ref. 14) because the production of PGE_2 and $PGF_{2\alpha}$ in tissues is thought to be brought about by cyclization of arachidonic acid liberated by the action of phospholipase A (ref. 15) and the incorporation of molecular oxygen[16,17].

Platelet-rich plasma was obtained from healthy donors, who had not taken aspirin for some days previously, by centrifuging blood containing 5.8 mM EDTA at 900g for 15 min at 18°C–20°C. For *in vitro* experiments the plasma was incubated with 0.5 μM 3^1-^{14}C-5-hydroxytryptamine creatinine sulphate (58 Ci/mol) for 1 h at 18°C–20°C, thereby incorporating radioactivity into the platelets[18]. The platelets were washed[19] once and

阿司匹林选择性抑制人血小板中前列腺素的合成

史密斯，威利斯

编者按

这篇文章是三篇连续发表的相关论文中的一篇，作者是药理学家史密斯和威利斯，他们在本文中提出了阿司匹林的一种作用机制。他们指出，在服用阿司匹林的志愿者的血小板中不再能合成前列腺素，前列腺素被认为是可调节炎症反应的复杂化学物质。其他两种消炎药——水杨酸钠和消炎痛也能产生类似的效应。这使他们提出上述三种药物的消炎作用都源于对前列腺素合成的抑制。

阿司匹林能降低柠檬酸钠血浆中血小板对玻璃的黏附性 [1]，减少由清洗过的结缔组织碎片造成的血小板聚集 [2]，并能抑制由二磷酸腺苷、肾上腺素和凝血酶诱导的血小板再次聚集 [3-5]。阿司匹林也能抑制一些与 5-羟色胺和组胺不同但能引起豚鼠回肠收缩的渗透性因子从清洗过的猪血小板或人血小板中释放出来 [6]。前列腺素 E_2（PGE_2）可能是其中的一个因子。这种化合物能增加血管的渗透性 [7,8]，并能使豚鼠的回肠收缩 [9]。当清洗过的人血小板与凝血酶共同孵育时，形成的前列腺素 E_2 和前列腺素 $F_{2\alpha}$（$PGF_{2\alpha}$）出现在了细胞外 [10]。

前列腺素有可能是炎症的重要介质，为了验证阿司匹林和其他消炎药能否抑制前列腺素的合成，我们设计了如下几个实验，这些实验与前一篇和后一篇文章中 [11,12] 描述的实验是独立展开的。

我们研究了阿司匹林及其他药物对前列腺素合成和由凝血酶诱导的"释放反应"（即 5-羟色胺、腺嘌呤核苷酸和溶酶体酶的释放 [13]）的作用效果。一般认为，组织中的 PGE_2 和 $PGF_{2\alpha}$ 是在磷脂酶 A 作用下（参考文献 15）释放的花生四烯酸发生环化并结合分子氧而形成的 [16,17]，因此我们检测了溶酶体磷脂酶 A 的释放情况（参考文献 14）。

取数天内未服用过阿司匹林的健康供者的血液，在 18℃~20℃、相对离心力为 900g 下将含 5.8 mM 乙二胺四乙酸的这些血液离心 15 min，即得到富含血小板的血浆。在体外实验中，将血浆和 0.5 μM 3^1-^{14}C-5-羟色胺肌酐硫酸盐（58 Ci/mol）于 18℃~20℃下孵育 1 h 从而使放射性掺入到血小板中 [18]。对血小板进行一次清洗 [19]，然后使其重新悬浮于缓冲盐溶液（134 mM NaCl；5 mM D-葡萄糖；15 mM 三羟甲

resuspended in a buffered saline solution (134 mM NaCl; 5 mM D-glucose; 15 mM Tris-HCl, pH 7.4). Portions (6 ml.) of this suspension, containing $5–20 \times 10^8$ platelets/ml., were shaken at 37°C for 5 min with 0.06 ml. thrombin (bovine thrombin, Leo Laboratories, Hayes, Middlesex; 500 NIH units/ml. buffered saline), 2 min after the addition of solutions of various drugs (0.001–0.120 ml.). These samples were then cooled to 0°C and centrifuged at $2,400g$ for 10 min. A portion (4.5 ml.) of the supernatant was adjusted to pH 2.5–3.0, extracted for prostaglandins with ethyl acetate and, after drying under reduced pressure, the prostaglandins were bioassayed as described before[20]. The rest of the supernatant was assayed for its content of ^{14}C-5-hydroxytrypt-amine[18], β-N-acetyl glucosamidase[21] and phospholipase A_1 (ref. 22) (pH optimum 4.0).

Aspirin considerably reduced the production of prostaglandins by platelets although it had no effect on the release of 5-hydroxytryptamine, β-N-acetyl glucosamidase or phospholipase A_1 (Table 1). The dose of aspirin causing a 50% inhibition of the production of prostaglandins was 0.3 μg/ml. (1.7 μM). In two experiments indomethacin was about ten times more potent than aspirin, whereas sodium salicylate was about ten times less potent. Hydrocortisone (100 μg/ml.) had no effect. Aspirin (10 and 100 μg/ml.) did not interfere with the extraction and assay of prostaglandins nor did it increase the small amounts of prostaglandin which remained in the platelet pellet[10].

Table 1. Effects of Aspirin Added to Resuspended Platelets on the Thrombin-induced Formation of Prostaglandins and on "the Release Reaction"

Aspirin addition and concentration (μg/ml.)	Second addition and concentration	^{14}C-5-HT release %	Activity/ml. supernate		
			β-N-Acetyl glucosamidase (mU)	Phospholipase A_1 (mU)	PG* (ng)
—	—	4.7	0.44	0.13	ND
—	PGE$_2$ 20 ng/ml.	4.9	0.37	0.06	15
—	Thrombin 5 U/ml.	77.7	2.53	1.12	53
0.01	Thrombin 5 U/ml.	74.1	2.55	0.91	48
0.10	Thrombin 5 U/ml.	75.5	2.55	0.98	35
1.00	Thrombin 5 U/ml.	73.2	2.32	0.91	18
10.00	Thrombin 5 U/ml.	74.7	2.77	1.01	7

*Assayed as PGE$_2$ on the rat fundus strip.
ND, Not detectable (less than 0.1 ng PGE$_2$/ml.).
There were 4.78×10^8 platelets/ml. of suspension and the platelets contained 3.2 mU β-N-acetyl glucosaminidase per 10^8 cells. Phospholipase A_1 was not assayed in platelets because endogenous substrate may effect the results (Silver, Smith and Webster, unpublished observations).

The effects of aspirin given to three volunteers were then examined. Blood was taken before and 1 h after taking two tablets (600 mg) of aspirin. Platelets were washed and re-suspended as soon as the second sample of platelet-rich plasma had been obtained. The two suspensions from each donor were incubated simultaneously with 5 U/ml. thrombin

基氨基甲烷盐酸盐，pH 值 7.4）中。将这个每 ml 含 5×10^8 个 ~ 20×10^8 个血小板的悬液分成若干份（每份 6 ml），与 0.06 ml 凝血酶（牛凝血酶，由米德尔塞克斯郡海斯的利奥实验室提供；每 ml 缓冲盐水中含 500 NIH 单位凝血酶〔译者注：1 NIH 单位相当于 7.5 国际单位 U，1 U 指在特定条件下每 min 催化 1 mmol 底物转化为产物的酶量〕）在 37℃ 下振荡 5 min，随后加入各种药液（0.001 ml ~ 0.120 ml）再振荡 2 min。然后将这些样品冷却到 0℃ 并在相对离心力 2,400 g 下离心 10 min。将一部分上清液(4.5 ml)的 pH 值调至 2.5 ~ 3.0，随即用乙酸乙酯抽提出前列腺素，在减压干燥后用以前报道过的方法 [20] 对前列腺素进行生物鉴定。用余下的上清液检测 ^{14}C–5– 羟色胺 [18]、β–N– 乙酰氨基葡萄糖苷酶 [21] 和磷脂酶 A$_1$ 的含量（参考文献 22）（最适 pH 值为 4.0）。

虽然阿司匹林对 5– 羟色胺、β–N– 乙酰氨基葡萄糖苷酶和磷脂酶 A$_1$ 的释放没有影响（表 1），但它可以使血小板合成的前列腺素明显减少。抑制 50% 前列腺素合成所需的阿司匹林的剂量是 0.3 μg/ml（1.7 μM）。在两次实验中，消炎痛的抑制作用都大约是阿司匹林的 10 倍，而阿司匹林的抑制作用又是水杨酸钠的 10 倍左右。氢化可的松（100 μg/ml）没有抑制作用。阿司匹林（10 μg/ml 和 100 μg/ml）对提取和检测前列腺素并无影响，也不会增加残留在血小板团中的少量前列腺素的量 [10]。

表 1. 加入到血小板悬液中的阿司匹林对由凝血酶诱导的前列腺素合成的影响和对"释放反应"的影响

是否添加阿司匹林及其添加浓度（μg/ml）	二次添加物及其添加浓度	^{14}C–5– 羟色胺肌酐硫酸盐的释放率 %	每 ml 上清液的活性		
			β–N– 乙酰氨基葡萄糖苷酶 (mU)	磷脂酶 A$_1$ (mU)	前列腺素 * (ng)
—	—	4.7	0.44	0.13	ND
—	PGE$_2$ 20 ng/ml	4.9	0.37	0.06	15
—	凝血酶 5 U/ml	77.7	2.53	1.12	53
0.01	凝血酶 5 U/ml	74.1	2.55	0.91	48
0.10	凝血酶 5 U/ml	75.5	2.55	0.98	35
1.00	凝血酶 5 U/ml	73.2	2.32	0.91	18
10.00	凝血酶 5 U/ml	74.7	2.77	1.01	7

* 检测的是用大鼠胃底条测出的 PGE$_2$。
ND 代表未检出（少于 0.1 ng PGE$_2$/ml）。
悬液中血小板的浓度为每 ml 4.78×10^8 个，血小板中每 10^8 个细胞包含 3.2 mU β–N– 乙酰氨基葡萄糖苷酶。
我们没有检测血小板中磷脂酶 A$_1$ 的含量，因为内源物质可能会影响结果（西尔弗、史密斯和韦伯斯特，尚未发表的观察结果）。

随后我们又检测了阿司匹林对三个志愿者所起的作用。分别在志愿者服用 2 片阿司匹林（600 mg）之前和之后 1 h 采血。在采集到第二个富含血小板的血浆样品后，立即清洗并重悬血小板样品。将每位供血者的两份血小板悬液在 37℃ 下与

for 5 min at 37°C; supernatants were prepared as described and portions were taken for the assay of prostaglandins, β-N-acetyl glucosamidase, β-glucuronidase, phospholipase A_1 and nucleotide[23].

Aspirin administration inhibited the production of prostaglandins in the platelet in a system in which "the release reaction" was unimpaired (Table 2). Indomethacin (50 mg) taken by two volunteers, and indomethacin (100 mg) taken by one volunteer, inhibited the production of prostaglandins by 51, 64 and 83% respectively without affecting the release of lysosomal enzymes. No change in the production of prostaglandins was found in platelets of one individual who took codeine (60 mg) or two individuals who took nothing.

Table 2. Comparison of the Supernatants of Thrombin Treated Platelet Suspensions from Three People before and after taking 600 mg Aspirin

Donor	R. S. (female)		B. S. (male)		J. W. (male)	
Aspirin	Before	After	Before	After	Before	After
Platelet count ($\times 10^8$/ml. platelet suspension)	13.8	15.3	11.5	11.4	9.64	9.94
Activity/ml. supernatant after thrombin treatment						
Nucleotide (n mol)	106	110	134	111	83	90
β-N-Acetyl glucosamidase (mU)	13.0	15.8	17.2	13.1	11.2	13.0
β-Glucuronidase (mU)	1.10	1.58	2.48	1.91	0.81	1.04
Phospholipase A_1 (mU)	0.99	1.22	3.57	2.97	0.99	1.49
Prostaglandin (ng)*	160	16	168	5	103	20

*Assayed as PGE_2 on the rat fundus strip and uncorrected for loss (estimated to be 20 to 30%) during extraction.

Aspirin inhibits "the release reaction" when induced by collagen or low concentrations of thrombin[24]. In our experiments relatively high concentrations of thrombin (5 U/ml.) were used and as a result aspirin did not inhibit the release of the platelet constituents. The production of prostaglandins is thus dissociated from "the release reaction". The finding that the production of prostaglandins is inhibited by aspirin, while the release of phospholipase A is unaffected, strongly suggests that one action of aspirin on platelets is inhibition of the conversion of arachidonic acid into prostaglandins. Other work from this laboratory (see preceding article[11]) has shown that aspirin inhibits the synthesis of prostaglandins from arachidonic acid in guinea-pig lung homogenates.

In these experiments, prostaglandin production in human platelets was inhibited by sodium salicylate, aspirin and indomethacin and this effect occurred after oral administration of the latter two drugs. Prostaglandins have been identified in exudate during the second phase of inflammation induced by carageenin in the rat[20] and in inflamed skin of patients with allergic contact eczema[25]. Low concentrations of PGE_1 or PGE_2 cause pronounced erythema when injected intradermally in man[5,26]. If the prostaglandins are indeed important mediators of inflammation[27], the clinical effectiveness of aspirin and indomethacin as anti-inflammatory agents could be explained by the inhibition of the production of prostaglandins.

5 U/ml 凝血酶共孵育 5 min。根据前述方法制备上清液，然后分成若干份，分别用于检测前列腺素、β–N– 乙酰氨基葡萄糖苷酶、β– 葡萄糖苷酸酶、磷脂酶 A_1 和核苷酸 [23]。

服用阿司匹林会抑制一个系统内血小板中前列腺素的合成，但并不影响血小板的"释放反应"（表 2）。对于两位服用 50 mg 消炎痛的志愿者和一位服用 100 mg 消炎痛的志愿者，其前列腺素合成的抑制率分别为 51%、64% 和 83%，但都没有影响到溶酶体酶的释放。在一位服用过可待因（60 mg）的志愿者和两位没有服用过任何药物的志愿者的血小板中，未发现前列腺素的合成有什么变化。

表 2. 三个志愿者服用 600 mg 阿司匹林之前和之后，其血小板经凝血酶处理后悬液的上清液的结果对比

供血者	R.S.（女性）		B.S.（男性）		J.W.（男性）	
阿司匹林	服用前	服用后	服用前	服用后	服用前	服用后
血小板计数（×10⁸/ml 血小板悬液）	13.8	15.3	11.5	11.4	9.64	9.94
每 ml 经凝血酶处理后的上清液的活性						
核苷酸（n mol）	106	110	134	111	83	90
β–N– 乙酰氨基葡萄糖苷酶（mU）	13.0	15.8	17.2	13.1	11.2	13.0
β– 葡萄糖苷酸酶（mU）	1.10	1.58	2.48	1.91	0.81	1.04
磷脂酶 A_1（mU）	0.99	1.22	3.57	2.97	0.99	1.49
前列腺素（ng）*	160	16	168	5	103	20

* 检测的是用大鼠胃底条测出的 PGE_2，未对抽提过程中的损失（大约 20% ~ 30%）进行修正。

阿司匹林能抑制胶原蛋白或者低浓度凝血酶诱导的"释放反应"[24]。我们在实验中使用了浓度较高的凝血酶（5 U/ml），所以阿司匹林没有抑制血小板中各种成分的释放。由此可见，前列腺素的合成与"释放反应"并无关联。研究结果显示，阿司匹林能抑制前列腺素的合成但不影响凝脂酶 A 的释放，这充分说明阿司匹林在血小板中所起的一个作用是抑制花生四烯酸向前列腺素的转化。我们实验室中的其他研究工作（见前一篇文章 [11]）也表明：阿司匹林能抑制豚鼠肺匀浆物中的花生四烯酸合成前列腺素。

在这些实验中，水杨酸钠、阿司匹林和消炎痛都可以抑制人血小板中的前列腺素合成，并且这种抑制作用会发生在口服后两种药物之后。在由角叉菜胶诱导的大鼠炎症的第二阶段渗出物中 [20] 以及在过敏性接触性湿疹患者的发炎皮肤中 [25] 均发现了前列腺素。给人皮下注射低浓度 PGE_1 或 PGE_2 会使人出现非常明显的红斑 [5,26]。如果前列腺素确实是炎症的重要介质 [27]，那么作为消炎药的阿司匹林和消炎痛的治疗作用就可以用抑制前列腺素的合成来解释。

We thank Professor G. R. Webster for advice on the assay of phospholipase A and the Medical Research Council for financial support.

(*Nature New Biology*, **231**, 235-237; 1971)

J. B. Smith and A. L. Willis[*]: Department of Pharmacology, Institute of Basic Medical Sciences, Royal College of Surgeons of England, Lincoln's Inn Fields, London WC2A 3PN.

Received May 6, 1971.

References:

1. Morris, C. D. W., *Lancet*, i, 279 (1967).

2. Weiss, W. J., and Aledort, L. M., *Lancet*, ii, 495 (1967).

3. Zucker, M. B., and Peterson, J., *Proc. Soc. Exp. Biol. Med.*, **127**, 547 (1968).

4. O'Brien, J. R., *Lancet*, i, 779 (1968).

5. Macmillan, D. C., *Lancet*, i, 1151 (1968).

6. Packham, M. A., Nishizawa, E. E., and Mustard, J. F., *Biochem. Pharmacol.*, Suppl., 171 (1968).

7. Crunkhorn, P., and Willis, A. L., *Brit. J. Pharmacol.*, **36**, 216P (1969).

8. Crunkhorn, P., and Willis, A. L., *Brit. J. Pharmacol.*, **41**, 49 (1971).

9. Piper, Priscilla J., and Vane, J. R., *Nature*, **223**, 29 (1969).

10. Smith, J. B., and Willis, A. L., *Brit. J. Pharmacol.*, **40**, 545P (1970).

11. Vane, J. R., *Nature New Biology*, **231**, 232 (1971).

12. Ferreira, G. H., Moncada, S., and Vane, J. R., *Nature New Biology*, **231**, 237 (1971).

13. Holmsen, H., Day, H. J., and Stormorken, H., *Scand. J. Haematol.*, Suppl., 8 (1969).

14. Smith, A. D., and Winkler, M., *Biochem. J.*, **108**, 867 (1968).

15. Bartels, J., Kunze, H., Vogt, W., and Wille, G., *Naunyn-Schmiedeberg's Arch. Pharmak. Exp. Pathol.*, **266**, 199 (1970).

16. Bergström, S., Danielsson, H., and Samuelsson, B., *Biochim. Biophys. Acta*, **90**, 207 (1964).

17. Samuelsson, B., *Progr. Biochem. Pharmacol.*, **3**, 59 (1967).

18. Mills, D. C. B., Robb, I. A., and Roberts, G. C. K., *J. Physiol.*, **195**, 715 (1968).

19. Haslam, R. J., *Nature*, **202**, 765 (1964).

20. Willis, A. L., *J. Pharm. Pharmacol.*, **21**, 126 (1969).

21. Holmsen, H., and Day, H. J., *J. Lab. Clin. Med.*, **75**, 840 (1970).

22. Cooper, M. F., and Webster, G. R., *J. Neurochem.*, **17**, 1543 (1970).

23. Davey, M. G., and Lüscher, E. F., *Biochim. Biophys. Acta*, **165**, 490 (1968).

24. Evans, G., Packham, M. A., Nishizawa, E. E., Mustard, J. F., and Murphy, E. A., *J. Exp. Med.*, **128**, 877 (1968).

25. Greaves, M. W., Sondergaard, J., and McDonald-Gibson, W., *Brit. Med. J.*, **2**, 258 (1971).

26. Juhlin, L., and Michaëlsson, G., *Acta Dermatol. Venet.*, **49**, 251 (1969).

27. Willis, A. L., in *Prostaglandins, Peptides and Amines* (edit. by Mantegazza, P., and Horton, E. W.), 31 (Academic Press, London, 1969).

---

[*] Present address: Department of Physiology, Stanford University, Stanford, California 94305.

感谢韦伯斯特教授在磷脂酶 A 的检测方面为我们提出了宝贵的建议，还要感谢医学研究理事会为我们提供了经费上的支持。

（彭丽霞 翻译；莫韫 孙军 审稿）

Indomethacin and Aspirin Abolish Prostaglandin Release from the Spleen

S. H. Ferreira *et al.*

Editor's Note

Prostaglandins are lipid molecules produced by cells for a wide variety of reasons, and are particularly associated with pain signals and tissue inflammation. Suppression of prostaglandins can therefore reduce pain and inflammation, and in this and an accompanying paper in the same issue of *Nature New Biology* British pharmacologist John Vane and his coworkers propose that this is essentially the mechanism by which aspirin (and the related drug indomethacin) exerts its analgesic effects. They show here that the two substances suppress prostaglandin synthesis in dog spleens. This discovery of the mode of action of one of the most important drugs in widespread use won Vane the 1982 Nobel Prize in medicine or physiology.

PROSTAGLANDIN release can be evoked from many tissues by physiological, pathological or mechanical stimulation[1]. In one of the accompanying articles Vane[2] has demonstrated that prostaglandin synthesis by lung homogenate is blocked by anti-inflammatory acids. In the other, Smith and Willis[3] have shown that prostaglandin production induced by the action of thrombin on platelets is also inhibited by these agents.

The dog spleen releases large amounts of prostaglandins, identified as a mixture of PGE_2 and $PGF_{2\alpha}$, when stimulated either by sympathetic nerve excitation, by adrenaline or by noradrenaline[4-6]. The release of prostaglandin can be reproduced consistently by repetition of the same stimulus and in much greater amounts than can be extracted from the tissues[4,6]. Therefore the release represents fresh synthesis of prostaglandins rather than mobilization from an intracellular reservoir. Thus the spleen affords an experimental model in which anti-inflammatory drugs can be tested as inhibitors of prostaglandin synthesis when the organ and cell integrity are preserved and prostaglandin synthesis is promoted by a naturally occurring mediator.

Spleens were perfused with Krebs-dextran solution by the method described earlier[6,7]. Eleven dogs of either sex (6–18 kg) were anaesthetized with pentobarbitone sodium (30 mg/kg given intravenously). The abdomen was opened and the splenic pedicle was carefully dissected to separate the splenic artery, vein and nerves. The dog was given heparin (1,000 IU/kg intravenously) and the splenic artery and vein were cannulated with polyethylene tubing. The spleen was then removed and perfused through the artery with Krebs solution containing 2–3% dextran (molecular weight 110,000; Fisons) warmed to 37°C and delivered from a constant output pump at a rate of 15–30 ml./min. A second roller pump took part of the splenic outflow at 10 ml./min to superfuse a series of isolated

930

消炎痛和阿司匹林阻断脾脏前列腺素的释放

费雷拉等

编者按

前列腺素是由细胞产生的脂质分子，产生的原因有很多，并且与疼痛信号和组织炎症有着特殊的联系。因此，对前列腺素的抑制能够缓解疼痛和炎症。在这篇以及前面一篇发表于同一期《自然·新兴生物学》杂志上的相关文章中，英国药理学家约翰·文及其同事们指出这就是阿司匹林（以及相关药物消炎痛）能够发挥镇痛作用的基本机制。他们在本文中证明上述两种药物可以抑制狗脾脏中前列腺素的合成。文因为发现了一种被人们广泛使用的最重要的药物——阿司匹林的作用机理而获得了 1982 年的诺贝尔生理学暨医学奖。

在生理学、病理学或机械刺激下，许多组织会释放出前列腺素 [1]。在一系列相关文章的其中一篇中，文 [2] 已经提出消炎酸可以抑制肺匀浆中前列腺素的合成。在另一篇文章中，史密斯和威利斯 [3] 指出这些药物也能抑制由血小板中凝血酶诱导的前列腺素合成。

在受到交感神经兴奋、肾上腺素或去甲肾上腺素的刺激时，狗的脾脏会释放出大量前列腺素——被认为是 PGE_2 和 $PGF_{2\alpha}$ 的混合物 [4-6]。反复进行相同的刺激，前列腺素的释放过程会持续不断地出现，其数量远远超过从这些组织中可以萃取出来的前列腺素的量 [4,6]。由此可见，释放意味着前列腺素的新鲜合成，而不是细胞内储存的前列腺素的排出。因此，如果能保证器官和细胞的完整性，并且前列腺素的合成能够被一种天然的介导物所诱导，那么脾就可以成为一个鉴定消炎药物是否会抑制前列腺素合成的实验模型。

按照以前报道过的方法 [6,7] 用克雷布斯 – 葡聚糖溶液灌注脾脏。将 11 只雌狗或雄狗（6 kg ~ 18 kg）用戊巴比妥钠麻醉（按 30 mg/kg 的剂量静脉注射）。然后打开腹腔，小心切开脾蒂，分离脾动脉、脾静脉和脾神经。给狗注射肝素（静脉注射 1,000 IU/kg）后，用聚乙烯管插入脾动脉和脾静脉。随后取出脾脏并用加热至 37℃、含 2% ~ 3% 葡聚糖的克雷布斯溶液（分子量为 110,000；费森斯公司）从动脉灌注脾脏，用一台恒功率泵以 15 ml/min ~ 30 ml/min 的速度输送溶液。用另一台滚柱泵以 10 ml/min 的速度抽取部分脾脏流出液以灌注一系列分离的受试组织。在

assay tissues; excess effluent was allowed to run to waste against a hydrostatic pressure of 6–12 cm. Four assay tissues were used, usually including a rat stomach strip[8], a rat colon[9] and a chick rectum[10], and sometimes including a gerbil colon[11]. This combination of tissues allows differentiation between prostaglandin E_2 and $F_{2\alpha}$ (refs. 5 and 6). The assay tissues were rendered insensitive to catecholamines, acetylcholine, histamine and 5-hydroxytryptamine by infusing a mixture of antagonists[6] into the Krebs solution superfusing them. In some experiments a smaller amount (1 ml./min or 5 ml./min) of the splenic effluent was used to superfuse the assay tissues; the superfusion rate was made up to 10 ml./min with fresh Krebs solution from another channel in the roller pump.

The spleen was supported on a plastic tray hanging from a strain gauge, the output from which was displayed on one channel of a pen recorder. Thus, contractions of the spleen were recorded as decreases in weight. The splenic perfusion pressure and the reactions of the assay tissues were displayed on other channels of the pen recorder. The spleen was kept warm by a lamp, the output of which was electronically controlled to maintain a thermistor placed on the surface of the spleen at 37°C.

Substances used were adrenaline bitartrate, indomethacin, sodium acetyl salicylate, sodium salicylate and hydrocortisone dihydrogen succinate. To make sure that the antagonist did not change the sensitivity of the assay tissues, it was infused from the beginning of the experiment into the splenic outflow. To test the effect of the antagonist on the spleen the infusion was changed from the splenic outflow to the splenic inflow.

There was usually a basal output of prostaglandins from the spleen, as shown by contraction of the assay tissues when the splenic outflow was first used to superfuse them. This basal output was estimated in three experiments by intermittently applying the splenic outflow to the assay tissues for periods of 3 min. Assayed as PGE_2 there was an output of 10–20 ng/ml. of splenic effluent. The relative contractions of the assay tissues showed that this output was a mixture of E and F prostaglandins.

When adrenaline (200 ng/min or 6–10 ng/ml. during 3 min) was infused intra-arterially splenic perfusion pressure increased. The spleen contracted, as shown by a loss in weight and by the increase in splenic effluent running to waste. There was an increase in prostaglandin output which, from the reactions of the assay tissues, was chiefly PGE_2. Estimated as PGE_2 there was an increase in prostaglandin output of 10–80 ng/ml. After the adrenaline infusion was stopped the spleen gradually returned to its initial weight. During this period the output was assessed by infusions of prostaglandins directly to the assay tissues. Repetition of the adrenaline infusion at 20–40 min intervals had similar effects on perfusion pressure, splenic weight, and prostaglandin output, even though there was a gradual increase in perfusion pressure throughout the experiment. After 3–5 h of perfusion the splenic weight also tended to increase and when this occurred there was a smaller reduction in weight induced by adrenaline infusions.

When indomethacin (0.37–4 µg/ml.) was infused into the spleen the assay tissues gradually

6 cm ~ 12 cm 的静压下放掉多余的流出液。受试组织有 4 种，通常是大鼠胃条 [8]、大鼠结肠 [9] 和小鸡直肠 [10]，有时也包括沙鼠结肠 [11]。联合采用这些组织可以使我们区分出前列腺素 E_2 和 $F_{2\alpha}$（参考文献 5 和 6）。我们在用于灌注的克雷布斯溶液中加入了一种由多种拮抗剂组成的混合物 [6]，目的是使这些受试组织失去对儿茶酚胺、乙酰胆碱、组胺和 5- 羟色胺的敏感性。在有些实验中，用于灌注受试组织的脾脏流出液会更少一些（1 ml/min 或 5 ml/min）；通过滚柱泵的另外一个通道灌注新鲜克雷布斯溶液的速度被控制在 10 ml/min。

将脾脏置于一个悬挂在应变计上的塑料托盘中，应变计的输出信号在描笔式记录器的一个频道上显示出来。由此可以得到脾脏收缩与脾脏重量减少之间的关系。脾脏灌注压和受试组织的反应在描笔式记录器的其他频道上显示。用一盏灯使脾脏保温，并用电子器件控制灯的输出功率以便使放在脾脏表面的热敏电阻的温度保持在 37℃。

实验中所用的药物有：酒石酸肾上腺素、消炎痛、乙酰水杨酸钠、水杨酸钠和氢化可的松二氢琥珀酸酯。为了确保拮抗剂不会对受试组织的敏感性造成影响，在实验一开始，我们就将拮抗剂注入脾脏流出液。为了检测拮抗剂对脾脏的影响，药液先加入脾脏流出液，再加入脾脏流入液。

当开始用脾脏流出液浇盖受试组织时，由受试组织的收缩可以看出：脾脏通常会有一个基础的前列腺素输出。通过三次用脾脏流出液不连续地灌注受试组织长达 3 min 的实验估算出了这个基础输出量，如果以 PGE_2 衡量，则脾脏流出液中的输出量可以达到 10 ng/ml ~ 20 ng/ml。由受试组织的相对收缩情况可知：上述输出物是前列腺素 E 和 F 的混合物。

当在动脉内注入肾上腺素（200 ng/min 或 6 ng/ml ~ 10 ng/ml 持续 3 min）时，脾灌注压就会升高。脾脏重量减轻和脾脏流出液的增加表明脾脏发生了收缩。从受试组织的反应可以看出，前列腺素输出的增加部分主要是 PGE_2。以 PGE_2 估算，前列腺素输出的增加量为 10 ng/ml ~ 80 ng/ml。一旦停止注入肾上腺素，脾脏就会慢慢恢复到它原来的重量。在此期间，输出量用直接注入受试组织的前列腺素的量来评估。如果每隔 20 min ~ 40 min 就重复进行一次肾上腺素的注入，那么灌注压、脾脏重量和前列腺素输出不会有太大的改变，虽然灌注压在整个实验中是在缓慢增加的。灌注 3 h ~ 5 h 后，脾脏的重量还会趋于增加，并且当脾脏重量增加时，由注入肾上腺素所诱导的重量减轻会更少一些。

当消炎痛（0.37 μg/ml ~ 4 μg/ml）被注入脾脏时，受试组织会逐渐松弛。同时，

relaxed. At the same time their sensitivity to prostaglandin infusions increased, suggesting that the basal output of prostaglandin from the spleen was being inhibited and that the assay tissues had previously been desensitized to additional prostaglandin infusions by the presence of this basal output. This was confirmed by intermittent application of the splenic outflow to the assay tissues. The sensitivity of the tissues to prostaglandin infusions was diminished during superfusion with splenic outflow but returned when normal Krebs solution was used. In later experiments the interference with the prostaglandin assay by the basal output was minimized by using only a small proportion of the splenic outflow to superfuse the tissues.

Indomethacin infusions into the spleen did not affect the splenic weight or the splenic perfusion pressure. The increase in prostaglandin output induced by adrenaline infusion, however, was abolished. In the first experiment indomethacin (4 µg/ml. for 15 min) completely abolished prostaglandin output, as did a concentration of 750 ng/ml. for 15 min (four experiments). In two other experiments indomethacin at 370 ng/ml. (1 µM) was used. In one there was a partial inhibition (60%) of prostaglandin output after 15 min of indomethacin infusion. In the other experiment the infusion was made for a further 15 min and there was then complete abolition of prostaglandin output. In all these experiments, after the indomethacin infusion was stopped, there was a gradual return of prostaglandin output induced by adrenaline infusion, which reached at least 60% of the initial output within 1–3 h and sometimes 100%.

Fig. 1 illustrates one of these experiments. The first panel shows the reaction of the assay tissues to prostaglandin E_2 (20 ng/ml.). When adrenaline was infused into the spleen there was a rise in perfusion pressure and a fall in spleen weight. The prostaglandin output, assayed as PGE_2, was about 20 ng/ml. After the adrenaline stimulation, the prostaglandin output, perfusion pressure and spleen weight gradually returned to initial levels. The indomethacin infusion (0.37 µg/ml.) which had so far been made into the splenic output was then made into the spleen. During the next 25 min the assay tissues, especially the rat stomach strip, gradually relaxed. They were then much more sensitive to PGE_2, 10 ng/ml. giving greater contractions than those produced by 20 ng/ml. previously. Adrenaline infusion into the spleen now induced a larger increase in perfusion pressure, a larger decrease in spleen weight but no prostaglandin release. The indomethacin was then given once more into the splenic outflow. The next adrenaline infusion shown was made 70 min after stopping the indomethacin treatment of the spleen. Although the effects on perfusion pressure were still greater than in the control period, the change in spleen weight had decreased to control values. There was also an output of prostaglandins which, assayed as PGE_2, was more than 10 ng/ml. In three of these experiments the solvent for the indomethacin (4% ethanol plus Krebs solution) was infused into the spleen. This did not change the prostaglandin output, the rise in perfusion pressure or fall in spleen weight induced by adrenaline infusion. In four of the six experiments with indomethacin the splenic contraction induced by adrenaline stimulation was clearly potentiated during the period in which prostaglandin release was abolished. In two experiments there was also a potentiation of the pressor response to adrenaline during the abolition of prostaglandin

934

它们对前列腺素注入的敏感性也会增加，这说明脾脏的基础前列腺素输出受到了抑制。之前由于这种基础输出的存在，受试组织对前列腺素的再注入失去了敏感性。这一点已经从用脾脏流出液不连续浇盖受试组织的实验中得到了证实。在用脾脏流出液灌注时，组织对前列腺素注入的敏感性下降，但当改用标准克雷布斯溶液时，组织的敏感性得到恢复。在后来的实验中，我们用很少量的脾脏流出液来灌注这些组织以便将基础输出对前列腺素分析的干扰降至最低。

将消炎痛注入脾脏后，脾重量和脾灌注压并未受到任何影响，但前列腺素的输出量也就不再会因为受到肾上腺素浸液的诱导而增加了。在第一个实验中，消炎痛（4 μg/ml，15 min）使前列腺素的输出完全被抑制，这与浓度为 750 ng/ml 消炎痛作用 15 min 的效果相同（四次实验）。在另外两个实验中，我们使用了 370 ng/ml（1 μM）的消炎痛。其中一个实验的结果是：在注入消炎痛 15 min 后，前列腺素的输出受到了部分抑制（60%）。在另一个实验中，我们又将注入时间延长了 15 min，这样前列腺素的输出就被完全抑制住了。在所有这些实验中，当消炎痛的注入停止时，由肾上腺素浸液所诱导的前列腺素输出会逐渐恢复，在 1 h ~ 3 h 内，输出量至少可以恢复到初始值的 60%，有时甚至能达到 100%。

图 1 显示了其中一次实验的结果。第一列数据表示受试组织对前列腺素 E_2（20 ng/ml）的反应。当用肾上腺素注入脾脏时，灌注压升高，脾脏重量下降。前列腺素的输出，以 PGE_2 计算，约为 20 ng/ml。肾上腺素的刺激完成后，前列腺素输出、灌注压和脾脏重量都会逐渐恢复到原来的水平。之前一直被加入到脾脏流出液中的消炎痛浸液(0.37 μg/ml)此时也被加入到脾脏中。在接下来的 25 min 之内，受试组织，尤其是大鼠胃条，会逐渐松弛。在这之后，受试组织变得对 PGE_2 更加敏感，10 ng/ml 浓度引起的收缩要大于之前由 20 ng/ml PGE_2 所引起的收缩。这时在脾脏中注入肾上腺素会使灌注压有更为显著的增加，脾脏重量的下降也更为明显，但并没有释放出前列腺素。随后消炎痛被再次加入到脾脏流出液中。图中显示的肾上腺素浸液的下一次刺激是在停止用消炎痛处理脾脏 70 min 以后进行的。虽然此时对灌注压的影响仍然高于对照期，但脾脏重量的变化已经减少到了对照值。另外，前列腺素还有一个输出量，以 PGE_2 计算，其值高于 10 ng/ml。在三次这样的实验中，消炎痛的溶剂（含 4% 乙醇的克雷布斯溶液）也被我们注入了脾脏，它们对由肾上腺素浸液诱导的前列腺素合成、灌注压升高和脾脏重量下降均没有任何影响。在六次使用消炎痛的实验中，有四次当前列腺素释放被阻断时，由肾上腺素刺激诱导的脾脏收缩在效力上明显增强。在两次实验中，当前列腺素释放受到抑制时，肾上腺素仍能导致压强的升高；但是其中只有一次在停止用消炎痛浸渍脾脏时，灌注压的升高能恢

release but in only one of these did the rise in perfusion pressure return to the previous levels when the indomethacin infusion to the spleen was stopped.

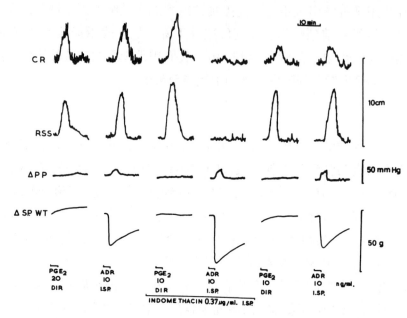

Fig. 1. A spleen from an 8.5 kg dog was perfused with Krebs-dextran solution at a rate of 20 ml./min. A continuous sample (10 ml./min) of the splenic outflow, with combined antagonists added, was used to superfuse the assay tissues. The figure shows the effects of prostaglandins on a chick rectum (CR; top) and a rat stomach strip (RSS). The next two tracings (bottom) show changes in perfusion pressure (ΔPP) and spleen weight ($\Delta SP.$ wt.). Except when infused into the spleen indomethacin was added to the splenic outflow to give a concentration of 0.37 μg/ml. The first panel shows contractions of CR and RSS induced by prostaglandin E_2 (20 ng/ml. DIR). Next an adrenaline infusion into the spleen (ADR 10 ng/ml. I. SP) induced a rise in perfusion pressure, a fall in spleen weight and an output of prostaglandins equivalent to PGF_2 at about 20 ng/ml. Indomethacin (0.37 μg/ml.) was then infused into the spleen. During the next 25 min the assay tissues relaxed (not shown) and were then more sensitive to PGE_2 (10 ng/ml. DIR). Adrenaline (40 min after start of indomethacin) now caused a greater increase in perfusion pressure, a greater decrease in spleen weight, but no output of prostaglandin. After stopping the indomethacin infusion into the spleen, the reactivity of the assay tissues gradually decreased and the output of prostaglandin induced by adrenaline infusion into the spleen gradually returned. The adrenaline stimulation shown was made 70 min after stopping the indomethacin. The rise in perfusion pressure was still larger, but the fall in spleen weight had returned to the original size.

Aspirin was much less effective as an inhibitor of prostaglandin release. At 15 μg/ml. (one experiment) there was 60% inhibition and at 40 μg/ml. (two experiments) there was 75% and 100% inhibition. Sodium salicylate (20 and 40 μg/ml.; two experiments) did not change the prostaglandin output induced by adrenaline infusion into the spleen. Hydrocortisone (2 μg/ml., one experiment; 25 μg/ml., two experiments) was also ineffective. In these experiments, after salicylic acid or hydrocortisone had been shown to be ineffective, prostaglandin output was reduced by indomethacin or aspirin.

复到原来的水平。

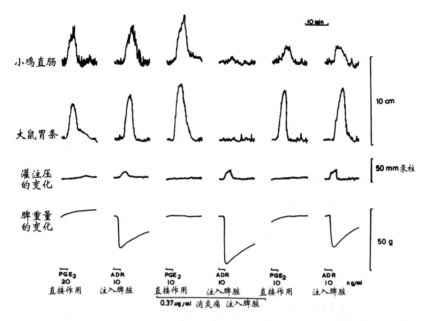

图 1. 我们用克雷布斯 – 葡聚糖溶液以 20 ml/min 的速度灌注一只体重为 8.5 kg 的狗的脾脏。用加入几种拮抗剂后脾脏的流出液样品（10 ml/min）持续灌注受试组织。这张图表明了前列腺素对小鸡直肠（CR，顶部）和大鼠胃条（RSS）的影响。最下面的两条曲线（底部）表示的是灌注压的变化（ΔPP）和脾重量的变化（ΔSP. wt.）。除了灌注脾脏时之外，消炎痛以 0.37 μg/ml 的浓度被加入到脾脏流出液中。第一列数据表示由前列腺素 E_2（20 ng/ml 直接作用）诱导的 CR 收缩和 RSS 收缩。第二列是由脾脏中注入肾上腺素（ADR 10 ng/ml 注入脾脏）所诱导的灌注压上升、脾重量下降和前列腺素输出（以 PGE_2 计算，约为 20 ng/ml）。随后将消炎痛（0.37 μg/ml）注入脾脏。在接下来的 25 min 时间里，受试组织变得松弛（未显示），并且对 PGE_2（10 ng/ml 直接作用）的敏感性更高。这时，肾上腺素（加入消炎痛 40 min后）会使灌注压的升高更加明显，而脾重量会降得更低，但是没有前列腺素输出。当我们停止在脾脏中注入消炎痛后，受试组织的敏感性逐渐降低，由在脾脏中注入肾上腺素而诱导的前列腺素输出将逐渐恢复。图中显示的肾上腺素刺激是在停止注入消炎痛 70 min 后进行的。灌注压的升高依然很明显，但脾重量的下降已经恢复到了原来的水平。

阿司匹林对前列腺素释放的抑制作用明显不及消炎痛。15 μg/ml 阿司匹林（一次实验）所产生的抑制率为 60%；40 μg/ml 阿司匹林（两次实验）所产生的抑制率为 75% 和 100%。水杨酸钠（20 μg/ml 和 40 μg/ml；两次实验）并不影响将肾上腺素注入脾脏所诱导的前列腺素合成。氢化可的松（2 μg/ml，一次实验；25 μg/ml，两次实验）也没有抑制效果。在这些实验中，首先发现的是水杨酸或氢化可的松对抑制前列腺素无效，随后才发现消炎痛或阿司匹林可使前列腺素的输出减少。

Both in human platelets[12] and dog spleen[4,6] prostaglandin output represents fresh prostaglandin synthesis. Thus we have now shown in cell-free homogenates[2], in isolated cells[3] and in a whole organ perfused *in vitro* (in this article), in experiments covering three species, that indomethacin is a potent inhibitor of prostaglandin synthesis. The concentrations used in the perfused spleen experiments were less than those reported in human plasma after an oral dose of 200 mg[13]. It is possible therefore that this activity of indomethacin may be the basis of some of its therapeutic actions. Neither hydrocortisone nor salicylate was effective in reducing prostaglandin output from the spleen. Hydrocortisone is an anti-inflammatory substance of a different nature but salicylate is from the same group as aspirin and indomethacin and contains the carboxy group. The lack of action of salicylate in these experiments does not necessarily mean that prostaglandin synthesis is unimportant in inflammation and pyrexia, for salicylate may be more effective against the synthesis of a prostaglandin such as PGE_1 from dihomo-γ-linolenic acid than it is against the synthesis of PGE_2 from arachidonic acid. PGE_1 is more effective than PGE_2 as a pyretic agent[14] and may also be important in inflammation[15]. In both these cases salicylate is about as potent as aspirin[16].

Hedqvist[17] has shown that prostaglandin E_2 release in the spleen may be a feedback mechanism, controlling the output of noradrenaline from sympathetic nerves. He measured the effects of PGE_2 on the increase in perfusion pressure induced by injections of noradrenaline; they ranged from potentiation to inhibition, depending on the concentration of prostaglandin used. These results may explain the variability in the effects of indomethacin on the rise in perfusion pressure induced by adrenaline in our experiments. Certainly the use of indomethacin in the perfused spleen preparation will allow further assessment of the hypothesis proposed by Hedqvist. The use of indomethacin as a prostaglandin synthesis inhibitor will also allow definition of the functions of prostaglandins in cells, tissues or in whole animals.

We thank Mr. P. Holgate for technical assistance and the Wellcome Trust for grants.

(*Nature New Biology*, **231**, 237-239; 1971)

S. H. Ferreira, S. Moncada and J. R. Vane: Department of Pharmacology, Institute of Basic Medical Sciences, Royal College of Surgeons of England, Lincoln's Inn Fields, London WC2A 3PN.

Received May 7, 1971.

References:

1. Piper, Priscilla J., and Vane, J. R., in *Prostaglandins* (edit. by Ramwell, P., and Shaw, J.), *Ann. NY Acad. Sci.*, **180**, 363 (1971).

2. Vane, J. R., *Nature New Biology*, **231**, 232 (1971).

3. Smith, J. B., and Willis, A. L., *Nature New Biology*, **231**, 235 (1971).

4. Davies, B. N., Horton, E. W., and Withrington, P. G., *J. Physiol.*, **188**, 38P (1967).

5. Ferreira, S. H., and Vane, J. R., *Nature*, **216**, 868 (1967).

在人血小板 [12] 和狗脾脏 [4,6] 中，前列腺素的输出意味着前列腺素的新鲜合成。因此我们分别用三个不同物种的无细胞匀浆物 [2]、分离出的细胞 [3] 以及体外灌注的整个器官（本文中）所进行的实验都验证了消炎痛是一种能强烈抑制前列腺素合成的物质。在灌注脾脏实验中所用的消炎痛浓度要低于人口服 200 mg 消炎痛后血浆中的报告浓度 [13]。因此，消炎痛的这种活性可能就是它具有某些治疗作用的基础。氢化可的松和水杨酸盐都不能减少脾脏中前列腺素的输出。氢化可的松是一种在化学结构上有所不同的消炎物质，而水杨酸盐与阿司匹林和消炎痛都属于同一类别，它们都含有羧基。水杨酸盐在上述实验中缺乏活性并不一定意味着前列腺素合成在发炎和发烧时是无关紧要的，因为水杨酸盐对合成 PGE$_1$ 等二高-γ-亚麻酸来源的前列腺素的抑制可能会强于对合成 PGE$_2$ 等花生四烯酸来源的前列腺素的抑制。PGE$_1$ 比 PGE$_2$ 更易引起发烧 [14]，在导致发炎上可能也很重要 [15]。在发烧和炎症的治疗上，水杨酸盐和阿司匹林的效果大致相同 [16]。

赫德奎斯特 [17] 曾指出脾脏释放前列腺素 E$_2$ 的过程可能是一个反馈机制，调控着交感神经内去甲肾上腺素的输出。他检测了 PGE$_2$ 对由注射去甲肾上腺素诱导的灌注压升高的影响。随着所用前列腺素浓度的不同，PGE$_2$ 对灌注压升高的影响可在增强和抑制之间转化。这些结果或许可以解释为什么在我们的实验中消炎痛会对由肾上腺素诱导的灌注压升高有不同的影响。很显然，用消炎痛灌注脾脏可以使我们更深入地评估赫德奎斯特提出的假设。此外，采用消炎痛作为前列腺素合成的抑制剂还能使我们明确前列腺素在细胞、组织甚或整个动物体中的功能。

我们衷心感谢霍尔盖特先生在技术上的帮助以及维康基金会在经费上的资助。

（彭丽霞 翻译；莫韫 审稿）

6. Gilmore, N., Vane, J. R., and Wyllie, J. H., *Nature*, **218**, 1135 (1968).

7. Gilmore, N., Vane, J. R., and Wyllie, J. H., in *Prostaglandins, Peptides and Amines* (edit. by Mantegazza, P., and Horton, E. W.), 21 (Academic Press, London, 1969).

8. Vane, J. R., *Brit. J. Pharmacol. Chemother.*, **23**, 360 (1964).

9. Regoli, D., and Vane, J. R., *J. Physiol.*, **183**, 513 (1966).

10. Mann, M., and West, G. B., *Brit. J. Pharmacol. Chemother.*, **5**, 173 (1950).

11. Ambache, N., Kavanagh, L., and Whiting, J., *J. Physiol.*, **176**, 378 (1965).

12. Smith, J. B., and Willis, A. L., *Brit. J. Pharmacol.*, **40**, 545P (1970).

13. Hucker, H. B., Zacchei, A. G., Cox, S. V., Brodie, D. A., and Cantwell, N. H. R., *J. Pharmacol. Exp. Ther.*, **153**, 237 (1960).

14. Milton, A. S., and Wendlandt, S. J., *J. Physiol.*, **207**, 76P (1970).

15. Willis, A. L., in *Prostaglandins, Peptides and Amines* (edit. by Mantegazza, P., and Horton, E. W.), 31 (Academic Press, London, 1969).

16. Collier, H. O. J., *Adv. Pharmacol. Chemother.*, 7, 333 (1969).

17. Hedqvist, P., *Life Sci.*, **9**, 269 (1970).

Possible Clonal Origin of Common Warts (*Verruca vulgaris*)

R. F. Murray, jun. *et al.*

Editor's Note

In the days before modern molecular approaches for proving clonality, researchers had to resort to more esoteric methods. Here Robert Murray and colleagues use a trick involving the cytosolic enzyme glucose-6-phosphate dehydrogenase to infer a possible clonal origin of common warts. The team realised that women heterozygous for the X-chromosome-linked gene have two populations of cells, each expressing a single phenotype corresponding to one or other of the two alleles. They found that verrucal tissue from such donors contained just one phenotype, a result consistent with a clonal origin. Although human papilloma virus had yet to be identified as the causative agent of warts, researchers were aware of the links between certain viruses and cancers, making this research all the more significant.

FEMALES heterozygous for the X-linked glucose-6-phosphate dehydrogenase (G6PD) locus have two populations of cells, each expressing a single phenotype corresponding to one or the other of the two alleles[1-3]. Where subjects are heterozygous for the A and B electrophoretic variants of G6PD[4], one cell population will show only the A electrophoretic variant and the other only the B. The study of these variants has been used to throw light on the origin of certain tumours. Thus, uterine leiomyomas from Negro females heterozygous for G6PD variants A and B are of only one phenotype, although the adjacent uninvolved tissue shows both phenotypes[5], which is consistent with a clonal or unicellular origin for this tumour. Similarly, chronic granulocytic leukaemia[6] is probably of clonal origin, like the metastatic lesions from one patient with chronic lymphocytic leukaemia[7]. On the other hand, studies of metastatic lesions from a colonic malignancy[7] and epidermoid carcinomas of the cervix have revealed[8] the presence of both G6PD phenotypes, which implies a multiclonal origin.

We have used this technique for the study of *verruca vulgaris* or common warts—thickenings or projections of the epidermis, classified as benign tumours of the skin, caused by a spherical virus 50 μm in diameter[9] which is structurally similar to the polyoma virus[10].

Verrucal tissue was obtained in a dermatology clinic from Negro females. Lesions were diagnosed as typical verruca vulgaris by J. H. The study included only lesions with typical appearance[11], and measuring 2–8 mm in diameter. They were excised surgically without removing adjacent normal tissue. The root of the wart was destroyed by cryotherapy and the specimen placed in a sterile tube containing about 2 cc of TC 199 tissue culture medium. Venous blood was taken from the patient (5–10 cm³) and placed in heparinized,

寻常疣为克隆性起源的可能性

小默里等

编者按

在还不能运用现代分子生物学方法来证明克隆性之前，研究人员不得不采取一些更为深奥难懂的方法。在本文中，罗伯特·默里及其同事们采用了一种与胞浆酶葡萄糖-6-磷酸脱氢酶有关的方法来推断寻常疣是否有可能为克隆性起源。该研究小组认为：X 染色体连锁基因的女性杂合子有两种类型的细胞，每一种都代表了一种单一的表型，对应于两个等位基因中的一个或者另一个。他们发现来自这类患者的疣组织只有一种表型，这一结果与克隆性起源是一致的。虽然那时候人们还没有确定人类乳头状瘤病毒就是疣的病原体，但研究人员意识到某些病毒与癌症之间是有关联的，这使本文中的研究工作显得尤为重要。

X 连锁葡萄糖-6-磷酸脱氢酶（G6PD）位点的女性杂合子有两种类型的细胞，每一种都代表了一种单一的表型，对应于两个等位基因中的一个或者另一个 [1-3]。当被测对象是 G6PD 的 A 和 B 电泳变异体的杂合子时 [4]，它的一群细胞将仅仅表现为 A 电泳变异体，而另一群则只表现出 B 电泳变异体的特点。人们已经开始用对这些变异体的研究成果来解释某些肿瘤的起源。因此，来自黑人女性身上的 G6PD A 和 B 变异体杂合子的子宫平滑肌瘤只有一种表型，尽管邻近肿瘤的未受累组织可以表现出两种表型 [5]，这与该肿瘤为克隆性起源或单细胞起源是一致的。同样，慢性粒细胞性白血病 [6] 可能也是克隆性起源，与一个慢性淋巴细胞性白血病患者的转移性病变是一样的 [7]。而另一方面，人们对结肠恶性肿瘤的转移性病变 [7] 和宫颈表皮样癌的研究结果表明 [8] 两种 G6PD 表型均存在，这间接说明了一种多克隆起源。

我们将这项技术应用于对寻常疣的研究——寻常疣指的是表皮的增厚或突起，它被归为皮肤的良性肿瘤，是由一种直径为 50 μm[9] 的球形病毒引起的，这种病毒在结构上类似于多瘤病毒 [10]。

疣组织是从一个皮肤病诊所的多位黑人女性患者身上获得的。詹姆斯·霍布斯将这种病诊断为典型的寻常疣。本研究所涉及的仅仅是具有典型表现 [11] 且直径为 2 mm ~ 8 mm 的病变区。把它们用手术切除的时候没有触及邻近的正常组织。用冷冻法破坏掉疣的根部，然后将样本置于一个含约 2 ml TC 199 组织培养基的无菌试管里。从患者身上采集静脉血（5 ml ~ 10 ml），并放入经肝素化处理的无菌容器

sterile containers and stored at 4°C until analysed (usually within 48 h). There was no significant loss of G6PD activity from the wart tissue in this time. Some samples were stored at −20°C until analysed. The G6PD phenotype of the red cells and corresponding wart specimens was determined on the same run. At the conclusion of each run the gel was sliced and stained[12] for zones of G6PD activity.

Only one specimen of verrucal tissue was obtainable from each individual, although if lesions were of sufficient size, they were cut into pieces and two or more runs made.

Adjacent normal skin tissue was not used to determine the patient's phenotype to minimize possible scarring after removal of the verrucae. Other evidence[13,14], however, indicates that the erythrocyte G6PD phenotype in heterozygotes is usually an adequate qualitative reflexion of the phenotype of skin or other tissues. In exceptional cases, the red cell phenotype is falsely homozygous although false homozygosity in skin in the presence of a heterozygous red cell phenotype has not been described. Thus, any resulting misclassification will probably result in a loss of information, but not misleading information.

Tissue and blood from twelve Afro-American females were tested. On starch gel electrophoresis wart tissue showed a single principal zone of G6PD activity and a slower minor zone. The former zone has the same mobility as that visible in mammalian liver and epidermal tissue and was found to have substrate specificity for galactose-6-phosphate dehydrogenase activity[15,16]. The mobility of this zone did not vary with the phenotype of the wart tissue.

Six of twelve patients studied were heterozygous for G6PD electrophoretic variants as determined from their red blood cell phenotypes (Table 1). In two of the six heterozygotes, the phenotype of the wart tissue was A only and in four cases B only. The blood of the remaining six patients showed only a single phenotype (one A and five B). The G6PD phenotype in the wart and blood was the same in each case. Three large warts, one from a heterozygote and two from homozygotes, were each cut into four approximately equal pieces, but the phenotype was the same in each piece.

Table 1. G6PD Phenotype of Blood and Wart Tissue of Twelve Afro-American Females Determined by Starch Gel Electrophoresis

No.	G6PD phenotype, blood	No.	G6PD phenotype, wart
5	B+	5	B+
1	A+	1	A+
6	A+ B+	4	B+
		2	A+

Verrucal tissue from individuals heterozygous for the common A and B electrophoretic variants of G6PD showed only one phenotype in each case tested, which is consistent with

内，然后在 4℃ 下保存直至用于分析（通常不超过 48 h）。在这段时间里，疣组织内的 G6PD 活性不会有明显的丧失。有些样本在使用前被保存于 –20℃ 下。在每一轮实验中，我们都会同时检测红细胞和对应疣样本的 G6PD 表型。每一轮实验结束后，对电泳胶进行切片并染色 [12] 以便观察具有 G6PD 活性的条带。

我们从每个人身上只取一个疣组织样本，但如果病变区足够大的话，可以把它们切成小块，用于 2 次或 2 次以上的实验。

我们不用邻近的正常皮肤组织来确定患者的表型，以减少疣摘除后可能出现的疤痕。但有其他证据 [13,14] 表明：杂合子中红细胞的 G6PD 表型常常足以定性反映皮肤或其他组织的表型。在一些特殊的病例中，红细胞的表型是假纯合子，不过现在还没有人描述过红细胞表型为杂合子时的皮肤假纯合子表型。这样一来，在分类上出现的任何错误都有可能会导致信息的丢失，但不会误导信息。

我们共检测了 12 名美国黑人女性的组织和血液。从淀粉凝胶电泳的结果来看，疣组织显示出一条主要的 G6PD 活性区带，还有一条较慢的次级区带。前一区带的移动能力与在哺乳动物肝脏和表皮组织中所看到的情况相同，而且其底物特异性与半乳糖 –6– 磷酸脱氢酶相同 [15,16]。无论疣组织的表型发生什么变化，这个区带的移动性都不会随之改变。

在我们所研究的 12 名患者中，有 6 名患者经红细胞表型分析之后被确定为 G6PD 电泳变异体的杂合子（表 1）。在 6 名杂合子个体中，有 2 名的疣组织表型只表现为 A，而另外 4 名只表现为 B。剩余 6 名患者的血液只表现出单一的表型（1 名为 A 和 5 名为 B）。疣组织和血液的 G6PD 表型在每个病例中均相同。我们将三个较大的疣（一个来自杂合子，另两个来自纯合子）分别切成四个大致相等的块，但每个块的表型都是相同的。

表 1. 由淀粉凝胶电泳测得的 12 名美国黑人女性血液和疣组织的 G6PD 表型

患者数	血液的G6PD表型	患者数	疣的G6PD表型
5	B+	5	B+
1	A+	1	A+
6	A+B+	4	B+
		2	A+

在每一个被检测的病例中，普通 G6PD 电泳变异体 A 和 B 杂合子个体的疣组织都只有一种表型，这支持了寻常疣是克隆性起源的假说。可以由此认为疣病毒感染

the hypothesis that verruca vulgaris is of clonal origin. This suggests that the verrucal virus infects a patch of cells all of the same G6PD phenotype, or that it infects only a single epidermal cell. The available data (six heterozygotes) are too limited to differentiate clearly between origin from a single cell or from a patch, especially as a normal patch size may be of the order of a thousand cells or more[5]. It is also possible that cells with both G6PD phenotypes are infected, but that one type is lost during cell proliferation.

Viral infection would be expected to involve a rather large target with spread to thousands of adjacent cells assuming a reasonable titre of the virus. If the proliferative process involves the spread of the virus, this should result in a lesion of multicellular origin, showing both G6PD phenotypes. On the other hand, if the process involves a change in a single cell or a small group of cells, what we have found would be expected, namely, evidence for single cell or clonal origin. The acidophilic inclusion bodies that are characteristic of pathologic sections of verruca vulgaris might still represent virus particles that have replicated with the proliferating cells rather that spread from an initial point of infection.

With the recent implication of viruses in the aetiology of some animal malignancies like the Rous sarcoma, and the murine leukaemias, and evidence that a tumour, like Burkitt's African lymphoma, is of possible clonal origin[17] and may be caused by reovirus 3 (ref. 18), the behaviour of viruses in causing tumour-like lesions takes on considerable significance. The somatic mutation that has been hypothesized by many to lead to neoplastic change might, in some cases, be preceded by the viral invasion of a susceptible cell or clone of cells, or all cells may harbour in a symbiotic fashion this virus and a somatic mutation may induce its oncogenic properties.

We thank Dr. S. M. Gartler for advice and suggestions. This work was supported by a grant from the US Children's Bureau and by a general research support grant from the US Public Health Service.

(**232**, 51-52; 1971)

Robert F. Murray, jun., James Hobbs and Brownell Payne: Medical Genetics Unit, Department of Paediatrics, Howard University College of Medicine, Washington DC 20001.

Received January 14, 1971.

References:

1. Davidson, R. G., Nitowsky, H. N., and Childs, B., *Proc. US Nat. Acad. Sci.*, **50**, 481 (1963).

2. Beutler, E., and Baluda, M. C., *Lancet*, i, 189 (1964).

3. DeMars, R., and Nance, W. E., *Wistar Inst. Symp. Monog.*, No. 1, 35 (Wistar Institute Press, Philadelphia, 1964).

4. Boyer, S. H., Porter, I. H., and Weilbacher, R. G., *Proc. US Nat. Acad. Sci.*, **48**, 1868 (1962).

5. Linder, D., and Gartler, S. M., *Science*, **150**, 67 (1965).

946

了一片具有相同 G6PD 表型的细胞，或者仅仅感染了一个表皮细胞。现有的数据（6个杂合子）太少，以至于难以明确地区分到底是来源于一个细胞还是一片细胞，尤其是当正常细胞的数量可能达到了 1,000 个或者更多的时候 [5]。也有可能两种 G6PD 表型的细胞都被感染，但其中一种表型在细胞增殖的过程中丢失了。

如果病毒的滴度合适的话，它将传播到数千个邻近的细胞，造成相当大面积的感染。如果细胞增殖过程涉及病毒的传播，那么就会导致一种表现出两种 G6PD 表型的多细胞起源的病变。另一方面，如果该过程涉及单个细胞或者一小组细胞的改变，那么我们所发现的结果就可以作为支持单细胞或者克隆性起源的证据，这与预期的结果一致。寻常疣的病理学特征性嗜酸性包涵体或许仍然代表了随着细胞的增殖已经发生了复制的病毒颗粒，而不是从感染的初始部位传播过来的病毒颗粒。

近年来的迹象表明病毒是一些诸如劳斯肉瘤和鼠类白血病等动物恶性肿瘤的病因，以及得到了类似非洲伯基特淋巴瘤这样的肿瘤可能为克隆性起源 [17] 并有可能是由呼肠孤病毒 3 引起（参考文献 18）的证据，因而病毒引起肿瘤样病变的行为就具有了非常重要的意义。有不少人认为体细胞突变是引起肿瘤变化的原因。在某些情况下，体细胞突变有可能发生在病毒侵染易感细胞或细胞克隆之后，或者这种病毒是以共生的方式藏匿于所有细胞之中，而体细胞突变也许能诱发病毒的致癌作用。

感谢加特勒博士为我们提出了一些意见和建议。本工作得到了美国儿童局的一项基金以及美国公共卫生署的一项一般研究经费补助金的资助。

（毛晨晖 翻译；袁峥 审稿）

6. Fialkow, P. J., Gartler, S. M., and Yoshida, A., *Proc. US Nat. Acad. Sci.*, **58**, 1468 (1967).

7. Beutler, E., Collins, Z., and Irwin, L. E., *New Engl. J. Med.*, **276**, 389 (1967).

8. Park, I., and Jones, H. W., *Amer J. Obstet. Gynec.*, **102**, 106 (1968).

9. Allen, A. C., *The Skin, a Clinicopathological Treatise*, 775 (Grune and Stratton, New York, 1967).

10. Williams, M. G., Howatson, A. F., and Almeida, J. D., *Nature*, **189**, 895 (1961).

11. Lewis, G. M., and Wheeler, C. E., *Practical Dermatology*, 520 (Saunders, Philadelphia, London, 1967).

12. Fildes, R. A., and Parr, C. W., *Nature*, **200**, 890 (1963).

13. Nance, W. E., *Cold Spring Harbor Symp. Quant. Biol.*, **29**, 415 (1964).

14. Gandini, E., Gartler, S. M., Angiuni, N., Argiolas, N., and Dell'Acqua, G., *Proc. US Nat. Acad. Sci.*, **61**, 945 (1968).

15. Ohkawara, A., Halperin, J., Barber, P., and Halperin, K. M., *Arch. Dermatol.*, **95**, 412 (1967).

16. Ohno, S., Payne, H. W., Morrison, M., and Beutler, E., *Science*, **153**, 1015 (1966).

17. Fialkow, P. J., Klein, G., Gartler, S. M., and Clifford, P., *Lancet*, i, 348 (1970).

18. Stanley, N. F., *Lancet*, i, 961 (1966).

Establishment of Symbiosis between *Rhizobium* and Plant Cells *in vitro*

R. D. Holsten *et al.*

Editor's Note

Leguminous plants such as those of the major food crops are able to survive only because their roots have structures that turn nitrogen in the air into nitrogenous chemicals essential for the wellbeing of the plants. This experiment, from a commercial laboratory in the United States, demonstrated that the process is indeed accomplished by the invasion of structures formed in the roots of plants by bacteria of the genus *Rhizobium*. The demonstration that the symbiosis between the bacteria and the plants can be established in the laboratory was a confirmation of what biologists believed but of which they nevertheless needed proof.

SYMBIOTIC fixation of atmospheric nitrogen by leguminous plants is perhaps the most quantitatively important agency by which this element is incorporated into the biosphere, and is the principal process for maintaining nitrogen fertility in agriculture. The interaction of plants of the family Leguminoseae and specific bacteria of the genus *Rhizobium* in this symbiotic relationship is not yet well understood, probably because of the absence of a suitable defined experimental system. Other workers have used isolated plant roots in combination with the appropriate bacteria[1-4], and, more recently, plant cell and tissue culture[5]. Although stimulation of plant cell differentiation and lignification apparently resulted, no evidence for intercellular bacterial symbiosis could be found.

We have attempted to establish a symbiotic N_2-fixing relationship with plant cells and microorganisms *in vitro*, using the techniques of aseptic plant cell culture in conjunction with the sensitive acetylene–ethylene assay for nitrogenase activity. We consider that such a system can be used further to elucidate the factors which control nodulation processes and N_2-fixing activity *in vivo*.

Soybean Root Cell Cultures

Seeds of soybean (*Glycine max* var. Acme) were surface sterilized by immersion in a commercial hypochlorite solution ("Zonite", Chemway Corp., Wayne, New Jersey) for 15 min, and rinsed with sterile distilled water. Seed was placed on the surface of an agar (1% w/v) solidified medium composed of the mineral salts (MS medium)[6] supplemented with sucrose, 30 g/l., for germination in the dark at 26°±1°C. All media were sterilized by autoclaving at 121°C for 15 min.

When the primary root was approximately 4 cm long, it was excised aseptically, and a 5 mm explant was removed 2.5 cm behind the growing tip. This explant of root origin was

950

在体外建立根瘤菌和植物细胞间的共生关系

霍尔斯滕等

编者按

豆科植物（如主要粮食作物中的那些豆科植物）能够生存下来的原因仅仅在于它们的根部具有许多可以把空气中的氮转化为植物良好生长所必需的含氮化合物的结构。本文描述的实验来自美国的一个商业实验室，该实验表明，这一过程确实是由根瘤菌属的细菌入侵植物根内所形成的结构来实现的。细菌和植物之间的共生关系可以在实验室条件下建成的事实证实了生物学家们虽然一直非常确信但仍需要拿出证据的观点。

豆科植物对大气氮的共生固定在数量上也许可以算是将氮元素整合进入生物圈的一种最为重要的方式，并且也是农业上维护氮素肥力的主要过程。也许是因为缺乏一种合适且确定的实验体系，所以人们至今仍不能完全理解在这种共生关系中豆科植物与根瘤菌属的一些特殊细菌之间的相互作用。其他一些研究者利用离体的植物根与适当的细菌相结合[1-4]；最近还有人使用了植物细胞培养与组织培养的方法[5]。尽管显然产生了刺激植物细胞分化和木质化的效果，但没有发现可以说明细胞间细菌共生关系的证据。

我们尝试结合利用无菌植物细胞培养技术与可以灵敏测定固氮酶活性的乙炔－乙烯还原法，在体外建立植物细胞与微生物之间的一种共生固氮关系。我们认为，这样一个体系可以用来进一步阐释那些控制体内结瘤过程与固氮活性的因素。

大豆根细胞的培养

通过将大豆（阿克姆）种子在商用次氯酸盐溶液（商品名为"花碧玉"，沁威公司，韦恩市，新泽西州）中浸泡 15 分钟使种子表面消毒，然后用无菌蒸馏水清洗。将种子置于琼脂固化的培养基（重量体积比为 1%）表面，这种培养基由矿物盐（MS 培养基）[6]组成，每升添加 30 克蔗糖。种子在 26℃±1℃ 的黑暗中发芽。实验中的所有培养基均在 121℃ 下高压蒸汽灭菌 15 分钟。

当初生根长到大约 4 厘米长时，将它在无菌条件下切开，并在其生长锥后 2.5 厘米处切出一块 5 毫米的外植体。将这个来自根部的外植体置于 50 毫升琼脂固化的

951

placed on the surface of 50 ml. of agar-solidified MS medium supplemented with 15% whole coconut milk (CM, Grand Island Biological Co.) and 2 mg/l. 2,4-dichlorophenoxya -cetic acid (2,4-D) contained in a cotton stoppered 250 ml. Erlenmeyer flask. The medium was adjusted to pH 6.0 before autoclaving. The explanted root section was incubated at $26°\pm1°C$ in the dark.

An actively growing mass of undifferentiated cells (callus) was produced on the root explant in 10–14 days. The callus was then aseptically removed and portions subcultured to fresh MS medium supplemented with 15% CM and 2 mg/l. 2,4-D. In these and subsequent subcultures, we used a liquid medium similar to that described, with 225 ml. of medium in a modified 1,000 ml. flask (Blaessig Glass Co., Rochester, New York)[7], as well as 50 ml. agar-solidified medium in 250 ml. Erlenmeyer flasks. The liquid cultures were incubated on a klinostat rotating at 1 r.p.m.[8] in the dark at $26°\pm1°C$. An actively growing cell and callus population with a cell doubling time of the order of 18 h was established in the liquid environment over a 14–21 day period. Subculture of the liquid system was carried out at 20–30 day intervals to maintain an actively dividing cell population. Cotyledonary tissue was similarly treated to establish cell cultures from this source.

Bacterial Growth

Rhizobium japonicum strains 61A76 (supplied by Dr. Joe C. Burton of the Nitragin Co., Milwaukee, Wisconsin) and ATCC 10324 were grown in a liquid culture medium composed of (in g/l.): K_2HPO_4, 1.0; KH_2PO_4, 1.0; $FeCl_3 \cdot 6H_2O$, 0.005; $MgSO_4 \cdot 7H_2O$, 0.36; $CaSO_4 \cdot 2H_2O$, 0.17; KNO_3, 0.70; yeast extract, 1.0; mannitol, 3.0. The medium was adjusted to pH 6.4 before autoclaving at $121°C$ for 15 min. A bacterial inoculum was added to 40 ml. portions of the above medium contained in 125 ml. Erlenmeyer flasks and incubated at $30°\pm1°C$ with constant shaking. Subcultures of the bacterial cultures were carried out at 7 day intervals to maintain vigorous stocks. At regular intervals aliquots of the *R. japonicum* cultures were examined microscopically and checked for nodulating activity by application to seedlings of Acme or Kent variety soybeans maintained in a Sherer–Gilette CEL 255–6 growth chamber under a 16 h photoperiod.

Establishment of a Symbiotic Relationship

A large culture flask from the klinostat containing 225 ml. of MS medium with 15% CM, 2 mg/l. 2,4-D and containing an actively growing root cell and callus population of approximately 15 g fresh weight was inoculated with 0.1 ml. (260×10^6 cells/ml.) of a liquid suspension culture of actively dividing *R. japonicum* and returned to the klinostat for incubation. After 3–7 days of incubation, the flask contents were poured through sterile cheesecloth. Plant cells and callus pieces retained on the cheesecloth (designated stage 1) were washed with 250 ml. of sterile MS medium, transferred to fresh growth media of varying composition, and incubated in the conditions described. Concomitantly, portions of the cell population inoculated with bacteria, as well as control tissue without bacteria,

MS 培养基表面。之前在培养基中补充了 15% 的全脂椰奶（简称 CM，格兰德艾兰生物公司）和 2 毫克 / 升的 2,4–二氯苯氧基乙酸（简称 2,4–D），并将其装入塞有棉花塞的 250 毫升锥形瓶中。在高压蒸汽灭菌之前将此培养基的 pH 值调至 6.0。将这个根部外植体在温度为 26℃±1℃ 的黑暗中进行培育。

经过 10 ~ 14 天，在根部外植体上长出了一块生长旺盛的未分化细胞（愈伤组织）。在无菌条件下将此愈伤组织取下，并分成数份置于含有 15% CM 和 2 毫克 / 升 2,4–D 的新鲜 MS 培养基上进行传代培养。在这些以及随后的传代培养实验中，我们采用了一种与前面描述过的培养基类似的液体培养基，使用了放在 1,000 毫升改良烧瓶（布莱西希玻璃公司，纽约州罗切斯特市）中的 225 毫升培养基[7]，以及放在 250 毫升锥形瓶中的 50 毫升琼脂固化培养基。将液态培养体系置于缓转器上以 1 转 / 分的转速[8]在温度为 26℃±1℃ 的黑暗中进行培育。14 ~ 21 天后，在液体环境中形成了一个生长旺盛的细胞和愈伤组织群体，其细胞倍增时间可达 18 小时左右。为了保持住一个积极分裂的细胞群体，以 20 ~ 30 天为间隔对这个液态体系进行传代培养。采用类似方法处理大豆子叶组织，以建立同一来源的细胞培养体系。

细菌的生长

将大豆根瘤菌菌株 61A76（由威斯康星州密尔沃基市根瘤菌剂公司的乔·伯顿博士提供）和 ATCC 10324 放置于包含以下成分的液态培养基中进行培育（单位为克 / 升）：K_2HPO_4，1.0；KH_2PO_4，1.0；$FeCl_3 \cdot 6H_2O$，0.005；$MgSO_4 \cdot 7H_2O$，0.36；$CaSO_4 \cdot 2H_2O$，0.17；KNO_3，0.70；酵母提取物，1.0 和甘露醇，3.0。调节此培养基的 pH 值到 6.4，然后在 121℃ 下高压蒸汽灭菌 15 分钟。将细菌接种物加入到含有 40 毫升上述液态培养基的 125 毫升锥形瓶中，并在 30℃±1℃ 下持续摇动培育。为了保持种子菌株的旺盛活力，每隔 7 天进行一次细菌培养物的传代培养。每隔一定时间，用显微镜检查分份的大豆根瘤菌培养物，并检测其结瘤活性，方法是用它来感染在谢勒–希列特 CEL 255–6 型生长室以 16 小时为光照期培养出的阿克姆大豆或者肯特大豆幼苗。

共生关系的建立

在缓转器的大培养瓶中放入 225 毫升含 15% CM 和 2 毫克 / 升 2,4–D 的 MS 培养基以及鲜重约为 15 克的生长旺盛的根细胞和愈伤组织群体，再接种 0.1 毫升（260×10^6 个细胞 / 毫升）积极分裂的根瘤菌液态混悬培养液，然后将培养瓶重新放在缓转器上进行培育。培育 3 ~ 7 天后，用无菌纱布过滤培养瓶中的液态混悬物。再用 250 毫升的无菌 MS 培养液清洗留在纱布上的植物细胞和愈伤组织（指定为阶段 1），然后将残留物转移至含有不同组分的新鲜培养基中，并按前述条件进行培育。与此同时，用乙炔–乙烯还原法对细菌接种过的细胞群体和没有细菌接种的对照组

were assayed for nitrogenase activity using the acetylene–ethylene assay procedure[9]. Dry weight determinations were performed after drying the tissue for 48 h at 60°C.

Cytological Examination of Tissue

Random samples of the soybean root cell cultures from bacteria-inoculated and control flasks were prepared for paraffin sectioning by fixing in formalin : acetic acid : ethanol (5 : 5 : 90 v/v) or in Fleming's osmic acid fixative for 24 h at room temperature. The tissues were then dehydrated through the ethanol series before being embedded in "Paraplast" (MP 56°–57°C) for sectioning. Sections were cut at 8–10 μm, stained with safranin-fast green or Van Gieson's Nile blue and examined with a Zeiss microscope fitted with phase contrast optics.

At the same time similar cell samples were overlaid with 5% glutaraldehyde in Millonig's phosphate buffer (*p*H 7.2–7.4) and fixed for 60 min at room temperature. The fixed samples were rinsed for 20 min, post fixed for 60 min at room temperature with 1% OsO_4, washed with phosphate buffer for 20 min, and dehydrated through an ethanol series up to 100% propylene oxide. The samples were then embedded in "Epon 812", polymerized at 60°C for 24 h, and sectioned at 500 Å using a diamond knife. The resulting sections were stained with uranyl acetate followed by Karnowsky's lead and examined in an RCA EMU-3G electron microscope.

Invasion of Cultured Root Cells

Examination of paraffin sections prepared at the time the bacteria-inoculated liquid plant cell cultures were transferred to fresh growth media (stage 1, day 3–7) revealed the presence of structures resembling the infection threads found in *in vivo* nodulating soybean root systems. These "pseudo-infection threads" of the *in vitro* system were observed to penetrate the undifferentiated cell mass for a considerable distance before entering the cell cytoplasm (Fig. 1). *Rhizobia* were observed within the thread-like structures as they traversed the intercellular spaces within the cell mass. Whenever *Rhizobia* entered a plant cell from the terminus of a "pseudo-infection thread", they multiplied to displace the plant cell contents completely with bacteria (Fig. 2). No other morphologically differentiated cells (for example conducting elements) were found in any of the bacterially populated plant cell masses at this developmental stage.

954

织进行固氮酶活性分析[9]。将组织在 60℃ 下干燥 48 小时，然后测量干重。

组织的细胞学检查

取接种过细菌的大豆根细胞培养物随机样本和对照培养瓶中的随机样本，用福尔马林、乙酸和乙醇的混合液（体积比为 5：5：90）或者用弗莱明锇酸固定剂在室温下固定 24 小时，以备制作石蜡切片。然后先将组织通过梯度乙醇脱水，再将其包埋于塑化石蜡（熔点为 56℃ ~ 57℃）中，以便进行切片。随后以 8 微米 ~ 10 微米的厚度切片，并用蕃红 - 固绿或者范吉森尼罗蓝染色，再在配有相差光学装置的蔡司显微镜下观察切片。

同时，将类似的细胞样品表面涂覆含 5% 戊二醛的米隆尼氏磷酸盐缓冲液（pH 值为 7.2 ~ 7.4），并在室温下固定 60 分钟。将固定后的样品用清水冲洗 20 分钟，并用 1% 的 OsO_4（四氧化锇）在室温下再次固定 60 分钟，接着用磷酸盐缓冲液清洗 20 分钟，然后经过梯度乙醇脱水，再用 100% 的环氧丙烷处理。接着将样品包埋于"环氧树脂 812"中，并在 60℃ 下聚合 24 小时，然后用金刚石刀切成厚度为 500 Å 的切片。将得到的切片先用醋酸铀酰再用卡诺斯基铅进行着色，然后在 RCA EMU-3G 型电子显微镜下进行观察。

侵入培养的根细胞

对将接种细菌的液态植物细胞培养物转移至新鲜生长培养基（第 1 阶段，3 ~ 7 天）时制备的石蜡切片进行观察，发现存在一些结构，这些结构与体内结瘤的大豆根系中的侵染丝相类似。我们观察到，这些体外系统中的"伪侵染丝"在进入细胞的细胞质之前，就已经穿入未分化细胞群中很长一段距离了（图 1）。当这些丝状结构穿越细胞群中的细胞间隙时，我们观察到丝状结构中有根瘤菌。一旦根瘤菌从"伪侵染丝"的末端进入植物细胞，它们就会进行繁殖，将植物细胞中的物质完全替换成细菌（图 2）。在处于这一发展阶段的任何被细菌填充的植物细胞群中，还没有发现过其他形态的分化细胞（例如疏导细胞）。

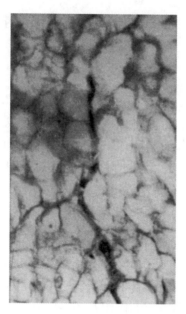

Fig. 1. Paraffin section of a callus culture showing *Rhizobium japonicum* penetration through an infection thread-like structure (×600).

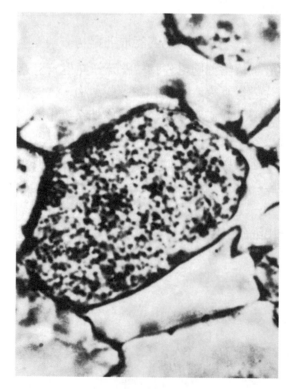

Fig. 2. Callus cell filled with *R. japonicum* after initial penetration.

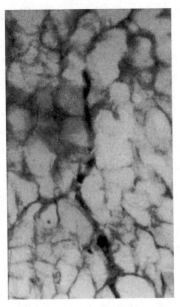

图 1. 愈伤组织培养物的石蜡切片。图中显示出大豆根瘤菌通过一个类似侵染丝的结构进行侵入（放大600倍）。

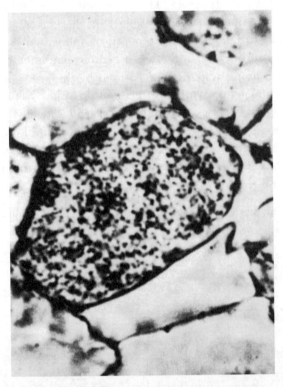

图 2. 在初步侵入之后，大豆根瘤菌充满愈伤组织细胞。

957

Continued Development of the Symbiotic System

Following the initial penetration of the plant cells by *Rhizobia* (3–7 days after bacterial inoculation of the cell culture flask), the cell cultures were transferred to fresh media, both liquid and agar-solidified, for development of stage 2. The following media were utilized: (1) MS, (2) MS plus 10–15% whole CM and (3) MS plus 10–15% whole CM and 2 mg/l. 2,4-D. Control cultures of soybean root cells (without bacteria) were transferred to similar media and incubated in the same conditions. All cultures were incubated for 7–20 days after transfer before being examined for bacterial penetration and for nitrogenase activity. Light-grown cultures received approximately 500 foot-candles of light from Duro-Test Optima lamps (Duro-Test Corp., North Bergen, New Jersey).

Fig. 3 shows that the degree of rhizobial invasion after transfer of the cultured soybean root cells is related to the growth factor additions to the MS medium. Most extensive invasion is found after transfer of bacterially populated plant cells to MS medium containing no exogenous growth supplements (Fig. 3*a*). With the addition of whole CM (Fig. 3*b*) and the combination of CM and 2,4-D (Fig. 3*c*) the number of cells inhabited by the *Rhizobia* is markedly reduced. Assays for nitrogenase activity under each of the tested culture conditions (Tables 1 and 2) show this activity to vary directly with the degree of bacterial invasion (Table 1). The tissue without rhizobial involvement shows little acetylene–ethylene reducing activity (Table 1) in these assays[11,12]. Table 2 shows that although there is wide variation in the response of individual Acme soybean root cell cultures to *R. japonicum* strain 61A76, there is a definite symbiotic relationship capable of fixing N_2 established in liquid systems, as measured by the acetylene–ethylene technique. Attempts to establish an active symbiotic relationship using Kent variety soybean root cells in liquid culture have been less successful. Even with Acme variety soybeans, the use of a different strain of *R. japonicum* (ATCC 10324) resulted in a lower over-all nitrogenase activity although electron microscopy showed bacterial penetration of the cultured cells. Cotyledonary tissue has shown no propensity to establish a symbiotic association in the conditions of this study.

Table 1. Acetylene to Ethylene Reduction by Soybean Root Cell Cultures inoculated with *Rhizobium japonicum* Strain 61A76

	nmol C_2H_4/g fresh weight ×24 h	
	Light	Dark
MS	72.8	318.6
MS+10% CM	36.6	129.0
MS+10% CM+2 p.p.m. 2,4-D	7.7	37.9
Uninoculated control MS+10% CM+2 p.p.m. 2,4-D	0.4	2.1

The assay was carried out 21 days after initial bacterial inoculation and after transfer to semi-solid agar medium for 14 days. Incubation was under continuous light or dark conditions following initial establishment of symbiosis.

共生体系的持续发展

在植物细胞开始被根瘤菌侵入之后（细胞培养瓶接种细菌后的 3～7 天），将细胞培养物转移至新鲜的培养基中，包括液体培养基和琼脂固化培养基，用来进行第 2 阶段的培育。使用的培养基如下：（1）MS；（2）MS 添加 10%～15% 全脂 CM；（3）MS 添加 10%～15% 全脂 CM 和 2 毫克/升 2,4–D。将大豆根细胞的对照培养物（不含细菌）转移至同样的培养基中，并在同样的条件下进行培育。在转移后将所有培养物培育 7～20 天，然后检测细菌侵入情况和固氮酶活性。光照下生长的培养物所接收的照度约为 500 英尺烛光，光源是杜罗试验最佳配置灯（杜罗试验公司，北伯根，新泽西州）。

由图 3 可以看出：在培养的大豆根细胞被转移后，根瘤菌的入侵程度与 MS 培养基中添加的生长因子之间存在着关联。将被细菌填满的植物细胞转移到不含任何外源生长添加剂的 MS 培养基中后，细菌的入侵范围最广（图 3*a*）。当添加全脂 CM（图 3*b*）和 CM 与 2,4–D 的混合物（图 3*c*）时，被根瘤菌占据的细胞数量明显减少。对上述每一种试验条件下的培养进行固氮酶活性检测的结果（表 1 和表 2）显示，酶活性与细菌入侵程度直接相关（表 1）。在这些分析测试中，没有根瘤菌介入的组织几乎不具有将乙炔还原为乙烯的活性（表 1）[11,12]。表 2 显示：尽管两个阿克姆大豆根细胞培养物对根瘤菌菌株 61A76 的反应有很大的差别，但乙炔–乙烯还原法的测定结果表明，在这些液态系统中形成了一种确定的能够固氮的共生关系。尝试用肯特大豆根细胞在液态培养体系中建立有效共生关系的成功率较低。即使是阿克姆大豆，在使用另一种大豆根瘤菌菌株（ATCC 10324）时得到的总固氮酶活性也会偏低一些，尽管通过电子显微镜术已经看到细菌侵入了培养细胞。在上述研究条件下，未曾发现子叶组织有建立共生关系的倾向。

表 1. 接种大豆根瘤菌菌株 61A76 的大豆根细胞培养物将乙炔还原为乙烯的活性

	纳摩尔 C_2H_4 每克鲜重 × 24 小时	
	光照	黑暗
MS	72.8	318.6
MS +10% CM	36.6	129.0
MS +10% CM + 2 p.p.m. 2,4–D	7.7	37.9
未经接种的对照物 MS +10% CM + 2 p.p.m. 2,4–D	0.4	2.1

这项分析测试是在初次接种细菌后 21 天和在转移至半固体琼脂培养基后 14 天进行的。在共生体系初步建立后开始培育，培育条件为连续光照或者避光。

Table 2. Acetylene to Ethylene Reduction by Two Separate Soybean Root Cell Cultures containing *R. japonicum* Strain 61A76 in Symbiotic Relationship

	nmol C$_2$H$_4$/g dry weight × 24 h	
	1	2
Tissue + 61A76 bacteria	638	88.0
Uninoculated control	307	25.7

Assays were made 3–5 days after addition of the microorganisms to liquid culture.

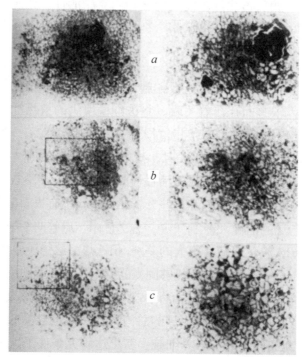

Fig. 3. Sections of callus tissue infected with *R. japonicum* showing the effects of growth supplements on development of the symbiotic association. See text for details of treatments. Left, ×350; right, ×700.

Electron Microscopic Examination of Symbiotic Cultures

Examination of *Rhizobium*-plant cell cultures with the electron microscope confirmed that *Rhizobia* have entered the cultured plant cells and displaced the normal cytoplasmic components (Fig. 4). In some cases the *Rhizobia* were enclosed within a vesicle which is morphologically analogous to the vesicles found in nodules obtained from whole soybean root systems (Fig. 5). It was further observed that the *Rhizobia* within the cultured root cells contained an inclusion which was similar in appearance to an inclusion of whole soybean nodule bacterioids designated as a polymer of β-hydroxybutyrate. This inclusion became more prominent with increasing time of incubation of the synthetic symbiotic system (up to 15–20 days after transfer). Microscopic examinations of uninoculated tissue showed that there were no microorganismal contaminants present. The media used for maintenance of the plant cell cultures were also found to be devoid of microorganisms and N$_2$-fixing activity; the bacterial inoculum contained only *Rhizobia* and exhibited no N$_2$-fixing activity. Thus it is clear that free-living N$_2$-fixing bacteria are not responsible for our observations.

960

表 2. 含有存在共生关系的大豆根瘤菌菌株 61A76 的两个独立大豆根部细胞培养物将乙炔还
原为乙烯的活性

	纳摩尔 C$_2$H$_4$ 每克干重 × 24 小时	
	1	2
组织 + 61A76 细菌	638	88.0
未经接种的对照物	307	25.7

分析测试是在将微生物加入到液体培养基中后 3 ~ 5 天进行的。

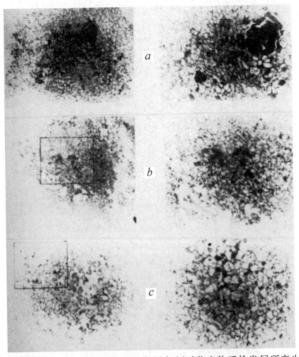

图 3. 感染大豆根瘤菌的愈伤组织切片，显示出生长添加剂对共生体系的发展所产生的影响。实验处理
细节参见正文中的叙述。左图：放大350倍；右图：放大700倍。

用电子显微镜观察共生培养物

对根瘤菌－植物细胞培养物的电子显微镜观察结果证实：根瘤菌进入了培养的植物细胞内部，并替换了正常的细胞质成分（图 4）。在有些样本中，根瘤菌被封入了一个囊泡，这个囊泡在形态上类似于来自完整大豆根系统的根瘤中的囊泡（图 5）。进一步的观察发现：培养的根细胞中的根瘤菌含有一种内含物，这种内含物在表观上类似于完整大豆根瘤类菌体中的内含物，它的成分被认定为 β– 羟基丁酸的聚合物。这种内含物会随着在人造共生体系中培育时间的增加(一直到转移后 15 ~ 20 天)而变得更加显著。用显微镜对未接种的组织进行观察,结果显示不存在微生物的污染。用于维持植物细胞培养物的培养基也被证明不具有微生物和固氮活性；细菌接种体只含有根瘤菌而没有表现出任何固氮活性。因此很明显，自生固氮菌与我们观察到的结果无关。

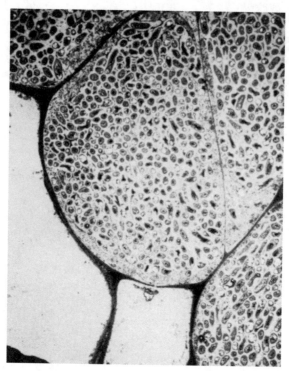

Fig. 4. Electron micrograph of cultured soybean callus infected with *R. japonicum* (×4,160).

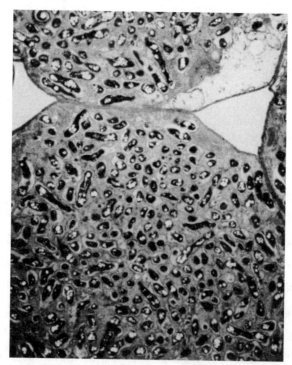

Fig. 5. Electron micrograph of a soybean root nodule infected with *R. japonicum* showing accumulation of β-hydroxybutyrate (×4,160).

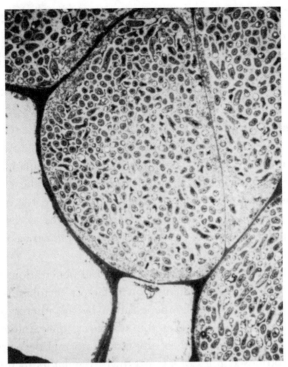

图 4. 培养的大豆愈伤组织被大豆根瘤菌感染后的电子显微镜照片（放大4,160倍）。

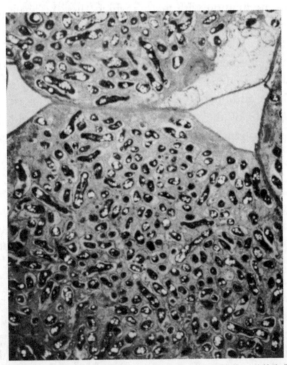

图 5. 大豆根瘤被根瘤菌感染后的电子显微镜照片。图中显示出 β– 羟基丁酸的聚集（放大 4,160 倍）。

Functionality of the Symbiotic Relationship

The functional system we have described provides a new tool for the elucidation of the mechanisms controlling nodulation and subsequent N_2-fixation in symbiotic associations.

The morphology of symbiosis *in vitro* is analogous to that of intact soybean root nodules[13,14]. *Rhizobia* within the cytoplasm of the cultured cells, as in root nodules, are at times enclosed within a vesicle. The intracellular bacteria (bacterioids?) are shown to contain an inclusion which has been tentatively identified as poly-β-hydroxybutyrate[15]. A structure similar in appearance to the infection thread of nodulating systems *in vivo* has been observed in the early stages of bacterial penetration of the cultured plant cells in the liquid milieu. This apparent mode of rhizobial entry into the cultured cells and the subsequent multiplication of the intracellular microorganisms to displace completely the cell cytoplasm are again analogous to processes which are known to occur *in vivo*.

It has further been found that following initial bacterial penetration of the cultured cells, depletion of exogenous supplies of plant growth factors (cytokinins and auxins) in the environment leads to increased cellular penetration by the microorganisms and greater nitrogenase induction. This suggests that the symbiosis, once initiated, is self-sufficient and that any exogenous additions tend to suppress its normal development and function. This agrees with the proposed roles of plant growth regulators in the establishment and maintenance of nodulated symbiotic systems *in vivo*[16,17]. Thus it may be possible to control the process of effective nodulation with factors which alter the levels of endogenous plant growth regulators such as auxins during the critical phases of invasion.

The reduction of acetylene to ethylene as a measure of nitrogenase activity shows that the N_2-fixing potential of the symbiosis increases with the proportion of cellular invasion by microorganisms, and the addition of plant growth factors depresses nitrogenase activity as well as the development of the symbiotic morphology. It should therefore be possible to increase the overall N_2-fixing activity of the system by regulating the degree of cell penetration and subsequent nitrogenase induction through chemical means.

Light does not seem to stimulate the symbiosis *in vitro* once it has been established; but cells grown in the light resisted rhizobial challenge, and no working symbiosis could be established in these cultures. It is possible that cells grown in the light develop a natural defence mechanism against invasion by microorganisms *in vitro*, perhaps through the phytochrome system[18].

The acetylene to ethylene reducing activities (and nitrogenase activity by analogy) achieved so far in the *in vitro* symbiotic system are of the order of 1% of those found in soybean nodules. Microscopic examinations of the symbiotic cell masses in culture showed that only 1–10% of the total cell population was infected with intracellular microorganisms. The activity per cell may therefore be comparable with that found in whole root nodules where more than 90% of the cells appear to contain bacteria. It is also pertinent (Fig.

共生关系的功能性

我们所描述的功能体系为阐释共生体系中控制结瘤的机制以及随后发生的固氮作用的机制提供了一个新的工具。

体外共生在形态上类似于完整的大豆根瘤 [13,14]。存在于培养细胞的细胞质中的根瘤菌,与在根瘤中的根瘤菌一样,有时会被封闭在一个囊泡中。胞内细菌(类菌体?)被证明含有一种内含物,其成分暂时被认定为 β−羟基丁酸 [15]。当细菌刚开始侵入液态环境中培养的植物细胞时,我们观察到了一种在外观上与体内结瘤系统的侵染丝相类似的结构。根瘤菌侵入培养细胞的表观模式,以及胞内微生物随后能繁殖到完全取代细胞的细胞质的模式,也与已知在体内发生的过程类似。

进一步又发现:在细菌开始侵入培养细胞之后,培养环境中植物生长因子(细胞分裂素和植物生长素)的外源供应被耗尽将导致微生物侵入细胞加强,并诱导产生更强的固氮酶活性。这表明共生体系一旦启动,便可以自给自足;而且任何外源添加物都会抑制其正常的发展和功能。关于植物生长调节剂在体内建立和维持成瘤共生体系中所起的作用,早已有人提出看法 [16,17],以上实验结果恰好与这种看法一致。因此,也许可以利用一些能改变内源植物生长调节剂水平的因素(如入侵过程关键时期的植物生长素水平)来控制有效成瘤的过程。

利用乙炔还原成乙烯的量来表征的固氮酶活性数值表明,共生体系的固氮潜能随着被微生物入侵的细胞比例的增加而增强,而添加植物生长因子不但会抑制共生体系的形态发展,也会抑制固氮酶的活性。因此,通过化学手段来调节细菌侵入细胞的程度以及随后诱导的固氮酶活性,以增强这个共生体系的总体固氮活性,应当是有可能实现的。

体外共生体系一旦建立,光似乎就不能再对它产生刺激作用了;但在光照下生长的细胞能抵制根瘤菌的入侵,因此无法在这些培养物中建成有效的共生体系。在光照下生长的细胞有可能会形成一种对抗体外微生物入侵的天然防卫机制,这也许是通过光敏色素系统实现的 [18]。

到目前为止,从体外共生体系中获得的使乙炔转变成乙烯的还原活性(以及由此推出的固氮酶活性)是在大豆根瘤中发现的 1% 左右。用显微镜对培养物中共生细胞群体的观察结果显示:在整个细胞种群中,只有 1%～10% 的细胞被细胞内的微生物感染。但在根瘤中看似有超过 90% 的细胞含有细菌,因此每个细胞的活性有可能与完整根瘤中的细胞活性相当,这也与细菌感染只发生在愈伤组织表面下的细

3) that infection occurs only in the subsurface cells of the callus where physiological conditions might be expected to simulate more closely the cortex of intact roots. The initial infection of the tissue culture occurs through a modification of a peripheral cell of the callus mass and not through a root hair cell as in intact roots. This suggests that symbiosis *in vivo* might also be expected to occur through penetration of epidermal cells other than root hairs in suitable conditions.

We thank Mrs. Mary Ann Mattson and Mr. M. L. Van Kavelaar for technical assistance.

(**232**, 173-176; 1971)

R. D. Holsten, R. C. Burns, R. W. F. Hardy and R. R. Hebert: Central Research Department, E. I. du Pont de Nemours and Company, Wilmington, Delaware 19898.

Received January 25, 1971.

References:

1. Lewis, K. N., and McCoy, E., *Bot. Gaz.*, **95**, 316 (1933).

2. McGonagle, M. P., *Nature*, **153**, 528 (1944).

3. Raggio, N., Raggio, M., and Burris, R. H., *Biochim. Biophys. Acta*, **32**, 274 (1959).

4. Raggio, M., Raggio, N., and Torrey, J. G., *Amer. J. Bot.*, **44**, 325 (1957).

5. Veliky, I., and LaRue, T. A., *Naturwissenschaften*, **54**, 96 (1967).

6. Murashige, T., and Skoog, F., *Physiol. Plant.*, **15**, 473 (1962).

7. Steward, F. C., and Shantz, E. M., *Chemistry and Mode of Action of Growth Substances* (edit. by Wain, R. L., and Wightman, F.), 312 (Academic Press, New York, 1956).

8. Caplin, S. M., and Steward, F. C., *Nature*, **163**, 920 (1949).

9. Hardy, R. W. F., Holsten, R. D., Jackson, E. K., and Burns, R. C., *Plant Physiol.*, **43**, 1185 (1968).

10. Millonig, G., *J. Appl. Phys.*, **32**, 1637 (1961).

11. Gamborg, O. L., and LaRue, T. A. G., *Nature*, **200**, 604 (1968).

12. Stewart, E. P., and Freebairn, H. T., *Plant Physiol.*, **42**, Suppl., S-30 (1967).

13. Goodchild, D. J., and Bergersen, F. J., *J. Bact.*, **92**, 204 (1966).

14. Holsten, R. D., Hebert, R. R., Jackson, E. K., Burns, R. C., and Hardy, R. W. F., *Bact. Proc.*, 149 (1969).

15. Klucas, R. V., and Evans, H. J., *Plant Physiol.*, **43**, 1458 (1968).

16. Kefford, N. P., Brockwell, J., and Zwar J. A., *Austral. J. Biol. Sci.*, **13**, 456 (1960).

17. Thimann, K., *Proc. US Nat. Acad. Sci.*, **22**, 511 (1936).

18. Lie, T. A., *Plant and Soil*, **30**, 391 (1969).

胞相一致（图3），可以设想此处的生理条件更近似于完整根部的皮层。在组织培养中，初期感染是通过改变愈伤组织的外周细胞而发生的，并不像在完整根部中那样，是通过根毛细胞发生的。这表明体内共生或许也可以被设想为是在适当条件下通过侵入表皮细胞发生的，而不是通过侵入根毛发生的。

感谢玛丽·安·马特森夫人和范卡维拉先生为我们提供了技术上的支持。

（刘振明 翻译；顾孝诚 审稿）